MARINE BIOENERGY

Trends and Developments

MARINE BIOENERGY

Trends and Developments

Edited by
Se-Kwon Kim
Choul-Gyun Lee

CRC Press
Taylor & Francis Group
Boca Raton London New York

CRC Press is an imprint of the
Taylor & Francis Group, an **informa** business

CRC Press
Taylor & Francis Group
6000 Broken Sound Parkway NW, Suite 300
Boca Raton, FL 33487-2742

Printed on acid-free paper
Version Date: 20150220

International Standard Book Number-13: 978-1-4822-2237-1 (Hardback)

Library of Congress Cataloging-in-Publication Data

Marine bioenergy : trends and developments / edited by Se-Kwon Kim, Choul-Gyun Lee.
pages cm
Includes bibliographical references and index.
ISBN 978-1-4822-2237-1
1. Biomass energy. 2. Ocean energy resources. I. Kim, Se-Kwon. II. Lee, Choul-Gyun.

TP339.M36 2015
333.95'39--dc23 2014040863

Visit the Taylor & Francis Web site at
http://www.taylorandfrancis.com

and the CRC Press Web site at
http://www.crcpress.com

Contents

Section IV Production of Bioenergy

Section V Bioelectricity and Microbial Fuel Cells

Preface

Marine bioenergy is one of the most important components to mitigate greenhouse gas (GHG) emissions and for the substitution of fossil fuels. The study of marine algae for fuel has become a hot topic in recent years with energy prices fluctuating widely and GHG emissions increasingly becoming a cause for concern. Marine microalgae are emerging as one of the most promising resources of biodiesel, with a projected yield of 58,700–136,900 L ha^{-1} $year^{-1}$. Bioethanol, as a clean and renewable combustible, is considered as a good alternative to oil. Although the energy equivalent of ethanol is 68% lower than that of petroleum fuel, the combustion of ethanol is cleaner (because it contains oxygen). Hydrogen gas is thought to be the ideal fuel for a world in which air pollution has been alleviated, global warming has been arrested, and the environment has been protected in an economically sustainable manner. Marine microbial fuel cell (MFC) is a promising new technology for generating electricity directly from biodegradable compounds. More recently, generation of electricity using MFC is gaining important attention in the research fraternity. MFC is a type of fuel cell that converts the chemical energy contained in organic matter to electricity using microorganisms as a biocatalyst.

Marine biodiesel is a renewable, nontoxic, biodegradable, and CO_2-neutral energy source. Therefore, in recent years it has become a hot topic for the exploitation of renewable and environment-friendly energy forms. Compared to conventional oil crops, microalgae are more attractive as feedstock for biodiesel production due to their high photosynthesis efficiency and lipid content. The marine microalgal lipid productivity/biomass (dry weight) is about 15–300 times that of conventional crops. Besides that, marine microalgae have the function of removing nitrogen and phosphorus and fixing CO_2, making the coupling of bioenergy production and wastewater treatment based on microalgae a promising technology in the future. Therefore, marine microalgae–based biodiesel has attracted increasing attention worldwide. However, the high cost of biodiesel production is the main bottleneck in its commercial application. Increasing the lipid content per microalgal biomass is one of the efficient methods to reduce the total cost of biodiesel production.

Marine Bioenergy: Trends and Developments covers the key aspects of marine bioenergy, namely, marine biomass, techniques, bioethanol, biomethane, biomethanol, biohydrogen, biodiesel, bioelectricity, marine waste, commercialization, and the global market. Section I provides an introduction to marine bioenergy; Section II describes marine algae as a source of bioenergy; Section III describes the biotechnological techniques for biofuel production; Section IV deals with the production of bioenergy particularly bioethanol, biomethane, biomethanol, biohydrogen, and biodiesel; Section V discusses bioelectricity and MFCs by marine algae and microbes; Section VI covers marine waste for bioenergy; and Section VII deals with commercialization and the global market. The chapters in each section are a good collection of comprehensive research on these bioenergies carried out by proficient scientists from around the world. In addition, the preparation methodologies for these bioenergies have also been well depicted in their respective chapters. We are quite certain that the findings and latest information presented in this book will be helpful for upcoming researchers to establish phenomenal research from an intersection of multiple research areas.

We are grateful to all the chapter authors who have provided state-of-the-art contributions in the field of marine bioenergy; their relentless effort was the result of scientific attitude, drawn from the past history in this field. I also thank the staff of Taylor & Francis Group and CRC Press for the continual support, which was essential for the successful completion of this book. We hope that the fundamental ideas presented in this book serve as potential research and development options for marine bioenergy for the benefit of humankind.

We strongly recommend this book to marine bioenergy researchers/students and hope that it helps to enhance their understanding in this field.

Se-Kwon Kim
Choul-Gyun Lee

MATLAB® is a registered trademark of The MathWorks, Inc. For product information, please contact:

The MathWorks, Inc.
3 Apple Hill Drive
Natick, MA 01760-2098 USA
Tel: 508-647-7000
Fax: 508-647-7001
E-mail: info@mathworks.com
Web: www.mathworks.com

Acknowledgments

We would like to thank CRC Press and Taylor & Francis Group for their encouragement and suggestions to get this wonderful compilation published. We extend our sincere gratitude to all the contributors for providing help, support, and advice to accomplish this task. Furthermore, we thank Dr. Panchanathan Manivasagan and Dr. Jayachandran Venkatesan who worked with us throughout the course of this book project.

Se-Kwon Kim
Choul-Gyun Lee

Editors

Se-Kwon Kim, PhD, is presently working as a distinguished professor in the Department of Marine-Bio Convergence Science and is director of Marine Bioprocess Research Center (MBPRC) at Pukyong National University, Busan, South Korea.

He received his MSc and PhD from Pukyong National University and conducted his postdoctoral studies at the Laboratory of Biochemical Engineering, University of Illinois, Urbana–Champaign, Illinois, United States. Later, he became a visiting scientist at the Memorial University of Newfoundland in Canada.

Dr. Kim served as president of the Korean Society of Chitin and Chitosan in 1986–1990 and the Korean Society of Marine Biotechnology in 2006–2007. For his research, he won the best paper award from the American Oil Chemists' Society in 2002. He was also the chairman of the 7th Asia-Pacific Chitin and Chitosan Symposium, which was held in South Korea in 2006. He was the chief editor at the Korean Society of Fisheries and Aquatic Science during 2008–2009. He is also a board member of the International Society of Marine Biotechnology (IMB) and the International Society of Nutraceuticals and Functional Food (ISNFF).

His major research interests are investigation and development of bioactive substances from seafood processing wastes and other marine sources. His immense experience in marine bioprocessing and mass production technologies for marine bioindustry is the key asset of holding majorly funded marine bio projects in Korea. Furthermore, he expanded his research fields up to the development of bioactive materials from marine organisms for their applications in oriental medicine, cosmeceuticals, and nutraceuticals. To date, he has authored around 450 research papers and holds 76 patents.

Choul-Gyun Lee, PhD, is currently a professor in the Department of Biological Engineering, Inha University, Incheon, South Korea. Dr. Lee earned his PhD from the Department of Chemical Engineering, University of Michigan, Ann Arbor, Michigan in 1994, and he worked for NASA, Kennedy Space Center in Advance Life Support Team before joining Inha University. His field of expertise is high-density and/or large-scale cultures of microalgae in various types of photobioreactors and molecular/biotechnological studies of microalgae based on whole-cell genome-scale in silico modeling. He also serves as the leader (director) for the National Marine Bioenergy (MBE) Research and Development Consortium funded by the Korean government and is developing massive microalgal culture systems for ocean deployment.

In addition to his research work, he has held a number of administrative positions in the University. He also chairs the Interdisciplinary Program on Future Energy Engineering and is the head of the Institute of Industrial Biotechnology as well as the Lipidomics Research Center. Dr. Lee is the president of the Korean Society of Marine Biotechnology since 2012 and has worked as either an editorial board member or an associate editor for six different journals.

Contributors

Sarmidi Amin (retired)
Agency for Assessment and Application of Technology (BPPT)
Jakarta, Indonesia

H. Ann Suji
Faculty of Marine Sciences
Annamalai University
Parangipettai, India

Dumitru Bulgariu
Department of Geology and Geochemistry
"Alexandru Ioan Cuza" University of Iasi
and
Collective of Geography
Romanian Academy
Iasi, Romania

Laura Bulgariu
Department of Environmental Engineering and Management
Gheorghe Asachi Technical University of Iasi
Iasi, Romania

Anong Chirapart
Department of Fishery Biology
Kasetsart University
Bangkok, Thailand

Trunil Desai
Department of Biotechnology
International Centre for Genetic Engineering and Biotechnology
New Delhi, India

Vaishali Dutt
Department of Biotechnology
International Centre for Genetic Engineering and Biotechnology
New Delhi, India

Kelvii Wei Guo
Department of Mechanical and Biomedical Engineering
City University of Hong Kong
Kowloon, Hong Kong

Vishal Gupta
Discipline of Marine Biotechnology and Ecology
Central Salt and Marine Chemicals Research Institute
Bhavnagar, India

Patrick C. Hallenbeck
Department of Microbiology, Infectology, and Immunology
University of Montreal
Montréal, Québec, Canada

and

Department of Biology
United States Air Force Academy
Colorado Springs, Colorado

Yong-Ki Hong
Department of Biotechnology
Pukyong National University
Busan, South Korea

Amita Jain
School of Biotechnology and Bioinformatics
D.Y. Patil University Navi Mumbai
Maharashtra, India

Gwi-Taek Jeong
Department of Biotechnology
Pukyong National University
Busan, South Korea

Bhavanath Jha
Discipline of Marine Biotechnology and Ecology
Central Salt and Marine Chemicals Research Institute
Bhavnagar, India

Pavan P. Jutur
Center for Advanced Bioenergy Research
International Centre for Genetic Engineering and Biotechnology
New Delhi, India

L. Kannan
Faculty of Marine Sciences
Annamalai University
Parangipettai, India

Valliappan Karuppiah
Marine Biotechnology Laboratory
State Key Laboratory of Microbial Metabolism
and
School of Life Sciences and Biotechnology
Shanghai Jiao Tong University
Minhang, Shanghai, People's Republic of China

Se-Kwon Kim
Department of Marine-Bio Convergence Science
and
Marine Bioprocess Research Center
Pukyong National University
Busan, South Korea

Sung-Koo Kim
Department of Biotechnology
Pukyong National University
Busan, South Korea

Ozcan Konur
Department of Materials Engineering
Yildirim Beyazit University
Ankara, Turkey

Ramya Kumar
Department of Biotechnology
Indian Institute of Technology
Kharagpur, India

Choul-Gyun Lee
Marine Bioenergy Research Center
Inha University
Incheon, South Korea

Zhiyong Li
Marine Biotechnology Laboratory
State Key Laboratory of Microbial Metabolism
and
School of Life Sciences and Biotechnology
Shanghai Jiao Tong University
Minhang, Shanghai, People's Republic of China

Panchanathan Manivasagan
Department of Marine-Bio Convergence Science
Specialized Graduate School Science and Technology Convergence
and
Marine Bioprocess Research Center
Pukyong National University
Busan, South Korea

Bintang Marhaeni
Fisheries and Marine Faculty
Jenderal Soedirman University
Purwokerto, Indonesia

Patrick J. McGinn
Algal Carbon Conversion Flagship Program
National Research Council of Canada
Halifax, Nova Scotia, Canada

Maria Dyah Nur Meinita
Fisheries and Marine Faculty
Jenderal Soedirman University
Purwokerto, Indonesia

C. Naveen
Centre for Energy and Environmental Science and Technology
National Institute of Technology
Tiruchirappalli, India

Asha A. Nesamma
Center for Advanced Bioenergy Research
International Centre for Genetic Engineering and Biotechnology
New Delhi, India

Poonam Singh Nigam
Faculty of Life and Health Sciences
University of Ulster
Northern Ireland, United Kingdom

E.M. Nithiya
Centre for Energy and Environmental Science and Technology
National Institute of Technology
Tiruchirappalli, India

Masahiro Notoya
Research Institute of Applied Phycology
Okabe Co., Ltd.
Shimane, Japan

Kiran Paranjape
Department of Microbiology, Infectology, and Immunology
University of Montreal
Montréal, Québec, Canada

Kyoung C. Park
Algal Carbon Conversion Flagship Program
National Research Council of Canada
Halifax, Nova Scotia, Canada

Kurniadhi Prabandono
Fish Quarantine Inspection Agency
Indonesian Ministry of Marine Affairs and Fisheries
Tangerang, Indonesia

Jantana Praiboon
Department of Fishery Biology
Kasetsart University
Bangkok, Thailand

M. Premalatha
Centre for Energy and Environmental Science and Technology
National Institute of Technology
Tiruchirappalli, India

Chae Hun Ra
Department of Biotechnology
Pukyong National University
Busan, South Korea

Dheeraj Rathore
School of Environment and Sustainable Development
Central University of Gujarat
Gandhinagar, India

C.R.K. Reddy
Discipline of Marine Biotechnology and Ecology
and
Central Salt and Marine Chemicals Research Institute
Academy of Scientific & Innovative Research
Bhavnagar, India

Rapeeporn Ruangchuay
Department of Technology and Industry
Prince of Songkla University
Pattani, Thailand

K. Sankaran
Centre for Energy and Environmental Science and Technology
National Institute of Technology
Tiruchirappalli, India

Ramkrishna Sen
Department of Biotechnology
Indian Institute of Technology
Kharagpur, India

Dong Ho Seong
Marine Bioenergy Research Center
Inha University
Incheon, South Korea

Anoop Singh
Department of Scientific and Industrial Research
Ministry of Science and Technology
Government of India
New Delhi, India

V.L. Sirisha
Department of Biology
and
Department of Atomic Energy
University of Mumbai
Mumbai, India

K. Sivakumar
Faculty of Marine Sciences
Annamalai University
Parangipettai, India

V.T. Somasundaram
Trichy Distilleries & Chemicals Ltd.
Tiruchirappalli, India

Shireesh Srivastava
Department of Biotechnology
International Centre for Genetic Engineering and Biotechnology
New Delhi, India

P. Subramanian (retired)
Centre for Energy and Environmental Science and Technology
National Institute of Technology
Tiruchirappalli, India

Nitin Trivedi
Discipline of Marine Biotechnology and Ecology
and
Central Salt and Marine Chemicals Research Institute
Academy of Scientific & Innovative Research
Bhavnagar, India

K.K. Vasumathi
Centre for Energy and Environmental Science and Technology
National Institute of Technology
Tiruchirappalli, India

Jayachandran Venkatesan
Department of Marine-Bio Convergence Science
Specialized Graduate School Science and Technology Convergence
and
Marine Bioprocess Research Center
Pukyong National University
Busan, South Korea

M. Vijayasekaran
Trichy Distilleries & Chemicals Ltd.
Tiruchirappalli, India

Geetanjali Yadav
Department of Biotechnology
Indian Institute of Technology
Kharagpur, India

Section I

Introduction to Marine Bioenergy

1

Introduction to Marine Bioenergy

Panchanathan Manivasagan and Se-Kwon Kim

CONTENTS

1.1 Introduction

Marine microalgae are currently considered to be one of the most promising alternative sources for biodiesel (Sheehan et al. 1998). Since many microalgal strains can be cultivated on nonarable land in a saline water medium, their mass farming does not place additional strains on food production (Widjaja et al. 2009). Their high photosynthetic rates, often ascribed to their simplistic unicellular structures, enable microalgae not only to serve as an effective carbon sequestration platform but also to rapidly accumulate lipids in their biomass (up to 77% of dry cell mass). Even using a conservative scenario, microalgae are still predicted to produce about 10 times more biodiesel per unit area of land than a typical terrestrial oleaginous crop (Sheehan et al. 1998; Chisti 2007; Rosenberg et al. 2008; Schenk et al. 2008).

Marine algae are basically a large and diverse group of simple, typically autotrophic organisms, ranging from unicellular to multicellular forms. Their unicellular to multicellular structure allows them to easily convert solar energy into chemical energy. A growing range of studies have been conducted to explore the techniques, procedures, and processes of producing large quantities of microalgae biomass (Spolaore et al. 2006). There are two most commonly used techniques to cultivate microalgae. These are open raceway pond system and closed photobioreactor system. The open pond system is less favorable due to limitations in controlling contaminations from predators, while the photobioreactors provide an easy system of controlling nutrients for growth and cultivation parameters such as temperature, dissolved CO_2, and pH and preventing contamination (Ugwu et al. 2008).

However, photobioreactors have a high initial cost and are very specific to the physiology of the microalgae strain being cultivated. Therefore, microalgae production facility is an important factor to be considered for the optimum production of a specific microalgal species (spp.). Methods such as flocculation, centrifugation, and filtration are used unaided to dewater the algae biomass (Heasman et al. 2000; Hung and Liu 2006; Knuckey et al. 2006). An optimum dewatering technique should be applicable to a wide range of microalgal strains, have high biomass recovery, and also be cost effective. It is therefore important to understand various technologies in cultivating and dewatering microalgae in order to maximize the production of microalgae at low cost (Harun et al. 2010).

Microalgae biotechnology has been developed for different commercial applications. As photosynthetic organisms, microalgae contain chlorophyll that can be used for food and cosmetic purposes (Spolaore et al. 2006). They can also be used in pharmaceutical industries, as some species of microalgae produce bioactive compounds such as antioxidants, antibiotics, and toxins (García-Casal et al. 2009). Besides, microalgae are used as nutrient supplements for human consumption due to their high protein, vitamin, and polysaccharide contents (Carballo-Cárdenas et al. 2003). Some microalgae species contain high levels of lipids, which can be extracted and converted into biofuels. The common methods that have been employed to extract the lipids from microalgae include oil press, solvent extraction, supercritical fluid extraction, and ultrasound (Pernet and Tremblay 2003; Andrich et al. 2005; Demirbaş 2008). The extracted lipids can further be transesterified into biodiesel (Vergara-Fernández et al. 2008). Furthermore, the waste biomass that is left behind after the lipids have been extracted could be converted to produce different types of biofuels such as biomethane (Vergara-Fernández et al. 2008), bioethanol (Tsukahara and Sawayama 2005), biohydrogen (Hankamer et al. 2007), and biodiesel (Ahmad et al. 2011). The objective of this chapter is to give an overview of marine microalgae as a prospective source for potential future biofuels.

1.2 Development of Biofuel Resources

In recent years, the use of liquid biofuels in the transport sector has shown rapid global growth, driven mostly by policies focused on achievement of energy security and mitigation of greenhouse gas emissions (Brennan and Owende 2010). First-generation biofuels, which have now attained economic levels of production, have been mainly extracted from food and oil crops including rapeseed oil, sugarcane, sugar beet, and maize (Brennan and Owende 2010) as well as vegetable oils and animal fats using conventional technology. It is projected that the growth in production and consumption of liquid biofuels will continue, but their impacts toward meeting the overall energy demands in the transport sector will remain limited due to competition with food and fiber production for the use of arable land, regionally constrained market structures, lack of well-managed agricultural practices in emerging economies, high water and fertilizer requirements, and a need for conservation of biodiversity (Brennan and Owende 2010).

Typically, the use of first-generation biofuels has generated a lot of controversy, mainly due to their impact on global food markets and on food security, especially with regard to the most vulnerable regions of the world economy. This has raised pertinent questions on their potential to replace fossil fuels and sustainability of their production (Moore 2008). For example, apart from the risk that higher food prices may have severe negative implications on food security, the demand for biofuels could place substantial additional pressure on the natural resource base, with potentially harmful environmental and social consequences. Currently, about 1% (14 million ha) of the world's available arable land is used for the production of biofuels, providing 1% of global transport fuels. Clearly, increasing that share to anywhere near 100% is impractical owing to the severe impact on the world's food supply and the large areas of production land required. The advent of second-generation biofuels is intended to produce fuels from the whole plant matter of dedicated energy crops or agricultural residues, forest harvesting residues, or wood processing waste (Moore 2008), rather than from food crops. However, the technology for conversion for the most part has not reached the scale for commercial exploitation, which has so far inhibited any significant exploitation.

Conditions for a technically and economically viable biofuel resource are that (Wang et al. 2008) it should be competitive or cost less than petroleum fuels, should require low to no additional land use, should enable air quality improvement (e.g., CO_2 sequestration), and should require minimal water use. Judicious exploitation of microalgae could meet these conditions and therefore make a significant contribution to meeting the primary energy demand while simultaneously providing environmental benefits (Wang et al. 2008).

1.3 Products from Marine Algae

Cyanobacteria are a diverse group of prokaryotic photosynthetic microorganisms that can grow rapidly due to their simple structures. They have been investigated for the production of different biofuels including biohydrogen, biodiesel, bioethanol, and biomethane. Cyanobacterial biofuel production is potentially sustainable. To make biofuel production economically viable, we also need to use remaining algal biomass for coproducts of commercial interests. It is possible to produce adequate cyanobacterial biofuels to satisfy the fast growing demand within the restraints of land and water resources (Parmar et al. 2011).

1.3.1 Bioethanol

Cyanobacteria and algae are capable of secreting glucose and sucrose. These simple sugars by anaerobic fermentation under dark conditions produce ethanol. If ethanol can be extracted directly from the culture media, the process may be drastically less capital and energy intensive than competitive biofuel processes. The process would essentially eliminate the need to separate the biomass from water and extract and process the oils.

Bioethanol could be very important to foster energy independence and reduce greenhouse gas emissions. A very strong debate on gradual substitution of petroleum by use of renewable alternatives such as biofuels dominates the political and economic agenda worldwide (Demain 2009). Alternative bioethanol production methods from cyanobacteria and microalgae need to be developed so that the costs associated with the land, labor, and time of traditionally fermented crops can be circumvented.

Ueda et al. (1996) have patented a two-stage process for microalgae fermentation. In the first stage, microalgae undergo fermentation in anaerobic environment to produce ethanol. The CO_2 produced in the fermentation process can be recycled in algae cultivation as a nutrient. The second stage involves utilization of remaining algal biomass for production of methane, by anaerobic digestion process, which can further be converted to produce electricity. Bush and Hall (2006) pointed out that the patented process of Ueda et al. (1996) was not commercially scalable due to the limitations of single cell free floating algae. They patented a modified fermentation process wherein yeasts, *Saccharomyces cerevisiae* and *S. uvarum*, were added to the algae fermentation broth for ethanol production (Hirayama et al. 1996; Bush and Hall 2006).

1.3.2 Biohydrogen

Hydrogen gas is seen as a future energy carrier by virtue of the fact that it is renewable, does not evolve the greenhouse gas CO_2 in combustion, and liberates large amounts of energy per unit weight in combustion. Biological hydrogen production has several advantages over hydrogen production by photoelectrochemical or thermochemical processes. Biological hydrogen production by photosynthetic microorganisms, for example, requires the use of a simple solar reactor such as a transparent closed box, with low energy requirements, whereas electrochemical hydrogen production via solar battery-based water splitting requires the use of solar batteries with high energy requirements.

Cyanobacteria can be used for the production of molecular hydrogen (H_2), a possible future energy carrier, and has been the subject of several recent reviews (Levin et al. 2004; Sakurai and Masukawa 2007; Tamagnini et al. 2007). Cyanobacteria are able to diverge the electrons emerging from the two primary reactions of oxygenic photosynthesis directly into the production of H_2, making them attractive for the production of renewable H_2 from solar energy and water. In cyanobacteria, two natural pathways for H_2 production can be used: first, H_2 production as a by-product during nitrogen fixation by nitrogenases, and second, H_2 production directly by bidirectional hydrogenase (Angermayr et al. 2009). Nitrogenases require ATP, whereas bidirectional hydrogenases do not require ATP for H_2 production, hence making them more efficient and favorable for H_2 production with a much higher turnover.

1.3.3 Biodiesel

Replacing all the transport fuel consumed in the United States with biodiesel will require 0.53 billion m^3 of biodiesel annually at the current rate of consumption. Oil crops, waste cooking oil, and animal fat cannot realistically satisfy this demand. For example, meeting only half the existing U.S. transport fuel needs by biodiesel would require unsustainably large cultivation areas for major oil crops. This is demonstrated in Table 1.1. Using the average oil yield per hectare from various crops, the cropping area needed to meet 50% of the U.S. transport fuel needs is calculated in column 3 (Table 1.1). In column 4 (Table 1.1), this area is expressed as a percentage of the total cropping area of the United States. If oil palm, a high-yielding oil crop can be grown, 24% of the total cropland will need to be devoted to its cultivation to meet only 50% of the transport fuel needs. Clearly, oil crops cannot significantly contribute to replacing petroleum-derived liquid fuels in the foreseeable future. This scenario changes dramatically, if microalgae are used to produce biodiesel. Between 1% and 3% of the total U.S. cropping area would be sufficient for producing algal biomass that satisfies 50% of the transport fuel needs (Table 1.1). The microalgal oil yields given in Table 1.1 are based on experimentally demonstrated biomass productivity in photobioreactors, as discussed later in this chapter. Actual biodiesel yield per hectare is about 80% of the yield of the parent crop oil given in Table 1.1.

In view of Table 1.1, microalgae appear to be the only source of biodiesel that has the potential to completely displace fossil diesel. Unlike other oil crops, microalgae grow

TABLE 1.1

Comparison of Some Sources of Biodiesel

Crop	Oil Yield (L/ha)	Land Area Needed (M ha)[a]	Percent of Existing U.S. Cropping Area[a]
Corn	172	1540	846
Soybean	446	594	326
Canola	1,190	223	122
Jatropha	1,892	140	77
Coconut	2,689	99	54
Oil palm	5,950	45	24
Microalgae[b]	136,900	2	1.1
Microalgae[c]	58,700	4.5	2.5

Source: Chisti, Y., *Biotechnol. Adv.*, 25(3), 294, 2007.
[a] For meeting 50% of all transport fuel needs of the United States.
[b] 70% oil (by wt) in biomass.
[c] 30% oil (by wt) in biomass.

extremely rapidly and many are exceedingly rich in oil. Microalgae commonly double their biomass within 24 h. Biomass doubling times during exponential growth are commonly as short as 3.5 h. Oil content in microalgae can exceed 80% by weight of dry biomass (Metting 1996; Spolaore et al. 2006). Oil levels of 20%–50% are quite common. Oil productivity, that is, the mass of oil produced per unit volume of the microalgal broth per day, depends on the algal growth rate and the oil content of the biomass. Microalgae with high oil productivities are desired for producing biodiesel.

Depending on species, microalgae produce many different kinds of lipids, hydrocarbons, and other complex oils (Banerjee et al. 2002; Metzger and Largeau 2005; Guschina and Harwood 2006). Not all algal oils are satisfactory for making biodiesel, but suitable oils occur commonly. Using microalgae to produce biodiesel will not compromise production of food, fodder, and other products derived from crops.

Potentially, instead of microalgae, oil-producing heterotrophic microorganisms (Ratledge 1993; Wynn and Ratledge 2002) grown on a natural organic carbon source such as sugar can be used to make biodiesel; however, heterotrophic production is not as efficient as using photosynthetic microalgae. This is because the renewable organic carbon sources required for growing heterotrophic microorganisms are produced ultimately by photosynthesis, usually in crop plants.

1.3.4 Coproducts

To make biofuels economically viable, using appropriate technologies, all primary components of algal biomass—carbohydrates, fats (oils), proteins, and a variety of inorganic and complex organic molecules—must be converted into different products, either through chemical, enzymatic, or microbial conversion means. The nature of the end products and of the technologies to be employed will be determined, primarily by the economics of the system, and they may vary from region to region according to the cost of the raw material (Willke and Vorlop 2004).

A large number of different commercial products have been derived from cyanobacteria and microalgae. These include products for human and animal nutrition, polyunsaturated fatty acids, antioxidants, coloring substances, fertilizers and soil conditioners, and a variety of specialty products such as bioflocculants, biodegradable polymers, cosmetics, pharmaceuticals, polysaccharides, and stable isotopes for research purposes.

1.4 Microalgae for Environmental Applications

Algae serve an advantage for effluent treatment via increasing performance of degradation, improving CO_2 balance, and lowering energy demand for oxygen supply in aerobic treatment stages. The role of algae was both to assimilate plant nutrients and to support bacteria with oxygen. Bacteria, in turn, were involved in degradation of organic material in wastewater, the same process utilized in activated sludge. Cyanobacteria were reported to be effectively used for treatment of organic pollutants from paper industry wastewater (Kirkwood et al. 2003). Besides wastewater containing organic compounds, phenol was efficiently removed with the aid of algae culture (Pinto et al. 2003). Conventional phenol degradation, using bacteria, requires addition of external carbon sources for growth while microalgae use CO_2 as carbon sources (photoautotrophically); hence, no addition of carbon sources to the system is needed (Hirooka et al. 2003). However, microalgae have limitations to be used in removal of organic compounds due to slower growth rate compared to bacteria (Lee 2001). Furthermore, microalgae are able to be used for removal of heavy metals in industrial wastewater. Brown alga, *Ascophyllum nodosum*, has proven to be the most effective algal species to remove metals of cadmium, nickel, and zinc from monometallic solutions compared to green and red algae (Shi et al. 2007). A similar study proved brown alga *Fucus vesiculosus* gave the highest removal efficiency of chromium (III) at high initial metal concentrations (Murphy et al. 2008). In accordance with this, biochemical composition of cell wall (alginate and fucoidan) of brown algae was found to have promising benefits for a biosorption process. Another algal species, *Scenedesmus obliquus*, was examined for degrading cyanide from mining process wastewaters (Gurbuz et al. 2009). It was observed that cyanide was degraded up to 90% after introduction of algae into the system. *Spirogyra condensata* and *Rhizoclonium hieroglyphicum* also employed as biosorption substrates to remove chromium from tannery wastewater (Gurbuz et al. 2009). The pH and concentration of algae were concluded to have significant effect on removal of chromium thus indicating potential of algae for removal hazardous heavy metals in wastewater (Gurbuz et al. 2009; Harun et al. 2010).

1.5 Conclusion

Producing biodiesel from algae has been touted as the most efficient way to make biodiesel fuel. Algal oil processes into biodiesel as easily as oil derived from land-based crops. The difficulties in efficient biodiesel production from algae lie not in the extraction of the oil but in finding an algal strain with a high lipid content and fast growth rate that is not too difficult to harvest and a cost-effective cultivation system (i.e., type of photobioreactor) that is best suited to that strain.

Algae are the fastest-growing plants in the world. Microalgae have much faster growth rates than terrestrial crops. Algae are very important as a biomass source. Different species of algae may be better suited for different types of fuel. Algae can be grown almost anywhere, even on sewage or saltwater, and do not require fertile land or food crops, and processing requires less energy than the algae provides. Algae can be a replacement for oil-based fuels, one that is more effective. Algae consume carbon dioxide as they grow, so they could be used to capture CO_2 from power stations and other industrial plant that would otherwise go into the atmosphere.

Acknowledgment

This research was supported by a grant from Marine Bioprocess Research Center of the Marine Biotechnology Program funded by the Ministry of Oceans and Fisheries, R&D/2004-6002, Republic of Korea.

References

Ahmad, AL, NH Yasin, CJC Derek, and JK Lim. 2011. Microalgae as a sustainable energy source for biodiesel production: A review. *Renewable and Sustainable Energy Reviews* 15 (1):584–593.

Andrich, G, U Nesti, F Venturi, A Zinnai, and R Fiorentini. 2005. Supercritical fluid extraction of bioactive lipids from the microalga *Nannochloropsis* sp. *European Journal of Lipid Science and Technology* 107 (6):381–386.

Angermayr, SA, KJ Hellingwerf, P Lindblad, and MJ Teixeira de Mattos. 2009. Energy biotechnology with cyanobacteria. *Current Opinion in Biotechnology* 20 (3):257–263.

Banerjee, A, R Sharma, Y Chisti, and UC Banerjee. 2002. *Botryococcus braunii*: A renewable source of hydrocarbons and other chemicals. *Critical Reviews in Biotechnology* 22 (3):245–279.

Brennan, L and P Owende. 2010. Biofuels from microalgae—A review of technologies for production, processing, and extractions of biofuels and co-products. *Renewable and Sustainable Energy Reviews* 14 (2):557–577.

Bush, RA and KM Hall. 2006. Process for the production of ethanol from algae. US Patent 7,135,308.

Carballo-Cárdenas, EC, PM Tuan, M Janssen, and RH Wijffels. 2003. Vitamin E (α-tocopherol) production by the marine microalgae *Dunaliella tertiolecta* and *Tetraselmis suecica* in batch cultivation. *Biomolecular Engineering* 20 (4):139–147.

Chisti, Y. 2007. Biodiesel from microalgae. *Biotechnology Advances* 25 (3):294–306.

Demain, AL. 2009. Biosolutions to the energy problem. *Journal of Industrial Microbiology & Biotechnology* 36 (3):319–332.

Demirbaş, A. 2008. Production of biodiesel from algae oils. *Energy Sources, Part A: Recovery, Utilization, and Environmental Effects* 31 (2):163–168.

García-Casal, MN, J Ramirez, I Leets, AC Pereira, and MF Quiroga. 2009. Antioxidant capacity, polyphenol content and iron bioavailability from algae (*Ulva* sp., *Sargassum* sp. and *Porphyra* sp.) in human subjects. *British Journal of Nutrition* 101 (01):79–85.

Gurbuz, F, H Ciftci, and A Akcil. 2009. Biodegradation of cyanide containing effluents by *Scenedesmus obliquus*. *Journal of Hazardous Materials* 162 (1):74–79.

Guschina, IA and JL Harwood. 2006. Lipids and lipid metabolism in eukaryotic algae. *Progress in Lipid Research* 45 (2):160–186.

Hankamer, B, F Lehr, J Rupprecht, JH Mussgnug, C Posten, and O Kruse. 2007. Photosynthetic biomass and H_2 production by green algae: From bioengineering to bioreactor scale-up. *Physiologia Plantarum* 131 (1):10–21.

Harun, R, M Singh, GM Forde, and MK Danquah. 2010. Bioprocess engineering of microalgae to produce a variety of consumer products. *Renewable and Sustainable Energy Reviews* 14 (3):1037–1047.
Heasman, M, J Diemar, W O'connor, T Sushames, and L Foulkes. 2000. Development of extended shelf-life microalgae concentrate diets harvested by centrifugation for bivalve molluscs—A summary. *Aquaculture Research* 31 (8–9):637–659.
Hirayama, S, H Nakayama, K Sugata, and R Ueda. 1996. Process for the production of ethanol from microalgae. US Patent 5,578,472.
Hirooka, T, Y Akiyama, N Tsuji et al. 2003. Removal of hazardous phenols by microalgae under photoautotrophic conditions. *Journal of Bioscience and Bioengineering* 95 (2):200–203.
Hung, MT and JC Liu. 2006. Microfiltration for separation of green algae from water. *Colloids and Surfaces B: Biointerfaces* 51 (2):157–164.
Kirkwood, AE, C Nalewajko, and RR Fulthorpe. 2003. Physiological characteristics of cyanobacteria in pulp and paper waste-treatment systems. *Journal of Applied Phycology* 15 (4):325–335.
Knuckey, RM, MR Brown, R Robert, and DMF Frampton. 2006. Production of microalgal concentrates by flocculation and their assessment as aquaculture feeds. *Aquacultural Engineering* 35 (3):300–313.
Lee, Y-K. 2001. Microalgal mass culture systems and methods: Their limitation and potential. *Journal of Applied Phycology* 13 (4):307–315.
Levin, DB, L Pitt, and M Love. 2004. Biohydrogen production: Prospects and limitations to practical application. *International Journal of Hydrogen Energy* 29 (2):173–185.
Metting Jr, FB. 1996. Biodiversity and application of microalgae. *Journal of Industrial Microbiology* 17 (5–6):477–489.
Metzger, P and C Largeau. 2005. *Botryococcus braunii*: A rich source for hydrocarbons and related ether lipids. *Applied Microbiology and Biotechnology* 66 (5):486–496.
Moore, A. 2008. Biofuels are dead: Long live biofuels (?)–Part one. *New Biotechnology* 25 (1):6–12.
Murphy, V, H Hughes, and P McLoughlin. 2008. Comparative study of chromium biosorption by red, green and brown seaweed biomass. *Chemosphere* 70 (6):1128–1134.
Parmar, A, NK Singh, A Pandey, E Gnansounou, and D Madamwar. 2011. Cyanobacteria and microalgae: A positive prospect for biofuels. *Bioresource Technology* 102 (22):10163–10172.
Pernet, F and R Tremblay. 2003. Effect of ultrasonication and grinding on the determination of lipid class content of microalgae harvested on filters. *Lipids* 38 (11):1191–1195.
Pinto, G, A Pollio, L Previtera, M Stanzione, and F Temussi. 2003. Removal of low molecular weight phenols from olive oil mill wastewater using microalgae. *Biotechnology Letters* 25 (19):1657–1659.
Ratledge, C. 1993. Single cell oils—Have they a biotechnological future? *Trends in Biotechnology* 11 (7):278–284.
Rosenberg, JN, GA Oyler, L Wilkinson, and MJ Betenbaugh. 2008. A green light for engineered algae: Redirecting metabolism to fuel a biotechnology revolution. *Current Opinion in Biotechnology* 19 (5):430–436.
Sakurai, H and H Masukawa. 2007. Promoting R&D in photobiological hydrogen production utilizing mariculture-raised cyanobacteria. *Marine Biotechnology* 9 (2):128–145.
Schenk, PM, SR Thomas-Hall, E Stephens et al. 2008. Second generation biofuels: High-efficiency microalgae for biodiesel production. *Bioenergy Research* 1 (1):20–43.
Sheehan, J, T Dunahay, J Benemann, and P Roessler. 1998. A look back at the US Department of Energy's aquatic species program: Biodiesel from algae. Vol. 328. In: NREL (ed.), National Renewable Energy Laboratory: US Department of Energy, Golden, CO, pp. 1–100.
Shi, J, B Podola, and M Melkonian. 2007. Removal of nitrogen and phosphorus from wastewater using microalgae immobilized on twin layers: An experimental study. *Journal of Applied Phycology* 19 (5):417–423.
Spolaore, P, C Joannis-Cassan, E Duran, and A Isambert. 2006. Commercial applications of microalgae. *Journal of Bioscience and Bioengineering* 101 (2):87–96.
Tamagnini, P, E Leitão, P Oliveira et al. 2007. Cyanobacterial hydrogenases: Diversity, regulation and applications. *FEMS Microbiology Reviews* 31 (6):692–720.

Tsukahara, K and S Sawayama. 2005. Liquid fuel production using microalgae. *Journal of the Japan Petroleum Institute* 48 (5):251–259.
Ueda, R., Hirayama, S., Sugata, K., and Nakayama, H., 1996. Process for the production of ethanol from microalgae. US Patent 5,578,472.
Ugwu, CU, H Aoyagi, and H Uchiyama. 2008. Photobioreactors for mass cultivation of algae. *Bioresource Technology* 99 (10):4021–4028.
Vergara-Fernández, A, G Vargas, N Alarcón, and A Velasco. 2008. Evaluation of marine algae as a source of biogas in a two-stage anaerobic reactor system. *Biomass and Bioenergy* 32 (4):338–344.
Wang, B, Y Li, N Wu, and CQ Lan. 2008. CO_2 bio-mitigation using microalgae. *Applied Microbiology and Biotechnology* 79 (5):707–718.
Widjaja, A, C-C Chien, and Y-H Ju. 2009. Study of increasing lipid production from fresh water microalgae *Chlorella vulgaris*. *Journal of the Taiwan Institute of Chemical Engineers* 40 (1):13–20.
Willke, T and K-D Vorlop. 2004. Industrial bioconversion of renewable resources as an alternative to conventional chemistry. *Applied Microbiology and Biotechnology* 66 (2):131–142.
Wynn, JP and C Ratledge. 2002. The biochemistry and molecular biology of lipid accumulation in oleaginous microorganisms. *Advances in Applied Microbiology* 51:1–51.

Section II

Biofuel Sources

2

Sources of Marine Biomass

Anong Chirapart, Jantana Praiboon, Rapeeporn Ruangchuay, and Masahiro Notoya

CONTENTS

2.1 Introduction

The majority of biomass used for biofuel production is from terrestrial sources. Land-based biomass for fuel can displace other agricultural activities and food production as well. The increasing demand for land has caused deforestation and shortages of food and resulted in increased food prices and civil unrest. Solutions to the increased demand for plant products for food, fuel, etc., include an increase in yields of all crops and a greater utilization of marine biomass. Nowadays, the increasing energy demand for public transport and the rise in oil prices are an intensifying interest in using green fuel for a sustainable future. The most common biofuels are ethanol produced from corn or sugarcane and biodiesel produced from a variety of oil crops such as soybeans and oil palm. Another potential type of biomass is marine biomass, which has the additional benefit. At present, marine biomass is considered as a potential biofuel feedstock. Marine biomass comprises of macro- and microalgae. Microalgae are potential sources of bio-oils, while macroalgae are potential sources of carbohydrates for fermentation or thermochemical-based conversions (Adams et al. 2011a). It can be anaerobically digested to produce methane, which, in turn, can be used to generate electricity, for heat or for transport (RCEP 2004). Plants in general are efficient solar energy converters and can create large amounts of biomass in a short term (Kraan 2013). However, marine biomass or marine algae are often an overlooked source when they potentially represent a significant source of renewable energy. Other points considered in view of marine biomass are increase in marine animal habitat size and fishery resources, protection of natural seaweed beds, and widening of seaweed feedstock farming areas, which are all beneficial for the conservation of biodiversity on coastal ecosystems and the global environment (Notoya 2007).

In addition, nowadays, lignocellulosic plant materials are preferable feedstocks for biofuels worldwide. The current microbial technologies for fermentation of the simple sugars in lignocellulose have yet to overcome the cost of the complex processes needed to release these sugars from recalcitrant polysaccharides (Wargacki et al. 2013). Distinct strategies are required to develop scalable and sustainable nonlignocellulosic biomass resources such as marine algae for use as next-generation feedstocks.

In recent years, marine algae have become a focus in both academic and commercial biofuel researches. Daroch et al. (2013) stated that micro- and macroalgae are the third-generation biofuel feedstocks after the first-generation biofuels made from edible feedstocks (corn, soybean, sugarcane, and rapeseed) and the second-generation biofuels from waste and dedicated lignocellulosic feedstocks. The applications of algae rely on the production of a high-volume and low-value biomass, which open opportunities to develop the culture of new commercial species (de Paula Silva et al. 2013). Marine algae are important photosynthetic organisms known to produce high oil and biomass yields. They can be cultivated within nonfreshwater sources including salt and wastewater, can be grown on nonarable land, do not compete with common food resources, and they very efficiently use water and fertilizers for growth (Hannon et al. 2010). Many algal species produce large amounts of lipids as storage products, as high as 50%–60% dry weight (Jones and Mayfield 2012), making algae a very productive potential source of biodiesel. Marine algae are also an alternative renewable energy source that can generate biogas with a considerable amount of methane (Kelly and Dworjanyn 2008; Bruhn et al. 2011; Gurung et al. 2012; Wei et al. 2013). Marine algae offer a vast renewable energy source for countries around the world that have a suitable coastline. Several species of seaweed are already farmed on a massive scale in the Far East (Kelly and Dworjanyn 2008).

Algae, which are known to lack lignin and have relatively small amounts of cellulose, could also be a promising feedstock for cellulosic ethanol production. Recent breakthroughs in converting diverse carbohydrates from seaweed biomass into liquid biofuels (e.g., bioethanol) through metabolic engineering have demonstrated potential for seaweed biomass as a promising, although relatively unexplored, source for biofuels (Wei et al. 2013). Although marine algae have been considered for bioenergy production (Bruton et al. 2009; Roesijadi et al. 2010; Sayadi et al. 2011), they have only attracted little attention by researchers compared to terrestrial-based feedstocks. Some genera investigated included several marine algae such as the microalgae *Tretraselmis*, *Dunaliella*, and *Chlorella* and macroalgae (seaweed) *Laminaria*, *Sargassum*, *Macrocystis*, *Ulva*, *Caulerpa*, and *Gracilaria* (Bird et al. 1990; Yokoyama et al. 2008; Adams et. al. 2011a,b; Bruhn et al. 2011; Borines et al. 2013; Kumar et al. 2013; Moheimani 2013; Poespowati et al. 2013). These algae are fast-growing photosynthetic plants that can grow to considerable sizes up to tens of meters in length, particularly the brown seaweed kelp. Growth rates of marine macroalgae far exceed those of terrestrial biomass (Kraan 2013).

In general, the structural characteristics of seaweed are based on complex sulfated polysaccharides (Lahaye and Robic 2007; Hernández-Garibay et al. 2011). The green algae such as *Ulva* species are reportedly composed of water-soluble ulvan and a sulfated, anionic polysaccharide that is mainly made of rhamnose, sulfate groups, and uronic acids (Lahaye and Robic 2007). Red seaweeds synthesize a great variety of sulfated galactans, which are the major components of their extracellular matrix; they are known as the source of unique sulfated galactans, such as agar, agarose, and carrageenan (Usov 2011). In brown seaweed, the most abundant sugars are mannitol and glucan (glucose polymers in the form of laminarin or cellulose). Ethanol production from glucan and mannitol yields approximately 0.08–0.12 wt ethanol/wt dry macroalgae (Roesijadi et al. 2010). Recently, Kraan (2013) stated that the potential of ethanol production from seaweed can be calculated; based on assumption of a carbohydrate content of 60% dry weight and a 90% conversion ratio to ethanol, fermentation of 1 g of sugar can yield 0.4 g ethanol; this will yield 0.22 kg or 0.27 L ethanol from 1 kg dry weight seaweed biomass. The chemical compositions from different seaweeds listed in Table 2.1 show the seaweeds' potential to be converted to biobase for production of bioenergy. However, their conversion efficiency depends on a variety of algal species, species of microorganism used, and conditions during the fermentation process (Kim et al. 2011; Cho et al. 2013; Wargacki et al. 2013; Wei et al. 2013). In this chapter, we review the progress of using marine algal biomass as a potential feedstock for the production of bioenergy.

2.2 Marine Biomass Species

2.2.1 Marine Macroalgae: Seaweed

There is a very large number of seaweeds around the world, belonging to several phylogenic groups; this includes green algae (Division: Chlorophyta), brown algae (Ochrophyta), and red algae (Rhodophyta). Production of seaweeds mainly from aquaculture has been centered in Asia, dominated by China, which accounted for 72% of the total production and 73% of the total value of cultured macroalgae (Roesijadi et al. 2010). Seaweeds are one source of marine biomass and are a potentially significant source of renewable energy (Mishima and Nguyen 2007; Roesijadi et al. 2010). Mishima and Nguyen (2007) suggested that the utilization of aquatic biomass from biofiltration for food processing, fertilizer, and other uses could cause many problems relating to food security in Asian countries, but such problems do not arise for the use of macroalgae in bioenergy processing since seaweed farms are able to act as biofilters and are able to remove nitrates and phosphates from the surrounding eutrophic inshore waters. Thus, seaweed biomass from nearshore cultivation can be harvested for biofuel feedstocks. In addition aquaculture-based production of seaweed has been focused mainly on the genera *Laminaria, Saccharina, Undaria, Porphyra, Eucheuma, Kappaphycus,* and *Gracilaria* (Buschmann et al. 2001; Bruton et al. 2009; Roesijadi et al. 2010; Msuya et al. 2014). The various brown seaweeds have been used for industrial applications since the early twentieth century. Currently, in many regions, attention has been focused on brown seaweed resources for energy production (Bruton et al. 2009; Roesijadi et al. 2010). Moreover, green seaweeds, in particular *Ulva* spp., are

TABLE 2.1

Seaweed Composition for a Variety of Species

Class	Seaweed	Carbohydrate Composition	Carbohydrates (%)	Crude Lipid (%)	Protein (%)	Fiber (%)	Ash (%)	References
Chlorophyceae	*C. lentillifera*	Starch, cellulose	11.8 ± 0.8	7.2 ± 0.3	9.7 ± 0.4	n.d.	46.4 ± 0.2	McDermid and Stuercke (2003)
(green seaweed)	*C. lentillifera*		59.27	0.86 ± 0.10	12.49 ± 0.3	3.17 ± 0.21	24.21 ± 1.7	Ratana-arporn and Chirapart (2006)
	U. lactuca		54.3	6.2	20.6	n.d	18.9	Kim et al. (2011)
	U. reticulata		55.77	0.75 ± 0.05	21.06 ± 0.42	4.84 ± 0.33	17.58 ± 2.0	Ratana-arporn and Chirapart (2006)
	U. intestinalis		48.97	0.62 ± 0.05	10.59 ± 0.13	5.87 ± 0.29	20.65 ± 0.14	Chirapart et al. (2014)
	Rhizoclonium sp.		57.4	n.d.	8.9	n.d	15.9	Chao et al. (1999)
	R. riparium		29.53	0.28 ± 0.01	12.77 ± 0.94	11.93 ± 0.16	37.62 ± 0.15	Chirapart et al. (2014)
	Rhizoclonium implexum		9.0 ± 0.4	8.3 ± 0.2	13.9 ± 0.6	n.d	35.5 ± 0.1	McDermid et al. (2007)
Phaeophyceae	*L. japonica*	Alginate, fucoidan, laminarin, mannitol, cellulose	51.9	1.8	14.8	n.d	31.5	Kim et al. (2011)
(brown seaweed)	*U. pinnatifida*		48.5	1.8	18.2	3.5	28	Cho et al. (2013)
	Sargassum vulgare		67.8	0.45	15.76	7.73	14.2	Marinho-Soriano et al. (2006)
	Sargassum fulvellum		39.6	1.4	13.0	n.d.	46.0	Kim et al. (2011)
	Sargassum longifolium		16.8 ± 0.7	8.2 ± 1.57	18.65 ± 1.21		n.d.	Narasimman and Murugaiyan (2012)
Rhodophyceae	*Gelidium amansii*	Agar, cellulose	77.2	1.1	13.1	n.d.	8.6	Kim et al. (2011)
(red seaweed)	*G. tenuistipitata*		54.89	0.26 ± 0.01	6.11 ± 0.04	4.96 ± 0.07	22.91 ± 0.09	Chirapart et al. (2014)
	Gracilaria salicornia		46.22	1.69 ± 0.07	16.28 ± 0.10	9.21 ± 0.11	13.49 ± 0.20	Chirapart et al. (2014)
	G. salicornia		24.6 ± 0.6	1.5 ± 0.1	3.9 ± 0.4	n.d.	49.5 ± 1.4	McDermid et al. (2007)
	Gracilaria cervicornis		63.12	0.43	22.96	5.65	7.72	Marinho-Soriano et al. (2006)
	Eucheuma denticulatum	Carrageenan, cellulose	28.0 ± 0.7	2.2 ± 0.2	4.9 ± 0.3	n.d.	43.6 ± 0.5	McDermid and Stuercke (2003)

n.d., no data.

being researched as potential renewable fuel feedstocks (Bruton et al. 2009; Bruhn et al. 2011). Among the seaweeds, the *Laminaria* spp. and *Ulva* spp. are the most important prospects from an energy perspective (Bruton et al. 2009). The vast majority of the seaweeds are collected for human consumption and for hydrocolloid production. Seaweed exploitation in Europe and the United States is restricted to manual and mechanized harvestings of natural stocks (Bruton et al. 2009; Roesijadi et al. 2010), and the majority of Asian seaweed resources are cultivated. The most common system in Europe to obtain seaweed biomass is by harvesting natural stocks in coastal areas with rocky shores and a tidal system, being collected largely as drift either by hand or by using rudimentary tools (Briand 1991; Critchley and Ohno 1998). The natural population of seaweeds is yet a significant resource.

2.2.1.1 Seaweed for Bioethanol Feedstock

Over the last decade, several groups have attempted to produce bioethanol from seaweed (Horn et al. 2000a,b; Shahbazi and Li 2006; Goh and Lee 2010; Candra et al. 2011; Fasahati and Liu 2012). Seaweed extract from *Laminaria hyperborea* has been used for fermentation and ethanol production (Horn et al. 2000a). Several strains of fermented organisms, in Table 2.2, have been applied to produce high yields of ethanol from the seaweed, and the most common microorganism for ethanol production is the yeast *Saccharomyces cerevisiae*. However, an optimization study comparing different producers suggested the superiority of *Pichia angophorae* over *S. cerevisiae* (Adams et al. 2011b). Fasahati and Liu (2012) recently proposed a process simulation for evaluating the large-scale production of ethanol from the brown seaweed *Saccharina japonica* in Korea. Thermal acid hydrolysis pretreatment has been used to improve the yield of saccharification in the seaweed (Horn et al. 2000a; Jang et al. 2012; Cho et al. 2013). Horn et al. (2000b) reported that ethanol produced from *L. hyperborea* extracts yielded 0.38 g ethanol/g mannitol. Mannitol and glucose utilized by the ethanogenic recombinant *Escherichia coli* KO11, which was cultured in *Laminaria japonica* hydrolysate supplemented with Luria–Bertani medium and hydrolytic enzymes, were able to produce 0.4 g ethanol/g carbohydrate (Kim et al. 2011). A report on long-term bioethanol production stated that the use of repeated-batch operation of surface-aerated fermentor, from the hydrolysate of *Sargassum sagamianum*, yielded ethanol of 0.386 g/g sugar (Yeon et al. 2011). Another report of the brown seaweed *Alaria crassifolia* hydrolysate gave an ethanol yield of 0.281 g/g sugar (Yanagisawa et al. 2011). The highest ethanol concentrations of 7.7 g/L (9.8 mL/L) and 9.42 g/L were produced from the brown seaweeds *S. japonica* (Jang et al. 2012) and *Undaria pinnatifida*, respectively (Cho et al. 2013).

Several studies reported on ethanol production from red and green seaweeds (Candra et al. 2011; Yanagisawa et al. 2011; Meinita et al. 2012; Trivedi et al. 2013; Chirapart et al. 2014). The biomass of *Ulva pertusa* is also used for ethanol production; the algal biomass gave an ethanol yield of 0.381 g/g reducing sugar (Yanagisawa et al. 2011). In recent works, a different enzymatic hydrolysis was used for efficient saccharification, and the fermentation of *Ulva fasciata* hydrolysate gave an ethanol yield of 0.45 g/g reducing sugar (Trivedi et al. 2013). The ethanol yield obtained is higher than the acid hydrolysis of *Ulva intestinalis* (9.98 μg ethanol/g glucose) and *Rhizoclonium riparium* (33.84 μg ethanol/g glucose) (Chirapart et al. 2014).

The biomass of the red seaweed *Gelidium elegans* has been used for ethanol production and obtained an ethanol yield of 0.376 g/g sugar (Yanagisawa et al. 2011). Fermentation of *Kappaphycus alvarezii* hydrolysates produced an ethanol yield of 0.21 g/g (Meinita et al. 2012). The enzymatic hydrolysates of *Gracilaria verrucosa* on fermentation with the

TABLE 2.2

Production of Ethanol from a Variety of Seaweed Feedstocks

Seaweed Sources		Fermented Organism	Yield	References
Chlorophyta	*U. pertusa*	*S. cerevisiae* IAM 4178	0.381 g/g	Yanagisawa et al. (2011)
	U. fasciata	*S. cerevisiae* MTCC No. 180	0.45 g/g reducing sugar	Trivedi et al. (2013)
	U. intestinalis	*S. cerevisiae* TISTR No. 5339	9.98 μg ethanol/g glucose	Chirapart et al. (2014)
	R. riparium	*S. cerevisiae* TISTR No. 5339	33.84 μg ethanol/g glucose	Chirapart et al. (2014)
Ochrophyta	*L. hyperborea*	*P. angophorae* SHF	0.43 g ethanol/g sugar	Horn et al. (2000a)
	L. hyperborea	*Zymobacter palmae*	0.38 g ethanol/g mannitol	Horn et al. (2000b)
	L. japonica	*E. coli* KO11	0.4 g ethanol/g sugars	Kim et al. (2011)
	Laminaria digitata	*P. angophorae* CBS 5830	167 mL ethanol/kg algae	Adams et al. (2011a)
	S. sagamianum	*Pichia stipitis*	0.386 g/g reducing sugar	Yeon et al. (2011)
	A. crassifolia	*S. cerevisiae* IAM 4178	0.281 g/g reducing sugar	Yanagisawa et al. (2011)
	S. japonica	*P. angophorae* KCTC 17574	7.7 g/L (9.8 mL/L) 33.3% theoretical yield	Jang et al. (2012)
	U. pinnatifida	*P. angophorae* KCTC 17574	9.42 g/L 27.3% theoretical yield from total CHO	Cho et al. (2013)
Rhodophyta	*E. cottonii (K. alvarezii)*	*S. cerevisiae*	4.6% ethanol in fermentation broth	Candra et al. (2011)
	K. alvarezii	*S. cerevisiae*	0.21 g/g galactose	Meinita et al. (2012)
	K. alvarezii	*S. cerevisiae* NCIM 3523	92.3% theoretical conversion	Khambhaty et al. (2012)
	G. elegans	*S. cerevisiae* IAM 4178	0.376 g/g	Yanagisawa et al. (2011)
	G. verrucosa	*S. cerevisiae* HAU strain	0.43 g/g sugars	Kumar et al. (2013)
	G. tenuistipitata	*S. cerevisiae* TISTR No. 5339	139.12 μg ethanol/g glucose	Chirapart et al. (2014)
	G. salicornia	*S. cerevisiae* TISTR No. 5339	1.43 μg ethanol/g glucose	Chirapart et al. (2014)

S. cerevisiae HAU strain produced an ethanol yield of 0.43 g/g sugars (Kumar et al., 2013). The fermentation of acid hydrolysates of *G. tenuistipitata* gave an ethanol yield of 139.12 μg ethanol/g glucose, while that of *G. salicornia* gave 1.43 μg ethanol/g glucose (Chirapart et al. 2014). The ethanol production reported for different seaweed feedstocks is given in Table 2.2.

2.2.1.2 Seaweed for Biogas Feedstock

Research on conversion of seaweed to biogas was started by the gas industry in the United States in 1974 (Flowers and Bird, 1990). The early works focused on the production of methane gas from the giant kelp, *Macrocystis pyrifera.* The potential of seaweed biomass as a source of methane has been reviewed (Gunaseelan 1997). In recent years, many attempts have been taken to improve the efficient conversion of seaweed biomass for biogas production (Vergara-Ferńandez et al. 2008; Nkemka and Murto 2010; Bruhn et al. 2011; Lee et al. 2012; Jard et al. 2013). An anaerobic digestion system has been approached as the best means of converting marine biomass feedstocks into energy (Flowers and Bird 1990). Bruton et al. (2009) stated that to achieve an energy output of the seaweed biomass from anaerobic digestion, it is assumed that one wet ton of seaweed yields 22 m^3 of methane with a gross calorific value of 39.8 MJ/m^3. The anaerobic digestion of *M. pyrifera*, *Durvillaea antarctica*, and their blend 1:1 (w/w) has been taken in a two-phase anaerobic digestion system, consisting of an anaerobic sequencing batch reactor and an upflow anaerobic filter (Vergara-Ferńandez et al. 2008). This system produced 70% of the total biogas and obtained the production of 180.4(71.5) mL/g dry algae/day. In addition, Tanisho and Notoya (2010) have studied on the production of fermentative biohydrogen from the mannitol of *S. japonica* (as *L. japonica*). They have been calculated obtaining ca. 80 kg mannitol and 16 Nm^3–H_2 from 1 ton wet weight of the cultured *S. japonica*.

A pilot-scale plant has been developed for biogas production from mixed *Laminaria* sp. and *Ulva* sp. with other organic waste (milk) and used as fermentation materials; the methane gas yield obtained is 22 m^3/ton for *Laminaria* sp. and 17 m^3/ton for *Ulva* sp. at a rate of about 0.6 ton seaweed (TS, 3%)/day (Matsui and Koike 2010). Biogas production from fresh and macerated *Ulva lactuca* can be yielded up to 271 mL CH_4/g VS, which is in the range of the methane production from cattle manure and land-based energy crops (Bruhn et al. 2011). Lee et al. (2012) reported that biogas production from *L. japonica* was increased by using a mixed culture with *Clostridium butyricum* and *Erwinia tasmaniensis*, with the total hydrogen and methane levels of 327.47% and 354.99% higher, respectively. Recently, a two-stage fermentation system with recycling of methane-fermented effluent was developed to produce H_2 and CH_4 from *L. japonica* (Jung et al. 2012). In another study, Gurung et al. (2012) evaluated energy recovery from brown algae, green algae, and fish viscera as substrates for methane production; the results showed that the CH_4 content of the biogas was approximately 70% for both substrates. The red seaweed *Palmaria palmata* has a methane production of 279 mL CH_4/g VS in batch reactor and 320 mL CH_4/g VS in semicontinuous reactor; the yields were higher than that of *S. latissima* under the same system (Jard et al. 2012). Methane production rate can be improved by codigesting macroalgae (*Ulva* sp.) with manure and waste-activated sludge (Costa et al. 2012). The biogas yields can also be obtained up to 20% from codigestion of seaweed biomass with digester sludge, after mechanical pretreatment (Tedesco et al. 2013). A recent work suggested that the red seaweed *P. palmata* is suitable for anaerobic digestion because of its low fiber content, low mineral concentration, and high volatile solid and carbohydrate content, which is validated by its high biochemical methane potential of 0.279 LCH_4/g VS (Jard et al. 2013). They also stated that *U. lactuca* seems to be a good substrate for anaerobic digestion because of its high biochemical methane potential (0.239 L/g VS). However, many species of seaweed have high sulfate content such as *Ulva* sp., which can be problematic for anaerobic digestion because of the competition between methanogens and sulfate-reducing bacteria (Peu et al. 2011). The comparison yield of methane gas produced from different seaweeds is given in Table 2.3.

2.2.1.3 Seaweed for Other Biofuel Products

Other potential biofuel products from seaweed also include bio-oil and biodiesel. Experimental works for the production of biodiesel from the green seaweed *Chaetomorpha linum* have been reported (Aresta et al. 2005a). The hydrothermal liquefaction and supercritical CO_2 extraction processes were studied; these methods obtained low yield at

TABLE 2.3
Methane Yields Produced from Some Different Seaweed Species

Sources	Fermenter	Yield	References
U. lactuca	Anaerobic digestion and direct combustion	271 mL CH_4/g VS	Bruhn et al. (2011)
Mixed *Gracilaria longissima*, *C. linum* (60:40)	Anaerobic digestion	0.375 dm^3 CH_4/g VS add	Migliore et al. (2012)
P. palmata	Batch reactor	279 mL CH_4/g VS	Jard et al. (2012)
P. palmata	Semicontinuous reactor	320 mL CH_4/g VS	Jard et al. (2012)
S. latissima	Batch reactor	210 mL CH_4/g VS	Jard et al. (2012)
S. latissima	Semicontinuous reactor	270 mL CH_4/g VS	Jard et al. (2012)
L. japonica	Anaerobic sequencing batch reactor	0.6 L CH_4/L/d	Jung et al. (2012)
L. japonica	Upflow anaerobic sludge blanket reactor	1.95 L CH_4/L/d	Jung et al. (2012)

4.5–8 kg/100 kg dry algae. The extraction of bio-oil from seaweed requires dry feedstock, which can be achieved using solar energy or process heat (Aresta et al. 2005b). The green tide alga *Enteromorpha prolifera* (reclassified now as *Ulva prolifera*) has been used for conversion to bio-oil by hydrothermal liquefaction in a batch reactor at temperatures of 220°C–320°C (Zhou et al. 2010). In the batch reactor, moderate temperature of 300°C with 5 wt% Na_2CO_3 and reaction time of 30 min led to the highest bio-oil yield of 23.0 wt%, and the higher heating values (HHVs) of bio-oils were around 28–30 MJ/kg. The advantage of the works is the hydrothermal liquefaction of *U. fasciata* using subcritical H_2O (300°C) as well as supercritical organic solvents CH_3OH and C_2H_5OH (300°C). The results revealed that use of alcoholic solvents significantly increased the bio-oil yield, and the bio-oil yield was 44% and 40% in case of liquefaction with CH_3OH and C_2H_5OH, respectively, whereas the bio-oil yield was 11% with H_2O (Singh et al. 2014). Thermal cracking of *E. prolifera* with vacuum gas oil gave rise to the maximum bio-oil yield of 90.5% at 300°C with a reaction time of 30 min; this process has the potential for industrial production of bio-oil from this alga (Song et al. 2014). In another work, a high lipid content of 12.72 mg/g dry wt in *C. linum* from the Orbetello lagoon in Italy was considered for biodiesel production (Borghini et al. 2014).

2.2.2 Marine Microalgae

Microalgae are microscopic photosynthetic organisms that are found in both marine and freshwater environments. Their photosynthetic mechanism is similar to land-based plants, but due to a simple cellular structure, and being submerged in an aqueous environment where they have efficient access to water, CO_2, and other nutrients, they are generally more efficient in converting solar energy into biomass (Wellinger 2009).

The most frequently used microalgae are Cyanophyceae (blue-green algae), Chlorophyceae (green algae), Bacillariophyceae (including the diatoms), and Chrysophyceae (including golden algae). Many microalgal species are able to switch from phototrophic to heterotrophic growth. Microalgae find uses as food and as live feed in aquaculture for production of bivalve molluscs; for juvenile stages of abalone, crustaceans, and some fish species; and for zooplankton used in aquaculture food chains. Therapeutic supplements from microalgae comprise an important market in which compounds such as β-carotene, astaxanthin, polyunsaturated fatty acid (PUFA) such as DHA and EPA, and polysaccharides such as β-glucan dominate (Pulz and Gross 2004).

The ability of microalgae to make lipid (as feedstock for biodiesel) or carbohydrates (for fermentation into ethanol or anaerobic digestion for methane production) is of great interest for fuel production. In the late 1970s, there were many researches on algal oil as a fuel product. This was based on the known ability of some microalgal species to accumulate large amounts of algal lipid, in particular under condition of nutrient (mainly N and Si) limitation (Dananandа et al. 2007).

There are several main groups of microalgae, which differ primarily in pigment composition, biochemical constituents, ultrastructure, and life cycle. The five groups that were of primary importance for biofuel productions are diatoms (class Bacillariophyceae), green algae (class Chlorophyceae), golden brown algae (class Chrysophyceae), prymnesiophyte (class Prymnesiophyceae), and eustigmatophyte (class Eustigmatophyceae) (Dananandа et al. 2007).

2.2.2.1 Green Algae

Green algae, often referred to as chlorophytes, are also abundant; approximately 8000 species are estimated to be in existence. This group has chlorophyll a and chlorophyll b. These algae use starch as their primary storage component. However, N deficiency promotes the accumulation of lipid in certain species. Green algae are the evolutionary progenitor higher plant and, as such, have received more attention than other groups of algae (Dananandа et al. 2007).

2.2.2.1.1 *Chlorella*

Chlorella is a unicellular species with a cell size in the range of 3–10 μm, which is ideal for combustion in a diesel engine. The fat content of *Chlorella* differs with strains and conditions, and it has been found that the content of the alga varies in the range 11%–63% (w/w). The reduction in nitrogen in the medium increases the lipid content in *Chlorella* strains, and the reported values were 63%–56% for *C. emersonii* and *C. minutissima*, 57.9% for *C. vulgaris*, and 29.2% for *C. pyrenoidosa*, respectively (Illman et al., 2000).

2.2.2.1.2 *Dunaliella*

The unicellular microalga *Dunaliella* naturally occurs in saline habitats and is well known to produce useful compounds, such as β-carotene (3%–5% [w/w]), glycerol, and lipids. *Dunaliella* contains 20.5% of fat and is easier to cultivate on a large scale; outdoor mass production system for *Dunaliella* has already been developed (Dananandа et al. 2007).

2.2.2.2 Prymnesiophytes

This group of algae, known as the haptophytes, consists of approximately 500 species. They are primarily marine organisms and can account for a substantial proportion of the primary productivity of tropical oceans. As with the diatoms and chrysophytes, fucoxanthin imparts a brown color to the cell and lipids, and chrysolaminarin is the major storage product. This group includes the coccolithophorids, which are distinguished by calcareous scales surrounding the cell wall (Dananandа et al. 2007).

2.2.2.3 Eustigmatophytes

This group represents an important component of the *picoplankton*, which are very small cells (2–4 μm in diameter). The genus *Nannochloropsis* is one of the few marine species in this class and is common in the world's oceans. Chlorophyll a is the only main pigment present in the cell, although several xanthophylls serve as accessory photosynthetic pigments.

2.2.2.4 Diatoms

Diatoms are among the most common and widely distributed groups of algae in existence; about 100,000 species are known. This group tends to dominate the phytoplankton of the oceans and has high levels of fucoxanthin, a photosynthetic accessory pigment. Several other main storage compounds of diatoms are lipids (TAGs) and an α-1,3-linked carbohydrate known as chrysolaminarin. A distinguishing feature of diatoms is the presence of a cell wall that contains substantial quantities of polymerized Si. This has an implication for media costs in a commercial production facility, because silicate is a relatively expensive chemical. On the other hand, Si deficiency is known to promote storage lipid accumulation in diatoms and thus could provide a controllable means to induce lipid synthesis in a two-stage production process. Another characteristic of diatoms that distinguishes them from most other algal groups is that they are diploid (having two copies of each chromosome) during vegetative growth; most algae are haploid (with one copy of each chromosome) except for a brief period when the cells are reproducing sexually. The main ramification of this form of strain development perspective is that it makes producing improved strains via classical mutagenesis and selection/screening substantially more difficult. As a consequence, diatom strain development programs must rely heavily on genetic engineering approaches (Dananand a et al. 2007).

Many species of marine microalgae accumulate large amounts of oil as shown in Table 2.4. The percentages vary with the type of algae; there are algal types that are comprised of up to 70% oil.

2.3 Marine Algal Cultivation

2.3.1 Marine Macroalgae

World production of seaweed was totally around 16,900 metric tons fw/year: 14,800 metric tons come from aquaculture and 1,100 metric tons fw come from fisheries; this included brown seaweed with 6,500 metric tons fw/year, red seaweed with 5,900 metric tons fw/year, and green seaweed with 30 metric tons fw/year (Paul et al. 2012). According to the report of FAO (2012) to date, only aquatic algae have been recorded globally in farmed aquatic plant production statistics. Global production has been dominated by marine macroalgae (seaweeds), grown in both marine and brackish waters. Aquatic algal production by volume increased at average annual rates of 9.5% in the 1990s and 7.4% in the 2000s. Cultivation has overshadowed production of algae collected from the wild, which accounted for only 4.5% of the total algal production in 2010. The estimated total value of farmed algae worldwide has been reduced for a number of years. The total value of farmed aquatic algae in 2010 is estimated at US$5.7 billion, while that for 2008 is now estimated at US$4.4 billion. The main area of global seaweed farmed production reported in year 2010 is given in Table 2.5. A few species dominate seaweed culture, with 98.9% of world production in 2010 coming from Japanese kelp (*S./L. japonica*) (mainly in the

TABLE 2.4

Oil Content of Marine Microalgae Expressed on a Dry Matter Basis

Taxonomy	Species	Oil Content (%)
Division: Chlorophyta		
Class: Chlorophyceae		
Order: Chlamydomonadales	*Dunaliella primolecta*	23.0[a], 23.1[b]
	Dunaliella salina	6.0–25.0[b]
	Dunaliella tertiolecta	16.7–71.0[b]
	Dunaliella bioculata	8[e]
	Dunaliella sp.	17.5–67.0[b]
Order: Sphaeropleales	*Neochloris oleoabundans*	35.0–54.0[a], 29.0–65.0[b]
Class: Chlorodendrophyceae		
Order: Chlorodendrales	*Tetraselmis maculata*	3[e]
	Tetraselmis suecica	15.0–32.0[c]
	T. suecica sp. (F&M-M33)	8.5[d]
	T. suecica (F&M-M35)	12.9[d]
	Tetraselmis sp. (F&M-M34)	14.7[d]
Class: Trebouxiophyceae		
Order: Chlorellales	*Nannochloris* sp.	20.0–35.0[a], 20.0–56.0[b], 20.0–63.0[c]
Order: Prasiolales	*Stichococcus* sp.	33.0[c]
Division: Dinophyta		
Class: Dinophyceae		
Order: Dinotrichales	*Crypthecodinium cohnii*	20.0[a]
Division: Ochrophyta		
Class: Coscinodiscophyceae		
Order: Chaetocerotales	*Chaetoceros muelleri* (F&M-M43)	33.6[d]
	C. muelleri	33.6[b]
	Chaetoceros calcitrans (Cs 178)	39.8[d]
	C. calcitrans	14.6–16.4[b]
Order: Thalassiosirales	*Thalassiosira pseudonana* (CS 173)	20.6[d]
	T. pseudonana	21.0–31.0[c]
	Skeletonema costatum	21.0[d]
	Skeletonema sp. (CS 252)	31.8[d]
Class: Eustigmatophyceae		
Order: Eustigmatales	*Nannochloropsis oculata*	22.7–29.7[b]
	Nannochloropsis sp.	31.0–68.0[d], 12.0–53.0[b]
	Nannochloropsis sp. (CS 246)	29.2[d]
	Nannochloropsis sp.(F&M-M26)	29.6[d]
	Nannochloropsis sp.(F&M-M27)	24.4[d]
	Nannochloropsis sp.(F&M-M24)	30.9[d]
	Nannochloropsis sp.(F&M-M29)	21.6[d]
	Nannochloropsis sp.(F&M-M28)	35.7[d]
Division: Haptophyta		
Class: Pavlovophyceae	*Pavlova salina*	30.9[b]
Order: Pavlovales	*P. salina* (CS 49)	30.9[d]
	Pavlova lutheri	35.5[b]
	P. lutheri (CS 182)	35.5[d]

(Continued)

TABLE 2.4 (*Continued*)

Oil Content of Marine Microalgae Expressed on a Dry Matter Basis

Taxonomy	Species	Oil Content (%)
Class: Cocclithophyceae		
Order: Isochrysidales	*Isochrysis galbana*	7.0–40.0[b]
	Isochrysis sp.	25.0–33.0[a], 7.1–33.0[b,c]
	Isochrysis sp. ((T-ISO) CS 177)	22.4[d]
	Isochrysis sp. (F&M-M37)	27.4[d]
Order: Prymnesiales	*Prymnesium parvum*	22–38[e]
Division: Ochrophyta		
Class: Bacillariophyceae		
Order: Bacillariales	*Cylindrotheca* sp.	16.0–37.0[a,c]
	Hantzschia sp. (DI-160)	66.0[c]
	Nitzschia sp.	45.0–47.0[a,c]
Order: Naviculales	*Phaeodactylum tricornutum* (F&M-M40)	18.7[d]
	P. tricornutum	20.0–30.0[a], 18.0–57.0[b], 31.0[c]
Division: Rhodophyta		
Class: Porphyridiophyceae		
Order: Porphyridiales	*Porphyridium cruentum*	9.5[b], 9–14[e]

[a] Harun et al. (2010).
[b] Amaro et al. (2011).
[c] Demirbas (2011).
[d] Singh and Gu (2010).
[e] Singh et al. (2011).

TABLE 2.5

Main Areas of Global Seaweed Farmed Production Reported in Year 2010

Area	Percent of Global Cultivation	Production (Million Tons)
China	58.4	11.1
Indonesia	20.6	3.9
Philippines	9.5	1.8
Republic of Korea	4.7	0.90
Democratic People's Republic of Korea	2.3	0.44
Japan	2.3	0.43
Malaysia	1.1	0.21
United Republic of Tanzania	0.7	0.13
Total	99.6	18.92

Source: Adapted from FAO, *The State of World Fisheries and Aquaculture 2012*, FAO, Rome, Italy, 2012, 209pp.

coastal waters of China), *Eucheuma* seaweeds (a mixture of *K. alvarezii*, formerly known as *Eucheuma cottonii*, and *Eucheuma* spp.), *Gracilaria* spp., nori/laver (*Porphyra spp.*), wakame (*U. pinnatifida*), and unidentified marine macroalgal species (3.1 million tons, mostly from China). The remainder consists of marine macroalgal species farmed in small quantities (such as *Sargassum* and *Caulerpa* spp.). The cultivation of aquatic algae is practiced in far fewer countries. Only 31 countries and territories are recorded with algal farming production in 2010, and 99.6% of global cultivated algal production comes from just eight countries: China (58.4%, 11.1 million tons), Indonesia (20.6%, 3.9 million tons), the Philippines (9.5%, 1.8 million tons), the Republic of Korea (4.7%, 901,700 tons), Democratic People's Republic of Korea (2.3%, 444,300 tons), Japan (2.3%, 432,800 tons), Malaysia (1.1%, 207,900 tons), and the United Republic of Tanzania (0.7%, 132,000 tons).

However, marine biomass is often an overlooked source and potentially represents a significant source of renewable energy because it was a source of carbohydrate and the average photosynthetic efficiency of seaweeds is higher than the terrestrial biomass (Kraan 2013). The red algal species such as *Gelidium* is a raw material in the agar process, in which the waste products from processing are converted into bioethanol. Also, green algal species such as the *Ulva* sp. has high levels of the polysaccharide ulvan, while the brown alga kelp contains 50%–60% carbohydrates dry weight and have been used in ethanol and methane productions (Kraan 2013). However, cultivation of seaweed for biomass is mostly in Asia, which has a long history of commercial seaweed farming. In the area, intensive cultivation belongs to the species of the brown algae (*L. japonica/S. japonica* and *U. pinnatifida*), the red algae (*Porphyra, Eucheuma, Kappaphycus, Gracilaria*), and the green algae (*Monostroma* and *Ulva/Enteromorpha*) (Ohno 1993; Nang and Dinh 1998; Wikfors and Ohno 2001; Lüning and Pang 2003).

2.3.1.1 Cultivation Method

The methods of seaweed cultivation are greatly varied. A seaweed species for cultivation is chosen according to the location of a farm and cultivation facilities (in the open sea, on the land, in the cold waters of a temperate zone, or in warm waters of the tropics) and on the productivity and adaptability of a species (Titlyanov and Titlyanov 2010). At present, there are six major types of seaweed cultivation: net method, long-line or raft method, monoline method, bottom planting method, pond method, and tank method.

2.3.1.1.1 Net Method

The net method is suitable for the foliose or thin seaweed such as *Ulva* (as *Enteromorpha*), *Monostroma*, and *Porphyra*, which are cultured in opened areas (Ohno 1993; Oohusa 1993). The method mostly starts from spore collection from hatchery or in nature. It was called a two-step farming: nursing and cultivating (Santelices 1999). Cultivation net is made of synthetic fiber in the size of $1.2 \times 18\ m^2$ and with mesh size of 15 cm. Spores attach on the net by setting the environments and then raring in nursing ground until seedlings grow up. The seedling net is separated to culture in a cultivation area. There are two types of net cultivation: fixed and floating type. In the first type, the net is fixed with a bamboo pole in the intertidal zone, which will expose the net to the air during low tidal levels. In the latter type, the net is supported by buoys to float in near water surface. The net method is available to classify seed source into two types: artificial spore collection method and natural spore collection method.

2.3.1.1.1.1 Artificial Spore Collection Method The artificial spore collection can be done by stimulating the mature thalli to release spores in hatchery condition. This method is preferably done in cultures of the green algae *Ulva* and *Monostroma*. Mature frond is selected to stimulate spore release. The selected frond is placed on setting plates and kept in a dark condition in the summer. Then the algal materials on the setting plates are cultured under low light intensity of 20–50 μmol photons/m^2/s for 1 month. Stimulation of spore releasing is performed at a higher intensity, 100 μmol photons/m^2/s, using fluorescence lamp, and then spores will be released into a culture solution. The released spore solution is poured to the tank with the soaked net, and then after a day, the net is transferred to the nursing ground to culture until seedling grows to 1–2 cm. The nets are separated to suitable ground at salinity less than 30 ppt and are fixed by bamboo poles in shallow waters or floating systems in the deep water zone. *Monostroma* is usually harvested for three or four times a year (in Okinawa, Japan); the crop is 1–1.2 kg dw/m^2 of net/year, while *Ulva* is harvested twice a year in winter and spring. The total yield ranges from 0.5 to 1 kg dw/m^2 of net per season (Titlyanov and Titlyanov 2010).

In *Porphyra*, the microthalli of conchocelis and the sporophyte stage of *Porphyra* serve as planting material for this seaweed. Conchocelis is grown from carpospores inside oyster shells in special tanks. Mature conchocelis in oyster shells is stimulated by decreased temperature of seawater until conchospore is released and attached on the cultivation net and then transported to nursing ground with several layers overlapping near the coast. Several days later, the conchospores attach to nets and grow into germlings; the nets are transferred to cultivate in one layer in the sea for several months (Titlyanov and Titlyanov 2010).

2.3.1.1.1.2 Natural Spore Collection Method This method should be interacted with the preceding method. The method is to avoid complex multistep in the hatchery. In shallow waters, where most seaweed crops are grown, an overlapping net is placed to collect spores for several days, in which environmental conditions may change very rapidly. After the germlings grow 2–3 cm long, the net is separated to transfer in a cultivation area.

2.3.1.1.2 Long-Line or Raft Method

This long-line or raft method is preferred for growing large seaweeds such as the kelps *Laminaria* and *Undaria*, in which seedling lines of 1–4 m long hang on cultured lines 70–140 m long and place in deep water zones (Paul et al. 2012). The culture method consists of several steps (Santelices 1999). In the first step, the culture starts from stimulating mature sporophylls to release spores on the palm strings on the frames and attachment is delayed by 25 days. The next step comprises of small-size developmental stages, which are often farmed in nursery houses. Then the frames of cooled houses are rearing at 6°C–8°C, with a controlled irradiance initially at 20 μmol photons/m^2/s and then increased by 20% every week until seedlings grow up as young sporophytes, 2–8 cm long. The sporophytes are fixed to the strings by holdfasts; the strings are cut into pieces and tied to culture ropes. The lines are supported by buoys as distance between the lines. The harvesting is done 10 or 20 months after planting, depending on species and technique of cultivation (Titlyanov and Titlyanov 2010). However, Notoya (2007) has demonstrated that the branches of some *Sargassum* species are released from the upstream of the ocean current along the coasts of Japan Sea, and several weeks or months after, they are caught at the coast of the downstream, and that there is no need for a culture system. However, there is a need for a high-performance calculation and simulation system for the coastal ocean current.

2.3.1.1.3 Monoline Method

The monoline method is applied in shallow waters for red seaweeds such as *Eucheuma* and *K. alvarezii*, whose life cycle is rather complex. The propagation by thallus fragment is a commercial way. Reef areas, with depth of 0.6–1 m, are a good locality for growing those seaweeds. Stocking density of 50–100 g is tied to the monoline of 10.5 m long. Their growth rates are higher above 5% per day. In Asian countries, for example, Indonesia, Malaysia,

Japan, China, India, and Vietnam, the line is hung on the floating rafts in more deeper zones in open seas and the growth rate reaches to 10% day (Trono 1998, 1999; Bindu and Levine 2011).

In *Gracilaria* spp., plants are sometimes cultivated using racks made of bamboo poles, with ropes stretched between poles. Branches of *Gracilaria* are cut and tied to the main ropes. If *Gracilaria* is cultivated at the sites where a natural population of this seaweed grows, spores may be directly inoculated on special string structures. Later, large sporelings are transplanted onto cultivation ropes. The plants on a rope are distributed and are repaired by the time of harvesting (Santelices and Doty 1989; Titlyanov and Titlyanova 2010).

2.3.1.1.4 *Bottom Planting Method*

This method is suitable for seaweeds that have small shrubs and long branches such as *G. chilensis*. Tidal mudflats are useful for this type of method. The cultivation is done in shallow lagoons and bays that are protected from winds on sandy or muddy sand bottoms. The thalli of *Gracilaria* are inserted in the soil by prongs, with the ends separated like "Y." These are protected by covering with a 10 mm mesh nylon to promote optimal plant growth. These are also banded to pebbles and then spread out on the bottom and covered with a net until the shrub erects (Critchley 1998). Seaweeds can be all harvested several times after planting by season. For subtidal areas in southern Chile, it has been established that *Gracilaria* production can reach 91–149 tons dw/ha/year (Buschmann et al. 2001). This method is economically profitable due to its very low cost.

2.3.1.1.5 *Pond Method*

The best species or forms of unattached seaweeds for this method are those vegetating all year round such as *Gracilaria* and *Caulerpa*. A pond size that is no larger than about one hectare in area is easiest to manage. Pieces of the seaweed thalli are spread to the bottom; the number of the pieces is estimated so as to occupy the whole area of the culture site by the time of harvesting. The seaweeds are harvested several times a number of months after planting. In *Gracilaria*, the crop is collected completely or partially, one-third or one-half of the crops in every 4–6 weeks. What is left is used to provide material for the next crop (Titlyanov and Titlyanova 2010). However, only some species of *Gracilaria* can be cultured in the pond due to optimum condition for cultivation based on low water motion. *G. tenuistipitata*, *G. parvispora*, and *G. blodgettii* are species that thrive in pond conditions. The salinity of the seawater about 25 ppt, temperature of 20°C–25°C, and water depth of 20–30 cm during spring to early summer and 60–80 cm in summer with peak irradiance are maintained (Shang 1976). The production is average about 1.3 tons dw/ha/year (Titlyanov and Titlyanova 2010).

The majority of the biomass of *Caulerpa* is from Philippines. The production of *Caulerpa* in the country was reported to be 4252 metric tons in year 2004 (Roesijadi et al. 2010). *Caulerpa lentillifera* is the most popular species for pond culture using semi-intensive method. In the *Caulerpa* farming, changes in productions of the seaweed are influenced by many factors, primarily on the hydrological and hydrochemical conditions of a site. Ponds used for *Caulerpa* production are closed to the sea to maintain high salinity (preferred salinity ranges from 25 to 30 ppt), which have a mud substrate and should be away from pollution sources, and freshwater influence is recommended (Nana et al. 2013). The pond culture may be supplied with a slow running mixture of seawater and river water, or seawater from ponds or tanks, in which marine animals are cultivated (Titlyanov and Titlyanova 2010; Poespowati et al. 2013). Changing seawater in *Caulerpa* culture is performed more often than in systems for *Gracilaria* (Titlyanov and Titlyanova 2010). In case the nutrient content is low, organic and inorganic fertilizers are added. Light penetrated in the culture ponds may be controlled using plastic mesh to shade the cultivated seaweeds (Titlyanov and Titlyanova 2010).

2.3.1.1.6 *Tank Method*

This method is done only in special valuable seaweeds in pharmaceutical or premium food species of *Ulva*, *Gracilaria*, and *Porphyra* (Paul et al. 2012). This type of culture has not attracted private investors because it is not profitable due to higher cost of additional pumping, nutrients, and CO_2 (Martínez and Buschmann 1996). However, cultivation in the tank has role for reproductive cell collection and spore stimulation. This method is used to culture or keep the spore phase of *Ulva* spp., *Monostroma* spp., *Porphyra*, and kelps such as *Laminaria* spp. and *U. pinnatifida* (Paul et al. 2012). This system is highly productive, with biomass production almost and over 50 kg wet wt/m^2/year. It is possible to coculture with fish to reduce the negative impact of fish waste, and most of the costs for cultivating the algae are covered by the operational costs of the fish farm, which turns the whole system both economically and ecologically profitable (Martínez and Buschmann 1996).

2.3.1.2 Integrated Aquaculture of Seaweed and Marine Animal for Biofiltration and Biofuel Feedstock

In recent years, the idea of integrated aquaculture has been often considered a mitigation approach against the excess nutrients/organic matter generated by intensive aquaculture activities (Chopin and Robinson 2004; Barrington et al. 2009). Interestingly, this practice has been defined based on pilot studies in marine habitats involving joint aquaculture of fed species, usually fish, together with extractive species such as bivalves and/or macroalgae (Barrington et al. 2009). Integrated multitrophic aquaculture (IMTA) can also allow an increase in production capacity (for harvesting) of a particular site when regular options have established limitations. Sometimes, the more general term *integrated aquaculture* is used to describe the integration of monocultures through water transfer between organisms (Neori et al. 2004). For all intents and purposes, however, the terms *IMTA* and *integrated aquaculture* differ primarily in their degree of descriptiveness; these terms are sometimes interchanged (Barrington et al. 2009). The IMTA systems can be land-based or open water systems, marine or freshwater systems, and may comprise several species combinations (Neori et al. 2004). Some IMTA systems have included such combinations as shellfish/shrimp, fish/seaweed/shellfish, fish/shrimp, and seaweed/shrimp (Troell et al. 2003). Some seaweed species have been reported by our researches, Kimura et al. (2007), Carton et al. (2010), and Carton et al. (2011). Table 2.6 shows the genera of particular interest and those with high potential for development in the IMTA systems in the world's marine temperate waters.

In the tropical waters, the bulk of the production comes from farming seaweeds, mussels (clams), and oysters in shallow coastal waters and the rest from production in lagoons and in land-based ponds. Seaweeds and molluscs, which dominated tropical coastal aquacultures before, have now been accompanied by modern fish cage aquaculture and other open water practices. Traditional integrated open water mariculture systems, located principally in China, Japan, and South Korea, also have a long history (Troll 2009). These operations have consisted of fish net pens, shellfish, and seaweed placed next to each other in bays and lagoons (Neori et al. 2004). In the tropics, it is more common to find small-scale farmers practicing integration with seaweeds in ponds (Chirapart and Lewmanomont 2004; Dubi et al. 2005; Nana et al. 2013). Small-scale culture of the

TABLE 2.6

Genera of Interest and Those with High Potential for Development in IMTA Systems in the World's Marine Temperate Waters

Seaweed		Fish	Echinoderms	Molluscs	Polychaetes	Crustaceans
Green	*Ulva*	*Salmon*	*Strongylocentrotus*	*Haliotis*	*Nereis*	*Penaeus*
Brown	*Laminaria*	*Oncorhynchus*	*Paracentrotus*	*Crassostrea*	*Arenicola*	*Homarus*
	Saccharina	*Scophthalmus*	*Psammechinus*	*Pecten*	*Glycera*	
	Saccorhiza	*Dicentrarchus*	*Loxechinus*	*Argopecten*	*Sabella*	
	Undaria	*Gadus*	*Cucumaria*	*Placopecten*		
	Alaria	*Anoplopoma*	*Holothuria*	*Mytilus*		
	Ecklonia	*Hippoglossus*	*Stichopus*	*Choromytilus*		
	Lessonia	*Melanogrammus*	*Parastichopus*	*Tapes*		
	Durvillaea	*Paralichthys*	*Apostichopus*			
	Macrocystis	*Pseudopleuronectes*	*Athyonidium*			
Red	*Gigartina*	*Mugil*				
	Sarcothalia					
	Chondracanthus					
	Callophyllis					
	Gracilaria					
	Gracilariopsis					
	Porphyra					
	Chondrus					
	Palmaria					
	Asparagopsis					

Source: Adapted from Barrington, K. et al., Integrated multi-trophic aquaculture (IMTA) in marine temperate waters, in: Soto, D. (ed.), *Integrated Mariculture: A Global Review*, FAO Fisheries and Aquaculture Technical Paper. No. 529, FAO, Rome, Italy, 2009, pp. 7–46.

gracilarioid *Hydropuntia fisheri* (as *Gracilaria fisheri*) and *G. tenuistipitata* has been taken seasonally in earthen ponds using shrimp farm effluent (Chirapart and Lewmanomont 2004), while integrated pond culture of finfish, shellfish, and seaweed (*Ulva reticulata* and *Gracilaria crassa*) was developed in Tanzania (Dubi et al. 2005). The *Ulva* and *Gracilaria* have been considered the most likely candidates for large-scale seaweed biomass production; in particular, the integrated culture system of the *Ulva* has proved easy to grow (Bolton et al. 2009). For the *Gracilaria, G. salicornia* (previously identified as *G. crassa*) was used as a biofilter of fishpond effluent waters in an integrated finfish–shellfish–seaweed land-based system (Msuya and Neori 2002). In addition, the integrated aquaculture of seaweeds (*U. lactuca* and *G. arcuata*) and marine fish (*Oreochromis spilurus*) was reported in Saudi Arabia for biomass production and inorganic nutrient bioremediation capabilities (Al-Hafedh et al. 2012). Other experiments on *Gracilaria* farming have been conducted to evaluate the seaweed as a biofilter of fishpond effluents and as an alternative species for cultivation in Tanzania, to produce agar and for other uses in value addition (Msuya et al. 2014).

Recently, there have been several attempts to grow other seaweeds for an alternative species of biofuel production; this seaweed is of less interest in general. There have been reports on intensive culture of the filamentous green algae (*Cladophora coelothrix, C. patentiramea, C. indica,* and *C. linum*), the so-called *green tide algae* (de Paula Silva et al. 2008, 2012, 2013). The green tide algal species had an ability to utilize alternative carbon sources under intensive culture. In situ growth rates of the *green tide* algae have broad applications across the environmental variables that characterize tropical pond-based aquaculture (de Paula Silva et al. 2008). In the intensive culture system, the *green tide algae* provide a bioremediation for both air (CO_2) and waste water (nitrogen, phosphorous, metals, and trace elements) and an opportunity to utilize this biomass for bioenergy products (de Paula Silva et al. 2013). Another report on preliminary study of *Caulerpa taxifolia* cultured in indoor aquarium with ornamental saltwater fish farming showed high growth rate (361.86%) and had a potential as a feedstock for production of bioenergy (Poespowati et al. 2013).

2.3.2 Marine Microalgal Culture

Microalgae are currently cultivated commercially for human nutritional products around the world in several dozen small- to medium-scale production systems, producing a few tens to several hundreds of tons of biomass annually. The main algal genera currently cultivated photosynthetically for various nutritional products are *Spirulina, Chlorella, Dunaliella,* and *Haematococcus.* Of these, only *Dunaliella* is predominantly a marine species. These are generally cultivated for extraction of high-value components such as pigments or proteins. The advantages of culturing microalgae as a source of biomass are many (Vonshak 1990):

- Microalgae are considered to be a very effective biological system for harvesting solar energy for the production of organic compounds via photosynthetic process.
- These are nonvascular plants lacking complex reproductive organ, making the entire biomass available for harvest and use.
- Many species of microalgae can be induced to produce particularly high concentrations of chosen, commercially valuable compounds such as proteins, carbohydrates, lipids, and pigments.

- Microalgae undergo a simple cell division that enables them to complete their life cycle faster compared to higher plants. This allows a more rapid development and demonstration of production process than with agricultural crops.
- For many regions suffering low productivity due to poor soil or shortage of sweet water, farming of microalgae that can be grown using sea or brackish water may be almost the only way to increase productivity, securing a basic protein supply.

Microalgal biomass systems can be adapted to various levels of operational or technological skills, from simple, labor-intensive production units to fully automated systems requiring high investments.

The studies on the growth and chemical composition of microalgae have revealed that many microalgae produce lipids as the primary storage molecule (Cohen 1999). Microalgae, like higher plants, produce storage lipid in the form of triacylglycerols (TAGs). Although TAGs could be used to produce a wide variety of chemicals, many researchers focused on the production of fatty acid methyl esters (FAMEs), which can be used as a substitute for fossil-derived diesel fuel. Biodiesel can be synthesized from TAGs via a simple transesterification reaction in the presence of acid or base. Microalgae exhibit properties that make them well suited for use in a commercial-scale biodiesel production facility. Many species exhibit rapid growth and high productivity, and many microalgal species can be induced to accumulate substantial quantities of lipid, often greater than 60% of their biomass. Microalgae can grow in saline water that are not suitable for agricultural irrigation or consumption by human and animals. The growth requirements are very simple, primarily carbon dioxide (CO_2) and water, although the growth rates can be accelerated by sufficient aeration and the addition of inorganic and organic nutrients (Dananganda et al. 2007).

Candidate microalgal species that exhibited characteristics desirable for a commercial production could be used. The desert regions were attractive areas in which to locate microalgal-based biofuel production facilities. The commercial production strain should have certain qualities for its mass production and are as follows. These characteristics included the ability of the strains to grow rapidly and have high lipid productivity when growing under high light intensity, high temperature, and in saline waters indigenous to the area in which the commercial production facility is located. In addition, because it is not possible to control the weather in the area of the ponds, the best strain should have good productivity under fluctuating light intensity, temperature, and salinity (Dananganda et al. 2007).

For the commercial scale-up, the strain should be able to grow in varied ranges of cultural conditions (pH, temperature, light intensity, salinity, and different media) with a considerable production of biocrude. And the effect of environmental variables and culture conditions on the growth and lipid composition of the selected strains has to be examined. No one algal strain was identified that exhibited the optimal properties of rapid growth and high lipid production. Many microalgae can be induced to accumulate lipid under condition of nutrient deprivation (Dananganda et al. 2007). There are many reports on the higher levels of lipid accumulation in microalgae induced by nitrogen (N) starvation and silica (Si) depletion in diatoms. Unlike N, Si is not a major component of cellular molecules; therefore, it was thought that the Si effect on lipid production might be less complex than the N effect.

Cost-effective production of biodiesel requires not only development of microalgal strains with optimal properties of growth and lipid production but also a production system and a clear understanding of the available resources (land, water, power, etc.) required on outdoor microalgal mass culture for production of biodiesel, as well as supporting engineering, economic, and resource analyses.

Microalgae are grown commercially in different ways:

1. Open cultivation using natural sunlight (raceway pond)
2. Closed cultivation using natural light (photobioreactor [PBR])
3. Closed cultivation using artificial illumination (PBR/fermentor)

Each system has its own advantage and disadvantage. The section of the production system depends on the degree of controlling the parameters required to produce the desired metabolite and the value of the product.

2.3.2.1 Open Ponds

At present, commercial production of phototrophic microbial biomass is limited to a few microalgal species that are cultivated in open ponds by means of a selective environment or a high growth rate. Most microalgae cannot be maintained long enough in outdoor open systems because of the risk of contamination by fungi, bacteria, and protozoa and competition by other microalgae that tend to dominate regardless of the original species used as inoculums (Richmond 1999). The problem of contamination has been solved to some extent by culturing algae that require special growth conditions that would prevent the contaminating microorganisms (Richmond 1986); however, doing so (i.e., facilitating species-specific conditions) limits the usefulness of these systems to a small number of organisms. Microalgae that tolerate unique conditions include *Dunaliella*, which can be grown at very high salinity. Low productivities are often seen associated with open culture systems, which need to process large quantities of water to harvest the algae. Also, outdoor open systems are subjected to daily and seasonal changes in light intensity and temperature, which make it very difficult to control operating parameter and reproduce specific culture condition. However, for some microalgal metabolites, this technology has proved very successful, producing tons of dried biomass per year (Lele and Kumar 2008).

A number of types of ponds have been designed and experimented with for microalgal cultivation. They vary in size, shape, material used for construction, type of agitation, and inclination. Often, the construction design is essentially dictated by local condition and available material. Despite the many different kinds of ponds proposed, only three major designs have been developed and operated at a relatively large scale: (1) inclined systems where mixing is achieved through pumping and gravity flow, (2) circular pond with agitation provided by a rotating arm, and (3) raceway pond constructed as an endless loop, in which the culture is circulated by paddle wheels. Only the last two, together with natural ponds, are used for commercial production of microalgae.

2.3.2.2 Closed Systems

Closed systems for microalgal cultivation essentially consist of a typical bioreactor with a suitable arrangement for illumination. In closed systems, the alga is enclosed in a transparent glass or plastic bag and the vessels are placed outdoor for entrapping sunlight. The closure of the systems prevents the contamination of the culture with other algal species. These systems have higher surface to volume ratios as compared to either ponds or raceways (open system). This increased surface to volume ratios of the vessels ensures better light entrapment than in open systems. However, the reproducibility of either biomass or metabolite productivities is problematic in outdoor closed systems as fluctuations in light and temperature are seen with changes in season. In addition, outdoor closed systems have problems in removal of oxygen and provision of temperature control. Although these issues can be solved, the cost of such installations can increase the cost advantage of natural sunlight. In an attempt to counteract the problems associated with outdoor systems, several designs have been proposed and constructed for the indoor, closed cultivation of algae using artificial illumination. These vessels, which are frequently referred to as PBRs, are in principle akin to conventional fermentor. PBRs apart from sharing the advantage (preventing contamination) of outdoor enclosures provide better controlling atmosphere in terms of light intensity, temperature, aeration, pH, and optimization of other culture parameters. PBRs are generally more expensive to construct than outdoor enclosed systems. However, the cost can be justified when high-value products are targeted from the biomass. PBRs can be used for cultivation of different microalgae; they can be a vital tool for small-scale production of high-value metabolites, such as stable-isotope-labeled biochemicals and unsaturated fatty acids.

PBRs are different types of tanks or closed horizontal or vertical pipes in which algae are cultivated (Richmond 2004). Algal cultures consist of a single or several specific strains optimized for producing the desired product. Water, necessary nutrients, and CO_2 are provided in a controlled way, while oxygen produced during the photovoltaic process has to be removed to avoid product inhibition. Algae receive sunlight either directly through the transparent container walls or via light fibers or tubes that channel the light from sunlight collectors.

PBRs can be classified on the basis of both design and mode of operation. In design terms, the main categories of reactors are (1) flat or tubular; (2) horizontal, inclined, vertical, or spiral; and (3) manifold or serpentine. An operation classification of PBRs would include (4) air or pump mixed and (5) single-phase reactors (filled with media, with gas exchange taking place in a separate gas exchanger) or two-phase reactors (in which both gas and liquid are present and continuous gas mass transfer takes place in the reactor itself). Construction materials provide additional variation and subcategories, for example, (6) glass or plastic and (7) rigid or flexible PBRs (Tredici 2004).

Fermentors come in a wide range of sizes from 1 L to more than 500,000 L. Commercial fermentors are readily available, so it is not necessary to describe their design. Fermentor operation includes batch and continuous models, as well as liquid and solid systems. Batch fermentation is the most common. There are common features between PBRs and fermentors (pH and temperature control, harvesting, etc.) (Table 2.7). The significant difference between a fermentor and PBRs is energy source, circulation, O_2 supply, and sterility. Organic carbon catabolism requires adequate oxygen, often making oxygen the single largest operating constraint. Glucose is the most widely used source of organic carbon and it is relatively inexpensive; however, acetate, citrate, and other organics have been used (Behrens 2005).

Although both PBR and fermentor can be used for the production of algae, they are different systems, and the economics of each system are different. Several excellent references provide a detailed description of capital and operation cost for fermentor. Similar descriptions are not readily available for PBRs because they are much less common systems and are usually customized. Table 2.8 lists some of the major cost factors in the construction and operation of PBRs and fermentors. Because PBRs are built in-house, they are usually of relative small scale, and this precludes the economy of scale of both construction and operating costs.

TABLE 2.7

Comparison of PBR and Fermentor Features

Feature	PBR	Fermentor
Energy source	Light	Organic carbon
Cell density/dry weight	Low	High
Limiting factor for growth	Light	Oxygen
Harvestability	Dilute, more difficult	Denser, less difficult
Vessel geometry	Dependent on light penetration	Independent of energy source
Control of parameters	High	High
Sterility	Usually sanitized	Can be completely sterilized
Availability of vessels	Often made in-house	Commercially available
Technology base	Relatively new	Centuries old
Construction costs	High per-unit volume	Low per-unit volume
Operating costs	High per-kg biomass	Low per-kg biomass
Applicability to algae	Photosynthetic algae	Heterotrophic algae

Source: Behrens, P.W., Photobioreactors and fermentors: The light and dark side of growing algae, in: Anderson, R.A. (ed.), *Algal Culturing Techniques*, Elsevier Academic Press, Burlington, MA, 2005, pp. 189–204.

TABLE 2.8

Some of the Major Cost Factors for PBRs and Fermentors

Cost Factor	PBR	Fermentor
Construction method	Individually constructed	Mass produced by craftsmen
Scale	Relatively small scale	Up to 500,000 L
Algal concentration	Dilute	High
Energy source	Light	Organic carbon

Source: Behrens, P.W., Photobioreactors and fermentors: The light and dark side of growing algae, in: Anderson, R.A. (ed.), *Algal Culturing Techniques*, Elsevier Academic Press, Burlington, MA, 2005, pp. 189–204.

Carbon is a key requirement, as the composition of microalgae is about 45% carbon. This is generally supplied as CO_2. For each kilogram of microalgae, at least 1.65 kg of CO_2 is required based on a mass balance (Berg-Nilsen 2006). A two-step cultivation process has been developed that involves a combination of raceway and PBR designs. The first step is the fast cultivation of biomass in the PBRs, and the second step is stress cultivation in open ponds. A PBR first step allows good protection of the growing biomass during early stages by maximizing the CO_2 capture. After that, the microalgal suspension is transferred to open ponds with low nitrogen nutrients and high CO_2 levels are maintained. The open raceway in the second step has some problems because higher density of algal biomass is more resistant to external contamination, and this nutrient-deficient phase avoided the growth of contaminating species (Bruton et al. 2009). Table 2.9 makes a comparison between PBR and ponds for several culture conditions and growth parameters.

The various large-scale culture systems also need to be compared on their basic properties such as their light utilization efficiency, ability to control temperature, the hydrodynamic stress placed on the algae, the ability to maintain the unialgal and/or axenic culture, and how easy they are to scale up from laboratory scale to large scale.

TABLE 2.9

Comparison of Open and Closed Large-Scale Culture Systems for Microalgae

Culture Systems for Microalgae	Closed Systems (PBRs)	Open Systems (Ponds)
Contamination control	Easy	Difficult
Contamination risk	Reduced	High
Sterility	Achievable	None
Process control	Easy	Difficult
Species control	Easy	Difficult
Mixing	Uniform	Very poor
Operation regime	Batch or semicontinuous	Batch or semicontinuous
Space required	A matter of productivity	PBRs ~ ponds
Area/volume ratio	High (20–200/m)	Low (5–10/m)
Population (algal cell) density	High	Low
Investment	High	Low
Operation costs	High	Low
Capital/operating costs	Ponds 3–10 times lower cost	PBRs > ponds
Light utilization efficiency	High	Poor
Temperature control	More uniform temperature	Difficult
Productivity	3–5 times more productive	Low
Water losses	Depends upon cooling design	PBRs ~ ponds
Hydrodynamic stress on algae	Low–high	Very low
Evaporation of growth medium	Low	High
Gas transfer control	High	Low
CO_2 losses	Depends on pH, alkalinity, etc.	PBRs ~ ponds
O_2 inhibition	Greater problem in PBRs	PBRs > ponds
Biomass concentration	3–5 times in PBRs	PBRs > ponds
Scale-up	Difficult	Difficult

Source: Mata, T.M. et al., *Renew. Sustain. Energ. Rev.*, 14, 217, 2010.

The final choice of system is almost always a compromise between all of these considerations to achieve an economically acceptable outcome (Borowitzka 1999).

2.4 Conclusion

Renewable technologies to supplement or replace fossil fuels are still in their early developmental stage. Marine algae are the major source of marine biomass and currently are an attractive renewable source for marine bioenergy. Most works have been done for bioethanol production from micro- and macroalgae (or seaweed) and for biodiesel from microalgal oils. Some seaweed species have high nitrogen, mineral, and sulfur contents, which can be problematic for the anaerobic digestion. The presence of salt and sulfated polysaccharides needs to be carefully managed in order to avoid inhibition of the fermentation process and a lowering of yields. Higher sulfur content in seaweed requires a higher removal cost for biofuel production. Marine algae have high carbohydrate and polysaccharide content, which require a new commercial process to break down into their constituent monomers prior to fermentation. On the other hand, all the processing stages of biofuels from marine algal biomass need to be simplified without involvement of extensive energy input.

Marine algal biomass has several advantages over terrestrial plant biomass because of its wide geographic distribution, and it can be a promising feedstock for biofuel production. Although biofuel production from marine algae has been reported in several research studies, large-scale production at low cost still faces numerous challenges. Application of marine algal biofuels will require increasing marine algal biomass farming, by expansion of farming areas or increased intensity of current marine algal farming areas. Production of biofuels from marine algal feedstock has an approach to produce feedstocks at a scale and cost needed to positively impact the biofuel economy. The integrated aquaculture (or IMTA) of marine algae and marine animal can increase production of the marine biomass with low cost and also mitigate the excess organic matter generated by intensive aquaculture activities. In the integrated aquaculture system, the marine algae can also enhance CO_2 fixation through photosynthetic processes. In addition, further works, for example, conversion of algal biomass into biofuels and value-added chemicals, development of technologies allowing an economically feasible design of each step including pretreatment and hydrolysis, fermentation with high yields and productivity, and cost-effective utilization of by-products and biomass residues, are needed.

References

Adams, J.M.M., A.B. Ross, K. Anastasakis, E.M. Hodgson, J.A. Gallagher, J.M. Jones, and I.S. Donnison. 2011a. Seasonal variation in the chemical composition of the bioenergy feedstock *Laminaria digitata* for thermochemical conversion. *Bioresour. Technol.* 102: 226–234.

Adams, J.M.M., T.A. Toop, I.S. Donnison, and J.A. Gallagher. 2011b. Seasonal variation in *Laminaria digitata* and its impact on biochemical conversion routes to biofuels. *Bioresour. Technol.* 102: 9976–9984.

Al-Hafedh, Y.S., A. Alam, A.H. Buschmann, and K.M. Fitzsimmons. 2012. Experiments on an integrated aquaculture system (seaweeds and marine fish) on the Red Sea coast of Saudi Arabia: Efficiency comparison of two local seaweed species for nutrient biofiltration and production. *Rev. Aquaculture* 4(1): 21–31.

Amaro, H.M., A.C. Guedes, and F.X. Malcata. 2011. Advances and perspectives in using microalgae to produce biodiesel. *Appl. Energ.* 88: 3402–3410.

Aresta, M., A. Dibenedetto, M. Carone, T. Colonna, and C. Fragale. 2005a. Production of biodiesel from macroalgae by supercritical CO_2 extraction and thermochemical liquefaction. *Environ. Chem. Lett.* 3(3): 136–139.

Aresta, M., A. Dibenedetto, and G. Barberio. 2005b. Utilization of macro-algae for enhanced CO_2 fixation and biofuels production: Development of a computing software for an LCA study. *Fuel Process. Technol.* 86: 1679–1693.

Barrington, K., T. Chopin, and S. Robinson. 2009. Integrated multi-trophic aquaculture (IMTA) in marine temperate waters. In: *Integrated Mariculture: A Global Review*, ed. D. Soto. FAO Fisheries and Aquaculture Technical Paper. No. 529, FAO, Rome, Italy, pp. 7–46.

Behrens, P.W. 2005. Photobioreactors and fermentors: The light and dark side of growing algae. In: *Algal Culturing Techniques*, ed. R.A. Anderson. Elsevier Academic Press, Burlington, MA, pp. 189–204.

Berg-Nilsen, J. 2006. *Production of Micro-Algae Based Products*. Nordic Innovation Centre, Oslo, Norway.

Bindu, M.S. and A.I. Levine. 2011. The commercial red seaweed *Kappaphycus alvarezii*—An overview on farming and environment. *J. Appl. Phycol.* 23: 789–796.
Bird, K.T., D.P. Chynoweth, and D.E. Jerger. 1990. Effects of marine algal proximate composition on methane yields. *J. Appl. Phycol.* 2: 207–213.
Bolton J.J., D.V. Robertson-Andersson, D. Shuuluka, and L. Kandjengo. 2009. Growing *Ulva* (Chlorophyta) in integrated systems as a commercial crop for abalone feed in South Africa: A SWOT analysis. *J. Appl. Phycol.* 21: 575–583.
Borghini, F., L. Lucattini, S. Focardi, S. Focardi, and S. Bastianoni. 2014. Production of bio-diesel from macro-algae of the Orbetello lagoon by various extraction methods. *Int. J. Sustain. Energ.* 33: 695–703.
Borines, M.G., R.L. de Leon, and J.L. Cuello. 2013. Bioethanol production from the macroalgae *Sargassum* spp. *Bioresour. Technol.* 138: 22–29.
Borowitzka, M.A. 1999. Commercial production of microalgae: Ponds, tanks, tubes and fermenters. *J. Biotechnol.* 70: 313–321.
Briand, X. 1991. Seaweed harvest in Europe. In: *Seaweed Resources in Europe: Uses and potential*, eds. M. Guriy and G. Blunden. John Wiley & Sons Ltd., New York, pp. 261–308.
Bruhn, A., J. Dahl, H.B. Nielsen, L. Nikolaisen, M.B. Rasmussen, S. Markager, B. Olesen, C. Arias, and P.D. Jensen. 2011. Bioenergy potential of *Ulva lactuca*: Biomass yield, methane production and combustion. *Bioresour. Technol.* 102: 2595–2604.
Bruton, T., H. Lyons, Y. Lerat, M. Stanley, and M.B. Rasmussen. 2009. *A Review of the Potential of Marine Algae as a Source of Biofuel in Ireland*. Sustainable Energy Ireland, Dublin, Ireland, 88pp.
Buschmann, H.A., A.J. Correa, R. Westermeier, C.M. Herńandez-Gonźalez, and R. Norambuena. 2001. Red algal farming in Chile: A review. *Aquaculture* 194: 203–220.
Candra, K.P., Sarwono, and Sarinah. 2011. Study on bioethanol production using red seaweed *Eucheuma cottonii* from Bontang sea water. *J. Coast. Dev.* 15: 45–50.
Carton, R.J., C.M. Caipang, M. Notoya, and D. Fujita. 2011. Physiological responses of two seaweed biofilter candidates, *Gracilariopsis bailinae* Zhang et Xia and *Hydropuntia edulis* (S. Gmelin), to nutrient source and environmental factors. *AACL Bioflux* 4(5): 635–643.
Carton, R.J., Y. Okuyama, H. Kimura, D. Fujita, and M. Notoya. 2010. Nutrient uptake and reduction efficiency by the red alga *Gracilaria bursa-pastoris* (S. Gemlin) Silva integrated with the red-sea bream, *Pagrus major*. *Algal Resources* 3(2): 99–110.
Chao, K.P., Y.C. Su, and C.S. Chen. 1999. Chemical composition and potential for utilization of the marine alga *Rhizoclonium* sp. *J. Appl. Phycol.* 11: 525–533.
Chirapart, A. and K. Lewmanomont. 2004. Growth and production of Thai agarophyte cultured in natural pond using the effluent seawater from shrimp culture. *Hydrobiologia* 512: 117–126.
Chirapart, A., J. Praiboon, P. Puangsombat, C. Pattanapon, and N. Nunraksa. 2014. Chemical composition and ethanol production potential of Thai seaweed species. *J. Appl. Phycol.* 26: 979–986.
Cho Y., H. Kim, and S.-K. Kim. 2013. Bioethanol production from brown seaweed, *Undaria pinnatifida*, using NaCl acclimated yeast. *Bioproc. Biosyst. Eng.* 36: 713–719.
Chopin, T. and S. Robinson. 2004. Defining the appropriate regulatory and policy framework for the development of integrated multi-trophic aquaculture practices: Introduction to the workshop and positioning of the issues. *Bull. Aquac. Assoc. Can.* 104 (3): 4–10.
Cohen, Z. 1999. *Chemicals from Microalgae*. Taylor & Francies Ltd., London, U.K., 419pp.
Costa, J.C., P.R. Gonçalves, A. Nobre, and M.M. Alves. 2012. Biomethanation potential of macroalgae *Ulva* spp. and *Gracilaria* spp. and in co-digestion with waste activated sludge. *Bioresour. Technol.* 114: 320–326.
Critchley, A.T. 1998. *Gracilaria* (Gracilariales, Rhodophyta): An economically important agarophyte. In: *Seaweed Resources of the World*, eds. A.T. Critchley and M. Ohno. Japan International Cooperation Agency, Yokosuka, Japan, pp. 89–112.
Critchley, A.T. and M. Ohno. 1998. *Seaweed Resources of the World*. Japan International Cooperation Agency (JICA), Yokosuka, Japan.

Danananda, C., R. Sarada, and G.A. Ravishankar. 2007. Fueling the future: By microalgae as a source of renewable energy. In: *Advances in Applied Phycology*, eds. R.K. Gupta and V.D. Pandey. Daya Publishing House, New Delhi, India, pp. 56–74.

Daroch, M., S. Geng, and G. Wang. 2013. Recent advances in liquid biofuel production from algal feedstocks. *Appl. Energ.* 102: 1371–1381.

Demirbas, A. 2011. Biodiesel from oilgae, biofixation of carbon dioxide by microalgae: A solution to pollution problems. *Appl. Energ.* 88: 3541–3547.

de Paula Silva P.H., R. de Nys, and N.A. Paul. 2012. Seasonal growth dynamics and resilience of the green tide alga *Cladophora coelothrix* in high-nutrient tropical aquaculture. *Aquacult. Environ. Interact.* 2: 253–266.

de Paula Silva P.H., S. McBride, R. de Nys, and N.A. Paul. 2008. Integrating filamentous 'green tide' algae into tropical pond-based aquaculture. *Aquaculture* 284: 74–80.

de Paula Silva, P.H., N.A. Paul, R. de Nys, and L. Mata. 2013. Enhanced production of green tide algal biomass through additional carbon supply. *PLoS ONE* 8(12): e81164.

Dubi, A.M., A.J. Mmochi, N.S. Jiddawi, M.S. Kyewalyanga, F.E. Msuya, Z. Ngazy, and A.W. Mwandya. 2005. Development of integrated pond culture of finfish, shellfish, and seaweed in Tanzania. Report no: WIOMSA/MASMA 2005-01 http://www.wiomsa.org/index.php?option = com_j (Accessed April 30, 2014).

FAO. 2012. *The State of World Fisheries and Aquaculture 2012*. Rome, Italy, 209pp.

Fasahati, P. and J.J. Liu. 2012. Process simulation of bioethanol production from brown algae. In: *Eighth IFAC Symposium on Advanced Control of Chemical Processes*, Part 1, Singapore, pp. 597–602.

Flowers, A. and K.T. Bird. 1990. Methane production from seaweed. In: *Introduction to Applied Phycology*, ed. I. Akatsuka. SPB Academic Publisher BV, The Hague, the Netherlands, pp. 575–587.

Goh, C.S. and K.T. Lee. 2010. A visionary and conceptual macroalgae-based third-generation bioethanol (TGB) biorefinery in Sabah, Malaysia as an underlay for renewable and sustainable development. *Renew. Sustain. Energ. Rev.* 14: 842–848.

Gunaseelan, V.N. 1997. Anaerobic digestion of biomass for methane production: A review. *Biomass Bioenergy* 13: 83–114.

Gurung, A., S.W.V. Ginkel, W.-C. Kang, N.A. Qambrani, and S.-E. Oh. 2012. Evaluation of marine biomass as a source of methane in batch tests: A lab-scale study. *Energy* 43: 396–401.

Hannon, M., J. Gimpel, M. Tran, B. Rasala, and S. Mayfield. 2010. Biofuels from algae: Challenges and potential. *Biofuels* 1(5): 763–784.

Harun, R., M. Singh, G.M. Forde, and M.K. Danquah. 2010. Bioprocess engineering of microalgae to produce a variety of consumer products. *Renew. Sustain. Energ. Rev.* 14: 1037–1047.

Hernández-Garibay, E., J.A. Zertuche-González, and I. Pacheco-Ruíz. 2011. Isolation and chemical characterization of algal polysaccharides from the green seaweed *Ulva clathrata* (Roth) C. Agardh. *J. Appl. Phycol.* 23: 537–542.

Horn, S.J., I.M. Aasen, and K. Østgaard. 2000a. Ethanol production from seaweed extract. *J. Ind. Microbiol. Biotechnol.* 25: 249–254.

Horn, S.J., I.M. Aasen, and K. Østgaard. 2000b. Production of ethanol from mannitol by *Zymobacter palmae*. *J. Ind. Microbiol. Biotechnol.* 24: 51–57.

Illman, A.M., A.H. Scragg, and S.W. Shales. 2000. Increase in *Chlorella* strain calorific values when grown in low nitrogen medium. *Enzyme Microb. Technol.* 27: 631–635.

Jang, J.-S., Y. Cho, G.-T. Jeong, and S.-K. Kim. 2012. Optimization of saccharification and ethanol production by simultaneous saccharification and fermentation (SSF) from seaweed, *Saccharina japonica*. *Bioproc. Biosyst. Eng.* 35: 11–18.

Jard, G., D. Jackowiak, H. Carrère, J.P. Delgenes, M. Torrijos, J.P. Steyer, and C. Dumas. 2012. Batch and semi-continuous anaerobic digestion of *Palmaria palmata*: Comparison with *Saccharina latissima* and inhibition studies. *Chem. Eng. J.* 209: 513–519.

Jard, G., H. Marfaing, H. Carrère, J.P. Delgenes, J.P. Steyer, and C. Dumas. 2013. French Brittany macroalgae screening: Composition and methane potential for potential alternative sources of energy and products. *Bioresour. Technol.* 144: 492–498.

Jones, C.S and S.P. Mayfield. 2012. Algae biofuels: Versatility for the future of bioenergy. *Curr. Opin. Biotech.* 23: 346–351.
Jung, K.-W., D.-H. Kim, and H.-S. Shin. 2012. Continuous fermentative hydrogen and methane production from *Laminaria japonica* using a two-stage fermentation system with recycling of methane fermented effluent. *Int. J. Hydrogen Energ.* 37: 15648–15657.
Kelly, M.S. and S. Dworjanyn. 2008. The potential of marine biomass for anaerobic biogas production. The Crown Estate, 103pp. http://www.thecrownestate.co.uk (Accessed December 24, 2013).
Khambhaty, Y., K. Mody, M.R. Gandhi, S. Thampy, P. Maiti, H. Brahmbhatt, K. Eswaran, and P.K. Ghosh. 2012. *Kappaphycus alvarezii* as a source of bioethanol. *Bioresour. Technol.* 103: 180–185.
Kim, N.-J., H. Li, K. Jung, H.N. Chang, and P.C. Lee. 2011. Ethanol production from marine algal hydrolysates using *Escherichia coli* KO11. *Bioresour. Technol.* 102: 7466–7469.
Kimura, H., D. Fujita, and M. Notoya. 2007. Nutrient uptake by *Undaria undarioides* (Yendo) Okamura and its application as an algal partner of fish-alga integrated culture. *Bull. Fish. Res. Agen.* 19: 143–147.
Kraan, S. 2013. Mass-cultivation of carbohydrate rich macroalgae, a possible solution for sustainable biofuel production. *Mitig. Adapt. Strateg. Glob. Change* 18: 27–46.
Kumar, S., R. Gupta, G. Kumar, D. Sahoo, and R.C. Kuhad. 2013. Bioethanol production from *Gracilaria verrucosa*, a red alga, in a biorefinery approach. *Bioresour. Technol.* 135: 150–156.
Lahaye, M. and A. Robic. 2007. Structure and functional properties of ulvan, a polysaccharide from green seaweeds. *Biomacromolecules* 8: 1765–1774.
Lee, S.-M., G.H. Kim, and J.-H. Lee. 2012. Bio-gas production by co-fermentation from the brown algae, *Laminaria japonica*. *J. Ind. Eng. Chem.* 18: 1512–1514.
Lele, S.S. and J.K. Kumar. 2008. *Algal Bioprocess Technology*. New Age International (P) Limited, Publishers, New Delhi, India, 133pp.
Lüning, K. and S. Pang. 2003. Mass cultivation of seaweeds: Current aspects and approaches. *J. Appl. Phycol.* 15: 115–119.
Marinho-Soriano E., P.C. Fonseca, M.A.A. Carneiro, and W.S.C. Moreira. 2006. Seasonal variation in the chemical composition of two tropical seaweeds. *Bioresour. Technol.* 97: 2402–2406.
Martinez, L.A. and A.H. Buschmann. 1996. Agar yield and quality of *Gracilaria chilensis* (Gigartinales, Rhodophyta) in tank culture using fish effluents. *Hydrobiologia* 326/327: 341–345.
Mata, T.M., A.A. Martins, and S.C. Nidia. 2010. Microalgae for biodiesel production and other applications: A review. *Renew. Sustain. Energ. Rev.* 14: 217–232.
Matsui, T. and Y. Koike. 2010. Methane fermentation of a mixture of seaweed and milk at a pilot-scale plant. *J. Biosci. Bioeng.* 110: 558–563.
McDermid, K.J. and B. Stuercke. 2003. Nutritional composition of edible Hawaiian seaweeds. *J. Appl. Phycol.* 15: 513–524.
McDermid, K.J., B. Stuercke, and G.H. Balazs. 2007. Nutritional composition of marine plants in the diet of the green sea turtle (*Chelonia mydas*) in the Hawaiian Islands. *Bull. Mar. Sci.* 81: 55–71.
Meinita, M.D.N., J.-Y. Kang, G.-T. Jeong, H.M. Koo, S.M. Park, and Y.-K. Hong. 2012. Bioethanol production from the acid hydrolysate of the carrageenophyte *Kappaphycus alvarezii* (*cottonii*). *J. Appl. Phycol.* 24: 857–862.
Migliore, G., C. Alisi, A.R. Sprocati, E. Massi, R. Ciccoli, M. Lenzi, A. Wang, and C. Cremisini. 2012. Anaerobic digestion of macroalgal biomass and sediments sourced from the Orbetello lagoon, Italy. *Biomass Bioenergy* 42: 69–77.
Mishima, Y. and T.D. Nguyen. 2007. Possibility of aquatic biomass utilization in Asian countries. http://www.biomass-asia-workshop. jp/biomassws/04workshop/index.html (Accessed September 4, 2008).
Moheimani, N.R. 2013. Long-term outdoor growth and lipid productivity of *Tetraselmis suecica*, *Dunaliella tertiolecta* and *Chlorella* sp. (Chlorophyta) in bag photobioreactors. *J. Appl. Phycol.* 25: 167–176.
Msuya, E.F., A. Buriyo, I. Omar, B. Pascal, K. Narrain, J.M.J. Ravina, E. Mrabu, and G.J. Wakibia. 2014. Cultivation and utilisation of red seaweeds in the Western Indian Ocean (WIO) Region. *J. Appl. Phycol.* 26: 699–705.

Msuya, E.F. and A. Neori. 2002. *Ulva reticulata* and *Gracilaria crassa*: Macroalgae that can biofilter effluent from tidal fishponds in Tanzania. *Western Indian Ocean J. Mar. Sci.* 1: 117–126.

Nana, S.S.U.P., I. Lapong, M.A. Rimmer, and S. Raharjo. 2013. *Caulerpa* culture in South Sulawesi—An alternative for brackishwater pond culture. http://putranana.blogspot.com/2013/03/caulerpa-culture-in-south-sulawesi.html (Accessed April 30, 2014).

Nang, H.Q. and Dinh, N.H. 1998. The seaweed resource of Vietnam. In: *Seaweed Resources of the World*, eds. A.T. Critchley and M. Ohno. Japan International Cooperation Agency, Yokosuka, Japan, pp. 62–69.

Narasimman, S. and K. Murugaiyan. 2012. Proximate composition of certain selected marine macroalgae form Mandapam coastal region (Gulf of Mannar), southeast coast of Tamil Nadu. *Int. J. Pharm. Biol. Arch.* 3(4): 918–921.

Neori, A., T. Chopin, M. Troell, A.H. Buschmann, G.P. Kraemer, C. Halling, M. Shpigel, and C. Yarish. 2004. Integrated aquaculture: Rationale, evolution and state of the art emphasizing seaweed biofiltration in modern mariculture. *Aquaculture* 231: 361–391.

Nkemka, V.N. and M. Murto. 2010. Evaluation of biogas production from seaweed in batch tests and in UASB reactors combined with the removal of heavy metals. *J. Environ. Manage.* 91: 1573–1579.

Notoya, M. 2007. The challenges of seaweed bio-fuel production, preservation of environment and fishery resources. In: *Program and Abstract of XIXth International Seaweed Symposium*, Kobe, Japan, 58pp.

Ohno, M. 1993. Chapter 2: Cultivation of green algae, *Monostroma* and *Enteromorpha* "Aonori." In: *Seaweed Cultivation and Marine Ranching*, eds. M. Ohno and A.T. Critchley. Japan International Cooperation Agency, Yokosuka, Japan, pp. 7–15.

Oohusa, T. 1993. Chapter 7: The cultivation of *Porphyra* "Nori." In: *Seaweed Cultivation and Marine Ranching*, eds. M. Ohno and A.T. Critchley. Japan International Cooperation Agency, Yokosuka, Japan, pp. 57–73.

Paul, N.A., C.K. Tseng, and M. Borowitzka. 2012. Seaweed and microalgae. In: *Aquaculture: Farming Aquatic Animal and Plants*, eds. J.S. Lucus and P.C. Soughgate. Wiley-Blackwell, Hoboken, NJ, 629pp.

Peu, P., J.-F. Sassi, R. Girault, S. Picard, P. Saint-Cast, F. Béline, and P. Dabert. 2011. Sulphur fate and anaerobic biodegradation potential during co-digestion of seaweed biomass (*Ulva* sp.) with pig slurry. *Bioresour. Technol.* 102: 10794–10802.

Poespowati, T., Jimmy, and S. Noertjahjono. 2013. Cultivation of *Caulerpa taxifolia* as feedstock of bioenergy. *J. Energ. Technol. Policy* 3: 13–20.

Pulz, O. and W. Gross. 2004. Valuable products from biotechnology of microalgae. *Appl. Microbiol. Biotechnol.* 65: 635–648.

Ratana-arporn, P. and A. Chirapart. 2006. Nutritional evaluation of tropical green seaweeds *Caulerpa lentillifera* and *Ulva reticulata*. *Kasetsart J.* (*Nat. Sci.*) 40(Suppl.): 75–83.

Richmond A. 1986. *CRC Handbook of Microalgal Mass Culture*. CRC Press, Inc., Boca Raton, FL.

Richmond, A. 1999. Physiological principles and modes of cultivation in mass production of photoautotrophic microalgae. In: *Chemical from Microalgae*, ed. Z. Cohen. Taylor & Francies, London, U.K., pp. 353–386.

Richmond, A. 2004. *Handbook of Microalgal Culture: Biotechnology and Applied Phycology*. Blackwell Science Ltd., Oxford, U.K., 566pp.

Roesijadi, G., S.B. Jones, L.J. Snowden-Swan, and Y. Zhu. 2010. *Macroalgae as a Biomass Feedstock: A Preliminary Analysis*. The US Department of Energy under Contract DE-AC05-76RL01830. PNNL-19944, Pacific Northwest National Laboratory Richland, Washington, DC.

Royal Commission on Environmental Pollution (RCEP). 2004. *Biomass as a Renewable Energy Source*. Royal Commission on Environmental Pollution, Westminster, London, U.K., SW1P 3JS. ISBN 0-9544186-1-1.

Santelices, B. 1999. A conceptual framework for marine agronomy. *Hydrobiologia* 398/399: 15–23.

Santelices, B. and M.S. Doty. 1989. A review of *Gracilaria* farming. *Aquaculture* 78: 95–133.

Sayadi, M.H., S.D. Ghatnekar, and M.F. Kavian. 2011. Algae a promising alternative for biofuel. *Proc. Int. Acad. Ecol. Environ. Sci.* 1(2): 112–124. http//www.iaees.org (Accessed April 19, 2014).

Shahbazi, A. and Y. Li. 2006. Availability of crop residues as sustainable feedstock for bioethanol production in North Carolina. *Appl. Biochem. Biotech.* 129–132: 41–54.

Shang, Y.C. 1976. Economic aspects of *Gracilaria* culture in Taiwan. *Aquaculture* 8: 1–7.

Singh, A., Nigam, P.S. and J.D. Murphy. 2011. Mechanism and challenges in commercialization of algal biofuels. *Bioresour. Technol.* 102: 26–34.

Singh, J. and S. Gu. 2010. Commercialization potential of microalgae for biofuels production. *Renew. Sustain. Energ. Rev.* 14: 2596–2610.

Singh, R., T. Bhaskar, and B. Balagurumurthy. 2014. Effect of solvent on the hydrothermal liquefaction of macro algae *Ulva fasciata. Process Saf. Environ.* http://www.sciencedirect.com/science/journal/aip/09575820 (Accessed April 15, 2014).

Song, L., M. Hu, D. Liu, D. Zhang, and C. Jiang. 2014. Thermal cracking of *Enteromorpha prolifera* with solvents to bio-oil. *Energ. Convers. Manage.* 77: 7–12.

Tanisho, S. and M. Notoya. 2010. Fermentative hydrogen production by using cultured seaweed. *Hydrogen Energ. Syst.* 35: 22–26.

Tedesco, S., K.Y. Benyounis, and A.G. Olabi. 2013. Mechanical pretreatment effects on macroalgae-derived biogas production in co-digestion with sludge in Ireland. *Energy* 61: 27–33.

Titlyanov E.A. and T.V. Titlyanova. 2010. Seaweed cultivation: Methods and problems. *Russ. J. Mar. Biol.* 36 (4): 227–242.

Tredici M.R. 2004. Mass production of microalgae: Photobioreactors. In: *Handbook of Microalgal Culture: Biotechnology and Applied Phycology*, ed. A. Richmon. Blackwell Publishing, Oxford, U.K., pp. 178–214.

Trivedi, N., V. Gupta, C.R.K. Reddy, and B. Jha. 2013. Enzymatic hydrolysis and production of bioethanol from common macrophytic green alga *Ulva fasciata* Delile. *Bioresour. Technol.* 150: 106–112.

Troell, M. 2009. Integrated marine and brackishwater aquaculture in tropical regions: Research, implementation and prospects. In: *Integrated Mariculture: A Global Review*, ed. D. Soto. FAO Fisheries and Aquaculture Technical Paper. No. 529. FAO, Rome, Italy, pp. 47–131.

Troell, M., C. Halling, A. Neori, T. Chopin, A.H. Buschmann, N. Kautsky, and C. Yarish. 2003. Integrated mariculture: Asking the right questions. *Aquaculture* 226: 69–90.

Trono G.C., Jr. 1998. *Eucheuma* and *Kappaphycus*: Taxonomy and cultivation. In: *Seaweed Resources of the World*, eds. A.T. Critchley and M. Ohno. Japan International Cooperation Agency, Yokosuka, Japan, pp. 75–88.

Trono G.C., Jr. 1999. Diversity of the seaweed flora of the Philippines and its utilization. *Hydrobiologia* 398/399: 1–6.

Usov, A.I. 2011. Chapter 4: Polysaccharides of the red algae. *Adv. Carbohydr. Chem. Biochem.* 65: 115–217.

Vergara-Ferñandez A., G. Vargas, N. Alarcón, and A. Velasco. 2008. Evaluation of marine algae as a source of biogas in a two-stage anaerobic reactor system. *Biomass Bioenergy* 32: 338–344.

Vonshak, A. 1990. Recent advance in microalgal biotechnology. *Biotechnol. Adv.* 8: 709–727.

Wargacki, A.J., E. Leonard, M.N. Win, D.D. Regitsky, C.N.S. Santos, P.B. Kim, S.R. Cooper et al. 2013. An engineered microbial platform for direct biofuel production from brown macroalgae. *Science* 335: 308–313.

Wei, N., J. Quarterman, and Y.-S. Jin. 2013. Marine macroalgae: An untapped resource for producing fuels and chemicals. *Trends Biotechnol.* 31(2): 70–77.

Wellinger A. 2009. Algal biomass does it save the world? Short reflections. IEA Bioenergy.

Wikfors, G.H. and M. Ohno. 2001. Impact of algal research in aquaculture. *J. Phycol.* 37: 968–974.

Yanagisawa, M., K. Nakamura, O. Ariga, and K. Nakasaki. 2011. Production of high concentrations of bioethanol from seaweeds that contain easily hydrolysable polysaccharides. *Process Biochem.* 46: 2111–2116.

Yokoyama, S., K. Jonouchi, and K. Imou. 2008. Energy production from marine biomass: Fuel cell power generation driven by methane produced from seaweed. *Int. J. Eng. Appl. Sci.* 4: 169–172.

Yeon, J.-H., S.-E. Lee, W.Y. Choi, D.H. Kang, H.-Y. Lee, and K.-H. Jung. 2011. Repeated-batch operation of surface-aerated fermentor for bioethanol production from the hydrolysate of seaweed *Sargassum sagamianum*. *J. Microbiol. Biotechnol.* 21(3): 323–331.

Zhou, D., L. Zhang, S. Zhang, H. Fu, and J. Chen. 2010. Hydrothermal liquefaction of macroalgae *Enteromorpha prolifera* to bio-oil. *Energ. Fuel* 24: 4054–4061.

3

Marine Algae: An Important Source of Bioenergy Production

Panchanathan Manivasagan, Jayachandran Venkatesan, and Se-Kwon Kim

CONTENTS

3.1 Introduction

The beginning of this century is marked by the large incentive given to biofuel use in replacement of gasoline. Several countries worldwide, including Brazil, United States, Canada, Japan, India, China, and Europe, are interested in developing their internal biofuel markets and established plans for use of these biofuels. Such interests are mainly motivated by (1) the rising oil prices and recognizing that the global oil reserves are exhausting fast, (2) concern about fuel emissions, (3) the requirements of the Kyoto Protocol and the Bali Action Plan on carbon emissions, and (4) the provision of alternative outlets for agricultural producers (Mussatto et al. 2010).

Global warming has become one of the most serious environmental problems. To cope with the problem, it is necessary to substitute renewable energy for nonrenewable fossil fuel. Biomass, which is one of the renewable energies, is considered to be carbon-neutral, meaning that the net CO_2 concentration in the atmosphere remains unchanged provided the CO_2 emitted by biomass combustion and that fixed by photosynthesis are balanced. Biomass is also unique because it is the only organic matter among renewable energies. In other words, fuels and chemicals can be produced from biomass in addition to electricity and heat (Yokoyama et al. 2008).

Marine biomass has attracted less attention than terrestrial biomass for energy utilization so far but is worth considering especially for many countries like Brazil, United States, Canada, Japan, India, China, and Europe, which have long available coastlines. If the sea area is utilized efficiently, a vast amount of renewable energy could be produced. In addition, native seaweeds often form submarine forests that serve as habitats for fish and shellfish, and so if a marine biomass energy system is realized, it may boost the production of marine food and promote the marine energy industry leading to CO_2 mitigation. The use of marine biomass energy was investigated in the United States (Chynoweth 2002) and Japan as an alternative energy in the 1970s after the oil crises, but the studies were discontinued when oil prices stabilized. However, now that global warming has become one of the most serious problems to be solved, we should reconsider the use of marine biomass energy as a means to mitigate CO_2 emissions (Yokoyama et al. 2008).

Marine microalgal biodiesel production system involves the following process steps: cultivation, harvesting, dewatering, extraction, and transesterification (Harun et al. 2010). To achieve high oil yields and CO_2 fixation capacity during cultivation, the key process considerations are the choice of microalgal strain, cultivation conditions, and the cultivation system (photobioreactors [PBRs] or open ponds). Different technologies are available for harvesting, dewatering, extraction, and transesterification. However, high-efficiency, energy-saving, and low CO_2 emission technologies are the optimum targets for full-scale industrial application of microalgae biotechnology (Zeng et al. 2011).

3.2 Marine Microalgae

Marine microalgae are prokaryotic or eukaryotic photosynthetic microorganisms that can grow rapidly and live in harsh conditions due to their unicellular or simple multicellular structure. Examples of prokaryotic microorganisms are cyanobacteria (*Cyanophyceae*), and eukaryotic microalgae are, for example, green algae (*Chlorophyta*) and diatoms (*Bacillariophyta*) (Li et al. 2008a,b). A more in-depth description of microalgae is presented by Richmond (2008).

Microalgae are present in all existing earth ecosystems, not just aquatic but also terrestrial, representing a big variety of species living in a wide range of environmental conditions. It is estimated that more than 50,000 species exist, but only a limited number, of around 30,000, have been studied and analyzed (Richmond 2008).

3.3 Marine Microalgae as Important Sources of Fuel and Energy

In terms of bioenergy applications, microalgae can be considered a solar energy capture, conversion, and storage mechanism. Like more complex land plants, microalgae carry out photosynthesis whereby solar energy in the form of photons is biologically transduced to energy (ATP) and reductant (NADPH). Adenosine triphosphate (ATP) and nicotinamide adenine dinucleotide phosphate hydrogen (NADPH) are both required to convert highly oxidized carbon in the form of CO_2 to reduced (energy-containing) 3-carbon sugar compounds (C3 pathway) through the Calvin–Benson cycle. In some microalgae, there is some evidence to show that a C4-like mechanism operates in conjunction with a C3 cycle (Reinfelder et al. 2000; McGinn and Morel 2008). The overall efficiency of energy conversion in photosynthesis, that is to say the efficiency of solar energy conversion into chemical energy, is typically on the order of 3%–4% (Wijffels and Barbosa 2010) (Figure 3.1). The chemical energy stored in the biomass mainly occurs in one of three different carbon-containing macromolecular classes: lipid, carbohydrate, or protein. Determining precisely the quantity of stored energy in the biomass is fairly straightforward and is usually done by measuring the heat of combustion in a standard laboratory bomb calorimeter. Typically, a known quantity of dried biomass is combusted in a 100% O_2 atmosphere, and the resultant thermal energy release is detected in a suitable conducting medium, a water bath, for instance. Growing under optimal conditions, the energy density of microalgal biomass from a variety of species is remarkably consistent ranging between 19 and 25 GJ/t, which is similar to the energy density of bituminous coal. Generally speaking, the energy density of the biomass increases with its lipid content, due to the higher energy density of lipid compared to the other major biological macromolecules, carbohydrate and protein. This has been observed in a number of elegant laboratory experiments where cultivation conditions have been manipulated in order to promote lipid accumulation.

On the other hand, cyanobacteria show little potential as a source of oil, although other storage products such as glycogen may find uses as a source of carbohydrate for ethanol production through anaerobic fermentation. Certain physiological *triggers* of TAG accumulation have been observed including nitrogen and sulfur limitation, high light stress, or changes in external pH, among others (Tatsuzawa et al. 1996; Hu et al. 2008). Singh et al. (2011) have suggested that mixotrophic growth may promote TAG accumulation

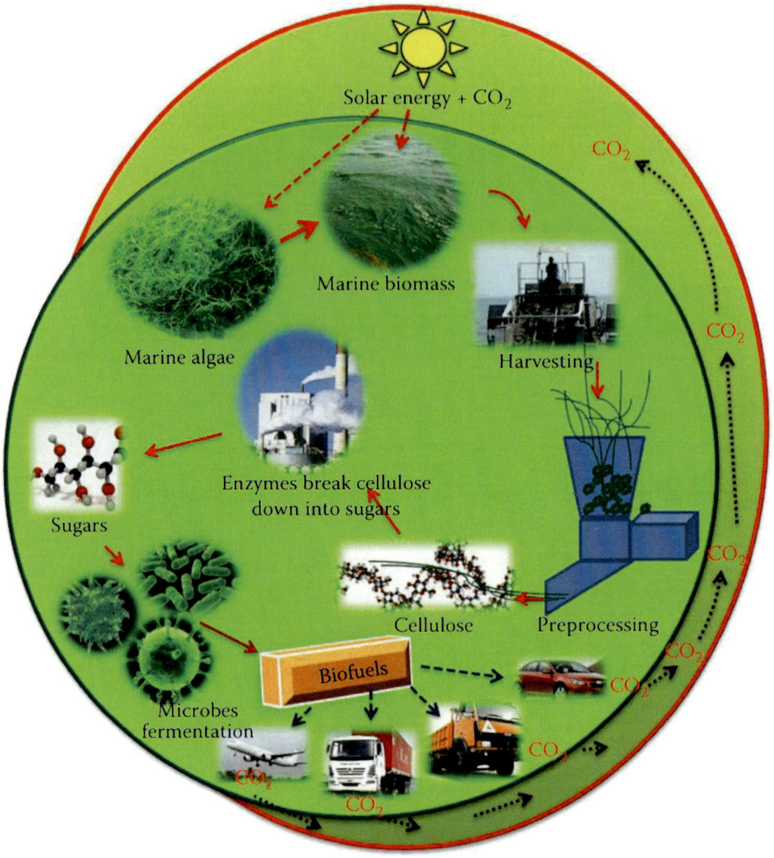

FIGURE 3.1
Potential pathways from microalgae to biofuels.

whereby an initial, photoautotrophic stage is used for bulk biomass production, followed by a second stage when organic carbon is supplied to the dense culture under nitrogen limitation. This approach has shown some promise for TAG accumulation in microalgae, in particular freshwater chlorophytes (Singh et al. 2011).

3.4 Microalgae Culture Systems

3.4.1 Open- versus Closed-Culture Systems

Microalgae cultivation can be done in open-culture systems such as lakes or ponds and in highly controlled closed-culture PBRs. A bioreactor is defined as a system in which a biological conversion is achieved. Thus, a PBR is a reactor in which phototrophs (microbial, algal, or plant cells) are grown or used to carry out a photobiological reaction. Although this definition may apply to both closed- and open-culture systems, for the purpose of this chapter, we limit the definition to the former ones.

Open-culture systems are normally less expensive to build and operate, are more durable than large closed reactors, and have a large production capacity when compared with closed systems. However, according to Richmond (2008), ponds use more energy to homogenize nutrients, and the water level cannot be kept much lower than 15 cm (or 150 L/m^2) for the microalgae to receive enough solar energy to grow. Generally, ponds are more susceptive to weather conditions, not allowing control of water temperature, evaporation, and lighting. Also, they may produce large quantities of microalgae but occupy more extensive land area and are more susceptible to contaminations from other microalgae or bacteria. Moreover, since the atmosphere only contains 0.03%–0.06% CO_2, it is expected that mass transfer limitation could slow down the cell growth of microalgae (Richmond 2008).

PBRs are flexible systems that can be optimized according to the biological and physiological characteristics of the algal species being cultivated, allowing one to cultivate algal species that cannot be grown in open ponds. On a PBR, direct exchange of gases and contaminants (e.g., microorganisms, dust) between the cultivated cells and atmosphere is limited or not allowed by the reactor's walls. Also, a great proportion of light does not impinge directly on the culture surface but has to cross the transparent reactor walls.

Depending on their shape or design, PBRs are considered to have several advantages over open ponds: PBRs offer better control over culture conditions and growth parameters (pH, temperature, mixing, CO_2, and O_2), prevent evaporation, reduce CO_2 losses, allow to attain higher microalgae densities or cell concentrations and higher volumetric productivities, and offer a more safe and protected environment, preventing contamination or minimizing invasion by competing microorganisms.

Despite their advantages, it is not expected that PBRs have a significant impact in the near future on any product or process that can be attained in large outdoor raceway ponds. PBRs suffer from several drawbacks that need to be considered and solved. Their main limitations include overheating; biofouling; oxygen accumulation; difficulty in scaling up; the high cost of building, operating, and algal biomass cultivation; and cell damage by shear stress and deterioration of material used for the photostage.

The cost of biomass production in PBRs may be one order of magnitude higher than in ponds. While in some cases, for some microalgal species and applications, it may be low enough to be attractive for aquaculture use, in other cases, the higher cell concentration and the higher productivity achieved in PBR may not compensate for its higher capital and operating costs.

3.4.2 Batch versus Continuous Operation

PBRs can be operated in batch or continuous mode. There are several advantages of using continuous bioreactors as opposed to the batch mode (Williams 2002):

- Continuous bioreactors provide a higher degree of control than do batch.
- Growth rates can be regulated and maintained for extended time periods and biomass concentration can be controlled by varying the dilution rate.
- Because of the steady state of continuous bioreactors, results are more reliable and easily reproducible and the desired product quality may be more easily obtained.
- Continuous reactions offer increased opportunities for system investigation and analysis.

3.4.3 Designs and Construction Materials of Culture Systems

Depending on the local conditions and available materials, it is possible to design different culture systems with variations in size, shape, construction materials, inclination, and agitation type, which influence their performance, cost, and durability (resistance to weathering).

Among the various sizes and shapes of ponds operated at a relatively large scale, the three major designs include (Richmond 2008; Schenk et al. 2008) (1) raceway ponds constructed as an endless loop, in which the culture is circulated by paddle wheels; (2) circular ponds with agitation provided by a rotating arm; and (3) inclined systems where mixing is achieved through pumping and gravity flow.

Raceway ponds and also natural ponds may be the most commonly used for commercial production of microalgae. Normally, open ponds are relatively economical, easy to clean up after cultivation, and good for mass cultivation of algae. However, they allow little control of culture conditions; their productivity is poor; they occupy a large land area; and cultures are easily contaminated, are limited to few strains of algae, and have difficulty in growing algal cultures for long periods (Ugwu et al. 2008; Mata et al. 2010).

PBRs can be classified on the basis of both design and mode of operation. Many different designs have been developed (serpentine, manifold, helical, and flat), where the main categories include (Richmond 2008) (1) flat or tubular; (2) horizontal, inclined, vertical, or spiral; and (3) manifold or serpentine. From these, elevated reactors can be oriented and tilted at different angles and can use diffuse and reflected light, which plays an important role in productivity.

3.5 Bioethanol Production

The principal fuel used as a petrol substitute for road transport vehicles is bioethanol. Bioethanol fuel is mainly produced by the algal fermentation process, although it can also be manufactured by the chemical process of reacting ethylene with steam.

The main sources of algae required to produce ethanol come from fuel or energy crops. These crops are grown specifically for energy use and include sugarcane, corn, maize and wheat crops, waste straw, willow and popular trees, sawdust, reed canary grass, cord grasses, Jerusalem artichoke, and miscanthus and sorghum plants. There is also ongoing research and development on the use of municipal solid wastes to produce ethanol fuel.

Ethanol or ethyl alcohol (C_2H_5OH) is a clear colorless liquid; it is biodegradable, is low in toxicity, and causes little environmental pollution if spilt. Ethanol burns to produce carbon dioxide and water. Ethanol is a high octane fuel and has replaced lead as an octane enhancer in petrol. By blending ethanol with gasoline, we can also oxygenate the fuel mixture, so it burns more completely and reduces polluting emissions. Ethanol fuel blends are widely sold in the United States. The most common blend is 10% ethanol and 90% petrol (E_{10}). Vehicle engines require no modifications to run on E_{10} and vehicle warranties are unaffected also. Only flexible fuel vehicles can run on up to 85% ethanol and 15% petrol blends (E_{85}).

3.5.1 Dry Grind

In the dry-grind ethanol process, the whole grain is processed, and the residual components are separated at the end of the process. There are five major steps in the dry-grind method of ethanol production.

Dry-grind ethanol processing steps are as follows:

1. Milling
2. Liquefaction
3. Saccharification
4. Fermentation
5. Distillation and recovery

3.5.1.1 Milling

Milling involves processing algae through a hammer mill to produce algae flour. This whole algae flour is slurried with water, and heat-stable enzyme (α-amylase) is added.

3.5.1.2 Liquefaction

This slurry is cooked, also known as *liquefaction*. Liquefaction is accomplished using jet cookers that inject steam into the algae flour slurry to cook it at temperatures above 100°C (212°F). The heat and mechanical shear of the cooking process break apart the starch granules present in the kernel endosperm, and the enzymes break down the starch polymer into small fragments. The cooked algae mash is then allowed to cool at 80°C–90°C (175°F–195°F), additional enzyme (α-amylase) is added, and the slurry is allowed to continue liquefying for at least 30 min.

3.5.1.3 Saccharification

After liquefaction, the slurry, now called *algae mash*, is cooled at approximately 30°C (86°F), and a second enzyme (glucoamylase) is added. Glucoamylase completes the breakdown of the starch into simple sugar (glucose). This step, called *saccharification*, often occurs, while the mash is filling the fermenter in preparation for the next step (fermentation) and continues throughout the next step.

3.5.1.4 Fermentation

In the fermentation step, yeast grown in seed tanks are added to the algae mash to begin the process of converting the simple sugars to ethanol. Other components of the algae kernel remain largely unchanged during the fermentation process. In most dry-grind ethanol plants, the fermentation process occurs in batches. A fermentation tank is filled, and the batch ferments completely before the tank is drained and refilled with a new batch.

The upstream processes (grinding, liquefaction, and saccharification) and downstream processes (distillation and recovery) occur continuously (grain is continuously processed through the equipment). Thus, dry-grind facilities of this design usually have three fermenters (tanks for fermentation) where, at any given time, one is filling, one is fermenting (usually for 48 h), and one is emptying and resetting for the next batch.

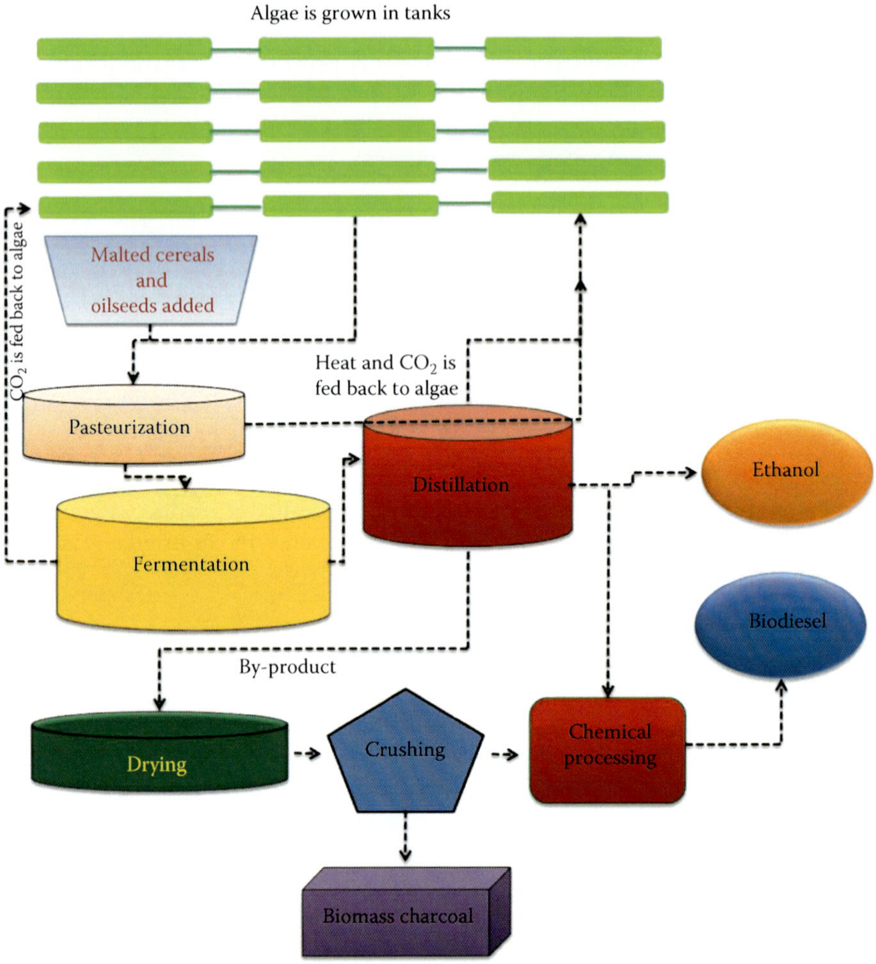

FIGURE 3.2
Main features of bioethanol production.

Carbon dioxide is also produced during fermentation. Usually, the carbon dioxide is not recovered and is released from the fermenters to the atmosphere. If recovered, this carbon dioxide can be compressed and sold for carbonation of soft drinks or frozen into dry ice for cold product storage and transportation.

After the fermentation is complete, the fermented algae mash is emptied from the fermenter into a beer well. The beer well stores the fermented beer between batches and supplies a continuous stream of material to the ethanol recovery steps, including distillation (Figure 3.2).

3.5.1.5 Distillation and Recovery

After fermentation, the liquid portion of the slurry has 8%–12% ethanol by weight. Because ethanol boils at a lower temperature than water does, the ethanol can be separated by a process called *distillation*. Conventional distillation/rectification systems can produce ethanol at 92%–95% purity. The residual water is then removed using molecular sieves that selectively adsorb the water from an ethanol/water vapor mixture, resulting in nearly pure ethanol (>99%).

3.5.2 Ethanol Production of Microalgae

Microalgae are a large group of fast-growing prokaryotic or eukaryotic photosynthetic microorganisms that can live in harsh conditions due to their unicellular or simple multicellular structure. Examples of prokaryotic microorganisms are cyanobacteria (*Cyanophyceae*), and eukaryotic microalgae include, for example, green algae (*Chlorophyceae*) and diatoms (*Bacillariophyceae*) (Li et al. 2008b).

These microorganisms convert sunlight, water, and CO_2 to algal biomass (composed mainly by carbohydrates, proteins, and oils) and can double their biomass in periods as short as 3.5 h presenting high growth rates in cheap culture media (Chisti 2007). While the mechanism of photosynthesis in microalgae is similar to that of higher plants, they are generally more efficient converters of solar energy due to their less complex structure. Furthermore, since microalgae are microscopic in size and grow in liquid culture, nutrients can be maintained at or near optimal conditions potentially providing the benefits of high levels of controlled continuous productivity, similar to those achieved in microbial fermentations (Walker et al. 2005).

Microalgae can provide feedstock for several different types of renewable fuels such as biodiesel, methane, hydrogen, and ethanol, among others (Figure 3.3). Using microalgae as a source of biofuels is not a new idea (Chisti 1980), but it is now being taken seriously

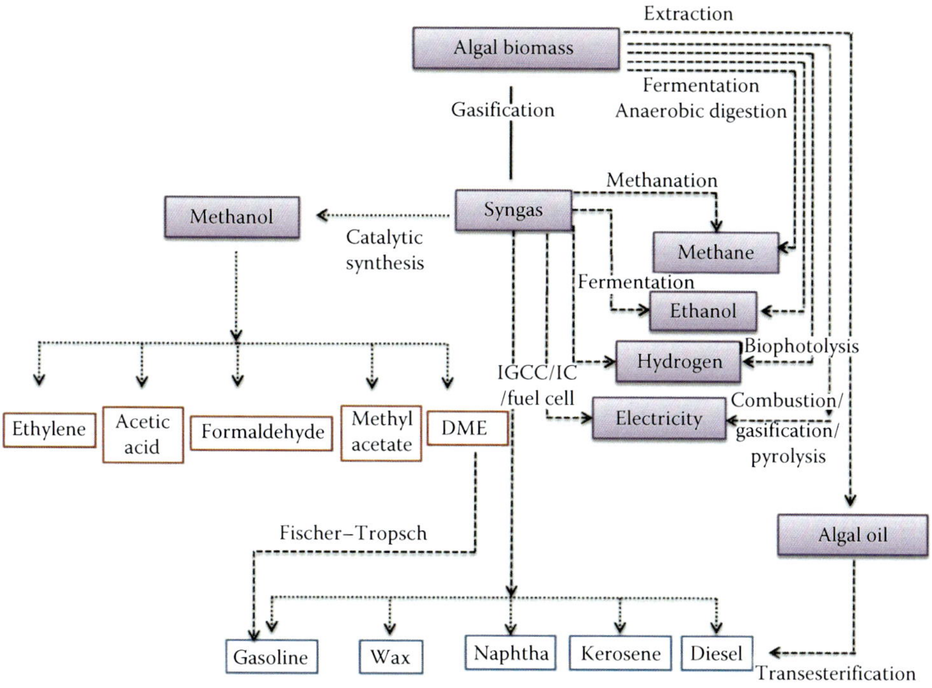

FIGURE 3.3
Main features of biomass energy technology.

because of the increasing cost of petroleum and, more significantly, the emerging concern about global warming arising from burning fossil fuels (Sawayama et al. 1995). The ability of microalgae to fix CO_2 has been proposed as a method of removing CO_2 from flue gases from power plants, thereby reducing the greenhouse gas (GHG) emissions (Abbasi and Abbasi 2010). Approximately half of the dry weight of microalgal biomass is carbon derived from CO_2 (Chisti 1980).

Other advantages of microalgal systems are the following:

1. Microalgae can be harvested batch-wise nearly all year round.
2. They grow in aqueous media but need less water than terrestrial crops, therefore reducing the load on freshwater sources.
3. Microalgae can be cultivated in brackish water on nonarable land and therefore may not incur land-use change, minimizing associated environmental impacts while not compromising the production of food, fodder, and other products derived from crops.
4. Microalgal biomass production can effect biofixation of waste CO_2 (1 kg of dry algal biomass utilizes about 1.83 kg of CO_2).
5. Nutrients for microalgae cultivation (especially nitrogen and phosphorus) can be obtained from wastewater; therefore, apart from providing growth medium, there is dual potential for treatment of industrial and domestic sewage.
6. Microalgae cultivation does not require herbicide or pesticide application.
7. They can also produce valuable coproducts such as proteins and residual biomass, which may be used as feed or fertilizer.
8. The biochemical composition of the microalgal biomass can be modulated by varying growth conditions; therefore, the oil or starch yields may be significantly enhanced (Brennan and Owende 2010).

Certain species of microalgae have the ability of producing high levels of carbohydrates instead of lipids as reserve polymers. These species are ideal candidates for the production of bioethanol as carbohydrates from microalgae can be extracted to produce fermentable sugars. It has been estimated that approximately 5,000–15,000 gal of ethanol/acre/year (46,760–140,290 L/ha) can be produced from microalgae. This yield is several orders of magnitude larger than yields obtained for other feedstocks. Even switch grass, considered as the cellulosic *second generation* of biofuels, achieves ethanol yields that are a fraction of microalgal yield.

The microalga *Chlorella vulgaris*, particularly, has been considered as a promising feedstock for bioethanol production because it can accumulate up to 37% (dry weight) of starch. However, higher starch contents can also be obtained for optimized culture conditions. Under favorable climate conditions, yields of 80–100 t dried *Chlorella* biomass per hectare area for a 300-day culture season can be reached (Doucha and Lívanský 2009). According to these authors, the *Chlorella* sp. strain is able to accumulate starch up to 70% of algal dry weight under conditions of protein synthesis suppression. *Chlorococum* sp. was also used as a substrate for bioethanol production under different fermentation conditions. Results showed a maximum bioethanol concentration of 3.83 g/L obtained from 10 g/L of lipid-extracted microalgae debris (Harun et al. 2010).

Production of ethanol by using microalgae as raw material can be performed according to the following procedure (Figure 3.4). In the first step, microalgae cultivation using

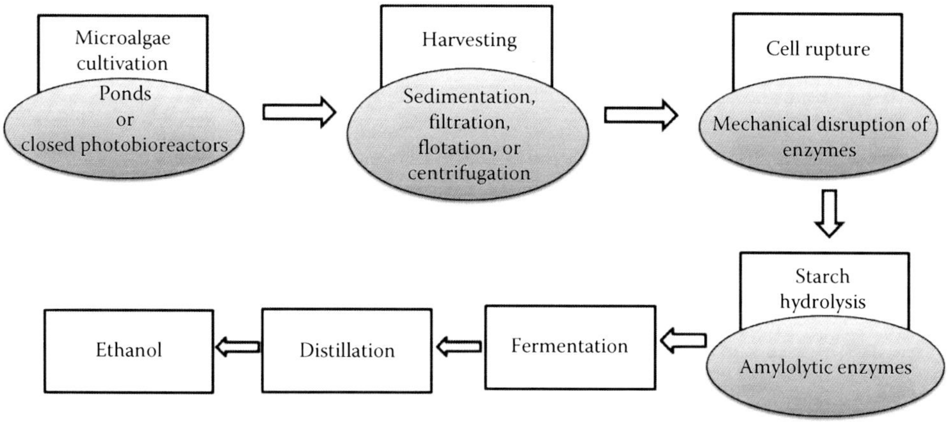

FIGURE 3.4
Procedure for bioethanol production from microalgae.

sunlight energy is carried out in open or covered ponds or closed PBRs, based on tubular, flat plate, or other designs. In the second step, the biomass needs to be concentrated by an initial factor of at least about 30-fold, requiring very-low-cost harvesting processes. After harvesting, microalgal starch is extracted from the cells with the aid of mechanical equipment or enzymes. Following starch extraction, amylolytic enzymes are used to promote formation of fermentable sugars. *Saccharomyces cerevisiae* is then added to begin alcoholic fermentation. At the end of fermentation, fermented broth containing ethanol is drained from the tank and pumped into a holding tank to be fed to a distillation unit (Amin 2009).

Apart from that, instead of extraction, there are also algal species able to conduct self-fermentation. Ueno et al. (1998) reported that dark fermentation in the marine green alga *Chlorococcum littorale* was able to produce 450 μmol/g dry-wt ethanol, at 30°C. Seambiotic, in collaboration with Inventure Chemicals, successfully demonstrated the production of ethanol by fermentation of microalgal polysaccharides. Seambiotic is the first company in the world that is utilizing flue gases from coal-burning power stations as a source of CO_2 for microalgae cultivation (Ueno et al. 1998).

3.6 Methane Production

3.6.1 What Is Biogas?

Biogas originates from marine bacteria in the process of biodegradation of organic material under anaerobic (without air) conditions. The natural generation of biogas is an important part of the biogeochemical carbon cycle. Methanogens (methane-producing bacteria) are the last link in a chain of microorganisms that degrade organic material and return the decomposition products to the environment. In this process, biogas is generated, a source of renewable energy.

3.6.2 Benefits of Biogas Technology

Well-functioning biogas systems can yield a whole range of benefits for their users, the society, and the environment in general:

- Production of energy (heat, light, and electricity)
- Transformation of organic waste into high-quality fertilizer
- Improvement of hygienic conditions through reduction of pathogens, worm eggs, and flies
- Reduction of workload, mainly for women, in firewood collection and cooking
- Environmental advantages through protection of soil, water, air, and woody vegetation
- Microeconomical benefits through energy and fertilizer substitution, additional income sources, and increasing yields of animal husbandry and agriculture
- Macroeconomical benefits through decentralized energy generation, import substitution, and environmental protection

Thus, biogas technology can substantially contribute to conservation and development, if the concrete conditions are favorable. However, the required high investment capital and other limitations of biogas technology should be thoroughly considered.

3.6.3 Three Steps of Biogas Production

Biogas microbes consist of a large group of complex and differently acting microbe species, notably the methane-producing bacteria. The whole biogas process can be divided into three steps: hydrolysis, acidification, and methane formation (Figure 3.5). Three types of bacteria are involved.

3.6.3.1 Hydrolysis

In the first step (hydrolysis), the organic matter is enzymolyzed externally by extracellular enzymes (cellulase, amylase, protease, and lipase) of microorganisms. Bacteria decompose the long chains of the complex carbohydrates, proteins, and lipids into shorter parts. For example, polysaccharides are converted into monosaccharides. Proteins are split into peptides and amino acids.

3.6.3.2 Acidification

Acid-producing bacteria, involved in the second step, convert the intermediates of fermenting bacteria into acetic acid (CH_3COOH), hydrogen (H_2), and carbon dioxide (CO_2). These bacteria are facultatively anaerobic and can grow under acid conditions. To produce acetic acid, they need oxygen and carbon. For this, they use the oxygen solved in the solution or bounded oxygen. Hereby, the acid-producing bacteria create an anaerobic condition, which is essential for the methane-producing microorganisms. Moreover, they reduce the compounds with a low molecular weight into alcohols, organic acids, amino acids, carbon dioxide, hydrogen sulfide, and traces of methane. From a chemical standpoint, this process is partially endergonic (i.e., only possible with energy input), since bacteria alone are not capable of sustaining that type of reaction.

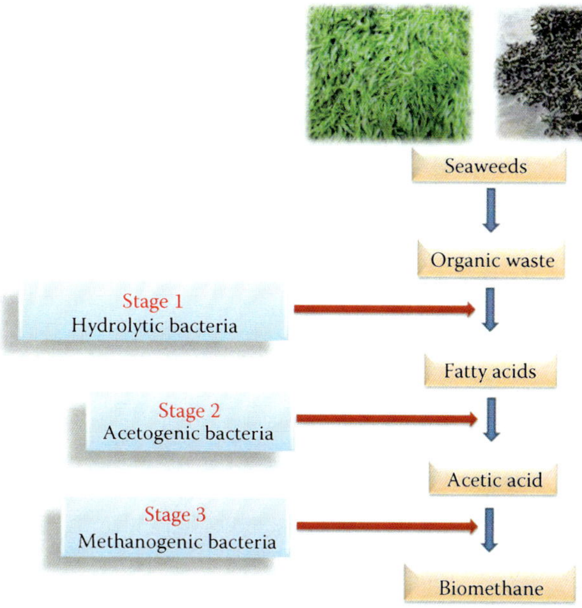

FIGURE 3.5
The three-stage anaerobic fermentation of biomass.

3.6.3.3 Methane Formation

Methane-producing bacteria, involved in the third step, decompose compounds with a low molecular weight. For example, they utilize hydrogen, carbon dioxide, and acetic acid to form methane and carbon dioxide. Under natural conditions, methane-producing microorganisms occur to the extent that anaerobic conditions are provided, for example, under water (e.g., in marine sediments), in ruminant stomachs, and in marshes. They are obligatory anaerobic and very sensitive to environmental changes. In contrast to the acidogenic and acetogenic bacteria, the methanogenic bacteria belong to the *Archaebacteria* genus, that is, to a group of bacteria with a very heterogeneous morphology and a number of common biochemical and molecular–biological properties that distinguish them from all other bacterial general. The main difference lies in the makeup of the bacteria's cell walls.

3.7 Production of Biohydrogen

Biohydrogen is defined as hydrogen produced biologically, most commonly by algae, bacteria, and archaea. Biohydrogen is a potential biofuel obtainable from both cultivation and waste organic materials (Demirbaş 2009).

3.7.1 Algaeic Biohydrogen

In 1939, a German researcher named Hans Gaffron, while working at the University of Chicago, observed that *Chlamydomonas reinhardtii* (a green algae), would sometimes switch from the production of oxygen to the production of hydrogen. Gaffron never

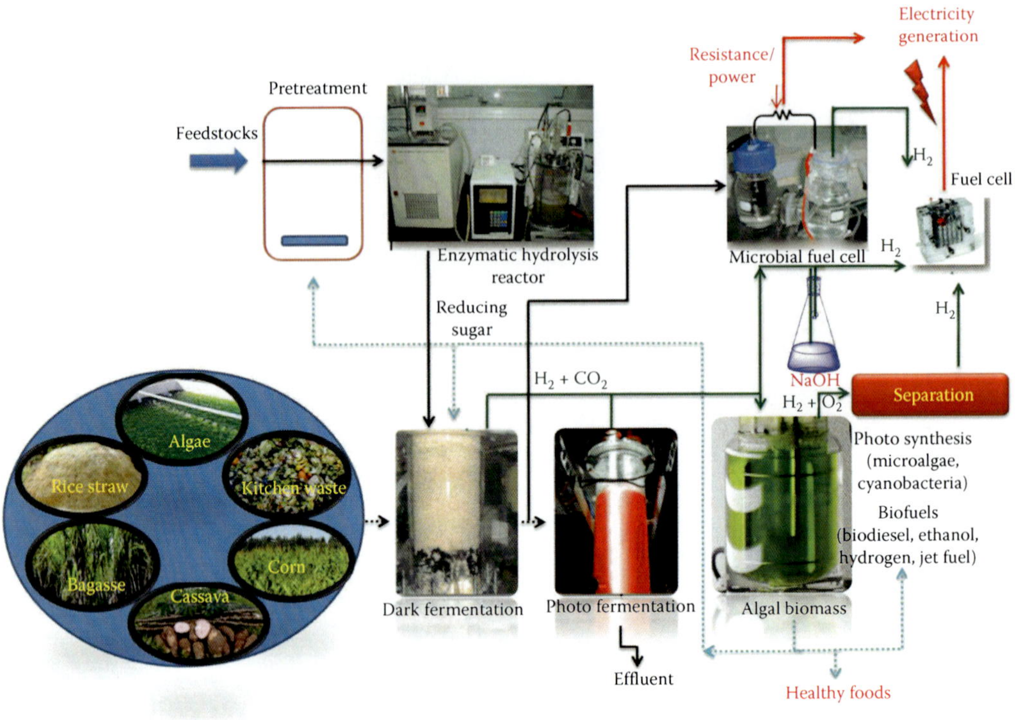

FIGURE 3.6
Procedure for biohydrogen production.

discovered the cause for this change, and for many years, other scientists failed in their attempts at its discovery. In the late 1990s, professor Anastasios Melis, a researcher at the University of California at Berkeley, discovered that if the algae culture medium is deprived of sulfur, it will switch from the production of oxygen (normal photosynthesis) to the production of hydrogen. He found that the enzyme responsible for this reaction is hydrogenase but that the hydrogenase lost this function in the presence of oxygen. Melis found that depleting the amount of sulfur available to the algae interrupted its internal oxygen flow, allowing the hydrogenase an environment in which it can react, causing the algae to produce hydrogen. *C. moewusii* is also a good strain for the production of hydrogen. Scientists at the U.S. Department of Energy's Argonne National Laboratory are currently trying to find a way to take the part of the hydrogenase enzyme that creates the hydrogen gas and introduce it into the photosynthesis process. The result would be a large amount of hydrogen gas, possibly on par with the amount of oxygen created (Figure 3.6) (Melis and Happe 2001).

3.7.2 Bacterial Biohydrogen

3.7.2.1 Process Requirements

If hydrogen by fermentation is to be introduced as an industry, the fermentation process will be dependent on organic acids as substrate for photofermentation. The organic acids are necessary for high hydrogen production rates (Tao et al. 2007).

The organic acids can be derived from any organic material source such as sewage wastewaters or agricultural wastes. The most important organic acids are acetic acid (HAc), butyric acid (HBc), and propionic acid (HPc). A huge advantage is that production of hydrogen by fermentation does not require glucose as substrate.

The fermentation of hydrogen has to be a continuous fermentation process, in order to sustain high production rates, since the amount of time for the fermentation to enter high production rates is in days.

3.7.2.2 Fermentation

Several strategies for the production of hydrogen by fermentation in lab scale have been found in literature. However, no strategies for industrial-scale productions have been found. In order to define an industrial-scale production, the information from lab-scale experiments has been scaled to an industrial-size production on a theoretical basis. In general, the method of hydrogen fermentation is referred to in three main categories. The first category is dark fermentation, which is fermentation that does not involve light. The second category is photofermentation, which is fermentation that requires light as the source of energy. The third is combined fermentation, which refers to the two fermentations combined.

3.7.2.3 Dark Fermentation

There are several bacteria with a potential for hydrogen production. The Gram-positive bacteria or the *Clostridium* genus is promising because it has a natural high hydrogen production rate. In addition, it is fast growing and capable of forming spores, which make the bacteria easy to handle in industrial application (Krupp and Widmann 2009).

Species of the *Clostridium* genus allow hydrogen production in mixed culture, under mesophilic or thermophilic conditions within a pH range of 5.0–6.5. Dark fermentation with mixed cultures seems promising since a mixed bacterial environment within the fermenter allows cooperation of different species to efficiently degrade and convert organic waste materials into hydrogen, accompanied by the formation of organic acids (Nath and Das 2004). The clostridia produce H_2 via a reversible hydrogenase (H_2ase) enzyme ($2H + 2e \Leftrightarrow H_2$), and this reaction is important in achieving the redox balance of fermentation. The rate of H_2 formation is inhibited as H_2 production causes the partial pressure of H_2 (pH 2.0) to increase. This can limit substrate conversion and growth, and the bacteria may respond by switching to a different metabolic pathway in order to achieve redox balance, energy generation, and growth—by producing solvents instead of hydrogen and organic acids.

Enteric bacteria such as *Escherichia coli* and *Enterobacter aerogenes* are also interesting for biohydrogen fermentation (Redwood et al. 2008). Unlike the clostridia, the enteric bacteria produce hydrogen primarily (or exclusively in the case of *E. coli*) by the cleavage of formate ($HCOOH \rightarrow H_2 + CO_2$), which serves to detoxify the medium by removing formate. Cleavage is not a redox reaction and it has no consequence on the redox balance of fermentation. This detoxification is particularly important for *E. coli* as it cannot protect itself by forming spores. Formate cleavage is an irreversible reaction; hence, H_2 production is insensitive to the partial pressure of hydrogen (pH 2.0) in the fermenter. *E. coli* has been referred to as the workhorse of molecular microbiology, and many workers have investigated metabolic engineering approaches to improve biohydrogen fermentation in *E. coli*.

Whereas oxygen kills clostridia, the enteric bacteria are facultative anaerobes; they grow very quickly when oxygen is available and transit progressively from aerobic to anaerobic metabolism as oxygen becomes depleted. Growth rate is much slower during anaerobic fermentation than during aerobic respiration because fermentation has less metabolic energy from the same substrate. In practical terms, facultative anaerobes are useful because they can be grown quickly to a very high concentration with oxygen and then used to produce hydrogen at a high rate when the oxygen supply is stopped (Redwood et al. 2012).

For fermentation to be sustainable at industrial scale, it is necessary to control the bacterial community inside the fermenter. Feedstocks may contain microorganisms, which could cause changes in the microbial community inside the fermenter. The enteric bacteria and most clostridia are mesophilic; they have an optimum temperature of around 30°C as do many common environmental microorganisms. Therefore, these fermentations are susceptible to changes in the microbial community unless the feedstock is sterilized, for example, where a hydrothermal pretreatment is involved, sterilization is a side effect (Orozco et al. 2012). A way to prevent harmful microorganisms from gaining control of the bacterial environment inside the fermenter could be through addition of the desired bacteria. Hyperthermophilic archaea such as *Thermotoga neapolitana* can also be used for hydrogen fermentation. Because they operate at around 70°C, there is little chance of feedstock contaminants becoming established.

Fermentations produce organic acids that are toxic to the bacteria. High concentrations inhibit the fermentation process and may trigger changes in metabolism and resistance mechanisms such as sporulation in different species. This fermentation of hydrogen is accompanied by the production of carbon dioxide that can be separated from hydrogen with a passive separation process.

The fermentation will convert some of the substrate (e.g., waste) into biomass instead of hydrogen. The biomass is, however, a carbohydrate-rich by-product that can be fed back into the fermenter to ensure that the process is sustainable. Fermentation of hydrogen by dark fermentation is restricted by incomplete degradation of organic material, into organic acids, and this is why we need the photofermentation.

The separation of organic acids from biomass in the outlet stream can be done with a settler tank in the outlet stream, where the sludge (biomass) is pumped back into the fermenter to increase the rate of hydrogen production.

In traditional fermentation systems, the dilution rate must be carefully controlled as it affects the concentration of bacterial cells and toxic end products (organic acids and solvents) inside the fermenter. A more complex *electrofermentation* technique decouples the retention of water and biomass and overcomes inhibition by organic acids.

3.7.2.4 Photofermentation

Photofermentation refers to the method of fermentation where light is required as the source of energy. This fermentation relies on photosynthesis to maintain the cellular energy levels. Fermentation by photosynthesis compared to other fermentations has the advantage of light as the source of energy instead of sugar. Sugars are usually available in limited quantities.

All plants, algae, and some bacteria are capable of photosynthesis: utilizing light as the source of metabolic energy. Cyanobacteria are frequently mentioned as capable of hydrogen production by oxygenic photosynthesis. However, the purple nonsulfur (PNS) bacteria (e.g., genus *Rhodobacter*) hold significant promise for the production of hydrogen by anoxygenic photosynthesis and photofermentation.

Studies have shown that *Rhodobacter sphaeroides* is highly capable of hydrogen production while feeding on organic acids, consuming 98%–99% of the organic acids during hydrogen production. Organic acids may be sourced sustainably from the dark fermentation of waste feedstocks. The resultant system is called combined fermentation.

Photofermentative bacteria can use light in the wavelength range 400–1000 nm (visible and near-infrared), which differs from algae and cyanobacteria (400–700 nm; visible).

Currently, there is limited experience with photofermentation at the industrial scale. The distribution of light within the industrial-scale photofermenter has to be designed to minimize self-shading. Therefore, any externally illuminated PBR must have a high ratio of high surface area to volume. As a result, PBR construction is material intensive and expensive.

A method to ensure proper light distribution and limit self-shading within the fermenter could be to distribute the light with an optic fiber where light is transferred into the fermenter and distributed from within the fermenter. Photofermentation with *R. sphaeroides* requires mesophilic conditions. An advantage of the optical fiber PBR is that radiant heat gain can be controlled by dumping excess light and filtering out wavelengths that cannot be used by the organisms.

3.7.2.5 Combined Fermentation

Combining dark and photofermentation has shown to be the most efficient method to produce hydrogen through fermentation. The combined fermentation allows the organic acids produced during dark fermentation of waste materials to be used as substrate in the photofermentation process. Many independent studies show this technique to be effective and practical.

For industrial fermentation of hydrogen to be economically feasible, by-products of the fermentation process have to be minimized. Combined fermentation has the unique advantage of allowing reuse of the otherwise useless chemical, organic acids, through photosynthesis.

Many wastes are suitable for fermentation, and this is equivalent to the initial stages of anaerobic digestion, now the most important biotechnology for energy from waste. One of the main challenges in combined fermentation is that effluent fermentation contains not only useful organic acids but excess nitrogenous compounds and ammonia, which inhibit nitrogenase activity by wild-type PNS bacteria. The problem can be solved by genetic engineering to interrupt downregulation of nitrogenase in response to nitrogen excess. However, genetically engineered bacterial strains may pose containment issues for application. A physical solution to this problem was developed at the University of Birmingham, United Kingdom, which involves selective electroseparation of organic acids from an active fermentation. The energetic cost of electroseparation of organic acids was found to be acceptable in a combined fermentation. *Electrofermentation* has the side effect of a continuous, high-rate dark hydrogen fermentation. As the method for hydrogen production, combined fermentation currently holds significant promise.

3.7.3 Hydrogen Production from Algae

Algae split water molecules to hydrogen ion and oxygen via photosynthesis. The generated hydrogen ions are converted into hydrogen gas by hydrogenase enzyme. *C. reinhardtii* is one of the well-known hydrogen-producing algae. Hydrogenase activity has been detected in green algae, *Scenedesmus obliquus*, in marine green algae *C. littorale* and *Platymonas subcordiformis*, and in *Chlorella fusca*. However, no hydrogenase activity was observed in *C. vulgaris* and *Dunaliella salina*. The hydrogenase activity of different algal species was compared by Winkler et al. (2002), and it was reported that enzyme activity of the *Scenedesmus* sp. (150 nmol/μg Chl a.h) is lower than *C. reinhardtii* (200 nmol/μg Chl a.h) (Winkler et al. 2002).

Cyanobacterial hydrogen gas evolution involves nitrogen fixing cultures such as nonmarine *Anabaena* sp., marine cyanobacterium *Oscillatoria* sp., *Calothrix* sp., and non-nitrogen fixing organisms such as *Synechococcus* sp. and *Gloeobacter* sp., and it was reported that *Anabaena* sp. has higher hydrogen evolution potential over the other cyanobacterial species. Heterocystous filamentous *Anabaena cylindrica* is a well-known hydrogen-producing cyanobacterium. However, *A. variabilis* has received more attention in recent years because of higher hydrogen production capacity. The growth conditions for *Anabaena* include nitrogen-free media, illumination, CO_2, and O_2. Since nitrogenase enzyme is inhibited by oxygen, hydrogen production is realized under anaerobic conditions. CO_2 is required for some cultures during the hydrogen evolution phase although an inhibition effect of CO_2 on the photoproduction of H_2 was also observed. Four percent to eighteen percent CO_2 concentrations were reported to increase cell density during the growth phase resulting in higher hydrogen evolution in the later stage. The use of simple sugars as supplement was reported to promote hydrogen evolution. Recent studies are concentrated on the development of hydrogenase and bidirectional hydrogenase-deficient mutant of *Anabaena* sp. in order to increase the rate of hydrogen production. At the present time, the rate of hydrogen production by *Anabaena* sp. is considerably lower than that obtained by dark or photofermentations (Masukawa et al. 2002).

The algal hydrogen production could be considered as an economical and sustainable method in terms of water utilization as a renewable resource and CO_2 consumption as one of the air pollutants. However, strong inhibition effect of generated oxygen on hydrogenase enzyme is the major limitation for the process. Inhibition of the hydrogenase enzyme by oxygen can be alleviated by cultivation of algae under sulfur deprivation for 2–3 days to provide anaerobic conditions in the light (Winkler et al. 2002). Low hydrogen production potential and no waste utilization are the other disadvantages of hydrogen production by algae. Therefore, dark and photofermentations are considered to be more advantageous due to simultaneous waste treatment and hydrogen gas production.

3.8 Microalgae for Biodiesel Production

During the past decades, extensive collections of microalgae have been created by researchers in different countries. An example is the freshwater microalgae collection of the University of Coimbra (Portugal), which is considered as one of the world's largest, having more than 4000 strains and 1000 species. This collection attests to the large variety of different microalgae available to be selected for use in a broad diversity of applications, such as value-added products for pharmaceutical purposes, food crops for human consumption, and as energy source.

A bit all over the world, other algae collections attest for the interest that algae have risen for many different production purposes. For example, the collection of Goettingen University, Germany (SAG), started in the early 1920s and has about 2213 strains and 1273 species. About 77% of all the strains in the SAG collection are green algae and about 8%

cyanobacteria (61 genera and 230 strains). Some of them are freshwater red algae and others from saline environments.

The University of Texas' Culture Collection of Algae is another very well-known collection of algae cultures that was founded in 1953. It includes 2300 different strains of freshwater algae (edaphic green algae and cyanobacteria) but includes representatives of most major algal taxa, including many marine macrophytic green and red algal species.

In the Asian continent, the National Institute for Environmental Studies (NIES) culture collection in Ibaraki, Japan, holds a collection of about 2150 strains, with around 700 species of different algae. The CSIRO Collection of Living Microalgae (CCLM), in Australia, holds about 800 strains of different algae, including representatives from the majority of classes of marine and some freshwater microalgae, being the majority of the strains isolated from Australian waters.

3.8.1 Advantages of Using Microalgae for Biodiesel Production

Many research reports and articles described many advantages of using microalgae for biodiesel production in comparison with other available feedstocks (Chisti 2007; Rosenberg et al. 2008). From a practical point of view, they are easy to cultivate, can grow with little or even no attention, using water unsuitable for human consumption and easy to obtain nutrients.

Microalgae reproduce themselves using photosynthesis to convert sun energy into chemical energy, completing an entire growth cycle every few days. Moreover, they can grow almost anywhere, requiring sunlight and some simple nutrients, although the growth rates can be accelerated by the addition of specific nutrients and sufficient aeration (Aslan and Kapdan 2006).

Different microalgal species can be adapted to live in a variety of environmental conditions. Thus, it is possible to find species best suited to local environments or specific growth characteristics, which is not possible to do with other current biodiesel feedstocks (e.g., soybean, rapeseed, sunflower, and palm oil).

They have much higher growth rates and productivity when compared to conventional forestry, agricultural crops, and other aquatic plants, requiring much less land area than other biodiesel feedstocks of agricultural origin, up to 49 or 132 times less when compared to rapeseed or soybean crops, for a 30% (w/w) of oil content in algal biomass (Chisti 2007). Therefore, the competition for arable soil with other crops, in particular for human consumption, is greatly reduced.

Microalgae can provide feedstock for several different types of renewable fuels such as biodiesel, methane, hydrogen, and ethanol, among others. Algal biodiesel contains no sulfur and performs as well as petroleum diesel while reducing emissions of particulate matter, CO, hydrocarbons, and SO_x. However, emissions of NO_x may be higher in some engine types (Delucchi 2003).

The utilization of microalgae for biofuel production can also serve other purposes. Some possibilities currently being considered are listed as follows:

- Removal of CO_2 from industrial flue gases by algae biofixation (Wang et al. 2008), reducing the GHG emissions of a company or process while producing biodiesel.
- Wastewater treatment by removal of NH_4^+, NO_3^-, and PO_4^{3-}, making algae grow using these water contaminants as nutrients (Wang et al. 2008).

- After oil extraction, the resulting algal biomass can be processed into ethanol, methane, and livestock feed, used as organic fertilizer due to its high N/P ratio, or simply burned for energy cogeneration (electricity and heat) (Wang et al. 2008).
- Combined with their ability to grow under harsher conditions, and their reduced needs for nutrients, they can be grown in areas unsuitable for agricultural purposes independently of the seasonal weather changes, thus not competing for arable land use, and can use wastewaters as the culture medium, not requiring the use of freshwater.
- Depending on the microalgal species, other compounds may also be extracted, with valuable applications in different industrial sectors, including a large range of fine chemicals and bulk products, such as fats, polyunsaturated fatty acids, oil, natural dyes, sugars, pigments, antioxidants, high-value bioactive compounds, and other fine chemicals and biomass (Raja et al. 2008).
- Because of this variety of high-value biological derivatives, with many possible commercial applications, microalgae can potentially revolutionize a large number of biotechnology areas including biofuels, cosmetics, pharmaceuticals, nutrition and food additives, aquaculture, and pollution prevention (Raja et al. 2008).

3.8.2 Microalgae Lipid Content and Productivities

Many microalgal species can be induced to accumulate substantial quantities of lipids, thus contributing to a high oil yield. The average lipid content varies between 1% and 70%, but under certain conditions, some species can reach 90% of dry weight (Chisti 2007).

Oil content in microalgae can reach 75% by weight of dry biomass but associated with low productivities (e.g., for *Botryococcus braunii*). Most common algae (*Chlorella*, *Crypthecodinium*, *Cylindrotheca*, *Dunaliella*, *Isochrysis*, *Nannochloris*, *Nannochloropsis*, *Neochloris*, *Nitzschia*, *Phaeodactylum*, *Porphyridium*, *Schizochytrium*, and *Tetraselmis*) have oil levels between 20% and 50%, but higher productivities can be reached.

Chlorella seems to be a good option for biodiesel production. Yet, as other species are so efficient and productive as this one, the selection of the most adequate species needs to take into account other factors, such as the ability of microalgae to develop using the nutrients available or under specific environmental conditions. All these parameters should be considered simultaneously in the selection of the most adequate species or strains for biodiesel production.

3.8.3 Microalgal Biodiesel Value Chain Stages

Although, in a simplistic view, microalgae may seem to not significantly differ from other biodiesel feedstocks, they are microorganisms living essentially in liquid environments, thus with particular cultivation, harvesting, and processing techniques that ought to be considered in order to efficiently produce biodiesel.

All existing processes for biodiesel production from microalgae include a production unit where cells are grown, followed by the separation of the cells from the growing media and subsequent lipid extraction. Then biodiesel or other biofuels are produced in a form akin to existing processes and technologies used for other biofuel feedstocks. Recently, other possibilities for biofuel production are being pursued instead of the transesterification reaction, such as the thermal cracking (or pyrolysis) involving the thermal decomposition or cleavage of triglycerides and other organic compounds presented in the feedstock

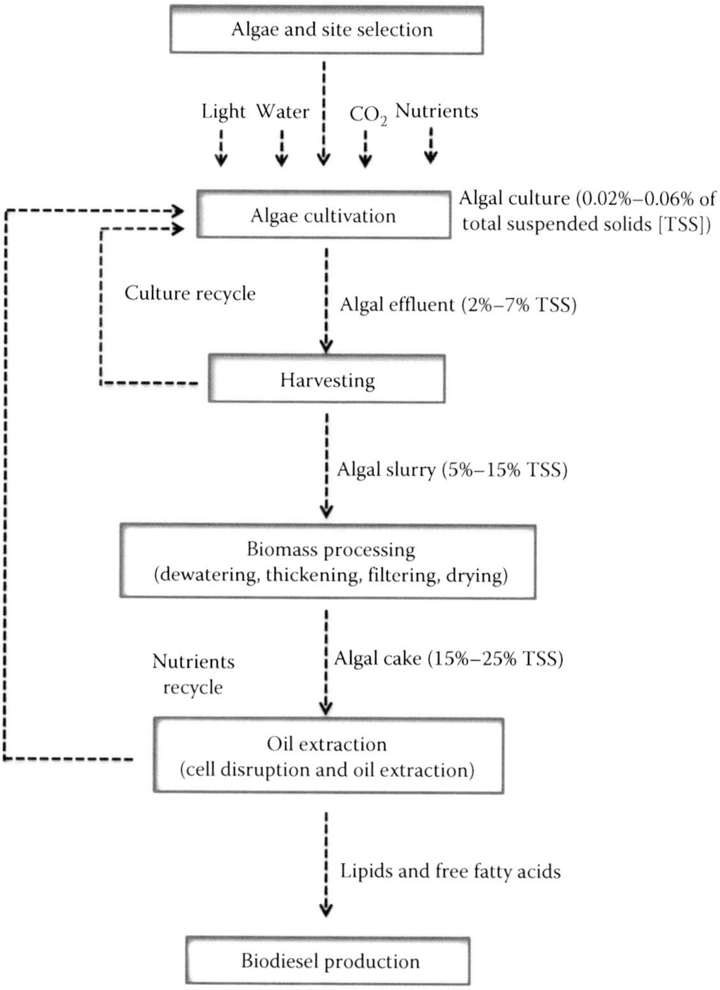

FIGURE 3.7
Microalgal biodiesel value chain stages.

and in simpler molecules, namely, alkanes, alkenes, aromatics, and carboxylic acids, among others (Bahadur et al. 1995).

Figures 3.7 and 3.8 show a schematic representation of the algal biodiesel value chain stages, starting with the selection of microalgal species depending on local specific conditions and the design and implementation of the cultivation system for microalgal growth. Then it follows the biomass harvesting, processing, and oil extraction to supply the biodiesel production unit.

3.8.4 Algae Cultivation

Microalgae are adapted to scavenge their environments for resources, to store them, or to increase their efficiency in resource utilization. In general, for biomass growth (consisting of 40%–50% carbon), microalgae depend on a sufficient supply of a carbon source and

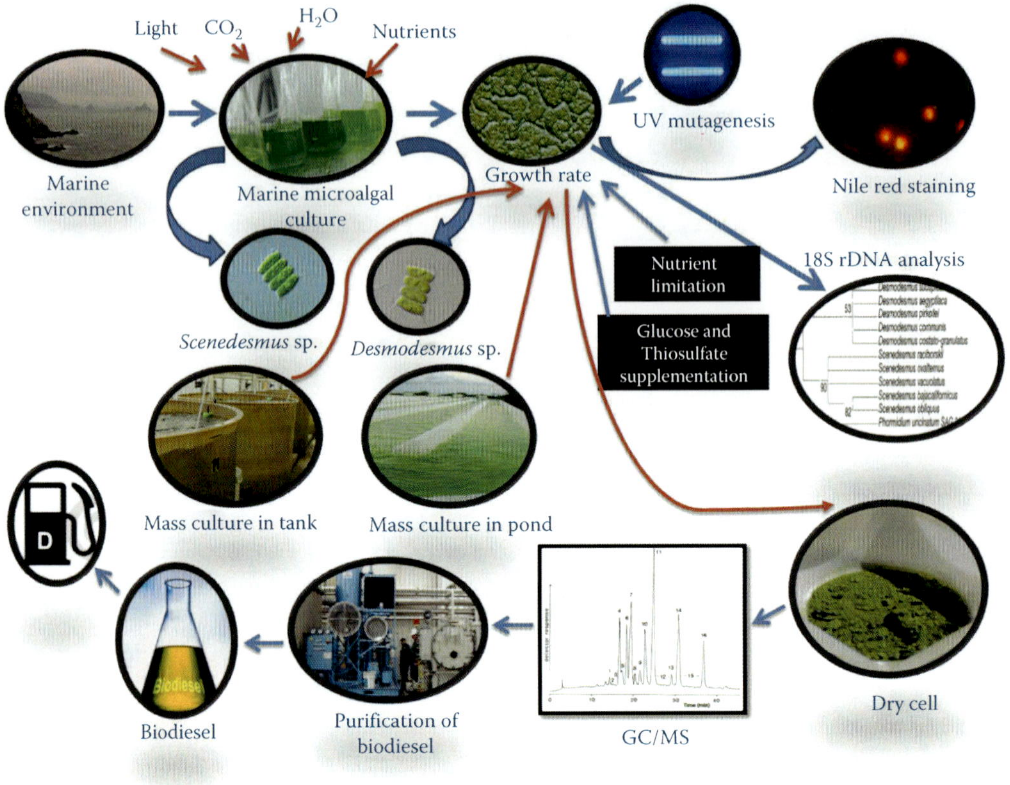

FIGURE 3.8
Biodiesel production of marine algae.

light to carry out photosynthesis (Chojnacka 2004). Yet they can adjust or change their internal structure (e.g., biochemical and physiological acclimation), while externally they can excrete a variety of compounds to among others, render nutrients available, or limit the growth of competitors (Richmond 2008).

Microalgae may assume many types of metabolisms (e.g., autotrophic, heterotrophic, mixotrophic, photoheterotrophic) and are capable of a metabolic shift as a response to changes in environmental conditions. For example, some organisms can grow

- Photoautotrophically, that is, using light as a sole energy source that is converted to chemical energy through photosynthetic reactions.
- Heterotrophically, that is, utilizing only organic compounds as carbon and energy source.
- Mixotrophically, that is, performing photosynthesis as the main energy source, though both organic compounds and CO_2 are essential. Amphitrophy, subtype of mixotrophy, means that organisms are able to live either autotrophically or heterotrophically, depending on the concentration of organic compounds and light intensity available.

- Photoheterotrophically, also known as photoorganotrophy, photoassimilation, or photometabolism, describes the metabolism in which light is required to use organic compounds as carbon source. The photoheterotrophic and mixotrophic metabolisms are not well distinguished; in particular, they can be defined according to a difference of the energy source required to perform growth and specific metabolite production.

3.8.5 Biodiesel Production

Biodiesel is a mixture of fatty acid alkyl esters obtained by transesterification (ester exchange reaction) of vegetable oils or animal fats. These lipid feedstocks are composed of 90%–98% (weight) of triglycerides and small amounts of mono- and diglycerides, free fatty acids (1%–5%), and residual amounts of phospholipids, phosphatides, carotenes, tocopherols, and sulfur compounds, and traces of water (Bozbas 2008).

Transesterification is a multiple-step reaction, including three reversible steps in series, where triglycerides are converted to diglycerides, and then diglycerides are converted to monoglycerides, and monoglycerides are then converted to esters (biodiesel) and glycerol (by-product). The overall transesterification reaction is described in Figure 3.9 where the radicals R1, R2, and R3 represent long-chain hydrocarbons, known as fatty acids.

For the transesterification reaction, oil or fat and a short-chain alcohol (usually methanol) are used as reagents in the presence of a catalyst (usually NaOH). Although the alcohol/oil theoretical molar ratio is 3:1, the molar ratio of 6:1 is generally used to complete the reaction accurately. The relationship between the feedstock mass input and biodiesel mass output is about 1:1, which means that theoretically, 1 kg of oil results in about 1 kg of biodiesel.

A homogeneous or heterogeneous, acid or basic catalyst can be used to enhance the transesterification reaction rate, although for some processes using supercritical fluids (methanol or ethanol), it may not be necessary to use a catalyst (Warabi et al. 2004). Most common industrial processes use homogeneous alkali catalysts (e.g., NaOH or KOH) in a stirred reactor operating in batch mode.

Recently, some improvements were proposed for this process, in particular, to be able to operate in continuous mode with reduced reaction time, such as reactors with improved mixing, microwave-assisted reaction, cavitation reactors, and ultrasonic reactors.

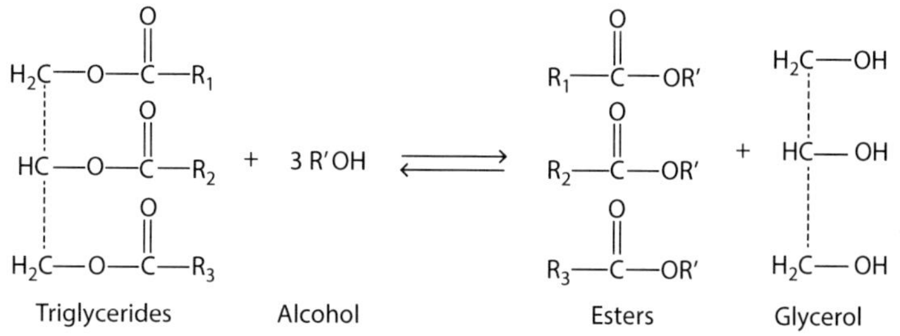

FIGURE 3.9
Transesterification of triglycerides (overall reaction).

3.9 Conclusions

Current efforts and business investment are driving attention and marketing efforts on the promises of producing algal biodiesel and superior production systems. A large number of companies are claiming that they are at the forefront of the technology and will be producing algal biodiesel economically within the next few years. However, most of these companies have limited technical expertise and few have actually made biodiesel from algae.

Producing algal biodiesel requires large-scale cultivation and harvesting systems, with the challenge of reducing the cost per unit area. At a large scale, the algal growth conditions need to be carefully controlled and optimum nurturing environment has to be provided. Such processes are most economical when combined with sequestration of CO_2 from flue gas emissions, with wastewater remediation processes, and/or with the extraction of high-value compounds for application in other process industries.

Current limitations to a more widespread utilization of this feedstock for biodiesel production concern the optimization of the microalgal harvesting, oil extraction processes, and supply of CO_2 for a high efficiency of microalgae production. Also, light, nutrients, temperature, turbulence, and CO_2 and O_2 levels need to be adjusted carefully to provide optimum conditions for oil content and biomass yield. It is therefore clear that a considerable investment in technological development and technical expertise is still needed before algal biodiesel is economically viable and can become a reality. This should be accomplished together with strategic planning and political and economic support.

Further efforts on microalgae production should concentrate in reducing costs in small-scale and large-scale systems. This can be achieved, for example, by using cheap sources of CO_2 for culture enrichment (e.g., from a flue gas), use of nutrient-rich wastewaters or inexpensive fertilizers, use of cheaper design culture systems with automated process control and with fewer manual labor, and use of greenhouses and heated effluents to increase algal yields.

Apart from saving costs of raw materials (nutrients and water use), these measures will help to reduce GHG emissions, waste amount, and the feed cost by using nitrogen fertilizers, will also raise the availability of microalgal biomass for different applications (e.g., food, agriculture, medicine, and biofuels, among others), and will contribute to the sustainability and market competitiveness of the microalgae industry.

Acknowledgment

This research was supported by a grant from the Marine Bioprocess Research Center of the Marine Biotechnology program funded by the Ministry of Oceans and Fisheries, Republic of Korea.

References

Abbasi, T. and S.A. Abbasi. 2010. Biomass energy and the environmental impacts associated with its production and utilization. *Renewable and Sustainable Energy Reviews* 14 (3):919–937.

Amin, S. 2009. Review on biofuel oil and gas production processes from microalgae. *Energy Conversion and Management* 50 (7):1834–1840.
Aslan, S. and I.K. Kapdan. 2006. Batch kinetics of nitrogen and phosphorus removal from synthetic wastewater by algae. *Ecological Engineering* 28 (1):64–70.
Bahadur, N.P., D.G.B. Boocock, and S.K. Konar. 1995. Liquid hydrocarbons from catalytic pyrolysis of sewage sludge lipid and canola oil: Evaluation of fuel properties. *Energy & Fuels* 9 (2):248–256.
Bozbas, K. 2008. Biodiesel as an alternative motor fuel: Production and policies in the European Union. *Renewable and Sustainable Energy Reviews* 12 (2):542–552.
Brennan, L. and P. Owende. 2010. Biofuels from microalgae—A review of technologies for production, processing, and extractions of biofuels and co-products. *Renewable and Sustainable Energy Reviews* 14 (2):557–577.
Chisti, Y. 1980. An unusual hydrocarbon. *Journal of the Ramsay Society* 81:27–28.
Chisti, Y. 2007. Biodiesel from microalgae. *Biotechnology Advances* 25 (3):294–306.
Chojnacka, K. 2004. Kinetic and stoichiometric relationships of the energy and carbon metabolism in the culture of microalgae. *Biotechnology* 3 (1):21–34.
Chynoweth, D.P. 2002. Review of biomethane from marine biomass. Department of Agricultural and Biological Engineering, University of Florida, Gainesville, FL, pp. 2–8.
Delucchi, M. 2003. A lifecycle emissions model (LEM): Lifecycle emissions from transportation fuels, motor vehicles, transportation modes, electricity use, heating and cooking fuels, and materials. Main Report, UCD-ITS-03-17, Institute of Transportation Studies, University of California, Davis. http://www.its.ucdavis.edu/people/faculty/delucchi/S.
Demirbaş, A. 2009. *Biohydrogen: For Future Engine Fuel Demands*. Springer, London, U.K.
Doucha, J. and K. Lívanský. 2009. Outdoor open thin-layer microalgal photobioreactor: Potential productivity. *Journal of Applied Phycology* 21 (1):111–117.
Harun, R., M.K. Danquah, and G.M. Forde. 2010. Microalgal biomass as a fermentation feedstock for bioethanol production. *Journal of Chemical Technology and Biotechnology* 85 (2):199–203.
Hu, Q., M. Sommerfeld, E. Jarvis et al. 2008. Microalgal triacylglycerols as feedstocks for biofuel production: Perspectives and advances. *The Plant Journal* 54 (4):621–639.
Krupp, M. and R. Widmann. 2009. Biohydrogen production by dark fermentation: Experiences of continuous operation in large lab scale. *International Journal of Hydrogen Energy* 34 (10):4509–4516.
Li, Y., M. Horsman, B. Wang, N. Wu, and C.Q. Lan. 2008a. Effects of nitrogen sources on cell growth and lipid accumulation of green alga *Neochloris oleoabundans*. *Applied Microbiology and Biotechnology* 81 (4):629–636.
Li, Y., M. Horsman, N. Wu, C.Q. Lan, and N. Dubois-Calero. 2008b. Biofuels from microalgae. *Biotechnology Progress* 24 (4):815–820.
Masukawa, H., M. Mochimaru, and H. Sakurai. 2002. Hydrogenases and photobiological hydrogen production utilizing nitrogenase system in cyanobacteria. *International Journal of Hydrogen Energy* 27 (11):1471–1474.
Mata, T.M., A.A. Martins, and N.S. Caetano. 2010. Microalgae for biodiesel production and other applications: A review. *Renewable and Sustainable Energy Reviews* 14 (1):217–232.
McGinn, P.J. and F.M.M. Morel. 2008. Expression and inhibition of the carboxylating and decarboxylating enzymes in the photosynthetic C4 pathway of marine diatoms. *Plant Physiology* 146 (1):300–309.
Melis, A. and T. Happe. 2001. Hydrogen production. Green algae as a source of energy. *Plant Physiology* 127 (3):740–748.
Mussatto, S.I., G. Dragone, P.M.R. Guimarães et al. 2010. Technological trends, global market, and challenges of bio-ethanol production. *Biotechnology Advances* 28 (6):817–830.
Nath, K. and D. Das. 2004. Improvement of fermentative hydrogen production: Various approaches. *Applied Microbiology and Biotechnology* 65 (5):520–529.
Orozco, R.L., M.D. Redwood, G.A. Leeke, A. Bahari, R.C.D. Santos, and L.E. Macaskie. 2012. Hydrothermal hydrolysis of starch with CO_2 and detoxification of the hydrolysates with activated carbon for bio-hydrogen fermentation. *International Journal of Hydrogen Energy* 37 (8):6545–6553.

Raja, R., S. Hemaiswarya, N.A. Kumar, S. Sridhar, and R. Rengasamy. 2008. A perspective on the biotechnological potential of microalgae. *Critical Reviews in Microbiology* 34 (2):77–88.

Redwood, M.D., I.P. Mikheenko, F. Sargent, and L.E. Macaskie. 2008. Dissecting the roles of *Escherichia coli* hydrogenases in biohydrogen production. *FEMS Microbiology Letters* 278 (1):48–55.

Redwood, M.D., R.L. Orozco, A.J. Majewski, and L.E. Macaskie. 2012. Electro-extractive fermentation for efficient biohydrogen production. *Bioresource Technology* 107:166–174.

Reinfelder, J.R., A.M.L. Kraepiel, and F.M.M. Morel. 2000. Unicellular C4 photosynthesis in a marine diatom. *Nature* 407 (6807):996–999.

Richmond, A. 2008. *Handbook of Microalgal Culture: Biotechnology and Applied Phycology*. John Wiley & Sons, Oxford.

Rosenberg, J.N., G.A. Oyler, L. Wilkinson, and M.J. Betenbaugh. 2008. A green light for engineered algae: Redirecting metabolism to fuel a biotechnology revolution. *Current Opinion in Biotechnology* 19 (5):430–436.

Sawayama, S., S. Inoue, Y. Dote, and S.-Y. Yokoyama. 1995. CO_2 fixation and oil production through microalga. *Energy Conversion and Management* 36 (6):729–731.

Schenk, P.M., S.R. Thomas-Hall, E. Stephens et al. 2008. Second generation biofuels: High-efficiency microalgae for biodiesel production. *Bioenergy Research* 1 (1):20–43.

Singh, A., P.S. Nigam, and J.D. Murphy. 2011. Mechanism and challenges in commercialisation of algal biofuels. *Bioresource Technology* 102 (1):26–34.

Tao, Y., Y. Chen, Y. Wu, Y. He, and Z. Zhou. 2007. High hydrogen yield from a two-step process of dark- and photo-fermentation of sucrose. *International Journal of Hydrogen Energy* 32 (2):200–206.

Tatsuzawa, H., E. Takizawa, M. Wada, and Y. Yamamoto. 1996. Fatty acid and lipid composition of the acidophilic green alga *Chlamydomonas* sp. 1. *Journal of Phycology* 32 (4):598–601.

Ueno, Y., N. Kurano, and S. Miyachi. 1998. Ethanol production by dark fermentation in the marine green alga, *Chlorococcum littorale*. *Journal of Fermentation and Bioengineering* 86 (1):38–43.

Ugwu, C.U., H. Aoyagi, and H. Uchiyama. 2008. Photobioreactors for mass cultivation of algae. *Bioresource Technology* 99 (10):4021–4028.

Walker, T.L., S. Purton, D.K. Becker, and C. Collet. 2005. Microalgae as bioreactors. *Plant Cell Reports* 24 (11):629–641.

Wang, B., Y. Li, N. Wu, and C.Q. Lan. 2008. CO_2 bio-mitigation using microalgae. *Applied Microbiology and Biotechnology* 79 (5):707–718.

Warabi, Y., D. Kusdiana, and S. Saka. 2004. Reactivity of triglycerides and fatty acids of rapeseed oil in supercritical alcohols. *Bioresource Technology* 91 (3):283–287.

Wijffels, R.H. and M.J. Barbosa. 2010. An outlook on microalgal biofuels. *Science* (*Washington*) 329 (5993):796–799.

Williams, J.A. 2002. Keys to bioreactor selections. *Chemical Engineering Progress* 98 (3):34–41.

Winkler, M., A. Hemschemeier, C. Gotor, A. Melis, and T. Happe. 2002. [Fe]-hydrogenases in green algae: Photo-fermentation and hydrogen evolution under sulfur deprivation. *International Journal of Hydrogen Energy* 27 (11):1431–1439.

Yokoyama, S., K. Jonouchi, and K. Imou. 2008. Energy production from marine biomass: Fuel cell power generation driven by methane produced from seaweed. *International Journal of Applied Science, Engineering & Technology* 4 (3):169–172.

Zeng, X., M.K. Danquah, X.D. Chen, and Y. Lu. 2011. Microalgae bioengineering: From CO_2 fixation to biofuel production. *Renewable and Sustainable Energy Reviews* 15 (6):3252–3260.

4

Significance of Harvesting in the Cultivation of Microalgae

K.K. Vasumathi, M. Premalatha, and P. Subramanian

CONTENTS

4.1 Introduction

Microalgae have the unique *cell factories* that convert carbon dioxide into biomass. They require mainly sunlight, CO_2, water, and inorganic nutrients for the growth. Cultivation of microalgae does have commercial interest, since it produces bioactive compounds such as cosmetic, pharmaceutical, food, feed, and biodiesel besides mitigating CO_2 emission. The overall economics lies in each of the following stages of the process (Figure 4.1).

Microalgae may vary from unicellular to long filamentous forms and in their origin from freshwater to marine. Strain selection is an important factor in overall success of microalgal technology, which is based on the industrial application and the targeted final products. It should be more robust, have a high growth rate, have higher photosynthetic efficiency, and be able to survive in shear stress conditions in photobioreactor and seasonal variations (Brennan and Owende, 2010). Site-specific adaptation of strain will make the technology viable (Sheehan et al., 1998).

Scale-up of the process is carried out generally by adding 10% of seed concentration. Seed concentration is another important factor that determines the biomass productivity on a large scale. According to the light availability at the cultivation site, the seed concentration has to be chosen to avoid photo damage to the pigments in the microalgae and corresponding to illuminated area of the photobioreactor. As the illuminated area is increased, so will the cell concentration.

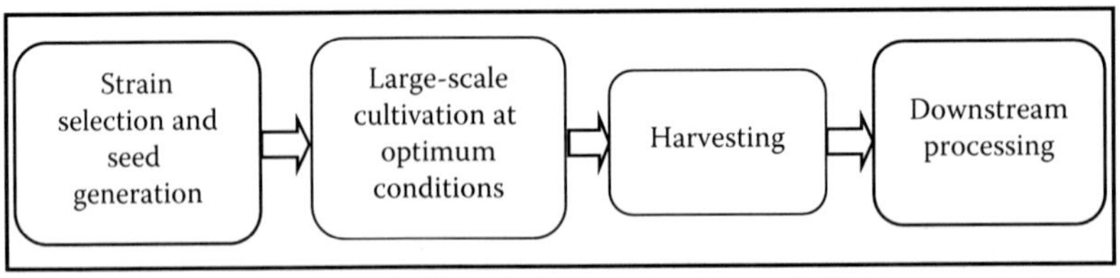

FIGURE 4.1
Microalgae cultivation, showing the major stages of the process.

In reality, synthesizing bulk products is very difficult (Eriksen, 2008). This may be due to the environmental factors in open-pond cultivation. If a closed photobioreactor is chosen, it will increase the operating cost. Investment cost per hectare for a raceway pond is 0.68 (M€), for a horizontal tubular reactor, 0.84 (M€), and for a flat panel photobioreactor, 1.2 (M€) (Norsker et al., 2011). The volume of the reactor is high; hence, the recovery cost of biomass seems higher that accounts for 20%–30% of the total costs (Gudin and Therpenier, 1986). If high-value products are produced from the biomass, it will help in revenue generation. But the product extraction process also adds cost to the system. Thus, design of photobioreactor, capital cost, nutrient consumption, and energy requirement for cultivation, harvesting, and product extraction are the major bottlenecks for implementing this technology on a large scale.

This chapter deals with the strategy of maintaining high cell concentration inside the reactor by harvesting at optimum cell densities corresponding to various process parameters such as pH, temperature, light intensity, and cell concentration, which directly helps in reducing the size of the reactor and thus the cost.

4.2 Large-Scale Cultivation at Optimum Conditions

4.2.1 Culture Methods

Photoautotrophic, heterotrophic, mixotrophic, and photoheterotrophic cultivation are the four major types of cultivation systems (Chojnacka and Marquez-Rocha, 2004). Table 4.1 elaborates the source of energy for cultivation condition, reactor type, cost, and issues associated with scale-up.

4.2.2 Photobioreactors

Open raceways are the common design employed for the growth of microalgae. The construction of open ponds is cheaper and easy to operate. Control of pH, temperature, and light is very difficult in open raceway ponds, which ultimately results in lower productivity. The productivity is very low (0.05–0.3 g DW/L) in open raceways (Brennan and Owende, 2010), whereas in closed photobioreactors, this value is in the range of 0.8–1.3 g DW/L day (Pulz, 2001). Land requirement is the major problem in commercializing these systems. Generally, for absorbing the CO_2 emitted from a 1 MW power plant, 0.44 km^2/MW is required (Vasumathi et al., 2012). However, 98% of commercial algae biomass production is carried out in open systems (Benemann, 2008). This is because their energy

TABLE 4.1

Comparison of Different Cultivation Conditions

Cultivation Conditions	Energy Source	Carbon Source	Cell Density	Reactor Scale-Up	Cost	Issues Associated with Scale-Up
Phototrophic	Light	Inorganic	Low	Open pond or photobioreactor	Low	Low cell density High condensation cost
Heterotrophic	Organic	Organic	High	Conventional fermentor	Medium	Contamination High substrate cost
Mixotrophic	Light and organic	Inorganic and organic	Medium	Closed photobioreactor	High	Contamination High equipment cost High substrate cost
Photoheterotrophic	Light	Organic	Medium	Closed photobioreactor	High	Contamination High equipment cost High substrate cost

Source: Chen, C.-Y. et al., *Bioresour. Technol.*, 102, 71, 2011.

costs and capital costs of open raceway ponds are 1 W/m^3 and US$9.4 m^{-2}, respectively (Fernandez et al., 2008; Hallenbeck and Benemann, 2002), whereas energy costs for closed photobioreactors have been estimated to range 50–300 W/m^3 (Fernandez et al., 2008) and capital cost of US$100 m^{-2} (Hallenbeck and Benemann, 2002).

Generally, closed photobioreactors provide better environmental control and higher volumetric productivities (Janssen et al., 2003) and are regarded as the best option for obtaining high-value-added products from microalgae.

Microalgal growth is measured as g/m^2/day. Based on the growth, the light utilization per square meter is calculated as photosynthetic efficiency. Theoretical photosynthetic efficiency for microalgal growth is 11%. But in actual cases, it is still lower (4%–6%). This is due to the poor supply of light to the individual cells. Design of the photobioreactor varies with reference to the way in which light is introduced to the cells. The reactor that has the better provision of light to the individual cells can maintain high cell concentration. Maintaining high cell concentration reduces the harvesting cost.

4.2.3 Optimum Process Conditions

Growth rates of microalgae are affected by diverse factors such as initial cell concentration, temperature, pH, irradiance, or different nutrient concentration, and they are interrelated (Perez et al., 2008). Each parameter has a saturation limit beyond which the growth rate decreases. Table 4.2 shows the optimum value at which maximum growth rate is obtained for different process parameters reported in literature for different species. The data indicate that all parameters showed the maximum growth rate at a particular value of the parameter.

Hence, the growth rate has to be maximized by studying the combined effect of all the parameters and subsequent optimization using the suitable optimization techniques like RSM. Maintaining the optimum parameters inside the reactor increases the cell concentration and hence reduces the harvesting cost.

TABLE 4.2
Optimum Process Parameters for Various Species

Species	Conditions	Range	Optimum	References
Phaeodactylum tricornutum	Temperature, °C	15–35	25	Fernandez et al. (2003)
P. tricornutum	pH	4–12	8	
	Temperature, °C	5–30	20	Perez et al. (2008)
	Light intensity, μE/m²/s	0–151	10.2	
P. tricornutum	Light intensity, μmol/m²/s	35–80	80	Sandnes et al. (2005)
		15–35	25.6	
	Temperature, °C			
Scenedesmus obliquus	pH	—	6.9	de Morais and Costa (2007)
Chlorella kessleri	pH	—	8.0	de Morais and Costa (2007)
—	Light intensity, μE/m²/s	0–2500	1000	Merchuk and Wu (2003)
Haematococcus pluvialis	Cell concentration, g/L	0.1–5.0	0.8	Wang et al. (2013)

4.3 Harvesting

Currently, research efforts are devoted to optimize traditional methods for microalgae harvesting. The method for harvesting microalgae is species specific, and it also depends on the final product. Due to the small particle size of the microalgal cells, low cell density, huge volume, and value of the target products, harvesting biomass remains a major challenge for its full scale.

Four important points to be considered to reduce the energy involved in harvesting are

1. Size of the species
2. Suitable harvesting technique, corresponding to the species and downstream products
3. Mode of harvesting
4. Time interval of harvesting

The economics of microalgal production lies in the choice of best harvesting technology (Schenk et al., 2008). Table 4.3 gives the physical, chemical, and biological methods used for harvesting different species (Fernandez and Ballesteros, 2013).

4.3.1 Mode of Operation

The production of microalgae on a large scale can be carried out either with a batch mode, semicontinuous mode, or continuous mode.

The general practice for maximal productivity (achieve maximum growth rate) is by increasing the cell density without mutual shading effect. It helps to improve the economics also (Richmond, 1992). *Pleurochrysis carterae* cells began to clump at cell densities of 8×10^5 cells/mL. To avoid problems due to sticking and self-shading, the culture was operated at a maximum cell concentration of $5.5–7 \times 10^5$ cells/mL (Moheimani and Borowitzka, 2006).

TABLE 4.3

Harvesting Techniques

Classification	Process	Types	Remarks
Physical	Gravity sedimentation	Sedimentation tanks, lamella	Physiology of strain. Applicable to high-cell-density cultures.
	Centrifugation	—	Effective separation—less volume. Time consuming. Cell damages due to high shear forces.
	Filtration	Screens, microstrainer, vibrating screen, submerged filtration	High-cell-density block screen. Cell damage due to pumping. Difficultly in membrane maintenance. Applying vacuum is energy intensive.
	Flotation	Dispersed air (DAF), dissolved air, suspended air (SAF), microbubble technology (MBT)	Separation is efficient but energy intensive. SAF requires less energy than DAF. MBT requires still less energy than SAF and DAF.
	Spiral plate technology	—	Energy requirement is less compared to centrifuge.
Chemical	Flocculation	Chemicals, metallic salts, polyelectrolytes	High pH will affect downstream processing. High cost of flocculants. The technique is not desirable when biomass is used as feed.
	Coagulation	Chemicals, electrocoagulation	Energy required for manufacturing coagulants. Electrocoagulation is a multistep process and energy consuming.
Biological	Microalgae	*Pediastrum* sp.	High efficiency. Manipulation of determined conditions promotes flocculation.
	Bacteria	Exopolysaccharides	The activity is dependent on the growth phase. Additional cost for bacterial growth.

Hase et al. (2000) also reported that the growth rate becomes slower at biomass concentration of 0.3 g dry biomass/L for *Chlorophyta* sp. and 0.5 g dry biomass/L for *Chlorella* sp. For maintaining steady state, these concentrations were found to be optimal under sunlight conditions. Hence, harvesting was done at periodical intervals. Sandnes et al. (2006) maintained the optical culture density of 1.5 g/L for *Nannochloropsis oceanica* in a biofence system and thus achieved 44 g/day.

Literature on harvesting recognizes the necessity of harvesting to avoid shading due to high cell densities, but does not recognize the impact of optimum biomass concentration on maximum growth rate. The harvesting has to be done in order to maintain the optimum biomass concentration in the photobioreactor.

Highest cell densities in any photobioreactor can be achieved by maintaining the highest microalgal growth rate. The nutrient concentration is higher during the log phase of growth, while the biomass concentration is lower. The cells start growing by utilizing the nutrients; thus the growth rate increases. It increases up to a certain point of optimum biomass concentration and then decreases as the nutrient concentration decreases, even though biomass concentration increases. Thus, microalgal growth is an autocatalytic process. Figure 4.2 depicts the autocatalytic nature of microalgal growth process.

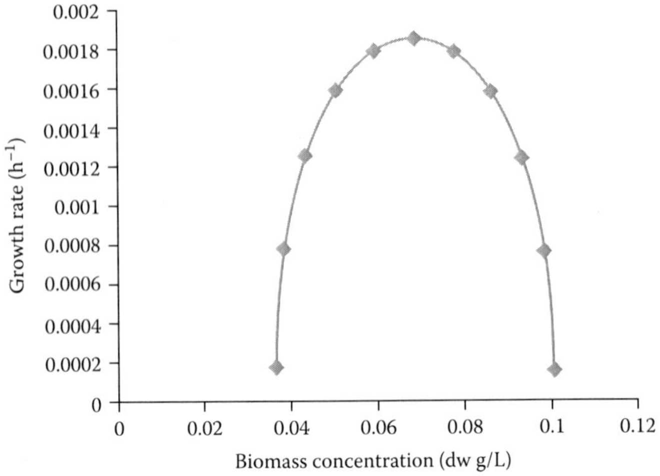

FIGURE 4.2
The inverted parabola resembles the autocatalytic nature of the microalgal growth. The growth rate increases up to a point and then decreases. (From Vasumathi, K.K. et al., *Renew. Sust. Energy Rev.*, 16, 5443, 2012.)

Therefore, an optimum concentration of biomass exists, offering the highest rate of growth. If this concentration is maintained, then the growth rate would be the highest, resulting in increased biomass yield and reduced volume of photobioreactor for the given rate of production (Vasumathi et al., 2012). The biomass concentration could be maintained at this optimum by removing the microalgae in excess of this optimum continuously. The volume of culture handling is less compared to batch and continuous modes; thus, the cost of harvesting is lower. The semicontinuous mode has been reported to yield higher productivity (Moheimani and Borowitzka, 2006). Similar observations have been reported that the population density must be maintained at its optimum, that is, the cell concentration that will result in the highest net yields (Moheimani and Borowitzka, 2007; Vonshak et al., 1982). Richmond (1992) reported that maintaining optimum population density results in highest net yields and the rate of O_2 evolution is also higher. Optimum pH, light intensity, and temperature are also to be maintained in photobioreactors for achieving high cell densities. The optimum biomass concentration will vary with species and process conditions. Hence, batch studies help to find the optimum biomass concentration for each species corresponding to any given process conditions.

In continuous culture, nutrient medium is added continuously to match the growth rate of microalgae and thus helps to maintain the culture at maximum growth rate, because the culture never runs out of the nutrients. Continuous culture systems differ from that of a batch system by supplying nutrient continuously to the cell culture at a constant rate, and an equal volume of cell culture is removed to maintain a constant volume. This makes the system to be at steady state. The main disadvantage for continuous system for microalgal technology is the requirement of automation that increases the cost.

In the semicontinuous reactor, the reactants are fed continuously, and the products are recovered batchwise. The harvesting is done once the system reaches certain cell densities to provide the light supply in the reactor (Moheimani and Borowitzka, 2006). Growth rates of *Chlorella* sp. were higher at high-cell-density cultures compared to low-density cultures. The high cell densities help to increase the CO_2 tolerance limits and subsequently reduce the batch time (Chiu et al., 2008). Batch studies are required to find out the optimum

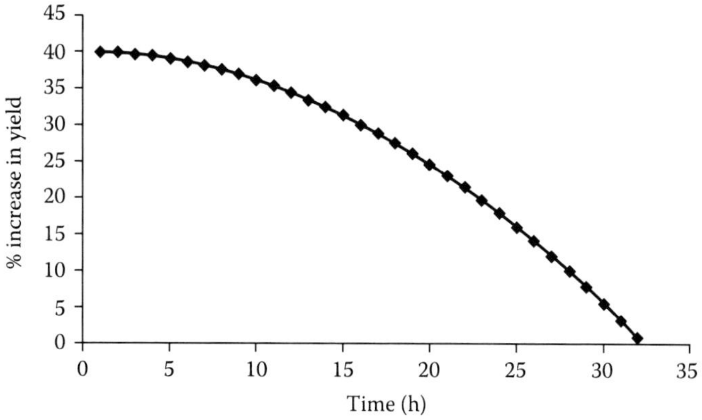

FIGURE 4.3
As the harvesting interval increases, the % increase in biomass yield decreases.

concentration to be maintained under semicontinuous mode. Figure 4.2 explains the optimum concentration obtained for *Scenedesmus* sp. carried out in a batch mode. Hence, to maintain high cell density and at the same time at an optimum level under semicontinuous mode, batch studies are required for a given species and process conditions.

4.3.2 Time of Harvesting

Increased cell densities with increased product formation decrease the cost of production and also decrease the harvesting and purification cost. The cost of a harvesting system is about \$33,000 ha^{-1}, and the cost varies with the efficiency of the system (Borowitzka, 1992). Hence, harvesting has to be done at optimum concentration, and, hence, study is required to find the interval of time at which harvesting has to be done to maintain the growth rate at the maximum level. For the *Scenedesmus* sp., the percentage increase in yield is calculated from the batch data for the given process condition. The percentage increase in biomass yield decreases with increasing harvesting interval (Vasumathi et al., 2013) and is shown in Figure 4.3. It varies with the species, growth characteristics, and doubling time. Depending upon the effect of harvesting interval on biomass production, the harvesting interval should be chosen. The growth characteristics of the *Scenedesmus* sp. studied under particular process condition, percentage increase in yield was not changing up to a harvesting interval of 5 h (Figure 4.3). Hence, harvesting could be done once every 5 h.

4.4 Conclusion

The size, shape of the species, and type of product expected from biomass decide the type of harvesting technique. Choosing the right technique without compromising quality will make the process energy efficient. Harvesting at the right interval is beneficial to maintain a higher growth rate inside the reactor. However, this optimum condition has to be found by conducting batch experiments for any species, given operating conditions and the design of the photobioreactor.

Acknowledgment

We sincerely thank the Department of Science and Technology (DST) for giving us an opportunity to work on the most important research area on microalgae for CO_2 mitigation.

References

Benemann, J.R. 2008. *Opportunities and Challenges in Algae Biofuel Production*. FAO, Algae World, Singapore, Singapore.

Borowitzka, M.A. 1992. Algal biotechnology products and processes—Matching science and economics. *Journal of Applied Phycology* 4:267–279.

Brennan, L. and P. Owende. 2010. Biofuels from microalgae—A review of technologies for production, processing, and extractions of biofuels and co-products. *Renewable and Sustainable Energy Reviews* 14:557–577.

Chen, C.-Y., K.-L. Yeh, R. Aisyah, D.-J. Lee, and J.-S. Chang. 2011. Cultivation, photobioreactor design and harvesting of microalgae for biodiesel production: A critical review. *Bioresource Technology* 102:71–81.

Chiu, S.-Y., C.-Y. Kao, C.-H. Chen, T.-C. Kuan, S.-C. Ong, and C.-S. Lin. 2008. Reduction of CO_2 by a high density culture of *Chlorella* sp. in a semi-continuous photobioreactor. *Bioresource Technology* 99:3389–3396.

Chojnacka, K. and F.J. Marquez-Rocha. 2004. Kinetic and stoichiometric relationships of the energy and carbon metabolism in the culture of microalgae. *Biotechnology* 3:21–34.

de Morais, M.G. and J.A.V. Costa. 2007. Isolation and selection of microalgae from coal fired thermoelectric power plant for biofixation of carbon dioxide. *Energy Conservation and Management* 48:2169–2173.

Eriksen, N.T. 2008. The technology of microalgal culturing. *Biotechnology Letters* 30:1525–1536.

Fernández, C.Z. and M. Ballesteros. 2013. Microalgae autoflocculation: An alternative to high-energy consuming harvesting methods. *Journal of Applied Phycology* 25:991–999.

Fernandez, F.C.A., D.O. Hall, E. Carizares Guerrero, K. Krishna Rao, and E. Molina Grima. 2003. Outdoor production of *Phaeodactylum tricornutum* biomass in a helical reactor. *Journal of Biotechnology* 103:137–152.

Fernandez, F.G.A., C.V. Gonzalez, J.M. Fernandez, M. García-González, J. Moreno, E. Sierra, M.G. Guerrero, and E. Molina. 2008. Removal of CO from flue gases coupled to the photosynthetic generation of biomass and exopolysaccharides by cyanobacteria. In: *Proceedings of the 11th International Conference on Applied Phycology*, Galway, Ireland, p. 2.

Gudin, C. and Therpenier, C. 1986. Bioconversion of solar energy into organic chemicals by microalgae. *Advances in Biotechnological Processes* 6:73–110.

Hallenbeck, P.C. and Benemann, J.R. 2002. Biological hydrogen production: Fundamentals and limiting processes. *International Journal of Hydrogen Energy* 27:1185–1193.

Hase, H., H. Oikawa, C. Sasao, M. Morita, and Y. Watanable. 2000. Photosynthetic production of microalgal biomass in a raceway system under greenhouse conditions in Sendai city. *Journal of Bioscience and Bioengineering* 89(2):157–163.

Janssen, M., J. Tramper, L.R. Mur, and R.H. Wijffels. 2003. Enclosed outdoor photobioreactors: Light regime, photosynthetic efficiency, scale-up, and future prospects. *Biotechnology and Bioengineering* 81(2):193–210.

Merchuk, J.C. and X. Wu. 2003. Modeling of photobioreactors: Application to bubble column simulation. *Journal of Applied Phycology* 15:163–169.

Mohemani, N.R. and M.A. Borowitzka. 2006. The long term culture of the coccolithophore *Pleurochrysis carterae* (Haptophyta) in outdoor raceway ponds. *Journal of Applied Phycology* 18:703–712.
Moheimani, N.R. and M.A. Borowitzka. 2007. Limits to productivity of the alga *Pleurochrysis carterae* (Haptophyta) grown in outdoor raceway ponds. *Biotechnology & Bioengineering* 96:27–36.
Norsker, N.H., M.J. Barbosa, M.H. Vermuë, and R.H. Wijffels. 2011. Microalgal production—A close look at the economics. *Biotechnology Advances* 29:4–27.
Perez, E.B., C.I. Pina, and L.P. Rodriguez. 2008. Kinetic model for growth of *Phaeodactylum tricornutum* in intensive culture photobioreactor. *Biochemical Engineering Journal* 20:520–525.
Pulz, O. 2001. Photobioreactors: Production systems for phototrophic microorganisms. *Applied Microbiology and Biotechnology* 57:287–293.
Richmond, A. 1992. Open systems for the mass production of photoautotrophic microalgae outdoors: Physiological principles. *Journal of Applied Phycology* 4:281–286.
Sandnes, J.M., T. Kallqvist, D. Wenner, and H.R. Gislerod. 2005. Combined influence of light and temperature on growth rates of *Nannochloropsis oceanica*: Linking cellular responses to large scale biomass production. *Journal of Applied Phycology* 17:515–525.
Sandnes, J.M., T. Ringstad, D. Wenner, P.H. Heyerdahl, T. Kallqvist, and H.R. Gislerod. 2006. Real-time monitoring and automatic density control of large scale microalgal cultures using near infrared (NIR) optical density sensor. *Journal of Biotechnology* 122:209–215.
Schenk, P., S. Thomas-Hall, E. Stephens, U. Marx, J. Mussgnug, C. Posten et al. 2008. Second generation biofuels: High-efficiency microalgae for biodiesel production. *Bioenergy Research* 1(1):20–43.
Sheehan, J., T. Dunahay, J.R. Benemann, and P. Roessler. 1998. A look back at the U.S. Department of Energy's Aquatic Species Program—Biodiesel from algae, NREL/TP-580-24190. National Renewable Energy Laboratory, Golden, CO.
Vasumathi, K.K., M. Premalatha, and P. Subramanian. 2012. Parameters influencing the design of photobioreactor for the growth of microalgae. *Renewable and Sustainable Energy Reviews* 16:5443–5450.
Vasumathi, K.K., M. Premalatha, and P. Subramanian. 2013. Experimental studies on the effect of harvesting interval on yield of *Scenedesmus arcuatus var. capitatus*. *Ecological Engineering* 58:13–16.
Vonshak, A., A. Abeliovich, Z. Boussiba, S. Arad, and A. Richmond. 1982. Production of *Spirulina* biomass: Effects of environmental factors and population density. *Biomass* 2:175–185.
Wang, J., D. Han, M.R. Sommerfeld, C. Lu, and Q. Hu. 2013. Effect of initial biomass density on growth and astaxanthin production of *Haematococcus pluvialis* in an outdoor photobioreactor. *Journal of Applied Phycology* 25:253–260.

5

Algal Photobioreactors

Ozcan Konur

CONTENTS

5.1 Introduction

5.1.1 Issues

Global warming, air pollution, and energy security have been some of the most important public policy issues in recent years (Jacobson 2009, Wang et al. 2008, Yergin 2006). With the increasing global population, food security has also become a major public policy issue (Lal 2004). The development of biofuels generated from biomass has been a long-awaited solution to these global problems (Demirbas 2007, Goldember 2007, Lynd et al. 1991). However, the development of early biofuels produced from agricultural plants such as sugar cane (Goldemberg 2007) and agricultural wastes such as corn stovers (Bothast and Schlicher 2005) has resulted in a series of substantial concerns about food security (Godfray et al. 2010). Therefore, the development of algal biofuels as a third-generation biofuel has been considered as a major solution for the global problems of global warming, air pollution, energy security, and food security (Chisti 2007, Demirbas 2007, Kapdan and Kargi 2006, Spolaore et al. 2006, Volesky and Holan 1995).

However, the production of algal biofuels in photobioreactors (PBRs) has been one of the most critical production factors from a technoeconomic perspective (e.g., Grima et al. 2003, Mata et al. 2010, Pulz 2001, Rodolfi et al. 2009). For example, Borowitzka (1999) argues that the main problem facing the commercialization of new microalgae and microalgal products is the need for closed culture systems and the fact that these are very capital intensive. The high cost of microalgal culture systems relates to the need for light and the relatively slow growth rate of the algae. Similarly, Pulz (2001) argues that the design of the technical and technological basis for PBRs is the most important issue for economic success in the field of phototrophic biotechnology. For future applications, he speculates that open-pond systems for large-scale production have a lower innovative potential than closed systems.

Although there have been many reviews on the use of marine algae for the production of algal PBRs and algal microbial fuel cells (MFCs) (e.g., Grima et al. 2003, Mata et al. 2010, Pulz 2001) and there have been a number of scientometric studies on algal biofuels (Konur 2011), there has not been any study on algal PBRs as in other research fields (Baltussen and Kindler 2004a,b, Dubin et al. 1993, Gehanno et al. 2007, Konur 2012a,b, 2013, Paladugu et al. 2002, Wrigley and Matthews 1986).

As North's new institutional theory suggests, it is important to have up-to-date information about current public policy issues to develop a set of viable solutions to satisfy the needs of all the key stakeholders (Konur 2000, 2002a,c, 2006a,b, 2007a,b, 2012c, North 1994).

Therefore, brief information on selected research on algal PBRs is presented in this chapter to inform the key stakeholders relating to the global problems of warming, air pollution, food security, and energy security about the production of algal biofuels in the algal PBRs for the solution of these problems in the long run, thus complementing a number of recent scientometric studies on biofuels and global energy research (Konur 2011, 2012d,e,f,g,h,i,j,k,l,m,n,o,p).

5.1.2 Methodology

A search was carried out in the Science Citation Index Expanded (SCIE) and Social Sciences Citation Index (SSCI) databases (version 5.11) in September 2013 to locate the papers relating to algal PBRs using the keyword set of Topic = (photobioreactor* or *photobio* reactor** or reactor* or *bioreactor** or *bio* reactor**) and Topic = (alga* or *macro* alga** or *micro* alga** or macroalga* or microalga* or cyanobacter* or seaweed* or diatom* or *sea* weed** or reinhardtii or braunii or chlorella or sargassum or gracilaria or spirulina) in the abstract pages of the papers. For this chapter, it was necessary to embrace the broad algal search terms to include diatoms, seaweeds, and cyanobacteria as well as to include many spellings of bioelectricity and MFCs rather than the simple keywords of bioelectricity, MFCs, and algae.

It was found that there were 2600 papers indexed by these keywords between 1980 and 2013. The key subject categories for the algal research were biotechnology applied microbiology (84 references, 34.3%), energy fuels (512 references, 19.1%), engineering chemical (411 references, 15.3%), and environmental sciences (345 references, 18.2%) altogether comprising 75.3% of all the references on algal PBRs. It was also necessary to focus on the key references by selecting articles and reviews.

The located highly cited 50 papers were arranged by decreasing number of citations for 2 groups of papers: 26 and 24 papers for nonexperimental research and experimental research, respectively. In order to check whether these collected abstracts correspond to the larger sample on these topical areas, new searches were carried out for each topical area.

The summary information about the located papers is presented under two major headings of nonexperimental and experimental research in the order of the decreasing number of citations for each topical area.

The information relating to the document type, author affiliation and nationality, the journal, the indexes, subject area, concise topic, total number of citations received, and total average number of citations received per year were given in the tables for each paper.

5.2 Nonexperimental Studies on Algal Photobioreactors

5.2.1 Introduction

The nonexperimental research on algal PBRs has been one of the most dynamic research areas in recent years. Twenty-six citation classics in the field of algal PBRs with more than sixty-one citations were located, and the key emerging issues from these papers were presented later in the decreasing order of the number of citations (Table 5.1). These papers give strong hints about the determinants of efficient algal PBR design and emphasize that efficient algal PBR design is possible in the light of the recent biotechnological advances in recent years.

The papers were dominated by the researchers from 18 countries, usually through the intracountry institutional collaboration, and they were multiauthored. France and Spain (four papers each) and England, the United States, Germany, Italy, Japan, and Portugal (two papers each) were the most prolific countries.

Similarly, all these papers were published in the journals indexed by the Science Citation Index (SCI) and/or SCIE. There were no papers indexed by the SSCI or Arts & Humanities Citation Index (A&HCI). The number of citations ranged from 61 to 396, and the average number of citations per annum ranged from 2.9 to 99.0. Sixteen of the papers were articles, while ten of them were reviews.

5.2.2 Nonexperimental Research on Algal Photobioreactors

Mata et al. (2010) discuss microalgae for biodiesel production and other applications in a review paper with 396 citations. They review the current status of microalgae use for biodiesel production, including their cultivation, harvesting, and processing. They present the microalgae species most used for biodiesel production and describe their main advantages in comparison with other available biodiesel feedstocks. They also describe the various aspects associated with the design of microalgae production units, giving an overview of the current state of development of algae cultivation systems (PBRs and open ponds).

Borowitzka (1999) discusses the commercial production of microalgae with a focus on the ponds, tanks, tubes, and fermenters in a paper with 272 citations. The commercial culture of microalgae is now over 30 years old, and the culture systems used in the 1990s to grow these algae are generally fairly unsophisticated. For example, *Dunaliella salina* is cultured in large (up to approx. 250 ha) shallow open-air ponds with no artificial mixing. Similarly, *Chlorella* and *Spirulina* also are grown outdoors in either paddle wheel mixed ponds or circular ponds with a rotating mixing arm of up to about 1 ha in area per pond. The production of microalgae for aquaculture is generally on a much smaller scale and in many cases is carried out indoors in 20–40 L carboys or in large plastic bags of up to approximately 1000 L in volume. More recently, a helical tubular PBR system, the Biocoil™, has been developed, which allows these algae to be grown reliably outdoors at high cell

TABLE 5.1

Nonexperimental Research on Algal PBRs

No.	Paper References	Year	Document	Affiliation	Country	No. of Authors	Journal	Index	Subjects	Topic	Total No. of Citations	Total Average Citations Per Annum	Rank
1	Mata et al.	2010	R	Univ. Porto, IPP	Portugal	3	*Renew. Sust. Energ. Rev.*	SCI, SCIE	Ener. Fuels	Algal photobioreactors general	396	99.0	1
2	Borowitzka	1999	A	Murdoch Univ.	Australia	1	*J. Biotechnol.*	SCI, SCIE	Biot. Appl. Microb.	Algal photobioreactors	272	18.1	3
3	Grima et al.	2003	R	Massey Univ.	New Zealand	5	*Biotechnol. Adv.*	SCI, SCIE	Biot. Appl. Microb.	Algal photobioreactors	251	20.9	4
4	Pulz	2001	R	IGV Inst. Cereal Proc.	Germany	1	*Appl. Microbiol. Biotechnol.*	SCI, SCIE	Biot. Appl. Microb.	Algal photobioreactors	205	22.8	5
5	Munoz and Guieysse	2006	R	Lund Univ., Univ. Valladolid	Sweden, Spain	2	*Water Res.*	SCI, SCIE	Eng. Env., Env. Sci., Water Res.	Algal photobioreactors	153	19.1	6
6	Carvalho et al.	2006	R	Univ. Catolica Portuguesa	Portugal	3	*Biotechnol. Prog.*	SCI, SCIE	Biot. Appl. Microb., Food Sci. Tech.	Algal photobioreactors	150	18.9	7
7	Ugwu et al.	2008	R	Univ. Tsukuba	Japan	3	*Bioresour. Technol.*	SCI, SCIE	Ener. Fuels, Biot. Appl. Microb.	Algal photobioreactors	134	22.3	8
8	Chen et al.	2011	R	Natl. Cheng Kung Univ., Natl. Taiwan Univ. Sci. & Technol.	Taiwan	5	*Bioresour. Technol.*	SCI, SCIE	Ener. Fuels, Biot. Appl. Microb.	Algal photobioreactors	118	39.3	9
9	Grima et al.	1999	A	Univ. Almeira	Spain	4	*J. Biotechnol.*	SCI, SCIE	Biot. Appl. Microb.	Algal photobioreactors	114	7.6	11
10	Greenwell et al.	2010	A	Univ. Durham, NREL, Univ. Swansea	England, Wales, United States	5	*J. R. Soc. Interface*	SCI, SCIE	Mult. Sci.	Algal photobioreactors	112	28.0	12
11	Lee	2001	A	Natl. Univ. Singapore	Singapore	1	*J. Appl. Phycol.*	SCI, SCIE	Biot. Appl. Microb., Mar. Fresh. Biol.	Algal photobioreactors	111	8.5	13
12	Del Campo et al.	2007	R	Univ. Seville	Spain	3	*Appl. Microbiol. Biotechnol.*	SCI, SCIE	Biot. Appl. Microb.	Algal photobioreactors	109	15.6	14
13	Hu et al.	1996	a	Ben Gurion Univ. Negev	Israel	3	*Biotechnol. Bioeng.*	SCI, SCIE	Biot. Appl. Microb.	Algal photobioreactors	107	5.9	16

(Continued)

TABLE 5.1 (*Continued*)

Nonexperimental Research on Algal PBRs

No.	Paper References	Year	Document	Affiliation	Country	No. of Authors	Journal	Index	Subjects	Topic	Total No. of Citations	Total Average Citations Per Annum	Rank
14	Tredici and Zittelli	1998	A	Univ. Florence	Italy	2	*Biotechnol. Bioeng.*	SCI, SCIE	Biot. Appl. Microb.	Algal photobioreactors	90	5.6	21
15	Chaumont	1993	A	Ctr. Cadarache	France	1	*J. Appl. Phycol.*	SCI, SCIE	Biot. Appl. Microb., Mar. Fresh Water	Algal photobioreactors	89	4.2	22
16	Tredici and Materassi	1992	A	Univ. Firenze	Italy	2	*J. Appl. Phycol.*	SCI, SCIE	Biot. Appl. Microb., Mar. Fresh Water	Algal photobioreactors	89	4.0	23
17	Posten	2009	R	Univ. Karlsruhe	Germany	1	*Eng. Life Sci.*	SCIE	Biot. Appl. Microb.	Algal photobioreactors	88	17.6	24
18	Gudin and Chaumont	1991	A	Cen Cadarache	France	2	*Bioresour. Technol.*	SCI, SCIE	Ener. Fuels, Biot. Appl. Microb.	Algal photobioreactors	79	3.4	30
19	Jorquera et al.	2010	A	NREL, Univ. Fed. Bahia	United States, Brazil	5	*Bioresour. Technol.*	SCI, SCIE	Ener. Fuels, Biot. Appl. Microb.	Algal photobioreactors	76	19.0	31
20	Eriksen	2008	R	Univ. Aalborg	Denmark	1	*Biotechnol. Lett.*	SCI, SCIE	Biot. Appl. Microb.	Algal photobioreactors	72	12.0	34
21	Stephenson et al.	2010	A	Univ. Cambridge	England	6	*Energy Fuels*	SCI, SCIE	Ener. Fuels, Eng. Chem.	Algal photobioreactors	70	17.5	35
22	Fernandez et al.	1997	A	Univ. Almeira	Spain	5	*Biotechnol. Bioeng.*	SCI, SCIE	Biot. Appl. Microb.	Algal photobioreactors	68	4.0	39
23	Benemann	1997	A	na	United States	1	*Int. J. Hydrog. Energy*	SCI, SCIE	Chem. Phys., Electroch. Ener. Fuels	Algal photobioreactors	66	3.9	40
24	Miyake et al.	1999	A	Natl. Inst. Adv. Interdisciplinary Res.	Japan	3	*J. Biotechnol.*	SCI, SCIE	Biot. Appl. Microb.	Algal photobioreactors	65	4.3	41
25	Cornet et al.	1992	A	Univ. Clermont Ferrand, Matra Espace Applicat Microgravite	France	3	*Biotechnol. Bioeng.*	SCI, SCIE	Biot. Appl. Microb.	Algal photobioreactors	63	2.9	44
26	Cornet et al.	1995	A	Matra Marconi Space, European Space Agcy.	France, Netherlands	5	*Chem. Eng. Sci.*	SCI, SCIE	Eng. Chem.	Algal photobioreactors	61	3.2	45

SCI, Science Citation Index; SCIE, Science Citation Index Expanded; SSCI, Social Sciences Citation Index; A, article; R, review.

densities in semicontinuous culture. Other closed PBRs such as flat panels are also being developed. They argue that the main problem facing the commercialization of new microalgae and microalgal products is the need for closed culture systems and the fact that these are very capital intensive. The high cost of microalgal culture systems relates to the need for light and the relatively slow growth rate of the algae.

Grima et al. (2003) discuss the recovery of microalgal biomass and metabolites with a focus on the process options and economics in a review paper with 251 citations. They argue that commercial production of intracellular microalgal metabolites requires the following: (1) large-scale monoseptic production of the appropriate microalgal biomass, (2) recovery of the biomass from a relatively dilute broth, (3) extraction of the metabolite from the biomass, and (4) purification of the crude extract. They examine the options available for recovery of the biomass and the intracellular metabolites from the biomass. They then discuss the economics of monoseptic production of microalgae in PBRs and the downstream recovery of metabolites using eicosapentaenoic acid (EPA) recovery as a representative case study.

Pulz (2001) discusses PBRs with a focus on the production systems for phototrophic microorganisms in a review paper with 205 citations. He argues that the design of the technical and technological basis for PBRs is the most important issue for economic success in the field of phototrophic biotechnology. For future applications, he speculates that open-pond systems for large-scale production have a lower innovative potential than closed systems. For high-value products in particular, he asserts that closed systems of PBRs are the more promising field for technical developments despite very different approaches in design.

Munoz and Guieysse (2006) discuss the algal–bacterial processes for the treatment of hazardous contaminants in a review paper with 153 citations. They note that when proper methods for algal selection and cultivation are used, it is possible to use microalgae to produce the O_2 required by acclimatized bacteria to biodegrade hazardous pollutants such as polycyclic aromatic hydrocarbons, phenolics, and organic solvents. They recommend well-mixed PBRs with algal biomass recirculation to protect the microalgae from effluent toxicity and optimize light utilization efficiency. The optimum biomass concentration to maintain in the system depends mainly on the light intensity and the reactor configuration. At low light intensity, the biomass concentration should be optimized to avoid mutual shading and dark respiration, whereas at high light intensity, a high biomass concentration can be useful to protect microalgae from light inhibition and optimize the light–dark cycle frequency. PBRs can be designed as open (stabilization ponds or high-rate algal ponds) or enclosed (tubular, flat plate) systems. The latter are generally costly to construct and operate but more efficient than open systems. The best configuration to select will depend on factors such as process safety, land cost, and biomass use.

Carvalho et al. (2006) discuss microalgal reactors with a focus on enclosed system designs and performances in a review paper with 150 citations. They argue that there is no such thing as *the best reactor system*—defined, in an absolute fashion, as the one able to achieve maximum productivity with minimum operation costs, irrespective of the biological and chemical system at stake. In fact, the choice of the most suitable system is situation dependent, as both the species of alga available and the final purpose intended will play a role. The need of accurate control impairs use of open-system configurations, so current investigation has focused mostly on closed systems. They outline several types of closed bioreactors described in the technical literature as able to support production of microalgae that are comprehensively discussed using transport phenomenon and process engineering methodological approaches with a focus on the reactor design, which includes tubular reactors, flat-plate reactors, and fermenter-type reactors and processing parameters, which include gaseous transfer, medium mixing, and light requirements.

Ugwu et al. (2008) discuss PBRs for mass cultivation of algae in a review paper with 134 citations. They note that in order to grow and tap the potentials of algae, efficient PBRs are required. Although a good number of PBRs have been proposed, they argue that only a few of them can be practically used for mass production of algae. One of the major factors that limits their practical application in algal mass cultures is mass transfer. Thus, a thorough understanding of mass transfer rates in PBRs is necessary for efficient operation of mass algal cultures. They then discuss various PBRs that are very promising for mass production of algae.

Chen et al. (2011) discuss the cultivation, PBR design, and harvesting of microalgae for biodiesel production in a review paper with 118 citations. They review recent advances in microalgal cultivation, PBR design, and harvesting technologies with a focus on microalgal oil (mainly triglycerides) production. They compare and critically discuss the effects of different microalgal metabolisms (i.e., phototrophic, heterotrophic, mixotrophic, and photoheterotrophic growth), cultivation systems (emphasizing the effect of light sources), and biomass harvesting methods (chemical/physical methods) on microalgal biomass and oil production.

Grima et al. (1999) discuss PBRs with a focus on light regime, mass transfer, and scale-up in a paper with 114 citations. They investigate the design and scale-up of tubular PBRs for outdoor culture of microalgae. They argue that the culture productivity is invariably controlled by availability of light, particularly as the scale of operation increases. Thus, they emphasize the light regime analysis with details of a methodology for computation of the internal culture illumination levels in outdoor systems. They also discuss the supply of carbon dioxide as another important feature of algal culture. Finally, they outline the potential scale-up approaches including promising novel concepts based on fundamentals of the unavoidable light–dark cycling of the culture.

Greenwell et al. (2010) discuss the technological challenges for the production of biofuels from microalgae in a recent paper with 112 citations. They review current designs of algal culture facilities, including PBRs and open ponds, with regard to photosynthetic productivity and associated biomass and oil production, and include an analysis of alternative approaches using models, balancing space needs, productivity, and biomass concentrations, together with nutrient requirements. They also evaluate the options for potential metabolic engineering of the lipid biosynthesis pathways of microalgae. They conclude that although significant literature exists on microalgal growth and biochemistry, significantly more work needs to be undertaken to understand and potentially manipulate algal lipid metabolism. Furthermore, with regard to chemical upgrading of algal lipids and biomass, they describe alternative fuel synthesis routes and discuss and evaluate the application of catalysts traditionally used for plant oils. They argue that simulations that incorporate financial elements, along with fluid dynamics and algae growth models, are likely to be increasingly useful for predicting reactor design efficiency and life-cycle analysis to determine the viability of the various options for large-scale culture. They assert that the greatest potential for cost reduction and increased yields most probably lies within closed or hybrid closed–open production systems.

Lee (2001) discusses microalgal mass culture systems and methods in a paper with 111 citations. He notes that the need to achieve higher productivity and to maintain monoculture of algae led to the development of enclosed tubular and flat-plate PBRs. Despite higher biomass concentration and better control of culture parameters, the illuminated areal, volumetric productivity, and cost of production in these enclosed PBRs are not better than those achievable in open-pond cultures. He argues that the technical difficulty in sterilizing these PBRs has hindered their application for the production of high-value pharmaceutical products, while the alternative of cultivating microalgae in heterotrophic mode in sterilizable fermenters has achieved some commercial success. However, the maximum specific growth rates of heterotrophic algal cultures are in general slower than those measured in photosynthetic cultures. Therefore, the biomass productivity of heterotrophic algal cultures has yet to achieve a level that is comparable to industrial production of yeast and other heterotrophic microorganisms. Mixotrophic cultivation of microalgae takes advantage of their ability to utilize organic energy and carbon substrates and perform photosynthesis concurrently. Moreover, production of some algal metabolites is light regulated. He recommends that future design of sterilizable bioreactors for mixotrophic cultivation of microalgae may have to consider the organic substrate the main source of energy and light the supplemental source of energy, a change in mindset.

Del Campo et al. (2007) discuss the outdoor cultivation of microalgae for carotenoid production in a review paper with 109 citations. They note that consumer demand for natural products favors development of pigments from biological sources, thus increasing opportunities for microalgae. The biotechnology of microalgae has gained considerable progress and relevance in recent decades, with carotenoid production representing one of its most successful domains. They review the most relevant features of microalgal biotechnology related to the production of different carotenoids outdoors, with a main focus on beta-carotene from *Dunaliella*, astaxanthin from *Haematococcus*, and lutein from chlorophycean strains. They compare the current state of the corresponding production technologies, based on either open-pond systems or closed PBRs. They also discuss the potential of scientific and technological advances for improvements in yield and reduction in production costs for carotenoids from microalgae.

Hu et al. (1996) discuss a flat inclined modular PBR for outdoor mass cultivation of photoautotrophs in a paper with 107 citations. It consists of flat glass reactors connected in cascade facing the sun with the proper tilt angles to assure maximal exposure to direct beam radiation. They evaluate the optimal cell density in reference to the length of the reactor light path and assess the effect of the tilt angle on utilization of both direct beam and diffuse sunlight. They optimize the mixing mode and extent in reference to productivity of biomass. The FIMP proved very successful in supporting continuous cultures of the tested species of photoautotrophs, addressing the major criteria involved in design optimization of PBRs. Made of fully transparent glass, inclined toward the sun and endowed with a high surface-to-volume ratio, it combines an optimal light path with a vigorous agitation system. They find that the maximal exposure to the culture to solar irradiance as well as the substantial control of temperature facilitate, under these conditions, a particularly high, extremely light-limited optimal cell density. They find that the integrated effects of these growth conditions resulted in record volumetric and areal output rates of *Monodus subterraneus*, *Anabana siamensis*, and *Spirulina platensis*.

Tredici and Zittelli (1998) discuss the efficiency of sunlight utilization for tubular and flat PBRs in a paper with 90 citations. They use a coiled tubular reactor and compare a near-horizontal straight tubular reactor and a near-horizontal flat panel in outdoor cultivation of the cyanobacterium (Spirulina) *Arthrospira platensis* under defined operating conditions for optimum productivity. The photosynthetic efficiency achieved in the tubular systems was significantly higher because their curved surface *diluted* the impinging solar radiation and thus reduced the light saturation effect. This interpretation was supported by the results of experiments carried out in the laboratory under continuous artificial illumination using both a flat and a curved chamber reactor. They also show that, when the effect of light saturation is eliminated or reduced, productivity and solar irradiance are linearly correlated even at very high diurnal irradiance values and supported findings that outdoor algal cultures are light limited even during bright summer days. They then observe that, besides improving the photosynthetic efficiency of the culture, spatial dilution of light also leads to higher growth rates and lowers the cellular content of accessory pigments; that is, it reduces mutual shading in the culture. They stress the inadequacy of using volumetric productivity as the sole criterion for comparing reactors of different surface-to-volume ratio and of the areal productivity for evaluating the performance of elevated PBRs operated outdoors. They argue that the photosynthetic efficiency achieved by the culture should also be calculated to provide a suitable parameter for comparison of different algal cultivation systems operated under similar climatic conditions.

Chaumont (1993) discusses the biotechnology of algal biomass production with a focus on systems for outdoor mass culture in an early paper with 89 citations. Industrial reactors for algal culture are all designed as open raceways (shallow open ponds where culture is circulated by a paddle wheel) in the early 1990s. Technical and biological limitations of these open systems have given rise to the development of enclosed photoreactors (made of transparent tubes, sleeves, or containers and where light source may be natural or artificial). He surveys advances in these two technologies for cultivation of microalgae. Starting from published results, he discusses the advantages and disadvantages of open systems and closed PBRs. He focuses on closed systems, which have been considered as capital intensive and are justified only when a fine chemical is to be produced.

Tredici and Materassi (1992) discuss the Italian experience in the development of reactors for the mass cultivation of phototrophic microorganisms in the early 1990s in an early paper with 89 citations. The need to develop new concepts in reactor design, and the growing interest in *Spirulina* prompted them to abandon open ponds in the 1970s and to focus interest mainly on closed systems. They develop two substantially different closed PBRs: the tubular PBR (made of rigid or collapsible tubes) and the recently devised vertical alveolar panel (VAP) made of 1.6 cm thick Plexiglas alveolar sheets. They discuss the technical characteristics of the two systems in relation to the main factors, which regulate the growth of oxygenic photosynthetic microorganisms in closed reactors.

Posten (2009) discusses the design principles of PBRs for cultivation of microalgae in a paper with 88 citations. He reviews present PBR designs and the basic limiting factors that include light distribution to avoid saturation kinetics, mixing along the light gradient to make use of light–dark cycles, aeration and mass transfer along the vertical or horizontal main axis for carbon dioxide supply and oxygen removal, and last but not least the energy demand necessary to fulfill these tasks. To make comparison of the performance of different designs easier, he develops a commented list of performance parameters. Based on these critical points, he discusses recent developments in the areas of membranes for gas transfer and optical structures for light transfer. He argues that the fundamental starting point for the optimization of photo-bioprocesses is a detailed understanding of the interaction between the bioreactor in terms of mass and light transfer as well as the microalgae physiology in terms of light and carbon uptake kinetics and dynamics.

Gudin and Chaumont (1991) discuss cell fragility as the key problem of microalgae mass production in closed PBRs in an early paper with 79 citations. They argue that the gap between the theoretical biological potential of microalgae and the biomass productivity obtained with algal culture in tubular PBRs is due to a reduced growth rate related to hydrodynamic stress of pumping. High levels of mixing are necessary to reach a turbulent flow of the culture, in order to optimize the light regime. However, the optimal conditions of pumping to produce this significant liquid mixing may produce some cell damage.

They discuss the factors affecting this hydrodynamic stress (geometry of the bioreactor involved, type of pump utilized, morphology of algal cells, physiological conditions of microalgae, etc.).

Jorquera et al. (2010) perform the comparative energy life-cycle analyses of microalgal biomass production in open ponds and tubular and flat-plate PBRs using the oil-rich microalgae *Nannochloropsis* sp. in a recent paper with 76 citations. They calculate the net energy ratio (NER) for each process. They find that the use of horizontal tubular PBRs is not economically feasible ([NER] < 1) and that the estimated NER for flat-plate PBRs and raceway ponds is >1. They argue that the NER for ponds and flat-plate PBRs could be raised to significantly higher values if the lipid content of the biomass were increased to 60% dw/cwd.

Eriksen (2008) discusses current status and recent developments in the technology of microalgal culturing in enclosed PBRs in a review paper with 72 citations. He argues that light distribution and mixing are the primary variables that affect productivities of photoautotrophic cultures and have strong impacts on PBR designs. Process monitoring and control, physiological engineering, and heterotrophic microalgae are additional aspects of microalgal culturing, which have gained considerable attention in recent years.

Stephenson et al. (2010) discuss the life-cycle assessment of potential algal biodiesel production in the United Kingdom with a focus on the comparison of raceways and air lift tubular bioreactors in a recent paper with 70 citations. They use the life-cycle assessment to investigate the global warming potential (GWP) and fossil-energy requirement of a hypothetical operation in which biodiesel is produced from the freshwater alga *Chlorella vulgaris*, grown using flue gas from a gas-fired power station as the carbon source. They consider the cultivation using a two-stage method, whereby the cells were initially grown to a high concentration of biomass under nitrogen-sufficient conditions, before the supply of nitrogen was discontinued, whereupon the cells accumulated triacylglycerides. They find that if the future target for the productivity of lipids from microalgae, such as *C. vulgaris*, of similar to 40 t/(ha year) could be achieved, cultivation in typical raceways would be significantly more environmentally sustainable than in closed air lift tubular bioreactors. While biodiesel produced from microalgae cultivated in raceway ponds would have a GWP similar to 80% lower than fossil-derived diesel (on the basis of the net energy content), if air lift tubular bioreactors were used, the GWP of the biodiesel would be significantly greater than the energetically equivalent amount of fossil-derived diesel. The GWP and fossil-energy requirement in this operation were particularly sensitive to (1) the yield of oil achieved during cultivation, (2) the velocity of circulation of the algae in the cultivation facility, (3) whether the culture media could be recycled or not, and (4) the concentration of carbon dioxide in the flue gas. They assert the crucial importance of using life-cycle assessment to guide the future development of biodiesel from microalgae.

Fernandez et al. (1997) discuss a model for light distribution and average solar irradiance inside outdoor tubular PBRs for the microalgal mass culture in an early paper with 68 citations. They propose a mathematical model to estimate the solar irradiance profile and average light intensity inside a tubular PBR under outdoor conditions, requiring only geographic, geometric, and solar position parameters. First, the length of the path into the culture traveled by any direct or disperse ray of light was calculated as the function of three variables: day of year, solar hour, and geographic latitude. Due to the existence of differential wavelength absorption, they find that none of the literature models are useful for explaining light attenuation by the biomass. Therefore, they propose an empirical hyperbolic expression. The equations to calculate light path length were substituted in the proposed hyperbolic expression, reproducing light intensity data obtained in the center of the loop tubes. The proposed model was also likely to estimate the irradiance accurately at any point inside the culture. Calculation of the local intensity was thus extended to the full culture volume in order to obtain the average irradiance, showing how the higher biomass productivities in a *Phaeodactylum tricornutum* UTEX 640 outdoor chemostat culture could be maintained by delaying light limitation.

Benemann (1997) discusses the feasibility analysis of photobiological hydrogen production in an early paper with 66 citations. He presents a preliminary analysis of a two-stage process in which microalgae are cultivated in large open ponds to produce a high carbohydrate biomass that then produces hydrogen in tubular PBRs. They propose PBRs constructed of inexpensive, commercially available, glass tubes for such applications. He argues that photobiological hydrogen production requires long-term research and development. However, it could be of lower cost than systems based on electrolysis of water using photovoltaic electricity—the current system of choice for solar hydrogen production.

Miyake et al. (1999) discuss the biotechnological hydrogen production with a focus on the research for efficient light energy conversion in a paper with 65 citations. They describe the subjects examined in the research for photosynthetic bacteria: analysis of the relationship between the penetration of light to PBR and hydrogen production and genetic engineering of photosynthetic bacteria to control the pigment content for making the light penetration easy. They present examples of bench-scale reactors.

Cornet et al. (1992) discuss a structured model for simulation of cultures of the cyanobacterium *Spirulina* (*platensis*) in PBRs with a focus on the coupling between light transfer and growth kinetics in an early paper with 63 citations. The study of the interactions between physical limitation by light and biological limitations in PBRs leads to very complex partial differential equations. Modeling of light transfer and kinetics and the assessment of radiant energy absorbed in PBRs require an equation including two parameters for light absorption and scattering in the culture medium. They discuss a simple model based on the simplified, monodimensional equation of Schuster for radiative transfer. This approach provides a simple way to determine a working illuminated volume in which growth occurs, therefore allowing identification of kinetic parameters. These parameters might then be extended to the analysis of more complex geometries such as cylindrical reactors. Moreover, they assert that this model allows the behavior of batch or continuous cultures of cyanobacteria under light and mineral limitations to be predicted.

Cornet et al. (1995) discuss a simplified monodimensional approach for modeling coupling between radiant light transfer and growth kinetics in PBRs in an early paper with 61 citations. Local information is essential to PBR modeling because of medium anisotropy in radiant light energy. Local available energy can be calculated using complex equations, applying the physical laws of radiative transfer and independently accounting for light absorption and scattering in the reactor. They simplify these equations postulating monodimensional approximation for the radiation field. This simplification is established for rectangular, cylindrical, and spherical coordinates, leading to simple analytical solutions for available radiant energy profiles inside the reactor. This approach provides a method of determining working illuminated volume defined as the PBR volume with sufficient radiant light energy for microorganism growth. This enables the coupling between radiant light transfer and growth kinetics to be easily studied. They use physical light transfer models to simulate volumetric biomass growth rates in a cylindrical PBR with kinetic parameters obtained from batch cultures of the cyanobacterium *S. platensis* in rectangular PBRs. They compare these calculations with experimental data obtained on continuous cultures in a wide range of incident radiant energy fluxes.

5.2.3 Conclusion

The nonexperimental research on algal PBRs has been one of the most dynamic research areas in recent years. Twenty-six citation classics in the field of algal PBRs with more than sixty-one citations were located, and the key emerging issues from these papers were presented in Table 5.1 in the decreasing order of the number of citations. These papers give strong hints about the determinants of efficient algal PBR design for efficient production of algal biofuels and biocompounds and emphasize that the design of efficient algal PBRs is possible in the light of the technological advances in recent years.

5.3 Experimental Studies on Algal Photobioreactors

5.3.1 Introduction

Experimental research on algal PBRs has been one of the most dynamic research areas in recent years. Twenty-four experimental citation classics in the field of algal PBRs with more than fifty-seven citations were located, and the key emerging issues from these papers are presented in the decreasing order of the number of citations (Table 5.2). These papers give strong hints about the determinants of efficient algal PBR design and emphasize that it is possible in the light of the recent biotechnological advances in recent years.

The papers were dominated by the researchers from 15 countries, usually through the intracountry institutional collaboration, and they were multiauthored. Spain (four papers); England (three papers); and China, Germany, Israel, Italy, Japan, and the United States (two papers each) were the most prolific countries.

Similarly, all these papers were published in the journals indexed by the SCI and/or SCIE. There were no papers indexed by the A&HCI or SSCI. The number of citations ranged from 57 to 337, and the average number of citations per annum ranged from 2.6 to 67.4. All of these papers were articles.

5.3.2 Experimental Research on Algal Photobioreactors

Rodolfi et al. (2009) study the strain selection, induction of lipid synthesis, and outdoor mass cultivation of microalgae in a low-cost PBR in a paper with 337 citations. Thirty microalgal strains were screened in the laboratory for their biomass productivity and lipid content. Four strains (two marine and two freshwater), selected because they are robust and highly productive and have a relatively high-lipid content, were cultivated under nitrogen deprivation in 0.6 L bubbled tubes. Only the two marine microalgae accumulated lipid under such conditions. One of them, the eustigmatophyte *Nannochloropsis* sp. F&M-M24, which attained 60% lipid content after nitrogen starvation, was grown in a 20 L flat alveolar panel PBR to study the influence of irradiance and nutrient (nitrogen or phosphorus) deprivation on fatty acid accumulation. They find that "fatty acid content increased with high irradiances (up to 32.5% of dry biomass) and following both nitrogen and phosphorus deprivation (up to about 50%)." To evaluate its lipid production potential under natural sunlight, the strain was grown outdoors in 110 L green-wall panel PBRs under nutrient-sufficient and nutrient-deficient conditions. They find that lipid productivity increased from 117 mg/L/day in nutrient-sufficient media (with an average biomass productivity of 0.36 g/L/day and 32% lipid content) to 204 mg/L/day (with an average

TABLE 5.2
Experimental Research on Algal PBRs

No.	Paper References	Year	Document	Affiliation	Country	No. of Authors	Journal	Index	Subjects	Topic	Total No. of Citations	Total Average Citations Per Annum	Rank
1	Rodolfi et al.	2009	A	Univ. Florence, CNR	Italy	7	*Biotechnol. Bioeng.*	SCI, SCIE	Biot. Appl. Microb.	Algal photobioreactors	337	67.4	2
2	Li et al.	2007	A	Tsing Hua Univ.	China	3	*Biotechnol. Bioeng*	SCI, SCIE	Biot. Appl. Microb.	Algal photobioreactors	115	16.4	10
3	Janssen et al.	2003	A	Univ. Wageningen & Res. Ctr., Univ. Amsterdam	Netherlands	4	*Biotechnol. Bioeng.*	SCI, SCIE	Biot. Appl. Microb.	Algal photobioreactors	108	9.0	15
4	Chiu et al.	2008	A	Natl. Chiao Tung Univ.	Taiwan	6	*Bioresour. Technol.*	SCI, SCIE	Ener. Fuels, Biot. Appl. Microb.	Algal photobioreactors	101	16.8	17
5	Miron et al.	1999	A	Univ. Almeira	Spain	5	*J. Biotechnol.*	SCI, SCIE	Biot. Appl. Microb.	Algal photobioreactors	99	6.6	18
6	Doucha et al.	2005	A	Acad. Sci. Czech Republ., Fuel Res. Inst.	Czech Republic	3	*J. Appl. Phycol.*	SCI, SCIE	Biot. Appl. Microb., Mar. Fresh Water	Algal photobioreactors	95	10.6	19
7	Olaizola	2000	A	Aquasearch Inc.	United States	1	*J. Appl. Phycol.*	SCI, SCIE	Biot. Appl. Microb., Mar. Fresh Water	Algal photobioreactors	94	6.7	20
8	Molina et al.	2001	A	Univ. Almeira	Spain	4	*J. Biotechnol.*	SCI, SCIE	Biot. Appl. Microb.	Algal photobioreactors	88	6.8	25
9	Richmond et al.	1993	A	Ben Gurion Univ. Negev	Israel	4	*J. Appl. Phycol.*	SCI, SCIE	Biot. Appl. Microb., Mar. Fresh Water	Algal photobioreactors	87	4.1	26
10	Tredici et al.	1991	A	CNR	Italy	4	*Bioresour. Technol.*	SCI, SCIE	Ener. Fuels, Biot. Appl. Microb.	Algal photobioreactors	81	3.5	27

(Continued)

TABLE 5.2 (*Continued*)
Experimental Research on Algal PBRs

No.	Paper References	Year	Document	Affiliation	Country	No. of Authors	Journal	Index	Subjects	Topic	Total No. of Citations	Total Average Citations Per Annum	Rank
11	Scragg et al.	2002	A	Univ. W England	England	4	*Biomass Bioenerg.*	SCIE	Agr. Eng., Biot. Appl. Microb., Ener. Fuels	Algal photobioreactors	79	6.1	28
12	Rubio et al.	1999	A	Univ. Almeria, Univ. Granada	Spain	5	*Biotechnol. Bioeng.*	SCI, SCIE	Biot. Appl. Microb.	Algal photobioreactors	79	5.3	29
13	de Morais and Costa	2007	A	Fed. Univ. Fdn. Rio Grande	Brazil	2	*J. Biotechnol.*	SCI, SCIE	Biot. Appl. Microb.	Algal photobioreactors	73	10.4	32
14	Javanmardian and Palsson	1991	A	Univ. Michigan	United States	2	*Biotechnol. Bioeng.*	SCI, SCIE	Biot. Appl. Microb.	Algal photobioreactors	73	3.2	33
15	Pruvost et al.	2009	A	Univ. Nantes	France	4	*Bioresour. Technol.*	SCI, SCIE	Ener. Fuels, Biot. Appl. Microb.	Algal photobioreactors	70	14.0	36
16	Gonzalez et al.	1997	A	Politecn. Nacl.	Mexico	3	*Bioresour. Technol.*	SCI, SCIE	Ener. Fuels, Biot. Appl. Microb.	Algal photobioreactors	69	4.1	37
17	Kern et al.	2005	A	Tech. Univ. Berlin, Free Univ. Berlin	Germany	7	*Biochim. Biophys. Acta—Bioenerg.*	SCI, SCIE	Bioch. Mol. Biol., Biophys.	Algal photobioreactors	68	7.6	38

(*Continued*)

TABLE 5.2 (*Continued*)
Experimental Research on Algal PBRs

No.	Paper References	Year	Document	Affiliation	Country	No. of Authors	Journal	Index	Subjects	Topic	Total No. of Citations	Total Average Citations Per Annum	Rank
18	Harker et al.	1996	A	Liverpool John Moores Univ.	England	3	*J. Ferment. Bioeng.*	SCI, SCIE	Biot. Appl. Microb., Food Sci. Tech.	Algal photobioreactors	65	3.6	42
19	Watanabe et al.	1995	A	Kings Coll., Univ. Laval	England, Canada	3	*Biotechnol. Bioeng.*	SCI, SCIE	Biot. Appl. Microb.	Algal photobioreactors	63	3.3	43
20	Matsunaga et al.	1991	A	STI Japan, Onoda Cement Co Ltd.	Japan	10	*Appl. Biochem. Biotechnol.*	SCI, SCIE	Bioch. Mol. Biol., Biot. Appl. Microb.	Algal photobioreactors	60	2.6	46
21	Degen et al.	2001	A	Fraunhofer Inst. Grenzflachen & Bioverfahrenstech, Fraunhofer Inst. Solar Energy Syst.	Germany	5	*J. Biotechnol.*	SCI, SCIE	Biot. Appl. Microb.	Algal photobioreactors	59	4.5	47
22	Sierra et al.	2008	A	Univ. Almeira	Spain	6	*Chem. Eng. J.*	SCI, SCIE	Eng. Env., Eng. Chem.	Algal photobioreactors	58	9.7	48
23	Richmond and Cheng-Wu	2001	A	Ben Gurion Univ. Negev, Nanjing Inst. Chem. Technol.	Israel, China	2	*J. Biotechnol.*	SCI, SCIE	Biot. Appl. Microb.	Algal photobioreactors	57	4.4	49
24	Hu et al.	1998	A	Marine Biotechnol. Inst.	Japan	5	*Appl. Microbiol. Biotechnol.*	SCI, SCIE	Biot. Appl. Microb.	Algal photobioreactors	57	3.6	50

SCI, Science Citation Index; SCIE, Science Citation Index Expanded; SSCI, Social Sciences Citation Index; A, article; R, review.

biomass productivity of 0.30 g/L/day and more than 60% final lipid content) in nitrogen-deprived media. In a two-phase cultivation process (a nutrient-sufficient phase to produce the inoculum followed by a nitrogen-deprived phase to boost lipid synthesis), they argue that the "oil production potential could be projected to be more than 90 kg per hectare per day." They obtain an increase of both lipid content and areal lipid productivity attained through nutrient deprivation in an outdoor algal culture. They estimate that this "marine eustigmatophyte has the potential for an annual production of 20 tons of lipid per hectare in the Mediterranean climate and of more than 30 tons of lipid per hectare in sunny tropical areas."

Li et al. (2007) study large-scale biodiesel production from microalga *C. protothecoides* through heterotropic cultivation in bioreactors in a paper with 115 citations. They find that "through substrate feeding and fermentation process controls, the cell density of *C. protothecoides* achieved 15.5 gL^{-1} in 5 L, 12.8 gL^{-1} in 750 L, and 14.2 gL^{-1} in 11,000 L bioreactors, respectively. Resulted from heterotrophic metabolism, the lipid content reached 46.1%, 48.7%, and 44.3% of cell dry weight in samples from 5 L, 750 L, and 11,000 L bioreactors, respectively." They next find that "transesterification of the microalgal oil was catalyzed by immobilized lipase from *Candidia* sp. 99–125. With 75% lipase 12,000 U g^{-1}, based on lipid quantity and 3:1 molar ratio of methanol to oil batch-fed at three times, 98.15% of the oil was converted to monoalkyl esters of fatty acids in 12 h." The expanded biodiesel production rates were 7.02, 6.12, and 6.24 g/L in 5, 750, and 11,000 L bioreactors, respectively. They assert that the "properties of biodiesel from *Chlorella* were comparable to conventional diesel fuel and comply with the U.S. Standard for Biodiesel (ASTM 6751)."

Janssen et al. (2003) study enclosed outdoor PBRs with a focus on the light regime, photosynthetic efficiency, scale-up, and future prospects in a paper with 108 citations. They analyze both light regime and photosynthetic efficiency in characteristic examples of state-of-the-art pilot-scale PBRs. They show that productivity of PBRs is determined by the light regime inside the bioreactors. In addition to light regime, oxygen accumulation and shear stress limit productivity in certain designs. In short light-path systems, they argue that "high efficiencies, 10% to 20% based on photosynthetic active radiation (PAR 400 to 700 nm), can be reached at high biomass concentrations (>5 kg [dry weight] m^{-3})." They show, however, that "these and other photobioreactor designs are poorly scalable (maximal unit size 0.1 to 10 m^3), and/or not applicable for cultivation of monocultures." Therefore, they propose a new PBR design in which light capture is physically separated from photoautotrophic cultivation. This system can possibly be scaled to larger unit sizes, 10 to >100 m^3, and the reactor liquid as a whole is mixed and aerated. They estimate that high photosynthetic efficiencies, 15% on a photosynthetically active radiant (PAR) basis, can be achieved.

Chiu et al. (2008) study the reduction of CO_2 by a high-density culture of *Chlorella* sp. in a semicontinuous PBR in a paper with 108 citations. The marine microalga, *Chlorella* sp., was cultured in a PBR to assess biomass, lipid productivity, and CO_2 reduction. They also determine the effects of cell density and CO_2 concentration on the growth of *Chlorella* sp. They find that during an 8-day interval, cultures in semicontinuous cultivation, the "specific growth rate and biomass of *Chlorella* sp. cultures in the conditions aerated 2–15% CO_2 were 0.58–0.66 d^{-1} and 0.76–0.87 g L^{-1}, respectively. At CO_2 concentrations of 2%, 5%, 10% and 15%, the rate of CO_2, reduction was 0.261, 0.316, 0.466 and 0.573 g h^{-1}, and efficiency of CO_2 removal was 58%, 27%, 20% and 16%, respectively." The efficiency of CO_2 removal was similar in the single PBR and in the six-parallel PBR. However, CO_2 reduction, production of biomass, and production of lipid were six times greater in the six-parallel PBR than those in the single PBR. They argue that the "inhibition of microalgal growth cultured in the system with high CO_2 (10–15%) aeration could be overcome via a high-density culture of microalgal inoculum that was adapted to 2% CO_2." Moreover, biological reduction of CO_2 in the established system could be parallely increased using the PBR consisting of multiple units.

Miron et al. (1999) evaluate the compact PBRs for large-scale monoculture of microalgae in a paper with 99 citations. They use engineering analyses combined with experimental observations in horizontal tubular PBRs and vertical bubble columns to demonstrate the potential of pneumatically mixed vertical devices for large-scale outdoor culture of photosynthetic microorganisms. Horizontal tubular PBRs and vertical bubble column–type units differ substantially in many ways, particularly with respect to the surface-to-volume ratio, the amount of gas in dispersion, the gas–liquid mass transfer characteristics, the nature of the fluid movement, and the internal irradiance levels. As shown for EPA production from the microalga *P. tricornutum*, they argue that a realistic commercial process cannot rely on horizontal tubular PBR technology. In bubble columns, the presence of gas bubbles generally enhances internal irradiance when the sun is low on the horizon. Near solar noon, the bubbles diminish the internal column irradiance relative to the ungassed state. The "optimal dimensions of vertical column photobioreactors are about 0.2 m diameter and 4 m column height. Parallel east–west oriented rows of such columns located at 36.8°N latitude need an optimal inter-row spacing of about 3.5 m." In vertical columns, the biomass productivity varies substantially during the year: the peak productivity during summer may be several times greater than in the winter. This seasonal variation occurs also in horizontal tubular units, but is much less pronounced. They estimate that under identical conditions, the "volumetric biomass productivity in a bubble column is similar to 60% of that in a 0.06 m diameter horizontal tubular loop, but there is substantial scope for raising this value."

Doucha et al. (2005) study the utilization of flue gas for cultivation of microalgae (*Chlorella* sp.) in an outdoor open thin-layer PBR in a paper with 95 citations. Flue gas generated by combustion of natural gas in a boiler was used for outdoor cultivation of *Chlorella* sp. in a 55 m^2 culture area PBR. A 6 mm thick layer of algal suspension continuously running down the inclined lanes of the bioreactor at 50 cm/s was exposed to sunlight. Flue gas containing 6%–8% by volume of CO_2 was substituted for more costly pure CO_2 as a source of carbon for autotrophic growth of algae. The degree of CO_2 mitigation (flue gas decarbonization) in the algal suspension was 10%–50% and decreased with increasing flue gas injection rate into the culture. A dissolved CO_2 partial pressure (pCO_2) higher than 0.1 kPa was maintained in the suspension at the end of the 50 m long culture area in order to prevent limitation of algal growth by CO_2. NO_X and CO gases (up to 45 mg/m^3 NO_X and 3 mg/m^3 CO in flue gas) had no negative influence on the growth of the alga. On summer days, the following daily net productivities of algae (g [dry weight]/m^2) were attained in comparative parallel cultures: flue gas = 19.4–22.8 and pure CO_2 = 19.1–22.6. Net utilization (eta) of the PAR energy was flue gas, 5.58%–6.94%, and pure CO_2, 5.49%–6.88%. They estimate that about 50% of flue gas decarbonization can be attained in the PBR and 4.4 kg of CO_2 is needed for production of 1 kg (dry weight) algal biomass. They propose a scheme of a combined process of farm unit size where this includes anaerobic digestion of organic agricultural wastes, production and combustion of biogas, and utilization of flue gas for production of microalgal biomass, which could be used in animal feeds.

Olaizola (2000) studies the commercial production of astaxanthin from *Haematococcus pluvialis* using 25,000 L outdoor PBRs in a paper with 94 citations. The core of astaxanthin production chain is the Aquasearch Growth Module (AGM), a 25,000 L enclosed and computerized outdoor PBR. At Aquasearch's newly expanded facility (dedicated January 1999), he uses three AGMs (total volume 75,000 L) to produce large amounts of clean, fast growing *H. pluvialis*. The *H. pluvialis* biomass produced in the AGMs is transferred daily to a pond culture system, where carotenogenesis and astaxanthin accumulation are induced. Following a 5-day induction period, the reddened *H. pluvialis* cells are harvested by gravitational settling. The harvested biomass, which averages >2.5 astaxanthin as percent of the dry weight, is transferred to a processing building where a high-pressure homogenizer is used to rupture the cells' walls. Once the biomass has been homogenized, it is dried to less than 5% moisture utilizing proprietary drying technology. The dried product is then ready to be packaged according to customer needs. He notes that the PBR research program has almost doubled the performance of the AGMs in the first 9 months of operations: standing biomass concentration increased from 50 to 90 g/m^2, and production increased from 9 to 13 g/(m^2 day) during this period.

Molina et al. (2001) study the tubular PBR design for algal cultures in a paper with 88 citations. They integrate principles of fluid mechanics, gas–liquid mass transfer, and irradiance controlled algal growth into a method for designing tubular PBRs in which the culture is circulated by an air lift pump. They test a 0.2 m^3 PBR designed using the proposed approach in continuous outdoor culture of the microalga *P. tricornutum*. The culture performance was assessed under various conditions of irradiance, dilution rates, and liquid velocities through the tubular solar collector. They find that a "biomass productivity of 1.90 g l^{-1} d^{-1} (or 32 g m^{-2} d^{-1}) could be obtained at a dilution rate of 0.04 h^{-1}." Photoinhibition was observed during hours of peak irradiance; the photosynthetic activity of the cells recovered a few hours later. "Linear liquid velocities of 0.50 and 0.35 m s^{-1} in the solar collector gave similar biomass productivities, but the culture collapsed at lower velocities." The effect of dissolved oxygen concentration on productivity was quantified in indoor conditions where dissolved oxygen levels higher or lower than air saturation values reduced productivity. Under outdoor conditions, for given levels of oxygen supersaturation, they find that "the productivity decline was greater outdoors than indoors," suggesting that "under intense outdoor illumination photooxidation contributed to loss of productivity in comparison with productivity loss due to oxygen inhibition alone." Dissolved oxygen values at the outlet of solar collector tube were up to 400% of air saturation.

Richmond et al. (1993) study a new tubular reactor for mass production of microalgae outdoors in an early paper with 83 citations. An air lift is used for circulation of the culture in transparent tubes lying on the ground and interconnected by a manifold. Dissolved O_2 is removed through a gas separator placed 2.0 m above the tubes, and water spray is used for cooling. The manifold permits short-run durations between leaving the gas separator and reentering it, preventing thereby damaging accumulation of dissolved oxygen. Day temperature control in summer is attained using water spray. In winter, temperature in the tubes rises rapidly in the morning, as compared to an open raceway even if placed in a greenhouse. The number of hours along which optimal temperature prevails in the culture throughout the year increased significantly. They obtain very high daily productivity computed on a volumetric basis (e.g., 550 mg dry weight/L culture), and they estimate that a "significantly higher output, e.g. 1500 mg dry wt l^{-1} d^{-1}, is attainable."

Tredici et al. (1991) study a VAP for outdoor mass cultivation of microalgae and cyanobacteria in an early paper. They construct VAP reactors with a surface area of 0.5–2.2 m^2 from Plexiglas alveolar sheets 1.6 cm in thickness and used from May 1989 to February 1990 to grow *Anabaena azollae* and *S. platensis* in the climatic conditions of central Italy.

They find that the VAP is well suited to the outdoor mass cultivation of cyanobacteria, allowing operation at high cell concentrations (4–7 g/L) and achieving high biomass productivity even in winter. They argue that the "high surface-to-volume ratio (80 m^{-1}), its flexible orientation with respect to the sun's rays, effective mixing and O_2 removal through air bubbling and a good control of environmental and nutritional conditions are the major advantages of the system."

Scragg et al. (2002) study the growth of microalgae with increased calorific values in a tubular bioreactor in a paper with 79 citations. They show that *C. vulgaris* and *C. emersonii* grow in a 2301 pumped tubular PBR in Watanabe's medium and a low-nitrogen medium. The low-nitrogen medium induces higher lipid accumulation in both algae, which increased their calorific value. They find that the "highest calorific value was obtained with *C. vulgaris* (28 kJ g^{-1}) grown in low nitrogen medium." However, the "biomass productivity was 24 mg dry wt 1^{-1} d^{-1} in the low nitrogen medium which was lower than in Watanabe's medium (40 mg dry wt 1^{-1} d^{-1}) and represents a reduced energy recovery."

Rubio et al. (1999) predict the dissolved oxygen and carbon dioxide concentration profiles in tubular PBRs for microalgal culture in a paper with 79 citations. They use experimental data to verify the model for continuous outdoor culture of *Porphyridium cruentum* grown in a 200 L reactor with 100 m long tubular solar receiver. The culture was carried out at a dilution rate of 0.05 h^{-1} applied only during a 10 h daylight period. They find that the "quasi-steady state biomass concentration achieved was 3.0 $g.L^{-1}$, corresponding to a biomass productivity of 1.5 $g.L^{-1}.d^{-1}$." The model could predict the dissolved oxygen level in both gas disengagement zone of the reactor and at the end of the loop, the exhaust gas composition, the amount of carbon dioxide injected, and the pH of the culture at each hour. In predicting the various parameters, the model took into account the length of the solar receiver tube, the rate of photosynthesis, the velocity of flow, the degree of mixing, and gas–liquid mass transfer. Because the model simulated the system behavior as a function of tube length and operational variables (superficial gas velocity in the riser, composition of carbon dioxide in the gas injected in the solar receiver, and its injection rate), they argue that it could potentially be applied to rational design and scale-up of PBRs.

de Morais and Costa (2007) study the biofixation of carbon dioxide by *Spirulina* sp. and *Scenedesmus obliquus* cultivated in a three-stage serial tubular PBR in a paper with 73 citations. They cultivate the *S. obliquus* and *Spirulina* sp. at 30°C in a temperature-controlled three-stage serial tubular PBR and determine the resistance of these organisms to limitation and excess of carbon dioxide and the capacity of the system to fix this greenhouse gas. They find that after 5 days of cultivation under conditions of carbon limitation, both organisms showed cell death, *Spirulina* sp. presenting better results for all parameters than *S. obliquus*. They note that for *Spirulina* sp., the "maximum specific growth rate and maximum productivity was 0.44 d^{-1}, 0.22 g L^{-1} d^{-1}, both with 6% (v/v) carbon dioxide and maximum cellular concentration was 3.50 g L^{-1} with 12% (v/v) carbon dioxide. Maximum daily carbon dioxide biofixation was 53.29% for 6% (v/v) carbon dioxide and 45.61% for 12% carbon dioxide to *Spirulina* sp. corresponding values for *S. obliquus* being 28.08% for 6% (v/v) carbon dioxide and 13.56% for 12% (v/v) carbon dioxide." The highest mean carbon dioxide fixation rates value was 37.9% to *Spirulina* sp. in the 6% carbon dioxide runs.

Javanmardian and Palsson (1991) study the high-density photoautotrophic algal cultures with a focus on the design, construction, and operation of a novel PBR system in an early paper with 73 citations. A fiber optic–based optical transmission system that is coupled to an internal light distribution system illuminates the culture volume uniformly, at light intensities of 1.7 mW/cm^2 over a specific surface area of 3.2 cm^2/cm^3. Uniform light distribution is achieved throughout the reactor without interfering with the flow pattern required to keep the cells in suspension. An online ultrafiltration unit exchanges spent with fresh medium and its use result in very high cell densities, up to 10^9 cells/mL (3% [w/v]) for eukaryotic green alga *C. vulgaris*. They find that online ultrafiltration influences the growth pattern. Prior to ultrafiltration, the cells seem to have halted at a particular point in the cell cycle where they contain multiple chromosomal equivalents. Following ultrafiltration, these cells divide, and the new cells are committed to division so that cell growth resumes. The prototype PBR system was operated both in batch and in continuous mode for over 2 months. They find that the "measured oxygen production rate of 4–6 mmol/L culture h under continuous operation is consistent with the predicted performance of the unit for the provided light intensity."

Pruvost et al. (2009) study the biomass and lipids production with *Neochloris oleoabundans* in PBR in a paper with 70 citations. They find, without nutrient limitation, a "maximal biomass areal productivity of 16.5 g m^{-2} day^{-1}." They then find that due to initial *N. oleoabundans* total lipids high content (23% of dry weight), "highest productivity was obtained without mineral limitation with a maximal total lipids productivity of 3.8 g m^{-2} day^{-1}." Regarding TAG, they find an almost similar productivity where "continuous production without mineral limitation (0.5 g m^{-2} day^{-1}) or batch production with either sudden or progressive nitrogen deprivation (0.7 g m^{-2} day^{-1})" took place. They argue that the decrease in growth rate reduces the benefit of the important lipids and TAG accumulation as obtained in nitrogen starvation (37% and 18% of dry weight, respectively).

Gonzalez et al. (1997) study the efficiency of ammonia and phosphorus removal from a Colombian agro-industrial wastewater of a dairy industry and pig farming by the microalgae *C. vulgaris* and *S. dimorphus* in an early paper with 69 citations. The microalgae were isolated from a wastewater stabilization pond. Batch cultures were made using both species in 4 L cylindrical glass bioreactors each containing 2 L of culture. *C. vulgaris* was also cultivated on wastewater in a triangular bioreactor where three 216 h experimental cycles were run for each microalga and in each bioreactor. They find that in the cylindrical bioreactor, "*S. dimorphus* was more efficient in removing ammonia than *C. vulgaris*." However, the "final efficiency of both microalgae at the end of each cycle was similar as both microalgae removed phosphorus from the wastewater to the same extent in a cylindrical bioreactor." They then find that using *C. vulgaris*, the triangular bioreactor was superior for removing ammonia, and the cylindrical bioreactor was superior for removing phosphorus.

Kern et al. (2005) study the purification, characterization, and crystallization of photosystem II (PSII) from *Thermosynechococcus elongatus* cultivated in a new type of PBR in a paper with 68 citations. The thermophilic cyanobacterium *T. elongatus* was cultivated under controlled growth conditions using a new tape of PBR, allowing us to optimize growth conditions and the biomass yield. A fast large-scale purification method for monomeric and dimeric PSII solubilized from thylakoid membranes of this cyanobacterium was developed using fast protein liquid chromatography (FPLC). They find "36 chlorophyll a (Chla), 2 pheophylin a (Pheoa), 9+/−1 beta-carotene (Car), 2.9+/−0.8 plastoquinone 9 (PQ9) and 3.8+/−0.5 Mn per active centre." For the monomeric and dimeric PSIIcc, they find "18 and 20 lipid as well as 145 and 220 detergent molecules in the detergent shell, respectively." The monomeric and dimeric complexes showed high oxygen evolution activity with 1/4 O_2 released per 37–38 Chla and flash in the best cases.

Harker et al. (1996) study the autotrophic growth and carotenoid production of *H. pluvialis* in a 30 L air lift PBR in an early paper with 65 citations. Due to the different culture requirements of the alga during the various stages of its development, a two-stage batch production process (effectively before and after addition of NaCl to the culture) was employed for (1) biomass and (2) astaxanthin by this alga. During the first stage, conditions within the reactor (light intensity, levels of nitrogen, and phosphate) were maintained so as to achieve high rates of algal growth. When the algae in the reactor reached the stationary phase of growth and levels of nitrogen and phosphate in the medium had become severely depleted, NaCl was added to stimulate the synthesis of ketocarotenoids (>95% astaxanthin, mainly in the form of mono- and especially diesters), partly overcoming the need to increase irradiance levels. They find that "*H. pluvialis* exhibited relatively high rates of growth in the air-lift and accumulated up to 2.7% astaxanthin (of the dry cell weight of the alga)." This was, however, lower than could be achieved under laboratory-scale conditions (>5.5%).

Watanabe et al. (1995) study the photosynthetic performance of a helical tubular PBR incorporating the cyanobacterium *Spirulina* (*platensis*) in an early paper with 63 citations. The PBR was constructed in a cylindrical shape (0.9 m high) with a 0.25 m^2 basal area and a photostage comprising 60 m of transparent PVC tubing of 1.6 cm inner diameter (volume = 12.1 L). The inner surface of the cylinder (area = 1.32 m^2) was illuminated with cool white fluorescent lamps; the energy input of PAR (400–700 nm) into the PBR was 2920 kJ/day. An air lift system incorporating 4% CO_2 was used to circulate the growth medium in the tubing. They find that the "maximum productivity achieved in batch culture was 7.18 g dry biomass per day [0.51 g d biomass/L day, or 5.44 g d biomass/m^2 (inner surface of cylindrical shape)/day] which corresponded to a photosynthetic (PAR) efficiency of 5.45%." They next find that the CO_2 was efficiently removed from the gaseous stream as monitoring the CO_2 in the outlet and inlet gas streams showed a 70% removal of CO_2 from the inlet gas over an 8 h period with almost maximum growth rate.

Matsunaga et al. (1991) study the glutamate production from CO_2 by marine cyanobacterium *Synechococcus* sp. using a novel biosolar reactor employing light-diffusing optical fibers (LDOFs) in an early paper with 60 citations. A PBR was constructed in the form of a Perspex column 900 mm tall with an internal diameter of 70 mm. The reactor volume was 1.8 L, and the light source consisted of a metal-halide lamp to reproduce sunlight. Light was distributed through the culture using a new type of optical fiber that diffuses light out through its surface, perpendicular to the fiber axis. A cluster of 661 LDOFs pass from the light source through the reactor column (60 cm culture depth) and are connected to a mirror at the top of the reactor. This biosolar reactor has been used for the production of glutamate from CO_2 by the marine cyanobacterium *Synechococcus* sp. NKBG040607. They find that the "maximum conversion ratio (28%) was achieved at a cell density of 3×10^8 cells/mL." A comparison of glutamate production using the LDOF biosolar reactor with production by batch culture using free or immobilized cells showed that "use of an optical-fiber biosolar reactor increased glutamate-production efficiency 6.75-fold." They argue that as a result of its high surface-to-volume ratio (692 m^{-1}), increased photoproduction of useful compounds may be achieved.

Degen et al. (2001) study a novel air lift PBR with baffles for improved light utilization through the flashing-light effect in a paper with 59 citations. The bioreactor exposed the cells to intermittent light to improve the efficiency of light utilization through the flashing-light effect. During batch cultures in the new PBR, they find that the "biomass productivity of *Chlorella vulgaris* was 1.7 times greater than in a randomly mixed bubble column of identical dimension whilst a reduction in light path from 30 to 15 mm increased the biomass productivity by 2.5-fold." Then, a "maximum dry biomass productivity of 0.11 g $l^{-1} h^{-1}$ was obtained at an artificial illumination of 980 muE $m^{-2} s^{-1}$."

Sierra et al. (2008) study the characterization of a flat-plate PBR for the production of microalgae in a paper with 58 citations. The gas holdup and mass transfer coefficient were consistent with referenced values for bubble columns observed in tubular PBR. A power supply of 53 W/m^3 promoted a mass transfer rate high enough to avoid the excessive accumulation of dissolved oxygen in this flat-panel PBR. This is similar to the 40 W/m^3 necessary in bubble columns and much lower than the 2000–3000 W/m^3 required in tubular PBRs. However, this power supply is in the order of magnitude of 100 W/m^3, which has been reported to damage some microalgal cells, whereas no damage has been referenced in tubular PBRs. Even at low-power supplies, the mixing time was shorter than 200 s, longer than the 60 s measured for bubble columns, but quite faster than the typical values found for tubular PBRs (1–10 h). The heat transfer coefficient of the internal heat exchanger (over 500 W/[m^2 K]) was much higher than the coefficient of the external surface of the reactor (30 W/[m^2 K]). They argue that the "major disadvantage of this reactor is the potential high stress damage associated with aeration whilst the main advantages are the low power consumption (53 W/m^3) and the high mass transfer capacity (0.007 1/s)."

Richmond and Chen-Wu (2001) study the optimization of a flat-plate glass reactor for mass production of *Nannochloropsis* sp. outdoors in a paper with 57 citations. They investigate the relationships between areal (g/m^2/day) and volumetric (g/L/day) productivity of *Nannochloropsis* sp. as affected by the light path (ranging from 1.3 to 17.0 cm) of a vertical flat-plate glass PBR. The areal productivity in relation to the light path, in contrast, yielded an optimum curve, and the highest areal productivity was obtained in a 10 cm LP reactor, which is regarded, therefore, optimal for mass production of *Nannochloropsis*. Two basic factors that relate to reactor efficiency and its cost-effectiveness were (a) the total illuminated surface required to produce a set quantity of product and (b) culture volume required to produce that quantity. They argue that the difference in light utilization efficiency between the two very different production systems involves three aspects—"first, the open raceway requires ca. 6 times greater volume than the 10 cm flat plate reactor to produce the same quantity of cell-mass. Second, the total ground area (i.e. including the ground area between reactors) for the vertical flat plate reactor is less than one half of that occupied by an open raceway, indicating the former is more efficient, photosynthetically, compared with the latter. Finally, the harvested cell density is close to one order of magnitude higher in the flat plate reactor, which carries economic significance." They obtain the "optimal population density (i.e. which results in the highest areal productivity) in the 10 cm plate reactor by a daily harvest of 10% of culture volume, yielding an annual average of ca. 12.1 g dry wt. m^{-2} per day or 240 mg l^{-1} per day."

Hu et al. (1998) study the ultrahigh-cell-density culture of a marine green alga *Chlorococcum littorale* in a flat-plate PBR to test the feasibility of CO_2 remediation by microalgal photosynthesis in a paper with 57 citations. The modified reactor has a narrow light path in which intensive turbulent flow is provided by streaming compressed air through perforated tubing into the culture suspension. The length of the reactor light path was optimized for the productivity of biomass. They find that by growing *C. littorale* cells in this reactor, a "CO_2 fixation rate of 16.7 g CO_2 l^{-1} 24 h^{-1} (or 200.4 g CO_2 m^{-2} 24 h^{-1}) could readily be sustained at a light intensity of 2000 μmol $m^{-2} s^{-1}$ at 25°C, and an ultrahigh cell density of well over 80 g l^{-1} could be maintained by daily replacing the culture medium."

5.4 Conclusion

The citation classics presented under the two main headings in this chapter confirm the predictions that marine algae have a significant potential to serve as a major solution for the global problems of warming, air pollution, energy security, and food security through the design of efficient algal PBRs for the production of algal biofuels and algal biocompounds.

Further research is recommended for the detailed studies in each topical area presented in this chapter including scientometric studies and citation classic studies to inform the key stakeholders about the potential of marine algae for the solution of the global problems of warming, air pollution, energy security, and food security in the form of algal PBRs.

References

Baltussen, A. and C.H. Kindler. 2004a. Citation classics in anesthetic journals. *Anesthesia & Analgesia* 98:443–451.

Baltussen, A. and C.H. Kindler. 2004b. Citation classics in critical care medicine. *Intensive Care Medicine* 30:902–910.

Benemann, J.R. 1997. Feasibility analysis of photobiological hydrogen production. *International Journal of Hydrogen Energy* 22:979–987.

Borowitzka, M.A. 1999. Commercial production of microalgae: Ponds, tanks, tubes and fermenters. *Journal of Biotechnology* 70:313–321.

Bothast, R.J. and M.A. Schlicher. 2005. Biotechnological processes for conversion of corn into ethanol. *Applied Microbiology and Biotechnology* 67:19–25.

Carvalho, A.P., L.A. Meireles, and F.X. Malcata. 2006. Microalgal reactors: A review of enclosed system designs and performances. *Biotechnology Progress* 22:1490–1506.

Chaumont, D. 1993. Biotechnology of algal biomass production—A review of systems for outdoor mass-culture. *Journal of Applied Phycology* 5:593–604.

Chen, C.Y., K.L. Yeh, R. Aisyah, D.J. Lee, and J.S. Chang. 2011. Cultivation, photobioreactor design and harvesting of microalgae for biodiesel production: A critical review. *Bioresource Technology* 102:71–81.

Chisti, Y. 2007. Biodiesel from microalgae. *Biotechnology Advances* 25:294–306.

Chiu, S.Y., C.Y. Kao, C.H. Chen, T.C. Kuan, S.C. Ong, and C.S. Lin. 2008. Reduction of CO_2 by a high-density culture of *Chlorella* sp in a semicontinuous photobioreactor. *Bioresource Technology* 99:3389–3396.

Cornet, J.F., C.G. Dussap, and G. Dubertret. 1992. A structured model for simulation of cultures of the cyanobacterium spirulina-platensis in photobioreactors. 1. Coupling between light transfer and growth-kinetics. *Biotechnology and Bioengineering* 40:817–825.

Cornet, J.F., C.G. Dussap, J.B. Gros, C. Binois, and C. Lasseur. 1995. A simplified monodimensional approach for modeling coupling between radiant light transfer and growth-kinetics in photobioreactors. *Chemical Engineering Science* 50:1489–1500.

Degen, J., A. Uebele, A. Retze, U. Schmid-Staiger, and W. Trosch. 2001. A novel airlift photobioreactor with baffles for improved light utilization through the flashing light effect. *Journal of Biotechnology* 92:89–94.

Del Campo, J.A., M. Garcia-Gonzalez, and M.G. Guerrero. 2007. Outdoor cultivation of microalgae for carotenoid production: Current state and perspectives. *Applied Microbiology and Biotechnology* 74:1163–1174.

Demirbas, A. 2007. Progress and recent trends in biofuels. *Progress in Energy and Combustion Science* 33:1–18.
de Morais, M.G. and J.A.V. Costa. 2007. Biofixation of carbon dioxide by *Spirulina* sp. and *Scenedesmus obliquus* cultivated in a three-stage serial tubular photobioreactor. *Journal of Biotechnology* 129:439–445.
Doucha, J., F. Straka, and K. Livansky. 2005. Utilization of flue gas for cultivation of microalgae (*Chlorella* sp.) in an outdoor open thin-layer photobioreactor. *Journal of Applied Phycology* 17:403–412.
Dubin, D., A.W. Hafner, and K.A. Arndt. 1993. Citation-classics in clinical dermatological journals—Citation analysis, biomedical journals, and landmark articles, 1945–1990. *Archives of Dermatology* 129:1121–1129.
Eriksen, N.T. 2008. The technology of microalgal culturing. *Biotechnology Letters* 30:1525–1536, C72.
Fernandez, F.G.A., F.G. Camacho, J.A.S. Perez, J.M.F. Sevilla, and E.M. Grima. 1997. A model for light distribution and average solar irradiance inside outdoor tubular photobioreactors for the microalgal mass culture. *Biotechnology and Bioengineering* 55:701–714.
Gehanno, J.F., K. Takahashi, S. Darmoni, and J. Weber. 2007. Citation classics in occupational medicine journals. *Scandinavian Journal of Work, Environment & Health* 33:245–251.
Godfray, H.C.J., J.R. Beddington, I.R. Crute et al. 2010. Food security: The challenge of feeding 9 billion people. *Science* 327:812–818.
Goldemberg, J. 2007. Ethanol for a sustainable energy future. *Science* 315:808–810.
Gonzalez, L.E., R.O. Canizares, and S. Baena. 1997. Efficiency of ammonia and phosphorus removal from a Colombian agroindustrial wastewater by the microalgae *Chlorella vulgaris* and *Scenedesmus dimorphus*. *Bioresource Technology* 60:259–262, C69.
Greenwell, H.C., L.M.L. Laurens, R.J. Shields, R.W. Lovitt, and K.J. Flynn. 2010. Placing microalgae on the biofuels priority list: A review of the technological challenges. *Journal of the Royal Society Interface* 7:703–726.
Grima, E.M., E.H. Belarbi, F.G.A. Fernandez, A.R. Medina, and Y. Chisti. 2003. Recovery of microalgal biomass and metabolites: Process options and economics. *Biotechnology Advances* 20:491–515.
Grima, E.M., F.G.A. Fernandez, F.G. Camacho, and Y. Chisti. 1999. Photobioreactors: Light regime, mass transfer, and scaleup. *Journal of Biotechnology* 70:231–247.
Gudin, C. and D. Chaumont. 1991. Cell fragility—The key problem of microalgae mass-production in closed photobioreactors. *Bioresource Technology* 38:145–151.
Harker, M., A.J. Tsavalos, and A.J. Young. 1996. Autotrophic growth and carotenoid production of *Haematococcus pluvialis* in a 30 liter air-lift photobioreactor. *Journal of Fermentation and Bioengineering* 82:113–118.
Hu, Q., H. Guterman, and A. Richmond. 1996. A flat inclined modular photobioreactor for outdoor mass cultivation of photoautotrophs. *Biotechnology and Bioengineering* 51:51–60.
Hu, Q., N. Kurano, M. Kawachi, I. Iwasaki, and S. Miyachi. 1998. Ultrahigh-cell-density culture of a marine green alga *Chlorococcum littorale* in a flat-plate photobioreactor. *Applied Microbiology and Biotechnology* 49:655–662.
Jacobson, M.Z. 2009. Review of solutions to global warming, air pollution, and energy security. *Energy & Environmental Science* 2:148–173.
Janssen, M., J. Tramper, L.R. Mur, and R.H. Wijffels. 2003. Enclosed outdoor photobioreactors: Light regime, photosynthetic efficiency, scale-up, and future prospects. *Biotechnology and Bioengineering* 81:193–210.
Javanmardian, M. and B.O. Palsson. 1991. High-density photoautotrophic algal cultures—Design, construction, and operation of a novel photobioreactor system. *Biotechnology and Bioengineering* 38:1182–1189.
Jorquera, O., A. Kiperstok, A. Sales, E.A. Embirucu, and M.L. Ghirardi. 2010. Comparative energy life-cycle analyses of microalgal biomass production in open ponds and photobioreactors. *Bioresource Technology* 101:1406–1413.
Kapdan, I.K. and F. Kargi. 2006. Bio-hydrogen production from waste materials. *Enzyme and Microbial Technology* 38:569–582.

Kern, J., B. Loll, C. Luneberg et al. 2005. Purification, characterisation and crystallisation of photosystem II from *Thermosynechococcus elongatus* cultivated in a new type of photobioreactor. *Biochimica et Biophysica Acta—Bioenergetics* 1706:147–157.

Konur, O. 2000. Creating enforceable civil rights for disabled students in higher education: An institutional theory perspective. *Disability & Society* 15:1041–1063.

Konur, O. 2002a. Assessment of disabled students in higher education: Current public policy issues. *Assessment and Evaluation in Higher Education* 27:131–152.

Konur, O. 2002b. Access to employment by disabled people in the UK: Is the disability discrimination act working? *International Journal of Discrimination and the Law* 5:247–279.

Konur, O. 2002c. Access to Nursing Education by disabled students: Rights and duties of nursing programs. *Nursing Education Today* 22:364–374.

Konur, O. 2006a. Participation of children with dyslexia in compulsory education: Current public policy issues. *Dyslexia* 12:51–67.

Konur, O. 2006b. Teaching disabled students in higher education. *Teaching in Higher Education* 11:351–363.

Konur, O. 2007a. A judicial outcome analysis of the Disability Discrimination Act: A windfall for the employers? *Disability & Society* 22:187–204.

Konur, O. 2007b. Computer-assisted teaching and assessment of disabled students in higher education: The interface between academic standards and disability rights. *Journal of Computer Assisted Learning* 23:207–219.

Konur, O. 2011. The scientometric evaluation of the research on the algae and bio-energy. *Applied Energy* 88:3532–3540.

Konur, O. 2012a. 100 citation classics in energy and fuels. *Energy Education Science and Technology Part A—Energy Science and Research* 2012(si):319–332.

Konur, O. 2012b. What have we learned from the citation classics in energy and fuels: A mixed study. *Energy Education Science and Technology Part A* 2012(si):255–268.

Konur, O. 2012c. The gradual improvement of disability rights for the disabled tenants in the UK: The promising road is still ahead. *Social Political Economic and Cultural Research* 4:71–112.

Konur, O. 2012d. Prof. Dr. Ayhan Demirbas' scientometric biography. *Energy Education Science and Technology Part A—Energy Science and Research* 28:727–738.

Konur, O. 2012e. The evaluation of the research on the biofuels: A scientometric approach. *Energy Education Science and Technology Part A—Energy Science and Research* 28:903–916.

Konur, O. 2012f. The evaluation of the research on the biodiesel: A scientometric approach. *Energy Education Science and Technology Part A—Energy Science and Research* 28:1003–1014.

Konur, O. 2012g. The evaluation of the research on the bioethanol: A scientometric approach. *Energy Education Science and Technology Part A—Energy Science and Research* 28:1051–1064.

Konur, O. 2012h. The evaluation of the research on the microbial fuel cells: A scientometric approach. *Energy Education Science and Technology Part A—Energy Science and Research* 29:309–322.

Konur, O. 2012i. The evaluation of the research on the biohydrogen: A scientometric approach. *Energy Education Science and Technology Part A—Energy Science and Research* 29:323–338.

Konur, O. 2012j. The evaluation of the biogas research: A scientometric approach. *Energy Education Science and Technology Part A—Energy Science and Research* 29:1277–1292.

Konur, O. 2012k. The scientometric evaluation of the research on the production of bio-energy from biomass. *Biomass and Bioenergy* 47:504–515.

Konur, O. 2012l. The evaluation of the global energy and fuels research: A scientometric approach. *Energy Education Science and Technology Part A—Energy Science and Research* 30:613–628.

Konur, O. 2012m. The evaluation of the biorefinery research: A scientometric approach. *Energy Education Science and Technology Part A—Energy Science and Research* 2012(si):347–358.

Konur, O. 2012n. The evaluation of the bio-oil research: A scientometric approach. *Energy Education Science and Technology Part A—Energy Science and Research* 2012(si):379–392.

Konur, O. 2012o. What have we learned from the citation classics in energy and fuels: A mixed study. *Energy Education Science and Technology Part A—Energy Science and Research* 2012(si): 255–268.

Konur, O. 2012p. The evaluation of the research on the biofuels: A scientometric approach. *Energy Education Science and Technology Part A—Energy Science and Research* 28:903–916.

Konur, O. 2013. What have we learned from the research on the International Financial Reporting Standards (IFRS)? A mixed study. *Energy Education Science and Technology Part D: Social Political Economic and Cultural Research* 5:29–40.

Lal, R. 2004. Soil carbon sequestration impacts on global climate change and food security. *Science* 304:1623–1627.

Lee, Y.K. 2001. Microalgal mass culture systems and methods: Their limitation and potential. *Journal of Applied Phycology* 13:307–315.

Li, X.F., H. Xu, and Q.Y. Wu. 2007. Large-scale biodiesel production from microalga *Chlorella protothecoides* through heterotropic cultivation in bioreactors. *Biotechnology and Bioengineering* 98:764–771.

Lynd, L.R., J.H. Cushman, R.J. Nichols, and C.E. Wyman. 1991. Fuel ethanol from cellulosic biomass. *Science* 251:1318–1323.

Mata, T.M., A.A. Martins, and N.S. Caetano. 2010. Microalgae for biodiesel production and other applications: A review. *Renewable & Sustainable Energy Reviews* 14:217–232.

Matsunaga, T., H. Takeyama, H. Sudo et al. 1991. Glutamate production from CO_2 by marine cyanobacterium *Synechococcus* sp. using a novel biosolar reactor employing light-diffusing optical fibers. *Applied Biochemistry and Biotechnology* 28–29:157–167.

Miron, A.S., A.C. Gomez, F.G. Camacho, E.M. Grima, and Y. Chisti. 1999. Comparative evaluation of compact photobioreactors for large-scale monoculture of microalgae. *Journal of Biotechnology* 70:249–270.

Miyake, J., M. Miyake, and Y. Asada. 1999. Biotechnological hydrogen production: Research for efficient light energy conversion. *Journal of Biotechnology* 70:89–101.

Molina, E., J. Fernandez, F.G. Acien, and Y. Chisti. 2001. Tubular photobioreactor design for algal cultures. *Journal of Biotechnology* 92:113–131.

Munoz, R. and B. Guieysse. 2006. Algal-bacterial processes for the treatment of hazardous contaminants: A review. *Water Research* 40:2799–2815.

North, D. 1994. Economic-performance through time. *American Economic Review* 84:359–368.

Olaizola, M. 2000. Commercial production of astaxanthin from *Haematococcus pluvialis* using 25,000-liter outdoor photobioreactors. *Journal of Applied Phycology* 12:499–506.

Paladugu, R., M.S. Chein, S. Gardezi, and L. Wise. 2002. One hundred citation classics in general surgical journals. *World Journal of Surgery* 26:1099–1105.

Posten, C. 2009. Design principles of photo-bioreactors for cultivation of microalgae. *Engineering in Life Sciences* 9:165–177.

Pruvost, J., G. Van Vooren, G. Cogne, and J. Legrand. 2009. Investigation of biomass and lipids production with *Neochloris oleoabundans* in photobioreactor. *Bioresource Technology* 100:5988–5995.

Pulz, O. 2001. Photobioreactors: Production systems for phototrophic microorganisms. *Applied Microbiology and Biotechnology* 57:287–293.

Richmond, A., S. Boussiba, A. Vonshak, and R. Kopel. 1993. A new tubular reactor for mass-production of microalgae outdoors. *Journal of Applied Phycology* 5:327–332.

Richmond, A. and Z. Cheng-Wu. 2001. Optimization of a flat plate glass reactor for mass production of *Nannochloropsis* sp outdoors. *Journal of Biotechnology* 85:259–269.

Rodolfi, L., G.C. Zittelli, N. Bassi et al. 2009. Microalgae for oil: Strain selection, induction of lipid synthesis and outdoor mass cultivation in a low-cost photobioreactor. *Biotechnology and Bioengineering* 102:100–112.

Rubio, F.C., F.G.A. Fernandez, J.A.S. Perez, F.G. Camacho, and E.G. Grima. 1999. Prediction of dissolved oxygen and carbon dioxide concentration profiles in tubular photobioreactors for microalgal culture. *Biotechnology and Bioengineering* 62:71–86.

Scragg, A.H., A.M. Illman, A. Carden, and S.W. Shales. 2002. Growth of microalgae with increased calorific values in a tubular bioreactor. *Biomass & Bioenergy* 23:67–73, doi: 10.1016/S0961-9534(02)00028-4.

Sierra, E., F.G. Acien, J.M. Fernandez, J.L. Garcia, C. Gonzalez, and E. Molina. 2008. Characterization of a flat plate photobioreactor for the production of microalgae. *Chemical Engineering Journal* 138:136–147.

Spolaore, P., C. Joannis-Cassan, E. Duran, and A. Isambert. 2006. Commercial applications of microalgae. *Journal of Bioscience and Bioengineering* 101:87–96.

Stephenson, A.L., E. Kazamia, J.S. Dennis, C.J. Howe, S.A. Scott, and A.G. Smith. 2010. Life-cycle assessment of potential algal biodiesel production in the United Kingdom: A comparison of raceways and air-lift tubular bioreactors. *Energy & Fuels* 24:4062–4077.

Tredici, M.R., P. Carlozzi, G.C. Zittelli, and R. Materassi. 1991. A vertical alveolar panel (VAP) for outdoor mass cultivation of microalgae and cyanobacteria. *Bioresource Technology* 38:153–159.

Tredici, M.R. and R. Materassi. 1992. From open ponds to vertical alveolar panels—The Italian experience in the development of reactors for the mass cultivation of phototrophic microorganisms. *Journal of Applied Phycology* 4:221–231.

Tredici, M.R. and G.C. Zittelli. 1998. Efficiency of sunlight utilization: Tubular versus flat photobioreactors. *Biotechnology and Bioengineering* 57:187–197.

Ugwu, C.U., H. Aoyagi, and H. Uchiyama. 2008. Photobioreactors for mass cultivation of algae. *Bioresource Technology* 99:4021–4028.

Volesky, B. and Z.R. Holan. 1995. Biosorption of heavy-metals. *Biotechnology Progress* 11:235–250.

Wang, B., Y.Q. Li, N. Wu, and C.Q. Lan. 2008. CO(2) bio-mitigation using microalgae. *Applied Microbiology and Biotechnology* 79:707–718.

Watanabe, Y., J. Delanoue, and D.O. Hall. 1995. Photosynthetic performance of a helical tubular photobioreactor incorporating the cyanobacterium *Spirulina platensis*. *Biotechnology and Bioengineering* 47:261–269.

Wrigley, N. and S. Matthews. 1986. Citation-classics and citation levels in geography. *Area* 18:185–194.

Yergin, D. 2006. Ensuring energy security. *Foreign Affairs* 85:69–82.

Section III

Biotechnological Techniques in Marine Bioenergy

6

Fermentation Techniques in Bioenergy Production

Geetanjali Yadav, Ramya Kumar, and Ramkrishna Sen

CONTENTS

6.1 Introduction

Fossil fuel combustion leads to increase in atmospheric CO_2 that poses great risk to worldwide sustainability. With the advent of the industrial revolution and increasing economic growth, the trend is going to increase (IPCC 2007). Available technologies for CO_2 removal or absorption rely heavily on chemical and physicochemical methods and geochemical storage into deep vents that always have a risk of leakage into the surroundings. Bioenergy is the energy from biomass or organic matter. It has the potential to greatly reduce our greenhouse gas emissions by using carbon dioxide from the atmosphere for its growth. The process is called photosynthesis. Biomass generates an equal amount of carbon dioxide as fossil fuels, but it utilizes CO_2 for its growth. Therefore, the net emission of carbon dioxide will be zero. Bioenergy can include energy from biomass in the form of biofuels, biopower, or other bioproducts. Biofuels are basically transportation fuels like bioethanol, biodiesel, biohydrogen, and biomethane. Biopower consists of direct biomass burning or converting it into gaseous fuel or oil to generate electricity. Among various biomass feedstocks, both micro- and macroalgae present a sustainable alternative (Singh and Sai 2010).

Fermentation is a process in which production is done by means of the mass culture of a microorganism. The product is either cell mass or various metabolites. In most of the fermentation processes, the end product depends on the type of substrate used. Also, the amount or quantity depends upon various fermentation techniques and modes of reactor operation. Optimization of process parameters for growth of microorganisms results in maximum product yield (Soccol et al. 2013).

This chapter presents an overview of bioenergy generation from marine biomass. Many aspects of marine algal biomass like biomass characteristics and pretreatment technologies are discussed in the following sections. The latter part of the chapter focuses mainly on various fermentation techniques and modes of fermenter operations, which can be applied to the bioconversion of algal biomass to biofuels.

6.2 Marine Bioenergy

Bioenergy is any form of energy that is derived from living or recently living biological organisms. Biological organisms are a rich source of organic material commonly termed as biomass. Recently, there have been many talks on biomass and biofuels, as we are facing a steep decline in our fossil fuel reserves. Every nation is exploring and evaluating newer routes to achieve petroleum independence. The overfamiliarity and overexploitation of terrestrial fuel sources have led countries to look toward marine ecosystems. Oceans have a tremendous amount of energy-rich reserves such as macroalgae, microalgae, and unique microbes that may be the key to providing a sustainable solution to the world's fossil fuel crisis. The most widely discussed biofuels around the globe are bioethanol, which can be produced from biomass rich in starch/cellulose, and biodiesel, a fatty acid alkyl ester (e.g., fatty acid methyl ester) that can be produced from vegetable oils, algal lipids, or microbial lipids by transesterification of lipids in the presence of alcohol (Brennan and Owende 2010). Biobutanol, biohydrogen, biogas, and syngas are a few other examples of biofuels produced by microorganisms through fermentation of carbohydrates as shown in Figure 6.1. Conventional biofuels were generated using sugars, starch, and vegetable oil, whereas second-generation biofuels such as algal biodiesel and cellulosic bioethanol are produced from sustainable feedstock. Sustainability of feedstock is based on its availability and impact on the environment (Roesijadi et al. 2010).

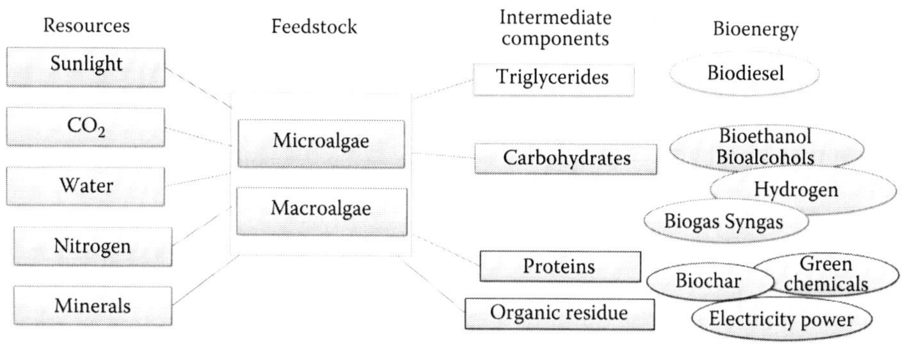

FIGURE 6.1
Bioenergy options from marine biomass. (From Adams, J.M. et al., *J. Appl. Phycol.*, 21, 569, 2009.)

6.3 Marine Feedstocks

Algae, both unicellular and multicellular, are a very large and diverse group of autotrophic organisms. Types of marine algae available for bioenergy generation are microalgae and macroalgae divided based on their morphology and size. The three most prominent categories based on pigment content of algae are green algae, brown algae, and red algae. Algae, like terrestrial plants, live through photosynthesis, a process whereby light energy is converted into chemical energy by fixing atmospheric CO_2 as in the following reaction:

$$6CO_2 + 6H_2O + \text{light energy} \rightarrow C_6H_{12}O_6(\text{sugars}) + 6O_2 \quad (6.1)$$

The sugars formed by photosynthesis are converted to other cellular components such as lipids, carbohydrates, and proteins that make up the biomass. Table 6.1 represents carbohydrate composition in algae

TABLE 6.1
Carbohydrate Composition in Algae

	Macroalgae		
Microalgae	**Green Algae**	**Brown Algae**	**Red Algae**
Cellulose	Cellulose	Laminarin	Carrageenan
Starch	Starch	Alginate	Agar
Rhamnose		Fucoidan	Cellulose
Mannose		Mannitol	Lignin
		Cellulose	

Source: Roesijadi, G. et al., *Macroalgae as a Biomass Feedstock: A Preliminary Analysis*, U.S. Department of Energy, Pacific Northwest Laboratory, Richland, WA, 2010.

6.3.1 Microalgae

Microalgae are unicellular, microscopic, photosynthetic algae. Generally, microalgal biomass contains three major macromolecular components: lipids, carbohydrates, and proteins. As shown in Table 6.1, cellulose is a major structural polysaccharide in the cell walls of microalgae depicted in Figure 6.2. Starch is also present in algal cells as a storage polysaccharide.

Microalgal biomass composition varies depending on the species. For example, *Chlorella* sp., *Dunaliella* sp., *Chlamydomonas* sp., and *Scenedesmus* sp. are reported to contain carbohydrates up to 50% of the dry weight of the cell (Singh and Sai 2010). *Nannochloris* sp., *Chlorella vulgaris*, and *Scenedesmus dimorphus* are a few examples that accumulate lipid content up to 60% of dry weight. The chemical components such as cellulose, starch, and triglycerides can be converted into a variety of fuel options such as alcohols, diesel, biogas, and hydrogen through appropriate conversion techniques. Selection of species should be made according to the desired biofuel route.

6.3.2 Macroalgae

Macroalgae represent a group of eukaryotic, photosynthetic marine organisms that possess plant-like characteristics. They are typically comprised of a blade or lamina, the stipe, and holdfast for anchoring the entire plant structure to hard surfaces in oceans. Brown macroalgae consist up to 65% carbohydrates on dry weight basis and contain high levels of complex carbohydrates such as alginate, laminarin, carrageenan, and agar (Roesijadi et al. 2010). In general, macroalgae have a low cellulose and lignin content. Polysaccharides and sugar alcohols in brown algae, for example, laminarin and mannitol, are candidate feedstocks for conversion to liquid fuels. Laminarin consists mainly of a linear β-(1 → 3)–linked glucan with some random β-(1 → 6)–linked side chains depending on the variety of seaweed shown in Figure 6.3 (Bruton et al. 2009).

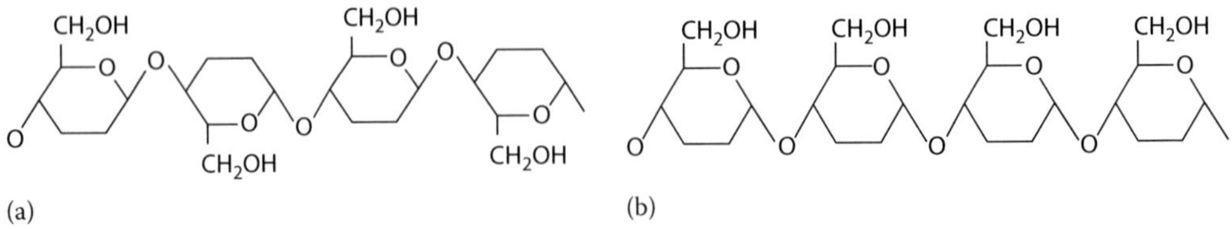

FIGURE 6.2
Backbone structures of (a) cellulose and (b) starch.

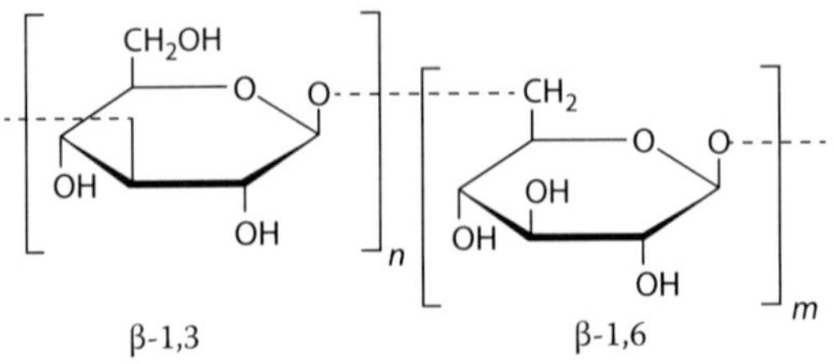

FIGURE 6.3
Backbone structure of laminarin.

TABLE 6.2
Composition of Algal Biomass

Algae	Carbohydrate Content (%)	Protein Content (%)	Lipid Content (%)
Microalgae			
Dunaliella bioculata	4	49	8
Dunaliella salina	32	57	6
Prymnesium parvum	25–33	28–45	22–38
Scenedesmus dimorphus	21–52	8–18	16–40
Porphyridium auentum	37–40	28–39	9–14
Chlamydomonas reinhardtii	17	48	21
Macroalgae			
Catenella repens	29–33	11–14	—
Ulva lactuca	35	8–11	—
Dictyota dichotoma	9–11	9–11	4
Enteromorpha intestinalis	26–30	13–15	5.2
T. ornata	14–16	14–16	3.5
Gracilaria verrucosa	16–18	8–10	3.2
H. musciformis	26–30	12–14	3.4
Saccharina japonica	60–66	10	1.6

Sources: Parthiban, C. et al., *Adv. Appl. Sci. Res.*, 4(3), 362, 2013; Singh, J. and Sai, G., *Renew. Sustain. Energ. Rev.*, 14, 2596, 2010; Banerjee, K. et al., *African J. Basic & Appl. Sci.* 1(5–6), 96, 2009.

Lipids in a variety of macroalgae are typically less than 5% of total dry weight, too low for conversion to biodiesel. Table 6.2 illustrates the biochemical composition of algae species.

6.4 Biomass Cultivation and Harvesting

Macroalgae are largely harvested from the wild. Cultivation of seaweed can be done in floating systems near the coastline as in some Asian countries. Although microalgae too can be harvested from the wild, they are cultivated, either in open-pond systems or closed photobioreactor systems in order to achieve high biomass concentrations. Cultivation of algal biomass is a critical step in obtaining desired biomass characteristics such as carbohydrate and lipid content. Changes in nutrient composition and culture conditions have a great effect on composition and characteristics of biomass (Bruton et al. 2009).

6.5 Pretreatment

6.5.1 Pretreatment of Algal Biomass for Utilization of Carbohydrates

Biomass contains complex polymers and compounds that may not be readily available for bioconversion either by enzymes or by microbes. Pretreatment entails disruption of the biomass in order to separate its components for further processing toward the final product.

The main obstacle in carbohydrates/glucose production is the rigid cell wall of algae. Pretreatment or carbohydrate extraction is an important step that liberates intercellular carbohydrates from the biomass. Table 6.3 enlists various pretreatment strategies employed for algal biomass conversion. Chemical and physicochemical pretreatment methods like alkali treatment and acid treatment in combination with application of high temperatures have been widely used in deconstruction of biomass (Harmsen et al. 2010). Studies on the effects of acid concentration on microalgal acid pretreatment confirm an increase in yield of sugars from biomass. Microwave-assisted extraction and ultrasonic-assisted extraction of carbohydrates have also been found to increase carbohydrate yield from biomass (Guili et al. 2003). Macroalgae have high levels of structural and storage complex such as alginate, laminarin, carrageenan, and agar. Polysaccharides and sugar alcohols in brown algae, for example, laminarin and mannitol, are potential feedstocks for conversion to liquid fuels and can be removed from the biomass by mild thermochemical acid pretreatment (Brennan and Owende 2010). Pretreatment method employed must be economical and suitable for scale-up.

6.5.1.1 Enzymatic Hydrolysis of Carbohydrates

Enzymes are proteins that enable many kinds of chemical reactions in living organisms. Cellulases are a group of enzymes that synergistically depolymerize cellulose fibril to

TABLE 6.3

List of Pretreatment Strategies for Algal Biomass Bioconversion

Pretreatment Options	Rationale
Physical Methods	
Washing/cleaning	Removal of foreign particles, removal of salts
Dewatering/drying	Removal of excess water from biomass
Milling/pulverizing	Size reduction, increasing surface area, release of biomolecules such as lipid
Microwave/microwave-assisted pyrolysis	Disruption of the cell walls, facilitates lipid extraction
Ultrasonication	Cell lysis and homogenization
Physicochemical Methods	
Steam explosion	Loosens up the biomass fibers and increases accessibility for further treatment
Thermal acid hydrolysis	Solubilization and releases fermentable sugars
Thermal alkaline hydrolysis	Increases cellulose digestibility
Liquid hot water	Improves enzymatic digestibility, removes soluble sugars
Chemical Methods	
Acid treatment	Improves enzymatic digestibility, removal of pentose sugars
Alkali treatment	Increases cellulose digestibility
Biological Methods	
Enzymatic	Biocatalytic conversion of polysaccharides to monomers, e.g., cellulases, amylases, xylanases
Whole cell	Biocatalysis; simultaneous saccharification and fermentation

Sources: Adapted from Harmsen, P.F.H. et al., *Literature Review of Physical and Chemical Pretreatment Processes for Lignocellulosic Biomass*, Wageningen UR, Food & Biobased Research, Wageningen, the Netherlands, 2010; Alvira, P. et al., *Bioresour. Technol.*, 101(13), 4851, 2010; Bruton, T.L.H. et al., *A Review of the Potential of Marine Algae as a Source of Biofuel in Ireland*, Sustainable Energy Ireland, Dublin, Ireland, 2009.

monomeric glucose units. They are produced by many fungi and bacteria. These enzymes consist of three main components: endoglucanases, exoglucanases, and β-glucosidases. Kinetics and mechanism of enzymatic hydrolysis of biomass from *Chloroccum* sp. by using cellulase from *Trichoderma reesei* have been investigated (Harun and Danquah 2011, Ueda et al. 1996). *Gracilaria verrucosa*, a red seaweed, is being used for the production of agar. The leftover pulp after agar extraction was found to contain 62%–68% cellulose, which, on enzymatic hydrolysis using cellulases, yielded 0.87 g sugars/g cellulose (Kumar 2013). Enzymatic hydrolysis using cellulases is an effective method of saccharification of algal biomass. A new generation of enzymes and production technologies is needed to cost-effectively convert complex carbohydrates into fermentable sugars.

6.5.2 Pretreatment of Algal Biomass for Lipid Extraction

Solvent extraction is most commonly used for lipid extraction. Approaches based on selective deconstruction of the microalgal cell wall, using ultrasound, microwave, enzymes, pressurized fluid extraction, and supercritical fluid extraction, are viable emerging lipid extraction methods. Since microalgae have rigid cell walls, effective extraction can be achieved only when combined with a pretreatment step. High-pressure homogenization is an effective method for cell disruption and has been widely investigated (Nguyen et al. 2009). Lipids obtained by microwave pretreatment followed by supercritical carbon dioxide extraction had a high concentration of fatty acids compared to conventional extraction without pretreatment in *Chlorella* sp. (Halim et al. 2012, Dejoye et al. 2011, Markou et al. 2012). The effect and efficacy of extraction techniques on microalgae can be observed by scanning electronic microscopy (SEM).

6.6 Fermentation

In the context of industrial biotechnology, the term *fermentation* refers to the growth of large quantities of cells under anaerobic or aerobic conditions within a vessel, called a fermenter or bioreactor. Fermented products have various applications in food, textile, and transportation industry. The basic function of a fermenter is to provide an environment suitable for the controlled growth of a pure culture or of a defined mixture of organisms. Although the fermentation process takes place in a fermenter or bioreactor, the whole production process comprises of many steps. The main process steps for a typical fermentation process involve media preparation, sterilization, inoculum development, fermentation or production stage, and downstream processing for product purification. A typical fermentation process is shown in Figure 6.4.

Although the main process step is fermentation, it is important to not ignore the upstream and downstream stages. The downstream step is concerned with production of desired product. The efficiency of a fermentation process could be assessed by finding the successful conversion of substrate into product. The productivity of fermentation is defined as

$$\text{Productivity (P)} = \frac{\text{Product concentration L}^{-1}}{\text{Fermentation time}} = \text{Units h}^{-1}$$

The process variable is a dynamic and most important feature of the process that may change rapidly. Hence, measurement of process parameters is important in

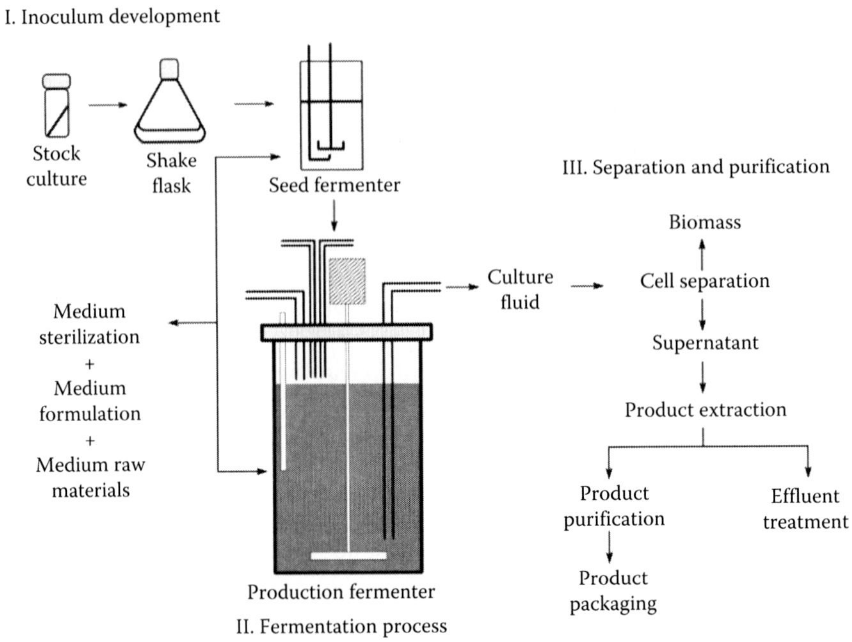

FIGURE 6.4
Generalized schematic representation of a fermentation process. (Adapted from Stanbury, P.F. et al., *Principles of Fermentation Technology*, Pergamon Press, Oxford, U.K., 1995.)

TABLE 6.4
Important Parameters for a Fermentation Process

S. No.	Process Parameters	Mathematical Characterization	Symbol Explanation
1	Temperature	°C	Celsius
2	pH	—	Negative log of hydronium ion concentration
3	Aeration rate (volume per volume per minute, vvm)	$AR = \frac{FG}{VR} (m^3/m^3 \text{ min})$	AR = aeration rate; FG = volumetric gas flow rate; VR = fermenter reaction volume
4	Agitation	rpm	Rpm (rotation per minute)
5	Mixing time (Tm)	$Tm = f(n, d_I, v) = V/Q = V/N_{fl} n\, d_I^3$ [s]	Tm = mixing time; n = stirrer speed; d_I = impeller diameter; v = kinematic viscosity; V = volume; Q = volumetric flow rate; N_{fl} = pumping number
6	Oxygen transfer rate (OTR)	$OTR = K_La(C_G - C_L)$ [kg O_2/m^3 h]	OTR = oxygen transfer rate; K_La = oxygen transfer coefficient; C_G = oxygen saturation concentration in the gas phase; C_L = measured oxygen saturation concentration in the liquid phase
7	Gas hold up (τ)	$\tau = \frac{VR}{FG}$	τ = gas hold up; VR = fermenter reaction volume; FG = volumetric gas flow rate

Sources: Modified from Hubbard, D.W., *Ann. NY Acad. Sci.*, 506, 600, 1987; Wang, D.I.C. and Cooney, C.L., Translation of laboratory, pilot, and plant scale data, in: Wang, D.L.C. et al. (eds.), *Fermentation and Enzyme Technology*, Wiley, New York, 1997, pp. 194–211.

controlling a process. There are mainly four along with other variables that affect any process, namely, temperature, pressure, flow rate, and pH of the system. The control of process parameters is very important for the optimal growth of microorganism, so whether it's a batch or continuous process, various factors important for a fermentation process are given in Table 6.4.

6.7 Classification of Fermentation Process

Fermentation can be largely subdivided into two categories, namely, solid-substrate fermentation and submerged fermentation. *Solid-state fermentation* (*SSF*) is defined as the growth of microorganisms on a solid-insoluble matrix in the absence of free water. Compared to SSF, *submerged fermentation* is a typical liquid fermentation process, where the cells are freely dispersed in the growth medium and interact as individual or flocculated units. Liquid fermentations are said to be perfectly homogeneous reactions due to its mixing ability, which is absent in SSF processes. Since algae are aquatic organisms, they usually require water for its growth; however, very few reports have been found for SSF for algae. Therefore, submerged or suspension cultures are primarily used for growing algae.

The fermentation process involves several chemical reactions like hydrolysis, oxidation, reduction, polymerization, biosynthesis, and formation of new cells. Fermentation is basically metabolism of microorganisms on carbon as a substrate. According to the requirement of air/oxygen, some fermentations are classified as aerobic and some anaerobic. We will briefly discuss these two processes:

1. Aerobic fermentation
2. Anaerobic fermentation

6.8 Types of Fermentation

6.8.1 Aerobic Fermentation

The term aerobic fermentation is a misnomer. More appropriate, aerobic respiration is a process in which organic compounds are degraded with the release of CO_2 and energy into simpler compounds. A similar reaction also takes place under anaerobic conditions when NO^{3-} or SO_4^{2-} can substitute for O_2 as electron acceptor. A comparison of carbon cycling in aerobic and anaerobic environment is depicted in Figure 6.5a. If no external electron acceptor is available, the degradation of organic material proceeds through the process of fermentation. A number of industrial processes, although called *fermentations*, are carried on by microorganisms under aerobic conditions. In older aerobic processes, it was necessary to furnish a large surface area by exposing fermentation media to air. In modern fermentation processes, aerobic conditions are maintained by passing air or oxygen in a closed fermenter with submerged cultures. The contents of the fermenter are agitated with the help of a stirrer and aerated by forcing sterilized air as shown in Figure 6.5b.

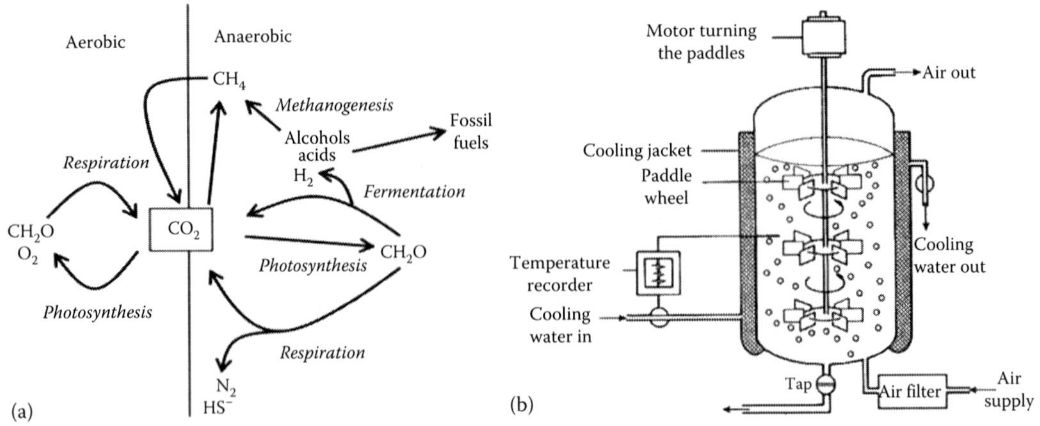

FIGURE 6.5
(a) Carbon cycling in aerobic environments. (b) An aerobic fermenter.

6.8.2 Anaerobic Fermentation

Anaerobic fermentation is a process of conversion of algal biomass into methane after lipid extraction (Sialve et al. 2009a). Fermentations are usually classified according to the main fermentation products, for instance, as alcohol, lactate, acetate, or methane fermentations. Both micro- and macroalgae act as suitable renewable substrates for the anaerobic digestion process (ADP). Algal biomass consists of a mixture of organic and inorganic matter. The organic part is composed of complex polymeric macromolecules, such as proteins, polysaccharides, lipids, and nucleic acids. The polymers appear in particulate or colloidal form. The ADP converts complex organic matter into final end products, that is, methane and carbon dioxide, new biomass, and inorganic residue.

Algae are advantageous in the sense they lack the recalcitrant lignin components in their cell wall. This gives them an upper hand for their use in ADP. Microalgae contain significant amounts of lipids and oils rather than sugars, which pose a detrimental effect on the ADP, along with high levels of proteins in the algal cells (Mata et al. 2010). Also, algal cells, especially macroalgae, contain high levels of fermentable carbohydrates and low lipid content, and they contain more carbohydrates after the lipid has been extracted in microalgae, which makes them an ideal substrate for anaerobic digestion and biogas production. Table 6.5 shows the difference in composition and productivity of micro- and macroalgae.

ADP generally leads to the production of methane and biohydrogen. Various products from algal biomass transformation processes are depicted in Figure 6.1.

6.8.2.1 Biomethane or Biogas

Algal biomass consists of a mixture of organic and inorganic matter. The organic part is composed of complex polymeric macromolecules, such as proteins, polysaccharides, lipids, and nucleic acids. ADP converts organic matter to the final products (methane and carbon dioxide), new biomass, and inorganic residue. Figure 6.6 describes mainly four key chemical and biological stages of anaerobic digestion, namely, hydrolysis, fermentation or acidogenesis, acetogenesis, and methanogenesis. Mixtures of bacteria are used to hydrolyze and break down the organic biopolymers (i.e., carbohydrates, lipids, and proteins) into monomers, which are then converted into a methane-rich gas via fermentation

TABLE 6.5

Data Illustrating Major Differences in Composition and Productivity between Microalgae and Macroalgae

Components	Microalgae (% dry wt.)	Macroalgae (% dry wt.)	References
Lipid	10–65	0–2	Adams et al. (2009)
Protein	20–60	12–19	Adams et al. (2009)
Total fermentable carbohydrates	11–47	20–60	Dismukes et al. (2008); Adams et al. (2009)
Lignin	0	0	Adams et al. (2009)
Dry matter	20	15	Adams et al. (2009)
Ash content	10	25	Adams et al. (2009)
Productivity (t ha^{-1} y^{-1})	20–75	11–45	Adams et al. (2009); Chynoweth, (2002)

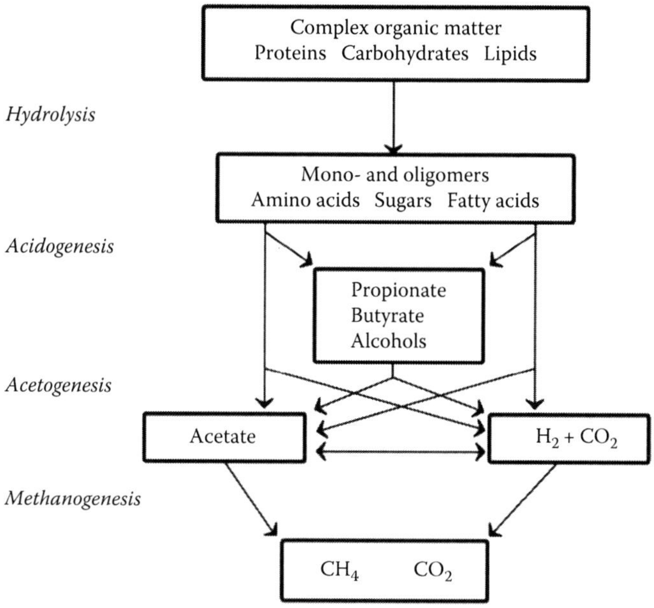

FIGURE 6.6
Simplified schematic representation of the anaerobic degradation process. (Adapted from de Mes, T.Z.D. et al., Chapter 4: Methane production by anaerobic digestion of wastewater of wastewater and solid wastes, in: Reith, J.H. et al. (eds.), *Biomethane and Biohydrogen. Status and Perspectives of Biological Methane and Hydrogen Production*, Dutch Biological Hydrogen Foundation, The Hague, the Netherlands, 2003, pp. 58–94.)

(typically 50%–75% CH_4). Carbon dioxide is the second main component found in biogas (approximately 25%–50%) and, like other interfering impurities, has to be removed before the methane is used for electricity generation.

1. *Hydrolysis*: The algal biomass is made up of complex organic polymers. These are first required to be broken down into smaller constituent parts so as to become accessible by anaerobic bacteria in the digester. The process of breaking these chains and dissolving the smaller molecules into solution is called hydrolysis.

2. *Acidogenesis or fermentation*: In this process, acidogenic (fermentative) bacteria convert the hydrolytic products mainly amino acids, sugars, and fatty acids into volatile fatty acids (VFAs) such as propionic acid, butyric acid, valeric acid, and various alcohols (Conrad 1999).
3. *Acetogenesis*: In the next step, the organic acids and alcohols are further digested to acetic acid, hydrogen, and carbon dioxide by acetogenic bacteria.
4. *Methanogenesis*: Is a terminal stage of anaerobic digestion in which methanogens utilize the intermediate products of the preceding stages and convert them into methane, carbon dioxide, and water. The end product is a combustible gas called biogas. A simplified generic chemical equation for the overall processes outlined earlier is as follows:

$$C_6H_{12}O_6 \rightarrow 3CO_2 + 3CH_4 \tag{6.2}$$

The theoretical methane and ammonium yields can be evaluated from the following formula (Symons and Buswell 1933).

$$C_aH_bO_cN_d + \left(\frac{4a-b-2c+3d}{4}\right)H_2O \rightarrow \left(\frac{4a+b-2c-3d}{8}\right)CH_4 + \left(\frac{4a-b+2c+3d}{8}\right)CO_2 + dNH_3 \tag{6.3}$$

where C, H, and O have their usual chemical significance and the subscripts refer to the number of the respective atoms. This equation gives a stoichiometric conversion of organic matter into methane, carbon dioxide, and ammonia.

The specific yield of methane produced can thus be expressed in terms of CH_4 per gram of volatile solids (VS) by the following:

$$B_o = \left(\frac{4a+b-2c-3d}{12a+b+16c+14d}\right) * V_m \tag{6.4}$$

where V_m is the normal molar volume of methane. There are various factors that have profound effect on biogas yield as shown in Table 6.6.

When considering biogas production from algae, two alternatives can be conceived: microalgae biodiesel production and further anaerobic digestion of microalgae residues for biogas production and anaerobic digestion of whole macroalgae or microalgae with biogas as sole biofuel.

6.8.2.2 Dark Fermentation for Biohydrogen Production

Biological hydrogen production is considered as one of the most promising green alternatives for sustainable energy production from biomass. Hydrogen is a clean fuel that does not produce carbon dioxide as a by-product when it is burnt for electricity generation in fuel cells (Park et al. 2009). Algal biomass (whole or after oil and/or starch removal) can be converted in bio-H_2 by dark fermentation as well as photofermentation using anaerobic

TABLE 6.6

Factors Affecting Biogas Yield in an Anaerobic Digester

S. No.	Methods	Range	References
1	C/N ratio	20:1–30:1	Parkin and Owen (1986)
2	Temperature	Psychrophilic: 10°C–20°C Mesophilic: 20°C–40°C Thermophilic: 50°C–60°C	Mes et al. (2003)
3	pH	6.5–7.5	Mes et al. (2003)
4	Toxicity	VFAs, ammonia, cations, namely, Na^+, K^+, Ca^{++}, heavy metals, sulfide, and xenobiotics	Mes et al. (2003)
5	Loading rate (kg VS/m^3 day)	1.4	Babaee and Shayegan (2011)
6	HRT (hydraulic retention time)	10–20 days	El-Mashad et al. (2004)

organisms for bio-H_2 production. Marine macroalgae, mainly brown seaweeds, are rich in biochemical components in their cell walls like cellulose, alginates, sulfated fucans, and protein (Kloareg et al. 1986). As a result, brown seaweeds are considered a suitable candidate for biohydrogen production by anaerobic fermentation due to rich carbohydrate content. Anaerobic bacteria use organic substrates for hydrogen production that can be depicted in the following two simple equations:

$$\text{Glucose} + 2H_2O \rightarrow 2\text{Acetate} + CO_2 + 4H_2 \quad (6.5)$$

$$\text{Glucose} \rightarrow \text{Butyrate} + 2CO_2 + 2H_2 \quad (6.6)$$

The enzyme hydrogenase carries out photoevolution of hydrogen in Cyanobacteria and algae by splitting of water molecule into oxygen and hydrogen (Adams 1990; Gaffron 1940). A light-dependent reaction is catalyzed by nitrogenase for hydrogen production in Cyanobacteria; hydrogenase does the same in dark anaerobic conditions (Hansel and Lindblad 1998; Rao and Hall 1996), while in green algae, hydrogen is produced photosynthetically by the ability to harness the solar energy resource, to drive H_2 production, from water (Ghirardi et al. 2000; Melis et al. 2000; Melis and Happe 2001; Ran et al. 2006; Yang et al. 2010).

Enterobacter and *Clostridium* strains of bacteria are well-known producers of bio-H_2 by utilizing various types of carbon sources (Angenent et al. 2004; Cantrell et al. 2008; Das 2009). It was reported that *Laminaria japonica* was the optimum substrate for hydrogen production among *Ulva lactuca, Porphyra tenera,* and *Undaria pinnatifida* producing about 4164 mL of hydrogen from 50 g L^{-1} of dry algae for 50 h under the optimum fermentation temperature, substrate concentration, and initial pH of 35°C, 5%, and 7.5, respectively (Park et al. 2009). Hydrogenase and nitrogenase enzymes are both capable of hydrogen production by two pathways:

1. *Hydrogenase-dependent hydrogen production*: The enzyme catalyzes the simplest of chemical reactions, the reversible reductive formation of hydrogen from protons and electrons (Tamagnini et al. 2002) as depicted in Figure 6.7.

$$2H^+ + 2e^- \leftrightarrow H_2 \quad (6.7)$$

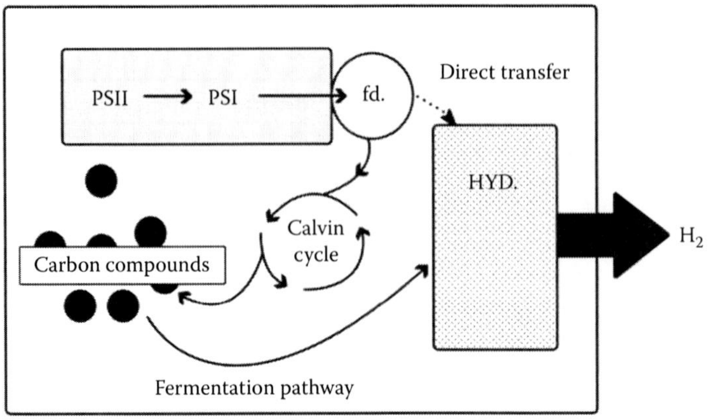

FIGURE 6.7
Hydrogenase-dependent hydrogen production. (Adapted from Miyamoto, K., Hydrogen production, In *Renewable Biological Systems for Alternative Sustainable Energy Production*, Miyamoto, K. (ed.), Issue 128, Food and Agriculture Organization, 108 pp., 1997.)

Hydrogenases are classified into three major groups: NiFe, Fe, and metal-free hydrogenases based on their metal composition of the active sites (Vignais et al. 2001). *Scenedesmus* have evolved molecular hydrogen under light conditions after being kept in anaerobic and dark conditions (Gaffron and Rubin 1942).

2. *Nitrogenase-dependent hydrogen production*: Nitrogenase is responsible for nitrogen fixation and is distributed mainly among prokaryotes, including Cyanobacteria, but does not occur in eukaryotes, under which microalgae are classified.

   Nitrogenase is a two-component protein system that uses MgATP (2ATP/e$^-$) and low potential electrons derived from reduced form of ferredoxin or flavodoxin to further reduce a variety of substrates (Hallenbeck and Benemann 2002). Molecular nitrogen is reduced to ammonium with consumption of reducing power (e$^-$ mediated by ferredoxin) and ATP. The reaction is substantially irreversible and produces ammonia:

$$N_2 + 6H^{1+} + 6e^- \rightarrow 2HN_3 \tag{6.8}$$

$$12ATP \rightleftharpoons 12(ADP + Pi) \tag{6.9}$$

However, nitrogenase catalyzes proton reduction in the absence of nitrogen gas (i.e., in an argon atmosphere) (Figure 6.8):

$$2H^+ + 2e^- \rightarrow H_2 \tag{6.10}$$

$$4ATP \rightleftharpoons 4(ADP + Pi) \tag{6.11}$$

Hydrogen production catalyzed by nitrogenase occurs as a side reaction at a rate of one-third to one-fourth that of nitrogen fixation, even in a 100% nitrogen gas atmosphere (Gouveia 2011). Reduction of nitrogen to ammonia by the enzyme nitrogenase was reported in *Anabaena cylindrica* (Masukawa et al. 2002).

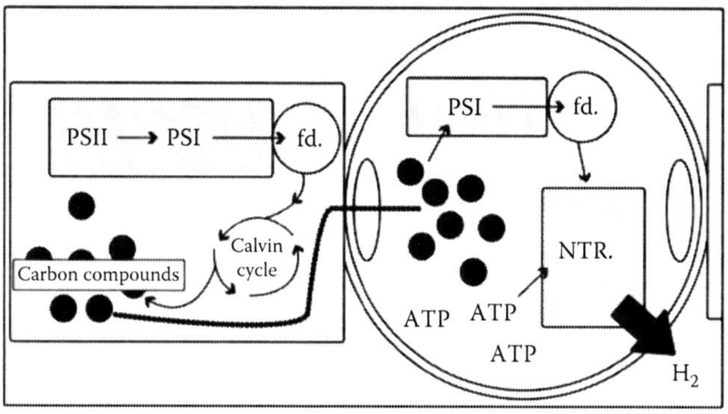

FIGURE 6.8
Nitrogenase-dependent hydrogen production. (Adapted from Miyamoto, K., Hydrogen production, In *Renewable Biological Systems for Alternative Sustainable Energy Production*, Miyamoto, K. (ed.), Issue 128, Food and Agriculture Organization, 108 pp., 1997.)

6.8.2.3 Bioethanol Production from Algae

Algal cell walls are typical in comprising particular biochemical components. Certain species of microalgae have the ability of producing high levels of carbohydrates instead of lipids as storage polymers. Hence, these storage carbohydrates can be extracted to produce fermentable sugars for the production of ethanol. Macroalgae or seaweeds are considered as ideal candidates for ethanol production since they contain no or very less amount of lignin. The most abundant sugars in brown seaweeds comprises of alginate, mannitol, and glucan (present as laminarin and cellulose) (Enquist-Newman et al. 2013). Alginate composes 30%–60% of the total sugars in brown macroalgae (Chapman 1970). Macroalgae contain a significant amount of sugars (at least 50%) that could be utilized in fermentation for bioethanol production (Wi et al. 2009). Some algae such as *Sargassum, Gracilaria, Prymnesium parvum,* and *Euglena gracilis* are promising candidates for ethanol production. The main microorganisms used to produce ethanol under anaerobic conditions are bacteria, yeast, or fungi (Harun et al. 2010). They metabolize carbohydrates and produce CO_2 and ethanol as metabolic end product in an anaerobic condition. The two most important microorganisms used for ethanol production are the yeast *Saccharomyces cerevisiae* and the bacterium *Zymomonas mobilis*. Brown seaweed produced higher amount of bioethanol as compared to other algal species (Moen 2008). Because brown macroalgae do not contain lignin, sugars can be released by simple operations such as milling or crushing. In the first step, the algal biomass is ground and hydrolyzed by enzymatic or chemical treatment to release its starch. In the subsequent steps, the hydrolyzed biomass is fermented with the addition of yeast *S. cerevisiae* and later distilled (Demirbas and Demirbas 2010). The final product after purification is ethanol. According to different ways of carrying out hydrolysis and fermentation in a system, there are three most important types:

6.8.2.3.1 *Separate Hydrolysis and Fermentation*

In separate hydrolysis and fermentation (SHF), the two steps, that is, enzymatic hydrolysis and fermentation, are performed separately in different vessels. This allows enzymes to operate at optimum activities and gives process better control. However, the main disadvantage in SHF is the accumulation of hydrolysis products that leads to substrate inhibition.

6.8.2.3.2 Simultaneous Saccharification and Fermentation

Unlike SHF, simultaneous saccharification and fermentation (SS&F) is a process step where the enzyme hydrolysis and fermentation are run in the same reactor vessel. Therefore, the end-product inhibition is prevented. This also leads to lower enzyme requirement and higher bioethanol yield (Lin and Tanaka 2006).

6.8.2.3.3 Simultaneous Saccharification and Cofermentation and Separate Hydrolysis and Cofermentation

Simultaneous saccharification and cofermentation (SSCF) is similar to SSF as in both processes, hydrolysis and fermentation steps are carried out in the same vessel, but here recombinant microbes are used. This is an improvement over SSF since microbes usually employed for bioethanol production cannot utilize all sugar sources, mainly pentoses, that reduce bioethanol yield.

In another similar bioprocess, separate hydrolysis and cofermentation (SHCF), processes take place in separate vessels so that each step can be performed at its optimal conditions. Besides, since the microbes utilize both pentoses and hexoses effectively in the cofermentation process in SHCF, the bioethanol yield is higher than SHF.

6.8.2.4 Biodiesel Production

Oil-rich biomass of algae can be largely produced through mass cultivation outdoors or even heterotrophic fermentation indoors by substrate feeding. These techniques put forward a novel pathway to produce oil feedstocks for biodiesel production. Biodiesel production mainly involves the following steps: biomass cultivation, harvesting, drying, cell disruption and lipid collection, and transesterification to convert triacylglycerides (TAGs) into fatty acid methyl esters (FAME). Transesterification is a chemical reaction between triglycerides and alcohol in the presence of a catalyst to produce monoesters that are termed as biodiesel.

6.9 Different Operating Fermentation Systems

Based on different modes for operation of fermentation process, there can be batch, continuous, or fed-batch fermentation systems depending on the goal of the process. We have already discussed how any fermentation process operates and the different types of products obtained; now we will discuss the important fermentation systems. A general abstract representation of these processes is given in Figure 6.9.

6.9.1 Batch System

Batch system refers to a partially closed system for growing microorganisms under optimized physiological conditions. During the whole process, nothing is added except for oxygen (in the form of air), antifoam agent, and acid or base to control pH variations. In batch system of fermentation, $Fi = Fo = 0$ that means the working volume in the reactor remains constant. Due to this, the medium composition, biomass concentration, and

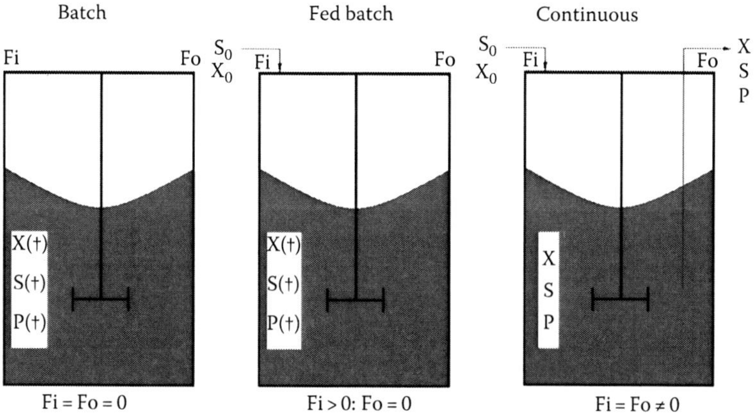

FIGURE 6.9
General diagram of operating fermentation systems. (Modified from Chen, F. and Zhang, Y., *Enzyme Microb. Technol.*, 20, 221, 1997.)

metabolite concentration change constantly with time. A typical mechanics of cell growth can be explained by four phases: lag phase, log phase, stationary phase, and death phase as shown in Figure 6.10. During the lag phase, microorganism acclimates itself to the new environment. It is an unproductive phase and hence generally recommended to avoid by inoculating the culture in active growth stage. When cells adapt to the environment, their

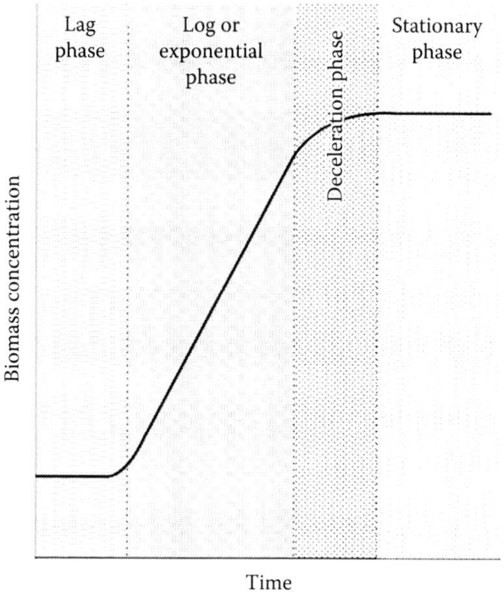

FIGURE 6.10
Growth of a typical microorganism under batch culture conditions. (Adapted from Stanbury, P.F. et al., *Principles of Fermentation Technology*, Pergamon Press, Oxford, U.K., 1995.)

growth follows typical first-order reaction kinetics where growth rate is proportional to biomass concentration:

$$\frac{dx}{dt} = \mu x \tag{6.12}$$

where μ is the specific growth rate, h^{-1}.

Next is the log or exponential phase, where cells actively grow in an exponential manner. Because a batch culture is a closed system, nutrients will become depleted, and cell growth will come to a halt in the course of an experiment. The duration of exponential phase in cultures depends upon the size of the inoculum, the growth rate, and the capacity of the medium and culturing conditions to support algal growth. In the stationary phase, cell number or biomass constantly increases as the cell nutrients deplete. According to the industrial point of view, log and stationary phases hold great biotechnological interest since most of the important primary and secondary metabolites are produced. When the substrate gets totally metabolized or the toxic substances have been formed, the growth rate slows down or completely stops, and the cell enters death phase where all the energy reserves are totally exhausted.

6.9.2 Continuous System

A continuous system is an open system where input–output interchange of materials is possible throughout the process. Here, $Fi = Fo \neq 0$. Fresh media are constantly added into the bioreactor and inside fluid is removed simultaneously. Hence, a steady state is achieved through regular mixing of the culture medium, and the reactor is called a chemostat. In this system under steady state, cell loss as a result of outflow must be balanced by growth of the organism, that is, $D = \mu$; D is volumetric flow rate (in and out) through the fermenter volume, where

$$\mu = \frac{1}{x}\frac{dx}{dt}, \quad \frac{\text{Growth rate}}{\text{Cell concentration in reactor}} \tag{6.13}$$

The biomass productivity of the microalgae *Scenedesmus* sp. was two times greater when growing in a continuous chemostat than in batch cultivation (McGinn et al. 2012), with the productivity in the latter being 130 mg L^{-1} d^{-1}.

6.9.3 Fed-Batch (Semicontinuous) System

Fed-batch (semicontinuous) system is another fermentation system that is a semiopen system. Here, $Fi > 0$ $Fo = 0$, which means fresh aliquot of medium is being continuously or periodically added without the removal of culture fluid. The substrate gets consumed as soon as it enters the control volume so that

$$\frac{ds}{dt} \approx 0 \quad \text{and} \quad \frac{dx}{dt} \approx 0 \tag{6.14}$$

thereby maintaining a quasi-steady state always. The *Spirulina platensis* biomass concentration in the fed-batch culture was found to be 4.25-fold than in the mixotrophic batch culture and 5.1-fold than in the photoautotrophic batch culture (Chen and Zhang 1997).

In the bioreactor, it is very important that maximum production of metabolites must be accomplished with emphasis on reliability of process and minimum capital investment and operating cost. There are several modes of reactor operation based on different criteria of classification and innovations particularly in mass transfer effects, agitation type or stirring, product inhibition, etc. (Gutiérrez-Correa and Villena 2010), as shown in Figure 6.11. The most commonly used bioreactor modes are as follows:

1. Continuous stirred tank reactor (CSTR): It is the most popular bioreactor configuration with internal agitation system that presents a homogeneous system. In CSTR, gas substrates are continuously fed into the reactor and mechanically stirred by baffled impellers into smaller bubbles, which have greater interfacial surface area for mass transfer. In addition, finer bubbles have a slower rising velocity and a longer retention time in the aqueous medium, resulting in higher gas-to-liquid mass transfer. The wall growth in stagnant zones like crevices and crannies are avoided due to continuous mixing.
2. Bubble column reactors (BCRs): BCRs differ from CSTR in the way that gas mixing and/or aeration is achieved by gas sparging and not by mechanical agitation using baffles or stirrers. This agitation is due to upward movement of gas (usually air) bubbles produced by the action of sparger mounted at the bottom. The reactor configuration has fewer moving parts and consequently has a lower associated capital and operational costs while exhibiting good heat and mass transfer efficiencies. However, one major drawback of these types of reactors is excess level of gas inflow for mixing, which may lead to heterogeneous flow and back mixing of the gas. The dependence of aeration rate on the growth rate of *S. platensis* and hence enhanced quantity of 69.4% gamma-linolenic acid (GLA) was observed for 0.2–2.5 vvm (Ronda et al. 2012).
3. Airlift reactors: CSTRs lack well-defined flow of air, whereas in airlift reactors, the air is pumped from below. This creates uniform bubbles in the medium that rises up through the draught tube by buoyancy and drags the surrounding fluid up. The air that lifts up stirs up the contents. Gas disengages at the top of the vessel leaving heavier bubble-free liquid to recirculate through the down comer. The growth profile of *C. sorokiniana* was reported better in an airlift reactor than in a BCR. The maximum biomass of 1.8 g L^{-1} was obtained corresponding to 3.24 g L^{-1} of CO_2 sequestered at 5% air–CO_2 concentration (Kumar and Das 2012).

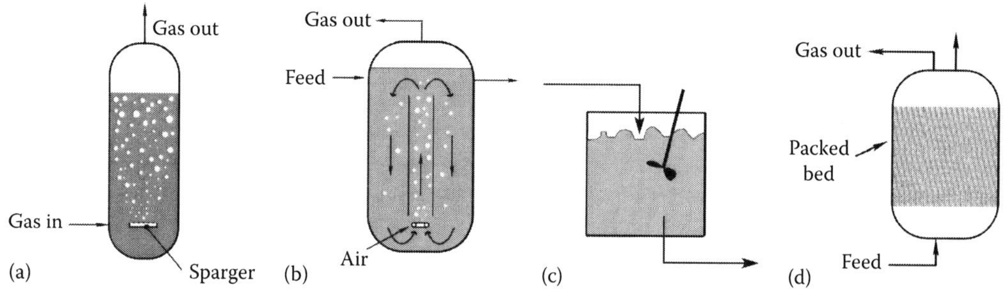

FIGURE 6.11
Principal bioreactor configurations. (a) Bubble column, (b) airlift, (c) stirred tank, and (d) packed bed.

4. Immobilized cell column reactor: Immobilized systems have great potential for large-scale production of fuels from microalgae. Microbes are immobilized through cross-linking or adsorption to insoluble biosupport materials, and then subsequent packing is done within the column. It offers several advantages over other systems like high cell densities, plug flow operation, high mass transfer rate via direct contact between microbe and gas, reduction of retention time, and operation without mechanical agitation. However, the major drawback is encountered when the microbe overgrows and completely fills the interstitial space. Cells of *C. vulgaris* were grown on immobilized *Clostridium butryicum* to produce hydrogen using H_2O as substrate following agar entrapment technique (Suzuki and Karube 1983).
    a. Packed bed reactor (PBR): PBRs are generally applied with attached biofilms especially in wastewater engineering. It consists of a cylindrical reactor body tightly packed with immobilized algal beads acting as a stationary phase (Preito et al. 2002). The reactor is fed with a nutrient-enriched medium kept in continuous circulation. This is used to remove toxicants from the wastewater stream. The disadvantage of packed bed lies in changed flow characteristics due to alterations in the bed porosity during operation.
    b. Trickle-bed reactor: Trickle-bed reactor is a gas- or liquid-continuous reactor consisting of packed bed; liquid culture trickles down through packing media containing suspended or immobilized cells. In this process, mechanical agitation is not required. Trickle-bed reactor was found to exhibit excellent gas conversion rates and higher productivities than CSTR and BCR.

6.10 Marine Microbes

Metagenomics, high throughput sequencing, and many other such advanced techniques have helped in revealing the rich microbial diversity in marine habitats. Many marine microbes have been reported to produce novel enzymes and bioactive compounds. *Alteromonas, Agarivorans, Alishewanella, Marinobacter, Marinobacterium, Microbulbifer,* and *Saccharophagus* are some marine microorganisms that have been identified for their ability to degrade many complex polysaccharides such as agar, alginate, laminarin, and cellulose (Bruton et al. 2009; Zhang et al. 2005). Hence, they can be exploited for bioenergy production from renewable sources.

6.11 Summary

One of the main goals for devising renewable options for energy production lies in the exploitation of alternate biomass sources. Bioenergy is the form of energy that comes from biomass. Due to food security and land scarcity, algae is a suitable candidate for bioenergy production. Algal biomass can be fermented to produce various products. In this chapter, various pretreatment methods for micro- as well as macroalgae cultures were presented. Biomass pretreatment results in breaking down of crystalline structures of cell wall and releases fermentable substrates. This helps in achieving greater product yields in less time.

This chapter discusses several fermentation systems and modes of reactor operations. Depending upon desired product, different fermentation systems are used in practice. To meet future energy demands, it is envisioned that extensive work on marine and freshwater algae production and fermentation systems could provide us some plausible sustainable solutions.

References

Adams, J.M., Gallagher, J.A., and Donnison, I.S. 2009. Fermentation study on *Saccharina latissima* for bioethanol production considering variable pretreatments. *Journal of Applied Phycology*. 21:569–574.

Adams, M.W.W. 1990. The structure and mechanism of iron-hydrogenases. *Biochimica et Biophysica Acta*. 1020:115–145.

Alvira, P., Tomás-Pejó, E., Ballesteros, M., and Negro, M.J. 2010. Pretreatment technologies for an efficient bioethanol production process based on enzymatic hydrolysis: A review. *Bioresource Technology*. 101(13):4851–4861.

Angenent, L.T., Karim, K., Al-Dahhan, M.H., Wrenn, B.A., and Rosa, D.E. 2004. Production of bioenergy and biochemicals from industrial and agricultural wastewater. *Trends in Biotechnology*. 22:477–485.

Babaee, A. and Shayegan, J. 2011. Effect of organic loading rates (OLR) on production of methane from anaerobic digestion of vegetables waste. In: *World Renewable Energy Congress 2011—Sweden*, Linkoping, Sweden, May 8–13, 2011.

Banerjee, K., Ghosh, R., Homechaudhuri, S., and Mitra, A. 2009. Biochemical composition of marine macroalgae from Gangetic delta at the apex of Bay of Bengal. *African Journal of Basic & Applied Sciences*. 1(5–6):96–104.

Brennan, L. and Owende, P. 2010. Biofuels from microalgae—A review of technologies for production, processing, and extractions of biofuels and co-products. *Renewable and Sustainable Energy Reviews*. 14:557–577.

Bruton, T.L.H., Lerat, Y., Stanley, M., and Rasmussen, M.B. 2009. *A Review of the Potential of Marine Algae as a Source of Biofuel in Ireland*. Sustainable Energy Ireland, Dublin, Ireland, pp. 1–88.

Cantrell, K.B., Ducey, T., Ro, K.S., and Hunt, P.G. 2008. Livestock waste-to-bioenergy generation opportunities. *Bioresource Technology*. 99:7941–7953.

Chapman, V.J. 1970. *Seaweeds and Their Uses*, 2nd edn. The Camelot Press, London, U.K.

Chen, F. and Zhang, Y. 1997. High cell density mixotrophic culture of *Spirulina platensis*. *Enzyme and Microbial Technology*. 20:221–224.

Chynoweth, D. 2002. *Review of Biomethane from Marine Biomass*. Department of Agricultural and Biological Engineering, University of Florida, Gainesville, FL.

Conrad, R. 1999. Contribution of hydrogen to methane production and control of hydrogen concentrations in methanogenic soils and sediments. *FEMS Microbiology Ecology*. 28(3):193–202.

Das, D. 2009. Advances in biological hydrogen production processes: An approach towards commercialization. *International Journal of Hydrogen Energy*. 34:7349–7357.

Dejoye, C., Vian, M.A., and Lumia, G. 2011. Combined extraction processes of lipid from *Chlorella vulgaris* microalgae: Microwave prior to supercritical carbon dioxide extraction. *International Journal of Molecular Sciences*. 12:9332–9341.

de Mes, T.Z.D., Stams, A.J.M., and Zeeman, G. 2003. Chapter 4: Methane production by anaerobic digestion of wastewater of wastewater and solid wastes. In: *Biomethane and Biohydrogen. Status and Perspectives of Biological Methane and Hydrogen Production*, eds. J.H. Reith, R.H. Wijffels, and H. Barten. Dutch Biological Hydrogen Foundation, The Hague, the Netherlands, pp. 58–94.

Demirbas, A. and Demirbas, M.F. 2010. *Algae Energy: Algae as a New Source of Biodiesel*. Springer-Verlag, London, U.K.

Dismukes, G.C., Carrieri, D., Bennette, N., Ananyev, G.M., and Posewitz, M.C. 2008. Aquatic phototrophs: Efficient alternatives to land-based crops for biofuels. *Current Opinion in Biotechnology*. 19(3):235–240.

El-Mashad, H.M., Zeeman, G., vanLoon, W.K., Bot, G.P., and Lettinga, G. 2004. Effect of temperature and temperature fluctuation on thermophilic anaerobic digestion of cattle manure. *Bioresource Technology*. 95(2):191–201.

Enquist-Newman, M., Faust, A.M.E., Bravo, D.D., Santos, C.N.S., Raisner, R.M., Hanel, A., Sarvabhowman, P. et al. 2013. Efficient ethanol production from brown macroalgae sugars by a synthetic yeast platform. *Nature*. 505(7482):239–243. doi:10.1038/nature12771.

Gaffron, H. 1940. Carbon dioxide reduction with molecular hydrogen in green algae. *American Journal of Botany*. 27:273–283.

Gaffron, H. and Rubin, J. 1942. Fermentative and photochemical production of hydrogen in algae. *The Journal of General Physiology*. 26:219–240.

Ghirardi, M.L., Kosourov, S., Tsygankov, A., and Seibert, M. 2000. Two-phase photobiological algal H2-production system. In: *Proceedings of the 2000 U.S. DOE Hydrogen Program Review*, National Renewable Energy Laboratory, Golden, CO, pp. 1–13.

Gouveia, L. 2011. *Microalgae as a Feedstock for Biofuels*. Springer Briefs in Microbiology. Springer, New York.

Gutierrez-Correa, M. and Villena, G.K. 2010. Chapter 7: Characteristics and techniques of fermentation systems. In: *Comprehensive Food Fermentation and Biotechnology*, eds. A. Pandey, C.R. Soccol, C. Larroche, E. Gnansounou, and P. Nigam-Singh, Vol. 1. Asiatech Publisher, Inc., New Delhi, India, pp. 183–227.

Halim, R., Harun, R., Danquah, M.K., and Webley, P.A. 2012. Microalgal cell disruption for biofuel development. *Applied Energy*. 91(1):116–121.

Hallenbeck, P.C. and Benemann, J.R. 2002. Biological hydrogen production; fundamentals and limiting processes. *International Journal of Hydrogen Energy*. 27:1185–1193.

Hansel, A. and Lindblad, P. 1998. Towards optimization of cyanobacteria as biotechnologically relevant producers of molecular hydrogen, a clean and renewable energy source. *Applied Microbiology and Biotechnology*. 50:153–160.

Harmsen, P.F.H., Huijgen, W.J.J., Bermúdez López, L.M., and Bakker, R.R.C. 2010. *Literature Review of Physical and Chemical Pretreatment Processes for Lignocellulosic Biomass*. Wageningen UR, Food & Biobased Research, Wageningen, the Netherlands.

Harun, R. and Danquah, M.K. 2011. Enzymatic hydrolysis of microalgal biomass for bioethanol production. *Chemical Engineering Journal*. 168(3):1079–1084.

Harun, R., Singh, M., Forde, G.M., and Danquah, M.K. 2010. Bioprocess engineering of microalgae to produce a variety of consumer products. *Renewable and Sustainable Energy Reviews*. 14:1037–1047.

Hubbard, D.W. 1987. Scale up strategies for bioreactors containing non-Newtonian broths. *Annals of the New York Academy of Sciences*. 506:600–607.

Intergovernmental Panel on Climate Change—IPCC (2007). Climate change 2007: Synthesis report, Geneva, Switzerland.

Kloareg, B., Demarty, M., and Mabeau, S. 1986. Polyanionic characteristics of purified sulphated homofucans from brown algae. *International Journal of Biological Macromolecules*. 8:380–386.

Kumar, K. and Das, D. 2012. Growth characteristics of *Chlorella sorokiniana* in airlift and bubble column photobioreactors. *Bioresource Technology*. 116:307–313.

Kumar, S. 2013. Bioethanol production from *Gracilaria verrucosa*, a red alga, in a biorefinery approach. *Bioresource Technology*. 135:150–156.

Lin, Y. and Tanaka, S. 2006. Ethanol fermentation from biomass resources: Current state and prospects. *Applied Microbiology and Biotechnology*. 69:627–642.
Markou, G., Angelidaki, I., and Georgakakis, D. (2012). Microalgal carbohydrates: An overview of the factors influencing carbohydrates production, and of main bioconversion technologies for production of biofuels. *Applied Microbiology and Biotechnology*. 96:631–645.
Masukawa, H., Mochimaru, M., and Sakurai, H. 2002. Disruption of the uptake hydrogenase gene, but not of the bi-directional hydrogenase gene, leads to enhanced photobiological hydrogen production by the nitrogen-fixing cyanobacterium *Anabaena* sp. PCC 7120. *Applied Microbiology and Biotechnology*. 58:618–624.
Mata, T.M., Martins, A.N.A., and Caetano, N.S. 2010. Microalgae for biodiesel production and other applications: A review. *Renewable and Sustainable Energy Reviews*. 14:217–232.
McGinn, P.J., Dickinson, K.E., Park, K.C., Whitney, C.G., MacQuarrie, S.P., Black, F.J., Frigon, J., Guiot, S.R., and O'Leary, S.J.B. 2012. Assessment of the bioenergy and bioremediation potentials of the microalga *Scenedesmus* sp. AMDD cultivated in municipal wastewater effluent in batch and continuous mode, *Algal Research*. 1:155–165.
Melis, A. and Happe, T. 2001. Hydrogen production. Green algae as a source of energy. *Plant Physiology*. 127:740–748.
Melis, A., Zhang, L., Forestier, M., Ghirardi, M.L., and Seibert, M. 2000. Sustained photobiological hydrogen gas production upon reversible inactivation of oxygen evolution in the green alga *Chlamydomonas reinhardtii*. *Plant Physiology*. 122:127–136.
Miyamoto, K., Hydrogen production. In: *Renewable Biological Systems for Alternative Sustainable Energy Production*, Miyamoto, K. (ed.), Issue 128, Food and Agriculture Organization, 108 pp., 1997.
Moen, E. 2008. *Biological Degradation of Brown Seaweeds. The Potential of Marine Biomass for Anaerobic Biogas Production*. Scottish Association for Marine Science Oban, Argyll, Scotland.
Nguyen, M.T., Choi, S.P., Lee, J., Lee, J.H., and Sim, S.J. 2009. Hydrothermal acid pretreatment of *Chlamydomonas reinhardtii* biomass for ethanol production. *Journal of Microbiology and Biotechnology*. 19:161–166.
Park, J., Lee, J., Sim, S.J., and Lee, J.H. 2009. Production of hydrogen from marine macro-algae biomass using anaerobic sewage sludge microflora. *Biotechnology and Bioprocess Engineering*. 14:307–315.
Parkin, G.F. and Owen, W.F. 1986. Fundamentals of anaerobic digestion of wastewater sludges. *Journal of Environmental Engineering*. 112(5):867–920.
Parthiban, C., Saranya, C., Girija, K., Hemalatha, A., Suresh, M., and Anantharaman, P. 2013. Biochemical composition of some selected seaweeds from Tuticorin coast. *Advances in Applied Science Research*. 4(3):362–366.
Prieto, G., Okumoto, M., Takashima, K., Katsura, S., Mizuno, A., Prieto, O., and Gay, C.R. 2003. Nonthermal plasma reactors for the production of light hydrocarbon olefins from heavy oil. *Brazilian Journal of Chemical Engineering*. 20(1):57–61. doi:10.1590/S0104–66322003000100011.
Ran, C.Q., Chen, Z.A., Zhang, W., Yu, X.J., and Jin, M.F. 2006. Characterization of photobiological hydrogen production by several marine green algae. *Wuhan Ligong Daxue Xuebao*. 28(Suppl 2):258–263.
Rao, K.K. and Hall, D.O. 1996. Hydrogen production by cyanobacteria: Potential, problems and prospects. *Journal of Marine Biotechnology*. 4:10–15
Roesijadi, G., Roesijadi, G., Jones, S.B., Snowden Swan, L., and Zhu, Y. 2010. *Macroalgae as a Biomass Feedstock: A Preliminary Analysis*. U.S. Department of Energy, Pacific Northwest Laboratory, Richland, WA.
Ronda, S.R., Bokka, C.S., Ketineni, C., Rijal, B., and Allu, P.R. 2012. Aeration effect on *Spirulina platensis* growth and γ-linolenic acid production. *Brazilian Journal of Microbiology*. 43(1):12–20.
Sialve, B., Bernet, N., and Bernard, O. 2009a. Anaerobic digestion of micro algae as a necessary step to make microalgal biodiesel sustainable. *Biotechnology Advances*. 27:409–416.
Singh, J. and Sai, G. 2010. Commercialization potential of microalgae for biofuels production. *Renewable and Sustainable Energy Reviews*. 14:2596–2610.
Soccol, C.R., Pandey, A., and Larroche, C. 2013. *Fermentation Processes Engineering in the Food Industry*. CRC Press, Boca Raton, FL, 510pp.

Stanbury, P.F., Whitaker, A., and Hallv, S.J. 1995. *Principles of Fermentation Technology*. Pergamon Press, Oxford, U.K.

Suzuki, S. and Karube, I. 1983. Energy production with immobilized cells. In: *Applied Biochemistry and Bioengineering*, eds. I. Chibata and L.B. Wingard, Vol. 4. Immobilized Microbial Cells. Academic Press, New York, pp. 281–310.

Symons, G.E. and Buswell, A.M. 1933. The methane fermentation of carbohydrates, *Journal of the American Chemical Society*. 55:2028–2036.

Tamagnini, P., Axelsson, R., Lindberg, P., Oxelfelt, F., Wünschiers, R., and Lindblad, P. 2002. Hydrogenases and hydrogen metabolism of cyanobacteria. *Microbiology and Molecular Biology Reviews*. 66:1–20.

Ueda, R., Hirayama, S., Sugata, K., and Nakayama, H. 1996. Process for the production of ethanol from microalgae. US Patent 5,578,472.

Vignais, P.M., Billoud, B., and Meyer, J. 2001. Classification and phylogeny of hydrogenases. *FEMS Microbiology Reviews*. 25:455–501.

Wang, D.I.C. and Cooney, C.L. 1997. Translation of laboratory, pilot, and plant scale data. In: *Fermentation and Enzyme Technology*, eds. D.L.C. Wang, C.L. Cooney, A.L. Demain, P. Dunnill, A.E. Humphrey, and M.D. Lilley. Wiley, New York, pp. 194–211.

Wi, S.G., Kim, H.J., Mahadevan, S.A., Yang, D.J., and Bae, H.J. 2009. The potential value of the seaweed Ceylon moss (*Gelidium amansii*) as an alternative bioenergy resource. *Bioresource Technology*. 100:6658–6660.

Yang, Z., Guo, R., Xu, X., Fan, X., and Li, X. 2010. Enhanced hydrogen production from lipid-extracted microalgal biomass residues through pretreatment. *International Journal of Hydrogen Energy*. 35:9618–9623.

Zhang, L., An, R., Wang, J., Sun, N., Zhang, S., Hu, J., and Kuai, J. 2005. Exploring novel bioactive compounds from marine microbes. *Current Opinion in Microbiology*. 8(3):276–281.

Zhao, G., Chen, X., Wang, L., Zhou, S., Feng, H., Chen, W.N., and Lau, R. 2003. Ultrasound assisted extraction of carbohydrates from microalgae as feedstock for yeast fermentation. *Bioresource Technology*. 128:337–344.

7

Nanotechnological Techniques in Bioenergy Production

Kelvii Wei Guo

CONTENTS

7.1 Introduction

Fossil fuels are well-established products that have served industry and consumers for a long time. For the foreseeable future, fuels will still be largely based on fossil fuels that however continue to be adjusted as they have been and are still being adjusted to meet changing demands from consumers; for example, traditional crude oil refining underwent increasing levels of sophistication to produce fuels of appropriate specifications, and increasing operating costs continuously put pressure on refining margins, but it remains problematic to convert all refinery streams into products with acceptable specifications at a reasonable return (Ashby et al. 2009a,b; Kostoff et al. 2007; Pitkethly 2004; Whitesides 2001).

However, fossil fuels once considered inexhaustible are now being depleted at a rapid rate. As the amount of available fossil fuels decreases, the need for alternate technologies that could potentially help prolong the traditional fuel culture and mitigate the forthcoming effects of the shortage of transportation fuels has been suggested to occur under the Hubbert peak oil theory (Ashby et al. 2009c; König et al. 1992; Palacios et al. 2000; Van Hove 2009; Zäch et al. 2006).

Moreover, today's energy supply is largely responsible for the anthropogenic greenhouse effect, acid rain, and other negative impacts on health and the environment. The current trend is clearly not sustainable, especially given the enormous demand for energy predicted for the future. Therefore, an adequate and secure supply of energy is essential to daily life, which needs to be achieved with minimum adverse environmental effects.

Voices are being raised for the establishment of an industry that produces and develops renewable energy (such as hydrogen fuel, solar cell, biotechnology, wind energy, and ocean energy, especially for the so-called green energy friendly to the environment) from nonconventional sources, but there is still a long way to go (Inman et al. 2010; Lashdaf et al. 2004; Moriarty and Honnery 2008; Reddy et al. 2010).

A growing number of scientists and engineers are exploring and tweaking material properties at the atomic scale to create designer materials, which might ultimately increase the efficiency of current energy sources or make new energy sources practical on a commercial scale. At the nanoscale, fundamental mechanical, electrical, optical, and other properties can significantly differ from their bulk material counterparts. Many nanoscale materials can spontaneously self-assemble into ordered structures. Nanostructured materials also have enormous surface areas, so that vastly more surface area is available and beneficial for interactions with other materials around them. That is useful because many important chemical and electrical reactions occur only at surfaces and are sensitive to the shape and texture of a surface as well as its chemical composition (Ashby et al. 2009a,b,c; Kostoff et al. 2007; Pitkethly 2004; Van Hove 2009; Zäch et al. 2006).

Aiming to advance the development of biotechnologies using nanotechnology, it is urgent to minimize potential environmental and human health risks associated with the manufacture and use of nanotechnology products in general, to apply nanophilosophy to solve legacy environmental problems, and to encourage replacement of existing products with new nanoproducts that are more environmentally friendly throughout their life cycles.

Definition of green chemistry:

1. Prevent waste: Design chemical syntheses to prevent waste, and no waste to treat or clean up.
2. Design safer chemicals and products: Design chemical products to be fully effective, yet have little or no toxicity.
3. Design less hazardous chemical syntheses: Design syntheses to use and generate substances with little or no toxicity to humans and the environment.
4. Use renewable feedstocks: Use raw materials and feedstocks that are renewable rather than depleting. Renewable feedstocks are often made from agricultural products or are the wastes of other processes; depleting feedstocks are made from fossil fuels (petroleum, natural gas, or coal) or are mined.
5. Use catalysts without stoichiometric reagents: Minimize waste by using catalytic reactions. Catalysts are used in small amounts and can carry out a single reaction many times. They are preferable to stoichiometric reagents, which are used in excess and work only once.
6. Avoid chemical derivatives: Avoid blocking or protecting groups or any temporary modifications if possible. Derivatives use additional reagents and generate waste.
7. Maximize atom economy: Design syntheses so that the final product contains the maximum proportion of the starting materials. There should be few, if any, wasted atoms.

8. Use safer solvents and reaction conditions: Avoid using solvents, separation agents, or other auxiliary chemicals. If these chemicals are necessary, use innocuous chemicals.
9. Increase energy efficiency: Run chemical reactions at ambient temperature and pressure whenever possible.
10. Design chemicals and products to degrade after use: Design chemical products to break down into innocuous substances after use so that they do not accumulate in the environment.
11. Analyze in real time to prevent pollution: Include in-process, real-time monitoring and control during syntheses to minimize or eliminate the formation of by-products.
12. Minimize the potential for accidents: Design chemicals and their forms (solid, liquid, or gas) to minimize the potential for chemical accidents, including explosions, fires, and releases to the environment.

Principles of green engineering:

1. Engineer process and products holistically, use systems analysis, and integrate environmental impact assessment tools.
2. Conserve and improve natural ecosystems while protecting human health and well-being.
3. Use life cycle thinking in all engineering activities.
4. Ensure that all material and energy inputs and outputs are as inherently safe and benign as possible.
5. Minimize depletion of natural resources.
6. Strive to prevent waste.
7. Develop and apply engineering solutions, while being cognizant of local geography, aspirations, and cultures.
8. Create engineering solutions beyond current or dominant technologies; improve, innovate, and invent (technologies) to achieve sustainability.
9. Actively engage communities and stakeholders in the development of engineering solutions.

As mentioned earlier, matter at the scale of 1–100 nm takes on new and interesting properties; for instance, bulk metals are not very chemically active, and nanoparticles of metals are often highly catalytic. Properties such as color, electrical conductivity, and magnetism can potentially be tuned by changing the size and shape of nanoparticles. Nanotechnology offers myriad new materials and methods for scientists and engineers to exploit in new applications.

Green nanophilosophy embodied in green chemistry includes such goals as preventing waste, maximizing the incorporation of raw materials, exploiting catalysis, and minimizing the use of toxic chemicals.

Green engineering likewise seeks to avoid harming the environment, but as research indicates, focuses more on the design of products and processes, such as making them more energy efficient and building them out of biodegradable materials. The green approach relies on life cycle assessment (LCA), a way of examining all of the impacts that a particular product has on the environment. This approach requires that the engineer consider the product's manufacture, its use over many years, and its ultimate resting place and decomposition. Ideally, an LCA looks at such things as the impacts of mining or manufacture of raw materials, factory emissions released during production, the waste materials disposed of, and the product's fate at a landfill, a recycling center, or elsewhere. Another approach to LCA would be to examine each step in the product's life span for opportunities to make better choices for the environment.

Green chemistry/engineering might seem like an odd mate for nanotechnology, but, in fact, both respects seek to emulate natural processes. The goal of green chemistry/engineering is to make industries function more like ecosystems or like cells, in which benign materials are used wisely, wastes are recycled, and energy is used efficiently. As it turns out, biological systems accomplish this feat by exploiting properties that occur in the nanodimension. Indeed, the cell is the quintessential green nanofactory. It uses natural ingredients at room temperature to assemble nanostructures, carries out its chemical reactions in water rather than in harmful solvents, employs smart controls with feedback loops, conserves energy, and reuses wastes. So, it should be no surprise that many researchers view nanotechnology and green chemistry/engineering as capable of working hand in hand to produce environmentally sustainable products and processes.

A marriage of nanotechnology with green chemistry/engineering serves two important purposes. First, emerging nanotechnologies could be made clean from the start. It would be foolhardy to build a new nanotechnology infrastructure from an old industrial model that would generate another set of environmental problems. While nanotechnology might never be as green as Mother Nature, adopting a green nano-approach to the technology's development ultimately promises to shift society into a new paradigm that is proactive, rather than reactive, when it comes to environmental problems.

Second, green technologies that benefit the environment could use nanotechnology to boost performance. In other words, nanotechnology could help us make every atom count—for example, by allowing us to create ultraefficient catalysts, detoxify wastes, assemble useful molecular machines, and efficiently convert sunlight into energy. It could potentially contribute to long-term sustainability for future generations, as more green products and green manufacturing processes replace the old harmful and wasteful ones.

Considering the remediation of environmental problems, in the development of green processing technologies, research that is looking into nanotechnology being related to biotechnology could be a possible solution to diminishing our dependence on depleting petroleum supplies (Dong and Dinu 2013; Khan et al. 2005; Krull et al. 2013; Lee et al. 2013; National Nanotechnology Initiative 2000; Retterer and Simpson 2012).

7.2 Nanotechnology Related to Bioenergy

The material of plants and animals, including their wastes and residues, is called biomass. It is an organic, carbon-based material that reacts with oxygen in combustion and natural metabolic processes to release heat. Such heat, especially if at temperatures >400°C, may be used to generate work and electricity. The initial material may be transformed by chemical and biological processes to produce biofuels, that is, biomass processed into a more convenient form, particularly liquid fuels for transport. Examples of biofuels include methane gas, liquid ethanol, methyl esters, oils, and solid charcoal.

The success of biomass systems is regulated by principles:

1. Every biomass activity produces a wide range of products and services. For instance, where sugar is made from cane, many commercial products can be obtained from the otherwise waste molasses and fiber. If the fiber is burnt, then any excess process heat can be used to generate electricity. Washings and ash can be returned to the soil as fertilizer.
2. Some high-value fuel products may require more low-value energy to manufacture than they produce, for example, ethanol from starch crops, hydrogen. Despite the energy ratio being >1, such an energy deficiency need not be an economic handicap provided that process energy can be available cheaply by consuming otherwise waste material, for example, straw, crop fiber, and forest trimmings.
3. The full economic benefit of agro-industries is likely to be widespread and yet difficult to assess. One of the many possible benefits is an increase in local *cash flow* by trade and employment.
4. Biofuel production is only likely to be economic if the production process uses materials already concentrated, probably as a by-product and so available at low cost or as extra income for the treatment and removal of waste. Thus, there has to be a supply of biomass already passing near the proposed place of production, just as hydropower depends on a natural flow of water already concentrated by a catchment. Examples are the wastes from animal enclosures, offcuts and trimmings from sawmills, municipal sewage, husks and shells from coconuts, and straw from cereal grains. It is extremely important to identify and quantify these flows of biomass in a national or local economy before specifying likely biomass developments. If no such concentrated biomass already exists as a previously established system, then the cost of biomass collection is usually too great and too complex for economic development. Some short-rotation crops may be grown primarily for energy production as part of intensive agriculture; however, within the widespread practice of agricultural subsidies, it is difficult to evaluate fundamental cost effectiveness.
5. The main dangers of extensive biomass fuel use are deforestation, soil erosion, and the displacement of food crops by fuel crops.
6. Biofuels are organic materials, so there is always the alternative of using these materials as chemical feedstock or structural materials. For instance, palm oil is an important component of soaps; many plastic and pharmaceutical goods are made from natural products; and much building board is made from plant fibers constructed as composite materials.
7. Poorly controlled biomass processing or combustion can certainly produce unwanted pollution, especially from relatively low-temperature combustion, wet fuels, and lack of oxygen supply to the combustion regions. Modern biomass processes require considerable care and expertise.
8. The use of sustainable biofuels in place of fossil fuels abates the emission of fossil CO_2 and so reduces the forcing of climate change. Recognition of this is a key aspect of climate change policies.

Moreover, in an attempt to alleviate fossil fuel usage and CO_2 emissions, fuels, heat, or electricity must be produced from biological sources in a way that is economic (and therefore efficient at a local scale), energetically (and greenhouse gas) efficient, environmentally friendly, and not competitive with food production.

In the past decade, many scientists apply biological organisms, systems, and processes related to DNA in industrial concerns, forming a new crosscutting technology platform known as bionanotechnology (Inman et al. 2010; Khan et al. 2005; Krull et al. 2013; Lashdaf et al. 2004; Lee et al. 2010, 2011, 2013; Moriarty and Honnery 2008; National Nanotechnology Initiative 2000; Reddy et al. 2010; Retterer and Simpson 2012; St-Pierre et al. 2010).

7.2.1 Biofuels

Finding sustainable feedstock alternatives to petroleum fuel is a key R&D goal, with biologically produced gasoline substitutes topping the list. Results already show environmental benefits from using plant-derived alcohol in engines. Unfortunately, biofuels are commercially produced from sugar, starch, and oil seed–based feedstocks currently. For example, bioalcohol is produced from corn starch, soybean, palm fruits, and rape and canola seeds are the common feedstocks for biodiesel production. The further expansion of biofuel production will trigger debate on food/feed versus fuel. Thus, for sustainable biofuel production, nonfood feedstock should be used. In the present methods, starch, which has various advantages over other fuels in terms of cost and convenience in handling and manufacture, is also investigated and used as a fuel or a fuel component for a combustor such as a boiler, kiln, dryer, or furnace (Anders et al. 2009; Copeland et al. 2007; Lewis 2010b; Munasinghe and Khanal 2010; Ramanavicius et al. 2008).

Therefore, for the use of biomass producing fuels, more efficient biomass conversion techniques would help make biofuels more cost-competitive. Land availability and crop selection are major issues in biomass fuel usage. Biomass alternatives can be expected to grow to a significantly larger scale for providing fuel.

Land availability may not be a major problem, but land use issues need to be coordinated. The long-term production of biofuels in substantial quantities will require a number of changes. Grain surpluses will not provide sufficient feedstocks for the fuel quantities needed. Producers need to switch to short-rotation woody plants and herbaceous grasses, which can sustain biofuel production in long-term, substantial quantities. Moreover, the increased use of municipal solid waste (MSW) could not only reduce feedstock costs and reserve agricultural land for food production but also reduce the environmental insult of incineration or landfill expansion.

To date, the more widespread use of alcohol could have some safety benefits since alcohol is water soluble and biodegradable and evaporates easily. However, a major drawback of alcohol compared to methanol is its price that can be almost twice as much as methanol.

Although most alcohol is now produced from corn, research has been done on producing this type of alcohol fuel from cellulosic biomass products including energy crops, forest and agricultural residues, and MSW, which would provide much cheaper feedstocks.

Conversion of cellulosic materials to alcohol consists of two main steps: hydrolysis and fermentation. Hydrolysis can be accomplished either with acids or with cellulase enzymes. Although the two process routes are currently about equal in cost, acid hydrolysis—a mature technology—is expected to yield only small cost improvements. The cost of the enzymes themselves dominates the enzymatic processes. Moreover, although cellulosic alcohol is the alternative to agriculturally derived alcohol, there are still some limitations associated with their use. A key issue in the case of cellulosic alcohol is the difficult and expensive derivation of fermentable sugars from lignocellulosic biomass. Beyond this, there are the limitations inherent in the fermentation process in terms of rate, efficiency, and the cost of isolating pure alcohol from a dilute aqueous solution. Additionally, alcohol is volatile, toxic, hydrophilic, potentially corrosive to engine components, and of relatively low energy content compared to gasoline or diesel fuel. While it would be expensive to harvest enough natural enzymes to convert biomass into alcohol on a large scale, developing an inexpensive artificial enzyme or other chemical change catalyst through biomimetic research may provide a solution (Bullen et al. 2006; Hennessey et al. 2010; Jia et al. 2009; Kim et al. 2006; Kubo and Nomoto 2010; Liu et al. 2010, 2012; Mascal 2010; Minteer et al. 2009; Roberts et al. 2009; Wang et al. 2012).

Although biomass has a large amount of advantage as substitutes for fossil fuels as well as electricity and heat, a large increase in biomass energy production has the potential to cause serious environmental problems. Land use issues and concerns about pollution are major concerns. Areas with fragile ecosystems and rare species would need to be preserved. Agricultural lands would also compete with food production. The loss of soil fertility from overuse is a concern. Biomass production would need to be varied and sustainable while preserving local ecosystems. Pollution problems could result from the expanded use of fertilizers and bioengineered organisms on energy farms. The introduction of hazardous chemicals from MSW into the agricultural system could result in increased air and water pollution.

7.2.2 Biomimetic Technology

Biomimetic materials research starts with the study of structure–function relationships in biological materials. Based on the strategies found in nature, bio-inspired materials will be developed. This technology is very useful in the design of composite materials, or material structures, which typically offer superior properties and functionality, such as hierarchical structures, superior assembly mechanisms, superior process control, multifunctionality and adaptability, and tremendous potential for improving the capabilities of a wide range of industrial components and systems (Chen et al. 2011; Jiang et al. 2011; Yin et al. 2012; Zhao et al. 2011).

7.2.2.1 Enzyme Catalyst

Enzyme catalysts illustrate the use of biomimetic chemistry to go beyond the range of biochemical catalysis. One of the distinct practical advantages is converting biomass to alcohol via biomimetic catalysis. Chemical reactions in natural systems are catalyzed by enzymes, which provide unparalleled efficiencies and specifications under near ambient conditions. Artificially produced enzymes may eventually match these high efficiencies and also accomplish reactions far more rapidly and under more aggressive process conditions. Because of the importance of catalytic processes in the chemical and manufacturing industries, research into the development of enzyme-based catalysts has potential for tremendous impact across a wide range of applications (Boeckh et al. 2010; Caimi et al. 2010; DiCosimo et al. 2011; Feron and Bergel 2011; Hughes and Symes 2010; Kalyanasundaram and Graetzel 2010; Kragl et al. 2010; Kubo et al. 2010; Martins et al. 2010; Sakai et al. 2010).

7.2.2.2 Mimicking Photosynthesis

Today, patterning or structuring of individual layers is performed at the process step of the specific layer, for example, when forming a solar cell. A solar or photovoltaic (PV) cell is a device that converts photons from the sun (solar light) into electricity using electrons. In general, a solar cell that includes the capacity to capture both solar and non-solar sources of light (such as photons from incandescent bulbs) is termed a PV cell. Fundamentally, the device needs to fulfill only two functions: photogeneration of charge carriers (electrons and holes) in a light-absorbing material and separation of the charge carriers to a conductive contact that will transmit the electricity. While plants then use the charges to electrochemically synthesize hydrocarbons for food, PV cells consolidate individual charges in a circuit to produce dc. Biomimetic approaches to light gathering and charge separation have shown attractive promise for advancing the efficiency and economy of PV cells (Argo et al. 2010; Catchpole and Polman 2009; Lagally and Liu 2010; Lee 2010; McConnell et al. 2010; Montemagno et al. 2007; Němec et al. 2011; Olson 2010; Sivaram et al. 2010; Wieting 2011).

Photosynthesis research has led to the device that has been incorporated into prototype solar cells assembled from a trimeric ruthenium complex adsorbed on films of nanometer-sized particles of titanium dioxide (Cao et al. 2011; Nakayama et al. 2010; Nam et al. 2011; Park et al. 2010; Parsons 2010). It is very attractive with higher conversion efficiencies and new cell combinations that are being studied for further improvement; exciting research progress in this area is being conducted by dedicated researchers (Holliman 2010; Jo et al. 2009; Kayama and Tanaka 2009; Lin et al. 2010; Park et al. 2011; Roscheisen et al. 2009; Wessels et al. 2010; Yamaguchi et al. 2010).

7.2.2.3 Water Decomposition

Biomimetic photocatalytic decomposition of water is of interest from the view point of photoenergy transformation. Further, a photocatalyst that shows activity of the photodecomposing reaction of water can be recognized as an excellent material that provides function such as photoabsorption, electrolytic separation, or oxidation–reduction reaction at the surface. Effective processes for splitting water into molecular hydrogen and oxygen are of special interest for hydrogen use as a noncarbonaceous fuel, either in direct combustion or as the primary consumable in fuel cells. It has been known for many years that zinc sulfide particles in aqueous suspension catalyze the photodecomposition of water. It should be noted, however, that the systems based on metal hydrides, and particularly complex metal hydrides, are rather costly and require preliminary preparation of reacting solutions and the use of expensive noble metal-based catalysts. Metal-based systems also suffer from a number of shortcomings. Metals or alloys, such as Li, Na, K, Ca, Al–Hg, Al–Li, Al–Na, and Si–Al–Ca–Na–Fe–Cu, that directly react with water at ambient temperature are either expensive, or hazardous, or present great safety concerns, including violent uncontrolled reaction with water. On the other hand, such inexpensive and readily available materials like aluminum (Al) and its alloys and iron (Fe) and its alloys do not react with water at ambient temperature. Al can release hydrogen from aqueous solutions only in the presence of substances such as alkali hydroxides, NaOH and KOH, that remove a protective oxide layer from Al surface and, as a result, are consumed in the process, since they are transformed into their respective aluminates. More recently, the biochemical approach is becoming more and more attractive. More efficient, simple, and inexpensive chemical compositions that exceed the performance characteristics, such as specific energy and power density, of the state-of-the-art hydrogen-generating systems to safely produce hydrogen using water have been studied. However, yields are still low (Domen et al. 2005; Lewis 2010a; Parida et al. 2010; Sepeur et al. 2008; Shelnutt et al. 2008; Smolyakov and Osinski 2011). In addition, research on biohydrogen with certain groups of microorganisms that possess functional nitrogenase and/or bidirectional hydrogenases shows that with oxygenic photosynthetic microbes, photobiological hydrogen can be produced at high rates. Hydrogen production in this experiment is mediated by an efficient nitrogenase system, which can be manipulated to convert solar energy into hydrogen at rates that are severalfold higher, compared with any previously described wild-type hydrogen-producing photosynthetic microbe (Bandyopadhyay et al. 2010). Furthermore, if a technique can indeed be developed for splitting water at industrially significant rates, novel process cycles may improve overall economics and provide additional benefits. For example, desalination of seawater—currently accomplished by high-temperature evaporation and condensation or by high-pressure reverse osmosis—could be integrated into the photoinduced decomposition process. The hydrogen split out of seawater could be combusted in a boiler or fed into a fuel cell to produce electricity. The water vapor resulting from combustion (or electrochemical reaction) could then be condensed and purified by conventional water-treatment technology to provide potable water. Such a cycle would, of course, need to be economically competitive with the conventional desalination approaches (Fukuzumi et al. 2010; Logan 2010; Wilkins et al. 2010).

7.2.2.4 Low-Temperature Biofuel Cell

It is well known that the fuel cells now under development must operate at medium to high temperatures—some as high as 900°C. An ambient-temperature fuel cell would represent an important advance of technology. Biomimetic research into ion transport in plants at the macromolecular level may make such a device possible (Jones 2001).

The electricity generation function of current fuel cells depends on catalytically stripping a hydrogen ion—a proton—from a hydrogen atom and inducing it to travel from the cell's anode, through an electrolyte, to its oxygen-bathed cathode. Electricity is generated when the electron remaining from each proton-stripped atom travels from the anode through an external circuit to reunite at the cathode. The process creates water and heat in addition to electricity (Ballantine et al. 2010; Kalyanasundaram and Graetzel 2010; Lee et al. 2010, 2011; Martins et al. 2010; Miyake et al. 2013; Tran 2011; Zhang et al. 2012).

The ability of some biological systems to transport ions across cellular and intracellular membranes provides a model for a lower-temperature, proton-pumping process (Pennisi et al. 2010). However, the process of proton pumping is not yet fully understood on a molecular level, and a more complete understanding of the photochemistry is required. Still, if biomimetic proton pumping could be applied to developing a low-temperature fuel cell, it would carry with it real advantages in cost, weight, life cycle, and construction materials.

7.3 Current and Future Developments

Oil and coal as the conventional energy resource will have a long-term role in providing energy diversity and security, providing ways that should be found to reduce CO_2 emissions in the longer term controlling SO_2, NO_x, and carbon in ash. As society moves from an economy based on fossil fuels to a more sustainable energy mix, scientists and engineers will be required not only to develop sustainable energy solutions but also to find more efficient ways of producing, refining, and using fossil fuels during the transition. At the same time, increasing the traditional power efficiency and recycling should be sought. The green bioenergy exploited by nanotechnology could diminish dependence on depleting petroleum supplies. Moreover, regardless of the specific nanotechnology or the specific energy application, finding either new sources of power or new efficiencies will require new breakthrough technologies. In the foreseeable future, research on the remediation of environmental problems and relevant green technologies will be realized to make the new clean eco-energy sources practical on a commercial scale.

7.4 Conclusion

Environmentally friendly nanotechnologies related to bioenergy have been reviewed for vital information about the growing field for green energy to minimize potential environmental risks. It shows that the significantly feasible world's eco-energy for the foreseeable future should not only be realized, but also methods for using the current energy and their by-products more efficiently should be found correspondingly, alongside technologies that will ensure minimal environmental impact due to unsustainable current fossil fuel usage and its successive greenhouse gas production. The green bioenergy exploited by nanotechnology will promise reliable solutions to reduce dependence on depleting petroleum supplies. At the same time, research on the remediation of environmental problems and relevant green technologies will be realized to make the new clean eco-energy sources practical on a commercial scale.

Acknowledgment

The support by City University of Hong Kong Strategic Research Grant (SRG) No. 7002669 is gratefully acknowledged.

References

Anders, V. N., A. Carsten, and J. Liu. 2009. Processes for production of a starch hydrolysate. EP2032713 (A2).

Argo, B., R. Vidu, P. Stroeve, J. Argo, S. Islam, J. R. Ku, and M. Chen. 2010. Nanostructure and photovoltaic cell implementing same. US7847180.

Ashby, M. F., P. J. Ferreira, and D. L. Schodek (eds.). 2009a. Nanomaterials and nanotechnologies: An overview. In: *Nanomaterials: Nanotechnologies and Design*. pp. 1–16. Butterworth-Heinemann, Elesevier: Oxford, UK.

Ashby, M. F., P. J. Ferreira, and D. L. Schodek. 2009b (eds.). Nanomaterials synthesis and characterization. In: *Nanomaterials: Nanotechnologies and Design*. pp. 257–290. Butterworth-Heinemann, Elesevier: Oxford, UK.

Ashby, M. F., P. J. Ferreira, and D. L. Schodek. 2009c (eds.). Nanomaterials and nanotechnologies in health and the environment. In: *Nanomaterials: Nanotechnologies and Design*. pp. 467–500. Butterworth-Heinemann, Elesevier: Oxford, UK.

Ballantine, A. W., D. C. Kirchhoff, J. F. McElroy, and M. P. Gordon. 2010. Separator plates, ion pumps, and hydrogen fuel infrastructure systems and methods for generating hydrogen. US7686937.

Bandyopadhyay, A., J. Stöckel, H. T. Min, L. A. Sherman, and H. B. Pakrasi. 2010. High rates of photobiological H2 production by a cyanobacterium under aerobic conditions. *Nature Communications* 1: 139.

Boeckh, D., B. Hauer, and D. Haring. 2010. Enzymatic synthesis of sugar acrylates. US7767425.

Bullen, R. A., T. C. Arnot, J. B. Lakeman, and F. C. Walsh. 2006. Biofuel cells and their development. *Biosensors and Bioelectronics* 21: 2015–2045.

Caimi, P. G., Y. C. Chou, M. A. Franden, K. Knoke, L. Tao, P. V. Viitanen, M. Zhang, and Y. Y. Zhang. 2010. Zymomonas with improved ethanol production in medium containing concentrated sugars and acetate. US7803623.

Cao, G. Z., X. Y. Zhou, J. Liu, Z. M. Nie, and Q. F. Zhang. 2011. Aggregate particles of titanium dioxide for solar cells. WO2011005440 (A2).

Catchpole, K. R. and A. Polman. 2009. Photovoltaic cell with surface plasmon resonance generating nano-structures. EP2109147 (A1).

Chen, S., N. Hirota, M. Okuda, M. Takeguchi, H. Kobayashi, N. Hanagata, and T. Ikoma. 2011. Microstructures and rheological properties of tilapia fish-scale collagen hydrogels with aligned fibrils fabricated under magnetic fields. *Acta Biomaterialia* 7: 644–652.

Copeland, B., G. S. Coil, R. A. Latta Jr., and T. A. Lenaghan. 2007. Methods for generating energy using agricultural biofuel. US7263934.

DiCosimo, R., A. Panova, S. D. Arthur, and H. K. Chenault. 2011. Process for enzymatically converting glycolonitrile to glycolic acid. US7875443.

Domen, K., M. Hara, T. Takata, and G. Hitoki. 2005. Photocatalysts for decomposition of water by visible light. US6864211.

Dong, C. B. and C. Z. Dinu. 2013. Molecular trucks and complementary tracks for bionanotechnological applications. *Current Opinion in Biotechnology* 24(4): 612–619.

Feron, D. and A. Bergel. 2011. Fuel cell, using oxidoreductase type enzymes in the cathodic compartment and possibly in the anodic compartment. US7875394.

Fukuzumi, S., T. Kishi, H. Kotani, Y. M. Lee, and W. W. Nam. 2010. High efficient photocatalytic oxygenation reactions using water as an oxygen. *Nature Chemistry* 3: 38–41.

Hennessey, S. M., M. Seapan, R. T. Elander, and M. P. Tucker. 2010. Process for concentrated biomass saccharification. US7807419.

Holliman, P. 2010. Low temperature sintering of dye-sensitised solar cells. WO20101304476 (A1).

Hughes, J. and K. C. Symes. 2010. Biocatalytic manufacturing of (meth) acrylylcholine or 2-(N, N-dimethylamino) ethyl (meth) acrylate. US7754455.

Inman, I. A., P. S. Datta, H. L. Du, C. Kübel, and P. D. Wood. 2010. High temperature tribocorrosion. *Shreir's Corrosion* 1: 331–398.

Jia, H. F., R. Narayanan, D. H. Reneker, P. Wang, and S. T. Wu. 2009. Nanofibers with high enzyme loading for highly sensitive biosensors. WO2009029180 (A1).

Jiang, Z. X., L. Geng, Y. D. Huang, S. A. Guan, W. Dong, and Z. Y. Ma. 2011. The model of rough wetting for hydrophobic steel meshes that mimic *Asparagus setaceus* leaf. *Journal of Colloid and Interface Science* 354(2): 866–872.

Jo, S. M., D. Y. Kim, S. Y. Jang, N. G. Park, and B. H. Yi. 2009. Dye-sensitized solar cell with metal oxide layer containing metal oxide nanoparticles produced by electrospinning and method for manufacturing same. EP2031613 (A2).

Jones, R. 2001. What can biology teach us? *Nature Nanotechnology* 1: 85–86.

Kalyanasundaram, K. and M. Graetzel. 2010. Artificial photosynthesis: Biomimetic approaches to solar energy conversion and storage. *Current Opinion in Biotechnology* 21: 298–310.

Kayama, S. and J. Tanaka. 2009. Ultrafine particulate titanium oxide with low chlorine and low rutile content, and production process thereof. US7591991.

Khan, F. I., K. Hawboldt, and M. T. Iqbal. 2005. Life cycle analysis of wind-fuel cell integrated system. *Renewable Energy* 30: 157–177.
Kim, J. B., H. F. Jia, and P. Wang. 2006. Challenges in biocatalysis for enzyme-based biofuel cells. *Biotechnology Advances* 24: 296–308.
König, H. P., R. F. Hertel, W. Koch, and G. Rosner. 1992. Determination of platinum emissions from a three-way catalyst-equipped gasoline engine. *Atmospheric Environment. A: General Topics* 26: 741–745.
Kostoff, R. N., R. G. Koytcheff, and G. Y. Lau. 2007. Global nanotechnology research literature overview. *Technological Forecasting and Social Change* 74: 1733–1747.
Kragl, U., N. Kaftzik, S. Fliege, and P. Wasserscheid. 2010. Enzyme catalysis in the presence of ionic liquids. US7754462.
Krull, R., T. Wucherpfennig, M. E. Esfandabadi, R. Walisko, G. Melzer, D. C. Hempel, I. Kampen, A. Kwade, and C. Wittmann. 2013. Characterization and control of fungal morphology for improved production performance in biotechnology. *Journal of Biotechnology* 163: 112–123.
Kubo, W. and T. Nomoto. 2010. Enzyme electrode, enzyme electrode producing method, sensor and fuel cell each using enzyme electrode. US7816025.
Kubo, W., T. Nomoto, and T. Yano. 2010. Enzyme electrode, and sensor and biofuel cell using the same. US7687186.
Lagally, M. and F. Liu. 2010. Graphite-based photovoltaic cells. US7858876.
Lashdaf, M., J. Lahtinen, M. Lindblad, T. Venäläinen, and A. O. I. Krause. 2004. Platinum catalysts on alumina and silica prepared by gas- and liquid-phase deposition in cinnamaldehyde hydrogenation. *Applied Catalysis A: General* 276: 129–137.
Lee, J. H., J. H. Lee, Y. J. Lee, and K. T. Nam. 2013. Protein/peptide based nanomaterials for energy application. *Current Opinion in Biotechnology* 24(4): 599–605.
Lee, J. Y., H. Y. Shin, S. W. Kang, C. Park, and S. W. Kim. 2010. Use of bioelectrode containing DNA-wrapped single-walled carbon nanotubes for enzyme-based biofuel cell. *Journal of Power Sources* 195(3): 750–755.
Lee, J. Y., H. Y. Shin, S. W. Kang, C. Park, and S. W. Kim. 2011. Improvement of electrical properties via glucose oxidase-immobilization by actively turning over glucose for an enzyme-based biofuel cell modified with DNA-wrapped single walled nanotubes. *Biosensors and Bioelectronics* 26(5): 2685–2688.
Lee, W. H. 2010. Method and structure for thin film tandem photovoltaic cell. WO2010039727 (A1).
Lewis, A. E. 2010a. Review of metal sulphide precipitation. *Hydrometallurgy* 104: 222–224.
Lewis, L. T. 2010b. Starch as a fuel or fuel component. US7815695.
Lin, M. C., Y. C. Tzeng, S. M. Lan, C.-S. Lee, T.-N. Yang, T.-Y. Wei, J.-P. Chiu, L.-F. Lin, D.-J. Shieh, and M.-C. Kuo. 2010. ln N/ln P/TiO.sub.2 photosensitized electrode. US7655575.
Liu, C., S. Alwarappan, Z. F. Chen, X. X. Kong, and C. Z. Li. 2010. Membraneless enzymatic biofuel cells based on grapheme nanosheets. *Biosensors and Bioelectronics* 25: 1829–1833.
Liu, J., X. H. Zhang, H. L. Pang, B. Liu, Q. Zou, and J. H. Chen. 2012. High-performance bioanode based on the composite of CNTs-immobilized mediator and silk film-immobilized glucose oxidase for glucose/O_2 biofuel cells. *Biosensors and Bioelectronics* 31: 170–175.
Logan, B. 2010. Desalination devices and methods. WO2010124079 (A2).
Martins, V. A., C. Bonfim, W. C. da Silva, and F. N. Crespilho. 2010. Iron (III) nanocomposites for enzynme-less biomimetic cathode: A promising material for use in biofuel cells. *Electrochemistry Communications* 12(11): 1509–1512.
Mascal, M. 2010. High-yield conversion of cellulosic biomass into furanic biofuels and value-added products. US7829732.
McConnell, I., G. H. Li, and G. W. Brudvig. 2010. Energy conversion in natural and artificial photosynthesis. *Chemistry & Biology* 17: 434–447.
Minteer, S. D., N. L. Akers, and C. M. Moore. 2009. Enzyme immobilization for use in biofuel cells and sensors. US7638228.
Miyake, T., K. Haneda, S. Yoshino, and M. Nishizawa. 2013. Flexible, layered biofuel cells. *Biosensors and Bioelectronics* 40(1): 45–49.

Montemagno, C. D., J. J. Schmidt, and S. P. Tozzi. 2007. Biomimetic membranes. US7208089.

Moriarty, P. and D. Honnery. 2008. The prospects for global green car mobility. *Journal of Cleaner Production* 16: 1717–1726.

Munasinghe, P. C. and S. K. Khanal. 2010. Biomass-derived syngas fermentation into biofuels: Opportunities and challenges. *Bioresource Technology* 101(13): 5013–5022.

Nakayama, K., T. Kubo, Y. Nishikitani, and H. Masuda. 2010. Nanotube-shaped titania and process for producing the same. US7687431.

Nam, W. H., K. S. Lee, D. H. Kim, S. J. Kim, N. H. Lee, and H. J. Oh. 2011. Method of manufacture Ni-doped TiO.sub.2 nanotube-shaped powder and sheet film comprising the same. US7867437.

National Nanotechnology Initiative. 2000. *Leading to the Next Industrial Revolution*. Committee on Technology, National Science, and Technology Council, Washington, DC.

Němec, H., E. Galoppini, H. Imahori, and V. Sundstrom. 2011. Solar energy conversion—Natural to artificial. *Comprehensive Nanoscience and Technology* 2: 325–359.

Olson, J. M. 2010. Multi-junction photovoltaic cell with nanowires. WO2010120233 (A2).

Palacios, M. A., M. M. Gómez, M. Moldovan, G. Morrison, S. Rauch, C. Mcleod, R. Ma et al. 2000. Platinum-group elements: Quantification in collected exhaust fumes and studies of catalyst surfaces. *The Science of the Total Environment* 257(1): 1–15.

Parida, K. M., K. H. Reddy, S. Martha, D. P. Das, and N. Biswal. 2010. Fabrication of nanocrystalline $LaFeO_3$: An efficient sol-gel auto-combustion assisted visible light responsive photocatalyst for water decomposition. *International Journal of Hydrogen Energy* 35: 12161–12168.

Park, H., W. R. Kim, H. T. Jeong, J. J. Lee, H. G. Kim, and W. Y. Choi. 2011. Fabrication of dye-sensitized solar cells by transplanting highly ordered TiO_2 nanotube arrays. *Solar Energy Materials and Solar Cells* 95(1): 184–189.

Park, J. T., D. Y. Roh, R. Patel, K. J. Son, W. G. Koh, and J. H. Kim. 2010. Fabrication of hole-patterned TiO_2 photoelectrodes for solid-state dye-sensitized solar cells. *Electrochimica Acta* 56: 68–73.

Parsons, G. 2010. Nano-structured photovoltaic solar cell and related methods. US7655860.

Pennisi, C. P., E. Greenbaum, and K. Yoshida. 2010. Analysis of light-induced transmembrane ion gradients and membrane potential in photosystem I proteoliposomes. *Biophysical Chemistry* 146: 13–24.

Pitkethly, M. J. 2004. Nanomaterials—The driving force. *Materials Today* 7: 20–29.

Ramanavicius, A., A. Kausaite, and A. Ramanaviciene. 2008. Enzymatic biofuel cell based on anode and cathode powered by ethanol. *Biosensors and Bioelectronics* 24: 761–766.

Reddy, B. R., B. Raju, J. Y. Lee, and H. K. Park. 2010. Process for the separation and recovery of palladium and platinum from spend automobile catalyst leach liquor using LIX 84I and Alamine 336. *Journal of Hazardous Materials* 180: 253–258.

Retterer, S. T. and M. L. Simpson. 2012. Microscale and nanoscale compartments for biotechnology. *Current Opinion in Biotechnology* 23: 522–528.

Roberts IV, W. L., H. H. Lamb, L. F. Stikeleather, and T. L. Turner. 2009. Process for conversion of biomass to fuel. EP2097496 (A2).

Roscheisen, M. R., B. M. Sager, and K. Pichler. 2009. Photovoltaic devices fabricated from nanostructured template. US7605327.

Sakai, H., T. Nakagawa, M. Kakuta, and Y. Tokita. 2010. New fuel cell, and power supply device and electronic device using the fuel cell. EP2216847 (A1).

Sepeur, S., G. Frenzer, and P. W. Oliveira. 2008. Particles or coating for splitting water. WO2008154894 (A2).

Shelnutt, J. A., J. E. Miller, Z. C. Wang, and C. J. Medforth. 2008. Water-splitting using photocatalytic porphyrin-nanotube composite devices. US7338290.

Sivaram, S., A. Agarwal, S. B. Herner, and C. J. Petti. 2010. Method to form a photovoltaic cell comprising a thin lamina. US7842585.

Smolyakov, G. A. and M. A. Osinski. 2011. Efficient hydrogen production by photocatalytic water splitting using surface plasmons in hybrid nanparticles. WO2011011064 (A2).

St-Pierre, P., N. Masri, M. C. Fournier, and T. C. White. 2010. Modified cellulases with increased thermostability, thermophilicity, and alkalophilicity. US7785854.

Tran, B. 2011. Nano-electronic array. US7864560.

Van Hove, M. A. 2009. Atomic-scale structure: From surface to nanomaterials. *Surface Science* 603: 1301–1305.

Wang, K. Q., J. Yang, L. G. Feng, Y. W. Zhang, L. Liang, W. Xing, and C. P. Liu. 2012. Photoelectrochemical biofuel cell using porphyrin-sensitized nanocrystalline titanium dioxide mesoporous film as photoanode. *Biosensors and Bioelectronics* 32: 177–182.

Wessels, K., M. Wark, and T. Oekermann. 2010. Efficiency improvement of dye-sensitized solar cells based on electrodeposited TiO_2 films by low temperature post-treatment. *Electrochimica Acta* 55: 6352–6357.

Whitesides, G. M. 2001. The once and future nanomachine. *Scientific American* 9: 79.

Wieting, R. D. 2011. Patterning electrode materials free from berm structures for thin film photovoltaic cells. US7863074.

Wilkins, F. C., A. D. Jha, and G. C. Ganzi. 2010. Method and apparatus for desalination. US7744760.

Yamaguchi, T., N. Tobe, D. Matsumoto, T. Nagai, and H. Arakawa. 2010. Highly efficient plastic-substrate dye-sensitized solar cells with validated conversion efficiency of 7.6%. *Solar Energy Materials and Solar Cells* 94: 812–816.

Yin, Y. Z., Z. Y. Dong, Q. Luo, and J. Q. Liu. 2012. Biomimetic catalysts designed on macromolecular scaffolds. *Progress in Polymer Science* 37: 1476–1509.

Zäch, M., C. Hägglund, D. Chakarov, and B. Kasemo. 2006. Nanoscience and nanotechnology for advanced energy systems. *Current Opinion in Solid State & Materials Science* 10: 132–143.

Zhang, L. L., M. Zhou, D. Wen, L. Bai, B. H. Lou, and S. J. Dong. 2012. Small-size biofuel cell on paper. *Biosensors and Bioelectronics* 35: 155–159.

Zhao, Q. B., T. X. Fan, J. Ding, D. Zhang, Q. X. Guo, and M. Kamada. 2011. Super black and ultrathin amorphous carbon film inspired by anti-reflection architecture in butterfly wing. *Carbon* 49(3): 877–883.

8

Marine Microalgae: Exploring the Systems through an Omics *Approach for Biofuel Production*

Pavan P. Jutur and Asha A. Nesamma

CONTENTS

8.1 Introduction

The decrease of fossil fuels and its impact on global warming led to an ever-increasing demand for its replacement by sustainable renewable biofuels. Marine microalgae may offer a potential feedstock for renewable biofuels capable of converting atmospheric CO_2 to substantial biomass and valuable biofuels, which is of great importance for the food and energy industries. Integrated *omics* research is a powerful tool in understanding the behavior of biological systems as a whole, where the metabolic pathways are often highly regulated and connected with a number of both feedforward and feedback mechanisms that can act positively and/or negatively ultimately affecting the systems output. Understanding the entire system through integrated omics research will identify relevant enzyme-encoding genes and reconstruct the metabolic pathways involved in the biosynthesis and degradation of precursor molecules that may have potential for biofuel production, aiming toward the vision of tomorrow's bioenergy needs.

8.1.1 Importance of Marine Microalgae

Marine photosynthesis is dominated by a diverse group of unicellular eukaryotic photosynthetic organisms like microalgae and cyanobacteria that are responsible for about 40%–50% of photosynthesis that occurs on Earth, despite the fact that photosynthetic biomass represents only 0.2% of that on land (Falkowski et al. 1998; Parker et al. 2008). They are primarily classified into two groups: (a) microalgae, which seems to be a more promising source of renewable energy to replace diminishing oil reserves as a source of lipids—a biodiesel feedstock, due to their high photosynthetic efficiency—and (b) macroalgae (or seaweed), which generally do not have lipids and are considered as rich sources for natural sugars and other carbohydrates, which can be further fermented to produce either biogas or bioalcohol-based fuels.

The major concerns and/or need for the production of biofuels from microalgae is mainly due to the following: (1) Fossil fuel reserves are limited and non-regenerable. (2) Global energy demands are ever-increasing. (3) CO_2 released from fossil fuel oxidation causes climatic change and leads to global warming. (4) Currently, liquid petroleum transportation fuels are hard to replace by other available renewable energy sources like wind, solar, and hydro. Alternatively, the microalgal-based biofuels, which produce carbon-neutral fuels, are reliable and regenerable resources.

Marine microalgae as the preferred option for the biofuel production over other biofuel feedstocks is advantageous due to the following reasons:

- They grow very fast, thus provide higher biomass yield.
- A high amount of oil productivity per acre (Chisti 2007).
- Use of non-arable land and a wide variety of water sources like marine/brackish waters.
- They can mitigate greenhouse gas emissions.
- A good potential for use in producing both biofuels and other valuable coproducts.
- These microalgal-based biodiesel derived fuels can be integrated into the current scenario of transportation infrastructure (Fang 2014).

Challenges that are relevant for biofuel production are (1) large-scale production of marine microalgae through either open systems or photobioreactors, which should be optimized using parameters like nutrient supply and recycling, photosynthetically active radiation (PAR) delivery, gas transfer and exchange, environmental conditions, and land and water availability; (2) optimization of harvesting techniques of microalgae through chemical coagulation, flocculation, filtering, sedimentation, etc.; and (3) reduction in the high cost for biofuel production to make biodiesel industrially competitive. All these emphases put forward together for developing techniques lead to the initiation of hypothesis-driven strain improvement strategies in marine microalgae for sustainable biofuel production.

8.1.2 Significance of Marine Microalgae in Biofuel Production

For economic viability and sustainability of microalgal biofuels, it is essential to understand the basic biology of these marine microalgae, particularly the effects of environmental stresses on cellular metabolisms and the regulation of biosynthetic pathways of fatty acids and triacylglycerols (TAGs), as well as carbon fixation and allocation. Neutral lipid biosynthesis and turnover have been studied in higher eukaryotes, particularly yeast, mice, and the model plant *Arabidopsis thaliana*, where extensive knowledge exists on lipid metabolism and several key enzymes identified. However, these pathways are not fully documented in marine microalgae and little is known about

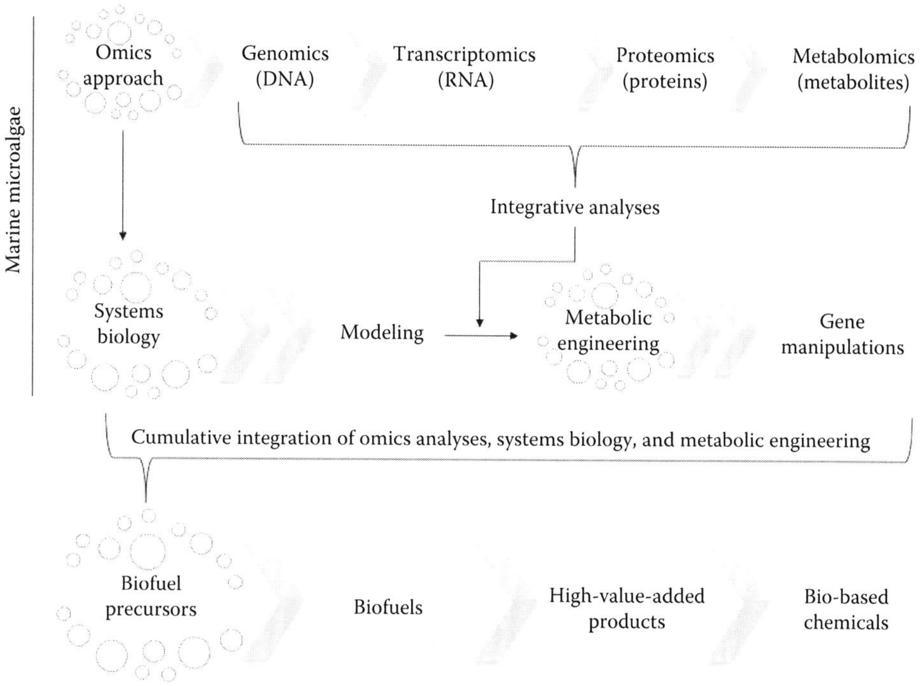

FIGURE 8.1
Schematic representation of cumulative integration of omics analyses and systems biology modeling in coherence with metabolic engineering that leads to production of sustainable biofuel precursors from marine microalgae.

the enzymes involved in the formation, accumulation, or degradation of TAGs. More research is needed to understand signal perception and transduction under stress and different molecular mechanisms leading to TAG and starch accumulations in microalgae. In this context, rapidly developing systems-level omics analyses (transcriptomics, proteomics, metabolomics), which are both sensitive and quantitative, and the availability of annotated genomes help us to investigate changes in response to environmental stress. Integrative omics research will provide useful hypotheses for microalgal strain improvement through metabolic engineering and systems biology to optimize their biofuel production (Figure 8.1).

8.2 Genomics and Transcriptomics

Understanding the basis of microalgal biology is important for establishing the foundation for innovative strategies for the development of ultimate fuel surrogates (Fang 2014). The algal genome sequences revealed that these organisms possess unique features in their patterns of primary sequence structures and gene compositions, showing that each system is useful for understanding basic physiological aspects and the evolution of photosynthetic eukaryotes (Shanmugam et al. 2013).

8.2.1 Genomics of Marine Microalgae

Algal genomics is the understanding of the systems through obtained DNA sequence information and assessing genetic variation at the DNA sequence and genomic architecture levels. The first sequenced genome in the field of emerging marine algal genomics started with the nuclear genome of the cryptophyte alga *Guillardia theta* (Douglas et al. 2001; Curtis et al. 2012), and *Cyanidioschyzon merolae* was the first ultrasmall unicellular red alga to provide a complete genome sequence (Matsuzaki et al. 2004). Later, it continued with the whole genome for each of the diatoms *Thalassiosira pseudonana* (Armbrust et al. 2004) and *Phaeodactylum tricornutum* (Bowler et al. 2008), the pelagophyte *Aureococcus anophagefferens* (Gobler et al. 2011), the haptophyte *Emiliania huxleyi* (Read et al. 2013), and four picoeukaryotes of green algal descent, the prasinophytes—*Ostreococcus tauri* (Derelle et al. 2006), *Ostreococcus lucimarinus* (Palenik et al. 2007), *Micromonas pusilla* strain NOUM17, and *M. pusilla* strain CCMP1545 (Worden et al. 2009). The completed genome sequences provide vital information about the series of events that occurred during the evolution of microalgae (Parker et al. 2008). Studies on genomic data showed that similar morphological species like *O. tauri* and *O. lucimarinus* were distinctively divergent, with an average amino acid identity of only 70% (Palenik et al. 2007). Analyzing such facts on microalgal genomes will provide important clues for algal research, which illustrates not only the similarity among nucleotides sequences but also morphogenesis, physiology, environmental adaptations, and evolutionary phenomenon among these organisms (Shanmugam et al. 2013). The approach of whole-genome sequencing will also provide leads toward identification of novel synthesis pathways and target sites for mutation or transgenic studies, facilitating genetic engineering.

8.2.2 Transcriptomics Profiling

Transcriptomics, also known as global analysis of gene expression or genome-wide expression profiling, is one of the major tools for measuring gene expression levels and uncovering differential expression of the whole set of all mRNA/miRNA molecules, or transcripts, produced in a single cell and/or a population of cells (Zhang et al. 2010). Unlike genome sequencing and comparative genomics technologies that mainly focus on DNA, it seems to be static information for any given species and normally does not change significantly in response to short-term external environmental changes. Transcriptomics has enabled quantitative measurements of the dynamic expression of mRNA/miRNA molecules and their variation at the genome scale between different states, thus reflecting the genes that are being actively expressed at any given time, with the exception of mRNA degradation phenomena (Ye et al. 2001; Horak and Snyder 2002). Such studies play a pivotal role in the identification of individual transcriptional units, the expression of unique transcripts restricted to cell conditions or cell types, the determination of the level of expression for each gene expressed, the precise assessment of transcript diversity at each transcriptional locus (i.e., alternative transcription start sites, splice isoforms, and polyadenylation sites), and the protein factors that control the transcriptional cassettes of a cell and their mechanisms of action (Shanmugam et al. 2013). Currently, there are technologies that are separated into approaches that primarily assess levels of expression, that provide precise transcript boundaries and other transcript-focused characteristics, and that look at the transcription factor–DNA interactions. Transcriptomics technologies have been used in various microalgal systems to explore genome-wide transcriptional activity (Miller et al. 2010; Rismani-Yazdi et al. 2011, 2012; Corteggiani Carpinelli et al. 2014; Dong et al. 2013; Liang et al. 2013; Lv et al. 2013; Yang et al. 2013).

High-throughput transcriptomics strategies are involved in the following:

1. Identifying mRNAs that differ in their expression patterns under different experimental conditions and further define the identity of those respective genes by differential display or produce the snapshot of mRNAs that correspond to fragments of those transcripts, also known as serial analysis of gene expression (SAGE). Long SAGE analyses in *E. huxleyi* revealed many new differentially regulated gene sequences and also assigned regulation data to EST sequences with no database homology and unknown function, which highlighted uncharacterized aspects of *E. huxleyi* N and P physiology (Dyhrman et al. 2006).
2. Alternatively assessing changes in the expression pattern of previously defined genes using cDNA or oligonucleotide microarrays (Kagnoff and Eckmann 2001) or chip-based nanoliter-volume reverse-transcription polymerase chain reaction (RT-PCR), which measures gene expression for several thousands of genes simultaneously at higher sensitivity and accuracy than microarrays (Stedtfeld et al. 2008).
3. A more recent approach in microalgae employs next-generation sequencers involving Roche 454 sequencer, applied biosystems SOLiD, and Illumina Genome Analyzer systems for direct sequencing of the cDNA converted from whole transcriptomes (Miller et al. 2010; Rismani-Yazdi et al. 2011, 2012; Corteggiani Carpinelli et al. 2014; Liang et al. 2013; Lv et al. 2013; Yang et al. 2013).
4. An alternative advanced technology that was employed in microbial systems not yet reported in microalgal systems is direct RNA sequencing (DRS) technology, which allows massively parallel sequencing of RNA molecules directly without prior synthesis of cDNA or the need for ligation/amplification steps developed by Helicos Genetic Analysis System (Ozsolak et al. 2009), where both the abundance and the identity of mRNA molecules can be determined in one analytical process. The major advantage of next-generation sequencing over the traditional sequencing method is the dramatically increased degree of parallelism, which can be represented by the number of reads (i.e., the number of DNA templates that can be sequenced simultaneously) in a single sequencing run and the number of sequenced bases per day (Zhang et al. 2010).

Expressed sequence tags (ESTs) are small pieces of the DNA sequence (usually ~200–500 nucleotides long) that are generated by sequencing either one or both ends of an expressed gene. The EST studies (Scala et al. 2002; Maheswari et al. 2005), together with the first whole-genome sequences from diatoms, *T. pseudonana* (Armbrust et al. 2004) and *P. tricornutum* (Bowler et al. 2008), revealed less than 50% of diatom genes can be assigned a putative function using homology-based methods. The diatom EST database (Maheswari et al. 2005, 2009) enabled comparative studies of eukaryotic algal genomes and revealed some interesting differences in the genes involved in basic cell metabolism (Montsant et al. 2005; Herve et al. 2006), along with the key signaling and regulatory pathways (Montsant et al. 2007), carbohydrate metabolism (Maheswari et al. 2005; Kroth et al. 2008), silica metabolism (Lopez et al. 2005; Montsant et al. 2005), and nitrogen metabolism (Allen et al. 2006). Patterns in transcript (ESTs) and protein abundance revealed the fact that *T. pseudonana* evolved with a sophisticated response to phosphorous (P) deficiency that involves multiple biochemical strategies that are essential for its ability to respond to variations in environmental P availability (Dyhrman et al. 2012). Transcriptome analysis in marine microalga *E. huxleyi* revealed that it retains the ability to reprogram its gene expression in response to reduced sulfate availability (Bochenek et al. 2013). Studies on the transcriptome and lipid biosynthetic pathways in marine *Nannochloropsis* species using Illumina HiSeq 2000 showed a total of 29,203 unigenes that were differentially expressed under conditions of low- and high-lipid-producing phases, wherein among those 195 unigenes were involved in lipid metabolism and 315 unigenes were putatively transcription factors (Zheng et al. 2013).

Next-generation DNA pyrosequencing technology analyses in *Dunaliella tertiolecta* identified that the majority of lipid and starch biosynthesis and catabolism pathways through assembled transcriptome provide a direct metabolic engineering methodology to enhance the quantity and quality of microalgae-based biofuel feedstock (Rismani-Yazdi et al. 2011). Metabolic measurements revealed by genes and pathway expression under the nitrogen limitation in *Neochloris oleoabundans* showed that carbon is portioned toward TAG production and the overexpression of the fatty acid synthesis pathway was bolstered by repression of the β-oxidation pathway along with upregulation of genes encoding for the pyruvate dehydrogenase complex that funnels acetyl-CoA to lipid biosynthesis (Rismani-Yazdi et al. 2012). Chromosome scale genome assembly and transcriptome profiling of marine *Nannochloropsis gaditana* under nitrogen depletion revealed traits of genes involved in DNA recombination, RNA silencing, and cell wall synthesis (Corteggiani Carpinelli et al. 2014). This study also showed that the content of lipids increased drastically, but without detectable major changes in expression of the genes involved in their biosynthesis and at the same time significant downregulation of mitochondrial gene expression, suggesting that the acetyl-CoA and NAD(P)H, normally oxidized through the mitochondrial respiration, are available for fatty acid synthesis, thus increasing the flux through the lipid biosynthetic pathway (Corteggiani Carpinelli et al. 2014).

8.3 Proteomics

Proteomics is a qualitative and quantitative determination of protein expression and modification, as proteins are the major components for building the cellular structure and they also serve as catalytic enzymes in metabolic pathways and as signal transduction molecules in regulatory pathways of cells (Graham et al. 2007). Proteomics mainly relies on two major strategies for separation and visualization of proteins: (1) Two-dimensional polyacrylamide gel electrophoresis (2D-PAGE), where proteins are separated based on their isoelectric points (pI) and mass, followed by mass spectrometric identification, that was established in marine microalga *D. salina* (Katz et al. 2007; Jia et al. 2010) and (2) gel-free profiling procedures with multidimensional separations coupling microscale separation with automated tandem mass spectrometry (LC-MS/MS) (Baggerman et al. 2005; Nie et al. 2008). For more advancement and precision, some accurate proteome measurements of quantitative proteomic methods, such as stable isotope labeling–based isotope-coded affinity tag (ICAT), isobaric tags for relative and absolute quantification (iTRAQ) (Yan et al. 2008), or label-free comparative quantitative proteomics, are needed to be employed (Haqqani et al. 2008).

A metaproteogenomic strategy to integrate proteomics and metabolomics data for systems-level analysis in the unicellular green algae *Chlamydomonas reinhardtii* was analyzed under different growth conditions and revealed metabolic pathways comprising of Calvin cycle, photosynthetic apparatus, starch synthesis, glycolysis, tricarboxylic acid (TCA) cycle, carbon-concentrating mechanism (CCM), and other pathways by targeted proteomics (mass Western). From the same samples, metabolite concentrations and metabolic fluxes by stable isotope incorporation were analyzed where the differences were found in growth-dependent crosstalk of chloroplastidic and mitochondrial metabolism. A few carbonic anhydrases that were partially involved in CCM revealed highest internal cell concentrations for a specific low-CO_2-inducible mitochondrial CAH isoform indicating its role as one of the strongest CO_2-responsive proteins in the crosstalk of air-adapted mixotrophic chloroplast and mitochondrial metabolism in *C. reinhardtii* (Wienkoop et al. 2010).

Oil bodies are sites of energy and carbon storage in many organisms including microalgae; proteomics profiling of these purified oil bodies from the model microalga *C. reinhardtii* grown under nitrogen deprivation revealed around 248 proteins (Z2 peptides) identified by LC-MS/MS, 33 were putatively involved in metabolism of lipids that are mostly acyl lipids and sterols (Nguyen et al. 2011). Among these, 19 new proteins of lipid metabolism were identified, spanning the key steps of TAG synthesis pathway including a glycerol-3-phosphate acyltransferase (GPAT), a lysophosphatidic acid acyltransferase (LPAT), and a putative phospholipid–diacylglycerol acyltransferase (PDAT). Some other proteins were putatively involved in deacylation/reacylation, sterol synthesis, lipid signaling, and lipid trafficking, and were also found to be associated with the oil bodies. These data provide evidence that *Chlamydomonas* oil bodies not only are storage compartments but also are dynamic structures likely to be involved in processes such as oil synthesis, degradation, and lipid homeostasis. Such proteomic profiling will provide useful insights in understanding TAG synthesis and the role of oil bodies in microalgal cell factories (Nguyen et al. 2011). Studies with random mutations (designated as CR12 and CR48) in *C. reinhardtii* with altered lipid productivities were generated, and comparative proteomics was analyzed to identify proteins that were significantly upregulated or downregulated and revealed some putative genes or pathways directly or indirectly linked to microalgal lipid production (Choi et al. 2013).

To optimize biofuel production in microalgae, it is desirable to induce lipid accumulation without compromising cell growth and survival. For this purpose, an 8-plex iTRAQ-based proteomic approach was carried out in the model alga *C. reinhardtii* CCAP 11/32CW15+ under nitrogen starvation. First-dimension fractionation was conducted using hydrophilic interaction liquid chromatography (HILIC) and strong cation exchange (SCX). A total of 587 proteins were identified (=3 peptides), of which 71 and 311 were differentially expressed at significant levels ($p < 0.05$), during nitrogen stress-induced carbohydrate and lipid production, respectively (Longworth et al. 2012). Also, a few other observations were carried out to determine the increase in energy metabolism, decrease in translation machinery, increase in cell wall production, and a change of balance between photosystems I and II. These findings suggest that a severely compromised system prevails where lipid is accumulated at the expense of normal functioning of the organism, whereas a more informed and controlled method of lipid induction than overall nutrient manipulation would be needed for the development of sustainable processes (Longworth et al. 2012).

Label-free comparative shotgun proteomic analyses have identified a number of novel targets, including previously unidentified transcription factors and proteins via a transcriptome-to-proteome pipeline, in the oleaginous microalga, *Chlorella vulgaris*, under nitrogen deprivation response. Among these identified targets are proteins involved in transcriptional regulation, lipid biosynthesis, cell signaling, and cell cycle progression (Guarnieri et al. 2013). Herein, the identified potential targets for strain-engineering strategies targeting enhanced lipid accumulation for microalgal biofuels applications. Protein profiling using a novel subtraction method in identifying the oil body–associated proteins in the diatoms *Fistulifera* sp. strain JPCC DA0580 in oleaginous microalgae that are promising producers of biodiesel has revealed 15 proteins as oil body–associated protein candidates. Among them, two proteins predicted to have transmembrane domains were confirmed to specifically localize to the oil bodies and were identified by using green fluorescent protein (GFP) fusion proteins. One protein that seems to be predicted through a potassium channel was also detected from the endoplasmic reticulum (ER), suggesting that oil bodies might originate from the ER (Nojima et al. 2013). Nitrogen deprivation in the marine diatom *P. tricornutum* led to an increase in the expression of genes involved in nitrogen assimilation and fatty acid biosynthesis and a concomitant decrease in photosynthesis and lipid catabolism enzymes. These changes at the molecular level are consistent with the physiological changes as observed in photosynthesis rate and saturated lipid content, which provides vital information at the proteomic level, the key enzymes involved in carbon flux toward lipid accumulation, and the candidates for genetic manipulation in microalgae for biodiesel production (Yang et al. 2014).

8.4 Metabolomics

Metabolomics is the study of global metabolite profiles of a cell under a given set of conditions (Tang 2011), where the comprehensive and quantitative analysis of all small molecules in a biological system and their concentration levels that differ as a consequence of genetic or physiological changes (Fiehn 2001; Raamsdonk et al. 2001). Metabolomic analysis is carried out in microbial and/or microalgal systems using high-throughput instrumentation like gas chromatography–time-of-flight mass spectrometry (GC-TOF-MS), high-performance liquid chromatography–mass spectrometry (LC-MS) or capillary electrophoresis–mass spectrometry (CE-MS), nuclear magnetic resonance (NMR) spectroscopy, and more recently vibrational spectroscopy (Zhang et al. 2010; Ito et al. 2013b). Metabolomic analyses usually involve two major approaches: (1) discovery metabolomics includes profiling, identification, and interpretation of metabolites with significant variations in abundance within a set of experimental and control conditions, and (2) targeted metabolomics is a quantitative methodology where analyses mainly focus on validation of hypotheses from discovery metabolomics experiments.

Using the GC-TOF-MS technique, some major metabolite changes during early induction of the CCM in the unicellular green alga *C. reinhardtii* revealed a metabolic pattern where photorespiration is increasing (Renberg et al. 2010). Comparative sequencing datasets among green, red, and brown microalgae have provided considerable insight into a number of important clues concerning their evolution, physiology, and metabolism. Further, the combinatorial application of metabolomics in understanding both the function of individual genes and metabolic processes in conjunction with sequencing data will provide greater insight into the metabolic hierarchies underpinning the function of individual organisms such as unicellular marine diatoms (Fernie et al. 2012). Studies on metabolites of central carbon metabolism in the diatom *Skeletonema marinoi* were comprehensively analyzed showing the changes during different growth phases (Vidoudez and Pohnert 2012).

Oil-rich algae *Pseudochoricystis ellipsoidea* MBIC 11204, a novel unicellular green algal strain, which accumulates a large amount of oil (lipids) in nitrogen-deficient (–N) conditions, was analyzed for its metabolome and lipidome profiles, respectively, using CE-MS and LC-MS (Ito et al. 2013b). Relative quantities of >300 metabolites were systematically compared between both these experimental (nitrogen-rich [+N; rapid growth] and –N) conditions. Studies revealed that there are relatively decreased levels of amino acids in nitrogen assimilation and N-transporting metabolisms in –N conditions, whereas in lipid metabolism, the quantities of neutral lipids increased drastically; however, amounts of all other lipids neither decreased nor changed slightly. This is one of the unique approaches to understand the novel microalgal strain's metabolism using a combination of wide-scale metabolome analysis and morphological analysis (Ito et al. 2013b).

Studies on characterization of global metabolism in microalga *P. ellipsoidea* using unlabeled and uniformly $^{(13)}$C-labeled metabolites were quantified using a CE-MS and LC-MS for ionic primary metabolites and lipids, respectively, and revealed four hierarchical characteristic clusters, two of which represented rapidly labeled metabolites, mainly consisting of primary metabolites, while the two other clusters represented slowly labeled metabolites, composed of lipids (Ito et al. 2013a).

8.5 From Omics to Systems Biology

In the future, the integrated omics biology of various cellular molecules that includes the transcriptome, proteome, and metabolome, and their interactions in the cells would lead to a systematic quantified model of cellular metabolism at a genome scale that can serve as a foundation for hypothesis-driven research (Ishii and Tomita 2009; Zhang et al. 2010) and will eventually lead new insights into microalgal cellular metabolism. Systems biology is an interdisciplinary science that studies the complex interactions and collective behavior of a cell or an organism (Liu et al. 2013). From omics to systems biology, the primary goal is to develop a comprehensive and consistent knowledge of a biological system by evaluating the behavior and interaction between its individual components (Jamers et al. 2009). One of the key steps in this process involves modeling, where the structure of the system is unraveled by mathematical algorithms that allow its dynamics. Thus, these mathematical models describe the system but also allow the prediction of the system's response to perturbations. A framework of studies in systems biology involves initially understanding the structure and identifying key elements in the system, such as gene networks, protein interactions, and metabolic pathways (Ideker et al. 2001), thus constructing an initial model of systems behavior. Secondly, the system is perturbed (genetically or using environmental stimuli), and corresponding responses are measured using high-throughput measurement tools, wherein the data generated at different levels of biological organization are integrated with each other and with the current model of the system. Obviously, the model is adapted in such a way that the experimentally observed phenomena correspond best with the model's predictions. These steps are continually repeated, thereby expanding and refining the model until the model's predictions reflect biological reality. For algae, no systems biology studies with extensive computational modeling efforts have been reported so far to our knowledge. Studies on metabolic, genomic, and transcriptomic data in *C. reinhardtii* provide genome-wide insights into the regulation of the metabolic networks under anaerobic conditions associated with H_2 production (Mus et al. 2007). Metabolic network reconstruction in model alga *C. reinhardtii* encompasses the organism's metabolism and genome annotation, providing a platform for omics data analysis and phenotype prediction. Integrating biological and optical data and reconstructing a genome-scale metabolic network offers insight into algal metabolism and potential for genetic engineering and efficient light source design, a pioneering resource for studying light-driven metabolism and quantitative systems biology (Chang et al. 2011).

The morphological characteristics and metabolome profiles of the oil-rich alga *P. ellipsoidea* exposed to +N and −N conditions were analyzed to determine how lipids synthesize and the mechanisms in which they accumulate in −N conditions. This study revealed few hypothetical metabolisms where advanced systems biology approaches such as metabolic flux analysis, turnover analysis, and pulse-chase experiments are required for more effective understanding of the physiological phenomenon occurring among these microalgal strains (Ito et al. 2013b). Systems biology will also make available synthetic biology tools more reliable, enabling the precise control of transcription and translation regardless of the undercontrolled gene (Mutalik et al. 2013). Similarly, simplified understanding of genetic systems created in synthetic biology will provide systems biology with new insights into the fundamentals of native gene regulation among microalgae, thus allowing simultaneous integration of omics data at global scales in living cells and forming a direct networking between systems and synthetic biologists that will hasten rapid progress in areas pertaining to genomics, transcriptomics, proteomics, and metabolomics (Liu et al. 2013).

8.6 Conclusions

Recent analysis on transcriptome, proteome, and metabolome studies has significantly contributed toward the understanding of microalgae cellular responses and has unraveled many of the underlying molecular mechanisms that are required for the carbon acquisition and accumulation (Winck et al. 2013). Certainly, the proof-of-principle studies on model strains and upcoming genome sequences of marine metagenomes will provide essential leads for metabolic engineering of marine microalgae. Metabolic engineering to increase yields of biofuel-related lipids in marine microalgae *T. pseudonana* without compromising growth provides insights for targeted metabolic manipulations to enhance lipid productivity in eukaryotic microalgae (Trentacoste et al. 2013). Subsequently, all metabolic pathways show certain differences in photosynthesis, carbon fixation and processing, carbon storage, and compartmentation of cellular and metabolic processes that seems to be substantial and likely to transcend into the efficiency of various processes involved in biofuel molecule production (Hildebrand et al. 2013). The omics approach studies identified differentially expressed enzyme encoding genes that represent few classes of genes, which may serve on future metabolic engineering with focus on enhancing the overall biomass and lipid accumulation. These omics datasets provide new insights and evidences on the cellular responses and serve as novel constraints for the improvement of metabolic models by system biology approach. Understanding the dynamics of microalgal systems from integrative omics analyses and systems biology in coherence with known functional aspects of essential genes in metabolic pathways provides an effective strategy for improved production of industrially feasible biofuel-based precursors from marine microalgae for tomorrow's energy needs.

References

Allen, A. E., A. Vardi, and C. Bowler. 2006. An ecological and evolutionary context for integrated nitrogen metabolism and related signaling pathways in marine diatoms. *Curr Opin Plant Biol* 9(3):264–273.

Armbrust, E. V., J. A. Berges, C. Bowler et al. 2004. The genome of the diatom *Thalassiosira pseudonana*: Ecology, evolution, and metabolism. *Science* 306(5693):79–86.

Baggerman, G., E. Vierstraete, A. De Loof, and L. Schoofs. 2005. Gel-based versus gel-free proteomics: A review. *Comb Chem High Throughput Screen* 8(8):669–677.

Bochenek, M., G. J. Etherington, A. Koprivova et al. 2013. Transcriptome analysis of the sulfate deficiency response in the marine microalga *Emiliania huxleyi*. *New Phytol* 199(3):650–662.

Bowler, C., A. E. Allen, J. H. Badger et al. 2008. The *Phaeodactylum* genome reveals the evolutionary history of diatom genomes. *Nature* 456(7219):239–244.

Chang, R. L., L. Ghamsari, A. Manichaikul et al. 2011. Metabolic network reconstruction of *Chlamydomonas* offers insight into light-driven algal metabolism. *Mol Syst Biol* 7:518.

Chisti, Y. 2007. Biodiesel from microalgae. *Biotechnol Adv* 25(3):294–306.

Choi, Y. E., H. Hwang, H. S. Kim, J. W. Ahn, W. J. Jeong, and J. W. Yang. 2013. Comparative proteomics using lipid over-producing or less-producing mutants unravels lipid metabolisms in *Chlamydomonas reinhardtii*. *Bioresour Technol* 145:108–115.

Corteggiani Carpinelli, E., A. Telatin, N. Vitulo et al. 2014. Chromosome scale genome assembly and transcriptome profiling of *Nannochloropsis gaditana* in nitrogen depletion. *Mol Plant* 7(2):323–335.

Curtis, B. A., G. Tanifuji, F. Burki et al. 2012. Algal genomes reveal evolutionary mosaicism and the fate of nucleomorphs. *Nature* 492(7427):59–65.

Derelle, E., C. Ferraz, S. Rombauts et al. 2006. Genome analysis of the smallest free-living eukaryote *Ostreococcus tauri* unveils many unique features. *Proc Natl Acad Sci USA* 103(31):11647–11652.

Dong, H.-P., E. Williams, D.-Z. Wang et al. 2013. Responses of *Nannochloropsis oceanica* IMET1 to long-term nitrogen starvation and recovery. *Plant Physiol* 162(2):1110–1126.

Douglas, S., S. Zauner, M. Fraunholz et al. 2001. The highly reduced genome of an enslaved algal nucleus. *Nature* 410(6832):1091–1096.

Dyhrman, S. T., S. T. Haley, S. R. Birkeland, L. L. Wurch, M. J. Cipriano, and A. G. McArthur. 2006. Long serial analysis of gene expression for gene discovery and transcriptome profiling in the widespread marine coccolithophore *Emiliania huxleyi*. *Appl Environ Microbiol* 72(1):252–260.

Dyhrman, S. T., B. D. Jenkins, T. A. Rynearson et al. 2012. The transcriptome and proteome of the diatom *Thalassiosira pseudonana* reveal a diverse phosphorus stress response. *PLoS One* 7(3):e33768.

Falkowski, P. G., R. T. Barber, and V. Smetacek. 1998. Biogeochemical controls and feedbacks on ocean primary production. *Science* 281(5374):200–206.

Fang, S.-C. 2014. Metabolic engineering and molecular biotechnology of microalgae for fuel production, Chapter 3. In *Biofuels from Algae*, A. Pandey, D.-J. Lee, Y. Chisti, and C. R. Soccol (eds.). pp. 47–65, Amsterdam, the Netherlands: Elsevier.

Fernie, A. R., T. Obata, A. E. Allen, W. L. Araujo, and C. Bowler. 2012. Leveraging metabolomics for functional investigations in sequenced marine diatoms. *Trends Plant Sci* 17(7):395–403.

Fiehn, O. 2001. Combining genomics, metabolome analysis, and biochemical modelling to understand metabolic networks. *Comp Funct Genomics* 2(3):155–168.

Gobler, C. J., D. L. Berry, S. T. Dyhrman et al. 2011. Niche of harmful alga *Aureococcus anophagefferens* revealed through ecogenomics. *Proc Natl Acad Sci USA* 108(11):4352–4357.

Graham, R. L. J., C. Graham, and G. McMullan. 2007. Microbial proteomics: A mass spectrometry primer for biologists. *Microb Cell Fact* 6:26.

Guarnieri, M. T., A. Nag, S. Yang, and P. T. Pienkos. 2013. Proteomic analysis of *Chlorella vulgaris*: Potential targets for enhanced lipid accumulation. *J Proteomics* 93:245–253.

Haqqani, A. S., J. F. Kelly, and D. B. Stanimirovic. 2008. Quantitative protein profiling by mass spectrometry using label-free proteomics. *Methods Mol Biol* 439:241–256.

Herve, C., T. Tonon, J. Collen, E. Corre, and C. Boyen. 2006. NADPH oxidases in eukaryotes: Red algae provide new hints! *Curr Genet* 49(3):190–204.

Hildebrand, M., R. M. Abbriano, J. E. Polle et al. 2013. Metabolic and cellular organization in evolutionarily diverse microalgae as related to biofuels production. *Curr Opin Chem Biol* 17(3):506–514.

Horak, C. E. and M. Snyder. 2002. Global analysis of gene expression in yeast. *Funct Integr Genomics* 2(4–5):171–180.

Ideker, T., T. Galitski, and L. Hood. 2001. A new approach to decoding life: Systems biology. *Annu Rev Genomics Hum Genet* 2:343–372.

Ishii, N. and M. Tomita. 2009. Multi-omics data-driven systems biology of *E. coli*. In *Systems Biology and Biotechnology of Escherichia coli*, S. Lee (ed.). pp. 41–57, Dordrecht, the Netherlands: Springer.

Ito, T., M. Sugimoto, Y. Toya et al. 2013a. Time-resolved metabolomics of a novel trebouxiophycean alga using $^{(13)}CO_2$ feeding. *J Biosci Bioeng* 116(3):408–415.

Ito, T., M. Tanaka, H. Shinkawa et al. 2013b. Metabolic and morphological changes of an oil accumulating trebouxiophycean alga in nitrogen-deficient conditions. *Metabolomics* 9(Suppl. 1):178–187.

Jamers, A., R. Blust, and W. De Coen. 2009. Omics in algae: Paving the way for a systems biological understanding of algal stress phenomena? *Aquat Toxicol* 92(3):114–121.

Jia, Y., L. Xue, J. Li, and H. Liu. 2010. Isolation and proteomic analysis of the halotolerant alga *Dunaliella salina* flagella using shotgun strategy. *Mol Biol Rep* 37(2):711–716.

Kagnoff, M. F. and L. Eckmann. 2001. Analysis of host responses to microbial infection using gene expression profiling. *Curr Opin Microbiol* 4(3):246–250.

Katz, A., P. Waridel, A. Shevchenko, and U. Pick. 2007. Salt-induced changes in the plasma membrane proteome of the halotolerant alga *Dunaliella salina* as revealed by blue native gel electrophoresis and nano-LC-MS/MS analysis. *Mol Cell Proteomics* 6(9):1459–1472.

Kroth, P. G., A. Chiovitti, A. Gruber et al. 2008. A model for carbohydrate metabolism in the diatom *Phaeodactylum tricornutum* deduced from comparative whole genome analysis. *PLoS One* 3(1):e1426.

Liang, C., S. Cao, X. Zhang et al. 2013. *De novo* sequencing and global transcriptome analysis of *Nannochloropsis* sp. (eustigmatophyceae) following nitrogen starvation. *Bioenerg Res* 6(2):494–505.

Liu, D., A. Hoynes-O'Connor, and F. Zhang. 2013. Bridging the gap between systems biology and synthetic biology. *Front Microbiol* 4:211.

Longworth, J., J. Noirel, J. Pandhal, P. C. Wright, and S. Vaidyanathan. 2012. HILIC- and SCX-based quantitative proteomics of *Chlamydomonas reinhardtii* during nitrogen starvation induced lipid and carbohydrate accumulation. *J Proteome Res* 11(12):5959–5971.

Lopez, P. J., J. Descles, A. E. Allen, and C. Bowler. 2005. Prospects in diatom research. *Curr Opin Biotechnol* 16(2):180–186.

Lv, H., G. Qu, X. Qi, L. Lu, C. Tian, and Y. Ma. 2013. Transcriptome analysis of *Chlamydomonas reinhardtii* during the process of lipid accumulation. *Genomics* 101(4):229–237.

Maheswari, U., T. Mock, E. V. Armbrust, and C. Bowler. 2009. Update of the diatom EST database: A new tool for digital transcriptomics. *Nucleic Acids Res* 37(Database issue):D1001–D1005.

Maheswari, U., A. Montsant, J. Goll et al. 2005. The diatom EST database. *Nucleic Acids Res* 33(Suppl. 1):D344–D347.

Matsuzaki, M., O. Misumi, I. T. Shin et al. 2004. Genome sequence of the ultrasmall unicellular red alga *Cyanidioschyzon merolae* 10D. *Nature* 428(6983):653–657.

Miller, R., G. Wu, R. R. Deshpande et al. 2010. Changes in transcript abundance in *Chlamydomonas reinhardtii* following nitrogen deprivation predict diversion of metabolism. *Plant Physiol* 154(4):1737–1752.

Montsant, A., A. E. Allen, S. Coesel et al. 2007. Identification and comparative genomic analysis of signaling and regulatory components in the diatom *Thalassiosira pseudonana*. *J Phycol* 43(3):585–604.

Montsant, A., K. Jabbari, U. Maheswari, and C. Bowler. 2005. Comparative genomics of the pennate diatom *Phaeodactylum tricornutum*. *Plant Physiol* 137(2):500–513.

Mus, F., A. Dubini, M. Seibert, M. C. Posewitz, and A. R. Grossman. 2007. Anaerobic acclimation in *Chlamydomonas reinhardtii*: Anoxic gene expression, hydrogenase induction, and metabolic pathways. *J Biol Chem* 282(35):25475–25486.

Mutalik, V. K., J. C. Guimaraes, G. Cambray et al. 2013. Precise and reliable gene expression via standard transcription and translation initiation elements. *Nat Methods* 10(4):354–360.

Nguyen, H. M., M. Baudet, S. Cuine et al. 2011. Proteomic profiling of oil bodies isolated from the unicellular green microalga *Chlamydomonas reinhardtii*: With focus on proteins involved in lipid metabolism. *Proteomics* 11(21):4266–4273.

Nie, L., G. Wu, and W. Zhang. 2008. Statistical application and challenges in global gel-free proteomic analysis by mass spectrometry. *Crit Rev Biotechnol* 28(4):297–307.

Nojima, D., T. Yoshino, Y. Maeda, M. Tanaka, M. Nemoto, and T. Tanaka. 2013. Proteomics analysis of oil body-associated proteins in the oleaginous diatom. *J Proteome Res* 12(11):5293–5301.

Ozsolak, F., A. R. Platt, D. R. Jones et al. 2009. Direct RNA sequencing. *Nature* 461(7265):814–818.

Palenik, B., J. Grimwood, A. Aerts et al. 2007. The tiny eukaryote *Ostreococcus* provides genomic insights into the paradox of plankton speciation. *Proc Natl Acad Sci USA* 104(18):7705–7710.

Parker, M. S., T. Mock, and E. V. Armbrust. 2008. Genomic insights into marine microalgae. *Annu Rev Genet* 42:619–645.

Raamsdonk, L. M., B. Teusink, D. Broadhurst et al. 2001. A functional genomics strategy that uses metabolome data to reveal the phenotype of silent mutations. *Nat Biotechnol* 19(1):45–50.

Read, B. A., J. Kegel, M. J. Klute et al. 2013. Pan genome of the phytoplankton *Emiliania* underpins its global distribution. *Nature* 499(7457):209–213.

Renberg, L., A. I. Johansson, T. Shutova et al. 2010. A metabolomic approach to study major metabolite changes during acclimation to limiting CO_2 in *Chlamydomonas reinhardtii*. *Plant Physiol* 154(1):187–196.

Rismani-Yazdi, H., B. Z. Haznedaroglu, K. Bibby, and J. Peccia. 2011. Transcriptome sequencing and annotation of the microalgae *Dunaliella tertiolecta*: Pathway description and gene discovery for production of next-generation biofuels. *BMC Genomics* 12:148.

Rismani-Yazdi, H., B. Z. Haznedaroglu, C. Hsin, and J. Peccia. 2012. Transcriptomic analysis of the oleaginous microalga *Neochloris oleoabundans* reveals metabolic insights into triacylglyceride accumulation. *Biotechnol Biofuels* 5(1):74.

Scala, S., N. Carels, A. Falciatore, M. L. Chiusano, and C. Bowler. 2002. Genome properties of the diatom *Phaeodactylum tricornutum*. *Plant Physiol* 129(3):993–1002.

Shanmugam, H., R. Rathinam, R. Ramanujam, K. Annamalai Yogesh, and S. C. Isabel. 2013. Microalgal omics and their applications. In *OMICS: Applications in Biomedical, Agricultural, and Environmental Sciences*, D. Barh, V. Zambare, and V. Azevedo (eds.). pp. 439–450. Hoboken, NJ: CRC Press.

Stedtfeld, R. D., S. W. Baushke, D. M. Tourlousse et al. 2008. Development and experimental validation of a predictive threshold cycle equation for quantification of virulence and marker genes by high-throughput nanoliter-volume PCR on the OpenArray platform. *Appl Environ Microbiol* 74(12):3831–3838.

Tang, J. 2011. Microbial metabolomics. *Curr Genomics* 12(6):391–403.

Trentacoste, E. M., R. P. Shrestha, S. R. Smith et al. 2013. Metabolic engineering of lipid catabolism increases microalgal lipid accumulation without compromising growth. *Proc Natl Acad Sci USA* 110(49):19748–19753.

Vidoudez, C. and G. Pohnert. 2012. Comparative metabolomics of the diatom *Skeletonema marinoi* in different growth phases. *Metabolomics* 8(4):654–669.

Wienkoop, S., J. Weiss, P. May et al. 2010. Targeted proteomics for *Chlamydomonas reinhardtii* combined with rapid subcellular protein fractionation, metabolomics and metabolic flux analyses. *Mol Biosyst* 6(6):1018–1031.

Winck, F. V., D. O. Paez Melo, and A. F. Gonzalez Barrios. 2013. Carbon acquisition and accumulation in microalgae *Chlamydomonas*: Insights from "omics" approaches. *J Proteomics* 94:207–218.

Worden, A. Z., J. H. Lee, T. Mock et al. 2009. Green evolution and dynamic adaptations revealed by genomes of the marine picoeukaryotes *Micromonas*. *Science* 324(5924):268–272.

Yan, W., D. Hwang, and R. Aebersold. 2008. Quantitative proteomic analysis to profile dynamic changes in the spatial distribution of cellular proteins. *Methods Mol Biol* 432:389–401.

Yang, S., M. Guarnieri, S. Smolinski, M. Ghirardi, and P. Pienkos. 2013. *De novo* transcriptomic analysis of hydrogen production in the green alga *Chlamydomonas moewusii* through RNA-Seq. *Biotechnol Biofuels* 6(1):118.

Yang, Z.-K., Y.-H. Ma, J.-W. Zheng, W.-D. Yang, J.-S. Liu, and H.-Y. Li. 2014. Proteomics to reveal metabolic network shifts towards lipid accumulation following nitrogen deprivation in the diatom *Phaeodactylum tricornutum*. *J Appl Phycol* 26(1):73–82.

Ye, R. W., T. Wang, L. Bedzyk, and K. M. Croker. 2001. Applications of DNA microarrays in microbial systems. *J Microbiol Methods* 47(3):257–272.

Zhang, W., F. Li, and L. Nie. 2010. Integrating multiple "omics" analysis for microbial biology: Application and methodologies. *Microbiology* 156(Pt. 2):287–301.

Zheng, M., J. Tian, G. Yang et al. 2013. Transcriptome sequencing, annotation and expression analysis of *Nannochloropsis* sp. at different growth phases. *Gene* 523(2):117–121.

9

Systems Biology and Metabolic Engineering of Marine Algae and Cyanobacteria for Biofuel Production

Trunil Desai, Vaishali Dutt, and Shireesh Srivastava

CONTENTS

9.1 Introduction

Our growing appetite for energy has made us deeply reliant on fossil fuels. Any sign of uncertainty in supply of petroleum products, whether due to production or supply, leads to an escalation in the price of this commodity, which increases overall inflation. In fact, the prices of crude petroleum have increased about 267% in the last 10 years (Dahiya et al., 2011) and have a direct effect on inflation. There is also awareness that petroleum products are nonrenewable sources with dwindling supplies. Additionally, burning of fossil fuels is a major cause of increment in atmospheric CO_2 levels. CO_2 is a greenhouse gas and a possible cause of global warming. Therefore, alternative, renewable sources of energy are being sought. While wind and solar energies are renewable and nonpolluting, they are geographically constrained and require high initial capital investment.

Biofuels, which are fuel analogs derived from biological sources, are other renewable sources of energy. Ethanol derived from fermentation of excess molasses was a first-generation biofuel, which was added to petrol (gasoline). This technology found commercial success in Brazil, which is the largest producer of sugarcane and has surplus sugarcane. Similarly, there were also efforts in the United States to make bioethanol from (excess) corn. For other countries, there are not enough extra sugar-rich crops to produce ethanol from. Additionally, ethanol being hygroscopic was found to be corrosive. Thus, there was a need to make other biofuels with improved characteristics. Another first-generation biofuel was biodiesel prepared by transesterification of vegetable oils with simple alcohols. However, spent vegetable oils contain many oxidized lipids, which produce less efficient fuel. Additionally, the viscosity of biodiesel is higher than that of diesel. Because vegetable oils, sugar, and corn are used as food, diverting these to produce biofuel will lead to reduced availability and increased prices of food. These constraints led to research on second-generation biofuels, which were biofuels that did not compete with food, such as cellulosic ethanol and biodiesel produced from nonfood oil plants such as Jatropha. A few demonstration and pilot plants have come up that produce cellulosic ethanol, though the price of cellulosic ethanol is higher than that of corn-based ethanol. Similarly, the yields of nonfood crops such as Jatropha are not high enough to be commercially viable.

Recently, there has been a significant interest in converting carbon dioxide directly into biofuels using fast-growing photosynthetic organisms such as cyanobacteria and algae as biological hosts (Lan and Liao, 2011; Liu et al., 2011a,b). The known fastest-growing algae and cyanobacteria have doubling times as low as 3.5 and 4 h, respectively (Hu et al., 2008; Sakamoto et al., 1998). Benefits of marine algae and cyanobacteria for biofuel production are that they have simple media requirements and do not compete for resources with food crops such as arable land and freshwater. Additionally, biofuels derived through carbon fixation are considered carbon neutral, that is, their use does not add CO_2 to the environment. In this chapter, cyanobacteria are the major focus, though we present some examples of studies employing algae. The freshwater cyanobacterial species employed in most recent studies are *Synechocystis* sp. PCC 6803 (hereafter *Syn* 6803) or *Synechococcus elongatus* PCC 7942 (hereafter *Syn* 7942), while the marine species used is *Synechococcus* sp. PCC 7002 (hereafter *Syn* 7002). All of these three cyanobacterial species are unicellular, have fast growth rates, and are amenable to genetic manipulations, making them suitable for metabolic engineering for diverse products.

In order to use organisms for commercially viable production of biofuels, a clear understanding of metabolic processes responsible for their growth and synthesis efficiency is needed (Rupprecht, 2009). A system-level understanding of microbial metabolism, rather than analysis of isolated pathways of biofuel synthesis, is necessary for this purpose (Mukhopadhyay et al., 2008). Production of the next-generation biofuel molecules such as butanol or triacylglycerols at commercially viable levels would require nonintuitive genetic and metabolic manipulations (Schmidt et al., 2010). In silico metabolic modeling of potential biofuel-producing organisms can help predict such nonintuitive gene targets for production of biofuel or other desired metabolic products. This approach presents an efficient alternative to the tortuous traditional method of strain improvement, which includes several rounds of mutagenesis, selection, mating, and hybridization (Nevoigt, 2008). Below, we present one of the most popular techniques of in silico metabolic modeling—flux balance analysis (FBA). A primary requirement of FBA is knowledge of metabolic reactions undergoing in the cell. While analyses could also be performed with core metabolic models, the best utility of FBA is in its ability to handle genome scale metabolic models (GSMMs) and provide reasonable predictions of phenotype under different environmental conditions and gene knockouts. Therefore, we first discuss the steps needed to create a GSMM from genome sequence and how to apply FBA to GSMMs. We discuss some published examples of applications of FBA in understanding cyanobacterial metabolism and cyanobacterial responses to various conditions. Later, we provide examples of metabolic engineering of cyanobacteria for production of biofuels and related compounds.

9.1.1 Reconstruction of GSMM

A GSMM represents all the metabolic reactions occurring in an organism and also provides gene, protein, and reaction associations for all annotated genes. GSMMs provide means to explore metabolic potential of organisms on a system level. For example, GSMM of *Escherichia coli* has been used to identify gene knockouts for improved 1,4-butanediol production (Yim et al., 2011). The engineered strain is now being used for commercial production of 1,4-butanediol. Genome sequence of the organism of interest serves as the starting point for model reconstruction. Genome sequences of various algae and cyanobacteria are available (e.g., *Chlamydomonas reinhardtii* [Merchant et al., 2007], *Acaryochloris marina* [Swingley et al., 2008], *Synechocystis* sp. PCC 6803 [Kaneko et al., 1996, 2003]). We will explain the reconstruction of GSMM based on methods used for reconstruction of *Syn* 6803 GSMM (Montagud et al., 2010). The process is outlined in Figure 9.1. The genome is searched for open reading frames (ORFs) and plausible protein-coding sequences using sequence comparison software like BLAST. This process is called annotation. Based on the annotation of the genome sequence, an organism-specific database of genes, enzymes, proteins, and metabolites is prepared using software such as *Pathway Tools* (Karp et al., 2002).

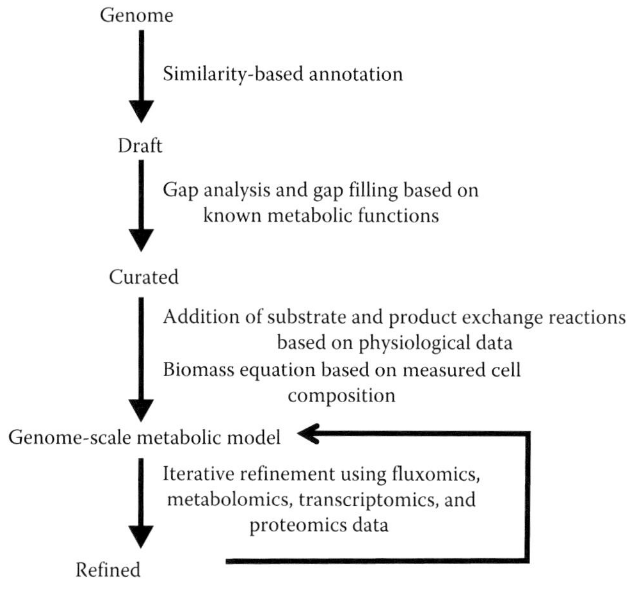

FIGURE 9.1
Iterative steps involved in construction of genome scale metabolic models.

The enzyme commission (EC) numbers of reactions and stoichiometry are checked with the help of enzyme nomenclature database (Bairoch, 2000) and KEGG pathway database (Kanehisa et al., 2008). Reversibility of reactions is checked in specific databases such as BRENDA. In case of lack of information, the reactions are assumed to be reversible.

In a network constructed through the method earlier, synthetic pathways for some metabolites are usually missing. These metabolites form gaps in the network. Reactions without any gene associations are added to provide for these metabolites. Additional reactions for exchange of metabolite with the environment and between compartments are included, and reactions for consumption of metabolites like ATP in cellular maintenance are also added. A biomass formation reaction, based on experimentally measured biomass composition (amount of proteins, carbohydrates, lipids, DNA, RNA, etc., in a given dry mass of cells) of the organism, is included so as to reflect the flux going toward growth. For detailed procedure of construction of GSMM, refer to Feist et al. (2009) and Thiele and Palsson (2010). The network thus constructed needs to be converted into a computable format. Systems Biology Markup Language (SBML) is one of the standard formats used to represent GSMMs (Hucka et al., 2003). The final step is evaluation and refinement of the network by comparison with experimental data. Metabolomics and transcriptomic data, if available or generated, provide additional information for addition or validation of the network and also adding constraints on reactions depending on growth conditions. Metabolomics data for *Syn* 7002 and *Syn* 7942 are available (Baran et al., 2010; Schwarz et al., 2011). Iterations of refinement and evaluation steps improve accuracy of the model. Bernhardt Palsson's group has developed a *constraint-based reconstruction and analysis (COBRA)* toolbox (Becker et al., 2007; Schellenberger et al., 2011), which runs on MATLAB platform and has many features such as identification as well as filling up of gaps, performing FBA, as well as simulating the effects of knockouts.

GSMM of some algae and cyanobacteria has been constructed. Shastri and Morgan (2005) developed the first metabolic model of a cyanobacterium, that of *Syn* 6803, and applied the modeling approach to photosynthetic organisms. Boyle and Morgan (2009) constructed GSMM of *C. reinhardtii*, the first algal model and the first GSMM with three metabolic compartments (cytoplasm, chloroplast, and mitochondria). In recent years, improved GSMMs of *Syn* 6803 (Knoop et al., 2010, 2013; Nogales et al. 2012; Saha et al., 2012), *Syn* 7002 (Hamilton and Reed, 2012), *Cyanothece* (Hamilton and Reed, 2012), and *C. reinhardtii* (Dal'Molin et al., 2011) have been constructed.

9.1.2 Metabolic Structural Modeling

Metabolic models can be kinetic or nonkinetic. Kinetic models require information about kinetic parameters (V_{max}, K_m, etc.) of enzymes and isozymes involved in pathways as well as concentrations of metabolites, which are often not available, especially for algal and cyanobacterial species. Generating such information is fairly tedious. Even if one generates such information, it is not static and varies with culture conditions. An alternative to kinetic models are structural models that are based on the stoichiometric information of metabolic network and can be conducted in the absence of detailed information of the kinetic parameters. Due to their inherent simplicity, one can solve a much larger system of equations. Therefore, most of the studies using GSMMs have employed the simpler stoichiometric/structural models. Stoichiometric models are capable of quantitatively predicting growth, product formation, and intracellular flux distributions. These predictions vary depending on the physiological constraints included during model setup.

9.1.3 Optimization of the GSMM

The GSMM once prepared can be used to predict possible metabolic routes (flux distribution) for growth and/or production of one or more metabolites. In a typical GSMM, the number of metabolites is less than the number of reactions. This forms an underdetermined system of linear equations, which can have many possible solutions. Therefore, optimization techniques are employed to solve this system of equation. The objective criterion chosen varies with the conditions used. Typically, maximization of biomass is used as objective function, which may be a valid objective under conditions of exponential growth. Some of the other objective functions used are minimization of total (moles or mass) intake of nutrients, minimization of ATP consumption, or minimization of nicotinamide adenine dinucleotide (reduced) (NADH) consumption (Savinell and Palsson, 1992). One of the widely used methods based on optimization is FBA (Orth et al., 2010) through which fluxes can be predicted based on stoichiometry of reactions, biomass composition, and limits on uptake/excretion rates of metabolites. FBA assumes a steady-state mass balance on intracellular metabolites, that is, the rate of formation of any intracellular metabolite is equal to its rate of consumption. The model thus forms a system of linear equations that can be represented mathematically as follows:

$$S \cdot v = 0$$

where

S is the stoichiometric matrix with rows representing the metabolites and columns representing the reactions

v is the vector of length equal to the number of reactions whose elements represent flux through reactions

Accuracy of FBA predictions depends on the quality of network reconstruction and availability of information about regulatory constraints. For example, if a reaction is experimentally determined to be blocked in aerobic conditions, the flux through that reaction can be constrained to zero while simulating growth under aerobic conditions.

9.1.4 Understanding Photosynthesis

Photosystems are the doorways for entry of solar energy into biological system. Increasing the efficiency of conversion of solar energy into biochemical energy is the most crucial step toward cost-effective production of desired metabolic products. Therefore, most modeling studies with cyanobacteria (and algae) focus on understanding photosynthesis.

Photosynthetic organisms show diurnal cycle regulation. Modeling the effect of light on reaction fluxes is a challenge. Recently developed cyanobacterial models incorporate diurnal cycle regulation (Knoop et al., 2013; Saha et al., 2012). Knoop et al. (2013) modeled one diurnal cycle of synchronized cells using dynamic FBA (Mahadevan et al., 2002) to calculate growth (biomass accumulation) with time. To mimic known variations in biomass components with time, the biomass components were represented in the biomass objective function (biomass equation) by a factor that varied during the diurnal cycle. For example, the factor for pigments increased 2 h before sunrise and reduced after noon. Biomass growth was optimized in both light as well as dark phases. In dark phase, stored glycogen utilization was set to be only slightly higher than the requirement for cellular maintenance.

Nogales et al. (2012) modeled alternate electron flow (AEF) pathways and showed that they act cooperatively to maintain the ATP/NADPH ratio according to metabolic demand. It has been shown recently in multiple studies employing FBA (Knoop et al., 2010, 2013; Nogales et al., 2012) and experimental gene knockouts (Eisenhut et al., 2008) that photorespiration is essential for survival of cyanobacteria. Optimizations using biomass maximization as the objective function also showed that the oxygenase efficiency of RuBisCO increased with the intensity of light (Knoop et al., 2010, 2013). The reason for this observation was the absence of stoichiometrically efficient pathways for producing amino acids serine, glycine, and cysteine in the model for *Syn* 6803. *Syn* 6803 is not yet known to have homologues for phosphoserine transaminase (EC 2.6.1.52) and phosphoserine phosphatase (EC 3.1.3.3). When simulations were run after adding these reactions for synthesis of serine, the photorespiration flux went to zero under autotrophic growth (Knoop et al., 2013). Glycine and cysteine were produced from the newly introduced pathway. This was consistent with experimental results that showed 13C enrichment of serine was substantially higher than glycine during transient labeling (Young et al., 2011). Indications of the presence of phosphoserine pathway and requirement of photorespiratory flux for growth suggests the possibility of as yet undiscovered gene homologues in *Syn* 6803.

9.1.5 Studies Employing Structural Metabolic Modeling for Producing Biofuels

Despite the availability of algal and cyanobacterial GSMMs, few studies have been done till date where the GSMMs are actually used for the prediction of metabolic capabilities of the organisms to produce biofuels. One such study models the stoichiometric potential of *Syn* 6803 for the production of ethanol, propane, ethylene, butanol, octadecanoic acid, heptadecane, and octadecanol (Kamarainen et al., 2012). The demand for NADP and ATP synthesis and carbon fixation reaction generally increases with increased chain length biofuel. However, the energy yield per mole of photon uptake is more or less constant (52.8–57.1 kJ/mol photon) for all biofuels except ethylene (23.1 kJ/mol photon) as the synthesis pathway of ethylene involves loss of carbon as carbon dioxide. Trade-offs are observed between synthesis of product and growth, meaning the production of biofuel would reduce the growth rate of the organism.

C. reinhardtii is a model alga. It is considered as a potential host organism for producing hydrogen (Dal'Molin et al., 2010). It is the only algal species whose GSMM (AlgaGEM) is available. This GSMM derives information from its genome sequence and compartmentalization data from Arabidopsis GSMM, AraGEM (Dal'Molin et al., 2010). AlgaGEM successfully predicted important pathways during anaerobic growth and predicted the effects of known mutations on hydrogen production. *C. reinhardtii* mutant, *stm6*, has impaired cyclic electron flow. It is known to produce 5–13-fold more hydrogen in mixotrophic conditions compared to wild type (Kruse et al., 2005). AlgaGEM, when optimized for production of hydrogen under mixotrophic conditions with growth rate fixed to wild-type growth rate and blocked cyclic electron flow, predicted fivefold increase in hydrogen production as compared to wild type.

9.1.6 Application of FBA in Marine Cyanobacteria

Although GSMMs of a few freshwater photosynthetic organisms have been constructed, very few marine algae and cyanobacteria are studied through this aspect. A GSMM (iSyp611) of marine cyanobacteria *Syn* 7002 is available. This model was used to identify metabolic differences between two cyanobacteria, namely, between *Syn* 7002 and *Cyanothece* 51142 (iCce806) (Hamilton and Reed, 2012). Gene essentiality was checked in both the models. This is done by blocking all reactions associated with one gene (or two genes in case of double gene deletion study) and optimizing for maximum biomass production. Gene deletions for which the growth rate is lower than a specified threshold were considered lethal. Hamilton and Reed found seven unique gene deletion sets lethal only in one of the cyanobacteria even though the genes were present in both of these cyanobacteria.

Production of ethanol, ethylene, 2-methyl-1-butanol, 3-methyl-1-butanol, and 1-butanol was simulated using GSMM of *Syn* 7002 (Vu et al., 2013). Theoretical yields of the chemicals were similar to yields previously obtained using GSMM of *Syn* 6803 as described earlier (Kamarainen et al., 2012). Three methods, minimization of metabolic adjustments (MOMA) (Segre et al., 2002), OptORF (Kim and Reed, 2010), and relative change (RELATCH) (Kim and Reed, 2012), were employed to identify gene deletion strategies for production of these biofuels. All these methods simulate adaptation of the mutant organisms with selection pressure to produce specific chemicals. These methods differ in calculation of fluxes in mutants with respect to flux distribution in wild-type organism's GSMM. The authors found that the mutations suggested by different algorithm were different and in some cases were exclusive. For example, MOMA and OptORF suggested that mutations in one of the NADH dehydrogenase (A0195, A0196, or A0197) would increase 1-butanol production, while RELATCH suggested single gene deletions in chlorophyll and heme biosynthesis pathways (A0707, A1023, or A2508) would increase 1-butanol synthesis. Experimental testing of these mutually exclusive sets of gene deletions predicted by the different algorithms will help identify the better one. Nonetheless, these simulations identify somewhat unintuitive targets and reduce the number of targets for further validation to the six listed earlier.

Optimization studies done using GSMMs suggest that metabolic models can be used to design strategies for improved production of biofuels. However, two major challenges remain. First, engineering light harvesting and carbon fixing complexes so as to increase the efficiency of conversion of light energy into biochemical energy and productivity; and second, increasing tolerance of host cyanobacteria/algae to higher biofuel concentrations.

9.2 Metabolic Engineering of Cyanobacteria for Biofuel Production

Considering the advantages offered by cyanobacteria for biofuel production, many researchers have employed metabolic engineering to introduce heterologous enzymes and pathways in order to make diverse products, many of which are not naturally produced in these organisms. To synthesize these biofuels and biochemicals, heterologous pathways from heterotrophic high producers were cloned into model cyanobacterial species *Syn* 7942, *Syn* 6803, or *Syn* 7002. Herein, we present these studies and their major findings. Figure 9.2 shows the pathways that have been engineered to produce different biofuels in cyanobacteria.

9.2.1 Natural Transformation and Homologous Recombination in Cyanobacteria

The unicellular freshwater species *Syn* 7942 and *Syn* 6803 as well as the marine species *Syn* 7002 all have been shown to be naturally transformable. Of the three transformation techniques studied in *Syn* 6803, electroporation, ultrasonic transformation, and natural

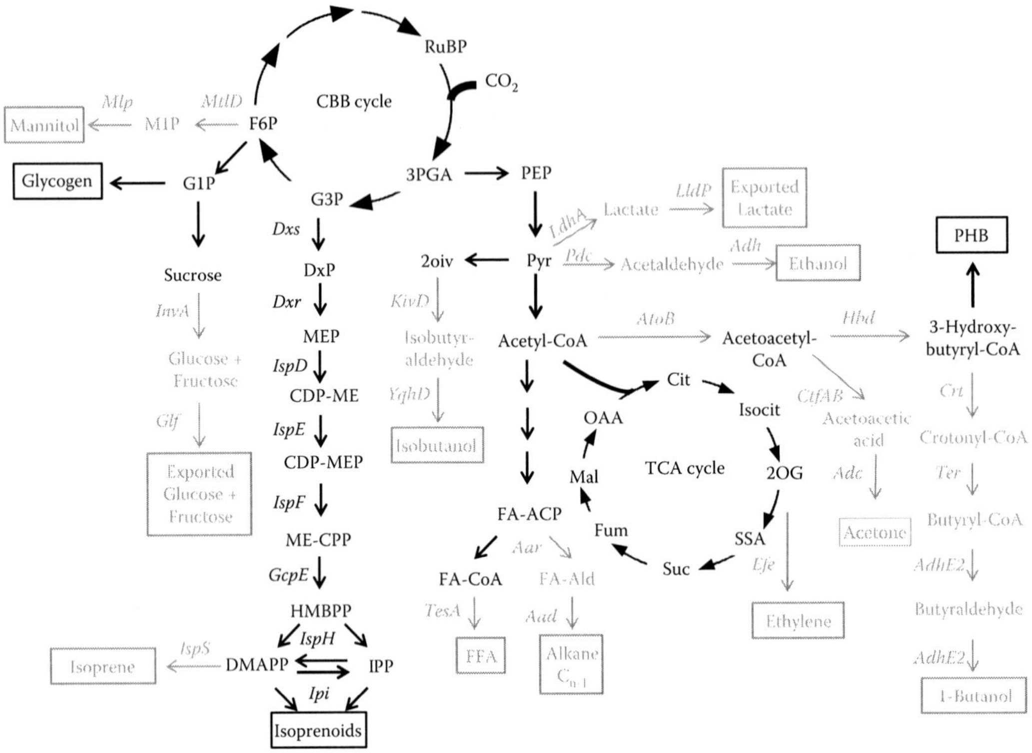

FIGURE 9.2
Metabolic pathways to biofuels and biochemicals from cyanobacteria. *Abbreviations for pathways*: CBB cycle, Calvin–Benson–Bassham cycle; TCA cycle, tricarboxylic acid cycle. *Abbreviations for metabolites*: RuBP, ribulose-1,5-P2; CO_2, carbon dioxide; 3PGA, 3-phosphoglycerate; PEP, phosphoenolpyruvate; Pyr, pyruvate; OAA, oxaloacetate; Cit, citrate; Isocit, isocitrate; OG, oxaloketoglutarate; SSA, succinic semialdehyde; Suc, succinate; Fum, fumarate; Mal, malate; G3P, glyceraldehyde-3-phosphate; DxP, deoxyxylulose-5-phosphate; MEP, methylerythritol-4-phosphate; CDP-ME, diphosphocytidyl methylerythritol; CDP-MEP, CDP-ME-2-phosphate; ME-CPP, methylerythritol-2,4-cyclodiphosphate; HMBPP, hydroxymethylbutenyl diphosphate; IPP, isopentenyl diphosphate; DMAPP, dimethylallyl diphosphate; F6P, fructose-6-phosphate; G1P, glucose-1-phosphate; M1P, mannitol-1-phosphate; FA-ACP, fatty acyl-ACP; FA-CoA, fatty acyl-CoA; FA-Ald, fatty acyl aldehyde; and PHB, polyhydroxybutyrate. *Abbreviations for enzymes*: *Efe*, ethylene-forming enzyme (*Pseudomonas syringae pv. phaseolicola PK2); IspS*, isoprene synthase (*Pueraria montana*); *CtfAB*, coenzyme A transferase; and *Adc*, acetoacetate decarboxylase (*C. acetobutylicum DSM* 1731); *LdhA*, lactate dehydrogenase, and *LldP*, lactate transporter (*E. coli DH5α*); *InvA*, invertase, and *Glf*, glucose or fructose transporter (*Z. mobilis*); *MtlD*, mannitol-1-phosphate dehydrogenase (*E. coli*); *Mlp*, mannitol-1-phosphatase (*E. tenella*); *Pdc*, pyruvate decarboxylase, and *Adh*, alcohol dehydrogenase (*Z. mobilis*); *KivD*, oxoacid decarboxylase (*Lactococcus lactis*); *YqhD*, NADPH-dependent alcohol dehydrogenase (*E. coli*); *AtoB*, acetyl-CoA acetyltransferase (*E. coli*); *Hbd*, 3-hydroxybutyryl-CoA dehydrogenase; *Crt*, crotonase, and *AdhE2*, bifunctional aldehyde/alcohol dehydrogenase (*C. acetobutylicum*); *Ter*, trans-2-enoyl-CoA reductase (*Treponema denticola*); *TesA*, Thioesterase (*E. coli*); *Aar*, Acyl-ACP Reductase; *Aad*, Aldehyde Decarbonylase. *Note:* Enzymes from heterologous sources, the associated metabolites and products are shown in gray.

transformation, natural transformation showed the highest efficiency (Zang et al., 2007). However, in the thermophilic marine cyanobacteria *Thermosynechococcus elongatus* BP-1, electroporation gave higher transformation efficiency compared to natural transformation (Rosgaard et al., 2012). The external DNA is incorporated into the genome through homologous recombination, which produces stable transformants. Golden et al. (1987) have described methods for direct genetic engineering of the cyanobacterial chromosome. A typical method employs flanking the heterologous sequences by cyanobacterial DNA on either side. The cyanobacterial DNA on either side helps in inserting the heterologous sequences at desired sites in the genome. For good expression of a heterologous gene, the choice of promoter is critical. Since the structure of cyanobacterial holopolymerase is different from that found in most bacteria, a systematic investigation is required as to which promoters would work the best in the given cyanobacterial system (Heidorn et al., 2011). The cyanobacterial promoters commonly used are of genes involved in photosynthesis like $P_{rbc\,L}$, which is the promoter of large subunit of RuBisCO, P_{psbA1}, and P_{psbA2} whose genes encode for proteins of photosystem II.

Unlike the unicellular cyanobacteria, multicellular filamentous cyanobacteria are difficult to transform, which has hindered their biotechnological applications. Research is underway to identify efficient transformation strategies in order to utilize these organisms.

9.2.2 Production of Biofuels in Cyanobacteria

9.2.2.1 Production of Shorter-Chain (C2 and C4) Biofuels

Photosynthetic production of up to 230 mg/L ethanol has been reported using genetically engineered *Syn* 7942, in which *Zymomonas mobilis pdc–adh* operon was expressed under a *Plac* and *Prbc* promoters via a shuttle vector pCB4 (Deng and Coleman, 1999). Recently, the *pdc–adh* expression cassette was integrated into the chromosome of *Syn* 6803 at the psbA2 locus. Driven by the strong psbA2 promoter, the expression of *pdc–adh* resulted in ~550 mg/L ethanol production by the engineered *Syn* 6803 under high light (~1000 μE/m^2/s) conditions (Dexter and Fu, 2009).

Compared to ethanol, isobutanol and 1-butanol have higher energy densities (30 MJ/kg for ethanol vs. 36.1 MJ/kg for (iso)butanol). Additionally, (iso)butanol is much less hygroscopic compared to ethanol. (Iso)butanol is, therefore, a better gasoline substitute (which has an energy density of approximately 46 MJ/kg). Photosynthetic production of 1-butanol in oxygenic cyanobacteria by expression of *C. acetobutylicum* 1-butanol synthesis pathway has proven difficult because of the intrinsic oxygen sensitivity and NADH dependence of the *C. acetobutylicum* pathway. This is in conflict with the photo-oxygenesis and the lack of NADH cofactors in cyanobacteria (Lan and Liao, 2011). While only ~1 mg/L 1-butanol was detected after 2 weeks cultivation under photosynthetic condition, cultivation in anoxic condition led to accumulation of 1-butanol up to a titer of 14.5 mg/L (Lan and Liao, 2011). In 2012, Lan and Liao reported 1-butanol production up to ~30 mg/L under the same photosynthetic condition by introducing several improvements. The strain had acetoacetyl-CoA synthase from *Streptomyces* sp. strain CL190 overexpressed in it. This enzyme, in an ATP-driven reaction, condenses malonyl-CoA and acetyl-CoA leading to the production of acetoacetyl-CoA. Also, AdhE2 in this strain was replaced with separate aldehyde and alcohol dehydrogenases, NADPH-dependent butyraldehyde dehydrogenase from *Clostridium saccharoperbutylacetonicum* N1-4 and NADPH-dependent alcohol dehydrogenase from *E. coli*, respectively (Lan and Liao, 2012). To photosynthetically produce isobutanol, an artificial isobutanol biosynthesis pathway was introduced into *Syn* 7942. Productivities were increased by overexpression of RuBisCO and in situ product recovery due to high vapor pressure of isobutyraldehyde. The engineered strain was able to photosynthetically produce isobutyraldehyde and isobutanol at titers of 1100 and 450 mg/L, respectively, under optimal conditions (Atsumi et al., 2009). The authors claim that these levels are commercially viable.

9.2.2.2 Production of Longer-Chain Biofuel-Related Molecules

Recently, *Syn* 6803 strains were engineered to produce and secrete FFAs to up to 197 mg/L at a cell density of 1.0 × 10^9 cells/mL (Liu et al., 2011b). The acetyl-CoA carboxylase (ACC) was overexpressed to drive the metabolic flux toward FFAs. The fatty acid activation gene *aas* (*slr1609*) was deleted to inactivate the FFAs degradation. Poly-β-hydroxybutyrate (PHB) synthesis genes (*slr1993* and *slr1994*) and the phosphotransacetylase gene *pta* (*slr2132*) were deleted. These deletions blocked the competitive pathways of FFA metabolism. Overexpression of thioesterase from *E. coli*, coupled with the aforementioned deletions, led to increased FFA production and secretion. While FFA themselves are not direct biofuels, they can be easily esterified with alcohol to produce biodiesel.

In order to produce long-chain alcohols, fatty acyl-CoA reductases from different sources were heterologously expressed in *Syn* 6803 and the resultant strains achieved production of fatty alcohols, including hexadecanol (C16) and octadecanol (C18) up to a titer of about 0.2 mg/L (Tan et al., 2011).

The lipid content of *Syn* 7002 was increased by antisense expression of PEPC-coding gene (*ppc*) (Song et al., 2008).

The alkane or alkene biosynthetic pathway has two successive biochemical reactions catalyzed by an acyl-ACP reductase and an aldehyde decarbonylase, respectively (Schirmer et al., 2010). Heterologous expression of acyl-ACP reductase and aldehyde decarbonylase genes from *Syn* 7942 in *Syn* 7002 resulted in *n*-alkane levels of up to 5% of the dry cell weight in the engineered cells (Reppas and Ridley, 2010). In *Syn* 6803, a total hydrocarbon (including pentadecane, hexadecane, heptadecane, 8-heptadecene, and 8-methyl heptadecane) yield of up to 150 μg/L/O.D.730 nm has been reported (Tan et al., 2011).

9.2.2.3 Production of Hydrogen

Many cyanobacteria naturally produce hydrogen. It is produced as a secondary metabolite to balance the redox state. Photosynthetic *Rhodospirillum rubrum* produces 4, 7, and 6 mol of H_2 per mol of acetate, succinate, and malate, respectively (Nandi and Sengupta, 1998). Cyanobacteria, namely, *Anabaena*, *Synechococcus*, and *Oscillatoria* sp., have been studied for photoproduction of H_2. Immobilized *Anabaena cylindrica* produced H_2 (20 mL/g dry wt/h) continually for 1 year (Nandi and Sengupta, 1998). *Cyanothece* sp. Miami BG 043511 has both dark- and light-induced H_2 effluxes. Dark, anoxic H_2 production occurs via hydrogenase utilizing reductant from glycolytic catabolism of carbohydrate (autofermentation). Photo-H_2 production occurs via nitrogenase and requires illumination of PSI. The production of O_2 by co-illumination of PSII is inhibitory to the nitrogenase above a threshold pO_2 (Skizim et al., 2011). In order to reduce the redox ratio (NAD^+/NADH) to drive the H_2 production, an ldhA mutant of the cyanobacterium *Syn* 7002 was constructed. It lacked the enzyme for the NADH-dependent reduction of pyruvate to L-lactate. The measured intracellular $NAD(P)^+$/NAD(P)H ratio in the ldhA-strain decreased appreciably during autofermentation. This was associated with an increase of up to fivefold in H_2 production via an NADH-dependent, bidirectional (NiFe) hydrogenase (McNeely et al., 2010).

9.2.3 Production of Other Compounds in Cyanobacteria

Ethylene is important to the synthetic chemical industry and its production almost exclusively relies on petroleum. The *efe* gene from *Pseudomonas syringae pv. phaseolicola* PK2, which encodes for the ethylene-forming enzyme (Efe), was introduced into *Syn* 7942 using the pUC303 shuttle vector. In pUC303-EFE03, the *efe* gene had its native promoter and terminator (Sakai et al., 1997). The amount of carbon incorporated into ethylene was 5.84% of the total C fixed in the recombinant cyanobacteria harboring pUC303-EFE03.

Isoprene is a potential biofuel. The *IspS* gene, which encodes for isoprene synthase, was isolated from *P. montana* and integrated into the genome of *Syn* 6803 under the *psbA2* promoter. Isoprene was produced at a rate of ~50 μg/g dry cell/day under high light flux, that is, ~500 $\mu E/m^2/s$ (Lindberg et al., 2010).

Acetone is an important industrial chemical. Recently, *Syn* 6803 was engineered to produce acetone. The engineering involved coexpressing acetoacetate decarboxylase (Adc) and coenzyme A transferase (CtfAB) and deleting PHB polymerase (PhaEC) and phosphotransacetylase (Pta). Acetone was produced at a titer of 36 mg/L in the culture (Zhou et al., 2012).

Cyanobacteria produce PHB naturally, which is one of the storage polymers in some of the species. *Syn* 6803 has been shown to accumulate PHB. The pathway and regulation of PHB in this organism has been studied. Similar to lipid accumulation in algae, the PHB content in cyanobacteria can be increased by nutrient deprivation. Phosphate limitation and acetate addition increased PHB content in *Syn* 6803 to 29% (Panda et al., 2006) or 38% (Panda and Mullick, 2007) by weight of dry cells. These values are about six- to eightfold higher as compared with the accumulation under photoautotrophic growth under nutrient sufficiency (Panda and Mullick, 2007). In addition to manipulation of culture conditions, genetic engineering of cyanobacteria has been tried to improve PHB synthesis. In *Syn* 6803, the mutation of the ADP-glucose pyrophosphorylase (*agp*) gene is linked to higher accumulation of PHB (14.9% of the dry cell weight) compared to the wild type during photoautotrophic growth (Wu et al., 2002). ADP-glucose pyrophosphorylase controls the synthesis of glycogen. Thus, mutation of *agp* gene led to reduced activity of this enzyme, which channeled the C-flux away from glycogen toward PHB synthesis. Introduction of PHB biosynthesis genes from *Ralstonia eutropha* (a heterotrophic producer of PHB that is used to produce PHB commercially) into *Syn* 7942 followed by culture of the cyanobacteria under nitrogen starvation and acetate supplementation led to PHB levels of 25.6% of the dry cell weight (Takahashi et al., 1998). Therefore, until now, genetic engineering approach has not yielded greater concentration of PHB compared to manipulation of culture conditions.

Lactic acid is a chemical, which is important to a variety of industries including biodegradable polyester and food. In *E. coli* cells, lactic acid is produced from pyruvate by the action of lactate dehydrogenase (Ldh) whose major isoform is LdhA. Lactic acid is transported out of *E. coli* using the transporter LldP. Production of lactate from pyruvate consumes NADH, which can be replenished by NADPH/NADH transhydrogenase (encoded by *udhA*). When the genes *ldhA*, *lldP*, and *udhA* from *E. coli* were heterologously expressed in *Syn* 7942 at the neutral site, lactic acid was produced at a titer of ~56 mg/L under photoautotrophic culture condition (Niederholtmeyer et al., 2010).

When exposed to salt stress, freshwater cyanobacteria accumulate solutes such as glucosylglycerol and sucrose (Hagemann, 2011). Overexpression of *invA*, *glf*, and *galU* genes in *Syn* 7942 resulted in up to 45 mg/L total hexose production (including glucose and fructose) in the culture supplemented with 200 mM NaCl. InvA catalyzes the conversion of sucrose to glucose and fructose; expression of the glucose or fructose transporter GLF (encoded by the *glf* gene) was essential for glucose or fructose secretion. Additional expression of *galU* enhanced the biosynthesis of intracellular precursors and thus further increased the hexose sugar production by over 30% in the culture (Niederholtmeyer et al., 2010).

Marine cyanobacteria have been engineered to produce medium-value compounds like mannitol and omega-3 fatty acids such as eicosapentaenoic acid (EPA). Mannitol biosynthesis from fructose-6-phosphate was introduced in the marine cyanobacterium *Syn* 7002 by expressing genes encoding mannitol-1-phosphate dehydrogenase (*mtlD*) from *E. coli* and codon-optimized mannitol-1-phosphatase (*mlp*) from the protozoan chicken parasite *E. tenella*. The concentration of mannitol in the culture reached 0.5 g/L after approximately 1 week and more than 1 g/L after prolonged incubation (Rosgaard et al., 2012).

EPA biosynthesis gene cluster from the EPA-producing bacterium *Shewanella* sp. SCRC-2738 was expressed in the marine cyanobacterium *Synechococcus* sp. NKBG15041c. A content of 7.5% EPA of the total fatty acids was obtained when the cells were cultured at 23°C and a light intensity of 40 lux. Higher light intensity (of 1000–1500 lux) reduced the content of EPA to 3.7% of the total fatty acids (Yu et al., 2000).

9.3 Future Directions

The initial studies have demonstrated the possibilities of synthesizing useful biofuel and related compounds by genetic engineering of cyanobacteria. As most of these studies were proof-of-concept in nature, not much emphasis was paid to identify the economic feasibility of using cyanobacteria for biofuel production. It is likely that, at this point, many of these products are not commercially viable. Multiple improvements are needed to boost productivity in order to make the processes commercially viable. This would include selecting the most suitable cyanobacteria depending on geographical conditions for cultivation, which affects important parameters such as duration and intensity of sunlight, temperature, and salinity. While initial large-scale cultivations employed ponds and raceways due to their low cost and ease of setup, it is being realized that such a process is sensitive to contamination and provides little control of important parameters. Thus, ponds and raceway systems are likely suitable for extremophiles (particularly organisms growing at normal temperatures but extreme pH). For mesophiles growing at normal pH (6–8), closed photobioreactors (PBRs) are better options. PBRs also provide the possibility for round-the-clock growth by providing artificial light. Significant research is being conducted to improve the light, energy, and gas transfer characteristics of PBRs in order to efficiently cultivate algae and cyanobacteria. A few start-up companies have also come up to utilize genetically modified cyanobacteria for biofuel production. Joule Unlimited and Algenol are two companies that are utilizing engineered cyanobacteria to commercially produce ethanol. Algenol also employs hydrothermal liquefaction to produce biodiesel from cyanobacterial biomass. Matrix Genetics is using engineered cyanobacteria to produce fatty acid aldehydes, which they suggest are flexible precursors to a variety of diesel substitutes (Kaiser et al., 2013). Along with higher productivity, technologies to efficiently (and cheaply) purify the compounds of interest are needed to reduce production costs. Extracellular products obviate the cell-harvesting step, which is a major cost factor. Roy Curtiss's group in Arizona State University is working on *suicidal* cyanobacteria, which will lyse on induction and release contents into medium, making it easier to isolate desired products.

The studies until now, most of which employed freshwater, model cyanobacterial strains, have shown promise in using cyanobacteria for a variety of biofuels. A few start-up companies that are currently working on demonstration-scale projects are claiming to produce biofuels (bioethanol and diesel mimics) at or below the cost of petroleum-based biofuels in the future. The use of marine/estuarine cyanobacteria (some of which have very fast growth rates) will likely further enhance the productivities and provide options to produce biofuels without using freshwater.

References

Atsumi S, Higashide W, Liao JC (2009). Direct photosynthetic recycling of carbon dioxide to isobutyraldehyde. *Nat. Biotechnol.* 27:1177–1180.

Bairoch A (2000). The ENZYME database in 2000. *Nucleic Acids Res.* 28:304–305.

Baran R, Bowen B, Bouskill N, Brodie EL et al. (2010). Metabolite identification in *Synechococcus* sp. PCC 7002 using untargeted stable isotope assisted metabolite profiling. *Anal. Chem.* 82:9034–9042.

Becker SA, Feist AM, Mo ML, Hannum G, Palsson BØ, Herrgard MJ (2007). Quantitative prediction of cellular metabolism with constraint-based models: The COBRA toolbox. *Nat Protoc.* 2(3):727–738.

Boyle N, Morgan J (2009). Flux balance analysis of primary metabolism in *Chlamydomonas reinhardtii*. *BMC Syst. Biol.* 3:4.

Dahiya A, Berwal A, Khan B (2011). Jatropha oil as a feed stock for bio diesel production in India. *Int. J. of Adv. Eng. Technol.* 2(2):218–221.

Dal'Molin CQL, Palfreyman R, Brumbley S et al. (2010). AraGEM, a genome-scale reconstruction of the primary metabolic network in Arabidopsis. *Plant Physiol.* 152:579–589.

Dal'Molin LQ, Palfreyman R, Nielsen L (2011). AlgaGEM—A genome-scale metabolic reconstruction of algae based on the *Chlamydomonas reinhardtii* genome. *BMC Genom.* 12(Suppl. 4):S5.

Deng MD, Coleman JR (1999). Ethanol synthesis by genetic engineering in cyanobacteria. *Appl. Environ. Microbiol.* 65:523–528.

Dexter J, Fu P (2009). Metabolic engineering of cyanobacteria for ethanol production. *Energ. Environ. Sci.* 2:857–864.

Eisenhut M, Ruth W, Haimovich M, Bauwe H et al. (2008). The photorespiratory glycolate metabolism is essential for cyanobacteria and might have been conveyed endosymbiotically to plants. *PNAS* 105:17199–17204.

Feist A, Herrgard M, Thiele I, Reed J et al. (2009). Reconstruction of biochemical networks in microorganisms. *Nat. Rev. Microbiol.* 7:129–143.

Golden S, Brusslan J, Haselkorn R (1987). Genetic engineering of the cyanobacterial chromosome. *Methods Enzymol.* 153:215–231.

Hagemann M (2011). Molecular biology of cyanobacterial salt acclimation. *FEMS Microbiol. Rev.* 35:87–123.

Hamilton J, Reed J (2012). Identification of functional differences in metabolic networks using comparative genomics and constraint-based models. *PLoS One* 7:e34670.

Heidorn T, Camsund D, Huang H, Lindberg P et al. (2011). Synthetic biology in cyanobacteria engineering and analyzing novel functions. *Methods Enzymol.* 497:539–579.

Hu Q, Sommerfeld M, Jarvis E, Ghirardi M et al. (2008). Microalgal triacylglycerols as feedstocks for biofuel production: Perspectives and advances. *Plant J.* 54:621–639.

Hucka M, Finney A, Sauro HM, Bolouri H et al. (2003). The systems biology markup language (SBML): A medium for representation and exchange of biochemical network models. *Bioinformatics* (Oxford, England) 19:524–531.

Kaiser BK, Carleton M, Hickman JW, Miller C et al. (2013). Fatty aldehydes in cyanobacteria are a metabolically flexible precursor for a diversity of biofuel products. *PLoS ONE* 8(3):e58307.

Kamarainen J, Knoop H, Natalie S, Guerrero F et al. (2012). Physiological tolerance and stoichiometric potential of cyanobacteria for hydrocarbon fuel production. *J. Biotechnol.* 162:67–74.

Kanehisa M, Araki M, Goto S, Hattori M et al. (2008). KEGG for linking genomes to life and the environment. *Nucleic Acids Res.* 36:D480–D484.

Kaneko T, Nakamura Y, Sasamoto S, Watanabe A et al. (2003). Structural analysis of four large plasmids harboring in a unicellular cyanobacterium, *Synechocystis* sp. PCC 6803. *DNA Res.* 10:221–228.

Kaneko T, Sato S, Kotani H, Tanaka A et al. (1996). Sequence analysis of the genome of the unicellular cyanobacterium *Synechocystis* sp. strain PCC6803. II. Sequence determination of the entire genome and assignment of potential protein-coding regions (supplement). *DNA Res.* 3:185–209.

Karp D, Paley S, and Romero P (2002). The pathway tools software. *Bioinformatics* 18:S225–S232.

Kim J, Reed J (2010). OptORF: Optimal metabolic and regulatory perturbations for metabolic engineering of microbial strains. *BMC Syst. Biol.* 4:53.

Kim J, Reed J (2012). RELATCH: Relative optimality in metabolic networks explains robust metabolic and regulatory responses to perturbations. *Genome Biol.* 13:R78.

Knoop H, Grundel M, Zilliges Y, Lehmann R et al. (2013). Flux balance analysis of cyanobacterial metabolism: The metabolic network of *Synechocystis* sp. PCC 6803. *PLOS Comp. Biol.* 9:e1003081.

Knoop H, Zilliges Y, Lockau W, Steuer R (2010). The metabolic network of *Synechocystis* sp. PCC 6803: Systemic properties of autotrophic growth. *Plant Physiol.* 154(1):410–422.

Kruse O, Rupprecht J, Bader KP, Thomas-Hall S et al. (2005). Improved photobiological H-2 production in engineered green algal cells. *J. Biol. Chem.* 280(40):34170–34177.

Lan EI, Liao JC (2011). Metabolic engineering of cyanobacteria for 1-butanol production from carbon dioxide. *Metab. Eng.* 13:353–363.

Lan EI, Liao JC (2012). ATP drives direct photosynthetic production of 1-butanol in cyanobacteria. *PNAS* 109:6018–6023.

Lindberg P, Park S, Melis A (2010). Engineering a platform for photosynthetic isoprene production in cyanobacteria, using *Synechocystis* as the model organism. *Metab. Eng.* 12:70–79.

Liu X, Fallona S, Sheng J, Curtiss R (2011a). CO_2-limitation-inducible green recovery of fatty acids from cyanobacterial biomass. *Proc. Natl. Acad. Sci. U.S.A.* 108(17):6905–6908.

Liu X, Sheng J, Curtiss R III (2011b). Fatty acid production in genetically modified cyanobacteria. *Proc. Natl. Acad. Sci. U.S.A.* 108:6899–6904.

Mahadevan R, Edwards J, Doyle F (2002). Dynamic flux balance analysis of diauxic growth in *Escherichia coli*. *Biophys. J.* 83:1331–1340.

McNeely K, Xu Y, Bennette N, Bryant DA et al. (2010). Redirecting reductant flux into hydrogen production via metabolic engineering of fermentative carbon metabolism in a cyanobacterium. *Appl. Environ. Microbiol.* 76:5032–5038.

Merchant SS, Prochnik SE, Vallon O, Harris EH et al. (2007). The *Chlamydomonas* genome reveals the evolution of key animal and plant functions. *Science* 318:245–250.

Montagud A, Zelezniak A, Navarro E, De Cordoba P et al. (2010) Reconstruction and analysis of genome-scale metabolic model of a photosynthetic bacterium. *BMC Syst. Biol.* 4:156.

Mukhopadhyay A, Redding A, Rutherford B, Keasling JD (2008). Importance of systems biology in engineering microbes for biofuel production. *Curr. Opin. Biotechnol.* 19:228–234.

Nandi R, Sengupta S (1998). Microbial production of hydrogen: An overview. *Crit. Rev. Microbiol.* 24:61–84.

Nevoigt E (2008). Progress in metabolic engineering of *Saccharomyces cerevisiae*. *Microbiol. Mol. Biol. Rev.* 72:379–412.

Niederholtmeyer H, Wolfstädter BT, Savage DF, Silver PA et al. (2010). Engineering cyanobacteria to synthesize and export hydrophilic products. *Appl. Environ. Microbiol.* 76:3462–3466.

Nogales J, Gudmundsson S, Knight E, Palsson B et al. (2012). Detailing the optimality of photosynthesis in cyanobacteria through systems biology analysis. *PNAS* 109:2678–2683.

Orth J, Thiele I, Palsson B (2010). What is flux balance analysis? *Nat. Biotechnol.* 28:245–248.

Panda B, Jain P, Sharma L, Mallick N (2006). Optimization of cultural and nutritional conditions for accumulation of poly-β-hydroxybutyrate in *Synechocystis* sp. PCC 6803. *Bioresour. Technol.* 97:1296–1301.

Panda B, Mallick N (2007). Enhanced poly-β-hydroxybutyrate accumulation in a unicellular cyanobacterium, *Synechocystis* sp. PCC 6803. *Lett. Appl. Microbiol.* 44:194–198.

Reppas NB, Ridley CP (2010). Methods and compositions for the recombinant biosynthesis of n-alkanes. Joule Unlimited Inc., Bedford, MA, US Patent No. 7794969.

Rosgaard L, de Porcellinis AJ, Jacobsen JH, Frigaard N-U et al. (2012). Bioengineering of carbon fixation, biofuels, and biochemicals in cyanobacteria and plants. *J. Biotechnol.* 162:134–147.

Rupprecht J (2009). From systems biology to fuel—*Chlamydomonas reinhardtii* as a model for a systems biology approach to improve biohydrogen production. *J. Biotechnol.* 142:10–20.

Saha R, Verseput AT, Berla BM, Mueller TJ et al. (2012). Reconstruction and comparison of the metabolic potential of cyanobacteria *Cyanothece* sp. ATCC 51142 and *Synechocystis* sp. PCC 6803. *PLoS ONE* 7:e48285.

Sakai M, Ogawa T, Matsuoka M, Fukuda H (1997). Photosynthetic conversion of carbon dioxide to ethylene by the recombinant cyanobacterium, *Synechococcus* sp. PCC 7942, which harbours a gene for the ethylene-forming enzyme of *Pseudomonas syringae*. *J. Ferment. Bioeng.* 84:434–443.

Sakamoto T, Delgaizo VB, Bryant DA (1998). Growth on urea can trigger death and peroxidation of the cyanobacterium *Synechococcus* sp. strain PCC 7002. *Appl. Environ. Microbiol.* 64(7):2361–2366.

Savinell JM, Palsson B (1992). Network analysis of intermediary metabolism using linear optimization. I. Development of mathematical formalism. *J. Theor. Biol.* 154(4):421–454.

Schellenberger J, Que R, Fleming RM, Thiele I et al. (2011) Quantitative prediction of cellular metabolism with constraint-based models: The COBRA Toolbox v2.0. *Nat Protoc.* 6(9):1290–1307.

Schirmer A, Rude MA, Xuezhi L, Popova E et al. (2010). Microbial biosynthesis of alkanes. *Science* 329:559–562.

Schmidt B, Schmidt X, Chamberlin A, Salehi-Ashtiani K et al. (2010) Metabolic systems analysis to advance algal biotechnology. *Biotechnol. J.* 5:660–670.

Schwarz D, Nodop A, Hüge J, Purfürst S et al. (2011). Metabolic and transcriptomic analysis of Synechococcus 7942. *Plant Physiol.* 155(4):1640–1655.

Segre D, Vitkup D, Church GM (2002). Analysis of optimality in natural and perturbed metabolic networks. *PNAS* 99:15112–15117.

Shastri AA, Morgan JA (2005). Flux balance analysis of photoautotrophic metabolism. *Biotechnol. Prog.* 21:1617–1626.

Skizim NJ, Ananyev GM, Krishnan A, Dismukes GC (2011). Metabolic pathways for photobiological hydrogen production by nitrogenase- and hydrogenase-containing unicellular cyanobacteria *Cyanothece*. *J. Biol. Chem.* 287(4):2777–2786.

Song D, Hou L, Shi D (2008) Exploitation and utilization of rich lipids-microalgae, as new lipids feedstock for biodiesel production—A review. *Sheng Wu Gong Cheng Xue Bao* 24:341–348.

Swingley W, Chen M, Cheung P, Conrad A et al. (2008). Niche adaptation and genome expansion in the chlorophyll-d producing cyanobacterium *Acaryochloris marina*. *PNAS* 105:2005–2010.

Takahashi H, Miyake M, Tokiwa Y, Asada Y (1998). Improved accumulation of poly-3-hydroxybutyrate by a recombinant cyanobacterium. *Biotechnol. Lett.* 20:183–186.

Tan X, Yao L, Gao Q, Wang W et al. (2011). Photosynthesis driven conversion of carbon dioxide to fatty alcohols and hydrocarbons in cyanobacteria. *Metab. Eng.* 13:169–176.

Thiele I, Palsson BØ (2010). A protocol for generating a high-quality genome-scale metabolic reconstruction. *Nat. Protoc.* 5(1):93–121.

Vu T, Hill E, Kusek L, Konopka A et al. (2013). Computational evaluation of *Synechococcus* sp. PCC 7002 metabolism for chemical production. *Biotechnol. J.* 8:619–630.

Wu GF, Shen ZY, Wu QY (2002). Modification of carbon partitioning to enhance PHB production in *Synechocystis* sp PCC6803. *Enzyme Microb. Technol.* 30:710–715.

Yim H, Haselbeck R, Niu W, Pujol-Baxley C et al. (2011). Metabolic engineering of *Escherichia coli* for direct production of 1,4-butanediol. *Nat. Chem. Biol.* 7:445–452.

Young JD, Shastri AA, Stephanopoulos G, Morgan JA (2011). Mapping photoautotrophic metabolism with isotopically nonstationary (13)C flux analysis. *Metab. Eng.* 13:656–665.

Yu R, Yamada A, Watanabe K, Yazawa K et al. (2000). Production of eicosapentanoic acid by a recombinant marine cyanobacterium, *Synechococcus* sp. *Lipids* 35:1061–1064.

Zang X, Liu B, Liu S, Arunakumara KK et al. (2007). Optimum conditions for transformation of *Synechocystis* sp. PCC 6803. *J. Microbiol.* 45:241–245.

Zhou J, Zhang H, Zhang Y, Li Y et al. (2012). Designing and creating a modularized synthetic pathway in cyanobacterium *Synechocystis* enables production of acetone from carbon dioxide. *Metab. Eng.* 14:394–400.

10

Biorefinery Concept for a Microalgal Bioenergy Production System

Dheeraj Rathore, Poonam Singh Nigam, and Anoop Singh

CONTENTS

10.1 Introduction

The concern for exhausting availability of fossil fuels for fulfilling future energy demands and considering changes in global climate by conventional energy resources has diverted research toward environmentally safe and sustainable energy resources. Total global primary energy use in 2010 was reported to be 12,002.4 million tons of oil equivalent (Mtoe), which, at a standardized energy density of 41.868 GJ T^{-1}, is equal to ~0.5 ZJ and was provided by oil, gas, coal, nuclear, hydroelectric, and other renewables (BP 2011). In the last few decades, intensive works from experts in various fields focused on finding ecologically secure biofuel options.

The term biofuel is used for any liquid, solid, or gaseous fuel that is derived from any organism or its parts. Photosynthesis is the first step in the conversion of light to chemical energy and is, therefore, ultimately responsible for the production of feedstock required for all biofuels: synthesis of protons and electrons (for bio-H_2), sugars and starch (for bioethanol), lipids/oils (for biodiesel), and biomass (for BTL products and biomethane) (Costa and Morais 2011; Hankamer et al. 2007).

Currently, all biofuels are commercially produced from food crops, developing serious ecological and socioeconomical anxiety such as land-use changes and food–fuel competition issue. Presently, about 1% (14 million ha) of the world's available arable land is used for the production of biofuels providing 1% of global transport fuels. Clearly, increasing the share will be impractical due to the severe impact on the world's food supply and the large areas of production land required (IEA 2006). This is manifested by the recent increase in grain prices due to the utilization of maize at large scale as a feedstock for the production of ethanol fuel in the United States. This caused riots in Mexico due to the increase in the price of tortillas, a staple food (Saraf and Hastings 2011). Furthermore, the requirement for greenhouse gas (GHG) reduction is another constraint for developing biofuels. The Intergovernmental Panel for Climate Change has calculated that reductions of 25%–40% of CO_2 emissions by 2020 and up to 80% by 2050 are required to avoid dangerous climate changes worldwide, as emphasized in the Copenhagen Climate Conference in 2009 (Kruse and Hankamer 2010).

The technical potential of microalgae for GHG abatement has been recognized for many years, given their ability to use carbon dioxide and their possibility of achieving higher productivities than land-based crops. Biofuel production from these marine resources, whether the use of biomass or the potential of some species to produce high levels of oil, is now an increasing topic of interest (EPOBIO 2007). The development of algal strains to better suit production scenarios, harvesting of other bioactives in the fuel extraction process (through biorefining), and synthetic biology through the engineering of microbes to facilitate extraction of other products (Wargacki et al. 2012) offer great potential. Algal lipid biorefineries, although at a very early stage of development, may provide a means for large-scale commercialization of algal biofuels (Federal Government of Germany 2012). This chapter is an effort to explore the opportunities of microalgal biorefinery for energy and nonenergy products.

10.2 Advantages of Using Microalgae for Biofuel

Algae are basically a large and diverse group of simple, typically autotrophic organisms, ranging from unicellular to multicellular forms. There is no distinct line between microalgae and macroalgae. Generally, microscopic photosynthetic organisms, which are found in both marine and freshwater environments, are considered as microalgae. Their body is simple, made up of one or more cells, and not differentiated into roots and stems. The cellular composition of microalgae is both eukaryotic and prokaryotic. Their photosynthetic mechanism is similar to terrestrial plants, but due to a simple cellular structure and because they are submerged in an aqueous environment where they have efficient access to water, CO_2, and other nutrients, they are generally more efficient in converting solar energy into biomass. Due to higher photosynthetic efficiency, algal species are capable of producing more biofuel from less land use in comparison to other terrestrial oil crops; however, the water footprint of algal species is very high (Table 10.1). Higher water footprint could not be a limitation in cultivation of algal biomass because they are able to grow in marine water and also have the efficiency to grow in industrial wastewater and reduce the chances of contaminations to other water bodies (Singh et al. 2011b, 2012).

The four main components of algal biomass are carbohydrates, protein, lipids, and nucleic acids (Table 10.2), which are considered suitable sources of biofuel, food, and medicine

TABLE 10.1

Water Footprint, Land Use, and Biofuel Yield of Various Energy Crops

	Water Footprint ($m^3\,GJ^{-1}$)	Land Use ($m^2\,GJ^{-1}$)	Biofuel Yield ($L\,ha^{-1}\,annum^{-1}$)
Bioethanol			
Cassava	148	79	6,000
Wheat	93	305	1,560
Paddy rice	85	212	2,250
Corn grain	50	133	3,571
Potatoes	105	114	4,167
Sugar cane	50	81	5,882
Sugar beet	46	95	5,000
Sorghum	180	386	1,235
Soybean (ethanol)	383	386	1,235
Biodiesel			
Soybean (biodiesel)	383	689	446
Jatropha	396	162	1,896
Rapeseed	383	258	1,190
Cotton	135	945	325
Sun flower	61	323	951
Oil palm	75	52	5,906
Coconut	49	128	2,399
Groundnut	58	220	1,396
Microalgae	<379	2–13	24,355–136,886

Sources: Chisti Y., *Biotechnol. Adv.*, 25, 294, 2007; Clarens, A.F. et al., *Environ. Sci. Technol.*, 44(15), 1813, 2010; Department of Agriculture, Coconut annual report, The Ministry of Agriculture and Cooperatives, Bangkok, Thailand; Dominguez-Faus, R. et al., *Environ. Sci. Technol.*, 43(9), 3005, 2009; Gerbens-Leenes, P. et al., *Ecol. Econ.*, 68, 1052, 2009; Jansson, C. et al., *Appl. Energy*, 86(1), S95, 2009; Rajvanshi, A.K. et al., Biofuels—Promise/prospects, in *National Oilseeds Conference*, Hyderabad, India, Nimbkar Agricultural Research Institute (NARI), Phaltan, India, 2007; Schenk, P.M. et al., *Bioenergy Res.*, 1, 20, 2008; Singh, A. and Nigam, P.S., *Bioresour. Technol.*, 2010.

production. For all of these reasons, algal biomass can be considered as a third-generation biomass feedstock for biorefinery with a high outlook in terms of energy requirements.

Some of the advantages of using microalgae over other renewable energy resources, highlighted by several researchers (Banerjee et al. 2002; Chisti 2007; Giselrød et al. 2008; Guschina and Harwood 2006; Hsueh et al. 2007; Hu et al. 2008; Nigam and Singh 2011; Shilton et al. 2008; Singh et al. 2011a,c), are as follows:

- No direct competition with the food/feed chain. Algae are considered as the only alternative to current biofuel crops such as corn and soybean as they do not require arable land.
- Higher photosynthetic efficiency than other biomass sources.
- Algae can utilize waste CO_2 streams (e.g., from power plants).
- Their capability to grow in industrial, municipal, and agricultural wastewaters and marine water.
- Possible utilization of growth nutrients such as nitrogen and phosphorus from a variety of wastewater sources.

TABLE 10.2

Chemical Composition (% Dry Matter Basis) of Selected Freshwater and Marine Microalgae

Algal Species	Protein	Carbohydrate	Lipids	Nucleic Acid
Freshwater algal species				
Scenedesmus obliquus	50–56	10–17	12–14	3–6
Scenedesmus quadricauda	47	—	1.9	—
Chlamydomonas reinhardtii	48	17	21	—
Chlorella vulgaris	51–58	12–17	14–22	4–5
Chlorella pyrenoidosa	57	26	2	—
Euglena gracilis	39–61	14–18	14–20	—
Spirulina platensis	46–63	8–14	4–9	2–5
Spirulina maxima	60–71	13–16	6–7	3–4.5
Anabaena cylindrica	43–56	25–30	4–7	—
Marine algal species				
Dunaliella salina	57	32	6	—
Prymnesium parvum	28–45	25–33	22–38	1–2
Tetraselmis maculata	52	15	3	—
Porphyridium cruentum	28–39	40–57	9–14	—
Synechococcus sp.	63	15	11	5

Sources: Becker, E.W., Microalgae in human and animal nutrition, in: Richmond, A., ed., *Handbook of Microalgal Culture*, Blackwell Publishing, Oxford, U.K., pp. 312–351, 2004; Bruton, T. et al., A review of the potential of marine algae as a source of biofuel in Ireland, *Sustainable Energy Ireland*, 2009; Singh, A. et al., *Bioresour. Technol.*, 102, 26, 2011a; Singh, A. et al., *J. Chem. Technol. Biotechnol.*, 86, 1349, 2011b; Singh, A. and Olsen, S.I., *Appl. Energy*, 88, 3548, 2011; Singh, A. et al., *Energy Educ. Sci. Technol. A Energy Sci. Res.*, 29, 687, 2012.

- Biomass doubling times during exponential growth are commonly as short as 3.5 h.
- Oil content in microalgae can be up to 80% of dry biomass. Depending on the species, microalgae produce many different kinds of lipids, hydrocarbons, and other complex oils as by-products.

10.3 Biorefinery Concept of Microalgal Biofuel

The term *bio-based products* refers to three different product categories: biofuels (e.g., biodiesel and bioethanol), bioenergy (heat and power), and bio-based chemicals and materials (e.g., succinic acid and polylactic acid). They are produced by a biorefinery that integrates the biomass conversion processes. The biorefinery concept is thus analogous to today's petroleum refineries that produce multiple fuels, power, and chemical products from petroleum (WEF 2010), which makes the refinery process more profitable as well as practical (Singh and Gu 2010).

Algae offer a varied range of valuable products (Figure 10.1) such as food, nutritional compounds, omega-3 fatty acids, animal feed, organic fertilizers, biodegradable plastics, recombinant proteins, pigments, medicines, pharmaceuticals, and vaccines along with energy sources (including jet fuel, aviation gas, biodiesel, gasoline, and bioethanol) (Pienkos and Darzins 2009; Pulz 2004; Singh and Olsen 2011).

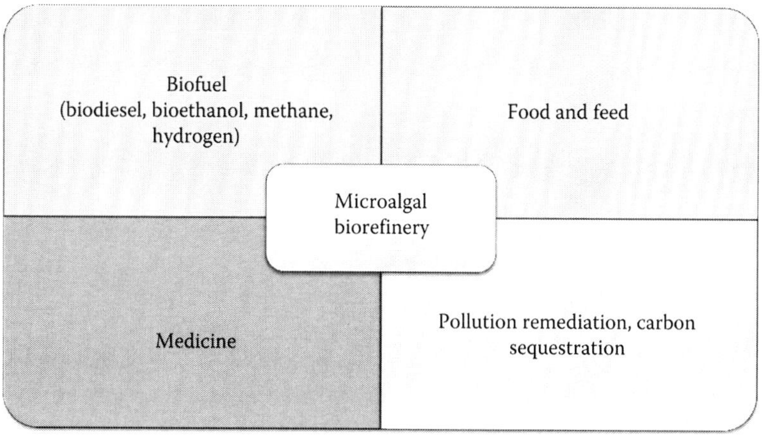

FIGURE 10.1
Main products of microalgal biorefinery.

10.3.1 Biofuels

Although the largest existing algal farms are for health food production (e.g., *Spirulina* production in China) and natural products (e.g., *Dunaliella* in Australia for β-carotene), those undergoing the most rapid expansion are currently aimed at biofuel production and associated R&D (Stephens et al. 2012). The whole algal biomass or algal oil extracts can be converted into different fuel forms, such as biomethane, biodiesel, bioethanol, kerosene, aviation fuel, and biohydrogen, through the implementation of appropriate processing technologies (Subhadra 2010).

10.3.1.1 Biodiesel

Biodiesel is typically a mixture of fatty acid alkyl esters, obtained by the transesterification (or ester exchange) of oils or fats and composed of 90%–98% triglycerides and smaller amounts of mono- and diglycerides and free fatty acids, besides residual amounts of phospholipids, phosphatides, carotenes, tocopherols, sulfur compounds, and water (Warabi et al. 2004). Many algae are exceedingly rich in oil, since their algal cells have been found heavily enriched with oil globules, which can be converted to biodiesel (Giselrød et al. 2008). Algae, when starved by a suitable source of nitrogen, produces mainly oil, whereas in the presence of sunlight, algae produce sugars and proteins from carbon dioxide. When grown under autotrophic and heterotrophic conditions, the microalgae *Chlorella protothecoides* accumulates lipids, which can be used for biodiesel production (Miao and Wu 2006). Nitrogen limitation is the most effective method of improving microalgal lipid accumulation, which not only results in the accumulation of lipids but also results in a gradual change of lipid composition from free fatty acids to triacylglycerol (Meng et al. 2009), a more useful lipid for biodiesel conversion (Tsukahara and Sawayama 2005).

Kaur et al. have studied the algal strains *Ankistrodesmus* spp., *Scenedesmus* spp., *Euglena* spp., *Chlorella* spp., and *Chlorococcus* spp. isolated from northeast India, which were found to accumulate a high intracellular lipid content as their energy storage product (Kaur et al. 2009). Amaro et al. reviewed the lipid content from microalgae and found 100% in freshwater species of *Botryococcus* and about 90% in marine alga *Dunaliella tertiolecta* (Amaro et al. 2012). The most common microalgae (viz., *Chlorella, Dunaliella, Isochrysis, Nannochloris, Nannochloropsis, Neochloris, Nitzschia, Phaeodactylum,* and *Porphyridium* spp.) possess oil levels between 20% and 50%, along with interesting productivities; *Chlorella* appears in particular to be a good option for biodiesel production. Greater lipid productivities of marine microalgae make them more prone to mass production, coupled with a realization that a high salinity prevents extensive contamination while allowing seawater to be directly used instead of depleting freshwater resources (Amaro et al. 2011a).

10.3.1.2 Bioethanol

One of the well-known first-generation biofuels is ethanol made by fermenting sugar extracted from crop plants and starch contained in maize kernels or other starchy crops. Carbohydrates are one of the most important sources of energy. Indian researchers explained that the biomass containing sugars can be directly fermented to ethanol by the least complex method in producing ethanol (Singh et al. 1995; Verma et al. 2000). Algae have a relatively high photoconversion efficiency and are able to accumulate higher carbohydrate content (higher than 50% of its dry weight) (Ho et al. 2012). In general, the algal carbohydrate includes starch, glucose, cellulose, hemicelluloses, and several other kinds of polysaccharides. Branyikova and coworkers studied starch production efficiency of freshwater alga *Chlorella* and reported 45% of starch content in the studied alga biomass prior to cell division, which is reduced to 13% and 4% during cell division under light and dark cycles, respectively. Sulfur deprivation further increased starch content by 50%. They suggested that for the production of biomass with high starch content, it is necessary to suppress cell division events, but not to disturb the synthesis of starch in the chloroplast (Branyikova et al. 2011). With these, algal starch/glucose can be conventionally used for biofuel production, especially for bioethanol (John et al. 2011) and hydrogen (Chochois et al. 2009).

10.3.1.3 Biomethane

Algal biomass is rich in nutrients, especially nitrogen and phosphorus, for which the use and potential loss may not be environmentally and economically sustainable (Sialve et al. 2009). The anaerobic digestion of algal waste not only recycles nutrients but also provides biomethane, a renewable energy. Anaerobic digestion involves the breakdown of organic matter to produce biogas (Brennan and Owende 2010). Sialve et al. reviewed methane production efficiency of microalgae and stated that the proportion of methane in the biogas produced is in a similar range (69%–75%) for the majority of the studies, regardless of species and operating conditions (Sialve et al. 2009). This reveals a good quality of conversion of algal organic matter into methane.

Besides carbon, nitrogen, and phosphorus, which are major components in microalgal composition, oligonutrients such as iron, cobalt, and zinc are also found (Grobbelaar 2004) and are known to stimulate methanogenesis (Speece 1996).

10.3.1.4 Biohydrogen

Biological hydrogen production from biomass is considered as one of the most promising alternatives for sustainable green energy production. With the development and commercialization of fuel cells, hydrogen production from biomass is being considered as an alternative energy source for decentralized power generation (Park et al. 2009). Among all photosynthetic organisms, only green microalgae and cyanobacteria have been shown to sustain hydrogen production (Schutz et al. 2004).

Hans Gaffron discovered hydrogen metabolism in unicellular green algae, which are eukaryotic organisms of oxygenic photosynthesis (Gaffron 1939, 1940). The first scientific investigation of H_2 evolution by microalgae was reported by Gaffron and Rubin (1942), who demonstrated that after a period of dark anaerobic adaptation, the green alga *Scenedesmus obliquus* produces H_2 in the dark at low rates, with H_2 production is greatly stimulated in the light, though only for relatively brief periods. The production pathway of H_2 production using water and sunlight in microalgal photobiolysis is given in detail by Rathore and Singh (2013). Laboratory-scale H_2 production from microalgae such as *Anabaena* (Jackson and Ellms 1896), *Chlamydomonas* mutant (Ghirardi et al. 1997), and *Chlamydomonas reinhardtii* (Melis et al. 2000) has been demonstrated. However, more study is needed for industrial biohydrogen production from microalgae.

10.3.2 Bio-Based Chemicals and Materials

10.3.2.1 Food and Feed

Microalgae are one of the most promising sources for new food and functional food products and can be used to enhance the nutritional value of foods, due to their well-balanced chemical composition (Batista et al. 2013). Besides application of microalgal pigments for coloring purposes, the novel use of microalgal biomass as a healthy and attractive food is an interesting field for providing protein, vitamins, minerals, and other nutritional supplements with biologically active compounds (e.g., antioxidants, polyunsaturated fatty acid [PUFA]-ω3, ω6).

Spirulina maxima is a filamentous cyanobacterium largely used as a feed and food supplement. This alga grows profusely in certain alkaline lakes in Mexico and Africa, forming massive blooms, and has been used as a staple food by local populations since ancient times (Yamaguchi 1997). Carotenoids such as canthaxanthin, astaxanthin, and lutein from *Chlorella* have been in regular use as pigments and have accordingly been included as feed ingredients for salmonid fish and trout, as well as poultry to enhance the reddish color of fish or the yellowish color of egg yolk (Cysewski and Lorenz 2004; Guerin et al. 2003; Lorenz and Cysewski 2000; Plaza et al. 2009). *Haematococcus pluvialis* has been identified as the organism that can accumulate the highest level of astaxanthin in nature (1.5%–3.0% dry weight), which is currently the prime natural source of astaxanthin for commercial exploitation (Lorenz and Cysewski 2000). Phycocyanin, a blue-colored protein pigment, constitutes about 15%–25% of the dry weight of the microalgae (Bermejo et al. 1997; Romay et al. 2003), which can be considered as a safe natural food colorant in nonacidic foodstuffs such as chewing gum, confectionaries, and dairy products (Bermejo et al. 1997; Downham and Collins 2000). In a study, van Harmelen and Oonk (2006) mentioned that the rapidly growing algae industry in Japan, the United States, India, China, and other countries is currently producing over 10,000 ton/annum of microbial biomass, mostly in open ponds and mainly for use as nutritional supplements.

Other than direct food or food supplements, antimicrobial compounds from microalgae used as a food preservative are increasing in demand due to the avoidance of synthetic preservatives by consumers. A study conducted by Guedes et al. (2011b) concluded that extracts of cyanobacteria and microalgae can be used as antibacterial agents against common food-borne pathogens. In this study, Guedes and coworkers observed that cyanobacteria possess wider antibacterial spectra than microalgae and intracellular extracts are more powerful than their extracellular counterparts. *Nostoc* sp. was effective against *Escherichia coli* and *Pseudomonas aeruginosa*, whereas *Gloeothece* sp. was able to inhibit also *Staphylococcus aureus* and *Salmonella* sp. Strain M2-1 of *S. obliquus* exhibited the strongest antibacterial activity, except against *Salmonella* sp., thus appearing to be the most promising antibacterial extract for eventual use as a preservative in food.

According to Pulz and Gross (2004), the market size of products from microalgae was estimated to have a retail value of US$5–6.5 billion. US$1.25–2.5 billion was generated by the health food sector, US$1.5 billion from the production of docosahexaenoic acid (DHA), and US$700 million from aquaculture.

10.3.2.1.1 Protein

The protein content of *Spirulina* (50%–70% of dried weight), which exceeds that of normal proteinous food sources, namely, meat, eggs, dried milk, grains, and soybeans, contains all the essential amino acids especially leucine, valine, and isoleucine. However, it seems somewhat deficient in methionine, cysteine, and lysine in comparison to standard alimentary protein sources, while it is superior to all plant proteins including proteins from legumes (Belay 2008; Henrikson 2009). In a study, Barbino and Lourenco quantified the protein content of some microalgae and reported 4%–15.6% in the tested microalgae (Barbino and Lourenco 2005).

10.3.2.1.2 Lipids

Many lipids are accumulated by algae (normally accounting for 30%–50% of their content by weight) under several specific cultural conditions, such as in a high C/N medium or under stress conditions. According to their carbon numbers, microalgal lipids are of two types, namely, (1) those with fatty acids containing 14–20 carbons used for biodiesel production and (2) PUFAs (with over 20 carbons) used as health food supplements (Yen et al. 2013). Because of their high levels of bioactivity, microalgal PUFAs, eicosapentaenoic acid (EPA, 20:5, ω3), and DHA (22:6, ω6) are of particular interest. Current food sources of omega-3 (ω3) and omega-6 (ω6) (PUFAs) are fish and shellfish, flaxseed (linseed), hemp oil, soya oil, canola (rapeseed) oil, chia seeds, pumpkin seeds, sunflower seeds, leafy vegetables, and walnuts; however, the major sources of EPA and DHA, on a worldwide basis, are still marine fish (Guedes et al. 2011a). Fishes have a limited capacity for synthesis of PUFAs, so most of them are simply accumulated from their microalgal diet. This realization has turned microalgae into one of the most important feed items in aquaculture, as they are de novo producers of PUFAs and can accumulate them to relatively high levels (Tonon et al. 2002), which, in turn, have been shown to trigger particularly high rates of growth of aquacultured fish (Guedes et al. 2011a).

A widely stated claim is that microalgae are capable of producing several times more oil per unit land area than terrestrial oilseed crops. The actual global oil production from oilseed crop was 0.592 t ha^{-1} in 2007–2008 (FAPRI 2008). If one assumes an oil concentration in algae of ~42% (Chica et al. 2005) and productivity of 365 t ha^{-1} $year^{-1}$ in AlgaeLink bioreactors, this equates to a 153.3 t ha^{-1} $year^{-1}$ oil production, which is about 259 times higher than the oilseed crops (Packer 2009). Even if these estimates are optimistic, the potential benefits are obvious (Singh et al. 2011a).

Nitrogen limitation is the most widely used and reliable strategy for increasing the lipid content of microalgae (Yen et al. 2013). Hu and coworkers collected the data for the lipid content of various classes of microalgae and cyanobacteria under normal growth and nitrogen stress conditions. Their results clearly showed that the lipid content of green microalgae, diatoms, and some other species under stress conditions was 10%–20% higher; however, the lipid content of cyanobacteria produced under stress conditions was usually less than 10%, making it unsuitable for lipid production (Hu et al. 2008). Although the overall fatty acid composition of a few microalgae is reasonably well documented, several native strains are still lacking.

10.3.2.2 Medicine

The therapeutic use of microalgal compounds can be traced back a long time. However, their systematic screening for biologically active principles began in the 1950s. In the last decade, microalgae have become the focus of extensive research efforts, aimed at finding novel compounds that might lead to therapeutically useful agents (Amaro et al. 2011b, 2013; Cardozo et al. 2007; Mayer and Hamann 2005; Mendes et al. 2003). Large numbers of microalgal extracts and/or extracellular products have been proven to have antibacterial, antifungal, antiprotozoal, antiplasmodial, antioxidants, antiviral, anti-inflammatory, and antitumor effects (Amaro et al. 2013; Ghasemi et al. 2004; Herrero et al. 2006; Kellan and Walker 1989; Ozemir et al. 2004). Chlorellin, a mixture of fatty acids found to be responsible for the inhibitory activity against bacteria, was first isolated by Pratt et al. (1944) from a microalga *Chlorella*. Antimicrobial activity detected in several pressurized extracts from *D. salina* may be explained not only by several fatty acids but also by compounds such as α- and β-ionone, β-cyclocitral, neophytadiene, and phytol (Herrero et al. 2006). PUFAs such as DHA (EPA) and arachidonic acid, obtained from microalgae, have been implicated in the prevention of coronary heart diseases, hypertriglyceridemia, blood platelet aggregation, atherosclerosis, general inflammation, and several carcinomas (Guil-Guerrero et al. 2001; Navarro-Pérez et al. 2001).

Microalgal compounds respond to a wide diversity of molecular targets with a marked selectivity, thus raising a potential pharmaceutical interest for tumor healing (Newman and Cragg 2006); for example, astaxanthin, a microalgal-derived carotenoid, possesses potent cancer chemopreventive features and is much more persuasive than β-carotene and other carotenoids (Ip and Chen 2005; Palozza et al. 2009). Amaro et al. reviewed the possibility of therapeutic use of algal metabolites for gastric cancer and associated *Helicobacter pylori* infection and concluded that microalgal carotenoids, sulfated polysaccharides, and PUFAs are potent chemoprevention agents for cancer (Amaro et al. 2013). Additionally, sulfated polysaccharides and carotenoids are useful for prophylaxis against *H. pylori* adhesion and prevention of infection. In a review article, Amaro et al. listed selective microalgal species and their bioactive metabolites (Table 10.3) that could be used as antitumor drugs (Amaro et al. 2013).

TABLE 10.3

Algal Species and Bioactive Metabolites of Antitumor Features

Algal Species	Bioactive Metabolites
Chlorella vulgaris	Peptide VECYGPNRPQF
Odontella aurita, *Chaetoceros* sp., *Isochrysis aff. galbana*	Fucoxanthin
Chlorella ellipsoidea	Violaxanthin
Haematococcus pluvialis	Astaxanthin
Gymnodinium sp.	Extracellular D-galactan sulfate, with/or without L-(+)-lactic acid

Source: Amaro, H.M. et al., *Trends Biotechnol.*, 31, 92, 2013.

10.4 Carbon Sequestration

The ability of algae to fix CO_2 can also be an interesting method of removing gases from power plants, and thus they can be used to reduce GHGs with a higher production microalgal biomass and consequently higher biodiesel yield. Algal biomass production systems can be easily adapted to various levels of operational and technological skills (Saraf and Hostings 2011). Biological CO_2 fixation as done by algae is considered to be a promising way of fixing CO_2, thus helping to prevent global warming (Douškova et al. 2010). Therefore, CO_2 sequestration can be regarded as an added benefit for growing algae (Benemann 1997; Brown 1996; Doucha and Lıvansky 2006). Douškova et al. reported that a microalga *Chlorella* is able to utilize flue gases and biogas as CO_2 source in a closed chamber (Douškova et al. 2010). Exploitation of microalgae for biofuel production and CO_2 mitigation, by which CO_2 is captured and sequestered, is under research (Chisti 2007; Hankamer et al. 2007; Li et al. 2007; Scragg et al. 2003; Tsukahara and Sawayama 2005).

10.5 Pollution Remediation

Algae can utilize nutrients (N and P) from a variety of wastewater sources (e.g., agricultural runoff, concentrated animal feed operations, industrial effluent, and municipal wastewater), thus providing sustainable bioremediation of these wastewaters for environmental and economic benefits (Shilton et al. 2008). In a review, Singh et al. concluded that the integration of microalgal cultivation with fish farms, food processing facilities, wastewater treatment plants, etc., will offer the possibility for waste remediation through recycling of organic matter and at the same time low-cost nutrient supply required for algal biomass cultivation (Singh et al. 2011a,b). A study conducted by Raposo et al. using brewery effluent as growth nutrient medium for culturing *C. vulgaris* and other autochthonous flora reported 27% and 15% reduction of BOD and COD, respectively, by autochthonous flora. In this study of autochthonous, they found removal of nitrogen up to 5855 g kg^{-1} biomass day^{-1} and removal of phosphate up to 805 g kg^{-1} biomass day^{-1} from the effluent (Raposo et al. 2010).

10.6 Limitations

High-value coproducts, such as nutritional supplements, currently the main commercial microalgal products, are not of interest in large-scale biofuels production, as their markets are too small to be relevant. Commodity animal feeds, with prices similar to, though somewhat higher than, those of liquid biofuels, provide potential synergies, and most current schemes for algal biofuel production rely on some types of commodity feed by-product. However, such coproduction of biofuels and animal feeds is problematic in that producers may prefer to sell the entire algal biomass for feeds, without oil extraction (Lundquist et al. 2010). Hossain and coworkers used common species such as *Oedogonium* and *Spirogyra* and found that biomass (after oil extraction) and sediments (glycerine, water, and pigments) were higher in *Spirogyra* but *Oedogonium* sp. has higher biodiesel content (Hossain et al. 2008). The selection of algal species (those should have higher oil content, higher biomass productivity, and also higher protein and medicinal products) and a robust technology (which should not degrade the quality of other products during production of biofuel from algal biomass) need to strengthen for the setup of a commercial-scale algal biorefinery.

10.7 Conclusion

The concept of algal biorefinery promises several benefits like better-quality biofuel, reducing energy dependency, reducing climate change, increasing biofuel production from less land use, reducing environmental pollution by utilizing nutrients from wastewater and CO_2 from flue gases, and also producing food and medicine. The integration of biofuel production and production of food and medicine at one unit is a challenge because algal species and a robust technology that suits for biorefinery still need further research.

References

Amaro HM, Barros R, Guedes AC, Sousa-Pinto I, and Malcata FX. 2013. Microalgal compounds modulate carcinogenesis in the gastrointestinal tract. *Trends in Biotechnology* 31:92–98.

Amaro HM, Guedes AC, and Malcata FX. 2011a. Advances and perspectives in using microalgae to produce biodiesel. *Applied Energy* 88:3402–3410.

Amaro HM, Guedes AC, and Malcata FX. 2011b. Antimicrobial activities of microalgae: An invited review. In: Vilas AM, ed. *Science Against Microbial Pathogens: Communicating Current Research and Technological Advances*. Formatex, Badajoz, Spain, pp. 1272–1280.

Amaro HM, Macedo AC, and Malcata FX. 2012. Microalgae: An alternative as sustainable source of biofuels? *Energy* 44:158–166.

Banerjee A, Sharma R, Chisti Y, and Banerjee UC. 2002. *Botryococcus braunii*: A renewable source of hydrocarbons and other chemicals. *Critical Reviews in Biotechnology* 22:245–279.

Barbino E and Lourenco SO. 2005. An evaluation of methods for extraction and quantification of protein from marine macro- and microalgae. *Journal of Applied Phycology* 17:447–460.

Batista AP, Gouveia L, Bandarra NM, Franco JM, and Raymundo A. 2013. Comparison of microalgal biomass profiles as novel functional ingredient for food products. *Algal Research* 2:164–173.

Becker EW. 2004. Microalgae in human and animal nutrition. In: Richmond A, ed. *Handbook of Microalgal Culture*. Blackwell Publishing, Oxford, U.K., pp. 312–351.

Belay A. 2008. *Spirulina arthrospira* production and quality assurance. In: Gershwin ME and Belay A, eds. *Spirulina in Human Nutrition and Health*. CRC Press, Taylor & Francis Group, Boca Raton, FL, pp. 1–27.

Benemann JR. 1997. CO_2 mitigation with microalgae systems. *Energy Conversion and Management* 38:S475–S479.

Bermejo R, Talavera EM, Alvarez-Pez JM, and Orte JC. 1997. Chromatographic purification of phycobiliproteins from *Spirulina platensis*. High-performance liquid chromatographic separation of their alpha and beta subunits. *Journal of Chromatography A* 778:441–450.

BP. 2011. *Statistical Review of World Energy*. British Petroleum, London, U.K. http://www.bp.com/liveassets/bp_internet/globalbp/globalbp_uk_english/reports_and_publications/statistical_energy_review_2011/STAGING/local_assets/pdf/statistical_review_of_world_energy_full_report_2011.pdf (accessed January 04, 2012).

Branyikova I, Marsalkova B, Doucha J, Branyik T, Bisova K, Zachleder V, and Vıtova M. 2011. Microalgae-novel highly efficient starch producers. *Biotechnology and Bioengineering Symposium* 108:766–776.

Brennan L and Owende P. 2010. Biofuels from microalgae—A review of technologies for production, processing, and extractions of biofuels and co-products. *Renewable and Sustainable Energy Reviews* 14:557–577.

Brown LM. 1996. Uptake of carbon dioxide from flue gas by microalgae. *Energy Conversion and Management* 37:1363–1367.

Bruton T, Lyons H, Lerat Y, Stanley M, and BoRasmussen M. 2009. A review of the potential of marine algae as a source of biofuel in Ireland. *Sustainable Energy Ireland*, Dublin, Ireland.

Cardozo KHM, Guaratini T, Barros MP, Falcão VR, Tonon AP, Lopes NP, Campos S, Torres MA, Souza AO, Colepicolo P, and Pinto E. 2007. Metabolites from algae with economical impact. *Comparative Biochemistry and Physiology Part C, Toxicology & Pharmacology* 146:60–78.

Chica RA, Doucet N, and Pelletier JN. 2005. Semi-rational approaches to engineering enzyme activity: Combining the benefits of directed evolution and rational design. *Current Opinion in Biotechnology* 16:378–384.

Chisti Y. 2007. Biodiesel from microalgae. *Biotechnology Advances* 25:294–306.

Chochois V, Dauvillee D, Beyly A, Tolleter D, Cuine S, Timpano H, Ball S, Cournac L, and Peltier G. 2009. Hydrogen production in chlamydomonas: Photosystem II-dependent and -independent pathways differ in their requirement for starch metabolism. *Plant Physiology* 151:631–640.

Clarens AF, Resurreccion EP, White MA, and Colosi LM. 2010. Environmental life cycle comparison of algae to other bioenergy feedstocks. *Environmental Science & Technology* 44(5): 1813–1819.

Costa JAV and Morais MG. 2011. The role of biochemical engineering in the production of biofuels from microalgae. *Bioresource Technology* 102:2–9.

Cysewski GR and Lorenz RT. 2004. Industrial production of microalgal cell-mass and secondary products-species of high potential: *Haematococcus*. In: Richmond A, ed. *Handbook of Microalgal Culture, Biotechnology and Applied Phycology*. Blackwell Science, Oxford, U.K., pp. 281–288.

Department of Agriculture. 2001. Coconut annual report. The Ministry of Agriculture and Cooperatives, Bangkok, Thailand.

Dominguez-Faus R, Powers S, Burken J, and Alvarez P. 2009. The water footprint of biofuels: A drink or drive issue? *Environmental Science and Technology* 43(9):3005–3010.

Doucha J and Lıvansky K. 2006. Productivity, CO_2/O_2 exchange and hydraulics in outdoor open high density microalgal (*Chlorella* sp.) photobioreactors operated in a Middle and Southern European climate. *Journal of Applied Phycology* 18:811–826.

Douškova I, Kaštanek F, Maleterova Y, Kaštanek P, Doucha J, and Zachleder V. 2010. Utilization of distillery stillage for energy generation and concurrent production of valuable microalgal biomass in the sequence: Biogas-cogeneration-microalgae-products. *Energy Conversion and Management* 51:606–611.

Downham A and Collins P. 2000. Colouring our foods in the last and next millennium. *Journal of Food Science and Technology* 35:5–22.

EPOBIO. 2007. Micro and macro-algae: Utility for industrial application. EPOBIO project report. CNAP, University of York, York, U.K.

FAPRI. 2008. *World Oilseeds and Products. FAPRI 2008 US and World Agricultural Outlook*. The Food and Agricultural Policy Research Institute, Iowa State University, Ames, IA and the University of Missouri-Columbia, Columbia, MO, pp. 224–283.

Federal Government of Germany. 2012. Biorefineries roadmap (as part of the German Federal Government action plans for the material and energetic utilisation of renewable raw materials). The Federal Ministry of Food, Agriculture & Consumer Protection, Berlin, Germany. www.bmbf.de/pub/BMBF_Roadmap-Bioraffinerien_en_bf.pdf (accessed on December 22, 2014).

Gaffron H. 1939. Reduction of CO_2 with H_2 in green plants. *Nature* 143:204–205.

Gaffron H. 1940. Carbon dioxide reduction with molecular hydrogen in green algae. *American Journal of Botany* 27:273–283.

Gaffron H and Rubin J. 1942. Fermentative and photochemical production of hydrogen in algae. *Journal of General Physiology* 26:219–240.

Gerbens-Leenes P, Hoekstra AY, and van der Meer TH. 2009. The water footprint of energy from biomass: A quantitative assessment and consequences of an increasing share of bio-energy in energy supply. *Ecological Economics* 68:1052–1060.

Ghasemi Y, Yazdi MT, Shafiee A, Amini M, Shokravi S, and Zarrini G. 2004. Parsiguine, a novel antimicrobial substance from *Fischerella ambigua*. *Pharmaceutical Biology* 42:318–322.

Ghirardi ML, Togasaki RK, and Seibert M. 1997. Oxygen sensitivity of algal hydrogen production. *Applied Biochemistry and Biotechnology* 63–65:141–151.

Giselrød HR, Patil V, and Tran K. 2008. Towards sustainable production of biofuels from microalgae. *International Journal of Molecular Sciences* 9:1188–1195.

Grobbelaar JU. 2004. Algal nutrition. In: Richmond A, ed. *Handbook of Microalgal Culture: Biotechnology and Applied Phycology*. Wiley-Blackwell, Ames, IA.

Guedes AC, Amaro HM, Barbosa CR, Pereira RD, and Malcata FX. 2011a. Fatty acid composition of several wild microalgae and cyanobacteria, with a focus on eicosapentaenoic, docosahexaenoic and α-linolenic acids for eventual dietary uses. *Food Research International* 44:2721–2729.

Guedes AC, Barbosa CR, Amaro HM, Pereira CI, and Malcata FX. 2011b. Microalgal and cyanobacterial cell extracts for use as natural antibacterial additives against food pathogens. *International Journal of Food Science and Technology* 46:862–870.

Guerin M, Huntley ME, and Olaizola M. 2003. Haematococcus astaxanthin: Applications for human health and nutrition. *Trends in Biotechnology* 21:210–215.

Guil-Guerrero JL, Belarbi H, and Rebolloso-Fuentes MM. 2001. Eicosapentaenoic and arachidonic acids purification from the red microalga *Porphyridium cruentum*. *Bioseparation* 9:299–306.

Guschina IA and Harwood JL. 2006. Lipids and lipid metabolism in eukaryotic algae. *Progress in Lipid Research* 45:160–186.

Hankamer B, Lehr F, Rupprecht J, Mussgnug J, Posten C, and Kruse O. 2007. Photosynthetic biomass and H_2 production by green algae: From bioengineering to bioreactor scale-up. *Physiologia Plantarum* 131:10–21.

Henrikson R. 2009. *Earth Food Spirulina*. Ronore Enterprises, Inc, Hana, HI. http://www.spirulinasource.com/PDF.cfm/EarthFoodSpirulina.pdf (accessed December 14, 2012).

Herrero M, Ibañez E, Cifuentes A, Reglero G, and Santoyo S. 2006. *Dunaliella salina* microalga pressurized liquid extracts as potential antimicrobials. *Journal of Food Protection* 69:2471–2477.

Ho S-H, Chen C-Y, and Chang J-S. 2012. Effect of light intensity and nitrogen starvation on CO_2 fixation and lipid/carbohydrate production of an indigenous microalga *Scenedesmus obliquus* CNW-N. *Bioresource Technology* 113:244–252.

Hossain ABMS, Salleh A, Boyce AN, Chowdhury P, and Naqiuddin M. 2008. Biodiesel fuel production from algae as renewable energy. *American Journal of Biochemistry and Biotechnology* 4:250–254.

Hsueh HT, Chu H, and Yu ST. 2007. A batch study on the bio-fixation of carbon dioxide in the absorbed solution from a chemical wet scrubber by hot spring and marine algae. *Chemosphere* 66:878–886.

Hu Q, Sommerfeld M, Jarvis E, Ghirardi M, Posewitz M, Seibert M, and Darzins A. 2008. Microalgal triacylglycerols as feedstocks for biofuel production: Perspectives and advances. *Plant Journal* 54:621–639.

IEA. 2006. *World Energy Outlook 2006*. International Energy Agency, Paris, France.

Ip P-F and Chen F. 2005. Production of astaxanthin by the green microalga *Chlorella zofingiensis* in the dark. *Process Biochemistry* 40:733–738.

Jackson DD and Ellms JW. 1896. On odors and tastes of surface waters with special reference to Anabaena, a microscopical organism found in certain water supplies of Massachusetts. The 1896 Report of the Massachusetts State Board of Health. pp. 410–420.

Jansson C, Westerbergh A, Zhang J, Hu X, and Sun C. 2009. Cassava, a potential biofuel crop in (the) People's Republic of China. *Applied Energy* 86(1):S95–S99.

John RP, Anisha GS, Nampoothiri KM, and Pandey A. 2011. Micro and macroalgal biomass: A renewable source for bioethanol. *Bioresource Technology* 102:186–193.

Kaur S, Gogoi HK, Srivastava RB, and Kalita MC. 2009. Algal diversity as a renewable feedstock for biodiesel. *Current Science* 96:182.

Kellan SJ and Walker JM. 1989. Antibacterial activity from marine microalgae. *British Journal of Phycology* 23:41–44.

Kruse O, Hankamer B. 2010. Microalgal hydrogen production. *Current Opinion in Biotechnology* 21:238–243.

Li J, Li B, Zhu G, Ren N, Bo L, and He J. 2007. Hydrogen production from diluted molasses by anaerobic hydrogen producing bacteria in an anaerobic baffled reactor (ABR). *International Journal of Hydrogen Energy* 32(15):3274–3283.

Lorenz TR and Cysewski GR. 2000. Commercial potential for Haematococcus microalgae as a natural source of astaxanthin. *Trends in Biotechnology* 18:160–167.

Lundquist T, Woertz I, Quinn N, and Benemann JR. 2010. *A Realistic Technology and Engineering Assessment of Algae Biofuel Production*. Energy Biosciences Institute, University of California, Berkely, CA.

Mayer AMS and Hamann MT. 2005. Marine pharmacology in 2001–2002: Marine compounds with antihelmintic, antibacterial, anticoagulant, antidiabetic, antifungal, anti-inflammatory, antimalarial, antiplatelet, antiprotozoal, antituberculosis, and antiviral activities; affecting the cardiovascular, immune and nervous systems and other miscellaneous mechanisms of action. *Comparative Biochemistry and Phycology, Part C* 140:265–286.

Melis A, Zhang L, Forestier M, Ghirardi ML, and Seibert M. 2000. Sustained photobiological hydrogen gas production upon reversible inactivation of oxygen evolution in the green alga *Chlamydomonas reinhardtii*. *Plant Physiology* 122:127–136.

Mendes RL, Nobre BP, Cardoso MT, Pereira AP, and Palabra AF. 2003. Supercritical carbon dioxide extraction of compounds with pharmaceutical importance from microalgae. *Inorganica Chimica Acta* 356:328–334.

Meng X, Yang J, Xu X, Zhang L, Nie Q, and Xian M. 2009. Biodiesel production from oleaginous microorganisms. *Renewable Energy* 34(1):1–5.

Miao X and Wu Q. 2006. Biodiesel production from heterotrophic microalgal oil. *Bioresource Technology* 97(6):841–846.

Navarro-Pérez A, Rebolloso-Fuentes MM, Ramos-Miras JJ, and Guil-Guerrero JL. 2001. Biomass nutrient profiles of the microalga *Phaeodactylum tricornutum*. *Journal of Food Biochemistry* 25:57–76.

Newman DJ and Cragg GM. 2006. Natural products from marine invertebrates and microbes as modulators of antitumor targets. *Current Drug Targets* 7:279–304.

Nigam PS and Singh A. 2011. Production of liquid biofuels from renewable resources. *Progress in Energy and Combustion Science* 37:52–68.

Ozemir G, Karabay NU, Dalay MC, and Pazarbasi B. 2004. Antibacterial activity of volatile components and various extracts of *Spirulina platensis*. *Phytotherapy Research* 18:754–757.

Packer M. 2009. Algal capture of carbon dioxide; biomass generation as a tool for greenhouse gas mitigation with reference to New Zealand energy strategy and policy. *Energy Policy* 37(9):3428–3437.

Palozza P, Torelli C, Boninsegna A, Simone R, Catalano A, Mele MC, and Picci N. 2009. Growth-inhibitory effects of the astaxanthinrich alga *Haematococcus pluvialis* in human colon cancer cells. *Cancer Letters* 283:108–117.

Park J, Lee J, Sim SJ, and Lee JH. 2009. Production of hydrogen from marine macro-algae biomass using anaerobic sewage sludge microflora. *Biotechnology and Bioprocess Engineering* 14:307–315.

Pienkos PT and Darzins A. 2009. The promise and challenges of micro-algal derived biofuels. *Biofuels, Bioproducts and Biorefining* 3:431–440.

Plaza M, Herrero M, Cifuentes A, and Ibáñez E. 2009. Innovative natural functional ingredients from microalgae. *Journal of Agricultural and Food Chemistry* 57:7159–7170.

Pratt R, Daniels TC, Eiler JB, Gunnison JB, Kumler WD, Oneto JF, Strait LA et al. 1944. Chlorellin, an antibacterial substance from *Chlorella*. *Science* 99:351–352.

Pulz O and Gross W. 2004. Valuable products from biotechnology of microalgae. *Applied Microbiology and Biotechnology* 65:635–648.

Rajvanshi AK, Singh V, and Nimbkar N. 2007. Biofuels—Promise/prospects. In: *National Oilseeds Conference*, Hyderabad (India). Nimbkar Agricultural Research Institute (NARI), Phaltan, India.

Raposo MFdJ, Oliveira SE, Castro PM, Bandarra NM, and Morais RM. 2010. On the utilization of microalgae for brewery effluent treatment and possible applications of the produced biomass. *Journal of The Institute of Brewing* 116:285–292.

Rathore D and Singh A. 2013. Biohydrogen production from microalgae. In: Gupta VK and Tuohy MG, eds. *Biofuel Technologies*. Springer-Verlag, Berlin, Germany, pp. 317–333.

Romay C, Gonzalez R, Ledon N, Remirez D, and Rimbau VC. 2003. Phycocyanin: A biliprotein with antioxidant, anti-inflammatory and neuroprotective effects. *Current Protein & Peptide Science* 4:207–216.

Saraf M and Hastings A. 2011. Biofuels, the role of biotechnology to improve their sustainability and profitability. In: Lichtfouse E, ed. *Biodiversity, Biofuels, Agroforestry and Conservation Agriculture*. Series: Sustainable Agriculture Reviews. Springer Science+Business Media B.V., New York, pp. 123–147.

Schenk PM, Thomas-Hall SR, Stephens E, Marx UC, Mussgnug JH, Posten C, Kruse O, and Hankamer B. 2008. Second generation biofuels: High-efficiency microalgae for biodiesel production. *BioEnergy Research* 1:20–43.

Schutz K, Happe T, Troshina O, Lindblad P, Leitao E, Oliveira P, and Tamagnini P. 2004. Cyanobacterial H_2 production: A comparative analysis. *Planta* 218:350–359.

Scragg AH, Morrison J, and Shales SW. 2003. The use of a fuel containing *Chlorella vulgaris* in a diesel engine. *Enzyme and Microbial Technology* 33(7):884–889.

Shilton AN, Powell N, Mara DD, and Craggs R. 2008. Solar-powered aeration and disinfection, anaerobic co-digestion, biological CO_2 scrubbing and biofuel production: The energy and carbon management opportunities of waste stabilization ponds. *Water Science and Technology* 58:253–258.

Sialve B, Bernet N, and Bernard O. 2009. Anaerobic digestion of microalgae as a necessary step to make microalgal biodiesel sustainable. *Biotechnology Advances* 27(4):409–416.

Singh A, Nigam PS, and Murphy JD. 2011c. Renewable fuels from algae: An answer to debatable land based fuels. *Bioresource Technology* 102:10–16.

Singh A, Nigam PS, and Murphy JD. 2011a Mechanism and challenges in commercialisation of algal biofuels. *Bioresource Technology* 102:26–34.

Singh A and Olsen SI. 2011. Critical analysis of biochemical conversion, sustainability and life cycle assessment of algal biofuels. *Applied Energy* 88:3548–3555.

Singh A, Olsen SI, and Nigam PS. 2011b. A viable technology to generate third generation biofuel. *Journal of Chemical Technology and Biotechnology* 86:1349–1353.

Singh A, Pant D, Olsen SI, and Nigam PS. 2012. Key issues to consider in microalgae based biodiesel. *Energy Education Science and Technology Part A: Energy Science and Research* 29:687–700.

Singh D, Dahiya JS, and Nigam P. 1995. Simultaneous raw starch hydrolysis and ethanol fermentation by glucoamylase from *Rhizoctonia solani* and *Saccharomyces cerevisiae*. *Journal of Basic Microbiology* 35:117–121.

Singh J and Gu S. 2010. Commercialization potential of microalgae for biofuels production. *Renewable and Sustainable Energy Reviews* 14:2596–2610.

Speece RE. 1996. *Anaerobic Biotechnology for Industrial Wastewaters*. Archae press, Nashville, TN.

Stephens E, Wagner L, Ross IL, and Hankamer B. 2012. Microalgal production systems—Global impact of industry scale up. In: Posten C and Walter C, eds. *Microalgal Biotechnology: Integration and Economy*. Walter de Gruyter GmbH, Berlin, Germany, pp. 267–307.

Subhadra BG. 2010. Sustainability of algal biofuel production using integrated renewable energy park (IREP) and algal biorefinery approach. *Energy Policy* 38:5892–5901.

Tonon T, Harvey D, Qing R, Larson TR, and Graham IA. 2002. Long chain polyunsaturated fatty acid production and partitioning to triacylglycerols in four microalgae. *Phytochemistry* 61:15–24.

Tsukahara K and Sawayama S. 2005. Liquid fuel production using microalgae. *Journal of The Japan Petroleum Institute* 48:251–259.

van Harmelen T and Oonk H. 2006. Microalgae biofixation processes: Applications and potential contributions to greenhouse gas mitigation options. TNO Built Environment Geosciences, Apeldoorn, the Netherlands.

Verma G, Nigam P, Singh D, and Chaudhary K. 2000. Bioconversion of starch to ethanol in a single-step process by co-culture of amylolytic yeasts and *Saccharomyces cerevisiae* 21. *Bioresource Technology* 72:261–266.

Warabi Y, Kusdiana D, and Saka S. 2004. Reactivity of triglycerides and fatty acids of rapeseed oil in supercritical alcohols. *Bioresource Technology* 91:283–287.

Wargacki AJ, Leonard E, Win MN, Regitsky DD, Santos CNS, Kim PB, Cooper SR et al. 2012. An engineered microbial platform for direct biofuel production from brown macroalgae. *Science* 335:308–313.

WEF. 2010. The future of industrial biorefineries. World Economy Forum. www3.weforum.org/docs/WEF_FutureIndustrialBiorefineries_Report_2010.pdf (accessed on December 22. 2014).

Yamaguchi K. 1997. Recent advances in microalgal bioscience in Japan, with special reference to utilization of biomass and metabolites: A review. *Journal of Applied Phycology* 8:487–502.

Yen HW, Hu IC, Chen CY, Ho SH, Lee DJ, and Chang JS. 2013. Microalgae-based biorefinery—From biofuels to natural products. *Bioresource Technology* 135:166–174.

Section IV

Production of Bioenergy

11

Marine Macroalgal Biomass as a Renewable Source of Bioethanol

Nitin Trivedi, Vishal Gupta, C.R.K. Reddy, and Bhavanath Jha

CONTENTS

11.1 Introduction

Depletion of fossil fuels, climate change, concern over energy securities, volatile energy costs, and negative impact on the environment, particularly greenhouse gas emissions, ignited efforts for finding renewable fuel alternatives. The concern over fossil fuel supply is clearly reflected by an increase in crude oil price from US$80 per barrel in 2006 to US$108.01 per barrel in 2013. As an initiative, the European Commission set a goal to substitute 5.75% of fossil fuels with alternate fuels in 2010 and subsequently aimed the same at 20% substitution by 2020 (Hahn-Hagerdal et al. 2006). Similarly, United States has also attained 10% blend of ethanol in gasoline in 2011 and is steadily planning to increase the blend to E15. Ethanol is the most widely used transport biofuel across the globe and showed a significant increase in production from less than a billion liters in 1975 to more than 86 billion liters in 2010, and it is expected to reach 170 billion liters by 2020 (Licht 2006; ISO World Fuel Ethanol Outlook 2012). The United States, Brazil, and the European Union (EU) are the pivotal markets for bioethanol production and consumption and accounted for 90% of the world's ethanol market share (Table 11.1). The rise in transport fuel demand globally has made it mandatory for other countries also to think over other alternatives like bioethanol. Canada, Colombia, China, Thailand, and India are among other countries involved in ethanol production currently at a small scale but with rise in the future. Bioethanol is largely derived from starch/cellulose-based

TABLE 11.1

Fuel Ethanol Production and Consumption across the Globe

	Production (Million Liters)		Consumption (Million Liters)	
Country	**2007**	**2011**	**2007**	**2011**
United States	24,552	52,805	25,917	48,685
Brazil	20,196	19,935	16,204	19,194
EU	1,803	4,450	2,298	5,647
China	1,700	2,100	1,700	2,100
Canada	640	1,350	1,118	1,948
Colombia	272	337	283	351
Thailand	192	510	176	420
India	170	430	170	430
Australia	50	298	50	288
Argentina	—	210	—	207
Others	301	992	758	1,510
Total	49,876	83,417	48,674	80,780

Sources: Adjusted from *ISO Ethanol Yearbook*, International Sugar Organization, London, U.K., 2012; MECAS(12)19, *World Fuel Ethanol Outlook to 2020*, International Sugar Organization, London, U.K., 2012.

biomass, that is, plants, crops, and agricultural wastes. The world leaders United States, Brazil, and EU utilize two major sources, namely, corn and sugarcane, as raw materials for ethanol production (Chiaramonti 2007). As an estimate, 70% of the world's ethanol has been produced from corn and sugarcane. The high cost involved in the production of ethanol from sugarcane and the food versus fuel debate over corn ethanol give a negative perception in governing bodies leading to a search for new feedstock for ethanol production. The lignocellulose-based feedstocks are nowadays looked as an alternative to food crops in most of the biorefineries to produce ethanol and other high-value chemicals (Cherubini 2010). Agriculture waste and fast-growing nonedible terrestrial plants such as *Populus* and *Eucalyptus* constitute the major lignocellulosic biomass. Another concern on availability of arable lands for agriculture limits the propagation of aforementioned trees for fuel purposes. It is therefore necessary to shift our focus from terrestrial to marine-based renewable biomass such as macroalgae (John et al. 2011; Subhadra and Grinson 2011; Kraan 2013). Marine macroalgae are a group of large photoautotrophic multicellular, sessile, benthic plants commonly called as seaweeds and consist of as many as 25,000 species worldwide with considerable morphological and functional diversity (Holdt and Kraan 2011). The continuous exploration of seaweeds for chemicals has resulted in a spurt of seaweed production from 3.8 million tons in 1990 to whopping 15.8 million tons fresh in 2008, ranking them as one of the major mariculture crops with a market value of over US$7.4 billion (FAO 2010). Macroalgae have wider industrial applications from conventional food, feed, and hydrocolloids to recently explored sources of fertilizer, bioactive compounds, and energy molecules (McHugh 2003) (Figure 11.1). Seaweeds are a rich source of easily depolymerizable polysaccharides with little or no impregnation of lignin, which have better potential as renewable biomass for bioethanol production over terrestrial resources. This chapter discusses the potential of seaweeds as renewable feedstock for bioethanol production with an overview on seaweed biomass characteristics, its saccharification, and fermentation. Further, global initiatives on seaweed bioethanol production and the challenges faced are also summarized.

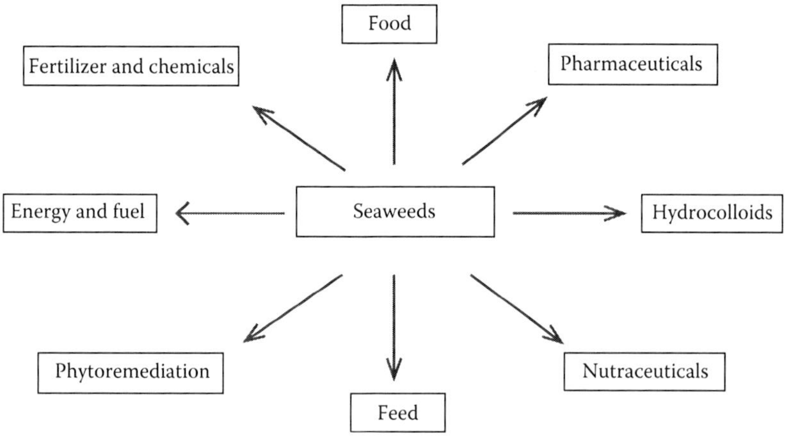

FIGURE 11.1
Potential uses of seaweeds.

11.2 Biofuel

Biofuels are any solid, liquid, or gaseous fuels derived from biomass or biological materials such as crops and/or crop residues. Liquid biofuels are utilized as an alternative transportation fuel and are classified mainly as bioethanol and biodiesel. Bioethanol is used as a replacement of gasoline, and biodiesel is a replacement of diesel. Depending upon the feedstock, biofuels are categorized as first-, second-, and third-generation biofuels. The first-generation biofuels are produced by the conversion of storage carbohydrates such as sugars and starch (Schubert 2006). These two are the major part of human dietary requirements and eventually initiated the food versus fuel debate. Resultantly, the second-generation biofuels are introduced, aiming at the conversion of the structural carbohydrates of the plant cell wall (Yuan et al. 2008). Agricultural waste rich in lignocellulose and nonedible fast-growing short-rotation forest trees such as *Populus* and *Eucalyptus* were considered as potential alternative feedstock. However, these materials are recalcitrant toward depolymerization because of the presence of complex impregnation of lignin and hemicellulose over cellulose. Different chemical pretreatment methods are therefore required to remove those hindrances so as to increase cellulose accessibility (Girio et al. 2010). The production of fermentable sugars from lignocellulosic biomass therefore tends to be complex and capital intensive and also has associated inherent environmental concerns. Therefore, it is necessary to develop a third-generation biofuel technology that is mainly based on algae. In recent years, algae have become the center of attraction in the area of academic and commercial biofuel research. Single-celled photoautotrophic microalgal strains are sourced for biodiesel, while multicellular, macroscopic, marine algae in common seaweeds are a renewable source of bioethanol production. Fourth-generation biofuels are the most recent introduction that includes metabolic engineering in photosynthetic microorganisms to produce and secrete biofuel. So far, this field is restricted to prokaryotic microbe, cyanobacteria, and no information is available on genetic engineering of micro- and macroalgae for biofuel production (Lu et al. 2011; Melis 2012).

11.3 Seaweeds as Renewable Source of Biofuel

The characteristics of biomass play a crucial role toward the feasibility of the process designed and developed for biofuel production. Henry (2009) reviewed the major criteria for selecting a plant group as renewable energy crop. These characteristics include (1) high biomass production, (2) optimal proximate composition, (3) high harvest index, (4) suitable storage of harvested material in the field, (5) nutrient partition to nonharvested parts, (6) high fraction of biofuel in harvested biomass, (7) high bulk density, (8) high water and nutrient uptake efficiency, (9) high coproduct potential, (10) low potential as a weed, (11) high scalability, (12) low harvesting cost, and (13) high suitability for genetic modification. Among these, higher growth rate, higher polysaccharide content, and no intervention in human food chain are the primary selection criteria for a source preferred for sustainable bioethanol production. Seaweeds fulfill these criteria and recently gained global attention as feedstock for bioethanol production. The major advantage of using seaweeds as feedstock for bioethanol is their growth rate, which is estimated to be higher than those of other terrestrial ethanol crops like wheat, maize, sugarcane, and sugar beet (Table 11.2). Species belonging to genus *Ulva* are regarded as opportunistic species with potentials to form blooms (green tide) when there is a sudden outburst of nutrients in the aquatic streams. The best energy crops, that is, sugarcane and maize, known for commercial ethanol productions, were shown to produce 6756 and 2010 L ha^{-1} year^{-1} volume of ethanol, respectively. On the contrary, the availability of marine macroalgae over the large expansion of sea around the globe accounts for a much higher yield (fivefold) of ethanol (23,400 L ha^{-1} year^{-1}) in comparison to any other land crops (Adams et al. 2009). The crop productivity and ethanol yields for various crops together with seaweeds are detailed in Table 11.2.

The bio-based fuels are considered carbon neutral and eventually can mitigate the negative impact on the environment arising due to excessive use of fossil fuels. Seaweeds, benthic photoautotrophic organisms, consume a large amount of CO_2 during photosynthesis and act as a CO_2 sink (Gao and McKinley 1994; Wang et al. 2008; Chung et al. 2013). The photosynthetic capacity of aquatic biomass (average of 6%–8%) is three to four times higher than terrestrial biomass (average of 1.8%–2.2%) indicating higher CO_2 fixation

TABLE 11.2

Potential of Macroalgae and Major Bioenergy Crop for Ethanol Production

Sources	Average World Yield (kg ha^{-1} year^{-1})	Potential Volume of Ethanol (L ha^{-1} year^{-1})
Wheat	2,800	1,010
Sorghum	1,300	1,235
Maize	4,815	2,010
Paddy rice	4,200	2,250
Potatoes	—	3,571
Sugar beet	47,070	5,150
Cassava	12,000	6,000
Sugar cane	68,260	6,756
Switch grass	—	10,760
Macroalgae	730,000	23,400

Sources: Adjusted from Adams, J.M. et al., *J. Appl. Phycol.*, 21, 569, 2009; Thi Hong Minh, N. and Van Hanh, V., *J. Vietnam. Environ.*, 3, 25, 2012.

and biomass production. The photosynthetic conversion efficiency (PCE) of macroalgae was reported to be 3.0%, while the same for a C4 plant like sugarcane was only 0.4% (Stephans et al. 2013). In 2010, the global production rate of aquatic plants was around 19.2 $\times$ 10^6 t, and a majority of these were accounted for by red seaweeds (9.0 $\times$ 10^6 t) followed by brown seaweeds (6.8 $\times$ 10^6 t), green seaweeds (0.2 $\times$ 10^6 t), and miscellaneous aquatic plants (3.2 $\times$ 10^6 t). Considering these values, Muraoka (2004) assumed that around 1000 t of carbon can be sequestered by aquatic plants. A higher share of seaweed productivity over other aquatic plants thereby reflects their potential as a major source of CO_2 sequestration as well. In another study by Kaladharan et al. (2009), seaweed biomass present all along the Indian coast was shown to utilize 9052 t CO_2 day^{-1} by photosynthesis against the emission of 365 t CO_2 day^{-1} during respiration, indicating a net carbon credit of 8687 t day^{-1}. Cultivated seaweed biomass removes 0.7 million tons of carbon each year (Turan and Neori 2010). Therefore, seaweed biomass may be considered as a potential bioenergy crop with the gain in reducing the CO_2 debt and other ancillary advantages, which are discussed further.

11.4 Proximate Composition of Seaweeds

Seaweeds are diverse in their chemical composition. Among the primary metabolic constituents, carbohydrate or polysaccharide is the major contributor toward bioethanol production. Seaweeds have a heterogeneous polysaccharide composition, which also varied among different divisions, that is, Chlorophyta, Rhodophyta, and Phaeophyta. The polysaccharides of seaweeds that hold potential to produce fermentable sugars after hydrolysis mainly include the galactans, carrageenan and agar from red seaweeds, alginate and laminarin from brown seaweeds, and cellulose and ulvan from green seaweeds. Carrageenan is a linear, sulfated polysaccharide composed of repeating units of alternatively linked D-galactose and anhydrogalactose, and agar is a polysaccharide composed of β-D-galactose and α-L-galactose derivatives. Unlike red seaweed, brown seaweeds are rich in alginate, a major structural polysaccharide of β-D-mannuronic acid and α-L-guluronic acid units (Jung et al. 2013). Apart from alginate, brown seaweeds also have other polysaccharides such as laminarin, a linear β-1,3-linked glucose residues with small amounts of β-1,6 linkages (Nobe et al. 2003), and fucoidan, a sulfated complex polysaccharide of fucose and a linear backbone of sulfate monosaccharides (Holtkamp et al. 2009; Adams et al. 2011). Green seaweeds are a rich source of structural polysaccharide cellulose along with a heterogeneous polysaccharide ulvan present mainly in *Ulva* species. Ulvan is a water-soluble sulfated polysaccharide composed of D-glucuronic acid, D-xylose, L-rhamnose, and sulfate (Lahaye and Robic 2007). These hydrocolloids are a valuable part of various industries and showed an increase in market value from US$644 million to US$1018 million in the past decade (Bixler and Porse 2011; Jung et al. 2013). In general, the carbohydrate content of green, red, and brown seaweeds was found to be in the range of 33%–58% DW, 31%–61% DW, and 21%–33% DW, respectively (Kumar et al. 2011). Along with the commercial values of polysaccharides, studies have shown that seaweed proteins can be consumed as human diet supplements. Data have been generated on the digestibility of seaweed proteins revealing their potential for human consumption (Fleurence 1999). The nutritional value of seaweeds has been known for centuries, but recent studies have reported proximate composition

TABLE 11.3
Proximate Composition of Seaweeds

Component (% DW)	Seaweeds		
	Green	Red	Brown
Carbohydrates	30–60	30–50	20–30
Polysaccharides	Ulvan, starch cellulose, mannan	Agar, carrageenan cellulose, lignin	Laminarin, alginate, mannitol, fucoidan, cellulose
Monosaccharides	Glucose, mannose, rhamnose, xylose, galactose, uronic acid, glucuronic acid	Glucose, galactose	Glucose, galactose, xylose, uronic acid, glucuronic acid, fucose, mannuronic acid, guluronic acid
Proteins	10–20	6–15	10–15
Lipids	1–3	0.5–1.5	1–2
Ash	13–22	5–15	14–28

Sources: Adjusted from Kumar, M. et al., *J. Appl. Phycol.*, 23, 797, 2011; Jung, K.A. et al., *Bioresour. Technol.*, 135, 182, 2013.

of seaweeds comparable to terrestrial vegetables and other food ingredients (Buchholz et al. 2012). The carbohydrate content and proximate compositions of seaweeds are summarized in Table 11.3.

11.5 Status of Bioethanol Production from Seaweeds

The potential of seaweed as a source of biofuel has been proposed by Howard Wilcox in the late 1960s. During the 1970s and 1980s, the United States carried out extensive research on macroalgae as alternative source of natural gas by anaerobic digestion of biomass with an aim of replacing the entire natural gas supply of the United States. However, this field remained abandoned afterward due to political disposition and low level of funding availabilities. In the past few years, Japan, Canada, and Europe restarted the use of macroalgae as a source of biofuel. The most common fuels produced from seaweeds are methane, ethanol, and butanol (Rothe et al. 2012). Seaweed being a renewable source of various energy molecules, this review mainly focused on illustrating the potential of seaweeds as source of ethanol. Seaweeds fulfill all the necessary requirements for an energy crop. The major advantages offered by seaweeds over terrestrial biomass are that they (1) have higher biomass production rate per unit area; (2) do not compete with agricultural plants for land; (3) require no agricultural input such as fertilizer, pesticides, and water; (4) have easier depolymerization as they do not contain lignin in their cell wall; (5) have low cost of harvesting; and (6) have high scalability (Jones and Mayfield 2012). Also, seaweeds are not an integral part of the human food chain and therefore do not possess the risk of food versus fuel concern originated from land-based biomass (Singh et al. 2011). Some of the seaweed biomass hydrolysates possess a higher hexose sugar content than that of some terrestrial lignocellulosic biomass feedstock (Chynoweth 2005; Sluiter 2006; Kim et al. 2011). These foregoing facts provide all circumstantial evidence for considering seaweeds as an effective renewable source of

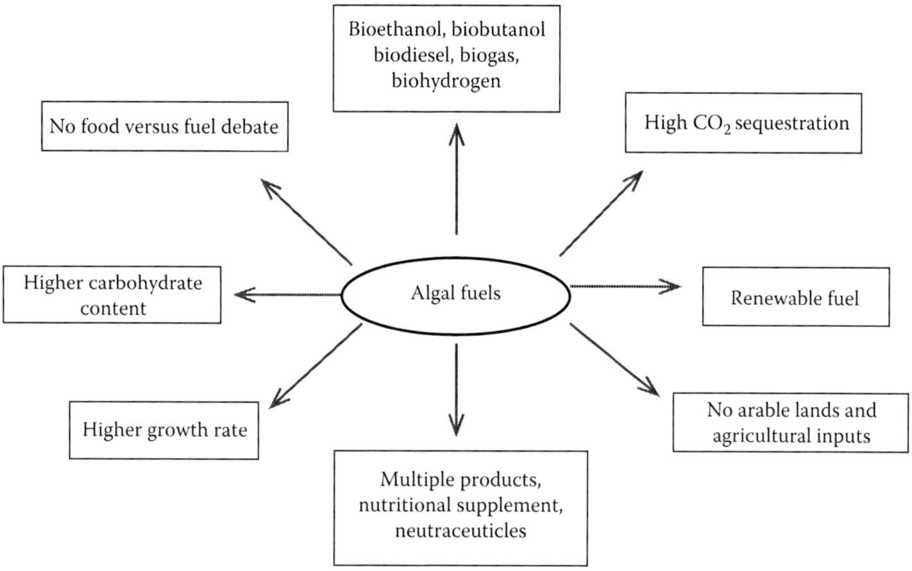

FIGURE 11.2
Algal feedstock as a source of biofuels.

bioenergy molecules. The characteristics of algal biomass as a feedstock for biofuel are summarized in Figure 11.2.

Recently, marine macroalgal species have regained considerable global attention as a source of third-generation liquid biofuel, mainly ethanol. Bioethanol production from macroalgae has been reported from all three groups of seaweeds: green seaweeds (*Ulva fasciata, U. lactuca, U. pertusa, U. armoricana*); red seaweeds (*Gracilaria verrucosa, Kappaphycus alvarezii, Gelidium amansii, G. elegans, G. salicornia, Chondracanthus tenellus, G. linoides, G. incurvata, G. vermiculophylla, Hypnea charoides, Prionitis angusta, P. divaricata, Pterocladia capillacea*); and brown seaweeds (*Laminaria japonica, L. hyperborea, Saccharina latissima, Sargassum fulvellum, S. sagamianum, Undaria pinnatifida, Alaria crassifolia, Dilophus okamurae, Eisenia bicyclis, Hizikia fusiformis, Ishige okamurae, Padina arborescens, S. ringgoldianum*) (Khambhaty et al. 2012; Daroch et al. 2013; Trivedi et al. 2013a). The production of bioethanol from seaweeds includes (1) hydrolysis of biomass into fermentable sugars followed by (2) fermentation by microbes. The saccharification and fermentation studies so far carried out for seaweeds are described in the next section and are also summarized in Table 11.4.

11.5.1 Conversion of Seaweed Carbohydrate to Fermentable Sugars

With the consideration that seaweeds can be a valuable feedstock for bioethanol production, the most daunting task is to design strategies for efficient saccharification of seaweed polysaccharides because of their chemical structure that differs from land-based crops. Saccharification of algal biomass can be achieved by chemical and enzymatic hydrolysis. Chemical hydrolysis (acid hydrolysis) is one of the viable methods largely being employed as a promising means of producing fermentable sugars. Chemical hydrolysis of seaweed polysaccharide was optimized for different acids, their concentrations, period

TABLE 11.4

Bioethanol Production from Macroalgal Feedstock

Seaweed	Type	Hydrolysis Conditions	Fermenting Strain	Ethanol Yield	References
Ulva fasciata	Green	Hot buffer + enzymatic	*Saccharomyces cerevisiae* MTCC No. 180	0.45 g g^{-1} reducing sugar	Trivedi et al. (2013a)
Gracilaria verrucosa	Red	Enzymatic	*Saccharomyces cerevisiae* HAU	0.43 g g^{-1} reducing sugar	Kumar et al. (2013)
Palmaria palmata	Red	Acidic	*Saccharomyces cerevisiae*	17.3 mg g^{-1} seaweed	Mutripah et al. (2014)
Undaria pinnatifida	Brown	Thermal acid hydrolysis	*Pichia angophorae* KCTC 17574	9.42 g L^{-1}	Cho et al. (2013)
Undaria pinnatifida	Brown	Acid + enzymatic	*Pichia angophorae* KCTC 17574	12.98 g L^{-1}	Kim et al. (2013a)
Sargassum spp.	Brown	Acid + enzymatic	*Saccharomyces cerevisiae*	—	Borines et al. (2013)
Eucheuma cottonii	Red	Amberlyst TM-15 (catalyst)	*Saccharomyces cerevisiae*	0.39 g g^{-1} sugar	Tan et al. (2013)
Ulva reticulata	Green	Enzymatic	*Saccharomyces cerevisiae* WLP099	90 L t^{-1} dried biomass	Yoza and Masutani (2013)
Gelidium amansii	Red	Thermal acid hydrolysis + enzymatic	*Scheffersomyces stipitis* KCTC 7228	20.5 g L^{-1}	Ra et al. (2013)
Kappaphycus alvarezii	Red	Acid + enzymatic	*Saccharomyces cerevisiae*	64.3 g L^{-1} (105 L t^{-1} dried biomass)	Hargreaves et al. (2013)
Eucheuma cottonii	Red	Acidic	*Saccharomyces cerevisiae*	9.6% v/v ethanol	Mansa et al. (2013)
Saccharina japonica	Brown	Acidic + enzymatic	*Saccharomyces cerevisiae* DK 410362	6.65 g L^{-1}	Lee et al. (2013a)
Kappaphycus alvarezii	Red	Acidic	*Saccharomyces cerevisiae*	0.369 g g^{-1} reducing sugar	Meinita et al. (2012)
Gelidium amansii	Red	Dilute acid hydrolysis	*Brettanomyces custersii* KCTC 18154P.	0.38 g g^{-1} reducing sugar	Park et al. (2012)
Kappaphycus alvarezii	Red	Acidic	*Saccharomyces cerevisiae* NCIM 3523	0.147 L ethanol kg^{-1} granules	Khambhaty et al. (2012)
Saccharina japonica	Brown	Thermal acid hydrolysis	*Pichia angophorae* KCTC 17574	0.169 g g^{-1} reducing sugar	Jang et al. (2012)
Saccharina japonica	Brown	Engineered microbial enzyme	Engineered BAL1611	0.41 g g^{-1} reducing sugar	Wargacki et al. (2012)
Laminaria japonica	Brown	Thermal liquefaction	*Pichia stipitis* KCTC7228	2.9 g L^{-1} using 100 g L^{-1} algae	Lee and Lee (2012)
Sargassum sagamianum	Brown	Thermal liquefaction	*Pichia stipitis* CBS 7126	0.43–0.44 g g^{-1} reducing sugar	Lee et al. (2012)
Gelidium elegans	Red	Enzymatic	*Saccharomyces cerevisiae* IAM 4178	55 g L^{-1} (5.5%)	Yanagisawa et al. (2011)
Laminaria japonica	Brown	Acid + enzymatic	Ethanologenic strain *E. coli* KO11	0.41 g g^{-1} reducing sugar	Kim et al. (2011)

(Continued)

TABLE 11.4 (*Continued*)

Bioethanol Production from Macroalgal Feedstock

Seaweed	Type	Hydrolysis Conditions	Fermenting Strain	Ethanol Yield	References
Sargassum sagamianum	Brown	Thermal hydrolysis	*Pichia stipitis CBS 7126*	0.386 g g^{-1} reducing sugar	Yeon et al. (2011)
Laminaria digitata	Brown	Shredding and enzymatic	*Pichia angophorae*	167 mL kg^{-1} algae	Adams et al. (2011)
Laminaria japonica	Brown	Acid + enzymatic	*Saccharomyces cerevisiae*	143 mL kg^{-1} floating residues	Ge et al. (2011)
Sargassum sagamianum	Brown	—	Pichia stipitis	0.133–0.233 g g^{-1} reducing sugar	Yeon et al. (2010)
Saccharina latissima	Brown	Shredding and enzymatic	*Saccharomyces cerevisiae* Ethanol red yeast	0.45% (v/v)	Adams et al. (2009)
Dilophus okamurae	Brown	Enzymatic	*Mixture of B5201 (Lactobacillus), Y5201 (Debaryomyces I) and Y5206 (Candida I)*	0.04 g 100 mL^{-1}	Uchida and Murata (2004)
Eisenia bicyclis				0.03 g 100 mL^{-1}	
Hizikia fusiformis				0.24 g 100 mL^{-1}	
Ishige okamurae				0.10 g 100 mL^{-1}	
Laminaria japonica				0.15 g 100 mL^{-1}	
Padina arborescens				0.08 g 100 mL^{-1}	
Sargassum ringgoldianum				0.04 g 100 mL^{-1}	
Undaria pinnatifida				0.38 g 100 mL^{-1}	
Undaria pinnatifida				0.12 g 100 mL^{-1}	
Ulva spp.	Green			0.16 g 100 mL^{-1}	
Ulva spp.				0.41 g 100 mL^{-1}	
Chondracanthus tenellus	Red	Enzymatic	*Mixture of B5201 (Lactobacillus), Y5201 (Debaryomyces I), and Y5206(Candida I)*	0.18 g 100 mL^{-1}	Uchida and Murata (2004)
Gelidium linoides				0.12 g 100 mL^{-1}	
Gracilaria incurvata				0.12 g 100 mL^{-1}	
Gracilaria vermiculophylla				0.23 g 100 mL^{-1}	
Hypnea charoides				0.16 g 100 mL^{-1}	
Prionitis angusta				0.17 g 100 mL^{-1}	
Prionitis divaricata				0.41 g 100 mL^{-1}	
Pterocladia capillacea				0.08 g 100 mL^{-1}	
Laminaria hyperborea	Brown	Thermal	*Pichia angophorae*	0.43 g g^{-1} sugar	Horn et al. (2000)

of hydrolysis, and reaction temperature in order to realize higher yields of fermentable sugars from red seaweed (*K. alvarezii, Palmaria palmata, Eucheuma cottonii*), brown seaweed (*S. japonica, U. pinnatifida, Sargassum* spp.), and green seaweed (*U. lactuca*) (Jang et al. 2012; Khambhaty et al. 2012; Meinita et al. 2012; Potts et al. 2012; Borines et al. 2013; Cho et al. 2013; Mutripah et al. 2014). The combination of high temperatures and strong acids for hydrolysis sometimes leads to the degradation of products and accumulation of nonsugar by-products such as 5-HMF and organic acids (formic acid and levulinic acid), which then inhibits the downstream fermentation of resultant saccharides (Kim et al. 2011; Mathew et al. 2011). Therefore, acid hydrolysis is required to be optimized for realizing the best yield of desired product. Environmental hazards are another inherent problem associated with the chemical hydrolysis that therefore necessitated alternative greener approaches.

Enzymatic hydrolysis represents a greener approach for the saccharification of algal biomass. The diversity and heterogeneity among polysaccharides of seaweeds demands for substrate-specific enzyme(s). Nevertheless, studies were conducted on saccharification of algal biomass using various commercial enzyme cocktails, mainly cellulases (Celluclast 1.5, Cellulase 22119, Novoprime 959, Novoprime 969, Viscozyme L) (Kim et al. 2011; Trivedi et al. 2013a) and a few on laminarinase (Adams et al. 2009), amyloglucosidase (AMG 300L) (Kim et al. 2011), and meicelase (Yanagisawa et al. 2011). The carbohydrate profiles of seaweeds are different from land-based crops as they contained special types of polysaccharide such as carrageenan, laminarin, alginate, agar, and ulvan. The nonavailability of polysaccharide lyases for complex seaweed heteropolysaccharides encouraged various research groups to isolate microorganisms capable of hydrolyzing these uncommon polysaccharides. As a result, marine bacteria *Pseudoalteromonas* sp. QY 203 and *Pseudoalteromonas carrageenovora* have been reported for κ-carrageenase activity (Lemoine et al. 2009; Shangyong et al. 2013). They cleave the β-1,4 linkage of carrageenan and resultantly produce oligogalactans, that is, neocarrabiose or neoagarobiose. Agarase is the most common group of enzymes studied for hydrolysis of algal biomass. Gupta et al. (2013) reported exo-β-agarase from an endophytic marine bacterium *Pseudomonas* spp. This enzyme on incubation with *G. dura* biomass yielded 20% galactose on DW basis. Recently, Jeng et al. (2011) studied the crystal structures of a laminarinase extracted from *Thermotoga maritima* MSB8 to get insights into its catalytic domain. Laminarinase catalyzes the β-1,3 linkage in laminarin to glucose. Alginate is the main polysaccharide of brown seaweed. Various microorganisms have been studied for alginate degradation (Huang et al. 2013; Kim et al. 2013b). Recently, Thomas et al. (2013) characterized two marine alginate lyase systems from *Zobellia galactanivorans* revealing their distinct modes of action. Alginate lyase via an endolytic β-elimination reaction catalyzes the depolymerization of alginate into oligomers that are further degraded to monomers via oligoalginate lyase (Oal) (Wargacki et al. 2012). Fucoidan is another polysaccharide found in brown seaweed, and a few reports have been available on the fucoidanase enzyme from marine microbes. Silchenko et al. (2013) reported hydrolysis of fucoidan by fucoidanase isolated from marine bacterium *Formosa algae*. Research on seaweed bioethanol is not only restricted to isolate and characterize extracellular enzymes but also investigated the means of improving the catalytic efficiencies of the enzymes by increasing the substrate accessibilities. For example, Trivedi et al. reported ionic liquid-stable, organic solvent–tolerant, and alkali-halotolerant cellulase from marine microbes isolated from green and brown seaweeds (Trivedi et al. 2011a,b, 2013b). ILs impart higher solubility to cellulose that as a result improved the substrate accessibility for enzyme and ultimately improved the catalytic efficiency. Recently, several microorganisms have also been genetically engineered for different enzyme production such as agarase (Xie et al. 2012; Lee et al. 2013b), κ-carrageenase (Liu et al. 2013), and cellulase (Ibrahim et al. 2013).

In addition to direct enzymatic hydrolysis, various pretreatment strategies were also investigated in order to realize improvement in reducing sugar yields. These pretreatment methods can be categorized as physical, chemical, and biological. Different pretreatments such as dilute acid, liquid ammonia, dry heat, and hot buffer were used to unlock the cellular components. Subsequent enzymatic hydrolysis converted the polysaccharides into fermentable sugar (Trivedi et al. 2013a). Most commonly, pretreatment with dilute acid (0.1%–0.3% H_2SO_4, 0.1 N HCl) at higher temperature of 120°C–130°C for 15–120 min was employed on various seaweed biomasses (Nguyen et al. 2009; Kim et al. 2011; Lee and Lee 2011; Meinita et al. 2012). Also, reduction in size by cutting (Adams et al. 2009), milling (Horn et al. 2000; Lee and Lee 2011; Yanagisawa et al. 2011), crushing (Kim et al. 2011), and homogenization (Wang et al. 2011) has also been the integral part for efficient saccharification. For the very first time, Tan et al. (2013) explored Amberlyst (TM)-15 as a potential heterogeneous catalyst to hydrolyze carbohydrates from red seaweed *E. cottonii* to simple sugars.

11.5.2 Fermentable Sugars to Bioethanol

The schematic diagram on ethanol production from seaweed biomass including hydrolysis and fermentation is given in Figure 11.3. Seaweed-housed polysaccharides on hydrolysis yield glucopyranose (cellulose and laminarin), galactopyranose (agar and carrageenan), and mannitol in species-specific manner as described earlier. Though glucose can be easily fermented into ethanol, technologies for fermentation of other monosugars are required to be established to achieve higher yields of ethanol. The most common organism to convert sugars obtained from hydrolysis of seaweed biomass into ethanol is *Saccharomyces cerevisiae*. Different strains of *S. cerevisiae* such as MTCC 180, HAU, *WLP099*, DK 410362, NCIM 3523, and IAM 4178 have been used for the production of bioethanol from different seaweeds (Yanagisawa et al. 2011; Khambhaty et al. 2012; Kumar et al. 2013; Lee et al. 2013a; Mutripah et al. 2014; Trivedi et al. 2013a). *Pichia angophorae* and *P. stipitis* are the other strains used for fermentation in addition to the most common fermentation strain *S. cerevisiae* (Horn et al. 2000; Adams et al. 2011; Yeon et al. 2011; Jang et al. 2012; Lee and Lee 2012; Cho et al. 2013).

Trivedi et al. (2013a) and Yoza and Masutani (2013) reported bioethanol production from cellulosic-rich green seaweed *U. fasciata* and *U. reticulata* by enzymatic hydrolysis of biomass followed by fermentation using *S. cerevisiae* MTCC No. 180 and *S. cerevisiae* WLP099, respectively. The former research group showed an interesting fact that there is no need of nitrogen source supplementation (yeast extract and peptone) in fermentation medium as ethanol yields were the same in both cases of with and without nitrogen source supplementation. Also, they showed a strategy for reusing the enzyme two times for hydrolysis of seaweed biomass. Lee et al. (2012) also used a cost-effective corn steep liquor (CSL) medium, replacing yeast extract and peptone, for bioethanol production from the hydrolysate of *S. sagamianum* in surface-aerated fermenter by repeated-batch operation.

The major polysaccharide of brown seaweeds, laminarin, on hydrolysis produced easily fermentable sugar glucose, while mannitol is another major sugar present in brown seaweeds that cannot be readily fermented by conventional yeast or ethanogenic bacteria. Nevertheless, a facultative anaerobe *Zymobacter palmae* isolated from palm sap was shown as a novel ethanogenic bacteria capable of fermenting mannitol (Horn et al. 2000).

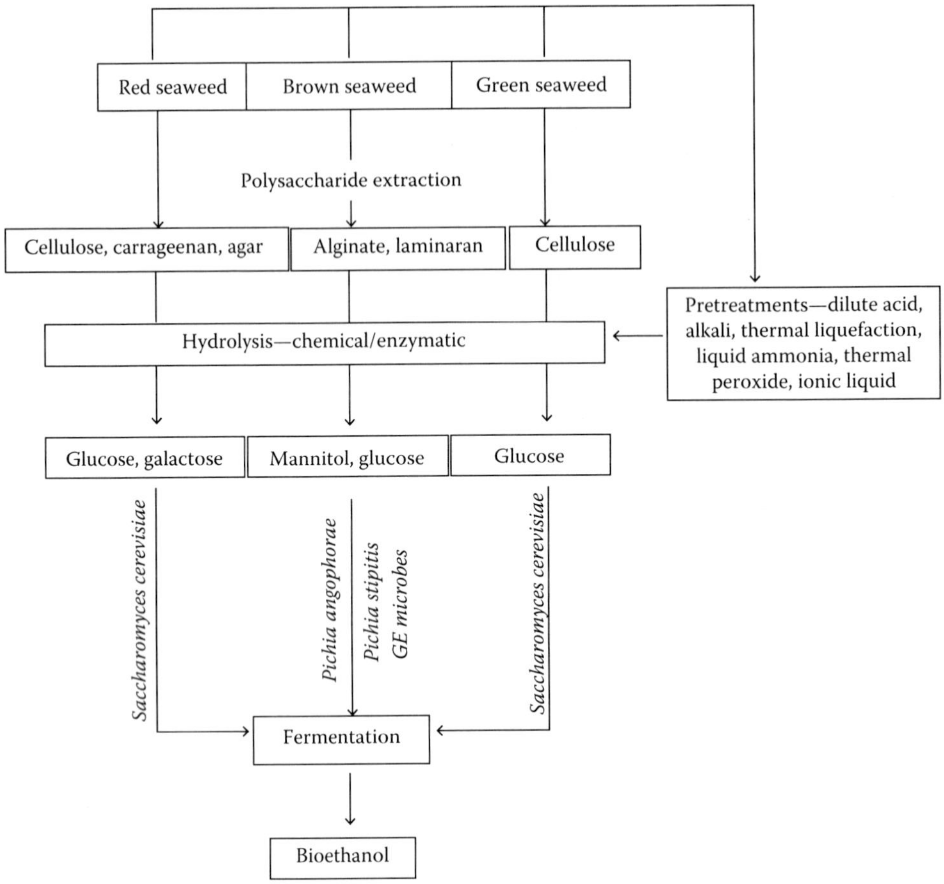

FIGURE 11.3
General scheme for bioethanol production from macroalgae.

Kim et al. (2011) used recombinant *Escherichia coli* KO11 to ferment both mannitol and glucose for bioethanol production from brown seaweed *L. japonica* and estimated a higher ethanol yield than that of the same obtained from the fermentation with *S. cerevisiae*. In another study, Wargacki et al. (2012) designed a consolidated bioprocessing platform by transforming a cluster of genes and transporter involved in alginate, mannitol, and glucose metabolism in *E. coli* to produce ethanol from a brown macroalga *S. japonica*. The strain BAL1611 engineered in this way utilized alginate, mannitol, and glucan as a substrate and gave ethanol yield of ~0.281 g ethanol g^{-1} biomass, which corresponds to a yield of ~0.41 g ethanol g^{-1} total sugars. Saccharification of red seaweed galactans (agar, carrageenan, and cellulose) gave galactose and glucose as fermentable sugars that subsequently fermented into bioethanol using different strain of *S. cerevisiae* such as *S. cerevisiae* HAU (Kumar et al. 2013), NCIM 3523 (Khambhaty et al. 2012), and IAM 4178 (Yanagisawa et al. 2011). Recently, Park et al. (2012) for the first time used *Brettanomyces custersii* KCTC 18154P for bioethanol production from red seaweeds *G. amansii*. This strain has the ability to use both galactose and glucose as a substrate for ethanol production (Yoon et al. 2011).

Few studies are also available on simultaneous saccharification and fermentation but are mainly restricted to brown seaweeds. Jang et al. (2012) developed a platform for simultaneous saccharification and fermentation after optimization of pretreatment conditions and enzymatic hydrolysis. They performed thermal acid hydrolysis of *S. japonica* with and without *Bacillus* sp. JS-1. The corresponding reducing sugar yields obtained were 21% and 69.1%, respectively, on total carbohydrate basis. For simultaneous saccharification and fermentation, both hydrolytic strain *Bacillus* sp. JS-1 and yeast *P. angophorae* were cocultured. Resultantly, an ethanol yield of 7.7 g L^{-1} was obtained that corresponds to 33.3% of theoretical yield. Lee and Lee (2011) isolated microorganisms from nuruk (mixture of yeast, bacteria, and fungi used to make Korean traditional alcohol) that showed the ability to produce ethanol by metabolizing alginate from *L. japonica*. Apart from bioethanol, seaweeds with higher carbohydrate contents provide suitable conditions to microbes for the production of other high-value chemicals and fuels such as lactic acid (Hwang et al. 2011; Mazumdar et al. 2014); 2,3-butanediol (Mazumdar et al. 2013); 5-hydroxymethyl furfural (HMF); levulinic acid and formic acid (Mondal et al. 2013); biobutanol (Potts et al. 2012); acetone, butanol, and ethanol (ABE) (Wal et al. 2012); acetic acid; citric acid; succinic acid; adipinic acid, butyraldehyde, propionic acid; and glycerol (Kraan 2013).

Seaweeds are historically considered as a nutrient-rich resource with edibility as human food and as a source of hydrocolloids. Proteins are the major component for seaweed nutrition value along with lipids, fatty acids, dietary fibers, vitamins, and macro- and micronutrients. The studies conducted to evaluate the digestibility of seaweed proteins showed their sensitivity to human intestinal juices adding value to their prospect as food ingredients. This has led to an impetus to design strategies for extracting a stream of products from seaweed biomass. The other reason for designing such a strategy is to avoid another foreseen debate of fuel versus hydrocolloid that may arise if these multibillion industrial constituents are used for bioethanol production. Studies by Kumar et al. (2013) and Khambhaty et al. (2012) demonstrated extraction of agar and liquid sap with fertilizer potential from *G. verrucosa* and *K. alvarezii*, respectively, and utilized the residues for the production of bioethanol.

11.6 Research Activities on Seaweed Ethanol across the Globe

In this review, an overview on the global research activities on seaweed-based ethanol production is given to indicate the exhilaration in this field globally. Currently, several companies, agencies, and universities are involved in different projects on seaweed-based ethanol production (Oilgae 2010; Rothe et al. 2012.). A few examples are as follows:

- In Denmark, the Danish Strategic Research Council started the Macro Algae Biorefinery (MAB3) project (2012–2016) on conversion of brown seaweeds *S. latissima* and *L. digitata* into bioethanol, biobutanol, and biogas. Another targeted seaweed species is *U. lactuca* from which bioethanol is produced following the method by Horn et al. (2000) (Huesemann et al. 2010).
- In Norway, a 3-year collaborative project between Statoil Company and California-based Bio Architecture Lab, Inc. (BAL) has been started on ethanol production from seaweed using BAL's consolidated bioprocessing technology.
- In Ireland, the National University of Ireland, Galway (NUIG), has initiated work on the development of an enzymatic process for seaweed fermentation.

- In the United Kingdom, the Supergene project has been started in collaboration with Irish researchers for the production of ethanol from seaweed as well as hydroliquefaction processes.
- A collaborative project (INTERREG IVA) between Scottish and Irish researchers is focused on large-scale biofuel production via sustainable production of algae.
- In Indonesia, the Korea Institute of Technology (KIT) and collaborators started seaweed cultivation for bioethanol production.
- In India, Central Salt and Marine Chemicals Research Institute (CSMCRI) has produced ethanol from red seaweed *K. alvarezii* and green seaweed *U. fasciata.*
- In the United States, the University of Maine completed a project on the life-cycle assessment of macroalgae for biofuel production and studied the conversion of Algefiber® from *E. spinosum* (red algae) into carboxylic acids and alcohol fuels.
- In Germany, offshore cultivation of brown seaweed *L. saccharina* has been carried out for food as well as fuel production (Buck and Buchholz 2004).
- In Japan, the Ocean Sunrise Project started with the aim of cultivation of brown seaweed *S. fulvellum* and its conversion to bioethanol (Aizawa et al. 2007). Tohoku University and Tohoku Electric Power Company discovered a natural yeast that converts macroalgae into ethanol with a rate of 200 mL (1 kg macroalgae)$^{-1}$.
- In Brazil, the State of Rio Grande do Norte Agricultural Research Company (EMPARN) is developing a technique for the production of fuel from seaweed.
- In South Korea, the South Korea National Energy Ministry plans to make a 35,000 ha offshore seaweed forest for the production of macroalgae-based ethanol. Recently, KIT has produced bioethanol from red algae *G. amansii* (Yoon et al. 2010).
- Seaweed Energy Solutions is involved in offshore cultivation and conversion to biogas and bioethanol (http://www.seaweedenergysolutions.com).
- Butamax Advance Fuels–DuPont–BAL–Statoil is involved in offshore kelp cultivation and conversion to ethanol and butanol (http://www.butamax.com/; http://www.ba-lab.com/).
- Blue Sun Energy, a Colorado-based company, is involved in jet fuel production (http://www.gobluesun.com/index.php).
- Holmfjord AS8, a Norway-based company, is involved in biofuel production from seaweed (www.holmfjord.no).
- Green Gold Algae and Seaweed Sciences, Inc. is involved in land-based ponds and conversion to ethanol (http://www.goldgreen.com).

11.7 Challenges and Future Plans in Seaweed Fuel

Marine macrophytic algae hold promising potential as a renewable source of the production of bioethanol. Though progress has been made in the development of strategies for seaweed bioethanol production, many challenges are yet to be solved. Commonly employed strategies of acid-based saccharification pose some potential problems such as intermediate production of furfurals, 5-hydroxymethylfurfural, and organic acids such as formic acid and levulinic acid, which inhibit the downstream fermentation process besides posing environmental hazards. Removal of these inhibitors thereby put forward hurdles in the success story of ethanol production. Enzymatic hydrolysis indeed presents the green approach but suffers from the high cost of commercial hydrolyzing enzyme. U.S. Department of Energy awarded $17 million each to two big industries: Genencor International (http://www.genencor.com/) and Novozymes, Inc. (http://www.novozymes.com/), with a goal to reduce enzyme cost by 10-fold (Hahn-Hagerdal et al. 2006).

Moreover, saccharification of seaweed polysaccharides and the fermentation of resultant sugars require specific microbes distinct from the ones employed for terrestrial biomass. The studies of the development of recombinant strains to utilize sugars from hydrolysis of macroalgal carbohydrates such as agar, carrageenan, and alginate are in their infancy (Daroch et al. 2013). The fuel versus hydrocolloid debate cannot be avoided if these polysaccharides are used for ethanol production. However, understanding the system biology and specialized metabolism of seaweeds for these polysaccharides may help reconstruct the metabolic pathways in vitro to get the desired product through a synthetic biology approach. Therefore, the success in this direction depends largely on the interdisciplinary collaboration of biologists, chemists, and engineers that will optimize the different parameters of different avenues. Beside these major technical challenges, yet another face-off is the technical constraint and economic feasibility of biomass production and harvest either by off- or onshore cultivation (Rothe et al. 2012; Kraan et al. 2013). Therefore, the development of cost-effective methods to grow, harvest, transport, and process macroalgae for various industrial applications is the most crucial need in the future.

11.8 Conclusion

The biomass characteristics of seaweed hold potential to overcome the economic hurdles and the life-cycle challenges associated with terrestrial biomass in renewable energy production. High photosynthesis efficiency aiding in the reduction in carbon debt, bioremediation, and controlling eutrophication are other deliverables from seaweed biomass if cultivated at large scale. Design of biorefineries for extracting a stream of multiple valuable products along with ethanol from seaweed will further strengthen their holding as a promising renewable energy crop. Seaweed as a source of alternative renewable fuel will substantially reduce the utility of fossil fuel. Further, macroalgal farming in the ocean will spare the land for production of food crops to mitigate the global concerns over food security.

Acknowledgment

Nitin Trivedi and Vishal Gupta gratefully acknowledge the Council of Scientific and Industrial Research (CSIR), New Delhi, for awarding senior research fellowships.

References

Adams, J. M., J. A. Gallagher, and I. S. Donnison. 2009. Fermentation study on *Saccharina latissima* for bioethanol production considering variable pre-treatments. *Journal of Applied Phycology* 21:569–574.

Adams, J. M. M., T. A. Toop, I. S. Donnison, and J. A. Gallagher. 2011. Seasonal variation in *Laminaria digitata* and its impact on biochemical conversion routes to biofuels. *Bioresource Technology* 102:9976–9984.

Aizawa, M., K. Asaoka, M. Atsumi, and T. Sakou. 2007. Seaweed bioethanol production in Japan—The Ocean Sunrise Project. OCEANS 2007, IEEE, Vancouver, Canada, 1–5.

Bixler, H. J. and H. Porse. 2011. A decade of change in the seaweed hydrocolloids industry. *Journal of Applied Phycology* 23:321–335.

Borines, M. G., R. L. de Leon, and J. L. Cuello. 2013. Bioethanol production from the macroalgae *Sargassum* spp. *Bioresource Technology* 138:22–29.

Buchholz, C. M., G. Krause, and B. H. Buck. 2012. Seaweed and man. In *Seaweed Biology, Ecological Studies*, eds. C. Wiencke and K. Bischof. Berlin, Germany: Springer, pp. 471–493.

Buck, B. H. and C. M. Buchholz. 2004. The offshore-ring: A new system design for the open ocean aquaculture of macroalgae. *Journal of Applied Phycology* 16:355–368.

Cherubini, F. 2010. The biorefinery concept: Using biomass instead of oil for producing energy and chemicals. *Energy Conversion and Management* 51:1412–1421.

Chiaramonti, D. 2007. Bioethanol: Role and production technologies. In *Improvement of Crop Plants for Industrial End Uses*, ed. P. Ranalli. Dordrecht, the Netherlands: Springer, pp. 209–251.

Cho, Y., H. Kim, and S.-K. Kim. 2013. Bioethanol production from brown seaweed, *Undaria pinnatifida*, using NaCl acclimated yeast. *Bioprocess and Biosystems Engineering* 36:713–719.

Chung, I. K., J. H. Oak, J. A. Lee, J. A. Shin, J. G. Kim, and K.-S. Park. 2013. Installing kelp forests/seaweed beds for mitigation and adaptation against global warming: Korean project overview. *ICES Journal of Marine Science* 70:1038–1044.

Chynoweth, D. P. 2005. Renewable biomethane from land and ocean energy crops and organic wastes. *Hortscience* 40:283–286.

Daroch, M., S. Geng, and G. Wang. 2013. Recent advances in liquid biofuel production from algal feedstocks. *Applied Energy* 102:1371–1381.

Fleurence, J. 1999. Seaweed proteins: Biochemical, nutritional aspects and potential uses. *Trends in Food Science and Technology* 10:25–28.

Food and Agriculture Organisation of the United Nations. 2010. The state of world fisheries and aquaculture. Rome, Italy: FAO.

Gao, K. and K. R. McKinley. 1994. Use of macroalgae for marine biomass production and CO_2 remediation: A review. *Journal of Applied Phycology* 6:45–60.

Ge, L., P. Wang, and H. Mou. 2011. Study on saccharification techniques of seaweed wastes for the transformation of ethanol. *Renewable Energy* 36:84–89.

Girio, F. M., C. Fonseca, F. Carvalheiro, L. C. Duarte, S. Marques, and R. Bogel-Lukasik. 2010. Hemicelluloses for fuel ethanol: A review. *Bioresource Technology* 101:4775–4800.

Gupta, V., N. Trivedi, M. Gupta, C. R. K. Reddy, and B. Jha. 2013. Purification and characterization of exo-β-agarase from an endophytic marine bacterium and its catalytic potential in bioconversion of red algal cell wall polysaccharides into galactans. *Biomass and Bioenergy* 49:290–298.

Hahn-Hagerdal, B., M. Galbe, M. F. Gorwa-Grauslund, G. Liden, and G. Zacchi. 2006. Bio-ethanol—The fuel of tomorrow from the residues of today. *Trends in Biotechnology* 24:549–556.

Hargreaves, P. L., C. A. Barcelos, A. C. Augusto da Costa, and N. Pereira, Jr. 2013. Production of ethanol 3G from *Kappaphycus alvarezii*: Evaluation of different process strategies. *Bioresource Technology* 134:257–263.

Henry, R. J. 2009. *Plant Resources for Food Fuel and Conservation*. London, U.K.: Earthscan.

Holdt, S. L. and S. Kraan. 2011. Bioactive compounds in seaweed: Functional food applications and legislation. *Journal of Applied Phycology* 23:543–597.

Holtkamp, A. D., S. Kelly, R. Ulber, and S. Lang. 2009. Fucoidans and fucoidanases-focus on techniques for molecular structure elucidation and modification of marine polysaccharides. *Applied Microbiology and Biotechnology* 82:1–11.

Horn, S. J., I. M. Aasen, and K. Ostgaard. 2000. Ethanol production from seaweed extract. *Journal of Industrial Microbiology and Biotechnology* 25:249–254.

Huang, L., J. Zhou, X. Li, Q. Peng, H. Lu, and Y. Du. 2013. Characterization of new alginate lyase from newly isolated *Flavobacterium* sp. S20. *Journal of Industrial Microbiology and Biotechnology* 40:113–122.

Huesemann, M., G. Roesjadi, J. Benemann, and F. B. Metting. 2010. Biofuels from microalgae and seaweeds. In *Biomass to Biofuels: Strategies for Global Industries*, eds., A. A. Vertès, N. Qureshi, H. P. Blaschek, and H. Yukawa. West Sussex, U.K.: John Wiley & Sons Ltd, pp. 165–184.

Hwang, H. J., S. Y. Lee, S. M. Kim, and S. B. Lee. 2011. Fermentation of seaweed sugars by *Lactobacillus* species and the potential of seaweed as a biomass feedstock. *Biotechnology and Bioprocess Engineering* 16:1231–1239.

Ibrahim, E., K. D. Jones, E. N. Hosseny, and J. Escudero. 2013. Molecular cloning and expression of cellulase and polygalacturonase genes in *E. coli* as a promising application for biofuel production. *Journal of Petroleum and Environmental Biotechnology* 4:147.

ISO Ethanol Yearbook. 2012. London, U.K.: International Sugar Organization.

Jang, J.-S., Y. Cho, G.-T. Jeong, and S.-K. Kim. 2012. Optimization of saccharification and ethanol production by simultaneous saccharification and fermentation (SSF) from seaweed, *Saccharina japonica*. *Bioprocess and Biosystems Engineering* 35:11–18.

Jeng, W.-Y., N.-C. Wang, C.-T. Lin, L.-F. Shyur, and A. H.-J. Wang. 2011. Crystal structures of the laminarase catalytic domain from *Thermotoga maritima* MSB8 in complex with inhibitors. Essential residues for b-1,3- and b-1,4-glucan selection. *The Journal of Biological Chemistry* 286:45030–45040.

John, R. P., G. S. Anisha, K. M. Nampoothiri, and A. Pandey. 2011. Micro and macroalgal biomass: A renewable source for bioethanol. *Bioresource Technology* 102:186–193.

Jones, C. S. and S. P. Mayfield. 2012. Algae biofuels: Versatility for the future of bioenergy. *Current Opinion in Biotechnology* 23:346–351.

Jung, K. A., S.-R. Lim, Y. Kim, and J. M. Park. 2013. Potentials of macroalgae as feedstocks for biorefinery. *Bioresource Technology* 135:182–190.

Kaladharan, P., S. Veena, and E. Vivekanandan. 2009. Carbon sequestration by a few marine algae: Observation and projection. *Journal of the Marine Biological Association of India* 51:107–110.

Khambhaty, Y., K. Mody, M. R. Gandhi, S. Thampy, P. Maiti, H. Brahmbhatt, K. Eswaran, and P. K. Ghosh. 2012. *Kappaphycus alvarezii* as a source of bioethanol. *Bioresource Technology* 103:180–185.

Kim, H., C. H. Ra, and S.-K. Kim. 2013a. Ethanol production from seaweed (*Undaria pinnatifida*) using yeast acclimated to specific sugars. *Biotechnology and Bioprocess Engineering* 18:533–537.

Kim, E. J., A. Fathoni, G.-T. Jeong, H. D. Jeong, T.-J. Nam, I.-S. Kong, and J. K. Kim. 2013b. *Microbacterium oxydans*, a novel alginate- and laminarin-degrading bacterium for the reutilization of brown-seaweed waste. *Journal of Environmental Management* 130:153–159.

Kim, N.-J., H. Li, K. Jung, H. N. Chang, and P. C. Lee. 2011. Ethanol production from marine algal hydrolysates using *Escherichia coli* KO11. *Bioresource Technology* 102:7466–7469.

Kraan, S. 2013. Mass-cultivation of carbohydrate rich macroalgae, a possible solution for sustainable biofuel production. *Mitigation and Adaptation Strategies for Global Change* 18:27–46.

Kumar, M., P. Kumari, N. Trivedi, M. K. Shukla, V. Gupta, C. R. K. Reddy, and B. Jha. 2011. Minerals, PUFAs and antioxidant properties of some tropical seaweeds from Saurashtra coast of India. *Journal of Applied Phycology* 23:797–810.

Kumar, S., R. Gupta, G. Kumar, D. Sahoo, and R. C. Kuhad. 2013. Bioethanol production from *Gracilaria verrucosa*, a red alga, in a biorefinery approach. *Bioresource Technology* 135:150–156.

Lahaye, M. and A. Robic. 2007. Structure and functional properties of Ulvan, a polysaccharide from green seaweeds. *Biomacromolecules* 8:1765–1774.

Lee, J. Y., P. Li, J. Lee, H. J. Ryu, and K. K. Oh. 2013a. Ethanol production from *Saccharina japonica* using an optimized extremely low acid pretreatment followed by simultaneous saccharification and fermentation. *Bioresource Technology* 127:119–125.

Lee, S.-E., J.-E. Lee, G.-Y. Shin, W. Y. Choi, D. H. Kang, H.-Y. Lee, K.-H. Jung. 2012. Development of a practical and cost-effective medium for bioethanol production from the seaweed hydrolysate in surface-aerated fermenter by repeated-batch operation. *Journal of Microbiology and Biotechnology* 22:107–113.

Lee S.M. and J.H. Lee. 2011. The isolation and characterization of simultaneous saccharification and fermentation microorganisms for *Laminaria japonica* utilization. *Bioresource Technology* 102:5962–5967.

Lee, S.-M. and J.-H. Lee. 2012. Ethanol fermentation for main sugar components of brown-algae using various yeasts. *Journal of Industrial and Engineering Chemistry* 18:16–18.

Lee, Y., C. Oh, M. De Zoysa, H. Kim, W. D. Wickramaarachchi, I. Whang, D. H. Kang, and J. Lee. 2013b. Molecular cloning, overexpression, and enzymatic characterization of glycosyl hydrolase family 16 β-agarase from marine bacterium *Saccharophagus* sp. AG21 in *Escherichia coli*. *Journal of Microbiology and Biotechnology* 23:913–922.

Lemoine, M., P. N. Collen, and W. Helbert. 2009. Physical state of kappa-carrageenan modulates the mode of action of kappa-carrageenase from *Pseudoalteromonas carrageenovora*. *Biochemical Journal* 419:545–553.

Li, S., P. Jia, L. Wang, W. Yu, and F. Han. 2013. Purification and characterization of a new thermostable kappa-carrageenase from the marine bacterium *Pseudoalteromonas* sp. QY203. *Journal of Ocean University of China* 12:155–159.

Licht, F. O. 2006. *World Ethanol Markets: The Outlook to 2015*. Tunbridge Wells, Agra Europe special report, U.K.

Liu, Z., G. Li, Z. Mo, and H. Mou. 2013. Molecular cloning, characterization, and heterologous expression of a new κ-carrageenase gene from marine bacterium *Zobellia* sp. ZM-2. *Applied Microbiology and Biotechnology* 97:10057–10067.

Lu, J., C. Sheahan, and P. C. Fu. 2011. Metabolic engineering of algae for fourth generation biofuels production. *Energy and Environmental Science* 4:2451–2466.

Mansa, R. F., W.-F. Chen, S.-J. Yeo, Y.-Y. Farm, and H. A. Bakar. 2013. Fermentation study on macroalgae *Eucheuma cottonii* for bioethanol production via varying acid hydrolysis. In *Advances in Biofuels*, eds. R. Pogaku and R. Hj. Sarbatly. New York: Springer Science + Business Media, pp. 219–240.

Mathew, A. K., K. Chaney, M. Crook, and A. C. Humphries. 2011. Dilute acid pre-treatment of oilseed rape straw for bioethanol production. *Renewable Energy* 36:2424–2432.

Mazumdar, S., J. Bang, and M.-K. Oh. 2014. L-Lactate production from seaweed hydrolysate of *Laminaria japonica* using metabolically engineered *Escherichia coli*. *Applied Biochemistry and Biotechnology* 172:1938–1952.

Mazumdar, S., J. Lee, and M.-K. Oh. 2013. Microbial production of 2,3 butanediol from seaweed hydrolysate using metabolically engineered *Escherichia coli*. *Bioresource Technology* 136:329–336.

McHugh, D. J. 2003. A guide to the seaweed industry. FAO Fisheries Technical Paper.

MECAS(12)19. 2012. *World Fuel Ethanol Outlook to 2020*. London, U.K.: International Sugar Organization.

Meinita, M. D. N., J.-Y. Kang, G.-T. Jeong, H. M. Koo, S. M. Park, and Y.-K. Hong. 2012. Bioethanol production from the acid hydrolysate of the carrageenophyte *Kappaphycus alvarezii* (cottonii). *Journal of Applied Phycology* 24:857–862.

Melis, A. 2012. Photosynthesis-to-fuels: From sunlight to hydrogen, isoprene, and botryococcene production. *Energy and Environmental Science* 5:5531–5539.

Mondal, D., M. Sharma, P. Maiti, K. Prasad, R. Meena, A. K. Siddhanta, P. Bhatt et al. 2013. Fuel intermediates, agricultural nutrients and pure water from *Kappaphycus alvarezii* seaweed. *RSC Advances* 3:17989–17997.

Muraoka, D. 2004. Seaweed resources as a source of carbon fixation. *Bulletin of Fisheries Research Agency* 1:59–63.

Mutripah, S., M. D. N. Meinita, J.-Y. Kang, G.-T. Jeong, A. B. Susanto, R. E. Prabowo, and Y.-K. Hong. 2014. Bioethanol production from the hydrolysate of Palmaria palmate using sulfuric acid and fermentation with brewer's yeast. *Journal of Applied Phycology* 26:687–693.

Nguyen, M. T., S. P. Choi, J. Lee, J. H. Lee, and S. J. Sim. 2009. Hydrothermal acid pretreatment of *Chlamydomonas reinhardtii* biomass for ethanol production. *Journal of Microbiology and Biotechnology* 19:161–166.

Nobe, R., Y. Sakakibara, N. Fukuda, N. Yoshida, K. Ogawa, and M. Suiko. 2003. Purification and characterization of laminaran hydrolases from *Trichoderma viride*. *Bioscience Biotechnology and Biochemistry* 67:1349–1357.

Oilgae. 2010. *Oilgae Guide to Fuels from Macroalgae*. Tamil Nadu, India.

Park, J.-H., J.-Y. Hong, H. C. Jang, S. G. Oh, S.-H. Kim, J.-J. Yoon, and Y. J. Kim. 2012. Use of *Gelidium amansii* as a promising resource for bioethanol: A practical approach for continuous dilute-acid hydrolysis and fermentation. *Bioresource Technology* 108:83–88.

Potts, T., J. Du, M. Paul, P. May, R. Beitle, and J. Hestekin. 2012. The production of butanol from Jamaica bay macro algae. *Environmental Progress and Sustainable Energy* 31:29–36.

Ra, C. H., G.-T. Jeong, M. K. Shin, and S.-K. Kim. 2013. Biotransformation of 5-hydroxymethylfurfural (HMF) by *Scheffersomyces stipitis* during ethanol fermentation of hydrolysate of the seaweed *Gelidium amansii*. *Bioresource Technology* 140:421–425.

Rothe, J., D. Hays, and J. Benemann. 2012. *New Fuels: Macroalgae*. Advancing Technology for America's Transportation, National Petroleum Council, California.

Schubert, C. 2006. Can biofuels finally take center stage? *Nature Biotechnology* 24:777–784.

Shangyong, L. I., J. I. A. Panpan, W. Linna, Y. U. Wengong, and H. Feng. 2013. Purification and characterization of a new thermostable κ-carrageenase from the marine bacterium *Pseudoalteromonas* sp. QY203. *Journal of Ocean University of China* 12:155–159.

Silchenko, A. S., M. I. Kusaykin, V. V. Kurilenko, A. M. Zakharenko, V. V. Isakov, T. S. Zaporozhets, A. K. Gazha, and T. N. Zvyagintseva. 2013. Hydrolysis of fucoidan by fucoidanase isolated from the marine bacterium, *Formosa algae*. *Marine Drugs* 11:2413–2430.

Singh, A., P. S. Nigam, and J. D. Murphy. 2011. Renewable fuels from algae: An answer to debatable land based fuels. *Bioresource Technology* 102:10–16.

Sluiter, A. 2006. Determination of structural carbohydrates and lignin in biomass. National Renewable Energy Laboratory, Golden, CO.

Stephens, E., R. D., Nys, I. L. Ross, and B. Hankamer. 2013. Algae fuels as an alternative to petroleum. *Journal of Petroleum & Environmental Biotechnology* 4:148.

Subhadra, B. and G. Grinson. 2011. Algal biorefinery-based industry: An approach to address fuel and food insecurity for a carbon-smart world. *Journal of the Science of Food and Agriculture* 91:2–13.

Tan, I. S., M. K. Lam, and K. T. Lee. 2013. Hydrolysis of macroalgae using heterogeneous catalyst for bioethanol production. *Carbohydrate Polymers* 94:561–566.

Thi Hong Minh, N. and V. Van Hanh. 2012. Bioethanol production from marine algae biomass: Prospect and troubles. *Journal of Vietnamese Environment* 3:25–29.

Thomas, F., L. C. E. Lundqvist, M. Jam, A. Jeudy, T. Barbeyron, C. Sandström, G. Michel, and M. Czjzek. 2013. Comparative characterization of two marine alginate lyase from *Zobellia galactanivorans* reveals distinct modes of action and exquisite adaptation to their natural substrate. *The Journal of Biological Chemistry* 288:23021–23037.

Trivedi, N., V. Gupta, M. Kumar, P. Kumari, C. R. K. Reddy, and B. Jha. 2011a. An alkali-halotolerant cellulase from *Bacillus flexus* Isolated from green seaweed *Ulva lactuca*. *Carbohydrate Polymers* 83:891–897.

Trivedi, N., V. Gupta, M. Kumar, P. Kumari, C. R. K. Reddy, and B. Jha. 2011b. Solvent tolerant marine bacterium *Bacillus aquimaris* secreting organic solvent stable alkaline cellulase. *Chemosphere* 83:706–712.

Trivedi, N., V. Gupta, C. R. K. Reddy, and B. Jha. 2013a. Enzymatic hydrolysis and production of bioethanol from common macrophytic green alga *Ulva fasciata* Delile. *Bioresource Technology* 150:106–112.

Trivedi, N., V. Gupta, C. R. K. Reddy, and B. Jha. 2013b. Detection of ionic liquid stable cellulase produced by the marine bacterium *Pseudoalteromonas* sp isolated from brown alga *Sargassum polycystum* C. Agardh. *Bioresource Technology* 132:313–319.

Turan, G. and A. Neori. 2010. Intensive seaweed aquaculture: A potent solution against global warming. In *Seaweeds and Their Role in Globally Changing Environments*, eds. A. Israel, R. Einav, and J. Seckbach. Dordrecht, the Netherlands: Springer, pp. 359–372.

Uchida, M. and M. Murata. 2004. Isolation of a lactic acid bacterium and yeast consortium from a fermented material of *Ulva* spp. (Chlorophyta). *Journal of Applied Microbiology* 97:1297–1310.

Wal, H., B. L. H. M. Sperber, B. H. Tan, R. B. Bakker, W. Brandenburg, and A. M. L. Contreras. 2012. Production of acetone, butanol, and ethanol from biomass of the green seaweed *Ulva lactuca*. *Bioresource Technology* 128:431–437.

Wang, B., Y. Li, N. Wu, and C. Q. Lan. 2008. CO(2) bio-mitigation using microalgae. *Applied Microbiology and Biotechnology* 79:707–718.

Wang, X., X. Liu, and G. Wang. 2011. Two-stage hydrolysis of invasive algal feedstock for ethanol fermentation. *Journal of Integrative Plant Biology* 53:246–252.

Wargacki, A. J., E. Leonard, M. N. Win, D. D. Regitsky, C. N. S. Santos, P. B. Kim, S. R. Cooper et al. 2012. An engineered microbial platform for direct biofuel production from brown macroalgae. *Science* 335:308–313.

Xie, W., B. Lin, Z. Zhou, G. Lu, J. Lun, C. Xia, S. Li, and Z. Hu. 2012. Characterization of a novel β-agarase fromanagar-degrading bacterium *Catenovulum* sp. X3. *Applied Microbiology and Biotechnology* 97:4907–4915.

Yanagisawa, M., K. Nakamura, O. Ariga, and K. Nakasaki. 2011. Production of high concentrations of bioethanol from seaweeds that contain easily hydrolyzable polysaccharides. *Process Biochemistry* 46:2111–2116.

Yeon, J.-H., S.-E. Lee, W. Y. Choi, D. H. Kang, H.-Y. Lee, and K.-H. Jung. 2011. Repeated-batch operation of surface-aerated fermentor for bioethanol production from the hydrolysate of seaweed *Sargassum sagamianum*. *Journal of Microbiology and Biotechnology* 21:323–331.

Yeon, J.-H., H.-B. Seo, S.-H. Oh, W.-S. Choi, D. H. Kang, H.- Y. Lee, and K.-H. Jung. 2010. Bioethanol production from hydrolysate of seaweed *Sargassum sagamianum*. *Korea Society for Biotechnology and Bioengineering* 25:283–288.

Yoon, J. J., S. H. Kim, Y. J. Kim, and J. S. Kim. 2011. Korea patent number 10-1075602.

Yoon, J. J., Y. J., Kim, S. H. Kim, H. J. Ryu, J. Y. Choi, G. S. Kim, and M. K. Shin. 2010. Production of polysaccharides and corresponding sugars from red seaweed. *Advanced Materials Research* 93–94:463–466.

Yoza, B. A. and E. M. Masutani. 2013. The analysis of macroalgae biomass found around Hawaii for bioethanol production. *Environmental Technology* 34:1859–1867.

Yuan, J. S., K. H. Tiller, H. Al-Ahmad, N. R. Stewart, and C. N. Stewart, Jr. 2008. Plants to power: Bioenergy to fuel the future. *Trends in Plant Science* 13:421–429.

12

Current State of Research on Algal Bioethanol

Ozcan Konur

CONTENTS

12.1 Introduction

12.1.1 Issues

Global warming, air pollution, and energy security have been some of the most important public policy issues in recent years (Jacobson 2009; Wang et al. 2008; Yergin 2006). With the increasing global population, food security has also become a major public policy issue (Lal 2004). The development of biofuels generated from biomass has been a long-awaited solution to these global problems (Demirbas 2007; Goldember 2007; Lynd et al. 1991). However, the development of early biofuels produced from agricultural plants such as sugar cane (Goldemberg 2007) and agricultural wastes such as corn stovers (Bothast and Schlicher 2005) has resulted in a series of substantial concerns about food security (Godfray et al. 2010). Therefore, the development of algal biofuels as a third-generation biofuel has been considered as a major solution for the global problems of global warming, air pollution, energy security, and food security (Chisti, 2007; Demirbas 2007; Kapdan and Kargi 2006; Spolaore et al. 2006; Volesky and Holan 1995).

Although there have been many reviews on the use of marine algae for the production of algal bioethanol (e.g., Amin 2009; Nigam and Singh 2011; Stal and Moezelaar 1997) and there have been a number of scientometric studies on algal biofuels (Konur 2011), there has not been any study of algal bioethanol research as in other research fields (Baltussen and Kindler 2004a,b; Dubin et al. 1993; Gehanno et al. 2007; Konur 2012a,b, 2013; Paladugu et al. 2002; Wrigley and Matthews 1986).

As North's new institutional theory suggests, it is important to have up-to-date information about current public policy issues to develop a set of viable solutions to satisfy the needs of all the key stakeholders (Konur 2000, 2002a–c, 2006a,b, 2007a,b, 2012c; North 1994).

Therefore, brief information from a selected set algal bioethanol studies is presented in this chapter to inform key stakeholders relating to the global problems of warming, air pollution, food security, and energy security about the production of algal bioethanol for the solution of these problems in the long run, thus complementing a number of recent scientometric studies on biofuels and global energy research (Konur 2011, 2012d–p).

12.1.2 Methodology

A search was carried out in the Science Citation Index Expanded (SCIE) and Social Sciences Citation Index (SSCI) databases (version 5.11) in September 2013 to locate papers relating to algal bioethanol using the keyword set of Topic = (ethanol* or *bio* ethanol** or *bioethanol** or *microbio* ethanol** or *ethyl* alcohol**) AND Topic = (alga* or *macro* alga** or *micro* alga** or macroalga* or microalga* or cyanobacteria or seaweed*) in the abstract pages of papers. For this chapter, it was necessary to embrace the broad algal search terms to include diatoms, seaweeds, and cyanobacteria as well as to include many spellings of ethanol rather than the simple keywords of bioethanol and algae.

It was found that there were 1042 papers indexed by these keywords between 1956 and 2013. The key subject categories for algal research were biotechnology and applied microbiology (304 references, 29.2%), energy and fuels (172 references, 16.5%), food science and technology (124 references, 11.2%), and marine freshwater biology (106 references, 10.2%), altogether comprising 68% of all the references on algal bioethanol. It was also necessary to focus on the key references by selecting articles and reviews.

The located papers were arranged by decreasing number of citations for two groups of papers: 25 papers each for nonexperimental research and experimental research. In order to check whether these collected abstracts correspond to the larger sample on these topical areas, new searches were carried out for each topical area.

The summary information about the located papers is presented under two major headings of nonexperimental and experimental research in the order of the decreasing number of citations for each topical area.

The information relating to the document type, author affiliation and location, the journal, the indexes, subject area of the journal, the concise topic, total number of citations, and total average number of citations received per year were given in the tables for each paper.

12.2 Nonexperimental Studies on Algal Bioethanol

12.2.1 Introduction

The nonexperimental research on algal bioethanol has been one of the most dynamic research areas in recent years. Twenty-five citation classics in the field of algal bioethanol with more than 25 citations were located, and the key emerging issues from these papers are presented in the decreasing order of the number of citations (Table 12.1). These papers give strong hints about the determinants of efficient algal bioethanol production and emphasize that marine algae are efficient bioethanol feedstocks in comparison with first- and second-generation biofuel feedstocks.

The papers were dominated by the researchers from 16 countries, usually through the intercountry institutional collaboration, and they were multiauthored. The United States (seven papers); England, Northern Ireland, and Ireland (three papers each); and the Netherlands, Canada, Brazil, and Turkey (two papers each) were the most prolific countries.

Similarly, all papers were published in the journals indexed by the Science Citation Index (SCI) and/or SCIE. There were no papers indexed by the Arts & Humanities Citation Index (A&HCI) or SSCI. The number of citations ranged from 26 to 127, and the average number of citations per annum ranged from 1.4 to 42.3. Sixteen of the papers were articles, while nine of them were reviews.

12.2.2 Algal Bioethanol Production Processes

12.2.2.1 Introduction

The nonexperimental research on algal bioethanol has been one of the most dynamic research areas in the recent years. Eight citation classics in the field of algal bioethanol with more than 35 citations were located, and the key emerging issues from these papers are presented in Table 12.1 in the decreasing order of the number of citations. These papers give strong hints about the determinants of the efficient algal bioethanol production and emphasize that marine algae are efficient bioethanol feedstocks in comparison with first- and second-generation biofuel feedstocks.

The papers were dominated by the researchers from 11 countries, usually through the intercountry institutional collaboration, and they were multiauthored. The Netherlands was the most prolific country with two papers.

Similarly, all papers were published in the journals indexed by the SCI and/or SCIE. There were no papers indexed by the A&HCI or SSCI. The number of citations ranged from 35 to 127, and the average number of citations per annum ranged from 1.7 to 42.3. Five of the papers were reviews, while three of them were articles.

12.2.2.2 Research on Algal Bioethanol Production Processes

Nigam and Singh (2011) review the research on the production of biofuels from biomass-based renewable resources including the production of algal bioethanol in a recent paper

TABLE 12.1
Nonexperimental Research on Algal Bioethanol

No.	Paper References	Year	Document	Affiliation	Country	No. of Authors	Class	Journal	Index	Subjects	Topic	Total No. of Citations	Total Average Citations per Annum	Rank
1	Nigam and Singh	2011	R	Univ. Ulster, Univ. Coll. Cork	N. Ireland, Ireland	2	P	*Prog. Energ. Combust. Sci.*	SCI, SCIE	Ther., En. Fuels, Eng. Chem., Eng. Mech.	Biofuel production	127	42.3	1
2	Stal and Moezelaar	1997	R	Neths. Inst. Ecol.	Netherlands	2	P	*Fems Microbiol. Rev.*	SCI, SCIE	Microbiol.	Algal fermentation	127	7.5	2
3	Groom et al.	2008	A	Univ. Wash. Nature Cons.	United States	2	Z	*Conserv. Biol.*	SCI, SCIE	Biodiv. Cons., Ecol., Env. Sci.	Biodiversity-friendly biofuels	101	16.8	3
4	John et al.	2011	A	M. Govt. Coll., Inst. Natl. Rech. Sci.+1	India, Canada	4	P	*Bioresour. Technol.*	SCI, SCIE	Agr. Eng., Biotech. Appl. Microb., En. Fuels	Algal bioethanol production	85	28.3	4
5	Singh et al.	2011a	A	Univ. Ulster, Univ. Coll. Cork	N. Ireland, Ireland	3	Z	*Bioresour. Technol.*	SCI, SCIE	Agr. Eng., Biotech. Appl. Microb., En. Fuels	Algal biofuels	85	28.3	5
6	Amin	2009	R	B.P.P.T., Tech. Univ. S.T.T. D.B.B.	Indonesia	1	P	*Energ. Conv. Manag.*	SCIE	Ther., En. Fuels, Mechs., Phys. Nucl.	Algal biofuel production	85	17.0	6
7	Harun et al.	2010a	R	Monash Univ., Univ. Pertanian Malaysia	Australia, Malaysia	4	P	*Renew. Sust. Energ. Rev.*	SCI, SCIE	En. Fuels	Algal bioprocess engineering	74	8.5	7

(Continued)

TABLE 12.1 (*Continued*)

Nonexperimental Research on Algal Bioethanol

No.	Paper References	Year	Document	Affiliation	Country	No. of Authors	Class	Journal	Index	Subjects	Topic	Total No. of Citations	Total Average Citations per Annum	Rank
8	Singh and Cu	2010	R	Univ. Southampton, Indian Inst. Petr.	England, India	2	Z	*Renew. Sust. Energ. Rev.*	SCI, SCIE	En. Fuels	Algal biofuel economics	70	17.5	8
9	Mussatto et al.	2010	R	Rev. Univ. Minho, Univ. Sao Paulo	Portugal, Brazil	9	Z	*Biotechnol. Adv.*	SCI, SCIE	Biotech. Appl. Microb.	Bioethanol production policies	62	15.5	9
10	Demirbas	2010	A	Sirnak Univ., Sila Sci.	Turkey	1	Z	*Energ. Conv. Manag.*	SCIE	Ther., En. Fuels, Mechs., Phys. Nucl.	Algal biofuels	56	14.0	10
11	Williams et al.	2009	R	Rev. E. Risk Sci. LLP, Natl. Renew. Energy Lab.	United States	4	Z	*Environ. Sci. Technol.*	SCI, SCIE	Eng. Env., Env. Sci.	Biofuel sustainability and environment	56	11.2	11
12	Fargione et al.	2009	A	Michigan Tech. Univ., Univ. Minnesota, Stanford Univ. +4	United States	10	Z	*Bioscience*	SCI, SCIE	Biology	Bioenergy and wildlife	55	11.0	12
13	Singh et al.	2011b	A	Univ. Ulster, Univ. Coll. Cork	N. Ireland, Ireland	2	Z	*Bioresour. Technol.*	SCI, SCIE	Agr. Eng., Biotech. Appl. Microb., En. Fuels	Algal biofuel economics	51	17.0	13
14	Angermayr et al.	2009	R	Univ. Amster. Uppsala Univ.	Netherlands, Sweden	4	P	*Curr. Opin. Biotechnol.*	SCI, SCIE	Bioch. Res. Meth., Biotech. Appl. Microb.	Algal biofuel production	49	9.8	14

(*Continued*)

TABLE 12.1 (*Continued*)
Nonexperimental Research on Algal Bioethanol

No.	Paper References	Year	Document	Affiliation	Country	No. of Authors	Class	Journal	Index	Subjects	Topic	Total No. of Citations	Total Average Citations per Annum	Rank
15	Rawat et al.	2011	A	Durban Univ. Technol.	S. Africa	4	Z	*Appl. Energ.*	SCI, SCIE	En. Fuels, Eng. Chem.	Algal biofuels and biosorption	47	15.7	15
16	Costa and de Morais	2011	A	Fed. Univ. Rio Grande	Brazil	2	P	*Bioresour. Technol.*	SCI, SCIE	Agr. Eng., Biotech. Appl. Microb., En. Fuels	Algal biofuels	40	13.3	16
17	Singh and Olsen	2011	A	Tech. Univ. Denmark	Denmark	2	Z	*Appl. Energ.*	SCI, SCIE	En. Fuels, Eng. Chem.	Algal biofuels	40	13.3	17
18	Konur	2012p	A	Sirnak Univ.	Turkey	1	Z	*Energ. Educ. Sci. Technol.-Part A*	SCIE	En. Fuels, Eng. Env., Eng. Chem.	Biofuel research evaluation	38	19.0	18
19	Wyman	1994	A	Natl. Renew. Energy lab.	United States	1	Z	*Appl. Biochem. Biotechnol.*	SCI, SCIE	Bioch. Mol. Biol., Bitech. Appl. Microb.	Algal biofuels and environment	38	1.9	19
20	Wyman and Goodman	1993	A	Natl. Renew. Energy Lab.	United States	2	P	*Appl. Biochem. Biotechnol.*	SCI, SCIE	Bioch. Mol. Biol., Bitech. Appl. Microb.	Biomass-based biotechnology	35	1.7	20

(*Continued*)

TABLE 12.1 (*Continued*)

Nonexperimental Research on Algal Bioethanol

No.	Paper References	Year	Document	Affiliation	Country	No. of Authors	Class	Journal	Index	Subjects	Topic	Total No. of Citations	Total Average Citations per Annum	Rank
21	Phalan	2009	A	Univ. Cambr.	England	1	Z	*Appl. Energ.*	SCI, SCIE	En. Fuels, Eng. Chem.	Biofuel impact in Asia	31	6.2	21
22	Taylor	2008	A	Univ. Southampon, UK Energy Res. Ctr.	England	1	Z	*Energ. Policy*	SCIE, SSCI	En. Fuels, Env. Sci., Env. Stud.	Algal biofuels	30	5.0	22
23	Cheng and Timilsina	2011	A	N. Carol. St. Univ., World bank	United States	2	Z	*Renew. Energ.*	SCIE	En. Fuels	Biofuel technologies	28	9.3	23
24	Sivakumar et al.	2010	R	Arkansas St. Univ., USDA ARS, Univ. Arkansas, Worcester P.I.	United States	7	Z	*Eng. Life Sci.*	SCIE	Biotech. Appl. Microb.	Biofuels	27	6.8	24
25	Kosaricj and Velikonja	1995	A	Univ. W. Ontario	Canada	2	Z	*Fems Microbiol. Rev.*	SCI, SCIE	Microbiol.	Bioethanol	26	1.4	25

SCI, Science Citation Index; SCIE, Science Citation Index Expanded; SSCI, Social Sciences Citation Index; P, production processes; Z, production policies; A, article; R, review.

with 127 citations. They note that renewable bioresources are available globally in the form of residual agricultural biomass and wastes, which can be transformed into liquid biofuels. However, they argue that the process of chemical conversion could be very expensive and not worthwhile to use for an economical large-scale commercial supply of biofuels, hinting at the need for further research for an effective, economical, and efficient conversion process.

Stal and Moezelaar (1997) review the research on fermentation processes in algae (cyanobacteria) to produce algal bioethanol and summarize relevant studies with a focus on the energetics of dark fermentation in a number of species in an early paper with 127 citations. They note that there are a variety of different fermentation pathways in cyanobacteria including home and heterolactic acid fermentation, mixed acid fermentation, and homoacetate fermentation. They find that all enzymes of the fermentative pathways are present in photoautotrophically grown cells, and many cyanobacteria are also capable of using elemental sulfur as an electron acceptor. The yield of ATP during fermentation exceeds the amount that is likely to be required for maintenance, which is very low in these cyanobacteria.

John et al. (2011) review the research on algal bioethanol. They argue that algae are sustainable green alternatives to food crops for bioethanol production in a recent paper with 85 citations. Dark, anaerobic fermentation is one of the most used methods for bioethanol production for some species of algae. Some algal species generate high starch biomass wastes after oil extraction, which are hydrolyzed to generate a sugary syrup used as substrate for ethanol production. They recommend further research for the efficient utilization of algal biomass and their industrial wastes to produce environmentally friendly algal bioethanol.

Amin (2009) presents a brief review on the main conversion processes for the production of algal biofuels including algal bioethanol in a paper with 85 citations. Bioethanol and biodiesel are produced by using biochemical processes, whereas bio-oil and biogas are produced by using thermochemical processes. They argue that as the properties of algal biofuels are almost similar to those of fish and vegetable oils, they are real-life substitutes for fossil oil and petroleum oil.

Harun et al. (2010a) review the research on bioprocess engineering of microalgae for the production of algal biofuels including algal bioethanol as well as high-value algal consumer products in a recent paper with 74 citations. They argue that the production process is moderately economically viable and the market is developing and assert that algae represent one of the most promising sources for new products and applications in the market. They further argue that with the development of detailed culture and screening techniques, algal biotechnology can meet the high demands of food, energy, and pharmaceutical industries.

Angermayr et al. (2009) review redirecting cyanobacterial intermediary metabolism by channeling (Calvin cycle) intermediates into fermentative metabolic pathways via the biosynthesis of fermentation end products, like alcohols (bioethanol) and biohydrogen, driven by solar energy, from water (and CO_2) in a paper with 49 citations. In this way, fortifying photosynthetic organisms with the ability to produce biofuels directly would bypass the need to synthesize all the complex chemicals of biomass.

Costa and de Morais (2011) present the main biofuels that can be derived from microalgae including algal bioethanol in a paper with 40 citations. They note that microalgae are an alternative energy source without the drawbacks of the first- and second-generation biofuels and, depending upon their growing conditions, microalgae can produce biocompounds that are easily converted into biofuels. The biofuels from microalgae are an alternative that can keep the development of human activity in harmony with the environment. The advantages of using microalgae to produce biofuels are their continuous production, simple cell division cycle, acquisition of organic compounds through photosynthesis, tolerance to varying environmental conditions, use of waste or brackish water, and use of land not used for agriculture, and, when subjected to physical and chemical stress, they can be induced to produce high concentrations of specific compounds.

Wyman and Goodman (1993) review the research on the biotechnology for production of fuels, chemicals, and materials from biomass in one of the earliest papers with 35 citations. Bioethanol and other products are derived from starch crops, such as corn. Enzyme-based technology is under development for conversion of lignocellulosic biomass (e.g., wood, grasses, and agricultural and municipal wastes) into bioethanol. The simultaneous saccharification and fermentation (SSF) process is employed to convert the cellulose fraction into bioethanol at improved rates, higher yields, and higher ethanol concentrations than using sequential processing through careful selection of improved cellulase enzymes and fermentative microorganisms. Medium-BTU gas can be derived from lignocellulosic biomass by anaerobic digestion and refined into a pipeline-quality gas. A high-solids fermenter achieves higher gas generation rates than conventional devices and promises to help make such gas economical. An extensive collection of more than 500 productive strains of microalgae has been established to produce lipid oils for diesel fuel and other compounds from carbon dioxide. They assert that acetyl coenzyme A (acetyl CoA) carboxylase (ACC) is a key enzyme in lipid oil synthesis, and genetic engineering approaches are being applied to enhance the rates and yields of product formation.

12.2.2.3 Conclusion

Algal bioethanol has been one of the most dynamic research areas in the recent years. Eight nonexperimental papers in the field of algal bioethanol production processes with more than 35 citations were located, and the key emerging issues from these papers were presented in Table 12.1 in the decreasing order of the number of citations. These papers gave strong hints about the determinants of efficient algal bioethanol production processes and emphasized that marine algae were more efficient biofuel feedstocks in comparison with the first- and second-generation biofuel feedstocks.

12.2.3 Algal Bioethanol Production Policies

12.2.3.1 Introduction

The experimental research on algal bioethanol has been one of the most dynamic research areas in the recent years. Seventeen papers in the field of algal bioethanol with more than 26 citations were located, and the key emerging issues from these papers are presented in Table 12.2 in the decreasing order of the number of citations. These papers give strong hints about the determinants of the efficient algal bioethanol production policies and emphasize that marine algae are efficient bioethanol feedstocks in comparison with the first- and second-generation biofuel feedstocks.

The papers were dominated by the researchers from 11 countries, usually through the intercountry institutional collaboration, and they were multiauthored. The United States (six papers), England (three papers), and Northern Ireland, Ireland, and Turkey (two papers each) were the most prolific countries.

TABLE 12.2

Experimental Research on Algal Bioethanol

No.	Paper References	Year	Document	Affiliation	Country	No. of Authors	Journal	Index	Subjects	Topic	Total No. of Citations	Total Average Citations per Annum	Rank
1	Deng and Coleman	1999	A	Univ. Toronto	Canada	2	*Appl. Environ. Microbiol.*	SCI, SCIE	Biotech. Appl. Microb., Microbiol.	Genetic engineering	76	5.1	1
2	Horn et al.	2000a	A	Norw. Univ. Sci. Technol., SINTEF Appl. Chem.	Norway	2	*J. Ind. Microbiol. Biotechnol.*	SCI, SCIE	Biotech. Appl. Microb.	Algal type	55	3.9	2
3	Wargacki et al.	2012	A	Bio Architecture Lab., BAL Chile SA, Univ. Wash.	Chile, United States	14	*Science*	SCI, SCIE	Multp. Sci.	Genetic engineering	47	23.5	3
4	Adams et al.	2009	A	Aberystwyth Univ.	Wales	3	*J. Appl. Phycol.*	SCI, SCIE	Biotech. Appl. Microb., Mar. Fresh. Biol.	Pretreatment	43	8.6	4
5	Dexter and Pengcheng	2009	A	China Univ. Petr., Univ. Hawaii	China, United States	2	*Energ. Environ. Sci.*	SCI, SCIE	Chem. Multp., En. Fuels, Eng. Chem., Env. Sci.	Genetic engineering	41	8.2	5
6	Harun et al.	2010b	A	Monash Univ., Univ. Putra Malaysia	Australia, Malaysia	3	*J. Chem. Technol. Biotechnol.*	SCI, SCIE	Biotech. Appl. Microb., Chem. Multp., Eng. Chem.	Algal type	39	9.8	6
7	Hirano et al.	1997	A	Mitsubishi Heavy Ind. Co. Ltd.	Japan	4	*Energy*	SCI, SCIE	Ther., En. Fuels	Intracellular fermentation	35	2.1	7
8	Ueno et al.	1998	A	Kamaishi Labs.	Japan	3	*J. Ferment. Bioeng.*	NA	Biot. Appl. Microb., Food Sci. Tech.	Intracellular fermentation	28	1.8	8
9	Nguyen et al.	2009	A	Sungkyunkwan Univ., Sogang Univ., Silla Univ.	S. Korea	5	*J. Microbiol. Biotechnol.*	SCIE	Biotech. Appl. Microb., Microbiol.	Pretreatment	27	5.4	9

(*Continued*)

TABLE 12.2 (*Continued*)

Experimental Research on Algal Bioethanol

No.	Paper References	Year	Document	Affiliation	Country	No. of Authors	Journal	Index	Subjects	Topic	Total No. of Citations	Total Average Citations per Annum	Rank
10	Tan et al.	2011	A	Chinese Acad. Sci.	China	6	*Metab. Eng.*	SCIE	Biotech. Appl. Microb.	Genetic engineering	26	8.7	10
11	Horn et al.	2000b	A	Norw. Univ. Sci. Technol., SINTEF Appl. Chem.	Norway	2	*J. Ind. Microbiol. Biotechnol.*	SCI, SCIE	Biotech. Appl. Microb.	Algal type	26	1.9	11
12	Choi et al.	2010	A	Sungkyunkwan Univ.	S. Korea	3	*Bioresour. Technol.*	SCI, SCIE	Agr. Eng., Biotech. Appl. Microb., En. Fuels	Pretreatment	24	6.0	12
13	Kim et al.	2011	A	KASIST, Ajou Univ.	S. Korea	5	*Bioresour. Technol.*	SCI, SCIE	Agr. Eng., Biotech. Appl. Microb., En. Fuels	Pretreatment	22	7.3	13
14	Ge et al.	2011	A	Ocean Univ. China	China	3	*Renew. Energ.*	SCIE	En. Fuels	Other production topics	19	6.3	14
15	Harun and Danquah	2011a	A	Monash Univ., Univ. Putra Malaysia	Australia, Malaysia	2	*Process Biochem.*	SCI, SCIE	Bioch. Mol. Biol., Biot. Appl. Microb., Eng. Chem.	Pretreatment	18	6.0	15
16	Matsumoto et al.	2003	A	Tokyo Univ. Agr. & Technol., Toshiba Co. Ltd., Elect. Power Dev. Co. Ltd.	Japan	5	*Appl. Biochem. Biotechnol.*	SCI, SCIE	Bioch. Mol. Biol., Biot. Appl. Microb.	Other production topics	16	1.5	16
17	Ueno et al.	2002	A	Tokyo Univ. Fisheries, RIKEN	Japan	4	*Arch. Microbiol.*	SCI, SCIE	Microbiol.	Algal type	15	1.3	17

(*Continued*)

TABLE 12.2 (*Continued*)

Experimental Research on Algal Bioethanol

No.	Paper References	Year	Document	Affiliation	Country	No. of Authors	Journal	Index	Subjects	Topic	Total No. of Citations	Total Average Citations per Annum	Rank
18	Hirayama et al.	1998	A	Mitsubishi Heavy. Ind. Ltd., Tokyo Elect. Power Co. Ltd.	Japan	7	*Stud. Surf. Sci. Catal.*	NA	Chem. Appl., Chem. Phys., Eng. Chem, Env. Sci.	Intracellular fermentation	15	0.9	18
19	Wang et al.	2011	A	Univ. Hawaii, Tianjin Univ., China. Peking Univ.	United States, China	3	*J. Integr. Plant Biol.*	SCIE	Bioch. Mol. Biol., Plant Sci.	Algal type	14	4.7	19
20	Harun et al.	2011	A	Monash Univ., Univ. Putra Malaysia	Australia, Malaysia	4	*Appl. Energ.*	SCI, SCIE	En. Fuels, Eng. Chem.	Pretreatment	13	4.3	20
21	Carrieri et al.	2010	A	Princ. Univ., Humboldt Univ., Penn St. Univ., Rutgers St. Univ.	United States, Germany	7	*Appl. Environ. Microbiol.*	SCI, SCIE	Biot. Appl. Microb., Microb.	Other production topics	13	3.3	21
22	Harun and Danquah	2011b	A	Monash Univ., Univ. Putra Malaysia	Australia, Malaysia	2	*Chem. Eng. J.*	SCI, SCIE	Eng. Env., Eng. Chem.	Pretreatment	12	4.0	22
23	Hellingwerf and de Mattos	2009	A	Univ. Amsterdam	Netherlands	2	*J. Biotechnol.*	SCI, SCIE	Biot. Appl. Microb.,	Other production topics	12	2.4	23
24	Powell and Hill	2009	A	Univ. Saskatchewan	Canada	2	*Chem. Eng. Res. Des.*	SCI, SCIE	Eng. Chem.	Other production topics	12	2.4	24
25	Miura et al.	1993	A	Kansai Elect. Power	Japan	5	*Appl. Biochem. Biotechnol.*	SCI, SCIE	Bioch. Mol. Biol., Biot. Appl. Microb.	Intracellular fermentation	11	0.5	25

SCI, Science Citation Index; SCIE, Science Citation Index Expanded; SSCI, Social Sciences Citation Index; A, article; R, review.

Similarly, all these papers were published in journals indexed by the SCI and/or SCIE. There were no papers indexed by the A&HCI or SSCI. The number of citations ranged from 26 to 101, and the average number of citations per annum ranged from 1.4 to 28.3. Only 4 of the papers were reviews, while 13 of them were articles.

12.2.3.2 Research on Algal Bioethanol Production Policies

Groom et al. (2008) develop principles that should be used in developing guidelines for certifying biodiversity-friendly biofuels in a paper with 101 citations. First, biofuel feedstocks should be grown with environmentally safe and biodiversity-friendly agricultural practices. The sustainability of any biofuel feedstock depends on good growing practices and sound environmental practices throughout the fuel production life cycle. Second, the ecological footprint of a biofuel, in terms of the land area needed to grow sufficient quantities of the feedstock, should be minimized. They argue that the best alternatives are fuels of the future, especially algal biofuels. Third, biofuels that can sequester carbon or that have a negative or zero carbon balance when viewed over the entire production life cycle should be given high priority. They then assert that corn-based bioethanol is the worst among the alternatives, although this is the biofuel that is most advanced for commercial production in the United States. They recommend aggressive pursuit of alternatives to corn as a bioethanol feedstock.

Singh et al. (2011a) review the research on algal bioethanol in comparison with first- and second-generation biofuel feedstocks as the sustainable alternatives to depleting fossil fuels in a recent paper with 85 citations. They argue that food prices increased because of the increased use of arable land for the cultivation of biomass used for the production of first- and second-generation biofuels. They assert that the use of non-arable land for the cultivation of algal biomass for the generation of third-generation bioethanol is a real-life solution. They focus on the cultivation of algal biomass and their use for algal bioethanol.

Singh and Cu (2010) review the research on the production of algal biofuels including algal bioethanol as well as high-value algal products and provide a critical appraisal of the commercialization potential of algal biofuels in a paper with 70 citations. The issues relating to the cultivation, life-cycle assessment (LCA), and conceptualization of an algal biorefinery are included in the review, and a critical analysis has been presented. They argue that the economic viability of the process in terms of minimizing the operational and maintenance cost along with maximization of oil-rich microalgae production is the key factor for successful commercialization of microalgal-based fuels.

Mussatto et al. (2010) review the research on the technological trends, global market, and challenges of bioethanol production including algal bioethanol with a focus on the raw materials, processes, and engineered strains development in a paper with 62 citations. They note that almost all bioethanol is produced from grain or sugarcane. They also present the main producer and consumer nations and future perspectives for the bioethanol market and discuss promising strategies like the use of microalgae and continuous systems with immobilized cells.

Demirbas (2010) investigates algal production technologies such as open, closed, and hybrid systems, production costs, and algal energy conversions in a paper with 56 citations. He finds that algal biofuel production costs can vary widely by feedstock, conversion process, scale of production, and region. He argues that algae are the only source of renewable biodiesel that is capable of meeting the global demand for transport fuels. Algae can also be converted to bio-oil, bioethanol, biohydrogen, and biomethane via thermochemical and biochemical methods.

Williams et al. (2009) assess what is known or anticipated about environmental and sustainability factors associated with next-generation biofuels (algal ethanol or algal biodiesel) relative to the primary conventional biofuels (i.e., corn-grain-based bioethanol and soybean-based biodiesel) in the United States during feedstock production and conversion processes in a paper with 56 citations. They consider greenhouse gas (GHG) emissions, air pollutant emissions, soil health and quality, water use and water quality, wastewater and solid waste streams, and biodiversity and land-use changes. They find that the production of next-generation feedstocks in the United States (e.g., municipal solid waste, forest residues, dedicated energy crops, microalgae) would fare better than corn-grain or soybean production on most of these factors, although the magnitude of these differences may vary significantly among feedstocks. Bioethanol produced using a biochemical or thermochemical conversion platform would result in fewer GHG and air pollutant emissions, but would have similar or potentially greater water demands and solid waste streams than conventional bioethanol biorefineries in the United States. However, they expect that these conversion-related differences would be small, particularly relative to those associated with feedstock production.

Fargione et al. (2009) examine the threats and opportunities for grassland conservation due to the production of the conventional and algal biofuels in a paper with 55 citations as they note that demand for land to grow corn for bioethanol increased in the United States by 4.9 million hectares between 2005 and 2008, with wide-ranging effects on wildlife, including habitat loss. They present a framework for assessing the impacts of biofuels on wildlife, and they use this framework to evaluate the impacts of conventional and algal biofuel feedstocks on grassland wildlife. They argue that meeting the growing demand for biofuels while avoiding negative impacts on wildlife will require either biomass sources that do not require additional land (e.g., wastes, residues, cover crops, algae) or crop production practices that are compatible with wildlife.

Singh et al. (2011b) review the research on the mechanism and challenges in commercialization of algal biofuels including algal bioethanol in a recent paper with 51 citations. They find that the efficient lipid-producer algal cell mass contains more than 30% of cell weight as lipids, and algal biomass has the potential to produce 100 times more oil per acre land than any terrestrial plants. They focus on the composition of algae, mechanism of oil droplets, triacylglycerol (TAG) production in algal biomass, cultivation of algal biomass, harvesting strategies, and recovery of lipids from algal mass. They then discuss the economical challenges in the production of biofuels from algal biomass in view of the future prospects in the commercialization of algal fuels.

Rawat et al. (2011) investigate phycoremediation of domestic wastewater and biomass production for sustainable biofuel production based on the dual role of microalgae in a recent paper with 47 citations. They discuss current knowledge regarding wastewater treatment using high-rate algal ponds (HRAPs) and microalgal biomass production techniques using wastewater streams. They then discuss the biomass harvesting methods and lipid extraction protocols in detail. They finally discuss biodiesel production via transesterification of the lipids and other biofuels such as algal biomethane and algal bioethanol using the biorefinery approach.

Singh and Olsen (2011) provide a critical review of biochemical conversion, sustainability, and LCA of algal biofuels including algal bioethanol in a recent paper with 40 citations. They note that algae sequester a significant quantity of carbon from atmosphere and industrial gases and utilize the nutrients from industrial effluents and municipal wastewater. Therefore, cultivation of algal biomass provides a dual benefit as it provides biomass for the production of biofuels and also saves our environment from air and water pollution. The LCA of algal biofuels shows that algal biomass is environmentally better than the fossil fuels, but it is not yet so attractive in economic terms.

Konur (2012p) explores the characteristics of the literature on the biofuels including algal bioethanol published during the last three decades, based on the databases of SCIE and SSCI, and its implications using the scientometric techniques in a recent paper with 38 citations. He finds that the literature on the biofuels has grown exponentially during this period, reaching 6770 papers in total, paralleling enormous changes in the research landscape. Papers mostly have been journal articles, reviews, and proceedings, predominantly in English. The United States, China, and Germany have been the three most prolific countries. The University of Illinois has been the most prolific institution. The most prolific authors have been Demirbas and Minter. *Biomass & Bioenergy* has been the most prolific journal, while *energy fuels* has been the most prolific subject area. The total number of citations is 79,304, giving a ratio for the *average citations per item* as 11.71 and *H-index* as 101. Ragauskas et al. have had the highest impact on the literature. Both the research output and the citations have thrived spectacularly after 2005.

Wyman (1994), in an early paper with 38 citations, reviews the research on the impact of alternative biofuels on carbon dioxide accumulation in the environment, presents an analysis of energy flows for algal bioethanol production, and provides a carbon dioxide balance for fossil fuels used. This analysis includes consideration of fuel utilization performance and assignment of carbon dioxide to coproduct. Biofuel technologies require little, if any, fossil fuel inputs. As a result, most or all of the carbon is recycled through their use, reducing substantially the net release of carbon dioxide to the atmosphere. He notes that the major fractions of lignocellulosic biomass, cellulose and hemicellulose can be broken down into sugars that can be fermented into ethanol. Biomass can also be gasified to a mixture of carbon monoxide and hydrogen for catalytic conversion into methanol. Algae could consume carbon dioxide from power plants and other sources to produce lipid oil that can be converted into a diesel fuel substitute. The costs have been reduced significantly for biofuels, and the potential exists for them to be competitive with conventional fuels.

Phalan (2009) provides a broad overview of the social and environmental costs and benefits of biofuels including algal bioethanol in Asia in a paper with 31 citations. He argues that the major factors determining the impacts of biofuels would be their contribution to land-use change, the feedstocks used, and issues of technology and scale. Biofuels offer economic benefits and in the right circumstances can reduce emissions and make a small contribution to energy security. Feedstocks that involve the conversion of agricultural land would affect food security and would cause indirect land-use change, while those that replace forests, wetlands, or natural grasslands would increase emissions and damage biodiversity. He contends that biofuels from cellulose, algae, or waste would avoid some of these problems, but come with their own set of uncertainties and risks.

Taylor (2008) identifies a range of research and development priorities regarding development of efficient algal biofuels including algal bioethanol in a paper with 30 citations. She notes that there are now a range of targets for biofuel use in the United Kingdom, although their environmental effects are disputed. Possible outputs include biodiesel and bioethanol, both of which can be used as transport fuel. Other potential products include hydrogen, polymers, and a wide range of value-added chemicals, making this technology important in a post-petrochemical world. She also suggests that biorefineries could use cogeneration to produce electricity.

Cheng and Timilsina (2011) review the research on the advanced biofuel technologies and summarize the current status of second-generation biofuel technologies including bioethanol from lignocellulosic materials and algal biodiesel in a recent paper with 28 citations. They describe the technologies, their advantages and challenges, feedstocks for the second-generation biofuels, the key barriers to their commercial applications, and future perspectives of the advanced technologies.

Sivakumar et al. (2010) review the research on bioethanol and biodiesel. They note that new technologies are needed for fuel extraction using feedstocks that do not threaten food security and cause minimal or no loss of natural habitat and soil carbon in a paper with 27 citations. At the same time, waste management has to be improved, and environmental pollution should be minimized or eliminated. They argue that liquid biofuels such as lignocellulosic-based bioethanol from plant biomass and algal-based biodiesel are sustainable alternative biofuels that could stabilize national security and provide clean energy for future generations.

Kosaric and Velikonja (1995) present challenging opportunities for production of liquid and gaseous fuels by biotechnology including algal bioethanol in an early paper with 26 citations. From the liquid fuels, bioethanol production has been widely researched and implemented. The major obstacle for large-scale production of bioethanol for fuel is the cost, whereby the substrate represents one of the major cost components. They present various scenarios for a critical assessment of cost distribution for production of ethanol from various substrates by conventional and high-rate processes. They then focus on recent advances in the research and application of biotechnological processes and methods for the production of liquid transportation fuels other than bioethanol (other oxygenates, diesel fuel extenders, and substitutes) as well as gaseous fuels (biogas, biomethane, reformed syngas). They also describe the construction and performance of microbial fuel cells for the direct high-efficiency conversion of chemical fuel energy to electricity.

12.2.3.3 Conclusion

The research on algal bioethanol has been one of the most dynamic research areas in the recent years. Seventeen nonexperimental papers in the field of algal bioethanol production policies with more than 26 citations were located, and the key emerging issues were presented in the decreasing order of the number of citations. These papers gave strong hints about the determinants of efficient algal bioethanol production policies and emphasized that marine algae were more efficient biofuel feedstocks in comparison with the first- and second-generation biofuel feedstocks.

12.3 Experimental Studies on Algal Bioethanol

12.3.1 Introduction

The experimental research on algal bioethanol has been one of the most dynamic research areas in recent years. Twenty-five experimental citation classics in the field of algal bioethanol with more than 11 citations were located, and the key emerging issues from these papers are presented in the decreasing order of the number of citations (Table 12.2). These papers give strong hints about the determinants of the efficient algal bioethanol production and emphasize that marine algae are efficient bioethanol feedstocks in comparison with the first- and second-generation biofuel feedstocks.

The papers were dominated by the researchers from 12 countries, usually through the intercountry institutional collaboration, and they were multiauthored. Japan (six papers); the United States (five papers); Australia and Malaysia (four papers each); S. Korea (three papers); and England, Northern Ireland, and Ireland (three papers each) were the most prolific countries.

Similarly, all these papers were published in the journals indexed by the SCI and/or SCIE. There were no papers indexed by the A&HCI or SSCI. The number of citations ranged from 26 to 76, and the average number of citations per annum ranged from 0.5 to 23.5. All of these papers were articles.

There were a number of subtopical areas: algal type (five papers), genetic engineering of algae (four papers), intracellular fermentation of algae (four papers), other production topics relating to algal bioethanol production (five papers), and pretreatment of algae (seven papers).

12.3.2 Experimental Research on Algal Bioethanol

12.3.2.1 Research on Genetic Engineering of Algae

Deng and Coleman (1999) investigate the production of algal bioethanol from cyanobacteria and introduce new genes into a cyanobacterium in order to create a novel pathway for fixed carbon utilization resulting in the synthesis of algal bioethanol in an early paper with 76 citations. The coding sequences of pyruvate decarboxylase (pdc) and alcohol dehydrogenase II (adh) from the bacterium *Zymomonas mobilis* were cloned into the shuttle vector pCB4 and then used to transform the cyanobacterium *Synechococcus* sp. strain PCC 7942. They find that "under control of the promoter from the rbcLS operon encoding the cyanobacterial ribulose-1,5-bisphosphate carboxylase/oxygenase, the pdc and adh genes were expressed at high levels." The transformed cyanobacterium then synthesized algal bioethanol, which diffused from the cells into the culture medium. They argue that as cyanobacteria have simple growth requirements and use light, CO_2, and inorganic elements efficiently, and that production of algal bioethanol by cyanobacteria is a potential system for bioconversion of solar energy and CO_2 into a valuable resource.

Wargacki et al. (2012) present the discovery of a 36 kilo base pair DNA fragment from *Vibrio splendidus* encoding enzymes for alginate transport and metabolism in a recent paper with 47 citations. They find that the genomic integration of this ensemble, together with an engineered system for extracellular alginate depolymerization, generated a microbial platform that can simultaneously degrade, uptake, and metabolize alginate. When further engineered for algal bioethanol synthesis, this platform "enables algal bioethanol production directly from macroalgae via a consolidated process, achieving a titer of 4.7% volume/volume and a yield of 0.281 weight ethanol/weight dry macroalgae (equivalent to similar to 80% of the maximum theoretical yield from the sugar composition in macroalgae)."

Dexter and Pengcheng (2009) report the creation of a *Synechocystis* sp. PCC 6803 strain that can photoautotrophically convert CO_2 to algal bioethanol in a paper with 41 citations. Transformation was performed using a double homologous recombination system to integrate the pdc and adh genes from obligately ethanol-producing *Z. mobilis* into the *Synechocystis* PCC 6803 chromosome under the control of the strong, light-driven psbAII promoter. Their system showed "an average yield of 5.2 mmol OD(730) unit^{-1} L^{-1} day^{-1} with no required antibiotic/selective agent."

Tan et al. (2011) investigate the photosynthesis-driven conversion of carbon dioxide to fatty alcohols and hydrocarbons in cyanobacteria for the production of high-value biochemicals and high-energy algal biofuels including algal bioethanol in a recent paper with 26 citations. They describe the biosynthesis of fatty alcohols in a genetically engineered cyanobacterial system through heterologously expressing fatty acetyl CoA reductase and the effect of environmental stresses on the production of fatty alcohols in the mutant strains. They evaluate hydrocarbon production in three representative types of native cyanobacterial model strains and the mutant strain overexpressing ACC. They find the potential for direct production of high-value chemicals and high-energy fuels in a single biological system that utilizes solar energy as the energy source and carbon dioxide as the carbon source.

12.3.2.2 Research on Algal Type

Horn et al. (2000a) test four algal microorganisms to carry out fermentation in a paper with 55 citations. They find that only *Pichia angophorae* was able to utilize both laminaran and mannitol for bioethanol production where laminaran and mannitol were consumed simultaneously, but with different relative rates. In batch fermentations, "mannitol was the preferred substrate whilst its share of the total laminaran and mannitol consumption rate increased with oxygen transfer rate (OTR) and pH." In continuous fermentations, "laminaran was the preferred substrate at low OTR, whereas at higher OTR, laminaran and mannitol were consumed at similar rates." They conclude that "optimization of bioethanol yield required a low OTR, and the best yield of 0.43 g bioethanol ethanol (g substrate)$^{-1}$ was achieved in batch culture at pH 4.5 and 5.8 mmol O-2 L^{-1} h^{-1}."

Harun et al. (2010b) study microalgal biomass as a fermentation feedstock for bioethanol production in a paper with 39 citations. They explore the suitability of microalgae (*Chlorococurn* sp.) as a substrate for bioethanol production via yeast (*Saccharomyces bayanus*) under different fermentation conditions. They find that a "maximum ethanol concentration of 3.83 g L^{-1} was obtained from 10 g L^{-1} of lipid-extracted microalgae debris." They assert that this productivity level (similar to 38% w/w) endorses microalgae as a promising substrate for bioethanol production.

Horn et al. (2000b) investigate the production of bioethanol from mannitol by *Zymobacter palmae* in a paper with 26 citations. They find that "bacterial growth as well as bioethanol yield depended on the amount of oxygen present." However, strictly anaerobic growth on mannitol was not observed. At excessive aeration, a change in the fermentation pattern was observed with high production of acetate and propionate. Under oxygen-limiting conditions, "the bacteria grew and produced bioethanol in a synthetic mannitol medium with a yield of 0.38 g ethanol (g mannitol)$^{-1}$." *Z. palmae* was also successfully applied for fermentation of mannitol from *Laminaria hyperborea* extracts.

Ueno et al. (2002) study the isolation, characterization, and fermentative pattern of a novel thermotolerant *Prototheca zopfii* var. *hydrocarbonea* strain producing bioethanol and CO_2 from glucose at 40°C in a paper with 15 citations. A novel thermotolerant strain of the achlorophyllous microalga *Prototheca* was isolated from a hot spring. The isolate produced an appreciable amount of bioethanol and CO_2 from glucose under anoxic conditions at both 25°C and 40°C. Moreover, it also evolved gas from sucrose after a time lag at 40°C. Its taxonomic characteristics coincided with those of *Prototheca zopfii* var. *hydrocarbonea*, and there was a close relationship between the two strains. "D-lactic acid, ethanol, CO_2 and a trace of acetic acid were produced from glucose, but L-lactic acid, formic acid, and H were not. At 25°C, D-lactic acid and ethanol were produced in approximately equimolar amounts under $N_2/H_2/CO_2$, whereas ethanol production was predominant under N_2." They then find that more bioethanol was produced at 40°C than at 25°C irrespective of the gas composition in the atmosphere.

Wang et al. (2011) study the two-stage hydrolysis of invasive algal feedstock for algal bioethanol fermentation, and they develop a saccharification method for the production of third-generation biofuel (i.e., bioethanol) using feedstock of the invasive marine macroalga *Gracilaria salicornia* in a paper with 14 citations. Under optimum conditions (120°C and 2% sulfuric acid for 30 min), "dilute acid hydrolysis of the homogenized invasive plants yielded a low concentration of glucose (4.1 mM or 4.3 g glucose/kg fresh algal biomass). However, two-stage hydrolysis of the homogenates (combination of dilute acid hydrolysis with enzymatic hydrolysis) produced 13.8 g of glucose from one kilogram of fresh algal feedstock." They then find that batch fermentation analysis produced 79.1 g EtOH from one kilogram of dried invasive algal feedstock using the ethanologenic strain *Escherichia coli* KO11. Furthermore, they find that the invasive algal feedstock contained different types of sugar, including C_5 sugar.

12.3.2.3 Research on Algal Pretreatment

Adams et al. (2009) carried out a fermentation study on *Saccharina latissima* for bioethanol production considering variable pretreatments in a paper with 43 citations. They showed that the pretreatments (pretreatment was at 65°C, pH 2 for 1 h prior to fermentation) were not required for the fermentations conducted, with higher algal bioethanol yields being achieved in untreated fermentations than in those with altered pH or temperature pretreatments. This result was seen in fresh and defrosted macroalgal samples using *S. cerevisiae* and 1 U kg^{-1} laminarinase. They note that macroalgae such as *Laminaria* spp. grow in abundance around the United Kingdom, reaching >4 m in length and containing up to 55% dry weight of the carbohydrates laminarin and mannitol. Using enzymes, they argue that these can be hydrolyzed and converted to glucose and fructose, which in turn can be utilized by yeasts to produce algal bioethanol.

Nguyen et al. (2009) study the hydrothermal acid pretreatment of *Chlamydomonas reinhardtii* biomass for algal bioethanol production in a paper with 27 citations. They focus upon dilute acid hydrothermal pretreatments at low cost and high efficiency to compete with current methods and employ *C. reinhardtii* UTEX 90 as the feedstock. With dry cells of 5% (w/v), the algal biomass was pretreated with sulfuric acid (1%–5%) under temperatures from 100°C to 120°C, from 15 to 120 min. They find that "the glucose release from the biomass was maximum at 58% (w/w) after pretreatment with 3% sulfuric acid at 110°C for 30 min." This method enabled not only starch but also the hydrolysis of other oligosaccharides in the algal cell in high efficiency. An Arrhenius-type of model equation enabled extrapolation of some yields of glucose beyond this range. They then find that the "pretreated slurry was fermented by yeast, *Saccharomyces cerevisiae* S288C, resulting in a bioethanol yield of 29.2% from algal biomass." They conclude that the pretreated algal biomass is a suitable feedstock for ethanol production and can have a positive impact on large-scale applied systems.

Choi et al. (2010) study the enzymatic pretreatment of *C. reinhardtii* biomass for algal bioethanol production in a paper with 24 citations. They convert algal biomass, *Chlamydomonas reinhardtii* UTEX 90, into a suitable fermentable feedstock by two commercial hydrolytic enzymes. They find that almost all starch was released and converted into glucose without steps for the cell wall disruption. They investigate the various conditions in the liquefaction and saccharification processes, such as enzyme concentration, pH, temperature, and residence time, to obtain an optimum combination using the orthogonal analysis. They find that "approximately 235 mg of bioethanol was produced from 1.0 g of algal biomass by a separate hydrolysis and fermentation (SHF) method." They assert that the main advantages of this process include the low cost of chemicals, short residence time, and simple equipment system, all of which promote its large-scale application.

Kim et al. (2011) treat biomass of the marine algae, *Ulva lactuca, Gelidium amansii, L. japonica,* and *Sargassum fulvellum,* with acid and commercially available hydrolytic enzymes in a recent paper with 22 citations. The hydrolysates contained glucose, mannose, galactose, and mannitol, among other sugars, at different ratios. The *L. japonica* hydrolysate contained up to 30.5% mannitol and 6.98% glucose in the hydrolysate solids. They find that "ethanogenic recombinant *Escherichia coli* KO11 was able to utilize both mannitol and glucose and produced 0.4 g ethanol per g of carbohydrate when cultured in *L. japonica* hydrolysate supplemented with Luria-Bertani medium and hydrolytic enzymes." They argue that the strategy of acid hydrolysis followed by simultaneous enzyme treatment and inoculation with *E. coil* KO11 could be a viable strategy to produce ethanol from marine alga biomass.

Harun and Danquah (2011a) study the influence of acid pretreatment on microalgal biomass for bioethanol production in a paper with 18 citations. They explore the influence of acid exposure as a microalgal pretreatment strategy for bioethanol production. Different parameters were investigated: acid concentration, temperature, microalgae loading, and pretreatment time. They find that the "highest bioethanol concentration obtained was 7.20 g/L and this was achieved when the pretreatment step was performed with 15 g/L of microalgae at 140°C using 1% (v/v) of sulphuric acid for 30 min." In terms of bioethanol yield, "similar to 52 wt% (g bioethanol/g microalgae) maximum was obtained using 10 g/L of microalgae and 3% (v/v) of sulphuric acid under 160°C for 15 min." They argue that temperature is the most critical factor during acid pretreatment of microalgae for bioethanol production.

Harun et al. (2011) study the alkaline pretreatment of microalgal biomass for algal bioethanol production from the species *Chlorococcum infusionum,* using NaOH in a paper with 13 citations. They examine three parameters of the concentration of NaOH, temperature, and pretreatment time. The bioethanol concentration, glucose concentration, and cell size were studied in order to determine the effectiveness of the pretreatment process. They find that the "highest glucose yield was 350 mg/g, and the maximum bioethanol yield obtained was 0.26 g ethanol/g algae using 0.75% (w/v) of NaOH and 120°C for 30 min." They assert that the alkaline pretreatment method was a promising option to pretreat microalgal biomass for bioethanol production.

Harun and Danquah (2011b) study the enzymatic hydrolysis of microalgal biomass for algal bioethanol production in a paper with 12 citations and examine the enzymatic hydrolysis of *Chloroccum* sp. by using cellulase obtained from *Trichoderma reesei,* ATCC 26921. They find that the "highest glucose yield of 64.2% (w/w) was obtained at a temperature of 40°C, pH 4.8, and a substrate concentration of 10 g/L of microalgal biomass." Comparative kinetic studies on glucose and cellobiose formation showed twice as fast glucose production than cellulobiose. They then find that the "value of $K_{m,app}$ was higher for the hydrolysis of cellobiose ($K_{m,app}$ = 15.18 g/L) compared to that of the substrate ($K_{m,app}$ = 1.48 g/L), thus displaying a competitive type of inhibition." They assert that the enzymatic hydrolysis process was an effective mechanism to enhance the saccharification process of microalgal biomass.

12.3.2.4 Research on Intracellular Bioethanol Production

Hirano et al. (1997) investigate CO_2 fixation and algal bioethanol production with microalgal photosynthesis and intracellular anaerobic fermentation in an early paper with 35 citations. Microalgae were screened from seawater where more than 250 strains were isolated, and some of the isolated strains and two strains from culture collections were tested to examine bioethanol productivity. Some strains had high growth rates of 20–30 g dry biomass/m^2/day and high starch content of more than 20% (dry base). A strain of *Chlorella vulgaris* (IAM C-534) had a high starch content of 37%. Starch was extracted from the cells of the *Chlorella*, saccharified and fermented with yeasts; 65% of the ethanol conversion rate was obtained as compared to the theoretical rate from starch. The algal starch was a good source for bioethanol production using the conventional process. As an example of another type of bioethanol production process, intracellular starch fermentation under dark and anaerobic conditions was examined. They find that "all of the tested strains showed intracellular starch degradation and ethanol production, but the levels of ethanol production were significantly different from each other." They then find that "higher bioethanol productions were obtained with *Chlamydomonas reinhardtii* (UTEX2247) and Sak-1 isolated from seawater." These showed a maximum ethanol concentration of 1 (w/w)%. The characteristics of intracellular ethanol production were examined with the *Chlamydomonas*. They conclude that intracellular ethanol production is simpler and less energy intensive than the conventional ethanol-fermentation process.

Ueno et al. (1998) investigate dark fermentation in the marine green alga, *C. littorale*, with emphasis on algal bioethanol production in an early paper with 28 citations. They find that under dark, anaerobic conditions, 27% of cellular starch was consumed within 24 h at 25°C, the cellular starch decomposition being accelerated at higher temperatures. Bioethanol, acetate, hydrogen, and carbon dioxide were obtained as fermentation products. The "maximum productivity of bioethanol was 450 μmol/g-dry wt. at 30°C." They propose the fermentation pathway for cellular starch from the yields of the end products and the determined enzyme activities. Bioethanol was formed from pyruvate by pdc and adh. The "change in fermentation pattern that varied with cell concentration in the reaction vials suggested that the hydrogen partial pressure affected the consumption mode of reducing equivalents under dark fermentation." Finally, they find that bioethanol productivity was improved by adding methyl viologen, while hydrogen production decreased.

Hirayama et al. (1998) study the bioethanol production from carbon dioxide by fermentative microalgae in an early paper with 15 citations. Microalgae were screened from seawater for CO, fixation, and ethanol by self-fermentation and tested for their growth rate, starch content, and conversion rate from starch to ethanol. More than 200 strains were isolated. One of the excellent strains, *Chlamydomonas* sp. YA-SH-1, which was isolated from the Red Sea showed (1) a growth rate of 30 g dry biomass/m^2/day, (2) a starch content of 30% (dry base), and (3) a conversion rate from intracellular starch to ethanol of 50% in the dark and anaerobic condition. New bioethanol production system consists of microalgal cultivation, algal cells' harvest, self-fermentation of algae, and bioethanol extraction processes. They assert that the system is "more simple and less energy consuming compared with the conventional one and if the microalgal productivity, starch content, and ethanol conversion rate are improved, the system should be an effective means for CO_2 fixation and energy production."

Miura et al. (1993) study the stimulation of hydrogen production in algal cells grown under high-CO_2 concentration and low temperature in one of the earliest papers with 11 citations. When cells of *Chlamydomonas* sp. MGA 161 were cultivated at a high-CO_2

concentration (15% CO_2) and low temperature (15°C), the growth lag time was much longer, but the starch accumulated was two times higher than under the basal conditions (5% CO_2 30°C). When the cells grown in the high-CO_2/low-temperature conditions were incubated under dark anaerobic conditions, they find that the "degradation of starch and production of hydrogen and bioethanol were remarkably higher than those grown under the basal conditions." The lag time of cell growth was shortened, whereas the high capacity of starch accumulation and hydrogen production was maintained, by cultivating the cells alternately every 12 h under the basal and high-CO_2/low-temperature conditions.

12.3.2.5 Research on Other Bioethanol Production Topics

Carrieri et al. (2010) examine sodium concentration cycling as a new strategy for redistributing carbon storage products and increasing autofermentative product yields following photosynthetic carbon fixation in the cyanobacterium Spirulina (*Arthrospira maxima*) in a paper with 13 citations. They find that for cells grown in 1.24 M NaCl, the "fermentative yields of acetate, bioethanol, and formate increase substantially to 1.56, 0.75, and 1.54 mmol/(g [dry weight] of cells. day), respectively (36-, 121-, and 6-fold increases in rates relative to cells grown in 0.24 M NaCl)." Catabolism of endogenous carbohydrate increased by approximately twofold upon hypoionic stress. For cultures grown at all salt concentrations, hydrogen was produced, but its yield did not correlate with increased catabolism of soluble carbohydrates. Instead, they argue that "ethanol excretion becomes a preferred route for fermentative NADH reoxidation, together with intracellular accumulation of reduced products of acetyl coenzyme A (acetyl-CoA) formation when cells are hypoionically stressed." In the absence of hypoionic stress, hydrogen production is a major beneficial pathway for NAD^+ regeneration without wasting carbon intermediates such as ethanol derived from acetyl CoA. They assert that this switch improves the overall cellular economy by retaining carbon within the cell until aerobic conditions return and the acetyl unit can be used for biosynthesis or oxidized via respiration for a much greater energy return.

Powell and Hill (2009) carry out an economic assessment of an integrated bioethanol–biodiesel–microbial fuel cell facility utilizing yeast and photosynthetic algae in a paper with 12 citations and present a strategy for the integration of a novel CO_2 photosynthetic culture and power generation system into a commercial bioethanol plant. Photosynthetic microalgal column photobioreactors, acting as cathodic half cells, are coupled with existing yeast fermentors at a bioethanol plant, acting as anodic half cells, to create coupled microbial fuel cells. The microalgal photobioreactors also sequester CO_2 emitted by the yeast fermentors. Incorporating microbial fuel cells into an existing bioethanol plant generates some of the power used in bioethanol production, and the microalgae species *C. vulgaris* contains oil, which acts as a by-product for the production of biodiesel. They determine the required design specifications of novel, air lift Business-Process Reengineering (BPR) cathodes to make the integrated system economically feasible at an existing bioethanol plant, and they develop the optimum integration strategy. These parameters include "PBR size, number of integrated MFCs, fuel cell outputs, oil (for biodiesel) production rate, and CO_2 consumption rate."

Hellingwerf and de Mattos (2009) discuss the light-driven biofuel formation from CO_2 and water based on the *photanol* approach in a paper with 12 citations and describe an approach in which basic reactions from phototrophy are combined in single organisms with key metabolic routes from chemotrophic organisms, with C_3 sugars as glyceraldehyde 3-phosphate as the central linking intermediate. Because various metabolic routes that lead to the formation of a range of short-chain alcohols can be used in this approach, they refer to it as the photanol approach. They explore the various strategies to optimize this biofuel production strategy.

Ge et al. (2011) study the saccharification techniques of seaweed wastes for the transformation of algal bioethanol and investigate the technical feasibility of floating residue (FR) utilization as a resource of renewable energy in a recent paper with 19 citations. They study the production of yeast-fermentable sugars (glucose) from FR by dilute sulfuric acid pretreatment and further enzymatic hydrolysis. They conduct dilute sulfuric acid pretreatment by using sulfuric acid at concentration of 0, 0.1, 0.2, 0.5, and 1.0% (w/v) for 0.5, 1.0, and 1.5 h, respectively, at 121°C. The system of enzymatic hydrolysis consisted of cellulase and cellobiase. They find that "FR might be a perfect bioenergy resource, containing high content of cellulose (30.0 +/− 0.07%) and little hemicellulose (2.2 +/− 0.86%)." They then find that the "acid pretreatment improved the hydrolysis efficiency of cellulase and cellobiase by increasing the reaction surface area of FR and enhanced the final yield of glucose for fermentation. The maximum yield of glucose reached 277.5 mg/g FR under the optimal condition of dilute sulfuric acid pretreatment (0.1% w/v, 121°C. 1.0 h) followed by enzymatic hydrolysis (50°C, pH 4.8, 48 h)." After fermentation by *S. cerevisiae* at 30°C for 36 h, the "bioethanol conversion rate of the concentrated hydrolysates reached 41.2%, which corresponds to 80.8% of the theoretical yield." They assert that cellulose in seaweed processing wastes including FR is easily hydrolyzed to produce glucose in comparison with that in terrestrial plants and FR shows excellent prospects as a potential feedstock for the production of bioethanol.

Matsumoto et al. (2003) investigate the saccharification of marine microalgae using amylase from marine bacteria in saline conditions for bioethanol production in a paper with 16 citations. An amylase-producing bacterium, *Pseudoalteromonas undina* NKMB 0074, was isolated and identified. They find that the "green microalga NKG 120701 had the highest concentration of intracellular carbohydrate obtained from algal culture stocks." *P. undina* NKMB 0074 was inoculated into suspensions containing NKG 120701 cells and increasingly reduced suspended sugars with incubation time. Terrestrial amylase and glucoamylase were inactive in saline suspension. Therefore, they assert that "marine amylase is necessary in saline conditions for successful saccharification of marine microalgae."

12.3.3 Conclusion

The experimental research on algal bioethanol has been one of the most dynamic research areas in the recent years. Twenty-five papers in the field of algal bioethanol with more than 11 citations were located, and the key emerging issues were presented in the decreasing order of the number of citations. These papers gave strong hints about the determinants of the efficient algal bioethanol production and emphasized that marine algae were efficient bioethanol feedstocks in comparison with the first- and second-generation biofuel feedstocks.

The experimental research focuses on a number of thematic areas such as algal type (five papers), genetic engineering of algae (four papers), intracellular fermentation of algae (four papers), other production topics relating to algal bioethanol production (five papers), and pretreatment of algae (seven papers).

12.4 Conclusion

The papers presented under the two main headings in this chapter confirm the predictions that the marine algae have a significant potential to serve as a major solution for the global problems of warming, air pollution, energy security, and food security in the form of algal bioethanol.

Further research is recommended for the detailed studies in each topical area presented in this chapter including scientometric studies and citation classic studies to inform the key stakeholders about the potential of the marine algae for the solution of the global problems of warming, air pollution, energy security, and food security in the form of algal bioethanol.

References

Adams, J.M., J.A. Gallagher, and I.S. Donnison. 2009. Fermentation study on *Saccharina latissima* for bioethanol production considering variable pre-treatments. *Journal of Applied Phycology* 21:569–574.

Amin, S. 2009. Review on biofuel oil and gas production processes from microalgae. *Energy Conversion and Management* 50:1834–1840.

Angermayr, S.A., K.J. Hellingwerf, P. Lindblad, and M.J.T. de Mattos. 2009. Energy biotechnology with cyanobacteria. *Current Opinion in Biotechnology* 20:257–263.

Baltussen, A. and C.H. Kindler. 2004a. Citation classics in anesthetic journals. *Anesthesia & Analgesia* 98:443–451.

Baltussen, A. and C.H. Kindler. 2004b. Citation classics in critical care medicine. *Intensive Care Medicine* 30:902–910.

Bothast, R.J. and M.A. Schlicher. 2005. Biotechnological processes for conversion of corn into ethanol. *Applied Microbiology and Biotechnology* 67:19–25.

Carrieri, D., D. Momot, I.A. Brasg et al. 2010. Boosting autofermentation rates and product yields with sodium stress cycling: Application to production of renewable fuels by cyanobacteria. *Applied and Environmental Microbiology* 76:6455–6462.

Cheng, J.J. and G.R. Timilsina. 2011. Status and barriers of advanced biofuel technologies: A review. *Renewable Energy* 36:3541–3549.

Chisti, Y. 2007. Biodiesel from microalgae. *Biotechnology Advances* 25:294–306.

Choi, S.P., M.T. Nguyen, and S.J. Sim. 2010. Enzymatic pretreatment of *Chlamydomonas reinhardtii* biomass for ethanol production. *Bioresource Technology* 101:5330–5336.

Costa, J.A.V. and M.G. de Morais. 2011. The role of biochemical engineering in the production of biofuels from microalgae. *Bioresource Technology* 102:2–9.

Demirbas, A. 2007. Progress and recent trends in biofuels. *Progress in Energy and Combustion Science* 33:1–18.

Demirbas, A. 2010. Use of algae as biofuel sources. *Energy Conversion and Management* 51:2738–2749.

Deng, D.M. and J.R. Coleman. 1999. Ethanol synthesis by genetic engineering in cyanobacteria. *Applied and Environmental Microbiology* 65:523–528.

Dexter, J. and P.C. Pengcheng. 2009. Metabolic engineering of cyanobacteria for ethanol production. *Energy & Environmental Science* 2:857–864.

Dubin, D., A.W. Hafner, and K.A. Arndt. 1993. Citation-classics in clinical dermatological journals—Citation analysis, biomedical journals, and landmark articles, 1945–1990. *Archives of Dermatology* 129:1121–1129.

Fargione, J.E., T.R. Cooper, D.J. Flaspohler et al. 2009. Bioenergy and wildlife: Threats and opportunities for grassland conservation. *Bioscience* 69:767–777.

Ge, L.L., P. Wang, and H.J. Mou. 2011. Study on saccharification techniques of seaweed wastes for the transformation of ethanol. *Renewable Energy* 36:84–89.

Gehanno, J.F., K. Takahashi, S. Darmoni, and J. Weber. 2007. Citation classics in occupational medicine journals. *Scandinavian Journal of Work, Environment & Health* 33:245–251.

Godfray, H.C.J., J.R. Beddington, I.R. Crute et al. 2010. Food security: The challenge of feeding 9 billion people. *Science* 327:812–818.

Goldemberg, J. 2007. Ethanol for a sustainable energy future. *Science* 315:808–810.

Groom, M.J., E.M. Gray, and P.A. Townsend. 2008. Biofuels and biodiversity: Principles for creating better policies for biofuel production. *Conservation Biology* 22:602–609.

Harun, R. and M.K. Danquah. 2011a. Influence of acid pre-treatment on microalgal biomass for bioethanol production. *Process Biochemistry* 46:304–309.

Harun, R. and M.K. Danquah. 2011b. Enzymatic hydrolysis of microalgal biomass for bioethanol production. *Chemical Engineering Journal* 168:1079–1084.

Harun, R., M.K. Danquah, and G.M. Forde. 2010b. Microalgal biomass as a fermentation feedstock for bioethanol production. *Journal of Chemical Technology and Biotechnology* 85:199–203.

Harun, R., W.S.Y. Jason, T. Cherrington, and M.K. Danquah. 2011. Exploring alkaline pre-treatment of microalgal biomass for bioethanol production. *Applied Energy* 88:3464–3467.

Harun, R., M. Singh, G.M. Forde, and M.K. Danquah. 2010a. Bioprocess engineering of microalgae to produce a variety of consumer products. *Renewable & Sustainable Energy Reviews* 14:1037–1047.

Hellingwerf, K.J. and M.J.T. de Mattos. 2009. Alternative routes to biofuels: Light-driven biofuel formation from CO_2 and water based on the 'photanol' approach. *Journal of Biotechnology* 142:87–90.

Hirano, A., R. Ueda, S. Hirayama, and Y. Ogushi. 1997. CO_2 fixation and ethanol production with microalgal photosynthesis and intracellular anaerobic fermentation. *Energy* 22:137–142.

Hirayama, S., R. Ueda, Y. Ogushi et al. 1998. Ethanol production from carbon dioxide by fermentative microalgae. *Studies in Surface Science and Catalysis* 114:657–660.

Horn, S.J., I.M. Aasen, and K. Ostgaard. 2000a. Ethanol production from seaweed extract. *Journal of Industrial Microbiology & Biotechnology* 25:249–254.

Horn, S.J., I.M. Aasen, and K. Ostgaard. 2000b. Production of ethanol from mannitol by *Zymobacter palmae*. *Journal of Industrial Microbiology & Biotechnology* 24:51–57.

Jacobson, M.Z. 2009. Review of solutions to global warming, air pollution, and energy security. *Energy & Environmental Science* 2:148–173.

John, R.P., G.S. Anisha, K.M. Nampoothiri, and A. Pandey. 2011. Micro and macroalgal biomass: A renewable source for bioethanol. *Bioresource Technology* 102:186–193.

Kapdan, I.K. and F. Kargi. 2006. Bio-hydrogen production from waste materials. *Enzyme and Microbial Technology* 38:569–582.

Kim, N.J., H. Li, K. Jung, H.N. Chang, and P.C. Lee. 2011. Ethanol production from marine algal hydrolysates using *Escherichia coli* KO11. *Bioresource Technology* 102:7466–7469.

Konur, O. 2000. Creating enforceable civil rights for disabled students in higher education: An institutional theory perspective. *Disability & Society* 15:1041–1063.

Konur, O. 2002a. Assessment of disabled students in higher education: Current public policy issues. *Assessment and Evaluation in Higher Education* 27:131–152.

Konur, O. 2002b. Access to employment by disabled people in the UK: Is the Disability Discrimination Act working? *International Journal of Discrimination and the Law* 5:247–279.

Konur, O. 2002c. Access to nursing education by disabled students: Rights and duties of nursing programs. *Nursing Education Today* 22:364–374.

Konur, O. 2006a. Participation of children with dyslexia in compulsory education: Current public policy issues. *Dyslexia* 12:51–67.

Konur, O. 2006b. Teaching disabled students in higher education. *Teaching in Higher Education* 11:351–363.

Konur, O. 2007a. A judicial outcome analysis of the Disability Discrimination Act: A windfall for the employers? *Disability & Society* 22:187–204.

Konur, O. 2007b. Computer-assisted teaching and assessment of disabled students in higher education: The interface between academic standards and disability rights. *Journal of Computer Assisted Learning* 23:207–219.

Konur, O. 2011. The scientometric evaluation of the research on the algae and bio-energy. *Applied Energy* 88:3532–3540.

Konur, O. 2012a. 100 citation classics in energy and fuels. *Energy Education Science and Technology Part A—Energy Science and Research* 2012(si):319–332.

Konur, O. 2012b. What have we learned from the citation classics in energy and fuels: A mixed study. *Energy Education Science and Technology Part A—Energy Science and Research* 2012(si):255–268.

Konur, O. 2012c. The gradual improvement of disability rights for the disabled tenants in the UK: The promising road is still ahead. *Social Political Economic and Cultural Research* 4:71–112.

Konur, O. 2012d. Prof. Dr. Ayhan Demirbas' scientometric biography. *Energy Education Science and Technology Part A—Energy Science and Research* 28:727–738.

Konur, O. 2012e. The evaluation of the research on the biofuels: A scientometric approach. *Energy Education Science and Technology Part A—Energy Science and Research* 28:903–916.

Konur, O. 2012f. The evaluation of the research on the biodiesel: A scientometric approach. *Energy Education Science and Technology Part A—Energy Science and Research* 28:1003–1014.

Konur, O. 2012g. The evaluation of the research on the bioethanol: A scientometric approach. *Energy Education Science and Technology Part A—Energy Science and Research* 28:1051–1064.

Konur, O. 2012h. The evaluation of the research on the microbial fuel cells: A scientometric approach. *Energy Education Science and Technology Part A—Energy Science and Research* 29:309–322.

Konur, O. 2012i. The evaluation of the research on the biohydrogen: A scientometric approach. *Energy Education Science and Technology Part A—Energy Science and Research* 29:323–338.

Konur, O. 2012j. The evaluation of the biogas research: A scientometric approach. *Energy Education Science and Technology Part A—Energy Science and Research* 29:1277–1292.

Konur, O. 2012k. The scientometric evaluation of the research on the production of bio-energy from biomass. *Biomass & Bioenergy* 47:504–515.

Konur, O. 2012l. The evaluation of the global energy and fuels research: A scientometric approach. *Energy Education Science and Technology Part A—Energy Science and Research* 30:613–628.

Konur, O. 2012m. The evaluation of the biorefinery research: A scientometric approach. *Energy Education Science and Technology Part A—Energy Science and Research* 2012(si):347–358.

Konur, O. 2012n. The evaluation of the bio-oil research: A scientometric approach. *Energy Education Science and Technology Part A-Energy Science and Research* 2012(si):379–392.

Konur, O. 2012o. What have we learned from the citation classics in energy and fuels: A mixed study. *Energy Education Science and Technology Part A—Energy Science and Research* 2012(si): 255–268.

Konur, O. 2012p. The evaluation of the research on the biofuels: A scientometric approach. *Energy Education Science and Technology Part A—Energy Science and Research* 28:903–916.

Konur, O. 2013. What have we learned from the research on the International Financial Reporting Standards (IFRS)? A mixed study. *Energy Education Science and Technology Part D: Social Political Economic and Cultural Research* 5:29–40.

Kosaric, N. and J. Velikonja. 1995. Liquid and gaseous fuels from biotechnology—Challenge and opportunities. *FEMS Microbiology Reviews* 16:111–142.

Lal, R. 2004. Soil carbon sequestration impacts on global climate change and food security. *Science* 304:1623–1627.

Lynd, L.R., J.H. Cushman, R.J. Nichols, and C.E. Wyman. 1991. Fuel ethanol from cellulosic biomass. *Science* 251:1318–1323.

Matsumoto, M., H. Yokouchi, N. Suzuki, H. Ohata, and T. Matsunaga. 2003. Saccharification of marine microalgae using marine bacteria for ethanol production. *Applied Biochemistry and Biotechnology* 105:247–254.

Miura, Y., W. Yamada, K. Hirata, K. Miyamoto, and M. Kiyohara. 1993. Stimulation of hydrogen-production in algal cells grown under high CO_2 concentration and low-temperature. *Applied Biochemistry and Biotechnology* 39:753–761.

Mussatto, S.I., G. Dragone, P.M.R. Guimaraes et al. 2010. Technological trends, global market, and challenges of bio-ethanol production. *Biotechnology Advances* 28:817–830.

Nguyen, M.T., S.P. Choi, J. Lee, J.H. Lee, and S.J. Sim. 2009. Hydrothermal acid pretreatment of *Chlamydomonas reinhardtii* biomass for ethanol production. *Journal of Microbiology and Biotechnology* 19:161–166.

Nigam, P.S. and A. Singh. 2011. Production of liquid biofuels from renewable resources. *Progress in Energy and Combustion Science* 37:52–68.

North, D. 1994. Economic-performance through time. *American Economic Review* 84:359–368.

Paladugu, R., M.S Chein, S. Gardezi, and L. Wise. 2002. One hundred citation classics in general surgical journals. *World Journal of Surgery* 26:1099–1105.

Phalan, B. 2009. The social and environmental impacts of biofuels in Asia: An overview. *Applied Energy* 86:S21–S29.

Powell, E.E. and G.A. Hill. 2009. Economic assessment of an integrated bioethanol-biodiesel-microbial fuel cell facility utilizing yeast and photosynthetic algae. *Chemical Engineering Research and Design* 87:1340–1348.

Rawat, I., R.R. Kumar, T. Mutanda, and F. Bux. 2011. Dual role of microalgae: Phycoremediation of domestic wastewater and biomass production for sustainable biofuels production. *Applied Energy* 88:3411–3424.

Singh, A., P.S. Nigam, and J.D. Murphy. 2011a. Renewable fuels from algae: An answer to debatable land based fuels. *Bioresource Technology* 102:10–16.

Singh, A., P.S. Nigam, and J.D. Murphy. 2011b. Mechanism and challenges in commercialisation of algal biofuels. *Bioresource Technology* 102:26–34.

Singh, A. and S.I. Olsen. 2011. A critical review of biochemical conversion, sustainability and life cycle assessment of algal biofuels. *Applied Energy* 88:3548–3555.

Singh, J. and S. Cu. 2010. Commercialization potential of microalgae for biofuels production. *Renewable & Sustainable Energy Reviews* 14:2596–610.

Sivakumar, G., D.R. Vail, J.F. Xu et al. 2010. Bioethanol and biodiesel: Alternative liquid fuels for future generations. *Engineering in Life Sciences* 10:8–18.

Spolaore, P., C. Joannis-Cassan, E. Duran, and A. Isambert. 2006. Commercial applications of microalgae. *Journal of Bioscience and Bioengineering* 101:87–96.

Stal, L.J. and R. Moezelaar. 1997. Fermentation in cyanobacteria. *FEMS Microbiology Reviews* 21:179–211.

Tan, X.M., L. Yao, Q.Q. Gao, W.H. Wang, F.X. Qi, and X.F. Lu. 2011. Photosynthesis driven conversion of carbon dioxide to fatty alcohols and hydrocarbons in cyanobacteria. *Metabolic Engineering* 13:169–176.

Taylor, G. 2008. Biofuels and the biorefinery concept. *Energy Policy* 36:4406–4409.

Ueno, R., N. Urano, M. Suzuki, and S. Kimura. 2002. Isolation, characterization, and fermentative pattern of a novel thermotolerant *Prototheca zopfii* var. hydrocarbonea strain producing ethanol and CO_2 from glucose at 40 degrees C. *Archives of Microbiology* 177:244–250.

Ueno, Y., N. Kurano, and S. Miyachi. 1998. Ethanol production by dark fermentation in the marine green alga, *Chlorococcum littorale*. *Journal of Fermentation and Bioengineering* 86:38–43.

Volesky, B. and Z.R. Holan. 1995. Biosorption of heavy-metals. *Biotechnology Progress* 11:235–250.

Wang, B.Y., Q. Liu, N. Wan, and C.Q. Lan. 2008. CO_2 bio-mitigation using microalgae. *Applied Microbiology and Biotechnology* 79:707–718.

Wang, X., X.H. Liu, and G.Y. Wang. 2011. Two-stage hydrolysis of invasive algal feedstock for ethanol fermentation. *Journal of Integrative Plant Biology* 53:246–252.

Wargacki, A.J., E. Leonard, M.N. Win et al. 2012. An engineered microbial platform for direct biofuel production from brown macroalgae. *Science* 355:308–313.

Williams, P.R.D., D. Inman, A. Aden, and G.A. Heath. 2009. Environmental and sustainability factors associated with next-generation biofuels in the US: What do we really know? *Environmental Science & Technology* 43:4763–4775.

Wrigley, N. and S. Matthews. 1986. Citation-classics and citation levels in geography. *Area* 18:185–194.

Wyman, C.E. 1994. Alternative fuels from biomass and their impact on carbon-dioxide accumulation. *Applied Biochemistry and Biotechnology* 45–46:897–915.

Wyman, C.E. and Goodman, B.J. 1993. Biotechnology for production of fuels, chemicals, and materials from biomass. *Applied Biochemistry and Biotechnology* 39:41–59.

Yergin, D. 2006. Ensuring energy security. *Foreign Affairs* 85:69–82.

13

Seaweed Bioethanol Production: Its Potentials and Challenges

Maria Dyah Nur Meinita, Bintang Marhaeni, Gwi-Taek Jeong, and Yong-Ki Hong

CONTENTS

13.1 Introduction

Seaweed biomass that differs from terrestrial plant biomass has attracted global attention as a potential bioethanol source. Seaweed poses many advantages over terrestrial biomass sources. *Kappaphycus alvarezii* (cottonii) is a red seaweed and an important source of carrageenan that has numerous applications in the food and pharmaceutical industries. Since the 1970s, when *Eucheuma* farming was first introduced by the late phycologist Maxwell S. Doty, the cultivation of this species has spread throughout Central America, Caribbean, South Pacific, and Southeast Asia. However, thus far, successful commercialization of *K. alvarezii* has occurred only in Indonesia (Adam and Porse 1987). Currently, the Philippines and Indonesia are the biggest producers of this species. Production of *K. alvarezii* in Indonesia averages 1,500,000 tons wet weight/year from 1,110,900 ha of potentially cultivatable seafloor (Pambudi et al. 2010).

Several studies have indicated significant differences in chemical composition among the Rhodophyta. In some red seaweeds, such as *Palmaria palmata* and Porphyra, the protein contents are 35% and 45% dry weight DW respectively. These levels are comparable to those of terrestrial plants (Mannivannan et al. 2009). Compared to other red seaweeds, Indonesian *K. alvarezii* has a unique proximate composition. Indonesian *K. alvarezii* is composed of 0.7% protein WW, 0.2% fat WW, 11.6% dietary fiber WW, and several macrominerals (Mg, Ca, K, Na) and microminerals (Cu, Zn, Fe) (Santoso et al. 2006; Fayaz et al. 2005). Nutrient compositions can vary widely depending on multiple environmental factors as water temperature, salinity, light, and available nutrient.

Seaweeds with a high biomass are strongly preferred for the production of bioethanol. Cultivation is required to meet the growing bioethanol demand. The industry currently utilizes 7.5–8 million tons of wet seaweed annually, either from naturally wild seaweed or from cultivated one. One million tons of wet seaweed are harvested to produce hydrocolloids (alginate, agar, and carrageenan) (McHugh 2003).

Countries with coastlines amenable to *K. alvarezii*, such as China, India, Indonesia, Madagascar, Malaysia, the Philippines, and Tanzania are potential cultivation sites. Indonesia, in particular, is currently one of the highest producers of *K. alvarezii* in the world and also one of the best areas for cultivation (Tables 13.1 and 13.2). The potential area in Indonesia that could be used for *K. alvarezii* cultivation is shown in Table 13.3, and the annual total production of *K. alvarezii* is shown in Table 13.4. Papua, Moluccas, and Central Sulawesi also have large potential areas for cultivating *K. alvarezii*. The South Sulawesi province now boasts the highest annual production of *K. alvarezii* (Pambudi et al. 2010). The *K. alvarezii* cultivation in Sulawesi began in 1988 and spread throughout the in surrounding area (Luxton 1993). Bioethanol production in Indonesia, primarily

TABLE 13.1

Potency of Cultivation Area for *K. alvarezii* Cultivation

Country	'000 km (%) of Coast	Year Started	Sources
China	15.3 (1.81)	1985	Wu et al. (1988)
India	7.0 (0.83)	1989	Mairh et al. (1995)
Indonesia	54.7 (6.48)	1975	Adnan and Porse (1987)
Madagascar	4.8 (0.57)	1998	Ask and Corrales (2002)
Malaysia	4.7 (0.55)	1977	Doty (1980)
Philippines	36.3 (4.30)	1971	Doty and Alvarez (1973)
Tanzania	1.4 (0.17)	1989	Lirasan and Twide (1993)

Source: Neish, I.C., The ABC of *Eucheuma* seaplant production, SuriaLink 1-0703, www.surialink.com, 2003.

TABLE 13.2

Status of *Kappaphycus* and *Eucheuma* Cultivation in the World

Country	Kappaphycus mt/Year (%)	Eucheuma mt/Year (%)	Develop Status
China	800 (0.7)	ND	Expand
India	200 (0.2)	ND	Expand
Indonesia	48,000 (42.0)	8,000 (35.7)	Expand
Madagascar	300 (0.3)	400 (1.8)	Expand
Malaysia	4,000 (3.5)	Trace	Expand
Philippines	60,000 (52.5)	10,000 (44.6)	Contract
Tanzania	1.4 (0.9)	4,000 (17.9)	Static

Source: Neish, I.C., The ABC of *Eucheuma* seaplant production, SuriaLink 1-0703, www.surialink.com, 2003.
ND, not detectable.

TABLE 13.3

Potential Area of Seaweed Cultivation in Indonesia

Province	Prefecture	Potential Area (ha)
Aceh	Aceh Besar, Aceh Barat, Pidie, Aceh Utara, Aceh Timur, Aceh Selatan, Simeulue, Singkil, Sabang	104,100
North Sumatra	North Sumatra	2,000
Jakarta	Seribu islet	1,800
West Java	Sumur, Panjang islet	2,400
Central Java	Jepara, Kebumen, Rembang	5,200
East Java	Sumenep, Situbondo, Pacitan	29,500
Bali	Buleleng, Karangasem, Klungkung, Badung, Jembrana	18,100
West Nusa Tenggara	Lombok Barat, Lombok Timur, Lombok Tengah, Sumbawa, Dompu, Bima	12,000
East Nusa Tenggara	Kupang	1,000
North Sulawesi	Manado, Minahasa, Bitung, Sangihe Talaud, Bolaang Mongondow, Gorontalo	10,500
Central Sulawesi	Donggala, Banggai, Banggai islet	106,300
South Sulawesi	Sinjai, Selayar, Bulukumba, Jeneponto, Takalar, Maros, Pangkep, Bantaeng, Mamuju	26,500
Southeast Sulawesi	Kendari, Kolaka, Buton, Muna	83,000
Moluccas	Maluku Utara, Maluku Tengah, Maluku Tenggara, Halmahera, Ternate	206,600
Papua	Biak Numfor, Yapen Waropen, Manokwari, Sorong	501,900
Total		1,110,900

Source: Pambudi, L. et al., *Mar. Biosci. Biotechnol.*, 4(1), 6, 2010.

from sugarcane, has stimulated economic growth while reducing poverty and pollution. However, land available for sugarcane production is limited. Seaweed would be an excellent alternative as a sustainable resource for bioethanol production. To date, there is lack of information regarding Indonesian seaweed and its applicability to bioethanol (Adnan and Porse 1987; Santoso et al. 2006). Therefore, the current study examines Indonesian *K. alvarezii* as a potential source of bioethanol. The aims of this study are (1) to observe the proximate composition (lipids, proteins, carbohydrates, ash, and moisture) of Indonesian *K. alvarezii*, (2) to determine its sugar and by-product content, and (3) to compare the fermentability and the potential for bioethanol production of nine *K. alvarezii* species collected from different areas of Indonesia (Figure 13.1).

13.2 Material and Method

13.2.1 Seaweed Material

Nine *K. alvarezii* strains were collected from different areas in Indonesia (Karimunjawa, Kupang, Bali, Makassar, Lombok, Madura, Lampung, Bone, Papua) (Figure 13.2). Collected seaweeds were washed thoroughly with tap water to remove all unwanted impurity. The washed seaweeds were dried and then ground to make seaweed powder.

TABLE 13.4

Potential Production of *Eucheuma* by Province

	Potential Production (Tons)				
Province	**2005**	**2006**	**2007**	**2008**	**2009**
Aceh	0	29,474	35,361	42,421	50,000
Riau	9,821	11,789	14,144	16,968	20,000
West Sumatra	9,821	11,789	14,144	16,968	20,000
Lampung	24,553	29,474	35,361	42,421	50,000
Jakarta	1,964	2,358	2,829	3,394	4,000
Banten	1,473	1,768	2,122	2,545	3,000
Central Java	9,821	11,789	14,144	16,968	20,000
East Java	26,605	36,447	48,221	62,342	77,500
Bali	39,284	47,158	56,577	67,874	80,000
West Nusa Tenggara	83,743	91,368	109,618	131,505	155,000
East Nusa Tenggara	81,287	88,421	106,082	127,263	150,000
South Kalimantan	3,928	4,716	5,658	6,787	8,000
East Kalimantan	9,821	11,789	14,144	16,968	20,000
North Sulawesi	22,185	23,579	28,288	33,937	40,000
Gorontalo	39,691	47,158	56,577	67,874	80,000
Central Sulawesi	39,691	47,158	56,577	67,874	80,000
West Sulawesi	27,675	31,640	33,466	35,655	38,006
South Sulawesi	78,979	96,965	116,331	139,560	164,494
Southeast Sulawesi	61,469	70,737	84,865	101,811	120,000
Moluccas	39,284	47,158	56,577	67,874	80,000
North Moluccas	58,926	70,737	84,865	101,811	120,000
West Papua	22,185	23,579	28,288	33,937	40,000
Papua	44,371	47,158	56,577	67,874	80,000
Total	736,579	884,210	1,060,816	1,272,631	1,500,000

Source: Pambudi, L. et al., *Mar. Biosci. Biotechnol.*, 4(1), 6, 2010.

13.2.2 Proximate (Carbohydrate, Protein, Lipid, Ash, and Moisture) Analysis

The total protein was estimated using the modified Lowry method (Lowry et al. 1951). Lipid determination was determined gravimetrically after extraction by chloroform–methanol (2:1) mixture. Total carbohydrate was determined by phenol–sulfuric acid (Kochert 1978). Ash content was determined by heating the samples for 5 h at 575°C. Moisture content was measured by drying sample at 105°C during 18 h until they reached a stable weight.

13.2.3 Acid Hydrolysis

Acid hydrolysis was carried out in 250 mL flask. The samples consisted of 10 g seaweed powder/100 mL of 0.2 M H_2SO_4 hydrolyzed in autoclave at 130°C for 15 min. After hydrolysis was completed, the residues were separated from liquid solution by filtration. The liquid solutions were analyzed for sugar determination and fermented for ethanol production.

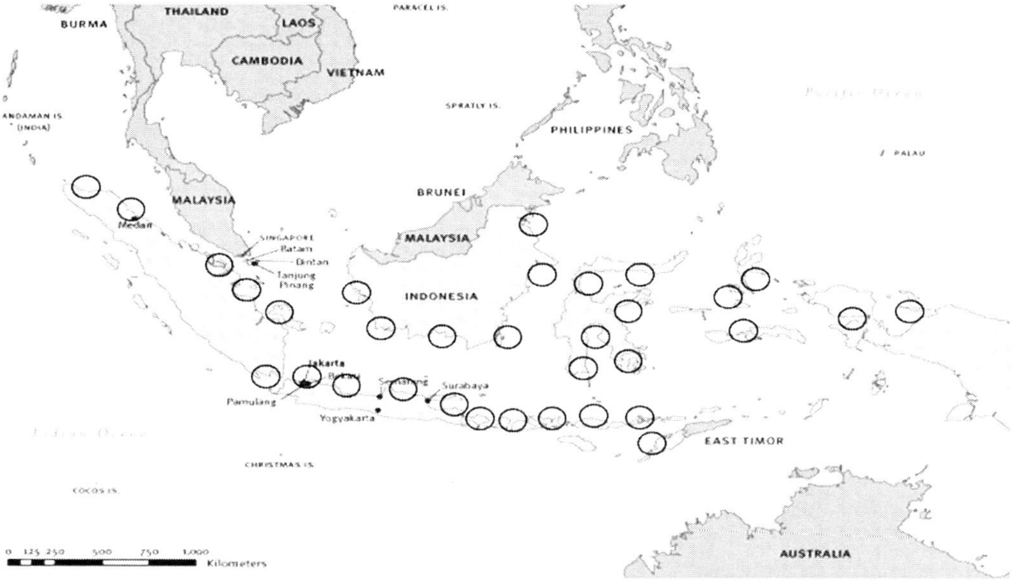

FIGURE 13.1
Location of potential area of seaweed cultivation in Indonesia. (From Pambudi, L. et al., *Mar. Biosci. Biotechnol.*, 4(1), 6, 2010.)

FIGURE 13.2
K. alvarezii materials from Bone (brown) (a), Kupang (b), Lampung (white) (c), Papua (d), Sumenep (e), Takalar (f), Lampung (brown) (g), Karimunjawa (h), Bali (i), Bone (white) (j), and Lombok (k).

13.2.4 Sugar and By-Product Determination

Monosaccharide (galactose, glucose), levulinic acid, and 5-hydroxymethyl-furfural (HMF) were determined by HPLC with Alltech IOA 1000 organic acid column (300 mm × 7.8 mm), which is equipped with RI detector and maintained at 60°C. The mobile phase used was 0.005 N of sulfuric acid with flow rate of 0.3 mL/min. The injection volume was 20 μL. Reducing sugar was determined using dinitrosalicylic acid (DNS) method (Chaplin 1986). The DNS solution was prepared by dissolving 3,5-dinitrosalicylic acid and potassium sodium tartrate (Rochelle salt) in 2 M NaOH (Chaplin 1986).

13.2.5 Fermentation, Microorganisms and Medium

Commercial yeast, *Saccharomyces cerevisiae* (JENICO, Korea), in basal medium was used in fermentation. The basal medium consisted of 0.02% $(NH_4)_2SO_4$ and 0.006% NaH_2PO_4 and neutralized at pH 5 (Prescott 1959). Fermentation was carried out in triplicate in 3 mL volume. The samples were incubated in shaking incubator at 30°C with gentle mixing at 120 rpm. Samplings were done for measurement of ethanol production.

13.3 Result and Discussion

13.3.1 Proximate Analysis

The comparison of proximate analysis was shown in Figure 13.3. The maximum carbohydrate content was recorded in Papua strain (52.2% ± 6.8% g/g), while the minimum one was

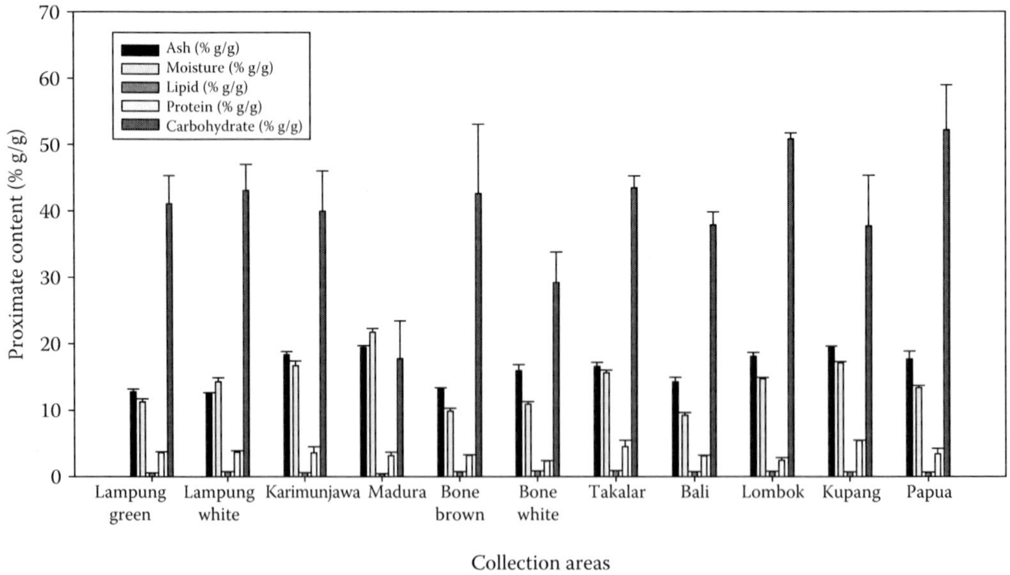

FIGURE 13.3
Carbohydrate, protein, lipid, ash, and moisture concentration of each *K. alvarezii* strains. (From Meinita, D.N.M. et al., *J. Appl. Phycol.*, 24, 857, 2012a.)

observed in Madura strain (17.8% ± 5.6% g/g); protein content was observed maximum in Kupang strain (5.4% ± 0.07% g/g) and minimum in Bone (white) strain (2.3% ± 0.07% g/g). The lipid content was acquired highest level in Lombok strain (0.79% ± 0.015% g/g) and minimum in Madura strain (0.39% ± 0.06% g/g). The range of ash in this study was 12.7%–19.7%. The moisture range in this study was between 9.2% ± 0.4% g/g and 21.7% ± 0.5% g/g (Figure 13.3).

13.3.2 Sugar, By-Product Content, and Ethanol Production

The comparison of sugar, by-product content, and ethanol production is shown in Figures 13.4 through 13.6. The highest galactose content was found in Bone (brown) strain (8.6 ± 0.14 g/L), and the lowest was found in Madura strain (5.0 ± 0.11 g/L), while glucose as the minor sugar was observed maximum also in Bone (brown) strain (0.6 ± 0.04 g/L) and minimum in Madura strain (0.2 ± 0.04 g/L). The range of levulinic acid and HMF content among the samples was between 0.32 ± 0.02 g/L and 0.64 ± 0.09 g/L and between 0.95 ± 0.06 g/L and 1.73 ± 0.14 g/L, respectively. The highest ethanol production was produced by Papua strain (0.70 ± 0.08 g/L) followed by Bone (brown) strain (0.65 ± 0.07 g/L) and Lombok strain (0.64 ± 0.07 g/L) (Meinita et al. 2012a).

13.3.3 Potentials of Indonesian *K. alvarezii* for Bioethanol Production

Indonesia and the Philippines are the main sources of k-carrageenan in the world market. The k-carrageenans are used as addition in food (jellies, ice cream, juice, jam, sausage) and nonfood (pharmaceutical). The high demand of carrageenan might be considered if we want to use this polysaccharide for bioethanol production. The raw *K. alvarezii*

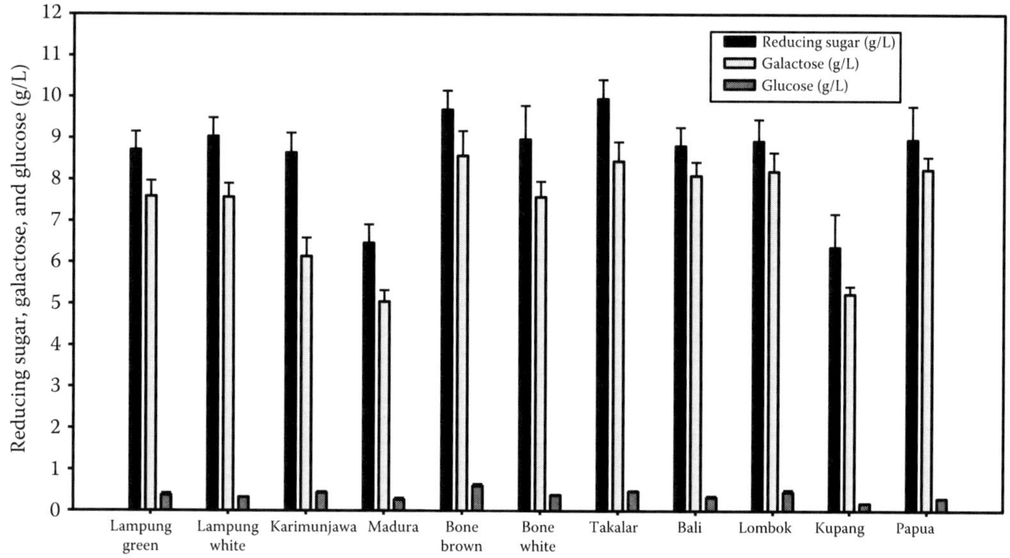

FIGURE 13.4
Reducing sugar, galactose, and glucose concentration of each *K. alvarezii* strains. (From Meinita, D.N.M. et al., *J. Appl. Phycol.*, 24, 857, 2012a.)

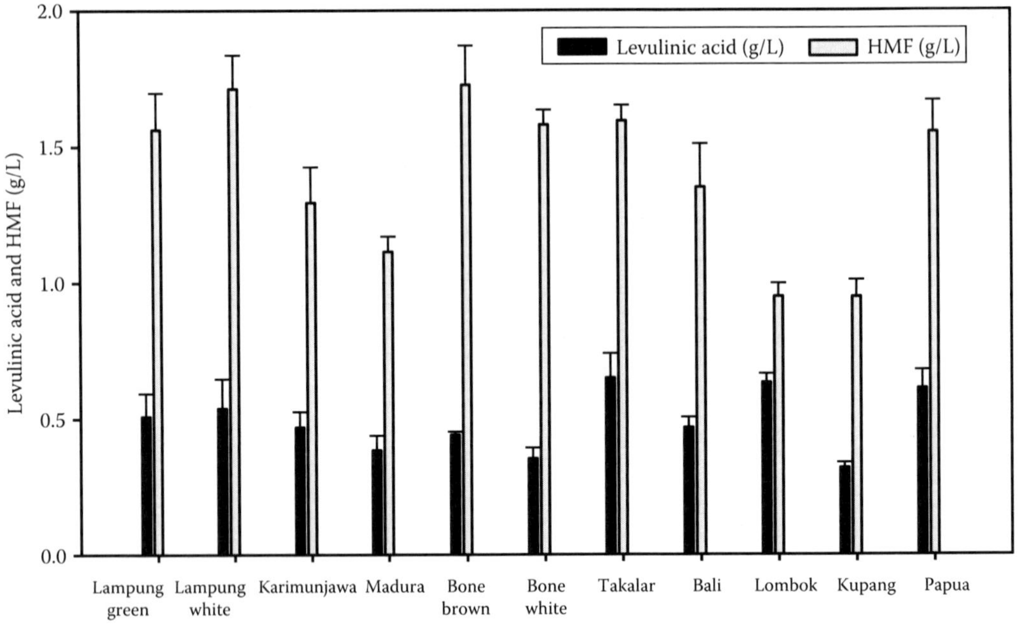

FIGURE 13.5
By-product concentration of each *K. alvarezii* strains. (From Meinita, D.N.M. et al., *J. Appl. Phycol.*, 24, 857, 2012a.)

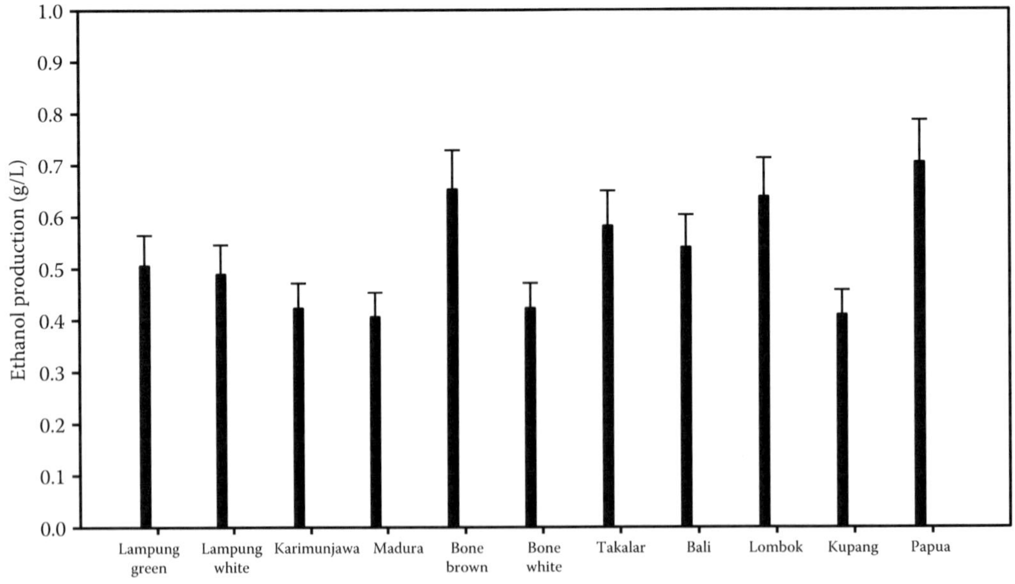

FIGURE 13.6
Ethanol production of each *K. alvarezii* strains. (From Meinita, D.N.M. et al., *J. Appl. Phycol.*, 24, 857, 2012a.)

price is about \$1.4–1.5/100 kg. The carrageenan industry consumes about 80,000 tons of seaweed per annum (McHugh 1991 in Luxton 1993). Indonesia as the world's largest archipelago country possesses great potential to develop seaweed bioethanol. Indonesia lies from 6 °N to 10 °S and from 95 °E to 142 °E with 18,110 islands and coastline about 54,700 km and can be a potential country for seaweed resources (Hutomo and Moosa 2005; Neish 2003). The total potential area in Indonesia for *Kappaphycus* cultivation is about 1,110,900 ha with the total production 975,000 tons dry weight/year (Table 13.5). According to our study, we found that Indonesian *K. alvarezii* contains 51% g/g of carbohydrate. The total carbohydrate was converted by acid hydrolysis into 26.5% ± 0.4% g/g of galactose and 2.15% ± 0.9% g/g of glucose. By assuming the sugar conversion based on our study, 131,771 tons dry weight of galactose and 10,690 tons dry weight of glucose can be extracted annually (Table 13.5). The total ethanol yield is about 53,814 ton dry weight/year based on the following Equation 13.1 (Goh and Lee 2010):

$$\text{Ethanol yield} = \text{galactose yield} \times \text{theoretical maximum ethanol yield} \times \text{fermentation efficiency} \qquad (13.1)$$

where the theoretical maximum ethanol yield is 0.51 g ethanol per gram sugar. The fermentation efficiency was calculated using the following Equation 13.2:

$$\text{Fermentation efficiency} = \text{ethanol produced}/\text{theoretical maximum ethanol yield} \qquad (13.2)$$

TABLE 13.5

Estimation of Ethanol Production from Indonesian *K. alvarezii*

Component	Quantity	Unit	References
A. Seaweed			
Total potential area in Indonesia	1,110,900	Ha/year	Pambudi et al. (2010)
Total potential production	1,500,000	Tons wet weight/year	Pambudi et al. (2010)
Total wet seaweed production	1.35	Tons wet weight/ha[a] year	
Total moisture	35	%	Luxton (1993)
Total dry seaweed production	975,000	Tons dry weight/year	
B. Galactose			
Carbohydrate	497,250	Tons dry weight/year	
Galactose	131,771	Tons dry weight/year	
Glucose	10,691	Tons dry weight/year	
C. Ethanol			
Theoretical maximum ethanol yield	51.1	%	
Fermentation efficiency	79.9	%	
Estimated ethanol yield			
By galactose	53,814	Tons dry weight/year	
By galactose and glucose	58,180	Tons dry weight/year	

[a] Carbohydrate, galactose, and glucose conversion were calculated using the result of our study. Estimated ethanol yield was calculated by assuming that fermentation efficiency is 79.9%.

13.3.4 Challenges of Indonesian *K. alvarezii* for Bioethanol Production

Despite recent studies showing significant progress, large scale commercialization of seaweed bioethanol remains challenging (Meinita et al. 2012b,c; Meinita et al. 2013). The primary constraints are technological concerns, economics, and sustainability. Technological constraints including pretreatment, saccharification, and fermentation must be solved to attain large scale production. Fermentation inhibitors such as HMF and levulinic acid that are liberated during the acid hydrolysis of terrestrial plants are also present in the hydrolysate of *K. alvarezii*. Detoxification treatments are required to remove these inhibitors. According to Taherzadeh et al. (1997) and Mussatto and Roberto (2004), there are four approaches that may be used the presence of inhibitors in hemicellulosic hydrolysates: (1) avoid formation of inhibitors during hydrolysis, (2) detoxify the hydrolysate before fermentation, (3) develop species of microorganisms that are able to resist inhibitors, and (4) convert toxic compounds into products that do not interfere with metabolism. The selection of an appropriate yeast strain also plays an important role in determining ethanol yields. Genetically engineered organisms that are able to ferment all of the available sugars in a given hydrolysate to ethanol would be a good choice. Further economic studies are required to estimate the total cost of bioethanol production from seaweed.

13.4 Conclusion

This study concludes that *K. alvarezii* contains high levels of carbohydrate; hence, it can be a good resource in bioethanol production, and some areas such as Bone, Papua, Lombok, and Makassar can be potential places to grow *K. alvarezii* in Indonesia. In this experiment, the samples were collected in one season. Therefore, further research is necessary to study more about seasonal variation on the nutrient composition. An economical analysis is also needed to investigate the feasibility of seaweed as a raw material in bioethanol production.

References

Adnan H, Porse H (1987) Culture of *Eucheuma cottonii* and *Eucheuma spinosum* in Indonesia. *Hydrobiologia* 151–152:355–358.

Ask EI, Azanza RV (2002) Advances in cultivation technology of commercial eucheumatoid species: A review with suggestions for future research. *Aquaculture* 206:257–277.

Chaplin MF (1986) Monosaccharide. In: Chaplin MF, Kennedy JF (eds.), *Carbohydrate Analysis: A Practical Approach*. IRC Press, Oxford, pp 1–36.

Doty MS, Alvarez, VB (1973) Seaweed farms: A new approach for U.S. industry. *Proceedings of the 9th Annual Conf. Proceedings*, University of Hawaii, Hawaii, pp. 701–708.

Fayaz M, Namitha K, Murthy KNC, Swamy MM, Sarada R, Khanam S, Subbarao PV, Ravishankar GA (2005) Chemical composition, iron bioavailability, and antioxidant activity of *Kappaphycus alvarezii* (Doty). *J Agric Food Chem* 53:792–797.

Goh CS, Lee KT (2010) A visionary and conceptual macroalgae-based third-generation bioethanol (TGB) biorefinery in Sabah, Malaysia as an underlay for renewable and sustainable development. *Re Sus En Re* 14:842–848.

Hutomo M, Moosa MK (2005) Indonesian marine and coastal biodiversity: Present status. *Ind J Mar Sci* 34(1):88–97.

Kochert G (1978) Carbohydrate determination by the phenol-sulfuric acid method . In: Hellebust JA, Craigie JS (eds.), *Handbook of Phycological Methods, Vol II, Physiological and Biochemical Methods.* Cambridge University Press, Cambridge, pp. 95–97.

Lirasan T, Twide P (1993) Farming *Eucheuma* in Zanzibar, Tanzania. *Hydrobiologia* 260–261:353–355.

Lowry OH, Rosebrough NJ, Farr AL, Randall RJ (1951) Protein measurement with the Folin phenol reagent. *J Biol Chem* 193:265–275.

Luxton DM (1993) Aspect of the farming and processing of *Kappaphycus* and *Eucheuma* in Indonesia. *Hydrobiologia* 260/261:365–371.

Mannivannan K, Thirumaran G, Devi GK, Anatharaman P, Balasubramanian (2009) Proximate composition of different group of seaweeds from Vedalai coastal waters (Gulf of Mannar): Southeast Coast of India. *Mid East J Sci R* 4(2):72–77.

Mairh OP, Zodape ST, Tewari A, Rajyaguru MR (1995) Culture of marine alga *Kappaphycus striatum* (Schmitz) Doty on the Suarashtra region, west coast of India. *Indian Journal of Marine Sciences* 24:24–31.

McHugh D (2003) A guide to the seaweed industry. Fisheries technical paper. FAO, Rome, Italy.

Meinita DNM, Kang JY, Jeong GT, Koo HM, Park SM, Hong YK (2012a) Bioethanol production from the acid hydrolysate of the carrageenophyte *Kappaphycus alvarezii (cottonii). J Appl Phycol* 24:857–862.

Meinita DNM, Jeong GT, Hong YK (2012b) Comparison of sulfuric and hydrochloric acids as catalysts in hydrolysis of *Kappaphycus alvarezii (cottonii). Bioproc Biosyst Eng* 35:123–128.

Meinita DNM, Jeong GT, Hong YK (2012c) Detoxification of acidic catalyzed hydrolysate of *Kappaphycus alvarezii (cottonii). Bioproc Biosyst Eng* 3:93–98.

Meinita DNM, Marhaeni B, Winanto JGT, Khan MNA, Hong YK (2013) Comparison of agarophytes (Gelidium, Gracilaria, and Gracilariopsis) as potential resources for bioethanol production. *J Appl Phycol.* doi:10.1007/s10811-013-0041-4.

Mussatto SI, Roberto IC (2004) Alternatives for detoxification of diluted-acid lignocellulosic hydrolyzates for use in fermentative processes: A review. *Bio Technol* 93:1–10.

Neish IC (2003) The ABC of *Eucheuma* seaplant production. SuriaLink 1-0703. www.surialink.com. Accessed on 5 December, 2013.

Pambudi L, Meinita MDN, Ariyati RW (2010) Seaweed cultivation in Indonesia: Recent status. *Mar Biosci Biotechnol* 4(1):6–10.

Prescott SC, Dun CG (1959) *Industrial Microbiology*. McGraw-Hill, New York.

Santoso J, Gunji S, Yoshie-Stark, Suzuki T (2006) Mineral contents of Indonesian seaweeds and mineral solubility affected by basic cooking. *Food Sci Technol Res* 12(1):59–66.

Taherzadeh MJ, Eklund R, Gustafsson L, Niklasson C, Lide G (1997) Characterization and fermentation of dilute-acid hydrolyzates from wood. *Ind Eng Chem Res* 36:4659–4665.

Wu CY, Li JJ, Xia EZ, Peng ZS, Tan SZ, Li J, Wen ZC, Huang XH, Cai ZL, Chen GJ (1988) Transplant and artificial cultivation of *Eucheuma striatum* in China. *Oceanol Limnol Sin* 19:410–417.

14

Bioethanol Production from Macroalgae and Microbes

Chae Hun Ra and Sung-Koo Kim

CONTENTS

14.1 Importance and Challenges of Macroalgae

Globally, fossil-fuel dependence resulted in considerable environmental problems, which include global warming, air quality deterioration, oil spills, and acid rain, among others. The combustion of fuel emits carbon dioxide (CO_2), one of the most significant greenhouse gases (GHGs) that trap heat in the earth's atmosphere. The level of atmospheric CO_2 concentration was reported to be around 350–380 ppm in 2010 and is predicted to increase to 450 ppm by 2020 if no action is taken (Kraan 2013). The heightened awareness of global warming as well as other environmental issues has increased interest in the development of methods to mitigate GHG emissions and lessen the production of pollutants. One of these abatement methods is the use of biofuels made from renewable sources of energy (Jang et al. 2012; Meinita et al. 2011).

Macroalgae (seaweed) has emerged as an alternative and promising feedstock to produce a myriad of renewable fuels. The use of seaweed as an energy feedstock for the production of biodiesel, bioethanol, biogas, and biohydrogen has been investigated (Adams et al. 2009; Park et al. 2012). Seaweed has a faster growth rate, lower land usage, higher CO_2 absorption and uptake rate, no need for fertilizers, and no competition for food or freshwater resources than lignocellulosic biomass (Singh et al. 2011). The seaweed has a high content of easily degradable carbohydrates, making it a potential substrate for the production of liquid fuels.

Macroalgae are divided into three major groups based on their photosynthetic pigments: green (*Chlorophyta* and *Charophyta*), red (*Rhodophyta*), and brown (*Phaeophyceae*) algae.

Green algae are composed of glucose, xylose, galactose, glucuronic acid, and rhamnose (Feng et al. 2011). The carbohydrates in red seaweed comprise a neutral polymer (agarose) and a sulfate polysaccharide (agaropectin). Galactose and glucose are obtained by agarose and agaropectin hydrolysis (Park et al. 2012). Brown algae are composed of alginate,

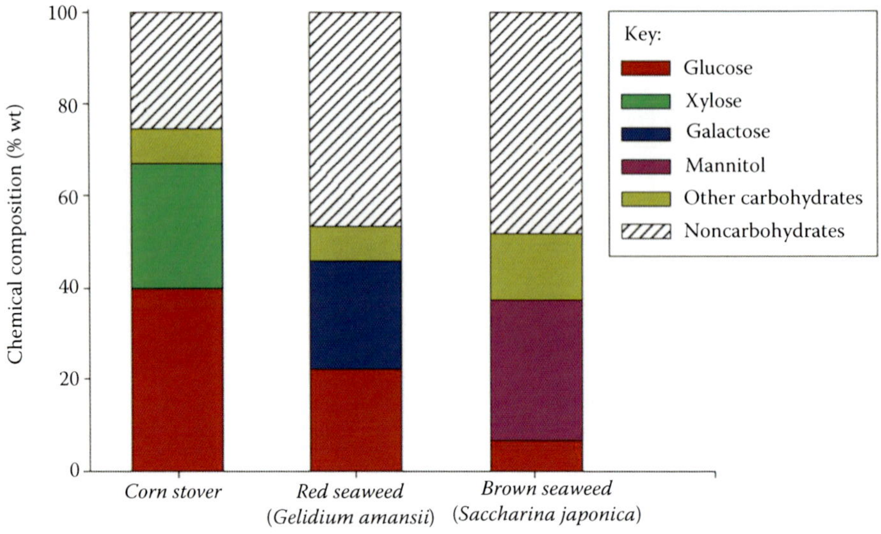

FIGURE 14.1
Chemical composition of hydrolysates obtained from corn stover, a red seaweed (*Gelidium amansii*), and a brown seaweed (*Saccharina japonica*). (From Kim, S.R. et al., *Trends Biotechnol.*, 30(5), 274, 2012.)

glucan, mannitol, and some other polysaccharides (Jang et al. 2012). Among them, the red algae are known for high carbohydrate content as one of the most abundantly available seaweed species.

The sugar composition of terrestrial plant biomass and macroalgae biomass is shown in Figure 14.1. As a result of the different chemical compositions of terrestrial plant biomass and macroalgae biomass, hydrolysates from these two sources are comprised of different types and amounts of fermentable sugars. Terrestrial plant hydrolysates obtained after pretreatment and enzyme saccharification contain 40% glucose and 30% xylose. By contrast, hydrolysates from red seaweed contain 22% glucose and 23% galactose as major sugars (Kim et al. 2012).

The significant difference in sugar composition suggests the importance of developing appropriate strategies for metabolic engineering of yeasts based on the target biomass resource. First, genes or networks of genes involved in the glucose signaling pathway to disable the glucose repression have been modified (Klein et al. 1999; Raamsdonk et al. 2001; Roca et al. 2004; Thanvanthri Gururajan et al. 2007). However, this approach alone might not resolve the problem because of complex and unknown interactions between the glucose signaling pathway and rapid glucose fermenting phenotypes. Second, heterologous expressions of secondary sugar transporters and metabolic genes have been explored to bypass the regulatory mechanisms of host *Saccharomyces cerevisiae* (Hector et al. 2008; Katajira et al. 2008; Runquist et al. 2009; Young et al. 2011).

14.2 Seaweed Cultivation and General Biofuel Production Processes

Currently, the macroalgae industry is primarily focused on products for human consumption, which account for 83%–90% of the global value of seaweed. The amount of macroalgae

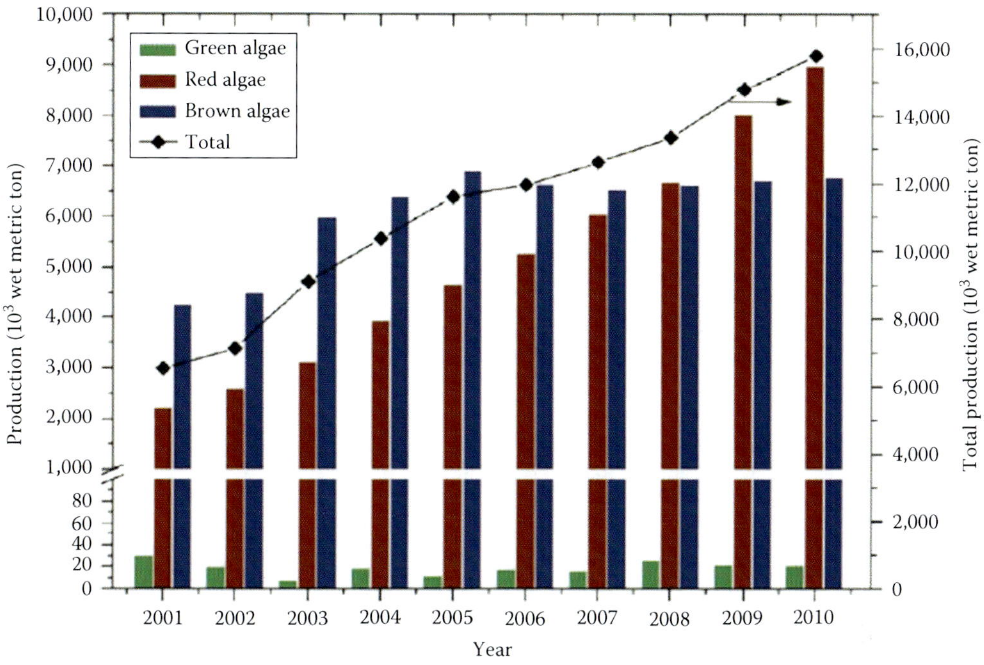

FIGURE 14.2
World production of farmed macroalgae from 2001 to 2010. (From Jung, K.A. et al., *Bioresour. Technol.*, 135, 182, 2013.)

mass cultivated in the world has continuously increased over the last 10 years at the average of 10% as shown in Figure 14.2 (Jung et al. 2013). Brown and red algae were cultivated more than green algae. It is noted that the red algae production dramatically increased, and the brown algae production attained 15.8 million wet tons, which were harvested from wild habitats and aquaculture farms in 2010.

As shown in Table 14.1, the amount of the mass cultivated macroalgae is four and six orders of magnitude greater than that of the microalgae and lignocellulosic biomass, respectively. Even though the macroalgae are two orders of magnitude less than the energy crops, this implies that macroalgae can be more mass cultivated to sufficiently supply biomass for bioethanol with current farming technology.

The use of bioethanol does not elevate carbon dioxide levels in the air. The following reaction mechanisms show the circulation system of carbon dioxide emitted by bioethanol as a fuel. Bioethanol can be produced using plants containing sugar, starch, and lignocelluloses (Hahn-Hägerdal et al. 2006; Saha and Cotta 2007). Bioethanol produced from materials containing sugar and starch is regarded as the first generation of bioethanol. However, many problems have arisen with increasing demand including food security and land substitution. The second generation of bioethanol uses cellulosic wastes, which are relatively abundant and low cost. However, the lignin contained in cellulosic wastes interferes with the ethanol production. Thus, a pretreatment process for removing lignin must be added. However, the use of lignin-free seaweed as a raw material is arising as a third generation of bioethanol production. Table 14.2 shows the typical contents of seaweed (Hong et al. 2013).

TABLE 14.1

World Production of Macroalgae, Microalgae, Energy Crops, and Lignocellulosic Biomass

Species	Group (or Phylum)	Production	% of Total
Macroalgae[a]			
Laminaria japonica	Brown algae	5,146,883	32.61
Eucheuma spp.	Red algae	3,489,388	22.11
Kappaphycus alvarezii	Red algae	1,875,277	11.88
Undaria pinnatifida	Brown algae	1,537,339	9.74
Gracilaria verrucosa[b]	Red algae	1,152,108	7.30
Porphyra spp.	Red algae	1,072,350	6.79
Gracilaria spp.[b]	Red algae	565,366	3.58
Porphyra tenera	Red algae	564,234	3.57
Eucheuma denticulatum	Red algae	258,612	1.64
Sargassum fusiforme	Brown algae	78,210	0.50
Phaeophyceae	Brown algae	21,747	0.14
Enteromorpha clathrata	Green algae	11,150	0.07
Monostroma nitidum	Green algae	4,531	0.03
Caulerpa spp.	Green algae	4,309	0.03
Codium fragile	Green algae	1,394	0.01
Gelidium amansii	Red algae	1,200	0.01
Total		15,784,098	100.00
Microalgae[c]			
Arthrospria sp.	Cyanophyta	3,000	
Chlorella sp.	Chlorophyta	2,000	
Dunaliella salina	Chlorophyta	1,200	
Haematococcus pluvialis	Chlorophyta	3,000	
Energy crops[d]			
Corn		844,405,181	
Palm oil		45,097,422	
Rapeseed		59,071,197	
Sugar cane		1,685,444,531	
Soybean		261,578,498	
Lignocellulosic biomass[e]			
Corn stover		12.6	
Switchgrass		9.0	

[a] Estimated in wet metric ton (FAO 2012a).
[b] Including macroalgae cultured in brackish water.
[c] Estimated in dry metric ton (Hejazi and Wijffels 2004; Lorenz and Cysewski 2000; Pulz and Gross 2004).
[d] Estimated in ton (FAO 2012b).
[e] Estimated in dry metric ton a hectare (Lemus et al. 2002; Shinners and Binversie 2007).

In general, bioethanol production from biomass involves pretreatment, enzymatic hydrolysis, fermentation, and distillation. When those are performed separately, the two processes of enzymatic saccharification and fermentation are referred to as separate hydrolysis and fermentation (SHF). Figure 14.3 shows a schematic diagram of ethanol production from macroalgae (seaweed) using SHF at pilot scale.

TABLE 14.2

Composition of Typical Seaweeds

	Species	Carbohydrate	Protein	Lipid	Ash
Green seaweed	*Codium fragile*	58.7	15.2	0.9	25.1
	Capsosiphon fulvescens	48.1	24.4	0.6	26.9
	Enteromorpha prolifera	53.3	22.9	0.8	22.9
	Ulva lactuca	50.4	26.8	0.6	22.2
	Caulerpa lentillifera	45.5	11.7	1.2	41.6
	Average	53.2 ± 1.8	19.8 ± 2.5	0.8 ± 0.1	26.1 ± 3.1
Red seaweed	*Gelidium amansii*	66.0	20.5	0.2	13.3
	Porphyra sp.	45.5	43.6	1.9	9.0
	Gigartina tenella	42.2	27.4	0.9	29.5
	Hypnea charoides	57.3	18.4	1.5	22.8
	Carpopeltis cornea	60.7	23.4	0.4	15.6
	Average	55.2 ± 2.8	23.1 ± 2.4	0.9 ± 0.2	20.6 ± 1.6
Brown seaweed	*Laminaria japonica*	51.5	8.4	1.3	38.8
	U. pinnatifida	43.2	23.8	3.5	29.5
	Hizikia fusiforme	47.5	9.8	1.2	41.5
	Eisenia bicyclis	72.7	8.2	0.2	18.8
	Ecklonia stolonifera	65.0	15.3	1.5	18.1
	Average	57.5 ± 4.2	12.1 ± 2.0	1.8 ± 0.6	28.6 ± 4.4

Source: Hong, I.K. et al., *J. Ind. Eng. Chem.*, 2013, http://dx.doi.org/10.1016/j.jiec.2013.10.056.

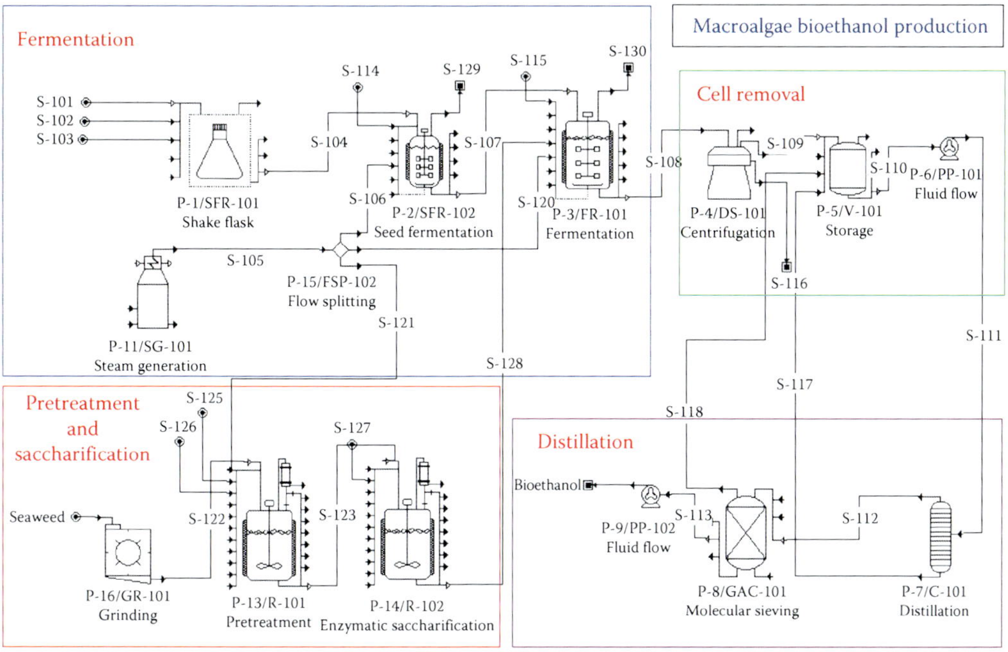

FIGURE 14.3
Schematic diagram of ethanol production from macroalgae (seaweed) using SHF at pilot scale (SuperPro Designer v.8.5).

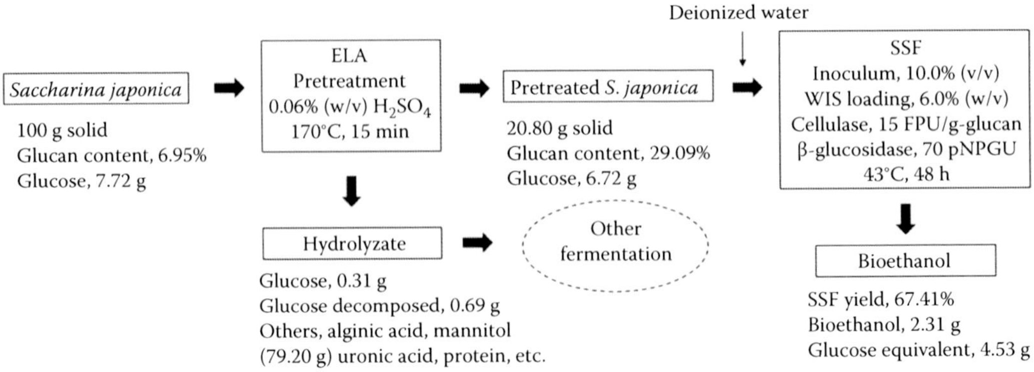

FIGURE 14.4
Detailed mass balance based on glucose for bioethanol production from *S. japonica* by 48 h using SSF. (From Lee, J.Y. et al., *Bioresour. Technol.*, 127, 119, 2013.)

Simultaneous saccharification and fermentation (SSF) is a single process that combines the saccharification and fermentation processes. Ethanol production from *Saccharina japonica* was carried out using an optimized extremely low acid (ELA) pretreatment followed by SSF (Lee et al. 2013). Six percent (w/v) of the water-insoluble solid (WIS) level in deionized water was prepared with 0.5, 0.025, and 1.0 g/L of $(NH_4)_2HPO_4$, $MgSO_4$ $7H_2O$, and yeast extracts. SSF was performed using enzyme loading of 15 FPU/g-glucan for cellulase and 70 pNPGU/g-glucan for beta-glucosidase with reaction temperature of 43°C and *S. cerevisiae* DK 410362. Figure 14.4 shows a detailed mass balance based on glucose amounts for bioethanol production from raw *S. japonica* by 48 h SSF.

When the ELA pretreatment was carried out with 0.06% (w/v) H_2SO_4 at 170°C for 15 min, conditions resulted in the highest glucan content in the pretreated *S. japonica*, and 79.20% of mass fraction of the macroalgae biomass was solubilized into the liquid hydrolysate. When treating 100 g of *S. japonica*, 20.80 g of WIS was obtained through the ELA treatment. A total of 2.31 g of bioethanol was produced from SSF at a loading of 20.80 g of pretreated *S. japonica*, with corresponding glucose amount of 6.72 g. This means that 67.41% of the glucan in the ELA-pretreated *S. japonica* was hydrolyzed and fermented into bioethanol with a theoretical maximum yield of 0.51 g/g glucose.

14.3 Pretreatment and Enzymatic Saccharification of Seaweed Biomass

Various physical, chemical, and biological pretreatments have been studied to increase the saccharification efficiency. Pretreatment techniques can be categorized into physical, chemical, biological, enzymatic, or a combination. Establishment of a pretreatment method to make polysaccharide more accessible to enzymes is the key aspect of effective fermentable sugar production. Energy and cost are also important factors. Acid pretreatment is one of the most popular methods to attain high sugar yields from lignocellulosic biomass for economic reasons. Borines et al. (2013) reported the pretreatment conditions employed for different terrestrial biomass and that of macroalgae (Table 14.3). Although pretreatment at extreme conditions is effective and high conversion of cellulose from terrestrial plants is possible, it entails high energy consumption and production cost (Ge et al. 2011).

TABLE 14.3

Pretreatment Conditions for Macroalgae and Various Terrestrial Biomass

Enzyme	T (°C)	Time (min)	H_2SO_4	References
Mixed wood (10% birch and 90% maple)	230	0.12	1.17 (w/w%)	Zheng (2007)
Wheat straw and aspen wood	140	60	0.50 (v/v)	Grohmann et al. (1985)
Olive tree biomass	170–210		0.2–1.4 (w/w%)	Kumar et al. (2009)
Corn stover	180–200	1	0.03–0.06 g acid /g dry biomass	Kumar et al. (2009)
Rye straw and Bermuda grass	121	90	1.5 (w/w%)	Sun and Cheng (2002)
Macroalgae (*Sargassum* spp.)	115	86–90	3.36–4.15 (w/v)	Borines et al. (2013)

Therefore, dilute acid hydrolysis is commonly used to prepare seaweed hydrolysates for enzymatic saccharification and fermentation.

At higher severe pretreatment conditions, it was reported that the glucose released from cellulose could be converted further to hydroxymethylfurfural (HMF), which might inhibit cell growth and decrease ethanol production. Ra et al. (2013a) reported that the conversion mechanism of furfural to furfural alcohol has been well established (Morimoto and Murakami 1967; Palmqvist et al. 1999; Taherzadeh et al. 2000). These reports indicated that HMF was converted to another compound by yeast and interpreted it as HMF alcohol since HMF has a structure similar to furfural (Nemirovskii et al. 1989). A similar UV/Vis spectrum was reported by Boopathy et al. (1993) during investigation of *Klebsiella pneumonia* and other enteric bacteria for HMF metabolism. Liu et al. (2004) later identified HMF conversion product with maximum absorbance at 222 nm as 2,5-*bis*-hydroxymethylfuran.

Figure 14.5 shows the decrease in HMF with concomitant increase in 2,5-*bis*-hydroxymethylfuran during ethanol fermentation by *Scheffersomyces stipitis* KCTC 7228. The biotransformation uses aldehyde as an electron accepter, simply converting HMF to HMF conversion product by $2e^-$ reduction (Belay 1997). Therefore, *S. stipitis* KCTC 7228 could convert HMF into a conversion product as 2,5-*bis*-hydroxymethylfuran. To decrease HMF concentration, the ethanol fermentation could be carried out without any damage to yeast.

After the acid hydrolysis, the seaweed is neutralized using sodium hydroxide. For enzymatic saccharification, commercial enzymes are usually used such as Lactozyme, Spirizyme, Viscozyme, Celluclast, and AMG as shown in Table 14.4.

Hydrolysis of seaweeds converts the stored carbohydrates into simple fermentable sugars, which can be easily used to produce ethanol by microorganisms (Table 14.5). Carbohydrate compounds are abundant in macroalgae. The carbohydrate contents of green, red, and brown algae are 50%–65%, 60%–77%, and 50%–60% dry weight, respectively (Jensen 1993; Ross et al. 2008).

Green algae have polysaccharides in the form of starch (α-1,4-glucan) and lipids; however, their proportions are small (Bruton et al. 2009). *Ulva* and *Enteromorpha* spp. have water-soluble ulvan and insoluble cellulose (38%–52% dry wt.) in their cell walls (Lahaye and Robic 2007). Ulvan, a distinctive feature of green algae, is composed mainly of D-glucuronic acid, D-xylose, L-rhamnose, and sulfate (Lobban and Wynne 1981).

The major polysaccharide constituents of red algae are galactans such as carrageenan (up to 75% dry wt.) and agar (up to 52%), which are the most commercially important polysaccharides from red algae (McHugh 2003). Carrageenan consists of repeating D-galactose unit and 3,6-anhydrogalactose. Carrageenans can be readily obtained by extracting from

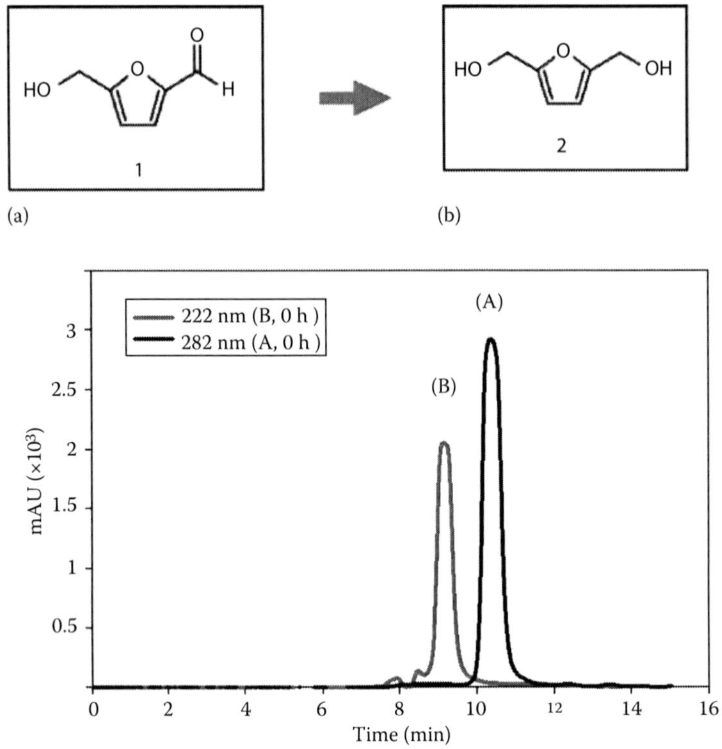

FIGURE 14.5
HPLC profile of HMF (a) and 2,5-*bis*-hydroxymethylfuran (b) during bioethanol fermentation using *Scheffersomyces stipitis* KCTC 7228. (From Ra, C.H. et al., *Bioresour. Technol.*, 140, 421, 2013a). The HMF peak was detected in seaweed hydrolysate at 282 nm (A, 0 h of fermentation). No HMF peak was detected at 282 nm after fermentation; however, 2,5-*bis*-hydroxymethylfuran was detected at 222 nm (B, 60 h of fermentation).

TABLE 14.4
Optimum Conditions of Enzymatic Saccharification

Enzyme	Temperature (°C)	pH	Note
Lactozyme	37	6.5	Glucoamylase
Spirizyme	60–63	4.0–4.5	Glucoamylase
Celluclast	25–55	3.0–5.0	Endoglucanase
Viscozyme	25–55	3.0–5.0	Beta-glucanase
AMG	60	4.5	Glucoamylase
Termamyl	90–110	5.5–7.5	Endoamylase

Sources: Hong, I.K. et al., *J. Ind. Eng. Chem.*, 2013, http://dx.doi.org/10.1016/j.jiec.2013.10.056; Kim, H.J. et al., *Biotechnol. Bioprocess Eng.*, 18, 533, 2013.

red seaweeds or dissolving them into an aqueous solution. Commercial carrageenans have been originated from *Chondrus*, *Gigartina*, and *Eucheuma* spp. (Vera et al. 2011). As another major constituent, agar is made up of alternating β-D-galactose and α-L-galactose with scarce sulfations. Agars are utilized as algal hydrocolloids in food, pharmaceutical, and biological industries (Bixler and Porse 2011). Agar is produced from *Gracilaria*, *Gelidium*, and *Pterocladia* spp. by treating them with thermal acid hydrolysis.

TABLE 14.5

Seaweed Composition and Sugars Released by Hydrolysis (% w/w Dry Biomass)

Seaweeds	Class	Carbohydrate Composition	Total Carbohydrate (%)	Sugars Released by Hydrolysis (%)
G. amansii *Gracilaria verrucosa* *Eucheuma spinosum*	Red	Agar, carrageenan Cellulose	64.7–77.2	34.6–56.6 (glucose, galactose)
S. japonica *Sargassum fulvellum* *U. pinnatifida*	Brown	Laminarin, mannitol Alginate, fucoidan Cellulose	51.9–59.5	34–37.6 (glucose, galactose, mannitol)
U. lactuca *U. pertusa*	Green	Starch, cellulose	54.3–65.2	19.4–59.6 (glucose)

A major polysaccharide of brown algae is alginate, which accounts for up to 40% dry wt. as a principal material of the cell wall. Alginate is composed of two different uronic acids: mannuronic acid blocks, guluronic acid blocks, and alternative blocks of mannuronic and guluronic units. Alginate tends to become gel due to its high affinity for divalent cations such as calcium, strontium, barium, and magnesium. Alginates have been used mostly in the textile (50%) and food (30%) industries. Structural information about polysaccharides abundant in seaweed biomass is listed (Figure 14.6).

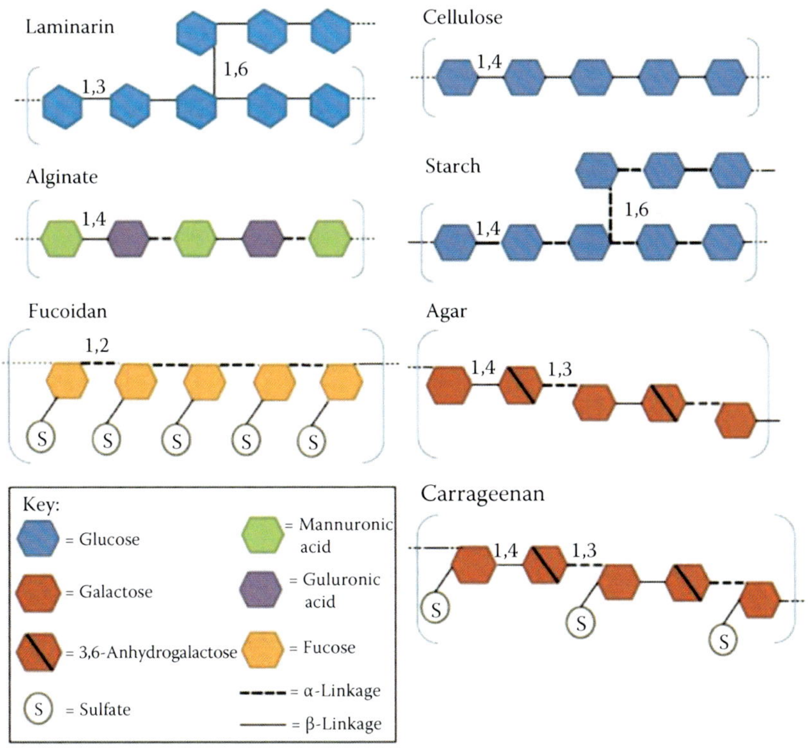

FIGURE 14.6
Structural information about polysaccharides abundant in seaweed biomass. (From Wei, N. et al., *Trends Biotechnol.*, 31(2), 74, 2013.)

14.4 Improving Ethanol Production and Related Microorganisms

Bioethanol can be produced from all kinds of macroalgae by converting their polysaccharides to simple sugars and by employing appropriate microorganisms. Since macroalgae have various carbohydrates such as starch, cellulose, laminarin, mannitol, and agar, carbohydrate conversion to sugars and the choice of appropriate microorganisms are pivotal for successful bioethanol fermentation. However, this research is now at the early stage. As shown in Table 14.6, there are some microorganisms that can accumulate high concentrations of ethanol.

Jang et al. (2012) reported that the saccharification of the seaweed *S. japonica* was performed using thermal acid hydrolysis with and without the addition of the isolated marine bacteria *Bacillus* sp. JS-1. SSF was carried out by the addition of 0.39 g dcw/L of *Bacillus* sp. JS-1 and 0.45 g dcw/L of yeasts to the thermal acid hydrolyzed *S. japonica* slurry. The highest amount of ethanol produced was 7.7 g/L (9.8 mL/L) using the yeast *Pichia angophorae* after 64 h of fermentation. Results of ethanol yield in this study showed 2–3 times higher yield than those of previous studies (Lee and Lee 2010). Further fermentation reduced the ethanol concentration due to the evaporation of ethanol. Fermentation with *P. stipitis*, *S. cerevisiae*, and *Pachysolen tannophilus* produced little ethanol, reaching only 0.9–1.8 g/L.

Cho et al. (2013a) reported that strain improvement of *P. angophorae* KCTC 17574 was successfully carried out for bioethanol fermentation of seaweed slurry with high salt concentration. *P. angophorae* KCTC 17574 was cultured under increasing salinity from 5 practical salinity unit (psu, ‰) to as high as 100 psu for 723 h. The seaweed *Undaria pinnatifida* (sea mustard, Miyuk) was fermented to produce bioethanol using high-salt acclimated yeast. The pretreatment of *U. pinnatifida* was optimized using thermal acid hydrolysis to obtain a high monosaccharide yield. Optimal pretreatment conditions of 75 mM H_2SO_4 and 13% (w/v) slurry at 121°C for 60 min were determined using response surface methodology. A maximum monosaccharide content of 28.65 g/L and the viscosity of 33.19 cP were obtained. The yeasts cultured under various salinity concentrations were collected and inoculated to the pretreated seaweed slurry after the neutralization using 5 N NaOH. The pretreated slurry was fermented with the inoculation of 0.1 g dcw/L of *P. angophorae* KCTC 17574 strains obtained at 90 psu. The maximum ethanol

TABLE 14.6

Yeast Species That Produce Ethanol as the Main Fermentation Product

Microorganisms	Temp (°C)	pH	Seaweed (Macroalgae)	Incubation Time (h)	Ethanol (g/L)	References
S. cerevisiae	30	4.0–6.0	*G. amansii*, *S. japonica*, *U. pinnatifida*, *Gracilaria verrucosa*	72–132	10–19	Cho et al. (2013a) Jang (2012)
P. stipitis	30	4.0–6.0	*G. amansii*	72	16–21	Cho et al. (2013b)
P. angophorae	30	4.0–6.0	*Gracilaria verrucosa* *S. japonica*, *U. pinnatifida*	132	8–12	Kim (2013) Cho et al. (2013a)
Kluyveromyces marxianus	30	4.0–6.0	*G. amansii*	72	16–20	Ra et al. (2013b)
Candida tropicalis	30	4.0–6.0	*G. amansii*	72	16–20	Ra et al. (2013b)
Recombinant *Escherichia. coli*	25 –30	6.0–8.0	*S. japonica* *U. pinnatifida*	48	47	Wargacki (2012)

concentration of 9.42 g/L with 27% yield of theoretical case of ethanol production from total carbohydrate of *U. pinnatifida* was obtained.

The preferential utilization of glucose over nonglucose sugars by yeast often results in low overall ethanol yield and production. Studies have been carried out to evaluate the mechanisms impeding fermentation with mixed sugars using engineered *S. cerevisiae* (Kubicek 1982; Ostergaard et al. 2000; Wisselink et al. 2009). However, these approaches were carried out by improving the utilization of mixed sugars, and further improvements in yields and productivities are required for ethanol fermentation by engineered yeasts to be adopted commercially. Kim et al. (2013) reported that ethanol production from *U. pinnatifida* was performed using yeast acclimated to specific sugars. Pretreatment conditions were optimized by thermal acid hydrolysis and enzyme treatment to increase the monosaccharide yield. Pretreatment by thermal acid hydrolysis was carried out using seaweed powder at 8%–17% (w/v) solid contents with a treatment time of 30–60 min. Enzyme treatment was carried out with 1% (v/v) Viscozyme L (1.2 FGU/mL), 1% (v/v) Celluclast 1.5 L (8.5 EGU/mL), 1% (v/v) AMG 300 L (3.0 AGU/mL), and 1% (v/v) Termamyl 120 L (0.72 KNU/mL). All enzymes except Termamyl 120 L, which was applied during pretreatment, were treated at 45°C for 24 h following pretreatment. Optimal pretreatment and enzyme conditions were determined to be 75 mM H_2SO_4, 13% (w/v) slurry, and 2.88 KNU/mL Termamyl 120 L at 121°C for 60 min. A maximum monosaccharide concentration of 33.1 g/L with 50.1% of theoretical yield was obtained. To increase the ethanol yield, *P. angophorae* KCTC 17574 was acclimated to a high concentration (120 g/L) of galactose and mannitol at 30°C for 24 h. Ethanol production of 12.98 g/L with 40.12% theoretical yield was obtained from *U. pinnatifida* through fermentation with 0.35 g dry cell weight/L *P. angophorae* KCTC 17574 acclimated to mannitol and galactose.

Cho et al. (2013b) reported that SHF was carried out by the addition of the galactose acclimated or nonacclimated *P. stipitis* or *S. cerevisiae*. Glucose was consumed first as the fermentation started because glucose was the preferred substrate to galactose. Glucose was consumed in 48 h, and then galactose was consumed for 24 h. However, galactose was not totally consumed until 96 h, and 7.1 g/L of galactose remained in the fermentation broth. The ethanol concentration after 96 h of fermentation with nonacclimated *P. stipitis* was 11.5 g/L with $Y_{EtOH} = 0.34$. By contrast, fermentation with acclimated *P. stipitis* totally consumed glucose during 60 h. The galactose was consumed until 84 h, and the final ethanol concentration of 16.6 g/L with $Y_{EtOH} = 0.5$ was produced.

The fermentation with nonacclimated *S. cerevisiae* to high concentration of galactose produced ethanol concentration of 6.9 g/L with $Y_{EtOH} = 0.21$ for 96 h. Therefore, the acclimation of yeasts to high concentration of galactose could make a simultaneous utilization of galactose and glucose for the production of ethanol from seaweed *Gelidium amansii*. Acclimated yeasts produced a higher ethanol concentration than nonacclimated yeasts.

14.5 Concluding Remarks and Future Perspectives

To commercialize the production of macroalgae-based fuels, the priority need is identifying microorganisms that can metabolize major and unique macroalgal carbohydrates. Some macroalgae of specific carbohydrates such as alginate and ulvan are not readily metabolized by commercially applied fermenting microorganisms as *S. cerevisiae* (Wegeberg and

Felby 2010). To overcome these problems, some researchers have developed macroalgae-specific enzymes to hydrolyze macroalgal carbohydrates using the SSF system (Jang et al. 2012). This SSF system has many advantages: low contamination, low initial osmotic stress of fermenting microorganisms, and high energy efficiency. However, the SSF system could inhibit enzyme activity and extracellular enzyme secretion. Therefore, various and complex carbohydrates of macroalgae can lead to low yield of biofuels in the SSF system.

Recently, advanced biotechnology has been applied to resolve the aforementioned challenges for the production of macroalgae-based fuels. Although there are natural microorganisms able to utilize seaweed sugar extracts, feasible biofuel production from seaweed requires efficient conversion of mixed sugars in seaweed hydrolysates. Therefore, research effort has been made to enhance mixed sugar fermentation to ethanol by engineered microorganisms. The wild-type yeast *S. cerevisiae* is capable of galactose fermentation; however, there are two major issues that limit the ethanol yield and productivity. Ethanol production rate and yield from galactose are considerably lower than those from glucose (Ostergaard et al. 2000). And the presence of glucose represses the utilization of galactose because of stringent transcriptional repression of *GAL* genes coding for enzymes for galactose metabolism (Johnston et al. 1994). These two issues of low ethanol yield and glucose repression to *GAL* genes lead to a diauxic consumption of glucose and galactose in red seaweed hydrolysates, which significantly reduces overall ethanol productivity (Ostergaard et al. 2001).

A recent breakthrough in the efficient use of brown seaweed biomass arose from an engineered *Escherichia coli* platform capable of alginate degradation and metabolism. Wargacki et al. (2012) reported that the genomic integration of alginate transport and metabolism, together with an engineered system for extracellular alginate depolymerization, generated a microbial platform that can simultaneously degrade, uptake, and metabolize alginate. This platform enables bioethanol production directly from macroalgae via a consolidated process, achieving a titer of 4.7% (v/v) and a yield of 0.281 weight ethanol/weight dry macroalgae (equivalent to 80% of the maximum theoretical yield from the sugar composition in macroalgae).

Even after the ethanol production, the leftover residue still contains a good amount of organic matter and useful minerals and eventually could be used as a biofertilizer. In addition, it has been observed that seaweeds can efficiently absorb CO_2 from seawater. Thus, carbon sequestration from the atmosphere will reduce the acidification of the ocean as well.

References

Adams, J. M., Gallagher, J. A., Donnison, I. S. 2009. Fermentation study on *Saccharina latissima* for bioethanol production considering variable pre-treatments. *J. Appl. Phycol.* 21: 569–574.

Belay, N. 1997. Anaerobic transformation of furfural by *Methanococcus deltae* (Delta) LH. *Appl. Environ. Microbiol.* 63: 2092–2094.

Bixler, H. J., Porse, H. 2011. A decade of change in the seaweed hydrocolloids industry. *J. Appl. Phycol.* 23: 321–335.

Boopathy, R., Bokang, H., Daniels, L. 1993. Biotransformation of furfural and 5-hydroxymethylfurfural by enteric bacteria. *J. Ind. Microbiol.* 11: 147–150.

Borines, M. G., de Leon, R. L., Cuello, J. L. 2013. Bioethanol production from the macroalgae *Sargassum* spp. *Bioresour. Technol.* 138: 22–29.

Bruton, T., Lyons, H., Lerat, Y., Stanley, M., Rasmussen, M. B. 2009. A review of the potential of marine algae as a source of biofuel in Ireland. Sustainable Energy Ireland, 88p, Available from: http://www.seai.ie/Publications/Renewables_Publications_/Bioenergy/Algaereport.pdf.

Cho, H. Y., Ra, C. H., Kim, S. K. 2013b. Ethanol production from the seaweed *Gelidium amansii*, using specific sugar acclimated yeasts. *J. Microbiol. Biotechnol.* 24: 264–269.

Cho, Y. K., Kim, H. J., Kim, S. K. 2013a. Bioethanol production from brown seaweed, *Undaria pinnatifida*, using NaCl acclimated yeast. *Bioprocess Biosyst. Eng.* 36: 713–719.

Draget, K. I., Smidsrød, O., Skjåk-Bræk, G. 2005. Alginates from algae. Biopolymers Online. 6, DOI: 10.1002/3527600035.bpol6008. Available from: http://www.wiley-vch.de/books/sample/3527313451_c01.pdf.

FAO (Food and Agriculture Organization of the United Nations). 2012a. 2010 Fishery and aquaculture statistics. Available from: ftp://ftp.fao.org/FI/CDrom/CD_yearbook_2010/booklet/ba0058t.pdf.

FAO (Food and Agriculture Organization of the United Nations). 2012b. *FAO Statistical Yearbook 2012: World Food and Agriculture*. Available from: http://reliefweb.int/sites/reliefweb.int/files/resources/i2490e.pdf.

Feng, D., Liu H., Li, F., Peng, J., Song, Q. 2011. Optimization of dilute acid hydrolysis of *Enteromorpha*. *Chin. J. Oceanol. Limnol.* 6: 1243–1248.

Ge, L., Wang, P., Mou, H. 2011. Study on saccharification techniques of seaweed wastes for the transformation of ethanol. *Renew. Energ.* 36: 84–89.

Grohmann, K., Torget, R., Himmel, M. 1985. Chemomechanical pretreatment of biomass. *Biotechnol. Bioeng. Symp.* 15: 59–80.

Hahn-Hägerdal, B., Galbe, M., Gorwa-Grauslund, M. F., Lidén, G., Zacchi, G. 2006. Bio-ethanol the fuel of tomorrow from the residues of today. *Trends Biotechnol.* 24: 549.

Hector, R. E., Qureshi, N., Hughes, S. R., Cotta, M. A. 2008. Expression of a heterologous xylose transporter in a *Saccharomyces cerevisiae* strain engineered to utilize xylose improves aerobic xylose consumption. *Appl. Microbiol. Biotechnol.* 80: 675–684.

Hejazi, M. A., Wijffels, R. H. 2004. Milking of microalgae. *Trends Biotechnol.* 22: 189–194.

Hong, I. K., Jeon, H., Lee, S. B. 2014. Comparison of red, brown and green seaweeds on enzymatic saccharification. *J. Ind. Eng. Chem.* 20(5): 2687–2691.

Jang, J. S., Cho, Y. G., Jeong, G. T., Kim, S. K. 2012. Optimization of saccharification and ethanol production by simultaneous saccharification and fermentation (SSF) from seaweed *Saccharina japonica*. *Bioprocess Biosyst. Eng.* 35: 11–18.

Jensen, A. 1993. Present and future needs for algae and algal products. *Hydrobiologia* 260–261: 15–23.

Johnston, M., Flick, J. S., Pexton, T. 1994. Multiple mechanisms provide rapid and stringent glucose repression of GAL gene expression in *Saccharomyces cerevisiae*. *Mol. Cell. Biol.* 14: 3834–3841.

Jung, K. A., Lim, S. R., Kim, Y. R., Park, J. M. 2013. Potentials of macroalgae as feedstocks for biorefinery. *Bioresour. Technol.* 135: 182–190.

Katajira, S., Ito, M., Takema, H., Fujita, Y., Tanaka, T., Fukuda, H., Kondo, A. 2008. Improvement of ethanol productivity during xylose and glucose co-fermentation by xylose-assimilating *S. cerevisiae* via expression of glucose transporter *Sut1*. *Enzyme Microb. Technol.* 43: 115–119.

Kim, H. J., Ra, C. H., Kim, S. K. 2013. Ethanol production from seaweed (*Undaria pinnatifida*) using yeast acclimated to specific sugars. *Biotechnol. Bioprocess Eng.* 18: 533–537.

Kim, S. R., Ha, S. J., Wei, N., Oh, E. J., Jin, Y. S. 2012. Simultaneous co-fermentation of mixed sugars: A promising strategy for producing cellulosic ethanol. *Trends Biotechnol.* 30(5): 274–282.

Klein, C. J. L., Rasmussen, J. J., Rønnow, B., Olsson, L., Nielsen, J. 1999. Investigation of the impact of MIG1 and MIG2 on the physiology of *Saccharomyces cerevisiae*. *J. Biotechnol.* 68: 197–212.

Kraan, S. 2013. Mass cultivation of carbohydrate rich macroalgae, a possible solution for sustainable biofuel production. *Mitig. Adap. Strat. Global Change* 18: 27–46.

Kubicek, C. P. 1982. β-glucosidase excretion by *Trichoderma pseudokoningii* correlation with cell wall bound β—1,3-glucanase activities. *Arch. Microbiol.* 132: 349–354.

Kumar, P., Barrett, D. M., Delwiche, M. J., Stroeve, P. 2009. Methods for pretreatment of lignocellulosic biomass for efficient hydrolysis and biofuel production. *Ind. Eng. Chem. Res.* 48: 3713–3729.

Lahaye, M., Robic, A. 2007. Structure and functional properties of ulvan, a polysaccharide from green seaweeds. *Biomacromolecules* 8: 1765–1774.

Lee, J. Y., Li, P., Lee, J., Ryu, H. J., Oh, K. K. 2013. Ethanol production from *Saccharina japonica* using an optimized extremely low acid pretreatment followed by simultaneous saccharification and fermentation. *Bioresour. Technol.* 127: 119–125.

Lee, S. M., Lee, J. H. 2010. Influence of acid and salt content on the ethanol production from Laminaria japonica. *Appl. Chem. Eng.* 21: 154–161.

Lemus, R., Brummer, E. C., Moore, K. L., Molstad, N. E., Burras, C. L., Barker, M. F. 2002. Biomass yield and quality of 20 switchgrass populations in southern Iowa, USA. *Biomass Bioenerg.* 23: 433–442.

Liu, Z. L., Slininger, P. J., Dien, B. S., Berhow, M. A., Kurtzman, C. P., Gorsich, S. W. 2004. Adaptive response of yeasts to furfural and 5-hydroxymethylfurfural and new chemical evidence for HMF conversion to 2,5-bis-hydroxymethylfuran. *J. Ind. Microbiol. Biotechnol.* 31: 345–352.

Lobban, C. S., Wynne, M. J. 1981. *The Biology of Seaweeds.* Blackwell Scientific Publications, Oxford, U.K.

Lorenz, R. T., Cysewski, G. R. 2000. Commercial potential for *Haematococcus* microalgae as a natural source of astaxanthin. *Trends Biotechnol.* 18: 160–167.

McHugh, D. J. 2003. A guide to the seaweed industry. FAO Fisheries Technical Paper 441.

Meinita, M. D., Hong, Y. K., Jeong, G. T. 2011. Comparison of sulfuric acid and hydrochloric acids as catalysts in hydrolysis of *Kappaphycus alvarezii* (cottonii). *Bioprocess Biosyst. Eng.* 35: 123–128.

Morimoto, S., Murakami, M. 1967. Studies on fermentation products from aldehyde by microorganisms: The fermentative production of furfural alcohol from furfural by yeasts (part I). *J. Ferment. Technol.* 45: 442–446.

Nemirovskii, V. G., Gusarova, L. A., Rakhmilevich, Y. D., Sizov, A. I., Kostenko, V. G. 1989. Furfural and hydroxymethylfurfural transformation route during culturing of nutrient yeasts. *Biotekhnologiya* 5: 285–289.

Ostergaard, S., Olsson, L., Johnston, M., Nielsen, J. 2000. Increasing galactose consumption by *Saccharomyces cerevisiae* through metabolic engineering of the GAL gene regulatory network. *Nat. Biotechnol.* 18: 1283–1286.

Ostergaard, S., WallÖe, K. O., Gomes, C. S. G., Olsson, L., Nielsen, J. 2001. The impact of *GAL6*, *GAL80*, and *MIG1* on glucose control of the *GAL* system in *Saccharomyces cerevisiae*. *FEMS Yeast Res.* 1: 47–55.

Palmqvist, E., Almeida, J. S., Hahn-Hagerdal, B. 1999. Influence of furfural on anaerobic glycolytic kinetics of *Saccharomyces cerevisiae* in batch culture. *Biotechnol. Bioeng.* 62: 447–454.

Park, J. H., Hong, J. Y., Jang, H. C., Oh, S. G., Kim, S. H., Yoon, J. J., Kim, Y. J. 2012. Use of *Gelidium amansii* as a promising resource for bioethanol: A practical approach for continuous dilute-acid hydrolysis and fermentation. *Bioresour. Technol.* 108: 83–88.

Pulz, O., Gross, W. 2004. Valuable products from biotechnology of microalgae. *Appl. Microbiol. Biotechnol.* 65: 635–648.

Ra, C. H., Jeong, G. T., Shin, M. K., Kim, S. K. 2013a. Biotransformation of 5-hydroxymethylfurfural (HMF) by *Scheffersomyces stipitis* during ethanol fermentation of hydrolysate of the seaweed *Gelidium amansii*. *Bioresour. Technol.* 140: 421–425.

Ra, C. H., Lee, H. J., Shin, M. K., Kim, S. K. 2013b. Bioethanol production from seaweed *Gelidium amansii* for Separate Hydrolysis and Fermentation (SHF). *KSBB J.* 28: 282–286.

Raamsdonk, L. M., Diderich, J. A., Kuiper, A., van Gaalen, M., Kruckeberg, A. L., Berden, J. A., van Dam, K. 2001. Co-consumption of sugars or ethanol and glucose in a *Saccharomyces cerevisiae* strain deleted in the *HXK2* gene. *Yeast* 18: 1023–1033.

Roca, C., Haack, M. B., Olsson, L. 2004. Engineering of carbon catabolite repression in recombinant xylose fermenting *Saccharomyces cerevisiae*. *Appl. Microbiol. Biotechnol.* 63: 578–583.

Ross, A. B., Jones, J. M., Kubacki, M. L., Bridgeman, T. 2008. Classification of macroalgae as fuel and its thermochemical behavior. *Bioresour. Technol.* 99: 6494–6504.

Runquist, D., Fonseca, C., Rådström, P., Spencer-Martins, I., Hahn-Häqerdal, B. 2009. Expression of the Gxf1 transporter from *Candida intermedia* improves fermentation performance in recombinant xylose-utilizing *Saccharomyces cerevisiae*. *Biotechnol. Biofuels* 3: 5.

Saha, B. C., Cotta, M. A. 2007. Enzymatic saccharification and fermentation of alkaline peroxide pretreated rice hulls to ethanol. *Enzyme Microb. Technol.* 41: 528.

Shinners, K. J., Binversie, B. N. 2007. Fractional yield and moisture of corn stover biomass produced in the Northern US Corn Belt. *Biomass Bioenerg.* 31: 576–584.

Singh, A., Nigam, P. S., Murphy, J. D. 2011. Renewable fuels from algae: An answer to debatable land based fuels. *Bioresour. Technol.* 102: 10–16.

Sun, Y., Cheng, J. 2002. Hydrolysis of lignocellulosic materials for ethanol production: A review. *Bioresour. Technol.* 83: 1–11.

Taherzadeh, M. J., Gustafsson, L., Niklasson, C. 2000. Physiological effects of 5-hydroxymethylfurfural on *Saccharomyces cerevisiae*. *Appl. Microbiol. Biotechnol.* 53: 701–708.

Thanvanthri Gururajan, V., Gorwa-Grauslund, M. F., Hahn-Häqerdal, B., Pretorius, I. S., Cordero Otero, R. R. 2007. A constitutive catabolite repression mutant of a recombinant *Saccharomyces cerevisiae* strain improves xylose consumption during fermentation. *Ann. Microbiol.* 57: 85–92.

Vera, J., Castro, J., Gonzalez, A., Moenne, A. 2011. Seaweed polysaccharides and derived oligosaccharides stimulate defense responses and protection against pathogens in plants. *Mar. Drugs* 9: 2514–2525.

Wargacki, A. J., Leonard, E., Win, M. N., Regitsky, D. D., Santos, C. N. S., Kim, P. B., Cooper, S. R. et al. 2012. An engineered microbial platform for direct biofuel production from brown macroalgae. *Science* 335: 308–313.

Wegeberg, S., Felby, C. 2010. Algae biomass for bioenergy in Denmark: Biological/technical challenges and opportunities. Available from: http://bio4bio.ku.dk/documents/nyheder/wegeberg-alage-biomass.pdf.

Wei, N., Quarterman, J., Jin, Y. S. 2013. Marine macroalgae: An untapped resource for producing fuels and chemicals. *Trends Biotechnol.* 31(2): 70–77.

Wisselink, H. W., Toirkens, M. J., Wu, Q., Pronk, J. T., van Maris, A. J. A. 2009. Novel evolutionary engineering approach for accelerated utilization of glucose, xylose, and arabinose mixtures by engineered *Saccharomyces cerevisiae* strains. *Appl. Metab. Microbiol.* 75: 907–914.

Young, E., Poucher, A., Comer, A., Bailey, A., Alper, H. 2011. Functional survey for heterologous sugar transport proteins, using *Saccharomyces cerevisiae* as a host. *Appl. Environ, Microbiol.* 77: 3311–3319.

Zheng, Y. 2007. Evaluation of different biomass materials as feedstock for fermentable sugar production. *Appl. Biochem. Biotechnol.* ABAB Symposium, part 3: 423–435.

15

Current State of Research on Algal Biomethane

Ozcan Konur

CONTENTS

15.1 Introduction

15.1.1 Issues

Global warming, air pollution, and energy security have been some of the most important public policy issues in recent years (Jacobson 2009, Wang et al. 2008, Yergin 2006). With increasing global population, food security has also become a major public policy issue (Lal 2004). The development of biofuels generated from biomass has been a long awaited solution to these global problems (Demirbas 2007, Goldember 2007, Lynd et al. 1991). However, the development of the early biofuels produced from the agricultural plants such as sugarcane (Goldemberg 2007) and agricultural wastes such as corn stovers (Bothast and Schlicher 2005) have resulted in a series of substantial concerns about food security (Godfray et al. 2010). Therefore, the development of algal biofuels as a third-generation biofuel has been considered as a major solution for the problems of global warming, air pollution, energy security, and food security (Chisti 2007, Demirbas 2007, Kapdan and Kargi 2006, Spolaore et al. 2006, Volesky and Holan 1995).

Although there have been many reviews on the use of marine algae for the production of algal biomethane (e.g., Sialve et al. 2009, Singh and Cu 2010, Yen and Brune 2007) and a number of scientometric studies on the algal biofuels (Konur 2011), there has not been any study of papers on algal biomethane as in other research fields (Baltussen and Kindler 2004a,b, Dubin et al. 1993, Gehanno et al. 2007, Konur 2012a,b, 2013, Paladugu et al. 2002, Wrigley and Matthews 1986).

As North's new institutional theory suggests, it is important to have up-to-date information about current public policy issues to develop a set of viable solutions to satisfy the needs of all key stakeholders (Konur 2000, 2002a–c, 2006a,b, 2007a,b, 2012c).

Therefore, the brief information on a selected set of papers on algal biomethane are presented in this chapter to inform the key stakeholders about the global problems of warming, air pollution, food security, and energy security and about the production of algal biomethane for the solution of these problems in the long run, thus complementing a number of recent scientometric studies on biofuel and global energy research (Konur 2011, 2012d–p).

15.1.2 Methodology

A search was carried out in Science Citation Index Expanded (SCIE) and Social Sciences Citation Index (SSCI) databases (version 5.11) in September 2013 to locate the papers relating to algal biomethane using the keyword set of Topic = (methane* or "bio* methane*" or biomethane* or "microbio* methane*" or "methyl* alcohol") AND Topic = (alga* or "macro* alga*" or "micro* alga*" or macroalga* or microalga* or cyanobacteria or seaweed* or diatom* or "sea* weed*" or reinhardtii or braunii or chlorella or sargassum or gracilaria) in the abstract pages of the papers. For this chapter, it was necessary to embrace the broad algal search terms to include diatoms, seaweeds, and cyanobacteria as well as to include many spellings of biomethane rather than the simple keywords of biomethane and algae.

It was found that there were 766 such papers indexed between 1956 and 2013. The key subject categories for algal research were environmental sciences (109 references, 14.2%), biotechnology applied microbiology (108 references, 14.1%), energy fuels (89 references, 11.6%), and geosciences multidisciplinary (66 references, 8.6%), altogether comprising 50% of all the references on algal biomethane. It was also necessary to focus on the key references by selecting articles and reviews.

The located 50 papers were arranged by decreasing number of citations for two groups of papers: 19 and 31 papers for nonexperimental research and experimental research, respectively. In order to check whether these collected abstracts correspond to the larger sample on these topical areas, new searches were carried out for each topical area.

The summary information about the located papers is presented under two major headings of nonexperimental and experimental research in the order of the decreasing number of citations for each topical area.

The information relating to the document type, the affiliation and location of the authors, the journal, the indexes, the subject area of the journal, the concise topic, the total number of citations and the total average number of citations received per year are given in the tables for each paper.

15.2 Nonexperimental Studies on Algal Biomethane

15.2.1 Introduction

The nonexperimental research on algal biomethane has been one of the most dynamic research areas in recent years. Nineteen papers in the field of algal biomethane with more than ten citations were located, and the key emerging issues are presented in the decreasing order of the number of citations in Table 15.1. These papers give strong hints about the determinants of efficient algal biomethane production and emphasize that marine algae are efficient biomethane feedstocks in comparison with the first- and second-generation biofuel feedstocks.

The papers were dominated by researchers from 10 countries, usually through the intercountry institutional collaboration, and they were multiauthored. The United States (eight papers), England and France (three papers each), and Australia (two papers) were the most prolific countries.

Similarly, all these papers were published in the journals indexed by the Science Citation Index (SCI) and/or SCIE. There were no papers indexed by the Arts & Humanities Citation Index (A&HCI) or SSCI. The number of citations ranged from 10 to 152 and the average number of citations per annum ranged from 0.6 to 25.3. Fifteen of the papers were articles, while four of them were reviews. Eight and eleven of these papers were related to algal biomethane production processes and algal biomethane production policies, respectively.

15.2.2 Nonexperimental Research on Algal Biomethane

15.2.2.1 Research on Algal Biomethane Production Processes

Sialve et al. (2009) discuss the anaerobic digestion of microalgae as a compulsory phase to make microalgal biodiesel and biomethane sustainable in a paper with 152 citations. They note that the conversion of algal biomass after lipid extraction into biomethane is a process that can recover more energy than the energy from the cell lipids. They identify three major bottlenecks to digest microalgae to produce biomethane. "First, the biodegradability of microalgae can be low depending on both the biochemical composition and the nature of the cell wall. Then, the high cellular protein content results in ammonia release which can lead to potential toxicity. Finally, the presence of sodium for marine species can also affect the digester performance." They argue that physicochemical pretreatment, codigestion, and control of gross composition are strategies that can significantly and efficiently increase the conversion yield of the algal biomass into biomethane. When the cell lipid content does not exceed 40%, anaerobic digestion of the whole algal biomass is the optimal strategy on an energy balance basis, for the energetic recovery of cell biomass.

Heaven et al. (2011) comment on Sialve et al. (2009) in a recent paper with 11 citations. They argue that there are a number of issues concerning the results and conclusions presented in the original paper. These include "the biomass energy values, which in some cases are unusually high; and the apparent production of more energy from processed biomass than is present in the original material." The main causes for these discrepancies, as they further argue, include "the choice of empirical formula for protein; confusion between values calculated on a total or volatile solid basis; and the lack of a mass balance approach." The choice of protein formula also affects predicted concentrations of ammonia in the digester. They assert that these and other minor errors contribute to some potentially misleading conclusions, which could affect subsequent interpretations of the overall process feasibility.

TABLE 15.1

Nonexperimental Research on Algal Biomethane

No.	Paper References	Year	Document	Affiliation	Country	No. of Authors	Journal	Index	Subjects	Topic	Total No. of Citations	Total Average Citations per Annum	Rank
1	Sialve et al.	2009	R	INRA, INRIA	France	3	*Biotechnol. Adv.*	SCI, SCIE	Biot. Appl. Microb.	Algal biomethane—prod. processes	152	25.3	1
2	Cantrell et al.	2008	R	ARS-USDA	United States	3	*Bioresour. Technol.*	SCI, SCIE	Agr. Eng., Biot. Appl. Microb., Ener. Fuels	Algal biomethane—prod. processes	90	15.0	3
3	Singh and Cu	2010	R	Univ. Southampton, Indian Inst. Petr.	England, India	2	*Renew. Sust. Energ. Rev.*	SCI, SCIE	Ener. Fuels	Algal biomethane—prod. policies	70	17.5	4
4	Kaddam	2002	A	NREL	United States	1	*Energy*	SCI, SCIE	Therm., Ener. Fuels	Algal biomethane—prod. policies	54	4.5	6
5	Rawat et al.	2011	A	Durban Univ. Technolç	S. Africa	4	*Appl. Energ.*	SCI, SCIE	Ener. Fuels, Eng. Chem.	Algal biomethane—prod. processes	47	15.7	7
6	Collet et al.	2011	A	INRA, Montpellier SupAgro, Naskeo	France	6	*Bioresour. Technol.*	SCI, SCIE	Agr. Eng., Biot. Appl. Microb., Ener. Fuels	Algal biomethane—prod. policies	46	15.3	8
7	Wyman	1994	A	NREL	United States	1	*Appl. Biochem. Biotechnol.*	SCI, SCIE	Bioch. Mol. Biol., Biot. Appl. Microb.	Algal biomethane—prod. policies	38	1.9	14

(*Continued*)

TABLE 15.1 (*Continued*)

Nonexperimental Research on Algal Biomethane

No.	Paper References	Year	Document	Affiliation	Country	No. of Authors	Journal	Index	Subjects	Topic	Total No. of Citations	Total Average Citations per Annum	Rank
8	Wyman and Goodman	1993	A	NREL	United States	2	*Appl. Biochem. Biotechnol.*	SCIE	Bioch. Mol. Biol., Biot. Appl. Microb.	Algal biomethane—prod. policies	35	1.7	15
9	Brune et al.	2009	A	Clemson Univ., Calif. Polytech State Univ. S.L.O., Inst. Environ. Management Inc.	United States	3	*J. Environ. Eng. ASCE*	SCI, SCIE	Eng. Env., Eng. Civil, Env. Sci.	Algal biomethane—prod. policies	32	6.4	17
10	Zamalloa et al.	2011	A	Univ. Ghent	Belgium	4	*Bioresour. Technol.*	SCI, SCIE	Agr. Eng., Biot. Appl. Microb., Ener. Fuels	Algal biomethane—prod. policies	31	2.4	18
11	Green et al.	1995a	A	Univ. Calif. Berkeley	United States	3	*Water Sci. Technol.*	SCIE	Eng. Env., Env. Sci., Water Res.	Algal biomethane—prod. policies	29	1.5	19
12	Green et al.	1995b	A	Univ. Calif. Berkeley	United States	6	*Water Sci. Technol.*	SCIE	Eng. Env., Env. Sci., Water Res.	Algal biomethane—prod. policies	26	1.4	23
13	Kosaric and Velikonja	1995	A	Univ. Western Ontario	Canada	2	*Fems Microbiol. Rev.*	SCI, SCIE	Microbiol.	Algal biomethane—prod. policies	26	1.4	24
14	Pearson	1996	A	Univ. Liverpool	England	1	*Water Sci. Technol.*	SCIE	Eng. Env., Env. Sci., Water Res.	Algal biomethane—prod. processes	21	1.7	29

(*Continued*)

TABLE 15.1 ***(Continued)***

Nonexperimental Research on Algal Biomethane

No.	Paper References	Year	Document	Affiliation	Country	No. of Authors	Journal	Index	Subjects	Topic	Total No. of Citations	Total Average Citations per Annum	Rank
15	Harun et al.	2011	A	Monash Univ., Univ. Putra Malaysia	Australia, Malaysia	6	*Biomass Bioenerg.*	SCIE	Agr. Eng., Biot. Appl. Microb., Ener. Fuels	Algal biomethane—prod. policies	13	4.3	32
16	Heaven et al.	2011	R	Univ. Southampton	England	3	*Biotechnol. Adv.*	SCI, SCIE	Biot. Appl. Microb.	Algal biomethane—prod. processes	11	3.7	40
17	Oswald et al.	1994	A	Univ. Calif B.	United States	3	*Water Sci. Technol.*	SCIE	Eng. Env., Env. Sci., Water Res.	Algal biomethane—prod. processes	11	0.6	43
18	Mairet et al.	2011	A	BIOCORE INRIA, Univ. Tecn. Federico Santa Maria, INRA	Chile, France	5	*Bioresour. Technol.*	SCI, SCIE	Agr. Eng., Biot. Appl. Microb., Ener. Fuels	Algal biomethane—prod. processes	10	3.3	45
19	Hodgson and Paspaliaris	1996	A	Univ. Melbourne	Australia	2	*Water Sci. Technol.*	SCIE	Eng. Env., Env. Sci., Water Res.	Algal biomethane—prod. processes	10	0.6	46

SCI, Science Citation Index; SCIE, Science Citation Index Expanded; SSCI, Social Sciences Citation Index; P, production processes; Z, production policies; A, article; R, review.

Cantrell et al. (2008) discuss the livestock waste-to-bioenergy generation opportunities by combining an algal CO_2-fixation treatment in a paper with 90 citations. They present established and emerging energy conversion opportunities that can transform the treatment of livestock waste from a liability to a profit center. While biological production of methanol and hydrogen is in early research stages, they argue that "anaerobic digestion is an established method of generating between 0.1 to 1.3 $m^3\ m^{-3}\ d^{-1}$ of biomethane-rich biogas." The TCC processes of pyrolysis, direct liquefaction, and gasification can convert waste into gaseous fuels, combustible oils, and charcoal. They further argue that integration of biological and thermal-based conversion technologies in a farm-scale hybrid design by combining an algal CO_2-fixation treatment requiring less than 27,000 m^2 of treatment area with the energy recovery component of wet gasification can drastically reduce CO_2 emissions and efficiently recycle nutrients. They assert that these designs have the potential to make future large-scale confined animal feeding operations sustainable and environmentally benign while generating on-farm renewable energy.

Rawat et al. (2011) discuss the phycoremediation of domestic wastewater and biomass production for sustainable biofuel production. They discuss current knowledge regarding wastewater treatment using HRAPs and microalgal biomass production techniques using wastewater streams. They also discuss the biomass harvesting methods and lipid extraction protocols. Finally, they discuss biodiesel production via transesterification of the lipids and other biofuels such as biomethane and bioethanol, which are described using the biorefinery approach.

Pearson (1996) discusses expanding the horizons of pond technology and application in an environmentally conscious world in an early paper with 21 citations. They note that ponds can be made to combine treatment and effluent storage and are the best and cheapest option for treating wastewaters for subsequent reuse in aquaculture and agriculture. They argue that ponds can be designed to produce large quantities of algal biomass for animal and potential human consumption and for energy production via biomethane. They assert that ponds are thus a modern wastewater reclamation and resource recovery technology in tune with current environment-friendly policies.

Oswald et al. (1994) discuss the performance of biomethane fermentation pits in advanced integrated wastewater pond systems (AIWPSs) in an early paper with 11 citations. They deal mainly with design and performance of advanced facultative ponds containing internally located fermentation pits. "Experiences with a 1,894 $m^3\ day^{-1}$ (0.5 MGD) AIWPS and a 7,576 $m^3\ day^{-1}$ (2.0 MGD) AIWPS show that primary facultative ponds with internal fermentation pits require less land than do conventional anaerobic ponds and that sludge removal is postponed for many years." They argue that new, more detailed, and controlled scientific studies on a 133 m(3) day^{-1} (0.035 MGD) demonstration AIWPS provide evidence that these simple pits remove suspended solids and biochemical oxygen demand more effectively than do comparably loaded conventional anaerobic ponds and produce much less odor. In addition, they improve removal of parasites, bacteria, viruses, heavy metals, and halogenated hydrocarbons. They then compare the reliability and cost-effectiveness of AIWPS with more conventional ponds and with mechanical wastewater treatment systems.

Mairet et al. (2011) model the anaerobic digestion of microalgae using ADM1 in a recent paper with 10 citations. They demonstrate the ability of the original ADM1 model and a modified version (based on Contois kinetics for the hydrolysis steps) to represent microalgal anaerobic digestion. They then compare simulations to experimental data of an anaerobic digester fed with *Chlorella vulgaris*. They find that the modified ADM1 fits adequately the data for the considered 140-day experiment encompassing a variety of influent load and flow rates. They assert that the ADM1 is a reliable predictive tool for optimizing the coupling of microalgae with anaerobic digestion processes.

Hodgson and Paspaliaris (1996) investigate wastewater treatment lagoons with a focus on the design modifications to reduce odors and enhance nutrient removal in an early paper with 10 citations. They describe some properties of 3 new style wastewater treatment lagoons 115E, 55E, and 25W at a wastewater treatment plant treating some 250 megaliters (ML) of untreated wastewater each day. "There is a potential residence time for each of 120 days and each consists of a sequence of up to 11 ponds. Pond 1 has an anaerobic reactor of 90, 150, and 150 ML, respectively, and Warmens floating aerators are installed on ponds 1 and 2 of 115E and 25W and pond 1 of 55E. BOD5 values of less than 50 are achieved by the end of pond 2 and these together with the installation of the HDPE cover on 115E have effectively reduced odor emissions." Nitrogen is removed by ammonification followed by either nitrification/denitrification or algal growth, which is grazed by zooplankton. Since the introduction of the aerators, chemolithotrophic ammonia-oxidizing bacteria are more frequently exposed to the inhibitory action of UV light, and therefore nitrification is more sporadic. "The lagoons have the potential to produce an effluent with inorganic-N levels of less than 2 mg/L, a BOD5 of less than 50 mg/L and low levels of algae" while the covered anaerobic reactor can in each case produce up to 20,000 cubic meters of gas each day comprising of 80% biomethane.

15.2.2.2 Research on Algal Biomethane Production Policies

Singh and Cu (2010) discuss the commercial potential of microalgae for biofuel production including biomethane in a recent paper with 70 citations. They scan the available literature on various aspects of microalgae (including its cultivation, life-cycle assessment [LCA], and conceptualization of an algal biorefinery) and present a critical analysis of these papers. They assert that the economic viability of the process in terms of minimizing the operational and maintenance cost along with maximization of oil-rich microalgal production is the key factor for successful commercialization of microalga-based fuels.

Kaddam (2002) discusses the environmental implications of power generation via coal–microalgal cofiring conducting an LCA to compare the environmental impacts of electricity production via coal firing versus coal/algal cofiring. The LCA results demonstrate that "there are potentially significant benefits to recycling CO_2 toward microalgal production. As it reduces CO_2 emissions by recycling it and uses less coal, there are concomitant benefits of reduced greenhouse gas emissions." However, there are also other energy and fertilizer inputs needed for algal production, which contribute to key environmental flows. "Lower net values for the algal cofiring scenario were observed for the following using the direct injection process: SO_x, NO_x, particulates, carbon dioxide, biomethane, and fossil energy consumption. Lower values for the algal cofiring scenario were also observed for greenhouse potential and air acidification potential." However, impact assessment for depletion of natural resources and eutrophication potential showed much higher values.

Collet et al. (2011) discuss the LCA of microalgal culture coupled to biogas production in a recent paper with 46 citations performing an LCA of biogas production including biomethane from the microalgae *C. vulgaris* and comparing the results to algal biodiesel and to first-generation biodiesels. They find that the impacts generated by the production of biomethane from microalgae are strongly correlated with the electric consumption. They assert that "improvements can be achieved by decreasing the mixing costs and circulation between different production steps, or by improving the efficiency of the anaerobic process under controlled conditions." They conclude that this new bioenergy generating process strongly competes with other biofuel production processes.

Wyman (1994) discusses the alternative fuels including biomethane from biomass and their impact on carbon dioxide accumulation in an early paper with 38 citations. He notes through anaerobic digestion, a consortium of bacteria can break down lignocellulosic biomass to generate a medium-energy-content gas that can be cleaned up for pipeline-quality biomethane. "Catalytic processing of pyrolytic oils from biomass produces a mixture of olefins that can be reacted with alcohols to form ethers, such as methyl tertiary butyl ether (MTBE), for use in reformulated gasoline to reduce emissions." He notes that the costs have been reduced significantly for biofuels and the potential exists for them to be competitive with conventional fuels. This analysis includes consideration of fuel utilization performance and assignment of carbon dioxide to coproduct. He asserts that most or all of the carbon is recycled through their use, reducing substantially the net release of carbon dioxide to the atmosphere.

Wyman and Goodman (1993) discuss the biotechnology for production of biofuels including biomethane, chemicals, and materials from biomass in an early paper with 35 citations. They note that medium-BTU gas can be derived from lignocellulosic biomass by anaerobic digestion and refined into a pipeline-quality biomethane gas. A high-solids fermenter achieves higher gas generation rates than conventional devices and promises to help make such gas economical. An extensive collection of more than 500 productive strains of microalgae has been established to produce lipid oils for diesel fuel and other compounds from carbon dioxide. "Acetyl CoA carboxylase (ACC) is a key enzyme in lipid oil synthesis, and genetic engineering approaches are being applied to enhance the rates and yields of product formation." In addition to fuels, they assert that a biorefinery could produce a wide range of chemicals and materials through microbial conversion of renewable resources and technology is being developed for production of chemicals and materials from biomass.

Brune et al. (2009) investigate the microalgal biomass for greenhouse gas (GHG) reductions with a focus on the potential for replacement of fossil fuels and animal feeds in a paper with 32 citations. They present an initial analysis of the potential for GHG avoidance using a proposed algal biomass production system coupled to recovery of flue-gas CO_2 combined with waste sludge and/or animal manure utilization. They construct a model around a 50 MW natural gas-fired electric generation plant operating at 50% capacity as a semibase-load facility. They project that "this facility produces 216 million k.Wh/240-day season while releasing 30.3 million kg-C/season of GHG-CO_2." They then design an algal system "to capture 70% of flue-gas CO_2 producing 42,400 t (dry wt) of algal biomass/season and requiring 880 ha of high-rate algal ponds operating at a productivity of 20 g-dry-wt/m^2-day." They assume that this algal biomass is "fractionated into 20% extractable algal oil, useful for biodiesel, with the 50% protein content providing animal feed replacement and 30% residual algal biomass digested to produce biomethane gas, providing gross GHG avoidances of 20, 8.5, and 7.8%, respectively." They estimate that "the total gross GHG avoidance potential of 36.3% results in a net GHG avoidance of 26.3% after accounting for 10% parasitic energy costs." At CO_2 utilization efficiencies predicted to range from 60% to 80%, they estimate that net GHG avoidances range from 22% to 30%.

Zamalloa et al. (2011) discuss the technoeconomic potential of biomethane through the anaerobic digestion of microalgae in a paper with 31 citations. They evaluate the potential of microalgae as feedstock for biomethane production from a process technical and economic point of view. Production of mixed culture algae in raceway ponds on nonagricultural sites, such as landfills, was identified as a preferred approach. They examine the potential of straightforward biomethanation, which includes "pre-concentration of microalgae and utilization of a high rate anaerobic reactor based on the premises of achievable up-concentration from 0.2–0.6 kg m^{-3} to 20–60 kg dry matter (DM) m^{-3} and an effective bio-methanation of the concentrate at a loading rate of 20 kg DM $m^{-3}d^{-1}$." They calculate the "costs of biomass available for biomethanation under such conditions in the range of €86–124 ton^{-1} DM." They estimate that the standardized cost of energy by means of the process line "algae biomass–biogas–total energy module" would be in the order of €0.170–0.087 kWh^{-1}, taking into account a carbon credit of about €30 ton^{-1} CO_{2eq}.

Green et al. (1995a) discuss the energetics of AIWPS for the biomethane production and present an energy balance for a second-generation AIWPS prototype in an early paper with 29 citations. Modifications were made to the existing 1.8 ML facultative pond in order to further optimize biomethane fermentation and to demonstrate the recovery of biomethane using a submerged gas collector. Biomethane production rates were determined over a range of in-pond digester loadings and temperatures. They find that "biomethane concentrations increased by more than 50% as the biogas emerged through the overlying water column and most of the carbon dioxide fraction was utilized by microalgae." Electric power requirements for mixing two 0.1 ha algal high-rate ponds (HRPs) were measured over a range of channel depths and velocities, and electric power requirements for daily recirculation pumping were also measured. Using preliminary biomethane production and recovery rates achieved at Richmond, CA, "the cogeneration potential was estimated and projected for larger second generation AIWPSs of 2 MLD and 200 MLD capacities." By incorporating biomethane recovery and electric power generation together with efficient HRP mixing using paddle wheels, they estimate that "full-scale second generation AIWPSs will be able to produce as much energy as they require for primary and secondary treatment."

Green et al. (1995b) discuss the biomethane fermentation, submerged gas collection, and the fate of carbon in AIWPSs in an early paper with 26 citations. They note that first-generation systems lack the facilities to recover and utilize the carbon-rich treatment by-products of biomethane and algal biomass. The recovery of biomethane using a submerged gas collector was demonstrated using a second-generation AIWPS prototype, and the optimization of in-pond biomethane fermentation, the growth of microalgae in high-rate foods, and the harvest of microalgae by sedimentation and dissolved air flotation were studied. They estimate that in the experimental system, "17% of the influent organic carbon was recovered as biomethane, and rut average of 6 g C/m^2/d were assimilated into harvestable algal biomass." They argue that in a full-scale second-generation AIWPS in a climate comparable to Richmond, California, located at 37°N latitude, these values would be significantly higher (as much as 30% of the influent organic carbon would be recovered as biomethane and as much as 10 g C m^{-2} day^{-1} would be assimilated).

Kosaric and Velikonja (1995) present challenging opportunities for production of liquid and gaseous fuels including biomethane by biotechnology in an early paper with 26 citations. They focus on recent advances in the research and application of biotechnological processes and methods for the production of liquid transportation fuels other than bioethanol (other oxygenates, diesel fuel extenders, and substitutes), as well as gaseous fuels (biogas, biomethane, reformed syngas). Potential uses of these biofuels are described, along with environmental concerns that accompany them. Emphasis is also put on microalgal lipids as diesel substitute and biogas/biomethane as a renewable alternative to natural gas. The capturing and use of landfill gases is also mentioned, as well as microbial coal liquefaction.

Harun et al. (2011) discuss the technoeconomic analysis of an integrated microalgal photobioreactor, biodiesel, and biogas production facility in a recent paper with 13 citations. They introduce a new concept as an option to reduce the total production cost of microalgal biodiesel. The integration of a biodiesel production system with biomethane production via anaerobic digestion is proved in improving the economics and sustainability of overall biodiesel stages. Anaerobic digestion of microalgae produces biomethane and is further converted to generate electricity. The generated electricity can surrogate the consumption of energy that is required in microalgal cultivation, dewatering, extraction, and transesterification processes. From theoretical calculations, they find that "the electricity generated from biomethane is able to power all of the biodiesel production stages and will substantially reduce the cost of biodiesel production (33% reduction)." The "carbon emissions of biodiesel production systems are also reduced by approximately 75% when utilizing biogas electricity compared to when the electricity is otherwise purchased from the Victorian grid." They assert that the approach of digesting microalgal waste to produce biogas will make the production of biodiesel from algae more viable by reducing the overall cost of production per unit of biodiesel and hence enable biodiesel to be more competitive with existing fuels.

15.2.3 Conclusion

The nonexperimental research on algal biomethane has been one of the most dynamic research areas in recent years. Nineteen papers in the field of algal biomethane with more than ten citations were located, and the key emerging issues were presented in the decreasing order of the number of citations. These papers give strong hints about the determinants of efficient algal biomethane production processes and policies and emphasize that marine algae are efficient biomethane feedstocks in comparison with the first- and second-generation biofuel feedstocks. Eight and eleven of these papers were related to the algal biomethane production processes and algal biomethane production policies, respectively.

15.3 Experimental Studies on Algal Biomethane

15.3.1 Introduction

The experimental research on algal biomethane has been one of the most dynamic research areas in recent years. Thirty-one experimental papers in the field of algal biomethane with more than eight citations were located, and the key emerging issues from these papers are presented in the decreasing order of the number of citations in Table 15.2. These papers give strong hints about the determinants of the efficient algal biomethane production and emphasize that marine algae are efficient biomethane feedstocks in comparison with the first- and second-generation biofuel feedstocks.

The papers were dominated by researchers from 19 countries, usually through the intercountry institutional collaboration, and they were multiauthored. The United States (seven papers); France (four papers); Norway and China (three papers each); and Belgium, New Zealand, Switzerland, and Wales (two papers each) were the most prolific countries.

Similarly, all these papers were published in the journals indexed by the SCI and/or SCIE. There were no papers indexed by the A&HCI or SSCI. The number of citations

TABLE 15.2
Experimental Research on Algal Biomethane

No.	Paper References	Year	Document	Affiliation	Country	No. of Authors	Journal	Index	Subjects	Topic	Total No. of Citations	Total Average Citations per Annum	Rank
1	Yen and Brune	2007	A	Tunghai Univ., Clemson Univ.	Taiwan, United States	2	*Bioresour. Technol.*	SCI, SCIE	Agr. Eng., Biot. Appl. Microb., Ener. Fuels	Algal biomethane—ingredients	94	13.4	2
2	Mussgnug et al.	2010	A	Univ. Bielefeld	Germany	4	*J. Biotechnol.*	SCI, SCIE	Biot. Appl. Microb.	Algal biomethane—algal type	56	14.0	5
3	Minowa and Sawayama	1999	A	Natl. Inst. Resources Environ.	Japan	2	*Fuel*	SCI, SCIE	Ener. Fuels, Eng. Chem.	Algal biomethane—nitrogen cycling	46	3.1	9
4	Duan and Savage	2011	A	Univ. Michigan	United States	2	*Ind. Eng. Chem. Res.*	SCI, SCIE	Eng. Chem.	Algal biomethane—catalyst effect	41	13.7	10
5	Stucki et al.	2009	A	Paul Scherrer Inst., Ecole Polytech. Fed. Lausanne	Switzerland	5	*Energ. Environ. Sci.*	SCI, SCIE	Chem. Mult., Ener. Fuels, Eng. Chem., Env. Sci.	Algal biomethane—catalyst effects	41	8.2	11
6	De Schamphelaire and Verstraete	2009	A	Univ. Ghent	Belgium	2	*Biotechnol. Bioeng.*	SCI, SCIE	Biot. Appl. Microb.	Algal biomethane—sunlight-to-biomethane energy conversion system	39	7.8	12
7	Briand and Morand	1997	A	Ctr. Etud. Valorisat Algues	France	2	*J. Appl. Phycol.*	SCI, SCIE	Biot. Appl. Microb., Mar. Fresh. Biol.	Algal biomethane—algal type	39	2.3	13
8	Gerhardt et al.	1991	A	Univ. Calif. Berkeley	United States	6	*Res. J. Water Pollut. C*	SCIE	Env. Sci., Limn., Water Res.	Algal biomethane—bioremoval and biomethane production	34	1.5	16

(*Continued*)

TABLE 15.2 (*Continued*)

Experimental Research on Algal Biomethane

No.	Paper References	Year	Document	Affiliation	Country	No. of Authors	Journal	Index	Subjects	Topic	Total No. of Citations	Total Average Citations per Annum	Rank
9	Haiduc et al.	2009	A	Ecole Polytech. Fed. Lausanne., Paul Scherrer Inst.	Switzerland	6	*J. Appl. Phycol.*	SCI, SCIE	Biot. Appl. Microb., Mar. Fresh. Biol.	Algal biomethane—catalyst effect	28	5.6	20
10	Grossi et al.	2001	A	CNRS, Netherlands Inst. Sea Res.	France, the Netherlands	3	*Org. Geochem.*	SCI, SCIE	Geoch. Geophys.	Algal biomethane—algal type	28	2.2	21
11	Chen and Oswald	1998	A	Natl. Chung Hsing Univ.	Taiwan	2	*Environ. Int.*	na	Env. Sci.	Algal biomethane—pretreatment	28	1.8	22
12	Chinnasamy et al.	2010	A	Univ. Georgia, Maharshi Dayanand Saraswati Univ.	United States, India	4	*Bioresour. Technol.*	SCI, SCIE	Agr. Eng., Biot. Appl. Microb., Ener. Fuels	Algal biomethane—ingredients	25	6.3	25
13	Bruhn et al.	2011	A	Aarhus Univ., Danish Technol. Inst., Tech. Univ. Denmark	Denmark	9	*Bioresour. Technol.*	SCI, SCIE	Agr. Eng., Biot. Appl. Microb., Ener. Fuels	Algal biomethane—algal type	24	8.0	26
14	Ras et al.	2011	A	INRA, Naskeo	France	5	*Bioresour. Technol.*	SCI, SCIE	Agr. Eng., Biot. Appl. Microb., Ener. Fuels	Algal biomethane—algal type	23	7.7	27
15	Ehimen et al.	2011	A	Univ. Otago	New Zealand	5	*Appl. Energ.*	SCI, SCIE	Ener. Fuels, Eng. Chem.	Algal biomethane—algal residues	22	7.3	28
16	Vergara-Fernandez et al.	2008	A	Univ. Catolica Temuco, Univ. Catolica Norte, CENICA INE	Chile, Mexico	4	*Biomass Bioenerg.*	SCIE	Agr. Eng., Biot. Appl. Microb., Ener. Fuels	Algal biomethane—stage effect	20	3.3	30

(*Continued*)

TABLE 15.2 (*Continued*)

Experimental Research on Algal Biomethane

No.	Paper References	Year	Document	Affiliation	Country	No. of Authors	Journal	Index	Subjects	Topic	Total No. of Citations	Total Average Citations per Annum	Rank
17	Gonzalez-Fernandez et al.	2011	A	Agr. Technol. Inst. Castilla Leon	Spain	3	*Appl. Energ.*	SCI, SCIE	Ener. Fuels, Eng. Chem.	Algal biomethane—manure couse	14	4.7	31
18	Nkemka and Murto	2010	A	Lund Univ.	Sweden	2	*J. Environ. Manage.*	SCI, SCIE	Env. Sci., Env. Stud.	Algal biomethane—algal type	13	3.3	34
19	Moen et al.	1997a	A	Norwegian Univ. Sci. Technol.	Norway	3	*J. Appl. Phycol.*	SCI, SCIE	Biot. Appl. Microb., Mar. Fresh. Biol.	Algal biomethane—algal type	13	0.8	33
20	Ostgaard et al.	1993	A	Norwegian Univ. Sci. Technol.	Norway	5	*J. Appl. Phycol.*	SCI, SCIE	Biot. Appl. Microb., Mar. Fresh. Biol.	Algal biomethane—algal type	13	0.6	35
21	Ungerfeld et al.	2005	A	Michigan State Univ., Univ. Hawaii Manoa	United States	5	*Anim. Feed Sci. Technol.*	SCI, SCIE	Agr. Dairy Anim. Sci.	Algal biomethane—algal type	12	1.3	36
22	Morand and Brian	1999	A	Ctr. Rech. Ecol. Marine Aquaculture, CNRS, Ctr. Etud. Valorisat. Algues	France	2	*J. Appl. Phycol.*	SCI, SCIE	Biot. Appl. Microb., Mar. Fresh. Biol.	Algal biomethane—algal type	12	0.8	37
23	Moen et al.	1997b	A	Norwegian Univ. Sci. Technol.	Norway	3	*J. Appl. Phycol.*	SCI, SCIE	Biot. Appl. Microb., Mar. Fresh. Biol.	Algal biomethane—algal type	12	0.7	38
24	Adams et al.	2011	A	Aberystwyth Univ.	Wales	4	*Bioresour. Technol.*	SCI, SCIE	Agr. Eng., Biot. Appl. Microb., Ener. Fuels	Algal biomethane—algal type	11	3.7	39

(*Continued*)

TABLE 15.2 (*Continued*)

Experimental Research on Algal Biomethane

No.	Paper References	Year	Document	Affiliation	Country	No. of Authors	Journal	Index	Subjects	Topic	Total No. of Citations	Total Average Citations per Annum	Rank
25	Yang et al.	2011	A	Chinese Acad. Sci.	China	5	*Int. J. Hydrogen Energ.*	SCI, SCIE	Chem. Phys., Electroch., Ener. fuels	Algal biomethane-stage effect	11	3.7	41
26	Ehimen et al.	2009	A	Univ. Otago, Waste Solut.	New Zealand	4	*GCB Bioenergy*	SCIE	Agronom., Ener. Fuels, Biot. Appl. Microb.	Algal biomethane—algal residues	11	2.2	42
27	Laws and Berning	1991	A	Hawaii Inst. Marine Biol., Elect. Power Res. Inst.	United States	2	*Bioresour. Technol.*	SCI, SCIE	Agr. Eng., Biot. Appl. Microb., Ener. Fuels	Algal biomethane—algal type	11	0.5	44
28	Yuan et al.	2011	A	Chinese Acad. Sci., Qingdao Technol. Univ.	China	6	*Energ. Environ. Sci.*	SCI, SCIE	Ener. Fuels, Eng. Chem., Env. Sci.	Algal biomethane—algal type	9	3.0	48
29	Lakaniemi et al.	2011	A	Tampere Univ. Technol., Bangor Univ., Finnish Environ. Inst., Ohio State Univ.	Finland, Wales, United States	5	*Biotechnol. Biofuels*	SCIE	Biot. Appl. Microb.	Algal biomethane—algal type	9	3.0	47
30	Zeng et al.	2010	A	Chinese Acad. Sci.	China	4	*J. Hazard. Mater.*	SCI, SCIE	Eng. Env., Eng. Civ., Env. Sci.	Algal biomethane—ISR	9	2.3	49
31	Zamalloa et al.	2012	A	Univ. Ghent	Belgium	3	*Appl. Energ.*	SCI, SCIE	Ener. Fuels, Eng. Chem.	Algal biomethane—algal type	8	4.0	50

SCI, Science Citation Index; SCIE, Science Citation Index Expanded; SSCI, Social Sciences Citation Index; A, article; R, review.

ranged from 8 to 94 and the average number of citations per annum ranged from 0.5 to 14.0. All of these papers were articles.

There were a number of subtopical areas: algal type (16 papers); catalyst effects (3 papers); ingredients, algal residues, stage effects (2 papers each); and other topics relating to algal biomethane production (6 papers).

15.3.2 Experimental Research on Algal Biomethane

15.3.2.1 Research on Algal Type

Mussgnug et al. (2010) investigate microalgae as substrates for fermentative biogas production including biomethane in a combined biorefinery concept in a paper with 56 citations. They focus on the suitability of six dominant microalgal species (freshwater and saltwater algae and cyanobacteria) as alternative substrates for biogas production. They demonstrate that biogas potential is strongly dependent on the species and on the pretreatment. They find that "fermentation of the green alga *Chlamydomonas reinhardtii* was efficient with a production of 587 ml (+/−8.8 SE) biogas g volatile solids^{-1} (VS^{-1}), whereas fermentation of *Scenedesmus obliquus* was inefficient with only 287 ml (+/−10.1 SE) biogas g VS^{-1} being produced." They then find that "drying as a pretreatment decreased the amount of biogas production to ca. 80%." The biomethane content of biogas from microalgae was 7%–13% higher compared to biogas from maize silage. To evaluate integrative biorefinery concepts, hydrogen production in *C. reinhardtii* prior to anaerobic fermentation of the algal biomass was measured and resulted in an increase of biogas generation to 123% (±3.7 SE). They assert that selected algal species can be good substrates for biogas production and that anaerobic fermentation can seriously be considered as the final step in future microalga-based biorefinery concepts.

Briand and Morand (1997) study the anaerobic digestion of *Ulva* sp. with a focus on the relationship between *Ulva* composition and mechanization in an early paper with 39 citations. They note that anaerobic degradation with a batch or completely stirred system is technically possible. However, "the biomethane yield reached only 0.20 m^3 kg^{-1} volatile solids and the epuration rate 50% volatile solids in experiments in batch or completely stirred reactors." More generally, "mechanization comes up against various practical obstacles such as seasonal growth of *Ulva*, low density of alga in suspension for loading the digester, high S concentration leading to the production of a biogas with a high H_2S content, and, finally, the existence of a refractory or slowly degradable part, which requires a compromise between productivity and biological yield."

Grossi et al. (2001) investigate the anaerobic biodegradation of lipids of marine microalga *Nannochloropsis salina* in a paper with 28 citations. In order to determine the susceptibility to anaerobic biodegradation of the different lipid biomarkers present in a marine microalga containing algaenan, portions of one large batch of cultured *N. salina* (Eustigmatophyceae) were incubated in anoxic sediment slurries for various times. They find that after 442 days, all lipids studied (mono-, di-, and triunsaturated hydrocarbons, long-chain unsaturated alcohols and alkyl diets, phytol, sterols, saturated and [poly]-unsaturated fatty acids) showed a significant decrease in concentration, which was accompanied by a strong production of sulfide and biomethane. However, the studied compounds showed a wide range of reactivity and different patterns and extent of degradation. "Polyunsaturated fatty acids, phytol and triunsaturated hydrocarbons were the most labile compounds and showed initially rapid degradation rates, followed by a substantial reduction in degradation rate during the later stages of incubation." Long-chain alkyl diols and unsaturated alkenols showed fluctuating concentrations with time clearly indicating their release from bound fractions in parallel with their degradation. Other lipids showed a continuous concentration decrease until the end of the incubation, with alkadienes and sterols being the most resistant compounds encountered. Besides providing an extended sequence of reactivity for lipids under anoxic conditions, they assert that "the presence of resistant algaenan in the outer cell wall of microalgae does not protect the other lipids of the cell from anaerobic microbial degradation."

Bruhn et al. (2011) study the bioenergy (biomethane) potential of *Ulva lactuca* with a focus on the biomass yield and biomethane production and combustion in a recent paper with 24 citations. They investigate the biomass production potential at temperate latitudes (56° N) and the quality of the biomass for energy production (anaerobic digestion to biomethane and direct combustion) for the green macroalgae *U. lactuca*. The algae were cultivated in a land-based facility demonstrating a production potential of 45 T (TS) ha^{-1} $year^{-1}$. They find that "biogas production from fresh and macerated *U. lactuca* yielded up to 271 ml CH_4 g^{-1} VS, which is in the range of the biomethane production from cattle manure and land based energy crops, such as grass-clover." Drying of the biomass resulted in a five- to ninefold increase in weight-specific biomethane production compared to wet biomass. Ash and alkali contents are the main challenges in the use of *U. lactuca* for direct combustion. They assert that application of a biorefinery concept could increase the economical value of the *U. lactuca* biomass as well as improve its suitability for production of bioenergy.

Ras et al. (2011) study the coupled process of production and anaerobic digestion of *C. vulgaris* in a recent paper with 23 citations as they investigate the feasibility of coupling algal production (*C. vulgaris*) to an anaerobic digestion unit for the biomethane production. An intermediate settling device was integrated in order to adapt the feed-flow concentration and the flow rate. Digestion of *C. vulgaris* was studied under 16- and 28-day hydraulic retention times (HRTs), with a corresponding organic loading rate (OLR) of 1 g_{COD} L^{-1}. They find that "increasing the HRT achieved 51% COD removal with a biomethane production measured at 240 mL $g_{VSS}{}^{-1}$." Performing different HRTs and dynamic monitoring during degradation highlighted differential hydrolysis of microalgal compartments. However, they then find that "50% of the biomass did not undergo anaerobic digestion, even under long retention times."

Nkemka and Murto (2010) evaluate biomethane production from seaweed in batch tests and in upflow anaerobic sludge blanket (UASB) reactors combined with the removal of heavy metals in a paper with 13 citations as they investigate the potential of seaweed and its leachate in the production of biomethane in batch tests. The effect of removing heavy metals from seaweed leachate was evaluated in both batch test and treatment in a UASB reactor. The heavy metals were removed from seaweed leachate using an iminodiacetic acid (IDA) polyacrylamide cryogel carrier. They find that "the biomethane yield obtained in the anaerobic digestion of seaweed was 0.12 N I CH_4/g VS_{added}. The same biomethane yield was obtained when the seaweed leachate was used for biomethane production." They argue that "the IDA-cryogel carrier was efficient in removing Cd^{2+}, Cu^{2+}, Ni^{2+} and Zn^{2+} ions from seaweed leachate. The removal of heavy metals in the seaweed leachate led to a decrease in the biomethane yield." They further find that "the maximum sustainable organic loading rate (OLR) attained in the UASB reactor was 20.6 g tCOD/l/day corresponding to a hydraulic retention time (HRT) of 12 h and with a total COD removal efficiency of about 81%." They conclude that hydrolysis and treatment with IDA cryogel reduced the heavy metal content in the seaweed leachate before biomethane production.

Moen et al. (1997a) investigate the alginate degradation during anaerobic digestion of *Laminaria hyperborea* stipes in an early paper with 13 citations. *L. hyperborea* stipes, harvested at 59° N off the Norwegian coast in autumn, were degraded at different concentrations of polyphenols in anaerobic batch reactors at 35°C and pH 7. This was done by removing or adding the mechanically peeled outer phenolic layer of the algae and using methanogenic and alginate-degrading inocula already adapted to *L. hyperborea* degradation. They find that initial alginate released from the algal particles was affected by NaOH titrations because the Ca/Na ratio was reduced. "After a rapid consumption of the mannitol, alginate lyases were induced, and guluronate lyases showed the highest extracellular activity. Then the microbes digested 0.12–0.23 g Na-alginate L^{-1} h^{-1}. Later the degradation rate of alginates declined almost to zero, and 13–50% of the alginate remained insoluble." They argue that the total solubilization of alginates was limited by both Ca-cross-linked guluronate residues and complexation with compounds such as polyphenols. They assert that the biomethane production had a lag phase that increased at higher amounts of soluble polyphenols and the total fermentation probably also became product inhibited if soluble compounds such as acetate, ethanol, and butyrate were accumulated.

Ostgaard et al. (1993) investigate the carbohydrate degradation and biomethane production during fermentation of *Laminaria saccharina* in an early paper with 13 citations. Anaerobic digestion of the brown alga *L. saccharina* (L.) Lamour harvested during spring and autumn was carried out at controlled laboratory conditions in stirred fermenter systems. Due to the normal seasonal variations, the autumn material had a much higher content of carbohydrates such as mannitol and laminaran. Both batch and semicontinuous feeding conditions were investigated for periods up to 800 h, with inoculum provided from previous kelp fermentations. They find that in batch cultures, "the biomethane yield from the autumn material was doubled compared to that of the spring material. Semi-continuous conditions gave more similar biomethane yields for both raw materials, 0.22 and 0.271 CH_4 per g VS for spring and autumn material, respectively." They then find that in all experiments, "mannitol and laminaran were reduced to less than 5% of the initial values within 24–48 hours after inoculation, whereas 30% of the alginate content was detectable even after 30 days." This material was severely depolymerized, and alginate lyase activity developed rapidly in all cultures. They assert that "although mannitol and laminaran were fermented much faster than alginate, the total accumulated biomethane yields were determined by the total carbohydrate content of the raw material during extended semi-continuous feeding."

Ungerfeld et al. (2005) investigate the effects of two lipids on in vitro ruminal biomethane production in a paper with 12 citations. The effects on mixed ruminal cultures of olive oil (OO) and a hexadecatrienoic acid (HA, cis-C-16:6.9.12) extracted from the Hawaiian algae *Chaetoceros* were studied in a 24 h batch fermentation. They find that "HA at 0.5, 1 and 2 ml/l linearly decreased CH_4 production by 25, 47 and 97%, respectively, while OO did not affect it. HA at 0.5, 1 and 2 ml/l increased H_2, accumulation by 2-, 2- and 5-fold, respectively. Release of CO_2, was linearly decreased by HA at 0.5, 1 and 2 ml/l by 10, 32 and 48%, respectively, while OO linearly increased it by 9, 2 and 17%, at the same concentrations." They argue that "fermented OM, as estimated through the VFA production stoichiometry, was linearly decreased by HA at 0.5, 1 and 2 ml/l by 9, 19 and 42%, respectively, while OO did not affect it." They further argue that "HA decreased acetate molar percentage, increased propionate, and tended to decrease butyrate. OO tended to decrease acetate molar percentage, and increased propionate and butyrate molar percentages. HA at 0.5, 1 and 2 ml/l linearly decreased NH_4^+ concentration by 5, 5 and 21%, respectively." They assert that HA was a strong inhibitor of methanogenesis but decreased fermentation and increased H-2 accumulation, while the addition of OO increased propionate production and did not seem to inhibit fermentation.

Morand and Briand (1999) study the anaerobic digestion of *Ulva* sp. with a focus on the study of *Ulva* degradation and mechanization of liquefaction juices in a paper with 12 citations. Observations on degradation of *Ulva* in such dumps led them to consider recovery of hydrolysis juice in order to methanize this rather than the entire alga. The hydrolysis step was then studied in the laboratory and under real conditions. They find that "the decomposition of *Ulva* was rapid (7.1% C d^{-1}), but its degradation incomplete (38% C remaining after 52 days). After 9 months in a dump, VFA contents in the flows were insignificant and N and C contents in the remaining material were due to the non-degradable fraction." Modifications of the physical or chemical conditions of hydrolysis didn't improve sufficiently significantly the results to be used on a large scale. On the other hand, they argue that the techniques that could allow a rapid recovery of the juice improved together the recovery of the chemical oxygen demand (COD). The hydrolysis juice is a very good substrate for mechanization as "the biomethane yield reached 330 L kg^{-1} VS, and the epuration rate 93%." They assert that the process combining the two steps, hydrolysis and juice mechanization, offers a reasonable compromise between biomethane output, productivity of the system, and treatment cost.

Moen et al. (1997b) study the biological degradation of *Ascophyllum nodosum* in an early paper with 12 citations. Aerobic and anaerobic batch reactors, operated at 35°C and pH 7, were fed milled *A. nodosum*, nutrients, and inocula adapted to seaweed degradation. They find that "the dominant factor for conversion of organic matter during anaerobic digestion was the inhibitory effect of the polyphenols on alginate lyases and biomethane production." They argue that the relative large fraction of high-molecular-weight polyphenols (>10 kDa) in this alga gave efficient binding of proteins during digestion. The anaerobic degradation was greatly stimulated when the polyphenols were fixed with low amounts of formaldehyde. An accumulated content of guluronate in the remaining alginate indicated that Ca cross-linking also limited the guluronate lyase access to the polymer. In contrast, the aerobic digestion of alga gave no increase in the guluronate content of the residual alginate. Compared to anaerobic conditions, they assert that the phenols had a much lower influence on the hydrolytic rate of organic matter during aerobic conditions.

Adams et al. (2011) study the seasonal variation in *Laminaria digitata* and its impact on biochemical conversion routes to biofuels in a recent paper with 11 citations. They show that it can be used as a feedstock in both bioethanol fermentation and anaerobic digestion for biomethane production. They optimize several parameters in the fermentation of *L. digitata* and investigate the suitability of the macroalgae through the year using samples harvested every month. For both biomethane and bioethanol productions, they find that "minimum yields were seen in material harvested in March when the carbohydrates laminarin and mannitol were lowest. July material contained the highest combined laminarin and mannitol content and maximum yields of 167 mL ethanol and 0.219 m^3 kg^{-1} *L. digitata*."

Laws and Berning (1991) carry out the photosynthetic efficiency (PE) optimization studies with the macroalga *Gracilaria tikvahiae* with a focus on the implications for CO_2 emission control from power plants in an early paper with 11 citations. The PE with which the macroalga *G. tikvahiae* converts visible light energy into chemical energy was studied as a function of irradiance, temperature, and salinity in tumble culture systems. The photosynthesis/irradiance curve exhibited a typical hyperbolic shape, the associated PEs being a maximum at a visible irradiance of about 500 kcal m^{-2} day^{-1}. They find that "the highest PEs were obtained in seawater diluted by 10% with freshwater; the maximum PEs under these conditions exceeded 7% in full sunlight and 12% in the region of optimal irradiance." They then find that "PEs were almost identical at 21°C and 25°C, but declined sharply as the temperature was reduced below 21°C." Conversion of the algal biomass to biomethane by anaerobic fermentation resulted in conversion efficiencies as high as 22% at a detention time of 15 days. They assert that to the extent that CO_2 emissions from electric power plants are reduced by scrubbing the stack gases, the growth of algae such as *G. tikvahiae* may be the most logical way to utilize the CO_2.

Lakaniemi et al. (2011) investigate the biohydrogen and biomethane production from *C. vulgaris* and *Dunaliella tertiolecta* biomass in a recent paper with nine citations. Utilization of *C. vulgaris* and *D. tertiolecta* biomass was tested as a feedstock for anaerobic biohydrogen and biomethane production. Anaerobic serum bottle assays were conducted at 37°C with enrichment cultures derived from municipal anaerobic digester sludge. They find that "low levels of H_2 were produced by anaerobic enrichment cultures, but H_2 was subsequently consumed even in the presence of 2-bromoethanesulfonic acid, an inhibitor of methanogens." Without inoculation, algal biomass still produced H_2 due to the activities of satellite bacteria associated with algal cultures. Biomethane was produced from both types of biomass with anaerobic enrichments. They find the presence of H_2-producing and H_2-consuming bacteria in the anaerobic enrichment cultures and the presence of H_2-producing bacteria among the satellite bacteria in both sources of algal biomass. They assert that "H_2 production by the satellite bacteria was comparable from *D. tertiolecta* (12.6 ml H_2/g volatile solids (VS)) and from *C. vulgaris* (10.8 ml H_2/g VS), whereas CH_4 production was significantly higher from *C. vulgaris* (286 ml/g VS) than from *D. tertiolecta* (24 ml/g VS)."

Yuan et al. (2011) investigate the biomethane production and microcystin (MC) biodegradation in anaerobic digestion of blue algae in a recent paper with nine citations. They study in more detail the biogas production, process stability, and the variation of MC concentration in anaerobic digestion of blue algae in order to demonstrate the potentials of both bioenergy production and MC biodegradation in methanogenic conditions. They obtain a "biomethane yield of 189.89 mL g^{-1} VS from the digester, and the average biomethane concentration in the biogas was 36.72%." During the digestion, the pH value was fairly constant (6.8–7.6), and soluble COD kept at a relatively stable level. The concentration of total volatile fatty acid (VFA) increased significantly in the first 8 days and then decreased, showing no inhibition on the digestion. They further find that "the concentration of MC could be significantly reduced from 1220.19 mg L^{-1} to 35.17 mg L^{-1} during the methanogenic process, which followed the first order kinetics well."

Zamalloa et al. (2012) investigate the anaerobic digestibility of *Scenedesmus obliquus* and *Phaeodactylum tricornutum* under mesophilic and thermophilic conditions in a recent paper with eight citations. They cultivate two types of nonaxenic algal cultures in two types of simple photobioreactor systems. The production rates, expressed on dry matter (DM) basis, were in the order of 0.12 and 0.18 g DM L^{-1} day^{-1} for *S. obliquus* and *P. tricornutum*, respectively. The biogas potential of algal biomass was assessed by performing standardized batch digestion as well as digestion in a hybrid flow-through reactor (combining a sludge blanket and a carrier bed), the latter under mesophilic and thermophilic conditions. They find that "the ultimate biomethane yield (B_0) of *P. tricornutum* biomass was a factor of 1.5 higher than that of *S. obliquus* biomass, i.e. 0.36 and 0.24 L CH_4 g^{-1} volatile solids (VS) added respectively." They then find that for *S. obliquus* biomass, "the hybrid flow-through reactor tests operated at volumetric organic loading rate (Bv) of 2.8 g VS L^{-1} d^{-1} indicated low conversion efficiencies ranging between 26–31% at a hydraulic retention time (HRT) of 2.2 days for mesophilic and thermophilic conditions respectively." When digesting *P. tricornutum* at a Bv of 1.9 g VS L^{-1} day^{-1} at either mesophilic or thermophilic conditions and at an HRT of 2.2 days, an overall conversion efficiency of about 50% was obtained. They assert that the hydrolysis of the algal cells is limiting the anaerobic processing of intensively grown *S. obliquus* and *P. tricornutum* biomass.

15.3.2.2 Research on Ingredient Effects

Yen and Brune (2007) investigate the anaerobic codigestion of algal sludge and waste paper to produce biomethane in a paper with 94 citations. They add carbon content of waste paper in algal sludge feedstock to have a balanced C/N ratio. They find that "adding 50% (based on volatile solid) of waste paper in algal sludge feedstock increased the biomethane production rate to 1170 +/− 75 ml/l day, as compared to 573 +/− 28 ml/l day of algal sludge digestion alone, both operated at 4 g VS/l day, 35°C and 10 days HRT." They then find that "the maximum biomethane production rate of 1,607 +/− 17 ml/l day was observed at a combined 5 g VS/l day loading rate with 60% (VS based) of paper adding in algal sludge feedstock." They assert that an optimum C/N ratio for codigestion of algal sludge and waste paper was in the range of 20–25 L^{-1}.

Chinnasamy et al. (2010) investigate the biomass and bioenergy production potential of microalgal consortium in open and closed bioreactors using untreated carpet industry effluent as growth medium in a recent paper with 25 citations. They evaluate the algal cultivation systems, namely, raceway ponds, vertical tank reactors (VTRs), and polybags for mass production of algal consortium using carpet industry (Cl) untreated wastewater. They find that "overall areal biomass productivity of polybags (21.1 g m^{-2} d^{-1}) was the best followed by VTR (8.1 g m^{-2} d^{-1}) and raceways (5.9 g m^{-2} d^{-1}). An estimated biomass productivity of 51 and 77 tons ha^{-1} $year^{-1}$ can be achieved using 20 and 30 L capacity polybags, respectively with triple row arrangement." They further find that biomass obtained from algal consortium was rich in proteins (near 53.8%) and low in carbohydrates (near 15.7%) and lipids (near 5.3%). They estimate that consortium cultivated in polybags has the potential to produce 12,128 m^3 of biomethane ha $year^{-1}$.

15.3.2.3 Research on Catalyst Effects

Duan and Savage (2011) investigate the hydrothermal liquefaction of a microalga with heterogeneous catalysts in a recent paper with 41 citations. They produce crude biooils from the microalga *Nannochloropsis* sp. via reactions in liquid water at 350°C in the presence of six different heterogeneous catalysts (Pd/C, Pt/C, Ru/C, Ni/SiO_2–Al_2O_3, CoMo/γ–Al_2O_3 [sulfided], and zeolite) under inert (helium) and high-pressure reducing (hydrogen) conditions. They find that in the absence of added H_2, "all of the catalysts produced higher yields of crude biooil from the liquefaction of *Nannochloropsis* sp., but the elemental compositions and heating values of the crude oil (about 38 MJ/kg) were largely insensitive to the catalyst used." The gaseous products were mainly H_2, CO_2, and CH_4, with lesser amounts of C_2H_4 and C_2H_6. The Ru and Ni catalysts produced the highest biomethane yields. Only the zeolite catalyst produced significant amounts of N_2. Typical H/C and O/C atomic ratios for the crude biooil are 1.7 and 0.09, respectively. On the other hand, in the presence of high-pressure H_2, the crude biooil yield and heating value were largely insensitive to the presence or identity of the catalyst. They argue that the presence of either the hydrogen or the higher pressure in the reaction system did suppress the formation of gas. The total gas yield was always lower in H_2 than it was in analogous experiments without H_2 and at lower pressure.

Stucki et al. (2009) investigate the catalytic gasification of algae in supercritical water for biofuel production and carbon capture in a paper with 41 citations. They propose a novel process based on microalgal cultivation using dilute fossil CO_2 emissions and the conversion of the algal biomass through a catalytic hydrothermal process. The resulting products are biomethane as a clean fuel and concentrated CO_2 for sequestration. The proposed gasification process mineralizes nutrient-bearing organics completely. They show that complete gasification of microalgae (*Spirulina platensis*) to a biomethane-rich gas is now possible in supercritical water using ruthenium catalysts. A percentage of 60–70 of the heating value contained in the algal biomass would be recovered as biomethane. They assert that such an efficient alga-to-biomethane process opens up an elegant way to tackle both climate change and dependence on fossil natural gas without competing with food production.

Haiduc et al. (2009) develop an integrated process for the hydrothermal production of biomethane from microalgae and CO_2 mitigation in a paper with 28 citations and describe a potential novel process (SunCHem) for the production of biomethane via hydrothermal gasification of microalgae where the nutrients, water, and CO_2 produced are recycled. The influence on the growth of microalgae of nickel was investigated. They find that "the growth was adversely affected by the nickel present (1, 5, and 10 ppm). At 25 ppm Ni, complete inhibition of cell division occurred." They show successful hydrothermal gasification of the microalgae *P. tricornutum* to a biomethane-rich gas with high carbon gasification efficiency (68%–74%) and C1–C3 hydrocarbon yields of 0.2 g_{C1-C3}/g_{DM} (DM). The biomass-released sulfur adversely affected Ru/C catalyst performance. They assert that liquefaction of *P. tricornutum* at short residence times around 360°C was possible without coke formation.

15.3.2.4 Research on Algal Residues

Ehimen et al. (2011) study the anaerobic digestion of microalgal residues resulting from the biodiesel production process in a recent paper with 22 citations. They deal with the anaerobic digestion of microalgal biomass residues (post transesterification) using semi-continuously fed reactors. They investigate the influence of substrate loading concentrations and HRTs on the specific biomethane yield of the anaerobically digested microalgal residues. They also examine the codigestion of the microalgal residues with glycerol as well as the influence of temperature. They find that the "hydraulic retention period was the most significant variable affecting biomethane production from the residues, with periods (>5 days) corresponding to higher energy recovery." The biomethane yield was also improved by a reduction in the substrate loading rates, with an optimum substrate carbon to nitrogen ratio of 12.44 seen to be required for the digestion process.

Ehimen et al. (2009) study the bioenergy recovery from lipid-extracted, transesterified, and glycerol-codigested microalgal biomass in a paper with 11 citations. They investigate the practical biomethane yields achievable from the anaerobic conversion of the microalgal residues (as well as codigestion with glycerol) after biodiesel production using both the conventional and in situ transesterification methods. They find that the "type of lipid extraction solvent utilized in the conventional transesterification process could inhibit subsequent biomethane production." On the basis of actual biomethane production, they obtain a recoverable energy of 8.7–10.5 MJ kg^{-1} of dry microalgal biomass residue using the lipid-extracted and transesterified microalgal samples. On codigesting the microalgal residues with glycerol, they observe a 4%–7% increase in biomethane production.

15.3.2.5 Research on Stage Effects

Vergara-Fernandez et al. (2008) investigate the marine algae as a source of biomethane in a two-stage anaerobic reactor system in a paper with 20 citations as they investigate the technical feasibility of marine algal utilization as a source of renewable energy to laboratory scale. They evaluate the anaerobic digestion of *Macrocystis pyrifera, Durvillaea antarctica*, and their blend 1:1 (w/w) in a two-phase anaerobic digestion system, which consisted of an anaerobic sequencing batch reactor and an upflow anaerobic filter (UAF). They find that "70% of the total biogas produced in the system was generated in the UAF, and both algae species have similar biogas productions of 180.4(+/− 1.5) mL g^{-1} dry algae d^{-1}, with a biomethane concentration around 65%." Further, they observe the same biomethane content in biogas yield of algal blend; however, they obtain a lower biogas yield. They assert that either algal species or their blend can be utilized to produce biomethane gas in a two-phase digestion system.

Yang et al. (2011) study the biohydrogen and biomethane production from lipid-extracted microalgal biomass residues (LMBRs) in a recent paper with 11 citations. They develop a two-stage process to produce hydrogen and biomethane from LMBRs. They compare the biogas production and energy efficiency between one- and two-stage processes. They find that the "two-stage process generated 46 +/− 2.4 mL H_2/g-volatile solid (VS), and 393.6 +/− 19.5 mL CH_4/g VS. The biomethane yield was 22% higher than the one in the one-stage process. Energy efficiency increased from 51% in the one-stage process to 65% in the two-stage process." Additionally, they find that repeated batch cultivation was a useful method to cultivate the cultures to improve the biomethane production rate and reduce the fermentation time. In the repeated batch cultivation, the biomethane yield slightly decreased if the ammonia levels rose, suggesting that the accumulation of ammonia could affect biomethane production.

15.3.2.6 Research on Other Topics

Minowa and Sawayama (1999) discuss a novel microalgal system for energy production with nitrogen cycling in an early paper with 46 citations. A microalga, *C. vulgaris*, could grow in the recovered solution from the low-temperature catalytic gasification of itself, by which biomethane-rich fuel gas was obtained. All nitrogen in the microalga was converted to ammonia during the gasification, and the recovered solution, in which ammonia was dissolved, could be used as nitrogen nutrient. They find that the novel microalgal system for energy production with nitrogen cycling could be coated.

De Schamphelaire and Verstraete (2009) investigate the revival of the biological sunlight-to-biogas (biomethane) energy conversion system in a paper with 39 citations. They investigate the use of algae for energy generation in a stand-alone, closed-loop system. The system encompasses an algal growth unit for biomass production, an anaerobic digestion unit to convert the biomass to biogas, and a microbial fuel cell to polish the effluent of the digester. The system continuously transforms solar energy into energy-rich biogas (biomethane) and electricity. They find that "algal productivities of 24–30 ton VS ha^{-1} $year^{-1}$ were reached, while 0.5 N m^3 biogas could be produced kg^{-1} algal VS." They then find that the system resulted in a power plant with a potential capacity of about 9 kW ha^{-1} of solar algal panel, with prospects of 23 kW ha^{-1}.

Gerhardt et al. (1991) investigate the removal of selenium using a novel algal bacterial process and the production of biomethane in an early paper with 34 citations. They study a process for removing selenium and nitrate from agricultural drainage water using algae and anaerobic bacteria in a field system. Algae were grown in HRPs containing drainage water, and the 178 ± 99 mg L^{-1} culture took up 18 ± 13 mg L^{-1} NO_3^--N. The algae and drainage water were then transferred to anoxic units where denitrifying and selenate-reducing bacteria, feeding on algae, reduced NO_3^--N from 100 ± 24 mg L^{-1} to <10 mg L^{-1} at times. They find that a "soluble selenium concentration, which was

200–400-μg/L in the influent, decreased only slightly in anoxic units, but speciation of effluent selenium showed that selenate was completely reduced to selenite and other reduced forms. Addition of 10–20 mg/L ferric chloride to the effluent reduced soluble selenium to 7–12-μg/L. Selenium reduction was not inhibited by 2000–4000 mg/L sulfate." Algae not used by denitrifying and selenate-reducing bacteria were fermented to biomethane in unmixed cylindrical digesters where biomethane production averaged 0.16 L g^{-1} VS was introduced.

Chen and Oswald (1998) investigate the thermochemical treatment for algal fermentation in a paper with 28 citations and determine the influence of the thermochemical pretreatment process of algal fermentation on the efficiency with which algal energy is converted microbiologically to the energy in biomethane. The variables studied were pretreatment temperature, duration, concentration, and dosage of sodium hydroxide (NaOH). They find that "pretreatment best efficiency was attained with a temperature of 100°C for 8 h at a concentration 3.7% solids and without NaOH." Compared with untreated algae, pretreatment improved the efficiency of biomethane fermentation a maximum at 33%. An orthogonal square design was selected to determine the mathematical model to describe the effects of algal thermochemical pretreatment and to determine the relative significance among the pretreatment parameters.

Gonzalez-Fernandez et al. (2011) evaluate the anaerobic codigestion of microalgal biomass and swine manure via response surface methodology in a recent paper with 14 citations. They evaluate the codigestion of this residue and microalgal biomass. They find that COD/VS and COD algal supplement presented a significant effect on biomethane yield. Nevertheless, biomethane yield values achieved were not expected. "Highest biomethane yield was exhibited by swine manure as a sole substrate, while algal biomass digestion reported the lowest." They argue that "biomethane production took place in a higher extent on samples with higher proportion of algae." As a result, nitrogen organic mineralization was low for those trials. They assert that even though biomethane production, hence breakage of the cells, was steadily occurring, the need of an algal biomass pretreatment is necessary for the feasibility of this integrated system.

Zeng et al. (2010) investigate the effect of inoculum/substrate ratio (ISR) on biomethane yield and orthophosphate release from anaerobic digestion of *Microcystis* spp. in a recent paper with nine citations. They conduct a batch anaerobic test to evaluate the effects of ISRs on the biomethane yield and orthophosphate release from the anaerobic digestion of *Microcystis* spp. They find an obvious influence on biomethane yield and orthophosphate release by ISR. "The maximum biomethane yield decreased from 140.48 to 94.42 mL/g VS when the ISR decreased from 2.0 to 0.5. The highest maximum biomethane yield calculated from Orskov equation was 153.66 mL/g VS at ISR value of 1.0." They assert that the values of pH, ammonia, and VFAs provided evidence for the appropriate stability of this anaerobic process.

15.3.3 Conclusion

The experimental research on algal biomethane has been one of the most dynamic research areas in recent years. Thirty-one experimental papers in the field of algal biomethane with more than eight citations were located, and the key emerging issues were presented in the decreasing order of the number of citations. These papers give strong hints about the determinants of efficient algal biomethane production and emphasize that marine algae are efficient biomethane feedstocks in comparison with the first- and second-generation biofuel feedstocks.

The experimental research on the algal biomethane focused on the algal type (16 papers); catalyst effects (3 papers); ingredients, algal residues, and stage effects (2 papers each); and other topics relating to algal biomethane production (6 papers).

15.4 Conclusion

The citation classics presented under the two main headings in this chapter confirm the predictions that marine algae have a significant potential to serve as a major solution for the global problems of warming, air pollution, energy security, and food security in the form of algal biomethane.

Further research is recommended for the detailed studies in each topical area presented in this chapter including scientometric studies and citation classic studies to inform key stakeholders about the potential of marine algae for the solution of the global problems of warming, air pollution, energy security, and food security in the form of algal biomethane.

References

Adams, J.M.M., T.A. Toop, I.S. Donnison, and J.A. Gallagher. 2011. Seasonal variation in *Laminaria digitata* and its impact on biochemical conversion routes to biofuels. *Bioresource Technology* 102:9976–9984.

Baltussen, A. and C.H. Kindler. 2004a. Citation classics in anesthetic journals. *Anesthesia & Analgesia* 98:443–451.

Baltussen, A. and C.H. Kindler. 2004b. Citation classics in critical care medicine. *Intensive Care Medicine* 30:902–910.

Bothast, R.J. and M.A. Schlicher. 2005. Biotechnological processes for conversion of corn into ethanol. *Applied Microbiology and Biotechnology* 67:19–25.

Briand, X. and P. Morand. 1997. Anaerobic digestion of *Ulva* sp. 1. Relationship between Ulva composition and methanisation. *Journal of Applied Phycology* 9:511–524.

Bruhn, A., J. Dahl, H.B. Nielsen, L. Nikolaisen, M.B. Rasmussen, S. Markager, B. Olesen, C. Arias, and P.D. Jensen. 2011. Bioenergy potential of *Ulva lactuca*: Biomass yield, methane production and combustion. *Bioresource Technology* 102:2595–2604.

Brune, D.E., T.J. Lundquist, and J.R. Benemann. 2009. Microalgal biomass for greenhouse gas reductions: Potential for replacement of fossil fuels and animal feeds. *Journal of Environmental Engineering—ASCE* 135:1136–1144.

Cantrell, K.B., T. Ducey, K.S. Ro, and P.G. Hunt. 2008. Livestock waste-to-bioenergy generation opportunities. *Bioresource Technology* 99:7941–7953.

Chen, P.H. and W.J. Oswald. 1998. Thermochemical treatment for algal fermentation. *Environment International* 24:889–897.

Chinnasamy, S., A. Bhatnagar, R. Claxton, and K.C. Das. 2010. Biomass and bioenergy production potential of microalgae consortium in open and closed bioreactors using untreated carpet industry effluent as growth medium. *Bioresource Technology* 101:6751–6760.

Chisti, Y. 2007. Biodiesel from microalgae. *Biotechnology Advances* 25:294–306.

Collet, P., A. Helias, L. Lardon, M. Ras, R.A. Goy, and J.P. Steyer. 2011. Life-cycle assessment of microalgae culture coupled to biogas production. *Bioresource Technology* 102:207–14.

De Schamphelaire, L. and W. Verstraete. 2009. Revival of the biological sunlight-to-biogas energy conversion system. *Biotechnology and Bioengineering* 103:296–304.

Demirbas, A. 2007. Progress and recent trends in biofuels. *Progress in Energy and Combustion Science* 33:1–18.

Duan, P.G. and P.E. Savage. 2011. Hydrothermal liquefaction of a microalga with heterogeneous catalysts. *Industrial & Engineering Chemistry Research* 50:52–61.

Dubin, D., A.W. Hafner, and K.A. Arndt. 1993. Citation-classics in clinical dermatological journals—Citation analysis, biomedical journals, and landmark articles, 1945–1990. *Archives of Dermatology* 129:1121–1129.

Ehimen, E.A., S. Connaughton, Z.F. Sun, and G.C. Carrington. 2009. Energy recovery from lipid extracted, transesterified and glycerol codigested microalgae biomass. *Global Change Biology Bioenergy* 1:371–381.

Ehimen, E.A., Z.F. Sun, C.G. Carrington, E.J. Birch, and J.J. Eaton-Rye. 2011. Anaerobic digestion of microalgae residues resulting from the biodiesel production process. *Applied Energy* 88:3454–3463.

Gehanno, J.F., K. Takahashi, S. Darmoni, and J. Weber. 2007. Citation classics in occupational medicine journals. *Scandinavian Journal of Work, Environment & Health* 33:245–251.

Gerhardt, M.B., F.B. Green, R.D. Newman, T.J. Lundquist, R.B. Tresan, and W.J. Oswald. 1991. Removal of selenium using a novel algal bacterial process. *Research Journal of the Water Pollution Control Federation* 63:799–805.

Godfray, H.C.J., J.R. Beddington, I.R. Crute et al. 2010. Food security: The challenge of feeding 9 billion people. *Science* 327:812–818.

Goldemberg, J. 2007. Ethanol for a sustainable energy future. *Science* 315:808–810.

Gonzalez-Fernandez, C., B. Molinuevo-Salces, and M.C. Garcia-Gonzalez. 2011. Evaluation of anaerobic codigestion of microalgal biomass and swine manure via response surface methodology. *Applied Energy* 88:3448–3453.

Green, F.B., L. Bernstone, T.J. Lundquist, J. Muir, R.B. Tresan, and W.J. Oswald. 1995b. Methane fermentation, submerged gas collection, and the fate of carbon in advanced integrated wastewater pond systems. *Water Science and Technology* 31:55–65.

Green, F.B., T.J. Lundquist, and W.J. Oswald. 1995a. Energetics of advanced integrated waste-water pond systems. *Water Science and Technology* 31:9–20.

Grossi, V., P. Blokker, and J.S.S. Damste. 2001. Anaerobic biodegradation of lipids of the marine microalga *Nannochloropsis salina*. *Organic Geochemistry* 32:795–808.

Haiduc, A.G., M. Brandenberger, S. Suquet, F. Vogel, R. Bernier-Latmani, and C. Ludwig. 2009. SunCHem: An integrated process for the hydrothermal production of methane from microalgae and CO_2 mitigation. *Journal of Applied Phycology* 21:529–541.

Harun, R., M. Davidson, M. Doyle, R. Gopiraj, M. Danquah, and G. Forde. 2011. Technoeconomic analysis of an integrated microalgae photobioreactor, biodiesel and biogas production facility. *Biomass & Bioenergy* 35:741–747.

Heaven, S., J. Milledge, and Y. Zhang. 2011. Comments on 'Anaerobic digestion of microalgae as a necessary step to make microalgal biodiesel sustainable'. *Biotechnology Advances* 29:164–167.

Hodgson, B. and P. Paspaliaris. 1996. Melbourne water's wastewater treatment lagoons: Design modifications to reduce odours and enhance nutrient removal. *Water Science and Technology* 33:157–164.

Jacobson, M.Z. 2009. Review of solutions to global warming, air pollution, and energy security. *Energy & Environmental Science* 2:148–173.

Kaddam, K.L. 2002. Environmental implications of power generation via coal-microalgae cofiring *Energy* 27:905–922.

Kapdan, I.K. and F. Kargi. 2006. Bio-hydrogen production from waste materials. *Enzyme and Microbial Technology* 38:569–582.

Konur, O. 2000. Creating enforceable civil rights for disabled students in higher education: An institutional theory perspective. *Disability & Society* 15:1041–1063.

Konur, O. 2002a. Assessment of disabled students in higher education: Current public policy issues. *Assessment and Evaluation in Higher Education* 27:131–152.

Konur, O. 2002b. Access to employment by disabled people in the UK: Is the Disability Discrimination Act working? *International Journal of Discrimination and the Law* 5:247–279.

Konur, O. 2002c. Access to nursing education by disabled students: Rights and duties of nursing programs. *Nursing Education Today* 22:364–374.

Konur, O. 2006a. Participation of children with dyslexia in compulsory education: Current public policy issues. *Dyslexia* 12:51–67.

Konur, O. 2006b. Teaching disabled students in higher education. *Teaching in Higher Education* 11:351–363.

Konur, O. 2007a. A judicial outcome analysis of the Disability Discrimination Act: A windfall for the employers? *Disability & Society* 22:187–204.

Konur, O. 2007b. Computer-assisted teaching and assessment of disabled students in higher education: The interface between academic standards and disability rights. *Journal of Computer Assisted Learning* 23:207–219.

Konur, O. 2011. The scientometric evaluation of the research on the algae and bio-energy. *Applied Energy* 88:3532–3540.

Konur, O. 2012a. 100 citation classics in energy and fuels. *Energy Education Science and Technology Part A—Energy Science and Research* 2012(si):319–332.

Konur, O. 2012b. What have we learned from the citation classics in energy and fuels: A mixed study. *Energy Education Science and Technology Part A* 2012(si):255–268.

Konur, O. 2012c. The gradual improvement of disability rights for the disabled tenants in the UK: The promising road is still ahead. *Social Political Economic and Cultural Research* 4:71–112.

Konur, O. 2012d. Prof. Dr. Ayhan Demirbas' scientometric biography. *Energy Education Science and Technology Part A—Energy Science and Research* 28:727–738.

Konur, O. 2012e. The evaluation of the research on the biofuels: A scientometric approach. *Energy Education Science and Technology Part A—Energy Science and Research* 28:903–916.

Konur, O. 2012f. The evaluation of the research on the biodiesel: A scientometric approach. *Energy Education Science and Technology Part A—Energy Science and Research* 28:1003–1014.

Konur, O. 2012g. The evaluation of the research on the bioethanol: A scientometric approach. *Energy Education Science and Technology Part A—Energy Science and Research* 28:1051–1064.

Konur, O. 2012h. The evaluation of the research on the microbial fuel cells: A scientometric approach. *Energy Education Science and Technology Part A—Energy Science and Research* 29:309–322.

Konur, O. 2012i. The evaluation of the research on the biohydrogen: A scientometric approach. *Energy Education Science and Technology Part A—Energy Science and Research* 29:323–338.

Konur, O. 2012j. The evaluation of the biogas research: A scientometric approach. *Energy Education Science and Technology Part A—Energy Science and Research* 29:1277–1292.

Konur, O. 2012k. The scientometric evaluation of the research on the production of bio-energy from biomass. *Biomass and Bioenergy* 47:504–515.

Konur, O. 2012l. The evaluation of the global energy and fuels research: A scientometric approach. *Energy Education Science and Technology Part A—Energy Science and Research* 30:613–628.

Konur, O. 2012m. The evaluation of the biorefinery research: A scientometric approach. *Energy Education Science and Technology Part A—Energy Science and Research* 2012(si):347–358.

Konur, O. 2012n. The evaluation of the bio-oil research: A scientometric approach. *Energy Education Science and Technology Part A—Energy Science and Research* 2012(si):379–392.

Konur, O. 2012o. What have we learned from the citation classics in energy and fuels: A mixed study. *Energy Education Science and Technology Part A—Energy Science and Research* 2012(si):255–268.

Konur, O. 2012p. The evaluation of the research on the biofuels: A scientometric approach. *Energy Education Science and Technology Part A—Energy Science and Research* 28:903–916.

Konur, O. 2013. What have we learned from the research on the International Financial Reporting Standards (IFRS)? A mixed study. *Energy Education Science and Technology Part D: Social Political Economic and Cultural Research* 5:29–40.

Kosaric, N. and J. Velikonja. 1995. Liquid and gaseous fuels from biotechnology—Challenge and opportunities. *FEMS Microbiology Reviews* 16:111–142.

Lakaniemi, A.M., C.J. Hulatt, D.N. Thomas, O.H. Tuovinen, and J.A. Puhakka. 2011. Biogenic hydrogen and methane production from *Chlorella vulgaris* and *Dunaliella tertiolecta* biomass. *Biotechnology for Biofuels* 4:34.

Lal, R. 2004. Soil carbon sequestration impacts on global climate change and food security. *Science* 304:1623–1627.

Laws, E.A. and J.L. Berning. 1991. Photosynthetic efficiency optimization studies with the macroalga *Gracilaria tikvahiae*—Implications for CO_2 emission control from power-plants. *Bioresource Technology* 37:25–33.

Lynd, L.R., J.H. Cushman, R.J. Nichols, and C.E. Wyman. 1991. Fuel ethanol from cellulosic biomass. *Science* 251:1318–23.

Mairet, F., O. Bernard, M. Ras, L. Lardon, and J.P. Steyer. 2011. Modeling anaerobic digestion of microalgae using ADM1. *Bioresource Technology* 102:6823–6829.

Minowa, T. and S. Sawayama. 1999. A novel microalgal system for energy production with nitrogen cycling. *Fuel* 78:1213–1215. C46.

Moen, E., S. Horn, and K. Ostgaard. 1997a. Alginate degradation during anaerobic digestion of *Laminaria hyperborea* stipes. *Journal of Applied Phycology* 9:157–166.

Moen, E., S. Horn, and K. Ostgaard. 1997b. Biological degradation of *Ascophyllum nodosum*. *Journal of Applied Phycology* 9:347–357.

Morand, P. and X. Briand. 1999. Anaerobic digestion of *Ulva* sp. 2. Study of Ulva degradation and methanisation of liquefaction juices. *Journal of Applied Phycology* 11:165–177.

Mussgnug, J.H., V. Klassen, A. Schluter, and O. Kruse. 2010. Microalgae as substrates for fermentative biogas production in a combined biorefinery concept. *Journal of Biotechnology* 150:51–56.

Nkemka, V.N. and M. Murto. 2010. Evaluation of biogas production from seaweed in batch tests and in UASB reactors combined with the removal of heavy metals. *Journal of Environmental Management* 91:1573–1579.

North, D. 1994. Economic-performance through time. *American Economic Review* 84:359–368.

Ostgaard, K., M. Indergaard, S. Markussen, S.H. Knutsen, and A. Jensen. 1993. Carbohydrate degradation and methane production during fermentation of *Laminaria saccharina* (laminariales, phaeophyceae). *Journal of Applied Phycology* 5:333–342.

Oswald, W.J., F.B. Green, and T.J. Lundquist. 1994. Performance of methane fermentation pits in advanced integrated waste-water pond systems. *Water Science and Technology* 30:287–295.

Paladugu, R., M.S. Chein, S. Gardezi, and L. Wise. 2002. One hundred citation classics in general surgical journals. *World Journal of Surgery* 26:1099–1105.

Pearson, H.W. 1996. Expanding the horizons of pond technology and application in an environmentally conscious world. *Water Science and Technology* 33:1–9.

Ras, M., L. Lardon, S. Bruno, N. Bernet, and J.P. Steyer. 2011. Experimental study on a coupled process of production and anaerobic digestion of *Chlorella vulgaris*. *Bioresource Technology* 102:200–206.

Rawat, I., R.R. Kumar, T. Mutanda, and F. Bux. 2011. Dual role of microalgae: Phycoremediation of domestic wastewater and biomass production for sustainable biofuels production. *Applied Energy* 88:3411–3424.

Sialve, B., N. Bernet, and O. Bernard. 2009. Anaerobic digestion of microalgae as a necessary step to make microalgal biodiesel sustainable. *Biotechnology Advances* 27:409–416.

Singh, J. and S. Cu. 2010 Commercialization potential of microalgae for biofuels production. *Renewable & Sustainable Energy Reviews* 14:2596–2610.

Spolaore, P., C. Joannis-Cassan, E. Duran, and A. Isambert. 2006. Commercial applications of microalgae. *Journal of Bioscience and Bioengineering* 101:87–96.

Stucki, S., F. Vogel, C. Ludwig, A.G. Haiduc, and M. Brandenberger. 2009. Catalytic gasification of algae in supercritical water for biofuel production and carbon capture. *Energy & Environmental Science* 2:535–541.

Ungerfeld, E.M., S.R. Rust, R.J. Burnett, M.T. Yokoyama, and J.K. Wang. 2005. Effects of two lipids on in vitro ruminal methane production. *Animal Feed Science and Technology* 119:179–185.

Vergara-Fernandez, A., G. Vargas, N. Alarcon, and A. Velasco. 2008. Evaluation of marine algae as a source of biogas in a two-stage anaerobic reactor system. *Biomass & Bioenergy* 32:338–344.

Volesky, B. and Z.R. Holan. 1995. Biosorption of heavy-metals. *Biotechnology Progress* 11:235–250.
Wang, B.Y., Q. Liu, N. Wan, and C.Q. Lan. 2008. CO_2 bio-mitigation using microalgae. *Applied Microbiology and Biotechnology* 79:707–718.
Wrigley, N. and S. Matthews. 1986. Citation-classics and citation levels in geography. *Area* 18:185–194.
Wyman, C.E. 1994. Alternative fuels from biomass and their impact on carbon-dioxide accumulation. *Applied Biochemistry and Biotechnology* 45–46:897–915.
Wyman, C.E. and B.J. Goodman. 1993. Biotechnology for production of fuels, chemicals, and materials from biomass. *Applied Biochemistry and Biotechnology* 39:41–59.
Yang, Z.M., R.B. Guo, X.H. Xu, X.L. Fan, and S.J. Luo. 2011. Hydrogen and methane production from lipid-extracted microalgal biomass residues. *International Journal of Hydrogen Energy* 36:3465–3470.
Yen, H.W. and D.E. Brune. 2007. Anaerobic co-digestion of algal sludge and waste paper to produce methane. *Bioresource Technology* 98:130–134.
Yergin, D. 2006. Ensuring energy security. *Foreign Affairs* 85:69–82.
Yuan, X.Z., X.S. Shi, D.L. Zhang, Y.L. Qiu, R.B. Guo, and L.S. Wang. 2011. Biogas production and microcystin biodegradation in anaerobic digestion of blue algae. *Energy & Environmental Science* 4:1511–1515.
Zamalloa, C., N. Boon, and W. Verstraete. 2012. Anaerobic digestibility of *Scenedesmus obliquus* and *Phaeodactylum tricornutum* under mesophilic and thermophilic conditions. *Applied Energy* 92:733–738.
Zamalloa, C., E. Vulsteke, J. Albrecht, and W. Verstraete. 2011. The techno-economic potential of renewable energy through the anaerobic digestion of microalgae. *Bioresource Technology* 102:1149–1158.
Zeng, S.J., X.Z. Yuan, X.S. Shi, and Y.L. Qiu. 2010. Effect of inoculum/substrate ratio on methane yield and orthophosphate release from anaerobic digestion of *Microcystis* spp. *Journal of Hazardous Materials* 178:89–93.

16

Production of Biomethane from Marine Microalgae

Kurniadhi Prabandono and Sarmidi Amin

CONTENTS

16.1 Introduction

One of the most important energy sources of renewable energy in the near future is biomass. And biofuel is one of the products of biomass, which can be used as a substitute for petroleum fuels. Recently, biomass has been considered as an alternative energy source because it is a renewable resource and has environmental consideration. Fuels from biomass have lower emission of sulfur oxide and nitrogen oxide and soot contents than fossil fuels. Because microalgae have higher photosynthetic efficiency, higher biomass production, and faster growth than other biomasses (e.g., trees), they are promising (Calvin and Taylor 1989). Moreover, to avoid competition between microalgae farming and food production, microalgae can be farmed using fresh or marine waters. If fuel is recovered efficiently from microalgae, the microalgae can be used as second-generation biofuels instead of using fossil fuels (Schenk et al. 2008).

Microalgae have been chosen as biofuel resources to produce biomethane, and their uses have grown in recent years (González-Fernández et al. 2012; Guiot and Frigon 2011). Biomethane can be obtained through purification of biogas from household waste, human waste, animal waste, and microalgae. The raw material is processed through anaerobic digestion, or the process by which biological material is broken down by bacteria in the absence of oxygen. Biogas can also be processed using other methods such as gasification and liquefaction.

In the process of anaerobic digestion, symbiotic microorganisms under oxygen-free condition are broken down into a weak acid. At hydrolysis stage, macromolecules (fats, proteins, and carbohydrates) are degraded into simpler compounds by bacteria, and at a later stage or stages of acidification, they are converted into simple compound acids. When acid is formed from degraded organic material, it is formed into biogas, carbon dioxide, and other gas nutrients, and additional cell matter, leaving salts, and refractory organic matter (Wilkie 2008).

In general, biogas consists of 55%–80% methane and 20%–45% CO_2. However, small amounts of other gases such as NH_3, H_2S, and H_2O may be present, depending on the source of the organic matter and the management of the anaerobic digestion process (Figure 16.1) (Ogejo et al. 2009). The typical biogas composition is shown in Table 16.1.

Generation of varying amounts of gas and gas of variable methane contents are caused by different energy content in each component of organic materials. Since microorganisms that are active during anaerobic decomposition use very small amounts of energy for their own growth, the majority of the available energy from the substrate becomes

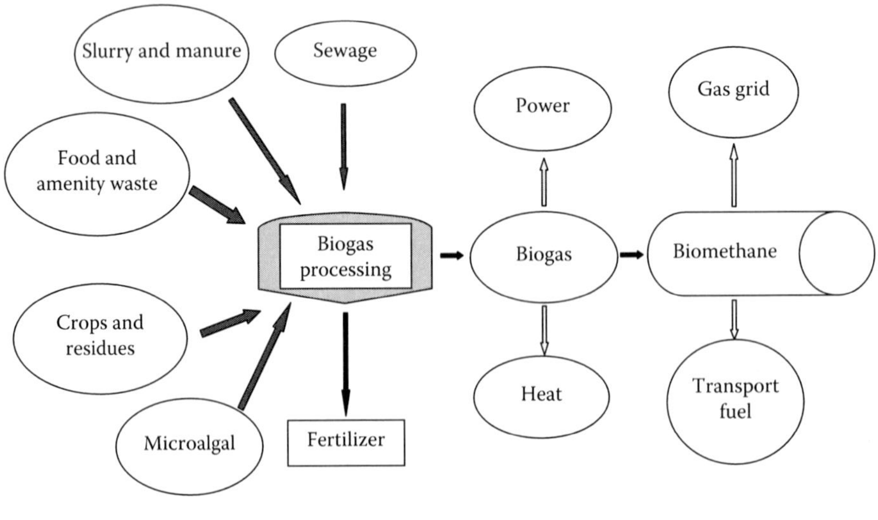

FIGURE 16.1
Flow diagram of biomethane processing.

TABLE 16.1
Biogas Composition

Component	Symbol	Concentration (Vol.)
Methane	CH_4	50%–75%
Carbon dioxide	CO_2	25%–45%
Water vapor	H_2O	2%–7%
Oxygen	O_2	<2%
Nitrogen	N_2	<2%
Ammonia	NH_3	<1%
Hydrogen	H_2	<1%
Sulfide	H_2S	20–20.000 ppm

Source: Alexopoulos, S., *J. Eng. Sci. Technol. Rev.*, 5, 48, 2012.

TABLE 16.2
Theoretical Quantity and Composition of Biogas

	Biogas Formed (m^3/kg VS)	Biogas Composition: CH_4:CO_2 (%)
Carbohydrates	0.38	50:50
Fat	1.0	70:30
Protein	0.53	60:40

Source: Berglund and Borjesson (2003) in (Schnürer, A. and Jarvis, A., Microbiological handbook for biogas plants, Swedish Gas Center Report 2007, 2010, pp. 9–50.)
VS, volatile solids, organic or carbon-containing fraction of total solids.

methane. Table 16.2 shows approximate biogas volumes and methane contents that can be formed from carbohydrates, protein, and fat. Using these values for a mixed material, it is possible to make a theoretical calculation of the amount of gas that can be formed (Schnürer and Jarvis 2010).

According to personal experience with household biogas processing, one house requires a minimum of four cattle such as cows or buffalos that will produce as much dung each day or agricultural waste, food processing waste, and others. According to Widodo's experience using of 10 cattle dung (about 20 kg/day/head or 200 kg dung per day) can produce biogas of 6 m^3 per day (Widodo et al. 2009). According to Sorathia et al. (2012), cattle dung can produce 0.037 m^3 of biogas per kg of cow dung; 0.07 m^3 (winter) and 0.1 m^3 (summer) per kg of pig dung; and 0.07 m^3 (winter) and 0.16 m^3 (summer) per kg of poultry dropping. The calorific value of gas is 21,000–23,000 kJ/m^3 or about 25,200 kJ/kg of gas. According to FAO (1996), cattle (cow and buffalo) dung can produce 0.023–0.040 m^3 of biogas per kg of dung, 0.04–0.059 m^3 of biogas per kg of pig dung, and 0.065–0.116 m^3 of biogas per kg of poultry (chicken) dung.

Animal manures may be adulterated or contaminated with other animal products, such as wool (shoddy and other hair), feathers, blood, and bone. Livestock feed can be mixed with the manure due to spillage. For example, chickens are often fed meat and bone meal, an animal product, which can end up becoming mixed with chicken litter.

A major limiting factor for the future growth of biomethane production as an example from plant sources is the availability of photosynthetically grown biomass. Currently, a 500 kW biomethane plant requires approx. 10,000–12,000 tons of biomass feedstock per year with maize currently being the major crop plant feedstock. Using cereals and sunflowers, typical yields between 2000 and 4500 m^3 biomethane per hectare per year have been reported (Amon et al. 2007; Weiland 2003). Yields from maize are higher and vary in dependence of the species and the time of harvesting between 5,700 and 12,400 m^3 of biomethane per year per hectare (Amon et al. 2007; Prochnow et al. 2005; Weiland 2003). For certain grass species like ryegrass, yields of up to 4000 m^3 of biomethane per year per hectare were reported (Weiland 2003). Microalgae are a major focus of interest as the efficiency of biomass production per hectare is estimated to reach 5–30 times that of crop plants (Sheehan et al. 1998). The relatively high lipid, starch and protein contents, and the absence of lignin that cannot be fermented easily make microalgae an ideal candidate for efficient biomethane production by fermentation in biogas plants.

16.1.1 Microalgae

Algae are simple organisms that are mainly aquatic and microscopic. There are two main populations of algae: filamentous and phytoplankton algae. They are categorized into four main classes: diatoms, green algae, blue-green algae, and golden algae. Algal organisms are photosynthetic macroalgae or microalgae growing in aquatic environments. Macroalgae are classified into three broad groups based on their pigmentation: (1) brown seaweed (*Phaeophyceae*), (2) red seaweed (*Rhodophyceae*), and (3) green seaweed (*Chlorophyceae*). Microalgae are unicellular photosynthetic microorganisms living in saline or freshwater environments that convert sunlight, water, and carbon dioxide to algal biomass (Ozkurt 2009). Biologists have categorized microalgae in a variety of classes, mainly distinguished by their pigmentation, life cycle, and basic cellular structure. The three most important classes of microalgae in terms of abundance are the diatoms (*Bacillariophyceae*), the green algae (*Chlorophyceae*), and the golden algae (*Chrysophyceae*). The cyanobacteria (blue-green algae) (*Cyanophyceae*) are also referred to as microalgae. This applies, for example, to Spirulina (*Arthrospira platensis* and *Arthrospira maxima*). Diatoms are the dominant life form in phytoplankton and probably represent the largest group of biomass producers on earth. Microalgae cultivation using sunlight energy can be carried out in open or covered ponds or closed photobioreactors (PBRs), based on tubular, flat plate, or other designs.

The concept of using algae as a fuel was first proposed by Meier in 1955 for the production of methane gas from the carbohydrate fractional cells. This idea was further developed by Oswald and Golueke in 1960, who introduced conceptual technoeconomic engineering analysis of digesting microalgal biomass grown in large raceway ponds to produce methane gas (Demirbas and Demirbas 2010).

Microalgae also have potential in the production of gaseous biofuel. According to Benemann, hydrogen can be produced via a number of photobiological processes either using hydrogenase through direct or indirect photolysis or using nitrogenase (in Liewellyn and Skill 2008). Microalgae also can be converted into synthesis gas by means of partial oxidation with air, oxygen, and/or steam at high temperature, typically in the range 800°C–900°C (Amin 2009).

Methane potential is the volume of methane biogas produced during anaerobic degradation in the presence of bacteria of a sample initially introduced and expressed under normal conditions of temperature and pressure (0°C, 1013 hPa) (Anonymous, Clarke Energy). The anaerobic digestion process indicates that the CH_4 content of anaerobic digesters is typically 55%–65% and cannot be much higher than 70%, even if the substrate is all fats

TABLE 16.3

Ranges of Biochemical Methane Potential Yield

Sample	Methane Yield (at Normal Condition) (L/g) of VS
Kelp (*Macrocystis*)	0.39–0.41
Sargassum	0.26–0.38
Laminaria	0.26–0.28
Hydrilla verticillata	58%–60% (Narayanaswami et al. 1986)

and vegetable oils. However, some standard anaerobic digesters have produced biogas with a CH_4 content higher than would be expected based on the anaerobic digestion process alone. Biogas methane contents of 65%–80% appear to be the result of absorption of excess CO_2 in the digester effluent. Higher CH_4 content than this is not likely, as it is not possible for digester effluents to absorb the additional CO_2 that would be needed to produce higher methane biogas (Krich et al. 2005). Table 16.3 gives ranges of methane potential yield for marine energy crops (Wilkie 2008).

Some macroalgae species like *Macrocystis pyrifera* and genera such as *Sargassum, Laminaria, Ascophyllum, Ulva, Cladophora, Chaetomorpha,* and *Gracilaria* have been explored as potential methane sources (Demirbas and Demirbas 2010).

16.1.2 Cultivation of Microalgae

Microalgae can be cultivated on land in large ponds, or in enclosed photobioreactors (PBRs), using enriched CO_2. Algae can reproduce very rapidly, faster than any other plants. There are three basic important components for microalgae cultivation: light, CO_2, and nutrients. The light comes from sunlight for outdoor culture systems or fluorescent lamps for indoor. Increasing the light intensity usually means better growth and faster division of algal cells. CO_2 can come in the form of flue gases from power plants or be obtained from other fossil fuel combustion and biological processes. Nutrients are the inorganic salts required for plant growth. The seawater containing algae must be clean from contaminants, for example, by using a filtration process.

The other important factor is water temperature, because most types of algae grow well at temperatures ranging from 17°C to 22°C. The lower temperatures will not usually kill the algae but will reduce their growth rate, and above 27°C, most types of algae will die (Laing 1991).

16.1.2.1 Open Ponds

Algae can be cultured in open ponds (such as raceway-type ponds and lakes) and PBRs. Raceway ponds and lakes are less expensive, but they are highly vulnerable to contamination by other microorganisms, such as other algal species or bacteria. Besides that, open systems also do not offer control over temperature and lighting. The growing season is largely dependent on location, and aside from tropical areas, it is limited to the warmer months.

The simple open algae cultivation systems are shallow, nonstirred ponds. The necessary carbon for their growth is received through atmospheric CO_2. The CO_2 transfer from the air to the water through naturally occurring dissolution restricts the significant growth of algae resulting in a low crop yield. Another disadvantage of these systems is the slow transfer of nutrients across the whole mass of the crop. Another way of transferring CO_2 to the pond water is to inject the gas at or near the bottom of a sump which spans all or part of the channel (Weissman and Goebel 1987).

Open ponds are relatively cheap to build compared with bioreactors, but what we are seeing is a real need for cost-effective PBRs that can address the capital expense issues while offering efficient cultivation in large volumes (Austin 2013). Another advantage of closed systems is that they open up sunny, dry areas such as the Southwest to biofuel production. Open ponds are unlikely to work in the Southwest because the water loss is going to be enormous (Austin 2013). Open ponds usually use paddle wheels for the circulation of the water.

Open pond refers to a simple open tank or natural pond. Algae are grown in suspension with additional fertilizer (Singh 2011). A raceway pond is made of a closed loop recirculation channel that is typically about 0.3 m deep. Mixing and circulation are produced by a paddlewheel (Christi 2007). The ponds are kept shallow because of the need to keep the algae exposed to sunlight (Demirbas 2010). These ponds are usually not more than 30 cm deep, and the water with nutrients (Janssen 2002).

16.1.2.2 Closed Bioreactors

The open ponds are highly vulnerable to contamination, thus the cultivators usually choose a closed system for monocultures such as in a PBR. A PBR is a bioreactor which incorporates a light source. Tubular PBRs consist of transparent tubes that are made of flexible plastic or glass. Tubes can be arranged vertically, horizontally, inclined, helical, or in a horizontal thin panel design. Tubes are generally placed in parallel to each other or

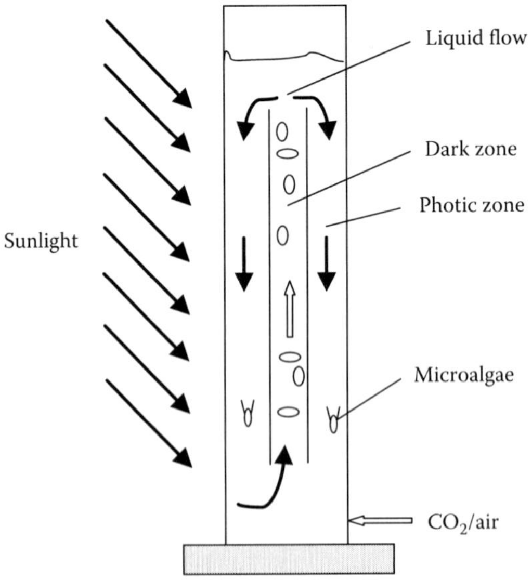

FIGURE 16.2
Outdoor air-lift PBR.

flat above the ground to maximize the illumination surface to volume ratio of the reactor (Demirbas 2010). There are two types of closed systems, that is, flat panel PBR and tubular PBR with air-lift column. Figure 16.2 shows an outdoor air lift with internal draught tube and transparent wall. Only the outer surface is exposed to sunlight, the so-called photic zone (Janssen 2002).

16.1.2.3 Open Ponds versus Closed Bioreactors

One of the major issues in the cultivation of microalgae is light limitation. The effective photosynthetic zone is within five centimeters of the surface of a pond. These small volumes allow light to penetrate better, but the problem is that it may lead to biofouling (the attachment of organisms to a surface in contact with water for a period of time) and the cost of pumping the algae around through the small volumes increases (Austin Biomass Magazine).

Although the problems of predators and weeds have been solved with bioreactors, such closed systems used to grow algae for other purposes have problems with virus susceptibility and/or bacteria attacks, which can take the whole system down in a matter of hours (Austin Biomass Magazine).

16.2 Environmental Factors of Methane Production

16.2.1 Temperature

Producing methane is strongly affected by the ambient temperature. While biogas fermentation can occur at temperature ranging from 0°C to 70°C, in Tatedo's website, the optimum temperature is 35°C (see Table 16.4) (Anonymous 2013). At that temperature, the organism will grow fastest and work most efficiently, but this will vary among species. Negligible temperature effect on methane production was reported by De Schamphelaire and Verstraete in Gonzales-Fernandes et al. (2012); the increase of temperature from 34° to 41°C did not have any effect on the anaerobic digestion of *Chlorella* ecosystem regardless of hydraulic retention time (14 and 25 days). Therefore, the increase in temperature in the mesophilic range did not favor the activity of degrading microorganisms.

TABLE 16.4

Biogas Yield at Various Temperatures

Temperature (°C)	Biogas Yield (m^3/Ton of Dung/Day)
15	0.150
20	0.300
25	0.600
30	1.000
35	2.000
40	0.700
45	0.320

TABLE 16.5
Process Temperature and Retention Time

Thermal Stage	Process Temperatures (°C)	Retention Time (Days)
Psychrophilic	<20	70–80
Mesophilic	30–42	30–40
Thermophilic	43–55	15–20

The temperature can be divided into three temperature ranges, that is, psychrophilic (below 25°C), mesophilic (25°C–45°C), and thermophilic (45°C–70°C). The relation between process temperature and retention time can be seen in Table 16.5 (Al Seadi et al. 2008). Other researchers called that temperature above 65°C as extreme thermophilic or hyperthermophilic (Schnürer and Jarvis 2010).

Many modern biogas plants operate at a thermophilic process temperature. This temperature provides many advantages compared to mesophilic and psychrophilic processes (Al Seadi et al. 2008):

- Effective destruction of pathogens
- Higher grow rate of methanogenic bacteria at higher temperature
- Reduced retention time, making the process faster and more efficient
- Improved digestibility and availability of substrate utilization
- Better possibility for separating liquid and solid fractions

The thermophilic process has also some disadvantages (Al Seadi et al. 2008):

- Larger degree of imbalance
- Larger energy demand due to high temperature
- Higher risk of ammonia inhibition. Ammonia toxicity increases with increasing temperature and can be relieved by decreasing the temperature; however, when decreasing to 50°C or below, the growth rate of the thermophilic microorganisms will drop drastically.

16.2.2 pH Value

Experience shows that methane formation takes place within a relatively narrow pH interval about 5.5–8.5, with an optimum interval between 7.0 and 8.0 for most methanogens (Al Seadi et al. 2008). Low pH will produce low yields of biogas and vice versa. Table 16.6 shows the correlation between pH and biogas yield (Tatedo 2013). According to Al Seadi et al. (2008), the optimum pH interval for mesophilic digestion is between 6.5 and 8.0, and the process is severely inhibited if the pH value decreases below 6.0 or rises above 8.3.

TABLE 16.6
pH and Biogas Production

	From	5	6	7	8	9	10
pH Value	**To**	**6**	**7**	**7**	**7.5**	**7**	**7**
Biogas yield		12.7	14.8	22.5	24.6	17.8	10.2

The value of pH can be increased by ammonia, produced during degradation of protein, or by the presence of ammonia in the feed stream, while accumulation of volatile fatty acids (VTAs) decreases the pH value (Al Seadi et al. 2008).

16.2.3 Carbon-to-Nitrogen Ratio

Carbon and nitrogen are the two most important nutrients for bacteria. Normally, fermentative bacteria will require 30 times more carbon than nitrogen. The optimum carbon–nitrogen (C/N) ratio is 30/1. Therefore, pig/cattle manure is more suitable for biogas (Tatedo 2013). If the C/N ratio is very high, nitrogen will be consumed rapidly by methanogens to meet their protein requirement and will no longer react on the leftover carbon content of the material. As a result, gas production will be low. On the other hand, if the C/N ratio is low, nitrogen will be liberated and accumulated in the form of ammonia (NH_4); NH_4 will increase the pH value of the content in the digester. A pH higher than 8.5 will start showing a toxic effect on methanogen population (FAO 1996). If the biogas feedstock has a low C/N ratio, increasing can be done by adding raw materials that have a higher C/N ratio.

16.2.4 Total Solid Concentration

Total solid (TS) concentration is a measure of the dilution ratio of the input material or ratio of the weight of the remaining material after drying at 105°C (to constant weight) and the original weight. A lower TS concentration will produce lower yield of biogas, but when TS concentration values exceed the optimal point, the yield also decreases. The optimum dilution ratio for cattle manure is one part of manure for five part of water (Tatedo 2013). Sorathia recommended that the proportion is cow dung + solid waste 1:1 by weight and forming to about 10% of solid content and 90% of water.

16.2.5 Retention Time

Retention time is the total time taken by the material to travel from the inlet to the outlet of the digester or reactor. Material is fed daily into the digester, and the fermented materials are pushed toward the system outlet. The complete digestion and gas extraction period is 50 days, but for an optimal cost–benefit ratio, retention time is 40 days (Tatedo 2013). According to Sorathia et al. (2012), the period ranges from 35 to 50 days depending upon the climatic condition and location of the digester. But according to Goeluke in Fernández (2012), a retention time longer than 30 days did not result in any methane conversion improvement. And according to Al Seadi et al. (2008), retention time is 70–80 days for a psychrophilic process temperature.

16.2.6 Toxicity

Toxics such as pesticides, ammonia, detergents, soap, heavy metal, and rain water influence the process; therefore, they are not allowed to be fed into the digester. Table 16.7 shows some harmful materials (Sorathia et al. 2012).

Stirring or agitation of the contents of the digester is not essential, but it is always advantageous, because if it is not stirred, slurry will tend to settle out and form a hard scum on the surface, which will prevent the release of biogas.

TABLE 16.7
Harmful Material in Biogas Digester

Harmful Material	Concentration
Sulfate (SO_4)	5000 ppm
Sodium Chloride (NaCl)	40,000 ppm
Copper (Cu)	100 mg/L
Chromium (Cr)	200 mg/L
Nickel (Ni)	200–500 mg/L
Cyanide (CN)	<25 mg/L
ABS (detergent compound)	20–40 ppm
Ammonia (NH_3)	1500–3000 mg/L
Sodium (Na)	3500–5500 mg/L
Potassium (K)	2500–4500 mg/L
Calcium (Ca)	2500–4500 mg/L
Magnesium (Mg)	1000–1500 mg/L

The use of ultrasound has proven to be successful at improving the disintegration and anaerobic biodegradability of *Chlorella vulgaris* (Park et al. 2013). Ultrasonic pretreatment in the range of 5–200 mJ/mL was applied to waste microalgal biomass, which was then used for batch digestion. Ultrasound techniques were successful to improve the solubility of the substrate. Disintegration reached up to 70% at 200 J/mL of energy.

16.3 Biogas Processing

There are some processes that can be used for producing biogas, that is, gasification, liquefaction, pyrolysis, and anaerobic digestion. In the process of anaerobic digestion, aside from the energy product, by-products such as sludge in the form of solids and liquids will be obtained.

The energy conversion reaction of biomass can be classified as biochemical, thermochemical, or direct combustion (Satin 2008; Tsukahara and Sawayama 2005). Biochemical conversion can be further subdivided into fermentation, anaerobic digestion, bioelectrochemical fuel cells, and other fuel-producing processes utilizing the metabolism of organisms. Thermochemical conversion can be subdivided into gasification, pyrolysis, and liquefaction. It is well known that microalgae have high water content (80%–90%) (Patil et al. 2008); therefore, not all energy conversion processes of biomass can be applied to microalgae. For example, direct combustion of microalgae is feasible only for biomass with content below 50%. High-moisture-content biomass is better suited to biological conversion processes (McKendry 2003).

16.3.1 Gasification

Gasification is the conversion of biomass into combustible gases such as CH_4, H_2, CO_2, and ammonia. The product gas has low calorific value (about 4–6 MJ/N m^3), and it can be burned directly or used as a fuel for gas engines and gas turbines or can be used

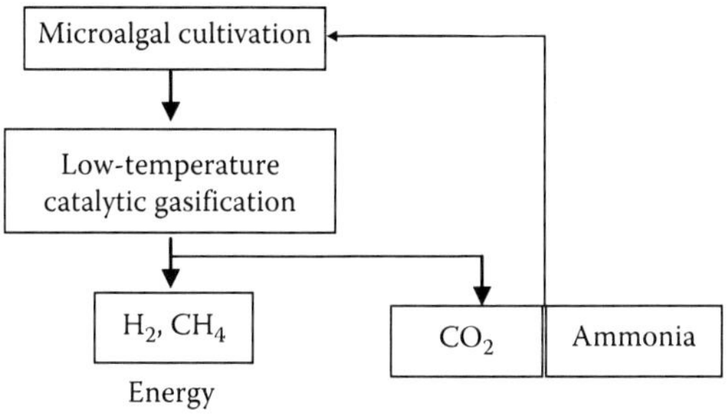

FIGURE 16.3
Flow diagram of microalgal system for fuel production by gasification.

as a feedstock (synthesis gas or syngas) in the production of chemicals (e.g., methanol) (McKendry 2003). Although syngas has a lower heating value than natural gas, it can still be used in highly efficient combined cycle electric power plants, or it can be used to make many products including ammonia fertilizers, methanol-derived chemicals, and clean burning synthetic fuels.

Gasification is a term that describes a chemical process by which carbonaceous materials (hydrocarbon) are converted into syngas by means of partial oxidation with air, oxygen, and/or steam at high temperature, typically in the range of 800°C–900°C. A flow diagram of a microalgal system for fuel production by low-temperature catalytic gasification of biomass is shown in Figure 16.3 (Amin 2009; Tsukahara and Sawayama 2005). A novel energy production system using microalgae with nitrogen cycling combined with low-temperature catalytic gasification of the microalgae has been proposed (Minowa and Sawayama 1999). Elliot and Sealock (1999) have also developed a low-temperature catalytic gasification of biomass with high moisture content. Biomass with high moisture is gasified directly to methane-rich fuel gas without drying. In addition, nitrogen in biomass is converted into ammonia during reaction (Minowa and Sawayama 1999).

16.3.2 Liquefaction

Microalgal cell precipitates derived from centrifugation, which are of high moisture content, are thus good raw materials for liquefaction (FAO 2007). Direct hydrothermal liquefaction in subcritical water condition is a technology that can be employed to convert wet biomass material into liquid fuel (Patil et al. 2008).

The separation scheme is presented in Figure 16.4 (Amin 2009; FAO 2007; Itoh et al. 1994; Minowa and Sawayama 1999; Minowa et al. 1995; Murakami et al. 1990; Yang et al. 2004). The liquefaction is performed in an aqueous solution of alkali or alkaline earth salt at about 300°C and 10 MPa without a reducing gas such as hydrogen and/or carbon monoxide (Minowa et al. 1995).

Liquefaction can be performed by using a stainless steel autoclave with mechanical mixing. The autoclave is charged with algal cells, after which nitrogen is introduced to purge the residual air. The reaction is initiated by heating the autoclave to a fixed temperature and by elevating nitrogen pressure. The temperature is maintained constant for a 5–6 min period, after which it is cooled with an electric fan.

The reaction is extracted with dichloromethane in order to separate the oil fraction. The dichloromethane extract is filtered from the reaction mixture, after which the residual

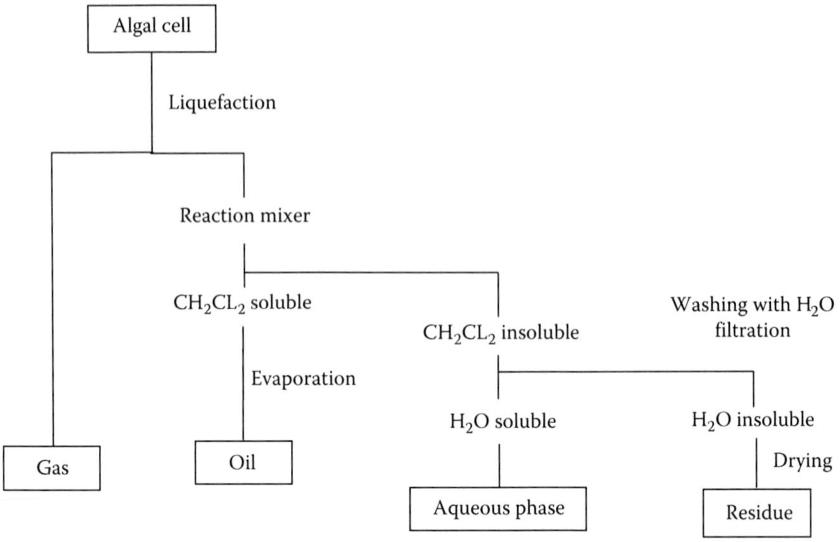

FIGURE 16.4
Separation scheme for liquefaction of microalgal cells.

dichloromethane is filtered and evaporated at 35°C under reduced pressure, yielding a dark brown viscous material (hereafter referred to as the oil). The aqueous phase resulting from dichloromethane extraction (insoluble fraction) is washed with water and filtered from the dichloromethane insoluble (FAO 2007; Minowa et al. 1995).

16.3.3 Fast Pyrolysis

Pyrolysis is the conversion of biomass into biofuel, charcoal, and gaseous fraction by heating the biomass in the absence of air to around 500°C (McKendry 2003; Miao et al. 2004) or by heating in the presence of a catalyst (Agarwal 2007) at high heating rate (103–104 K/s) and with short gas residence time to crack into short-chain molecules and then being cooled to liquid rapidly (Qi et al. 2007). Previous studies used slow pyrolysis processes; they were performing at a low heating rate and long residence time. The longer residence time can cause secondary cracking of the primary products, reducing yield and adversely affecting the biofuel properties. In addition, a low heating rate and long residence time may increase the energy input (Agarwal 2007). In recent years, fast pyrolysis processes for biomass have attracted a great deal of attention for maximizing liquid yields, and many researches have been performed (Miao et al. 2004).

The advantage of fast pyrolysis is that it can directly produce a liquid fuel (Bridgwater and Peacocke 2000) and also produce biogas. If flash pyrolysis is used, the conversion of biomass into biocrude with efficiency of up to 80% is enabled. A conceptual fluidized-bed fast pyrolysis system is shown in Figure 16.5 (Bridgwater and Peacocke 2000). Since microalgae usually have a high moisture content, a drying process requires much heating energy (Yang et al. 2004). Algae are subjected to pyrolysis in the fluidized-bed reactor. The result of the reaction then flows to a cyclone and is separated into char, biofuel, and gas. The resultant gas can be used for drying the raw material, or for heating of the process, or as a biomethane source.

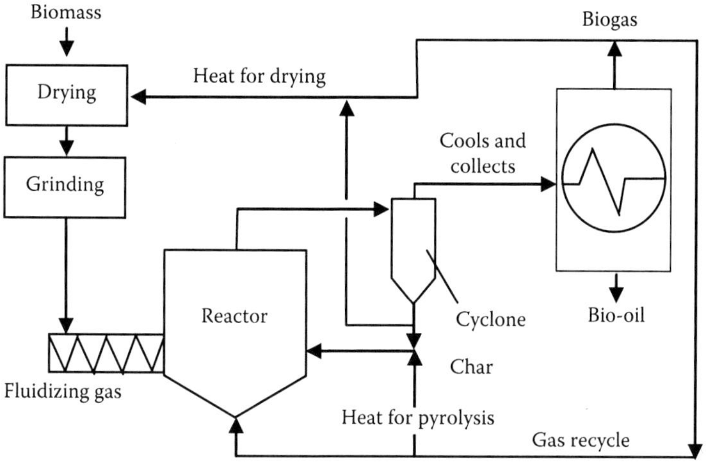

FIGURE 16.5
Fast pyrolysis principles.

16.3.4 Anaerobic Digestion

An anaerobic organism or anaerobe is any organism that does not require oxygen for growth. It may react negatively or even die if oxygen is present, which means that it can perform its bodily functions better in the absence of oxygen. Anaerobic digestion is a collection of processes by which microorganisms break down biodegradable material in the absence of oxygen.

Digestion of organic compounds such as carbohydrate, proteins, and lipids, occurs within microorganisms in a process that breaks the complex molecules into simpler compounds. There are two types of digestion processes: aerobic and anaerobic digestion (Suyog 2011). The digestion process occurring in the presence of O_2 is called aerobic digestion, while the digestion process occurring in the absence of O_2 is called anaerobic digestion.

Anaerobic digestion is an anaerobic process of degrading waste and biodegradable material and it produces fertilizer and biogas. Anaerobic digestion can occur at mesophilic (35°C–45°C) or thermophilic (50°C) temperatures. Both types of digestion typically require supplementary sources of heat to reach their optimal temperature (Anonymous, Clarke Energy). It is a natural process that takes place in the absence of air. It involves biochemical decomposition of complex organic material by biochemical processes: hydrolysis, acidification, and methanogenesis.

Hydrolysis is the first step of organic matter enzymolyzed externally by extracellular enzymes, cellulose, amylase, protease, and lipase. Bacteria decompose long chains of complex organic matter such as carbohydrates, proteins, and lipids into small chains. During hydrolysis, complex organic matters (polymers) are converted into glucose, glycerol, purines, and pyridines (Al Seadi et al. 2008). For example, polysaccharides are converted into monosaccharides. Proteins are split into peptides and amino acids (Sorathia et al. 2012). In the first step, complicated molecules like polymers, fat, protein, and carbohydrates are converted into monomers, fatty acids, amino acids, and saccharides (Alexopoulos 2012).

The second step is acidification, converting the intermediates of fermenting bacteria into acetic acid, hydrogen, and carbon dioxide. The acidification step has two parts: acidogenesis and acetogenesis. During acidogenesis, the product of the first step is converted by acidogenic (fermentative) bacteria into methanogenic substrates. Product from acidogenesis is converted into methanogenic.

The third step is decomposing compounds having low molecular weight. They utilize hydrogen, carbon dioxide, and acetic acid to form methane and carbon dioxide, with other small traces of H_2S, H_2, and N_2 (Sorathia et al. 2012). According to Al Seadi et al. (2008), 70% of the formed methane originates from acetate, while the remaining 30% is produced from conversion of hydrogen and carbon dioxide. Figures 16.5 and 16.6 are steps involved in the breakdown of organic material to produce biogas, according to Ogejo et al. (2009) and Al Seadi et al. (2008).

There are two types of processing for anaerobic fermentation: continuous and batch. In the continuous process, the raw material is charged into a digester, and the same volume of the fermented material overflows from it (Figure 16.7).

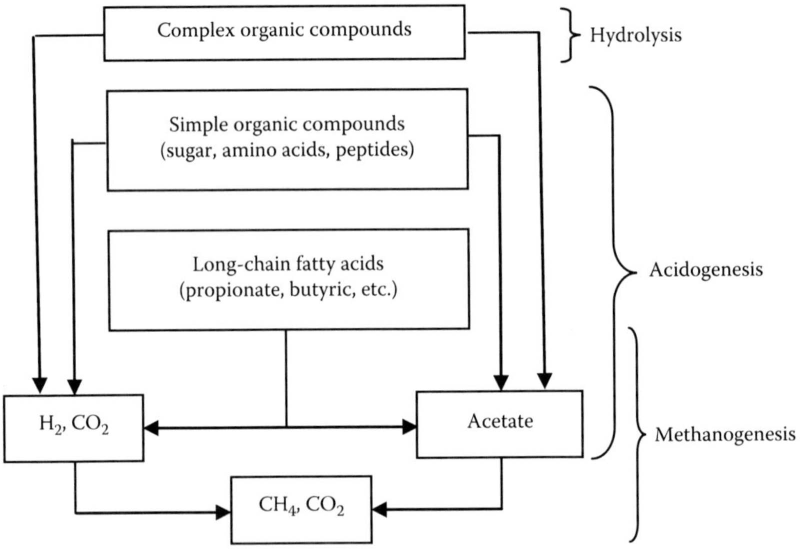

FIGURE 16.6
Steps involved in the breakdown of organic material to produce biogas. (From Ogejo, J.A. et al., Biomethane technology. *Virginia Cooperative Extension*, pp. 442–881.)

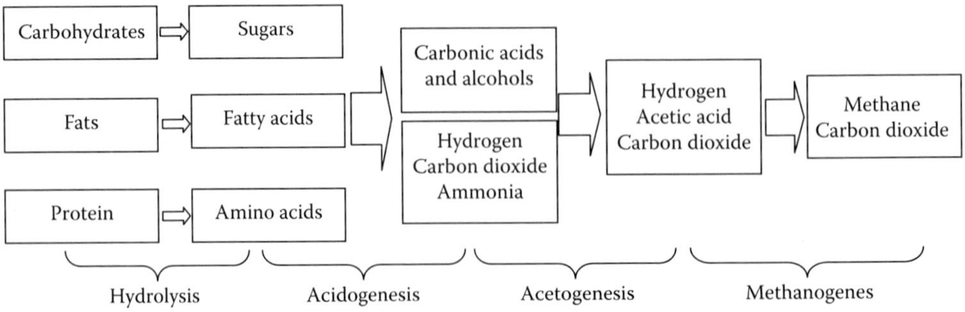

FIGURE 16.7
Steps involved in the breakdown of organic material to produce biogas. (From Al Seadi, T. et al., More about anaerobic digestion, in: Al Seadi, T. ed., *Biogas Handbook*, University of Southern Denmark Esbjerg, Odense, Denmark, pp. 21–29, 2008.)

16.3.4.1 Digester

There are three main types of digester used for the production of biogas: (1) the fixed-dome type, (2) the floating gas holder type, and (3) the plug-flow type. Other researchers also use the balloon-type digester. The balloon plant consists of a heat-sealed plastic or rubber bag, and it is used as a combined digester and gas holder. The gas is stored in the upper part of the balloon. Other digester types such as Deenbandhu model, bag digester, and upflow anaerobic sludge blanket are shown in FAO (1996).

16.3.4.1.1 Fixed-Dome Digester

The fixed-dome digester (Figure 16.8) is made of brick and cement structure, consisting of mixing tank, inlet chamber, digester, outlet chamber, and overflow tank. In this design, the fermentation and gas holder are combined as one unit. The gas holder is placed on top of the digester to collect the produced biogas from digester.

A complete mix digester is a controlled temperature, constant volume, mechanical mixed unit designed to process slurry manure with a solid concentration from 2% to 10% (Ogejo et al. 2009). Biomass as raw material is mixed with water in the mixing tanks, and then the slurry is fed into the digester and left unused for about 1–2 months. During that time, an anaerobic bacterium present in the slurry decomposes. As a result of anaerobic decomposition, biogas is formed and collected in the dome of digester. The gas control valve is opened when a supply of biogas is required.

16.3.4.1.2 Floating Drum Digester

Biomass as raw material is mixed with water in the mixing tanks, and then the slurry is fed into the digester and left unused for about 1–2 months. During that time, an anaerobic bacterium present in the slurry decomposes. As a result of anaerobic decomposition, biogas is formed. Biogas being lighter rises up and is collected in the gas holder. The gas holder now starts moving up. The digester chamber is made of brick masonry and cement

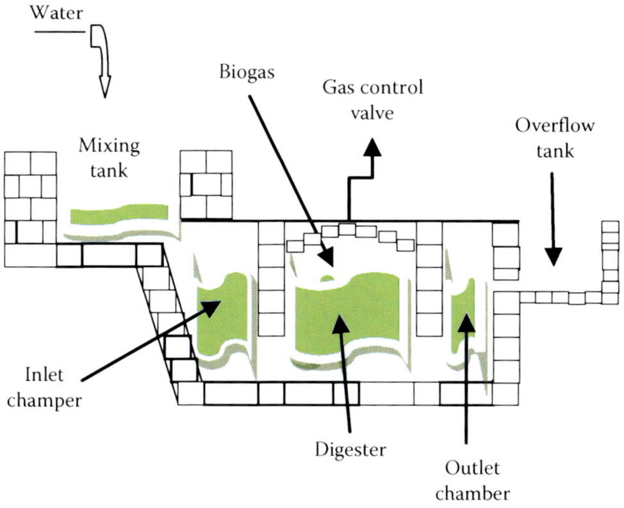

FIGURE 16.8
Fixed dome digester.

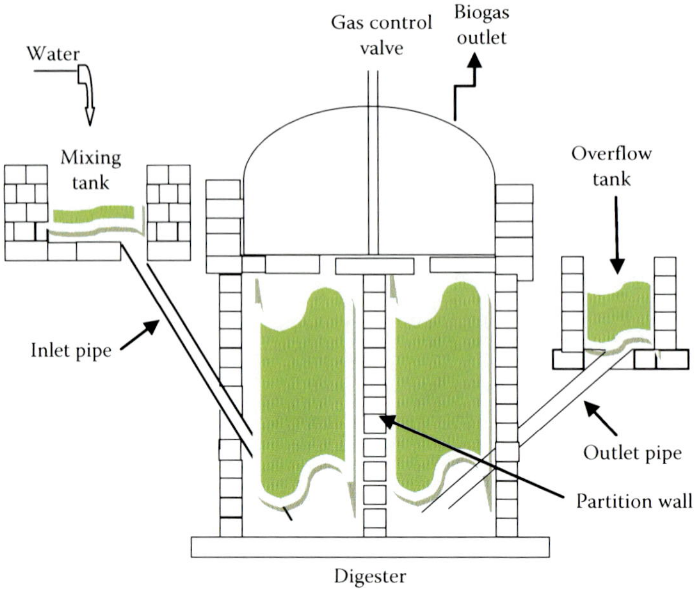

FIGURE 16.9
Floating gas holder digester.

mortar, and the gas holder is a steel drum; thus, there are two separate structures for gas production and collection. The floating drum digester (Figure 16.9) produces biogas at a constant pressure with variable volume, but the gas volume cannot rise up beyond a certain level. The control valve is opened when a supply of biogas is required.

16.3.4.1.3 Plug-Flow Digester

The disadvantage of the fixed-dome and floating drum model is that once installed, they are difficult to move. Hence, portable models built over the ground called tubular or plug-flow digester (Figure 16.10) were developed (Rajendran et al. 2012).

According to Ogejo et al. (2009), the plug-flow digester works best for dairy manure with 11%–14% TS. In this digester, raw material enters at one end of the reactor and exits at the opposite end. Manure is added daily, and an equal volume of manure will be forced out at the other end.

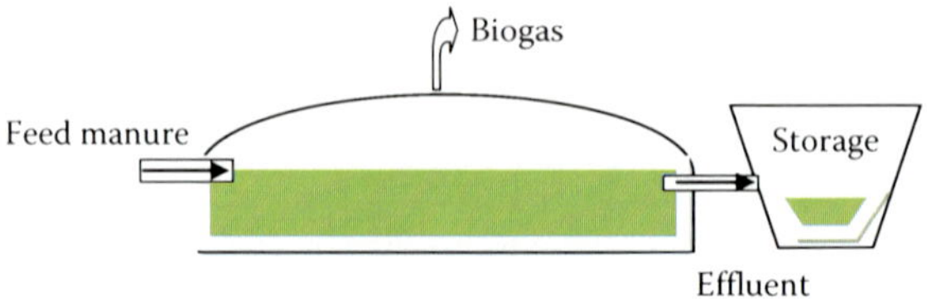

FIGURE 16.10
Plug flow digester.

16.4 Transformation of Biogas to Biomethane

Raw biogas consists mainly of CH_4 and contaminant such as CO_2, H_2O, H_2S, NH_3, oxygen, nitrogen, siloxanes, and particles. Table 16.1 illustrates the biogas composition when the raw gas leaves the fermenter. Contaminants should be removed from biogas. An illustration of the required separating and polishing step from the fermenter to CH_4 is presented in Figure 16.11 (Scholz et al. 2013). The most widely used technologies for biogas into biomethane or biogas upgrading are the following.

Depending on the production process and type of organic matter used, biogas requires treatment to remove the contaminants. Biogas will become biomethane and can be used in vehicle engines. H_2S should be removed because it contains toxins and also causes corrosion. When H_2S is burned in a gas engine, SO_2 or SO_3 is formed and can condense with water and become H_2SO_3 or H_2O_4.

Processes for removal of H_2S include activated carbon filters, low level oxygen dosing into digester heat space (typically <1%), external biological scrubber towers, and ferric chloride dosing into the digester (Anonymous, Clarke Energy).

CO_2 must also be removed, if the biogas is to be used as vehicle fuel. H_2O must be also removed because it can be corrode and increase the ignition point. Water can be removed from the gas by using gas dehumidification (drying) units or by using ground tube dewatering (Anonymous, Clarke Energy).

In some cases, biogas contains siloxanes that are formed from the anaerobic decomposition of materials commonly found in soaps and detergents. During the combustion process, silicon is released and can combine with free oxygen or various other elements in the combustion gas. Deposits are formed containing mostly silica or silicates. These deposits must be removed by chemical or mechanical means (Anonymous, Clarke Energy).

16.4.1 Pressure Swing Adsorption

The pressure swing adsorption (PSA) process primarily uses kinetic effect to separate undesirable gases from the raw material. Activated carbon, zeolites, or hydrocarbon molecular sieve generally act as absorbers (Biogas Association 2012). PSA is a technology used to separate certain components and has a reduced level of CO_2. The main disadvantage is that H_2S content in the biogas needs to be removed prior to PSA (de Hullu et al. 2008), because the adsorption material absorbs H_2S irreversibly and it is thus poisoned by H_2S.

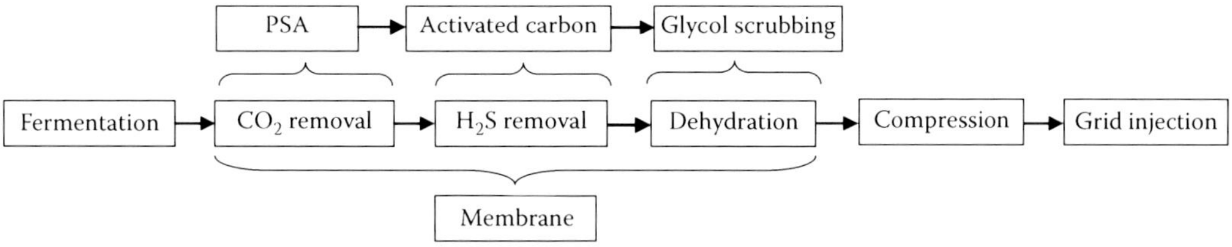

FIGURE 16.11
Various unit operations of biogas upgrading.

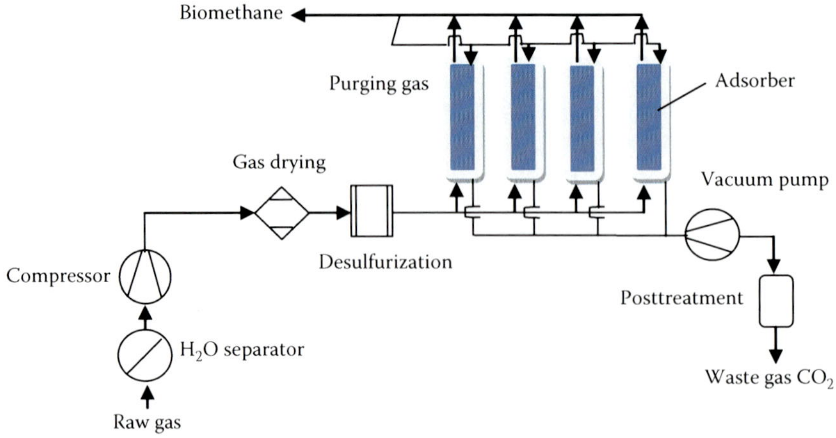

FIGURE 16.12
PSA. (From Biogas Association, Farm to fuel. Developers' guide to biomethane, www.cga.ca/wp-content/uploads/2012/08/Farm-to-Fuel-Developers-Guide-toBiomethane.pdf, accessed August 19, 2013, 2012.)

The biogas is first compressed with a high-pressure compressor, then cooled and desulfurized in the H_2S removal and then subjected to gas conditioning. During the cooled process, the gas is usually dehumidified to such an extent that separate gas drying is no longer necessary (Biogas Association 2012). In gas conditioning, condensate separated from the gas is then passed to the PSA. In the PSA, carbon molecular sieve will absorb off-gases such as CO_2, N_2, O_2, H_2O, and H_2S. The waste gas generated through this process has to be posttreated before being released into the atmosphere.

The main advantages of PSA are high CH_4 enrichment, low power demand, low level of emission, no use of chemicals, no occurrence of wastewater, no process heat required for regeneration, and generally no downstream gas drying (Figure 16.12).

16.4.2 High-Pressure Water Scrubbing

This process is called water scrubbing, pressurized water wash, or high-pressure water scrubbing (HPWS). Water is sprayed from the top of the column so that it flows down countercurrent to the gas. The column is usually filled with packing material for high-transfer surface gas–liquid contact. After the drying step, the obtained methane purity can reach 98%. The disadvantage of this technique is that it requires a large amount of water. According to the data from de Hullu et al. (2008), a 330 Nm^3/h wash of biogas requires around 1500 L/h of water (Figure 16.13).

16.4.3 Chemical Absorption

Chemical absorption is comparable to water absorption. A liquid such as amine is chemically bonded to CO_2. The biogas is dried and desulfurized prior to the separation of the CO_2. The absorber is an aqueous monoethanolamine (MEA) or diethanolamine (DEA) mixture that enables the reversible absorption of CO_2 (Figure 16.14).

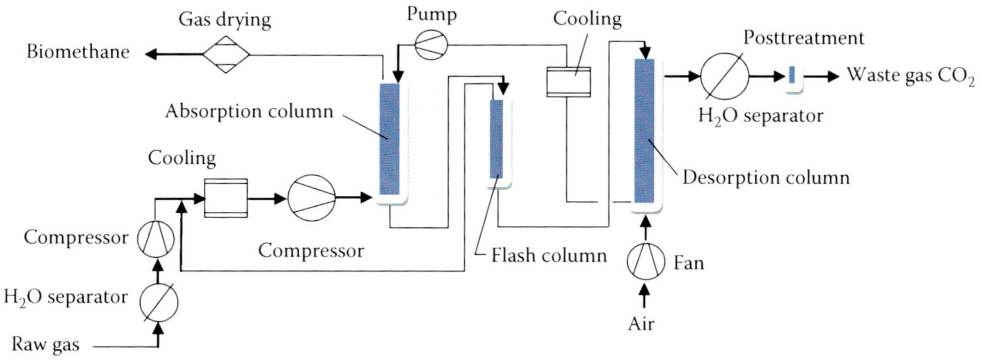

FIGURE 16.13
HPWS.

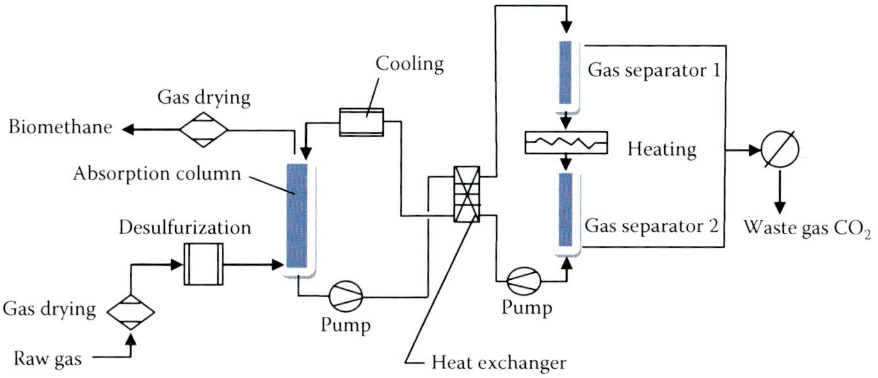

FIGURE 16.14
Amine scrubbing. (From Biogas Association, Farm to fuel. Developers' guide to biomethane, www.cga.ca/wp-content/uploads/2012/08/Farm-to-Fuel-Developers-Guide-toBiomethane.pdf, accessed August 19, 2013, 2012.)

The advantages of chemical absorption are (Biogas Association 2012)

- High CO_2 loading capacity
- Unpressurized process
- Very high methane purity
- Low methane loss
- No thermal posttreatment of the waste gas necessary

16.4.4 Membrane Separation

Methane can be separated from CO_2 using semipermeable membranes. The force can be a pressure difference, a concentration gradient, or electrical potential difference. Biogas upgrading can use the membrane process. The raw biogas has to be compressed to the membrane system to purify the gas. A fine desulfurization unit lowers the hydrogen sulfide level when the membrane system is not able to achieve the required hydrogen sulfide level.

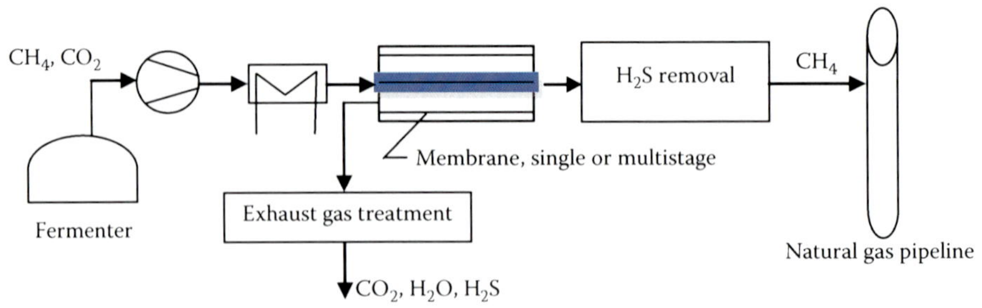

FIGURE 16.15
Process equipment for a membrane-based upgrading process.

Membrane process is an alternative to conventional biogas upgrading technology. Membrane process is usually used in natural gas treatment. Although the process condition of natural gas treatment and biogas upgrading is different, the membrane process can be used for it. The natural gas is under pressure when it leaves the natural gas field, whereas in biogas upgrading, the raw gas has to be compressed into the pipeline (Scholz et al. 2013). CH_4 and CO_2 can be separated by using a membrane; because of the difference in particle size, certain molecules pass through a membrane while others do not (de Hullu et al. 2008).

Various membrane materials are able to separate CO_2 and CH_4, and both polymeric and inorganic materials can be used. However, in industrial-scale gas separation, only polymeric membrane material is applied, due to its low manufacturing cost compared to inorganic material (Scholz et al. 2013).

Figure 16.15 presents the basic process equipment for a membrane-based upgrading process (Scholz et al. 2013; Biomethane Regions). The first raw biogas should be compressed from atmospheric pressure to 31 barg (Molino et al. 2013) and then cooled down for drying and removal of ammonia, and then H_2S is removed by means of adsorption on iron or zinc oxide. The gas is piped to a single-stage or multistage gas permeation unit.

The product gas is generated at the target methane purity, and it is dried and collected at about 30 barg pressure. To increase methane recovery, the permeate gas from the first-stage membrane module is processed through a second-stage membrane system. The first membrane module splits the feed gas stream into two gas streams: the biomethane product nonpermeate gas stream with a methane content higher than 95% by volume collected at high pressure of about 30 barg and the permeate gas that contains majority of CO_2, H_2O, and additional impurities collected at low pressure of 2 barg. With the second module, it is possible to recover additional product stream with methane concentration greater than 85%, and a second-stage permeate with low methane content can be used as a fuel (Monilo et al. 2013).

16.4.5 Cryogenic Separation

Cryogenic separation is the process of separation using low temperature close to –170°C and high pressure (approximately 80 bar). Trace gases and CO_2 are removed by cooling down the gas in various temperature steps. Because CO_2 and other biogas contaminants liquefy at different temperature–pressure domains, it is possible to separate them from biogas and produce pure biomethane. Figure 16.16 shows the process flow diagram of a cryogenic separation unit (de Hullu et al. 2008).

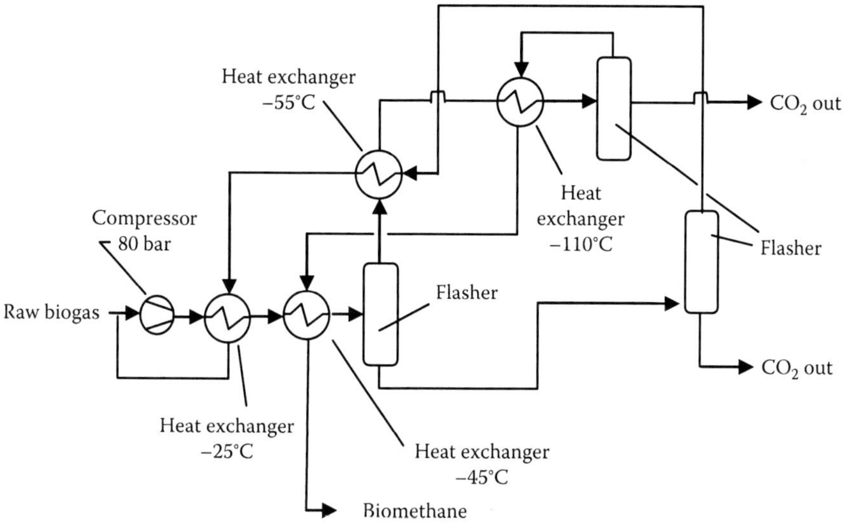

FIGURE 16.16
Flow diagram of cryogenic separation process.

16.5 Conclusion

Biogas can be produced by using raw materials such as household waste, human waste, animal waste, and microalgae, while biomethane is produced from purification of biogas. Raw biogas is processed through anaerobic digestion or other methods such as gasification, liquefaction, and fast pyrolysis. As a raw material, microalgae have potential in the production of biofuel because they can be relatively easy to cultivate using sunlight carried out in open or covered pond or closed PBR. Biogas consists of 55%–80% methane, 20%–45% CO_2, and a small amount of other contaminant gases such as NH_3, H_2S, and H_2O. The contaminants should be removed because they contain toxins and also cause corrosion. CO_2 content must also be removed, if the biogas is used as vehicle fuel. There are many processes for the removal of the contaminant in the biogas. The processes of transformation of biogas to biomethane among others are PSA, HPWS, chemical absorption, and membrane separation.

References

Agarwal, A.K. 2007. Biofuel (alcohol and biodiesel) application as fuel for internal combustion engine. *Prog Ener Comb Sci* 33: 233–271.

Al Seadi, T., Rutz, D., Prassl, H., Köttner, M., Finsterwalder, T., Volk, S., and Janssen, R. 2008. More about anaerobic digestion. In *Biogas Handbook*, ed. Al Seadi, T. University of Southern Denmark Esbjerg, Odense, Denmark, pp. 21–29.

Alexopoulos, S. 2012. Biogas system: Basic, biogas multifunction, principle of fermentation and hybrid application with a solar tower for the treatment of waste animal manure. *J Eng Sci Technol Rev* 5: 48–55.
Amin, S. 2009. Review on biofuel oil and gas production process from microalgae. *Energy Conv Manag* 50: 1834–1840.
Amon, T., Amon, B., Kryvoruchko, V., Machmüller, A., Hopfner-Sixt, K., Bodiroza, V., Hrbek, R. et al. 2007. Methane production through anaerobic digestion of various energy crops grown in sustainable crop rotations. *Bioresour Technol* 98: 3204–3212.
Anonymous. 2013. Biogas. http://www.clarke-energy.com/gas-type/biogas/, accessed July 20, 2013.
Austin, A. Open ponds versus closed bioreactors. http://biomassmagazine.com/articles/3618/open-ponds-versus-closed-bioreactors/, accessed September 29, 2013.
Biogas Association. 2012. Farm to fuel. Developer's guide to biomethane. www.cga.ca/wp-content/uploads/2012/08/Farm-to-Fuel-Developers-Guide-toBiomethane.pdf, accessed August 19, 2013.
Biomethane Regions. Introduction to the production of biomethane from biogas. A guide for England and Wales. Intelligent Energy.
Bridgwater, A.V. and Peacocke, G.V.C. 2000. Fast pyrolysis processes for biomass. *Renew Sust Energ Rev* 4: 1–73.
Calvin, M. and Taylor, S. 1989. Fuels from algae. In: *Algal and Cyanobacterial Biotechnology*, eds. Cresswell, RC, Rees, TAV, and Shah, N. Longman and John Wiley & Son, New York.
Christi Y. 2007. Biodiesel from microalgae. *Biotech Adv* 25: 294–306.
de Hullu, J., Maassen, J.I.W., van Meel, P.A., Shazad, S., Vaessen, J.M.P., Bini, L., and Reijenga, J.R. 2008. Comparing different biogas upgrading techniques. Interim report, Eindhoven University of Technology, Eindhoven, the Netherlands.
Demirbas, A. 2010. Use of algae as biofuel sources. *En Conv Management* 51: 2738–2749.
Demirbas, A. and Demirbas, M.F. 2010. *Algae Energy*. Springer, London, U.K., pp. 97–133.
Elliot, D.C. and Sealock, L.J. 1999. Chemical processing in high-pressure aqueous environments: Low temperature catalytic gasification. *Trans, IchemE* 74: 563–566.
FAO. 1996. *Biogas Technology: A Training Manual for Extension*. Consolidated Management Services Nepal (P) Ltd., Nepal, September.
FAO. 2007. Oil production, FAO Corp Doc Repository. www.fao.org/docrep/w7241e/w7241eOh.htm, accessed February 5, 2007.
González-Fernández, C., Sialve, B., Bernet, N., and Steyer, J.-P. 2012. Impact of microalgae characteristics on their conversion to biofuel. Part II: Focus on biomethane production. *Biofuel Bioprod Bioref* 6: 205–218.
Guiot, S.R. and Frigon, J.C. 2011. Anaerobic digestion as an effective biofuel production technology, Hallenbeck, P.C. (Ed.), *Microbial Technologies in Advanced Biofuels Production*, pp. 143–164.
Itoh, S., Suzuki, A., Nakamura, T., and Yokoyama, S. 1994. Production of heavy oil from sewage sludge by direct thermochemical liquefaction. *Desalination* 98: 127–133.
Janssen, M.G.J. 2002. Cultivation of microalgae: Effect of light/dark cycles on biomass yield. Thesis of Wageningen University, Ponsen & Looijen BV, the Netherlands.
Krich, K., Augenstein, D., Batmale, J.P., Benemann, J., Rutledge, B., and Salour, D. 2005. Biomethane from dairy waste. Prepare for Western United Dairymen, Modesto, CA.
Laing, I. 1991. Cultivation of marine unicellular algae. Laboratory Leaflet Number 67. Ministry of Agriculture, Fisheries and Food's Directorate of Fisheries Research, Lowestoft, U.K.
Liewellyn, C. and Skill, S. Biofuel production from marine algae. www.nbu.ac.uk/biota/Archieve_marineBD/9251.htm, accessed July 21, 2008.
McKendry, P. 2003. Energy production from biomass (part 2): Conversion technologies. *Biores Tech* 83: 47–54.
Miao, X., Wu, Q., and Yang, C. 2004. Fast pyrolysis of microalgae to produce renewable fuels. *J Anal Appl Pyrolysis* 71: 855–863.
Minowa, T. and Sawayama, S. 1999. A novel microalgal system for energy production with nitrogen cycling. *Fuel* 78: 1213–1215.

Minowa, T., Yokoyama, S., Kishimoto, M., and Okakurat, T. 1995. Oil production from algal cells of *Dunaliella tertiolecta* by direct thermochemical liquefaction. *J Fuel* 74(12): 1735–1738.
Molino, A., Nanna, F., Ding Y., Bikson, B., and Braccio, G. 2013. Biomethane production by anaerobic digestion of organic waste. *Fuel* 103: 1003–1009.
Murakami, M., Yokoyama, S., Ogi, T., and Koguchi, K. 1990. Direct liquefaction of activated sludge from aerobic treatment of effluents from the corn starch industry. *Biomass* 23: 215–228.
Narayanaswami, V., Sankar, K., Sekaran, P.M.C., and Lalitha, K. 1986. Biomethanation of *Leucaena leucocephala:* A potential biomass substrate. *Fuel* 65: 1129–1133.
Ogejo, J.A., Wen, Z., Ignosh, J., Bendfeldt, E., and Collins, E.R. 2009. Biomethane technology. *Virginia Cooperative Extension*, pp. 442–881.
Ozkurt, I. 2009. Qualifying of safflower and algae for energy. *Energy Edu Sci Technol Part A* 23: 145–151.
Park, K.Y., Kweon, J., Chantrasakdakul, P., Lee, K., and Cha, H.Y. 2013. Anaerobic digestion of microalgal biomass with ultrasonic disintegration. *Int Biodet Biodeg* 85: 598–602.
Patil, V., Tran, K.Q., and Giselrod, H.R. 2008. Toward sustainable production of biofuel from microalgae. *Int J Mol Sci* 9: 1188–1195.
Prohnow, A., Heiermann, M., Drenckhan, A., and Schelle, H. 2005. Seasonal pattern of biomethanisation of grass from landscape management. *CIGR E-Journal* 7: 1–17.
Qi, Z., Lie, C., Tiejun, W., and Ying, X. 2007. Review of biomass pyrolysis oil properties and upgrading research. *En Conv Mngt* 48: 87–92.
Rajendran, K., Aslanzadeh, S., and Taherzadeh, M.J. 2012. Household biogas digester—A review. *Energies* 5: 2911–2942.
Satin, M. Microalgae. www.fao.org/ag/ags/Agsi/MIRCOALG.htm, accessed June 16, 2008.
Schenk, P., Thomas-Hall, S., Stephens, E., Marx, U., Mussgnug, J., Posten, C., Kruse, O., and Hankamer, B. 2008. Second generation biofuels: High-efficiency microalgae for biodiesel production. *BioEnergy Rese* 1: 20–43.
Schnürer, A. and Jarvis, A. 2010. Microbiological handbook for biogas plants. Swedish Gas Center Report 2007, pp. 9–50.
Scholz, M., Melin, T., and Wessling, M. 2013. Transforming biogas into biomethane using membrane technology. *Renew Sustain Energ Rev* 17: 199–212.
Sheehan, J., Dunahay, T., Benemann, J., and Roessler, P. 1998. A look back at the U.S. Department of Energy's Aquatic Species Program—Biodiesel from algae. U.S. Department of Energy's Office of Fuels Development, Colorado.
Singh A. Nigam P.S., Murphy J.D. 2011. Mechanism and challenges in commercialisation of algal biofuels. *Bioresource Technology* 102: 26–34.
Sorathia, H.S., Rathod, P.P., and Sorathia, A.S. 2012. Biogas generation and factors affecting the biogas generation—A review study. *Int J Adv Eng Technol* III(III): 72–78.
Suyog, V.I.J. 2011. Biogas production from kitchen waste. Seminar report. National Institute of Technology, Rourkela, India.
Tatedo. Biogas Technology, Tanzania Traditional Energy Development and Environment Organization (Tatedo). www.tatedo.org, accessed July 23, 2013.
Tsukahara, K. and Sawayama, S. 2005. Liquid fuel production using microalgae. *J Jpn Petro Inst* 48(5): 251–259.
Weiland, P. 2003. Production and energetic use of biogas from energy crops and wastes in Germany. *Appl Biochem Biotechnol* 109: 263–274.
Weissman, J.C and Goebel, R.P. 1987. Design and analysis of microalgal open pond systems for the purpose of producing fuels. Solar Energy Research Institute.
Widodo, T.W., Asari, A., Ana, N., and Elita, R. 2009. Design and development of biogas reactor for farmer group scale. *Indon J Agri* 2(2): 121–128.
Wilkie, A. 2008. Biomethane from biomass, biowaste, and biofuels. In *Bioenergy*, Wall, J. ed. ASM Press, Washington, DC, pp. 195–205.
Yang, Y.F., Feng, C.P., Inamori, Y., and Maekawa, T. 2004. Analysis of energy conversion characteristics in liquefaction of algae. *Res Cons Recycl* 43: 21–33.

17

Current State of Research on Algal Biomethanol

Ozcan Konur

CONTENTS

17.1 Introduction

17.1.1 Issues

Global warming, air pollution, and energy security have been some of the most important public policy issues in recent years (Jacobson et al. 2009, Wang et al. 2008,

Yergin 2006). With the increasing global population, food security has also become a major public policy issue (Lal 2004). The development of biofuels generated from biomass has been a long-awaited solution to these global problems (Demirbas 2007, Goldember 2007, Lynd et al. 1991). However, the development of early biofuels produced from agricultural plants such as sugarcane (Goldemberg 2007) and agricultural wastes such as corn stovers (Bothast and Schlicher 2005) has resulted in a series of substantial concerns about food security (Godfray et al. 2010). Therefore, the development of algal biofuels as a third-generation biofuel has been considered as a major solution for the problems of global warming, air pollution, energy security, and food security (Chisti 2007, Demirbas 2007, Kapdan and Kargi 2006, Spolaore et al. 2006, Volesky and Holan 1995).

Although there have been many reviews on the use of marine algae for the production of algal biofuels (e.g., Chisti 2007, Mata et al. 2010, Meng et al. 2009) and methanol fuels in general (Klier 1982, Kung 1980, Liu et al. 2006, Pena et al. 1996, Wasmus and Kuver 1999, Wender 1996) and there have been a number of scientometric studies on algal biofuels (Konur 2011), there has not been any study on the citation classics on algal biomethanol as well as methanol fuels in general (Baltussen and Kindler 2004a,b, Dubin et al. 1993, Gehanno et al. 2007, Konur 2012a,b, 2013, Paladugu et al. 2004, Wrigley and Matthews 1986).

As North's new institutional theory suggests, it is important to have up-to-date information about current public policy issues to develop a set of viable solutions to satisfy the needs of all key stakeholders (Konur, 2000, 2002a,c, 2006a,b, 2007a,b, 2012c, North 1994).

Therefore, brief information on a selected set of citation classics on algal biomethanol as well as methanol fuels in general are presented in this chapter to inform key stakeholders about the global problems of warming, air pollution, food security, and energy security and about the potential of algal biomethanol production for the solution of these problems in the long run, thus complementing a number of recent scientometric studies on biofuels and global energy research (Konur 2011, 2012d–p).

This study shows that the research on algal biomethanol has been scarce and there is a great potential for the production and use of algal biomethanol in many industrial contexts, especially as algal methanol fuel cells (Kreuer 2001, Reddington et al. 1998, Wasmus and Kuver 1999) and algal methanol fuels in transportation vehicles (Brown 2001, Kumar et al. 2003, Trimm and Onsan 2001) as well as an ingredient for the production of hydrogen through stream reforming (Breen and Ross 1999, Peppley et al. 1999a,b). This study also shows that methanol has great potential as a future fuel that meets the renewable and sustainable energy needs in comparison with hydrogen fuels, based on both natural gas and biomass through a number of interdisciplinary studies on the use of methanol as a fuel in various industrial contexts (Hamelinck and Faaij 2006, Lange 2007, MacLean and Lave 2003, Olah 2005, Olah et al. 2009).

17.1.2 Methodology

A search was carried out in the Science Citation Index Expanded (SCIE) and Social Sciences Citation Index (SSCI) databases (version 5.11) in October 2013 to locate the papers relating to algal biomethanol and methanol in general using the keyword set of Topic = (methanol or biomethanol or MeOH or $CH3OH$) Refined by: Document Types = (article or review) Timespan = All years. (Databases = SCI-Expanded, SSCI) in the abstract pages of the papers.

It was found that there were 127,250 papers indexed between 1980 and 2013. The key subject categories for the methanol research were chemistry, while 6600 papers were indexed by the energy and fuels subject. It was also necessary to focus on the key references by selecting articles and reviews.

A further search was carried out for algal biofuels using the keyword set of Topic = (methanol or biomethanol or MeOH or CH_3OH) and Topic = ((alga* or *macro* alga** or *micro* alga** or macroalga* or microalga* or cyanobacter* or seaweed* or diatoms or sea* weed* or reinhardtii or braunii or *Chlorella* or *Sargassum* or *Gracilaria* or *Spirulina*)) Timespan = All years. (Databases = SCI-Expanded, SSCI). For this chapter, it was necessary to embrace the broad algal search terms to include diatoms, seaweeds, and cyanobacteria. This search resulted in 1000 references. However, it was realized that the found references were not mostly related to algal methanol biofuels.

Therefore, the larger set of 127,250 references was screened out for methanol biofuels in the decreasing order of the citations received for each paper. The found references were grouped under a number of topical headings. New searches were carried out for each topical area using the set of special keywords for each group. Four main groups of papers emerged from this grouping exercise: algal biomethanol, production of methanol as a fuel, use of methanol as a fuel, and economics and assessment of methanol as a fuel. Depending on the size of the highly cited papers, 2, 20, 40, 20 papers were selected for these topical groups, respectively, for this chapter.

The information relating to the document type, the affiliation and location of the authors, the journal, the indexes, the subject area, the concise topic, the total number of citations and the total average number of citations received per year were given in the tables for each paper.

17.2 Algal Biomethanol

17.2.1 Introduction

Although there were around 1000 references related to both algae and methanol, most of these papers were not related to algal methanol biofuels. However, there were only two papers on algal methanol biofuels (Table 17.1).

17.2.2 Research on Algal Biomethanol

Hirano et al. (1998) study the temperature effect on continuous gasification of microalgal biomass with a focus on the theoretical yield of methanol production and its energy balance in an early paper with 50 citations. A microalga, *Spirulina*, was partially oxidized at temperatures of 850°C, 950°C, and 1000°C, and the composition of produced gas was determined in order to evaluate the theoretical yield of methanol from the gas. They find that the "gas composition depended on the temperature, and the gasification at 1000°C gave the highest theoretical yield of 0.64 g methanol from 1 g of the biomass." Based on this yield, the total energy requirement for the whole process including microalgal biomass production and conversion into methanol was obtained. They find that "energy balance, the ratio of the energy of methanol produced to the total required energy, was 1.1, which

TABLE 17.1

Research on Algal Biomethanol

No.	Paper References	Year	Document	Affiliation	Country	Number of Authors	Journal	Index	Subjects	Topic	Total Number of Citations	Total Average Citations per Annum	Rank
1	Hirano et al.	1998	A	Tokyo Elect. Power Co. Ltd., Mitsubishi Heavy Ind. Ltd.	Japan	5	*Catal. Today*	SCI, SCIE	Chem. Appl., Chem. Phys., Eng. Chem.	Algal biomethanol	50	3.1	64
2	Liu and Ma	2009	A	S China Univ. Technol.	China	2	*Energ. Pol.*	SCIE, SSCI	Ener. Fuels, Env. Sci., Env. Stud.	Algal biomethanol	12	2.4	82

SCI, Science Citation Index; SCIE, Science Citation Index Expanded; SSCI, Social Sciences Citation Index; A, article; R, review.

indicates that this process was plausible as an energy producing process." They stress that the "greater part of the total required energy, almost four-fifths, was consumed with the microalgal biomass production, suggesting that more efficient production of microalgal biomass might greatly improve its energy balance."

Liu and Ma (2009) analyze the energy and environmental impacts of microalgal biomethanol in China by using a method of life cycle assessment (LCA) in a paper with 12 citations. They use LCA to identify and quantify the environment emissions and energy efficiency of the system throughout the whole life cycle, including microalgae cultivation, methanol conversion, transport, and end use. They find that "energy efficiency, defined as the ratio of the energy of methanol produced to the total required energy, is 1.24." They argue that this is plausible as an energy-producing process. The "environmental impact loading of microalgal methanol fuel is 0.187 $mPFT_{2000}$ in contrast to 0.828 $mPET_{2000}$ for gasoline." The effect of photochemical ozone formation is the highest of all the calculated categorization impacts of the two fuels. They argue that "utilization of microalgae as a raw material of producing methanol fuel is beneficial to both production of renewable fuels and improvement of the ecological environment." They assert that algal methanol is friendly to the environment, which should take an important role in automobile industry development and gasoline fuel substitute.

17.2.3 Conclusion

Although there were around 1000 references related to both algae and methanol, it was found that most of these papers were not related to algal methanol biofuels. There were only two papers on algal methanol biofuels. These two papers show that the production of algal biomethanol is efficient and environment friendly in comparison to gasoline and other fossil-based transportation fuels. These findings give support to carry out the research based on methanol fuels in general to explore the emerging research and policy issues in the production, use, and assessment of methanol fuels. The following sections of the chapter deal with these issues.

17.3 Studies on the Production of Methanol

17.3.1 Introduction

The production of methanol has been one of the most dynamic research areas in recent years. Twenty citation classics in the field of methanol production with more than eighty-one citations were located and the key emerging issues from these papers are presented in the decreasing order of the number of citations (Table 17.2).

These papers give strong hints about the determinants of efficient methanol production and emphasize that it is possible to optimize the production of methanol. The brief information on these papers was provided under the headings of synthesis gas (syngas) production and methanol synthesis (10 papers each), respectively.

Syngas production and the methanol synthesis are governed by Equations 17.1 and 17.2, respectively. In the first stage of the methanol production, syngas is produced, and in the second stage of methanol production, methanol synthesis takes place.

TABLE 17.2

Research on the Production of Methanol

No.	Paper References	Year	Document	Affiliation	Country	Number of Authors	Journal	Index	Subjects	Topic	Total Number of Citations	Total Average Citations per Annum	Rank
1	Klier	1982	R	Lehigh Univ.	United States	1	*Adv. Catal.*	SCI, SCIE	Chem. Phys.	Methanol synthesis	726	22.7	3
2	Hickman and Schmidt	1993	A	Univ. Minnesota	United States	2	*Science*	SCI, SCIE	Mult. Sci.	Synthesis gas production	607	28.9	6
3	Pena et al.	1996	R	CSIC, REPSOL Petr. SA	Spain	3	*Appl. Catal. A Gen.*	SCI, SCIE	Chem. Phys., Env. Sci.	Synthesis gas production	380	21.1	13
4	Chinchen et al.	1986	A	ICI PLC	England	3	*Adv. Catal.*	SCI, SCIE	Chem. Phys.	Methanol synthesis	348	12.4	15
5	Kung	1980	R	Northwestern Univ.	United States	1	*Catal. Rev. Sci. Eng.*	SCI, SCIE	Chem. Phys.	Methanol synthesis	347	10.3	16
6	Ichikawa	1978	A	Sagami Chem. Res. Ctr.	Japan	1	*Bull. Chem. Soc. Jpn.*	SCI, SCIE	Chem. Mult.	Methanol synthesis	220	6.1	21
7	Waugh	1992	A	ICI PLC	England	1	*Catal. Today*	SCI, SCIE	Chem. Appl., Chem. Phys., Eng. Chem.	Methanol synthesis	209	9.5	23
8	Bowker et al.	1981	A	ICI PLC	England	3	*J. Chem. Soc. Farad. Trans.*	SCI, SCIE	Chem. Mult., Chem. Phys.	Methanol synthesis	208	6.3	24
9	Grunwaldt et al.	2000	A	Haldor Topsoe Res. Labs.	Denmark	5	*J. Catal.*	SCI, SCIE	Chem. Phys., Eng. Chem.	Synthesis gas production	201	14.4	25
10	Wender	1996	R	Univ. Pittsburgh	United States	1	*Fuel Process. Technol.*	SCI, SCIE	Chem. Appl., Ener. Fuels, Eng. Chem.	Synthesis gas production	198	11.0	27

(Continued)

TABLE 17.2 (*Continued*)
Research on the Production of Methanol

No.	Paper References	Year	Document	Affiliation	Country	Number of Authors	Journal	Index	Subjects	Topic	Total Number of Citations	Total Average Citations per Annum	Rank
11	Bharadwaj and Schmidt	1995	A	Univ. Minnesota	United States	2	*Fuel Process. Technol.*	SCI, SCIE	Chem. Appl., Ener. Fuels, Eng. Chem.	Synthesis gas production	185	9.7	28
12	Wilhelm et al.	2001	A	SFA Pacific Inc.	United States	4	*Fuel Process. Technol.*	SCI, SCIE	Chem. Appl., Ener. Fuels, Eng. Chem.	Synthesis gas production	169	13	33
13	Graaf et al.	1986	A	State Univ. Groningen	Netherlands	4	*Chem. Eng. Sci.*	SCI, SCIE	Eng. Chem.	Methanol synthesis	151	5.4	38
14	Deutschmann and Schmidt	1998	A	Univ. Minnesota	United States	2	*AICHE J.*	SCI, SCIE	Eng. Chem.	Synthesis gas production	140	8.8	39
15	Askgaard et al.	1995	A	Tech. Univ. Denmark	Denmark	4	*J. Catal.*	SCI, SCIE	Chem. Phys., Eng. Chem.	Methanol synthesis	124	6.5	45
16	Centi and Perathoner	2009	A	Univ. Messina, CASPE INSTM	Italy	2	*Catal. Today*	SCI, SCIE	Chem. Appl., Chem. Phys., Eng. Chem.	Synthesis gas production	122	24.4	46
17	Ovesen et al.	1997	A	Tech. Univ. Denmark	Denmark	6	*J. Catal.*	SCI, SCIE	Chem. Phys., Eng. Chem.	Methanol synthesis	119	7.0	49
18	Fisher and Bell	1997	A	Univ. Calif. B.	United States	2	*J. Catal.*	SCI, SCIE	Chem. Phys., Eng. Chem.	Methanol synthesis	117	16.7	50
19	Ng et al.	1999	A	Univ. London Imperial Coll. Sci. Technol. & Med., Air Prod & Chem Inc.	England, United States	3	*Chem. Eng. Sci.*	SCI, SCIE	Eng. Chem.	Synthesis gas production	100	6.7	51
20	Ponec	1992	A	Leiden Univ.	Netherlands	1	*Catal. Today*	SCI, SCIE	Chem. Appl., Chem. Phys., Eng. Chem.	Synthesis gas production	81	3.7	57

SCI, Science Citation Index; SCIE, Science Citation Index Expanded; SSCI, Social Sciences Citation Index; A, article; R, Review.

$$\text{Production of syngas (synthesis gas)}: CH_4 + 0.5O_2 \rightarrow CO + 2H_2 \quad (17.1)$$

$$\text{Production of methanol (methanol synthesis)}: CO + 2H_2 \rightarrow CH_3OH \quad (17.2)$$

These papers were dominated by researchers from seven countries, usually through the intracountry institutional collaboration, and they were multiauthored. The United States (nine papers), England (four papers), Denmark (three papers), and the Netherlands (two papers) were the most prolific countries. Similarly, Imperial Chemical Industries PLC, University of Minnesota (three papers each), and Technical University of Denmark (two papers) were the most prolific institutions.

Similarly, all these papers were published in journals indexed by the SCI and SCIE. There were no papers indexed by the Arts & Humanities Citation Index (A&HCI) or SSCI. The number of citations ranged from 81 to 726 and the average number of citations per annum ranged from 3.7 to 28.9. Sixteen of these papers were articles while four of them were reviews.

17.3.2 Research on the Production of Methanol

17.3.2.1 Research on the Synthesis Gas

Hickman and Schmidt (1993) discuss the production of syngas by direct catalytic oxidation of methane in a paper with 607 citations. They argue that the reaction between methane and oxygen over platinum and rhodium surfaces in metal-coated ceramic monoliths can be made to produce mostly hydrogen and carbon monoxide (greater than 90% selectivity for both) with almost complete conversion of methane and oxygen at reaction times as short as 10^{-3} s. They next argue that this process has great promise for conversion of abundant natural gas into liquid products such as methanol and hydrocarbons (HCs), which can be easily transported from remote locations. Rhodium was considerably superior to platinum in producing more H_2 and less H_2O, which can be explained by the known chemistry and kinetics of reactants, intermediates, and products on these surfaces.

Pena et al. (1996) discuss new catalytic routes for syngas and hydrogen production starting from simple hydrogen-containing molecules in a review paper with 380 citations. They focus on the new direct catalytic alternatives in natural gas conversion and discuss the improvements in syngas technology, including partial oxidation (POX), autothermal reforming (ATR), combined reforming, and carbon dioxide reforming, and compare the energy efficiencies of direct and indirect methane conversion. They emphasize that most of the ongoing research related to direct processes is at the exploratory stage, while technology utilizing indirect approach has advanced to semiworks and initial commercialization plants. The new emerging processes based on POX features are unique for syngas generation. They anticipate further enhancement of such processes plus improvements in other second-generation technologies and advances in direct processes to provide additional, new attractive paths to the chemical conversion of natural gas. Similarly, they discuss onboard generation of hydrogen-rich gaseous fuels either for spark ignition (SI) engines or for coupled-fuel-cells electric engines within the scope of both POX and catalytic decomposition of methanol.

Grunwaldt et al. (2000) investigate structural changes in Cu/ZnO catalysts in a paper with 201 citations. They find that under typical mild reduction conditions, very small copper particles (10–15 Å) are formed. Upon change in the reduction potential of the methanol syngas, reversible changes of the Cu–Cu coordination number are observed by x-ray absorption spectroscopy (EXAFS). These structural changes are accompanied by changes in catalytic activity, and the highest activities were observed after exposure to the most reducing conditions. In this state, the catalyst exhibited low Cu–Cu coordination numbers. They propose a model that reversible changes in the wetting of ZnO by Cu may occur upon changes in the reaction conditions. They also find that such dynamical changes in Cu morphology may influence the catalytic properties. All the conditions used in these studies are less severe than those observed to result in bulk alloy formation. Only under severe reduction conditions, they observe significant alloying of copper and zinc in EXAFS in addition to the morphological changes.

Wender (1996) discusses the reactions of syngas in a review paper with 198 citations. There has been steady growth in the traditional uses of syngas. Almost all hydrogen gas is manufactured from syngas and there has been a tremendous spurt in the demand for this basic chemical; indeed, the chief use of syngas is in the manufacture of hydrogen for a growing number of purposes. Methanol not only remains the second largest consumer of syngas but has shown remarkable growth as part of the methyl ethers used as octane enhancers in automotive fuels. Fischer–Tropsch synthesis remains the third largest consumer of syngas, mostly for transportation fuels but also as a growing feedstock source for the manufacture of chemicals, including polymers. The hydroformylation of olefins (the oxo reaction), a complete chemical use of syngas, is the fourth largest use of carbon monoxide and hydrogen mixtures; research and industrial application in this field continue to grow steadily. A direct application of syngas as fuel (and eventually also for chemicals) that promises to increase is its use for integrated gasification combined cycle (IGCC) units for the generation of electricity (and also chemicals) from coal, petroleum coke, or heavy residuals.

Bharadwaj and Schmidt (1995) discuss the catalytic POX of natural gas to syngas in a paper with 185 citations. They trace the development of catalytic POX technology for the conversion of natural gas to syngas from steam reforming to ATR to direct oxidation. Syngas, which has applications in methanol, ammonia, and Fischer–Tropsch synthesis, has been conventionally produced by endothermic steam-reforming processes in fire tube furnaces. Catalytic POX is much faster, highly selective in a single reactor, and much more energy efficient. It could thus significantly decrease capital and operating costs of syngas production. They consider catalysts as well as reactors where they discuss further processes, issues, and practical difficulties with academic and industrial efforts presented in parallel. They assert that new millisecond contact time direct oxidation processes, which eliminate the use of steam and the use of autothermal reactors orders of magnitude smaller than those used for conventional steam reforming, hold promise for commercialization.

Wilhelm et al. (2001) discuss the syngas production for gas-to-liquid applications in a paper with 169 citations. The principal technologies for producing syngas from natural gas are catalytic steam methane reforming (SMR), two-step reforming, ATR, POX, and heat exchange reforming. As a frame of reference, in terms of syngas flow rates, they note that a 20,000 barrels/day F–T plant would be comparable to three 2500 mt/day methanol plants. Single-train methanol plants are now producing more than 2500 t/day and plants approaching 3000 mt/day have been announced. The projected relative economies of scale of the various syngas production technologies indicate that "two-step reforming and ultimately, ATR, should be the technologies of choice for large-scale GTL plants." Nevertheless, for a 20,000 barrels/day F–T liquids plant, capital charges still dominate the manufacturing costs. "Syngas production (oxygen plant and reforming) comprises half of the total capital cost of this size GTL plant." While air-blown reforming eliminates the expensive oxygen plant, air-blown reforming is unlikely to be competitive with, or offer the flexibility of, oxygen-blown reforming. They argue that the proposed and future gas-to-liquids (GTL) facilities should be substantially less costly than their very expensive predecessors—as the result of improvements in FT catalyst and reactor design, the most significant of which have been pioneered by Sasol. In the absence of a breakthrough technology, they argue that "economy of scale will be the only significant mechanism by which GTL can achieve greater economic viability." However, even with such further cost reductions, the economic viability of GTL plants will remain confined to special situations until crude price levels rise substantially. In the long term, if a ceramic membrane reactor (combining air separation and POX) can be developed that enables the 20% reduction in GTL investment costs that the R&D effort is targeting, they argue that "GTL could become economically viable at crude prices below US$20/b."

Deutschmann and Schmidt (1998) model the POX of methane in a short-contact-time reactor in a paper with 140 citations. POX of methane in monolithic catalysts at very short contact times offers a promising route to convert natural gas into syngas, which can then be converted to higher alkanes or methanol. Detailed modeling is needed to understand their complex interaction of transport and kinetics in these systems and for their industrial application. In this work, the POX of methane in noble-metal (Rh and Pt)–coated monoliths was studied numerically as an example of short-contact-time reactor modeling. A tube wall catalytic reactor was simulated as a model for a single pore of the monolithic catalyst using a 2D flow field description coupled with detailed reaction mechanisms for surface and gas-phase chemistry. The catalytic surface coverages of adsorbed species are calculated versus position. The reactor is characterized by competition between complete and POX of methane. At atmospheric pressure, CO_2 and H_2O are formed on the catalytic surface at the entrance of the catalytic reactor. At higher pressure, gas-phase chemistry becomes important, forming more complete oxidation products downstream and decreasing syngas selectivity by about 2% at 10 bar.

Centi and Perathoner (2009) discuss opportunities and prospects in the chemical recycling of carbon dioxide to fuels as a complementary technology to carbon sequestration and storage (CSS) in a paper with 122 citations. They argue that the requisites for this objective are (1) minimize as much as possible the consumption of hydrogen (or hydrogen sources), (2) produce fuels that can be easily stored and transported, and (3) use renewable energy sources. From this perspective, they argue that the preferable option is to produce alcohols (preferably $\geq C_2$) using solar energy to produce the protons and electrons necessary for the reaction of CO_2 reduction. They discuss (1) reverse water–gas shift and (2) hydrogenation to HCs, alcohols, dimethyl ether, and formic acid; (3) reaction with HCs to syngas; (4) photo- and electrochemical/catalytic conversion; and (5) thermochemical conversion.

Ng et al. (1999) study kinetics and modeling of dimethyl ether synthesis from syngas over a commercial $CuO/ZnO/Al_2O_3$ (methanol forming) and a gamma-alumina (dehydration) catalyst 250°C and 5 MPa using a gradientless, internal-recycle-type reactor in a paper with 100 citations. They test a kinetic model for the combined methanol + DME synthesis based on a methanol synthesis model using results obtained from a wide range of CO_2:CO feed ratios. They find results at different CO_x:H_2 ratio and catalyst loading ratios, while they observe catalyst deactivation during DME synthesis at high space velocities and a large ratio of dehydration catalyst.

Ponec (1992) discusses active centers for syngas reactions in a paper with 81 citations. He argues that for methanol and higher alcohol synthesis, an active center always requires the presence of a promoter within close proximity. In contrast to this, formation of HCs and aldehydes can, in principle, take place without any promoter. He discusses some aspects of the mechanism of the reactions and of the promoter function focusing on the discussion of the active centers.

17.3.2.2 Research on the Methanol Synthesis

Klier (1982) discusses methanol synthesis in an early review paper with 726 citations. He focuses on the binary copper–zinc oxide system and mechanisms of the methanol synthesis.

Chinchen et al. (1986) discuss the activity and state of the copper surface in methanol synthesis catalysts in a paper with 348 citations. The measurement of copper metal surface areas by monitoring nitrous oxide chemisorption is a well-established technique. They develop a frontal chromatographic version of this technique, which is very suitable for in situ measurements and for measuring apparent copper areas of various catalysts after exposures to methanol syngases of different compositions at typical industrial conditions in microreactors commonly used for assessing the methanol synthesis activity of such catalysts. Using such techniques, they show that first, "there is a linear relationship between the methanol synthesis activity of copper/zinc oxide/alumina catalysts and their total copper surface area." Second, "copper supported on other materials has approximately the same turnover number as copper/zinc oxide/alumina catalysts." Third, "under industrial conditions the copper surface of the catalyst is partially oxidized, to an extent which depends on the composition of the synthesis gas, particularly the CO_2/CO ratio."

Kung (1980) discusses methanol synthesis in an early review paper with 347 citations. He notes that methanol is used as a solvent in many industrial processes, as a starting material for the production of other compounds, notably formaldehyde, and as a freezing point suppressing agent for gasoline lines and window washing liquids, as well as for many other purposes. Traditionally, methanol has been produced by catalytic hydrogenation of carbon monoxide.

Ichikawa (1978) discusses catalysis by supported metal crystallites from carbonyl clusters with a focus on the catalytic methanol synthesis under mild conditions over supported rhodium, platinum, and iridium crystallites prepared from Rh, Pt, and Ir carbonyl cluster compounds deposited on ZnO and MgO in an early paper with 220 citations.

Waugh (1992) discusses methanol synthesis with a focus on the kinetics and mechanism of methanol synthesis over copper/zinc oxide/alumina and other oxide-supported catalysts in a paper with 209 citations. The kinetics and mechanism of methanol synthesis over a multicomponent catalyst have been elucidated by the following process. The adsorptive and reactive interactions of each of the gas-phase species (CO, CO_2, H_2) with each of the catalytic components (Cu, ZnO, Al_2O_3) separately have been studied. The interactions of combinations of the gas-phase species with the individual catalytic components and then with pairs of the catalytic components, were then studied, ultimately building up to a study of the full system. What has emerged is that it is the CO_2 component of the CO, CO_2 mixture which is the immediate precursor to methanol, being adsorbed on the partially oxidized copper as a symmetric carbonate. This carbonate is hydrogenated on the copper component of the catalyst to a formate species, the most stable and longest-lived intermediate to methanol, with hydrogenation of the formate being the rate-determining step (RDS) of the reaction. The specific activity of copper in synthesizing methanol is unaffected by the nature of the oxide support, indicating that no unique synergy attaches to the copper/zinc oxide combination. The role of the CO is to keep the copper in a more reduced (more active) state than could be achieved with hydrogen alone.

Bowker et al. (1981) discuss the mechanism and kinetics of methanol synthesis on zinc oxide in an early paper with 208 citations. The kinetics of the adsorption and surface reactions of hydrogen, water, carbon dioxide, carbon monoxide, formaldehyde, and methanol on what is mainly the prism face of zinc oxide have been studied using temperature-programmed desorption and reaction. Both hydrogen and carbon dioxide show a multiplicity of desorption energies, adsorption into those states showing the highest binding energies being activated. Temperature programming after the room-temperature adsorption of formaldehyde or methanol shows evidence of surface reaction, with the formation of a formate intermediate. The surface reaction mechanisms of this formate intermediate and their kinetics are identical regardless of whether it was formed from methanol adsorption or formaldehyde adsorption. The formate appears to be the pivotal intermediate in zinc oxide–catalyzed syngas chemistry, decomposing (1) to carbon monoxide and hydrogen or (2) (depending on the hydrogen coverage) to methanol. The same formate intermediate is formed by coadsorption of carbon dioxide and hydrogen, while hydrogen/carbon monoxide dosing does not result in its formation; indeed, carbon monoxide itself did not adsorb on the defected zinc oxide.

Graaf et al. (1986) discuss the chemical equilibria in methanol synthesis in an early paper with 151 citations. The chemical equilibria of the methanol reaction and the water–gas shift reaction, starting from carbon monoxide, carbon dioxide, and hydrogen, were studied in a fixed-bed catalytic reactor at P = 10–80 bar and T = 200°C–270°C. It was found that the chemical equilibria could be described very well by thermochemical data based on ideal gas behavior in combination with a correction for the nonideality of the gas mixture as predicted by the Soave–Redlich–Kwong equation of state. This correction for nonideality results in significantly better agreement with experimental data than a correction based on the original Redlich–Kwong equation of state, the Peng–Robinson equation of state, the virial equation truncated after the second virial coefficient, Lewis and Randall's rule, or not correcting at all for nonideality, thus assuming ideal gas behavior.

Askgaard et al. (1995) discuss a kinetic model of methanol synthesis in a paper with 124 citations. All parameters in the model are estimated from gas-phase thermodynamics and surface science studies. The rate-limiting step in the kinetic model was determined from Cu_{100} single-crystal experiments to be the hydrogenation of H_2COO star to methoxide and oxide. Calculated methanol rates from the model extrapolated to industrial working conditions were in good agreement with rates measured on a real catalyst.

Ovesen et al. (1997) describe a dynamic microkinetic model of the methanol synthesis reaction over Cu/ZnO catalysts in a paper with 117 citations. The model is based on surface science measurements and it includes the dynamic changes in particle shape and active surface area, which have recently been observed by in situ EXAFS measurements to take place upon change in the redox potential of the reaction gas. They argue that the change in particle morphology is related to a change in the number of oxygen vacancies at the Zn–O–Cu interface. Furthermore, the structure sensitivity of the methanol synthesis reaction is also taken into account. The dynamic microkinetic model gives a much better description of the kinetic measurements over a working methanol catalyst compared to a static microkinetic model. The new model also gives an explanation of the kinetic behavior during transient conditions. The dynamic aspects provide a basis for understanding the apparently conflicting reaction orders reported in the literature.

Fisher and Bell (1997) carry out an in situ infrared study of methanol synthesis from H_2/CO_2 over Cu/SiO_2 and $Cu/ZrO_2/SiO_2$ with the aim of understanding the nature of the species involved in methanol synthesis and the dynamics of the formation and consumption of these species in a paper with 117 citations. In the case of Cu/SiO_2, the only species observed during methanol synthesis are formate groups on Cu and methoxide groups on silica. When ZrO_2/SiO_2 or $Cu/ZrO_2/SiO_2$ is exposed to H_2/CO_2, the majority of the species observed are associated with ZrO_2. CO_2 adsorption on either ZrO_2/SiO_2 or $Cu/ZrO_2/SiO_2$ leads to the appearance of carbonate and bicarbonate species on the surface of ZrO_2. In the presence of H_2, these species are converted to formate and then methoxide species adsorbed on ZrO_2. The presence of Cu greatly accelerates the hydrogenation of bicarbonate to formate species and the hydrogenation of formate to methoxide species. On $Cu/ZrO_2/SiO_2$, methylenebisoxy species are observed as intermediates in the latter reaction. While Cu promotes the reductive elimination of methoxide species as methanol, it is observed that hydrolytic release of methoxide species from ZrO_2 occurs much more rapidly than reductive elimination. Thus, the methanol synthesis over $Cu/ZrO_2/SiO_2$ is envisioned to occur on ZrO_2, with the primary role of Cu being the dissociative adsorption of H_2. The spillover of atomic H onto ZrO_2 provides the source of hydrogen needed to hydrogenate the carbon-containing species. They propose that the formation of CO via the reverse-water–gas-shift reaction occurs on Cu.

17.4 Studies on the Use of Methanol

17.4.1 Introduction

The research on the use of methanol has been one of the most dynamic research areas in recent years. Forty citation classics in the field of methanol use as a fuel with more than twenty-nine citations were located and the key emerging issues from these papers are presented in the decreasing order of the number of citations (Table 17.3). These citation classics are presented under four topical headings starting with methanol fuel cells in general and followed by the specific use of methanol fuel cells in transportation vehicles, the use of liquid methanol fuels in car engines, and the use of methanol in hydrogen production by the method of steam reforming. Due to its importance, 10 papers were presented for each topical area (Table 17.3).

These papers give strong hints about the determinants of efficient use of methanol in four major industrial contexts and emphasize that it is possible to use methanol as a fuel in all these contexts optimally in relation to other fuels.

The papers were dominated by the researchers from 16 countries, usually through the intracountry institutional collaboration, and they were multiauthored. The United States (nine papers), China (six papers), Australia and Spain (five papers each), Germany (four papers), and Canada and Turkey (three papers each) were the most prolific countries. Similarly, Hokkaido University and Royal Institute of Technology–KTH, Princeton University, Royal Military College of Canada, University of California, University of New South Wales, CSIRO, University of Cambridge, and Xi'an Jiaotong University (two papers each) were the most prolific institutions.

Similarly, all these papers were published in the journals indexed by the SCI and/or SCIE or SSCI. There were no papers indexed by the A&HCI. The number of citations ranged from 29 to 1432 and the average number of citations per annum ranged from 2.8 to 110.1. All of these papers were articles. *J. Power Sourc.* (seven papers), *Appl. Catal. A Gen.* (six papers), and *Catal. Today*, *Energ. Pol.*, *Fuel*, and *Int. J. Hydrogen Energ.*, *J. Catal.* (two papers each) were the most prolific journals publishing papers on the use of methanol as a fuel.

TABLE 17.3
Research on the Use of Methanol

No.	Paper References	Year	Document	Affiliation	Country	Number of Authors	Journal	Index	Subjects	Topic	Total Number of Citations	Total Average Citations per Annum	Rank
1	Kreuer	2001	A	Max Planck Inst.	Germany	1	*J. Membr. Sci.*	SCI, SCIE	Eng. Chem., Poly. Sci.	Methanol fuel cells—membrane development	1432	110.1	1
2	Wasmus and Kuver	1999	R	Degussa AG	Germany	2	*J. Electroanal. Chem.*	SCI, SCIE	Chem. Anal., Electrochem.	Methanol fuel cells—review	748	49.9	2
3	Reddington et al.	1998	A	IIT, Penn State Univ., ICET Inc.	United States	7	*Science*	SCI, SCIE	Mult. Sci.	Methanol fuel cells—electrocatalyst selection	624	39.0	4
4	Li et al.	2003	A	Chinese Acad. Sci., Dalian Univ. Technol.	China	7	*J. Phys. Chem. B*	SCI, SCIE	Chem. Phys.	Methanol fuel cells—nanocomposite catalysts	622	56.5	5
5	Liu et al.	2006	R	Natl. Res. Council Canada, Univ. British Columbia	Canada	6	*J. Power Sourc.*	SCI, SCIE	Electrochem., Ener. Fuels	Methanol fuel cells—anode catalysts	584	73.0	7
6	Arico et al.	2001	R	Inst. CNR TAE, Princeton Univ.	Italy, United States	3	*Fuel Cells*	SCIE	Electrochem., Ener. Fuels	Methanol fuel cells— general	547	42.1	8
7	Heinzel and Barragan	1999	A	Fraunhofer Inst. Solar Energy Syst ISE, Univ. Complutense Madrid	Germany, Spain	2	*J. Power Sourc.*	SCI, SCIE	Electrochem., Ener. Fuels	Methanol fuel cells—methanol crossover	479	31.9	9
8	Xu et al.	2008	A	Nanjing Univ. Sci. & Technol.	China	3	*J. Phys. Chem. C*	SCI, SCIE	Chem. Phys., Nanosci. Nanotech., Mats. Sci. Mult.	Methanol fuel cells—nanocomposite catalysts	472	78.7	10

(*Continued*)

TABLE 17.3 (*Continued*)

Research on the Use of Methanol

No.	Paper References	Year	Document	Affiliation	Country	Number of Authors	Journal	Index	Subjects	Topic	Total Number of Citations	Total Average Citations per Annum	Rank
9	Wainright et al.	1995	A	Case Western Reserve Univ.	United States	5	*J. Electrochem. Soc.*	SCI, SCIE	Electrochem., Mats. Sci. Coat. Films	Methanol fuel cells—polymer electrolytes	457	24.1	11
10	Ren et al.	2000	A	Univ. Calif. Los Alamos Natl. Lab.	United States	5	*J. Power Sourc.*	SCI, SCIE	Electrochem., Ener. Fuels	Methanol fuel cells—methanol crossover	444	29.6	12
11	Trimm and Onsan	2001	R	Univ. New S Wales, Bogazici Univ.	Australia, Turkey	2	*Catal. Rev. Sci. Eng.*	SCI, SCIE	Chem. Phys.	Methanol fuel vehicles—comparative performance	358	27.5	14
12	Brown	2001	A	Univ. Calif. Los Alamos Natl. Lab.	United States	1	*Int. J. Hydrogen Energ.*	SCI, SCIE	Chem. Phys., Electrochem., Ener. Fuels	Methanol fuel vehicles—comparative performance	333	25.6	17
13	Peppley et al. (1999a)	1999a	A	Royal Mil. Coll. Canada	Canada	4	*Appl. Catal. A Gen.*	SCI, SCIE	Chem. Phys., Env. Sci.	Methanol–steam reforming	301	20.1	18
14	Peppley et al. (1999a)	1999a	A	Royal Mil. Coll. Canada	Canada	4	*Appl. Catal. A Gen.*	SCI, SCIE	Chem. Phys., Env. Sci.	Methanol–steam reforming	252	16.8	19
15	Breen and Ross	1999	A	Univ. Limerick	Ireland	2	*Catal. Today*	SCI, SCIE	Chem. Appl., Chem. Phys., Eng. Chem.	Methanol–steam reforming	248	16.5	20
16	Velu et al.	2000	A	Natl. Ind. Res. Inst. Nagoya, Osaka Natl. Res. Inst.	Japan	6	*J. Catal.*	SCI, SCIE	Chem. Phys., Eng. Chem.	Methanol–steam reforming	217	15.5	22
17	Takezawa and Iwasa	1997	A	Hokkaido Univ.	Japan	2	*Catal. Today*	SCI, SCIE	Chem. Appl., Chem. Phys., Eng. Chem.	Methanol–steam reforming	201	11.8	26
18	Palo et al.	2007	A	Pacific NW Natl. Lab.	United States	3	*Chem. Rev.*	SCI, SCIE	Chem. Mult.	Methanol–steam reforming	182	26	30

(*Continued*)

TABLE 17.3 (*Continued*)
Research on the Use of Methanol

No.	Paper References	Year	Document	Affiliation	Country	Number of Authors	Journal	Index	Subjects	Topic	Total Number of Citations	Total Average Citations per Annum	Rank
19	Ogden et al.	2004	A	Princeton Univ.	United States	3	*J. Power Sourc.*	SCI, SCIE	Electrochem., Ener. Fuels	Methanol fuel vehicles—comparative performance	180	45.0	31
20	Iwasa et al.	1995	A	Hokkaido Univ.	Japan	4	*Appl. Catal. A Gen.*	SCI, SCIE	Chem. Phys., Env. Sci.	Methanol–steam reforming	171	9.0	32
21	Jiang et al. (1993a)	1993a	A	Univ. New S Wales, Macquarie Univ.	Australia	4	*Appl. Catal. A Gen.*	SCI, SCIE	Chem. Phys., Env. Sci.	Methanol–steam reforming	168	8.0	34
22	Agrell et al.	2003	A	KTH-Royal Inst. Technol., CSIC	Sweden, Spain	6	*J. Catal.*	SCI, SCIE	Chem. Phys., Eng. Chem.	Methanol–steam reforming	159	14.5	35
23	Jiang et al. (1993b)	1993b	A	Univ. New S Wales, Macquarie Univ.	Australia	4	*Appl. Catal. A Gen.*	SCI, SCIE	Chem. Phys., Env. Sci.	Methanol–steam reforming	156	7.4	36
24	Fieweger et al.	1997	A	Rhein Westfal TH Aachen	Germany	3	*Combust. Flame*	SCI, SCIE	Eng. Mult., Eng. Chem.	Methanol transportation fuels—self-ignition	140	8.3	40
25	Kumar et al.	2003	A	Indian Inst. Technol.	India	3	*Biomass Bioenerg.*	SCIE	Agr. Eng., Biot. Appl. Microb., Ener. Fuels	Methanol transportation fuels—comparative performance	128	11.6	43
26	McNicol et al.	1999	A	CSIRO Energy Technol., Univ. Cambridge	England, Australia	3	*J. Power Sourc.*	SCI, SCIE	Electrochem., Ener. Fuels	Methanol fuel vehicles—review	124	8.3	44
27	Tzeng et al.	2005	A	Natl. Chiao Tung Univ., Univ. Belgrade	Taiwan, Serbia	3	*Energ. Pol.*	SCIE, SSCI	Ener. Fuels, Env. Sci., Env. Stud.	Methanol fuel vehicles—hybrid electric buses	120	13.3	48

(*Continued*)

TABLE 17.3 (*Continued*)
Research on the Use of Methanol

No.	Paper References	Year	Document	Affiliation	Country	Number of Authors	Journal	Index	Subjects	Topic	Total Number of Citations	Total Average Citations per Annum	Rank
28	Thomas et al.	2000	A	Directed Technol. Inc.	United States	4	*Int. J. Hydrogen Energ.*	SCI, SCIE	Chem. Phys., Electrochem., Ener. Fuels	Methanol fuel vehicles—comparative performance	95	6.8	53
29	McNicol et al.	2001	A	CSIRO Energy Technol., Univ. Cambridge	Australia, England	3	*J. Power Sourc.*	SCI, SCIE	Electrochem., Ener. Fuels	Methanol fuel vehicles—review	90	6.9	54
30	Lindstrom et al.	2002	A	Royal Inst. Technol.–KTH	Sweden	3	*Appl. Catal. A Gen.*	SCI, SCIE	Chem. Phys., Env. Sci.	Methanol fuel vehicles—methanol reforming	87	7.3	55
31	Ogden et al.	2004	A	Princeton Univ.	United States	3	*Energ. Pol.*	SCIE, SSCI	Ener. Fuels, Env. Sci., Env. Stud.	Methanol fuel vehicles—comparative performance	82	8.2	56
32	Zervas et al.	2002	A	Inst. Francais Petr., Inst. Chim. Sur. Int.	France	3	*Environ. Sci. Technol.*	SCI, SCIE	Eng. Env., Env. Sci.	Methanol transportation fuels—comparative emissions	75	6.3	60
33	Kawatsu	1998	A	Toyota Motor Corp.	Japan	1	*J. Power Sourc.*	SCI, SCIE	Electrochem., Ener. Fuels	Methanol fuel vehicles—comparative performance	73	4.6	61
34	Cheng et al.	2008	A	Tianjin Univ., Hong Kong Polytech. Univ.	China	6	*Fuel*	SCI, SCIE	Ener. Fuels, Eng. Chem.	Methanol transportation fuels—comparative performance	46	7.7	65

(*Continued*)

TABLE 17.3 (*Continued*)
Research on the Use of Methanol

No.	Paper References	Year	Document	Affiliation	Country	Number of Authors	Journal	Index	Subjects	Topic	Total Number of Citations	Total Average Citations per Annum	Rank
35	Qi et al.	2010	A	Changan Univ.	China	5	*Appl. Energ.*	SCI, SCIE	Ener. Fuels, Eng. Chem.	Methanol transportation fuels—comparative performance	38	9.5	69
36	Chao et al.	2001	A	Natl. Cheng Kung Univ., Chinese Petr. Corp.	Taiwan	5	*Sci. Total Environ.*	SCI, SCIE	Env. Sci.	Methanol transportation fuels—emission performance	37	2.8	71
37	Huang et al. (2004a)	2004a	A	Xi'an Jiaotong Univ.	China	7	*Bioresour. Technol.*	SCI, SCIE	Agr. Eng., Ener. Fuels, Biot. Appl. Microb.	Methanol transportation fuels—combustion behavior	36	3.6	72
38	Huang et al. (2004b)	2004b	A	Xi'an Jiaotong Univ.	China	7	*Proc. Inst. Mech. Eng. Part D–J. Automob. Eng.*	SCI, SCIE	Ng. mech., Trans. Sci. Tech.	Methanol transportation fuels—combustion behavior	35	3.5	73
39	Bayraktar	2008	A	Karadeniz Tech. Univ.	Turkey	1	*Fuel*	SCI, SCIE	Ener. Fuels, Eng. Chem.	Methanol transportation fuels—performance	32	5.3	75
40	Sayin et al.	2009	A	Kocaeli Univ., Marmara Univ., GM Acad.	Turkey	4	*Renew. Energ.*	SCIE	Ener. Fuels	Methanol transportation fuels—performance	29	5.8	78

SCI, Science Citation Index; SCIE, Science Citation Index Expanded; SSCI, Social Sciences Citation Index; A, article; R, review.

17.4.2 Research on the Use of Methanol

17.4.2.1 Research on Methanol Fuel Cells

Kreuer (2001) discuss the development of proton-conducting polymer membranes for hydrogen and methanol fuel cells in a paper with 1432 citations. He finds that the less pronounced hydrophobic/hydrophilic separation of sulfonated polyetherketones compared to NAFION corresponds to narrower, less connected hydrophilic channels and to larger separations between less acidic sulfonic acid functional groups. At high water contents, this significantly reduces electroosmotic drag and water permeation while maintaining high proton conductivity. Blending of sulfonated polyetherketones with other polyaryls even further reduces the solvent permeation, increases the membrane flexibility in the dry state, and leads to an improved swelling behavior. He asserts that polymers based on sulfonated polyetherketones are not only an interesting low-cost alternative membrane material for hydrogen-fuel-cell applications, they may also help to reduce the problems associated with high water drag and high methanol crossover in direct liquid methanol fuel cells (DMFC).

Wasmus and Kuver (1999) discuss methanol oxidation and direct methanol fuel cells (DMFCs) in a paper with 748 citations. They focus on strategies and approaches rather than on individual results as they describe the state of the art in the fields of methodology, catalysis, catalyst characterization, polymer electrolytes, and assessment and interpretation of cell and electrode performance.

Reddington et al. (1998) investigate the screening method for discovery of better electrocatalysts for DMFCs in a paper with 624 citations. By converting the ions generated in an electrochemical half-cell reaction to a fluorescence signal, they identify the most active compositions in a large electrode array. They use a fluorescent acid–base indicator to image high concentrations of hydrogen ions, which were generated in the electrooxidation of methanol. They screen a 645-member electrode array containing five elements (platinum, ruthenium, osmium, iridium, and rhodium), 80 binary, 280 ternary, and 280 quaternary combinations to identify the most active regions of phase space. They find several very active compositions, some in ternary and quaternary regions, that were bounded by rather inactive binaries. They argue that the best catalyst, platinum(44)/ruthenium(41)/osmium(10)/iridium(5) (numbers in parentheses are atomic percent), was significantly more active than platinum(50)/ruthenium(50) in a DMFC operating at 60°C, even though the latter catalyst had about twice the surface area of the former.

Li et al. (2003) prepare multiwalled carbon nanotube–supported Pt (Pt/MWNT) nanocomposites by both the aqueous solution reduction of a Pt salt (HCHO reduction) and the reduction of a Pt ion salt in ethylene glycol solution in comparison to a Pt/XC-72 nanocomposite in a paper with 622 citations. They find that the "Pt/MWNT catalyst prepared by the EG method has a high and homogeneous dispersion of spherical Pt metal particles with a narrow particle-size distribution" as the "Pt particle size is in the range of 2–5 nm with a peak at 2.6 nm." They next find that surface chemical modifications of MWNTs and water content in EG solvent are the key factors in depositing Pt particles on MWNTs. In the case of the DMFC test, the Pt/MWNT catalyst prepared by EG reduction is slightly superior to the catalyst prepared by aqueous reduction and displays significantly higher performance than the Pt/XC-72 catalyst. They assert that these differences in catalytic performance are due to "a greater dispersion of the supported Pt particles when the EG method is used, in contrast to aqueous HCHO reduction and to possible unique structural and higher electrical properties when contrasting MWNTs to carbon black XC-72 as a support."

Liu et al. (2006) discuss anode catalysis in DMFCs in a paper with 584 citations focusing on the three most active areas: (1) progress in preparation methods of Pt–Ru catalysts with respect to activity improvement and utilization optimization, (2) preparation of novel carbon materials as catalyst supports to create a highly dispersed and stably supported catalysts, and (3) exploration of new catalysts with a low noble-metal content and non-noble-metal elements through fast activity down-selection methods such as combinatorial methods.

Arico et al. (2001) describe recent developments in fundamental and technological aspects of DMFCs in a paper with 547 citations. They focus on the electrocatalysis of the methanol oxidation reaction and oxygen electroreduction where they discuss the promoting effect on Pt of additional elements, and some aspects of the electrocatalysis of oxygen reduction in the presence of methanol crossover have been treated. They then discuss the development of both components and devices focusing on the development of high-surface-area electrocatalysts and alternative electrolyte membranes to Nafion, as well as the fabrication methodologies for the membrane/electrode assembly (MEA). They finally discuss the recent efforts in developing DMFC stacks for both portable and electrotraction applications.

Heinzel and Barragan (1999) provide a review of the state of the art of the methanol crossover in DMFCs in a paper with 479 citations. They note that although DMFCs are attractive for several applications, there are several barriers that must be overcome before they can become an alternative to internal-combustion engines (ICEs). They argue that the methanol crossover from the anode to the cathode is the major limitation. They discuss the influence of methanol crossover in DMFC and the efforts to get a more methanol-impermeable polymer electrolyte.

Xu et al. (2008) discuss graphene–metal particle nanocomposites in a paper with 472 citations and present a general approach for the preparation of graphene–metal particle nanocomposites in a water–ethylene glycol system using graphene oxide as a precursor and metal nanoparticles (Au, Pt, and Pd) as building blocks. They argue that these metal nanoparticles are adsorbed on graphene oxide sheets and play a pivotal role in catalytic reduction of graphene oxide with ethylene glycol, leading to the formation of graphene–metal particle nanocomposites. They assert that the typical methanol oxidation of graphene–Pt composites in cyclic graphene voltammogram analyses indicated its potential application in DMFCs, bringing graphene–particle nanocomposites close to real technological applications.

Wainright et al. (1995) discuss acid-doped polybenzimidazoles, a new polymer electrolyte, as potential polymer electrolytes for use in hydrogen/air and DMFCs in a paper with 457 citations. They present experimental findings on proton conductivity, water content, and methanol vapor permeability of this material, as well as preliminary fuel cell results. They argue that the low methanol vapor permeability of these electrolytes significantly reduces the adverse effects of methanol crossover typically observed in direct methanol polymer electrolyte membrane fuel cells.

Ren et al. (2000) describe recent advances in the science and technology of DMFCs made at Los Alamos National Laboratory (LANL) in a paper with 444 citations. The effort on DMFCs at LANL includes work devoted to portable power applications and work devoted to potential transport applications. They describe recent results with a new type of DMFC stack hardware that allows to lower the pitch per cell to 2 mm while allowing low air flow and air pressure drops. Such stack technology lends itself to both portable power and potential transport applications. They find that power densities of 300 W/l and 1 kW/l are achievable under conditions applicable to portable power and transport applications, respectively. DMFC power system analysis based on the performance of this stack, under conditions applying to transport applications (joint effort with U.C. Davis), has shown that, in terms of overall system efficiency and system packaging requirements, a power source for a passenger vehicle based on a DMFC could compete favorably with a hydrogen-fueled fuel-cell system, as well as with fuel-cell systems based on fuel processing onboard. As part of more fundamental studies performed, we describe optimization of anode catalyst layers in terms of PtRu catalyst nature, loading, and catalyst layer composition and structure. We specifically show that optimized content of recast ionic conductor added to the catalyst layer is a sensitive function of the nature of the catalyst. Other elements of MEA optimization efforts are also described, highlighting our ability to resolve, to a large degree, a well-documented problem of polymer electrolyte DMFCs, namely, methanol crossover. This was achieved by appropriate cell design, enabling fuel utilization as high as 90% in highly performing DMFCs.

17.4.2.2 Research on Methanol-Fuel-Cell Vehicles

Trimm and Onsan (2001) discuss onboard fuel conversion for hydrogen-fuel-cell-driven vehicles in a review paper with 358 citations. Increasingly, stringent legislation controls emissions from ICEs to the point where alternative power sources for vehicles are necessary. The hydrogen fuel cell is one promising option, but the nature of the gas is such that the conversion of other fuels to hydrogen onboard the vehicle is necessary. They discuss the conversion of methanol, methane, propane, and octane to hydrogen. They argue that a combination of oxidation and steam reforming (indirect POX) or direct POX is the most promising process. Indirect POX involves combustion of part of the fuel to produce sufficient heat to drive the endothermic steam-reforming reaction. Direct POX is favored only at high temperatures and short residence times but is highly selective. However, indirect POX is shown to be the preferred process for all fuels. The product gases can be taken through a water–gas shift reactor but still retain similar to 2% carbon monoxide, which poisons fuel-cell catalysts. Selective oxidation is the preferred route to removal of residual carbon monoxide.

Brown (2001) carries out a comparative study of fuels for onboard hydrogen production for fuel-cell-powered automobiles in a paper with 33 citations. He compares seven common fuels (methanol, natural gas, gasoline, diesel fuel, aviation jet fuel, ethanol, and hydrogen) for their utility as hydrogen sources for proton-exchange-membrane fuel cells used in automotive propulsion. He notes that except for steam-reforming methanol and using pure hydrogen, all processes for generating hydrogen from these fuels require water–gas shift reactors of significant size. This occurs because their higher processing temperatures produce unacceptably large amounts of CO. All processes require low- or zero-sulfur fuels and this may add cost to some of them. "Fuels produced by pure steam-reforming contain similar to 70–80% hydrogen, those by pure partial oxidation similar to 35–45%." The lower percentages may adversely affect cell performance. Pure steam-reforming suffers from poor transient operation. Theoretical input energies do not differ markedly among the processes for generating hydrogen from organic-chemical fuels. Pure hydrogen has formidable distribution and storage difficulties. He recommends some combination of POX and steam reforming of methanol, technically the leading candidate for onboard generation of hydrogen for automotive propulsion. The possible use of methanol suffers from a lack of infrastructure and solubility in water combined with toxicity.

Ogden et al. (1999) compare hydrogen, methanol, and gasoline as fuels for fuel-cell vehicles (FCVs) with a focus on the implications for vehicle design and infrastructure development in a paper with 180 citations. They present modeling results comparing three leading options for fuel storage onboard FCVs: (1) compressed gas hydrogen storage, (2) onboard steam reforming of methanol, and (3) onboard POX of HC fuels derived from crude oil. They develop an FCV model, including detailed models of onboard fuel processors. This allowed them to compare vehicle performance; fuel economy, weight, and cost for various vehicle parameters; fuel storage choices; and driving cycles. They also compare the infrastructure requirements for gaseous hydrogen, methanol, and gasoline, including the added costs of fuel production, storage, distribution, and refueling stations. They then estimate delivered fuel cost, total life cycle cost of transportation, and capital cost of infrastructure development for each alternative.

McNicol et al. (1999) discuss DMFCs for road transportation in a paper with 124 citations. They discuss the history of the technology and assess the various problems associated with its commercial development, in particular the mechanisms of the electrode reactions, the development of effective catalysts, and the possible electrolytes that can be used. They then discuss the barriers to successful commercialization.

Tzeng et al. (2005) carry out the multicriteria analysis of alternative-fuel buses for public transportation in a paper with 120 citations. They consider several types of fuels as alternative-fuel modes, that is, electricity, fuel cell (hydrogen), and methanol. They argue that electric vehicles may be considered the alternative-fuel vehicles with the lowest air pollution while the hybrid electric vehicles provide an alternate mode, at least for the period of improving the technology of electric vehicles. A hybrid electric vehicle is a vehicle with the conventional ICE and an electric motor as its major sources of power. Experts from different decision-making groups performed the multiple attribute evaluation of alternative vehicles. They find that the "hybrid electric bus is the most suitable substitute bus for Taiwan urban areas in the short and median term" while "if the cruising distance of the electric bus extends to an acceptable range, the pure electric bus could be the best alternative."

Thomas et al. (2000) discuss the fuel options for the FCV for hydrogen, methanol, or gasoline in a paper with 90 citations. FCVs can be powered directly by hydrogen, with an onboard chemical processor, or other liquid fuels such as gasoline or methanol. They argue that hydrogen is the preferred fuel in terms of reducing vehicle complexity, but the cost of a hydrogen infrastructure would be excessive. They next argue that the automobile industry must develop complex onboard fuel processors to convert methanol, ethanol, or gasoline to hydrogen. They show that the total fuel infrastructure cost to society including onboard fuel processors may be less for hydrogen than for either gasoline or methanol, the primary initial candidates currently under consideration for FCVs. They also present the local air pollution and greenhouse gas advantages of hydrogen-fuel-cell vehicles compared to those powered by gasoline or methanol.

McNicol et al. (2001) discuss fuel cells for road transportation purposes as they focus on the advantages and disadvantages of the candidate fuel-cell systems and the various fuels together with the issue of whether the fuel should be converted directly in the fuel cell or should be first converted to hydrogen onboard the vehicle. They review developments in competing vehicle technologies, namely, internal-combustion-engine vehicles (ICEVs), pure-battery vehicles (EVs), and ICE-battery hybrid vehicles (HEVs). Finally, they examine the impact of the introduction of FCVs on industry and, in particular, on the oil and automotive industries. For FCVs to compete successfully with conventional ICEVs, they assert that direct-conversion fuel cells, using probably hydrogen, but possibly methanol, are the only realistic contenders for road transportation applications.

Lindstrom et al. (2002) investigate the influence of catalyst properties on the activity and selectivity of hydrogen generation by methanol reforming over copper-based catalysts impregnated on gamma-alumina pellets in a paper with 87 citations as they tested three sets of copper-based catalysts with various compositions: $Cu/Zn/Al_2O_3$, $Cu/Cr/Al_2O_3$, and $Cu/Zr/Al_2O_3$. They find a "correlation between the copper surface area and catalytic activity." The results of the activity tests indicate that the "choice of promoter and the catalyst composition greatly influence the activity as well as the selectivity for CO_2 formation" as the "highest conversions were achieved for the zinc-containing catalysts ($Cu/Zn/Al_2O_3$) for both steam reforming and the combined reforming process." Complete conversion of methanol was only obtained for the zinc-containing catalysts when running the steam-reforming process. The combined reforming process generally yielded a product stream containing lower carbon monoxide concentrations compared to steam reforming at the equivalent reactor temperature for all of the catalysts tested.

Ogden et al. (2004) discuss the societal life cycle costs of cars with alternative fuels/engines in a paper with 82 citations. They use the societal life cycle cost of transportation, including the vehicle first cost (assuming large-scale mass production), fuel costs (assuming a fully developed fuel infrastructure), externality costs for oil supply security, and damage costs for emissions of air pollutants and greenhouse gases calculated over the full fuel cycle as the basis for comparing alternative automotive engine/fuel options in evolving toward zero-pollutant goals. They consider several engine/fuel options—including current gasoline ICEs and a variety of advanced lightweight vehicles such as ICE vehicles fueled with gasoline or hydrogen; ICE/hybrid electric vehicles fueled with gasoline, compressed natural gas, diesel, and Fischer–Tropsch liquids or hydrogen; and FCVs fueled with gasoline, methanol, or hydrogen (from natural gas, coal, or wind power). They estimate the life cycle costs for a range of possible future conditions. Under base-case conditions, they find that "several advanced options have roughly comparable lifecycle costs that are lower than for today's conventional gasoline internal combustion engine cars, when environmental and oil supply insecurity externalities are counted-including advanced gasoline internal combustion engine cars, internal combustion engine/hybrid electric cars fueled with gasoline, diesel, Fischer-Tropsch liquids or compressed natural gas, and hydrogen fuel cell cars." They argue that the hydrogen-fuel-cell car stands out as having the lowest externality cost of any option and, when mass-produced and with high valuations of externalities, the least projected life cycle cost.

Kawatsu (1998) discusses advanced polymer electrolyte fuel cell (PEFC) development for fuel-cell-powered vehicles in Japan in a paper with 73 citations. He discusses a Toyota fuel-cell electric vehicle (FCEV) using hydrogen as the fuel, which was developed and introduced in 1996, followed by another Toyota FCEV using methanol as the fuel, developed and introduced in 1997. In those Toyota FCEVs, a fuel-cell system is installed under the floor of each RAV4L, a sports utility vehicle. He finds that the "CO concentration in the reformed gas of methanol reformer can be reduced to 100 ppm in wide ranges of catalyst temperature and gas flow rate, by using the ruthenium (Ru) catalyst as the CO selective oxidizer, instead of the platinum (Pt) catalyst." He then finds that a "fuel cell performance equivalent to that with pure hydrogen can be ensured even in the reformed gas with the carbon monoxide (CO) concentration of 100 ppm, by using the Pt-Ru (platinum ruthenium alloy) electrocatalyst as the anode electrocatalyst of a (PEFC), instead of the Pt electrocatalyst."

17.4.2.3 Research on Methanol Transportation Fuels

Fieweger et al. (1997) investigate the self-ignition of SI engine model fuels (isooctane, methanol, methyl tert-butyl ether [MTBE], and three different mixtures of isooctane and n-heptane), mixed with air with a focus on a shock tube investigation at high pressure in a paper with 140 citations. They find that for temperatures relevant to piston engine combustion, the "self-ignition process always starts as an inhomogeneous, deflagrative mild ignition." This instant is defined by the ignition delay time, T_{defl}. The deflagration process in most cases is followed by a secondary explosion (DDT). This transition defines a second ignition delay time, T_{DDT}, which is a suitable approximation for the chemical ignition delay time, if the change of the thermodynamic conditions of the unburned test gas due to deflagration is taken into account. For isooctane at p = 40 bar, they observe an "NTC (negative temperature coefficient), behavior connected with a two step (cool flame) self-ignition at low temperatures." This process was very pronounced for rich and less pronounced for stoichiometric mixtures. The results of the T_{DDT} delays of the stoichiometric mixtures were shortened by the primary deflagration process in the temperature range between 800 and 1000 K. They find a "strong influence of the n-heptane fraction in the mixture, both on the ignition delay time and on the mode of self-ignition." The self-ignition of methanol and MTBE is characterized by a very pronounced initial deflagration. For temperatures below 900 K (methanol: 800 K), no secondary explosion occurs. The measured delays T_{DDT} of the secondary explosion are shortened by up to one order of magnitude.

Kumar et al. (2003) compare the methods to use methanol and jatropha oil such as blending, transesterification, and dual-fuel operation in a compression ignition (CI) (diesel) engine in a paper with 128 citations. They carry out the tests at a constant speed of 1500 rev/min at varying power outputs. In dual-fuel operation, the methanol-to-jatropha-oil ratio was maintained at 3:7 on the volume basis, which is close to the fraction of methanol used to prepare the ester with jatropha oil. They find that "brake thermal efficiency was better in the dual fuel operation and with the methyl ester of jatropha oil as compared to the blend." It increased from 27.4% with neat jatropha oil to a maximum of 29% with the methyl ester and 28.7% in the dual-fuel operation. Smoke was reduced with all methods compared to neat vegetable oil operation. The values of smoke emission are 4.4 Bosch smoke units (BSU) with neat jatropha oil, 4.1 BSU with the blend, 4 BSU with methyl ester of jatropha oil, and 3.5 BSU in the dual-fuel operation. The nitric oxide (NO) level was lower with jatropha oil compared to diesel as it was further reduced in dual-fuel operation and the blend with methanol. Dual-fuel operation showed higher HC and carbon monoxide (CO) emissions than the ester and the blend. Ignition delay was higher with neat jatropha oil as it increased further with the blend and in dual-fuel operation. It was reduced with the ester. Peak pressure and rate of pressure rise were higher with all the methods compared to neat jatropha oil operation. jatropha oil and methyl ester showed higher diffusion combustion compared to standard diesel operation. However, dual-fuel operation resulted in higher premixed combustion. They assert that transesterification of vegetable oils and methanol induction can significantly enhance the performance of a vegetable oil fueled diesel engine.

Zervas et al. (2002) investigate the emission of alcohols and carbonyl compounds from an SI engine to study the impact of fuel composition and of the air/fuel equivalence (lambda) ratio on exhaust emissions of alcohols and aldehydes/ketones in a paper with 75 citations. Fuel blends contained eight HCs (n-hexane, 1-hexene, cyclohexane, n-octane, isooctane, toluene, o-xylene, and ethylbenzene [ET8]) and four oxygenated compounds (methanol,

ethanol, 2-propanol, and MTBE). Exhaust methanol is principally produced from fuel methanol and MTBE but also from ethanol, 2-propanol, isooctane, and hexane. Exhaust ethanol and 2-propanol are produced only from the respective fuel compounds. Exhaust formaldehyde is mainly produced from fuel methanol, acetaldehyde from fuel ethanol, and propionaldehyde from straight-chain HCs. Exhaust acrolein comes from fuel 1-hexene, acetone from 2-propanol, n-hexane, n-octane, isooctane, and MTBE. Exhaust crotonaldehyde comes from fuel 1-hexene, cyclohexane, n-hexane, and n-octane; methacrolein from fuel isooctane; and benzaldehyde from fuel aromatics. They argue that light pollutants (C_1–C_2) are most likely formed from intermediate species, which are quite independent of the fuel composition. An increase in λ increases the exhaust concentration of acrolein, crotonaldehyde, and methacrolein and decreases these three alcohols for the alcohol-blended fuels. The concentration of methanol, formaldehyde, propionaldehyde, and benzaldehyde is a maximum at stoichiometry. The exhaust concentration of acetaldehyde and acetone presents a complex behavior: it increases in some cases, decreases in others, or presents a maximum at stoichiometry. The concentration of four aldehydes (formaldehyde, acetaldehyde, propionaldehyde, and benzaldehyde) is also linked with the exhaust temperature and fuel H/C ratio.

Cheng et al. (2008) compare the emissions of a direct injection diesel engine operating on biodiesel with emulsified and fumigated methanol in a paper with 46 citations. They compare the effect of applying a biodiesel with either 10% blended methanol or 10% fumigation methanol. The biodiesel was converted from waste cooking oil. Experiments were performed on a four-cylinder naturally aspirated direct injection diesel engine operating at a constant speed of 1800 rev/min with five different engine loads. They find a "reduction of CO_2, NO_x, and particulate mass emissions and a reduction in mean particle diameter, in both cases, compared with diesel fuel." They then find that for the blended mode, "there is a slightly higher brake thermal efficiency at low engine load while the fumigation mode gives slightly higher brake thermal efficiency at medium and high engine loads." In the fumigation mode, an extra fuel injection control system is required, and there is also an increase in CO, HC, and NO_2 and particulate emissions in the engine exhaust, which are disadvantageous compared with the blended mode.

Qi et al. (2010) study the performance and combustion characteristics (engine performance, emissions, and combustion characteristics) of a biodiesel–diesel–methanol blend fueled engine to evaluate the effects of using methanol as additive to biodiesel–diesel blends in a direct injection diesel engine under variable operating conditions in a paper with 36 citations. BD50 (50% biodiesel and 50% diesel in vol.) was prepared as the baseline fuel. Methanol was added to BD50 as an additive by volume percent of 5% and 10% (denoted as BDM5 and BDM10). They find that the "combustion starts later for BDM5 and BDM10 than for BD50 at low engine load, but is almost identical at high engine load." They next find that "at low engine load of 1500 r/min, BDM5 and BDM10 show the similar peak cylinder pressure and peak of pressure rise rate to BD50, and higher peak of heat release rate than that of BD50. At low engine load of 1800 r/min, the peak cylinder pressure and the peak of pressure rise rate of BDM5 and BDM10 are lower than those of BD50, and the peak of heat release rate is similar to that of BD50. The crank angles at which the peak values occur are later for BDM5 and BDM10 than for BD50." On the other hand, at a high engine load, the "peak cylinder pressure, the peak of pressure rise rate, and peak of heat release rate of BDM5 and BDM10 are higher than those of BD50, and the crank angle of peak values for all tested fuels are almost same." The power and torque outputs of BDM5 and BDM10 are slightly lower than those of BD50. BDM5 and BDM10 show dramatic reduction of smoke emissions.

Chao et al. (2001) investigate the effects of methanol-containing additive (MCA) on emission characteristics of HCs, carbon monoxide (CO), nitrogen oxides (NO_x), particulate matter (PM), and unregulated carbon dioxide (CO_2) and polycyclic aromatic HCs (PAHs) from a heavy-duty diesel engine in a paper with 37 citations. The engine was tested on a series of diesel fuels blended with five additive levels (0%, 5%, 8%, 10%, and 15% of MCA by volume). Emissions tests were performed under both cold- and hot-start transient heavy-duty federal test procedure (HD-FTP) cycles and two selected steady-state modes. They find that "MCA addition slightly decreases PM emissions but generally increases both THC and CO emissions." Decrease in NO emissions was common in all MCA blends. As for unregulated emissions, CO_2 emissions did not change significantly for all MCA blends, while vapor-phase and particle-associated PAHs emissions in high load and transient cycle tests were relatively low compared to the base diesel when either 5% or 8% MCA was used, due to the lower PAHs levels in MCA blends. Finally, the particle-associated PAHs emissions also showed trends quite similar to that of the PM emissions in this study.

Huang et al. (2004a) study the combustion behaviors of a CI engine fueled with diesel/methanol blends under various fuel delivery advance angles in a paper with 36 citations. They find that "increasing methanol mass fraction of the diesel/methanol blends would increase the heat release rate in the premixed burning phase and shorten the combustion duration of the diffusive burning phase." Furthermore, the ignition delay increased with the advancing of the fuel delivery advance angle for both the diesel fuel and the diesel/methanol blends. For a specific fuel delivery advance angle, the ignition delay increased with the increase of the methanol mass fraction (oxygen mass fraction) in the fuel blends and the behaviors were more obvious at low engine load and/or high engine speed. The rapid burn duration and the total combustion duration increased with the advancing of the fuel delivery advance angle. The center of the heat release curve was close to top dead center with the advancing of the fuel delivery advance angle. Maximum cylinder gas pressure increased with the advancing of the fuel delivery advance angle, and the maximum cylinder gas pressure of diesel/methanol blends gave a higher value than that of diesel fuel. The maximum mean gas temperature remained almost unchanged or had a slight increase with the advancing of the fuel delivery advance angle, and it only slightly increased for diesel/methanol blends compared to that of diesel fuel.

Huang et al. (2004b) investigate the engine performance and emissions of a CI engine operating on diesel–methanol blends in a paper with 35 citations. They find that the "engine thermal efficiency increases and the diesel equivalent brake specific fuel consumption (BSFC) decreases with increase in the oxygen mass fraction (or methanol mass fraction) of the diesel-methanol blends due to an increased fraction of premixed combustion phase, oxygen enrichment and improvement in the diffusive combustion phase." They argue that "further increase in the fuel delivery advance angle will achieve a better engine thermal efficiency when the diesel engine is operated using the diesel-methanol fuel blends." A marked reduction in exhaust CO and smoke can be achieved when operating with diesel–methanol blend. There is not a large variation in the exhaust HC with the addition of methanol in diesel fuel. On the other hand, "NO_x increases with increase in the mass of methanol added whilst methanol addition to diesel fuel had a strong influence on the NO_x concentration at high engine loads rather than at low engine loads, and a flat NO_x-smoke trade-off curve exists when operating with the diesel-methanol fuel blends."

Bayraktar (2008) investigates the performance parameters of an experimental CI engine fueled with diesel–methanol–dodecanol blends in a paper with 32 citations. He finds that the "methanol concentration in the blend has been changed from 2.5% to 15% with the increments of 2.5%, and 1% dodecanol was added into each blend to solve the phase separation problem." An experimental study was conducted on a single-cylinder, water-cooled CI engine. The engine was operated at different compression ratios (19, 21, 23, and 25) and the engine speed varied from 1000 to 1600 rpm at each compression ratio. He calculates the performance parameters such as torque, effective power, specific fuel consumption, and effective efficiency for each blend at various conditions depending on the experimental data. He asserts that among the different blends, the blend including 10% methanol (DM10) is the most suited one for CI engines from the engine performance point of view. Improvements obtained up to 7% in performance parameters with this blend without any modification to engine design and fuel system are very promising.

Sayin et al. (2009) investigate the effect of injection timing on the exhaust emissions of a single-cylinder, naturally aspirated, four-stroke, direct injection diesel engine using diesel–methanol blends by using methanol-blended diesel fuel from 0% to 15% with an increment of 5% in a paper with 29 citations. They conduct tests for three different injection timings (15°, 20°, and 25° CA bottom top dead center [BTDC]) at four different engine loads (5, 10, 15, 20 Nm) at 2200 rpm. They find that "BSFC, NO_x and CO_2 emissions increased as BTE, smoke opacity, CO and UHC emissions decreased with increasing amount of methanol in the fuel mixture." When the results are compared to those of original injection timing, NO_x and CO_2 emissions decreased; smoke opacity, UHC, and CO emissions increased for the retarded injection timing (15 CA BTDC). On the other hand, with the advanced injection timing (25° CA BTDC), decreasing smoke opacity, UHC and CO emissions diminished, and NO_x and CO_2 emissions boosted at all test conditions. In terms of BSFC and BTE, retarded and advanced injection timings gave negative results for all fuel blends in all engine loads.

17.4.2.4 Research on Hydrogen Production from Methanol

Peppley et al. (1999a) investigate methanol–steam reforming on $Cu/ZnO/Al_2O_3$ with a focus on the reaction network in a paper with 301 citations. They show that "in order to explain the complete range of observed product compositions, rate expressions for all three reactions (methanol-steam reforming, water-gas shift and methanol decomposition) must be included in the kinetic analysis." Furthermore, variations in the selectivity and activity of the catalyst indicate that the "decomposition reaction occurs on a different type of active site than the other two reactions." Although the decomposition reaction is much slower than the reaction between methanol and steam, they assert that it must be included in the kinetic model since the small amount of CO that is produced can drastically reduce the performance of the anode electrocatalyst in low-temperature fuel cells.

Peppley et al. (1999b) investigate methanol–steam reforming on $Cu/ZnO/Al_2O_3$ catalysts with a focus on the comprehensive kinetic model in a paper with 252 citations. They develop surface mechanisms for methanol–steam reforming on $Cu/ZnO/Al_2O_3$ catalysts, which account for all three of the possible overall reactions: methanol and steam reacting directly to form H_2 and CO_2, methanol decomposition to H_2 and CO, and the water–gas shift reaction. The elementary surface reactions used in developing the mechanisms were chosen based on a review of the extensive literature concerning methanol synthesis on $Cu/ZnO/Al_2O_3$ catalysts and the more limited literature specifically dealing with methanol–steam reforming. They argue that the key features of the mechanism are "(i) that hydrogen adsorption does not compete for the active sites which the oxygen-containing species adsorb on, (ii) there are separate active sites for the decomposition reaction distinct from the active sites for the methanol-steam reaction and the water-gas shift reaction, (iii) the rate-determining step (RDS) for both the methanol-steam reaction and the methanol decomposition reaction is the dehydrogenation of adsorbed methoxy groups and (iv) the

RDS for the water-gas shift reaction is the formation of an intermediate formate species." They argue that the resultant kinetic model accurately predicted both the rates of production of hydrogen, carbon dioxide, and carbon monoxide for a wide range of operating conditions including pressures as high as 33 bar.

Breen and Ross (1999) investigate the methanol reforming for fuel-cell applications with a focus on the development of zirconia-containing Cu–Zn–Al catalysts in a paper with 248 citations. After outlining some of the constraints inherent in the use of the reaction and the types of catalysts that have been used by other investigators, they present results on the preparation and testing of a series of copper-containing catalysts for this reaction. They argue that the reaction sequence probably involves the formation of methyl formate, which then decomposes to give CO_2 as the primary product where CO is formed by the reverse water–gas shift reaction and this only occurs to an appreciable extent when the methanol is almost completely converted. They finally describe a number of different copper-containing catalysts and show that of these sequentially precipitated $Cu/ZnO/ZrO_2/Al_2O_3$ materials have the highest activities and stabilities for the steam-reforming reaction.

Velu et al. (2000) investigate the oxidative steam reforming of methanol over CuZnAl(Zr)-oxide catalysts for the selective production of hydrogen for fuel cells with a focus on the catalyst characterization and performance evaluation in a paper with 217 citations. X-ray diffraction (XRD) of the catalysts indicated the "presence of a mixture of poorly crystallized CuO and ZnO phases whose crystallinity increased with decreasing Al content. TPR results demonstrated that substitution of Zr for Al improved the copper reducibility and dispersion." UV–Vis DRS and EPR results revealed that isolated Cu^{2+} ions interacting with Al were formed in the Al-rich samples, while mostly bulk-like or cluster-like Cu^{2+} species were present in the Zr-rich samples. The oxidative steam reforming of methanol reaction was performed over these catalysts in the temperature range 180°C–290°C at atmospheric pressure using H_2O/CH_3OH; molar ratio = 3. Initially, the Cu:Zn:Al metallic composition was optimized, and catalytic performance in terms of methanol conversion and Hz production rate increased with decreasing Al content. Among CuZnAl-oxide catalysts, the one with Cu:Zn:Al = 37.6: 50.7:11.7 (wt%) was the most active. Replacement of Al either partially or completely by Zr further improved the catalytic performance.

Takezawa and Iwasa (1997) investigate the steam reforming and dehydrogenation of methanol and the difference in the catalytic functions of copper and Group VIII metals in a paper with 201 citations. They find that HCHO species were effectively involved in the steam reforming and the dehydrogenation of methanol over supported copper and Group VIII metal (Ni, Rh, Pd, and Pt) catalysts. They argue that the difference in the catalytic performances of copper and Group VIII metals was ascribed to the difference in the reactivity of HCHO species formed in the course of the reactions.

Palo et al. (2007) discuss methanol–steam reforming for hydrogen production based on a literature review in a paper with 182 citations.

Iwasa et al. (1995) investigate steam reforming of methanol over Pd/ZNo with a focus on the effect of the formation of PdZn alloys upon the reaction in a paper with 171 citations. They find that the "catalytic performance of Pd/ZnO for steam reforming of methanol was greatly improved by previously reducing the catalysts at higher temperatures." The original catalytic functions of metallic palladium were greatly modified as a result of the formation of PdZn alloys. Over the catalysts containing alloys, they argue that formaldehyde species formed in the reaction were effectively attacked by water, being transformed into carbon dioxide and hydrogen. By contrast, the formaldehyde species decomposed selectively to carbon monoxide and hydrogen over catalysts containing metallic palladium.

Jiang et al. (1993a) carry out the kinetic study of steam reforming of methanol over copper-based catalysts at temperatures from 443 to 533 K in a paper with 168 citations. Screening experiments showed that a coprecipitated $CuO–ZnO–Al_2O_3$ low-temperature methanol synthesis catalyst had the highest activity and did not deactivate with time online. It also exhibited 100% selectivity to carbon dioxide and hydrogen (the desired reaction products). Kinetic measurements made over the coprecipitated $CuO–ZnO–Al_2O_3$ fit the power law expression: $r_{SR} = k_0e^{-1.5\ kJ/mol/RT}P^{0.26}{}_{MeOH}P^{0.03}{}_{H2O}P^{0.2}{}_{H2}$. Carbon dioxide had no effect on the kinetics of steam reforming of methanol. When carbon monoxide was added to the feed, there was negligible influence on the steam-reforming reaction with an order of 0.016 being observed.

Agrell et al. (2003) investigate the production of hydrogen from methanol over Cu/ZnO catalysts promoted by ZrO_2 and Al_2O_3 in a paper with 159 citations. They observe distinct differences between the processes with respect to catalyst behavior. They find that ZrO_2-containing catalysts, especially $Cu/ZnO/ZrO_2/Al_2O_3$, exhibit the best performance in the steam-reforming reaction. During POX, however, a binary Cu/ZnO catalyst exhibits the lowest light-off temperature and the lowest level of CO by-product. They argue that the redox properties of the catalyst play a key role in determining the pathway for hydrogen production. In particular, the extent of methanol and/or hydrogen combustion at differential O_2 conversion is strongly dependent on the ease of copper oxidation in the catalyst.

Jiang et al. (1993b) investigate the kinetic mechanism for the reaction between methanol and water over a $Cu–ZnO–Al_2O_3$ catalyst in a paper with 156 citations. They find that the reaction yields carbon dioxide and hydrogen in the ratio of one to three, with small amounts of dimethyl ether and carbon monoxide being produced at high conversion. Comparison of the rates of methanol dehydrogenation and of steam reforming over the same catalyst indicates that "steam reforming proceeds via dehydrogenation to methyl formate." Methyl formate then hydrolyzes to formic acid, which decomposes to carbon dioxide and hydrogen. Detailed studies of the kinetics of the reactions show that methanol dehydrogenation controls the rate of steam reforming. Further, Langmuir–Hinshelwood modeling indicates that hydrogen extraction from adsorbed methoxy groups is rate determining to the overall processes.

17.5 Studies on the Economics and Assessment of Methanol

17.5.1 Introduction

The research on the social aspects of the use of methanol as a fuel in various industrial contexts such as a methanol fuel cell or as a pure methanol fuel in vehicles as well as in the production of hydrogen through steam reforming has also been one of the most dynamic research areas in recent years. Twenty citation classics in this field with more than fourteen citations were located and the key emerging issues are presented in the decreasing order of the number of citations.

It was found that there were four main research areas in this context: the research on LCA of methanol, the research on the exergetic analysis of methanol, the research on the economics of methanol, and the research on the optimization of methanol. Five citation classics for each topical area were located (Table 17.4).

TABLE 17.4

Research on the Economics and Assessment of Methanol

No.	Paper References	Year	Document	Affiliation	Country	Number of Authors	Journal	Index	Subjects	Topic	Total Number of Citations	Total Average Citations per Annum	Rank
1	Olah	2005	A	Univ. So. Calif.	United States	1	*Angew. Chem. Int. Edit.*	SCI, SCIE	Chem. Mult.	Methanol economics	183	20.3	29
2	Olah et al.	2009	A	Univ. So. Calif.	United States	3	*J. Org. Chem.*	SCI, SCIE	Chem. Org.	Methanol economics	155	31	37
3	Lange	2007	A	Shell Global Solut.	England	1	*Biofuels Bioprod. Biorefining*	SCIE	Biot. Appl. Microb., Ener. Fuels	Methanol economics	135	19.3	41
4	MacLean and Lave	2003	R	Univ. Toronto, Carnegie Mellon Univ.	Canada, United States	2	*Prog. Energ. Combust. Sci.*	SCI, SCIE	Therm., Ener. Fuels, Eng. Chem., Eng. Mech.	Methanol life cycle assessment	135	12.3	42
5	Hamelinck and Faaij	2006	A	Univ. Utrecht	Netherlands	2	*Energ. Pol.*	SCIE, SSCI	Ener. Fuels, Env. Sci., Env. Stud.	Methanol economics	121	15.1	47
6	Ogden	1999	A	Princeton Univ.	United States	1	*Int. J. Hydrogen Energ.*	SCI, SCIE	Chem. Phys., Electrochem., Ener. Fuels	Methanol economics	96	6.4	52
7	Arico et al.	1998	A	CNR, Univ. Reggio Seoul Natl. Univ.	Italy, S. Korea	6	*Electrochim. Acta*	SCI, SCIE	Electrochem.	Methanol optimization	79	4.9	58
8	Wahlund et al.	2004	A	Royal Inst. Technol.	Sweden	3	*Biomass Bioenerg.*	SCIE	Agr. Eng., Ener. Fuels, Biot. Appl. Microb.	Methanol optimization	77	7.7	59
9	Rakopoulos and Kyritsis	2001	A	Natl. Tech. Univ. Athens	Greece	2	*Energy*	SCI, SCIE	Therm., Ener. Fuels	Methanol exergetic analysis	61	4.7	62
10	Niu et al.	2003	A	Shandong Univ.	China	5	*Synth. Met.*	SCI, SCIE	Mats. Sci. Mult., Phys. Cond. Mat., Poly. Sci.	Methanol optimization	54	4.9	63
11	Meyers and Newman	2002	A	Univ. Calif. Berkeley	United States	2	*J. Electrochem. Soc.*	SCI, SCIE	Electrochem., mats. Sci. Coat. Films	Methanol optimization	45	3.8	66

(Continued)

TABLE 17.4 (*Continued*)

Research on the Economics and Assessment of Methanol

No.	Paper References	Year	Document	Affiliation	Country	Number of Authors	Journal	Index	Subjects	Topic	Total Number of Citations	Total Average Citations per Annum	Rank
12	Pehnt	2001	A	DLR–German Aerosp. Ctr.	Germany	1	*Int. J. Hydrogen Energ.*	SCI, SCIE	Chem. Phys., Electrochem., Ener. Fuels	Methanol life cycle assessment	43	3.3	67
13	Ishihara et al.	2004	A	Yokohama Natl. Univ., Japan Sci. & Technol. Corp.	Japan	4	*J. Power Sourc.*	SCI, SCIE	Electrochem., Ener. Fuels	Methanol exergetic analysis	41	4.1	68
14	Wright and Brown	2007	R	Iowa State Univ.	United States	2	*Biofuels Bioprod. Biorefining*	SCIE	Biot. Appl. Microb., Ener. Fuels	Methanol optimization	38	5.4	70
15	Aresta et al.	2001	A	Univ. Bari, Ctr. METEA	Italy	3	*Energ. Fuel*	SCI, SCIE	Ener. Fuels, Eng. Chem.	Methanol life cycle assessment	35	2.7	74
16	Hinderink et al.	1996	A	Stork Comprimo Bv.	Sweden	5	*Chem. Eng. Sci.*	SCI, SCIE	Eng. Chem.	Methanol exergetic analysis	32	1.8	76
17	Ptasinski et al.	2002	A	Eindhoven Univ. Technol.	Netherlands	3	*Energ. Conv. Manag.*	SCIE	Therm., Ener. Fuels, Mechs., Phys. Nucl.	Methanol exergetic analysis	31	2.6	77
18	Ptasinski	2008	A	Eindhoven Univ. Technol.	Netherlands	1	*Biofuels Bioprod. Biorefining*	SCIE	Biot. Appl. Microb., Ener. Fuels	Methanol exergetic analysis	25	4.2	79
19	Pehnt (2003a)	2003a	A	Inst. Energy & Environm. Res.	Germany	1	*Int. J. Life Cycle Assess.*	SCIE	Eng. Env., Env. Sci.	Methanol life cycle assessment	20	1.8	80
20	Pehnt (2003b)	2003b	A	Inst. Energy & Environm. Res.	Germany	1	*Int. J. Life Cycle Assess.*	SCIE	Eng. Env., Env. Sci.	Methanol life cycle assessment	14	1.3	81

SCI, Science Citation Index; SCIE, Science Citation Index Expanded; SSCI, Social Sciences Citation Index; A, article; R, review.

These papers give strong hints about the determinants of the efficient methanol production and use and emphasize that it is possible to produce and use methanol efficiently as a fuel in relation to its competitors such as hydrogen, gasoline, and diesel.

These papers were dominated by the researchers from 11 countries, usually through the intracountry institutional collaboration and they were multiauthored. The United States (six papers), Germany and Netherlands (three papers each), and Sweden (two papers) were the most prolific countries.

Similarly, all these papers were published in the journals indexed by the SCI and/or SCIE or SSCI. There were no papers indexed by the A&HCI. The number of citations ranged from 14 to 183 and the average number of citations per annum ranged from 1.3 to 31. All of these papers were articles. *Biofuels Bioprod. Biorefining* (three papers) and *Int. J. Hydrogen Energ.* and *Int. J. Life Cycle Assess.* (two papers each) were the most prolific journals.

17.5.2 Research on the Economics and Assessment of Methanol

17.5.2.1 Research on Life Cycle Assessment of Methanol

MacLean and Lave (2003) evaluate that automobile fuel/propulsion system technologies could power cars and light trucks in the United States and Canada over the next two to three decades in a paper with 135 citations: (1) reformulated gasoline and diesel, (2) compressed natural gas, (3) methanol and ethanol, (4) liquid petroleum gas, (5) liquefied natural gas, (6) Fischer–Tropsch liquids from natural gas, (7) hydrogen, and (8) electricity and (1) SI port injection engines, (2) SI direct injection engines, (3) CI engines, (4) electric motors with battery power, (5) hybrid electric propulsion options, and (6) fuel cells. They review recent studies to evaluate the environmental, performance, and cost characteristics of fuel/propulsion technology combinations that are currently available or will be available in the next few decades. The vehicle options likely to be competitive during the next two decades are those using improved ICES, including HEVs burning *clean* gasoline or diesel. An extensive infrastructure has been developed to locate, extract, transport, refine, and retail gasoline and diesel. The current infrastructure is a major reason for continuing to use gasoline and diesel fuels. Absent a breakthrough in electrochemistry, battery-powered vehicles will remain expensive and have an unattractive range. Fuel-cell propulsion systems are unlikely to be competitive before 2020, if they are ever competitive. Although, fuel cells have high theoretical efficiencies, and do not need a tailpipe and therefore have vehicle emissions benefits over conventional vehicles; generating the hydrogen and getting it to the vehicle requires large amounts of energy. The current well-to-wheel analyses show that using a liquid fuel and onboard reforming produces a system inferior to gasoline-powered ICEs on the basis of efficiency and environmental discharges. Storage of hydrogen onboard the vehicle is another challenge. Fischer–Tropsch liquids from natural gas and ethanol from biomass may become widespread. The Fischer–Tropsch liquids will penetrate if there are large amounts of stranded natural gas selling for very low prices at the same time that petroleum is expensive or extremely low sulfur is required in diesel fuel. Ethanol could become the dominant fuel if energy independence, sustainability, or very low carbon dioxide emissions become important or if petroleum prices double. Absent major technology breakthroughs, a doubling of petroleum prices, or stringent regulation of fuel economy or greenhouse gas emissions, the 2030 LDV will be powered by a gasoline ICE. The continuing progress in increasing engine efficiency, lowering emissions, and supplying inexpensive gasoline makes it extremely difficult for any of the alternative fuels or propulsion technologies to displace the gasoline (diesel)-fueled ICE. Extensive progress has been made by analysts in examining the life cycles of a range of fuels and propulsion systems for personal transportation vehicles. The most important contribution of these methods and studies is getting decision makers to focus on the important attributes and to avoid looking only at one aspect of the fuel cycle or propulsion system or at only one media for environmental burdens. The current state of knowledge should avoid the recurrence of the fiasco of requiring battery-powered cars on the grounds that they are good for the environment and will appeal to consumers.

Pehnt (2001) discusses the LCA of fuel-cell stacks in a paper with 43 citations. He investigates the production process of PEFC stacks, identifies the ecological contributions of various components and materials, and compares the results with impacts due to utilization of the stacks in a vehicle (i.e., hydrogen or methanol production and direct emissions). He argues that the production of fuel-cell stacks leads to environmental impacts, which cannot be neglected compared to the utilization of the stacks in a vehicle (the actual driving process). These impacts are mainly caused by the platinum group metals for the catalyst and, to a lesser degree, the materials and energy for the flow field plates.

Aresta et al. (2001) discuss the innovative synthetic technologies of industrial relevance based on carbon dioxide as raw material in a paper with 35 citations. The reduction of carbon dioxide emission into the atmosphere requires the implementation of several convergent technologies in different production sectors. The chemical industry can contribute to the issue with innovative synthetic technologies, which implement the principles of atom economy, dematerialization, energy saving, and raw material diversification with carbon recycling. Such methodologies based on carbon dioxide utilization merge two issues, namely, avoiding the production of CO_2 and carbon recycling. In this paper, some options are discussed, such as the synthesis of carboxylic acids, organic carbonates, and carbamates/isocyanates. Synthetic processes of methanol from either CO_2 or CO are compared. The results of application of LCA to selected cases are discussed.

Pehnt (2003a,b) assesses the future energy and transport systems for fuel cells with a focus on the methodological issues in a paper with 20 and 14 citations, respectively. Principally, the investigated forecasting methods are suitable for future energy system assessment. The selection of the best method depends on different factors such as required resources, quality of the results, and flexibility. In particular, the time horizon of the investigation determines which forecasting tool may be applied. Environmentally relevant process steps exhibiting a significant time dependency shall always be investigated using different independent forecasting tools to ensure stability of the results. The results of the LCA underline that principally, fuel cells offer advantages in the impact categories that are typically dominated by pollutant emissions, such as acidification and eutrophication, whereas for global warming and primary energy demand, the situation depends on a set of parameters such as driving cycle and fuel economy ratio in mobile applications and thermal/total efficiencies in stationary applications. For the latter impact categories, the choice of the primary energy carrier for fuel production (renewable or fossil) dominates the impact reduction. With increasing efficiency and improving emission performance of conventional systems, the competition regarding all impact categories in both mobile and stationary applications is getting even stronger. The production of the fuel-cell system is of low overall significance in stationary applications, whereas in vehicles, the lower lifetime of the vehicle leads to a much higher significance of the power train production.

17.5.2.2 Research on the Exergetic Analysis of Methanol

Rakopoulos and Kyritsis (2001) discuss the comparative second-law analysis of ICE operation for methane, methanol, and dodecane fuels in a paper with 61 citations. The results of the second-law analysis of engine operation with n-dodecane (n-$C_{12}H_{26}$) fuel are compared with the results of a similar analysis for cases where a light, gaseous (methane) and an oxygenated (methanol) fuel is used. They calculate the rate of entropy production during combustion as a function of the fuel reaction rate with the combined use of first- and second-law arguments and a chemical equilibrium hypothesis. They show theoretically that the decomposition of lighter molecules leads to less entropy generation compared to heavier fuels. They verify this computationally for particular fuels and calculate the corresponding decrease in combustion irreversibility. They focus on the effect of the lower mixing entropy of the exhaust gas of an oxygenated fuel (methanol) as a contribution to the discussion of the advantages and disadvantages of the use of such fuels.

Ishihara et al. (2004) carry out the exergy analysis of polymer electrolyte fuel-cell systems using methanol. An exergy (available energy) analysis has been conducted on a typical PEFC system using methanol in a paper with 41 citations. The material balance and enthalpy balance were calculated for the PEFC system using methanol–steam reforming, and the exergy flow was obtained. Based on these results, the exergy loss in each unit was obtained, and the difference between the enthalpy and exergy was discussed. The exergy loss in this system was calculated to be 193 kJ/mol methanol for the steam-reforming process of methanol. Although the enthalpy efficiency approached unity as the recovery rate of the waste heat from the cell approached unity, the exergy efficiency remained around 0.45 since the cell's operating temperature of 80°C is low. They find that the cell voltage should exceed 0.82 V in order to obtain the exergy efficiency of 0.5 or higher. A DMFC was analyzed using the exergy and compared with the methanol reforming PEFC. In order to obtain the exergy efficiency higher than that of PEFC with steam reforming, the cell voltage of the DMFC should be 0.50 V or larger at the current density of 600 mA/cm^2.

Hinderink et al. (1996) discuss the exergy analysis with a flowsheeting simulator with a focus on the syngas production from natural gas in a paper with 32 citations. Several processes, producing syngas from natural gas, have been analyzed by the exergy method, showing exergy analysis as a valuable diagnostic tool. In addition, a generally applicable and systematic way of performing exergy analyses and dealing with their results is illustrated. Exergy calculations have been carried out by user-defined subroutines, which are integrated with a flowsheeting simulator. They find that compared to the conventional steam-reforming process, giving an overall exergy loss of approximately 8.5 GJ/t methanol, the exergy loss can be reduced to about 4.9 GJ/t methanol by application of the convective reforming option in combination with POX. Secondly, by considering progressively smaller subsystems within the overall process, locations of major exergy loss are revealed and their potential for improvement can be indicated. Finally, the minus value of the standard Gibbs energy of the overall reaction of each process, denoted as available reaction exergy, is compared to the exergy loss associated with this overall reaction. This comparison demonstrates that available reaction exergies should always be minimized to reduce exergy losses associated with chemical reactions. They cannot however be eliminated completely when reactions are only thermally coupled. Further improvement can be attained by direct coupling of chemical reactions such that the overall Gibbs energy of reaction is still reasonably negative.

Ptasinski et al. (2002) carry out the exergy analysis of methanol from the sewage sludge process in a paper with 31 citations. This paper concerns a new method of sewage sludge treatment that contributes more than traditional methods to the sustainable technology by achieving a higher rational efficiency of sludge processing. This is obtained by preserving the chemical exergy present in the sludge and transforming it into a chemical one, methanol. The proposed method combines a sludge gasification process and a modified methanol plant. The sludge gasification produces a synthetic gas mainly containing CO and H_2, in which gas is next used as a reactant to make methanol.

The plant, with a capacity of 50,000 t (dry) solids per year, is simulated by a computer model using the flowsheeting program Aspen Plus (Aspen Technology). The exergy analysis is performed for various operating conditions and the optimal values of these conditions are found. The irreversibilities of different plant segments, thermal dryer, gasifier, gas cleaning, compressors, methanol reactor, distillation column, and purge gas combustion, are evaluated. It is shown that the rational efficiency of the overall process is much higher than that of traditional plants of thermal sludge treatment.

Ptasinski (2008) discusses the thermodynamic efficiency of biomass gasification and biofuels conversion in a review paper with 25 citations. He uses the exergy analysis, which is based on the second law of thermodynamics to analyze the biomass gasification and conversion of biomass to biofuels. He reviews thermodynamic efficiency of biomass gasification for air-blown as well as steam-blown gasifiers. Finally, he evaluates the overall technological chains biomass-to-biofuels including methanol, Fischer–Tropsch HCs, and hydrogen. He finally compares the efficiency of biofuel production with that of fossil fuels.

17.5.2.3 Research on the Economics of Methanol

Olah (2005) discusses the methanol economy in relation to the hydrogen and oil economy and asserts that the methanol economy is the future for the society in meeting the goals of sustainable energy requirements.

Olah et al. (2009) discuss the chemical recycling of carbon dioxide to methanol and dimethyl ether: from greenhouse gas to renewable, environmentally carbon-neutral fuels and synthetic HCs in a paper with 155 citations. They argue that chemical recycling of carbon dioxide from natural and industrial sources as well as varied human activities can be achieved via its capture and subsequent reductive hydrogenative conversion. They discuss this new approach and their research in the field over the last 15 years. They argue that carbon recycling represents a significant aspect of the methanol economy. Any available energy source (alternative energies such as solar, wind, geothermal, and atomic energy) can be used for the production of needed hydrogen and chemical conversion of CO_2. Improved new methods for the efficient reductive conversion of CO_2 to methanol and/or DME that they developed include bireforming with methane and ways of catalytic or electrochemical conversions. Liquid methanol is preferable to highly volatile and potentially explosive hydrogen for energy storage and transportation. Together with the derived DME, they are excellent transportation fuels for ICEs and fuel cells as well as convenient starting materials for synthetic HCs and their varied products. They assert that carbon dioxide thus can be chemically transformed from a detrimental greenhouse gas causing global warming into a valuable, renewable, and inexhaustible carbon source of the future allowing environmentally neutral use of carbon fuels and derived HC products.

Lange (1997) discusses lignocellulose conversion in a paper with 135 citations. He discusses the variety of chemistries and technologies that are being explored to valorize lignocellulosic biomass. He shows the need to *deoxygenate* biomass and reviews the main chemical routes for it, that is, the pyrolysis to char, biocrude, or gas; the gasification to syngas and its subsequent conversion, for example, to alkanes or methanol; and the hydrolysis to sugar and their subsequent upgrading to oxygenated intermediates via chemical or fermentation routes. He argues that the economics of biomass conversion also needs to be considered: the current production cost of biofuels is typically \$60–\$120/barrel of oil equivalent. Influential factors include the cost of the biomass at the plant gate, the conversion efficiency, the scale of the process, and the value of the product (e.g., fuel, electricity, or chemicals).

Hamelinck and Faaij (2006) discuss the outlook for advanced biofuels to assess which biofuels have the better potential for the short term or the longer term (2030) and what developments are necessary to improve the performance of biofuels; the production of four promising biofuels, methanol, ethanol, hydrogen, and synthetic diesel, is systematically in a paper with 121 citations. First, they model the key technologies for the production of these fuels, such as gasification, gas processing, synthesis, hydrolysis, and fermentation, and their improvement options. Then, they analyze the production facility's technological and economic performance, applying variations in technology and scale. Finally, they compare likely biofuel chains (including distribution to cars and end use) on an equal economic basis, such as costs per kilometer driven. They find that "production costs of these fuels range 16–22 Euro/GJ(HHV) now, down to 9–13 Euro/GJ(HHV) in future (2030)." This performance assumes both certain technological developments as well as the availability of biomass at 3 Euro/GJ(HHV). They argue that "feedstock costs strongly influence the resulting biofuel costs by 2–3 Euro/GJ(fuel) for each Euro/GJHHV feedstock difference." In biomass-producing regions such as Latin America or the Russian Federation, the "four fuels could be produced at 7–11 Euro/GJ(HHV) compared to diesel and gasoline costs of 7 and 8 Euro/GJ (excluding distribution, excise and VAT; at crude oil prices of similar to 35 Euro/bbl or 5.7 Euro/GJ)." The uncertainties in biofuel production costs of the four selected biofuels are 15%–30%. When applied in cars, biofuels have driving costs in ICEVs of about 0.18–0.24 Euro/km now (fuel excise duty and VAT excluded) and may be about 0.18 in future. The cars' contribution to these costs is much larger than the fuels' contribution. They note that large-scale gasification, thorough gas cleaning, and microbiological processes for hydrolysis and fermentation are key major fields for RD&D efforts, next to consistent market development and larger-scale deployment of those technologies.

Ogden (1999) discusses the development of an infrastructure for hydrogen vehicles in a paper with 96 citations. He examines the technical feasibility and economics of developing a hydrogen vehicle refueling infrastructure for a specific area where zero-emission vehicles are being considered in Southern California. He assesses in detail several near-term possibilities for producing and delivering gaseous hydrogen transportation fuel including (1) hydrogen produced from natural gas in a large, centralized steam-reforming plant and truck delivered as a liquid to refueling stations; (2) hydrogen produced in a large, centralized steam-reforming plant and delivered via small-scale hydrogen gas pipeline to refueling stations; (3) by-product hydrogen from chemical industry sources; (4) hydrogen produced at the refueling station via small-scale steam reforming of natural gas; and (5) hydrogen produced via small-scale electrolysis at the refueling station. He estimates the capital cost of infrastructure and the delivered cost of hydrogen for each hydrogen supply option. He then compares hydrogen to other fuels for FCVs (methanol, gasoline) in terms of vehicle cost, infrastructure cost, and life cycle cost of transportation.

17.5.2.4 Research on the Optimization of Methanol

Arico et al. (1998) discuss the optimization of operating parameters of a DMFC and physicochemical investigation of catalyst–electrolyte interface in a paper with 79 citations. They carry out a physicochemical investigation of catalyst–Nafion® electrolyte interface of a DMFC, based on a Pt–Ru/C anode catalyst, by XRD, SEM–EDAX, and TEM. They find no interaction between catalyst and electrolyte and they observe no significant interconnected network of Nafion micelles inside the composite catalyst layer. Optimal conditions were 2 M methanol, 5 atm cathode pressure, and 2–3 atm anode pressure. Power densities of 110 and 160 mW/cm^2 were obtained for operation with air and oxygen, respectively, at temperatures of 95°C–100°C and with 1 mg/cm^2 Pt loading.

Wahlund et al. (2004) carry out a comparative study of CO_2 reduction and cost for different bioenergy processing options in a paper with 77 citations. They systematically investigate several bioenergy processing options, quantify the reduction rate, and calculate the specific cost of reduction. They address the issue of which option Sweden should concentrate on to achieve the largest CO_2 reduction at the lowest cost. They find that the "largest and most long-term sustainable CO_2 reduction would be achieved by refining the woody biomass to fuel pellets for coal substitution, which have been done in Sweden." Refining to motor fuels, such as methanol, DME, and ethanol, gives only half of the reduction and furthermore at a higher specific cost. Biomass refining into pellets enables transportation over long distances and seasonal storage, which is crucial for further utilization of the woody biomass potential.

Niu et al. (2003) discuss the formation optimization of platinum-modified polyaniline films for the electrocatalytic oxidation of methanol in a paper with 54 citations. They find that the oxidation of methanol depends greatly on the nature of both polyaniline matrix and platinum particles, which can be optimized by electrochemical formation conditions, such as cycle numbers, sweep rates, and potential limits.

Meyers and Newman (2002) discuss the simulation of the DMFC in a paper with 45 citations. By mapping methanol concentration and potential profiles under steady-state operating conditions, they gain insight into the behavior and limitation of the cell performance. They examine methanol fuel efficiency, catalyst layer design, and the effect of methanol crossover on cell performance and discuss the optimization of membrane separator thickness and methanol feed concentration.

Wright and Brown (2007) discuss factors that influence the optimal size of biorefineries and the resulting unit cost of biofuels produced by them in a review paper with 38 citations. Technologies examined include dry grind corn to ethanol; lignocellulosic ethanol via enzymatic hydrolysis; gasification and upgrading to hydrogen, methanol, and Fischer–Tropsch liquids; gasification of lignocellulosic biomass to mixed alcohols; and fast pyrolysis of lignocellulosic biomass to bio-oil. On the basis of gallons of gasoline equivalent (gge) capacity, they find that "optimally sized gasification-to-biofuels plants were 50–100% larger than biochemical cellulosic ethanol plants. Biorefineries converting lignocellulosic biomass into transportation fuels were optimally sized in the range of 240–486 million gge per year compared to 79 million gge per year for a grain ethanol plant." Among the biofuel options, ethanol, whether produced biochemically or thermochemically, is the most expensive to produce. "Lignocellulosic biorefineries will require 4.7–7.8 million tons of biomass annually compared to 1.2 million tons of corn grain for a grain ethanol plant."

17.6 Conclusion

The citation classics presented under the four main headings in this chapter confirm the predictions through the extrapolation that marine algae have a significant potential to serve as a major solution for the global problems of warming, air pollution, energy security, and food security in the form of algal biomethanol based on the studies on the production, use, and economics and assessment of methanol from natural gas and biomass.

Further research is recommended for the detailed studies in each topical area presented in this chapter including scientometric studies and citation classic studies to inform the key stakeholders about the potential of marine algae for the solution of the global problems of warming, air pollution, energy security, and food security in the form of algal biomethanol.

References

Agrell, J., H. Birgersson, M. Boutonnet, I. Melian-Cabrera, R.M. Navarro, and J.L.G. Fierro. 2003. Production of hydrogen from methanol over Cu/ZnO catalysts promoted by ZrO2 and Al2O3. *Journal of Catalysis* 219:389–403.

Aresta, M., A. Dibenedetto, and I. Tommasi. 2001. Developing innovative synthetic technologies of industrial relevance based on carbon dioxide as raw material. *Energy & Fuels* 15:269–273.

Arico, A.S., P. Creti, P.L. Antonucci, J. Cho, H. Kim, and V. Antonucci. 1998. Optimization of operating parameters of a direct methanol fuel cell and physico-chemical investigation of catalyst-electrolyte interface. *Electrochimica Acta* 43:3719–3729.

Arico, A.S., S. Srinivasan, and V. Antonucci. 2001. DMFCs: From fundamental aspects to technology development. *Fuel Cells* 1:133–161.

Askgaard, T.S., J.K. Norskov, C.V. Ovesen, and P. Stoltze. 1995. A kinetic-model of methanol synthesis. *Journal of Catalysis* 156:229–242.

Baltussen, A. and C.H. Kindler. 2004a. Citation classics in anesthetic journals. *Anesthesia & Analgesia* 98:443–451.

Baltussen, A. and C.H. Kindler. 2004b. Citation classics in critical care medicine. *Intensive Care Medicine* 30:902–910.

Bayraktar, H. 2008. An experimental study on the performance parameters of an experimental CI engine fueled with diesel-methanol-dodecanol blends. *Fuel* 87:158–164.

Bharadwaj, S.S. and L.D. Schmidt. 1995. Catalytic partial oxidation of natural-gas to syngas. *Fuel Processing Technology* 42:109–127.

Bothast, R.J. and M.A. Schlicher. 2005. Biotechnological processes for conversion of corn into ethanol. *Applied Microbiology and Biotechnology* 67:19–25.

Bowker, M., H. Houghton, and K.C. Waugh. 1981. Mechanism and kinetics of methanol synthesis on zinc-oxide. *Journal of the Chemical Society—Faraday Transactions I* 77:3023–3036.

Breen, J.P. and J.R.H. Ross. 1999. Methanol reforming for fuel-cell applications: Development of zirconia-containing Cu-Zn-Al catalysts. *Catalysis Today* 51:521–533.

Brown, L.F. 2001. A comparative study of fuels for on-board hydrogen production for fuel-cell-powered automobiles. *International Journal of Hydrogen Energy* 26:381–397.

Centi, G. and S. Perathoner. 2009. Opportunities and prospects in the chemical recycling of carbon dioxide to fuels. *Catalysis Today* 148:191–205.

Chao, M.R., T.C. Lin, H.R. Chao, F.H. Chang, and C.B. Chen. 2001. Effects of methanol-containing additive on emission characteristics from a heavy-duty diesel engine. *Science of the Total Environment* 279:167–179.

Cheng, C.H., C.S. Cheung, T.L. Chan, S.C. Lee, C.D. Yao, and K.S. Tsang. 2008. Comparison of emissions of a direct injection diesel engine operating on biodiesel with emulsified and fumigated methanol. *Fuel* 87:1870–1879.

Chinchen, G.C., K.C. Waugh, and D.A. Whan. 1986. The activity and state of the copper surface in methanol synthesis catalysts. *Applied Catalysis* 25:101–107.

Chisti, Y. 2007. Biodiesel from microalgae. *Biotechnology Advances* 25:294–306.

Demirbas, A. 2007. Progress and recent trends in biofuels. *Progress in Energy and Combustion Science* 33:1–18.

Deutschmann, O. and L.D. Schmidt. 1998. Modeling the partial oxidation of methane in a short-contact-time reactor. *AICHE Journal* 44:2465–2477.
Dubin, D., A.W. Hafner, and K.A. Arndt. 1993. Citation-classics in clinical dermatological journals—Citation analysis, biomedical journals, and landmark articles, 1945–1990. *Archives of Dermatology* 129:1121–1129.
Fieweger, K., R. Blumenthal, and G. Adomeit. 1997. Self-ignition of SI engine model fuels: A shock tube investigation at high pressure. *Combustion and Flame* 109:599–619.
Fisher, I.A. and A.T. Bell. 1997. In-situ infrared study of methanol synthesis from H-2/CO_2 over Cu/SiO_2 and Cu/ZrO_2/SiO_2. *Journal of Catalysis* 172:222–237.
Gehanno, J.F., K. Takahashi, S. Darmoni, and J. Weber. 2007. Citation classics in occupational medicine journals. *Scandinavian Journal of Work, Environment & Health* 33:245–251.
Godfray, H.C.J., J.R. Beddington, I.R. Crute et al. 2010. Food security: The challenge of feeding 9 billion people. *Science* 327:812–818.
Goldemberg, J. 2007. Ethanol for a sustainable energy future. *Science* 315:808–810.
Graaf, G.H., P.J.J.M. Sijtsema, E.J. Stamhuis, and G.E.H. Joosten. 1986. Chemical-equilibria in methanol synthesis. *Chemical Engineering Science* 41:2883–2890.
Grunwaldt, J.D., A.M. Molenbroek, N.Y. Topsoe, H. Topsoe, and B.S. Clausen. 2000. In situ investigations of structural changes in Cu/ZnO catalysts. *Journal of Catalysis* 194:452–460.
Hamelinck, C.N. and A.P.C. Faaij. 2006. Outlook for advanced biofuels. *Energy Policy* 34:3268–3283.
Heinzel, A. and V.M. Barragan. 1999. A review of the state-of-the-art of the methanol crossover in direct methanol fuel cells. *Journal of Power Sources* 84:70–74.
Hickman, D.A. and L.D. Schmidt. 1993. Production of syngas by direct catalytic-oxidation of methane. *Science* 259:343–346.
Hinderink, A.P., F.P.J.M. Kerkhof, A.B.K. Lie, J.D. Arons, and H.J. vanderKooi. 1996. Exergy analysis with a flowsheeting simulator. 2. Application; Synthesis gas production from natural gas. *Chemical Engineering Science* 51:4701–4715.
Hirano, A., K. Hon-Nami, S. Kunito, M. Hada, and Y. Ogushi. 1998. Temperature effect on continuous gasification of microalgal biomass: Theoretical yield of methanol production and its energy balance. *Catalysis Today* 45:399–404.
Huang, Z.H., H.B. Lu, D.M. Jiang et al. 2004a. Combustion behaviors of a compression-ignition engine fuelled with diesel/methanol blends under various fuel delivery advance angles. *Bioresource Technology* 95:331–341.
Huang, Z.H., H.B. Lu, D.M. Jiang et al. 2004b. Engine performance and emissions of a compression ignition engine operating on the diesel-methanol blends. *Proceedings of the Institution of Mechanical Engineers Part D: Journal of Automobile Engineering* 218:435–447.
Ichikawa, M. 1978. Catalysis by supported metal crystallites from carbonyl clusters. 1. Catalytic methanol synthesis under mild conditions over supported rhodium, platinum, and iridium crystallites prepared from Rh, Pt, and Ir carbonyl cluster compounds deposited on ZnO and MgO. *Bulletin of the Chemical Society of Japan* 51:2268–2272.
Ishihara, A., S. Mitsushima, N. Kamiya, and K. Ota. 2004. Exergy analysis of polymer electrolyte fuel cell systems using methanol. *Journal of Power Sources* 126:34–40.
Iwasa, N., S. Masuda, N. Ogawa, and N. Takezawa. 1995. Steam reforming of methanol over Pd/ZnO—Effect of the formation of PdZn alloys upon the reaction. *Applied Catalysis A: General* 125:145–157.
Jacobson, M.Z. et al. 2009. Review of solutions to global warming, air pollution, and energy security. *Energy & Environmental Science* 2:148–173.
Jiang, C.J., D.L. Trimm, M.S. Wainwright, and N.W. Cant. 1993a. Kinetic-study of steam reforming of methanol over copper-based catalysts. *Applied Catalysis A: General* 93:245–255.
Jiang, C.J., D.L. Trimm, M.S. Wainwright, and N.W. Cant. 1993b. Kinetic mechanism for the reaction between methanol and water over a Cu-ZnO-Al2O3 catalyst. *Applied Catalysis A: General* 97:145–158.
Kapdan, I.K. and F. Kargi. 2006. Bio-hydrogen production from waste materials. *Enzyme and Microbial Technology* 38:569–582.

Kawatsu, S. 1998. Advanced PEFC development for fuel cell powered vehicles. *Journal of Power Sources* 71:150–155.

Klier, K. 1982. Methanol synthesis. *Advances in Catalysis* 31:243–313.

Konur, O. 2000. Creating enforceable civil rights for disabled students in higher education: An institutional theory perspective. *Disability & Society* 15:1041–1063.

Konur, O. 2002a. Assessment of disabled students in higher education: Current public policy issues. *Assessment and Evaluation in Higher Education* 27:131–152.

Konur, O. 2002b. Access to employment by disabled people in the UK: Is the Disability Discrimination Act working? *International Journal of Discrimination and the Law* 5:247–279.

Konur, O. 2002c. Access to nursing education by disabled students: Rights and duties of nursing programs. *Nursing Education Today* 22:364–374.

Konur, O. 2006a. Participation of children with dyslexia in compulsory education: Current public policy issues. *Dyslexia* 12:51–67.

Konur, O. 2006b. Teaching disabled students in higher education. *Teaching in Higher Education* 11:351–363.

Konur, O. 2007a. A judicial outcome analysis of the Disability Discrimination Act: A windfall for the employers? *Disability & Society* 22:187–204.

Konur, O. 2007b. Computer-assisted teaching and assessment of disabled students in higher education: The interface between academic standards and disability rights. *Journal of Computer Assisted Learning* 23:207–219.

Konur, O. 2011. The scientometric evaluation of the research on the algae and bio-energy. *Applied Energy* 88:3532–3540.

Konur, O. 2012a. 100 citation classics in energy and fuels. *Energy Education Science and Technology Part A: Energy Science and Research* 2012(si):319–332.

Konur, O. 2012b. What have we learned from the citation classics in Energy and Fuels: A mixed study. *Energy Education Science and Technology Part A: Energy Science and Research* 2012(si):255–268.

Konur, O. 2012c. The gradual improvement of disability rights for the disabled tenants in the UK: The promising road is still ahead. *Social Political Economic and Cultural Research* 4:71–112.

Konur, O. 2012d. Prof. Dr. Ayhan Demirbas' scientometric biography. *Energy Education Science and Technology Part A: Energy Science and Research* 28:727–738.

Konur, O. 2012e. The evaluation of the research on the biofuels: A scientometric approach. *Energy Education Science and Technology Part A: Energy Science and Research* 28:903–916.

Konur, O. 2012f. The evaluation of the research on the biodiesel: A scientometric approach. *Energy Education Science and Technology Part A: Energy Science and Research* 28:1003–1014.

Konur, O. 2012g. The evaluation of the research on the bioethanol: A scientometric approach. *Energy Education Science and Technology Part A: Energy Science and Research* 28:1051–1064.

Konur, O. 2012h. The evaluation of the research on the microbial fuel cells: A scientometric approach. *Energy Education Science and Technology Part A: Energy Science and Research* 29:309–322.

Konur, O. 2012i. The evaluation of the research on the biohydrogen: A scientometric approach. *Energy Education Science and Technology Part A: Energy Science and Research* 29:323–338.

Konur, O. 2012j. The evaluation of the biogas research: A scientometric approach. *Energy Education Science and Technology Part A: Energy Science and Research* 29:1277–1292.

Konur, O. 2012k. The scientometric evaluation of the research on the production of bio-energy from biomass. *Biomass and Bioenergy* 47:504–515.

Konur, O. 2012l. The evaluation of the global energy and fuels research: A scientometric approach. *Energy Education Science and Technology Part A: Energy Science and Research* 30:613–628.

Konur, O. 2012m. The evaluation of the biorefinery research: A scientometric approach. *Energy Education Science and Technology Part A: Energy Science and Research* 2012(si):347–358.

Konur, O. 2012n. The evaluation of the bio-oil research: A scientometric approach. *Energy Education Science and Technology Part A: Energy Science and Research* 2012(si):379–392.

Konur, O. 2012o. What have we learned from the citation classics in energy and fuels: A mixed study. *Energy Education Science and Technology Part A: Energy Science and Research* 2012(si):255–268.

Konur, O. 2012p. The evaluation of the research on the biofuels: A scientometric approach. *Energy Education Science and Technology Part A: Energy Science and Research* 28:903–916.

Konur, O. 2013. What have we learned from the research on the International Financial Reporting Standards (IFRS)? A mixed study. *Energy Education Science and Technology Part D: Social Political Economic and Cultural Research* 5:29–40.

Kreuer, K.D. 2001. On the development of proton conducting polymer membranes for hydrogen and methanol fuel cells. *Journal of Membrane Science* 185:29–39.

Kumar, M.S., A. Ramesh, and B. Nagalingam. 2003. An experimental comparison of methods to use methanol and Jatropha oil in a compression ignition engine. *Biomass & Bioenergy* 25:309–318.

Kung, H.H. 1980. Methanol synthesis. *Catalysis Reviews—Science and Engineering* 22:235–259.

Lal, R. 2004. Soil carbon sequestration impacts on global climate change and food security. *Science* 304:1623–1627.

Lange, J.P. 2007. Lignocellulose conversion: An introduction to chemistry, process and economics. *Biofuels Bioproducts & Biorefining: BIOFPR* 1:39–48.

Li, W.Z., C.H. Liang, W.J. Zhou et al. 2003. Preparation and characterization of multiwalled carbon nanotube-supported platinum for cathode catalysts of direct methanol fuel cells. *Journal of Physical Chemistry B* 107:6292–6299.

Lindstrom, B., L.J. Pettersson, and P.G. Menon. 2002. Activity and characterization of Cu/Zn, Cu/Cr and Cu/Zr on gamma-alumina for methanol reforming for fuel cell vehicles. *Applied Catalysis A: General* 234:111–125.

Liu, H.S., C.J. Song, L. Zhang, J.J. Zhang, H.J. Wang, and D.P. Wilkinson. 2006. A review of anode catalysis in the direct methanol fuel cell. *Journal of Power Sources* 155:95–110.

Liu, J. and X.Q. Ma. 2009. The analysis on energy and environmental impacts of microalgae-based fuel methanol in China. *Energy Policy* 37:1479–1488.

Lynd, L.R., J.H. Cushman, R.J. Nichols, and C.E. Wyman. 1991. Fuel ethanol from cellulosic biomass. *Science* 251:1318–1323.

MacLean, H.L. and L.B. Lave. 2003. Evaluating automobile fuel/propulsion system technologies. *Progress in Energy and Combustion Science* 29:1–69.

Mata, T.M., A.A. Martins, and N.S. Caetano. 2010. Microalgae for biodiesel production and other applications: A review. *Renewable & Sustainable Energy Reviews* 14:217–232.

McNicol, B.D., D.A.J. Rand, and K.R. Williams. 1999. Direct methanol-air fuel cells for road transportation. *Journal of Power Sources* 83:15–31.

McNicol, B.D., D.A.J. Rand, and K.R. Williams. 2001. Fuel cells for road transportation purposes—Yes or no? *Journal of Power Sources* 100:47–59.

Meng, X., J.M. Yang, X. Xu, L. Zhang, Q.J. Nie, and M. Xian. 2009. Biodiesel production from oleaginous microorganisms. *Renewable Energy* 34:1–5.

Meyers, J.P. and J. Newman. 2002. Simulation of the direct methanol fuel cell—III. Design and optimization. *Journal of the Electrochemical Society* 149:A729–A735.

Ng, K.L., D. Chadwick, and B.A. Toseland. 1999. Kinetics and modelling of dimethyl ether synthesis from synthesis gas. *Chemical Engineering Science* 54:3587–3592.

Niu, L., Q.H. Li, F.H. Wei, X. Chen, and H. Wang. 2003. Formation optimization of platinum-modified polyaniline films for the electrocatalytic oxidation of methanol. *Synthetic Metals* 139:271–276.

North, D. 1994. Economic-performance through time. *American Economic Review* 84:359–368.

Ogden, J.M. 1999. Developing an infrastructure for hydrogen vehicles: A Southern California case study. *International Journal of Hydrogen Energy* 24:709–730.

Ogden, J.M., Steinbugler, M.M., and Kreutz, T.G. 1999. A comparison of hydrogen, methanol and gasoline as fuels for fuel cell vehicles: Implications for vehicle design and infrastructure development. *Journal of Power Sources* 79:143–168.

Ogden, J.M., R.H. Williams, and E.D. Larson. 2004. Societal lifecycle costs of cars with alternative fuels/engines. *Energy Policy* 32:7–27.

Olah, G.A. 2005. Beyond oil and gas: The methanol economy. *Angewandte Chemie—International Edition* 44:2636–2639.

Olah, G.A., A. Goeppert, and G.K.S. Prakash. 2009. Chemical recycling off carbon dioxide to methanol and dimethyl ether: From greenhouse gas to renewable, environmentally carbon neutral fuels and synthetic hydrocarbons. *Journal of Organic Chemistry* 74:487–498.

Ovesen, C.V., B.S. Clausen, J. Schiotz, P. Stoltze, H. Topsoe, and J.K. Norskov. 1997. Kinetic implications of dynamical changes in catalyst morphology during methanol synthesis over Cu/ZnO catalysts. *Journal of Catalysis* 168:133–142.

Paladugu, R., M.S. Chein, S. Gardezi, and L. Wise. 2002. One hundred citation classics in general surgical journals. *World Journal of Surgery* 26:1099–1105.

Palo, D.R., R.A. Dagle, and J.D. Holladay. 2007. Methanol steam reforming for hydrogen production. *Chemical Reviews* 107:3992–4021.

Pehnt, M. 2001. Life-cycle assessment of fuel cell stacks. *International Journal of Hydrogen Energy* 26:91–101.

Pehnt, M. 2003a. Assessing future energy and transport systems: The case of fuel cells Part I: Methodological aspects. *International Journal of Life Cycle Assessment* 8:283–289.

Pehnt, M. 2003b. Assessing future energy and transport systems: The case of fuel cells—Part 2: Environmental performance. *International Journal of Life Cycle Assessment* 8:365–378.

Pena, M.A., J.P. Gomez, and J.L.G. Fierro. 1996. New catalytic routes for syngas and hydrogen production. *Applied Catalysis A: General* 17–57.

Peppley, B.A., J.C. Amphlett, L.M. Kearns, and R.F. Mann. 1999a. Methanol-steam reforming on Cu/ZnO/Al_2O_3. Part 1: The reaction network. *Applied Catalysis A: General* 179:21–29.

Peppley, B.A., J.C. Amphlett, L.M. Kearns, and R.F. Mann. 1999b. Methanol-steam reforming on Cu/ZnO/Al_2O_3 catalysts. Part 2. A comprehensive kinetic model. *Applied Catalysis A: General* 179:31–49.

Ponec, V. 1992. Active-centers for synthesis gas reactions. *Catalysis Today* 12:227–254.

Ptasinski, K.J. 2008. Thermodynamic efficiency of biomass gasification and biofuels conversion. *Biofuels Bioproducts & Biorefining: BIOFPR* 2:239–253.

Ptasinski, K.J., C. Hamelinck, and P.J.A.M. Kerkhof. 2002. Exergy analysis of methanol from the sewage sludge process. *Energy Conversion and Management* 43:1445–1457.

Qi, D.H., H. Chen, L.M. Geng, Y.Z. Bian, and X.C. Ren. 2010. Performance and combustion characteristics of biodiesel-diesel-methanol blend fuelled engine. *Applied Energy* 87:1679–1686.

Rakopoulos, C.D. and D.C. Kyritsis. 2001. Comparative second-law analysis of internal combustion engine operation for methane, methanol, and dodecane fuels. *Energy* 26:705–722.

Reddington, E., A. Sapienza, B. Gurau et al. 1998. Combinatorial electrochemistry: A highly parallel, optical screening method for discovery of better electrocatalysts. *Science* 280:1735–1737.

Ren, X.M., P. Zelenay, S. Thomas, J. Davey, and S. Gottesfeld. 2000. Recent advances in direct methanol fuel cells at Los Alamos National Laboratory. *Journal of Power Sources* 86:111–116.

Sayin, C., M. Ilhan, M. Canakci, and M. Gumus. 2009. Effect of injection timing on the exhaust emissions of a diesel engine using diesel-methanol blends. *Renewable Energy* 34:1261–1269.

Spolaore, P., C. Joannis-Cassan, E. Duran, and A. Isambert. 2006. Commercial applications of microalgae. *Journal of Bioscience and Bioengineering* 101:87–96.

Takezawa, N. and N. Iwasa. 1997. Steam reforming and dehydrogenation of methanol: Difference in the catalytic functions of copper and group VIII metals. *Catalysis Today* 36:5–56.

Thomas, C.E., B.D. James, F.D. Lomax, and I.F. Kuhn. 2000. Fuel options for the fuel cell vehicle: Hydrogen, methanol or gasoline? *International Journal of Hydrogen Energy* 25:551–567.

Trimm, D.L. and Z.I. Onsan. 2001. Onboard fuel conversion for hydrogen-fuel-cell-driven vehicles. *Catalysis Reviews—Science and Engineering* 43:31–84.

Tzeng, G.H., C.W. Lin, and S. Opricovic. 2005. Multi-criteria analysis of alternative-fuel buses for public transportation. *Energy Policy* 33:1373–1383.

Velu, S., K. Suzuki, M. Okazaki, M.P. Kapoor, T. Osaki, and F. Ohashi. 2000. Oxidative steam reforming of methanol over CuZnAl(Zr)-oxide catalysts for the selective production of hydrogen for fuel cells: Catalyst characterization and performance evaluation. *Journal of Catalysis* 194:373–384.

Volesky, B. and Z.R. Holan. 1995. Biosorption of heavy-metals. *Biotechnology Progress* 11:235–250.
Wahlund, B., J.Y. Yan, and M. Westermark. 2004. Increasing biomass utilisation in energy systems: A comparative study of CO_2 reduction and cost for different bioenergy processing options. *Biomass & Bioenergy* 26:531–544.
Wainright, J.S., J.T. Wang, D. Weng, R.F. Savinell, and M. Litt. 1995. Acid-doped polybenzimidazoles—A new polymer electrolyte. *Journal of the Electrochemical Society* 142:L121–L123.
Wang, B., Y.Q. Li, N. Wu, and C.Q. Lan. 2008. CO(2) bio-mitigation using microalgae. *Applied Microbiology and Biotechnology* 79:707–718.
Wasmus, S. and A. Kuver. 1999. Methanol oxidation and direct methanol fuel cells: A selective review. *Journal of Electroanalytical Chemistry* 461:14–31.
Waugh, K.C. 1992. Methanol synthesis. *Catalysis Today* 15:51–75.
Wender, I. 1996. Reactions of synthesis gas. *Fuel Processing Technology* 48:189–297.
Wilhelm, D.J., D.R. Simbeck, A.D. Karp, and R.L. Dickenson. 2001. Syngas production for gas-to-liquids applications: Technologies, issues and outlook. *Fuel Processing Technology* 71:139–148.
Wright, M. and R.C. Brown. 2007. Establishing the optimal sizes of different kinds of biorefineries. *Biofuels Bioproducts & Biorefining: BIOFPR* 1:191–200.
Wrigley, N. and S. Matthews. 1986. Citation-classics and citation levels in geography. *Area* 18:185–194.
Xu, C., X. Wang, and J.W. Zhu. 2008. Graphene-metal particle nanocomposites. *Journal of Physical Chemistry C* 112:19841–19845.
Yergin, D. 2006. Ensuring energy security. *Foreign Affairs* 85:69–82.
Zervas, E., X. Montagne, and J. Lahaye. 2002. Emission of alcohols and carbonyl compounds from a spark ignition engine. Influence of fuel and air/fuel equivalence ratio. *Environmental Science & Technology* 36:2414–2421.

18

Microalgal Production of Hydrogen and Biodiesel

Kiran Paranjape and Patrick C. Hallenbeck

CONTENTS

18.1 Introduction

Since the middle of the nineteenth century, fossil fuels have been the main energy source consumed by humanity. In the beginning, coal was first mined and consumed at an industrial level as a result of the First Industrial Revolution, from 1760 to around 1840, in Great Britain and around Europe. Petroleum or crude oil was also consumed during this period; however, it was not until the Second Industrial Revolution and the advent of the internal combustion engine that oil started to be produced and consumed at industrial levels. As of 2012, the U.S. Energy Information Administration (EIA) projected that the highest consumed energy source, in the United States, was petroleum (36%), followed by natural gas (26%), coal (18%), renewable energy (9%), and nuclear electric power (8%) (EIA, 2013). As fossil fuels are formed from fossilized remains of organisms over long geological periods of time, there is a limited amount of these fuels. Thus, the notion of *peak oil* has come about. Peak oil refers to the point in time when maximum rate of petroleum extraction will be reached, after which the rate of production will diminish eventually dropping down to zero. The point in time when peak oil will be reached is highly debated; however, most specialists can be grouped into two categories: the late-peak advocates and the early-peak advocates (Chapman, 2013). Some examples of late-peak advocates are the Cambridge Energy Research Associates (CERA), the EIA, and Shell Company who, respectively, believe peak oil will be reached by 2017, 2035, and around 2020 (Chapman, 2013). On the opposite side, early-peak advocates believe the peak has already been reached. For instance, some have stated that the peak was reached in 2005 and that that level of oil production will never again be surpassed (Deffeyes, 2010).

Furthermore, fossil fuels have been immensely criticized for their negative effects on the environment. Since fossil fuels are mainly formed of hydrocarbons, their consumption liberates carbon compounds (mainly CO_2) that have been trapped for millions of years in the Earth's crust. Carbon dioxide being a greenhouse gas, the increase of atmospheric CO_2 concentrations has been linked to climate change and the global warming phenomenon. In addition, the extraction and purification processes for these fuels are highly polluting, releasing toxic compounds and thus causing damage to nearby ecosystems. Oil spills have been one of the major problems of the oil industry causing much discord within the business. There have been many examples of oil spills throughout the past few decades; most notable is the BP oil spill during the summer of 2010. The BP oil spill began in April 2010, in the Gulf of Mexico, and lasted until July 15, 2010. This particular spill discharged between 62,500 and 84,000 bopd (barrels of oil per day) for 87 days, making it the largest oil spill yet (Joye et al., 2011). The spill mainly affected marine ecology with the discharge of toxic chemical compounds, such as polycyclic aromatic hydrocarbons and methane, harming many marine species. This in turn had a negative ripple effect both on economy (fishing and tourism) and human health in surrounding countries (mainly the United States).

As a result of all these negative effects of fossil fuel use, there has been a great interest in recent years in the development of greener and more sustainable energies. Hydroelectricity, solar power, and wind power are all examples of green technologies. Among these new technologies, microalgae have been in the spotlight in recent years. Microalgae are unicellular photosynthetic organisms found in freshwater or marine systems. Their diversity is highly debated, as some consider them solely to be in the eukaryotic realm and others consider them to be both in the prokaryotic (cyanobacteria) and the eukaryotic realms. However, commonly they are defined as oxygen-producing photosynthetic microorganisms containing chlorophyll *a* (Leite et al., 2012). Lately, microalgae have been extensively studied for their capacity to produce lipids and hydrogen. Indeed, microalgae have been known to accumulate large quantities of lipids within their cells under certain conditions rivaling or exceeding the lipid content of traditional crops, such as soya or corn. The lipids produced can be extracted and either converted into biodiesel or used as nutritional supplements, such as omega-3 fatty acids (FAs). Some microalgae are also known to produce hydrogen, which could be used as a potential biofuel.

The advantages of using microalgae for biofuel production are numerous. Since algae are phototrophic, they have the capacity to fix CO_2 using energy produced from light. Consequently, this would have an impact on atmospheric CO_2 sequestration and help to mitigate CO_2 emission levels from fossil fuel production and consumption. Furthermore, microalgae have a faster growth rate than traditional crops, with some species having a doubling time of less than 24 h, and, under certain conditions, could be grown year round (Leite et al., 2012). Another advantage is their capacity to be grown on nonarable land avoiding competition with food crops. This would help to diminish the high food prices for cultivated plants that are now used for biofuel production, such as corn and soya. Moreover, microalgae have the capacity of growing in many different types of media and thus could be used to treat certain polluted waters such as wastewater (Abdelaziz et al., 2013a,b; Makareviciene et al., 2013). Wastewater contains high amounts of nitrogen and phosphorus coming from excessive use of cleaning products and fertilizers; the algae would in principle be able to use up these elements for their own growth while removing pollutants from municipal and industrial wastewaters and other liquid wastes (Makareviciene et al., 2013).

Thus, algae have many advantages; however, each advantage has one or more drawbacks making algal biofuels still not commercially competitive. One of the biggest disadvantages for microalgal cultivation is the availability of carbon dioxide (Chisti, 2013). To produce 1 t of algal biomass, a minimum of 1.83 t of carbon dioxide is needed (Chisti, 2007). As a result, most pilot-scale algal cultures rely on purchased carbon dioxide, creating high cultivation costs (Chisti, 2013). Moreover, the high growth rates that were mentioned earlier depend largely on optimal conditions, easily produced in laboratories but not necessarily reproducible at an industrial scale. Therefore, the cultivation of microalgae is complex and still in the beginning phase of development. Yet advances have been made, and some start-up companies have been attempting to commercialize algal fuels. Some examples of such start-up companies are Algenol Biofuels, Aquaflow, LiveFuels Inc., Solazyme Inc., Joule Unlimited Inc., and Solix Biofuels Inc (Chisti, 2013). Thus, advances are being made in the algal biofuel field, yet much research still needs to be done in order to make biofuels from algae a competitive and abundant commodity. Thus, this chapter focuses on recent topics and research advances in the algal production of biodiesel and biohydrogen.

18.2 Microalgae

18.2.1 Microalgal Diversity

As a general rule, microalgae consist of any microscopic photosynthetic organism, usually unicellular, containing chlorophyll *a* and a thallus not differentiated into roots, stems, and leaves (Tomaselli, 2008). This definition being very general, microalgae form an extremely diverse polyphyletic group within the tree of life. Though sometimes contested, this definition takes into account species from the prokaryotic realm, such as cyanobacteria, and species from the eukaryotic realm, such as the green algae. As a result, most microalgae can be regrouped into five kingdoms: the Protozoan kingdom, the Plantae kingdom, the Chromista kingdom, the Fungi kingdom, and the Cyanobacteria phylum (Leite et al., 2012; Tomaselli, 2008) (Table 8.1).

Microalgae are mainly found in aquatic or highly wet environments, such as riverbeds, lakeshores, and freshwater or saltwater. However, since their phylogenetic diversity is extremely high, some species can be found in very different habitats with some known to grow under extreme conditions. *Chlamydomonas nivalis*, *Chlamydomonas brevispina*, and *Chloromonas granulosa* are all microalgae within the Chlorophyta phylum (green algae). These species are also commonly known as *snow algae* and are capable of growing on snow at very low temperatures (Raven et al., 2007). These species have a high tolerance for extreme temperatures, acidity, and exposure to sun (Raven et al., 2007). They also have

TABLE 18.1
Groups Where Hyper Oil Producers Are Found

Taxa	Rank	Kingdom	Habitats	Carbon Reserves	Examples
Bacillariophyceae	Class	Chromista	Marine, freshwater terrestrial	Fat, chrysolaminarin	Diatoms *Navicula*
Chlorophyta	Phylum	Plantae	Marine, freshwater terrestrial	Starch, inulin, fat	Green algae *Chlorella* sp.
Dinophyceae	Class	Protozoa	Marine and freshwater	Starch, fat	Dinoflagellates *Crypthecodinium cohnii*
Haptophyta	Phylum	Chromista	Mostly marine	Fat, chrysolaminarin	Golden brown *Pavlova lutheri*
Chrysophyceae	Class	Chromista	Marine freshwater	Fat, chrysolaminarin	Golden algae *Chrysocapsa* sp.
Xanthophyceae	Class	Chromista	Freshwater, damp soil	Fat, chrysolaminarin	Yellow-green *Pleurochloris* sp.

the capacity to grow with oligotrophic concentrations of minerals for growth (Raven et al., 2007). *C. nivalis* contains a red carotenoid pigment, which protects the chlorophyll from extreme sunlight exposure, creating a phenomenon known as *watermelon snow* or *red snow*, when the algae bloom (Raven et al., 2007). The great versatility of growth habitats can be an advantage for algal cultivation. As different locations have different specific geographical and meteorological conditions, each could use specific species or strains for algal cultivation, which are tailored to withstand and thrive under those conditions. For example, species capable of growing at freezing temperatures, such *C. nivalis*, could possibly be used in very cold locations.

As a result of this wide variety, microalgae are extremely diverse in terms of physiology, anatomical structures, and metabolism. Typically, each species, and even strain, has its own set of physiological, structural, and metabolic systems helping it to survive under specific environmental conditions.

18.2.2 Cell Wall

Structurally speaking, microalgae can be placed in either the prokaryotic realm or the eukaryotic realm. Prokaryotic microalgae are mainly composed of the cyanobacterial phylum, which is further subdivided into four main groups (Stanier and Cohen-Bazire, 1977). Thus, we can find the chroococcacean group, the pleurocapsalean group, and the oscillatorian and the heterocystous groups, which both form filaments, composed of mother and daughter cells, called trichomes (Stanier and Cohen-Bazire, 1977). Apart from the heterocystous group, the other three taxa are still highly debated and have not yet been validly added to the Bacteriological Code.

Cyanobacteria have a cell wall that is structurally and chemically homologous to Gram-negative bacteria (Hiczyk and Hansel, 2000; Stanier and Cohen-Bazire, 1977). This cell wall has four layers thus making it thicker than most Gram-negative bacteria. The structural part of the cell wall is formed of a murein layer, and the outside membrane is formed of lipopolysaccharides (Tomaselli, 2008). The cell wall also contains species-specific components, such as lipopolysaccharides, carotenoids, lipids, and proteins (Hoiczyk and Hansel, 2000). These specific components can be used as biomarkers or molecular markers for phylogenetic or genetic studies of cyanobacteria. Recently, hopanoids, such as 2-methylhopanoids, have been used as molecular fossil biomarkers for the study of ancient fossilized cyanobacteria and early oxygenic photosynthesis (Garby et al., 2013). Hopanoids are pentacyclic compounds, which play a role in membrane integrity, akin to the role of sterols in eukaryotes (Garby et al., 2013). The hopanoid composition can vary depending on the species of cyanobacterium and the environmental conditions.

Eukaryotic microalgal cell walls also have a high degree of variation. However, the cell wall composition and function are different from those of cyanobacteria. For one thing, as the eukaryotic algae can be divided into four kingdoms, the diversity of cell wall composition is colossal, while in terms of functionality, eukaryotic cell walls all have the same main function, protection of the cell from the environment. As a result, most cell wall components are formed of very strong and resistant biological molecules. Species found in the Chlorophyta phylum, the green algae, typically have similar components to vascular plants, since they share a relatively recent common ancestor. Cellulose is the major component found in these species; however, xyloglucans, mannans, glucuronan, and ulvans, all highly resistant polysaccharides, are also found (Popper et al., 2011). Other algae, such as diatoms, a group of microalgae belonging to the heterokonts, contain inorganic molecules, which give the cell wall even more resistance. This group has the particularity of containing silica within their cell wall. The biological silica contained in the cell wall is synthesized within the cell in the form of silicic acid monomers (Scheffel et al., 2011). The monomers are then extruded and added to the cell wall. More recently, it has been found that some diatom species, such as *Thalassiosira pseudonana*, also contain chitin within their cell walls (Brunner et al., 2009; Durkin et al., 2009). Chitin is a polysaccharide closely related to glucose and comparable to cellulose. This polysaccharide is abundant and can be found in arthropods, such as insects and crustaceans, and as a component of the cell wall of most fungal species. Thus, the diversity in cell wall structures and composition is very big in microalgae.

18.2.2.1 Chloroplasts

As microalgae are found in the prokaryotic and eukaryotic realm, their cellular components are organized accordingly. Thus, eukaryotic species have a nucleus, containing the genetic material, and several organelles inside the cell, whereas cyanobacteria do not have any organelles and have a single circular chromosome.

In order to carry out photosynthesis, eukaryotic algae use a specific organelle, the chloroplast. These organelles conduct photosynthesis with the help of specific pigments and an electron transport chain found in the membranes of these units. They also carry out most of the fatty acid synthesis (FAS). During photosynthesis, molecular pigments found in the membrane of the thylakoid of the chloroplast are excited when they absorb light, causing the loss of an electron. This electron passes through a series of complexes in the membrane, the electron transport chain, to ultimately reduce NADP to NADPH. At the same time, a proton gradient is created across the chloroplast membrane and used to drive phosphorylation of ADP by ATP synthase to produce ATP. The NADPH and ATP are then used as energy sources to fix CO_2 through the Calvin cycle and produce carbohydrates, such as sugars, and lipids, either for direct consumption or energy storage.

On the other hand, cyanobacteria do not have any organelles and carry out photosynthesis directly in cell membrane. The inner cell membrane forms complex folds and is called the thylakoid membrane. This thylakoid membrane contains the different complexes of the electron transport chain used in photosynthesis, such as ATP synthase and photosystems I and II. The folds created in the thylakoid membrane help to increase the surface area of the membrane. As a result, cyanobacteria have a higher concentration of photosynthetic complexes than other photosynthetic bacteria, such as purple sulfur bacteria, and so have a better efficiency of converting light to chemical energy. Interestingly enough, the current formed by the flow of electrons through the electron transport chain has been studied for its use in microbial fuel cells. Thus, it is possible to use cyanobacteria to produce electricity, with one study showing that some strains of *Synechocystis* can produce up to 6.7 $mW \cdot m^{-3}$ (Madiraju et al., 2012).

In addition, the photosynthetic pigments are very variable between species. As previously mentioned, microalgae all contain chlorophyll *a*, but many species also have additional pigments. For example, diatoms contain chlorophylls *a* and *c* and fucoxanthin, while green algae contain chlorophylls *a* and *b* and zeaxanthin (Brennan et al., 2010). Cyanobacteria also contain chlorophyll *a*; however, they also have a high concentration of phycobilins, such as the blue pigment phycocyanin or the red pigment phycoerythrin (Chang et al., 2012; Horton et al., 2006; Tomaselli, 2008). Cyanobacteria get their blue-green color from the phycocyanin pigment. Thus, the diversity in photosynthetic pigments is great that helps to capture different wavelengths of light. For instance, phycocyanin will absorb light in the orange range at around 620 nm, whereas chlorophyll *a* has an absorption maximum at around 664 and 430 nm (Chang et al., 2012). Secondary functions can include photoprotection (carotenoids), singlet oxygen scavenging, structure stabilization, and excess energy dissipation (Frank et al., 1996).

Finally, chloroplasts are thought to have arisen from a primary endosymbiosis between the ancestral eukaryote and cyanobacterium (Alberts et al., 2007). In this scenario, an ancestral eukaryote would have engulfed a cyanobacterium through phagocytosis. However, due to unknown circumstances, the cyanobacteria were not degraded and survived inside its host, producing energy by photosynthesis. This endosymbiotic theory is supported by many facts. Indeed, chloroplasts retain their own genome, however reduced it may be, a genome containing many genes resembling those of cyanobacteria (McFadden, 2001). Furthermore, this genome shows typical prokaryotic organization. Secondary endosymbiosis is an even more interesting phenomenon. In this event, some chloroplasts originated from the endosymbiosis between a heterotrophic eukaryote, where an algal species, already the product of a primary endosymbiosis, underwent a secondary endosymbiosis (McFadden, 2001). This secondary endosymbiosis would explain why some chloroplasts in some species have up to four cell walls (heterokonts and dinoflagellates) and nucleomorphs (remnants of a nucleus). Interestingly, some species of protists, *Hatena arenicola*, and animals, such as sacoglossan slugs, have the capacity to sequester plastids from algae and use them for energy production through photosynthesis (McFadden, 2001).

18.2.2.2 Cellular Division

For the most part, cyanobacteria reproduce asexually through binary fission. In this process, either all cellular membranes (wall and membranes) or just the inner membrane creates invaginations in the middle of a mother cell, which then constrict to split the cell in to two daughter cells of equal size (Tomaselli, 2008). Binary fission is the most common method of reproduction, but some species have been known to have other forms of reproduction, such as reproduction by fragmentation to form hormogonia, reproduction by budding (*Chamaesiphon*), and production of akinetes (Tomaselli, 2008). Though sexual reproduction is unknown in cyanobacteria, conjugation and horizontal gene transfer do exist. Recent studies on the *Leptolyngbya*, *Fischerella*, and *Chlorogloeopsis* genera have shown that foreign genes, such as a GFP reporter transgene, can be introduced through conjugation, electroporation, and biolistic DNA transfer methods (Stucken et al., 2012; Taton et al., 2012).

In contrast, eukaryotic algae can reproduce asexually but also have the capacity to reproduce sexually. For asexual reproduction, just like cyanobacteria, eukaryotic microalgae will reproduce by cell division, fragmentation, and production of spores (Tomaselli, 2008). In the case of sexual reproduction, individuals of the same species will combine their gametes to produce offspring. Thus, there are five major life-cycle types found in eukaryotic microalgae (Tomaselli, 2008). The first would be a predominantly diploid life cycle wherein meiosis occurs before the formation of the gametes. Opposite to this would be a predominantly haploid life cycle where meiosis occurs right after the zygote forms. Third, a generational alternation of life cycles can exist in some species where individuals of a species will alternate between being haploid (gametophytes) and diploid (sporophytes) every other generation. The fourth life cycle is called heteromorphy alternation of generation where individuals go through an generational alternation of life cycle; however, one of the phases (haploid or diploid) will be dominant over the other one, making either the haploid individuals dependent upon the diploid individual or vice versa. Finally, some species, such as red algae, have a triphasic lifestyle, also called a triphasic alternation of generation, where individuals will have one haploid phase and two distinct diploid phases.

18.2.2.3 Growth Modes and Metabolism

Microalgae are extremely diverse and as a result can grow under a variety of different conditions. Most microalgae are photoautotrophic, using photosynthesis to fix ambient carbon dioxide to create their own organic compound for cell growth or energy storage. Additionally, some species have the capacity to grow in a heterotrophic growth mode, using reduced carbon sources found in their environment, such as sugars. Consequently, a variety of carbon sources, such as glucose, xylose, glycerol, and acetate, can be used to grow some species of microalgae in a heterotrophic manner. Many experiments have shown that *Chlorella vulgaris* and *Chlorella protothecoides* are capable of using sugars, such as glucose, and glycerol for heterotrophic growth (Liang et al., 2009; O'Grady and Morgan, 2011; Perez-Garcia et al., 2010). Recently, mixotrophic growth has been studied in some species. For algae, mixotrophy, as the name implies, is the capacity to mix phototrophic growth with heterotrophic growth. Hence, *C. vulgaris* has been shown to grow on glucose in a mixotrophic manner, fixing CO_2 through photosynthesis while at the same time consuming glucose (Ogawa and Aiba, 1981).

These different growth modes are a result of the diversity in metabolic networks seen across the different groups of microalgae. Indeed, the metabolic diversity is quite large in microalgae and can be seen at the species level and even at the strain level. For example, some strains of *C. vulgaris* have been shown to grow only heterotrophically, whereas others can only grow autotrophically (Liang et al., 2009). This diversity may be advantageous for the algal industry since each species can offer one or more interesting characteristics (such as protein production, growth medium, growth rate, or productivity) for cultivation. Microalgae have been known to have different intracellular lipid profiles according to species and to environmental conditions. Depending on the lipids produced, different species of microalgae could be of interest for the biofuel industry. Thus, microalgae are a highly diverse group spread throughout the eukaryotic and prokaryotic realms. Their diversity can be seen at the morphological level, the physiological level, and the metabolic level.

18.3 Microalgae and Biofuel

18.3.1 Biofuel Generalities

Biofuels are fuels produced from energy-rich organic compounds usually produced by plants or microalgae through contemporary photosynthesis. First-generation biofuels are produced using vegetable oil, starches, and sugars from cultivated plants, such as corn and soya, to produce different types of fuels. Bioalcohols (biologically produced ethanol) and biodiesel are examples of such biofuels. Second-generation biofuels take into account the sustainability of the feedstock as well as the requirement to not compete with the food supply. The sustainability of a feedstock would in principal depend upon the abundance of the feedstock as well as the consequences of its utilization on cultivable land usage, greenhouse gas emissions, and general impact on the environment.

Thus, the major thrust in second-generation biofuel production is the use of a feedstock that does not compete with food crops and has a very low impact on the environment. The major criticism of first-generation biofuels is that they use arable land, which could otherwise have been used for food crops, or they directly use food crops. Consequently, first-generation biofuels have the potential for increasing food prices. This is seen in the bioethanol industry. Bioethanol is produced through the fermentation of sugar derived from corn, and ethanol plants already in use or under construction will within a few years burn up to half of the U.S. domestic corn supplies (Runge and Senauer, 2007). Already in March of 2007, corn prices rose over $4.38 a bushel, which was one of the highest prices recorded in 10 years (Runge and Senauer, 2007). This is bad news for consumers and especially in poor developing countries where marginal increases in the cost of staple grains could be devastating (Runge and Senauer, 2007). Therefore, the use of second-generation biofuels would help to solve this problem by using the inedible waste products of agriculture, such lignocellulose.

Third-generation biofuels use nonarable lands to produce biofuels. Thus, biodiesel produced from microalgae fits the description of a third-generation biofuel. Biodiesel is a mixture of FAs that have been transesterified with an alcohol (usually methanol but ethanol and propanol can be used) to produce FA alkyl esters. Among the stored products resulting from microalgal photosynthesis, lipids mainly triacylglycerols (TAGs), three FA molecules attached to glycerol, are made by some strains that can be used for biodiesel production. The FA composition can be different for different TAG molecules from different strains. For the production of biodiesel, FA composition is crucial since FA length and degree of saturation will greatly influence the resulting fuel properties (Leite et al., 2012).

FAs are carboxylic acids with a long aliphatic chain that can either be saturated or unsaturated. FAs differ from one another in the length of their hydrocarbon tails, the number of carbon–carbon double bonds, the positions of the double bonds in the chains, and the number of branches (Horton et al., 2006). Thus, over 100 different FA molecules have been identified in living organisms (Horton et al., 2006). In the case of biodiesel, FAs are evaluated for their hydrocarbon tails and their degree of unsaturation. These parameters will affect a number of general properties of the resulting biodiesel, such as the neutralization number, cold filter plugging point (CFPP, which reflects a fuel's performance in cold weather), cetane number (ignition characteristics), viscosity, and storage stability (Meher et al., 2006). Consequently, FAs used for gasoline would need hydrocarbon chains with a length of 6–12 carbons with a mixture of saturated chains, whereas FAs used for biodiesel would need chains of 12–18 carbons (Leite et al., 2012).

TABLE 18.2

FA Composition of Some Microalgal Species

Species	Lipid Content (% dry wt)	Lauric Acid C12:0 (%)	Palmitic Acid C16:0 (%)	Stearic Acid C18:0 (%)	Oleic Acid C18:1 (%)	ϒ-Linolenic Acid C18:3 (%)	Others (%)
Scenedesmus obliquus	29	11	29	17	20	23	0
Chlamydomonas pitschmannii	51	10	26	20	13	23	8
Chlorella vulgaris	26	5	22	5	53	8	7
Chlamydomonas mexicana	29	34	50	6	0	0	10

Fortunately, many species of microalgae are capable of producing FAs with tails of around the desired sizes (Table 18.2). For example, some species of microalgae (*Spirulina maxima, C. vulgaris, Scenedesmus obliquus*) have been shown to produce from 17% to 40% by weight of palmitic acid, an FA with 16 carbons in its tail (Gouveia and Oliveira, 2009). Another study showed that different species change their FA composition under nitrogen replete and nitrogen depleted conditions (Griffiths et al., 2012). Thus, the Chlorophyta species studied showed an increase in C18:1 FAs with nitrogen limitation, along with a decrease in C16:0, C18:2, and C18:3 FAs (Griffiths et al., 2012). Due to the variety of FAs produced in nature, certain standards have been put in place by industrial and governmental organizations detailing the exact requirements biodiesels must have in order to be accepted for use. In Europe, the EN 14214 gives specific requirements and test methods for fatty acid methyl ester (FAME) quality. Examples of requirements are the acceptability or not of dye or marker usage, of additives, and of stabilizing agents and even percentage of different FAs in the biodiesel (European Standard, 2008).

18.3.2 Lipid Metabolism

18.3.2.1 Photosynthetic Efficiency

As microalgae are photosynthetic organisms, one of the limits to biodiesel production is efficiency with which each species is capable of capturing light and using the captured energy to fix carbon dioxide, also known as the photosynthetic efficiency. Microalgae contain chlorophyll *a* as their major pigment, and thus, species can only capture up to a maximum of 45% of the whole solar spectrum (Leite et al., 2012). However, additional amounts of energy are said to be lost due to the reflection of light on the surface of the reactor (about 10%), to transform light energy into chemical energy at the reaction center, to respiration, to photosaturation, and to photoinhibition (Leite et al., 2012). Thus, the theoretical maximum photosynthetic efficiency is said to be between 4.6% and 6% for microalgae (Leite et al., 2012; Ort et al., 2011).

Numerous ideas, from cultivation techniques to genetic engineering, have been proposed in order to improve the photosynthetic efficiency. Among some of these ideas, the optimization of the light-harvesting complex or chlorophyll antenna to improve photosynthetic efficiency through genetic manipulation is one of the more interesting ones. The light-harvesting complex is a group of proteins and chlorophyll molecules fixed in the thylakoid membrane, which capture light and transfer the energy to a chlorophyll *a* molecule in the reaction center of a photosystem. In the case of microalgal cultivation, the density of cells and the intensity of light in a culture affect the photosynthetic efficiency of the species cultured. Thus, if the cell density in a culture is too high, the efficiency will be lowered, and although increasing the light intensity might help, the algae at the surface might become photosaturated. Reducing the size of the light-harvesting complex through genetic manipulation has been shown to be effective in lessening photosynthetic inefficiency related with the over absorption of light during mass culture (Ort et al., 2011). Microalgal strains with reduced chlorophyll antenna have been shown to improve solar to biomass conversion efficiencies in mass cultures (Ort et al., 2011). Recently, a *tla3* mutant (*truncated light-harvesting antenna size*) of *Chlamydomonas reinhardtii* was described (Kirst et al., 2012). The deletion of the *TLA3–CpSRP43* gene showed a reduction in the light-harvesting antenna size and a twofold increase in the light saturation maximum over the wild-type strain (Kirst et al., 2012).

18.3.2.2 TAG Synthesis

Microalgae are extremely variable in their lipid composition. Many species can produce a plethora of different types of lipids of value for the biodiesel industry, the pharmaceutical industry, the food industry, and the cosmetic industry. Thus, algae can produce a number of polyunsaturated FAs (PUFAs) (such as eicosapentaenoic acid, docosahexaenoic acid, and arachidonic acid), new fatty compounds (long-chained hydrocarbons with 35–40 carbons, unusual hydrocarbons, such as *n*-alkadienes and trienes, galactolipids, triterpenoids, tetraterpenoids, and lycopadienes), and oxylipins (Guschina and Hardwood., 2006). Consequently, FA and TAG synthesis is highly complex, and much of it still requires elucidation at the molecular level.

Recently, some species, mostly green algae, have been shown to have enzymes that are homologous to key enzymes used in TAG synthesis in higher plants and fungi (Khozin-Goldberg and Cohen, 2011; Merchant et al., 2012). For example, the acetyl coenzyme A carboxylase (ACCase), an enzyme used in the first step of FAS, found in most green algae and some red algae species is similar to those found in higher plants (Khozin-Goldberg and Cohen, 2011). ACCase is responsible for the carboxylation of acetyl-CoA to malonyl-CoA, which goes into the FAS pathway (Khozin-Goldberg and Cohen, 2011).

As very little is known about microalgal TAG synthesis, one approach for identifying genes involved in TAG synthesis involves analysis of orthologues of known enzymes from yeast and other animals. In yeast, TAG synthesis has been comprehensively studied in *Saccharomyces cerevisiae*. The pathway is separated into three parts: (1) the generation of diacylglycerol (DAG), (2) the esterification of DAG to generate TAG, and (3) the degradation of TAG (Merchant et al., 2012). Acetyl-CoA and NADPH are required for TAG synthesis. These are produced by a citrate/malate cycle where citrate is accumulated in the mitochondria via the tricarboxylic acid cycle and excreted out into the cytoplasm. It is then converted to acetyl-CoA and oxaloacetate by the key enzyme acyl-CoA ligase, which uses ATP (Hallenbeck, 2012; Hu et al., 2008). The oxaloacetate is converted into malate and enters the mitochondria, where it is converted into pyruvate, CO_2, and NADPH by the malic enzyme (Hallenbeck, 2012). The NADPH will help fuel the production of FAs. After the production of acetyl-CoA, ACCase will transform acetyl-CoA into malonyl-CoA, used in FAS. The products of FAS will be FAs with a CoA attached, forming molecules of acyl-CoA. After this, a first acyl-CoA molecule will be attached to a glycerol-3-phosphate (G-3-P) molecule by an acyltransferase via esterification, producing a lysophosphatidic acid (Merchant et al., 2012). A second acyltransferase then attaches, on the C2 position of the glycerol molecule of the lysophosphatidic acid, a second acyl-CoA molecule. This produces a phosphatidic acid. The phosphate group from the G-3-P is then removed by a phosphatase to produce a DAG molecule.

Two routes are known for the conversion of DAG into TAG: the acyl-CoA-dependant route (also known as the Kennedy pathway) and by transesterification. In the acyl-CoA-dependant route, a DAG acyltransferase (DAGAT) esterifies a third acyl-CoA onto the C3 position of the DAG to produce a TAG. In the transesterification route, a membrane phospholipid is transferred to the C3 position of a DAG to form a TAG (Bensheng and Benning, 2013; Hildebrand et al., 2013; Hu et al., 2008; Merchant et al., 2012). The transesterification is catalyzed by a phospholipid diacylglycerol acyltransferase (PDAT). Interestingly enough, in *S. cerevisiae,* the DAGAT route for TAG synthesis is used in the stationary phase, whereas the transesterification route is operative during the exponential growth phase (Merchant et al., 2012).

This general scheme for TAG synthesis is applicable to fungal species and plants. In the case of microalgae, this network may be applicable in at least some cases. Recent studies have shown that *Chlamydomonas* tends to have only a single copy of the genes for proteins used in FAS, suggesting that lipid metabolism in this genus is much simpler than that of higher plants, which tend to have multiple copies of genes for FAS (Liu and Benning, 2013). However, in the case of TAG synthesis, this same organism has been found to contain six-gene coding for DAGAT enzymes suggesting a more complicated metabolic network for TAG synthesis than plants, which only have two-gene coding for the same enzyme (Liu and Benning, 2013). Moreover, other significant differences are thought to occur in microalgal TAG synthesis. For instance, studies on *Chlamydomonas* have hypothesized that TAG synthesis occurs in the chloroplasts, whereas for higher plants, TAGs are assembled both in the endoplasmic reticulum and the chloroplasts (Liu and Benning, 2013). Other studies have shown that lipid droplet production inside *Chlamydomonas* uses a *major lipid droplet protein* (MLDP), which is specific to green algae (Khozin-Goldberg and Cohen, 2011). Obviously, there is still much research to be done on TAG synthesis in microalgae.

18.3.2.3 Lipid Accumulation

Environmental stress is known to affect lipid metabolism in microalgae (Guschina and Hardwood, 2006; Hu et al., 2008; Khozin-Goldberg and Cohen, 2011; Liu and Benning, 2013; Longworth et al., 2012; Merchant et al., 2012; Msanne et al., 2012). Numerous studies have shown that nutrient limitation, temperature changes, light intensity changes, salinity changes, and pH changes contribute to modifying the quantities and composition of algal FAs.

Nutrient limitation is the reduction of a nutrient in a medium to the point where it hinders growth. This has led to a two-system growth method where a strain is grown in a normal medium, containing all of the necessary nutrients until it reaches a certain biomass density. Then, the strain is left to grow in a second medium containing a limiting nutrient. The limiting nutrient can be any nutrient in the media; however, the most common ones are usually nitrogen, phosphorus, sulfur, and silicon for diatoms. Depending on the nutrient, different effects may occur on the lipid composition and quantity in the cell of a given strain. Thus, nitrogen starvation has shown to increase the lipid content of many species of microalgae. *S. obliquus, Neochloris oleoabundans, C. vulgaris,* and *C. zofingiensis* have been shown to accumulate up to 46% lipids by weight when starved for nitrogen (Breuer et al., 2012) (Table 18.3).

TABLE 18.3

Enhancement of Lipid Production in Different Microalgae

Species	Rich Media (%)	Nitrogen Deficient (%)
Chlamydomonas applanata	18	33
Chlorella emersonii	29	63
Chlorella minutissima	31	57
Chlorella vulgaris	18	40
Ettlia oleoabundans	36	42
Scenedesmus obliquus	12	27
Selenastrum gracile	21	28

Phosphorus starvation has been linked to an increase in lipid content in *Phaeodactylum tricornutum*, *Chaetoceros* sp., and *Pavlova lutheri* while giving a decreased lipid content in the green flagellates, *Nannochloris atomus* and *Tetraselmis* sp. (Guschina and Hardwood, 2006). On the other hand, very high levels of nutrients have also been shown to increase lipid content in some cases. For example, at 15 mM nitrate, *Ulva pertusa* showed an increase in crude lipid content as a percentage of biomass, and when nitrogen levels were augmented, PUFA (which cannot be used for biodiesel production) concentrations decreased and palmitate and lineolate (FAs acceptable for biodiesel production) levels increased in the same species (Guschina and Hardwood, 2006). Moreover, silicon starvation for *Stephanodiscus minutulus*, a freshwater diatom, was found to increase TAG content and decrease polar lipid content (Guschina and Hardwood, 2006). Finally, different studies have shown that physical growth conditions, such as light intensity, pH, and temperature, can have different effects on lipid accumulation when microalgae are grown under nitrogen starvation (Breuer et al., 2013). As a result, *S. obliquus* was shown to have an optimal biomass productivity at pH 7, at 27.5°C. When grown at 20°C or 35°C, the total FA and TAG content grew with increasing pH under nitrogen starvation (Breuer et al., 2013).

Efforts have been made to understand lipid accumulation under starvation at the cellular level, but there is still much to be learned. Most of the research has been focused on the effects of nitrogen starvation. With the absence of nitrogen, protein synthesis and consequently cell growth are hindered (Msanne et al., 2012). As a result, cells in a nitrogen-deprived environment are thought to channel excess fixed carbon from photosynthesis into storage molecules, such as TAGs and starch, instead of using the carbon and energy for growth. Recently, a study with *C. reinhardtii* showed that in the first 2 days of nitrogen starvation, cells accumulate up to 14-fold higher levels of starch. However, after 10 days, the FA content increased significantly, while starch levels declined (Msanne et al., 2012). These findings suggested that FAS is a *de novo* process and that FAS and TAG synthesis, in nitrogen-depleted environment, are unlikely to originate only from newly fixed carbon, since most likely most proteins used in carbon fixation, such as Rubisco, will be greatly reduced (Breuer et al., 2012; Msanne et al., 2012).

Thus, it is hypothesized that TAG accumulation under nitrogen deprivation comes from carbon already assimilated in other cellular components, such as proteins, ribosomal subunits, and starch. Furthermore, as mentioned earlier, *C. reinhardtii* contains six different genes coding for DAGAT enzymes grouped in two families: type 1 (*DGAT*) and type 2 (*DGTT*). *DGTT* are found in five forms in *Chlamydomonas*, denoted as *DGTT 1–5*, but there is only one form of *DGAT* discovered up until now. *DGAT1* and *DGTT1* are thought to play vital roles in TAG accumulation under nitrogen starvation. Studies have shown that the expression levels for these two genes increase when cells are put under nitrogen starvation conditions and even other types of starvation, such as sulfur, phosphorus, zinc, and iron (Boyle et al., 2012; Liu and Benning, 2013). However, although *DAGAT* genes are thought to play a crucial role in TAG accumulation, the molecular details are not fully understood. For instance, in a knockout and overexpression study of *DGTT1, 3,* and *4*, *C. reinhardtii* showed changes in TAG concentrations. However, another study involving the individual overexpression of *DGTT1, 2,* and *3* showed no changes in TAG accumulation (Liu and Benning, 2013). Thus, different acetyltransferases must play a crucial role in lipid accumulation under nitrogen starvation, but the exact mechanism is not yet elucidated. Interestingly, research on different species in the green algae taxon has shown similar trends in starch and TAG accumulation (*C. reinhardtii* and *Coccomyxa* sp.), suggesting that the metabolic response of lipid accumulation under nitrogen starvation may be shared by a large range of green microalgae, since most species are spread over a large portion of the tree of life (Msanne et al., 2012). Thus, the genetics of lipid accumulation in microalgae is still in its infancy and needs further research.

18.3.3 Algal Culture and Genetic Manipulation

One of the major factors in algal cultures for biofuel production is productivity. Productivity in terms of biodiesel consists in the amount of lipids produced per unit volume of growth medium per unit time. Consequently, productivity depends on three factors: the growth rate of the strain, the quantity of lipid produced, and the volume of growth media needed. Generally, the best microalgal strain to select for biofuel production would be the species or strains with the highest productivity (Table 18.4). Cultivation methods and genetic manipulation have been considered in order to make algal oil more competitive.

The photosynthetic growth of algae necessitates water, carbon dioxide, inorganic salts, and light. The essential inorganic salts required for algal growth must contain nitrogen, phosphorus, iron, and for diatoms silicon (Chisti, 2007). Large-scale algal production usually uses a continuous method of culture where fresh medium is fed at the same rate as culture is withdrawn. Agitation is important as it prevents settling of the biomass, which can lead to lowered photosynthetic activity and, in some cases, can help aerate the culture. Thus, there are two typical cultivation methods for algal culture: open-pond systems or photobioreactors. Open-pond systems are characterized by low-maintenance and low-energy requirements and can consist of tanks, circular ponds, raceway ponds, and shallow big ponds with machinery used for mixing (Abdelaziz et al., 2013b; Chisti, 2007; Huang et al., 2010; Leite et al., 2012; Ugwu et al., 2008). Raceway ponds are channels organized

TABLE 18.4

Some Desirable Algal Strain Characteristics for Large-Scale Culture

Characteristics	Advantages
Tolerance of culture conditions	Require less control of pH, temperature, and others
CO_2 uptake efficiency	Less cost required to supplement CO_2
Tolerance to contaminants	Potential growth on very eutrophic water or flue gases
Tolerance of shear force	Allow cheaper pumping and mixing methods to be used
No excretion of autoinhibitors	Higher cell density expected: higher biomass productivity
Naturally competitive	Harder to be overcome by invader species

into closed-loop systems typically about 0.3 m deep (Chisti, 2007). Paddle wheels are used for circulation and mixing of the culture, and the channels are usually made of concrete or compacted earth (Chisti, 2007). Open-pond systems can use a wide variety of media; however, attention has been focused on the use of wastewater as a growth medium, in particular coupling open-pond operations with wastewater plants (Abdelaziz et al., 2013b; Leite et al., 2012). Other ideas have been to couple algal growth operations with industrial complexes emitting CO_2 and using the emitted CO_2 to directly feed algal cultures (Chisti, 2007; Huang et al., 2010). The advantages of using open-pond systems are the generally low cost of construction and maintenance, the ease of scaling up growth operations, the relatively low-energy requirements, and the easy maintenance (Leite et al., 2012).

However, there are a number of significant disadvantages associated with this type of growth operation. Contamination is one of the major concerns with open-pond systems. Predatory or competitively commercially uninteresting organisms can overrun a culture and thus ruin entire cultivation runs. Water loss through evaporation, poor mixing of CO_2, inadequate control of light intensity and pH, and suboptimal temperatures are all major factors that are capable of slowing or even halting growth and inhibiting lipid production in cultures. All these represent major obstacles that must be overcome in order to make biodiesel production cost-effective.

Photobioreactors in principle offer solutions to some of the problems associated with open-pond systems; however, the major disadvantages with these systems are their high cost and high-maintenance requirements. Photobioreactors are closed transparent containers designed to increase control over a culture. Thus, factors such as temperature, light intensity, pH, contamination from other species, and aeration can be controlled to a larger degree than that possible with pond systems. A wide variety of different styles of bioreactors have been invented that are highly suitable for laboratory or small-scale industries. Photobioreactors include flat plate, tubular, vertical column, internally illuminated, and torus-shaped vessels (Ugwu et al., 2008). Among all these vessels, tubular photobioreactors are one of the more appropriate types for outdoor mass culture (Chisti, 2007; Ugwu et al., 2008). These reactors are generally made of a network of transparent tubes usually made of glass or plastic. The medium inside is circulated by a series of pumps, helping to create mixing and increase aeration. These systems generally use continuous culture methods (Chisti, 2007). Thus, the use of photobioreactors increases control, productivity, and biomass yields and minimizes contamination. However, their cost, the difficulty in scaling up operations, system fragility, and the high maintenance are all major disadvantages of using these systems.

A second option to increasing productivity is the use of genetic manipulation on selected species. Cellular function can be redirected to synthesize desired products through metabolic engineering. However, very little is known in general about algal genetics, and therefore, except for *C. reinhardtii* and *Volvox carteri*, few species, if any, have been used for genetic manipulation experiments. It has been hypothesized that the difficulty of transformation might be due to an innate defense mechanism that algae possess to suppress transposons or viral invasion (Rosenberg et al., 2008). Nevertheless, a few techniques, taken from yeast or plant transformation methods, exist to transform algae, including electroporation, agitation with glass beads and DNA, biolistic particle delivery systems (such as with a gene gun), and transfection that are all potentially effective ways to transform cells (Rosenberg et al., 2008).

Moreover, some advances in algal genetic engineering have been made relatively recently. Overexpression of acetyl-CoA carboxylase in order to augment lipid production was achieved with *Cyclotella cryptica* with, however, no difference in lipid production

(Courchesne et al., 2009; Hu et al., 2008; Radakovits et al., 2010; Rosenberg et al., 2008). *V. carteri* was the first microalga to be successfully transformed with the *Hup1* gene, hexose/H+ symporter gene, from *Chlorella*, and similar experiments have been done with *C. reinhardtii* and *P. tricornutum* (Beer et al., 2009; Rosenberg et al., 2008). As mentioned earlier, experiments have been undertaken in order to improve photosynthetic efficiency via truncation of the light-harvesting complex. Improvement in gene silencing strategies has been one of the most significant advances in algal genetics (Beer et al., 2009). New anti-RNA technology will most likely emerge and thus help to elucidate metabolic pathways utilized for the production of TAGs in microalgae. Other experiments have been focused on overexpression of enzymes used in FAS, FA catabolism, and lipid modification to suit biodiesel composition requirements (Radakovits et al., 2010).

Finally, one of the more interesting modifications would be to create strains capable of directly producing biodiesel (Hallenbeck, 2012; Radakovits et al., 2010). At present, microalgae produce lipids, which then need to be processed chemically to give biodiesel. Thus, producing genetically modified organisms, which directly produce biodiesel, would increase the competiveness of algal biodiesel, lower production costs, and in general be a boon to the biodiesel industry. Microalgal TAGs are the type of lipids required for biodiesel production. These are converted through transesterification to FAMEs, the main component of biodiesel. This reaction usually uses methanol in the presence of an acid or base, but ethanol and propanol can also be used. Thus, the creation of algal strains capable of directly producing FAMEs and bypassing TAG extraction and chemical conversion outside the cell altogether would be highly interesting. Most organisms produce ethanol; as a result, fatty acid ethyl esters (FAEEs) will be produced for biodiesel production using this type of scheme (Hallenbeck, 2012). This might be an extra advantage as FAEEs have better low-temperature characteristics than FAMEs (Hallenbeck, 2012).

This concept might seem highly imaginative; however, experiments in *Escherichia coli* have already been able to produce the FAEE ethyl oleate through genetic modification (Hallenbeck, 2012; Radakovits et al., 2010). A nonspecific acetyltransferase from *Acinetobacter* and a pyruvate decarboxylase and alcohol dehydrogenase from *Zymomonas mobilis* were inserted into *E. coli*; the end result was the production of ethyl oleate when glucose and oleic acid were supplied (Hallenbeck, 2012; Radakovits et al., 2010). The FAEE yield from this manipulation was not very impressive, so initiatives to see if higher production can be attained and how FA ester accumulation works in microalgae are highly desirable. Thus, the combination of algal cultivation and genetic manipulation will most certainly help increase algal productivity for the production of biodiesel. However, more research is needed in order to make algal biodiesel competitive with fossil fuels and first-generation biodiesel operations.

18.4 Microalgae and Biohydrogen Production

18.4.1 Hydrogen as a Fuel and Commodity

Hydrogen can be used as a fuel for the transportation sector, as it is highly combustible and flammable. Indeed, a mixture of hydrogen and oxygen is used for rocket fuel in spaceship engines. In addition, hydrogen is presently used for a number of different applications, such as in the petroleum industry, chemical industry, as coolant, and in the semiconductor industry. Global demand is expected to rise about 4% annually, creating a total demand in 2016 to 286 billion m^3 valued at $43.2 billion. The vast majority of hydrogen available on the present day market is derived from fossil fuels. Biological hydrogen production holds the promise of the sustainable production of hydrogen without the generation of greenhouse gases.

18.4.2 Hydrogen-Producing Enzymes

Hydrogen production by microalgae has been known since the beginning of the twentieth century. Since that time, much has been learned about this process at the molecular level, including details of the enzymatic machinery involved. Hydrogen is produced by hydrogen-producing enzymes that reduce protons with electrons generated either directly from the splitting of water by photosystem II or indirectly from the degradation of organic molecules such as starch (Hallenbeck and Benemann, 2002; Kruse et al., 2010). Three types of enzyme are known to produce hydrogen: nitrogenase, [FeFe]-hydrogenase, and [NiFe]-hydrogenase (Ghirardi et al., 2007; Hallenbeck and Benemann, 2002).

Nitrogenase fixes atmospheric N_2 to produce ammonium and, as a by-product, molecular hydrogen (Hallenbeck and Benemann, 2002; Koku et al., 2002; Zehr et al., 2003). A large quantity of energy in the form of ATP is needed to separately activate the triple bond of N_2. Furthermore, the turnover rate, the rate at which an enzyme cycles through its catalytic cycle, is low for this enzyme (6.4 s^{-1}), making it an inefficient candidate for hydrogen production (Hallenbeck and Benemann, 2002).

On the other hand, hydrogenases are more effective in hydrogen production because of their relatively more efficient use of energy. Two types of hydrogenases exist: [FeFe]-hydrogenase and [NiFe]-hydrogenase. [FeFe]-hydrogenase is mainly found in green algae, fungi, protist, and anaerobic bacteria and is highly conserved. [FeFe]-hydrogenase contains a metallocluster, where catalytic reactions occur (binding of a proton with an electron to form molecular hydrogen), constituted of a [4Fe–4S] cubane linked through a cysteine residue to a 2Fe subcluster (Ghirardi et al., 2007). A number of eukaryotic hydrogenase sequences have already been obtained. [FeFe]-hydrogenase is highly sensitive to oxygen and is irreversibly destroyed in its presence. Finally, the last class of hydrogenase is [NiFe]-hydrogenase, found mainly in cyanobacterial groups among the microalgae. Two types of [NiFe]-hydrogenase are found: an uptake and a bidirectional protein (Ghirardi et al., 2007; Rashid et al., 2013). Uptake [NiFe]-hydrogenase is usually found in nitrogen-fixing cyanobacteria and oxidizes hydrogen produced by nitrogenase. On the other hand, bidirectional [NiFe]-hydrogenase can either oxidize or produce molecular hydrogen. The oxidization of hydrogen is a well-understood process where hydrogen is transported by a hydrophobic channel to the [NiFe] cluster. The H_2 brought by the channel is thought to bind with the Ni atom of the cluster. Hydrogen cleavage produces protons and electrons, which are transferred to the protein environment and a redox partner, respectively (Ghirardi et al., 2007). The mechanism of the production of hydrogen is not well understood.

As oxygen irreversibly inactivated hydrogenase, a number of different processes have evolved to protect hydrogenases. Some species produce separate anaerobic compartments, thereby protecting hydrogen-producing enzymes from the destructive action of oxygen. Other species only produce hydrogen when oxygen levels are low enough to create an anaerobic environment within the cell. Finally, some species can have a mutualistic or symbiotic relationship with another organism helping to create a suitable environment for hydrogen production.

18.4.3 Hydrogen Production Mechanisms

Hydrogen production is mediated through various metabolic pathways. However, a general scheme for algae has been produced in recent years. Photosynthesis produces protons and molecular oxygen through the splitting of water. The protons and electrons that are produced are fed to a hydrogen-producing enzyme, such as nitrogenase or hydrogenase, to generate molecular hydrogen. However, the inactivation of hydrogenase by oxygen within a short period of the start of photosynthesis severely limits the amount of hydrogen produced. One way around this problem would be to develop a two-stage method where, in the first stage, microalgae are grown photosynthetically to produce carbohydrates and oxygen. Then, in the second stage, the algal cells are placed under anaerobic conditions and degrade the organic compounds generated in the first stage to produce hydrogen. The anaerobic stage can either be with illumination or without, also called dark fermentation. Obviously, maximum biomass production in the first stage requires optimum conditions, such as good lighting conditions, optimal pH, and ideal temperatures.

The second stage can either use dark fermentation or a light-dependent reaction. In dark fermentation, fixed carbon reserves produced in the first stage (starch, lipid, and glycogen) are converted into pyruvate, either through glycolysis or the respective catabolic pathway. The pyruvate is then used to produce hydrogen with the help of two enzymes: pyruvate formate lyase (PFL) and pyruvate ferredoxin oxidoreductase (PFOR). PFL will produce acetyl-CoA and formate, whereas PFOR will produce acetyl-CoA and ferredoxin (Hallenbeck and Benemann, 2002). Enteric bacteria derive hydrogen from formate, while strict anaerobes derive hydrogen from ferredoxin. The yield of hydrogen production is low in dark fermentation with around 1–2 mol of hydrogen produced from 1 mol of pyruvate (Hallenbeck and Benemann, 2002). Illumination could drive additional hydrogen production due to the energy captured by photosynthesis but also risks restarting oxygen evolution.

Sulfur starvation has been extensively used to augment hydrogen production by microalgae through creating an anaerobic cellular environment. Photosystem II contains an essential protein, D1, which is easily photodamaged. This protein cannot be repaired but needs to be resynthesized, a process requiring the sulfur-containing amino acids—methionine and cysteine. Obviously, sulfur deprivation causes methionine and cysteine biosynthesis to halt, and thus, illumination during sulfur deprivation will gradually decrease the amount of active D1 protein (Kruse et al., 2010; Melis and Happe, 2001). Consequently, oxygen levels are diminished in the culture, and hydrogenase can become active within the cells to produce hydrogen (Kruse et al., 2010).

Thus, hydrogen production relies on a number of different metabolic pathways and nutrient limitation process. Further research must be followed in order to make competitive biohydrogen industry.

18.5 Conclusion

In conclusion, microalgae represent a potential energy resource for the future. They have the advantage of potentially being sustainable and ecologically friendly. They are highly diverse and therefore highly suitable for different cultivation locations. Furthermore, algae have the capacity to produce a number of products of interest for the transportation industry, the cosmetic industry, the food industry, and the pharmaceutical industry. However, for the time being, algal cultivation is only associated with high end products due to their relatively low productivity and high cost of cultivation and harvesting. In the case of hydrogen and biodiesel production, these barriers must be overcome in order to make algal fuels cost-effective. Some recent advances are encouraging, and a significant number of algal product companies are active and nearing pilot-scale process demonstration. Further advances in genetic manipulation and algal cultivation will help increase the affordability of algal biofuels in the near future.

Acknowledgments

Algal biofuel research in the laboratory of PCH is supported by FQRNT (Fonds québécois de la recherche sur la nature et les technologies). Part of this work was performed while PCH held a National Research Council Research Associateship Award at the Life Sciences Research Center, Department of Biology, United States Air Force Academy.

References

Abdelaziz, A. E. M., Leite, G. B., and Hallenbeck, P. C. Addressing the challenges for sustainable production of algal biofuels: I. Algal strains and nutrient supply. *Environmental Technology* 34 (2013a): 1783–1805.

Abdelaziz, A. E. M., Leite, G. B., and Hallenbeck, P. C. Addressing the challenges for sustainable production of algal biofuels: II. Harvesting and conversion to biofuels. *Environmental Technology* 34 (2013b): 1807–1836.

Alberts, B., Johnson, A., Lewis, J., Raff, M., Roberts, K., and Walter, P. *Biologie Moléculaire de la Cellule*. Paris, France: Médecine-Science Flammarion, 2007.

Beer, L. L., Boyd, E. S., Peters, J. W., and Posewtiz, M. C. Engineering algae for biohydrogen and biofuel production. *Current Opinion in Biotechnology* 20 (2009): 264–271.

Bensheng, L. and Benning, C. Lipid metabolism in microalgae distinguishes itself. *Current Opinion in Biotechnology* 24 (2013): 300–309. Print

Boyle, N. R., Page, M. D., Liu, B., Blaby, I. K., Casero, D., Kropat, J., Cokus, S. J. et al. Three acyltransferases and nitrogen-responsive regulator are implicated in nitrogen starvation-induced triacylglycerol accumulation in *Chlamydomonas*. *Journal of Biological Chemistry* 287 (2012): 15811–15825.

Brennan, L. and Owende, P. Biofuels from microalgae—A review of technologies for production, processing, and extraction of biofuels and co-products. *Renewable and Sustainable Energy Reviews* 14 (2010): 557–577.

Breuer, G., Lamers, P. P., Martens, D. E., Draaisma, R. B., and Wijffels, R. H. The impact of nitrogen starvation on the dynamics of triacylglycerol accumulation in nine microalgae strains. *Bioresource Technology* 124 (2012): 217–226.

Breuer, G., Lamers, P. P., Martens, D. E., Draaisma, R. B., and Wijffels, R. H. Effect of light intensity, pH and temperature on triacylglycerol (TAG) accumulation induced by nitrogen starvation in *Scenedesmus obliquus*. *Bioresource Technology* 143 (2013): 1–9.

Brunner, E., Richthammer, P., Ehrlich, H., Paasch, S., Simon, P., Ueberlein, S., and Van Pée, K-H. Chitin-based organic networks: An integral part of cell wall biosilica in the diatom *Thalassiosira pseudonana*. *Angewandte Chemie International Edition* 48 (2009): 9724–9727.

Chang, D. W., Hobson, P., Burch, M., and Lin, T. F. Measurement of cyanobacteria using in vivo fluoroscopy—Effect of cyanobacterial species, pigments, and colonies. *Water Research* 46 (2012): 5037–5048.

Chapman, I. The end of peak oil? Why this topic is still relevant despite recent denials. *Elsevier Energy Policy* 64 (2014): 93–101.

Chisti, Y. Biodiesel from microalgae. *Biotechnology Advances* 25 (2007): 294–306.

Chisti, Y. Constraints to commercialization of algal fuels. *Journal of Biotechnology* 167 (2013): 201–214.

Courchesne, N. M. D., Parisien, A., Wang, B., and Lan, C. Q. Enhancement of lipid production using biochemical, genetic and transcription factor engineering approaches. *Journal of Biotechnology* 141 (2009): 31–41.

Deffeyes, K. S. *When Oil Peaked*. New York: Hill & Wang, 2010.

Durkin, C. A., Mock, T., and Armbrust, E. V. Chitin in diatoms and its association with the cell wall. *Eukaryotic Cell* 8 (2009): 1038–1050.

EIA. What are the major sources and users of energy in the United States. Eia.gov. U.S. Energy Information Administration. August 1, 2013. Web. September 10, 2013. http://www.eia.gov/energy_in_brief/article/major_energy_sources_and_users.cfm.

European Standard. EN 14214—Automotive fuels—Fatty acid methyl esters (FAME) for diesel engines—Requirements and test methods. Brussels, Belgium: Comité Européen de Normalisation, 2008.

Frank, H. A. and Cogdell, R. J. Carotenoids in photosynthesis. *Photochemistry and Photobiology* 63 (1996): 257–264.

Garby, T. J., Walter, M. R., Larkum, A. W. D., and Nellan, B. A. Diversity of cyanobacterial biomarker genes from the stromatolites of Shark Bay, Western Australia. *Environmental Microbiology* 15 (2013): 1464–1475.

Ghirardi, M. L., Posewitz, M. C., Maness, P. C., Dubini, A., Yu, J., and Seibert, M. Hydrogenases and hydrogen photoproduction in oxygenic photosynthetic organisms. *Annual Review of Plant Biology* 58 (2007): 71–91.

Gouveia, L. and Oliveira, A. C. Microalgae as a raw material for biofuels production. *Journal of Industrial Microbiology and Biotechnology* 36 (2009): 269–274.

Griffiths, M. J., Van Hille, R. P., and Harrison, S. T. L. Lipid productivity, settling potential and fatty acid profile of 11 microalgal species grown under nitrogen replete and limited conditions. *Journal of Applied Phycology* 24 (2012): 989–1001.

Guschina, I. A. and Hardwood, J. L. Lipids and lipid metabolism in eukaryotic algae. *Progress in Lipid Research* 45 (2006): 160–186.

Hallenbeck, P. C. Microbial production of fatty acid based biofuels. In Hallenbeck, P. C., ed., *Microbial Technologies in Advanced Biofuels Production*. New York: Springer Science + Business Media, 2012, pp. 213–230.

Hallenbeck, P. C. and Benemann, J. R. Biological hydrogen production; fundamentals and limiting processes. *International Journal of Hydrogen Energy* 27 (2002): 1185–1193.

Hildebrand, M., Abbriano, R. M., Polle, J. E. W., Traller, J. C., Trentacoste, E. M., Smith, S. R., and Davis, A. K. Metabolic and cellular organization in evolutionarily diverse microalgae as related to biofuels production. *Current Opinion in Chemical Biology* 17 (2013): 506–514.

Hoiczyk, E. and Hansel, A. cyanobacterial cell walls: News from an unusual prokaryotic envelope. *Journal of Bacteriology* 182 (2000): 1191–1199.

Horton, H. R., Moran, L. A., Scrimgeour, K. G., Perry, M. D., and Rawn, J. D. *Principles of Biochemistry*. Upper Saddle River, NJ: Pearson Prentice Hall, 2006. Print.

Hu, Q., Sommerfeld, M., Jarvis, E., Ghirardi, M., Posewitz, M., Seibert, M., and Darzins, A. Microalgal triacylglycerols as feedstocks for biofuel production: Perspectives and advances. *The Plant Journal* 54 (2008): 621–639.

Huang, G. H., Chen, F., Wei, D., Zhang, X. W., and Chen, G. Biodiesel production by microalgal biotechnology. *Applied Energy* 87 (2010): 38–46.
Joye, S. B., Macdonald, I. R., Leifer, I., Asper, V. Magnitude and oxidation potential of hydrocarbon gases released from the BP oil well blowout. *Nature Geoscience* 4 (2011): 161–164.
Khozin-Goldberg, I. and Cohen, Z. Unraveling algal lipid metabolism: Recent advances in gene identification. *Biochimie* 93 (2011): 91–100.
Kirst, H., Garcia-Cerdan, J. G., Zurbriggen, A., Ruehle, T., and Melis, A. Truncated photosystem chlorophyll antenna size in green microalga *Chlamydomonas reinhardtii* upon deletion of *TLA3-CpSRP43* gene. *Plant Physiology* 160 (2012): 2251–2260.
Koku, H., Eriglu, I., Gündüz, U., Yücel, M., and Türker, L. Aspects of the metabolism of hydrogen production by *Rhodobacter sphaeroides*. *International Journal of Hydrogen Energy* 27 (2002): 1315–1329.
Kruse, O. and Hankamer, B. Microalgal hydrogen production. *Current Opinion in Biotechnology* 21 (2010): 238–243.
Leite, G. B. and Hallenbeck, P. C. Algae oil. In Hallenbeck, P. C., ed., *Microbial Technologies in Advanced Biofuels Production*. New York: Springer Science + Business Media, 2012, pp. 231–259.
Liang, Y., Sarkany, N., and Cui, Y. Biomass and lipid productivities of *Chlorella vulgaris* under autotrophic, heterotrophic and mixotrophic growth conditions. *Biotechnology Letters* 31 (2009): 1043–1049.
Liu, B. and Benning, C. Lipid metabolism in microalgae distinguishes itself. *Current Opinion in Biotechnology* 24 (2013): 300–309.
Longworth, J., Noirel, J., Pandhal, J., Wright, P. C., and Vaidyanathan, S. HIIC- and SCX-based quantitative proteomics of *Chlamydomonas reinhardtii* during nitrogen starvation induced lipid and carbohydrate accumulation. *Journal of Proteome Research* 11 (2012): 5959–5971.
Madiraju, K. S., Lyew, D., Kok, R., and Raghavan V. Carbon neutral electricity production by *Synechocystis sp.* PCC6803 in a microbial fuel cell. *Bioresource Technology* 110 (2012): 214–218.
Makareviciene, V., Skorupskaite, V., and Andruleviciute, V. Biodiesel fuel from microalgae-promising alternative fuel for the future: A review. *Reviews in Environmental Science and Biotechnology* 12 (2013): 119–130.
Mcfadden, G. I. Primary and secondary endosymbiosis and the origin of plastids. *Journal of Phycology* 37 (2001): 951–959.
Meher, L. C., Vidya, D., and Naik, S. N. Technical aspects of biodiesel production by transesterification—A review. *Renewable and Sustainable Energy Reviews* 10 (2006): 248–268.
Melis, A. and Happe, T. Hydrogen production. Green algae as a source of energy. *Plant Physiology* 127 (2001): 740–748.
Merchant, S. S., Kropat, J., Liu, B., Shaw, J., and Warakanont, J. TAG, you're it! Chlamydomonas as a reference organism for understanding algal triacylglycerol accumulation. *Current Opinion in Biotechnology* 23 (2012): 352–363.
Msanne, J., Xu, D., Konda, A. R., Casas-Mollano, J. A., Awada, T., Cahoon, E. B., and Cerutti, H. Metabolic and gene expression changes triggered by nitrogen deprivation in photoautotrophically grown microalgae *Chlamydomonas reinhardtii* and *Coccomyxa* sp. *C-169*. *Phytochemistry* 75 (2012): 50–59.
Ogawa, T. and Aiba, S. Bioenergetic analysis of mixotrophic growth in *Chlorella vulgaris* and *Scenedesmus acutus*. *Biotechnology and Bioengineering* 23 (1981): 1121–1132.
O'Grady, J. and Morgan, J. A. Heterotrophic growth and lipid production of *Chlorella protothecoides* on glycerol. *Bioprocess and Biosystems Engineering* 34 (2011): 121–125.
Ort, D. R., Zhu, X., and Melis, A. Optimizing antenna size to maximize photosynthetic efficiency. *Plant Physiology* 155 (2011): 79–85.
Perez-Garcia, O., De-Bashan, L. E., Hernandez, J. P., and Bashan Y. Efficiency of growth and nutrient uptake from wastewater by heterotrophic, autotrophic, and mixotrophic cultivation of *Chlorella vulgaris* immobilized with *Azospirillum brasilense*. *Journal of Phycology* 46 (2010): 800–812.
Popper, Z. A., Michel, G., Hervé, C., Domozych, D. S., Willats, W. G. T., Tuohy, M. G., Kloareg, B., and Stengel, D. B. Evolution and diversity of plant cell walls: From algae to flowering plants. *The Annual Review of Plant Biology* 62 (2011): 567–590.

Radakovits, R., Jinkersin, R. E., Darzin, A., and Posewitz, M. C. Genetic engineering of algae for enhanced biofuel production. *Eukaryotic Cell* 9 (2010): 486–501.
Rashid, N., Rehman, M. S. U., Memon, S., Rahman, Z. U., Lee, K., and Han, J. I. Current status, barriers and developments in biohydrogen production by microalgae. *Renewable and Sustainable Energy Review* 22 (2013): 571–579.
Raven, P. H., Evert, R. F., and Eichhorn, S. E. *Biologie végétale*. Brussels, Belgium: De Boeck, 2007.
Rosenberg, J. N., Oyler, G. A., Wilkinson, L., and Betenbaugh, M. J. A green light for engineered algae: Redirecting metabolism to fuel a biotechnology revolution. *Current Opinion in Biotechnology* 19 (2008): 430–436.
Runge, C. F. and Senauer, B. How biofuels could starve the poor. *Foreign Affairs* 86 (2007): 41–53.
Scheffel, A., Poulsen, N., Shian, S., and Kröger, N. Nanopatterned protein microrings from a diatom that direct silica morphogenesis. *Proceedings of the National Academy of Science* 108 (2011): 3175–3180.
Stanier, R. Y. and Cohen-Bazire, G. Phototrophic Prokaryotes: The Cyanobacteria. *Ann. Rev. Microbiol.* 31 (1977): 225–274. Print.
Stucken, K., Ilhan, J., Roettger, M., Dagan, T., and Martin, W. F. Transformation and conjugal transfer of foreign genes into the filamentous multicellular cyanobacteria (subsection V) *Fischerella* and *Chlorogloeopsis*. *Current Microbiology* 65 (2012): 552–560.
Taton, A., Lis, E., Adin, D. M., Dong, G., Cookson, S., Kay, S. A., Golden, S. S., and Golden, J. W. Gene transfer in *Leptolyngbya* sp. strain BL0902, a cyanobacterium suitable for production of biomass and bioproducts. *PLoS ONE* 7 (2012): e30901.
Tomaselli, L. The microalgal cell. In Richmond, A. ed., *Handbook of Microalgal Culture Biotechnology and Applied Phycology*. Ames, IA: Blackwell Publishing, 2008, pp. 3–19.
Ugwu, C. U., Aoyagi, H., and Uchiyama, H. Photobioreactors for mass cultivation of algae. *Bioresource Technology* 99 (2008): 4021–4028.
Zehr, J. P., Jenkins, B. D., Short S. M., and Stewart, G. F. Nitrogenase gene diversity and microbial community structure: A cross-system comparison. *Environmental Microbiology* 5 (2003): 539–554.

19

Current State of Research on Algal Biohydrogen

Ozcan Konur

CONTENTS

19.1 Introduction

19.1.1 Issues

Global warming, air pollution, and energy security have been some of the most important public policy issues in recent years (Jacobson 2009, Wang et al. 2008, Yergin 2006). With increasing global population, food security has also become a major public policy issue (Lal 2004). The development of biofuels generated from biomass has been a long-awaited solution to these global problems (Demirbas 2007, Goldember 2007, Lynd et al. 1991). However, the development of early biofuels produced from agricultural plants such as sugarcane (Goldemberg 2007) and agricultural wastes such as corn stovers (Bothast and Schlicher 2005) has resulted in a series of substantial concerns about food security (Godfray et al. 2010). Therefore, the development of algal biofuels as a third-generation biofuel has been considered as a major solution for the global problems of global warming, air pollution, energy security, and food security (Chisti 2007, Demirbas 2007, Kapdan and Kargi 2006, Spolaore et al. 2006, Volesky and Holan 1995).

Although there have been many reviews on the use of marine algae for the production of algal biohydrogen (e.g., Das and Veziroglu 2001, Ghirardi et al. 2000, Nandi and Sengupta 1998) and there have been a number of scientometric studies on algal biofuels (Konur 2011), there has not been any study on the citation classics in algal biohydrogen as in the other research fields (Baltussen and Kindler 2004a,b, Dubin et al. 1993, Gehanno et al. 2007, Konur 2012a,b, 2013, Paladugu et al. 2002, Wrigley and Matthews 1986).

As North's new institutional theory suggests, it is important to have up-to-date information about current public policy issues to develop a set of viable solutions to satisfy the needs of all the key stakeholders (Konur 2000, 2002a,b,c, 2006a,b, 2007a,b, 2012c, North 1994).

Therefore, the brief information on a selected set of citation classics in algal biohydrogen are presented in this chapter to inform the key stakeholders relating to the global problems of warming, air pollution, food security, and energy security about the production of algal biohydrogen for the solution of these problems in the long run, thus complementing a number of recent scientometric studies on biofuel and global energy research (Konur 2011, 2012d–p).

19.1.2 Methodology

A search was carried out in the Science Citation Index Expanded (SCIE) and Social Sciences Citation Index (SSCI) databases (version 5.11) in September 2013 to locate the papers relating to algal biohydrogen using the keyword set of Topic = (hydrogen* or *bio* hydrogen* or biohydrogen** or *microbio* hydrogen** or h2 or *h-2 h*(2)) AND Topic = (alga* or *macro* alga** or *micro* alga** or macroalga* or microalga* or cyanobacteria or cyanobacterium or seaweed* or diatoms or *sea* weed** or reinhardtii or braunii or chlorella or sargassum or gracilaria) in the abstract pages of the papers. For this chapter, it was necessary to embrace the broad algal search terms to include diatoms, seaweeds, and cyanobacteria as well as to include many spellings of biohydrogen rather than the simple keywords of biohydrogen and algae.

It was found that there were 3945 papers indexed between 1980 and 2013. The key subject categories for algal research were biochemistry and molecular biology (678 references, 15.5%), biotechnology and applied microbiology (545 references, 12.4%), physical chemistry (421 references, 9.6%), and applied physics (417 references, 9.5%), altogether comprising 50% of all the references on algal biohydrogen. It was also necessary to focus on the key references by selecting articles and reviews.

The located highly cited 50 papers were arranged in order of decreasing number of citations for two groups of papers: 33 and 17 papers for nonexperimental research and experimental research, respectively. In order to check whether these collected abstracts correspond to the larger sample on these topical areas, new searches were carried out for each topical area.

The summary information about the located citation classics is presented under two major headings of nonexperimental and experimental research in the order of decreasing number of citations for each topical area.

The information relating to the document type, affiliation and location of the authors, the journal, the indexes, subject area of the journal, the concise topic, total number of citations, and total average number of citations received per year were given in the tables for each paper.

19.2 Nonexperimental Studies on Algal Biohydrogen

19.2.1 Introduction

The nonexperimental research on algal biohydrogen has been one of the most dynamic research areas in recent years. Thirty-three citation classics in the field of algal biohydrogen with more than seventy citations were located, and the key emerging issues from these papers are presented in decreasing order of the number of citations (Table 19.1). These papers give strong hints about the determinants of the efficient algal biohydrogen production and emphasize that marine algae are efficient biohydrogen feedstocks in comparison with the first- and second-generation biofuel feedstocks.

The papers were dominated by researchers from 13 countries, usually through the intercountry institutional collaboration, and they were multiauthored. The United States (17 papers), Germany (6 papers), Sweden (5 papers), and Portugal (4 papers) were the most prolific countries. Similarly, Uppsala University (6 papers), University of California at Berkeley (5 papers), and University of Porto and University of Bielefeld (3 papers each) were the most prolific institutions.

Similarly, all these papers were published in journals indexed by the SCI and/or SCIE. There were no papers indexed by the Arts and Humanities Citation Index (A&HCI) or SSCI. The number of citations ranged from 70 to 725, and the average number of citations per annum ranged from 2.5 to 55.8. Eleven of the papers were articles, while 22 of them were reviews. Seven and twenty-six of these papers were related to comparative biohydrogen production and algal biohydrogen production, respectively.

19.2.2 Nonexperimental Research on Algal Biohydrogen

19.2.2.1 Comparative Studies

Das and Veziroglu (2001) discuss biohydrogen production in a seminal paper with 725 citations. They note that biohydrogen is the fuel of the future mainly due to its high conversion efficiency, recyclability, and nonpolluting nature. Biohydrogen production processes are more environment friendly and less energy intensive as compared to thermochemical and electrochemical processes. These processes are mostly controlled by either photosynthetic or fermentative organisms where more emphasis has been given on the former processes. Nitrogenase and hydrogenase (H_2ase) play a very important role. Genetic manipulation of cyanobacteria (H_2ase-negative gene) improves biohydrogen generation. They present a survey of biohydrogen production processes. They then present the microorganisms and biochemical pathways involved in biohydrogen generation processes in some detail. They argue that an immobilized system is suitable for continuous biohydrogen production, while about 28% of energy can be recovered in the form of biohydrogen using sucrose as a substrate, and fermentative biohydrogen production processes have some edge over other biological processes.

Levin et al. (2004) discuss biohydrogen production with a focus on the prospects and limitations to practical application in a paper with 536 citations. They note that hydrogen may be produced by a number of processes, including electrolysis of water, thermocatalytic reformation of hydrogen-rich organic compounds, and biological processes. Hydrogen is produced almost exclusively by electrolysis of water or by steam reformation of methane. Biohydrogen technologies provide a wide range of approaches to generate

TABLE 19.1

Nonexperimental Research on Algal Biohydrogen

No.	Paper References	Year	Document	Affiliation	Country	No. of Authors	Journal	Index	Subjects	Topic	Total No. of Citations	Total Average Citations per Annum	Rank
1	Das and Veziroglu	2001	R	Indian Inst Technol, Univ Miami	India, United States	2	*Int. J. Hydrogen Energ.*	SCI, SCIE	Chem. Phys., Electrochem., En. Fuels	Algal biohydrogen production—comp. stud.	725	55.8	1
2	Levin et al.	2004	A	Univ. Victoria	Canada	3	*Int. J. Hydrogen Energ.*	SCI, SCIE	Chem. Phys., Electrochem., En. Fuels	Algal biohydrogen production—comp. stud.	536	53.6	2
3	Kapdan and Kargi	2006	R	Dokuz Eylul Univ.	Turkey	2	*Enzyme Microb. Technol.*	SCI, SCIE	Biotech., Appl. Microb.	Algal biohydrogen production from wastes—comp. stud.	394	49.3	3
4	Hallenbeck and Benemann	2002	A	Univ. Montreal	Canada	2	*Int. J. Hydrogen Energ.*	SCI, SCIE	Chem. Phys., Electrochem., En. Fuels	Algal biohydrogen production—comp. stud.	329	27.4	5
5	Nandi and Sengupta	1998	R	Indian Inst. Chem. Biol.	India	2	*Crit. Rev. Microbiol.*	SCI, SCIE	Microbiol.	Algal biohydrogen production—comp. stud.	294	18.4	6
6	Ghirardi et al.	2000	R	Natl. Renew. Energy Lab., Oak Ridge Natl. Lab.	United States	7	*Trends Biotechnol.*	SCI, SCIE	Biotech. Appl. Microb.	Algal biohydrogen production	229	16.4	7
7	Melis and Happe	2001	A	Univ. Calif. Berkeley, Univ. Bonn	United States, Germany	2	*Plant Physiol.*	SCI, SCIE	Plant Sci.	Algal biohydrogen production	207	15.9	8
8	Tamagnini et al.	2002	R	Univ. Uppsala, Univ. Porto	Sweden, Portugal	6	*Microbiol. Mol. Biol. Rev.*	SCI, SCIE	Microbiol.	Algal biohydrogen production	186	15.5	9
9	Vignais and Colbeau	2005	R	CNRS	France	2	*Curr. Issues Mol. Biol.*	SCIE	Bioch. Mol. Biol.	Algal biohydrogen production	165	18.3	10
10	Wu and Mandrand	1993	R	INSA	France	2	*FEMS Microb. Rev.*	SCI, SCIE		Algal biohydrogen production	147	7.0	13

(Continued)

TABLE 19.1 (*Continued*)

Nonexperimental Research on Algal Biohydrogen

No.	Paper References	Year	Document	Affiliation	Country	No. of Authors	Journal	Index	Subjects	Topic	Total No. of Citations	Total Average Citations per Annum	Rank
11	Magnuson et al.	2009	R	Uppsala Univ., Lund Univ., Univ. S Bohemia	Sweden, Czech Republic	11	*Accounts Chem. Res.*	SCI, SCIE	Chem. Mult.	Algal biohydrogen production	141	28.2	14
12	Houchins	1984	R	Univ. Minnesota	United States	1	*Biochim. Biophys. Acta*	SCI, SCIE	Bioch. Mol. Biol., Biophys.	Algal biohydrogen production	139	4.6	15
13	Ghirardi et al.	2007	R	Natl. Renew. Ener. Lab., Color. Sch. Mines	United States	6	*Annu. Rev. Plant Biol.*	SCI, SCIE	Bioch. Mol. Biol., Plant Sci.	Algal biohydrogen photoproduction	130	18.6	16
14	Stal and Moezelaar	1997	R	DLO	Netherlands	2	*Fems Microbiol. Rev.*	SCI, SCIE	Microbiol.	Algal biohydrogen production—dark fermentation	127	7.5	18
15	Melis	2002	A	Univ. Calif. Berk.	United States	1	*Int. J. Hydrogen Energ.*	SCI, SCIE	Chem. Phys., Electrochem., En. Fuels	Algal biohydrogen production	113	9.2	25
16	Kruse et al.	2005a	R	Univ. Bielefeld, Univ. Queensland, Princeton Univ.	Germany, Australia, United States	5	*Photochem. Photobiol. Sci.*	SCI, SCIE	Bioch. Mol. Biol., Biophys., Chem., Phys.	Algal biohydrogen production—photosynthesis	104	11.6	28
17	Hoehler et al.	2001	A	NASA	United States	3	*Nature*	SCI, SCIE	Mult. Sci.	Algal biohydrogen production—microbial mats	104	8.0	27
18	Tamagnini et al.	2007	R	Univ. Porto, Uppsala Univ.	Portugal, Sweden	8	*Fems Microbiol. Rev.*	SCIE, SCIE	Microbiol.	Algal biohydrogen production—H_2ases	98	14.0	31
19	Ghirardi et al.	2009	A	Natl. Renew. Ener. Lab.	United States	3	*Chem. Soc. Rev.*	SCI, SCIE	Chem. Mult.	Algal biohydrogen photoproduction	92	18.4	32
20	Melis	2007	R	Univ. Calif. Berk.	United States	1	*Planta*	SCI, SCIE	Plant Sci.	Algal biohydrogen production	91	13.0	34

(Continued)

TABLE 19.1 ***(Continued)***

Nonexperimental Research on Algal Biohydrogen

No.	Paper References	Year	Document	Affiliation	Country	No. of Authors	Journal	Index	Subjects	Topic	Total No. of Citations	Total Average Citations per Annum	Rank
21	Hansel and Lindblad	1998	R	Uppsala Univ.	United States	2	*Appl. Microbiol. Biotechnol.*	SCI, SCIE	Biot. Appl. Microb.	Algal biohydrogen production	91	5.7	33
22	Schutz et al.	2004	A	Univ. Porto, Ruhr Univ. Bochum, Uppsala Univ., Russian Acad. Sci.	Portugal, Sweden, Russia, Germany	7	*Planta*	SCI, SCIE	Plant Sci.	Algal biohydrogen production	90	9.0	36
23	Beer et al.	2009	R	Colorado Sch. Mines, Montana St. Univ.	United States	4	*Curr. Opin. Biotechnol.*	SCI, SCIE	Bioch., Res. Meth., Biot. Appl. Microb.	Algal biohydrogen production	88	17.6	37
24	Mus et al.	2007	A	Carnegie Inst. Wash., Natl. Renew. Energy Lab., Colorado Sch. Mines	United States	5	*J. Biol. Chem.*	SCI, SCIE	Bioch. Mol. Biol.	Algal biohydrogen production	88	6.8	38
25	Prince and Kheshgi	2005	R	ExxonMobil Res and Engn. Co.	United States	2	*Crit. Rev. Microbiol.*	SCI, SCIE	Microbiol.	Algal biohydrogen photoproduction—comp. stud.	86	9.6	39
26	Asada and Miyake	1999	R	AIST	Japan	2	*J. Biosci. Bioeng.*	SCI, SCIE	Biot. Appl. Microb., Fodo Sci. Tech.	Algal biohydrogen photoproduction	81	5.4	41

(Continued)

TABLE 19.1 (*Continued*)

Nonexperimental Research on Algal Biohydrogen

No.	Paper Ref.	Year	Doc.	Affiliation	Country	No. of Authors	Journal	Index	Subjects	Topic	Total No. of Citations	Total Average Citations per Annum	Rank
27	Lambert and Smith	1981	R	Australian Natl. Univ.	Australia	2	*Biol. Rev. Cambridge Philosophic. Soc.*	na	Biol.	Algal biohydrogen production	81	2.5	42
28	Dau and Zaharieva	2009	R	Free Berlin Univ.	Germany	2	*Accounts Chem. Res.*	SCI, SCIE	Chem. Multp.	Algal biohydrogen production—photosynthesis	80	16.0	43
29	Lindberg et al.	2010	A	Univ. Calif. B.	United States	3	*Metab. Eng.*	SCIE	Biot. Appl. Microb.	Algal biohydrogen production—photosynthesis	76	19.0	44
30	Rupprecht et al.	2006	R	Univ. Bielefeld, Univ. Queensland, Princeton Univ.	Germany, Australia, United States	6	*Appl. Microbiol. Biotechnol.*	SCI, SCIE	Biot. Appl. Microb.	Algal biohydrogen production—comp. stud.	76	9.5	45
31	Benemann	2000	A	Univ. Calif. B.	United States	1	*J. Appl. Phycol.*	SCI, SCIE	Biot. Appl. Microb., Mar. Fresh. Biol.	Algal biohydrogen production	72	5.1	47
32	Pinto et al.	2002	A	Uppsala Univ., Univ. Minho, Russian Acad. Sci.	Portugal, Russia, Sweden	3	*Int. J. Hydrogen Energ.*	SCI, SCIE	Chem. Phys., Electroch., Ener. Fuels	Algal biohydrogen production	70	5.8	50
33	Appel and Schulz	1998	R	Univ. Marburg	Germany	2	*J. Photochem. Photobiol. B Biol*	SCI, SCIE	Bioch. Mol. Biol., Biophys.	Algal biohydrogen production—photosynthesis	70	4.4	49

SCI, Science Citation Index; SCIE, Science Citation Index Expanded; SSCI, Social Sciences Citation Index. A, article. R, review.

hydrogen, including direct biophotolysis, indirect biophotolysis, photofermentations, and dark fermentation. They argue that the practical application of these technologies to every day energy problems is unclear. They compare biohydrogen production rates of various biohydrogen systems by first standardizing the units of biohydrogen production and then by calculating the size of biohydrogen systems that would be required to power proton exchange membrane (PEM) fuel cells of various sizes.

Kapdan and Kargi (2006) discuss biohydrogen production from waste materials in a paper with 394 citations. They note that electrolysis of water, steam reforming of hydrocarbons, and autothermal processes for hydrogen gas production are not cost-effective due to high energy requirements. Biohydrogen production has significant advantages over chemical methods. The major biological processes utilized for biohydrogen gas production are biophotolysis of water by algae and dark fermentation and photofermentation of organic materials, usually carbohydrates by bacteria. Sequential dark fermentation and photofermentation processes are rather new approaches for biohydrogen production. One of the major problems in dark and photofermentative hydrogen production is the raw material cost. They argue that carbohydrate-rich, nitrogen-deficient solid wastes such as cellulose and starch-containing agricultural and food industry wastes and some food industry wastewaters such as cheese whey, olive mill, and baker's yeast industry wastewaters can be used for hydrogen production by using suitable bioprocess technologies. Utilization of these wastes for hydrogen production provides inexpensive energy generation with simultaneous waste treatment. They summarize biohydrogen production from some waste materials. Types of potential waste materials, bioprocessing strategies, microbial cultures to be used, bioprocessing conditions, and the recent developments are discussed with their relative advantages.

Hallenbeck and Benemann (2002) discuss biohydrogen production with a focus on fundamentals and limiting processes in a paper with 329 citations. They review various approaches and identify critical limiting factors. The low energy content of solar irradiation dictates that photosynthetic processes operate at high conversion efficiencies and places severe restrictions on photobioreactor economics. They argue that conversion efficiencies for direct biophotolysis are below 1% and indirect biophotolysis remains to be demonstrated. Dark fermentation of biomass or wastes presents an alternative route to biological hydrogen production that has been little studied. In this case, the critical factor is the amount of hydrogen that can be produced per mole of substrate. Known pathways and experimental evidence indicate that at most 2–3 mol of hydrogen can be obtained from substrates such as glucose.

Nandi and Sengupta (1998) discuss microbial production of hydrogen in an early paper with 294 citations. They note that production of hydrogen by anaerobes, facultative anaerobes, aerobes, methylotrophs, and photosynthetic bacteria is possible. Anaerobic clostridia are potential producers, and immobilized *Clostridium butyricum* produces 2 mol H_2/mol glucose at 50% efficiency. Spontaneous production of H_2 from formate and glucose by immobilized *Escherichia coli* showed 100% and 60% efficiencies, respectively. Enterobacteriaceae produce H_2 at similar efficiency from different monosaccharides during growth. Among methylotrophs, methanogenes, rumen bacteria, and thermophilic archae, *Ruminococcus albus* is promising (2.37 mol/mol glucose). Immobilized aerobic *Bacillus licheniformis* optimally produces 0.7 mol H_2/mol glucose. Photosynthetic *Rhodospirillum rubrum* produces 4, 7, and 6 mol of H_2 from acetate, succinate, and malate, respectively. Excellent productivity (6.2 mol H_2/mol glucose) by cocultures of *Cellulomonas* with a H_2ase uptake (Hup) mutant of *Rhodopseudomonas capsulata* on cellulose was found. Cyanobacteria, namely, *Anabaena*, *Synechococcus*, and *Oscillatoria* spp., have been studied for photoproduction of H_2. Immobilized *Aegilops cylindrica* produces H_2 (20 mL/g dry wt/h) continually for 1 year. Increased H_2 productivity was found for Hup mutant of *Anabaena variabilis. Synechococcus* sp. has a high potential for H_2 production in fermenters and outdoor cultures. Simultaneous productions of oxychemicals and H_2 by *Klebsiella* sp. and by enzymatic methods were also attempted.

Prince and Kheshgi (2005) discuss photobiological production of hydrogen with a focus on potential efficiency and effectiveness as a renewable fuel in a paper with 86 citations. They note that photosynthetic microorganisms can produce hydrogen when illuminated, and there has been considerable interest in developing this to a commercially viable process. Its appealing aspects include the fact that hydrogen would come from water and that the process might be more energetically efficient than growing, harvesting, and processing crops. They review current knowledge about photobiological hydrogen production and identify and discuss some of the areas where scientific and technical breakthroughs are essential for commercialization. First, they describe the underlying biochemistry of the process and identify some opportunities for improving photobiological hydrogen production at the molecular level. Then, they address the fundamental quantum efficiency of various processes that have been suggested, technological issues surrounding large-scale growth of hydrogen-producing microorganisms, and the scale and efficiency on which this would have to be practiced to make a significant contribution to current energy use.

Rupprecht et al. (2006) discuss the perspectives and advances of biological H_2 production in microorganisms in a paper with 76 citations. They note that the rapid development of clean fuels for the future is a critically important global challenge for two main reasons. First, new fuels are needed to supplement and ultimately replace depleting oil reserves. Second, fuels capable of zero CO_2 emissions are needed to slow the impact of global warming. They summarize the development of solar powered bio-H_2 production processes based on the conversion of photosynthetic products by fermentative bacteria, as well as using photoheterotrophic and photoautotrophic organisms.

19.2.2.2 Algal Studies

Ghirardi et al. (2000) discuss microalgae as a green source of renewable hydrogen in a paper with 200 citations. They summarize recent advances in the field of algal hydrogen production. They note that two fundamental approaches are being developed. One involves the temporal separation of the usually incompatible reactions of O_2 and H_2 production in green algae, and the second involves the use of classical genetics to increase the O_2 tolerance of the reversible H_2ase enzyme.

Melis and Happe (2001) discuss hydrogen production from green algae as a source of energy in a paper with 207 citations. They note that hydrogen and electricity could team to provide attractive options in transportation and power generation. Interconversion between these two forms of energy suggests on-site utilization of hydrogen to generate electricity, with the electrical power arid serving in energy transportation, distribution utilization, and hydrogen regeneration as needed. A challenging problem in establishing H_2 as a source of energy for the future is the renewable and environmentally friendly generation of large quantities of H_2 gas. Thus, processes that are presently conceptual in nature, or at a developmental stage in the laboratory, need to be encouraged, tested for feasibility, and otherwise applied toward commercialization.

Tamagnini et al. (2002) discuss the H_2ases and hydrogen metabolism of cyanobacteria in a paper with 186 citations. They note that cyanobacteria may possess several enzymes that are directly involved in dihydrogen metabolism: nitrogenase(s) catalyzing the production of hydrogen concomitantly with the reduction of dinitrogen to ammonia, an uptake H_2ase (encoded by hupSL) catalyzing the consumption of hydrogen produced by nitrogenase, and a bidirectional H_2ase (encoded by hoxFUYH) that has the capacity to both take zip and produce hydrogen. They summarize the knowledge about cyanobacterial H_2ases, focusing on recent progress since the first molecular information was published in 1995. They present the molecular knowledge about cyanobacterial hupSL and hoxFUYH, their corresponding gene products, and their accessory genes before finishing with an applied aspect—the use of cyanobacteria in a biological, renewable production of the future energy carrier molecular hydrogen.

Vignais and Colbeau (2005) discuss the molecular biology of microbial H_2ases in a paper with 165 citations. They note that H_2ases are metalloproteins. The great majority of them contain iron–sulfur clusters and two metal atoms at their active center, either a Ni and an Fe atom, the [NiFe]-H_2ases, or two Fe atoms, the [FeFe]-H_2ases. Enzymes of these two classes catalyze the reversible oxidation of hydrogen gas ($H_2 \leftrightarrow 2H^+ + 2e^-$) and play a central role in microbial energy metabolism; in addition to their role in fermentation and H_2 respiration, H_2ases may interact with membrane-bound electron transport systems in order to maintain redox poise, particularly in some photosynthetic microorganisms such as cyanobacteria. Recent work has revealed that some H_2ases, by acting as H_2 sensors, participate in the regulation of gene expression and that H_2-evolving H_2ases, thought to be involved in purely fermentative processes, play a role in membrane-linked energy conservation through the generation of a proton-motive force. The Hmd H_2ases of some methanogenic archaea constitute a third class of H_2ases, characterized by the absence of Fe–S cluster and the presence of an iron-containing cofactor with catalytic properties different from those of [NiFe]- and [FeFe]-H_2ases. They emphasize recent advances that have greatly increased the knowledge of microbial H_2ases, their diversity, the structure of their active site, how the metallocenters are synthesized and assembled, how they function, how the synthesis of these enzymes is controlled by external signals, and their potential use in biological H_2 production.

Wu and Mandrand (1993) discuss microbial H_2ases with a focus on primary structure, classification, signatures, and phylogeny in an early paper. Thirty sequenced microbial H_2ases are classified into six classes according to sequence homologies, metal content, and physiological function. The first class contains nine H_2-uptake membrane-bound NiFe-H_2ases from eight aerobic, facultative anaerobic, and anaerobic bacteria. The second comprises four periplasmic and two membrane-bound H_2-uptake NiFe(Se)-H_2ases from sulfate-reducing bacteria. The third consists of four periplasmic Fe-H_2ases from strict anaerobic bacteria. The fourth contains eight methyl viologen (MV), factor F420 (F420), or NAD-reducing soluble H_2ases from methanobacteria and *Alcaligenes eutrophus* H16. The fifth is the H_2-producing labile H_2ase isoenzyme 3 of *E. coli*. The sixth class contains two soluble tritium-exchange H_2ases of cyanobacteria. They argue that the 30 H_2ases have evolved from at least three different ancestors. While those of class I, II, IV, and V H_2ases are homologous, both class III and VI H_2ases are related neither to each other nor to the other classes. They further find that class II falls into two distinct clusters composed of NiFe- and NiFeSe-H_2ases, respectively. They also find that class IV comprises three distinct clusters: MV-reducing, F420-reducing, and NAD-reducing H_2ases. They argue that all H_2ases, except those of class VI, must contain some common motifs probably participating in the formation of hydrogen activation domains and electron transfer domains. The regions of hydrogen activation domains are highly conserved and can be divided into two categories. One corresponds to the *nickel-active center* of NiFe(Se)-H_2ases. It consists of two possible specific nickel-binding motifs, RxCGxCxxxH and DPCxxCxxH, located at the N- and C-termini of the so-called large subunits in the dimeric H_2ases, respectively. The other is the H cluster of the Fe-H_2ases. It might comprise three motifs on the C-terminal half of the large subunits.

Magnuson et al. (2009) discuss the biomimetic and microbial approaches to solar fuel generation in a paper with 141 citations. They link photosensitizers and charge-separation motifs to potential catalysts in supramolecular assemblies. In photosynthesis, production of carbohydrates demands the delivery of multiple reducing equivalents to CO_2. In contrast, the two-electron reduction of protons to molecular hydrogen is much less demanding. They argue that the catalytic sites of H_2ases are now the center of attention of biomimetic efforts, providing prospects for catalytic hydrogen production with inexpensive metals. Thus, they might complete the water-to-fuel conversion. An alternative route to hydrogen from solar energy is therefore to engineer these organisms to produce hydrogen more efficiently. They describe their original approach to combine research in these two fields: mimicking structural and functional principles of both photosystem II (PSII) and H_2ases by synthetic chemistry and engineering cyanobacteria to become better hydrogen producers and ultimately developing new routes toward synthetic biology.

Houchins (1984) discusses the physiology and biochemistry of hydrogen metabolism in cyanobacteria in an early paper with 139 citations.

Ghirardi et al. (2007) discuss the H_2ases and hydrogen photoproduction in oxygenic photosynthetic organisms in a paper with 130 citations. They note that the photobiological production of H_2 gas, using water as the only electron donor, is a property of two types of photosynthetic microorganisms: green algae and cyanobacteria. In these organisms, photosynthetic water splitting is functionally linked to H_2 production by the activity of H_2ase enzymes. Interestingly, each of these organisms contains only one of two major types of H_2ases, [FeFe] or [NiFe] enzymes, which are phylogenetically distinct but perform the same catalytic reaction, suggesting convergent evolution. This idea is supported by the observation that each of the two classes of H_2ases has a different metallocluster, is encoded by entirely different sets of genes (apparently under the control of different promoter elements), and exhibits different maturation pathways. The genetics, biosynthesis, structure, function, and O_2 sensitivity of these enzymes have been the focus of extensive research in recent years.

Stal and Moezelaar (1997) discuss fermentation in cyanobacteria in a paper with 128 citations. They summarize the papers and give a critical consideration of the energetics of dark fermentation in a number of species. There are a variety of different fermentation pathways in cyanobacteria. These include home and heterolactic acid fermentation, mixed acid fermentation, and homoacetate fermentation. Products of fermentation include CO_2, H_2, formate, acetate, lactate, and ethanol. In all species investigated, fermentation is constitutive. All enzymes of the fermentative pathways are present in photoautotrophically grown cells. Many cyanobacteria are also capable of using elemental sulfur as electron acceptor. In most cases, it is unlikely that sulfur respiration occurs. The main advantage of sulfur reduction is the higher yield of ATP, which can be achieved during fermentation. Besides oxygen and elemental sulfur, no other electron acceptors for chemotrophic metabolism are known so far in cyanobacteria. Calculations show that the yield of ATP during fermentation, although it is low relative to aerobic respiration, exceeds the amount that is likely to be required for maintenance, which appears to be very low in these cyanobacteria. The possibility of a limited amount of biosynthesis during anaerobic dark metabolism is discussed.

Melis (2002) discusses green alga hydrogen production with a focus on the progress, challenges, and prospects in a paper with 113 citations.

Kruse et al. (2005a) discuss the photosynthesis as a blueprint for solar-energy capture and biohydrogen production technologies in a paper with 104 citations. They note that chlorophyll photochemistry within PSII drives the water-splitting reaction efficiently at room temperature, in contrast with the thermal dissociation reaction that requires a temperature of ca. 1550 K. The successful elucidation of the high-resolution structure of PSII, and in particular the structure of its Mn_4Ca cluster, provides an invaluable blueprint for designing solar-powered biotechnologies for the future. This knowledge, combined with new molecular genetic tools, fully sequenced genomes, and an ever-increasing knowledge base of physiological processes of oxygenic phototrophs, has inspired scientists from many countries to develop new biotechnological strategies to produce renewable CO_2-neutral energy from sunlight. They focus particularly on the potential of use of cyanobacteria and microalgae for biohydrogen production. Specifically, they review the predicted size of the global energy market and the constraints of global warming upon it, before detailing the complex set of biochemical pathways that underlie the photosynthetic process and how they could be modified for improved biohydrogen production.

Hoehler et al. (2001) discuss the role of microbial mats in the production of reduced gases on the early Earth in an early paper with 104 citations. They report that in modern, cyanobacteria-dominated mats from hypersaline environments in Guerrero Negro, Mexico, photosynthetic microorganisms generate H_2 and CO gases that provide a basis for direct chemical interactions with neighboring chemotrophic and heterotrophic microbes. They also observe an unexpected flux of CH_4, which is probably related to H_2-based alteration of the redox potential within the mats. These fluxes would have been most important during the nearly two-billion-year period during which photosynthetic mats contributed substantially to biological productivity—and, hence, to biogeochemistry—on Earth. In particular, the large fluxes of observed H_2 could, with subsequent escape to space, represent a potentially important mechanism for oxidation of the primitive oceans and atmosphere.

Tamagnini et al. (2007) discuss cyanobacterial H_2ases with a focus on diversity, regulation, and applications in a paper with 98 citations. They note that cyanobacteria may possess two distinct nickel–iron (NiFe)-H_2ases: an uptake enzyme found in N_2-fixing strains and a bidirectional one present in both non-N_2-fixing and N_2-fixing strains. The uptake H_2ase (encoded by hupSL) catalyzes the consumption of the H_2 produced during N_2 fixation, while the bidirectional enzyme (hoxEFUYH) probably plays a role in fermentation and/or acts as an electron valve during photosynthesis. HupSL constitutes a transcriptional unit and is essentially transcribed under N_2-fixing conditions. The bidirectional H_2ase consists of a H_2ase and a diaphorase part, and the corresponding five hox genes are not always clustered or cotranscribed. The biosynthesis/maturation of NiFe-H_2ases is highly complex, requiring several core proteins. In cyanobacteria, the genes that are thought to affect H_2ases pleiotropically (hyp), as well as the genes presumably encoding the H_2ase-specific endopeptidases (hupW and hoxW), have been identified and characterized. Furthermore, NtcA and LexA have been implicated in the transcriptional regulation of the uptake and the bidirectional enzyme, respectively. Recently, the phylogenetic origin of cyanobacterial and algal H_2ases was analyzed, and it was proposed that the current distribution in cyanobacteria reflects a differential loss of genes according to their ecological needs or constraints.

Ghirardi et al. (2009) discuss the photobiological hydrogen-producing systems in a paper with 92 citations. They note that hydrogen photoproduction by microorganisms combines the photosynthetic properties of oxygenic and nonoxygenic microbes with the activity of H_2-producing enzymes in nature: H_2ases and nitrogenases. The overall efficiency of the process depends on the separate efficiencies of photosynthesis and enzymatic catalysis. They discuss the biochemical pathways for H_2 production in different organisms, barriers to be overcome, and possible suggestions for integrating photobiological H_2 production with fermentative, anaerobic systems for a potentially more efficient process.

Melis (2007) discusses the photosynthetic H_2 metabolism in *Chlamydomonas reinhardtii* (unicellular green algae) in a paper with 91 citations. He notes that the H_2 evolution process was induced upon sulfate nutrient deprivation of the cells, which reversibly inhibits PSII and O_2 evolution in their chloroplast. In the absence of O_2, and in order to generate ATP, green algae resorted to anaerobic photosynthetic metabolism, evolved H_2 in the light, and consumed endogenous substrates. He summarizes recent advances on green algal hydrogen metabolism including the mechanism of a substantial tenfold starch accumulation in the cells, observed promptly upon S deprivation, and the regulated starch and protein catabolism during the subsequent H_2 evolution. He also discusses the function of a chloroplast envelope–localized sulfate permease and the photosynthesis–respiration relationship in green algae as potential tools by which to stabilize and enhance H_2 metabolism. He then discusses the biochemistry of anaerobic H_2 photoproduction, its genes, proteins, regulation, and communication with other metabolic pathways in microalgae. He argues that photosynthetic H_2 production by green algae may hold the promise of generating a renewable fuel from nature's most plentiful resources, sunlight and water.

Hansel and Lindbald (1998) discuss the optimization of cyanobacteria as biotechnologically relevant producers of molecular hydrogen, a clean and renewable energy source, in a paper with 91 citations. They argue that basic and applied research leading to biological production of molecular hydrogen utilizing cyanobacteria deserves serious attention. In these oxygenic phototrophic bacteria, hydrogen can be produced by the activity of either nitrogenases or reversible/bidirectional H_2ases. Knowledge of the physiological and molecular basis of some of the processes involved in hydrogen metabolism in these peculiar microorganisms has increased during the last decade. This information might then constitute the basis for optimizing the efficiency of hydrogen evolution by cyanobacteria. They assert that progress might be achieved by screening more cyanobacterial strains for their ability to produce and evolve hydrogen, by genetically manipulating specific strains, and by improving the conditions for cultivation in bioreactors.

Schutz et al. (2004) discuss cyanobacterial H_2 production with a focus on the comparative analysis in a paper with 90 citations. Several unicellular and filamentous, nitrogen-fixing, and non-nitrogen-fixing cyanobacterial strains have been investigated on the molecular and the physiological level in order to find the most efficient organisms for photobiological hydrogen production. These strains were screened for the presence or absence of hup and hox genes, and it was shown that they have different sets of genes involved in H_2 evolution. The uptake H_2ase was identified in all N_2-fixing cyanobacteria, and some of these strains also contained the bidirectional H_2ase, whereas the non-nitrogen-fixing strains only possessed the bidirectional enzyme. In N_2-fixing strains, hydrogen was mainly produced by the nitrogenase as a by-product during the reduction of atmospheric nitrogen to ammonia. Therefore, hydrogen production was investigated both under non-nitrogen-fixing conditions and under nitrogen limitation. It was shown that the hydrogen uptake activity is linked to the nitrogenase activity, whereas the hydrogen evolution activity of the bidirectional H_2ase is not dependent or even related to diazotrophic growth conditions. With regard to large-scale hydrogen evolution by N_2-fixing cyanobacteria, hydrogen uptake–deficient mutants have to be used because of their inability to reoxidize the hydrogen produced by the nitrogenase. On the other hand, fermentative H_2 production by the bidirectional H_2ase should also be taken into account in further investigations of biological hydrogen production.

Beer et al. (2009) discuss engineering of algae for biohydrogen and biofuel production in a paper with 88 citations. They note that there is currently a substantial interest in utilizing eukaryotic algae for the renewable production of several bioenergy carriers, including starches for alcohols, lipids for diesel fuel surrogates, and H_2 for fuel cells. Relative to terrestrial biofuel feedstocks, algae can convert solar energy into fuels at higher photosynthetic efficiencies and can thrive in saltwater systems. Recently, there has been considerable progress in identifying relevant bioenergy genes and pathways in microalgae, and powerful genetic techniques have been developed to engineer some strains via the targeted disruption of endogenous genes and/or transgene expression. Collectively, the progress that has been realized in these areas is rapidly advancing our ability to genetically optimize the production of targeted biofuels.

Mus et al. (2007) discuss the anaerobic acclimation in *C. reinhardtii* with a focus on anoxic gene expression, H_2ase induction, and metabolic pathways in a paper with 88 citations. Metabolite analyses, quantitative PCR, and high-density *Chlamydomonas* DNA microarrays were used to monitor changes in metabolite accumulation and gene expression during acclimation of the cells to anoxia. Elevated levels of transcripts encoding proteins associated with the production of H_2, organic acids, and ethanol were observed in congruence with the accumulation of fermentation products. The levels of over 500 transcripts increased significantly during acclimation of the cells to anoxic conditions. Among these were transcripts encoding transcription/translation regulators, prolyl hydroxylases, hybrid cluster proteins, proteases, transhydrogenase, catalase, and several putative proteins of unknown function. Overall, they use metabolite, genomic, and transcriptome data to provide genome-wide insights into the regulation of the complex metabolic networks utilized by *Chlamydomonas* under the anaerobic conditions associated with H_2 production.

Asada and Miyake (1999) discuss photobiological hydrogen production in an early paper with 81 citations as they review the principles and recent progress in the research and development (R&D) of photobiological hydrogen production. Cyanobacteria produce hydrogen gas using nitrogenase and/or H_2ase. Hydrogen production mediated by native H_2ases in cyanobacteria occurs in the dark under anaerobic conditions by degradation of intracellular glycogen. In vitro and in vivo couplings of the cyanobacterial photosynthetic system with a clostridial H_2ase via cyanobacterial ferredoxin were demonstrated in the presence of light. Genetic transformation of *Synechococcus* PCC7942 with the H_2ase gene from *C. pasteurianum* was successful; the active enzyme was expressed in PCC7942. The strong hydrogen producers among photosynthetic bacteria were isolated and characterized. Coculture of *Rhodobacter* and *Clostridium* was applied for hydrogen production from glucose. A mutant strain of *Rhodobacter sphaeroides* RV whose light-harvesting proteins were altered was obtained by UV irradiation. Hydrogen productivity by the mutant was improved when irradiated with monochromatic light of some wavelengths.

Lambert and Smith (1981) discuss hydrogen metabolism of cyanobacteria in one of the earliest papers with 81 citations.

Dau and Zaharieva (2009) discuss principles, efficiency, and blueprint character of solar-energy conversion in photosynthetic water oxidation in a paper with 80 citations. They describe and apply a rationale for estimating the solar-energy conversion efficiency (eta (SOLAR)) of PSII: the fraction of the incident solar energy is absorbed by the antenna pigments and eventually stored in the form of chemical products. For PSII at high concentrations, approximately 34% of the incident solar energy is used for creation of the photochemistry-driving excited state, P680*, with an excited-state energy of 1.83 eV. Subsequent electron transfer results in the reduction of a bound quinone (Q(A)) and oxidation of the Tyr(z) within 1 µds. This radical-pair state is stable against recombination losses for approximately 1 ms. At this level, the maximal eta (SOLAR) is 23%. After the essentially irreversible steps of quinone reduction and water oxidation (the final steps catalyzed by the PSII complex), a maximum of 50% of the excited-state energy is stored in chemical form; eta (SOLAR) can be as high as 16%. Extending their considerations to a photosynthetic organism optimized to use PSII and PSI to drive H_2 production, the theoretical maximum of the solar-energy conversion efficiency would be as high as 10.5%, if all electrons and protons derived from water oxidation were used for H_2 formation. The overpotential for catalysis of water oxidation at the Mn_4Ca complex of PSII may be as low as 0.3 V. To address the specific energetics of water oxidation at the Mn complex of PSII, they propose a new conceptual framework that will facilitate quantitative considerations on the basis of oxidation potentials and pK values. They conclude that photosynthetic water oxidation works at high efficiency and thus can serve as both an inspiring model and a benchmark in the development of future technologies for production of solar fuels.

Lindberg et al. (2010) discuss engineering a platform for photosynthetic isoprene production in cyanobacteria, using *Synechocystis* as the model organism in a paper with 76 citations. They note that the concept of *photosynthetic biofuels* envisions application of a single organism, acting both as photocatalyst and producer of ready-made fuel. They apply this concept upon genetic engineering of the cyanobacterium *Synechocystis*, conferring the ability to generate volatile isoprene hydrocarbons from CO_2 and H_2O. Heterologous expression of the *Pueraria montana* (kudzu) isoprene synthase (IspS) gene in *Synechocystis* enabled photosynthetic isoprene generation in these cyanobacteria, while codon-use optimization of the kudzu IspS gene improved expression of the isoprene synthase in *Synechocystis*. They find that use of the photosynthesis psbA2 promoter, to drive the expression of the IspS gene, resulted in a light-intensity-dependent isoprene synthase expression. They assert that oxygenic photosynthesis can be redirected to generate useful small volatile hydrocarbons, while consuming CO_2, without a prior requirement for the harvesting, dewatering, and processing of the respective biomass.

Benemann (2000) discusses hydrogen production by microalgae in a paper with 72 citations. He argues that the most plausible processes for future applied R&D are those that couple separate stages of microalgal photosynthesis and fermentations (*indirect biophotolysis*). These involve fixation of CO_2 into storage carbohydrates followed by their conversion to H_2 by the reversible H_2ase, both in dark- and possibly light-driven anaerobic metabolic processes. Based on a preliminary engineering and economic analysis, biophotolysis processes must achieve close to an overall 10% solar-energy conversion efficiency to be competitive with alternative sources of renewable H_2, such as photovoltaic–electrolysis processes. He further argues that "such high solar conversion efficiencies in photosynthetic CO_2 fixation could be reached by genetically reducing the number of light harvesting (antenna) chlorophylls and other pigments in microalgae." Similarly, greatly increased yields of H_2 from dark fermentation by microalgae could be obtained through application of the techniques of metabolic engineering. Another challenge is to scale up biohydrogen processes with economically viable bioreactors. He asserts that "solar energy driven microalgae processes for biohydrogen production are potentially large-scale, but also involve long-term and economically high-risk R&D."

Pinto et al. (2002) discuss the three decades of research on cyanobacterial hydrogen evolution in a paper with 70 citations. They note that hydrogen evolution by cyanobacteria is a potential way of biohydrogen production for the future. The basic and early applied research over the last 30 years has established the basis of present knowledge in the field and is a platform for future R&D directions. They briefly survey some of the progress made in the field of cyanobacterial hydrogen evolution during this time period.

Appel and Schulz (1998) discuss hydrogen metabolism in organisms with oxygenic photosynthesis and H_2ases as important regulatory devices for a proper redox poising in a paper with 70 citations. They note that H_2ases are well-characterized enzymes on the enzymatic, structural, and genetic level, especially in prokaryotic microorganisms. They can be classified regarding the metal composition of their active site (Fe only, NiFe- or metal-free), their preferential direction of reaction (uptake only or bidirectional/reversible), and their in vivo electron donors or acceptors. The main physiological role of the uptake H_2ase in cyanobacteria is probably recapturing the hydrogen produced by nitrogenase. They argue that the role of the bidirectional H_2ase in phototrophs is still a matter of debate. Based on recent results that showed it to be of the NAD(P)-reducing type, they suggest a model for its physiological function. This model includes that "this type of hydrogenase is linked to complex I of the respiratory electron-transport chain and might be an important electron valve during photosynthesis under rapidly changing light conditions."

19.2.3 Conclusion

The nonexperimental research on algal biohydrogen has been one of the most dynamic research areas in recent years. Thirty-three citation classics in the field of algal biohydrogen with more than seventy citations were located, and the key emerging issues from these papers were presented in decreasing order of the number of citations. These papers give strong hints about the determinants of efficient algal biohydrogen production processes and policies and emphasize that marine algae are efficient biohydrogen feedstocks in comparison with the first- and second-generation biofuel feedstocks. Seven and twenty-six of these papers were related to comparative biohydrogen production and algal biohydrogen production, respectively.

19.3 Experimental Studies on Algal Biohydrogen

19.3.1 Introduction

The experimental research on algal biohydrogen has been one of the most dynamic research areas in recent years. Seventeen experimental citation classics in the field of algal biohydrogen with more than seventy-one citations were located, and the key emerging issues from these papers are presented in decreasing order of the number of citations (Table 19.2). These papers give strong hints about the determinants of efficient algal biohydrogen production and emphasize that marine algae are efficient biohydrogen feedstocks in comparison with the first- and second-generation biofuel feedstocks.

The papers were dominated by researchers from six countries, usually through the intracountry institutional collaboration, and they were multiauthored. The United States

TABLE 19.2
Experimental Research on Algal Biohydrogen

No.	Paper References	Year	Document	Affiliation	Country	No. of Authors	Journal	Index	Subjects	Topic	Total No. of Citations	Total Average Citations per Annum	Rank
1	Melis et al.	2000	A	Univ. Calif B., NREL	United States	5	*Plant Physiol.*	SCI, SCIE	Plant Sci.	Algal biohydrogen photoproduction	353	25.2	4
2	Posewitz et al.	2004	A	Colorado Sch. Mines, NREL	United States	6	*J. Biol. Chem.*	SCI, SCIE	Bioch. Mol. Biol.	Algal biohydrogen production	161	16.1	11
3	Zhang et al.	2002	A	Univ. Calif B.	United States	3	*Planta*	SCI, SCIE	Plant Sci.	Algal biohydrogen production	160	13.3	12
4	Happe and Naber	1993	A	Ruhr Univ. Bochum	Germany	2	*Eur. J. Biochem.*[a]	SCI, SCIE	Bioch. Mol. Biol.	Algal biohydrogen production	127	6.1	17
5	Kruse et al.	2005b	A	Univ. Bielefeld, CNRS, Univ. Queensland	Germany, France, Australia	7	*J. Biol. Chem.*	SCI, SCIE	Bioch. Mol. Biol.	Algal biohydrogen photoproduction	122	13.6	19
6	Florin et al.	2001	A	Univ. Bonn	Germany	3	*J. Biol. Chem.*	SCI, SCIE	Bioch. Mol. Biol.	Algal biohydrogen photoproduction	120	6.3	20
7	King et al.	2006	A	NREL, Colorado Sch. Mines	United States	4	*J. Bacteriol.*	SCI, SCIE	Microbiol.	Algal biohydrogen production	117	14.6	21
8	Happe and Kaminski	2002	A	Univ. Bonn.	Germany	2	*Eur. J. Biochem.*[a]	SCI, SCIE	Bioch. Mol. Biol.	Algal biohydrogen production	115	8.9	22
9	Kosourov et al.	2002	A	NREL	United States	4	*Biotechnol. Bioeng.*	SCI, SCIE	Biot. Appl. Microb.	Algal biohydrogen photoproduction	114	9.5	24
10	Ghirardi et al.	1997	A	Indiana Univ.	United States	3	*Appl. Biochem. Biotechnol.*	SCI, SCIE	Biot. Appl. Microb., Bioch. Mol. Biol.	Algal biohydrogen production	114	6.7	23

(Continued)

TABLE 19.2 (*Continued*)

Experimental Research on Algal Biohydrogen

No.	Paper References	Year	Document	Affiliation	Country	No. of Authors	Journal	Index	Subjects	Topic	Total No. of Citations	Total Average Citations per Annum	Rank
11	Schmitz et al.	1995	A	Max Planck Inst. Biochem., Univ. Cologne	Germany	7	*Eur. J. Biochem.*[a]	SCI, SCIE	Bioch. Mol. Biol.	Algal biohydrogen production	106	5.6	26
12	Kosourov et al.	2003	A	NREL	United States	3	*Plant Cell Physiol.*	SCI, SCIE	Plant Sci., Cell Biol.	Algal biohydrogen production	102	9.3	29
13	Happe et al.	1994	A	Ruhr Univ.	Germany	3	*Eur. J. Biochem.*[a]	SCI, SCIE	Bioch. Mol. Biol.	Algal biohydrogen production	99	5.0	30
14	Takeda et al.	1995	A	Kinki Univ., Rite	Japan	3	*Plant Cell Physiol.*	SCI, SCIE	Plant Sci., Cell Biol.	Algal biohydrogen production	91	4.8	35
15	Antal et al.	2003	A	NREL, Moscow MV Lomonosov. St. Univ, RAS	Unite States, Russia	8	*Biochim. Biophys. Acta Bioenerg.*	SCI, SCIE	Bioch. Mol. Biol., Biophys.	Algal biohydrogen production—photosynthesis	83	7.5	40
16	Brown et al.	2010	A	Univ. Michig.	United States	3	*Energ. Fuel*	SCI, SCIE	Ener. Fuels, Eng. Chem.	Algal biohydrogen production	74	18.5	46
17	Mussgnug et al.	2007	A	Univ. Queensland, Univ. Bielefeld	United States, Australia	9	*Plant Biotechnol. J.*	SCIE	Biot. Appl. Microb., Plant Sci.	Algal biohydrogen production—photosynthesis	71	10.1	48

SCI, Science Citation Index; SCIE, Science Citation Index Expanded; SSCI, Social Sciences Citation Index; A, article; R, review.

[a] *Eur. J. Biochem.*, continues as *FEBS J.*

(10 papers), Germany (6 papers), and Australia (2 papers) were the most prolific countries. Similarly, the National Renewable Energy Laboratory (NREL) (6 papers), Ruhr University, University of Bonn, University of California Berkeley, University of Queensland, and University of Bielefeld (2 papers each) were the most prolific institutions.

Similarly, all these papers were published in journals indexed by the SCI and/or SCIE. There were no papers indexed by the A&HCI or SSCI. The number of citations ranged from 71 to 353, and the average number of citations per annum ranged from 4.8 to 25.2. All of these papers were articles.

19.3.2 Experimental Research on Algal Biohydrogen

Melis et al. (2000) study sustained photobiological hydrogen gas production upon reversible inactivation of oxygen evolution in the green alga *C. reinhardtii* in a paper with 353 citations as they describe a novel approach for sustained photobiological production of H_2 gas via the reversible H_2ase pathway in the green alga *C. reinhardtii*. They argue that "this single-organism, two-stage H_2 production method circumvents the severe O_2 sensitivity of the reversible hydrogenase by temporally separating photosynthetic O_2 evolution and carbon accumulation (stage 1) from the consumption of cellular metabolites and concomitant H_2 production (stage 2)." A transition from stage 1 to stage 2 was effected upon S deprivation of the culture, which reversibly inactivated PSII and O_2 evolution. Under these conditions, oxidative respiration by the cells in the light depleted O_2 and caused anaerobiosis in the culture, which was necessary and sufficient for the induction of the reversible H_2ase. Subsequently, sustained cellular H_2 gas production was observed in the light but not in the dark. They find that "the mechanism of H_2 production entailed protein consumption and electron transport from endogenous substrate to the cytochrome b_6-f and PSI complexes in the chloroplast thylakoids." Light absorption by PSI was required for H_2 evolution, suggesting that "photoreduction of ferredoxin is followed by electron donation to the reversible hydrogenase." The latter catalyzes the reduction of protons to molecular H_2 in the chloroplast stroma.

Posewitz et al. (2004) discuss the discovery of two novel radical S-adenosylmethionine (SAM) proteins required for the assembly of an active [Fe] H_2ase in a paper with 161 citations. To identify genes necessary for the photoproduction of H_2 in *C. reinhardtii*, random insertional mutants were screened for clones unable to produce H_2. Within several prokaryotic genomes, HydE, HydF, and HydG are found in putative operons with [Fe] H_2ase structural genes. Both HydE and HydG belong to the emerging radical SAM (commonly designated *radical SAM*) superfamily of proteins. They demonstrate that "HydEF and HydG function in the assembly of [Fe] hydrogenase." Northern blot analysis indicates that "mRNA transcripts for both the HydEF gene and the HydG gene are anaerobically induced concomitantly with the two *C. reinhardtii* [Fe] hydrogenase genes, HydA1 and HydA2." They find that "complementation of the bx; 1 *C. reinhardtii* hydEF-1 mutant with genomic DNA corresponding to a functional copy of the HydEF gene restores hydrogenase activity." Moreover, coexpression of *C. reinhardtii* HydEF, HydG, and HydA1 genes in *E. coli* results in the formation of an active HydA1 enzyme.

Zhang et al. (2002) study biochemical and morphological characterizations of sulfur-deprived and H_2-producing *C. reinhardtii* in a paper with 160 citations. Profile analysis of selected photosynthetic proteins showed "a precipitous decline in the amount of ribulose-1,5-bisphosphate carboxylase-oxygenase (Rubisco) as a function of time in S deprivation, a more gradual decline in the level of photosystem (PS) II and PSI proteins, and a change in the composition of the PSII light-harvesting complex (LHC-II)." An increase in the level of the enzyme Fe-H_2ase was noted "during the initial stages of S deprivation (0–72 h) followed by a decline in the level of this enzyme during longer (t > 72 h) S-deprivation times." Microscopic observations showed distinct morphological changes in *C. reinhardtii* during S deprivation and H_2 production. "Ellipsoid-shaped cells (normal photosynthesis) gave way to larger and spherical cell shapes in the initial stages of S deprivation and H_2 production, followed by cell- mass reductions after longer S-deprivation and H_2-production times." They argue that "under S-deprivation conditions, electrons derived from a residual, PSII H_2O-oxidation activity feed into the hydrogenase pathway, thereby contributing to the H_2-production process in *Chlamydomonas reinhardtii*."

Happe and Naber (1993) study isolation, characterization, and N-terminal amino acid sequence of H_2ase from *C. reinhardtii* in an early paper with 127 citations. H_2ase from *C. reinhardtii* was purified to homogeneity by five column-chromatography steps under strict anaerobic conditions. The enzyme was purified 6100-fold, resulting in a specific activity for H_2 evolution of 935 µmol/(min mg protein) at 25°C, using reduced MV as electron donor. The optimal temperature for hydrogen evolution is 60°C; the optimal pH value is 6.9. The K_m value for MV is 0.83 mM and for ferredoxin 35 µM. The protein was pure. On nondenaturing gels, run under nitrogen, a single band was detected after activity staining. This band corresponded to the single band observed on denaturing SDS gels, which had a molecular mass of 48 kDa. If the band was cut out of the native gel and incubated with reduced MV, hydrogen evolution could be measured. They find that the purified enzyme contains 4 Fe atoms/mol. The amino acid composition and the N-terminal amino acid sequence (24 residues) of the protein were determined. They assert that no significant amino acid sequence homologies could be found to any sequences from prokaryotic H_2ases.

Kruse et al. (2005b) study improved photobiological H_2 production in engineered green algal cells in a paper with 122 citations. To improve H_2 production in *Chlamydomonas,* they have developed a new approach to increase H^+ and e^- supply to the H_2ases. In a first step, mutants blocked in the state 1 transition were selected. These mutants are inhibited in cyclic e^- transfer around photosystem I, eliminating possible competition for e^- with H_2ase. Selected strains were further screened for increased H_2 production rates, leading to the isolation of Stm6. This strain has a modified respiratory metabolism, providing it with two additional important properties as follows: "large starch reserves (i.e. enhanced substrate availability), and a low dissolved O_2 concentration (40% of the wild type (WT)), resulting in reduced inhibition of H_2ase activation." They find that the "H_2 production rates of Stm6 were 5–13 times that of the control WT strain over a range of conditions (light intensity, culture time, +/– uncoupler)." Typically, "similar to 540 ml of H_2 liter^{-1} culture (up to 98% pure) were produced over a 10–14-day period at a maximal rate of 4 ml h^{-1}" (efficiency = similar to 5 times the WT). They conclude that stm6 represents an important step toward the development of future solar-powered H_2 production systems.

Florin et al. (2001) study whether a novel type of iron H_2ase in the green alga *Scenedesmus obliquus* is linked to the photosynthetic electron transport chain in a paper with 120 citations. They report biochemical and genetical characterizations of a new type of iron H_2ase (HydA) in this photosynthetic organism. The monomeric enzyme has a molecular mass of 44.5 kDa. "The complete hydA cDNA of 2609 base pairs comprises an open reading frame encoding a polypeptide of 448 amino acids." The protein contains a short transit peptide that routes the nucleus encoded H_2ase to the chloroplast. "Antibodies raised against the iron hydrogenase from *Chlamydomonas reinhardtii* react with both the isolated and in Escherichia coli overexpressed protein of *S. obliquus* as shown by Western blotting." By analyzing 5 kb of the genomic DNA, they find the transcription initiation site and five introns within hydA. Northern experiments suggest that "hydA transcription is induced during anaerobic incubation." They argue that "alignments of *S. obliquus* HydA with known iron hydrogenases and sequencing of the N terminus of the purified protein confirm that HydA belongs to the class of iron hydrogenases." The C terminus of the enzyme including the catalytic site (H cluster) reveals a high degree of identity to iron H_2ases. However, they argue that "the lack of additional Fe S clusters in the N-terminal domain indicates a novel pathway of electron transfer." Inhibitor experiments show that the "ferredoxin PetF functions as natural electron donor linking the enzyme to the photosynthetic electron transport chain and PetF probably binds to the hydrogenase through electrostatic interactions."

King et al. (2006) carry out functional studies of [FeFe] H_2ase maturation in an *E. coli* biosynthetic system in a paper with 117 citations. Two radical SAM proteins proposed to function in H cluster biosynthesis, HydEF and HydG, were recently identified in the hydEF-1 mutant of the green alga *C. reinhardtii* by Posewitz et al. (2004). A more stable [FeFe] H_2ase expression system has been achieved in *E. coli* by cloning and coexpression of hydE, hydF, and hydG from the bacterium *C. acetobutylicum*. Coexpression of the *C. acetobutylicum* maturation proteins with various algal and bacterial [FeFe] H_2ases in *E. coli* resulted in purified enzymes with specific activities that were similar to those of the enzymes purified from native sources. In the case of structurally complex [FeFe] H_2ases, they argue that maturation of the catalytic sites could occur in the absence of an accessory iron–sulfur cluster domain. Initial investigations of the structure and function of the maturation proteins HydE, HydF, and HydG showed that the "highly conserved radical-SAM domains of both HydE and HydG and the GTPase domain of HydF were essential for achieving biosynthesis of active [FeFe] hydrogenases." They assert that "the catalytic domain and a functionally complete set of Hyd maturation proteins are fundamental to achieving biosynthesis of catalytic [FeFe] hydrogenases."

Happe and Kaminski (2002) study the differential regulation of the Fe-H_2ase during anaerobic adaptation in *C. reinhardtii* in a paper with 115 citations. Using the suppression subtractive hybridization (SSH) approach, the differential expression of genes under anaerobiosis was analyzed. A PCR fragment with similarity to the genes of bacterial Fe-H_2ases was isolated and used to screen all anaerobic cDNA expression library of *C. reinhardtii*. The "cDNA sequence of hydA contains a 1494-bp ORF encoding a protein with an apparent molecular mass of 53.1 kDa." The transcription of the H_2ase gene is very rapidly induced during anaerobic adaptation of the cells. The deduced amino acid sequence corresponds to two polypeptide sequences determined by sequence analysis of the isolated native protein. The "Fe-hydrogenase contains a short transit peptide of 56 amino acids, which routes the hydrogenase to the chloroplast stroma." The isolated protein belongs to a new class of Fe-H_2ases. All four cysteine residues and 12 other amino acids, which are strictly conserved in the active site (H cluster) or Fe-H_2ases, have been identified. The "N-terminus of the *C. reinhardtii* protein is markedly truncated compared to other nonalgal Fe-hydrogenases." They argue that further conserved cysteines that coordinate additional Fe–S cluster in other Fe-H_2ases are missing. They assert that "Ferredoxin PctF, the natural electron donor, links the hydrogenase from *C. reinhardtii* to the photosynthetic electron transport chain" and the hydrogenase enables the survival of the green algae under anaerobic conditions by transferring the electrons from reducing equivalents to the enzyme.

Kosourov et al. (2002) study sustained hydrogen photoproduction by *C. reinhardtii* with a focus on the effects of culture parameters in a paper with 114 citations. Using a computer-monitored photobioreactor system, they investigate the behavior of sulfur-deprived algae and find that "(1) the cultures transition through five consecutive phases: an aerobic phase, an O_2-consumption phase, an anaerobic phase, a H_2-production phase and a termination phase; (2) synchronization of cell division during pre-growth with 14:10 h light: dark cycles leads to earlier establishment of anaerobiosis in the cultures and to earlier onset of the H_2-production phase; (3) re-addition of small quantities of sulfate (12.5–50 μM $MgSO_4$, final concentration) to either synchronized or unsynchronized cell suspensions results in an initial increase in culture density, a higher initial specific rate of H_2 production, an increase in the length of the H_2-production phase, and an increase in the total amount of H_2 produced; and (4) increases in the culture optical density in the presence of 50 μM sulfate result in a decrease in the initial specific rates of H_2 production and in an earlier start of the H_2-production phase with unsynchronized cells." They assert that the "effects of sulfur re-addition on H_2 production, up to an optimal concentration, are due to an increase in the residual water-oxidation activity of the algal cells." They also argue that, in principle, "cells synchronized by growth under light: dark cycles can be used in an outdoor H_2-production system without loss of efficiency compared to cultures that up until now have been pre-grown under continuous light conditions."

Ghirardi et al. (1997) study the oxygen sensitivity of algal H_2 production in an early paper with 114 citations. They design an experimental technique for the selection of O_2-tolerant, H_2-producing variants of *C. reinhardtii* based on the ability of wild-type cells to survive a short (20 min) exposure to metronidazole in the presence of controlled concentrations of O_2. They find that the "number of survivors depends on the metronidazole concentration, light intensity, preinduction of the hydrogenase, and the presence or absence of O_2." Finally, they demonstrate that some of the selected survivors in fact exhibit H_2-production capacity that is less sensitive to O_2 than the original wild-type population.

Schmitz et al. (1995) study the molecular biological analysis of a bidirectional H_2ase from cyanobacteria in a paper with 106 citations. An 8.9 kb segment with H_2ase genes from the cyanobacterium *A. variabilis* has been cloned and sequenced. The sequences show "homology to the methyl-viologen-reducing hydrogenases from archaebacteria and, even more striking, to the NAD^+-reducing enzymes from *Alcaligenes eutrophus* and *Nocardia opaca* as well as to the $NADP^+$-dependent protein from *Desulfovibrio fructosovorans*." The cluster from *A. variabilis* contains genes coding for both the H_2ase heterodimer (hoxH and hoxY) and for the diaphorase moiety (hoxU and hoxF) described for the *A. eutrophus* enzyme. In "*A. variabilis* the gene cluster is split by two open reading frames (between hoxY and hoxH and between hoxU and hoxY respectively), and a probably non-coding 0.9-kb segment in an unusual way." The hoxH partial sequence from *Anabaena* 7119 and *Anacystis nidulans* was amplified by PCR. Using the labeled segment from A. 7119 as probe, southern analysis revealed "homologous gene segments in the cyanobacteria A. 7119, *Anabaena cylindrica, Anacystis nidulans* and *A. variabilis.*" The bidirectional H_2ase from *A. nidulans* was purified and digests were sequenced. The amino acid sequences obtained showed "partial identities to the amino acid sequences deduced from the DNA data of the 8.9-kb segment from *A. variabilis.*" Therefore, they argue that "the 8.9-kb segment contains the genes coding for the bidirectional, reversible hydrogenase from cyanobacteria whilst crude extracts from *A. nidulans* perform NAD(P)H-dependent H_2 evolution corroborating the molecular biological demonstration of the $NAD(P)^+$-dependent hydrogenase in cyanobacteria."

Kosourov et al. (2003) study the effects of extracellular pH on the metabolic pathways in sulfur-deprived, H_2-producing *C. reinhardtii* cultures in a paper with 102 citations. They focus on (1) the effects of different initial extracellular pHs on the inactivation of PSII and O_2-sensitive H_2-production activity in sulfur-deprived algal cells and (2) the relationships among H_2 production and photosynthetic, aerobic, and anaerobic metabolisms under different pH regimens. They find that the "maximum rate and yield of H_2 production occur when the pH at the start of the sulfur deprivation period is 7.7 and decrease when the initial pH is lowered to 6.5 or increased to 8.2." The pH profile of hydrogen photoproduction correlates with that of the residual PSII activity (optimum pH 7.3–7.9), but not with the pH profiles of photosynthetic electron transport through photosystem I or of starch and protein degradation. In vitro H_2ase activity over this pH range is much higher than the actual in situ rates of H_2 production, indicating that H_2ase activity per se is not limiting. They then find that "starch and protein catabolisms generate formate, acetate and ethanol; contribute some reductant for H_2 photoproduction, as indicated by 3-(3,4-dichlorophenyl)1,1-dimethylurea and 2,5-dibromo-6-isopropyl-3-methyl-1,4-benzoquinone inhibition results; and are the primary sources of reductant for respiratory processes that remove photo-synthetically generated O_2." They assert that "alternative metabolic pathways predominate at different pHs, and these depend on whether residual photosynthetic activity is present or not."

Happe et al. (1994) study the induction, localization, and metal content of H_2ase in *C. reinhardtii* in a paper with 99 citations. They note that the H_2ase enzyme occurring in *C. reinhardtii* is induced by anaerobic adaptation of the cells. In aerobically growing cells, antibodies against the H_2ase failed to detect either active or inactive enzyme. However, already 10 min after the onset of anaerobic adaptation, the protein could be detected. They find that the "maximal amount of enzyme was reached after 2–3 hours anaerobiosis whilst addition of nickel or iron to the growth medium did not influence activity." In atomic absorption experiments, a Ni/Fe ratio of about 1:250 was measured. They, therefore, propose that the "hydrogenase from *C. reinhardtii* is of the Fe-only type." Adaptation in the presence of uncouplers of phosphorylation showed this process to be energy dependent. From protein synthesis inhibition experiments, they conclude that the "protein is synthesized on cytoplasmic ribosomes and, therefore, must be nuclear encoded." After isolation of intact chloroplasts from adapted cells, the active enzyme was located in the chloroplasts.

Takeda et al. (1995) discuss the resistance of photosynthesis to hydrogen peroxide in algae in an early paper with 91 citations. The effects of H_2O_2 on the photosynthetic fixation of CO_2 and on thiol-modulated enzymes involved in the photosynthetic reduction of carbon in algae were studied in a comparison with those in chloroplasts isolated from spinach leaves. In both systems, they find that H_2O_2-scavenging enzymes were inhibited by addition of 0.1 mM NaN_3 1 h prior to the addition of H_2O_2. A "concentration (10^{-4} M) of H_2O_2 caused strong inhibition of the CO_2 fixation by intact spinach chloroplasts," "but not that by Euglena and Chlamydomonas cells." The same results were also obtained with cells of the cyanobacteria *Synechococcus* PCC 7942 and *Synechocystis* PCC 6803 in the presence of 1 mM hydroxylamine. They argue that "algal photosynthesis is rather resistant to H_2O_2." They also observe the "insusceptibility to H_2O_2 of thiol-modulated enzymes, namely, fructose-1,6-bisphosphatase, NADP-glyceraldehyde-3-phosphate dehydrogenase, and ribulose-5-phosphate kinase, in the chloroplasts of Euglena and Chlamydomonas and in cyanobacterial cells." They assert that the "resistance of photosynthesis to H_2O_2 is due in part to the insusceptibility of the algal thiol-modulated enzymes to H_2O_2."

Antal et al. (2003) study the dependence of algal H_2 production on PSII and O_2 consumption activities in sulfur-deprived *C. reinhardtii* cells in a paper with 83 citations. They demonstrate that (1) the photochemical activity of PSII, monitored by in situ fluorescence, also decreases slowly during the aerobic period; (2) at the exact time of anaerobiosis, the remaining PSII activity is rapidly downregulated; and (3) electron transfer from PSII to PSI abruptly decreases at that point. Shortly thereafter, the PSII photochemical activity is partially restored, and H_2 production starts. Hydrogen production, which lasts for 3–4 days, is catalyzed by an anaerobically induced, reversible H_2ase. While most of the reductants used directly for H_2 gas photoproduction come from water, the remaining electrons must come from endogenous substrate degradation through the NAD(P)H plastoquinone (PQ) oxidoreductase pathway. They propose that the "induced hydrogenase activity provides a sink for electrons in the absence of other alternative pathways, and its operation allows the partial oxidation of intermediate photosynthetic carriers, including the PQ pool, between PSII and PSI." They conclude that the "reduced state of this pool, which controls PSII photochemical activity, is one of the main factors regulating H_2 production under sulfur-deprived conditions. Residual O_2 evolved under these conditions is probably consumed mostly by the aerobic oxidation of storage products linked to mitochondrial respiratory processes involving both the cytochrome oxidase and the alternative oxidase." They assert that these functions maintain the intracellular anaerobic conditions required to keep the hydrogenase enzyme in the active, induced form.

Brown et al. (2010) study the hydrothermal liquefaction and gasification of *Nannochloropsis* sp. in a recent paper with 74 citations. They convert the marine microalga *Nannochloropsis* sp. into a crude biooil product and a gaseous product via hydrothermal processing from 200°C to 500°C and a batch holding time of 60 min. A moderate temperature or 350°C led to the highest biooil yield of 43 wt%. They estimate the heating value of the biooil as 39 MJ/kg, which is comparable to that of a petroleum crude oil. They find that the "H/C and O/C ratios for the bio-oil decreased from 1.73 and 0.12, respectively, for the 200°C product to 1.04 and 0.05, respectively, for the 500°C product." Major biooil constituents include phenol and its alkylated derivatives, heterocyclic N-containing compounds, long-chain fatty acids, alkanes and alkenes, and derivatives of phytol and cholesterol. CO_2 was always the most abundant gas product. "H_2 was the second most abundant gas at all temperatures other than 500°C where its yield was surpassed by that of CH_4." They assert that "nearly 80% of the carbon and up to 90% of the chemical energy originally present in the microalga can be recovered as either biooil or gas products."

Mussgnug et al. (2007) study engineering photosynthetic light capture with a focus on the impacts on improved solar energy to biomass conversion in a paper with 71 citations. They apply RNAi technology to downregulate the entire LHC gene family simultaneously to reduce energy losses by fluorescence and heat. They find that the mutant Stm3LR3 had significantly reduced levels of LHCI and LHCII mRNAs and proteins, while chlorophyll and pigment synthesis was functional. The grana were markedly less tightly stacked, consistent with the role of LHCII. Stm3LR3 also exhibited reduced levels of fluorescence, a higher photosynthetic quantum yield, and a reduced sensitivity to photoinhibition, resulting in an increased efficiency of cell cultivation under elevated light conditions. Collectively, they argue that these properties offer three advantages in terms of algal bioreactor efficiency under natural high-light levels: "(i) reduced fluorescence and LHC-dependent heat losses and thus increased photosynthetic efficiencies under high-light conditions; (ii) improved light penetration properties; and (iii) potentially reduced risk of oxidative photodamage of PSII."

19.3.3 Conclusion

The experimental research on algal biohydrogen has been one of the most dynamic research areas in recent years. Seventeen experimental citation classics in the field of algal biohydrogen with more than 71 citations were located, and the key emerging issues from these papers were presented above in decreasing order of the number of citations. These papers give strong hints about the determinants of efficient algal biohydrogen production and emphasize that marine algae are efficient biohydrogen feedstocks in comparison with the first- and second-generation biofuel feedstocks.

19.4 Conclusion

The citation classics presented under the two main headings in this chapter confirm the predictions that marine algae have a significant potential to serve as a major solution for the global problems of warming, air pollution, energy security, and food security in the form of algal biohydrogen.

Further research is recommended for the detailed studies in each topical area presented in this chapter including scientometric studies and citation classic studies to inform the key stakeholders about the potential of marine algae for the solution of the global problems of warming, air pollution, energy security, and food security in the form of algal biohydrogen.

References

Antal, T.K., T.E. Krendeleva, T.V. Laurinavichene et al. 2003. The dependence of algal H-2 production on Photosystem II and O-2 consumption activities in sulfur-deprived *Chlamydomonas reinhardtii* cells. *Biochimica et Biophysica Acta—Bioenergetics* 1607:153–160.

Appel, J. and R. Schulz. 1998. Hydrogen metabolism in organisms with oxygenic photosynthesis: Hydrogenases as important regulatory devices for a proper redox poising? *Journal of Photochemistry and Photobiology B—Biology* 47:1–11.

Asada, Y. and J. Miyake. 1999. Photobiological hydrogen production. *Journal of Bioscience and Bioengineering* 88:1–6.

Baltussen, A. and C.H. Kindler. 2004a. Citation classics in anesthetic journals. *Anesthesia & Analgesia* 98:443–451.

Baltussen, A. and C.H. Kindler. 2004b. Citation classics in critical care medicine. *Intensive Care Medicine* 30:902–910.

Beer, L.L., E.S. Boyd, J.W. Peters, and M.C. Posewitz. 2009. Engineering algae for biohydrogen and biofuel production. *Current Opinion in Biotechnology* 20:264–271.

Benemann, J.R. 2000. Hydrogen production by microalgae. *Journal of Applied Phycology* 12:291–300.

Bothast, R.J. and M.A. Schlicher. 2005. Biotechnological processes for conversion of corn into ethanol. *Applied Microbiology and Biotechnology* 67:19–25.

Brown, T.M., P.G. Duan, and P.E. Savage. 2010. Hydrothermal liquefaction and gasification of *Nannochloropsis* sp. *Energy & Fuels* 24:3639–3646.

Chisti, Y. 2007. Biodiesel from microalgae. *Biotechnology Advances* 25:294–306.

Das, D. and T.N. Veziroglu. 2001. Hydrogen production by biological processes: A survey of literature. *International Journal of Hydrogen Energy* 26:13–28.

Dau, H. and I. Zaharieva. 2009. Principles, efficiency, and blueprint character of solar-energy conversion in photosynthetic water oxidation. *Accounts of Chemical Research* 42:1861–1870.

Demirbas, A. 2007. Progress and recent trends in biofuels. *Progress in Energy and Combustion Science* 33:1–18.

Dubin, D., A.W. Hafner, and K.A. Arndt. 1993. Citation-classics in clinical dermatological journals—Citation analysis, biomedical journals, and landmark articles, 1945–1990. *Archives of Dermatology* 129:1121–1129.

Florin, L., A. Tsokoglou, and T. Happe. 2001. A novel type of iron hydrogenase in the green alga *Scenedesmus obliquus* is linked to the photosynthetic electron transport chain. *Journal of Biological Chemistry* 276:6125–6132.

Gehanno, J.F., K. Takahashi, S. Darmoni, and J. Weber. 2007. Citation classics in occupational medicine journals. *Scandinavian Journal of Work, Environment & Health* 33:245–251.

Ghirardi, M.L., A. Dubini, J.P. Yu, and P.C. Maness. 2009. Photobiological hydrogen-producing systems. *Chemical Society Reviews* 38:52–61.

Ghirardi, M.L., M.C. Posewitz, P.C. Maness, A. Dubini, J.P. Yu, M. Seibert. 2007. Hydrogenases and hydrogen photoproduction in oxygenic photosynthetic organisms. *Annual Review of Plant Biology* 58:71–91.

Ghirardi, M.L., R.K. Togasaki, and M. Seibert. 1997. Oxygen sensitivity of algal H_2-production. *Applied Biochemistry and Biotechnology* 63–65:141–151.

Ghirardi, M.L., J.P. Zhang, J.W. Lee et al. 2000. Microalgae: A green source of renewable H_2. *Trends in Biotechnology* 18:506–511.

Godfray, H.C.J., J.R. Beddington, I.R. Crute et al. 2010. Food security: The challenge of feeding 9 billion people. *Science* 327:812–818.

Goldemberg, J. 2007. Ethanol for a sustainable energy future. *Science* 315:808–810.

Hallenbeck, P.C. and J.R. Benemann. 2002. Biological hydrogen production; fundamentals and limiting processes. *International Journal of Hydrogen Energy* 27:1185–1193.

Hansel, A. and P. Lindblad. 1998. Towards optimization of cyanobacteria as biotechnologically relevant producers of molecular hydrogen, a clean and renewable energy source. *Applied Microbiology and Biotechnology* 50:153–160.

Happe, T. and A. Kaminski. 2002. Differential regulation of the Fe-hydrogenase during anaerobic adaptation in the green alga *Chlamydomonas reinhardtii*. *European Journal of Biochemistry* 269:1022–1032.

Happe, T., B. Mosler, and J.D. Naber. 1994. Induction, localization and metal content of hydrogenase in the green-alga chlamydomonas-reinhardtii. *European Journal of Biochemistry* 222:769–774.

Happe, T. and J.D. Naber. 1993. Isolation, characterization and n-terminal amino-acid-sequence of hydrogenase from the green-alga *Chlamydomonas reinhardtii*. *European Journal of Biochemistry* 214:475–481.

Hoehler, T.M., B.M. Bebout, and D.J. Des Marais. 2001. The role of microbial mats in the production of reduced gases on the early Earth. *Nature* 142:324–327.

Houchins, J.P. 1984. The physiology and biochemistry of hydrogen metabolism in cyanobacteria. *Biochimica et Biophysica Acta* 768:227–255.

Jacobson, M.Z. 2009. Review of solutions to global warming, air pollution, and energy security. *Energy & Environmental Science* 2:148–173.

Kapdan, I.K. and F. Kargi. 2006. Bio-hydrogen production from waste materials. *Enzyme and Microbial Technology* 38:569–582.

King, P.W., M.C. Posewitz, M.L. Ghirardi, and M. Seibert. 2006. Functional studies of [FeFe] hydrogenase maturation in an *Escherichia coli* biosynthetic system. *Journal of Bacteriology* 188:2163–2172.

Konur, O. 2000. Creating enforceable civil rights for disabled students in higher education: An institutional theory perspective. *Disability & Society* 15:1041–1063.

Konur, O. 2002a. Assessment of disabled students in higher education: Current public policy issues. *Assessment and Evaluation in Higher Education* 27:131–152.

Konur, O. 2002b. Access to employment by disabled people in the UK: Is the Disability Discrimination Act working? *International Journal of Discrimination and the Law* 5:247–279.

Konur, O. 2002c. Access to Nursing Education by disabled students: Rights and duties of Nursing programs. *Nursing Education Today* 22:364–374.

Konur, O. 2006a. Participation of children with dyslexia in compulsory education: Current public policy issues. *Dyslexia* 12:51–67.

Konur, O. 2006b. Teaching disabled students in Higher Education. *Teaching in Higher Education* 11:351–363.
Konur, O. 2007a. A judicial outcome analysis of the Disability Discrimination Act: A windfall for the employers? *Disability & Society* 22:187–204.
Konur, O. 2007b. Computer-assisted teaching and assessment of disabled students in Higher Education: The interface between academic standards and disability rights. *Journal of Computer Assisted Learning* 23:207–219.
Konur, O. 2011. The scientometric evaluation of the research on the algae and bio-energy. *Applied Energy* 88:3532–3540.
Konur, O. 2012a. 100 citation classics in Energy and Fuels. *Energy Education Science and Technology Part A—Energy Science and Research* 2012(si):319–332.
Konur, O. 2012b. What have we learned from the citation classics in Energy and Fuels: A mixed study. *Energy Education Science and Technology Part A* 2012(si):255–268.
Konur, O. 2012c. The gradual improvement of disability rights for the disabled tenants in the UK: The promising road is still ahead. *Social Political Economic and Cultural Research* 4:71–112.
Konur, O. 2012d. Prof. Dr. Ayhan Demirbas' scientometric biography. *Energy Education Science and Technology Part A—Energy Science and Research* 28:727–738.
Konur, O. 2012e. The evaluation of the research on the biofuels: A scientometric approach. *Energy Education Science and Technology Part A—Energy Science and Research* 28:903–916.
Konur, O. 2012f. The evaluation of the research on the biodiesel: A scientometric approach. *Energy Education Science and Technology Part A—Energy Science and Research* 28:1003–1014.
Konur, O. 2012g. The evaluation of the research on the bioethanol: A scientometric approach. *Energy Education Science and Technology Part A—Energy Science and Research* 28:1051–1064.
Konur, O. 2012h. The evaluation of the research on the microbial fuel cells: A scientometric approach. *Energy Education Science and Technology Part A—Energy Science and Research* 29:309–322.
Konur, O. 2012i. The evaluation of the research on the biohydrogen: A scientometric approach. *Energy Education Science and Technology Part A—Energy Science and Research* 29:323–338.
Konur, O. 2012j. The evaluation of the biogas research: A scientometric approach. *Energy Education Science and Technology Part A—Energy Science and Research* 29:1277–1292.
Konur, O. 2012k. The scientometric evaluation of the research on the production of bio-energy from biomass. *Biomass and Bioenergy* 47:504–515.
Konur, O. 2012l. The evaluation of the global energy and fuels research: A scientometric approach. *Energy Education Science and Technology Part A—Energy Science and Research* 30:613–628.
Konur, O. 2012m. The evaluation of the biorefinery research: A scientometric approach. *Energy Education Science and Technology Part A—Energy Science and Research* 2012(si):347–358.
Konur, O. 2012n. The evaluation of the bio-oil research: A scientometric approach, *Energy Education Science and Technology Part A—Energy Science and Research* 2012(si):379–392.
Konur, O. 2012o. What have we learned from the citation classics in energy and fuels: A mixed study. *Energy Education Science and Technology Part A—Energy Science and Research* 2012(si):255–268.
Konur, O. 2012p. The evaluation of the research on the biofuels: A scientometric approach. *Energy Education Science and Technology Part A—Energy Science and Research* 28:903–916.
Konur, O. 2013. What have we learned from the research on the International Financial Reporting Standards (IFRS)? A mixed study. *Energy Education Science and Technology Part D: Social Political Economic and Cultural Research* 5:29–40.
Kosourov, S., M. Seibert, and M.L. Ghirardi. 2003. Effects of extracellular pH on the metabolic pathways in sulfur-deprived, H_2-producing *Chlamydomonas reinhardtii* cultures. *Plant and Cell Physiology* 44:146–155.
Kosourov, S., A. Tsygankov, M. Seibert, and M.L. Ghirardi. 2002. Sustained hydrogen photoproduction by *Chlamydomonas reinhardtii*: Effects of culture parameters. *Biotechnology and Bioengineering* 78:731–740.
Kruse, O., J. Rupprecht, K.P. Bader et al. 2005b. Improved photobiological H_2 production in engineered green algal cells. *Journal of Biological Chemistry* 280:34170–34177.

Kruse, O., J. Rupprecht, J.H. Mussgnug, G.C. Dismukes, and B. Hankamer. 2005a. Photosynthesis: A blueprint for solar energy capture and biohydrogen production technologies. *Photochemical & Photobiological Sciences* 4:957–970.

Lal, R. 2004. Soil carbon sequestration impacts on global climate change and food security. *Science* 304:1623–1627.

Lambert, G.R. and G.D. Smith. 1981. The hydrogen metabolism of cyanobacteria (blue-green-algae). *Biological Reviews of the Cambridge Philosophical Society* 56:589–660.

Levin, D.B., L. Pitt, and M. Love. 2004. Biohydrogen production: Prospects and limitations to practical application. *International Journal of Hydrogen Energy* 29:173–185.

Lindberg, P., S. Park, and A. Melis. 2010. Engineering a platform for photosynthetic isoprene production in cyanobacteria, using Synechocystis as the model organism. *Metabolic Engineering* 12:70–79.

Lynd, L.R., J.H. Cushman, R.J. Nichols, and C.E. Wyman. 1991. Fuel ethanol from cellulosic biomass. *Science* 251:1318–1323.

Magnuson, A., M. Anderlund, O. Johansson et al. 2009. Biomimetic and microbial approaches to solar fuel generation. *Accounts of Chemical Research* 42:1899–1909.

Melis, A. 2002. Green alga hydrogen production: Progress, challenges and prospects. *International Journal of Hydrogen Energy* 27:1217–1228.

Melis, A. 2007. Photosynthetic H_2 metabolism in *Chlamydomonas reinhardtii* (unicellular green algae). *Planta* 226:1075–1086.

Melis, A. and T. Happe. 2001. Hydrogen production. Green algae as a source of energy. *Plant Physiology* 127:740–748.

Melis, A., L.P. Zhang, M. Forestier, M.L. Ghirardi, and M. Seibert. 2000. Sustained photobiological hydrogen gas production upon reversible inactivation of oxygen evolution in the green alga *Chlamydomonas reinhardtii*. *Plant Physiology* 122:127–135.

Mus, F., A. Dubini, M. Seibert, M.C. Posewitz, and A.R. Grossman. 2007. Anaerobic acclimation in *Chlamydomonas reinhardtii*—Anoxic gene expression, hydrogenase induction, and metabolic pathways. *Journal of Biological Chemistry* 282:25475–25486.

Mussgnug, J.H., S. Thomas-Hall, J. Rupprecht et al. 2007. Engineering photosynthetic light capture: Impacts on improved solar energy to biomass conversion. *Plant Biotechnology Journal* 5:802–814.

Nandi, R. and S. Sengupta. 1998. Microbial production of hydrogen: An overview. *Critical Reviews in Microbiology* 24:61–84.

North, D. 1994. Economic-performance through time. *American Economic Review* 84:359–368.

Paladugu, R., M.S. Chein, S. Gardezi, and L. Wise. 2002. One hundred citation classics in general surgical journals. *World Journal of Surgery* 26:1099–1105.

Pinto, F.A.L., O. Troshina, and P. Lindblad. 2002. A brief look at three decades of research on cyanobacterial hydrogen evolution. *International Journal of Hydrogen Energy* 27:1209–1215.

Posewitz, M.C., P.W. King, S.L. Smolinski, L.P. Zhang, M. Seibert, M.L. Ghirardi. 2004. Discovery of two novel radical S-adenosylmethionine proteins required for the assembly of an active [Fe] hydrogenase. *Journal of Biological Chemistry* 279:25711–25720.

Prince, R.C. and H.S. Kheshgi. 2005. The photobiological production of hydrogen: Potential efficiency and effectiveness as a renewable fuel. *Critical Reviews in Microbiology* 31:19–31.

Rupprecht, J., B. Hankamer, J.H. Mussgnug, G. Ananyev, C. Dismukes, and O. Kruse. 2006. Perspectives and advances of biological H_2 production in microorganisms. *Applied Microbiology and Biotechnology* 72:442–449.

Schmitz, O., G. Boison, R. Hilscher et al. 1995. Molecular biological analysis of a bidirectional hydrogenase from cyanobacteria. *European Journal of Biochemistry* 233:266–276.

Schutz, K., T. Happe, O. Troshina et al. 2004. Cyanobacterial H_2 production—A comparative analysis. *Planta* 218:350–359.

Spolaore, P., C. Joannis-Cassan, E. Duran, and A. Isambert. 2006. Commercial applications of microalgae. *Journal of Bioscience and Bioengineering* 101:87–96.

Stal, L.J. and R. Moezelaar. 1997. Fermentation in cyanobacteria. *FEMS Microbiology Reviews* 21:179–211.

Takeda, T., A. Yokota, and S. Shigeoka. 1995. Resistance of photosynthesis to hydrogen-peroxide in algae. *Plant and Cell Physiology* 36:1089–1095.
Tamagnini, P., R. Axelsson, P. Lindberg, F. Oxelfelt, R. Wunschiers, and P. Lindblad. 2002. Hydrogenases and hydrogen metabolism of cyanobacteria. *Microbiology and Molecular Biology Reviews* 66:1–20.
Tamagnini, P., E. Leitao, P. Oliveira et al. 2007. Cyanobacterial hydrogenases: Diversity, regulation and applications. *FEMS Microbiology Reviews* 31:692–720.
Vignais, P.M. and A. Colbeau. 2005. Molecular biology of microbial hydrogenases. *Current Issues in Molecular Biology* 6:159–188.
Volesky, B. and Z.R. Holan. 1995. Biosorption of heavy-metals. *Biotechnology Progress* 11:235–250.
Wang, B., Y.Q. Liu, N. Wan, and C.Q. Lan. 2008. CO_2 bio-mitigation using microalgae. *Applied Microbiology and Biotechnology* 79:707–718.
Wrigley, N. and S. Matthews. 1986. Citation-classics and citation levels in geography. *Area* 18:185–194.
Wu, L.F. and M.A. Mandrand. 1993. Microbial hydrogenases—Primary structure, classification, signatures and phylogeny. *FEMS Microbiology Reviews* 104:243–270.
Yergin, D. 2006. Ensuring energy security. *Foreign Affairs* 85:69–82.
Zhang, L.P., T. Happe, and A. Melis. 2002. Biochemical and morphological characterization of sulfur-deprived and H_2-producing *Chlamydomonas reinhardtii* (green alga). *Planta* 214:552–561.

20

*Algal Biodiesel: Third-Generation Biofuel**

Amita Jain and V.L. Sirisha

CONTENTS

* *Note:* Both the authors Amita Jain and V.L. Sirisha contributed equally for the chapter.

20.1 Introduction

All nations have been confronted with the energy crisis due to depletion of finite fossil fuel reserves. Their continued consumption is at an alarming stage due to depletion of resources and emissions of greenhouse gases (GHGs) in the environment (Demirbas 2010). A constantly rising worldwide requirement of fossil fuels to satisfy demand of motor and power generation fuels causes increasing anthropogenic GHG emissions and depletion of fossil reserves. Therefore, it is highly important to think about various environmentally friendly alternate sources of energy to meet global demand. Finding sufficient supplies of clean energy for the future is one of the world's most frightening challenges and is intimately linked with global stability, economic prosperity, and quality of life. Fuel represents around 70% of the total global energy requirements, particularly in transportation, manufacturing, and domestic heating. Electricity accounts for only 30% of global energy consumption. In recent years, a lot of thrust has been put into the search for potential biomass feedstock from different sources, which can be converted to liquid as well as gas fuels for energy generations. Various biomasses from different alternate sources including forestry, agricultural, and aquatic sources have been identified for the production of different biofuels, including biodiesel, bioethanol, biohydrogen, bio-oil, and biogas.

These years, most industrial biodiesels are made from oil (triglycerides [TAGs]) of raw materials (rapeseed, sunflower, soybean, etc.). With the intent to change their physicochemical properties similar to petroleum-based diesel, TAGs are transesterified into fatty acid alkyl esters (FAAEs), which can be used in a conventional engine without modifications (Knothe 2010). On the ecological side, in addition to the ability of oleaginous plants to reduce pollutant emissions of GHGs by their capacity to trap and use the carbon dioxide (CO_2), using biodiesel also reduces net emissions of pollutants. Typically, the addition of 20% (v/v) of soybean-based biodiesel in petrodiesel reduces emissions of carbon monoxide (CO), CO_2, particulate matter (PM), and hydrocarbons (HCs) by 11%, 15.5%, 10%, and 21%, respectively (Sheehan et al. 1998a; United States Environmental Protection Agency 2002). Biodiesel production through oil-containing crops presents a route for renewable and carbon-neutral fuel production. However, current supplies from oil crops account for only 0.3% of the current demand for transport fuels. Increasing biofuel production on arable land could have severe consequences for the global food supply. On the contrary, producing biodiesel from algae is widely accepted and regarded as one of the most efficient ways of generating biofuels and also appears to represent the only current renewable source of oil that could meet the global demand for transport fuels.

20.1.1 Biofuel

A biofuel is any HC fuel that is produced from organic matter (living or once-living material) in a short period of time, days, weeks, or even months), and uses energy from carbon fixation. This contrasts with fossil fuels, which take millions of years to form. Biofuels are solid, liquid, or gaseous fuels primarily produced from biomass and can replace fossil fuels. Biofuels are gaining increased public and scientific attention, driven by factors such as oil price spikes, the need for increased energy security, concern over GHG emissions from fossil fuels, and government subsidies (Sameera et al. 2011). The increasing petroleum price and negative impact of fossil fuels on the environment are encouraging the use of lignocellulosic materials to help meet energy needs. From a theoretical viewpoint, biofuels represent a direct substitute for fossil fuels used in transportation because they are obtained from biomass, a renewable energy source. These can be integrated into already existing systems for fuel supply preparing thus the way to more performant fuels, such as hydrogen (Victor et al. 2010). Even though the majority of biofuels still are a lot more expensive than fossil fuels, their use is on the increase worldwide due to financial incentives both for processing and consumption. Encouraged by various economic measures and instruments (especially subventions), the global production of biofuels is estimated currently at over 35 million tons (Victor et al. 2010).

They are generally divided into primary and secondary biofuels (Figure 20.1): while primary biofuels such as wood and animal waste are used in an unprocessed form mainly for heating, cooking, or electricity production, secondary biofuels such as bioethanol, biomethanol, and biodiesel are produced by processing biomass and can be used in vehicles and various industrial processes (Dragone et al. 2010). The secondary biofuels can be categorized into three generations: first-, second-, and third-generation biofuel on the basis of various parameters, like type of feedstock, the type of processing technology, and their level of development (Nigam and Singh 2011).

20.1.1.1 First-Generation Biofuels

First-generation biofuels are produced directly from food crops by extracting the oils for use in biodiesel or producing bioethanol through fermentation (United Nations, 2007). Crops such as wheat and sugarcane are the most widely used feedstock for bioethanol, while oilseed rape has proven very effective for use in biodiesel. The three main types of first-generation biofuels used commercially are biodiesel, bioethanol, and biogas of which worldwide large quantities have been produced so far and for which the production process is considered established technology. Biodiesel is produced through transesterification of vegetable oils and residual oils and fats; it can serve as a full substitute for conventional diesel. Bioethanol is a substitute for gasoline, and it is used in flexi-fuel vehicles. It is derived from sugars or starch through fermentation. Bioethanol also serves as feedstock for ethyl tertiary butyl ethers (ETBEs) that blends more easily with gasoline. It can be produced through anaerobic digestion of liquid manure and other digestible feedstock. At present, biodiesel, bioethanol, and biogas are produced from commodities that are also used for food (Naik et al. 2010). Although biofuel processes have a great potential to provide a carbon-neutral route to fuel production, first-generation production systems have considerable economic and environmental limitations (Peer et al. 2008). There is hot debate over their actual benefit in reducing GHG and CO_2 emissions due to the fact that some biofuels can produce negative effects rather than energy gains, by releasing more carbon in their production than their feedstock's capture in their growth. However, the most contentious issue with first-generation biofuels is *fuel* vs. *food*. As the majority of biofuels are produced directly from food crops, the rise in demand for biofuels has led to an

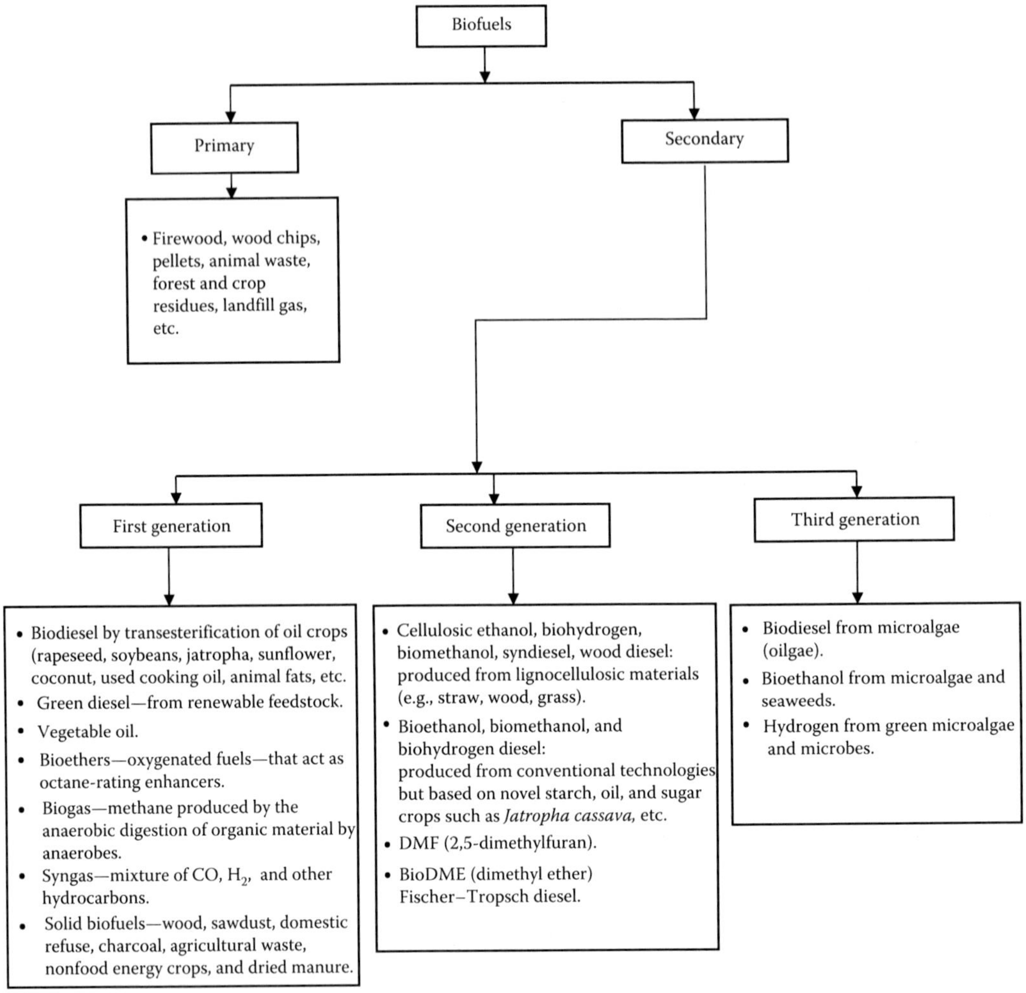

FIGURE 20.1
Classification of biofuels. (Modified from Sameera, V. et al., *J. Microbial. Biochem. Technol.*, R1, 002, 2011; Nigam, P.S. and Singh, A., *Prog. Energ. Combus. Sci.*, 37, 52, 2011.)

increase in the volumes of crops being diverted away from the global food market. This has been blamed for the global increase in food prices over the last couple of years.

First-generation biofuels are hence an expensive option for energy security taking into account total production costs excluding government grants and subsidies, as they provide only limited GHG reduction benefits, do not meet their claimed environmental benefits because the biomass feedstock may not always be produced sustainably, and are accelerating deforestation (Goldemberg and Guardabassi 2009). Hence, they have a potentially negative impact on biodiversity and compete for scarce water resources in some regions. It is proven that selected first-generation biofuels have contributed to the recent increases in world prices for food and animal feeds. However, much uncertainty exists in this regard, and estimates of the biofuel contribution in the literature range from 15% to 25% of the total price increase. Despite the culpability, competition with food crops dominates total biofuel production (IEA 2008). Therefore, now, research is focusing on other alternate renewable resources to produce biofuels to meet the required demands.

20.1.1.2 Second-Generation Fuel

Second-generation biofuels are advanced biofuels produced from lignocellulosic biomass in a more sustainable fashion, that is, carbon neutral or even sometimes carbon negative in terms of their impact on CO_2 concentrations. Biomass used to produce second-generation biofuels can be any source of organic carbon that is renewed rapidly as part of the carbon cycle. It can be plant material that refers to lignocellulosic, cheap and abundant nonfood material, agricultural residues or waste available from plants, and animal materials. Plant biomass represents one of the most abundant and underutilized biological resources on the earth and is seen as a promising source of material for fuels and raw materials. By burning it, heat and electricity can be produced. Most importantly, through production of second-generation biofuel, the amount can be increased sustainably by using biomass consisting of the residual nonedible parts of cash crops, such as stems, leaves, straw, and husks, which are left behind once the food crop has been extracted, as well as nonfood crops such as jatropha, whole-crop maize, switch grass, grass, and cereals that bear little grain, and also industrial waste like woodchips and skins and pulp from fruit pressing (Oliver et al. 2009). There is great potential in the use of plant biomass to produce liquid biofuels, and these are mainly comprised of plant cell walls, that is, composed of 75% polysaccharides (Pauly and Keegstra 2008). These polysaccharides are an important pool of latent sugars locked in by lignin, hemicelluloses, and cellulose; moreover, even in potential food crops like wheat, rice, ragi, and other monocots, sugar is tied up in straw, that is, in stems, as well as in the starch of grains.

Lignocellulosic ethanol or fuel is made by freeing the sugar molecules from cellulose using various enzymes, stem heating, and other pretreatments. These sugar molecules are then fermented as in first-generation or any other ethanol production and give lignin as by-product that can be recycled easily as a carbon-neutral fuel to produce heat and electricity. These fluids include bioalcohols, bio-oil, 2,5-dimethylfuran (BioDMF), biohydrogen, Fischer–Tropsch (FT) diesel, and wood diesel (Figure 20.2) (Demirbras 2009; Roman-Leshkov et al. 2007). FT diesel or biomass to liquid (BTL) is a full substitute of diesel. Lignocellulosic biomass is gasified to produce syngas that is in turn transformed into liquid HCs, mostly diesel and kerosene. Syngas can also be transformed into methane, that is, Bio-DME (dimethyl ether), that can replace diesel after slight adaptations (Balat 2011a). Bio-SNG (synthetic natural gas) can be used in gasoline vehicles with some modifications. However, commercial utilization for these fuels is still under development. To date, the potential of many crop residues to supply sugar feedstocks for biofuel production has not been realized. Though biofuel production from agricultural residues could only satisfy a segment of the ever-increasing demand for liquid fuel, this has generated great interest in making use of dedicated biomass crops as feedstock for biofuel production.

Producing second-generation biofuel offers greater GHG emission savings than those obtained by first-generation biofuels. Lignocellulosic biofuels can reduce GHGs by around 90% when compared with fossil petroleum; in contrast, first-generation biofuels offer savings of only 20%–70% (http://ies.jrc.ec.europa.eu/wtw.html). However, converting woody biomass into fermentable sugars and subsequently biofuel is not cost-effective, as there are many technical barriers that need to be overcome before their potential can be realized, meaning that second-generation biofuels cannot yet be produced economically on a large scale to fulfill present demand of biofuels. Therefore, third-generation biofuels derived

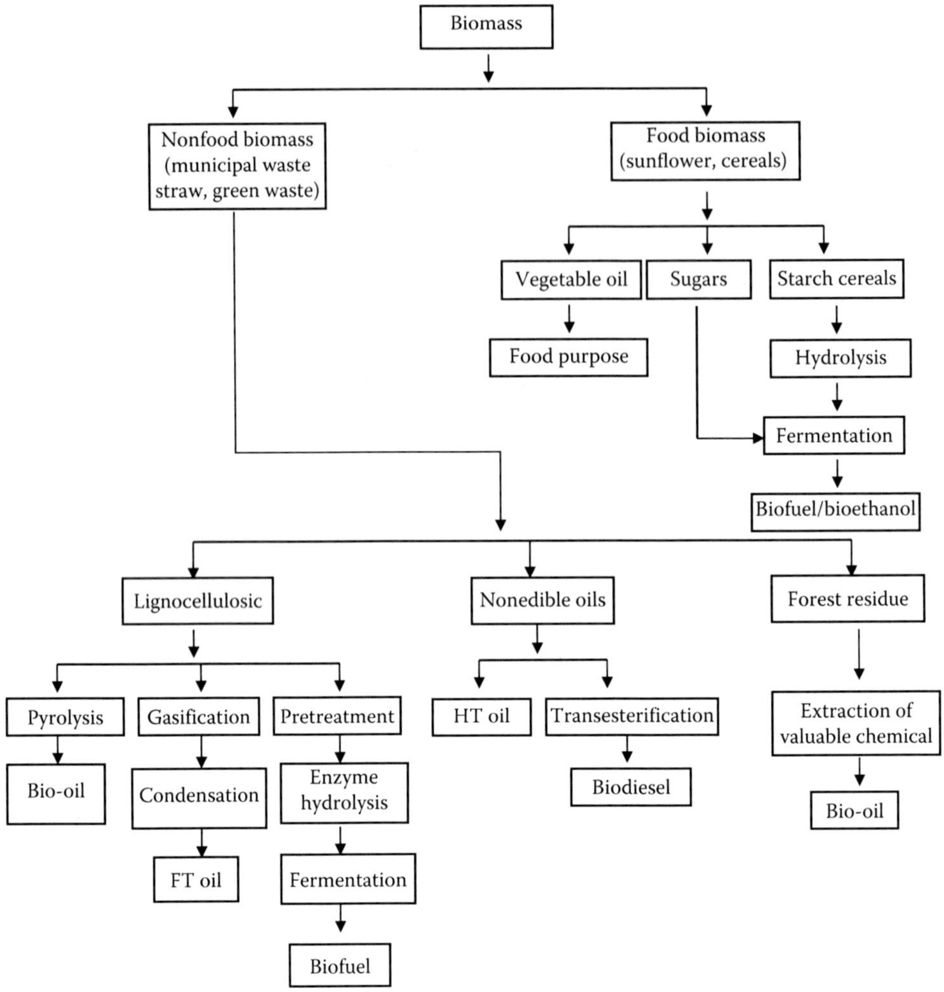

FIGURE 20.2
Second-generation biofuel production from edible and nonedible biomass. (Modified from Naik, S.N. et al., *Renew. Sustain. Energ. Rev.*, 14, 578, 2010.)

from microalgae are considered to be a viable alternative energy resource that is devoid of the major drawbacks associated with first- and second-generation biofuels (Chisti 2007; Li et al. 2008; Nigam and Singh 2011).

20.1.1.3 Third-Generation Biofuels: Algal Biodiesel

Third-generation biofuels are produced from extracting oil from algae—also known as *oilgae*. Its production is supposed to be low cost and high yielding, giving up to around 30 times the energy per unit area as can be realized from current, conventional *first-generation* and *second-generation* biofuel feedstocks. The cultivation of algal biomass for biofuel production has great promise because algae generate high energy yield and require much less space to grow than conventional feedstocks, and also algal biomass does not require fertile land. Hence, it would not compete with food and could be grown with minimal inputs and a variety of nutrients, and carbon sources can be used (Singh et al. 2012). Microalgae are able to produce 15–300 times more oil for biodiesel production than first- and second-generation biomass. When compared with conventional crops that are usually harvested annually once or twice, microalgae have a short harvesting cycle (1–10 days vary with the process), allowing multiple or continuous harvests with significantly increased yields (Schenk et al. 2008). Therefore, third-generation biofuels derived from microalgae are considered as the only source of renewable biodiesel that is capable of meeting the global demand for transport fuels.

20.1.2 Algae: As an Energy Resource

Algae are a very large and diverse group of simple, typically autotrophic organisms, ranging from unicellular to multicellular forms. Although the term algae originally referred to aquatic plants, it is now broadly used to include a number of different groups of unrelated organisms. There are seven divisions of organisms that make up the algae. They are grouped according to the types of pigments they use for photosynthesis, the makeup of their cell walls, the types of carbohydrate compounds they store for energy, and the types of flagella (whiplike structures) they use for movement. The colors of the algal types are due to their particular mixtures of photosynthetic pigments, which typically include a combination of one or more of the green-colored chlorophylls as their primary pigments (http://www.scienceclarified.com). Algae are classified into different groups and its classification is shown in Figure 20.3.

Algae are photosynthetic microorganisms that convert sunlight, water, and CO_2 to sugars, from which biological macromolecules, such as lipids and triacylglycerol, can be obtained and converted to energy. It is a carbon-neutral source of energy. Microalgae are present in all existing earth ecosystems, not just aquatic but also terrestrial, representing a big variety of species living in a wide range of environmental conditions. The most recent estimate suggests a total number of 72,500 algal species worldwide (Guiry 2012), but only a limited number, of around 30,000, have been studied and analyzed (Richmond 2004). The principal energy processes being considered for aquatic biomass are shown in Figure 20.4.

The main component of typical algal feedstock are proteins, carbohydrates, lipids, and valuable components, for example, pigments, antioxidants, fatty acids, and vitamins.

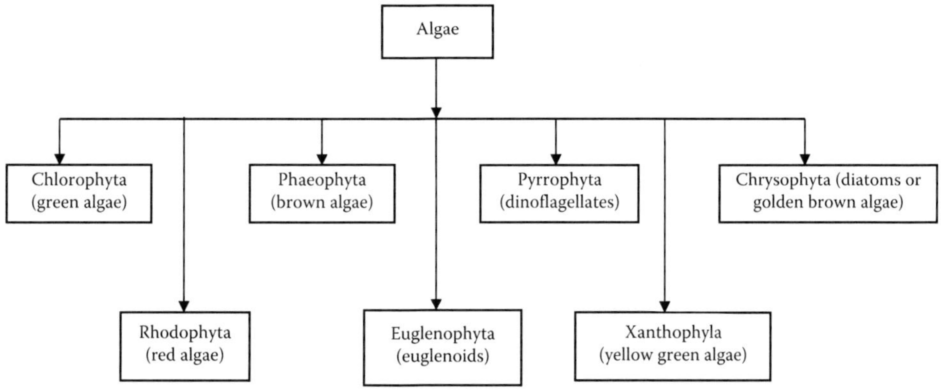

FIGURE 20.3
Classification of algae.

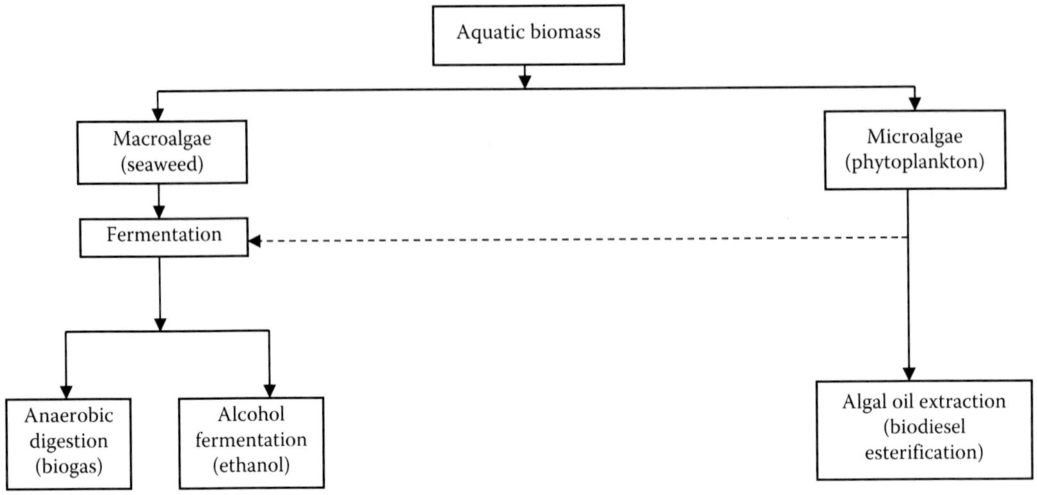

FIGURE 20.4
Various energy processes.

The protein and carbohydrate contents in various strains of microalgae are up to 50% of its dry weight, and the maximum lipid contents are around 40% on weight basis, which is quite high to consider microalgae as a potential candidate for biofuel production (Table 20.1). The triacylglycerol content of some microalgae is so high that they are promising sustainable feedstock for biodiesel production. In order to decrease CO_2 emissions coming from road transport and industrial combustions, biodiesel demand is constantly increasing, but oil crops are not able to satisfy it, and however, full replacement of diesel fuel with biodiesel coming from oil crops would not be sustainable. The use of algae for producing biofuel as a major source can be a sustainable alternative. Algae grown in ponds can be far more efficient in capturing solar energy than higher plants especially when grown in bioreactors. If industrial scale production of algae is achieved, less than six million hectares would be needed worldwide to meet the current fuel demand. This consists of less than 0.4% of arable land, which would be an achievable goal from global agriculture. The use of microalgae as feedstocks (biofuels and biomaterials) can be a suitable alternative because algae are the most efficient biological producer of oil on the planet and a versatile biomass source. In the near future, it will become one of the earth's most important renewable fuel crops (Campbell 1997) due to higher photosynthetic efficiency, higher biomass productivities, and greater lipids (20%–50% DCW) per hectare than any kind of terrestrial biomass.

20.2 Understanding Microalgae as a Potential Source of Biodiesel

Algae are the important food sources for many animals and belong to the bottom of the food chain. Moreover, they are the principal producers of oxygen on earth (Khan et al. 2009) and are one of the best sources of biodiesel production (Shay 1993). They are the highest yielding feedstock for biodiesel. Algae have the ability to produce up to 250 times the amount of oil per acre as soybeans and 7–31 times greater oil than

TABLE 20.1

Different Microalgal Species with Their Lipid Content and Productivities

Marine and Freshwater Microalgal Species	Lipid Content (% Dry Weight Biomass)	Lipid Productivity (mg/L/Day)	Volumetric Productivity of Biomass (g/L/Day)	Areal Productivity of Biomass (g/m²/Day)
Ankistrodesmus sp.	24.0–31.0	—	—	11.5–17.4
Botryococcus braunii	25.0–75.0	—	0.02	3.0
Chaetoceros muelleri	33.6	21.8	0.07	—
Chaetoceros calcitrans	14.6–16.4/39.8	17.6	0.04	—
Chlorella emersonii	25.0–63.0	10.3–50.0	0.036–0.041	0.91–0.97
Chlorella protothecoides	14.6–57.8	1214	2.00–7.70	—
Chlorella sorokiniana	19.0–22.0	44.7	0.23–1.47	—
Chlorella vulgaris	5.0–58.0	11.2–40.0	0.02–0.20	0.57–0.95
Chlorella sp.	10.0–48.0	42.1	0.02–2.5	1.61–16.47/25
Chlorella pyrenoidosa	2.0	—	2.90–3.64	72.5/130
Chlorella	18.0–57.0	18.7	—	3.50–13.90
Chlorococcum sp.	19.3	53.7	0.28	—
Crypthecodinium cohnii	20.0–51.1	—	10	—
Dunaliella salina	6.0–25.0	116.0	0.22–0.34	1.6–3.5/20–38
Dunaliella primolecta	23.1	—	0.09	14
Dunaliella tertiolecta	16.7–71.0	—	0.12	—
Dunaliella sp.	17.5–67.0	33.5	—	—
Ellipsidion sp.	27.4	47.3	0.17	—
Euglena gracilis	14.0–20.0	—	7.70	—
Haematococcus pluvialis	25.0	—	0.05–0.06	10.2–36.4
Isochrysis galbana	7.0–40.0	—	0.32–1.60	—
Isochrysis sp.	7.1–33	37.8	0.08–0.17	—
Monodus subterraneus	16.0	30.4	0.19	—
Monallanthus salina	20.0–22.0	—	0.08	12
Nannochloris sp.	20.0–56.0	60.9–76.5	0.17–0.51	—
Nannochloropsis oculata	22.7–29.7	84.0–142.0	0.37–0.48	—
Nannochloropsis sp.	12.0–53.0	37.6–90.0	0.17–1.43	1.9–5.3
Neochloris oleoabundans	29.0–65.0	90.0–134.0	—	—
Nitzschia sp.	16.0–47.0	—	—	8.8–21.6
Oocystis pusilla	10.5	—	—	40.6–45.8
Pavlova salina	30.9	49.4	0.16	—
Pavlova lutheri	35.5	40.2	0.14	—
Phaeodactylum tricornutum	18.0–57.0	44.8	0.003–1.9	2.4–21
Porphyridium cruentum	9.0–18.8/60.7	34.8	0.36–1.50	25
Scenedesmus obliquus	11.0–55.0	—	0.004–0.74	—
Scenedesmus quadricauda	1.9–18.4	35.1	0.19	—
Scenedesmus sp.	19.6–21.1	40.8–53.9	0.03–0.26	2.43–13.52
Skeletonema sp.	13.3–31.8	27.3	0.09	—
Skeletonema costatum	13.5–51.3	17.4	0.08	—
Spirulina platensis	4.0–16.6	—	0.06–4.3	1.5–14.5/24–51
Spirulina maxima	4.0–9.0	—	0.21–0.25	25
Thalassiosira pseudonana	20.6	17.4	0.08	—
Tetraselmis suecica	8.5–23.0	27.0–36.4	0.12–0.32	19
Tetraselmis sp.	12.6–14.7	43.4	0.30	—

Source: Mata, T.M. et al., *Renew. Sustain. Energ. Rev.*, 14, 217, 2010.

palm oil. It is very simple and convenient to extract oil from algae. Microalgae are the best algae for the production of biodiesel. Recent research has proven that oil production from microalgae is clearly superior to that of terrestrial plants such as rapeseed, soybean, palm, or jatropha.

Microalgae are prokaryotic or eukaryotic photosynthetic microorganisms having simple structures, which can grow rapidly and live under severe conditions due to their unicellular or simple multicellular structure (Mata et al. 2010). They can function as sunlight-driven cell factories that convert CO_2 to potential biofuels, food, feeds and high-value bioactives (Akkerman et al. 2002; Banerjee et al. 2002; Borowitzka 1999a; Ghirardi et al. 2000; Kay 1991; Melis 2002; Metting and Pyne 1986; Schwartz 1990; Shimizu 1996, 2003; Spolaore et al. 2006; Walter et al. 2005). They are capable of producing several different types of renewable biofuels and by-products. These include renewable biofuel (Demirbas et al. 2007; Subramaniam et al. 2010); biohydrogen (Chisti 2007; Fedorov et al. 2005; Illman et al. 2000; Schenk et al. 2008); HC (Bajhaiya et al. 2010; Barupal et al. 2010; Kojima and Zhang 1999); methane (Mata et al. 2010); ethanol (Bush and Hall 2009); phycocolloids and carotenoids (Azocar et al. 2011); minerals, vitamins, polyunsaturated fatty acids (PUFAs), α-linolenic, eicosapentaenoic, and doco succinic acids, belonging to the w-3 group (Khan et al. 2009; Shimizu 2003); propylene glycol, acetol, and butanol (Verma et al. 2010); biogas (Costa and De Morais 2011; Gunaseelan 1997; Ras et al. 2011); neutral lipid (Chen et al. 2011); polar lipid and carbohydrates (Moreno Garido et al. 2008); sterols, carotenoids, tocopherols, quinines, terpenes, and phytylated pyrrole derivatives such as the chlorophylls (Bisen et al. 2010); β-carotene (Singh et al. 2011); and antioxidants, antibiotics, and astaxanthin and pigments (Ratledge and Cohen 2008).

Algal biomass is one emerging source of sustainable energy. Therefore, introducing large-scale biomass could contribute to sustainable development environmentally, socially, and economically (Turkenburg 2000). Moreover microalgae commonly double their biomass within 24 h (Bajhaiya et al. 2010). During exponential growth, biomass doubling time is commonly as short as 3.5 h (Chisti et al. 2008; Patil et al. 2008), which can double their biomass in less than 2–5 days, achieving large yields, without the need for applying any pesticides, fungicides, or herbicides (Costa and De Morais 2011). They are capable of synthesizing more oil per acre than the terrestrial plants that are currently used for the fabrication of biofuels. Moreover, using microalgae for biodiesel production will not compromise the production of food, fodder, and other products derived from crops (Kozlovska et al. 2012) (Table 20.1). More than 50,000 microalgae species exist in the world, but only 30,000 species have been studied and analyzed (Richmond 2004). They can also be used as a source of wastewater remediation by removal of NH_4^+, NO_3^-, and PO_4^{3-} from a variety of wastewater runoff, for example, industrial and municipal wastewaters and concentrated animal feed operations. It produces value-added coproducts or by-products that can be used in different industrial sectors (e.g., biopolymers, proteins, polysaccharides, pigments, animal feed, fertilizer, and H_2). Algae can be grown in a suitable culture vessel (photobioreactor [PBR]) throughout the year with an annual biomass production (Naik et al. 2010).

The idea of using microalgae as a source of fuel is not new (Chisti 1980; Nagle and Lemke 1990; Sawayama et al. 1995); because of the escalating price of petroleum and, more significantly, the emerging concern about global warming that is associated with burning fossil fuels (Gavrilescu and Chisti 2005), it is now being taken seriously. The biodiesel generated from biomass from various plants and animal oils is a mixture of monoalkyl ester, which is currently obtained from transesterification of TAGs and monohydric alcohols. But this

TABLE 20.2

Comparison of Some Sources of Biodiesel

Crop	Oil Yield (L/ha)	Land Area Needed (Mha)[a]	Percent of Existing U.S. Cropping Area[a]
Corn	172	1540	846
Soybean	446	594	326
Canola	1,190	223	122
Jatropha	1,892	140	77
Coconut	2,689	99	54
Oil palm	5,950	45	24
Microalgae[b]	136,900	2	1.1
Microalgae[c]	58,700	4.5	2.5

Source: Mata, T.M. et al., *Renew. Sustain. Energ. Rev.*, 14, 217, 2010.

[a] For meeting 50% of all transport fuel needs of the US.

[b] 70% oil (by weight) in biomass.

[c] 30% oil (by weight) in biomass.

trend is changing as several companies are attempting to generate large-scale algal biomass as there are several advantages associated with it, which can be used for commercial production of algal biodiesel.

Table 20.2 clearly shows that microalgal strains with high oil content are of great interest in the search for sustainable feedstock for biodiesel production (Chisti 2007; Spolaore et al. 2006) and have the capacity to replace fossil fuels. Algae have 20%–80% of oil by weight of dry mass (Table 20.1). Moreover, microalgae grow extremely rapidly and many are exceedingly rich in oil unlike other oil crops. Oil content in some microalgae can exceed 80% by weight of dry biomass (Metting 1996; Spolaore et al. 2006). Quite common are 20%–50% of oil levels (Table 20.2). The amount of oil produced per unit volume of the microalgal broth per day depends on the growth rate of algae and the oil content of the biomass. Moreover, lipid accumulation in algae typically occurs during environmental stress, including nutrient-inadequate conditions. Biochemical studies have proposed that acetyl-CoA carboxylase (ACC), a biotin-containing enzyme that catalyzes an early step in biosynthesis of fatty acids, may be involved in the control of this lipid accumulation process. Therefore, for enhanced lipid production, this enzyme activity can be increased by genetic engineering.

As microalgal industrial culture does not directly compete with food and wood production, it can represent a great potential economic development. In fact, the use of microalgae for biofuel production would reduce deforestation, preserving the forest heritage. Moreover, the development of microalgae could favor the energetic autonomy of all countries, including developing countries. Thus, the industrial production of microalgae could be a sustainable solution to energetic, environmental, and food problems. There are multiple disadvantages associated with using microalgae for fuel production. One disadvantage is the low biomass concentration in microalgal culture (a result of limit of light penetration). Another is the small size of algal cells that makes their harvesting relatively costly. And finally, microalgal farming needs greater care than conventional agricultural crops. Nevertheless, despite these disadvantages, microalgae have vast potential as efficient primary producers of biomass if proper technology is developed, thereby helping solve global demand for transport fuels.

20.3 Technological Development

20.3.1 Strain Selection

Microalgae can be found in a large range of places where light and water are present including ocean, soils, lakes, rivers, and ice (Deng et al. 2009). Approximately 22,000–26,000 species of microalgae exist, of which only a few have been identified for successful commercial application (Norton et al. 1996). The best performing microalgal strain can be obtained by screening of a large range of naturally available isolates, and their efficiency can be improved by selection, adaptation, and genetic engineering (Singh et al. 2011a). Isolation of appropriate microalgae from the natural environment is the first critical step in developing oil-rich strains and can then be further exploited in engineered systems for the production of biodiesel feedstock (Doan et al. 2011). Depending on species, microalgae produce many different kinds of lipids, HCs, and other complex oils (Banerjee et al. 2002; Guschina and Harwood 2006; Metzger and Largeau 2005). Among algal strains, the amount of lipid produced varies within a vast range (4%–80% dry weight basis), and the variation is as a result of the environmental conditions. For example, some microalgae species like *Botryococcus braunii* or *Schizochytrium* sp. can contain up to 80% of their dry weight of lipids (Deng et al. 2009). These species can produce lipid yields by acre up to 770 times higher than oleaginous plants (colza, sunflower, etc.). A list of protein, carbohydrate, and lipid and composition of some microalgae are shown in Table 20.3.

TABLE 20.3

Chemical Composition of Some Microalgae on the Basis of % Dry Matter

Algal Species	Proteins	Carbohydrates	Lipids
Anabaena cylindrica	43–56	25–30	4–7
Botryococcus braunii	8–17	8–20	25–75
Chlamydomonas reinhardtii	48	17	21
Chlorella pyrenoidosa	57	26	2
Chlorella vulgaris	51–58	12–17	14–22
Dunaliella bioculata	49	4	8
Dunaliella salina	57	32	6
Euglena gracilis	39–61	14–18	14–20
Isochrysis sp.	31–51	11–14	20–22
Neochloris oleoabundans	20–60	20–60	35–54
Porphyridium cruentum	28–39	40–57	9–14
Prymnesium parvum	28–45	25–33	22–38
Scenedesmus dimorphus	8–18	21–52	16–40
Scenedesmus obliquus	50–56	10–17	12–14
Scenedesmus quadricauda	48	17	21
Spirogyra sp.	6–20	33–64	11–21
Spirulina maxima	60–71	13–16	6–7
Spirulina platensis	46–63	8–14	4–9
Synechococcus sp.	63	15	11
Tetraselmis maculata	52	15	3

Sources: Um, B.H. and Kim, Y.S., *J. Ind. Eng. Chem.*, 15, 1, 2009; Sydney, E.B. et al., *Bioresour. Technol.*, 101, 5892, 2010.

Microalgae, which can grow heterotrophically, use exogenous carbon source as chemical energy that is stored as lipid droplets (Konur 2012). For example, *Chlorella protothecoides*, which was cultivated heterotrophically, has accumulated higher lipids (about 55% dry weight) when compared to the one grown photoautotrophically (14% of dry weight). Nitrogen (N) starvation is another mechanism to alter the lipid content in microalgae. In the green alga *Haematococcus pluvialis*, because of N deficiency, there is inhibition of cell cycle and production of cellular components. However, the rate of lipid accumulation is higher, which leads to the accumulation of oil in starved cells and also enhances the accumulation of antioxidant pigment astaxanthin (Boussiba 2000). Both these responses help the alga to survive under stress conditions.

Another important approach while selecting microalgal strains is to combine their growth with CO_2 bio-mitigation. This is due to the fact that microalgae have much higher CO_2 fixing abilities as compared to agricultural and aquatic plants and conventional forestry (Borowitzka 1999b; Li et al. 2008). This approach of microalgal biofuel production becomes more attractive when combined with fixing industrial exhaust gases (flue gas) and integrating the cultivation of algae with wastewater treatment.

For profitable commercialization of microalgae, apart from species selection, genetic and metabolic engineering tools can be adapted. By manipulating metabolic pathways, cellular function can be redirected for the production of desired products and even expand the processing capabilities of microalgae. Metabolic engineering allows direct control over the organism's cellular machinery through mutagenesis or the introduction of transgenes. The development of a number of transgenic algal strains boasting recombinant protein expression, engineered photosynthesis, and enhanced metabolism encourages the prospects of designer microalgae (Rosenberg et al. 2008). Production of algal oils requires an ability to inexpensively produce large quantities of oil-rich microalgal biomass.

20.3.2 Production of Microalgal Biomass

Microalgal cultivation can be done in open-culture systems such as lakes or ponds and in closed-culture systems that are highly controlled, such as PBRs. The photosynthetic growth of microalgae requires light, CO_2, water, organic salts, and temperature of 20°C–30°C. As large quantities of algal biomass are required for the production of microalgal biodiesel, it is better to produce biomass using free sunlight in order to minimize the expense.

Large-scale production of microalgal biomass generally uses continuous culture during daylight. In this method, fresh culture medium is fed continuously at a constant rate, while the same quantity of microalgal broth is withdrawn continuously (Molina Grima 1999). As feeding ceases during night because of low light, the mixing of broth should continue to prevent the settling of the biomass (Molina Grima 1999). Around 25% of the biomass produced during the day may be lost during the night because of respiration. This loss of biomass mainly depends on the light level, growth temperature, and temperature during night. The only feasible methods for large-scale production of microalgal biomass can be achieved by growing them in raceway ponds (Terry and Raymond 1985) or PBRs (Molina Grima 1999; Sánchez Mirón et al. 1999; Tredici 1999).

20.3.2.1 Raceway Ponds

A raceway pond is a shallow artificial pond used in the cultivation of algae. The pond is divided into a rectangular grid, with each rectangle containing one channel in the shape

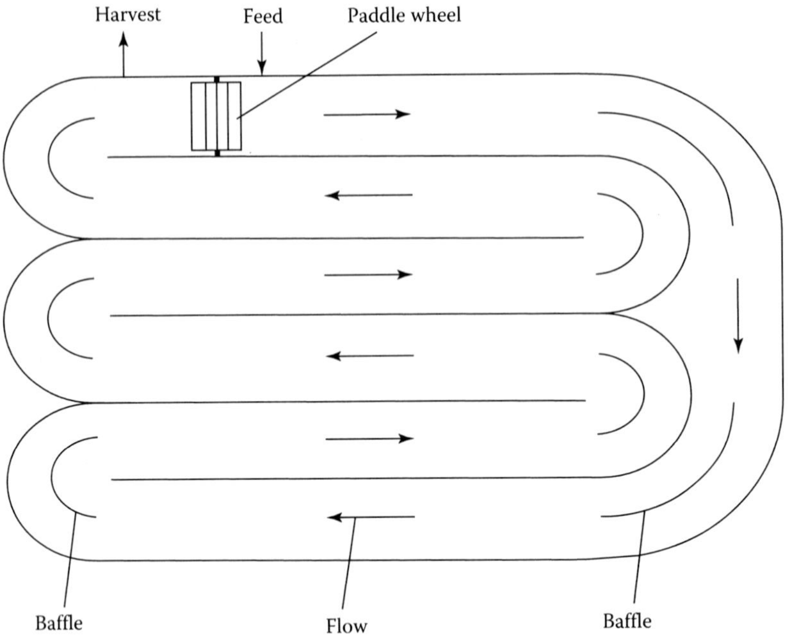

FIGURE 20.5
Schematic representation of raceway pond. (From Surendhiran, D. and Vijay, M., *Res. J. Chem. Sci.*, 2(11), 71, 2012.)

of an oval, like an automotive raceway circuit. In this, the algae, water, and nutrients circulate around a track. A raceway pond is a closed-loop recirculation channel that is about 0.3 m deep (Figure 20.5). In order to prevent sedimentation, a paddle wheel is used for mixing and circulating the algal biomass. Flow is guided around bends by baffles placed in the channel. Raceway channels are built in compacted earth or concrete and may be lined with white plastic. During daytime, the culture is fed continuously in front of the paddle wheel where the flow begins (Figure 20.5). Broth is harvested behind the paddle wheel, on completion of the circulation loop. The paddle wheel operates continuously to prevent sedimentation. Raceway ponds have been used since the 1950s for mass culture of microalgae. Open raceway pond is a cost-effective method of growing microalgae, and in commercial raceway ponds, the biomass production is in excess of 0.5 g/L (Rawat et al. 2010). The advantage of this method is that municipal wastewater can be used as a medium for cultivation with added benefit of bioremediation. If the raceway pond is located near a power plant, cheaply available flue gas can be used to speed up the photosynthetic rates in the pond, or pure CO_2 can be bubbled into the pond (Rawat et al. 2010; Tsukahara and Sawayama 2005). Some drawbacks of cultivation in raceways are: temperature fluctuates within a diurnal cycle and seasonally, and cooling is achieved only by evaporation; the water loss by evaporation is significant. As there is significant losses to atmosphere, raceways use CO_2 less efficiently than PBRs. Contamination with unwanted algae and microorganisms that feed on algae will affect the productivity. In raceways, the biomass concentration remains low because it is poorly mixed and cannot sustain dark zone. Terry and Raymond (1985) further discussed about raceway ponds and other open-culture systems for producing microalgae. Although raceways are low cost, they have a low biomass productivity compared with PBRs.

20.3.2.2 Photobioreactors

Fully closed PBRs provide opportunities for production of monoseptic culture of a greater variety of algae than is possible in open systems. PBRs have been successfully used for producing large-scale microalgal biomass (Carvalho et al. 2006; Molina Grima 1999; Pulz 2001; Tredici 1999). PBRs with tubular solar collectors are the most promising (Molina Grima 1999; Tredici 1999) for producing algal biomass on a large scale needed for biofuel production.

A tubular PBR consists of an array of straight transparent tubes that are usually made of plastic or glass. This tubular array, or solar collector, is where the sunlight is captured for photosynthesis (Figure 20.6). The solar collector tubes are generally 0.1 m or less in diameter. Tube diameter is limited because light does not penetrate too deeply in the dense culture broth that is necessary for ensuring a high biomass productivity of the PBR. Microalgae, required nutrients, and water are circulated from a reservoir such as a feeding vessel to the solar collector and back to the reservoir. It is usually operated as a continuous culture during daylight. In continuous culture, fresh medium is fed continuously at a constant rate, and the same quantity of microalgal broth is removed. PBRs require cooling during daytime. The loss of biomass because of respiration can be reduced by decreasing the temperature at night. To utilize sunlight to the maximum extent, the tubes of the solar collector are generally placed horizontally flat on the surface (Figure 20.6). The ground underneath the solar collectors is pointed with white sheets of plastic to increase the reflectance (Banerjee et al. 2002; Chisti 1999; Miron et al. 1999), which in turn increases the total light received by the tubes. To prevent sedimentation of biomass, a highly turbulent flow is maintained. This flow is produced either using a gentle air lift pump (Garcia et al. 2001, 2007; Mazzuca et al. 2006; Sanchez et al. 2003) or a mechanical pump. The typical biomass produced by a PBR is nearly 30 times the concentration of biomass produced by raceway ponds.

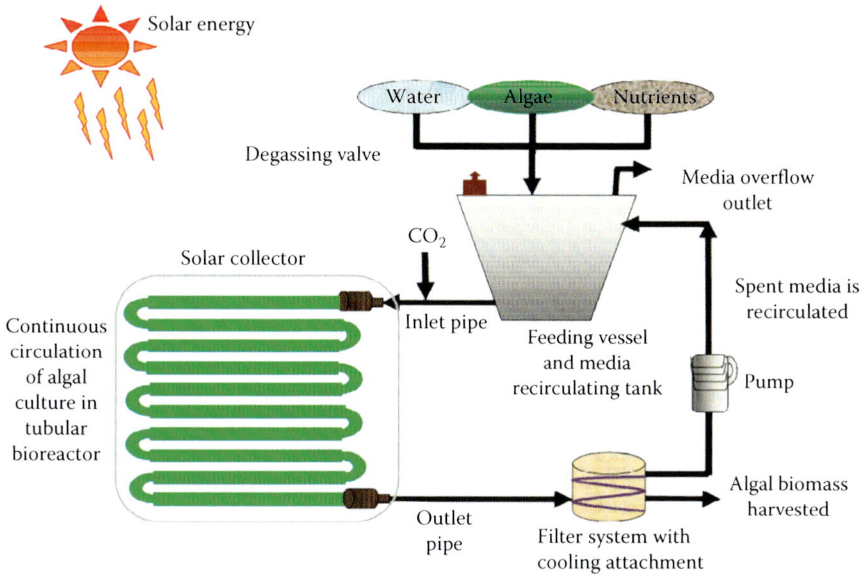

FIGURE 20.6
Schematic representation of PBR. (From Bajhaiya, A.K. et al., *Asian J. Exp. Biol. Sci.*, 1, 728, 2010.)

20.3.2.3 Comparison of Raceways and Tubular Photobioreactors

The microalgal biomass produced from PBR and raceway methods is shown in Table 20.4. This comparison is for an annual production level of 100 tons of biomass in both cases. The amounts of CO_2 consumed by both methods are almost identical (Table 20.4). They are compared for optimal combinations of biomass productivity and concentration that have been actually achieved in large-scale PBRs and raceways. The amount of oil produced per hectare was significantly higher in PBRs as compared to raceway ponds (Table 20.4). This is because the volumetric biomass productivity of PBRs is more than 13-fold greater in comparison with raceway ponds (Table 20.4). Both raceway and PBR production methods are technically feasible. Production facilities using PBRs and raceway units of dimensions similar to those in Table 20.4 have indeed been used extensively in commercial operations (Lorenz and Cysewski 2003; Molina Grima 1999; Pulz 2001; Spolaore et al. 2006; Terry and Raymond 1985; Tredici 1999).

20.3.3 Harvesting Techniques to Increase Algal Biomass Yield

Highly efficient harvesting biomass from cultivation broth is most important for mass production of biodiesel from microalgae. The selection of harvesting technique depends on the type of microalgae used and their properties like density, size, and value of the desired products. The harvesting of algal biomass could be the most energy-demanding process due to its concentration, small size, and surface charge. In pond cultures, the cells are very dilute as compared to bioreactor cultures and natural oceanic cultures. Out of many methods that are used for harvesting, the major techniques that are being used are centrifugation, flocculation, filtration, screening, gravity sedimentation, immobilization, flotation, and electrophoresis (Chen et al. 2011; Mutanda et al. 2011). The choice of method

TABLE 20.4

Comparison of PBR and Raceway Production Methods

	PBR Facility	Raceway Ponds
Annual biomass production (kg)	100	100,000
Volumetric productivity (kg/m^3/day)	1.535	0.117
Areal productivity (kg/m^2/day)	0.048[a] 0.072[c]	0.035[b]
Biomass concentration in broth (kg/m^3)	4.00	0.14
Dilution rate (day^{-1})	0.384	0.250
Area needed (m^2)	5681	7828
Oil yield (m^3/ha)	136.9[d] 58.7[e]	99.4[d] 42.6[e]
Annual CO_2 consumption (kg)	183,333	183,333
System geometry	132 parallel tubes/unit; 80 m long tubes; 0.06 m tube diameter	978 m^2/pond; 12 m wide, 82 m long; 0.30 m deep
Number of units	6	8

Source: Chisti, Y., *Biotechnol. Adv.*, 25, 294, 2007.
[a] Based on facility area.
[b] Based on actual pond area.
[c] Based on projected area of PBR tubes.
[d] Based on 70% by wt oil in biomass.
[e] Based on 30% by wt oil in biomass.

depends on the size and density of algae, target product, and the production process used to get the final product.

20.3.3.1 Centrifugation

For commercial-scale algal biomass production, centrifugation is quite feasible and is the process that could separate the algal biomass from the full volume of liquid medium in a single step, hence increasing the efficiency of harvest. Centrifugation is the most rapid, efficient, and universal technique of harvesting most of the microalgal biomass (Chen et al. 2011; Rawat et al. 2010). A report has shown that laboratory centrifugation tests were conducted on pond effluent at 500–1000*g*, which recovered 80%–90% of microalgae within 2–5 min (Chen et al. 2011). Nevertheless, the amount of energy needed to power such a colossal centrifuge is significant, and as the algal cells are very delicate, the speed applied to the algae during centrifugation must be accurate to concentrate the mass but not burst the cells that would release the valuable components found inside the algae into the liquid (Sanchez et al. 2003). However, for large-scale production, this method is not economically feasible as it is energy intensive and time consuming (Rawat et al. 2010).

20.3.3.2 Flocculation

Flocculation is the process in which scattering particles are aggregated together to form large particles for settling. It is the means by which the algal cells gather and form a mass by the addition of a chemical or organic substance. There are two types of flocculation such as autoflocculation, which occurs as a result of precipitation of carbonate salts with algal cells, and chemical flocculation in which adding chemicals to microalgal culture induces flocculation (Chen et al. 2011). Care should be exercised when picking the substance to induce flocculation, since some of the substances can contaminate the final product. For large-scale harvesting, flocculation is the first step to aggregate the cells, so increasing their particle size eases subsequent centrifugation, filtration, or sedimentation steps. In wastewater industry, alum ($Al_2(SO_4)_3$) in particular has been widely used for this purpose. Autoflocculation is also termed as bioflocculation (Pittman et al. 2011). Using cationic polymers or the addition of an alkali substance to increase the pH (Rawat et al. 2010), flocculation may be achieved. Li et al. showed that the addition of organic carbon substances, such as glucose, acetate, or glycerine, induces significant flocculation and increases efficiency of algal biomass recovery as seen in Table 20.5. The lipid content

TABLE 20.5

Comparison of Effects of Three Different Additives on Algal Flocculation

	Substrate Concentration (c)	0.1 g/L		0.5 g/L	
Substrate	**Mixing Time (t)**	**6 h**	**24 h**	**6 h**	**24 h**
Acetate	Recovery efficiency (%)	52 (51–53)	88 (75–96)	52 (50–53)	90 (81–95)
	Concentration factor	149 (146–151)	240 (215–264)	133 (106–150)	100 (89–106)
Glucose	Recovery efficiency (%)	53 (44–70)	90 (83–96)	48 (35–60)	93 (89–97)
	Concentration factor	167 (146–182)	236 (232–239)	181 (141–242)	109 (89–121)
Glycerine	Recovery efficiency (%)	45 (36–54)	94 (91–96)	50 (38–61)	93 (93–97)
	Concentration factor	179 (143–216)	204 (192–228)	171 (152–210)	131 (100–185)

Source: Li, Y. et al., *Biotechnol. Prog.*, 24, 815, 2008.

of the algal cells was not affected by adding these additives. Moreover, these are relatively low cost and induce flocculation in a reliable way.

20.3.3.3 Flotation

Flotation is a gravity separation process that uses air to bring the algal population to the surface of the liquid medium. A compressor is used to dissolve the air flotation that supersaturates the liquid with air and then moves the medium into a pressurized chamber where the precipitation of air into small bubble occurs, which adheres algal cells to bring them to surface. Flocculants neutralize or reduce the negative charge on the algal surface and prevent their sticking together in the suspension (Molina Grima 1999). Particles with a diameter of less than 500 µm can be captured using flotation by collision between a bubble and a particle and the subsequent adhesion of bubble and particle. This process is believed to be the best until Wiley et al. showed that suspended air flotation is most efficient in bringing the cells to the surface of the liquid medium. In suspended air flotation, air bubbles are created using surfactants, making the process less expensive with fewer mechanical components, and use less energy than dissolved air flotation. The algal biomass can be skimmed or collected easily on the top of the liquid medium if flocculation and flotation are combined, thereby not affecting the cultivation process, and the liquid media can be reused to reduce the cost of nutrients and water demands. Kim et al. (2011) screened a number of flocculants to harvest *Scenedesmus* sp. and concluded that flocculation using consecutive treatment with calcium chloride and ferric chloride and a bioflocculant from the culture broth of *Paenibacillus polymyxa* AM49 was found to be effective for the flocculation of high-density *Scenedesmus* sp. and also suggested that a flocculated medium can be effectively reused as a growth-supporting medium without compromising the algal growth and biomass yield, thereby significantly reducing the cost of biodiesel production from algae.

20.3.3.4 Gravity Sedimentation

Gravity sedimentation is a basic method of harvesting biomass. It is highly energy efficient and works for various types of algae (Rawat et al. 2010). It is effective in separating large and small organisms (Mutanda et al. 2011). Increased microalgal harvesting using sedimentation can be achieved using lamella separators and sedimentation tanks (Chen et al. 2011).

20.3.3.5 Filtration

Filtration is the most commonly used method for solid–liquid separation (Rawat et al. 2010). In this technique, the liquid medium is filtered by using vacuum force through a porous membrane while collecting the algal biomass. This process is only suitable for large microalgal strains (>70 µm) and unsuitable for strains of a diameter less than 30 µm. Mohn et al. showed that filtration processes can achieve a concentration factor of 245 times the original concentration for *Coelastrum proboscideum* to produce a sludge with 27% solids. The disadvantage of using this process is that small microalgae cannot be recovered. Moreover, in this process, the entire volume of liquid medium must be passed through the filter, and for a commercial-scale algae farm, that is, a massive volume of liquid. The process is also very tedious since the filtration must be carefully monitored and the algal biomass must be scraped off at intervals to avoid clogging the filter (Kanellos 2010).

20.3.3.6 Electrolytic Method

This is another potential approach to separate algae without the need to add any chemicals. In this process, an electric field drives charged algae to move out of the solution. Water electrolysis generates hydrogen that adheres to the microalgal floc and carries them to the surface. There are several advantages using electrochemical methods, including environmental compatibility, versatility, energy efficiency, safety, selectivity, and effectiveness (Chen et al. 2011).

20.3.3.7 Immobilization

Many microorganisms, including some of the microalgae, have the tendency to attach to surfaces and grow on them (Moreno Garido et al. 2008). Immobilization is the artificial attachment or encapsulation in alginates or similar substances. Immobilization of microalgal cultures provides a ready-to-retrieve alternative for biomass retrieval (Rawat et al. 2010). Immobilized biomass can be used for biofuel conversion by thermal or fermentative means. For example, HC-rich microalgae, *B. braunii* and *B. protuberance*, which are immobilized in alginate beads, yielded a significant increase in chlorophyll, carotenoids, dry weight, and lipids during the stationary and resting growth phases compared to free living cells. In addition, as compared to free cells, immobilized cells are more stable.

20.3.3.8 Ultrasonic Aggregation

Another method of harvesting microalgal biomass is ultrasonic aggregation. This method was successfully used to achieve 92% separation efficiency and a concentration factor of 20 times (Tsukahara and Sawayama 2005).

20.3.4 Oil Extraction from Algal Biomass for Biodiesel Production

Under favorable conditions, algae synthesize fatty acids for esterification into glycerol-based membrane lipids, which constitute about 5%–20% of their cell dry weight. However, under unfavorable conditions, many algae change their lipid biosynthetic pathways to the formation of neutral lipids (20%–50% DCW), mostly in the form of TAGs. For biodiesel production from microalgae, these neutral lipids have to be extracted from their biomass. Algal oil extraction is one of the expensive processes that determine the sustainability of microalgal-based biodiesel.

For biodiesel production, algal biomass is an interesting sustainable feedstock. As compared to many oil-producing crops, algae can produce 30 times more oil per acre. Algal cell has a thick cell wall that makes extraction of oil complicated. Before oil extraction, algae have to be dried out completely. There are various approaches available for extraction of algal oil, such as chemical extraction through different organic solvents, enzymatic extraction, mechanical extraction using hydraulics or a screw, ultrasonic extraction, expeller/press, and supercritical extraction using CO_2 above its standard temperature and pressure.

20.3.4.1 Chemical Extraction Methods

Algal oil extraction using chemical methods can be done using various solvents like hexane, benzene, and ether. Hexane is the most popular and inexpensive solvent used for algal oil extraction only for lipids of low polarity. Benzene is highly carcinogenic and hence cannot be used and may be replaced by toluene. The commonly used methods are described as follows.

20.3.4.1.1 Hexane Solvent Method

Hexane solvent is used along with mechanical extraction methods, first pressing the oil. After the oil is extracted using an expeller, the remaining product is mixed with hexane to extract all the oil content. The oil and hexane can be separated by distillation. Various solvents like ethanol (96%) and a hexane–ethanol (96%) mixture can be used. About 98% quantitative extraction of purified fatty acids can be obtained using these solvents.

20.3.4.1.2 Soxhlet Method

The Soxhlet method is the most commonly used solvent extraction method for the extraction of oil from various plants and algal strains. By this method, oil and fat can be extracted from solid material by repeated washings using an organic solvent. Usually n-hexane or petroleum ether is used under reflux in a special glassware Soxhlet extractor. This method has several benefits in that a large amount of oil can be extracted using limited solvent, and it is cost-effective and becomes very economical if used on a large scale. Though this process has many advantages, there are some drawbacks like poor extraction of polar lipids, a long time for extraction, and hazards of boiling solvents. Even then, this is the commonly used method of oil extraction in most laboratories.

20.3.4.1.3 Folch Method

In this method, the tissue is homogenized with chloroform/methanol (2:1 v/v) for 1.5 h. The aqueous phase is recovered by centrifugation or filtration, and the solvent is washed with 0.9% NaCl solution. The lipids present in the lower chloroform phase are evaporated using vacuum in a rotary evaporator.

20.3.4.2 Enzymatic Extraction

In enzymatic extraction, water is used as solvent with the cell wall–degrading enzymes to facilitate an easy and mild fractionation of oil, proteins, and hulls. In an algal cell, content is surrounded by a sturdy cell wall that has to be opened to release proteins and oil. So when the cell is opened by enzymatic degradation, downstream processing makes fractionation of the components possible to a degree that cannot be reached when using a conventional technique like mechanical pressing. This is the biggest advantage of enzymatic extraction over other extraction methods. But this method is expensive when compared with solvent-based extraction processes. The high cost of extraction serves as a limitation factor for large-scale utilization of this process.

20.3.4.3 Expeller Press

In this process, algae are dried to retain their oil content and can be pressed out with an oil press. Commercial manufacturers use a combination of chemical solvents and mechanical press for oil extraction.

20.3.4.4 Ultrasonic Extraction

This method involves intense sound waves that propagate into the liquid medium resulting in alternating high and low pressure cycles. These ultrasonic waves create bubbles in a solvent material; when these bubbles collapse near the cell walls, it creates shock waves and liquid jets that cause the cell walls to break and release their contents into the solvent. After the oil is dissolved in cyclohexane, the pulp/tissue is filtered out. The solution is distilled to separate the oil from the hexane. This method can be used with dry or wet algae. For wet algae, it is necessary to extract part of the water from the mash before extraction. Ultrasonication not only improves the extraction of oil from the algal cells but also helps in the conversion into biodiesel. The drawback of this method is that it is only cost-effective for large-scale application.

20.3.4.5 CO_2 Supercritical Extraction

In supercritical fluid/CO_2 extraction, CO_2 is liquefied under pressure and heated to a point where it has the properties of both gas and liquid (supercritical point 31°C 74 bar). This liquefied fluid then acts as a solvent for extracting the oil. The algal material is brought in contact with this supercritical solvent in an extraction vessel. Due to its high diffusion rates and gas-like low viscosity, CO_2 penetrates into the smallest pores of the material. In a separate vessel, CO_2 is depressurized, and substances of interest are collected with less solvent residues as compared to other extraction methods. CO_2 is the most used supercritical solvent because the compounds can be obtained without contamination by toxic organic solvents and without thermal degradation. Advantages associated with $SCCO_2$ extraction include lineable solvating power, low toxicity of the supercritical fluid, favorable mass transfer equilibrium due to the intermediate diffusion/viscosity properties of the fluid, and the production of solvent-free extract. It has many advantages in that the biomass residues that remain after extraction of the oil could be used as high-protein animal feed. The algal biomass residue that remains can also be used to produce biomass by anaerobic digestion. In addition, the remaining biomass fraction can be used as a protein feed for livestock. This gives further value to the process and reduces waste. Though CO_2 supercritical extraction is the most advanced oil extraction method, it has some disadvantages including an elevated pressure requirement and high capital investment for equipment.

20.3.5 Converting Algal Oil to Biodiesel

Oils derived from algae have high viscosity, higher flash point (>200°C), high molecular weight, and low volumetric heating values compared to diesel fuels. In conventional diesel engines, using such bio-oils leads to a number of problems due to significant difference in the automation, injection, and combustion characteristics of these diesel engines. The major problem in substituting diesel fuels with TAGs is mostly associated with their low viscosity, high viscosities, and polyunsaturated character (Balat 2011b). Hence, refinement of bio-oils into quality fuels is essential (Lin et al. 2011). Various efforts have been made to convert vegetable oil derivatives that have properties similar to that of the hydrocarbon based diesel fuels. Currently, four methods used for production of biodiesel include pyrolysis, thermal liquefication, microemulsification, and transesterification (chemical and enzymes).

20.3.5.1 Pyrolysis

The method of pyrolysis of microalgae was developed in Germany in 1986. It is the thermal decomposition of algal biomass in an oxygen-deprived environment and high temperature or in the presence of much less oxygen than required for thermal combustion (Mutanda et al. 2011).

The pyrolysis process consists of both simultaneous and successive reactions when organic material is heated in a nonreactive atmosphere. Thermal decomposition of organic components in biomass starts at 350°C–550°C and goes up to 700°C–800°C in the absence of air/oxygen. The long chains of carbon, hydrogen, and oxygen compounds in biomass break down into smaller molecules in the form of gases, condensable vapors (tars and oils), and solid charcoal under pyrolysis conditions. Rate and extent of decomposition of each of these components depends on the process parameters of the reactor temperature, biomass heating rate, pressure, reactor configuration, feedstock, etc.

Depending on the thermal environment and the final temperature, pyrolysis will yield mainly biochar at low temperatures less than 450°C when the heating rate is quite slow, and mainly gases at high temperatures greater than 800°C with rapid heating rates. At an intermediate temperature and under relatively high heating rates, the main product is bio-oil. So this process involves reducing TAGs chemically to FAAEs by the application of extreme heat. This method is suitable for microalgae oil extraction because of lower temperature and high quality of oil obtained.

There are two types of pyrolysis process, namely slow and fast. In slow pyrolysis, the algal biomass is associated with liquid fuels at low temperature (675–775 K), and/or gas at high temperature (Mutanda et al. 2011). The drawbacks of slow pyrolysis include expensive equipment for separation of various fractions. Also the product obtained similar to gasoline containing sulfur, which makes it less eco-friendly (Bisen et al. 2010). The process yields an average particle size of 0.8 mm (Olguin 2003; Walter et al. 2005).

Compared to slow pyrolysis, fast pyrolysis is a new technology that produces biofuels in an air-deprived environment at atmospheric pressure with a relatively low temperature (450°C–550°C) and high heating rate (10^3°C/s–10^4°C/s) as well as a short gas residence time to crack into short-chain molecules and be cooled to liquid quickly. Using fast pyrolysis, biodiesel can be produced in less time and requires less energy than slow pyrolysis.

20.3.5.2 Thermal Liquefaction

In thermal liquefaction, biodiesel can be produced directly from wet algal biomass (Mutanda et al. 2011; Olguin 2003). Thermal liquefaction is a high pressure (5–20 MPa), low temperature (300°C–350°C) process aided by a catalyst in the presence of hydrogen to yield bio-oil. The microalgae are converted into oily substances under the influence of high pressure and temperature. Liquefaction is performed using a conventional stainless steel autoclave with mechanical stirring. The autoclave is charged with algal cells, and then N is introduced to purge the residual air. Pressurized N at 2–3 MPa is required to control evaporation of water. The reaction starts by heating the autoclave at 250°C–400°C. This temperature is kept constant for 60 min and then it is cooled. Once the reaction is finished, the mixture is extracted with dichloromethane in order to separate the oil fraction from the aqueous phase. The final product is dark-brown viscous oil. Liquefaction of microalgae has resulted in the production of 30%–50% of dry weight of oil depending on the species used (Walter et al. 2005). In microalgae species like *B. braunii*, *Dunaliella tertiolecta*, and *Spirulina platensis* by thermochemical liquefaction, around 64%, 42%, and 30%–80% of dry weight basis of oil and fuel properties of bio-crude oil (30–45.9 MJ/kg) have been produced. This indicates that the thermal conversion of biomass to biofuel is an attractive method for liquid fuel production. The most important advantage of this technology is conversion of wet biomass into useful energy (Harun et al. 2010; Huber et al. 2006; Mutanda et al. 2011; Rawat et al. 2010) as there is no drying process involved. However, the drawbacks are that reactors for thermochemical liquefaction and fuel-feed systems are complex and expensive.

20.3.5.3 Transesterification of Algal Oil

Out of the various methods that are developed for extraction of algal oil, transesterification is the most promising solution to the high viscosity problem, and the biodiesel produced by this method is almost similar to conventional diesel that can be blended in any proportion. Crude microalgal oil is high in viscosity, thus requiring conversion to lower-molecular-weight constituents in the form of FAAEs. Transesterification has been demonstrated as the simplest and most efficient route for biodiesel production in large quantities, versus less eco-friendly, costly, and eventual low yield methods of pyrolysis and micro-emulsification. Therefore, transesterification has become the production method of choice (Ghaley et al. 2010).

Transesterification is a multiple step process, including three reversible steps, where TAGs are converted to diglyceride, and then diglycerides are converted to monoglycerides, and monoglycerides are then converted to esters (biodiesel) and glycerol (by-product) (Mata et al. 2010). Each conversion step yields one FAAE molecule, giving a total of three FAAEs per TAG molecule. The process is described by the equation given in Figure 20.7 (Ghaley et al. 2010).

Transesterification is a reversible reaction of oil or fat (which is composed of TAG) with an alcohol to form FAAEs and glycerol. The stoichiometry ratio of alcohol and oil for the reaction is 3:1 (Bajhaiya et al. 2010; Mishra et al. 2012; Rawat et al. 2010). Methanol, ethanol, propanol, butanol, and amyl alcohol can be used in the transesterification process. Among these alcohols, methanol and ethanol are frequently used. Methanol is preferred for commercial development because of its low cost and its physical and chemical advantages (polar and shortest-chain alcohol) (Bisen et al. 2010). Transesterification of algal oil to biodiesel can be done in a number of ways such as using an alkali catalyst, acid catalyst, enzyme catalyst, or heterogeneous catalyst or using alcohol in its supercritical state (Bajhaiya et al. 2010). The reaction can be catalyzed both by homogeneous and heterogeneous catalysts. Homogenous catalysts include alkalis and acids.

20.3.5.3.1 Acid Transesterification of Microalgal Oil

Acid transesterification is mostly suitable for oils containing high levels of fatty acids. The acid catalysis is carried out in flasks and heated to the reaction temperature in a water bath. The mixture consists of microalgal oil, methanol, and concentrated sulfuric acid. The reaction is depicted in Figure 20.8. The mixture was heated for a specific time, allowed to cool, and left separated in the settling vessel to obtain two layers. The upper oil layer (biodiesel) was separated, washed with petroleum ether, and then washed with hot water (50°C). The biodiesel is obtained by evaporating the ether solution.

The green alga *Scenedesmus obliquus* was studied using an acid catalysis procedure with 60:1 molar ratio of methanol to oil, at 30°C and 100% sulfuric acid catalyst concentration. It was found that the accumulation of lipid started at the early growth phase, and maximum accumulation (12.7% dcw) was observed at stationary phase. Using methanol and acid catalysis transesterification, the most abundant composition of microalgal oil transesterified is $C_{19}H_36O_2$, which is suggested to accord with the standards of biodiesel. However, the reaction is slow. Speeding up the acid-catalyzed reaction requires an increase in

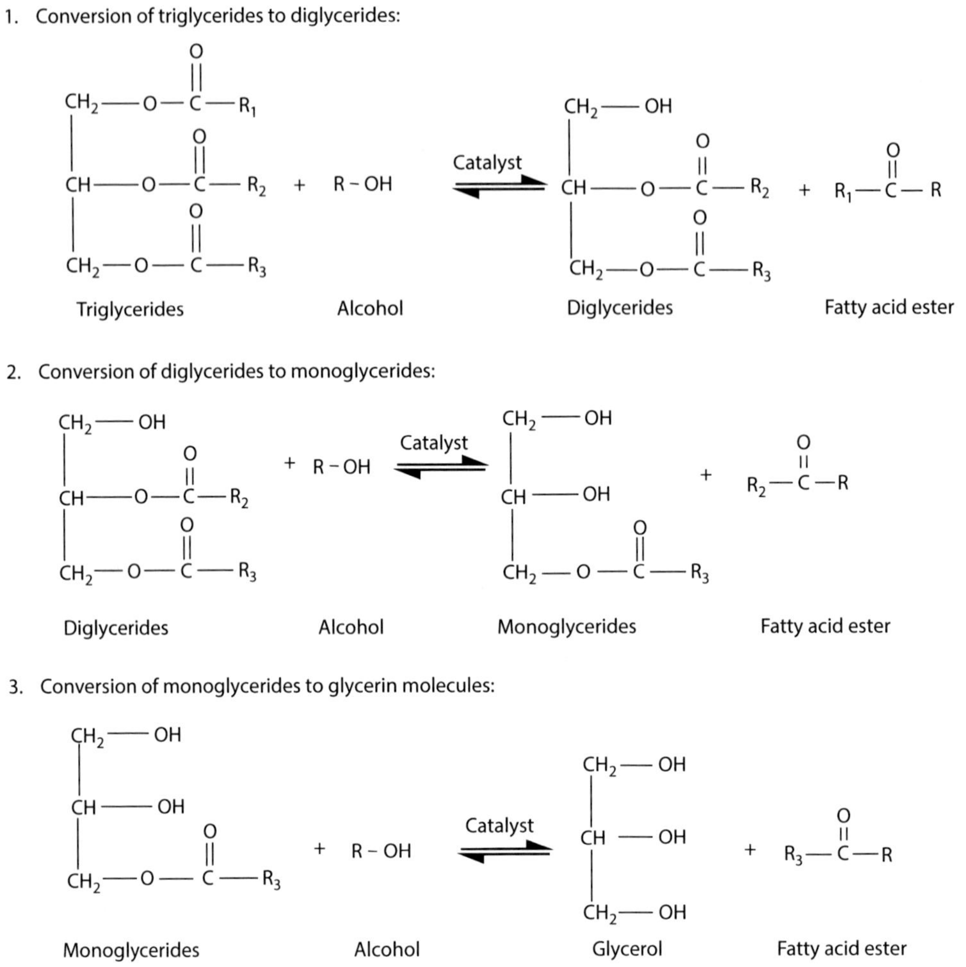

FIGURE 20.7
Transesterification stepwise reaction. (From Surendhiran, D. and Vijay, M., *Res. J. Chem. Sci.*, 2(11), 71, 2012.)

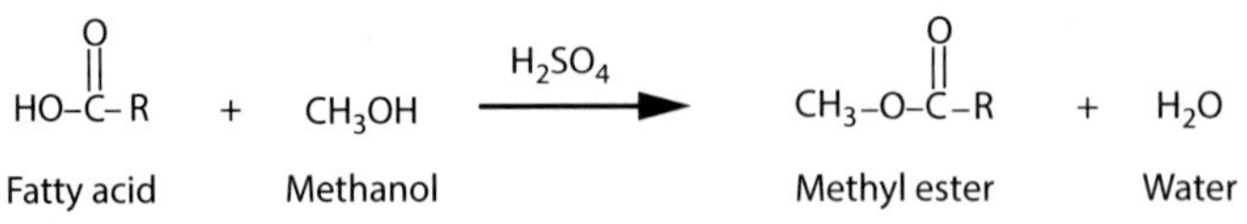

FIGURE 20.8
Acid catalyst transesterification reaction.

temperature and pressure, making it prohibitively expensive at large scale (Rawat et al. 2010). Sulfuric, sulfonic, and hydrochloric acids are usually used as catalysts in the acid-catalyzed reaction.

20.3.5.3.2 Alkyl Transesterification of Microalgal Oil

The alkali-catalyzed transesterification is about 4000 times faster than the acid-catalyzed reaction (Fukuda et al. 2001). Alkali-catalyzed transesterification is carried out at approximately 60°C under atmospheric pressure, as methanol boils off at 65°C at atmospheric pressure. Under these conditions, the reaction takes about 90 min to complete. A higher temperature can be used in combination with higher pressure. As methanol and oil do not mix, the reaction mixture contains two liquid phases. As methanol is least expensive, it has been used widely for this reaction. To prevent yield loss due to saponification reactions (i.e., soap formation), the oil and alcohol must be dry, and the oil should have a minimum of free fatty acids. Biodiesel is recovered by repeated washing with water to remove glycerol and methanol (Banerjee et al. 2002). This process of biodiesel production is found to be most efficient and least corrosive of all the processes as the reaction rate is reasonably high even at a low temperature of 60°C. This alkyl catalysis is limited by the free fatty acid content. Free fatty acid content in the region of 20%–50% is responsible for saponification during the base-catalyzed transesterification. Saponification is responsible for consumption of the base catalyst as well as making downstream recovery difficult. Alkalis such as sodium and potassium hydroxide are commonly used as commercial catalysts at a concentration of about 1% by weight of oil. Other alkaline catalysts include carbonates, methoxides, sodium ethoxide, sodium propoxide, and sodium butoxide (Ghaley et al. 2010). An advantage of this method is the higher conversion rate in short reaction time.

20.3.5.3.3 Enzymatic Transesterification of Microalgal Oil

Biocatalysts are becoming increasingly important in biodiesel preparation as it is believed that these catalysts will eventually have the ability to outperform chemical catalysts. Enzymes are biological catalysts that allow many chemical reactions to occur within the homeostasis constraints of a living system (Ghaley et al. 2010). Biocatalysts are naturally occurring enzymes like lipases that are used to catalyze some reactions such as hydrolysis of glycerol, acidolysis, and alcoholysis, and they have also been used as catalysts for transesterification and esterification reactions (Bisen et al. 2010). A number of lipases have been studied from microorganisms such as *Arthrobacter, Achromobacter, Alcaligenes, Burkholderia, Bacillus, Chromobacterium,* and *Pseudomonas* (Sharma et al. 2012). Enzyme-catalyzed transesterification reactions have been extensively used in production of drug intermediates, biosurfactants, and designer fats (Ghaley et al. 2010).

Lipases are found in all living organisms and are broadly classified as intracellular and extracellular and also classified based on their sources such as microorganisms, animal as pancreatic lipases, and plant as papaya latex, oat seed lipase, and castor seed lipase. The selection of a lipase for lipid modification is based on the nature of modification desired, for example, position-specific modification of triacylglycerol, fatty acid–specific modification, modification by hydrolysis, and modification by synthesis (direct synthesis and transesterification). The microorganisms, *Candida antarctica, C. rugosa, Pseudomonas cepacia, P. fluorescens, Rhizomucor miehei, Rhizopus chinensis, R. oryzae,* and *Thermomyces lanuginosa,* have produced the most effective lipases for transesterification (Ghaley et al. 2010).

20.3.5.3.4 Esterification Using Immobilized Extracellular Enzyme

Microbial lipases are mostly intracellular, produced by solid-state fermentation or submerged fermentation. For producing extracellular lipase, an important purification step is required, which is a very complex process, and it depends on the origin and structure of lipase. Moreover, for the large-scale production of extracellular lipases, they should be economical, fast, easy, and efficient. The majority of immobilized lipases that are commercially available are extracellular. The most commonly used ones are Novozym 435, which is from *C. antarctica*; Lipozyme RM IM, lipase produced by *R. miehei*; and Lipozyme TL IM, from *T. lanuginosus* (Ghaley et al. 2010).

20.3.5.3.5 Esterification Using Immobilized Whole-Cell Biocatalyst (Intracellular Lipases)

The most important challenge in generating biodiesel is the cost of enzymes. Hence, use of whole cells is being implemented for the production. The intracellular lipases are the ones that can be utilized while still present in the cells. This is advantageous, for this process does not involve the purification of the enzyme, which is cost consuming. For bulk production of biodiesel, direct use of compact cells for intracellular production of lipases or fungal cells immobilized within porous biomass support particles as a whole biocatalyst represents an attractive process. Immobilization techniques can be classified under four general techniques: (1) adsorption, (2) cross-linking, (3) entrapment, and (4) encapsulation. Use of intracellular lipases as compared to extracellular lipases slows down the transesterification process but increases the conversion efficiency. Using intracellular lipases has a lot of advantages like easy handling, its enhanced thermal and chemical stability, and reusability (Sharma and Kanwar 2012). Some of the microorganisms that are used as whole-cell biocatalysts are *C. antarctica*, *R. chinensis*, *R. oryzae*, and rarely *Saccharomyces cerevisiae*. Both whole cells and extracellular lipases should be immobilized so that they resemble ordinary solid phase catalysts that are conventionally used in chemical reactions (Ghaley et al. 2010).

20.3.5.3.5.1 Advantages of Using Lipase Enzyme in Biodiesel Production

1. The immobilized residue can possibly be reused and regenerated, as it can be left in the reactor unaffected.
2. High concentration of the enzymes is to be used in the reactors that makes for longer activation of lipases.
3. Enzymes have high thermal stability due to the native state.
4. Immobilization of lipase protects it from the solvent that could be used in the reaction and that will prevent all the enzyme particles getting together.
5. Product separation will be much easier using this catalyst.
6. No soap formation and the ability to esterify both TAGs and FFAs in one step without the need of a washing step.
7. Ability to handle large variation in raw material quality.
8. Second-generation raw materials like waste cooking oils, animal fat, and similar waste fractions, with high FFA and water content, can be catalyzed with complete conversion to alkyl esters with a significantly condensed amount of wastewater.
9. Works under milder conditions (which lead to less energy consumption) with lower alcohol to oil ratio than chemical catalysts, which is quite economical.
10. They are robust and versatile enzymes that can be produced in bulk because of their extracellular nature in most producing system.
11. Most of the lipases show significant activity to catalyze transesterification with long- or branched-chain alcohols, which are very hard to convert to fatty acid esters in the presence of conventional alkaline catalyst.
12. Product and by-product separation in downstream process is extremely easier.

13. Immobilization of lipases on a carrier has facilitated the reuse of enzymes after removal from the reaction mixture, and when the lipase is in a packed bed reactor, no separation is necessary after transesterification.
14. Higher thermostability and short-chain alcohol-tolerant abilities of lipase make it very convenient for use in biodiesel production.

20.3.6 Biodiesel Purification and Resultant By-Products

The by-products and biodiesel obtained from microalgae have to be purified to increase the biodiesel production. The main separation processes use hot water (50°C) (Li et al. 2007), organic solvents such as hexane (Halim et al. 2010; Wiltshire et al. 2000), and water–organic solvent for a liquid–liquid separation (Couto et al. 2010; Lewis et al. 2000; Samorì et al. 2010). There are three main means of purifying microalgal biodiesel: *water washing*, *dry washing*, and *membrane extraction* (Leung et al. 2010). The main by-products produced by microalgae and vegetable oil during biodiesel production (Leung et al. 2010) include unreacted lipids, water, alcohol, chlorophyll, metals, and glycerol.

Glycerol is the most interesting by-product of biodiesel production. Biodiesel production will generate about 10% (w/w) glycerol as the main by-product. In other words, every gallon of biodiesel produced generates around 1.05 lb of glycerol. It was projected that the world biodiesel market would reach 37 billion gallons by 2016, which implied that approximately 4 billion gallons of crude glycerol would be produced (Anand and Saxena 2011). Since purified glycerol is a high-value commercial chemical with thousands of uses, the crude glycerol presents great opportunities for new applications. For that reason, more attention is being paid to the utilization of crude glycerol from biodiesel production in order to defray the production cost of biodiesel and to promote biodiesel industrialization on a large scale. Hence, the glycerol by-product during biodiesel production can be transformed into value-added products using many paths including chemical, thermochemical, or biological conversion.

Glycerol can be oxidized or reduced and can produce various chemical added-value products like propylene glycol, acrylic acid, propionic acid, isopropanol, propanol, allyl alcohol, and acrolein, but only some of these products are used in terms of market or profitability (Johnson and Taconi 2007).

The anaerobic fermentative production of 1,3-propanediol is the most promising option for biological conversion of glycerol. Mu et al. (2006) showed that in fed-batch cultures of *Klebsiella pneumoniae*, crude glycerol could be used directly for the production of 1,3-propanediol. The differences between the final concentrations of 1,3-propanediol were small for crude glycerol from the methanolysis of soybean oil by alkali (51.3 g/L) and lipase catalysis (53 g/L). This implied that the composition of crude glycerol had little effect on the biological conversion, and a low fermentation cost could be expected. Additionally, an incorporated bioprocess that combined biodiesel production by lipase with microbial production of 1,3-propanediol by *K. pneumoniae* was developed in a hollow-fiber membrane. The bioprocess avoided glycerol inhibition on lipase and reduced the production cost (Mu et al. 2008).

Furthermore, FT fuel can be produced from glycerol at low temperature (225°C–300°C) by catalytic processes (Soares et al. 2006) or transformed into hydrogen (H_2) by catalytic (generally nickel, platinum, or ruthenium) or noncatalytic reforming (Vaidya and Rodrigues 2009).

20.4 Major Limiting Factors in Commercialization of Microalgal Biodiesel

Oil crops are a potential renewable and carbon-neutral source of biodiesel that can replace petroleum fuels. But it did not find its way to become a top competitor for petroleum fuels because of very limited supply in comparison to global demand as well as high cost. Therefore, biodiesel is a bright and attractive hope to both investors and consumers. Because biodiesel from oil crops, waste cooking oil, and animal fat cannot fulfill a small fraction of the existing demand for transport fuels, microalgae appears to be the only source of renewable biodiesel that is capable of meeting the global demand for transport fuels (Hossain et al. 2008). But, at the same time, it has to be thought upon that algal fuels can be produced in sufficient quantity to genuinely replace petroleum fuels. Obstacles to commercialization of algal fuels need to be understood and addressed for any future commercialization (Chisti 2013). First of all, CO_2 availability is the major impediment to any significant production of biodiesel (Chisti 2013). Production of each ton of algal biomass requires at least 1.83 tons of CO_2 (Chisti 2007). Almost all commercial algal culture setups buy CO_2 that contribute substantially (~50%) to the cost of producing the biomass. Hence, the amount of CO_2 available is a major cause that is hampering large scale biodiesel production and commercialization. Algal culture for biodiesel production is not feasible unless CO_2 is available and free (Chisti 2007). Moreover, nutrient supply in the form of nitrogen (N) and phosphorus (P) to grow algal biomass is another hindrance in algal biodiesel commercialization. Although P is in ample supply (Cordell et al. 2009; Gilbert 2009) there isn't sufficient nitrogen even for agricultural food crops. Hence nitrogen cannot be used for any significant scale of algal biodiesel production. Fixation of environmental N by the Haber–Bosch process (Travis 1993) requires an extensive amount of energy.

In addition to enabling reclamation and reuse of nutrients, the biogas produced by anaerobic digestion can be burnt to supply all the electrical power that is needed for production of algal biomass and its separation from the water (Chisti 2008; Harun et al. 2011). Algae have to be dried for efficient extraction of oil, but the process of drying is not feasible, as net energy recovery in the oil would be very less or can be even negative. Presently, techniques used for anaerobic digestion are only suited to algal biomass of low salinity and not to marine (high salinity) microalgal biomass. Therefore, researchers have to invent a new anaerobic digestion technology for biogas production and nutrient recovery for growth of the algal biofuel or biodiesel market.

Another constraint that has to be considered here is the supply of freshwater. At present, algal fuel would cost about $8 a gallon. That same gallon would also require 350 gallons of water to be produced (Keune 2012). Therefore, the use of marine algae is the only option for making biofuels, but even cultivation of these does not avoid the requirement of freshwater as it is needed to compensate some evaporative losses that depend on climatic conditions. Production of algal biofuels requires inputs of energy obtained from fossil fuels (Chisti 2013). Energy ratio, the ratio of the energy contained in the oil to the fossil energy required for making it, is an important measure that determines whether production of oil is worthwhile (Chisti 2008). An energy ratio of 1:1 means a nil recovery of energy in the oil. Ideally, an energy ratio of at least 7 is recommended to make algal biodiesel commercial (Chisti 2008). This can be possible by an efficient recovery process of N and P fertilizer by anaerobic digestion, through minimizing freshwater and fossil energy input (Sompech et al. 2012; Wongluanget al. 2013). Algal biomass production process is the next important strategy;

it has to be economical and easier. On the basis of several available reports, the Algae 2020 study has reported the estimated costs of producing algal oils and algal biodiesel today are between \$9 and \$25 per gallon in ponds and \$15–\$40 in PBRs. Algal oil production has many processes, that is, production, harvesting, extraction, and drying systems, and reducing the number of processes in algal biofuel production is essential to provide easier, better, and lower-cost systems. This can be addressed with the advent of cheaper PBRs in the next few years (Singh and Gu 2010).

Finally, but importantly, the coproduction of some more valuable fractions and their marketing is also important for the success. Even with algal species with up to 50% oil content, the additional 50% of the biomass remains. These biomass fractions contain valuable protein for livestock, poultry, and fresh-feed additives (Singh and Gu 2010). Remaining fractions of algal biomass still contain some important compounds that can be used to produce by-products (cleaners, detergents, plastic, etc.). This coproduct marketing will be a way to successful commercialization of algal biodiesel. Manipulation of metabolic pathways through genetic engineering can redirect cellular function toward the synthesis of products and even expand the processing capabilities of microalgae (Singh et al. 2011). The algal microrefineries can avoid the harvesting, extraction, and refining systems by excreting lipids directly from the cells using nonlethal extraction known as milking. Apart from these, biofuels like bioethanol, biomethane and other valuable products can be cogenerated to make commercialization process a profitable venture. Such methods have the capability to reduce the production cost significantly, and the future of algal biodiesel will be bright. Interest of consumers and investors in addition to the present-hour need in commercial production of biodiesel from microalgae is so strong that economical and simplified methods of viable production at large scale will be possible in a certain quantum of time. As Al Darzins, coauthor of a Department of Energy report on algae's viability, said, "The path of algal biofuels commercialization will not be totally dependent on any one unit operation" or technology but rather on the industry's ability to string together or "integrate robust and scalable technology solutions."

20.5 Conclusion

Microalgae comprise a vast group of photosynthetic, auto-/heterotrophic cell factories that convert CO_2 to potential biofuels, foods, feeds, and high-value additives that are also considered as energy crops. Microalgae offer great potential as a sustainable feedstock for the production of third-generation biofuels, such as biodiesel and bioethanol. Microalgal biodiesel has the potential to replace petroleum, and it is technically feasible and economically competitive.

However, several important scientific and technical barriers remain to be overcome before the large-scale production of microalgal-derived biofuels can become commercial reality. Microalgal strain selection, biomass production, harvesting, drying, processing, water sources, nutrient and growth inputs, and advances in engineering of PBRs are important areas that must be optimized. The use of the biorefinery concept and genetically manipulated algal strains for more lipid production, less water requirement, etc., will further make algal biofuel sustainable. The cultivation of algal biomass also saves our environment from air and water pollution and minimizes the waste disposal problems by utilizing wastewater and fuel gases for algal growth. The short and simple life cycle, energy balance, biofuel yield per unit area, carbon balance, N and P balance, water availability, land use, and nutrient sources are very important factors to decide the sustainability of algal biofuels including biodiesel. Considerations of these areas may lead enhanced cost-effectiveness and, therefore, successful commercial implementation of microalgal biofuel.

References

Akkerman, I., M. Janssen, J. Rocha, R.H. Wijffels. 2002. Photobiological hydrogen production: Photochemical efficiency and bioreactor design. *Int J Hydrogen Energ* 27:1195–1208.

Anand, P., R.K. Saxena. 2011. A comparative study of solvent-assisted pretreatment of biodiesel derived crude glycerol on growth and 1,3-propanediol production from *Citrobacter freundii*. *N Biotechnol* 29:199–205.

Azocar, L., G. Ciudad, H.J. Heipieper, R. Munoz, R. Navia. 2011. Lipase-catalyzed process in an anhydrous medium with enzyme reutilization to produce biodiesel with low acid value. *J Biosci Bioeng* 112:583–589.

Bajhaiya, A.K., S.K. Mandotra, M.R. Suseela, K. Toppa, S. Ranade. 2010. Algal biodiesel: The next generation biofuel for India: Review article. *Asian J Exp Biol Sci* 1:728–739.

Balat, M. 2011a. Production of bioethanol from lignocellulosic materials via the biochemical pathway: A review. *Energ Convers Manag* 52(2):858–875.

Balat, M. 2011b. Challenges and opportunities for large-scale production of biodiesel. *Energ Educ Sci Tech A* 27:427–434.

Banerjee, A., R. Sharma, Y. Chisti, U.C. Banerjee. 2002. *Botryococcus braunii*: A renewable source of hydrocarbons and other chemicals. *Crit Rev Biotechnol* 22:245–279.

Barupal, D.K., T. Kind, S.L. Kothari, D.Y. Lee, O. Fiehn. 2010. Hydrocarbon phenotyping of algal species using pyrolysis-gas chromatography mass spectrometry. *BMC Biotechnol* 10:40.

Bisen, P.S., B.S. Sanodiya, G.S. Thakur, R.K. Baghel, G.B.K.S. Prasad. 2010. Biodiesel production with special emphasis on lipase-catalyzed transesterification. *Biotechnol Lett* 32:1019–1030.

Borowitzka, M.A. 1999a. Commercial production of microalgae: Ponds, tanks, tubes and fermenters. *J Biotechnol* 70:313–321.

Borowitzka, M.A. 1999b. Pharmaceuticals and agrochemicals from microalgae. In: Cohen, Z., ed., *Chemicals from Microalgae*. Taylor & Francis, London, U.K., pp. 313–352.

Boussiba, S. 2000. Carotenogenesis in the green alga *Haematococcus pluvialis*: Cellular physiology and stress response. *Physiol Plant* 108:111–117.

Bush, R.A., K.M. Hall. 2009. Process for the production of ethanol from algae. US patent no. 7,507,554 B2, Issued March 24.

Campbell, C.J. 1997. *The Coming Oil Crisis*. Multi-Science Publishing Company and Petroconsultants, Essex, U.K.

Carvalho, A.P., L.A. Meireles, F.X. Malcata. 2006. Microalgal reactors: A review of enclosed system designs and performances. *Biotechnol Prog* 22:1490–1506.

Chen, C.Y., K.L. Yeh, R. Aisyah, D.J. Lee, J.S. Chang. 2011. Cultivation, photobioreactor design and harvesting of microalgae for biodiesel production: A critical review. *Bioresour Technol* 102:71–81.

Chisti, Y. 1980. An unusual hydrocarbon. *J Ramsay Soc* 27–28:24–26.

Chisti, Y. 1999. Shear sensitivity. In: *Encyclopedia of Bioprocess Technology: Fermentation. Biocatalysis, and Bioseparation*, Flickinger, M.C. and Drew, S.W., (eds.), Vol. 5, Wiley, pp. 2379–2406.

Chisti, Y. 2007. Biodiesel from microalgae: A review. *Biotechnol Adv* 25:294–306.

Chisti, Y. 2008. Response to Reijnders: Do biofuels from microalgae beat biofuels from terrestrial plants? *Trends Biotechnol* 26:351–352.

Chisti, Y. et al. 2008. Biodiesel from microalgae beats bioethanol. *Trends Biotechnol* 26:126–131.
Chisti, Y.J. 2013. Constraints to commercialization of algal fuels. *J Biotechnol* 167:201–214.
Cordell, D., J.O. Drangert, S. White. 2009. The story of phosphorus: Global food security and food for thought. *Global Environ Change* 19:292–305.
Costa, J.A.V., M.G. De Morais. 2011. The role of biochemical engineering in the production of biofuel from microalgae. *Bioresour Technol* 102:2–9.
Couto, R.M., P.C. Simões, A. Reis, T.L. Da Silva, V.H. Martins, Y. Sánchez-Vicente. 2010. Supercritical fluid extraction of lipids from the heterotrophic microalga *Crypthecodinium cohnii*. *Eng Life Sci* 10:158–164.
Demirbas, A. 2007. Progress and recent trends in biofuels. *Prog Energ Combust Sci* 33:1–18.
Demirbas, A. 2009. Political economic and environmental impacts of biofuels: A review. *App Energ* 86(1):108–117.
Demirbas, A. 2010. Use of algae as biofuel sources. *Energ Convers Manage* 51:2738–2749.
Deng, X., Y. Li, X. Fei. 2009. Microalgae: A promising feedstock for biodiesel. *Afr J Microbiol Res* 3:1008–1014.
Doan, T.T.Y., B. Sivaloganathan, J.P. Obbard. 2011. Screening of marine microalgae for biodiesel feedstock. *Biomass Bioenergy* 35:2534–2544.
Dragone, G., B. Fernandes, A.A. Vicente, J.A. Teixerira. 2010. Third generation biofuels from microalgae. *Curr Res Technol Educ Topics Appl Microbiol Microbial Biotechnol* 2:1355–1366.
Fedorov, A., S. Kosourov, M.L. Ghirardi, M. Seibert. 2005. Continuous hydrogen photoproduction by *Chlamydomonas reinhardtii* using a novel two-stage, sulfate-limited chemostat system. *Appl Biochem Biotechnol* 121–124:403–412.
Fukuda, H., A. Kondo, H. Noda. 2001. Biodiesel fuel production by transesterification of oils. *J Biosci Bioeng* 92:405–416.
Garcia Camacho, F., E. Molina Grima, A. Sanchez Miron, V. Gonzalez Pascual, Y. Chisti. 2001. Carboxymethyl cellulose protects algal cells against hydrodynamic stress. *Enzyme Microb Technol* 29:602–610.
Garcia Camacho, F., J.G. Rodriguez, A.S. Miron, M.C.C. Garcia, E.H. Belarbi, Y. Chisti, E.M. Grima. 2007. Biotechnological significance of toxic marine dinoflagellates. *Biotechnol Adv* 25:176–194.
Gavrilescu, M., Y. Chisti. 2005. Biotechnology—A sustainable alternative for chemical industry. *Biotechnol Adv* 23:471–499.
Ghaley, A.E., D. Dave, M.S. Brooks, S. Budge. 2010. Production of biodiesel by enzymatic transesterification: Review. *Am J Biochem Biotechnol* 6:54–76.
Ghirardi, M.L., J.P. Zhang, J.W. Lee, T. Flynn, M. Seibert, E. Greenbaum. 2000. Microalgae: A green source of renewable H_2. *Trends Biotechnol* 18:506–511.
Gilbert, N. 2009. The disappearing nutrient. *Nature* 461:716–718.
Goldemberg, J., P. Guardabassi. 2009. Are biofuels a feasible option? *Energ Policy* 37(1):10–14.
Guiry, M. 2012. How many species of algae are there? *J Phycol* 48(5):1057–1063.
Gunaseelan, V.N. 1997. Anaerobic digestion of biomass for methane production: A review. *Biomass Bioenergy* 13:83–114.
Guschina, I.A., J.L. Harwood. 2006. Lipids and lipid metabolism in eukaryotic algae. *Prog Lipid Res* 45:160–186.
Halim, R., B. Gladman, M.K. Danquah, P.A. Webley. 2010. Oil extraction from microalgae for biodiesel production. *Bioresour Technol* 102:178–185.
Harun, R., M. Davidson, M. Doyle, R. Gopiraj, M. Danquah, G. Forde. 2011. Techno economic analysis of an integrated microalgae photobioreactor, biodiesel and biogas production facility. *Biomass Bioenergy* 35:741–747.
Harun, R., M. Singh, G.M. Forde, M.K. Danquah. 2010. Bioprocess engineering of microalgae to produce a variety of consumer products. *Renew Sustain Energ Rev* 14:1037–1047.
Hossain, S., A. Sallen, A. Wasrulhag Boyce, P. Chowdhury, M. Naqiuddin. 2008. Biodiesel fuel production from algae as a renewable energy. *Am J Biochem Biotechnol* 4(3):320–254. http://www.scienceclarified.com.

Huber, G.W., S. Iborra, A. Corma. 2006. Synthesis of transportation fuels from biomass: Chemistry, catalysts, and engineering. *Chem Rev* 106:4044–4098.

Illman, A.M., A.H. Scragg, S.W. Shales. 2000. Increase in Chlorella strains calorific values when grown in low nitrogen medium. *Enzyme Microbial Technol* 27:631–635.

International Energy Agency. 2008. *World Energy Outlook*. OECD/IEA, Paris, France, pp. 1–569.

Johnson, D.T., K.A. Taconi. 2007. The glycerin glut: Options for the value-added conversion of crude glycerol resulting from biodiesel production. *Environ Prog* 26:338–348.

Kanellos, M. 2010. Fermentation or photosynthesis: The debate in algae fuel. *CBS Interactive*, January 28, 2008. April 10, 2010.

Kay, R.A. 1991. Microalgae as food and supplement. *Crit Rev Food Sci Nutr* 30:555–573.

Keune, N. 2012. Algae: Fuel of the future. *National Review online*.

Khan, S.A., Rashmi, Z. Mir Hussain, S. Prasad, U.C. Banerjee. 2009. Prospects of biodiesel production from microalgae in India. *Renew Sustain Energ Rev* 13:2361–2372.

Kim, D.G., H.J. La, C.Y. Ahn, Y.H. Park, H.M. Oh. 2011. Harvest of *Scenedesmus* sp. with bioflocculant and reuse of culture medium for subsequent high-density cultures. *Bioresour Technol* 102:3163–3168.

Knothe, G. 2010. Biodiesel and renewable diesel: A comparison. *Prog Energ Combust Sci* 36(3):364–373.

Kojima, E., K. Zhang. 1999. Growth and hydrocarbon production of microalga *Botryococcus braunii* in bubble column photobioreactors. *J Biosci Bioeng* 87:811–815.

Konur, O. 2012. The evaluation of the research on the biodiesel: A scientometric approach. *Energ Educ Sci Technol A* 28:1003–1014.

Kozlovska, J., K. Valancius, E. Petraitis. 2012. Sapropel use as a biofuel feasibility studies. *Res J Chem Sci* 2:29–34.

Leung, D.Y.C., X. Wu, M.K.H. Leung. 2010. A review on biodiesel production using catalyzed transesterification. *Appl Energ* 87:1083–1095.

Lewis, T., P.D. Nichols, T.A. McMeekina. 2000. Evaluation of extraction methods for recovery of fatty acids from lipid-producing microheterotrophs. *J Microbiol Methods* 42:107–116.

Li, X., H. Xu, Q. Wu. 2007. Large-scale biodiesel production from microalga *Chlorella protothecoides* through heterotrophic cultivation in bioreactors. *Biotechnol Bioeng* 98:764–771.

Li, Y., M. Horsman, N. Wu, C.Q. Lan, N. Dubois Calero. 2008. Biofuels from microalgae. *Biotechnol Prog* 24:815–820.

Lin, L., Z. Cunshan, S. Vittayapadung, S. Xiangqian, D. Mingdong. 2011. Opportunities and challenges for biodiesel fuel. *Appl Energ* 88:1020–1031.

Lorenz, R.T., G.R. Cysewski. 2003. Commercial potential for *Haematococcus microalga* as a natural source of astaxanthin. *Trends Biotechnol* 18:160–167.

Mata, T.M., A.A. Martins, N.S. Caetano. 2010. Microalgae for biodiesel production and other application: A review. *Renew Sustain Energ Rev* 14:217–232.

Mazzuca Sobczuk, T., F. Garcia Camacho, E. Molina Grima, Y. Chisti. 2006. Effects of agitation on the microalgae *Phaeodactylum tricornutum* and *Phaeodactylum cruentum*. *Bioprocess Biosyst Eng* 28:243–250.

Melis, A.2002. Green alga hydrogen production: Progress, challenges and prospects. *Int J Hydrogen Energ* 27:1217–1228.

Metting, B., J.W. Pyne. 1986. Biologically-active compounds from microalgae. *Enzyme Microb Technol* 8:386–394.

Metting, F.B. 1996. Biodiversity and application of microalgae. *J Ind Microbiol* 17:477–489.

Metzger, P., C. Largeau. 2005. *Botryococcus braunii*: A rich source for hydrocarbons and related ether lipids. *Appl Microbiol Biotechnol* 66:486–496.

Miron, A.S., A.C. Gomez, F.G. Camacho, E. Molina Grima, Y. Chisti. 1999. Comparative evaluation of compact photobioreactors for large scale monoculture of microalgae. *J Biotechnol* 70:249–270.

Mishra, S.R., M.K. Mohanty, S.P. Das, A.K. Pattanaik. 2012. Production of bio-diesel (methyl ester) from simarouba glauca oil. *Res J Chem Sci* 2:66–71.

Molina Grima, E. 1999. Microalgae, mass culture methods. In: Flickinger, M.C., Drew, S.W., eds., *Encyclopedia of Bioprocess Technology: Fermentation, Biocatalysis and Bioseparation*, Vol. 3, Wiley, New York, pp. 1753–1769.

Moreno Garido, I. 2008. Review-microalgae immobilization: Current techniques and cases. *Bioresour Technol* 99:3949–3964.

Mu, Y., H. Teng, D.J. Zhang, W. Wang, Z.L. Xiu. 2006. Microbial production of 1,3-propanediol by *Klebsiella pneumonia* using crude glycerol from biodiesel preparations. *Biotechnol Lett* 28:1755–1759.

Mu, Y., Z.L. Xiu, D.J. Zhang. 2008. A combined bioprocess of biodiesel production by lipase with microbial production of 1,3-propanediol by *Klebsiella pneumonia*. *Biochem Eng J* 40:537–541.

Mutanda, T., D. Ramesh, S. Karthikeyan, S. Kumari, A. Anandraj, F. Bux. 2011. Bioprospecting for hyper-lipid producing microalgal strains for sustainable biofuel production. *Bioresour Technol* 102:57–70.

Nagle, N., P. Lemke. 1990. Production of methyl-ester fuel from microalgae. *Appl Biochem Biotechnol* 24–25:355–361.

Naik, S.N., V.V. Goud, P.K. Raut, A.K. Dalai. 2010. Production of first and second generation biofuels. A comprehensive review. *Renew Sustain Energ Rev* 14:578–597.

Nigam, P.S., A. Singh. 2011. Production of liquid biofuels from renewable resources. *Prog Energ Combus Sci* 37:52–68.

Norton, T.A., M. Melkonian, R.A. Andersen. 1996. Algal biodiversity. *Phycologia* 35:308–326.

Olguin, E.J. 2003. Phycoremediation: Key issues for cost-effective nutrient removal processes. *Biotechnol Adv* 22:81–91.

Oliver, R.I., A.K. David. 2009. Quo vadis biofuels. *Energ Environ Sci* 2:343. doi:10.1039/b82295ic.

Patil, V., K.Q. Tran, H.R. Giselrod. 2008. Towards sustainable production of biofuels from microalgae: A review. *Int J Mol Sci* 9:1188–1195.

Pauly M., K. Keegstra. 2008. Cell-wall carbohydrates and their modification as a resource for biofuels. *Plant J* 54:559–568.

Peer, M.S., T.H. Skye, S. Evan, M.C. Vte, M.H. Jon, P. Clenens, K. Olaf, B. Hankamer. 2008. Second generation biofuels; High efficiency microalgae for biodiesel production. *Bioenerg Res* 1:20–43.

Pittman, J.K., A.P. Dean, O. Osundeko. 2011. The potential of sustainable algal biofuel production using wastewater resources. *Bioresour Technol* 102:17–25.

Pulz, O. 2001. Photobioreactors: Production systems for phototrophic microorganisms. *Appl Microbiol Biotechnol* 57:287–293.

Ras, M., L. Lardon, S. Bruno, N. Bernet, J.P. Steyer. 2011. Experimental study on a coupled process of production and anaerobic digestion of *Chlorella vulgaris*. *Bioresour Technol* 102:200–206.

Ratledge, C., Z. Cohen. 2008. Feature microbial and algal oils: Do they have a future for biodiesel or as commodity oils. *Lipid Technol* 20:155–160.

Rawat, I., R. Kumar, T. Mutanda, F. Bux. 2010. Dual role of microalgae: Phycoremediation of domestic wastewater and biomass production for sustainable biofuels production. *Appl Energ* 88:3411–3424.

Richmond, A. 2004. *Handbook of Microalgal Culture: Biotechnology and Applied Phycology*. Blackwell Science Ltd, Oxford, U.K.

Román-Leshkov, Y., C.J. Barrett, Z.Y. Liu, J.A. Dumesic. 2007. Production of dimethylfuran for liquid fuels from biomass-derived carbohydrates. *Nature* 447:982–998.

Rosenberg, J.N., G.A. Oyler, L. Wilkinson, M.J. Betenbaugh. 2008. A green light for engineered algae: Redirecting metabolism to fuel a biotechnology revolution. *Curr Opin Biotechnol* 19:430–436.

Sameera, V., C. Sameera, Y. Ravi Teja. 2011. Current strategies involved in biofuel production from plants and algae. *J Microbial Biochem Technol* R1:002. doi:10.4172/1948-5948.R1-002.

Samorì, C., C. Torri, G. Samorì, D. Fabbri, P. Galletti, F. Guerrini, R. Pistocchi, E. Tagliavini. 2010. Extraction of hydrocarbons from microalga *Botryococcus braunii* with switchable solvents. *Bioresour Technol* 101:3274–3279.

Sanchez Miron, A., M.C. Ceron Garcia, A. Contreras Gomez, F. Garcia Camacho, E. Molina Grima, Y. Chisti. 2003. Shear stress tolerance and biochemical characterization of *Phaeodactylum tricornutum* in quasi steady-state continuous culture in outdoor photobioreactors. *Biochem Eng J* 16:287–297.

Sánchez Mirón, A., A. Contreras Gómez, F. García Camacho, E. Molina Grima, Y. Chisti. 1999. Comparative evaluation of compact photobioreactors for large-scale monoculture of microalgae. *J Biotechnol* 70:249–270.

Sawayama, S., S. Inoue, Y. Dote, S.-Y. Yokoyama. 1995. CO_2 fixation and oil production through microalga. *Energ Convers Manag* 36:729–731.

Schenk, P.M., S.R. Thomas-Hall, E. Stephens, U.C. Marx, J.H. Mussgnug, C. Posten, O. Kruse, B. Hankamer. 2008. Second generation biofuels: High-efficiency microalgae for biodiesel production. *Bioenerg Res* 1:20–43.

Schwartz, R.E. 1990. Pharmaceuticals from cultured algae. *J Ind Microbiol* 5:113–123.

Sharma, C.K., S.S. Kanwar. 2012. Synthesis of methyl cinnamate using immobilized lipase from *B. licheniformis* MTCC-10498. *Res J Recent Sci* 1(3):68–71.

Sharma, C.K., P.K. Sharma, S.S. Kanwar. 2012. Optimization of production conditions of lipase from *B. licheniformis* MTCC-10498. *Res J Recent Sci* 1(7):25–32.

Shay, E.G. 1993. Diesel fuel from vegetable oils: Status and opportunities. *Biomass Bioenergy* 4:227–242.

Sheehan, J., V. Combreco, J. Duffield, M. Graboski, H. Shapouri. 1998a. *An Overview of Biodiesel and Petroleum Diesel Life Cycles*. National Renewable Energy Laboratory, Golden, CO, pp. 1–47.

Shimizu, Y. 1996. Microalgal metabolites: A new perspective. *Annu Rev Microbiol* 50:431–465.

Shimizu, Y. 2003. Microalgal metabolites. *Curr Opin Microbiol* 6:236–243.

Singh, A., P.S. Nigam, J.D. Murphy. 2011a. Renewable fuels from algae: An answer to debatable land based fuels. *Bioresour Technol* 102:10–16.

Singh, A., P.S. Nigam, J.D. Murphy. 2011b. Mechanism and challenges in commercialisation of algal biofuels. *Bioresour Technol* 102:26–34.

Singh, A., D. Pant, S.I. Olesen, P.S. Nigam. 2012. Key issues to consider in microalgae based biodiesel production. *Energ Educ Sci Technol A Energ Sci Res* 29(1):687–700.

Singh, J., S. Gu. 2010. Commercialization potential of microalgae for biofuels production. *Renew Sustain Energ Rev* 14:2596–2610.

Soares, R.R., D.A. Simonetti, J.A. Dumesic. 2006. Glycerol as a source for fuels and chemicals by low-temperature catalytic processing. *Angew Chem Int Ed* 45:3982–3985.

Sompech, K., Y. Chisti, T. Srinophakun. 2012. Design of raceway ponds for producing microalgae. *Biofuels* 3:387–397.

Spolaore, P., C. Joannis-Cassan, E. Duran, A. Isambert. 2006. Commercial applications of microalgae. *J Biosci Bioeng* 101:87–96.

Subramaniam, R., S. Dufreche, M. Zappi, R. Bajpai. 2010. Microbial lipids from renewable resources: Production and characterization. *J Ind Microbial Biotechnol* 37:1271–1287.

Surendhiran D. and Vijay M. 2012. Microalgal biodiesel—A comprehensive review on the potential and alternative biofuel. *Res J Chem Sci* 2(11): 71–82.

Sydney, E.B. Sturm, W. de Carvalho, J.C. Thomaz-Soccol, V. Larroche, C. Pandey, A. and Soccol, C.R. 2010. Potential carbon dioxide fixation by industrially important microalgae. *Bioresour Technol* 101: 5892–5896.

Terry, K.L., L.P. Raymond. 1985. System design for the autotrophic production of microalgae. *Enzyme Microb Technol* 7:474–487.

Travis, T. 1993. The Haber–Bosch process-exemplar of 20th-century chemical industry. *Chem Ind (Lond)* 15:581–585.

Tredici, M.R. 1999. Bioreactors, photo. In: Flickinger, M.C., Drew, S.W., eds., *Encyclopedia of Bioprocess Technology: Fermentation, Biocatalysis, and Bioseparation*. Wiley, New York, pp. 395–419.

Tsukahara, K., S. Sawayama. 2005. Review: Liquid fuel production from microalgae. *J Jpn Petrol Inst* 48(5):251–259.

Turkenburg, W.C. 2000. Renewable energy technologies. In: Goldemberg, J., ed., *World Energy Assessment*. Preface. United Nations Development Programme, New York.

Um, B.H. and Kim, Y.S. 2009. Review: A chance for Korea to advance algal-biodiesel technology. *J Ind Eng Chem* 15:1–7.

United Nations. 2007. Sustainable bioenergy: A framework for decision makers. UN Report, April.

United States Environmental Protection Agency. 2002. A comprehensive analysis of biodiesel impacts on exhaust emissions, pp. 1–118.

Vaidya, P.D., A.E. Rodrigues. 2009. Glycerol reforming for hydrogen production: A review. *Chem Eng Technol* 32:1463–1469.

Verma, N.M., S. Mehrotra, A. Shukla, B. Mishra. 2010. Review: Prospective of biodiesel production utilizing microalgae as the cell factories: A comprehensive discussion. *Afr J Biotechnol* 9(10):1402–1411.

Victor, P., F. Simona, C. Andreea, J. Sorina. 2010. Current perspectives and challer biofuel production and consumption. *Roman J Econ* 31:107–111.

Walter, T.L., S. Purton, D.K. Becker, C. Collet. 2005. Review: Microalgae as bioreactors. *Plant Cell Rep* 24:629–641.

Wiltshire, K.H., M. Boersma, A. Möller, H. Buhtz. 2000. Extraction of pigments and fatty acids from the green alga *Scenedesmus obliquus* (Chlorophyceae). *Aqua Ecol* 34:119–126.

Wongluang, P., Y. Chisti, T. Srinophakun. 2013. Optimal hydro dynamic design of tubular photobioreactors. *J Chem Technol Biotechnol* 88:55–61. www.task39.org.

21

Biodiesel Production from Marine Macroalgae

Laura Bulgariu and Dumitru Bulgariu

CONTENTS

21.1 Introduction

Accelerated development of humanity has the effect of increasing energy demand worldwide. The annual reports of International Energy Agency (IEA) show that the total energy consumption increased from 196 EJ in 1973 to more than 350 EJ in 2009 (1 EJ = 10^{18} J) (Schlagermann et al. 2012), and the tendency is upward. More important is that about 80% of energy demand is obtained from oil and other fossil fuels, whose use has two serious consequences:

1. The fossil fuel supplies are constantly diminishing—It is estimated that in Europe, oil resources will be exhausted in 40 years, natural gas in 60 years, and coal in 200 years, and this means that in about 20 years, Europe will have to import 70% of energy needs (Balat 2010).
2. The utilization of fossil fuels resulted in an increase of greenhouse gas emission that provokes serious climate change by global warming—Only the energy transport activities generate, at global level, over 60% of anthropogenic emissions of greenhouse gas (Ulusarslan et al. 2009).

Negative consequences of environmental pollution caused by the use of fossil fuels (oil, natural gas, or coal) and their imminent exhaustion that can generate important policy issues have determined the orientation of scientific research in order to identify new sources of clean energy, using renewable resources. Fortunately, technological developments are making transitions possible (Demirbas 2009).

Renewable energy is a promising alternative solution that can be considered clean and environmentally safe, mainly because it produces lower or negligible levels of greenhouse gas and other pollutants, in comparison with fossil energy (Dincer 2008). The major sources of renewable energy are different kinds of biomass, hydro, solar, wind, geothermal, etc., each of them having their own advantages and disadvantages, including political, economical, and practical issues (Demirbas and Demirbas 2011). It is therefore difficult to say which of these energy sources will constitute the energy sources of the future.

Among all mentioned renewable energy resources, only biomass is suitable for the production of alternative fuels for vehicles, because it is totally dependent on fossil oil. Thus, various kinds of biomasses (such as soybean, palm oil, rape seeds, sunflower, mustard seeds) (Atabani et al. 2012) can be used for the extraction of vegetable oils that by suitable chemical processes can be transformed into an ecological fuel called biodiesel. Unfortunately, there are some barriers to the development of biodiesel production. First, the finding of adequate biomass with high oil content, for the biodiesel production that to be efficient from an economic perspective, but this will be not acceptable if vegetable oils will be more expensive than petroleum fuels. Second, the intensive cultivation of biomass for biodiesel production does not significantly reduce the share of agricultural crops, because this will generate other significant problems such as the rise of vegetable costs and even starvation especially in the developing countries. Therefore, such biomasses based on cultivated terrestrial plants seem to be unsuitable, at least now for biodiesel production as it requires substantial arable lands and consumes large amounts of freshwater (Chisti 2007), due to negative environmental as well as economical impacts. In consequence, it is necessary to find other suitable biomass resources that do not involve the use of agricultural land, but that meet the expected increasing demand for biofuels. This necessity has led to an interest into the use of aquatic biomasses, such as marine macroalgae (John et al. 2011).

Marine macroalgae also called seaweeds seem to be one of the potential alternatives to fully displace terrestrial biomass in the production of sustainable energy. The most important advantages of macroalgae use as feedstock for biodiesel production are the following: (1) marine macroalgae are considered fast-growing plants that are available in large quantities in many regions of the world; (2) marine macroalgae do not need fertile land and freshwater for their cultivation (Demirbas and Karslioglu 2007); (3) marine macroalgae can convert solar energy into chemical energy with higher efficiency (6%–8%) than terrestrial biomass (1.8%–2.2%) (Jung et al. 2013); and (4) marine macroalgae have a lower risk for the competition for food and energy than other crops, because only in a few countries from East Asia, marine macroalgae are used for food, fertilizer, and animal feed (Bixler and Porse 2011). Unfortunately, besides the environmental and economical advantages, the utilization of marine macroalgae for biodiesel production also has a major drawback. Due to their unique carbohydrates, which are distinctively different from those of terrestrial biomass, the existing technologies for biodiesel production cannot be directly applied to marine macroalgae biomass.

In this chapter, the utilization of marine macroalgae as potential feedstock for biodiesel production is presented. Basic information, useful for marine macroalgae characterization (classification, chemical composition, habitat environment, etc.), is reviewed. Also, the main steps of macroalgae-based technologies developed to convert various macroalgae biomasses to biodiesel and the economical considerations associated with production are summarized and discussed.

21.2 Basic Characterization of Marine Macroalgae

Marine macroalgae (or seaweeds) are multicellular photosynthetic biological organisms that are included in the lower plant category, because they have a thallus-type leaf instead of roots, stems, and leaves. The size of marine macroalgae varies in a wide range, from tiny microscopic forms (3–10 μm) to large macroscopic forms (up to 70 m long). Because they are photosynthetic biological organisms, most marine macroalgae have chlorophyll, but they also contain additional pigments that are the basis of their classification (Guiry 2012). The mixture of pigments in their chloroplasts lends characteristic colors to marine macroalgae, and many of their scientific names are based on these colors (Peterfi and Ionescu 1976). As a function of the color of additional pigments, marine macroalgae are classified into three categories:

1. Brown algae (Phaeophyta division)—Their principal photosynthetic pigments are chlorophyll a and c, b-carotene, and other xanthophylls.
2. Red algae (Rhodophyta division)—Red color is derived from chlorophyll a, phycoerythrin, and phycocyanin.
3. Green algae (Chlorophyta division)—They have the same ratio of chlorophyll a to b as land plants.

There are about 1500–2000 species of brown macroalgae (Phaeophyceae class under phylum Chrysophyta); 4000–6000 species of red macroalgae, most of which exist in tropical marine environment (Florideophycidae and Bangiophycidae subclasses); and at least 1500 species of seawater-favorable green macroalgae (Bryopsidophyceae, Dasycladophyceae, Siphoncladophyceae, and Ulvophyceae classes) (Yu et al. 2002; Guiry 2012).

Each category of marine macroalgae has a particular cell wall structure. Thus, the cell walls of brown algae generally contain three main components: cellulose (that is the structural support), alginic acid (a polymer of mannuronic and guluronic acids, and the corresponding salts of sodium, potassium, magnesium, and calcium and represent 14%–40% of the dry weight of the biomass), and sulfated polysaccharides (Lodeiro et al. 2005; Romera et al. 2007). Red macroalgae also contain cellulose as structural support, but their biosorptive properties are mainly determined by the sulfated polysaccharides made from galactanes (agar and carragenates) (Romera et al. 2006, 2007). Green algae are mainly cellulose, and a high percentage of their cell walls are proteins bonded to polysaccharides to form glycoproteins (Romera et al. 2007; Wang et al. 2009).

TABLE 21.1

Chemical Composition of Some Marine Macroalgae

Marine Macroalgae		Dry Weight (%)			
		Ash	Protein	Carbohydrates	Lipids
Brown algae	*Dictyota ciliolata*	47.2	4.1	15.2	7.8
	Hydroclathrus clathratus	49.4	4.2	18.3	2.9
	Fucus virsoides	17.7	12.3	15.8	3.2–4.7
	Cystoseira barbata	20.4	13.5	17.4	1.3–2.4
	Padina boryana	33.5	10.6	18.4	5.2
	Rosenvingea nhatrangensis	56.6	6.6	8.4	3.1
	Turbinaria conoides	34.4	5.9	19.7	2.3
Red algae	*Gracilaria compressa*	23.5	17.7	20.2	—
	Gracilaria salicornia	49.3	6.0	24.4	1.3
	Laurencia majuscula	42.2	12.5	18.8	5.1
	Portieria hornemannii	37.4	9.8	21.8	5.3
	Hypnea sp.	34.7	6.9	31.7	3.4
Green algae	*Enteromorpha intestinalis*	49.5	3.2	18.7	1.8
	Halimeda macroloba	64.4	4.6	2.7	2.5
	Halimeda opuntia	89.7	3.2	2.5	2.9
	Caulerpa racemosa	47.7	6.9	14.7	4.4
	Ulva lactuca	16.6	13.9	29.5	1.8

Sources: Peterfi, S. and Ionescu, A., *Algal Handbook* (in Romanian), Romanian Academy Publisher, Bucharest, Romania, Vol. 2, pp. 25–42, 1976; Renaud, S.M. and Luong-Van, J.T., *J. Appl. Phycol.*, 18, 381, 2006.

Table 21.1 summarizes the main constituents that are present in the structure of some marine macroalgae.

Besides these main constituents, marine macroalgae also contain several compounds (such as fat, fatty acids, fatty alcohols, phenols, terpenes, steroids, resin acids, and waxes) (Kumar et al. 2007) that are easily extracted in organic and/or aqueous solvents, most of them being the basis for biofuel production.

From a chemical composition point of view, marine macroalgae are significantly different from other biomasses and terrestrial plants. The main differences are as follows: (1) marine macroalgae have unique carbohydrates (such as manan, ulvan, agar, carrageenan, mannitol, alginate, and fucoidan) (Guiry 2012) that are the most abundant compounds in marine macroalgae (25–60 wt%) and that are not included in the microalgae and terrestrial biomass. The monosaccharides hydrolyzed from these carbohydrates are the target for biofuel production using this kind of biomass; (2) marine macroalgae have low content of proteins (7–20 wt%) and lipids (1–10 wt%); (3) marine macroalgae almost do not include lignin, and therefore, the cell walls of marine macroalgae are structurally flexible (Wegeberg and Felby 2010). The low lignin content of marine macroalgae can provide many advantages for biofuel production, because it is not necessary that the lignin be removed by complex processes. However, like terrestrial biomass, marine macroalgae do not have high starch and oil content (Roesijadi et al. 2010), which is the main difference in comparison with microalgae, and they also have high water (70–90 wt%) and mineral content (10–15 wt%) (Peterfi and Ionescu 1976; Renaud and Luong-Van 2006).

Since the performance of biofuel production depends on the chemical composition of the feedstock, information about the carbohydrate contents of marine macroalgae are necessary and will be helpful to evaluate if the conversion of a given marine macroalgae to biofuels is economically feasible.

21.3 Habitat and Production of Marine Macroalgae

Experimental studies have shown that the chemical composition, growing rate, size, and pigment intensity of marine macroalgae are significantly affected by environmental conditions such as light, temperature, salinity, nutrient contents, and pollution. Among all these conditions, light has the most important role.

In their natural environment, marine macroalgae grow on rocky substrate and form stable, multilayer, perennial vegetation and are generally vertically distributed from the upper zone (close to the sea surface) to the lower sublittoral zone, so that their pigments can selectively absorb the light of specific wavelengths (Guiry 2012). Although most marine macroalgae are found in the littoral zone near the coast, the *Gelidium* sp. red macroalgae inhabit the deep sea (over 25 m below surface), where sunlight intensity is lower (Yu et al. 2002). This behavior is determined by the fact that the *Gelidium* sp. red macroalgae have particular phycoerythrin and phycocyanin pigments that can efficiently absorb the photosynthetically active radiation, which can penetrate seawater to the deep zone (Wegeberg and Felby 2010).

From the sublittoral zone, marine macroalgae are mechanically collected, dried in air, and used as feedstock in industrial production. Although huge amounts of marine macroalgae are produced on the coasts of seas and oceans, their utilization as feedstock in the biodiesel production has two main disadvantages: (1) the quality and quantity of biomass that can be obtained are dependent significantly on climatic conditions, and this determines the quantity of marine macroalgae that can be taken depending on the season, and (2) in seawater, more species of marine macroalgae grow together, which do not have the same efficiency as feedstock in the biodiesel production.

In order to minimize these drawbacks, the cultivation of marine macroalgae is more feasible in specialized farms using open-pond technology. Whether they are located on the seashore or in special places, the commercial marine macroalgae farms have a long history, especially in Asia.

Although few species of marine macroalgae are cultivated for commercial purpose, over 2000 species are reported worldwide. More important is that the amount of marine macroalgae cultivated in the world continuously increased (Figure 21.1). The reports of the Food and Agriculture Organization of the United Nations from 2012 note that brown and red marine macroalgae were cultivated in aquaculture farms more than green algae, their production attaining 15.8 million tons of wet biomass in 2010 (Table 21.2). So the farming technology permits the cultivation of marine macroalgae on a large scale, in order to supply sufficient feedstock for biofuel production.

The most promising marine macroalgae species that are used as feedstock for biofuel production and most intensively cultivated are brown macroalgae *Laminaria japonica* and *Undaria pinnatifida* (over 40% from the total); *Porphyra*, *Eucheuma*, *Kappaphycus*, and *Gracilaria* sp. from the red macroalgae category (over 40% from total); and *Monostroma* and

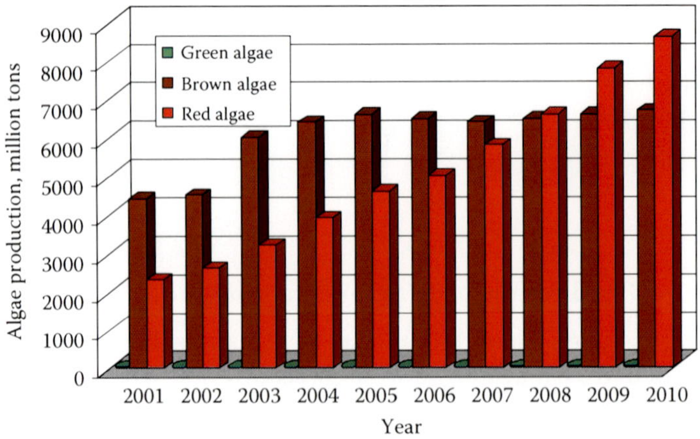

FIGURE 21.1
Global production of marine macroalgae in specialized farms. (According to the data from Food and Agriculture Organization of the United Nations, 2010 Fishery and Aquaculture Statistics, 2012, ftp://ftp.fao.org/FI/CDrom/CD_yearbook_2010/index.htm.)

TABLE 21.2
Global Production of Selected Marine Macroalgae, in Specialized Farms

Marine Macroalgae		Global Production (Million Tons)	% of Total
Brown algae	*Laminaria japonica*	5.14	32.16
	Undaria pinnatifida	1.54	9.74
	Sargassum fusiforme	0.08	0.50
	Phaeophyceae	0.02	0.14
Red algae	*Eucheuma* sp.	3.49	22.11
	Kappaphycus sp.	1.87	11.89
	Gracilaria sp.	1.15	7.30
	Porphyra sp.	1.07	6.79
	Porphyra tenera	0.56	3.57
	Gelidium sp.	1.20×10^{-3}	0.01
Green algae	*Enteromorpha clathrata*	1.12×10^{-2}	0.07
	Monostroma nitidum	4.53×10^{-3}	0.03
	Caulerpa sp.	4.31×10^{-3}	0.03
	Codium fragile	1.39×10^{-3}	0.01

Source: Jung, K.A. et al., *Bioresour. Technol.*, 135, 182, 2013.

Enteromorpha sp. from the green macroalgae category (Luning and Pang 2003), but their production is much lower than those of brown and red marine macroalgae.

Considering the officially reported data and taking into account the chemical composition of marine macroalgae, it seems that the most efficient species for biodiesel production are brown and red macroalgae. Under these conditions, technology development should be focused on this kind of feedstock.

Regardless of the category to which they belong, the cultivation of marine macroalgae for biodiesel production must be highly productive, easily harvestable by simple mechanical operations, able to withstand water motion in the open system, and the production cost should be equal to or even lower than other available biomasses (Carlsson et al. 2007). At this moment, the main drawback of marine macroalgae cultivation is that the production cost is still too high.

21.4 Stages of Biodiesel Production Processes

The utilization of marine macroalgae as feedstock for biodiesel production could provide a key tool for reducing the emission of greenhouse gas from coal-fired power plants and other carbon-intensive industrial processes. Moreover, algae biomass can play an important role in solving the problem between food production and biodiesel in the near future. However, biodiesel production from marine macroalgae biomass is still discussed both from technological and economical considerations. According to its definition, biodiesel is a mixture of fatty acid alkyl esters, produced from triacylglycerols, diacylglycerols, free fatty acids, and phospholipids (Patil et al. 2012), that, in comparison with conventional diesel, generally contains a higher level of oxygen and lower levels of sulfur and nitrogen (responsible for the emission of SO_x and NO_x). In consequence, the production of biodiesel from marine macroalgae will mainly suppose the obtaining of lipids from this biomass that will be subsequently converted into biodiesel.

From a technological point of view, the obtaining of biodiesel from marine macroalgae involves several processes that can be included into three major categories. A schematic representation of the main processes is illustrated in Figure 21.2. All these processes, from algae growing to biodiesel purification require high energy consumption and high costs, and for this reason, rigorous research is still needed to reduce them.

21.4.1 Preparation of Marine Macroalgae Biomass

Although the growing of marine macroalgae is the starting point of biodiesel production, the achievement of this process significantly affects the quality of biomass. In favorable environmental conditions, marine macroalgae grow very fast and have a high lipid content. These lipids are the active compounds in biodiesel production processes, and their synthesis at the cellular level is based on the photoautotrophic property of marine macroalgae.

Thus, in adequate conditions (sunlight and nutrients), the cells of marine macroalgae fix CO_2 from air during the day by photophosphorylation and produce carbohydrates through the Calvin cycle, which are then metabolically converted into various products including lipids, as a function of the macroalgae species and growing conditions (Liu and Benning 2012). A schematic representation of lipid biosynthesis by a photoautotrophic mechanism is given in Figure 21.3.

The biological studies performed in this area have shown that most marine macroalgae directly uptake HCO_3^- rather than CO_2 for their growth, mainly because the diffusion of CO_2 in seawater is very low (Drechsler et al. 1993; Beer and Koch 1996). Even though there are few marine macroalgae that can utilize CO_2 as a direct substrate, in most cases, the interconversion between CO_2 and HCO_3^- takes place, and this process is mediated by specific enzymes such as *RuBP carboxylase* and *carbonic anhydrase* (Drechsler et al. 1993; Zhou and Gao 2013).

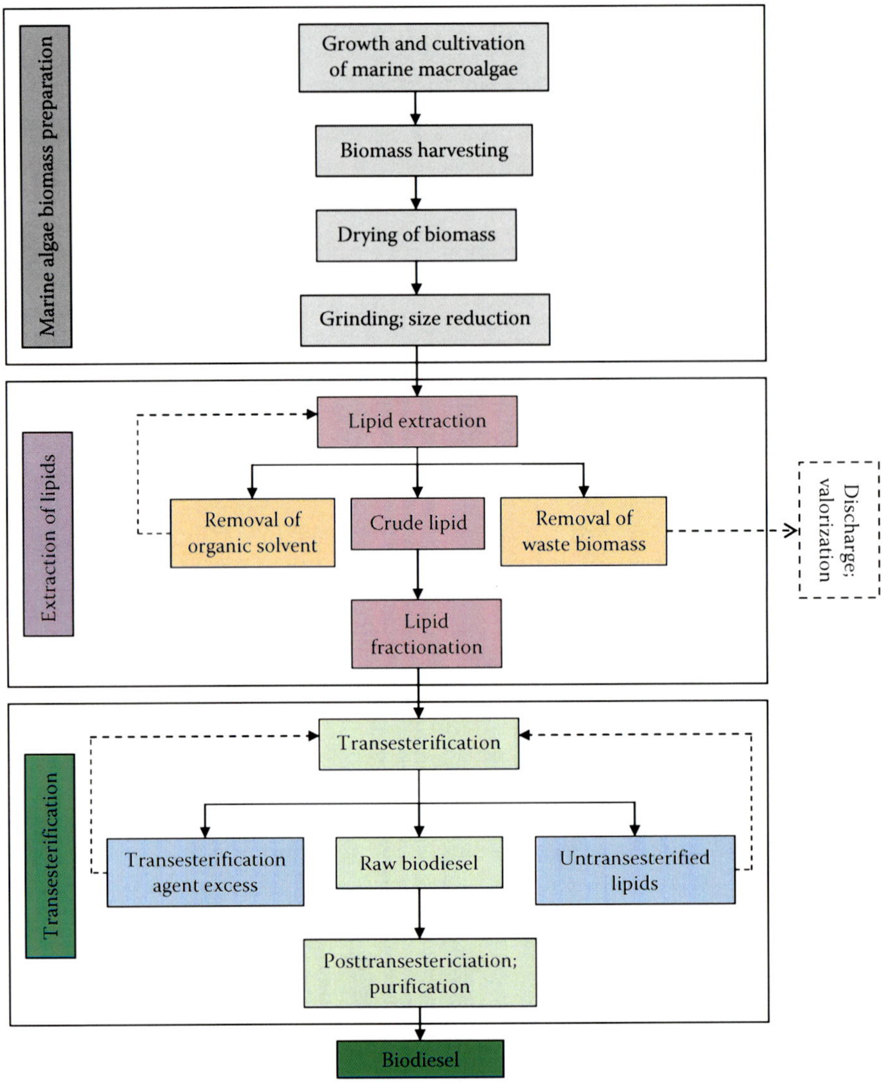

FIGURE 21.2
Schematic representation of the main processes involved in biodiesel production from marine macroalgae biomass.

Due to their photoautotrophic characteristic, marine macroalgae produce and store significant amounts of organic carbon, which represent the carbon source for biofuel production. Photosynthetic rates for several species of marine macroalgae are presented in Table 21.3.

The studies performed by Chung et al. (2011) have shown that marine macroalgae cultivated along the coastline can retain, annually, around one billion tons of carbon, which is an important benefit for the environment. Even if the photosynthetic rates highly depend on the algae species (even in the same category), the high values of this parameter combined with productivity rates, which are higher than terrestrial biomass, make marine macroalgae a suitable feedstock for biodiesel production.

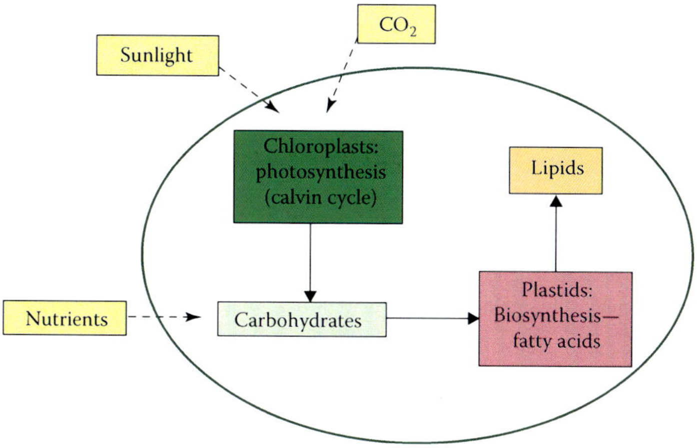

FIGURE 21.3
Schematic representation of lipid synthesis in marine macroalgae cells.

TABLE 21.3
Photosynthetic Rates for Selected Marine Macroalgae

Macroalgae Category	Macroalgae Specie	Photosynthetic Rate (mmol CO_2/[h·g])
Brown algae	*Alaria marginata*	0.109
	Fucus sp.	0.561
	Laminaria sp.	0.124
	Sargassum sp.	0.415
Red algae	*Gracilaria* sp.	0.085
	Iridaea cordata	0.029
	Porphyra sp.	1.809
Green algae	*Cladophora rupestris*	0.031
	Ulva sp.	0.048
	Enteromorpha sp.	1.786

Source: Jung, K.A. et al., *Bioresour. Technol.*, 135, 182, 2013.

The growth of marine macroalgae is mostly achieved based on open-pond technology, in specialized farms, as it was shown in the previous paragraph. After growing, marine macroalgae undergo some operations (see Table 21.4) and are transformed into raw material for the next steps of the technological process. All these operations can be performed in a single step or in multiple steps.

The last two operations, drying and grinding, have a particular importance for marine macroalgae biomass, because those operations eliminate residual water, are known to inhibit the efficiency of mass transfer during the lipid extraction step, and result in the formation of dried powder with high specific surface. However, these operations are energy intensive, and a high number of drying and grinding cycles should be only carried out if they substantially enhance the efficiency of macroalgae lipid extraction.

TABLE 21.4

Main Operations Involved in the Preparation of Marine Macroalgae Biomass

Operation	Observation
Harvesting	Is done mechanically using special devices with which algae are brought to the surface.
Washing	Is using clean water, and the main goal of this process step is to remove solid impurities (from harvesting step) and the sea salts from macroalgae leaves; this is the most pollutant operation, due to the large quantities of resulted wastewater.
Drying	Requires hot air (50°C–60°C) and is performed in several steps, for a few hours, until drying of biomass is complete.
Grinding, size reduction	Dry macroalgae biomass is comminuted until a suitable particle size (fine powder) is obtained by milling with specific sieve or by crushing.

The operation included in the preparation step of macroalgae biomass is not significantly different for those required for other terrestrial biomass, so from this perspective, biodiesel production has similar costs. Nevertheless, the development of cost-effective and energy-efficient technological processes for biomass preparation is currently an active field of research.

21.4.2 Extraction of Lipids

Extraction is one of the most important processing steps used for the recovery of lipids (commonly called *oil*) from biomass for the production of biodiesel. In case of macroalgae, the extraction of oil is not as simple as it would be from other crop seeds, which is usually done by mechanical pressing or simple solvent extraction methods, due to their rigid structure of cell walls. For this reason, in order to extract the *unusual* lipid classes and fatty acids localized at the cellular level, in many cases, a pretreatment step is necessary—cell disruption—which can be done by ultrasonication, microwave, enzymatic degradation, etc. (Lee et al. 1998; Gutierrez et al. 2008; Halim et al. 2012). Such pretreatment methods reduce the rigidity of cell walls, allowing thus more efficient removal of lipids by extraction.

Thus, Suganya and Renganathan (2012) recommend the use of ultrasonication as a pretreatment method, because it highly improves the extraction of lipids (in case of *Ulva lactuca* green marine algae, the extraction efficiency increases with 42% after 5 min of ultrasonication at 24 kHz), reduces the solvent requirement and extraction time, permits a more efficient penetration of organic solvent into cellular material, and improves release of cell contents into the bulk medium. In addition, pretreatment has also some economical and environmental benefits due to low moderate cost and negligible added toxicity.

Regardless of the pretreatment method chosen for the cell disruption, this adds significant cost to macroalgae biodiesel products; therefore, the development of new cost-viable pretreatment methods is still in the attention of the researchers.

It is not always necessary that the macroalgae biomass should be pretreated. In case of some marine macroalgae species (with low content of cellulose, so with a lower rigid structure of cell walls), no pretreatment is performed, and the dry biomass is directly processed for lipid extraction.

In lipid extraction, the macroalgae biomass comes in contact with an adequate extraction solvent, which extracts the lipids from the cellular matrices. After the crude lipids are separated from the extraction system (which contains, besides extracted lipids and cell debris, extraction solvent and water), they are fractionated and purified.

From an applicative point of view, the solvent extraction processes that can be used for lipid extraction must display a high specificity toward lipids, in order to minimize the coextraction of nonlipid compounds (like proteins or carbohydrates) (in this way, it significantly reduced the costs of purification/fractionation processes); the extracted lipids should not interact with extraction solvent during the extraction process (such secondary processes will decrease the yield of biodiesel production); and the extraction process should be efficient, relatively cheap, and safe (Halim et al. 2012; Suganya and Renganathan 2012).

Thus, several extraction techniques (such as solvent extraction, wet lipid extraction, hydrothermal liquefaction, ultrasonic extraction, enzymatic extraction, and supercritical fluid extraction) (Kumari et al. 2011; Eline et al. 2012; Borghini et al. 2014) have been developed for lipid extraction, and these have proven to be useful especially in the case of microalgae. The extraction of lipids is frequently performed by solvent extraction or supercritical fluid extraction, because only these techniques have reasonable operating costs.

21.4.2.1 Solvent Extraction

Solvent extraction is a common and efficient technique for lipid extraction, which involves the transfer of a soluble fraction from a solid material (macroalgae biomass) to a liquid solvent (organic solvent) (Amin et al. 2010), accordingly with the basic chemistry concept of *like dissolves like*. Therefore, the neutral lipids that have long hydrophobic chains are generally extracted using nonpolar organic solvents (such as n-hexane and chloroform), through van der Waals' interactions, while the polar lipids require a polar organic solvent and their extraction involves stronger interactions.

In solvent extraction, the nonpolar organic solvent comes close to the macroalgae biomass particles and covers them with a continuous thin film of organic solvent. This continuous thin film is formed due to the interactions between organic solvent and macroalgae cell walls and remains undisturbed by any solvent flow or agitation (Halim et al. 2012). In addition, the presence of the continuous thin organic solvent film is the driving force that governs the lipid extraction at cellular level and can be described through five elementary steps as schematically illustrated in Figure 21.4:

1. Diffusion of organic solvent across macroalgae cell—This elementary step depends on the cell membrane permeability that ensures the penetration, and here, the cell disruption can have a significant importance.
2. Organic solvent inside of macroalgae cell will interact with lipids, through van der Waals' forces—The efficiency of this process is strictly determined by the adequate selection of organic solvent.
3. Formation of organic solvent—Lipid complex inside of macroalgae cell.
4. Diffusion of organic solvent—Lipid complex outside of macroalgae cell, under the action of concentration gradient.
5. Diffusion of extracted lipids into the bulk organic solvent.

In this way, the lipids (most of them in neutral form) are extracted from solid biomass and remain dissolved in the nonpolar organic solvent, where they are then properly processed.

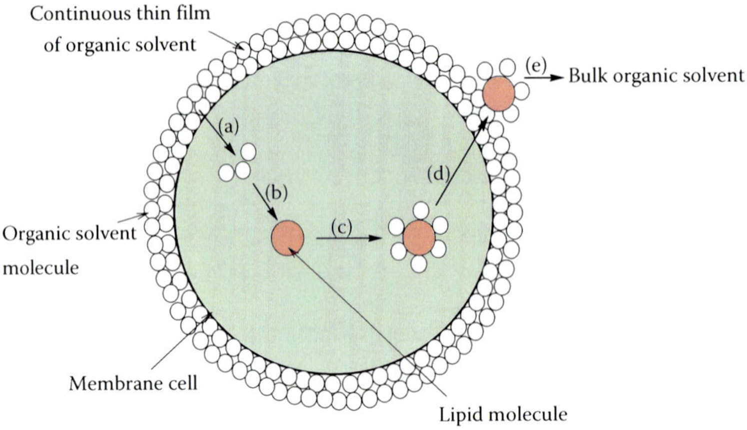

FIGURE 21.4
Schematic illustration of the solvent extraction of lipids, at cellular level.

Unfortunately, nonpolar organic solvents extract only the nonpolar lipids from marine macroalgae, which represent a fraction of total lipid content. Another important part of lipids is strongly bonded by hydrogen bonds to proteins in the cell membrane, and the polar lipid–protein associations cannot be disrupted by the van der Waals' forces, induced by the presence of nonpolar organic solvent in the extraction system (Suganya et al. 2013; Im et al. 2014). Therefore, the utilization of polar organic solvents (such as methanol, ethanol, and isopropanol) that can break the bond from lipid–protein associations is necessary for the extraction of polar lipids and to form new hydrogen bonds with the polar lipids. Many experimental studies (El-Moneim et al. 2010; Stabili et al. 2012; Suganya and Renganathan 2012) have mentioned that the utilization of a mixture of nonpolar and polar organic solvents ensures a higher extraction of lipids, both in freestanding form and protein-associated complex form. After lipid extraction from solid biomass, a new volume of the same nonpolar organic solvent and water is added to the obtained organic extract, and two liquid (one organic and another aqueous) phases are formed. The neutral and polar lipids will mainly partition in the organic phase (formed from nonpolar and polar organic solvent mixture), while the aqueous phase will contain primarily nonlipid compounds (such as proteins and carbohydrates).

In order to design an efficient solvent extraction system of lipids, the selection of some experimental parameters should be done. The main such parameters that influenced the extraction efficiency of lipids are as follows (Siddiquee and Rohan 2011; Suganya and Renganathan 2012):

- *Nature of organic solvent system*—The selection of organic solvents used for lipid extraction from macroalgae biomass is, probably, the most important experimental factor, because the choice of a suitable organic solvent system significantly reduces the biodiesel production cost (especially in the product purification step). Therefore, an adequate organic solvent system for the lipid extraction process should have the following characteristics:
    - A good extraction capacity
    - Low viscosity, that is, to enhance free circulation

- The same polarity as the targeted lipids
- Easy penetration of the biomass cells
- Higher solubility with lipids, without interacting with them
- Low volatility and toxicity

In Table 21.5, the influence of organic solvent systems on lipid extraction efficiency from various marine macroalgae biomasses is illustrated.

Hexane is the most extensively used organic solvent for lipid extraction from macroalgae biomass, because of its high stability, low greasy residual effects, boiling point, and low corrosiveness.

- *Humidity content of biomass*—Experimental studies have shown that the presence of high humidity in macroalgae biomass (>5%) decreases the efficiency of lipid extraction (Suganya and Renganathan 2012). This is mainly because a high level of humidity content creates a barrier between organic solvent system and macroalgae particle, which may restrict the penetration of solvent into biomass cell. From this reason, before the extraction process, the macroalgae biomass should be very well dried, and this operation is performed several times.
- *Particle size of biomass*—Since lipid extraction efficiency depends on the interfacial area between solid particles of biomass and organic solvent, it is expected that a decrease of biomass particle size increases the values of lipid extraction percentage. This is because a high size of biomass particles creates difficulties for the organic solvent to cross the membrane from each macroalgae cell. In most cases, a particle size around 0.1 mm can be considered suitable to obtain a maximum of lipid extraction in a given organic solvent system.
- *Temperature*—In general, the lipid extraction efficiency from macroalgae biomass is enhanced with an increase of temperature, mainly determined by (1) the increase of solvent diffusion system rate inside of macroalgae cell and (2) the increase of organic solvent system capacity to solubilize the lipids from macroalgae cell. It

TABLE 21.5

Effect of Various Organic Solvent Systems on the Lipid Extraction from Marine Macroalgae Biomasses

Macroalgae Category	Marine Macroalgae	Organic Solvent System	Lipid Extraction Efficiency (%)	References
Brown algae	*Colpomenia sinuosa*	$CHCl_3$:CH_3OH (2:1)	3.50	El-Moneim et al. (2010)
		Hexane/ether (1:1)	2.30	
Red algae	*Jania rubens*	$CHCl_3$:CH_3OH (2:1)	4.40	El-Moneim et al. (2010)
		Hexane/ether (1:1)	2.80	
	Gelidium latifolium	$CHCl_3$:CH_3OH (2:1)	3.00	El-Moneim et al. (2010)
		Hexane/ether (1:1)	3.10	
Green algae	*Ulva lactuca*	Hexane	8.09	Suganya and Renganathan (2012)
		Hexane/ether (1:1)	3.50	El-Moneim et al. (2010) and
		Hexane/ether (3:1)	8.16	Suganya and Renganathan (2012)
		$CHCl_3$:CH_3OH (2:1)	4.20	El-Moneim et al. (2010)
		$CHCl_3$/2-propanol (2:1)	7.12	Suganya and Renganathan (2012)

should be noted that the increase of temperature must be done in relatively low temperature range (between 50°C and 60°C, where the organic solvents are not destroyed or degraded), but even in these conditions, the extraction of lipids can be complete (Amin et al. 2010).

- *Extraction time*—Its optimization is important in the selection of the optimum contact time between macroalgae biomass and organic solvent system for the extraction process. Lipid extraction efficiency always increases with the increase of the extraction time. The experimental studies have shown that the extraction time depends on the nature of organic solvent systems, nature and particle size of macroalgae biomass, and on temperature and can range from several hours to several tens of hours (El-Moneim et al. 2010; Suganya and Renganathan 2012; Suganya et al. 2013).
- *Organic solvent/solid biomass ratio*—Almost each experimental study indicates a different value of organic solvent/solid biomass ratio, depending on the lipid content of biomass and intrinsic properties of organic solvent system. Generally, the increase of this parameter determines the increase of the lipid extraction efficiency. But an organic solvent/biomass ratio that is too high results in excessive consumption of organic solvent, while if this value is too low, it leads to handling difficulties and incomplete extraction.

Solvent extraction was the first technique used for the lipid extraction from marine macroalgae biomass and is still widely used in biodiesel production technologies due to its advantages. The possibility to utilize a high number of organic solvent systems, easy recuperation of organic solvents, easy adaptability of the method to various experimental conditions, cheap equipments, etc., makes solvent extraction a powerful method, which still requires further research.

Unfortunately, the solvent extraction of lipids from marine macroalgae biomass has several important disadvantages that drastically limit its practical applicability. The most important disadvantages are that (1) it has a limited selectivity that can be enhanced by an adequate selection of the organic solvent system; (2) the rate of lipid extraction process is slow and requires a long work time for completion; (3) it requires large amounts of organic solvents that are aggressive to the environment and harmful for human health due to their high volatility and toxicity; and (4) even if the energy consumption is low because the process occurs near to ambient conditions, the removal of organic solvent from lipid extracts and the recovery require energy-intensive processes that significantly increase the production costs.

Some of these disadvantages can be significantly minimized by using modified solvent extraction methods such as microwave-assisted organic solvent extraction and accelerated or subcritical organic solvent extraction (Patil et al. 2012).

21.4.2.2 Supercritical Fluid Extraction

In comparison with solvent extraction, the supercritical fluid extraction of lipids is considered an emerging green technology that does not require the utilization of toxic and volatile organic solvents, being thus most friendly to the environment.

In general terms, supercritical fluid extraction involves the utilization of a certain fluid (most frequently used is CO_2) in its supercritical region (over critical values of temperature and pressure) that has the same role as the organic solvent in traditional solvent extraction systems (Macias-Sanchez et al. 2007). The wide utilization of CO_2 as supercritical fluid for extraction is mainly due to practical ease to reach the supercritical region without degrading the lipid molecules. The critical temperature and pressure of CO_2 are 31.1°C and 72.9 atm, respectively (Lang and Wai 2001). In addition, CO_2 has low toxicity, low flammability, and low reactivity, which make its utilization in extraction processes to be considered safe.

The utilization of CO_2 supercritical fluid extraction can be considered a viable alternative because

1. The extraction capacity of supercritical fluid can be continuously adjusted by changing the experimental conditions (temperature and pressure) of the extraction process and improving its selectivity
2. Supercritical fluid has physical properties intermediate to a liquid and a gas, can easily and rapidly cross the membrane cells of biomass, and the result is a higher efficiency of lipid extraction in a shorter contact time
3. The extracted lipids are free from organic solvents, which means that additional operations are not necessary for their removal and recovery. Therefore, the energy consumption is significantly reduced in comparison with solvent extraction

From a practical point of view, the supercritical fluid extraction of lipids from marine macroalgae biomass is done in the following manner (Figure 21.5): the biomass is placed inside of the extraction vessel, equipped with a heating element. From the reservoir tank, CO_2 at a pressure higher than its critical value (72.9 atm) is introduced in the extraction vessel. When the extraction vessel is heated above the critical temperature of CO_2, it reaches its critical state and behaves as an extraction agent, and the lipid extraction from macroalgae biomass takes place.

At the molecular level, the supercritical extraction process of lipids can be described through three elementary steps (Figure 21.6) (Pourmortazavi and Hajimirsadeghi 2007; Halim et al. 2012):

1. Lipids from macroalgae biomass are desorbed from macroalgae biomass into the static layer of supercritical CO_2.
2. Here, the lipid molecules are surrounded by supercritical CO_2 molecules, forming a specific complex (lipid–supercritical CO_2).
3. The formed complex will diffuse from the static supercritical CO_2 layer into bulk supercritical CO_2 flow, which takes it to the collection vessel where it is precipitated.

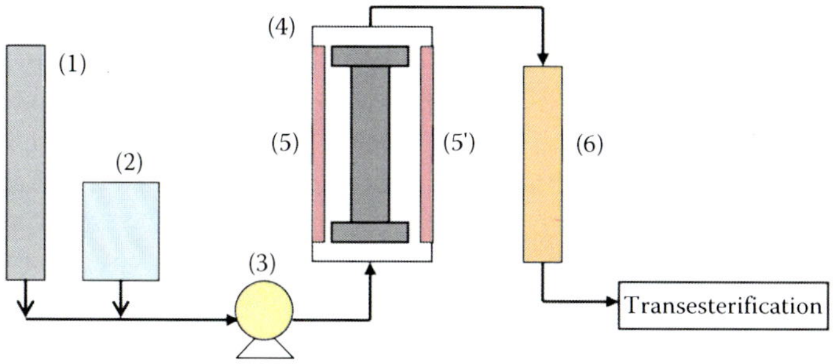

FIGURE 21.5
Schematic representation of the supercritical fluid extraction system. (1) CO_2 tank; (2) cosolvent tank; (3) peristaltic pump; (4) extractor; (5), (5′) heating elements; (6) lipid fractionation element.

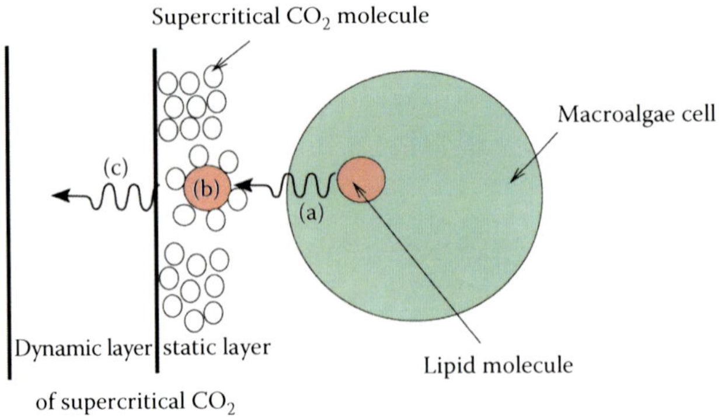

FIGURE 21.6
Elementary steps involved in the supercritical fluid extraction of lipids from marine macroalgae.

TABLE 21.6
Values of Experimental Parameters Required for the Supercritical Lipid Extraction from Marine Macroalgae Biomass

Marine Macroalgae	Extraction Temperature (°C)	Extraction Pressure (atm)	Supercritical CO_2 Flow Rate	Cosolvent	References
Hypnea charoides	40–50	238	4.8 mL/min	—	Cheung (1999)
Undaria pinnatifida	140	260	2.0 mL/min	—	Quitain et al. (2013)
Sargassum muticum	60	150	2.2 mL/min	12% ethanol	Esteban et al. (2013)
	40–55	197.5–395.0	140 g/min	Ethanol	Pérez-López et al. 2014
Sargassum hemiphyllum	40–50	237–375	—	—	Cheung et al. (1998)
Laminaria japonica	40	300–400	54.55 g/min	—	Lee and Chun (2013)

The performance of supercritical CO_2 extraction of lipids from marine macroalgae is influenced by several experimental parameters (pressure, temperature, flow rate, etc.), which mainly affect the supercritical behavior of the fluid (CO_2). In Table 21.6, the experimental conditions used for lipid extraction from some marine macroalgae are summarized using supercritical fluid extraction method.

In comparison with the traditional solvent extraction method, the supercritical fluid extraction has several important advantages (Halim et al. 2012; Azmir et al. 2013): (1) it can be considered a green extraction method, because it does not require the use of toxic and volatile organic solvents; (2) due to its intermediate liquid–gas properties, supercritical CO_2 has higher diffusion coefficient, lower viscosity, and surface tension and can more rapidly penetrate the membrane cell, which makes the lipid extraction more efficient and requires a short period of time; (3) it has a higher selectivity in comparison with solvent extraction, and this can be tuned by changing either temperature or pressure; (4) the extracted lipids are free from organic solvents and do not require further process for solvent removal; (5) because the lipid extraction occurs close to room temperature, the thermal degradation of lipids is minimized; (6) due to the low reactivity of CO_2, it will not interact with lipids during the extraction process, which makes, after extraction, the lipid separation easy to perform; (7) supercritical CO_2 can be easily recycled and reused, which minimizes the waste generation; and (8) supercritical fluid extraction can be easily adapted from laboratory sampling to an industrial scale.

However, it should be mentioned that supercritical fluid extraction has also several disadvantages; the most important of these are: (1) the equipment for supercritical fluid extraction is very expensive and needs rigorous precautions; (2) it is considered a highly energy-intensive method, as fluid compression and heating are needed to bring CO_2 in the supercritical region; and (3) in most of cases, it requires the addition of supplementary organic solvents (such as methanol) to obtain a selective extraction of lipids and to minimize the coextraction of other compounds. This is because supercritical CO_2 is a nonpolar extracting agent, which cannot interact with polar lipids, both in free or complex form. The addition of a polar organic solvent (named cosolvent), besides, will diminish the fluid viscosity, allow the rapid penetration of membrane cell, and also enhance the affinity of supercritical CO_2 toward polar lipids as well as lipid complexes. In this way, the selective extraction of lipids with a higher efficiency will be facilitated (Pourmortazavi and Hajimirsadeghi 2007).

All experimental studies performed in this area have shown that supercritical fluid extraction can successfully replace traditional solvent extraction of lipids from marine macroalgae biomass. However, for marine macroalgae where the lipids strongly interact with other compounds from the cell matrix, a pretreatment could be required. Unfortunately, the high cost of supercritical extraction method reduces its applicability for biodiesel production at industrial scale.

After lipid extraction, the obtained mixture can contain:

- *Lipids*—which are isolated and represent the valuable compounds
- *Excess of organic solvent used in the extraction process*—which is recovered by distillation, vacuum evaporation, or adsorption on solid adsorbent
- *Residual water*—which is removed by biphasic separation and decantation
- *Cell debris*—which is removed by filtration

In function of the work methodology used for the lipid extraction, some of these components do not always need to be separated. Thus, if the lipid extraction is performed by Soxhlet method, the removal of cell debris is not required. Also, in the case of supercritical fluid extraction, further extra steps for the removal of organic solvent and residual water are not necessary, because in working conditions, CO_2 and water are in a gaseous state and are eliminated by decompression. The lipids isolated in this way are called *crude lipids*, or *total lipids* representing the total content of lipids from a given marine macroalgae biomass that are then further processed to obtain the biodiesel.

Often, the lipid fractions that are in liquid state at room temperature are called *oil*. Unfortunately, this term is not very rigorous especially in the case of lipids extracted from marine macroalgae, because these are extracted as a complex mixture that contains various fractions, and not all lipid fractions are in the liquid state.

21.4.2.3 Fractionation of Lipids

In general terms, lipids are defined as any biological molecules that are soluble in a given organic solvent, and most of them contain fatty acids (Magnusson et al. 2014). As a function of the polarity of their head groups, the lipids can be classified in two categories (Chen et al. 2007; Halim et al. 2012):

- *Neutral lipids*—Have the role to store the energy in macroalgae cell and include
    - *Free fatty acids*—Are composed by a hydrophilic carboxyl group attached to one end of a hydrophilic hydrocarbon chain and are bonded to a hydrogen atom
    - *Acylglycerols*—Comprise fatty acid ester bonded to a glycerol skeleton and, in function of its number of fatty acids, can be classified into monoacylglycerols, diacylglycerols, and triacylglycerols
    - *Lipids without fatty acids*—Include hydrocarbons, sterols, ketones, and pigments that are soluble in organic solvents but cannot be transformed into biodiesel
- *Polar lipids*—Comprise phospholipids and glycolipids; these lipids are located in the membrane cells

The lipid content varies within large limits from one species of marine macroalgae to another, and some experimental studies have demonstrated that some species are richer in neutral lipids than others (Huerlimann et al. 2010; Gosch et al. 2012). However, it should be mentioned that the high content of lipids does not always determine the high oil productivity per mass unit of dry macroalgae biomass. For this reason, the suitability of marine macroalgae lipids for biodiesel production is difficult to assess and needs to be examined in each case.

In the crude lipids extracted from marine macroalgae, the acylglycerols generally have a lower degree of unsaturation than other lipid fractions, and in consequence, they are more suitable for biodiesel production, because they are more easily transesterified and the obtained fatty acid methyl esters have higher oxidation stability. In addition, the transesterification process is more efficient with acylglycerols than other lipid fractions (polar lipids and free fatty acids) and can be easily designed for industrial-scale applications. But besides acylglycerols, the crude lipids obtained from marine macroalgae contain free fatty acids, lipids without fatty acids, and polar lipids, which are considered contaminants from a biodiesel production point of view and must be removed. Therefore, the crude lipids obtained after extraction process often need to be fractioned, before transesterification, and this can be done using different fractionation methods, such as liquid chromatography, acid precipitation, and urea crystallization (Ruberto et al. 2001).

In addition, the nature and content of fatty acids from acylglycerol composition, for a given marine macroalgae species, are not only influenced by the macroalgae life cycle as well as its growing conditions (temperature, light intensity, nutrients content, etc.) but also the working conditions used in extraction process. In general, marine macroalgae fatty acids range from 12 to 22 carbons in length and can be either saturated or unsaturated. The number of double bonds in the fatty acid chains is no more than 6, and almost all of the unsaturated fatty acids are *cis* isomers (Kumari et al. 2013). In Table 21.7, the fatty acid composition of lipids extracted from different species of marine macroalgae is illustrated.

TABLE 21.7

Fatty Acid Composition of Lipids Extracted from Different Species of Marine Macroalgae

Fatty Acid	Carbon Atom Number	*Himanthalia elongata* (Brown Algae)[a]	*Porphyra* sp. (Red Algae)[a]	*Ulva lactuca* (Green Algae)[b]
Lauric acid	12:0	ud	ud	3.08
Tridecanoic acid	13:0	ud	ud	3.72
Myristic acid	14:0	9.57	0.53	5.85
Pentadecanoic acid	15:0	ud	ud	1.77
Palmitic acid	16:0	36.73	63.19	50.16
Heptadecanoic acid	17:0	ud	ud	0.75
Stearic acid	18:0	0.59	1.23	11.07
Oleic acid	18:1	22.64	6.70	16.57
Linoleic acid	18:2	5.80	1.17	3.23
Linolenic acid	18:3	6.77	0.23	0.38
Nonadecanoic acid	19:0	ud	ud	0.02
Arachidic acid	20:0	9.78	6.80	2.05
Heneicosanoic acid	21:0	ud	ud	0.23
Behenic acid	22:0	ud	ud	1.12

ud, undetermined in mentioned study.
[a] In extraction conditions mentioned in reference (Plaza et al. 2008).
[b] In extraction conditions mentioned in reference (Suganya and Renganathan 2012).

It can be observed that for the mentioned marine macroalgae species, the saturated fatty acid content is relatively low compared to the total unsaturated content. This is an advantage for cold flow properties of obtained biodiesel, because at low temperature, the saturated chains of fatty acids will pack rapidly and will form a semicrystalline structure and impede its flow. Also the red and green marine macroalgae (*Porphyra* sp. and *U. lactuca*) have lower contents of polyunsaturated fatty acids with two or three double bonds (C18:2 and C18:3) in comparison with brown macroalgae (*H. elongate*). This is desirable because the high content of polyunsaturated fatty acids will determine a poor volatility, low oxidation stability, and gum formation tendency, observed in case of some biodiesels obtained from oilseeds (Lang and Wai 2001).

Therefore, several additional solvent extraction steps can be used to remove the undesirable fatty acids and thus to improve the quality of biodiesel obtained after transesterification.

21.4.3 Transesterification

The transesterification process supposes the reaction between fatty acid containing lipid fraction and alcohols (such as methanol, ethanol, isopropanol, and butanol); in the presence of a catalyst, glycerol and monoalkyl fatty acid esters are obtained (Harrison et al. 2012). Methanol is widely applied for the transesterification of lipid fractions extracted from marine macroalgae, mainly because it is a cheaper reagent and offers some chemical and physical advantages over other alcohols. When methanol is used for transesterification, fatty acid methyl ester (or biodiesel) and glycerol as by-products are obtained accordingly with the following reaction:

CH_2-OCOR_1			$R_1-COOCH_3$		CH_2-OH
|					|
$CH-OCOR_2$	$+ 3\ CH_3OH$ methanol	$\xrightleftharpoons{\text{catalyst}}$	$R_2-COOCH_3$	+	$CH-OH$
|					|
CH_2-OCOR_3 triacylglycerides			$R_3-COOCH_3$ fatty acid methyl esters		CH_2-OH glycerol

The transesterification process occurs step by step (triacylglycerides are transformed into diacylglycerides, then into monoacylglycerides, and finally into fatty acid methyl esters), and further, this process reduced viscosity of mixture in comparison with crude lipids, without that fatty acid composition to be altered.

The transesterification process can be performed using acid (such as H_2SO_4 or HCl), base (such as NaOH or KOH), or enzyme as catalysts, and the type of transesterification reaction is determined by the nature of the catalyst. The transesterification process can also be performed in the absence of catalyst, using a supercritical methanol, at high temperature (200°C–300°C) and pressure (200–400 atm). Under these conditions, the transesterification process requires shorter time (<5 min) and can be successfully applied for the conversion of vegetable oils obtained from oilseeds and animal fats (Kusdiana and Saka 2001).

In function of the nature of catalyst, the transesterification processes can be classified into four categories (Mohan et al. 2014):

- *Direct transesterification*—involves combining extraction and transesterification in a single step (Sathish and Sims 2012); the main advantages of this process are high efficiency of process, minimization of organic solvent requirement, and less reaction time.
- *Acid-catalyzed transesterification*—involves the use of acid catalysts, such as H_2SO_4 or HCl in high concentration, and occurs at high alcohol/oil ratio and relatively low temperature and pressure. The main advantage of this transesterification method is that the acid catalysts are less susceptible to the presence of free fatty acids from crude lipids, but the transesterification process occurs very slowly, and the biodiesel obtained after acid transesterification has higher acidity (Helwani et al. 2009).
- *Base-catalyzed transesterification*—the most frequently used method for the conversion of marine macroalgae crude lipids to biodiesel that involves the presence of a base (NaOH, KOH) as catalyst for transesterification. The base-catalyzed transesterification process occurs at atmospheric pressure and relatively low temperatures (60°C–70°C), using a relatively high excess of methanol (to ensure the quantitative conversion), and in general supposes two successive steps (Meher et al. 2006):
  - Catalyst cleaves the ester bonds between fatty acids and the glycerol skeleton.
  - Liberated fatty acids react then with methanol and form fatty acid methyl esters.

Since base catalysts have faster reaction rates and higher conversion efficiency than acid catalysts, they are used for obtaining biodiesel from plants and animal oils at the industrial scale (Kusdiana and Saka 2001). The main drawback of this process is the formation of soap at high free fatty acid concentrations, which means that the free fatty acids and water should be removed from macroalgae crude lipids prior to transesterification.

- *Enzyme-catalyzed transesterification*—involves the use of enzymes as catalyst, when all lipid fractions can be converted to biodiesel (Gerpen 2005; Bisen et al. 2010). Unfortunately, the high cost of enzyme catalysts and complex processing equipment limits the applicability of enzyme-catalyzed transesterification.

Once the transesterification process is complete, the obtained mixture (that contains biodiesel, glycerol, catalyst, excess of methanol, and untransesterified lipids) is purified in order to remove the by-product contaminants, and this purification process can be performed in two steps (Gerpen 2005; Wahlen et al. 2011):

- The mixture obtained after transesterification is left in standby, which under gravity is separated in two phases: in the top phase, the biodiesel and untransesterified lipids are partitioned, and in the bottom phase, the glycerol is concentrated.
- After phase separation (usual by decantation), the mixture is washed several times with doubly distilled water in order to remove the catalyst and excess methanol.

21.4.4 Characterization of Marine Macroalgae Biodiesel

The macroalgae oil obtained after transesterification and purification should be rigorously characterized in order to show the possibility of its utilization as biodiesel. The most important parameters used for the characterization of marine macroalgae oil are (Mohan et al. 2014):

- Acid number—indicates the corrosiveness of the oil
- Iodine number—is connected with the unsaturation degree
- Specific gravity and density—show the energetic efficiency of macroalgae oil as fuel
- Flash point—expresses the lowest temperature where the oil vaporization occurs to form an ignitable mixture
- Pour point—represents the lowest temperature over that the macroalgae oil becomes semisolid and loses its flow properties
- Viscosity—indicates the macroalgae oil resistance to flow
- Heating value—is the energy released as heat during combustion
- Cetane number—refers to the ignition quality of the diesel engines in efficient operating conditions

In Table 21.8, the quality characteristics of biodiesel obtained from some species of marine green macroalgae are presented, in comparison with the conventional fuel and quality standards used for their evaluation. It can be observed that most

TABLE 21.8

Quality Parameters of Some Marine Green Macroalgae Biodiesel

Quality Characteristic	EN 14212:2008	ASTM D6751-09	Conventional Fuel (Mohan et al. 2014)	Marine Green Macroalgae Biodiesel		
				Enteromorpha compressa (Suganya et al. 2013)	*Caulerpa peltata* (Tamilarasan and Sahadevan 2014)	*Ulva lactuca* (Ciubota-Rosie et al. 2010)
Acid number (mg KOH/g of oil)	≤0.5	≤0.5	0.7–1.0	0.43	0.40	0.38
Iodine value (g I/100 g of oil)	≤120	≤120	120	N/A	N/A	93.22
Density (kg/m^3)	860–900	860–900	880	848.47	885.47	872
Kinematic viscosity (mm^2/s)	3.5–5.0	1.9–6.0	3.5	4.35	4.25	4.0–4.8
Free glycerin (% mass)	0.020	0.020	N/A	0.0034	0.0020	0.00031
Carbon residue (% mass)	0.3	0.050	N/A	0.01	0.0096	0.012
Sulfur (% mass)	0.01	0.05	0.05–0.06	0.00056	0.00064	0.00061
Phosphorus (% mass)	0.001	0.001	0.0001	0.0004	0.0001	0.0002
Flash point (°C)	≥101	≥93	≥62	166	178	≥143
Pour point (°C)	−15	−11.6	−16	−2	−1	−1
Distillation temperature (°C)	N/A	360	N/A	346	352	N/A
Cold point (°C)	N/A	N/A	N/A	3	3	N/A
Cetane number	≥51	≥47	60	58.50	57.90	58.23

Observation: EN, European Normative; ASTM, American Society for Testing and Materials; N/A, not available.

physicochemical properties of marine green macroalgae biodiesel are similar to those of fossil diesel fuels.

All these characteristics of obtained macroalgae biodiesel are mainly dependent on the nature of the marine macroalgae feedstock and also of the extraction and transesterification method used for the biodiesel production. The quality of biodiesel is also influenced by the nature of triacylglycerides and fatty acids from biological feedstock. Thus, the fatty acid composition (carbon chain length and unsaturation degree) from fatty acid methyl esters influenced the biodiesel properties, such as oxidative stability or ignition quality—affected by the unsaturation degree—while the cold flow properties are influenced by the carbon chain length (Ramos et al. 2009). Fatty acid methyl esters that contain in their composition only saturated fatty acids have high oxidative stability with high cetane number but also poor cold flow properties (Harrison et al. 2012). Likewise, the fatty acids from the composition of fatty acid methyl esters have a higher unsaturation degree, so cold flow properties of the obtained biodiesel are significantly enhanced but decrease the oxidative stability and cetane number. Marine macroalgae oil in general differs from vegetable oil being quite rich in polyunsaturated fatty acids with two or more double bonds (see Table 21.7).

Such fatty acids and fatty acid methyl esters are susceptible to oxidation during storage, and this reduces their applicability for the use in biodiesel especially for vehicle use. In consequence, the European standard limits their concentration to 12% from total mass (El-Moneim et al. 2010). This is not an important impediment, because the content of polyunsaturated fatty acids can be easily reduced by partial catalytic hydrogenation of the marine macroalgae oil (Jang et al. 2005). Nevertheless, the experimental studies (Jang et al. 2005; Ramos et al. 2009; Harrison et al. 2012) have shown that higher concentrations of some fatty acids, such as palmitic acid (C16:0) and oleic acid (C18:1), in the oil obtained from macroalgae biomass have a positive influence on the obtained biodiesel properties.

21.5 Conclusion and Final Remarks

Marine macroalgae biomass can have an important role in solving the conflict between the production of food and that of biodiesel in the near future and can help to satisfy the marked requirements, while reducing greenhouse gas emissions. However, before using marine macroalgae biodiesel in large-scale commercial applications, several challenges must be overcome.

In the last few years, numerous studies have been performed to obtain biodiesel from different kinds of marine macroalgae, until now, it is not clear what marine macroalgae species would be most appropriate for the production of commercially viable biofuels. In many studies, locally available marine macroalgae or those that can be easily cultivated in specific regional climatic conditions have been used, without a rigorous comparison to be made regarding their efficiency in the production of biodiesel. Even so, productivity of marine macroalgae culture, cultivation system design, growth conditions and nutrient uptake, lipid extraction, and many other aspects are important issues that still require considerable attention before marine macroalgae can be considered a feasible feedstock for biorefineries. Even if most of the technological processes have been optimized from the efficiency point of view, the main drawbacks that still need to be solved are economic. At this point in time, almost each technological process requires expensive reagents or high energy consumption, so a high production cost increases the sale price of the obtained biodiesel over that of fossil fuels, and thus, this commercial appeal is relatively low. At present, 70%–90% of the total energy consumption is used for harvesting marine macroalgae and extracting lipids. A solution for this problem can be the selection in each technological step of the processes that require minimum energy consumption. Unfortunately, it is not always possible, more adequately from the sustainable development point of view, to offset the cost of marine macroalgae biodiesel by the revenues generated from other coproducts of the macroalgae biomass.

Since the experimental research has proven that marine macroalgae contain significant quantities of proteins and carbohydrates, as well as other compounds (such as carotenes, chlorophylls, and omega-3 and omega-6 free fatty acids), the adequate utilization of these can cogenerate high-value useable products. Thus, by an adequate fractionation of crude lipids extracted from macroalgae biomass, the mentioned high-value compounds can be quantitatively separated from lipids for biodiesel and used for different (pharmaceutics, cosmetics, or food) purposes. On the other hand, the carbohydrates separated from marine macroalgae biomass can be fermented to produce bioethanol. Therefore, the integration of marine macroalgae biodiesel production with simultaneous production of valuable coproducts will have a positive impact on the overall process economics. Unfortunately, such combined technologies are still in the early stages of development. But the considerable interest, highlighted by the high amount of research, gives us hope that biodiesel production from marine macroalgae biomass will be economically viable and will be able to replace in some proportion the utilization of fossil fuels in the near future.

References

Amin, S.K., S. Hawash, G. El Diwani, and S. El-Rafei. 2010. Kinetics and thermodynamics of oil extraction from *Jatropha curcas* in aqueous acidic hexane solutions. *Journal of American Science* 6: 293–300.

Atabani, A.E., A.S. Silitonga, I.A. Badruddin, T.M.I. Mahlia, H.H. Masjuki, and S. Mekhilef. 2012. A comprehensive review on biodiesel as an alternative energy resource and its characteristics. *Renewable & Sustainable Energy Review* 16: 2070–2093.

Azmir, J., I.S.M. Zaidul, M.M. Rahman, K.M. Sharif, A. Mohamed, F. Sahena, M.H.A. Jahurul, K. Ghafoor, N.A.N. Norulaini, and A.K.M. Omar. 2013. Techniques for extraction of bioactive compounds from plant materials: A review. *Journal of Food Engineering* 117: 426–436.

Balat, H. 2010. Prospects of biofuels for a sustainable energy future: A critical assessment. *Energy Education Science and Technology Part A* 24: 85–111.

Beer, S. and E. Koch. 1996. Photosynthesis of marine macroalgae and seagrasses in globally changing CO_2 environments. *Marine Ecology Progress Series* 141: 199–204.

Bisen, P.S., B.S. Sanodiya, G.S. Thakur, R.K. Baghel, and G.B.K.S. Prasad. 2010. Biodiesel production with special emphasis on lipase-catalyzed transesterification. *Biotechnology Letter* 32: 1019–1030.

Bixler, H.J. and H. Porse. 2011. A decade of change in the seaweed hydrocolloids industry. *Journal of Applied Phycology* 23: 321–335.

Borghini, F., L. Lucattini, S. Focardi, S. Focardi, and S. Bastianoni. 2014. Production of bio-diesel from macroalgae of the *Orbetello lagoon* by various extraction methods. *International Journal of Sustainable Energy* 33(3): 697–703.

Carlsson, A.S., J.B. van Beilen, R. Moller, and D. Clayton. 2007. Micro- and macro-algae utility for industrial applications. Outputs from the EPOBIO project, University of York, Heslington, U.K.

Chen, G.Q., Y. Jiang, and F. Chen. 2007. Fatty acid and lipid class composition of the eicosapentaenoic acid-producing microalga, *Nitzschia laevis*. *Food Chemistry* 104: 1580–1585.

Cheung, P.C.K. 1999. Temperature and pressure effects on supercritical carbon dioxide extraction of n-3 fatty acids from red seaweed. *Food Chemistry* 65(3): 399–403.

Cheung, P.C.K., A.Y.H. Leung, and P.O. Ang. 1998. Comparison of supercritical carbon dioxide and Soxhlet extraction of lipids from a brown seaweed, *Sargassum hemiphyllum* (Turn.). *Journal of Agricultural and Food Chemistry* 46: 4228–4232.

Chisti, Y. 2007. Research review paper: Biodiesel from microalgae. *Biotechnology Advances* 25: 294–306.

Chung, I., J. Beardall, S. Mehta, D. Sahoo, and S. Stojkovic. 2011. Using marine macroalgae for carbon sequestration: A critical appraisal. *Journal of Applied Phycology* 23: 877–886.

Ciubota-Rosie, C., L. Bulgariu, C. Catrinescu, I. Volf, and M. Macoveanu. 2010. Study of biodiesel obtaining through valorification of biomass. Cinquieme Edition du Colloque Francophone COFRET 2010 sur l'energy environment, Iaşi, Romania (personal communication).

Demirbas, A. 2009. Future energy sources: Part I. *Future Energy Sources* 1: 1–95.

Demirbas, A. and M.F. Demirbas. 2011. Importance of algae oil as a source of biodiesel. *Energy Conversion and Management* 52: 163–170.

Demirbas, A. and S. Karslioglu. 2007. Biodiesel production facilities from vegetable oils and animal fats. *Energy Sources Part A* 29: 133–141.

Dincer, K. 2008. Lower emissions from biodiesel combustion. *Energy Sources Part A* 30: 963–968.

Drechsler, Z., R. Sharkia, Z.I. Cabantchik, and S. Beer. 1993. Bicarbonate uptake in the marine macroalgae *Ulva* sp. Is inhibited by classical probes of anion exchange by red blood cells. *Planta* 191(1): 34–40.

Eline, R., M. Koenraad, and F. Imogen. 2012. Optimization of an analytical procedure for extraction of lipids from microalgae. *Journal of American Oil Chemistry Society* 89(2): 189–198. doi: http://dx.doi.org/10.1007/s11746-011-1903-z.

El-Moneim, A., M.R. Afify, E.A. Shalaby, and S.M.M. Shanab. 2010. Enhancement of biodiesel production from different species of algae. *Grasas Y Aceites* 61(4): 416–422.

Esteban, L.S., L. Vandanjon, E. Ibanez, J.A. Mendiola, S. Cerantola, N. Kervarec, S. LaBarre, L. Marchal, and V. Stiger-Pouvreau. 2013. Green improved processes to extract bioactive phenolic compounds from brown macroalgae using *Sargassum muticum* as model. *Talanta* 104: 44–52.

Food and Agriculture Organization of the United Nations. 2012. 2010 Fishery and Aquaculture Statistics. ftp://ftp.fao.org/FI/CDrom/CD_yearbook_2010/index.htm.

Gerpen, V.J. 2005. Biodiesel processing and production. *Fuel Processing Technology* 86: 1097–1107.

Gosch, B.J., M. Magnusson, N.A. Paul, and R. de Nys. 2012. Total lipid and fatty acid composition of seaweeds for the selection of species for oil-based biofuel and bioproducts. *Global Change Biology Bioenergy* 4: 919–930.

Guiry, M.D. 2012. The Seaweed Site: Information on marine algae. Available from: http://www.seaweed.ie/algae/index.html. Accessed on December 22, 2014.

Gutierrez, L.F., C. Ratti, and K. Belkacemi. 2008. Effects of drying method on the extraction yields and quality of oils from Quebec sea buckthorn (*Hippophae rhamnoides* L.) seeds and pulp. *Food Chemistry* 106: 896–904.

Halim, R., M.K. Danquah, and P.A. Webley. 2012. Extraction of oil from microalgae for biodiesel production: A review. *Biotechnology Advances* 30: 709–732.

Harrison, B.B., E.B. Marc, and J.M. Anthony. 2012. Chemical and physical properties of algal methyl ester biodiesel containing varying levels of methyl eicosapentaenoate and methyl docosahexaenoate. *Algal Research* 1: 57–69.

Helwani, Z., M.R. Othman, N. Aziz, W.J.K. Fernando, and J. Kim. 2009. Technologies for production of biodiesel focusing on green catalytic techniques: A review. *Fuel Processing Technology* 90: 1502–1514.

Huerlimann, R., R. de Nys, and K. Heimann. 2010. Growth, lipid content, productivity, and fatty acid composition of tropical microalgae for scale-up production. *Biotechnology and Bioengineering* 107: 245–257.

Im, H., H. Lee, M.S. Park, J.W. Yang, and J.W. Lee. 2014. Concurrent extraction and reaction for the production of biodiesel from wet microalgae. *Bioresource Technology* 152: 534–537.

Jang, E.S., M.Y. Jung, and D.B. Min. 2005. Hydrogenation for low trans and high conjugated fatty acids. *Comprehensive Review in Food Science* 4: 22–30.

John, R.P., G.S. Anisha, K.M. Nampoothiri, and A. Pandey. 2011. Micro- and macro-algal biomass: A renewable source for bioethanol. *Bioresource Technology* 102(1): 186–193.

Jung, K.A., S.R. Lim, Y. Kim, and J.M. Park. 2013. Potentials of macroalgae as feedstock for biorefinery. *Bioresource Technology* 135: 182–190.

Kumar, Y.P., P. King, and V.S.R.K. Prasad. 2007. Adsorption of zinc from aqueous solution using marine green algae-*Ulva fasciata* sp. *Chemical Engineering Journal* 129: 161–166.

Kumari, P., A.J. Bijo, V.A. Mantri, C.R.K. Reddy, and B. Jha. 2013. Fatty acid profiling of tropical marine macroalgae: An analysis from chemotaxonomic and nutritional perspectives. *Phytochemistry* 86: 44–56.

Kumari, P., C.R.K. Reddy, and B. Jha. 2011. Comparative evaluation and selection of a method for lipid and fatty acid extraction from macroalgae. *Analytical Biochemistry* 415(2): 134–144.

Kusdiana, D. and S. Saka. 2001. Kinetics of transesterification in rapeseed oil to biodiesel fuel as treated in supercritical methanol. *Fuel* 80: 693–698.

Lang, Q. and C.M. Wai 2001. Supercritical fluid extraction in herbal and natural product studies—A practical review *Talanta* 53(4): 771–782.

Lee, J.H. and B.S. Chun. 2013. Effect of antioxidant activity of mixture obtained from brown seaweed and wheat germ oils using different extraction methods. *Food Science Biotechnology* 22(S): 9–17.

Lee, S.J., B.D. Yoon, and H.M. Oh. 1998. Rapid method for the determination of lipid from the green alga *Botryococcus braunii*. *Biotechnology Techniques* 12: 553–556.

Liu, B. and C. Benning. 2012. Lipid metabolism in microalgae distinguishes itself. *Current Opinion in Biotechnology* 24: 300–309.

Lodeiro, P., B. Cordero, J.L. Barriada, R. Herrero, and M.E. Sastre de Vicente. 2005. Biosorption of cadmium by biomass of brown marine macroalgae. *Bioresource Technology* 96: 1796–1803.

Luning, K. and S.J. Pang. 2003. Mass cultivation of seaweeds: Current aspects and approaches. *Journal of Applied Phycology* 15: 115–119.

Macias-Sanchez, M.D., C. Mantell, M. Rodriguez, E.M. de la Ossa, L.M. Lubian, and O. Montero. 2007. Supercritical fluid extraction of carotenoids and chlorophyll a from *Synechococcus* sp. *Journal of Supercritical Fluids* 39: 323–329.

Magnusson, M., L. Mata, R. De Nys, and N.A. Paul. 2014. Biomass, lipid and fatty acid production in large-scale cultures of the marine macroalga *Derbesia tenuissima* (Chlorophyta). *Marine Biotechnology* 16(4): 456–464.

Meher, L.C., D.V. Sagar, and S.N. Naik. 2006. Technical aspects of biodiesel production by transesterification: A review. *Renewable & Sustainable Energy Reviews* 10: 248–268.

Mohan, S.V., M.P. Devi, G.V. Subhash, and R. Chandra. 2014. Algae oils as fuels, in *Biofuels from Algae*. A. Pandey, D.J. Lee, Y. Chisti, C. Soccol (eds.), Elsevier, Amsterdam, the Netherlands, Chapter 8, pp. 155–187.

Patil, P.D., V.G. Gude, A. Mannarswamy, P. Cooke, N. Nirmalakhandan, P. Lammers, and S. Deng. 2012. Comparison of direct transesterification of algal biomass under supercritical methanol and microwave irradiation conditions. *Fuel* 97: 822–831.

Pérez-López, P., E.M. Balboa, S. González-García, H. Domínguez, G. Feijoo, and M.T. Moreira. 2014. Comparative environmental assessment of valorization strategies of the invasive macroalgae *Sargassum muticum*. *Bioresource Technology* 161: 137–148.

Peterfi, S. and A. Ionescu. 1976. *Algal Handbook* (in Romanian). Romanian Academy Publisher, Bucharest, Romania, Vol. 2, pp. 25–42.

Plaza, M., A. Cifuentes, and E. Ibanez. 2008. In the search of new functional food ingredients from algae. *Trends in Food Science & Technology* 19: 31–39.

Pourmortazavi, S.M. and S.S. Hajimirsadeghi. 2007. Supercritical fluid extraction in plant essential and volatile oil analysis—Review. *Journal of Chromatography A* 1163: 2–24.

Quitain, A.T., T. Kai, M. Sasaki, and M. Goto. 2013. Microwave-hydrothermal extraction and degradation of fucoidan from supercritical carbon dioxide deoiled *Undaria pinnatifida*. *Industrial Engineering Chemical Research* 52(23): 7940–7946.

Ramos, M.J., C.M. Fernandez, A. Casas, L. Rodriguez, and A. Perez. 2009. Influence of fatty acid composition of raw materials on biodiesel properties. *Bioresource Technology* 100: 261–268.

Renaud, S.M. and J.T. Luong-Van. 2006. Seasonal variation in the chemical composition of tropical Australian marine macroalgae. *Journal of Applied Phycology* 18: 381–387.

Roesijadi, G., S.B. Jones, L.J. Snowden-Swan, and Y. Zhu. 2010. Macroalgae as a biomass feedstock: A Preliminary Analysis (PNNL-19944). Available from: http://www.pnl.gov/main/publications/external/technical_reports/PNNL-19944.pdf. Accessed on December 22, 2014.

Romera, E., F. Gonzalez, A. Ballester, M.L. Blasquez, and J.A. Munoz. 2006. Biosorption with algae: A statistical review. *Critical Review in Biotechnology* 26: 223–235.

Romera, E., F. Gonzalez, A. Ballester, M.L. Blazquez, and J.A. Munoz. 2007. Comparative study of biosorption of heavy metals using different types of algae. *Bioresource Technology* 98: 3344–3353.

Ruberto, G., M.T. Baratta, D.M. Biondi, and V. Amico. 2001. Antioxidant activity of extracts of the marine algal genus Cystoseira in a micellar model system. *Journal of Applied Phycology* 13: 403–407.

Sathish, A. and R.C. Sims. 2012. Biodiesel from mixed culture algae via a wet lipid extraction procedure. *Bioresource Technology* 118: 643–647.

Schlagermann, P., G. Gottlicher, R. Dillschneider, R. Rosello-Sastre, and C. Posten. 2012. Composition of algal oil and its potential as biofuel. *Journal of Combustion* (Open Access Journal): Article ID 285185, 14 pp, doi:10.1155/2012/285185.

Siddiquee, M.N. and S. Rohan. 2011. Lipid extraction and biodiesel production from municipal sewage sludges: A review. *Renewable & Sustainable Energy Review* 15: 1067–1072.

Stabili, L., M.I. Acquaviva, F. Biandolino, R.A. Cavallo, S.A. De Pascali, F.P. Fanizzi, M. Narracci, A. Petrocelli, and E. Cecere. 2012. The lipidic extract of the seaweed *Gracilariopsis longissima* (Rhodophyta, Gracilariales): A potential resource for biotechnological purposes. *New Biotechnology* 29(3): 443–450.

Suganya, T., N.N. Gandhi, and S. Renganathan. 2013. Production of algal biodiesel from marine macroalgae *Enteromorpha compressa* by two step process: Optimization and kinetic study. *Bioresource Technology* 198: 392–400.

Suganya, T. and S. Renganathan. 2012. Optimization and kinetic studies on algal oil extraction from marine macroalgae *Ulva lactuca*. *Bioresource Technology* 107: 319–326.

Tamilarasan, S. and R. Sahadevan. 2014. Ultrasonic assisted acid base transesterification of algal oil from marine macroalgae *Caulerpa peltata*: Optimization and characterization studies. *Fuel* 128: 347–355.

Ulusarslan, D., Z. Gemici, and I. Teke. 2009. Currency of district cooling systems and alternative energy sources. *Energy Education Science and Technology Part A* 23: 31–53.

Wahlen, B.D., R.M. Willis, and L.C. Seefeldt. 2011. Biodiesel production by simultaneous extraction and conversion of total lipids from microalgae, cyanobacteria, and wild mixed cultures. *Bioresource Technology* 102: 2724–2730.

Wang, X.S., Z.Z. Li, and C. Sun. 2009. A comparative study of removal of Cu(II) from aqueous solutions by locally low-cost materials: Marine macroalgae and agricultural by-products. *Desalination* 235: 146–159.

Wegeberg, S. and C. Felby. 2010. Algae biomass for bioenergy in Denmark: Biological/technical challenges and opportunities. Available from: http://www.bio4bio.dk/~/media/Bio4bio/publications/Review_of_algae_biomass_for_energy_SW_CF_April2010.ashx. Accessed on December 22, 2014.

Yu, S., A. Blennow, M. Bojko, F. Madsen, C.E. Olsen, and S.B. Engelsen. 2002. Physico-chemical characterization of Floridean starch of red algae. *Starch* 54: 66–74.

Zhou, D. and K. Gao. 2013. Thermal acclimation of respiration and photosynthesis in the marine macroalga *Gracilaria lemaneiformis* (Gracilariales, Rhodophyta). *Journal of Phycology* 49(1): 61–68.

22

Current State of Research on Algal Biodiesel

Ozcan Konur

CONTENTS

22.1 Introduction

22.1.1 Issues

Global warming, air pollution, and energy security have been some of the most important public policy issues in recent years (Jacobson 2009, Wang et al. 2008, Yergin 2006). With increasing global population, food security has also become a major public policy issue (Lal 2004). The development of biofuels generated from biomass has been a long-awaited solution to these global problems (Demirbas 2007, Goldember 2007, Lynd et al. 1991). However, the development of early biofuels produced from agricultural plants such as sugar cane (Goldemberg 2007) and agricultural wastes such as corn stovers (Bothast and Schlicher 2005) has resulted in a series of substantial concerns about food security (Godfray et al. 2010). Therefore, the development of algal biofuels as a third-generation biofuel has been considered as a major solution for the global problems of global warming, air pollution, energy security, and food security (Chisti, 2007, Demirbas 2007, Kapdan and Kargi 2006, Spolaore et al. 2006, Volesky and Holan 1995).

Although there have been many reviews on the use of the marine algae for the production of algal biodiesel (e.g., Chisti 2007, Mata et al. 2010, Meng et al. 2009) and there have been a number of scientometric studies on the algal biofuels (Konur 2011), there has not been any study on the citation classics in algal biodiesel as in the other research fields as highly cited papers (Baltussen and Kindler 2004a,b, Dubin et al. 1993, Gehanno et al. 2007, Konur 2012a,b, 2013, Paladugu et al. 2002, Wrigley and Matthews 1986).

As North's new institutional theory suggests, it is important to have up-to-date information about the current public policy issues to develop a set of viable solutions to satisfy the needs of all the key stakeholders (North 1994, Konur, 2000, 2002a–c, 2006a,b, 2007a,b, 2012c).

Therefore, brief information on a selected set of citation classics on algal biodiesel is presented in this chapter to inform the key stakeholders relating to the global problems of warming, air pollution, food security, and energy security about the production of algal biodiesel for the solution of these problems in the long run, thus complementing a number of recent scientometric studies on biofuels and global energy research (Konur, 2011, 2012d–p).

22.1.2 Methodology

A search was carried out in the Science Citation Index Expanded (SCIE) and Social Sciences Citation Index (SSCI) databases (version 5.11) in September 2013 to locate the papers relating to the algal biodiesel using the keyword set of Topic = (diesel* or *bio* diesel** or biodiesel* or *microbio* diesel** or *mono-alkyl ester**) AND Topic = (alga* or *macro* alga** or *micro* alga** or macroalga* or microalga* or cyanobacter* or seaweed* or diatoms or reinhardtii or braunii or chlorella or sargassum or gracilaria) in the abstract pages of the papers. For this chapter, it was necessary to embrace the broad algal search terms to include diatoms, seaweeds, and cyanobacteria as well as to include many spellings of biodiesel rather than the simple keywords of biodiesel and algae.

It was found that there were 1293 papers indexed between 1980 and 2013. The key subject categories for algal research were biotechnology applied microbiology (659 references, 49.5%), energy fuels (603 references, 45.3%), agricultural engineering (330 references, 24.8%), and engineering chemical (260 references, 19.5%), altogether comprising 80% of all the references on algal biodiesel. It was also necessary to focus on the key references by selecting articles and reviews.

The located highly cited 50 papers were arranged by decreasing number of citations for two groups of papers: 31 and 19 papers for nonexperimental research and experimental research, respectively. In order to check whether these collected abstracts correspond to the larger sample on these topical areas, new searches were carried out for each topical area.

The summary information about the located citation classics is presented under two major headings of nonexperimental and experimental researches in the order of the decreasing number of citations for each topical area.

The information relating to the document type, affiliation and location of the authors, the journal, the indexes, subject area of the journal, the concise topic, total number of citations received, and total average number of citations received per year is given in the tables for each paper.

22.2 Nonexperimental Studies on Algal Biodiesel

22.2.1 Introduction

The nonexperimental research on algal biodiesel has been one of the most dynamic research areas in recent years. Thirty-one citation classics in the field with more than

64 citations were located, and the key emerging issues from these papers were presented in decreasing order of the number of citations (Table 22.1). These papers give strong hints about the determinants of efficient algal biodiesel production and emphasize that marine algae are efficient biodiesel feedstocks in comparison with first- and second-generation biofuel feedstocks.

The papers were dominated by the researchers from the 19 countries, usually through the intercountry institutional collaboration, and they were multiauthored. The United States (six papers); India, China, and England (three papers each); and Canada, Wales, France, Ireland, Northern Ireland, New Zealand, Portugal, and Taiwan (two papers each) were the most prolific countries.

Similarly, all these papers were published in journals indexed by the SCI and/or SCIE. There were no papers indexed by the Arts & Humanities Citation Index (A&HCI) or SSCI. The number of citations ranged from 64 to 1312, and the average number of citations per annum ranged from 10.7 to 187.4. Ten of the papers were articles, while twenty-one of them were reviews.

22.2.2 Nonexperimental Research on Algal Biodiesel

Chisti (2007) discusses algal biodiesel in a review paper with 1312 citations. He argues that renewable, carbon-neutral, transport fuels are necessary for environmental and economic sustainability. Biodiesel derived from oil crops is a potential renewable and carbon-neutral alternative to petroleum fuels. However, biodiesel from oil crops, waste cooking oil, and animal fat cannot realistically satisfy even a small fraction of the existing demand for transport fuels. He asserts that microalgae are the only source of renewable biodiesel that is capable of meeting the global demand for transport fuels. Like plants, microalgae use sunlight to produce oils, but they do so more efficiently than crop plants. Oil productivity of many microalgae greatly exceeds the oil productivity of the best producing oil crops.

Mata et al. (2010) discuss algal biodiesel production and other applications in a review paper with 396 citations. They argue that the first-generation biofuels, primarily produced from food crops and mostly oil seeds, are limited in their ability to achieve targets for biofuel production, climate change mitigation, and economic growth. These concerns have increased the interest in developing second-generation biofuels produced from nonfood feedstocks such as microalgae, which potentially offer greatest opportunities in the longer term. They review the current status of microalgae use for biodiesel production, including their cultivation, harvesting, and processing. They present the microalgal species most used for biodiesel production and their main advantages described in comparison with other available biodiesel feedstocks. They describe the various aspects associated with the design of microalgae production units, giving an overview of the current state of development of algae cultivation systems (photobioreactors and open ponds).

Chisti (2008) discusses algal biodiesel in a review paper with 387 citations. He argues that renewable biofuels are needed to displace petroleum-derived transport fuels, which contribute to global warming and are of limited availability. Biodiesel and bioethanol are the two potential renewable fuels that have attracted the most attention. However, biodiesel and bioethanol produced from agricultural crops using existing methods cannot sustainably replace fossil-based transport fuels. He asserts that biodiesel from microalgae is the only renewable biofuel that has the potential to completely displace petroleum-derived transport fuels without adversely affecting supply of food and other crop products. Most productive oil crops, such as oil palm, do not come close to microalgae in being able to

TABLE 22.1

Nonexperimental Research on Algal Biodiesel

No.	Paper References	Year	Document	Affiliation	Country	No. of Author(s)	Journal	Index	Subjects	Topic	Total No. of Citations	Total Average Citations (Pa)	Rank
1	Chisti	2007	R	Massey Univ.	New Zealand	1	*Biotechnol. Adv.*	SCI, SCIE	Biot. Appl. Microb.	Algal biodiesel	1312	187.4	1
2	Mata et al.	2010	R	Univ. Porto, IPP	Portugal	3	*Renew. Sust. Energ. Rev.*	SCI, SCIE	Ener. Fuels	Algal biodiesel	396	99.0	2
3	Chisti	2008	R	Massey Univ.	New Zealand	1	*Trends Biotechnol.*	SCI, SCIE	Biot. Appl. Microb.	Algal biodiesel	387	64.5	3
4	Schenk et al.	2008	A	Univ. Queensland, Univ. Bielefeld, Univ. Karlsruhe	Germany, Australia	8	*BioEnergy Res.*	SCIE	Ener. Fuels, Env. Sci.	Algal biodiesel production	305	50.8	6
5	Gouveia and Oliveira	2009	A	Inst. Nacl. Engn.	Portugal	2	*J. Ind. Microbiol. Biotechnol.*	SCI, SCIE	Biot. Appl. Microb.	Algal biodiesel production	197	39.4	8
6	Lardon et al.	2009	A	INRA, Montpellier SupAgro	France	5	*Environ. Sci. Technol.*	SCI, SCIE	Eng. Env., Env. Sci.	Algal biodiesel production LCA	193	38.6	9
7	Li et al.	2008a	A	Univ. Ottawa, S. China Normal Univ.	China, Canada	5	*Biotechnol. Prog.*	SCI, SCIE	Biot. Appl. Microb. Food Sci. Tech.	Algal biodiesel production	178	29.7	10
8	Meng et al.	2009	R	Chinese Acad. Sci., Qingdao Agr. Univ.	China	6	*Renew. Energ.*	SCIE	Ener. Fuels	Algal biodiesel production	177	35.4	11
9	Griffiths and Harrison	2009	A	Univ. Cape Town	South Africa	2	*J. Appl. Phycol.*	SCI, SCIE	Biot. Appl. Microb. Mar. Fresh. Biol.	Algal biodiesel production—algal selection	174	34.8	12
10	Vasudevan and Briggs	2008	R	Univ New Hampshire	United States	2	*J. Ind. Microbiol. Biotechnol.*	SCI, SCIE	Biot. Appl. Microb.	Algal biodiesel production	174	29.0	13
11	Wang et al.	2008	R	Univ. Ottawa	Canada	4	*Appl. Microbiol. Biotechnol.*	SCI, SCIE	Biot. Appl. Microb.	Algal biodiesel—CO_2 mitigation	158	26.3	14

(*Continued*)

TABLE 22.1 (*Continued*)

Nonexperimental Research on Algal Biodiesel

No.	Paper References	Year	Document	Affiliation	Country	No. of Author(s)	Journal	Index	Subjects	Topic	Total No. of Citations	Total Average Citations (Pa)	Rank
12	Sialve et al.	2009	R	INRA, INRIA	France	3	*Biotechnol. Adv.*	SCI, SCIE	Biot. Appl. Microb.	Algal biodiesel production	154	25.7	15
13	Li et al.	2008	R	Tsinghua Univ.	China	3	*Appl. Microbiol. Biotechnol.*	SCI, SCIE	Biot. Appl. Microb.	Algal biodiesel production	145	24.2	18
14	Huang et al.	2010	A	China Univ. Min. Technol., Univ. Hong Kong, S. China Univ. Technol.	China	5	*Appl. Energ.*	SCI, SCIE	Ener. Fuels, Eng. Chem.	Algal biodiesel production	143	35.8	19
15	Nigam and Singh	2011	R	Univ. Ulster, Univ. Coll. Cork	Ireland, Northern Ireland	2	*Prog. Energ. Combust. Sci.*	SCI, SCIE	Eng. Chem., Eng. Mech.	Algal biodiesel production	132	44.0	20
16	Gressel	2008	R	Weizmann Inst. Sci., Assif Strategies Ltd.	Israel	1	*Plant Sci.*	SCI, SCIE	Bioch. Mol. Biol., Plant Sci.	Algal biodiesel production—transgenics	131	21.8	21
17	Chen et al.	2011	R	Natl. Cheng Kung Univ., Natl. Taiwan Univ. Sci. Technol.	Taiwan	5	*Bioresour. Technol.*	SCI, SCIE	Agr. Eng., Biot. Appl. Microb., Ener. Fuels	Algal biodiesel production	118	39.3	23
18	Greenwell et al.	2010	R	Univ. Durham, England. NREL, Univ. Swansea	England, United States, Wales	5	*J. R. Soc. Interface*	SCI, SCIE	Mult. Sci.	Algal biodiesel production	112	28.0	25
19	Radakovits et al.	2010	R	Col. Sch. Mines, NREL	United States	4	*Eukaryot. Cell*	SCI, SCIE	Microbiol.	Algal biodiesel production—genetics	111	27.8	27
20	Williams and Laurens	2010	R	Univ. Bangor	Wales	2	*Energ. Environ. Sci.*	SCI, SCIE	Chem. Mult., Ener. Fuels, Chem. Eng., Env. Sci.	Algal biodiesel production	106	26.5	30

(Continued)

TABLE 22.1 (*Continued*)

Nonexperimental Research on Algal Biodiesel

No.	Paper References	Year	Document	Affiliation	Country	No. of Author(s)	Journal	Index	Subjects	Topic	Total No. of Citations	Total Average Citations (Pa)	Rank
21	Akoh et al.	2007	A	Natl. Chung Hsing Univ., Univ. Georgia, Chung Chou Univ. Technol, Natl. Taiwan Normal Univ., Acad. Sinica	Taiwan, United States	4	*J. Agric. Food Chem.*	SCI, SCIE	Agr. Mult., Chem. Appl., Food Sci. Tech.	Algal biodiesel production	101	14.4	31
22	Demirbas	2010	A	Sila Sci.	Turkey	1	*Energ. Educ. Sci. Technol. Part A*	SCIE	Ener. Fuels, Eng. Env., Eng. Chem.	Algal biodiesel production	100	25.0	32
23	Vyas et al.	2010	R	Nirma Univ.	India	3	*Fuel*	SCIE	Enr. Fuels, Eng. Chem.	Algal biodiesel production	96	24.0	34
24	Pienkos and Darzins	2009	A	NREL	United States	2	*Biofuels Bioprod. Biorefining*	SCIE	Biot. Appl. Microb., Ener. Fuels	Algal biodiesel	94	18.8	35
25	Beer et al.	2009	R	Colorado Sch. Mines, Montana St. Univ.	United States	4	*Curr. Opin. Biotechnol.*	SCI, SCIE	Bioch. Res. Meth., Biot. Appl. Microb.	Algal biodiesel	89	17.8	38
26	Scott et al.	2010	R	Univ. Cambridge	England	7	*Curr. Opin. Biotechnol.*	SCI, SCIE	Bioch. Res. Meth., Biot. Appl. Microb.	Algal biodiesel	87	21.8	39

(*Continued*)

TABLE 22.1 (*Continued*)

Nonexperimental Research on Algal Biodiesel

No.	Paper References	Year	Document	Affiliation	Country	No. of Author(s)	Journal	Index	Subjects	Topic	Total No. of Citations	Total Average Citations (Pa)	Rank
27	Singh et al.	2011	A	Univ. Ulster, Univ. Coll. Cork	Ireland, Northern Ireland	3	*Bioresour. Technol.*	SCI, SCIE	Agr. Eng., Biot. Appl. Microb., Ener. Fuels	Algal biodiesel	86	28.7	40
28	Amin	2009	R	BPPT, Tech. Univ. STT Duta Bangsa Bekasi	Indonesia	1	*Energ. Conv. Manag.*	SCIE	Therm., Ener. Fuels, Mechs., Phys. Nucl.	Algal biodiesel	85	17.0	41
29	Patil et al.	2008	R	Norwegian Univ. Life Sci.	Norway	3	*Int. J. Mol. Sci.*	SCIE	Chem. Mult.	Algal biodiesel production	73	12.2	44
30	Singh and Cu	2010	R	Univ. Southampton, Indian Inst Petr.	England, India	2	*Renew. Sust. Energ. Rev.*	SCI, SCIE	Ener. Fuels	Algal biodiesel	70	17.5	47
31	Khan et al.	2009	R	Indian Agr Res Inst, NIPER	India	4	*Renew. Sust. Energ. Rev.*	SCI, SCIE	Ener. Fuels	Algal biodiesel production	64	10.7	50

SCI, Science Citation Index; SCIE, Science Citation Index Expanded; SSCI, Social Sciences Citation Index; A, article; R, review.

sustainably provide the necessary amounts of biodiesel. Similarly, he argues that bioethanol from sugarcane is no match for microalgal biodiesel.

Schenk et al. (2008) discuss the high-efficiency microalgae for biodiesel production in a paper with 305 citations. They note that biodiesel is currently produced from oil synthesized by conventional fuel crops that harvest the sun's energy and store it as chemical energy. This presents a route for renewable and carbon-neutral fuel production. However, current supplies from oil crops and animal fats account for only approximately 0.3% of the current demand for transport fuels. Increasing biofuel production on arable land could have severe consequences for the global food supply. In contrast, they argue that producing biodiesel from algae is one of the most efficient ways of generating biofuels and also represents the only current renewable source of oil that could meet the global demand for transport fuels. The main advantages of second-generation microalgal systems are that they (1) have a higher photon conversion efficiency (as evidenced by increased biomass yields per hectare); (2) can be harvested batchwise nearly all year round, providing a reliable and continuous supply of oil; (3) can utilize salt and waste water streams, thereby greatly reducing freshwater use; (4) can couple CO_2-neutral fuel production with CO_2 sequestration; and (5) produce nontoxic and highly biodegradable biofuels. Current limitations exist mainly in the harvesting process and in the supply of CO_2 for high-efficiency production.

Gouveia and Oliveira (2009) discuss microalgae as a raw material for biofuel production including algal biodiesel in a paper with 197 citations. The screening of microalgae (*Chlorella vulgaris*, *Spirulina maxima*, *Nannochloropsis* sp., *Neochloris oleabundans*, *Scenedesmus obliquus*, and *Dunaliella tertiolecta*) was done in order to choose the best one(s), in terms of quantity and quality as oil source for biofuel production. They find that *N. oleabundans* (fresh water microalga) and *Nannochloropsis* sp. (marine microalga) are suitable as raw materials for biofuel production, due to their high oil content (29.0% and 28.7%, respectively). Both microalgae, when grown under nitrogen shortage, show a great increase (50%) in oil quantity. If the purpose is to produce biodiesel only from one species, they assert that *S. obliquus* has the most adequate fatty acid profile, namely, in terms of linolenic and other polyunsaturated fatty acids. However, the microalgae *N. oleabundans*, *Nannochloropsis* sp., and *D. tertiolecta* can also be used if associated with other microalgal oils and/or vegetable oils.

Lardon et al. (2009) discuss the life-cycle assessment (LCA) of biodiesel production from microalgae in a paper with 193 citations. They provide an analysis of the potential environmental impacts of biodiesel production from microalgae. A comparative LCA study of a virtual facility has been undertaken to assess the energetic balance and the potential environmental impacts of the whole process chain, from the biomass production to the biodiesel combustion. Two different culture conditions, nominal fertilizing or nitrogen starvation, as well as two different extraction options, dry or wet extraction, have been tested. The best scenario has been compared to first-generation biodiesel and oil diesel. Their results confirm the potential of microalgae as an energy source but also highlight the imperative necessity of decreasing energy and fertilizer consumption. Therefore, control of nitrogen stress during the culture and optimization of wet extraction are valuable options.

Li et al. (2008b) discuss algal biofuels including algal biodiesel in a paper with 178 citations. Microalgae have been investigated for the production of a number of different biofuels including biodiesel, bio-oil, biosyngas, and biohydrogen. The production of these biofuels can be coupled with flue gas CO_2 mitigation, wastewater treatment, and the production of high-value chemicals. Microalgal farming can also be carried out with seawater using marine microalgal species as the producers. They expect that developments in microalgal cultivation and downstream processing (e.g., harvesting, drying, and thermochemical processing) further enhance the cost-effectiveness of the biofuel from microalgae strategy.

Meng et al. (2009) discuss biodiesel production from oleaginous microorganisms in a review paper with 177 citations. They note that biodiesel, a mixture of fatty acid methyl esters (FAMEs) derived from animal fats or vegetable oils, is rapidly moving toward the mainstream as an alternative source of energy. However, biodiesel derived from conventional petrol or from oilseeds or animal fat cannot meet realistic needs and can only be used for a small fraction of existing demand for transport fuels. In addition, expensive large acreages for sufficient production of oilseed crops or cost to feed animals are needed for raw oil production. Therefore, oleaginous microorganisms are available for substituting conventional oil in biodiesel production. Most of the oleaginous microorganisms like microalgae, bacillus, fungi, and yeast are all available for biodiesel production. They discuss the regulation mechanism of oil accumulation in microorganism and approach of making microbial diesel economically competitive with petrodiesel.

Griffiths and Harrison (2009) discuss lipid productivity as a key characteristic for choosing algal species for biodiesel production in a paper with 174 citations. They review information available in the literature on microalgal growth rates, lipid content, and lipid productivities for 55 species of microalgae, including 17 Chlorophyta, 11 Bacillariophyta, and 5 Cyanobacteria, as well as other taxa. They find that the data available in the literature are far from complete and rigorous comparison across experiments carried out under different conditions is not possible. However, the collated information provides a framework for decision making and a starting point for further investigation of species selection. They show the importance of lipid productivity as a selection parameter over lipid content and growth rate individually.

Vasudevan and Briggs (2008) discuss biodiesel production with a focus on current state of the art and challenges in a review paper with 174 citations. They note that since biodiesel is made entirely from vegetable oil or animal fats, it is renewable and biodegradable. The majority of biodiesel today is produced by alkali-catalyzed transesterification with methanol, which results in a relatively short reaction time. However, the vegetable oil and alcohol must be substantially anhydrous and have low free fatty acid content, because the presence of water or free fatty acid or both promotes soap formation. They examine different biodiesel sources (edible and nonedible), virgin oil versus waste oil, algae-based biodiesel that is gaining increasing importance, role of different catalysts including enzyme catalysts, and the current state of the art in biodiesel production.

Wang et al. (2008) discuss CO_2 biomitigation using microalgae in a review paper with 158 citations. They note that microalgae are a group of unicellular or simple multicellular photosynthetic microorganisms that can fix CO_2 efficiently from different sources, including the atmosphere, industrial exhaust gases, and soluble carbonate salts. They assert that combination of CO_2 fixation, biofuel production, and wastewater treatment may provide a very promising alternative to current CO_2 mitigation strategies.

Sialve et al. (2009) discuss the anaerobic digestion of microalgae as a necessary step to make microalgal biodiesel sustainable in a review paper with 154 citations. They argue that anaerobic digestion is a key process that can solve the waste issue as well as the economical and energetic balance of such a promising technology. Indeed, the conversion of algal biomass after lipid extraction into methane is a process that can recover more energy than that from cell lipids. They identify three main bottlenecks to digest microalgae. First, the biodegradability of microalgae can be low depending on both the biochemical composition and the nature of the cell wall. Then, the high cellular protein content results in ammonia release, which can lead to potential toxicity. Finally, the presence of sodium in marine species can also affect the digester performance. They assert that physicochemical pretreatment, codigestion, and control of gross composition are strategies that can significantly and efficiently increase the conversion yield of the algal organic matter into methane. They conclude that when the cell lipid content does not exceed 40%, anaerobic digestion of the whole biomass is the optimal strategy on an energy balance basis for the energetic recovery of cell biomass.

Li et al. (2008a) discuss the perspectives of microbial oils for biodiesel production in a review paper with 145 citations. They note that recently, much attention has been paid to the development of microbial oils, and it has been found that many microorganisms, such as algae, yeast, bacteria, and fungi, have the ability to accumulate oils under some special cultivation conditions. Compared to other plant oils, microbial oils have many advantages, such as short life cycle; less labor required; less affection by venue, season, and climate; and easier to scale up. With the rapid expansion of biodiesel, they argue that microbial oils might become one of the potential oil feedstocks for biodiesel production in the future, though there are many works associated with microorganisms producing oils that need to be carried out further. They review related research about different oleaginous microorganisms producing oils and discuss the prospects of such microbial oils used for biodiesel production.

Huang et al. (2010) discuss the biodiesel production by microalgal biotechnology in a paper with 143 citations. They note that microalgal biotechnology possesses high potential for biodiesel production because a significant increase in lipid content of microalgae is now possible through heterotrophic cultivation and genetic engineering approaches. They provide an overview of the technologies in the production of biodiesel from microalgae, including the various modes of cultivation for the production of oil-rich microalgal biomass, as well as the subsequent downstream processing for biodiesel production. They then discuss the advances and prospects of using microalgal biotechnology for biodiesel production.

Nigam and Singh (2011) discuss the production of liquid biofuels including algal biodiesel from renewable resources in a recent review paper with 132 citations. They note that researchers have been redirecting their interests in biomass-based fuels, which currently seem to be the only logical alternative for sustainable development in the context of economical and environmental considerations. Renewable bioresources are available globally in the form of residual agricultural biomass and wastes, which can be transformed into liquid biofuels. However, the process of conversion, or chemical transformation, could be very expensive and not worthwhile to use for an economical large-scale commercial supply of biofuels. Hence, they argue that there is still need for much research to be done for an effective, economical, and efficient conversion process.

Gressel (2008) discusses the transgenics for algae as a feedstock for algal biodiesel in a review paper with 131 citations. They note that algae and cyanobacteria for third-generation biodiesel need transgenic manipulation to deal with weeds, light penetration, photoinhibition, carbon assimilation, etc. The possibilities of producing fourth-generation biohydrogen and bioelectricity using photosynthetic mechanisms are being explored. There seem to be no health or environmental impact study requirements when undomesticated biofuel crops are grown; they note that yet there are illogically stringent requirements, they should transgenically be rendered less toxic and more efficient as biofuel crops.

Chen et al. (2011) discuss the cultivation, photobioreactor design, and harvesting of microalgae for biodiesel production in a recent review paper with 118 citations. They present recent advances in microalgal cultivation, photobioreactor design, and harvesting technologies with a focus on microalgal oil (mainly triglycerides) production. They compare and critically discuss the effects of different microalgal metabolisms (i.e., phototrophic, heterotrophic, mixotrophic, and photoheterotrophic growth), cultivation systems (emphasizing the effect of light sources), and biomass harvesting methods (chemical/physical methods) on microalgal biomass and oil production.

Greenwell et al. (2010) discuss microalgae on the biofuel priority list with a focus on the technological challenges in a review paper with 112 citations. They critically review current designs of algal culture facilities, including photobioreactors and open ponds, with regard to photosynthetic productivity and associated biomass and oil production, and include an analysis of alternative approaches using models, balancing space needs, productivity, and biomass concentrations, together with nutrient requirements. In light of the current interest in synthetic genomics and genetic modifications, they also evaluate options for potential metabolic engineering of the lipid biosynthesis pathways of microalgae. They conclude that although significant literature exists on microalgal growth and biochemistry, significantly more work needs to be undertaken to understand and potentially manipulate algal lipid metabolism. Furthermore, with regard to chemical upgrading of algal lipids and biomass, they describe alternative fuel synthesis routes and discuss and evaluate the application of catalysts traditionally used for plant oils. They argue that simulations that incorporate financial elements, along with fluid dynamics and algae growth models, are likely to be increasingly useful for predicting reactor design efficiency and life cycle analysis to determine the viability of the various options for large-scale culture. They assert that the greatest potential for cost reduction and increased yields most probably lies within closed or hybrid closed–open production systems.

Radakovits et al. (2010) discuss the genetic engineering of algae for enhanced biofuel production in a review paper with 111 citations. They argue that although the application of genetic engineering to improve energy production phenotypes in eukaryotic microalgae is in its infancy, significant advances in the development of genetic manipulation tools have recently been achieved with microalgal model systems and are being used to manipulate central carbon metabolism in these organisms. It is likely that many of these advances can be extended to industrially relevant organisms. They focus on potential avenues of genetic engineering that may be undertaken in order to improve microalgae as a biofuel platform for the production of biohydrogen, starch-derived alcohols, diesel fuel surrogates, and/or alkanes.

Williams and Laurens (2010) discuss microalgae as biodiesel and biomass feedstocks with a focus on the biochemistry, energetics, and economics in a review paper with 106 citations. They show that the maximum conversion efficiency of total solar energy into primary photosynthetic organic products falls in the region of 10%. Biomass biochemical composition further conditions this yield: 30% and 50% of the primary product mass is lost on producing cell protein and lipid. Obtained yields are 1/3rd to 1/10th of the theoretical ones. Wasted energy from captured photons is a major loss and a major challenge in maximizing mass algal production. Using irradiance data and kinetic parameters derived from reported field studies, they produce a simple model of algal biomass production and its variation with latitude and lipid content. An economic analysis of algal biomass production considers a number of scenarios and the effect of changing individual parameters. They find that (1) the biochemical composition of biomass influences the economics (in particular, increased lipid content reduces other valuable compounds in the biomass); (2) the *biofuel only* option is unlikely to be economically viable; and (3) among the hardest problems in assessing the economics are the cost of the CO_2 supply and uncertain nature of downstream processing.

Akoh et al. (2007) discuss the enzymatic approach to biodiesel production in a paper with 101 citations. They note that enzymatic reactions involving lipases can be an excellent alternative to produce biodiesel through a process commonly referred to alcoholysis, a form of transesterification reaction, or through an interesterification (ester interchange) reaction. Protein engineering can be useful in improving the catalytic efficiency of lipases as biocatalysts for biodiesel production. They argue that the use of recombinant DNA technology to produce large quantities of lipases and the use of immobilized lipases and immobilized whole cells may lower the overall cost, while presenting less downstream processing problems, to biodiesel production. In addition, they assert that the enzymatic approach is environmentally friendly, is considered a *green reaction*, and needs to be explored for industrial production of biodiesel.

Demirbas (2010) discusses microalgae as a feedstock for biodiesel in a paper with 100 citations. He notes that the oil productivity of many microalgae exceeds the best producing oil crops. In recent years, use of microalgae as an alternative biodiesel feedstock has gained renewed interest from researchers, entrepreneurs, and the general public. Biodiesel produced from microalgae is being investigated as an alternative. The lipid and fatty acid contents of microalgae vary in accordance with culture conditions. The average fatty acid contents of the algal oils are 36% oleic (18:1), 15% palmitic (16:0), 11% stearic (18:0), 8.4% iso (17:0), and 7.4% linoleic (18:2).

Vyas et al. (2010) discuss FAME production processes for biodiesel production in a review paper with 96 citations. They explore the various production processes, few of which are applied at an industrial level also, to produce basically FAME (later can be utilized as biodiesel after purification). They argue that transesterification of vegetable oils/fats and extraction from algae are the leading process options for biodiesel production on large scale. They review briefly the literature on transesterification reaction using homogeneous, heterogeneous, and enzyme catalysts.

Pienkos and Darzins (2009) discuss the promise and challenges of microalgal-derived biofuels including algal biodiesel in a paper with 94 citations. They note that the interest in algal biofuels has been growing recently due to increased concern over peak oil, energy security, greenhouse gas emissions, and the potential for other biofuel feedstocks to compete for limited agricultural resources. The high productivity of algae suggests that much of the U.S. transportation fuel needs can be met by algal biofuels at a production cost competitive with the cost of petroleum seen during the early part of 2008. Development of algal biomass production technology, however, they argue remains in its infancy. They provide a brief overview of past algal research sponsored by the DoE, the potential of microalgal biofuels, and a discussion of the technical and economic barriers that need to be overcome before production of microalgal-derived diesel-fuel substitutes can become a large-scale commercial reality.

Beer et al. (2009) discuss the engineering algae for biohydrogen and biofuel production including algal biodiesel in a review paper with 89 citations. They note that recently, there has been considerable progress in identifying relevant bioenergy genes and pathways in microalgae, and powerful genetic techniques have been developed to engineer some strains via the targeted disruption of endogenous genes and/or transgene expression. Collectively, the progress that has been realized in these areas is rapidly advancing our ability to genetically optimize the production of targeted biofuels.

Scott et al. (2010) discuss the challenges and prospects relating to algal biodiesel production in a review paper with 87 citations. They note that microalgae offer great potential for exploitation, including the production of biodiesel, but the process is still some way from being carbon neutral or commercially viable. Part of the problem is that there is little established background knowledge in the area. They argue that we should look both to achieve incremental steps and to increase our fundamental understanding of algae to identify potential paradigm shifts. In doing this, integration of biology and engineering will be essential. They present an overview of a potential algal biofuel pipeline and focus on recent work that tackles optimization of algal biomass production and the content of fuel molecules within the algal cell.

Singh et al. (2011) discuss the renewable fuels from algae, including algal biodiesel in a recent paper with 86 citations. They review the utilization of first- and second-generation biofuels as the suitable alternatives to depleting fossil fuels. Then the concern is presented over a debate on the most serious problem arising from the production of these biofuels, which is the increase of food market prices because of the increased use of arable land for the cultivation of biomass used for the production of first- and second-generation biofuels. The solution to this debate has been suggested with the use of nonarable land for the cultivation of algal biomass for the generation of third-generation biofuels.

Amin (2009) discusses biofuel oil and gas production processes from microalgae in a paper with 85 citations. This paper presents a brief review on the main conversion processes of microalgae into energy. Since microalgae have high water content, not all biomass energy conversion processes can be applied. By using thermochemical processes, oil and gas can be produced, and by using biochemical processes, ethanol and biodiesel can be produced. The properties of the microalgae product are almost similar to those of fish and vegetable oils, and therefore, it can be considered as a substitute of fossil oil.

Patil et al. (2008) discuss the sustainable production of biofuels from microalgae in a review paper with 83 citations. They note that the viability of first-generation biofuel production is questionable because of the conflict with food supply, while the microalgal biofuels are a viable alternative. The oil productivity of many microalgae exceeds the best producing oil crops. They analyze and promote integration approaches for sustainable microalgal biofuel production to meet the energy and environmental needs of the society. The emphasis is on hydrothermal liquefaction technology for direct conversion of algal biomass to liquid fuel.

Singh and Cu (2010) discuss the commercialization potential of microalgae for biofuel production in a review paper with 70 citations as they present a critical appraisal of the commercialization potential of microalgae biofuels. The available literature on various aspects of microalgae, for example, its cultivation, LCA, and conceptualization of an algal biorefinery, has been scanned, and a critical analysis has been presented. A critical evaluation of the available information suggests that the economic viability of the process in terms of minimizing the operational and maintenance cost along with maximization of oil-rich microalgae production is the key factor for successful commercialization of microalgae-based fuels.

Khan et al. (2009) discuss the prospects of biodiesel production from microalgae in India in a review paper with 64 citations. Energy is essential and vital for development, and the global economy literally runs on energy. They argue that the production of biodiesel using microalgae biomass is a viable alternative. The oil productivity of many microalgae exceeds the best producing oil crops. Microalgae are photosynthetic microorganisms that convert sunlight, water, and CO_2 to sugars, from which macromolecules, such as lipids and triacylglycerols (TAGs), can be obtained. These TAGs are the promising and sustainable feedstock for biodiesel production. A microalgal biorefinery approach can be used to reduce the cost of making microalgal biodiesel. They finally argue that microalgal-based carbon sequestration technologies cover the cost of carbon capture and sequestration.

22.2.3 Conclusion

The nonexperimental research on algal biodiesel has been one of the most dynamic research areas in recent years. Thirty-one citation classics in the field of algal biodiesel with more than 64 citations were located, and the key emerging issues from these papers were presented in the decreasing order of the number of citations. These papers give strong hints about the determinants of efficient algal biodiesel production processes and policies and emphasize that marine algae are efficient biodiesel feedstocks in comparison with first- and second-generation biofuel feedstocks.

22.3 Experimental Studies on Algal Biodiesel

22.3.1 Introduction

The experimental research on algal biodiesel has been one of the most dynamic research areas in recent years. Nineteen experimental citation classics in the field with more than 65 citations were located, and the key emerging issues from these papers are presented in the decreasing order of the number of citations (Table 22.2). These papers give strong hints about the determinants of efficient algal biodiesel production and emphasize that marine algae are efficient biodiesel feedstocks in comparison with first- and second-generation biofuel feedstocks.

The papers were dominated by the researchers from eight countries, usually through the intracountry institutional collaboration, and they were multiauthored. The United States (seven papers), China (five papers), and Italy and South Korea (two papers each) were the most prolific countries. Similarly, Tsinghua University (four papers) and Virginia Polytechnic Institute and State University (three papers) were the most prolific institutions.

Similarly, all these papers were published in the journals indexed by the SCI and/or SCIE. There were no papers indexed by the A&HCI or SSCI. The number of citations ranged from 65 to 337, and the average number of citations per annum ranged from 10.8 to 67.4. All of these papers were articles.

22.3.2 Experimental Research on Algal Biodiesel

Rodolfi et al. (2009) study the strain selection, induction of lipid synthesis, and outdoor mass cultivation in a low-cost photobioreactor for algal biodiesel production in a paper with 337 citations. Thirty microalgal strains were screened in the laboratory for their biomass productivity and lipid content. Four strains (two marine and two freshwater), selected because they are robust and highly productive and have a relatively high lipid content, were cultivated under nitrogen deprivation in 0.6 L bubbled tubes. They find that only the two marine microalgae accumulated lipid under such conditions. One of them, the eustigmatophyte *Nannochloropsis* sp. F&M-M24, which attained 60% lipid content after nitrogen starvation, was grown in a 20 L flat alveolar panel photobioreactor to study the influence of irradiance and nutrient (nitrogen or phosphorus) deprivation on fatty acid accumulation. They find that "fatty acid content increased with high irradiances (up to 32.5% of dry biomass) and following both nitrogen and phosphorus deprivation (up to about 50%)." To evaluate its lipid production potential under natural sunlight,

TABLE 22.2

Experimental Research on Algal Biodiesel

No.	Paper References	Year	Document	Affiliation	Country	No. of Author(s)	Journal	Index	Subjects	Topic	Total No. of Citations	Total Average Citations (Pa)	Rank
1	Rodolfi et al.	2009	A	Univ. Florence, CNR	Italy	7	*Biotechnol. Bioeng.*	SCI, SCIE	Biot. Appl. Microb.	Algal biodiesel production	337	67.4	4
2	Miao and Wu	2006	A	Tsing Hua Univ., Ningde Teachers Coll.	China	2	*Bioresour. Technol.*	SCI, SCIE	Agr. Eng., Biot. Appl. Microb., Ener. Fuels	Algal biodiesel production	306	38.3	5
3	Xu et al.	2006	A	Tsing Hua Univ.	China	3	*J. Biotechnol.*	SCI, SCIE	Biot. Appl. Microb.	Algal biodiesel production	221	27.6	7
4	Li et al.	2008b	A	Univ. Ottawa, S. China Normal Univ.	Canada, China	4	*Appl. Microbiol. Biotechnol.*	SCI, SCIE	Biot. Appl. Microb.	Algal biodiesel production	153	25.5	16
5	Schirmer et al.	2010	A	LS9 Inc	United States	5	*Science*	SCI, SCIE	Mult. Sci.	Algal biodiesel production	150	37.5	17
6	Miao et al.	2004	A	Tsing Hua Univ., Ningde Teachers Coll., Chinese Acad. Sci.	China	3	*J. Anal. Appl. Pyrolysis*	SCI, SCIE	Chem. Anal., Spectroscopy	Algal biodiesel production—Fast pyrolysis	120	12.0	22
7	Li et al.	2007	A	Tsing Hua Univ.	China	3	*Biotechnol. Bioeng.*	SCI, SCIE	Biot. Appl. Microb.	Algal biodiesel production	115	16.4	24
8	Converti et al.	2009	A	Univ. Genoa	Italy	5	*Chem. Eng. Process.*	SCI, SCIE	Ener. Fuels, Chem. Eng.	Algal biodiesel production	112	22.4	26
9	Chiu et al.	2009	A	Natl. Chiao Tung Univ.	Taiwan	6	*Bioresour. Technol.*	SCI, SCIE	Agr. Eng., Biot. Appl. Microb., Ener. Fuels	Algal biodiesel production	109	21.8	28
10	Xiong et al.	2008	A	Tsing Hua Univ.	China	4	*Appl. Microbiol. Biotechnol.*	SCI, SCIE	Biot. Appl. Microb.	Algal biodiesel production	109	18.2	29
11	Liang et al.	2009	A	So Illinois Univ.	United States	3	*Biotechnol. Lett.*	SCI, SCIE	Biot. Appl. Microb.	Algal biodiesel production	97	19.4	33

(Continued)

TABLE 22.2 (*Continued*)

Experimental Research on Algal Biodiesel

No.	Paper References	Year	Document	Affiliation	Country	No. of Author(s)	Journal	Index	Subjects	Topic	Total No. of Citations	Total Average Citations (Pa)	Rank
12	Chi et al.	2007	A	Virginia Polytech. Inst. St. Univ., Wash. St. Univ.	United States	5	*Process Biochem.*	SCI, SCIE	Bioch. Mol. Biol., Biot. Appl. Microb., Eng. Chem.	Algal biodiesel production	94	13.4	36
13	Lee et al.	2010	A	KRIBB	South Korea	5	*Bioresour. Technol.*	SCI, SCIE	Agr. Eng., Biot. Appl. Microb., Ener. Fuels	Algal biodiesel production	89	22.3	37
14	McNeff et al.	2008	A	SorTec Corp., Augsburg Coll., Univ. Minnesota	United States	9	*Appl. Catal. A Gen.*	SCI, SCIE	Chem. Phys., Env. Sci.	Algal biodiesel production	77	12.8	42
15	Chinnasamy et al.	2010	A	Univ. Georgia, Maharshi Dayanand Saraswati Univ.	United States, India	4	*Bioresour. Technol.*	SCI, SCIE	Agr. Eng., Biot. Appl. Microb., Ener. Fuels	Algal biodiesel production	73	18.3	43
16	Johnson and Wen	2009	A	Virginia Polytech. Inst. St. Univ.	United States	2	*Energ. Fuel*	SCI, SCIE	Ener. Fuels, Eng. Chem.	Algal biodiesel production	72	14.4	46
17	Stephenson et al.	2010	A	Univ. Cambridge	England	6	*Energ. Fuel*	SCI, SCIE	Ener. Fuels, Eng. Chem.	Algal biodiesel—LCA	70	17.5	45
18	Yoo et al.	2010	A	KRIBB	South Korea	6	*Bioresour. Technol.*	SCI, SCIE	Agr. Eng., Biot. Appl. Microb., Ener. Fuels	Algal biodiesel production	68	17.0	48
19	Pyle et al.	2008	A	Virginia Polytech. Inst. St. Univ., USDA	United States	3	*J. Agric. Food Chem.*	SCI, SCIE	Agr. Mult., Chem. Appl., Food Sci. Tech.	Algal biodiesel production	65	10.8	49

SCI, Science Citation Index; SCIE, Science Citation Index Expanded; SSCI, Social Sciences Citation Index; A, article; R, review.

the strain was grown outdoors in 110 L green wall panel photobioreactors under nutrient-sufficient and nutrient-deficient conditions. They find that "lipid productivity increased from 117 mg/L/day in nutrient sufficient media (with an average biomass productivity of 0.36 g/L/day and 32% lipid content) to 204 mg/L/day (with an average biomass productivity of 0.30 g/L/day and more than 60% final lipid content) in nitrogen deprived media." They argue that in a two-phase cultivation process (a nutrient-sufficient phase to produce the inoculum followed by a nitrogen-deprived phase to boost lipid synthesis), the oil production potential could be projected to be more than 90 kg/ha/day. They assert that "this marine eustigmatophyte has the potential for an annual production of 20 tons of lipid per hectare in the Mediterranean climate and of more than 30 tons of lipid per hectare in sunny tropical areas."

Miao and Wu (2006) study biodiesel production from heterotrophic microalgal oil in a paper with 306 citations as they introduce an integrated method for the production of biodiesel from microalgal oil. Heterotrophic growth of *C. protothecoides* resulted in the accumulation of high lipid content (55%) in cells. A large amount of microalgal oil was efficiently extracted from these heterotrophic cells by using n-hexane. Biodiesel comparable to conventional diesel was obtained from heterotrophic microalgal oil by acidic transesterification. They find that the "best process combination was 100% catalyst quantity (based on oil weight) with 56:1 molar ratio of methanol to oil at temperature of 30°C, which reduced product specific gravity from an initial value of 0.912 to a final value of 0.8637 in about 4 h of reaction time." They assert that the new process, which combined bioengineering and transesterification, was a feasible and effective method for the production of high-quality biodiesel from microalgal oil.

Xu et al. (2006) study high-quality biodiesel production from a microalga *C. protothecoides* by heterotrophic growth in fermenters in a paper with 221 citations. The technique of metabolic controlling through heterotrophic growth of *C. protothecoides* was applied, and the heterotrophic *C. protothecoides* contained the crude lipid content of 55.2%. To increase the biomass and reduce the cost of alga, corn powder hydrolysate instead of glucose was used as organic carbon source in heterotrophic culture medium in fermenters. They find that "cell density significantly increased under the heterotrophic condition, and the highest cell concentration reached 15.5 g L^{-1}." A large amount of microalgal oil was efficiently extracted from the heterotrophic cells by using n-hexane and then transmuted into biodiesel by acidic transesterification. The "biodiesel was characterized by a high heating value of 41 MJ kg^{-1}, a density of 0.864 kg L^{-1}, and a viscosity of 5.2×10^{-4} Pas (at 40°C)." They assert that this method has great potential in the industrial production of liquid fuel from microalga.

Li et al. (2008c) study the effects of nitrogen sources on cell growth and lipid accumulation of green alga *N. oleoabundans* in a paper with 153 citations. While the highest lipid cell content of 0.40 g/g was obtained at the lowest sodium nitrate concentration (3 mM), a "remarkable lipid productivity of 0.133 g l^{-1} day^{-1} was achieved at 5 mM with a lipid cell content of 0.34 g/g and a biomass productivity of 0.40 g l^{-1} day^{-1}." Then they find that the "highest biomass productivity was obtained at 10 mM sodium nitrate, with a biomass concentration of 3.2 g/l and a biomass productivity of 0.63 g l^{-1} day^{-1}." Cell growth continued after the exhaustion of the external nitrogen pool, hypothetically supported by the consumption of intracellular nitrogen pools such as chlorophyll molecules.

Schirmer et al. (2010) study the microbial biosynthesis of alkanes in a paper with 150 citations. They describe the discovery of an alkane biosynthesis pathway from cyanobacteria. The pathway consists of an acyl–acyl carrier protein reductase and an aldehyde decarbonylase, which together convert intermediates of fatty acid metabolism to alkanes and alkenes. "The aldehyde decarbonylase is related to the broadly functional nonheme diiron enzymes. Heterologous expression of the alkane operon in *Escherichia coli* leads to the production and secretion of C_{13} to C_{17} mixtures of alkanes and alkenes." They argue that these genes and enzymes can now be leveraged for the simple and direct conversion of renewable raw materials to fungible hydrocarbon fuels.

Miao et al. (2004) study the fast pyrolysis of microalgae to produce renewable fuels in a paper with 120 citations. Fast pyrolysis tests of microalgae were performed in a fluid bed reactor. The experiments were completed at a temperature of 500°C with a heating rate of 600°/s and a sweep gas (N_2) flow rate of 0.4 m^3/h and a vapor residence time of 2–3 s. In comparison with the previous studies on slow pyrolysis from microalgae in an autoclave, a greater amount of high-quality bio-oil can be directly produced from continuously processing microalgae feeds at a rate of 4 g/min in their work, which has a potential for commercial application of large-scale production of liquid fuels. The liquid product yields of 18% and 24% from fast pyrolysis of *C. protothecoides* and *Microcystis aeruginosa*, respectively, were obtained. The saturated and polar fractions account for 1.14% and 31.17% of the bio-oils of microalgae on average, which are higher than those of bio-oil from wood. The H/C and O/C molar ratios of microalgae bio-oil are 1.7 and 0.3, respectively. The gas chromatograph analyses showed that the "distribution of straight-chain alkanes of the saturated fractions from microalgae biooils were similar to diesel fuel." They find that bio-oil product from fast pyrolysis microalgae is characterized by "low oxygen content with a higher heating value of 29 MJ/kg, a density of 1.16 kg l^{-1} and a viscosity of 0.10 Pa s." They argue that these properties of bio-oil of microalgae make it more suitable for fuel oil use than fast pyrolysis oils from lignocellulosic materials.

Li et al. (2007) discuss large-scale biodiesel production from microalga *C. protothecoides* through heterotrophic cultivation in bioreactors in a paper with 115 citations. They focused on scaling up fermentation in bioreactors. Through substrate feeding and fermentation process controls, the "cell density of *C. protothecoides* achieved 15.5 g L^{-1} in 5 L, 12.8 g L^{-1} in 750 L, and 14.2 g L^{-1} in 11,000 L bioreactors, respectively." Resulting from heterotrophic metabolism, the "lipid content reached 46.1%, 48.7%, and 44.3% of cell dry weight in samples from 5 L, 750 L, and 11,000 L bioreactors, respectively. Transesterification of the microalgal oil was catalyzed by immobilized lipase from *Candida* sp. 99–125. With 75% lipase 12,000 U g^{-1}, (based on lipid quantity) and 3:1 molar ratio of methanol to oil batch-fed at three times, 98.15% of the oil was converted to monoalkyl esters of fatty acids in 12 h. The expanded biodiesel production rates were 7.02 g L^{-1}, 6.12 g L^{-1}, and 6.24 g L^{-1} in 5 L, 750 L, and 11,000 L bioreactors, respectively." They argue that the properties of biodiesel from *Chlorella* were comparable to conventional diesel fuel and comply with the U.S. Standard for Biodiesel (ASTM 6751) and assert that it is feasible to expand heterotrophic *Chlorella* fermentation for biodiesel production at the industry level.

Converti et al. (2009) study the effect of temperature and nitrogen concentration on the growth and lipid content of *Nannochloropsis oculata* and *C. vulgaris* for biodiesel production in a paper with 112 citations. In addition, they investigate various lipid extraction methods. The extracted lipids were quantitatively and qualitatively analyzed by gravimetric and gas chromatographic methods, respectively, in order to check their suitability according to the European standards for biodiesel. They find that the lipid content of microalgae was strongly influenced by the variation of tested parameters; indeed, "an increase in temperature from 20 to 25°C practically doubled the lipid content of *N. oculata* (from 7.90 to 14.92%), while an increase from 25 to 30°C brought about a decrease of the lipid content of *C. vulgaris* from 14.71 to 5.90%." On the other hand, they find that "a 75% decrease of the nitrogen concentration in the medium with respect to the optimal values for growth, increased the lipid fractions of *N. oculata* from 7.90 to 15.31% and of *C. vulgaris* from 5.90 to 16.41%, respectively."

Chiu et al. (2009) study the lipid accumulation and CO_2 utilization of *N. oculata* in response to CO_2 aeration in a paper with 109 citations. They also explore the lipid content of *N. oculata* cells at different growth phases. The lipid accumulation from logarithmic phase to stationary phase of *N. oculata* NCTU-3 was significantly increased from 30.8% to 50.4%. In the microalgal cultures aerated with 2%, 5%, 10%, and 15% CO_2, they find that the "maximal biomass and lipid productivity in the semicontinuous system were 0.480 and 0.142 g L^{-1} d^{-1} with 2% CO_2 aeration, respectively. Even the *N. oculata* NCTU-3 cultured in the semicontinuous system aerated with 15% CO_2, the biomass and lipid productivity could reach to 0.372 and 0.084 g L^{-1} d^{-1}, respectively." In the comparison of productive efficiencies, the semicontinuous system was operated with two culture approaches over 12 days. They find that the "biomass and lipid productivity of *N. oculata* NCTU-3 were 0.497 and 0.151 g L^{-1} d^{-1} in one-day replacement (half broth was replaced each day), and were 0.296 and 0.121 g L^{-1} d^{-1} in three-day replacement (three fifth broth was replaced every 3 d), respectively." To optimize the condition for long-term biomass and lipid yield from *N. oculata* NCTU-3, they recommend that this microalga be grown in the semicontinuous system aerated with 2% CO_2 and operated by 1-day replacement.

Xiong et al. (2008) study the high-density fermentation of microalga *C. protothecoides* in bioreactor for microbiodiesel production in a paper with 109 citations. They find that "cell density achieved was 16.8 g 1^{-1} in 184 h and 51.2 g 1^{-1} in 167 h in a 5-1 bioreactor by performing preliminary and improved fed-batch culture strategy, respectively. The lipid content was 57.8, 55.2, and 50.3% of cell dry weight from batch, primary, and improved fed-batch culture in 5-1 bioreactor." Transesterification was catalyzed by immobilized lipase, and the conversion rate reached up to 98%. They argue that the properties of biodiesel from *Chlorella* were comparable to conventional diesel fuel and comply with U.S. Standard for Biodiesel. They assert that the approach including high-density fermentation of *Chlorella* and enzymatic transesterification process was a promising alternative for biodiesel production.

Liang et al. (2009) study the biomass and lipid productivities of *C. vulgaris* under autotrophic, heterotrophic, and mixotrophic growth conditions for biodiesel production in a paper with 97 citations. They find that while "autotrophic growth did provide higher cellular lipid content (38%), the lipid productivity was much lower compared with those from heterotrophic growth with acetate, glucose, or glycerol whilst the optimal cell growth (2 g l^{-1}) and lipid productivity (54 mg l^{-1} day^{-1}) were attained using glucose at 1% (w/v) whereas higher concentrations were inhibitory." Growth of *C. vulgaris* on glycerol had similar dose effects as those from glucose. Overall, they argue that *C. vulgaris* is mixotrophic.

Chi et al. (2007) study the production of the docosahexaenoic acid (DHA, 22:6 n – 3) from biodiesel-waste glycerol by microalgal fermentation (*Schizochytrium limacinum*) in a paper with 94 citations. Crude glycerol is the primary by-product in the biodiesel industry, which is too costly to be purified into higher-quality products used in the health and cosmetics industries. They find that "crude glycerol supported alga growth and DHA production, with 75–100 g/L concentration being the optimal range." Among other medium and environmental factors influencing DHA production, "temperature, trace metal (PI) solution concentration, ammonium acetate, and NH_4Cl had significant effects ($P < 0.1$)." They then find that "their optimal values were determined 30 mL/L of PI, 0.04 g/L of NH_4Cl, 1.0 g/L of ammonium acetate, and 19.2°C. A highest DHA yield of 4.91 g/L with 22.1 g/L cell dry weight was obtained." They assert that biodiesel-derived crude glycerol is a promising feedstock for production of DHA from heterotrophic algal culture.

Lee et al. (2010) study the effective lipid extraction from microalgae in a paper with 89 citations. They test various methods, including autoclaving, bead beating, microwaves, sonication, and a 10% NaCl solution, to identify the most effective cell disruption method. The total lipids from *Botryococcus* sp., *C. vulgaris*, and *Scenedesmus* sp. were extracted using a mixture of chloroform and methanol (1:1). They find that the "lipid contents from the three species were 5.4–11.9, 7.9–8.1, 10.0–28.6, 6.1–8.8, and 6.8–10.9 g L^{-1} when using autoclaving bead-beating, microwaves, sonication, and a 10% NaCl solution, respectively." Additionally, "*Botryococcus* sp. showed the highest oleic acid productivity at 5.7 mg L^{-1} d^{-1} when the cells were disrupted using the microwave oven method." They assert that the microwave oven method was the most simple, easy, and effective for lipid extraction from microalgae.

McNeff et al. (2008) study a continuous catalytic system (fixed-bed reactor process) using a metal oxide–based catalyst for biodiesel production in a paper with 77 citations. They show that "porous zirconia, titania and alumina micro-particulate heterogeneous catalysts are capable of continuous rapid esterification and transesterification reactions under high pressure (ca. 2500 psi) and elevated temperature (300–450°C)." They then describe the continuous transesterification of triglycerides and simultaneous esterification of free fatty acids with residence times as low as 5.4 s. They find that "biodiesel produced from soybean oil, acidulated soapstock, tall oil, algae oil, and corn oil with different alcohols to make different alkyl esters using this new process pass all current ASTM testing specifications." Furthermore, they argue that the economics of this novel process is much more cost competitive due to the use of inexpensive lipid feedstocks that often contain high levels of free fatty acids. They show that the process easily scales up to a factor of 49 for more than 115 h of continuous operation without loss of conversion efficiency.

Chinnasamy et al. (2010) study the microalgae cultivation in a wastewater dominated by carpet mill effluents (85%–90%) for biodiesel applications in a paper with 73 citations. Native algal strains were isolated from carpet wastewater. Preliminary growth studies indicated both freshwater and marine algae showed good growth in wastewaters. They find that a "consortium of 15 native algal isolates showed >96% nutrient removal in treated wastewater. Furthermore, biomass production potential and lipid content of this consortium cultivated in treated wastewater were similar to 9.2–17.8 tons ha^{-1} $year^{-1}$ and 6.82%, respectively." They estimate that about 63.9% of algal oil obtained from the consortium could be converted into biodiesel. However, they caution that further studies on anaerobic digestion and thermochemical liquefaction are required to make this consortium approach economically viable for producing algae biofuels.

Johnson and Wen (2009) study the production of biodiesel fuel from the microalga *S. limacinum* by direct transesterification of algal biomass in a paper with 72 citations. When freeze-dried biomass was used as feedstock, they find that the "two-stage method resulted in 57% of crude biodiesel yield (based on algal biomass) with a fatty acid methyl ester (FAME) content of 66.37%." Then the "one-stage method (with chloroform, hexane, or petroleum ether used in transesterification) led to a high yield of crude biodiesel, whereas only chloroform-based transesterification led to a high FAME content." When wet biomass was used as feedstock, the one-stage method resulted in a much-lower biodiesel yield. The biodiesel prepared via the direct transesterification of dry biomass was subjected to ASTM standard tests. They find that "parameters such as free glycerol, total glycerol, acid number, soap content, corrosiveness to copper, flash point, viscosity, and particulate matter met the ASTM standards, while the water and sediment content, as well as the sulfur content did not pass the standard." They assert that the alga *S. limanicum* is a suitable feedstock for producing biodiesel via the direct transesterification method.

Stephenson et al. (2010) discuss the LCA of potential algal biodiesel production with a focus on a comparison of raceways and air-lift tubular bioreactors in a paper with 70 citations. LCA has been used to investigate the global warming potential (GWP) and fossil-energy requirement of a hypothetical operation in which biodiesel is produced from the freshwater alga *C. vulgaris*, grown using flue gas from a gas-fired power station as the carbon source. Cultivation using a two-stage method was considered, whereby the cells were initially grown to a high concentration of biomass under nitrogen-sufficient conditions, before the supply of nitrogen was discontinued, whereupon the cells accumulated triacylglycerides. Cultivation in typical raceways and air-lift tubular bioreactors was investigated, as well as different methods of downstream processing. They find that "if the future target for the productivity of lipids from microalgae, such as *C. vulgaris*, of similar to 40 tons ha^{-1} $year^{-1}$ could be achieved, cultivation in typical raceways would be significantly more environmentally sustainable than in closed air-lift tubular bioreactors." On the other hand, "while biodiesel produced from microalgae cultivated in raceway ponds would have a GWP similar to 80% lower than fossil-derived diesel (on the basis of the net energy content), if air-lift tubular bioreactors were used, the GWP of the biodiesel would be significantly greater than the energetically equivalent amount of fossil-derived diesel." The GWP and fossil-energy requirement in this operation were particularly sensitive to (1) the yield of oil achieved during cultivation, (2) the velocity of circulation of the algae in the cultivation facility, (3) whether the culture media could be recycled or not, and (4) the concentration of carbon dioxide in the flue gas. They recommend using LCA to guide the future development of biodiesel from microalgae.

Yoo et al. (2010) study the selection of microalgae for lipid production under high levels of carbon dioxide in a paper with 68 citations. To select microalgae with a high biomass and lipid productivity, *Botryococcus braunii*, *C. vulgaris*, and *Scenedesmus* sp. were cultivated with ambient air containing 10% CO_2 and flue gas. They find that the "biomass and lipid productivity for *Scenedesmus* sp. with 10% CO_2 were 217.50 and 20.65 mg L^{-1} d^{-1} (9% of biomass), while those for *B. braunii* were 26.55 and 5.51 mg L^{-1} d^{-1} (21% of biomass). Additionally, with flue gas, the lipid productivity for *Scenedesmus* sp. and *B. braunii* was increased 1.9-fold (39.44 mg L^{-1} d^{-1}) and 3.7-fold (20.65 mg L^{-1} d^{-1}), respectively." Oleic acid, a main component of biodiesel, occupied 55% among the fatty acids in *B. braunii*. They assert that *Scenedesmus* sp. is appropriate for mitigating CO_2, due to its high biomass productivity and C-fixation ability, whereas *B. braunii* is appropriate for producing biodiesel, due to its high lipid content and oleic acid proportion.

Pyle et al. (2008) study the production of DHA-rich algae from biodiesel-derived crude glycerol with a focus on the effects of impurities on DHA production and algal biomass composition in a paper with 65 citations. All of the glycerol samples contained methanol, soaps, and various elements including calcium, phosphorus, potassium, silicon, sodium, and zinc. They find that both methanol and soap negatively influenced algal DHA production; in these two, impurities can be removed from culture medium by evaporation through autoclaving (for methanol) and by precipitation through pH adjustment (for soap). The "glycerol-derived algal biomass contained 45–50% lipid, 14–20% protein, and 25% carbohydrate, with 8–13% ash content. Palmitic acid (C16:0) and DHA were the two major fatty acids in the algal lipid." The algal biomass was rich in lysine and cysteine, relative to many common feedstuffs. Boron, calcium, copper, iron, magnesium, phosphorus, potassium, silicon, sodium, and sulfur were present in the biomass, whereas no heavy metals (such as mercury) were detected in the algal biomass. They assert that crude glycerol was a suitable carbon source for algal fermentation. The crude glycerol-derived algal biomass had a high level of DHA and a nutritional profile similar to that of commercial algal biomass, suggesting a great potential for using crude glycerol-derived algae in omega-3-fortified food or feed.

22.3.3 Conclusion

The experimental research on algal biodiesel has been one of the most dynamic research areas in recent years. Nineteen experimental citation classics in the field of algal biodiesel with more than 65 citations were located, and the key emerging issues from these papers were presented in the decreasing order of the number of citations. These papers give strong hints about the determinants of efficient algal biodiesel production and emphasize that marine algae are efficient biodiesel feedstocks in comparison with first- and second-generation biofuel feedstocks.

22.4 Conclusion

The citation classics presented under the two main headings in this chapter confirm the predictions that marine algae have a significant potential to serve as a major solution for the global problems of warming, air pollution, energy security, and food security in the form of algal biodiesel.

Further research is recommended for the detailed studies in each topical area presented in this chapter including scientometric studies and citation classic studies to inform key stakeholders about the potential of the marine algae for the solution of the global problems of warming, air pollution, energy security, and food security in the form of algal biodiesel.

References

Akoh, C.C., S.W. Chang, G.C. Lee, and J.F. Shaw. 2007. Enzymatic approach to biodiesel production. *Journal of Agricultural and Food Chemistry* 55:8995–9005.

Amin, S. 2009. Review on biofuel oil and gas production processes from microalgae. *Energy Conversion and Management* 50:1834–1840.

Baltussen, A. and C.H. Kindler. 2004a. Citation classics in anesthetic journals. *Anesthesia & Analgesia* 98:443–451.

Baltussen, A. and C.H. Kindler. 2004b. Citation classics in critical care medicine. *Intensive Care Medicine* 30:902–910.

Beer, L.L., E.S. Boyd, J.W. Peters, and M.C. Posewitz. 2009. Engineering algae for biohydrogen and biofuel production. *Current Opinion in Biotechnology* 20:264–271.

Bothast, R.J. and M.A. Schlicher. 2005. Biotechnological processes for conversion of corn into ethanol. *Applied Microbiology and Biotechnology* 67:19–25.

Chen, C.Y., K.L. Yeh, R. Aisyah, D.J. Lee, and J.S. Chang. 2011. Cultivation, photobioreactor design and harvesting of microalgae for biodiesel production: A critical review. *Bioresource Technology* 102:71–81.

Chi, Z.Y., D. Pyle, Z.Y. Wen, C. Frear, and S.L. Chen. 2007. A laboratory study of producing docosahexaenoic acid from biodiesel-waste glycerol by microalgal fermentation. *Process Biochemistry* 42:1537–1545.
Chinnasamy, S., A. Bhatnagar, R.W. Hunt, and K.C. Das. 2010. Microalgae cultivation in a wastewater dominated by carpet mill effluents for biofuel applications. *Bioresource Technology* 101:3097–3105.
Chisti, Y. 2007. Biodiesel from microalgae. *Biotechnology Advances* 25:294–306.
Chisti, Y. 2008. Biodiesel from microalgae beats bioethanol. *Trends in Biotechnology* 26:126–131.
Chiu, S.Y., C.Y. Kao, M.T. Tsai, S.C. Ong, C.H. Chen, and C.S. Lin. 2009. Lipid accumulation and CO(2) utilization of *Nannochloropsis oculata* in response to CO(2) aeration. *Bioresource Technology* 100:833–838.
Converti, A., A.A. Casazza, E.Y. Ortiz, P. Perego, and M. Del Borghi. 2009. Effect of temperature and nitrogen concentration on the growth and lipid content of *Nannochloropsis oculata* and *Chlorella vulgaris* for biodiesel production. *Chemical Engineering and Processing* 48:1146–1151.
Demirbas, A. 2007. Progress and recent trends in biofuels. *Progress in Energy and Combustion Science* 33:1–18.
Demirbas, M.F. 2010. Microalgae as a feedstock for biodiesel. *Energy Education Science and Technology Part A—Energy Science and Research* 25:31–43.
Dubin, D., A.W. Hafner, and K.A. Arndt. 1993. Citation-classics in clinical dermatological journals—Citation analysis, biomedical journals, and landmark articles, 1945–1990. *Archives of Dermatology* 129:1121–1129.
Gehanno, J.F., K. Takahashi, S. Darmoni, and J. Weber. 2007. Citation classics in occupational medicine journals. *Scandinavian Journal of Work, Environment & Health* 33:245–251.
Godfray, H.C.J., J.R. Beddington, I.R. Crute et al. 2010. Food security: The challenge of feeding 9 billion people. *Science* 327:812–818.
Goldemberg, J. 2007. Ethanol for a sustainable energy future. *Science* 315:808–810.
Gouveia, L. and A.C. Oliveira. 2009. Microalgae as a raw material for biofuels production. *Journal of Industrial Microbiology & Biotechnology* 36:269–274.
Greenwell, H.C., L.M.L. Laurens, R.J. Shields, R.W. Lovitt, and K.J. Flynn. 2010. Placing microalgae on the biofuels priority list: A review of the technological challenges. *Journal of the Royal Society Interface* 7:703–726.
Gressel, J. 2008. Transgenics are imperative for biofuel crops. *Plant Science* 174:246–263.
Griffiths, M.J. and S.T.L. Harrison. 2009. Lipid productivity as a key characteristic for choosing algal species for biodiesel production. *Journal of Applied Phycology* 21:493–507.
Huang, G.H., F. Chen, D. Wei, X.W. Zhang, and G. Chen. 2010. Biodiesel production by microalgal biotechnology. *Applied Energy* 87:38–46.
Jacobson, M.Z. 2009. Review of solutions to global warming, air pollution, and energy security. *Energy & Environmental Science* 2:148–173.
Johnson, M.B. and Z.Y. Wen. 2009. Production of biodiesel fuel from the microalga *Schizochytrium limacinum* by direct transesterification of algal biomass. *Energy & Fuels* 23:5179–5183.
Kapdan, I.K. and F. Kargi. 2006. Bio-hydrogen production from waste materials. *Enzyme and Microbial Technology* 38:569–582.
Khan, S.A., M.Z. Hussain, S. Prasad, and U.C. Banerjee. (2009). Prospects of biodiesel production from microalgae in India. *Renewable & Sustainable Energy Reviews* 13:2361–2372.
Konur, O. 2000. Creating enforceable civil rights for disabled students in higher education: An institutional theory perspective. *Disability & Society* 15:1041–1063.
Konur, O. 2002a. Assessment of disabled students in higher education: Current public policy issues. *Assessment and Evaluation in Higher Education* 27:131–152.
Konur, O. 2002b. Access to employment by disabled people in the UK: Is the Disability Discrimination Act working? *International Journal of Discrimination and the Law* 5:247–279.
Konur, O. 2002c. Access to Nursing Education by disabled students: Rights and duties of Nursing programs. *Nursing Education Today* 22:364–374.

Konur, O. 2006a. Participation of children with dyslexia in compulsory education: Current public policy issues. *Dyslexia* 12:51–67.
Konur, O. 2006b. Teaching disabled students in higher education. *Teaching in higher education* 11:351–363.
Konur, O. 2007a. A judicial outcome analysis of the Disability Discrimination Act: A windfall for the employers? *Disability & Society* 22:187–204.
Konur, O. 2007b. Computer-assisted teaching and assessment of disabled students in higher education: The interface between academic standards and disability rights. *Journal of Computer Assisted Learning* 23:207–219.
Konur, O. 2011. The scientometric evaluation of the research on the algae and bio-energy. *Applied Energy* 88:3532–3540.
Konur, O. 2012a. 100 citation classics in energy and fuels. *Energy Education Science and Technology Part A—Energy Science and Research* 2012(si):319–332.
Konur, O. 2012b. What have we learned from the citation classics in energy and fuels: A mixed study. *Energy Education Science and Technology—Part A* 2012(si):255–268.
Konur, O. 2012c. The gradual improvement of disability rights for the disabled tenants in the UK: The promising road is still ahead. *Social Political Economic and Cultural Research* 4:71–112.
Konur, O. 2012d. Prof. Dr. Ayhan Demirbas' scientometric biography. *Energy Education Science and Technology Part A—Energy Science and Research* 28:727–738.
Konur, O. 2012e. The evaluation of the research on the biofuels: A scientometric approach. *Energy Education Science and Technology Part A—Energy Science and Research* 28:903–916.
Konur, O. 2012f. The evaluation of the research on the biodiesel: A scientometric approach. *Energy Education Science and Technology Part A—Energy Science and Research* 28:1003–1014.
Konur, O. 2012g. The evaluation of the research on the bioethanol: A scientometric approach. *Energy Education Science and Technology Part A—Energy Science and Research* 28:1051–1064.
Konur, O. 2012h. The evaluation of the research on the microbial fuel cells: A scientometric approach. *Energy Education Science and Technology Part A—Energy Science and Research* 29:309–322.
Konur, O. 2012i. The evaluation of the research on the biohydrogen: A scientometric approach. *Energy Education Science and Technology Part A—Energy Science and Research* 29:323–338.
Konur, O. 2012j. The evaluation of the biogas research: A scientometric approach. *Energy Education Science and Technology Part A—Energy Science and Research* 29:1277–1292.
Konur, O. 2012k. The scientometric evaluation of the research on the production of bio-energy from biomass. *Biomass and Bioenergy* 47:504–515.
Konur, O. 2012l. The evaluation of the global energy and fuels research: A scientometric approach. *Energy Education Science and Technology Part A—Energy Science and Research* 30:613–628.
Konur, O. 2012m. The evaluation of the biorefinery research: A scientometric approach. *Energy Education Science and Technology Part A—Energy Science and Research* 2012(si):347–358.
Konur, O. 2012n. The evaluation of the bio-oil research: A scientometric approach. *Energy Education Science and Technology Part A—Energy Science and Research* 2012(si):379–392.
Konur, O. 2012o. What have we learned from the citation classics in energy and fuels: A mixed study. *Energy Education Science and Technology Part A—Energy Science and Research* 2012(si):255–268.
Konur, O. 2012p. The evaluation of the research on the biofuels: A scientometric approach. *Energy Education Science and Technology Part A—Energy Science and Research* 28:903–916.
Konur, O. 2013. What have we learned from the research on the International Financial Reporting Standards (IFRS)? A mixed study. *Energy Education Science and Technology Part D: Social Political Economic and Cultural Research* 5:29–40.
Lal, R. 2004. Soil carbon sequestration impacts on global climate change and food security. *Science* 304:1623–1627.
Lardon, L., A. Helias, B. Sialve, J.P. Steyer, and O. Bernard. 2009. Life-cycle assessment of biodiesel production from microalgae. *Environmental Science & Technology* 43:6475–6481.
Lee, J.Y., C. Yoo, S.Y. Jun, C.Y. Ahn, and H.M. Oh. 2010. Comparison of several methods for effective lipid extraction from microalgae. *Bioresource Technology* 101:S75–S77.

Li, Q., W. Du, and D.H. Liu. 2008a. Perspectives of microbial oils for biodiesel production. *Applied Microbiology and Biotechnology* 80:749–756.

Li, X.F., H. Xu, and Q.Y. Wu. 2007. Large-scale biodiesel production from microalga *Chlorella protothecoides* through heterotrophic cultivation in bioreactors. *Biotechnology and Bioengineering* 98:764–771.

Li, Y., M. Horsman, N. Wu, C.Q. Lan, and N. Dubois-Calero. 2008b. Biofuels from microalgae. *Biotechnology Progress* 24:815.

Li, Y.Q., M. Horsman, B. Wang, N. Wu, and C.Q. Lan. 2008c. Effects of nitrogen sources on cell growth and lipid accumulation of green alga *Neochloris oleoabundans*. *Applied Microbiology and Biotechnology* 81:629–636.

Liang, Y.N., N. Sarkany, and Y. Cui. 2009. Biomass and lipid productivities of *Chlorella vulgaris* under autotrophic, heterotrophic and mixotrophic growth conditions. *Biotechnology Letters* 31:1043–1049.

Lynd, L.R., J.H. Cushman, R.J. Nichols, and C.E. Wyman. 1991. Fuel ethanol from cellulosic biomass. *Science* 251:1318–1323.

Mata, T.M., A.A. Martins, and N.S. Caetano. 2010. Microalgae for biodiesel production and other applications: A review. *Renewable & Sustainable Energy Reviews* 14:217–232.

McNeff, C.V., L.C. McNeff, B. Yan et al. 2008. A continuous catalytic system for biodiesel production. *Applied Catalysis A—General* 343:39–48.

Meng, X., J.M. Yang, X. Xu, L. Zhang, Q.J. Nie, and M. Xian. 2009. Biodiesel production from oleaginous microorganisms. *Renewable Energy* 34:1–5.

Miao, X.L. and Q.Y. Wu. 2006. Biodiesel production from heterotrophic microalgal oil. *Bioresource Technology* 97:841–846.

Miao, X.L., Q.Y. Wu, and C.Y. Yang. 2004. Fast pyrolysis of microalgae to produce renewable fuels. *Journal of Analytical and Applied Pyrolysis* 71:855–863.

Nigam, P.S. and A. Singh. 2011. Production of liquid biofuels from renewable resources. *Progress in Energy and Combustion Science* 37:52–68.

North, D. 1994. Economic-performance through time. *American Economic Review* 84:359–368.

Paladugu, R., M.S. Chein, S. Gardezi, and L. Wise. 2002. One hundred citation classics in general surgical journals. *World Journal of Surgery* 26:1099–1105.

Patil, V., K.Q. Tran, and H.R Giselrod. 2008. Towards sustainable production of biofuels from microalgae. *International Journal of Molecular Sciences* 9:1188–1195.

Pienkos, P.T. and A. Darzins. 2009. The promise and challenges of microalgal-derived biofuels. *Biofuels Bioproducts & Biorefining—BIOFPR* 3:431–440.

Pyle, D.J., R.A. Garcia, and Z.Y. Wen. 2008. Producing docosahexaenoic acid (DHA)-rich algae from biodiesel-derived crude glycerol: Effects of impurities on DHA production and algal biomass composition. *Journal of Agricultural and Food Chemistry* 56:3933–3939.

Radakovits, R., R.E. Jinkerson, A. Darzins, and M.C. Posewitz. 2010. Genetic engineering of algae for enhanced biofuel production. *Eukaryotic Cell* 9:486–501.

Rodolfi, L., G.C. Zittelli, N. Bassi et al. 2009. Microalgae for oil: Strain selection, induction of lipid synthesis and outdoor mass cultivation in a low-cost photobioreactor. *Biotechnology and Bioengineering* 102:100–112.

Schenk, P.M., S.R. Thomas-Hall, E. Stephens et al. 2008. Second generation biofuels: High-efficiency microalgae for biodiesel production. *Bioenergy Research* 1:20–43.

Schirmer, A., M.A. Rude, X.Z. Li, E. Popova, and S.B. del Cardayre. 2010. Microbial biosynthesis of alkanes. *Science* 329:559–562.

Scott, S.A., M.P. Davey, J.S. Dennis et al. 2010. Biodiesel from algae: Challenges and prospects. *Current Opinion in Biotechnology* 21:277–286.

Sialve, B., N. Bernet, and O. Bernard. 2009. Anaerobic digestion of microalgae as a necessary step to make microalgal biodiesel sustainable. *Biotechnology Advances* 27:409–416.

Singh, A., P.S. Nigam, and J.D. Murphy. 2011. Renewable fuels from algae: An answer to debatable land based fuels. *Bioresource Technology* 102:10–16.

Singh, J. and S. Cu. 2010. Commercialization potential of microalgae for biofuels production. *Renewable & Sustainable Energy Reviews* 14:2596–2610.
Spolaore, P., C. Joannis-Cassan, E. Duran, and A. Isambert. 2006. Commercial applications of microalgae. *Journal of Bioscience and Bioengineering* 101:87–96.
Stephenson, A.L., Kazamia, J.S. Dennis, C.J. Howe, S.A. Scott, and A.G. Smith. 2010. Life-cycle assessment of potential algal biodiesel production in the united kingdom: A comparison of raceways and air-lift tubular bioreactors. *Energy & Fuels* 24:4062–4077.
Vasudevan, P.T. and M. Briggs. 2008. Biodiesel production-current state of the art and challenges. *Journal of Industrial Microbiology & Biotechnology* 35:421–430.
Volesky, B. and Z.R. Holan. 1995. Biosorption of heavy-metals. *Biotechnology Progress* 11:235–250.
Vyas, A.P., J.L. Verma, and N. Subrahmanyam. 2010. A review on FAME production processes. *Fuel* 89:1–9.
Wang, B., Y.Q. Li, N. Wu, and C.Q. Lan. 2008. CO(2) bio-mitigation using microalgae. *Applied Microbiology and Biotechnology* 79:707–718.
Williams, P.J.L. and L.M.L. Laurens. 2010. Microalgae as biodiesel & biomass feedstocks: Review & analysis of the biochemistry, energetics & economics. *Energy & Environmental Science* 3:554–590.
Wrigley, N. and S. Matthews. 1986. Citation-classics and citation levels in geography. *Area* 18:185–194.
Xiong, W., X.F. Li, J.Y. Xiang, and Q.Y. Wu. 2008. High-density fermentation of microalga *Chlorella protothecoides* in bioreactor for microbio-diesel production. *Applied Microbiology and Biotechnology* 78:29–36.
Xu, H., X.L. Miao, and Q.Y. Wu. 2006. High quality biodiesel production from a microalga *Chlorella protothecoides* by heterotrophic growth in fermenters. *Journal of Biotechnology* 126:499–507.
Yergin, D. 2006. Ensuring energy security. *Foreign Affairs* 85:69–82.
Yoo, C., S.Y. Jun, J.Y. Lee, C.Y. Ahn, and H.M. Oh. 2010. Selection of microalgae for lipid production under high levels carbon dioxide. *Bioresource Technology* 101:S71–S74.

Section V

Bioelectricity and Microbial Fuel Cells

23

Bioelectricity Production by Marine Bacteria, Actinobacteria, and Fungi

K. Sivakumar, H. Ann Suji, and L. Kannan

CONTENTS

23.1 Introduction

Microbial fuel cells (MFCs) can be best defined as fuel cells where microbes act as catalysts in degrading the organic content to generate electricity (Muralidharan et al., 2011). MFCs can convert biodegradable and reduced compounds, such as glucose, acetate, and lactate, directly into electricity, offering a clean and renewable source of energy (Zhang et al., 2012). In addition, MFCs may also assist environmental protection, for example, through wastewater treatment. In MFC devices, microbial cells in the anode chamber play a key role in catalyzing the oxidation of an organic substrate (the fuel) and transferring the electrons, derived from the metabolic processes to the electrode (Zhang, 2011). Recently, MFC technology has been developed as a novel biotechnology for the harvest of energy from biodegradable materials, and the interest in MFCs is steadily increasing mainly because they offer the possibility of directly harvesting electricity from organic wastes and renewable biomass.

Present-day thinking about energy around the globe is heading toward the search for alternatives to fossil fuels. Increasing energy consumption creates unbalanced energy management and requires power sources that are able to sustain for longer periods. In this regard, MFC technology represents a new form of renewable energy by generating electricity from what would otherwise be considered waste. The technology can use microbes already present in the wastewater as a catalyst to generate electricity while simultaneously treating the wastewater.

In recent times, several modifications in the design of MFCs have been adopted to improve current and power densities. A dual-chamber configuration is the most common one applied in research (Logan et al., 2006). The anode and cathode materials for the reactor are selected based on several properties, including high surface area, improved chemical stability, biocompatibility (anode), and good conductivity. Choice of materials and conditions and the understanding of the mechanism by which the microbes attach and grow on these electrodes are important to obtain better yields of current (Paul, 2009).

In this context, microbes, especially marine microbes, are attracting global attention as new biotechnological resources for MFCs. These microbes also serve as a potential source of new bioactive compounds for industrial, agricultural, environmental, pharmaceutical, medical, and other uses (Lam, 2006; Debnath et al., 2007; Sivakumar et al., 2007).

23.2 Marine Bacteria as MFCs

Marine bacterial MFCs have been studied for their application as a potential long-term energy supply in remote areas. Biofilm growth on the cathodes of such MFCs cannot be avoided, because the cathodes are exposed in an aqueous environment, containing a diverse microbial community. During a long-term project with an abiotic seawater fuel cell, consisting of a magnesium alloy anode and a carbon fiber cathode, Hasvold et al. (1997) found that a biofilm on the cathode improved the oxygen reduction rates, indicating that the cathode biofilm functioned as a biocatalyst. Bergel et al. (2005) tested a seawater fuel cell with an abiotic platinum anode and stainless steel cathode with a biofilm separated by a PEM and observed that the maximum power density decreased from 270 to 2.8 mW/m^2 after the biofilm was removed from the cathode. The authors believed that the microorganisms in the biofilm were responsible for catalyzing the oxygen reduction and electrochemically active bacteria can grow as a biofilm attached to the anode. Bacterial species, feasible for MFC are given in Table 23.1.

Microbial communities associated with the electrodes from underwater fuel cells, for harvesting electricity from five different aquatic sediments, were investigated. Analysis of 16S rRNA gene sequences after 3–7 months of incubation demonstrated that all of the energy-harvesting anodes were highly enriched with the microorganisms (δ-Proteobacteria) when compared to the control electrodes, not connected to a cathode. Quantitative PCR analysis of 16S rRNA genes and culture studies have also indicated that Geobacteraceae was 100-fold more abundant on the marine-deployed anodes (Holmes et al., 2004).

Zhang et al. (2012) isolated 20 different species (74 strains) of bacteria from a consortium under anaerobic conditions and cultured in the laboratory; of which 34% was found to be exoelectrogens in single-species studies. So, future studies should be designed to help optimize the harvesting of electricity from aquatic sediments or waste organic matters, with a focus on the electrode interactions of the bacteria that are most competitive in colonizing anodes and cathodes (Holmes et al., 2004). The two most important models for the investigation of electron transfer mechanisms today are *Shewanella* and *Geobacter* (Karin, 2010).

TABLE 23.1

Bacterial Species, Feasible for MFC

Transfer Type	Bacterial Species	Mediator	External Mediator
Oxidative metabolism			
Membrane driven	*Rhodoferax ferrireducens*	Unknown	—
	Geobacter sulfurreducens	89 kDa c-type cytochrome	—
	Aeromonas hydrophila	c-type cytochrome	—
Mediator driven	*Escherichia coli*	Hydrogenase	Neutral red
	Shewanella putrefaciens	Quinones	
	Pseudomonas aeruginosa	Pyocyanin and phenazine	—
	Erwinia dissolvens	Unknown	Fe(III) CyDTA (an iron chelator)
	Desulfovibrio desulfuricans	S^{2-}	—
Fermentative metabolism			
Membrane driven	*Clostridium butyricum*	Cytochromes	—
Mediator driven	*Enterococcus faecium*	Unknown	Pyocyanin

Source: Shah Chirag, K. and Yagnik, B.N., *Res. J. Biotechnol.*, 8(3), 84, 2013.

23.2.1 *Shewanella*

Shewanella species are Gram-negative, proteobacteria that are facultative anaerobes and can respire on a tremendous variety of inorganic and organic electron acceptors. One such electron acceptor is Fe_2O_3, which is commonly found in clay hematite. These facultative anaerobes are often found in marine sediments and can swim with the aid of a single polar flagellum (Venkateswaran et al., 1999). Since the modern characterization of *Shewanella* in 1988, DNA: DNA hybridization and 16S rRNA sequencing have been used to identify more than 40 distinct species. The features that characterize this genus include psychrotolerance, mild halophilicity, and the capacity to reduce an unparalleled array of inorganic and organic compounds for respiration (Gralnick and Newman, 2007). Their capacity to respire on various metals as well as their production of endogenous hydrocarbons has ignited tremendous interest in the characterization and potential applications of these microorganisms (Camiodio, 2011). C-type cytochromes were identified to be among the most significantly upregulated genes of *Shewanella oneidensis* (Figure 23.1), when respiring metals.

Model species of the genus is *S. oneidensis* MR-1 and it showed the capacity to reduce manganese. Interestingly, oxidized manganese is an insoluble substance, which suggests that the bacterium has developed a way to transfer electrons to metals outside of their cells. Moreover, some species of *Shewanella* rely heavily on the soluble electron shuttle riboflavin for electron transport to extracellular acceptors. *S. oneidensis* MR-1 has been shown to transport electrons across distances greater than 50 μm using conductive nanowires that are developed in electron acceptor limited conditions.

In-depth analysis of the process of extracellular electron transfer became possible with the development of a genetic system for *Geobacter sulfurreducens* and *S. oneidensis*. Mutation studies of *Shewanella* have identified a number of genes that play a role in extracellular electron transfer to insoluble electron acceptors. While genetic studies have not been conducted on the structural components of the *Shewanella* nanowires (specialized pili), mutation of pilD, a type IV prepilin peptidase, caused a reduction in power production in an MFC. Other genes identified through knockout mutants in *S. oneidensis* that caused a decrease in power and cellular mass attached to the electrode included mtrA, mtrB,

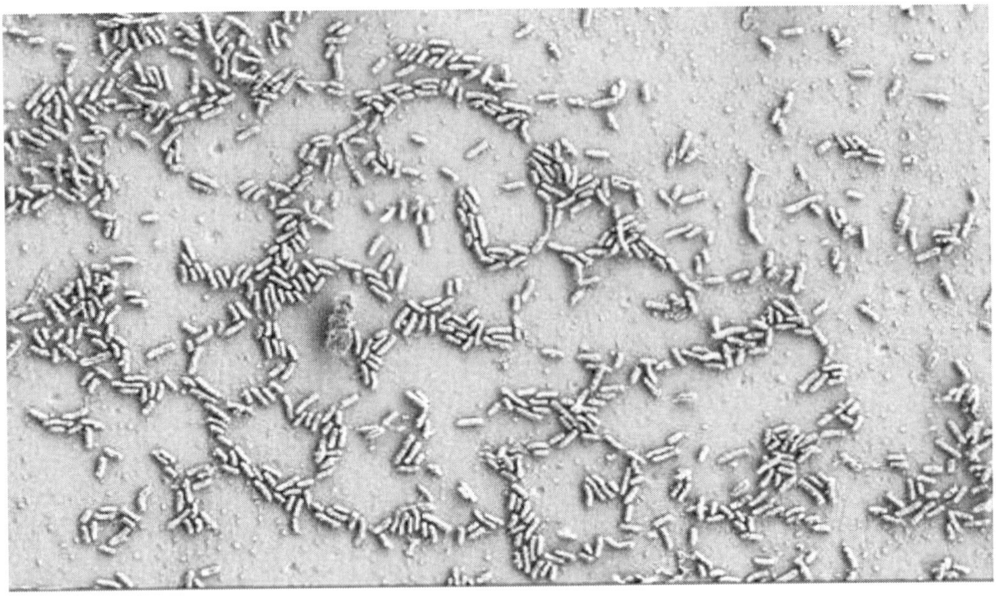

FIGURE 23.1
Shewanella oneidensis. (From Science and Tech, 26th March 2013 online.)

omcA, mtrC, cyma, and gspG. It is proposed to form a complex of MtrA, MtrB, and MtrC, in which MtrC is an extracellular element that mediates extracellular electron transfer and MtrB is a trans outer membrane b-barrel protein that serves as a sheath within which MtrA and MtrC exchange electrons. OmcA and MtrC have been shown to directly interact with an electrode and type II secretion has been found to be important for the correct transport and localization of cytochromes out of the cell (Franks et al., 2010).

Interestingly, antibody recognition force microscopy of *Shewanella* grown on insoluble Fe(III) has indicated that OmcA is localized between the cell and the mineral, while MtrC is displayed more uniformly around the cell surface. These results may question the original speculation that MtrA and OmcA are found localized on the nanowires directly (Franks et al., 2010). *Shewanella* has a protein on its surface that works like an electrical wire between the interior and exterior of the bacterium. The protein is called deca-heme c-class cytochrome and can bond to the surface molecules of the rock allowing in this way the transfer of electrons through the membrane. This process also releases chemical elements as iron and manganese, altering the rock (Franks et al., 2010).

Recently, researchers have created a synthetic version of the marine bacterium, *S. oneidensis*, using just the proteins thought to shuttle the electrons from the inside of the microbe to a rock. They inserted them into the layers of vesicles, which are small capsules of lipids, or fats, such as the ones that make up a bacterial membrane. Then, they tested how well electrons traveled between an electron donor on the inside and an iron-bearing mineral on the outside (Daily Mail, 2013). Scientists have also found how the rock-dwelling bacterium, *Shewanella*, can transform minerals by zapping them with tiny electrical currents. This could be the discovery of the century, as it could lead to the production of a new type of fuel cells that will generate electricity (Daily Mail, 2013).

Shewanella is the key to using such bacteria not only in cheap electricity production but also in oil-spill cleanup. According to the biochemist, David Richardson, of the University of East Anglia in the United Kingdom, *Shewanella* is the ideal candidate for environmental-cleanup tasks as it lives underground: "Understanding their biochemistry could help develop strategies to stimulate their activities [at the cleanup sites]." Laboratory tests have shown that other types of bacteria can also generate electricity making the discovery far more important (Cristi, 2009).

Scientists have begun to utilize the respiring ability of *Shewanella* species to develop new methods of obtaining energy. One such method is the development of fuel cells that can be used for wastewater treatment as well as to power data-monitoring devices.

23.2.2 *Geobacter*

In sediment fuel cells, natural populations of microorganisms in the anoxic sediments are responsible for electron transfer to the current-harvesting anode. Previous molecular and culture studies have indicated that a specific group of Fe(III)-reducing microorganisms of *Geobacteraceae* are primarily responsible for electricity production by sediment fuel cells. Further analysis of the pure cultures of *Geobacteraceae* has shown that these organisms can conserve energy to support growth by oxidizing organic compounds to carbon dioxide, with an electrode serving as the sole electron acceptor. Therefore, a likely explanation for the functioning of sediment fuel cells is that microorganisms in the sediments could convert the complex sediment organic matter to fermentation products, most notably acetate, and that *Geobacteraceae* members colonizing the anode would oxidize these fermentation products to carbon dioxide, with the transfer of electrons to the current-harvesting electrode. These electrons would then flow to the cathode in the overlying aerobic water, where they react with oxygen (Holmes et al., 2009).

Nanowires have been identified in *G. sulfurreducens* (Figure 23.2) using atomic force microscopy fitted with a conductive tip. Pili in *G. sulfurreducens* have been proposed as a conduit for the extracellular transfer of electrons to insoluble Fe(III) oxide particles due to

FIGURE 23.2
Geobacter sulfurreducens. (From Wilson, Blog on the future of biohybrid and biomimetic technology, http://www.robotcompanions.eu/blog/2012/01/a-bit-of-bacteria-goes-a-long-way/geobacter-sulfurreducens/, 2012.)

the close association of pili and insoluble Fe(III) oxide particles. It is due to the fact that pili are highly expressed at optimal growth temperatures when using Fe(III) oxides but not soluble electron acceptors and because Fe(III) oxide reduction is lowered due to deletion of pilA, a structural component of the pilin. Investigations of pilA regulation have identified a two-component regulatory protein PilR. Disruption of PilR not only caused a decrease in pilA and other pilin-related genes but also resulted in differential expressions of many c-type cytochromes involved in Fe(III) reduction, in some cases, causing mislocalization of cytochromes. The pilA mutant produced considerably less power than the wild type in the poised MFC, forming a monolayer of cells on the anode surface. Only deletion of pilA and omcZ inhibited current production in the high-current density MFC. Furthermore, these mutants had no impairment when grown on a graphite electrode with fumigate as an electron acceptor. This study has thus implicated that pilA did not have a structural role within the biofilm under these conditions but was essential for long-range electron transfer (Alves et al., 2010).

G. sulfurreducens is a biocatalyst, able to convert organic compounds into a flow of electrons (Bond and Lovley, 2003). This depends upon a suite of intracellular enzymes, linked to macromolecules at the outer membrane that have an ability to interface with insoluble Fe(III) minerals and conducive surfaces (Holmes et al., 2006; Lovley, 2006; Reguera et al., 2006). The goal of this initial investigation was to prove that cells entrapped near electrode surfaces could establish electrical contact. This has led to a series of electrochemical observations regarding the coupling between cytoplasmic oxidation and electron transport at the outer surface of *G. sulfurreducens* (Srikanth et al., 2007), and the organism seems to be unique in its ability to produce a biofilm greater than 50 μm thick in pure culture. Formation of a relatively thick biofilm occurred within a single passage of a poised MFC grown in batch mode with concentrations of acetate at 10 mM. Analysis of the biofilm formed on the anode surface at different current values indicated a direct correlation between the biomass/biofilm thickness and the current produced (He and Angenent, 2006). *G. sulfurreducens* does not produce an electron shuttle as most of this biofilm is not in direct contact with the anode surface, and it is thought that *G. sulfurreducens* cells are able to transfer electrons directly to the anode across the biofilm thickness through a conductive network (Franks et al., 2010).

23.2.3 *Desulfuromonas*

Like *Geobacter, Desulfuromonas* species perform complete oxidation of their carbon sources and are present in more than half of the population on the energy-harvesting electrodes. *Desulfuromonas acetoxidans* is dominant in marine studies (Alves et al., 2010).

23.3 Marine Actinobacteria as MFCs

Actinobacteria belonging to the Phylum Actinobacteria are Gram-positive bacteria with higher G + C content in the DNA. They are metabolically active (Moran et al., 1995) and physiologically adapted to the salt concentration encountered in the sea (Jensen et al., 1991) and can be recovered from deep-sea sediments (Weyland, 1969). *Rhodococcus marinonascens* was the first actinobacterial species that was described and accepted as an autochthonous marine species.

Studies on marine actinobacteria are very limited and they have been mentioned incidentally on the microbial community of marine habitats (Sivakumar, 2001). There is no report on the conversion of organic matter into electricity by marine actinobacteria as MFCs. Electricity can be directly generated by bacteria in MFCs from many different biodegradable substrates. When cellulose is used as the substrate, electricity generation would require a microbial community with both cellulolytic and exoelectrogenic activities (Farzaneh et al., 2009). Hence, identification of the effective cellulose-degrading actinobacteria that are electrochemically active is the present-day need to overcome the problems of electricity shortage and industrial pollution. In this context, Sethubathi (2012) carried out a study that is the first of its kind on the marine actinobacterial strain to obtain bioelectricity from the industrial effluents, and the present effort would pave way to improve the power generation to fulfill the energy needs exponentially worldwide at a lower cost. Results have indicated the possibility of bioelectricity generation from industrial effluents, using the marine actinobacterial strain *Streptomycetes* sp., as MFC, fabricated with low-cost anode material, without any toxic mediators. Apart from power generation, the fuel cell is effective in the removal of COD and heavy metals from the industrial effluent as a measure of bioremediation.

23.4 Marine Fungi as MFCs

Among the three major habitats of the biosphere, the marine realm that covers 70% of the earth's surface provides the largest habitable space for living organisms, particularly microbes including fungi. The best working definition for a marine fungus was proposed by Kohlmeyer and Kohlmeyer (1979): obligate marine fungi are those that grow and sporulate exclusively in a marine or estuarine habitat and the facultative marine fungi are those from fresh water and terrestrial environments, able to grow and possibly sporulate in the marine environment.

Four marine-derived fungi belonging to the phyla Ascomycota (NIOCC #C3 and NIOCC #16V) and Basidiomycota (NIOCC #2a and NIOCC #15V) have been used for the remediation of two raw textile-mill effluents (Verma, 2011). But yet, they have to be fully explored for their MFC potential, and several techniques are needed to improvise the isolation of culture-independent fungi from different marine habitats. A high-throughput screening strategy for lignin-degrading marine fungi is also required to tap these vast resources from marine ecosystems. Assessments of their biological activity, using functional genomics, will be immensely helpful in targeting such fungi for an array of bioactive compounds and MFC potential.

23.5 Yeasts as MFCs

A wide range of organisms have been isolated that can produce power individually in an MFC. These include representatives of Firmicutes, Acidobacteria, and four of the five classes of Proteobacteria, as well as the yeast strains *S. cerevisiae* and *H. anomala*. Yeast is a heterotroph, found in a wide range of natural habitats, and it is a facultative anaerobe (growing rapidly under both aerobic and anaerobic conditions) and *Saccharomyces* has now been isolated from seawater sediments and estuary (Cheng and Lin, 1977; Bhat et al., 1995; Takami et al., 1998). It has simple nutritional requirements and so it can utilize a wide variety of substrates that make it ideal for an MFC (Mathuriya and Sharma, 2010).

Each microorganism involved in anaerobic degradation has a specific pH optimum in which it grows and performs its best and generally microbial activity is lower at sub-optimal pH than the optimal pH. At pH 6.0, *S. cerevisiae* generates comparatively better electricity than *Clostridium acetobutylicum*. The culture of *S. cerevisiae* started current generation after 24 h and gave maximum current output of 5.18 and 8.54 mA, respectively, after 4 and 8 days of operation. This was higher than that of *C. acetobutylicum*, which produced 4.89 and 6.51 mA current, respectively, after 5 and 9 days (Mathuriya and Sharma, 2010).

Rapid current fall was observed due to substrate exhaustion, which was recovered after replacement of 50% fresh substrate, and the current output of *S. cerevisiae* reached a peak value of 10.45 mA after 9 days of operation (Mathuriya and Sharma, 2010). *S. cerevisiae* started current generation after about 24 h, which reached a peak value of 6.13 mA after 4 days of operation. The current fall observed after 5 days was recovered when 50% AW was replaced with fresh AW.

23.6 Sediment MFCs

Energy can be harvested from aquatic sediments by burying a graphite electrode in anoxic sediments and making an electrical connection between this electrode and a similar

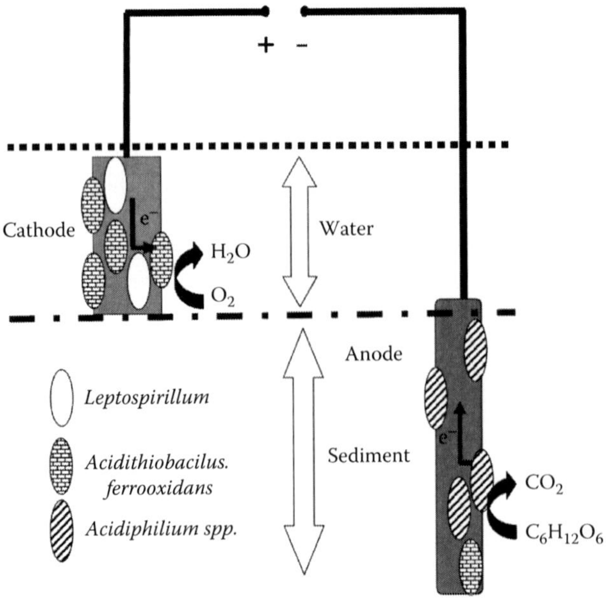

FIGURE 23.3
Scheme of the anodic and cathodic main reactions in the sediment microbial fuel cell. (From Munoz, J.G. et al., *Int. Microbiol.*, 14(2), 73, 2011.)

electrode in the overlying aerobic water. Recovery of electricity from these sediments will be analogous to that from previously described biological fuel cells (Holmes et al., 2009).

By placing one electrode into the marine sediments rich in organic matter and sulfides and the other in the overlying oxic water, electricity can be generated at sufficient levels to power some marine devices. Protons conducted by the seawater can produce a power density of up to 28 mW/m.

In sediment MFCs, graphite disks can be used for the electrodes, although platinum mesh electrodes have been used. Bottle brush cathodes used for seawater batteries may hold the most promise for long-term operation of unattended systems as these electrodes provide a high surface area and are made of noncorrosive materials. Sediments have also been placed into H-tube-configured two-chamber systems to allow investigation of the bacterial community (Singh et al., 2010). Munoz et al. (2011) have given the scheme of the anodic and cathodic main reactions in the sediment MFC (Figure 23.3).

Benthic MFC (BMFC) is a field-deployable and uniquely configured MFC that relies on the natural redox processes in the aqueous sediments. These fuel cells could be developed as long-term power sources for autonomous sensors and acoustic communication devices deployed in fresh- and saltwater environments (Reimers et al., 2006).

23.7 Future Applications of MFCs

As MFCs have the following operational and functional advantages over the technologies currently used for generating energy from organic matter, they can be pursued to the fullest extent in the near future for both economic and environmental benefits.

- Direct conversion of substrate energy to electricity enables high conversion efficiency.
- Efficient operation at ambient temperature.
- No requirement of gas treatment because the off-gases of MFCs are enriched with carbon dioxide and normally have no useful energy content.
- No need for energy input for aeration, provided the cathode is passively aerated.
- Potential for widespread application at locations, lacking electrical infrastructures.
- Acting on diverse substrate ranges including environmental wastes.

23.8 Conclusion

MFCs have gained significance in the last few decades due to their capability to produce energy, either as electricity or hydrogen production, from renewable resources, sewage wastes, and other similar waste sources (Muralidharan et al., 2011). MFCs run without any expensive catalysts—just on the microbes plus organic materials as food for the microbes under the anaerobic conditions (Karin, 2010). We can believe that many new types of bacteria, capable of anodophilic electron transfer (electron transfer to an anode) or even interspecies electron transfer (electrons transferred between bacteria in any form) will be discovered. We can even produce clean energy by using MFCs and the benefits would include clean, safe, and quiet performance, low emissions, high efficiency, and direct electricity recovery (Singh et al., 2010).

Lovley (2006) has stated that MFC is a very clever approach, and it would help open our eyes to how we might use microbes and electrodes for practical applications. Understanding of how these microbes function would enable the scientists to develop biobatteries that could store energy for sensors in the remote environments (Lewis, 2013). Biobatteries have now taken a giant leap toward becoming a reality.

British scientists have made an important breakthrough in the quest to generate clean electricity from bacteria (Sinha, 2013). "The bacteria are capable of continuously generating electricity at levels that could be used to operate small electronic devices," says Charles Milliken of the Medical University of South Carolina, who conducted research with his colleague, Harold May. As long as the bacteria are fed with fuel, they will be able to produce electricity 24 h a day (Live Science, 2005).

Harnessing electricity and transporting it over long distances is arguably one of the most important innovations that humans have made. The ability to transmit electric energy and an impulse through wires has enabled basically every form of technology and communication at our disposal. As it turns out, however, humans were not the first creatures to find a means to conduct electric current. A recent report in the journal *Nature* describes a newly discovered species of bacterium that long ago beat us to the punch when it came to long-distance electron transport (Jesse, 2013). Scientists are excited because this discovery may lead to the creation of clean electricity from bacteria, otherwise known as *biobatteries*. Biobatteries would be perfect for use underground, where solar power is unavailable (Hutchens, 2013). MFCs could one day become a viable source of sustainable energy. Batteries made from bacteria could soon be powering our gadgets (Daily Mail, 2013).

Acknowledgments

Authors thank the Annamalai University authorities and dean, Faculty of Marine Sciences, for providing them with necessary facilities and encouragement.

References

Alves, A.S., B.M. Fonseca, and R.O. Louro, 2010. Sludge oomph: Harnessing the power of sediment microbiota, A. Mendez-Vilas (ed.), In *Current Research Applied Microbiology and Microbial Biotechnology*, 64–73.

Bergel, A., D. Féron, and A. Mollica, 2005. Catalysis of oxygen reduction in PEM fuel cell by seawater biofilm. *Electrochem. Commun.*, 7: 900–904.

Bhat, J.V., N. Kachwala, and B.N. Moody, 1995. Marine yeast off the Indian coast. *J. SciInd. Res.*, 4: 9–15.

Bond, D.R. and D.R. Lovley, 2003. Electricity production by *Geobacter sulfurreducens* attached to electrodes. *Appl. Environ. Microbiol.*, 69: 1548–1555.

Camiodio, 2011. *Shewanella oneidensis* MR-1: Background and applications, Retrieved July 23, 2011, from http://microbewiki.kenyon.edu/index.php?title=Shewanella_oneidensis_MR1:_Background_and_Applications&oldid=65066.

Cheng, Y.C. and L.P. Lin, 1977. Microbiological studies on western coast of Taiwan. Enumeration, isolation and identification of marine-occurring yeasts. *Acta Occeanogr. Taiwan*, 7: 216–228.

Cristi, 2009. New discovery: Generating electricity with ground bacteria, *Shewanella. Energy News, Green News*, Retrieved December 22, 2009/01.

Daily Mail, 2013. Fuel cell that uses bacteria to generate electricity. *Science Daily*. Retrieved November 1, 2013, from http://www.sciencedaily.com/releases/2008/01/080103101137.htm.

Debnath, M., A.K. Paul, and P.S. Bisen, 2007. Natural bioactive compounds and biotechnological potential of marine bacteria. *Curr. Pharm. Biotechnol.*, 8: 253–260.

Franks, A.E., N. Malvankar, and K.P. Nevin, 2010. Bacterial biofilms: The use of a microbial fuel cell. *Biofuels*, 1(4): 589–604.

Gralnick, J.A. and D.K. Newman, 2007. Extracellular respiration. *Mol. Microbiol.* 65: 1–11.

Hasvold, O., H.H. Melvaer, G. Citi, B.O. Johansen, T. Kjonigsen, and R. Galetti, 1997. Sea-water battery for subsea control systems. *J. Power Sources*, 65: 253–261.

He, Z. and L.T. Angenent, 2006. Application of bacterial biocathodes in microbial fuel cells. *Electroanalysis*, 18: 2009–2015.

Holmes, D.E., S.K. Chaudhuri, and K.P. Nevin, 2006. Microarray and genetic analysis of electron transfer to electrodes in *Geobacter sulfurreducens. Env. Microbiol.*, 8(10): 1805–1815.

Holmes, D.E., J.S. Nicoll, D.R. Bond, and D.R. Lovley, 2004. Potential role of a novel psychrotolerant member of the family Geobacteraceae, *Geopsychrobacter electrodiphilus* gen. nov., sp. nov., in electricity production by a marine sediment fuel cell. *Appl. Environ. Microbiol.*, 70(10): 6023–6030.

Holmes, K., Tavender, T.J., Winzer, K., Wells, J.M., and Hardie, K.R. 2009. A1-2 does not function as a quorum sensing molecule in *Campylobacter jejuni* during exponential growth in vitro. *BMC Microbiol.* 9, 214.

Hutchens, L., 2013. Bio-batteries produce clean electricity from bacteria. *Energy News*, Retrieved March 28, 2013/01.

Jensen, P.R., R. Dwight, and W. Fenical, 1991. Distribution of actinomycetes in near-shore tropical marine sediments. *Appl. Environ. Microbiol.*, 57: 1102–1108.

Jesse, L., 2013. Electricity-conducting bacteria from living wires on ocean floor. Research response, MIT and Massachusetts General Hospital, Boston, MA.

Karin, H., 2010. Waste to Watts: With a little help from tiny power horses. *Analysis*, 5: 22–26.

Kohlmeyer, J. and E. Kohlmeyer, 1979. *Marine Mycology: The Higher Fungi.* Academic Press, New York.

Lam, 2006. Discovery of novel metabolites from marine actinomycetes. *Curr. Opin. Microbiol.*, 9: 245–251.

Lewis, T., 2013. Electric bacteria could be used for bio battery. *Live Science* Retrieved March 25, 2013, from http://www.livescience.com/28163-bio-batteries-onestep-closer.html.

Liu, H., S.A. Cheng, and B.E. Logan, 2005. Power generation in fed-batch microbial fuel cells as a function of ionic strength, temperature and reactor configuration. *Environ. Sci. Technol.*, 39: 5488–5493.

Live Science, 2005. Double bonus. Bacteria eat pollution, generate electricity. *Live Science*, Retrieved June 07, 2005, from http://www.livescience.com/01.

Logan, B.E., B. Hamelers, R. Rozendal, U. Schroder, J. Keller, W. Verstraete, and K. Rabaey, 2006. Microbial fuel cells: Methodology and technology. *Environ. Sci. Technol.* 40: 5181–5192.

Mail online, 2013. The clean energy that's produced by grime: Batteries which use bacteria to make electricity will be on sale "in ten years". *Science and Tech*, Retrieved on March 26, 2013.

Mathuriya, A.S. and V.N. Sharma, 2010. Electricity generation by *Saccharomyces cerevisiae* and *Clostridium acetobutylicum* via, microbial fuel cell technology: A comparative study. *Adv. Biol. Res.*, 4(4): 217–223.

Moran, M.A., L.T. Rutherford, and R.E. Hodson, 1995. Evidence for indigenous *Streptomyces* populations in a marine environment determined with a 16S rRNA probe. 61(10): 3695–3700.

Munoz, J.G., R. Amits, V.M. Fernandez, A.L. De Lacey, and M. Malki, 2011. Electricity generation by microorganisms in the sediment—Water interface of an extreme acidic microcosm. *Int. Microbiol.* 14(2): 73–81.

Muralidharan, A., O.K. Ajay Babu, N. Raman, and M. Ramya, 2011. Impact of salt concentration on electricity production in microbial hybrid fuel cells. *Ind. J. Fund. Appl. Life Sci.*, 1(2): 178–184.

Paul, V., 2009. Electricity generation and ethanol production using iron reducing haloalkaliphilic bacteria. MSc dissertation, Missouri University of Science and Technology, Rolla, MO, 65pp.

Reguera, G., K.P. Nevin, J.S. Nicoll, S.F. Covalla, T.L. Woodard, and D.R. Lovley, 2006. Biofilm and nanowire production leads to increased current in *Geobacter sulfurreducens* fuel cells. *Appl. Environ. Microbiol.*, 72(11): 7345–7348.

Reimers, C.E., P. Gigruis, H. Stecher, L.M. Tender, N. Ryckelynck, and P. Whaling, 2006. Microbial fuel cell energy from an ocean cold seep. *Geobiology*, 4: 123–136.

Rezaei, F., D. Xing, R. Wagner, J.M. Regan, T.L. Richard, and B.E. Logan, 2009. Simultaneous cellulose degradation and electricity production by enterobactercloacae in a microbial fuel cell. *Appl. Environ. Microbiol.*, 75(11): 3673–3678.

Sethubathi, G., 2012. Evaluation of marine actinobacteria from the coral reef environment of the little Andaman Island (India) for microbial fuel cell potential. PhD thesis, Annamalai University, Tamil Nadu, India.

Shah Chirag, K. and B.N. Yagnik, 2013. Review paper: Bioelectricity production using microbial fuel cell. *Res. J. Biotech.*, 8(3): 84–90.

Singh, D., D. Pratap, Y. Baranwal, B. Kumar, and R.K. Chaudhary, 2010. Microbial fuel cells: A green technology for power generation. *Annals Biol. Res.*, 1(3): 128–138.

Sinha, K., 2013. Bacteria could generate clean electricity, TNN, March 28.

Sivakumar, K., 2001. Actinomycetes of an Indian mangrove (Pichavaram) environment: An inventory. PhD thesis, Annamalai University, Tamil Nadu, India, 91pp.

Sivakumar, K., M.K. Sahu, T. Thankaradjou, and L. Kannan, 2007. Research on marine actinobacteria in India. *Ind. J. Microbiol.*, 47(3): 186–196.

Srikanth, S., E. Marsili, M.C. Flickinger, and D.R. Bond, 2007. Electrochemical characterization of *Geobacter sulfurreducens* cells immobilized on graphite paper electrodes. *Biotechnol. Bioeng.*, 99: 1065–1073. Wiley InterScience.

Takami, H., T. Nagahama, F. Fuji, A. Inoue, F. Abe, and K. Horikoshi, 1998. Microbial diversity in the deep sea environment. *American Geophysical Union Spring Meeting* 171, Boston, MA.

Venkateswaran, K., D. Moser, M. Dollhopf, D. Lies, and D. Saffarini, 1999. Polyphasic taxonomy of the genus *Shewanella* and description of *Shewanella oneidensis* sp. nov. *Int. J. Syst. Bacteriol.*, 49: 705–724.

Verma, A.K., 2011. Potential of marine-derived fungi and their enzymes in bioremediation of industrial pollutants. PhD thesis, Goa University, Bambolim, Goa, p. 297.

Weyland, H., 1969. Actinomycetes in North Sea and Atlantic Ocean sediment. *Nature*, 223: 858.

Wilson, 2012. Blog on the future of biohybrid and biomimetic technology. A Bit of Bacteria Goes a Long Way. January 26.

Zhang, G., 2011. Improved performance of microbial fuel cell using combination biocathode of graphite fiber brush and graphite granules. *J. Power Sources*, 196: 6036–6041.

Zhang, G., K. Wang, Q. Zhao, Y. Jiao, and D.J. Lee, 2012. Effect of cathode types on long-term performance and anode bacterial communities in microbial fuel cells. *Bioresource Technol.*, 118: 249–256.

24

Current State of Research on Algal Bioelectricity and Algal Microbial Fuel Cells

Ozcan Konur

CONTENTS

24.1 Introduction

24.1.1 Issues

Global warming, air pollution, and energy security have been some of the most important public policy issues in recent years (Jacobson 2009, Wang et al. 2008, Yergin 2006). With increasing global population, food security has also become a major public policy issue (Lal 2004). The development of biofuels generated from biomass has been a long-awaited solution to these global problems (Demirbas 2007, Goldember 2007, Lynd et al. 1991). However, the development of early biofuels produced from agricultural plants such as sugar cane (Goldemberg 2007) and agricultural wastes such as corn stovers (Bothast and Schlicher 2005) has resulted in a series of substantial concerns about food security (Godfray et al. 2010). Therefore, the development of algal biofuels as a third-generation biofuel has been considered as a major solution for the global problems of global warming, air pollution, energy security, and food security (Chisti 2007, Demirbas 2007, Kapdan and Kargi 2006, Spolaore et al. 2006, Volesky and Holan 1995).

Although there have been many reviews on the use of marine algae for the production of algal bioelectricity and algal microbial fuels cells (MFCs) (e.g., Ghirardi 2006, Rosenbaum and Schroeder 2010, Satyanaranaya et al. 2011) and there have been a number of scientometric studies on algal biofuels (Konur 2011), there has not been any study on the citation classics in algal bioelectricity and algal MFCs as in the other research fields (Baltussen and Kindler 2004a,b, Dubin et al. 1993, Gehanno et al. 2007, Konur 2012a,b, 2013, Paladugu et al. 2002, Wrigley and Matthews 1986).

As North's new institutional theory suggests, it is important to have up-to-date information about the current public policy issues to develop a set of viable solutions to satisfy the needs of all the key stakeholders (Konur 2000, 2002a,b,c, 2006a,b, 2007a,b, 2012c, North 1994).

Therefore, brief information on a selected set of citation classics in algal bioelectricity and algal MFCs is presented in this chapter to inform the key stakeholders relating to the global problems of warming, air pollution, food security, and energy security about the production of algal bioelectricity and their use in algal MFCs for the solution of these problems in the long run, thus complementing a number of recent scientometric studies on the biofuels and global energy research (Konur 2011, 2012d,e,f,g,h,i,j,k,l,m,n,o,p).

24.1.2 Methodology

A search was carried out in the Science Citation Index Expanded (SCIE) and Social Sciences Citation Index (SSCI) databases (version 5.11) in September 2013 to locate the papers relating to algal bioelectricity and algal MFCs using the keyword set of Topic = (electricity or *bio* electricity* or bioelectricity or *microbio* electricity* or *microbial* fuel* cell** or *fuel* cell** or *Photomicrobial* fuel* cell**) AND Topic = (alga* or *macro* alga** or *micro* alga** or macroalga* or microalga* or cyanobacter* or seaweed* or diatoms or *sea* weed** or reinhardtii or braunii or chlorella or sargassum or gracilaria) in the abstract pages of the papers. For this chapter, it was necessary to embrace the broad algal search terms to include diatoms, seaweeds, and cyanobacteria as well as to include many spellings of bioelectricity and MFCs rather than the simple keywords of bioelectricity, MFCs, and algae.

It was found that there were 245 papers indexed by these indexes between 1980 and 2013. The key subject categories for algal research were energy fuels (84 references, 34.3%), biotechnology applied microbiology (66 references, 27.0%), environmental sciences (55 references, 22.4%), and engineering chemical (32 references, 13.1%), altogether comprising 86% of all the references on algal bioelectricity and algal MFCs. It was also necessary to focus on the key references by selecting articles and reviews.

The located highly cited 50 papers were arranged by decreasing number of citations for 2 groups of papers: 15 and 35 papers for nonexperimental research and experimental research, respectively. In order to check whether these collected abstracts correspond to the larger sample on these topical areas, new searches were carried out for each topical area.

The summary information about the located citation classics are presented under two major headings of nonexperimental and experimental research in the order of decreasing number of citations for each topical area.

The information relating to the document type, the affiliation and location of the authors, the journal, the indexes, the subject area, the concise topic, the total number of citations received, and the total average number of citations received per year is given in the tables for each paper.

24.2 Nonexperimental Studies on Algal Bioelectricity and Algal MFCs

24.2.1 Introduction

The nonexperimental research on algal bioelectricity and algal MFCs has been one of the most dynamic research areas in recent years. Fifteen citation classics in the field of algal bioelectricity and algal MFCs with more than ten citations were located and the key emerging issues from these papers are presented in the decreasing order of the number of citations (Table 24.1). These papers give strong hints about the determinants of efficient algal bioelectricity production and algal MFCs processes and emphasize that marine algae are efficient algal bioelectricity and algal MFC feedstocks in comparison with first- and second-generation biofuel feedstocks.

The papers were dominated by the researchers from nine countries, usually through the intracountry institutional collaboration and they were multiauthored. The United States (eight papers) was the most prolific country.

Similarly, all these papers were published in journals indexed by the Science Citation Index (SCI) and/or SCIE. There was also one paper indexed by the SSCI. The number of citations ranged from 10 to 212 and the average number of citations per annum ranged from 1.1 to 16.3. Twelve of the papers were articles, while three of them were reviews.

24.2.2 Nonexperimental Research on Algal Bioelectricity and Algal MFCs

24.2.2.1 Nonexperimental Research on Algal Bioelectricity

24.2.2.1.1 Biohydrogen

Melis and Happe (2001) discuss the production of biohydrogen and bioelectricity from algae in a paper with 212 citations. They argue that hydrogen and electricity could team to provide attractive options in transportation and power generation. Interconversion between these two forms of energy suggests on-site utilization of hydrogen to generate electricity, with the electrical power grid serving energy transportation, distribution utilization, and hydrogen regeneration as needed. They suggest that a challenging problem in establishing H_2 as a source of energy for the future is the renewable and environmentally friendly generation of large quantities of H_2 gas. Thus, they assert that processes that are presently conceptual in nature, or at a developmental stage in the laboratory, need to be encouraged, tested for feasibility, and otherwise applied toward commercialization.

Ghirardi (2006) discusses the production of biohydrogen and bioelectricity by photosynthetic green algae in a review paper with 22 citations. She notes that oxygenic photosynthetic organisms such as cyanobacteria, green algae, and diatoms are capable of absorbing light and storing up to 10%–13% of its energy into the H–H bond of hydrogen gas. This process, which takes advantage of the photosynthetic apparatus of these organisms to convert sunlight into chemical energy, could conceivably be harnessed for production of significant amounts of energy from a renewable resource, water. The harnessed energy could then be coupled to a fuel cell for electricity generation and recycling of water molecules. She discusses current biochemical understanding of this reaction in green algae and some of the major challenges facing the development of future commercial algal photobiological systems for H_2 production.

24.2.2.1.2 Integrated Systems

Subhadra and Edwards (2010) discuss an integrated renewable energy park (IREP) approach for algal biofuel production in United States in a paper with 20 citations.

TABLE 24.1

Nonexperimental Research on Algal Bioelectricity and Algal Microbial Fuel Cells

No.	Paper References	Year	Document	Affiliation	Country	No. of Authors	Journal	Index	Subjects	Topic	Total No. of Citations	Total Average Citations per Annum	Rank
1	Melis and Happe	2001	A	Univ. Calif. B., Univ. Bonn	United States	2	*Plant Physiol.*	SCI, SCIE	Plant Sci.	Algal electricity—hydrogen	212	16.3	2
2	Kadam	2002	A	NREL	United States	1	*Energy*	SCI, SCIE	Ener. Fuels	Algal electricity—cofiring in power plants	54	4.5	4
3	Mansfeld	2007	A	Univ. So. Calif.	United States	1	*Electrochim. Acta*	SCI, SCIE	Electroch.	Algal electricity—MFCs	31	4.4	15
4	Ghirardi	2006	R	NREL	United States	1	*Indian J. Biochem. Biophys.*	SCIE	Bioch. Mol. Biol., Biophys.	Algal electricity—hydrogen	22	2.8	18
5	Clarens et al.	2011	A	Univ. Virginia	United States	5	*Environ. Sci. Technol.*	SCI, SCIE	Eng. Env., Env. Sci.	Algal electricity—biodiesel	20	6.7	20
6	Subhadra and Edwards	2010	A	Univ. New Mexico, Arizona St. Univ.	United States	2	*Energy Policy*	SCIE, SSCI	Ener. Fuels, Env. Sci., Env. Stud.	Algal electricity—biofuels	20	5.0	21
7	Satyanarayana et al.	2011	R	NPDEAS, Univ. Fed. Parana	Brazil	3	*Int. J. Energy Res.*	SCI, SCIE	Ener. Fuels, Nucl. Sci. Tech.	Algal electricity—general	19	6.3	22
8	Subhadra	2010	A	Univ. New Mexico	United States	1	*Energy Policy*	SCIE, SSCI	Ener. Fuels, Env. Sci., Env. Stud.	Algal electricity—biofuels	16	4.0	26

(*Continued*)

TABLE 24.1 (*Continued*)

Nonexperimental Research on Algal Bioelectricity and Algal Microbial Fuel Cells

No.	Paper References	Year	Document	Affiliation	Country	No. of Authors	Journal	Index	Subjects	Topic	Total No. of Citations	Total Average Citations per Annum	Rank
9	Rosenbaum and Schroder	2010	R	Tech. Univ. Carolo Wilhelmina Braunschweig, Cornell Univ.	United States, Germany	2	*Electroanalysis*	SCI, SCIE	Chem. Anal., Electroch.	Algal electricity—MFCs, solar cells	15	3.8	29
10	Wunschiers and Lindblad	2002	A	Uppsala Univ.	Sweden	2	*Int. J. Hydrog. Energy*	SCI, SCIE	Chem. Phys., Electroch., Ener. Fuels	Algal electricity—MFCs hydrogen	15	1.3	30
11	Pisciotta et al.	2010	A	Univ. Maryland	United States	3	*PLoS One*	SCIE	Mult. Sci.	Algal electricity—MFCs	14	3.5	31
12	Strik et al.	2010	A	Wageningen Univ.	Netherlands	3	*Environ. Sci. Technol.*	SCI, SCIE	Eng. Env., Env. Sci.	Algal electricity—MFCs	14	3.5	32
13	Harun et al.	2011	A	Monash Univ., Univ. Putra Malaysia	Australia, Malaysia	6	*Biomass Bioenerg.*	SCIE	Agr. Eng., Biot. Appl. Microb., Ener. Fuels	Algal electricity—biodiesel	13	4.3	35
14	Powell and Hill	2009	A	Univ. Saskatchewan	Canada	2	*Chem. Eng. Res. Des.*	SCI, SCIE	Eng. Chem.	Algal electricity—MFCs	12	2.4	37
15	Dante	2005	A	Inst. Tecnol. Estudios Super Monterrey Mexico City	Mexico	1	*Int. J. Hydrog. Energy*	SCI, SCIE	Chem. Phys., Electroch., Ener. Fuels	Algal electricity—MFCs	10	1.1	39

SCI, Science Citation Index; SCIE, Science Citation Index Expanded; SSCI, Social Sciences Citation Index; A, article; R, review.

They argue that fossil energy inputs and intensive water usage diminish the positive aspects of algal energy production. They propose the IREP approach for aligning renewable energy industries in resource-specific regions in the United States for synergistic electricity and liquid biofuel production from algal biomass with net zero-carbon emissions. They then discuss the benefits, challenges, and policy needs of this approach.

Subhadra (2010) discusses the sustainability of algal biofuel production using an IREP and algal biorefinery approach in a paper with 16 citations. He focuses on three integrated approaches to meet challenges facing algal biofuel sector. Firstly, he delineates an integrated algal biorefinery for sequential biomass processing for multiple high-value products to bring in the financial sustainability to the algal biofuel production units. Secondly, he proposes an IREP approach for amalgamating various renewable energy industries established in different locations. This would aid in synergistic and efficient electricity and liquid biofuel production with zero net carbon emissions while obviating numerous sustainability issues such as productive usage of agricultural land, water, and fossil fuel usage. A *renewable energy corridor* rich in multiple energy sources needed for algal biofuel production for deploying IREPs in the United States is also illustrated. Finally, he argues that the integration of various industries with the algal biofuel sector can bring a multitude of sustainable deliverables to society, such as renewable supply of cheap protein supplements, health products, and aquafeed ingredients. He discusses the benefits, challenges, and policy needs of the IREP approach.

24.2.2.1.3 Cofiring Power Plants

Kadam (2002) discusses the environmental implications of power generation via coal–microalgae cofiring in a paper with 54 citations. He argues that power-plant flue gas can serve as a source of CO_2 for microalgae cultivation, and algae can be cofired with coal. He conducts a life cycle assessment (LCA) to compare the environmental impacts of electricity production via coal firing versus coal/algae cofiring. He finds that "there are potentially significant benefits to recycling CO_2 toward microalgal production." As it reduces CO_2 emissions by recycling it and uses less coal, there are concomitant benefits of reduced greenhouse gas emissions. However, "there are also other energy and fertilizer inputs needed for algal production, which contribute to key environmental flows." Lower net values for the algal cofiring scenario were observed for the following using the direct injection process (in which the flue gas is directly transported to the algae ponds): SO_x, NO_x, particulates, carbon dioxide, methane, and fossil energy consumption. Lower values for the algal cofiring scenario were also observed for greenhouse potential and air acidification potential. However, he argues that "impact assessment for depletion of natural resources and eutrophication potential showed much higher values." He asserts that the LCA can help in the decision-making process for implementation of the algal scenario.

24.2.2.1.4 Biodiesel

Clarens et al. (2011) discuss the environmental impacts of algae-derived biodiesel and bioelectricity for transportation in a paper with 20 citations. They note that large-scale deployment of algal bioenergy systems could have unexpectedly high environmental burdens. They undertake a *well-to-wheel* LCA to evaluate algae's potential use as a transportation energy source for passenger vehicles. They assess four algal conversion pathways resulting in combinations of bioelectricity and biodiesel for several relevant nutrient procurement scenarios. They find that "algae-to-bioenergy systems can be either net energy positive or negative depending on the specific combination of cultivation and conversion processes used." For example, conversion pathways involving direct combustion for bioelectricity production generally outperformed systems involving anaerobic digestion and biodiesel production, and they generated 4 and 14 times as many vehicle kilometers traveled (VKT) per hectare as switchgrass or canola, respectively. Despite this, they argue that algal systems exhibited mixed performance for environmental impacts (energy use, water use, and greenhouse gas emissions) on a *per km* basis relative to the benchmark crops. They recommend that both cultivation and conversion processes must be carefully considered to ensure the environmental viability of algae-to-bioenergy processes.

Harun et al. (2011) discuss the technoeconomic analysis of an integrated microalgae photobioreactor (PBR), biodiesel, and biogas production facility in a recent paper with 13 citations. They note that the production of microalgal biodiesel is not economically viable in the current environment because it costs more than conventional fuels. Therefore, they introduce a new concept as an option to reduce the total production cost of microalgal biodiesel. They find that the "integration of biodiesel production system with methane production via anaerobic digestion improved the economics and sustainability of overall biodiesel stages." Anaerobic digestion of microalgae produces methane, which is further converted to generate electricity. The generated electricity can surrogate the consumption of energy that is required in microalgal cultivation, dewatering, extraction, and transesterification process. From theoretical calculations, they find that the "bioelectricity generated from methane is able to power all of the biodiesel production stages and will substantially reduce the cost of biodiesel production (33% reduction)." The carbon emissions of biodiesel production systems are also reduced by approximately 75% when utilizing biogas electricity compared to when the electricity is otherwise purchased from the Victorian grid. They assert that the approach of digesting microalgal waste to produce biogas will make the production of biodiesel from algae more viable by reducing the overall cost of production per unit of biodiesel and hence enable biodiesel to be more competitive with existing fuels.

24.2.2.1.5 Biomass

Satyanarayana et al. (2011) discuss microalgae as a versatile source for sustainable energy and materials in a recent review paper with 19 citations. They note that microalgae possess a high growth rate, need abundantly available solar light and CO_2, and thus are more photosynthetically efficient than oil crops. Also, they tolerate a high concentration of salts, allowing the use of any type of water for the agriculture and the possibility of production using innovative compact PBRs. In addition, microalgae are a potential source of biomass, which may have great biodiversity and consequent variability in their biochemical composition. They present an overview on microalgae with particular emphasis as a source for energy (biofuel/bioelectricity) and new biomaterials. They discuss the critical issues involved in production of microalgae and their use and future R&D to overcome these.

24.2.2.2 Nonexperimental Research on Algal MFCs

Mansfeld (2007) discusses the production of bioelectricity from algae in MFCs in a paper with 31 citations with a focus on the different examples for the interaction of bacteria and metal surfaces based on work reported previously. He notes that certain strains of *Shewanella* can prevent pitting of Al 2024 in artificial seawater, tarnishing of brass, and rusting of mild steel. The corrosion started again when the biofilm was killed by adding antibiotics. The mechanism of corrosion protection is different for different bacteria since the corrosion potential E_{corr} becomes more negative in the presence of *Shewanella ana* and algae, but more positive in the presence of *Bacillus subtilis*. These findings have been used in an initial study of a bacterial battery in which *Shewanella oneidensis* MR-1 was added to a cell containing Al 2024 and Cu in a growth medium. They find that the "power output of this cell continuously increased with time. In the microbial fuel cell (MFC) bacteria oxidize the fuel and transfer electrons directly to the anode." In initial studies, electrochemical impedance spectroscopy (EIS) has been used to characterize the anode, cathode, and membrane properties for different operating conditions of an MFC that contained *S. oneidensis* MR-1. Cell voltage (V)–current density (i) curves were obtained using potentiodynamic sweeps.

Rosenbaum and Schroder (2010) discuss photomicrobial solar and fuel cells in a review paper with 15 citations. Microbial solar cells and photomicrobial fuel cells (PFCs) exploit the energy of light and the activity of phototrophic microorganisms to produce electricity. Whereas microbial solar cells use light as the sole energy source, PFCs degrade organic matter in the presence of light. They provide an overview about the recent developments in the field of microbial-based photobiological fuel cells and microbial solar cells with a focus on the possible electron transfer mechanisms and elementary photobiological reaction pathways.

Wunschiers and Lindblad (2002) discuss the development of educational instruments for teaching the production of biohydrogen and bioelectricity from algae in a paper with 15 citations. They argue that an integral part of current research and development must be the inclusion of new technology into public education. By means of a model bioreactor for light-dependent (photobiological) production of hydrogen gas with green algae, they serve this goal in biological education. Various simple PBR types (closed batch, open batch) were analyzed for their capability to produce hydrogen under different conditions. The focus laid on functionality and simplicity rather than on high efficiency. Easy-to-handle systems that can be used in the classroom are presented. In a more sophisticated version, a proton exchange membrane (PEM) fuel cell was connected to a continuous gas flow tube bioreactor. They developed a software interface, designed to read light intensity, temperature, and power generation by the bioreactor and the connected fuel cell, respectively.

Pisciotta et al. (2010) discuss the light-dependent electrogenic activity of cyanobacteria in a paper with 14 citations. They show that diverse genera of cyanobacteria including biofilm-forming and pelagic strains have a conserved light-dependent electrogenic activity, that is, the ability to transfer electrons to their surroundings in response to illumination. Naturally growing biofilm-forming photosynthetic consortia also displayed light-dependent electrogenic activity, demonstrating that this phenomenon is not limited to individual cultures. Treatment with site-specific inhibitors revealed that "electrons originate at the photosynthetic electron transfer chain (P-ETC). Moreover, electrogenic activity was observed upon illumination only with blue or red but not green light confirming that P-ETC is the source of electrons." They find that the "yield of electrons harvested by extracellular electron acceptor to photons available for photosynthesis ranged from 0.05% to 0.3%, although the efficiency of electron harvesting likely varies depending on terminal electron acceptor." They illustrate that cyanobacterial electrogenic activity is an important microbiological conduit of solar energy into the biosphere. They argue that the mechanism responsible for electrogenic activity in cyanobacteria is fundamentally different from the one exploited in previously discovered electrogenic bacteria, such as *Geobacter*, where electrons are derived from oxidation of organic compounds and transported via a respiratory electron transfer chain (R-ETC). They assert that the electrogenic pathway of cyanobacteria might be exploited to develop light-sensitive devices or future technologies that convert solar energy into limited amounts of electricity in a self-sustainable, CO_2-free manner.

Strik et al. (2010) discuss the solar energy–powered MFC with a reversible bioelectrode in a paper with 14 citations. They note that a general problem with MFCs is the pH membrane gradient that reduces cell voltage and power output. This problem is caused by acid production at the anode, alkaline production at the cathode, and the nonspecific proton exchange through the membrane. They report a solution for a new kind of solar energy–powered MFC via development of a reversible bioelectrode responsible for both biocatalyzed anodic and cathodic electron transfer. Anodic produced protons were used for the cathodic reduction reaction that held the formation of a pH membrane gradient. They find that the MFC continuously generated electricity and repeatedly reversed polarity dependent on aeration or solar energy exposure. Identified organisms within the biocatalyzing biofilm of the reversible bioelectrode were algae, cyanobacteria, and protozoa.

Powell and Hill (2009) discuss the economic assessment of an integrated bioethanol–biodiesel–MFC facility utilizing yeast and photosynthetic algae in a paper with 12 citations. They present a strategy for the integration of a novel CO_2 photosynthetic culture and power generation system into a commercial bioethanol plant. Photosynthetic microalgae column PBRs, acting as cathodic half cells, are coupled with existing yeast fermenters at a bioethanol plant, acting as anodic half cells, to create coupled MFCs. The microalgae PBRs also sequester CO_2 emitted by the yeast fermenters. Incorporating MFCS into an existing bioethanol plant generates some of the power used in bioethanol production, and the microalgae species *Chlorella vulgaris* contains oil, which acts as a by-product for the production of biodiesel. They determine the required design specifications of novel, airlift PBR cathodes to make the integrated system economically feasible at an existing bioethanol plant. Data from previous experimental studies was used to develop the optimum integration strategy. The reported parameters include PBR size, number of integrated MFCs, fuel cell outputs, oil (for biodiesel) production rate, and CO_2 consumption rate.

Dante (2005) discusses hypotheses for direct PEM fuel cells applications of photobio-produced hydrogen by *Chlamydomonas reinhardtii* in a paper with 10 citations. He investigates the possibility to exploit the energy content of hydrogen released by green algae directly by a fuel cells stack. The power output can vary widely depending on the use of flow rate controls, since excessive dilution of hydrogen and power peaks can be avoided provoking a temporal delay between hydrogen production and the fuel cell stack feed. They estimate that a 100 m^3 algae culture can give an average 240 W power for 100 h, also taking into account fuel cell efficiency. He asserts that there are possibilities to increase the power output, overall raising algae efficiency to transform solar energy into chemical hydrogen energy.

24.2.3 Conclusion

The nonexperimental research on algal bioelectricity and algal MFCs has been one of the most dynamic research areas in recent years. Fifteen citation classics in the field of algal bioelectricity and algal MFCs with more than ten citations were located and the key emerging issues from these papers were presented in the decreasing order of the number of citations. These papers give strong hints about the determinants of the efficient algal bioelectricity production and algal MFC processes and policies and emphasize that marine algae are efficient algal bioelectricity and algal MFC feedstocks in comparison with first- and second-generation biofuel feedstocks.

24.3 Experimental Studies on Algal Bioelectricity and Algal MFCs

24.3.1 Introduction

The experimental research on algal bioelectricity and algal MFCs has been one of the most dynamic research areas in recent years. Thirty-five experimental citation classics in the field of algal bioelectricity and algal MFCs with more than five citations were located and the key emerging issues from these papers are presented in the decreasing order of the number of citations (Table 24.2). These papers give strong hints about the determinants of efficient algal bioelectricity production and algal MFCs processes and emphasize that marine algae are efficient biodiesel feedstocks in comparison with first- and second-generation biofuel feedstocks.

The papers were dominated by the researchers from 14 countries, usually through the intercountry institutional collaboration, and they were multiauthored. The United States (10 papers), Japan (9 papers), Canada (5 papers), England (4 papers), India (3 papers), and South Korea, China, and Taiwan (2 papers each) were the most prolific countries.

Similarly, all these papers were published in journals indexed by the SCI and/or SCIE. There were no papers indexed by the Arts & Humanities Citation Index (A&HCI) or SSCI. The number of citations ranged from 5 to 520 and the average number of citations per annum ranged from 0.5 to 65.0. All of these papers were articles.

24.3.2 Experimental Research on Algal Bioelectricity and Algal MFCs

24.3.2.1 Experimental Research on Algal Bioelectricity

Chen et al. (2010) develop strategies to enhance cell growth and achieve high-level oil production of a *C. vulgaris* isolate in a paper with 15 citations. The autotrophic growth of an oil-rich indigenous microalgal isolate, *C. vulgaris* C–C, was promoted by using engineering strategies to obtain the microalgal oil for biodiesel synthesis. "Illumination with a light/dark cycle of 14/10 (i.e., 14 h light-on and 10 h light-off) resulted in a high overall oil production rate (v_{oil}) of 9.78 mg/L/day and a high electricity conversion efficiency (E_c) of 23.7 mg cell/kw h" They then find that "when using a $NaHCO_3$ concentration of 1500 mg/L as carbon source, the v_{oil} and E_c were maximal at 100 mg/L/day and 128 mg/kw h, respectively." They describe the "microalgal growth kinetics with an estimated maximum specific growth rate (μ_{max}) of 0.605 day^{-1} and a half saturation coefficient (K_s) of 124.9 mg/L." They assert that an optimal nitrogen source (KNO_3) concentration of 625 mg/L could further enhance the microalgal biomass and oil production, leading to a nearly 6.19-fold increase in v_{oil} value.

24.3.2.2 Experimental Research on Algal MFCs

Gorby et al. (2006) study the electrically conductive bacterial nanowires produced by *S. oneidensis* strain MR-1 and other microorganisms in a paper with 520 citations. *S. oneidensis* MR-1 produced electrically conductive pilus-like appendages called bacterial nanowires in direct response to electron-acceptor limitation. Mutants deficient in genes for c-type decaheme cytochromes MtrC and OmcA and those that lacked a functional Type II secretion pathway displayed nanowires that were poorly conductive. These mutants were also deficient in their ability to reduce hydrous ferric oxide and in their ability to generate current in an MFC. They assert that "nanowires produced by the oxygenic phototrophic

TABLE 24.2

Experimental Research on Algal Bioelectricity and Algal Microbial Fuel Cells

No.	Paper References	Year	Document	Affiliation	Country	No. of Authors	Journal	Index	Subjects	Topic	Total No. of Citations	Total Average Citations per Annum	Rank
1	Gorby et al.	2006	A	Pacific NW Natl. Lab., Univ. Guelph, Korea Inst. Sci. Technol., Gwangju Inst. Sci. Technol., Univ. Wisconsin, Univ. Missouri, Maryland Biotechnol. Inst., Penn. St. Univ., Univ. So Calif.	Canada, USA, S. Korea, Japan	24	*Proc. Natl. Acad. Sci. U.S.A.*	SCI, SCIE	Mult. Sci.	Algal electricity—MFCs—nanosystems	520	65.0	1
2	El-Naggar et al.	2010	A	Univ. So Calif., J. Craig Venter Inst., Univ. Western Ontario	United States, Canada	9	*Proc. Natl. Acad. Sci. U.S.A.*	SCI, SCIE	Mult. Sci.	Algal electricity—MFCs—nanosystems	77	19.3	3
3	Velasquez-Orta et al.	2009	A	Univ. Newcastle, Penn. St. Univ.	England, United States	3	*Biotechnol. Bioeng.*	SCI, SCIE	Biot. Appl. Microb.	Algal electricity—MFCs.	51	10.2	5
4	Rodrigo et al.	2007	A	Univ. Castilla La Mancha	Spain	6	*J. Power Sources*	SCI, SCIE	Electroch., Ener. Fuels	Algal electricity—MFCs.	50	7.1	6
5	Chiao et al.	2006	A	Univ. British Columbia, Univ. Calif. Berkeley	Canada, United States	3	*J. Micromech. Microeng.*	SCI, SCIE	Eng. Elect. Elect., Nano Sci. Tech., Inst. Inst., Mats. Sci. Mult., Mechs	Algal electricity—MFCs.	47	5.9	7
6	Tanaka et al.	1985	A	Inst. Phys. Chem. Res. Riken	Japan	3	*J. Chem. Technol. Biot.*	SCI, SCIE	Biot. Appl. Microb.	Algal electricity—MFCs.	43	1.5	8
7	De Schamphelaire and Verstraete	2009	A	Univ. Ghent	Belgium	2	*Biotechnol. Bioeng.*	SCI, SCIE	Biot. Appl. Microb.	Algal electricity—MFCs.	39	7.8	9

(*Continued*)

TABLE 24.2 (*Continued*)

Experimental Research on Algal Bioelectricity and Algal Microbial Fuel Cells

No.	Paper References	Year	Document	Affiliation	Country	No. of Authors	Journal	Index	Subjects	Topic	Total No. of Citations	Total Average Citations per Annum	Rank
8	Mohan	2009a	A	Indian Inst. Chem. Technol.	India	1	*Int. J. Hydrog. Energy*	SCI, SCIE	Chem. Phys., Electroch., Ener. Fuels	Algal electricity—MFCs.	37	7.4	10
9	He et al.	2009	A	Univ. So Calif., Cornell Univ.	United States	5	*Environ. Sci. Technol.*	SCI, SCIE	Eng. Env., Env. Sci.	Algal electricity—MFCs.	35	7.0	11
10	Tsujimura et al.	2001	A	Kyoto Univ., Univ. Osaka Prefecture	Japan	4	*Enzyme Microb. Technol.*	SCI, SCIE	Biot. Appl. Microb.	Algal electricity—MFCs	35	2.7	12
11	Strik et al.	2008	A	Univ. Wageningen & Res. Ctr.	Netherlands	4	*Appl. Microbiol. Biotechnol.*	SCI, SCIE	Biot. Appl. Microb.	Algal electricity—MFCs.	34	5.7	13
12	Rosenbaum et al.	2005	A	Univ. Greifswald	Germany	3	*Appl. Microbiol. Biotechnol.*	SCI, SCIE	Biot. Appl. Microb.	Algal electricity—MFCs.	34	3.8	14
13	Powell et al.	2009	A	Univ. Saskatchewan	Canada	4	*Bioresour. Technol.*	SCI, SCIE	Agr. Eng., Biot. Appl. Microb., Ener. Fuels	Algal electricity—MFCs.	27	5.4	16
14	Yagishita et al.	1993	A	Inst. Phys. & Chem. Res. Riken, Tokyo Univ. Agr. & Technol.	Japan	3	*J. Chem. Technol. Biotechnol.*	SCI, SCIE	Biot. Appl. Microb., Chem. Mult., Eng. Chem.	Algal electricity—MFCs.	27	1.3	17
15	Yagishita et al.	1997a	A	Inst. Phys. & Chem. Res.	Japan	4	*Sol. Energy*	SCI, SCIE	Ener. Fuels	Algal electricity—MFCs.	22	1.3	19
16	Mohan et al.	2009b	A	Indian Inst. Chem. Technol.	India	6	*Bioresour. Technol.*	SCI, SCIE	Agr. Eng., Biot. Appl. Microb., Ener. Fuels	Algal electricity—MFCs.	19	3.8	23

(Continued)

TABLE 24.2 (*Continued*)

Experimental Research on Algal Bioelectricity and Algal Microbial Fuel Cells

No.	Paper References	Year	Document	Affiliation	Country	No. of Authors	Journal	Index	Subjects	Topic	Total No. of Citations	Total Average Citations per Annum	Rank
17	Zou et al.	2009	A	Univ. Maryland	United States	4	*Biotechnol. Bioeng.*	SCI, SCIE	Biot. Appl. Microb.	Algal electricity—MFCs.	19	3.8	24
18	Ryu et al.	2010	A	Stanford Univ., Yonsei Univ.	United States, S. Korea	9	*Nano Letters*	SCI, SCIE	Chem. Mult., Chem. Phys., Nano S.T., Mats. Sci. Mult., Phys. Appl., Phys. Cond. M.	Algal electricity—MFCs, nanosystems	16	4.0	25
19	McCormick et al.	2011	A	Univ. Cambridge	England	9	*Energy Environ. Sci.*	SCI, SCIE	Chem. Mult., Ener. Fuels, Eng. Chem., Env. Sci.	Algal electricity—MFCs—photovoltaic cells	15	5.0	27
20	Chen et al.	2010	A	Natl. Cheng Kung Univ., Taiwan Fisheries Res. Inst., Natl. Kaohsiung Marine Univ.	Taiwan	6	*Biotechnol. Prog.*	SCI, SCIE	Biot. Appl. Microb., Food Sci. Tech.	Algal electricity—biodiesel	15	3.8	28
21	Wang et al.	2010	A	Harbin Inst. Technol.	China	7	*Biosens. Bioelectron.*	SCI, SCIE	Biophys., Biot. Appl. Microb., Chem. Anal., Electroch., Nano ST	Algal electricity—MFCs	14	3.5	33
22	Yagishita et al.	1999	A	Natl. Inst. Resources Environm.	Japan	4	*J. Biosci. Bioeng.*	SCI, SCIE	Biot. Appl. Microb., Food Sci. Tech.	Algal electricity—MFCs	14	0.9	34

(*Continued*)

TABLE 24.2 (*Continued*)

Experimental Research on Algal Bioelectricity and Algal Microbial Fuel Cells

No.	Paper References	Year	Document	Affiliation	Country	No. of Authors	Journal	Index	Subjects	Topic	Total No. of Citations	Total Average Citations per Annum	Rank
23	Zou et al.	2010	A	Univ. Maryland	Untied States	3	*Bioelectrochemistry*	SCI, SCIE	Bioch. Mol. Biol., Biol., Biophys., Electroch.	Algal electricity—MFCs	13	3.3	36
24	Zhang et al.	2011	A	Tech. Univ. Denmark	Denmark	3	*Energy Environ. Sci.*	SCI, SCIE	Chem. Mult., Ener. Fuels, Eng. Chem., Env. Sci.	Algal electricity—MFCs	11	3.7	38
25	Yagishita et al.	1997b	A	Natl. Inst. Resources & Environm.	Japan	4	*Bioelectrochem. Bioenerg.*	SCI, SCIE	Bioch. Mol. Biol., Biophys.	Algal electricity—MFCs	10	0.6	40
26	Yagishita et al.	1998	A	Natl. Inst. Resources & Environm.	Japan	4	*J. Ferment. Bioeng.*[a]	SCI, SCIE	Biot. Appl. Microb., Food Sci. Tech.	Algal electricity—MFCs	10	0.6	41
27	Martens and Hall	1994	A	Univ. Cambridge	England	2	*Photochem. Photobiol.*	SCI, SCIE	Bioch. Mol. Biol., Biophys.	Algal electricity—MFCs	10	0.5	42
28	Bombelli et al.	2011	A	Univ. Cambridge, Univ. Bath	England	14	*Energy Environ. Sci.*	SCI, SCIE	Chem. Mult., Ener. Fuels, Eng. Chem., Env. Sci.	Algal electricity—MFCs—photovoltaic cells	9	3.0	43
29	Fu et al.	2009	A	Natl. Cheng Kung Univ., Taiwan. Univ. Calif. Davis	Taiwan, United States	6	*Bioresour. Technol.*	SCI, SCIE	Agr. Eng., Biot. Appl. Microb., Ener. Fuels	Algal electricity—MFCs.	9	1.8	44

(Continued)

TABLE 24.2 (*Continued*)

Experimental Research on Algal Bioelectricity and Algal Microbial Fuel Cells

No.	Paper References	Year	Document	Affiliation	Country	No. of Authors	Journal	Index	Subjects	Topic	Total No. of Citations	Total Average Citations per Annum	Rank
30	Croisetiere et al.	2001	A	Univ. Quebec	Canada	3	*Appl. Microbiol. Biotechnol.*	SCI, SCIE	Biot. Appl. Microb.	Algal electricity—MFCs	9	0.7	45
31	Dawar et al.	1998	A	Maharshi Dayanand Univ.	India	3	*Int. J. Energy Res.*	SCI, SCIE	Ener. Fuels, Nucl. Sci. Tech.	Algal electricity—MFCs	9	0.6	46
32	Nishio et al.	2010	A	Univ. Tokyo, ERATO JST	Japan	3	*Appl. Microbiol. Biotechnol.*	SCI, SCIE	Biot. Appl. Microb.	Algal electricity—MFCs	8	2.0	47
33	Martins et al.	2010	A	Univ. Minho	Portugal	6	*Bioelectrochemistry*	SCI, SCIE	Bioch. Mol. Biol., Biol., Biophys. Electroch.	Algal electricity—MFCs	7	1.8	48
34	Gong et al.	2011	A	Oregon St. Univ., Harvard Univ., Teledyne Benthos Inc	United States	6	*Environ. Sci. Technol.*	SCI, SCIE	Envir. Sci., Eng. Envir.	Algal electricity—MFCs	5	1.7	49
35	Yuan et al.	2011	A	Guangdong Inst. Ecoenvironm. Soil Sci., Sichuan Normal Univ.	China	6	*J. Hazard. Mater.*	SCI, SCIE	Eng. Env., Eng. Civ., Env. Sci.	Algal electricity—MFCs	5	1.7	50

SCI, Science Citation Index; SCIE, Science Citation Index Expanded; SSCI, Social Sciences Citation Index; A, article; R, Review.

[a] *Journal of Fermentation and Bioengineering* continues as *Journal of Bioscience and Bioengineering*.

cyanobacterium *Synechocystis* PCC6803 and the thermophilic, fermentative bacterium *Pelotomaculum thermopropionicum* reveal that electrically conductive appendages are not exclusive to dissimilatory metal-reducing bacteria and may, in fact, represent a common bacterial strategy for efficient electron transfer and energy distribution."

El-Naggar et al. (2010) study the electrical transport along bacterial nanowires from *S. oneidensis* MR-1. They report electron transport measurements along individually addressed bacterial nanowires derived from electron-acceptor-limited cultures of the dissimilatory metal-reducing bacterium *S. oneidensis* MR-1. Transport along the bacterial nanowires was independently evaluated by two techniques: (1) nanofabricated electrodes patterned on top of individual nanowires and (2) conducting probe atomic force microscopy at various points along a single nanowire bridging a metallic electrode and the conductive atomic force microscopy tip. They find that the "*S. oneidensis* MR-1 nanowires were electrically conductive along micrometer-length scales with electron transport rates up to 10(9)/s at 100 mV of applied bias and a measured resistivity on the order of 1 Omega.cm." Mutants deficient in genes for c-type decaheme cytochromes MtrC and OmcA produce appendages that are morphologically consistent with bacterial nanowires, but they were nonconductive.

Velasquez-Orta et al. (2009) study the production of energy from algae using MFCs in a paper with 51 citations. They examine bioelectricity production from a phytoplankton, *C. vulgaris*, and a macrophyte, *Ulva lactuca*, in single-chamber MFCs. MFCs were fed with the two algae (as powders), obtaining differences in energy recovery, degradation efficiency, and power densities. They find that "*C. vulgaris* produced more energy generation per substrate mass (2.5 k Wh/kg), but *U. lactuca* was degraded more completely over a batch cycle (73 +/− 1%, COD)." Additionally, "maximum power densities obtained using other single cycle or multiple cycle methods were 0.98 W/m^2 (277 W/m^3) using *C. vulgaris*, and 0.76 W/m^2 (215 W/m^3) using *U. lactuca*." Polarization curves obtained using a common method of linear sweep voltammetry (LSV) overestimated maximum power densities at a scan rate of 1 mV/s. At 0.1 mV/s, however, the LSV polarization data were in better agreement with single- and multiple-cycle polarization curves. The fingerprints of microbial communities developed in reactors had only 11% similarity to inocula and clustered according to the type of bioprocesses used.

Rodrigo et al. (2007) study the production of bioelectricity from the treatment of urban wastewater using an MFC in a paper with 50 citations. By using an anaerobic pretreatment of the activated sludge of an urban wastewater treatment plant, they obtain the electricity generation in a WC after a short acclimatization period of less than 10 days. They find that the "power density generated depend mainly on the organic matter contain (COD) but not on the wastewater flow-rate with maximum power densities of 25 mW m^{-2} (at a cell potential of 0.23 V)." The rate of consumption of oxygen in the cathodic chamber was very low. They note that as the oxygen reduction is coupled with the chemical oxygen demand (COD) oxidation in the anodic chamber, the COD removed by the electricity-generating process is very small. Thus, taking into account the oxygen consumption, they conclude that "only 0.25% of the removed COD was used for the electricity-generation processes" and argue that "the remaining COD should be removed by anaerobic processes." They also note that the presence of oxygen in the anodic chamber leads to a deterioration of the MFC performance, and this deterioration of the MFC process occurs rapidly after the appearance of nonnegligible concentrations of oxygen. Hence, they recommend that the growth of algae should be avoided to assure a good performance of this type of MFC.

Chiao et al. (2006) study the micromachined microbial and photosynthetic fuel cells in a paper with 47 citations. They present two types of fuel cells: a miniature MFC (μMFC) and a miniature photosynthetic electrochemical cell (μPEC). A bulk micromachining process is used to fabricate the fuel cells, and the prototype has an active PEM area of 1 cm^2. Two different microorganisms are used as biocatalysts in the anode: (1) *Saccharomyces cerevisiae* (baker's yeast) is used to catalyze glucose and (2) phylum Cyanophyta (blue-green algae) is used to produce electrons by a photosynthetic reaction under light. In the dark, they find that the "μPEC continues to generate power using the glucose produced under light. In the cathode, potassium ferricyanide is used to accept electrons and electric power is produced by the overall redox reactions." They next find that the "bio-electrical responses of μMFCs and μPECs are characterized with the open-circuit potential measured at an average value of 300–500 mV." They finally find that under a 10 Ω load, the power density is measured as 2.3 and 0.04 nW/cm^2 for μMFCs and μPECs, respectively.

Tanaka et al. (1985) study the bioelectrochemical fuel cells operated by the cyanobacterium, *Anabaena variabilis*, in a paper with 43 citations. Substantial electric output was delivered from the bioelectrochemical fuel cells operated under anaerobic conditions by *A. variabilis* strain M-2 and 2-hydroxy-1,4-naphthoquinone (HNQ). They find that "there was a linear relationship between the coulombic output of the fuel-cells operated in the dark and the glycogen content of the organisms." The coulombic output was increased substantially in the light as the increase was observed even in the absence of CO_2. Oxygen was evolved by *Anabaena* cells under the operating conditions of the fuel cells. They assert that the source of bioelectricity obtained from the fuel cells is endogenous glycogen in the dark and both glycogen and electrons produced by photosynthetic oxidation of water in the light.

De Schamphelaire and Verstraete (2009) study the biological sunlight-to-biogas energy conversion system in a paper with 39 citations. They note that algal biomass can be used as a substrate for anaerobic digestion and investigate the use of algae for energy generation in a stand-alone, closed-loop system. The system encompasses an algal growth unit for biomass production, an anaerobic digestion unit to convert the biomass to biogas, and an MFC to polish the effluent of the digester. Nutrients set free during digestion can accordingly be returned to the algal growth unit for a sustained algal growth. Hence, they present a "system that continuously transforms solar energy into energy-rich biogas and bioelectricity where algal productivities of 24–30 ton VS ha^{-1} $year^{-1}$ were reached, while 0.5 N m^3 biogas could be produced kg^{-1} algal VS." This system resulted in "a power plant with a potential capacity of about 9 kW ha^{-1} of solar algal panel, with prospects of 23 kW ha^{-1}."

Mohan (2009) studies the production of biohydrogen and bioelectricity wastewater treatment using mixed fermentative consortia in a paper with 37 citations. They review and summarize the work carried out in their laboratory on dark fermentation process of biohydrogen production utilizing wastewater as primary substrate under acidogenic mixed microenvironment toward optimization of dynamic process. They evaluate the process based on the nature and composition of wastewater, substrate loading rates, reactor configuration, operation mode, pH microenvironment, and pretreatment procedures adopted for mixed anaerobic culture to selectively enrich acidogenic H_2-producing consortia. The fermentative conversion of the substrate to H_2 is possible by a series of complex biochemical reactions manifested by selective bacterial groups. In spite of striking advantages, they argue that the main challenge of fermentative H_2 production is that relatively low energy from the organic source was obtained in the form of H_2. They argue that further utilization of unutilized carbon sources present in wastewater for additional H_2 production will sustain the practical applicability of the process. They discuss in this context "enhancing H_2 production by adapting various strategies, viz., self-immobilization of mixed consortia (onto mesoporous material and activated carbon), integration with terminal methanogenic and photo-biological processes and bioaugmentation with selectively enriched acidogenic consortia." They also present application of acidogenic microenvironment for in situ production of bioelectricity through wastewater treatment employing MFC.

He et al. (2009) study the self-sustained phototrophic MFCs based on the synergistic cooperation between photosynthetic microorganisms and heterotrophic bacteria in a paper with 35 citations. They develop a sediment-type self-sustained phototrophic MFC to generate bioelectricity through the synergistic interaction between photosynthetic microorganisms and heterotrophic bacteria. Under illumination, the MFC continuously produced electricity without the external input of exogenous organics or nutrients. The current increased in the dark and decreased with the light on, possibly because of the negative effect of the oxygen produced via photosynthesis. Continuous illumination inhibited the current production, while the continuous dark period stimulated the current production. Extended darkness resulted in a decrease of current probably because of the consumption of the organics accumulated during the light phase. Using color filters or increasing the thickness of the sediment resulted in a reduction of the oxygen-induced inhibition. Based on the molecular taxonomic analysis, they find that "photosynthetic microorganisms including cyanobacteria and microalgae predominated in the water phase, adjacent to the cathode and on the surface of the sediment." In contrast, the sediments were dominated by heterotrophic bacteria, becoming less diverse with increasing depth. In addition, "results from the air-cathode phototrophic MFC confirmed the light-induced current production, while the test with the two-chamber MFC (in the dark) indicated the presence of electricigenic bacteria in the sediment."

Tsujimura et al. (2001) study a photosynthetic bioelectrochemical cell utilizing cyanobacteria and water-generating oxidase in a paper with 35 citations. They construct a novel photosynthetic bioelectrochemical cell that utilizes biocatalysts in both anode and cathode compartments in the anodic half cell. Some of the electrons produced by the oxidation of water in the photosystem of cyanobacteria are transferred to the carbon-felt anode through quinonoid electron transfer mediators. The electron is passed to dioxygen to regenerate water in the cathodic half cell reaction with an aid of bilirubin oxidase reaction via a mediator. They find that the "maximum bioelectric power was about 0.3–0.4 W m^{-2} for the projective electrode surface area at an apparent efficiency of the light energy conversion of 2–2.5%."

Strik et al. (2008) study renewable sustainable biocatalyzed electricity production in a photosynthetic algal MFC (PAMFC) in a paper with 34 citations. They describe the proof of principle of a PAMFC based on naturally selected algae and electrochemically active microorganisms in an open system and without addition of instable or toxic mediators. They develop solar-powered PAMFC produced continuously over 100 days renewable biocatalyzed electricity. They find that the "sustainable performance of the PAMFC resulted in a maximum current density of 539 mA/m^2 projected anode surface area and a maximum power production of 110 mW/m^2 surface area photobioreactor." They argue that the energy recovery of the PAMFC can be increased by optimization of the PBR, by reducing the competition from nonelectrochemically active microorganisms and by increasing the electrode surface and establishment of a further-enriched biofilm. They recommend that future research should also focus on the development of low-energy input PAMFCs as the current algae production systems have energy inputs similar to the energy present in the outcoming valuable products.

Rosenbaum et al. (2005) study the utilization of the green alga *C. reinhardtii* for microbial electricity generation in a paper with 34 citations. By employing living cells of the green alga *C. reinhardtii*, they demonstrate the possibility of direct bioelectricity generation from microbial photosynthetic activity. The "underlying concept is based on an in situ oxidative depletion of hydrogen, photosynthetically produced by *C. reinhardtii* under sulfur-deprived conditions, by polymer-coated electrocatalytic electrodes."

Powell et al. (2009) study the growth kinetics of *C. vulgaris* and its use as a cathodic half cell in a paper with 27 citations. The kinetics of growth of the algal species *C. vulgaris* has been investigated using CO_2 as the growth substrate. They find that the growth rate increased as the dissolved CO_2 increased to 150 mg/L but fell dramatically at higher concentrations. Increasing the radiant flux also increased growth rate. With a radiant flux of 32.3 mW falling directly on the 500 mL culture media, the growth rate reached 3.6 mg of cells/L. Both pH variation (5.5–7.0) and mass transfer rate of CO_2 (K_La between 6 and 17 h^{-1}) had little effect on growth rate. Growing on glucose, the yeast *S. cerevisiae* produced a stable 160 mV potential difference when acting as an MFC anode with ferricyanide reduction at the cathode. The algal culture was a workable electron acceptor in a cathodic half cell. They assert that "using an optimum methylene blue mediator concentration, a net potential difference of 70 mV could be achieved between the growing *C. vulgaris* culture acting as a cathode and a 0.02 M potassium ferrocyanide anodic half cell." They measure the surge current and power levels of 1.0 μA/mg of cell dry weight and 2.7 mW/m^2 of cathode surface area between these two half cells.

Yagishita et al. (1993) study the effects of light, CO_2, and inhibitors on the current output of biofuel cells containing the photosynthetic organism *Synechococcus* sp. in an early paper with 27 citations. The current output of the biofuel cells containing a marine alga, *Synechococcus* sp., and an electron transport mediator, HNQ, was increased under illumination and in the presence of CO_2. The inhibitory effects of carbonyl cyanide *m*-chlorophenylhydrazone (CCCP), 3-(3,4-dichlorophenyl)-1,1-dimethylurea (DCMU), 2,5-dibromomethylisopropyl-*p*-benzoquinone (DBMIB), phenylmercury acetate (PMA), and *N,N'*-dicyclohexylcarbodiimide (DCCD) on the output current of fuel cells run in the light suggested that "HNQ accepts electrons mainly at the site of ferredoxin-NADP+ reductase (FNR) in the electron transfer chain." The inhibition of light-induced generation of current output by CCCP indicates that the "current is derived from photosynthetic oxidation of water." They assert that endogenous glycogen in algae is required to sustain a steady current output from the fuel cells.

Yagishita et al. (1997a) study the effects of intensity of incident light and concentrations of *Synechococcus* sp. and HNQ on the current output of PEC operated under illumination in an early paper with 22 citations. They find that "though the current density was higher with increasing concentration of *Synechococcus* sp., increase in the current density was suppressed at more than 24 mu g chlorophyll/ml." The current density reached a maximum value at 1 mM HNQ. They argue that the "current density was saturated at 50 W/m^2 probably due to reoxidation of HNQ by oxygen photosynthetically evolved." The conversion efficiency of light energy to bioelectrical energy was 3.3%. For the analysis of energy loss for four steps in the electrochemical cell reaction, they assert that the "energy loss remarkably arose from the reaction of the electron transfer chain and the high internal resistance of the electrochemical cell."

Mohan et al. (2009) study the potential of various aquatic ecosystems in harnessing bioelectricity through a benthic fuel cell with a focus on the effect of electrode assembly and water characteristics using 6 different types of ecological water bodies in a paper with 19 citations. Experiments were designed with various combinations of electrode assemblies, surface area of anode, and anodic materials. Among the 32 experiments conducted, they find that 9 combinations evidenced stable electron discharge/current.

"Nature, flow conditions and characteristics of water bodies showed significant influence on the power generation apart from electrode assemblies, surface area of anode and anodic material." Stagnant water bodies showed comparatively higher power output than the running water bodies. They then find that "placement of cathode on algal mat (as bio-cathode) documented several folds increment in power output." They finally find that "electron-discharge started at 1,000 Ω resistance in polluted water bodies (Nacaharam Cheruvu, Hussain Sagar Lake, Musi River), whereas, in relatively less polluted water bodies (Uppal Pond/Stream, Godavari River) electron-discharge was observed at low resistances (500/750 Ω)."

Zou et al. (2009) study photosynthetic MFCs with positive light response in a paper with 19 citations. They introduce an aerobic single-chamber photosynthetic MFC (PMFC). Evaluation of PMFC performance using naturally growing freshwater photosynthetic biofilm revealed a weak-positive light response, that is, an increase in cell voltage upon illumination. When the PMFC anodes were coated with electrically conductive polymers, they found that "the rate of voltage increased and the amplitude of the light response improved significantly." The rapid immediate positive response to light was consistent with a mechanism postulating that the photosynthetic electron transfer chain is the source of the electrons harvested on the anode surface. They argue that "this mechanism is fundamentally different from the one exploited in previously designed anaerobic MFCs, sediment MFCs, or anaerobic PMFCs, where the electrons are derived from the respiratory electron-transfer chain." They next find that the "power densities produced in PMFCs, were substantially lower than those that are currently reported for conventional MFC (0.95 mW/m^2 for polyaniline-coated and 1.3 mW/m^2 for polypyrrole-coated anodes)." However, they argue that the PMFC did not depend on an organic substrate as an energy source and was powered only by light energy while its operation was CO_2 neutral and did not require buffers or exogenous electron transfer shuttles.

Ryu et al. (2010) study the direct extraction of photosynthetic electrons from single algal cells by nanoprobing system in a paper with 16 citations. To more efficiently use the solar energy harvested by photosynthetic organisms, they evaluate the feasibility of generating bioelectricity by directly extracting electrons from the photosynthetic electron transport (PET) chain before they are used to fix CO into sugars and polysaccharides. From a living algal cell, "*C. reinhardtii*, photosynthetic electrons (1.2 pA at 6000 mA/m^2) were directly extracted without a mediator electron carrier by inserting a nanoelectrode into the algal chloroplast and applying an overvoltage." They speculate that this finding may represent an initial step in generating *high-efficiency* bioelectricity by directly harvesting high-energy photosynthetic electrons.

McCormick et al. (2011) study photosynthetic biofilms in pure culture in a mediatorless biophotovoltaic cell (BPV) system in a recent paper with 15 citations. They report on light-driven electrical power generated with biofilms grown from photosynthetic freshwater or marine species without the addition of an artificial electron-shuttling mediator. Green alga (*C. vulgaris, Dunaliella tertiolecta*) or cyanobacterium (*Synechocystis* sp. PCC 6803, *Synechococcus* sp. WH5701) strains were grown directly on a transparent, conductive anode (indium tin oxide–coated polyethylene terephthalate), and power generation under light and dark conditions was evaluated using a single-chamber BPV system. They observe "increased power outputs for all strains upon illumination, with the largest light effect observed for *Synechococcus* (maximum 10.3 mW m^{-2} total power output recorded under 10 W m^{-2} white light)." Further experiments conducted with *Synechococcus* and *C. vulgaris* indicated that "photosynthetic oxygen evolution rates were consistent with BPV power outputs under different light regimes (red, green and blue light), indicating a direct link between power output and photosynthetic activity." Biofilm power generation in these BPV devices was self-sustained for several weeks and was highly sensitive to ambient light levels.

Wang et al. (2010) study the sequestration of CO_2 discharged from anode by algal cathode in microbial carbon capture cells (MCCs) in a paper with 14 citations. They construct new MCCs by introducing anodic off-gas into an algae grown cathode (*C. vulgaris*) and demonstrate this approach as an effective technology for CO_2 emission reduction with simultaneous voltage output without aeration (610 ± 50 mV, 1000 Ω). They find that "maximum power densities increased from 4.1 to 5.6 W/m^3 when the optical density (OD) of cathodic algae suspension increased from 0.21 to 0.85 (658 nm)." Compared to a stable voltage of 706 ± 21 mV (1000 Ω) obtained with cathodic dissolved oxygen (DO) of 6.6 ± 1.0 mg/L in MCC, they next find that "voltage outputs decreased from 654 to 189 mV over 70 h in the control reactor (no algae) accompanied with a decrease in DO from 7.6 to 0.9 mg/L, indicating that cathode electron acceptor was oxygen." Gas analysis showed that all the CO_2 generated from anode was completely eliminated by catholyte, and the soluble inorganic carbon was further converted into algal biomass. They assert that their approach showed the possibility of a new method for simultaneous carbon fixing, power generation, and biodiesel production during wastewater treatment without aeration.

Yagishita et al. (1999) study the effects of glucose addition and light on current outputs in PECs using *Synechocystis* sp. PCC6714 under photo- and chemoheterotrophical conditions in a paper with 14 citations. They find that the addition of glucose to the anode solutions of the electrochemical cells resulted in a rapid increase in the current outputs under both light and dark conditions. Although the coulombic outputs were almost the same under light and dark conditions, the rate of glucose consumption was faster under illumination than in the dark. The total sugar content in the cells of strain PCC6714 increased with the addition of glucose and the total sugar accumulated remained intact during the discharge under illumination, while it decreased gradually in the dark. They next find that when the light was switched off after the addition of glucose, the current output markedly increased.

> The coulombic outputs obtained after darkening were 10 to 80 times larger than that obtained by the addition of glucose under the continuous light or dark conditions. *Synechocystis* sp. completely incorporated 0.14 mM and 0.42 mM glucose for 1 h and 3 h, respectively, under illumination. There was no difference in the coulombic outputs between 1 h and 12 h illumination times in the electrochemical cells with 0.14 mM glucose. When the light was switched off after 1 h illumination in the electrochemical cells with 0.42 mM glucose, the coulombic output obtained from the electrochemical cell was lower than that in the electrochemical cell with 12 h illumination.

They assert that the current output was produced with higher efficiency with glucose incorporated under illumination than that in the case of glucose incorporated after darkening as the highest coulombic yield of 54% in this experiment was obtained by darkening in the electrochemical cell with 0.14 mM glucose.

Zou et al. (2010) study the nanostructured polypyrrole-coated anode for sun-powered MFCs in a paper with 13 citations. They show that "nanostructured electrically conductive polymer polypyrrole substantially improved the efficiency of electron collection from photosynthetic biofilm in photosynthetic MFCs (PMFCs)." Nanostructured fibrillar polypyrrole showed better performance than granular polypyrrole. "Cyclic voltammetry and impedance spectroscopy analyses revealed that better performance of nanostructured anode materials was due to the substantial improvement in electrochemical properties including higher redox current and lower interface electron-transfer resistance." They then find that "at loading density of 3 mg/cm^2, coating of anode with fibrillar polypyrrole resulted in a 450% increase in the power density" compared to the previous studies on PMFCs that used the same photosynthetic culture.

Zhang et al. (2011) study the simultaneous organic carbon, nutrients removal, and energy production in a PFC in a recent paper with 11 citations. They develop a sediment-type PFC, based on the synergistic interaction between microalgae (*C. vulgaris*) and electrochemically active bacteria, to remove carbon and nutrients from wastewater, and produce bioelectricity and algal biomass simultaneously. Under illumination, they find that "a stable power density of 68 +/− 5 mW m^{-2} and a biomass of 0.56 +/− 0.02 g L^{-1} were generated at an initial algae concentration of 3.5 g L^{-1}." Accordingly, the removal efficiency of organic carbon, nitrogen, and phosphorus was 99.6%, 87.6%, and 69.8%, respectively. Mass balance analysis suggested the "main removal mechanism of nitrogen and phosphorus was the algae biomass uptake (75% and 93%, respectively), while the nitrification and denitrification process contributed to a part of nitrogen removal (22%)." In addition, they investigate the effect of illumination period on the performance of PFC. They find that "except notable fluctuation of power generation, carbon and nutrients removal was not significantly affected after changing the light/dark photoperiod from 24 h/0 h to 10 h/14 h." They assert that this effective bacteria–algae coupled system was capable for extracting energy and removing carbon, nitrogen, and phosphorus from wastewater in one step.

Yagishita et al. (1997b) study the behavior of glucose degradation in *Synechocystis* sp. M-203 in bioelectrochemical fuel cells in a paper with 10 citations. By the addition of glucose, they find that "the voltage output rapidly increased under illumination, and in the dark substantial voltage output was sustained stably for 20 h of discharge. The addition of the glucose analogue 3-O-methyl-D-glucose did not change the current output both in the light and the dark, indicating that added glucose was transported into Synechocystis cells and was consumed to produce bioelectricity." An inhibitor, CCCP, induced the increase in the current output under illumination, while DCMU partially inhibited the current output.

Yagishita et al. (1998) study the performance of PECs using immobilized *A. variabilis* M-3 immobilized within alginate beads during repeated discharge and culture cycles of 10 h each in a paper with 10 citations. They examine the effect of the light intensity during the culture periods on the duration of the current output and the recovery of endogenous total sugar in the *Anabaena* cells. They find that compared with continuous discharge operation, the "duration of the current output was extended by 10 times when the electrochemical cell was operated under 10-h discharge (dark)/culture (light) cycles with a light intensity of 100 W/m^2 during the culture periods." The conversion efficiency of light energy into electricity was about 0.2% under these conditions.

Martens and Hall (1994) study the diaminodurene (DAD) as a mediator of a photocurrent using intact cells of cyanobacteria in an early paper with 10 citations. They compare the efficiency of DAD as an electron shuttle between the PET chain within the cells and the electrode with ferrocyanide and *p*-benzoquinone; they construct a simple *Synechococcus*-carbon-paste electrode containing DAD as the mediator in the electrode material and the photosynthetic cells immobilized in a polymer matrix. The device required no addition of further reagents to the bulk solution for the generation of a photocurrent. Using cyclic voltammetry, they determine the second-order rate constant for the reaction of DAD with the PET. They find that "diaminodurene compared favorably with p-benzoquinone in that the overpotential was reduced to 0.2 V." They next find that the deterioration of the PET by the DAD was significantly decreased compared with *p*-benzoquinone, resulting in a prolonged lifetime of the electrode.

Bombelli et al. (2011) study the quantitative analysis of the factors limiting solar power transduction by *Synechocystis* sp. PCC 6803 in biological photovoltaic devices in a paper with nine citations. They describe the fabrication of an MFC-inspired photovoltaic (BPV) device with multiple microchannels. This allows a direct comparison between subcellular photosynthetic organelles and whole cells and quantitative analysis of the parameters affecting power output. They study electron transfer within the photosynthetic materials using the metabolic inhibitors DCMU and methyl viologen (1,1′-dimethyl-4,4′-bipyridinium dichloride). They find that the "electrons that cause an increase in power upon illumination leave the photosynthetic electron transfer chain from the reducing end of photosystem I."

Fu et al. (2009) study the effects of biomass weight and light intensity on the performance of photosynthetic microbial fuel cells (PMFCs) with *Spirulina platensis* in a paper with nine citations. They attach microalgae *S. platensis* to the anode of a membrane-free and mediator-free MFC to produce bioelectricity through the consumption of biochemical compounds inside the microalgae. They observe an increase in open-circuit voltage (OCV) with decreasing light intensity and optimal biomass area density. "The highest OCV observation for the MFC was 0.39 V in the dark with a biomass area density on the anode surface of 1.2 g cm^{-2}. Additionally, they observe that the MFC with 0.75 g cm^{-2} of biomass area density produced 1.64 mW m^{-2} of electrical power in the dark, which is superior to the 0.132 mW m^{-2} produced in the light." They assert that the MFC can be applied to generate electrical power under both day and night conditions.

Croisetiere et al. (2001) discuss a simple mediatorless amperometric method using the cyanobacterium *Synechococcus leopoliensis* for the detection of phytotoxic pollutants in a paper with nine citations. They use the unicellular cyanobacterium *S. leopoliensis* in a microelectrochemical cell to generate photocurrents. The photocurrent is dependent on PET and is mediated by hydrogen peroxide formation following the reduction of oxygen on the acceptor side of photosystem I. This was the first known application of cyanobacteria in an electrochemical device where no artificial electroactive mediator is needed. They then discuss the potential for the development of this microelectrochemical cell for the detection of phytotoxic pollutants, such as herbicides and toxic metal cations, using the photosynthetic system of the cyanobacteria without interference from added electron acceptor.

Dawar et al. (1998) study the development of a low-cost oxyhydrogen biofuel cell for generation of bioelectricity using *Nostoc* as a source of hydrogen in a paper with nine citations. They fabricate an oxyhydrogen biofuel cell, based on a carbon–carbon electrode. The electrode pellets were prepared by taking carbon powder mixed with polyvinyl alcohol as a binder. The anode was charged with Co–Al spinel mixed oxide at 700°C, 30% KOH acted as an electrolyte. For the cyanobacterial bioreactor, a potential heterocystous blue-green alga of *Nostoc* spp. has been used for hydrogen production and electrical energy generation. Various nutrient enrichment techniques are employed to increase the hydrogen generation efficiency of the algae. They find that "one liter free cell algal reactor attached to the fuel cell, at the anode end for hydrogen gas input, generated about 300 mV of voltage and 100 mA of current." They assert that the development of a low-cost fuel cell with high efficiency of current output may be helpful in commercializing this technology.

Nishio et al. (2010) study the light/electricity conversion by a self-organized photosynthetic biofilm in a single-chamber reactor in a paper with eight citations. They show a biofilm-based light/electricity-conversion system that was self-organized from a natural microbial community. A bioreactor equipped with an air cathode and graphite-felt anode was inoculated with a green hot-spring microbial mat. When the reactor was irradiated with light, electric current was generated between the anode and cathode in accordance with the formation of green biofilm on the anode. Fluorescence microscopy of the green biofilm revealed the "presence of chlorophyll-containing microbes of similar to 10 A μm in size, and these cells were abundant close to the surface of the biofilm." The biofilm community was also analyzed by sequencing of polymerase chain reaction–amplified small-subunit rRNA gene fragments, showing that sequence types affiliated with Chlorophyta, Betaproteobacteria, and Bacteroidetes were abundantly detected. They assert that green algae and heterotrophic bacteria cooperatively convert light energy into bioelectricity.

Martins et al. (2010) study the implementation of a benthic MFC (BMFC) in Lake Furnas (Azores) with a focus on the phylogenetic affiliation and electrochemical activity of sediment bacteria. They examine the composition and electrochemical activity of the bacterial community inhabiting this lake. Fingerprinting analysis of the bacterial 16S rRNA gene fragment was done by denaturing gradient gel electrophoresis. The sequences retrieved from this lake sediments were affiliated to Bacteroidetes/Chlorobi group; Chloroflexi; alpha, delta, and gamma subclasses of Proteobacteria; Cyanobacteria; and Gemmatimonadetes. A cyclic voltammetric study was carried out with an enriched sediment bacterial suspension in a standard two-chamber electrochemical cell using a carbon paper anode cyclic voltammograms (scan rate of 50 mV/s) and showed the "occurrence of oxidation-reduction reactions at the carbon anode surface." They then find that the "benthic MFC operated with Lake Furnas sediments presented a low power density (1 mW/m^2) indicating that further work is required to optimize its power generation." They assert that sediment bacteria, probably from the delta and gamma subclasses of Proteobacteria, were electroactive under tested conditions.

Gong et al. (2011) study the BMFC as direct power source for an acoustic modem and seawater oxygen/temperature sensor system in a recent paper with five citations. They use a chambered BMFC with a 0.25 m^2 footprint to power an acoustic modem interfaced with an oceanographic sensor that measures DO and temperature. The experiment was conducted in Yaquina Bay, Oregon, over 50 days. Several improvements were made in the BMFC design and power management system based on lessons learned from earlier prototypes. The energy was harvested by a dynamic gain charge pump circuit that maintains a desired point on the BMFC's power curve and stores the energy in a 200 F supercapacitor. The system also used an ultralow power microcontroller and quartz clock to read the oxygen/temperature sensor hourly, store data with a time stamp, and perform daily polarizations. Data records were transmitted to the surface by the acoustic modem every 1–5 days after receiving an acoustic prompt from a surface hydrophone. After jump-starting energy production with supplemental macroalgae placed in the BMFC's anode chamber, the average power density of the BMFC adjusted to 44 mW/m^2 of seafloor area, which is better than past demonstrations at this site. They find that the "highest power density was 158 mW/m^2, and the useful energy produced and stored was ≥1.7 times the energy required to operate the system."

Yuan et al. (2011) study bioelectricity generation and microcystins removal in a blue-green alga–powered single-chamber tubular MFC in a recent paper with five citations. They find that the "blue-green algae powered MFC produced a maximum power density of 114 mW/m^2 at a current density of 0.55 mA/m^2." Coupled with the bioenergy generation, high removal efficiencies of COD and nitrogen were also achieved in MFCs. "Over 78.9% of total chemical oxygen demand (TCOD), 80.0% of soluble chemical oxygen demand (SCOD), 91.0% of total nitrogen (total-N) and 96.8% ammonium-nitrogen (NH(3)-N) were removed under closed circuit conditions in 12 days, which were much more effective than those under open circuit and anaerobic reactor conditions." Most importantly, "the MFC showed great ability to remove microcystins released from blue-green algae. Over 90.7% of MC-RR and 91.1% of MC-LR were removed under closed circuit conditions (500 Ω)." They assert that the MFC could provide a potential means for bioelectricity production from blue-green algae coupling algae toxins removal.

24.4 Conclusion

The citation classics presented under the two main headings in this chapter confirm the predictions that marine algae has a significant potential to serve as a major solution for the global problems of warming, air pollution, energy security, and food security in the form of algal bioelectricity and algal MFCs.

Further research is recommended for the detailed studies in each topical area presented in this chapter including scientometric studies and citation classic studies to inform the key stakeholders about the potential of marine algae for the solution of the global problems of warming, air pollution, energy security, and food security in the form of algal bioelectricity and algal MFCs.

References

Baltussen, A. and C.H. Kindler. 2004a. Citation classics in anesthetic journals. *Anesthesia & Analgesia* 98:443–451.

Baltussen, A. and C.H. Kindler. 2004b. Citation classics in critical care medicine. *Intensive Care Medicine* 30:902–910.

Bombelli, P., R.W. Bradley, A.M. Scott et al. 2011. Quantitative analysis of the factors limiting solar power transduction by *Synechocystis* sp. PCC 6803 in biological photovoltaic devices. *Energy & Environmental Science* 4:4690–4698.

Bothast, R.J. and M.A. Schlicher. 2005. Biotechnological processes for conversion of corn into ethanol. *Applied Microbiology and Biotechnology* 67:19–25.

Chen, C.Y., K.L. Yeh, H.M. Su, Y.C. Lo, W.M. Chen, and J.S. Chang. 2010. Strategies to enhance cell growth and achieve high-level oil production of a *Chlorella vulgaris* isolate. *Biotechnology Progress* 26:679–686.

Chiao, M., K.B. Lam, and L.W. Lin. 2006. Micromachined microbial and photosynthetic fuel cells. *Journal of Micromechanics and Microengineering* 16:2547–2553.

Chisti, Y. 2007. Biodiesel from microalgae. *Biotechnology Advances* 25:294–306.

Clarens, A.F., H. Nassau, E.P. Resurreccion, M.A. White, and L.M. Colosi. 2011. Environmental impacts of algae-derived biodiesel and bioelectricity for transportation. *Environmental Science & Technology* 45:7554–7560.

Croisetiere, L., R. Rouillon, and R. Carpentier. 2001. A simple mediatorless amperometric method using the cyanobacterium *Synechococcus leopoliensis* for the detection of phytotoxic pollutants. *Applied Microbiology and Biotechnology* 56:261–264.

Dante, R.C. 2005. Hypotheses for direct PEM fuel cells applications of photobioproduced hydrogen by *Chlamydomonas reinhardtii*. *International Journal of Hydrogen Energy* 30:421–424.

Dawar, S., B.K. Behara, and P. Mohanty. 1998. Development of a low-cost oxy-hydrogen bio-fuel cell for generation of electricity using Nostoc as a source of hydrogen. *International Journal of Energy Research* 22:1019–1028.

De Schamphelaire, L. and W. Verstraete. 2009. Revival of the biological sunlight-to-biogas energy conversion system. *Biotechnology and Bioengineering* 103:296–304.

Demirbas, A. 2007. Progress and recent trends in biofuels. *Progress in Energy and Combustion Science* 33:1–18.

Dubin, D., A.W. Hafner, and K.A. Arndt. 1993. Citation-classics in clinical dermatological journals—Citation analysis, biomedical journals, and landmark articles, 1945–1990. *Archives of Dermatology* 129:1121–1129.

El-Naggar, M.Y., G. Wanger, K.M. Leung et al. 2010. Electrical transport along bacterial nanowires from *Shewanella oneidensis* MR-1. *Proceedings of the National Academy of Sciences of the United States of America* 107:18127–18131.

Fu, C.C., C.H. Su, T.C. Hung, C.H. Hsieh, D. Suryani, and W.T. Wu. 2009. Effects of biomass weight and light intensity on the performance of photosynthetic microbial fuel cells with *Spirulina platensis*. *Bioresource Technology* 100:4183–4186.

Gehanno, J.F., K. Takahashi, S. Darmoni, and J. Weber. 2007. Citation classics in occupational medicine journals. *Scandinavian Journal of Work, Environment & Health* 33:245–251.

Ghirardi, M.L. 2006. Hydrogen production by photosynthetic green algae. *Indian Journal of Biochemistry & Biophysics* 43:201–210.

Godfray, H.C.J., J.R. Beddington, I.R. Crute et al. 2010. Food security: The challenge of feeding 9 billion people. *Science* 327:812–818.

Goldemberg, J. 2007. Ethanol for a sustainable energy future. *Science* 315:808–810.

Gong, Y.M., S.E. Radachowsky, M. Wolf, M.E. Nielsen, P.R. Girguis, and C.E. Reimers. 2011. Benthic microbial fuel cell as direct power source for an acoustic modem and seawater oxygen/temperature sensor system. *Environmental Science & Technology* 45:5047–5053.

Gorby, Y.A., S. Yanina, J.S. McLean et al. 2006. Electrically conductive bacterial nanowires produced by *Shewanella oneidensis* strain MR-1 and other microorganisms. *Proceedings of the National Academy of Sciences of the United States of America* 103:11358–11363.

Harun, R., M. Davidson, M. Doyle, R. Gopiraj, M. Danquah, and G. Forde. 2011. Technoeconomic analysis of an integrated microalgae photobioreactor, biodiesel and biogas production facility. *Biomass & Bioenergy* 35:741–747.

He, Z., J. Kan, F. Mansfeld, L.T. Angenent, and K.H. Nealson. 2009. Self-sustained phototrophic microbial fuel cells based on the synergistic cooperation between photosynthetic microorganisms and heterotrophic bacteria. *Environmental Science & Technology* 43:1648–1654.

Jacobson, M.Z. 2009. Review of solutions to global warming, air pollution, and energy security. *Energy & Environmental Science* 2:148–173.

Kadam, K.L. 2002. Environmental implications of power generation via coal-microalgae cofiring. *Energy* 27:905–922.

Kapdan, I.K. and F. Kargi. 2006. Bio-hydrogen production from waste materials. *Enzyme and Microbial Technology* 38:569–582.

Konur, O. 2000. Creating enforceable civil rights for disabled students in higher education: An institutional theory perspective. *Disability & Society* 15:1041–1063.

Konur, O. 2002a. Assessment of disabled students in higher education: Current public policy issues. *Assessment and Evaluation in Higher Education* 27:131–152.

Konur, O. 2002b. Access to employment by disabled people in the UK: Is the Disability Discrimination Act working? *International Journal of Discrimination and the Law* 5:247–279.

Konur, O. 2002c. Access to nursing education by disabled students: Rights and duties of nursing programs. *Nursing Education Today* 22:364–374.

Konur, O. 2006a. Participation of children with dyslexia in compulsory education: Current public policy issues. *Dyslexia* 12:51–67.

Konur, O. 2006b. Teaching disabled students in higher education. *Teaching in Higher Education* 11:351–363.

Konur, O. 2007a. A judicial outcome analysis of the Disability Discrimination Act: A windfall for the employers? *Disability & Society* 22:187–204.
Konur, O. 2007b. Computer-assisted teaching and assessment of disabled students in higher education: The interface between academic standards and disability rights. *Journal of Computer Assisted Learning* 23:207–219.
Konur, O. 2011. The scientometric evaluation of the research on the algae and bio-energy. *Applied Energy* 88:3532–3540.
Konur, O. 2012a. 100 Citation classics in energy and fuels. *Energy Education Science and Technology Part A—Energy Science and Research* 2012(si):319–332.
Konur, O. 2012b. What have we learned from the citation classics in energy and fuels: A mixed study. *Energy Education Science and Technology Part A* 2012(si):255–268.
Konur, O. 2012c. The gradual improvement of disability rights for the disabled tenants in the UK: The promising road is still ahead. *Social Political Economic and Cultural Research* 4:71–112.
Konur, O. 2012d. Prof. Dr. Ayhan Demirbas' scientometric biography. *Energy Education Science and Technology Part A—Energy Science and Research* 28:727–738.
Konur, O. 2012e. The evaluation of the research on the biofuels: A scientometric approach. *Energy Education Science and Technology Part A—Energy Science and Research* 28:903–916.
Konur, O. 2012f. The evaluation of the research on the biodiesel: A scientometric approach. *Energy Education Science and Technology Part A—Energy Science and Research* 28:1003–1014.
Konur, O. 2012g. The evaluation of the research on the bioethanol: A scientometric approach. *Energy Education Science and Technology Part A—Energy Science and Research* 28:1051–1064.
Konur, O. 2012h. The evaluation of the research on the microbial fuel cells: A scientometric approach. *Energy Education Science and Technology Part A—Energy Science and Research* 29:309–322.
Konur, O. 2012i. The evaluation of the research on the biohydrogen: A scientometric approach. *Energy Education Science and Technology Part A—Energy Science and Research* 29:323–338.
Konur, O. 2012j. The evaluation of the biogas research: A scientometric approach. *Energy Education Science and Technology Part A—Energy Science and Research* 29:1277–1292.
Konur, O. 2012k. The scientometric evaluation of the research on the production of bio-energy from biomass. *Biomass and Bioenergy* 47:504–515.
Konur, O. 2012l. The evaluation of the global energy and fuels research: A scientometric approach. *Energy Education Science and Technology Part A—Energy Science and Research* 30:613–628.
Konur, O. 2012m. The evaluation of the biorefinery research: A scientometric approach. *Energy Education Science and Technology Part A—Energy Science and Research* 2012(si):347–358.
Konur, O. 2012n. The evaluation of the bio-oil research: A scientometric approach. *Energy Education Science and Technology Part A—Energy Science and Research* 2012(si):379–392.
Konur, O. 2012o. What have we learned from the citation classics in energy and fuels: A mixed study. *Energy Education Science and Technology Part A—Energy Science and Research* 2012(si): 255–268.
Konur, O. 2012p. The evaluation of the research on the biofuels: A scientometric approach. *Energy Education Science and Technology Part A—Energy Science and Research* 28:903–916.
Konur, O. 2013. What have we learned from the research on the International Financial Reporting Standards (IFRS)? A mixed study. *Energy Education Science and Technology Part D: Social Political Economic and Cultural Research* 5:29–40.
Lal, R. 2004. Soil carbon sequestration impacts on global climate change and food security. *Science* 304:1623–1627.
Lynd, L.R., J.H. Cushman, R.J. Nichols, and C.E. Wyman. 1991. Fuel ethanol from cellulosic biomass. *Science* 251:1318–1323.
Mansfeld, F. 2007. The interaction of bacteria and metal surfaces. *Electrochimica Acta* 52:7670–7680.
Martens, N. and E.A.H. Hall. 1994. Diaminodurene as a mediator of a photocurrent using intact-cells of cyanobacteria. *Photochemistry and Photobiology* 59:91–98.
Martins, G., L. Peixoto, D.C. Ribeiro, P. Parpot, A.G. Brito, and R. Nogueira. 2010. Towards implementation of a benthic microbial fuel cell in Lake Furnas (Azores): Phylogenetic affiliation and electrochemical activity of sediment bacteria. *Bioelectrochemistry* 78:67–71.

McCormick, A.J., P. Bombelli, A.M. Scott et al. 2011. Photosynthetic biofilms in pure culture harness solar energy in a mediatorless bio-photovoltaic cell (BPV) system. *Energy & Environmental Science* 4:4699–4709.

Melis, A. and Happe, T. 2001. Hydrogen production: Green algae as a source of energy. *Plant Physiology* 127:740–748.

Mohan, S.V. 2009. Harnessing of biohydrogen from wastewater treatment using mixed fermentative consortia: Process evaluation towards optimization. *International Journal of Hydrogen Energy* 34:7460–7474.

Mohan, S.V., S. Srikanth, S.V. Raghuvulu, G. Mohanakrishna, A.K. Kumar, and P.N. Sarma. 2009. Evaluation of the potential of various aquatic eco-systems in harnessing bioelectricity through benthic fuel cell: Effect of electrode assembly and water characteristics. *Bioresource Technology* 100:2240–2246.

Nishio, K., K. Hashimoto, and K. Watanabe. 2010. Light/electricity conversion by a self-organized photosynthetic biofilm in a single-chamber reactor. *Applied Microbiology and Biotechnology* 86:957–964.

North, D. 1994. Economic-performance through time. *American Economic Review* 84:359–368.

Paladugu, R., M.S. Chein, S. Gardezi, and L. Wise. 2002. One hundred citation classics in general surgical journals. *World Journal of Surgery* 26:1099–1105.

Pisciotta, J.M., Y. Zou, and I.V. Baskakov. 2010. Light-dependent electrogenic activity of cyanobacteria. *PLoS ONE* 5:e10821.

Powell, E.E. and G.A. Hill. 2009. Economic assessment of an integrated bioethanol–biodiesel–microbial fuel cell facility utilizing yeast and photosynthetic algae. *Chemical Engineering Research & Design* 87:1340–1348.

Powell, E.E., M.L. Mapiour, R.W. Evitts, and G.A. Hill. 2009. Growth kinetics of *Chlorella vulgaris* and its use as a cathodic half cell. *Bioresource Technology* 100:269–274.

Rodrigo, M.A., P. Canizares, J. Lobato, R. Paz, C. Saez, and J.J. Linares. 2007. Production of electricity from the treatment of urban waste water using a microbial fuel cell. *Journal of Power Sources* 169:198–204.

Rosenbaum, M. and U. Schroder. 2010. Photomicrobial solar and fuel cells. *Electroanalysis* 22:844–855.

Rosenbaum, M., U. Schroder, and F. Scholz. 2005. Utilizing the green alga *Chlamydomonas reinhardtii* for microbial electricity generation: A living solar cell. *Applied Microbiology and Biotechnology* 68:753–756.

Ryu, W., S.J. Bai, J.S. Park et al. 2010. Direct extraction of photosynthetic electrons from single algal cells by nanoprobing system. *Nano Letters* 10:1137–1143.

Satyanarayana, K.G., A.B. Mariano, and J.V.C. Vargas. 2011. A review on microalgae, a versatile source for sustainable energy and materials. *International Journal of Energy Research* 35:291–311.

Spolaore, P., C. Joannis-Cassan, E. Duran, and A. Isambert. 2006. Commercial applications of microalgae. *Journal of Bioscience and Bioengineering* 101:87–96.

Strik, D.P.B.T.B., H.V.M. Hamelers, and C.J.N. Buisman. 2010. Solar energy powered microbial fuel cell with a reversible bioelectrode. *Environmental Science & Technology* 44:532–537.

Strik, D.P.B.T.B., H. Terlouw, H.V.M. Hamelers, and C.J.N. Buisman. 2008. Renewable sustainable biocatalyzed electricity production in a photosynthetic algal microbial fuel cell (PAMFC). *Applied Microbiology and Biotechnology* 81:659–668.

Subhadra, B. and M. Edwards. 2010. An integrated renewable energy park approach for algal biofuel production in United States. *Energy Policy* 38:4897–4902.

Subhadra, B.G. 2010. Sustainability of algal biofuel production using integrated renewable energy park (IREP) and algal biorefinery approach. *Energy Policy* 38:5892–5901.

Tanaka, K., R. Tamamushi, and T. Ogawa. 1985. Bioelectrochemical fuel-cells operated by the cyanobacterium, anabaena-variabilis. *Journal of Chemical Technology and Biotechnology B—Biotechnology* 35:191–197.

Tsujimura, S., A. Wadano, K. Kano, and T. Ikeda. 2001. Photosynthetic bioelectrochemical cell utilizing cyanobacteria and water-generating oxidase. *Enzyme and Microbial Technology* 29:225–231.

Velasquez-Orta, S.B., T.P. Curtis, and B.E. Logan. 2009. Energy from algae using microbial fuel cells. *Biotechnology and Bioengineering* 103:1068–1076.

Volesky, B. and Z.R. Holan. 1995. Biosorption of heavy-metals. *Biotechnology Progress* 11:235–250.

Wang, B., Y.Q. Li, N. Wu, and C.Q. Lan. 2008. CO(2) bio-mitigation using microalgae. *Applied Microbiology and Biotechnology* 79:707–718.

Wang, X., Y.J. Feng, J. Liu, H. Lee, C. Li, N. Li, and N.Q. Ren. 2010. Sequestration of CO_2 discharged from anode by algal cathode in microbial carbon capture cells (MCCs). *Biosensors & Bioelectronics* 25:2639–2643.

Wrigley, N. and S. Matthews. 1986. Citation-classics and citation levels in geography. *Area* 18:185–194.

Wunschiers, R. and P. Lindblad. 2002. Hydrogen in education—A biological approach. *International Journal of Hydrogen Energy* 27:1131–1140.

Yagishita, T., T. Horigome, and K. Tanaka. 1993. Effects of light, CO_2 and inhibitors on the current output of biofuel cells containing the photosynthetic organism *Synechococcus* sp. *Journal of Chemical Technology and Biotechnology* 56:393–399.

Yagishita, T., S. Sawayama, K. Tsukahara, and T. Ogi. 1999. Effects of glucose addition and light on current outputs in photosynthetic electrochemical cells using *Synechocystis* sp. PCC6714. *Journal of Bioscience and Bioengineering* 88:210–214.

Yagishita, T., S. Sawayama, K.I. Tsukahara, and T. Ogi. 1997a. Effects of intensity of incident light and concentrations of *Synechococcus* sp. and 2-hydroxy-1,4-naphthoquinone on the current output of photosynthetic electrochemical cell. *Solar Energy* 61:347–353.

Yagishita, T., S. Sawayama, K.I. Tsukahara, and T. Ogi. 1997b. Behavior of glucose degradation in *Synechocystis* sp. M-203 in bioelectrochemical fuel cells. *Bioelectrochemistry and Bioenergetics* 43:177–180.

Yagishita, T., S. Sawayama, K.I. Tsukahara, and T. Ogi. 1998. Performance of photosynthetic electrochemical cells using immobilized *Anabaena variabilis* M-3 in discharge/culture cycles. *Journal of Fermentation and Bioengineering* 85:546–549.

Yergin, D. 2006. Ensuring energy security. *Foreign Affairs* 85:69–82.

Yuan, Y., Q. Chen, S.G. Zhou, L. Zhuang, and P. Hu. 2011. Bioelectricity generation and microcystins removal in a blue-green algae powered microbial fuel cell. *Journal of Hazardous Materials* 187:591–595.

Zhang, Y.F., J.S. Noori, and I. Angelidaki. 2011. Simultaneous organic carbon, nutrients removal and energy production in a photomicrobial fuel cell (PFC). *Energy & Environmental Science* 4:4340–4346.

Zou, Y.J., J. Pisciotta, and I.V. Baskakov. 2010. Nanostructured polypyrrole-coated anode for sun-powered microbial fuel cells. *Bioelectrochemistry* 79:50–56.

Zou, Y.J., J. Pisciotta, R.B. Billmyre, and I.V. Baskakov. 2009. Photosynthetic microbial fuel cells with positive light response. *Biotechnology and Bioengineering* 104:939–946.

25

Marine Microbial Fuel Cells

Valliappan Karuppiah and Zhiyong Li

CONTENTS

25.1 Introduction

The requirement of energy in one form or another is important for all human activities. The U.K. Energy Research Center reported that the production of oil will reach its peak in 2030, and after that, it will start to decline (Sorrel et al. 2010). Similarly, the production of gas reaches its peak in 2020 (Bentley 2002). Further rise in global temperature results in melting glaciers and rising sea levels (Kerr 2007; Voiland 2009). Hence, the search for alternative renewable energy sources is facilitated (Faaij 2006; Rodrigo et al. 2007). The renewable energy sources are based on solar, wind, tidal, bio, piezoharvester, and fuel cells (Tsuneo 2001). Presently, biofuels such as ethanol from starch or sugarcane and biodiesel from vegetable oil or animal fats are widely used (Fortman et al. 2008). However, ethanol is not a better fuel, because it is not compatible with the present fuel transportation such as distribution and storage due to its corrosivity and high hygroscopicity (Atsumi et al. 2008; Stephanopoulos 2007). Apart from that, it holds only about 70% of the energy content. Similarly, biodiesel also cannot be transported in pipelines due to its cloud and pour points that are higher than those for petroleum diesel (petrodiesel), and its energy content is approximately 11% lower than that of petrodiesel.

The increasing awareness toward the requirement of cleaner and renewable energy resources made an important change in the area of energy research. The availability of biomass and bio-related waste in great quantity assures an eco-friendly solution to the increasing demand for alternative energy. Microbial fuel cells (MFCs) are devices capable of directly transforming chemical to electrical energy by the electrochemical reactions involved in the biochemical pathways of microorganism. MFCs can possibly produce current from a wide range of complex organic wastes and renewable biomass. MFCs are unique in their ability to utilize microorganisms, rather than an enzyme or inorganic molecule, as catalysts for converting the chemical energy of feedstock directly into electricity. MFCs utilize biomass and directly produce electricity from biomass with high competence. MFCs are not so large as to require storage outside a building or in a separate room. They can be very small and can run on organic matter. There is presently a worldwide niche market for stationary fuel cells, about the size of a one-car garage with 300–200 kW fuel-cell power plants.

MFCs make use of bacteria attaching to the electrode surfaces as catalyst of the electrochemical reactions. Microorganisms present in different natural environments such as marine sediments (Reimers et al. 2001; Tender et al. 2002), wastewater (Liu et al. 2004; Min and Logan 2004), and soils (Dylan et al. 2007) that can form biofilms on graphite anodes, oxidize the dissolved organic matter contained in the environment, and use the electrode as the final electron acceptor. MFCs show promising advantages with value to standard abiotic fuel cells (Lovley 2006). They suggest the opportunity of producing electricity from organic waste and renewable biomass, as the catalyzing bacteria can adapt to different organic materials contained in a wide range of environments such as wastewaters or sediments. The costly catalysts essential in abiotic fuel cells are replaced by naturally growing microorganisms.

Several investigators have recommended that production of electricity by microbes may become a significant form of bioenergy. The power densities of MFCs are too low for most envisioned applications. While sediment MFCs extract electrons from organic matter present in marine sediments to power electronic monitoring devices (Tender et al. 2008), probably sediment fuel cells in a pot can provide power for a light source or battery charger in off-grid areas (Lovely 2008). Marine MFCs have been mainly considered for low-power-consuming marine instrumentations (Lovley 2006), such as oceanographic sensors, monitoring devices, and telemetry systems (Liu et al. 2007).

25.2 General Principle and Techniques of MFCs

25.2.1 Principle

To know how an MFC generates electricity, we must know how bacteria capture and process energy. Bacteria grow by catalyzing chemical reactions and harnessing and storing energy in the form of adenosine triphosphate (ATP). In some bacteria, reduced substrates are oxidized and electrons are transferred to respiratory enzymes by NADH, the reduced form of nicotinamide adenine dinucleotide. These electrons flow down a respiratory chain, a series of enzymes that function to move protons across an internal membrane, creating a proton gradient. The protons flow back into the cell through the enzyme ATPase, creating one ATP molecule from one adenosine diphosphate for every three to four protons. The electrons are finally released to a soluble terminal electron acceptor, such as nitrate, sulfate, or oxygen.

The operating principle of a typical dual chamber MFC is demonstrated in Figure 25.1 (Ren et al. 2012). It comprised of two chambers: anode and cathode chambers, which are divided by an ion-exchange membrane (proton-exchange membrane [PEM]). The PEM splits the electrolyte in the anode (anolyte) and the cathode (catholyte) chambers. Bacteria in the anode chamber execute a respiration and break down organic substrates to generate carbon dioxide, protons, and electrons, and these electrons are transferred to the anode via extracellular electron transfer. Then these electrons pass through the external circuit to the cathode driven by the potential difference between the two electrodes and reduced at the cathode by electron acceptors. At the same moment, an unbalanced charge distribution occurs between the anode and cathode in an electrical field gradient. This electrical field gradient transfers cations from anode to cathode chamber and anions from cathode to anode chamber, respectively. This process results in the conversion of biomass into electricity. Oxygen is consumed in the cathode chamber; the half reactions at the anode and cathode can be written as

$$CH_3COO^- + 2H_2O \rightarrow 2CO_2 + 8e^- + 7H^+ \text{ for anode} \tag{25.1}$$

$$O_2 + 4H^+ + 4e^- \rightarrow 2H_2O \text{ for cathode} \tag{25.2}$$

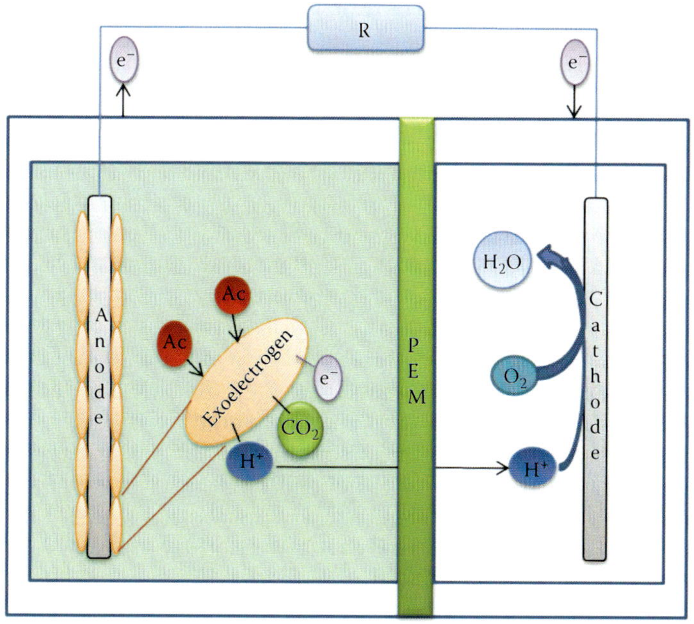

FIGURE 25.1
Operating principle of an MFC.

The electric power (measured in watts) generated by the MFC is based on the amount of electrons moving through the circuit (current, measured in amps) and electrochemical potential difference (volts) across the electrodes. At the cathode, the electrons and protons mingle to reduce the terminal electron acceptor with the help of oxygen.

In the absence of oxygen, microorganisms utilize the substrate such as sugar/glucose in anaerobic conditions, thereby producing carbon dioxide and electrons as given in the following equation (Chaudhuri and Lovley 2003):

$$C_{12}H_{22}O_{11} + 13H_2O \rightarrow 12CO_2 + 48H^+ + 48e^- \qquad (25.3)$$

The release of electrons refers that the mediator proceeds to its original oxidized state to repeat the process. It is important to note that this reaction will take place under anaerobic conditions because in the aerobic condition, microorganisms gather all the electrons and have higher electronegativity than the mediator.

Electricity generation in MFCs is influenced by a lot of factors such as substrate concentration, nature of the electrodes, membrane type, bacterial substrate oxidation rate, presence of alternative electron acceptors, and microbial growth. On the other hand, electrochemical potential depends on the potential difference between the bacterial respiratory enzyme or electron carrier and the potential of the anode and the terminal electron acceptor.

The MFC device should be able to oxidize the substrate at the anode, either continuously or intermittently; otherwise, the system is considered as a battery. Electrons can be moved to the anode by electron mediators or shuttles (Rabaey et al. 2004, 2005a), by direct membrane–associated electron transfer (Bond and Lovley 2003), or by nanowires (Gorby et al. 2005, 2006; Reguera et al. 2005) produced by the bacteria. Chemical mediators, such as neutral red or anthraquinone-2,6-disulfonate, can be added to the device to produce electricity by bacteria unable to use the electrode (Bond et al. 2002; Park and Zeikus 1999). If no exogenous mediators are added to the device, the MFC is called a *mediator-less* MFC even though the mechanism of electron transfer is unknown (Logan 2004).

In majority of MFCs, the electrons reach the cathode and it combines with protons (diffuse from the anode through a separator) and oxygen (from air) to produce water as an end product (Bond et al. 2002; Kim et al. 1999, 2002; Min and Logan 2004). Chemical oxidizers, such as ferricyanide or Mn(IV), can also be used although these must be replaced or regenerated (Rabaey et al. 2003, 2004; Rhoads et al. 2005; Shantaram et al. 2005). In the case of metal ions, bacteria can help to catalyze the oxidation of the metal using dissolved oxygen (Rhoads et al. 2005; Shantaram et al. 2005).

MFCs operated by means of mixed cultures are currently achieving higher power densities than those with pure cultures (Rabaey et al. 2004, 2005a). Community study of the microorganisms that exist in MFCs has so far discovered a great diversity in composition (Aelterman et al. 2006; Bond et al. 2002; Logan et al. 2005; Phung et al. 2004; Rabaey et al. 2004). Based on the existing data, it is believed that many new types of bacteria are capable of anodophilic electron transfer (electron transfer to an anode) or even interspecies electron transfer (electrons transferred between bacteria in any form).

MFCs are being built with a range of materials. MFCs are operated in various conditions including differences in temperature, pH, electron acceptor, electrode surface areas, reactor size, and operation time.

25.2.2 Materials of Construction

25.2.2.1 Anode

Anodic materials used in MFCs should be conductive, biocompatible, and chemically stable in the reactor solution. Metal anodes consisting of noncorrosive stainless steel mesh have been developed (Logan et al. 2006), but copper is not a suitable anode due to its toxicity against bacteria. The most adaptable electrode substance is carbon; it can be available as compact graphite plates, rods, or granules, fibrous material (felt, cloth, paper, fibers, foam), and glassy carbon.

The simplest anode electrodes are graphite plates or rods, because they are relatively cheap and easy to handle and have a defined surface area. High surface areas are attained with graphite felt electrodes (Gil et al. 2003; Park and Zeikus 1999). Even though, the indicated surface area is not available to bacteria. Hence, carbon fiber, paper, foam, and cloth (Toray) have been broadly used as electrodes. It has been revealed that current density increases with the overall internal surface area in the order carbon felt > carbon foam > graphite (Chaudhuri and Lovley 2003). Considerably larger surface areas are attained either by using a compact material such as reticulated vitreous carbon (RVC; ERG Materials and Aerospace Corp., Oakland, CA) (He et al. 2005; Kim et al. 2000) or by using layers of packed carbon granules or beads (Rabaey et al. 2005b; Sell et al. 1989). In both situations, maintaining the high porosity is vital to avoid clogging.

To improve the performance of anodes, various chemical and physical stratagems have been applied. The transfer of electrons to anodes has been improved by (Park and Zeikus 2003) using Mn(IV), Fe(III), and covalently linked neutral red. The electricity production is also increased by the addition of electrocatalytic materials such as polyanilines/Pt composites, which increase the current generation by supporting the direct oxidation of microbial metabolites (Lowy et al. 2006; Niessen et al. 2004; Schroder et al. 2003).

25.2.2.2 Cathode

The materials used in anodes can also be used in the cathode. The materials such as carbon without platinum catalyst, plain carbon, metals other than platinum, and biocathodes are mostly used as the cathode.

The selection of cathode material greatly distresses performance and it differs based on its function. In the sediment fuel cells, plain graphite disk electrodes dipped in seawater over the sediment have been used (Tender et al. 2002). Due to the very low kinetics of the oxygen reduction in plain carbon and the resultant of large overpotential, the utilization of such cathodes limits the use of this noncatalyzed material to systems that can tolerate low performance. In seawater, oxygen reduction on carbon cathodes has been supported by microbes (Rhoads et al. 2005; Shantaram et al. 2005). Such microbial-mediated reduction in stainless steel cathodes has also been noted, which rapidly reduces oxygen with the help of bacterial biofilm (Bergel et al. 2005).

25.2.2.3 Catholyte and Anolyte

The effectiveness of fuel cells depends upon the redox reaction taking place in cathodic and anodic chambers. Some of the important catalysts such as Pt, MnO_2, CNT, and TiO_2 are used in MFCs. Ferricyanide ($K_3[Fe(CN)_6]$) is an important electron acceptor in MFCs (Park and Zeikus 2003). The greatest advantage of using ferricyanide is due to the low overpotential. The greatest disadvantage is the inadequate reoxidation by oxygen and the frequent change of catholyte (Rabaey et al. 2005b). In addition, the long-term production of electricity can be influenced by the flow of ferricyanide into the anode chamber by CEM. Oxygen is the best electron acceptor for the MFC, because of its high oxidation potential, availability, low cost (it is free), sustainability, and the lack of a chemical waste product (water is formed as the only end product).

To improve the amount of oxygen reduction, Pt catalysts are frequently applied for dissolved oxygen (Reimers et al. 2001) or open-air (gas diffusion) cathodes (Liu et al. 2004; Sell et al. 1989). To reduce the costs of MFCs, the Pt load can be reduced as 0.1 mg cm^{-2} (Cheng et al. 2006a). The long-term stability of Pt needs further investigation, and there remains a need for new types of inexpensive catalysts. For example, recently, noble-metal-free catalysts that use pyrolyzed iron(II) phthalocyanine or CoTMPP have been used as MFC cathodes (Cheng et al. 2006a; Zhao 2005).

Biological incident of MFCs is generally concerned with the anodic chamber. Different types of anolytes are used in MFCs, which contain different exoelectrogens of pure or mixed culture. Research on MFCs reveals that the interaction of biofilm electrodes and diverse substrates increase the power generation. Different types of substrates such as glucose, acetate, lactate, sodium formate, starch, urban wastewater, and artificial wastewater are used in MFCs. The microorganisms interact with these substrates and decompose it to produce protons and electrons. Transfer of electrons from a microorganism to the electrode can be enhanced by the addition of electron-shuttling compounds such as neutral red, thionin, methyl viologen, and phenazine ethosulfate. These compounds accept electrons from intracellular, membrane-bound redox proteins and transmit the electrons to the electrode surface (Logan et al. 2006).

25.2.2.4 Membrane

The majority of MFC devices require a partition between anode and cathode compartment by a membrane or salt bridge. But there are some exceptions regarding naturally separated systems such as sediment MFCs (Reimers et al. 2001) or specially designed single-compartment MFCs (Cheng et al. 2006a; Liu et al. 2004). The selection of a membrane should represent a choice between two opposing interests such as high selectivity and high stability. Large numbers of membranes are being studied based upon the application of MFCs. Some of the membranes that are being used are Ralex, Ultrex, Fumatech, PEO, PEG, etc. (Mo et al. 2009; Mohan et al. 2008; Zuo et al. 2008). The most commonly used ion exchange membrane (IEM) is Nafion (Dupont Co., United States), which is accessible from several suppliers such as Aldrich and Ion Power, Inc. Nafion is made up of a perfluorinated sulfonic acid polymer (pop), which contains the continuous skeleton of $-(CF_2)_n-$ groups attached to the numerous hydrophilic segments of a sulfonic acid group $-SO_3H$. Its super acidity is recognized by the electron-withdrawing effect of the perfluorocarbon chain acting on the sulfonic acid group. Nafion is capable of catalyzing various reactions, such as alkylation, disproportionation, and esterification. The other membrane such as Ultrex CMI-7000 (Membranes International Inc., Glen Rock, NJ) is also used for MFC applications. But it is more expensive than Nafion. IEM used in an MFC should be permeable to chemicals such as oxygen, ferricyanide, other ions, or organic matter present in the substrate. The market for ion-exchange membranes is budding continually, and more research is required to analyze the effect of the membrane on electricity production and long-term stability (Rozendal et al. 2006).

25.2.3 MFC Design

Many different designs are available for MFCs (Figures 25.2 and 25.3). A widely used and inexpensive configuration is a two-chamber MFC built in a traditional *H* shape, consisting of two bottles connected by a tube with a separator, that is, usually a cation-exchange membrane (CEM) (also called as proton-exchange membrane) (Bond et al. 2002; Logan et al. 2005; Min et al. 2005; Park and Zeikus 1999) or Ultrex (Rabaey et al. 2003) or a plain salt bridge (Min et al. 2005) (Figure 25.2a and b). The membrane is the key to this design, which allows protons to pass between the chambers. In the H-configuration, the membrane is fixed in the center of the tubes connecting the bottle (Figure 25.2a). However, the tube itself is not required. The two chambers can be compressed together by pressing up onto either side of the membrane to form a huge surface (Figure 25.2b). An inexpensive way to join the bottles is to use a glass tube that is heated and bent into a U-shape, filled with agar and salt to serve as a CEM, and inserted through the lid of each bottle (Figure 25.2b). The salt bridge MFC, however, produces little power due to the high internal resistance.

H-shape configurations are applicable for basic research on examining the power production using new materials or different types of biodegrading microbial communities, but they typically produce low power densities. The amount of current produced in these systems is affected by the surface area of the cathode relative to that of the anode (Oh et al. 2004; Oh and Logan 2006) and the surface of the membrane (Oh and Logan 2006). The power density (P) created by these systems is typically inadequate by high internal resistance and electrode-based losses.

It is not essential to place the cathode in water or in a separate chamber when using oxygen at the cathode. The cathode can be placed in direct contact with air (Figures 25.2e and 25.3c and d), packed-bed reactor that uses ferricyanide as a catholyte (Figure 25.2c), and photoheterotrophic-type MFC (Figure 25.2d), which have been designed (Liu et al. 2004). In one design, a kaolin clay-based separator and graphite cathode were coupled together to form a joint separator–cathode structure (Park and Zeikus 2003). Much larger power densities have been attained via oxygen as the electron acceptor when aqueous cathodes are replaced with air cathodes. In the simplest design, the anode and cathode are located on either side of a tube, where the anode is sealed against a flat plate and the cathode is exposed to air on one side and water on the other (Figure 25.2e). When a membrane is used in this air-cathode system, it mainly prevents water from leaking through the cathode, and also it reduces the diffusion of oxygen into the anode chamber. The oxygen consumption by bacteria in the anode chamber leads to lower coulombic efficiency (Liu et al. 2004). Hydrostatic pressure on the cathode will create water leakage, but it can be reduced by applying coatings, such as polytetrafluoroethylene, to the outer surface of the cathode that permit oxygen diffusion but prevent the water loss (Cheng et al. 2006a).

A number of discrepancies on these crucial designs have appeared in an attempt to raise power density or to give a continuous flow through the anode chamber. The apparatus has been designed with an outer cylindrical reactor with a concentric inner tube as the cathode (Habermann and Pommer 1991; Liu et al. 2004) (Figure 25.3d) and with an inner cylindrical reactor (anode consisting of granular media) with the cathode on the outside (Rabaey et al. 2005b) (Figure 25.3a). Another variation in design is the upflow fixed-bed biofilm reactor, with the fluid flowing continuously through porous anodes toward a membrane separating the anode from the cathode chamber (He et al. 2005) (Figure 25.3b). Systems have been designed to resemble hydrogen fuel cells, where a CEM is sandwiched between the anode

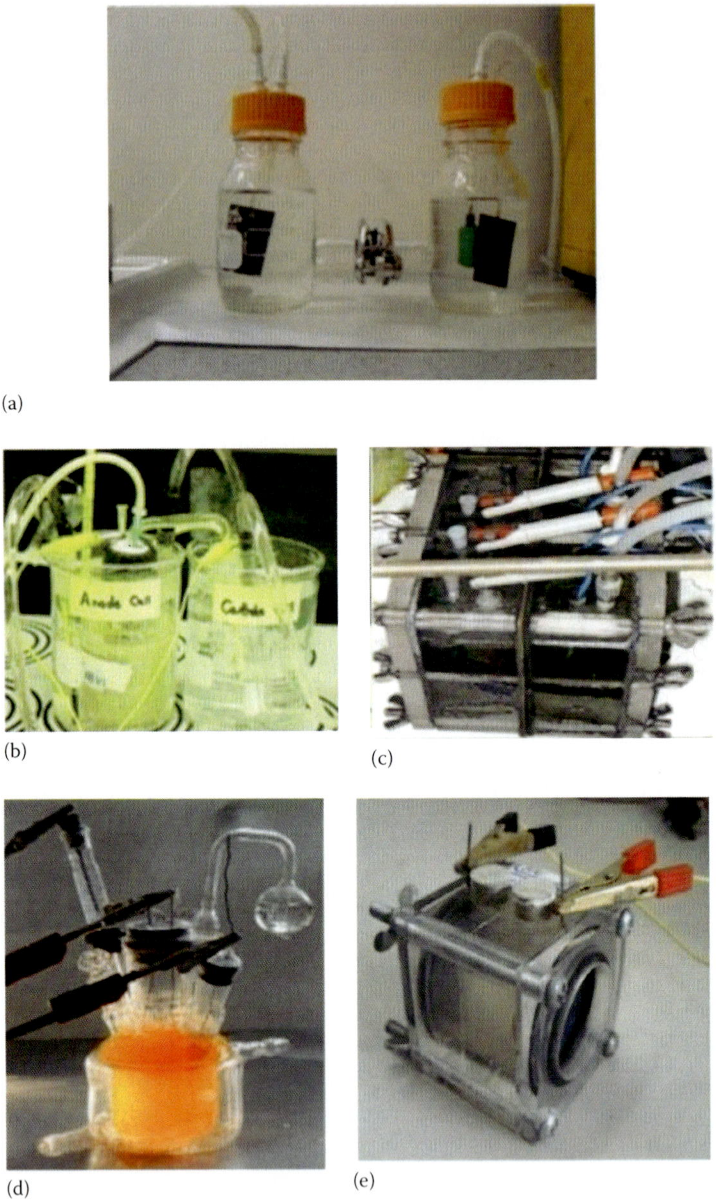

FIGURE 25.2
Different types of MFC: (a) two-chamber H-type system showing anode and cathode chambers equipped for gas sparging (From Logan, B.E. et al., *Water Res.*, 39, 942, 2005), (b) easily constructed system containing a salt bridge (From Min, B. et al., *Water Res.*, 39, 1675, 2005), (c) four batch-type MFCs where the chambers are separated by the membrane (without a tube) and held together by bolts (From Rabaey, K. et al., *Environ. Sci. Technol.*, 39, 3401, 2005a), (d) photoheterotrophic type MFC (76), and (e) single-chamber, air-cathode system in a simple *tube* arrangement. (From Liu, H. et al., *Environ. Sci. Technol.*, 38, 2281, 2004.)

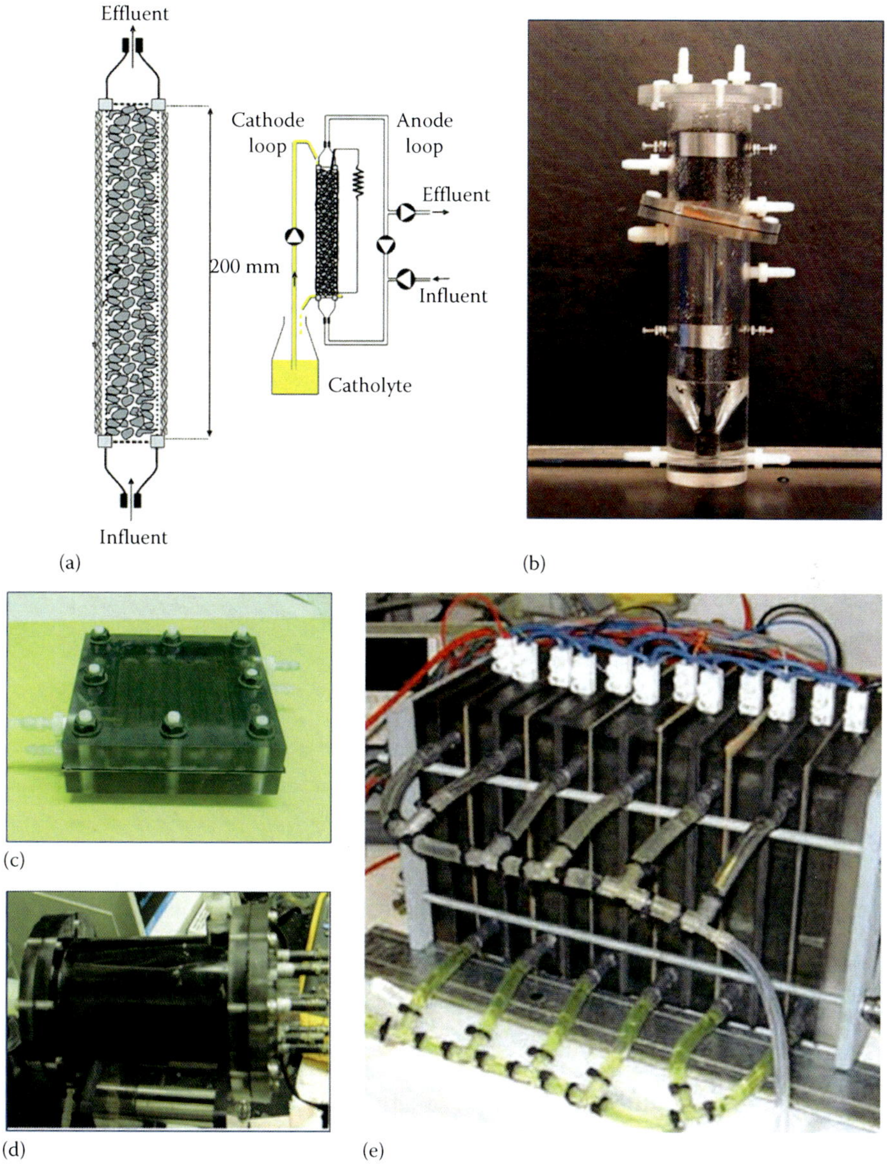

FIGURE 25.3
MFCs used for continuous operation: (a) upflow, tubular-type MFC with an inner graphite bed anode and an outer cathode (From Rabaey, K. et al., *Environ. Sci. Technol.*, 39, 8077, 2005b); (b) upflow, tubular-type MFC with an anode below and a cathode above, where the membrane is inclined (From He, Z. et al., *Environ. Sci. Technol.*, 39, 5262, 2005); (c) flat plate design where a channel is cut in the blocks so that liquid can flow in a serpentine pattern across the electrode (From Min, B. and Logan, B.E., *Environ. Sci. Technol.*, 38, 5809, 2004); (d) single-chamber system with an inner concentric air cathode surrounded by a chamber containing graphite rods as anode (From Liu, H. et al., *Environ. Sci. Technol.*, 38, 2281, 2004); (e) stacked MFC, in which six separate MFCs are joined in one reactor block. (From Aelterman, P. et al., *Environ. Sci. Technol.*, 40, 3388, 2006.)

and cathode (Figure 25.3c). To increase the overall system voltage, MFCs can be stacked with the systems shaped as a series of flat plates or linked together in series (Aelterman et al. 2006) (Figure 25.3e).

25.2.4 Characterization Techniques for MFCs

MFC experiments require specific electrochemical instruments to measure the generated electricity (Liu et al. 2004; Rabaey et al. 2004). Generally, cell voltages and electrode potentials are effectively calculated with normal existing voltage meters, multimeters, and data acquisition systems attached in parallel with the circuit. Cell voltages can be measured directly from the voltage difference between the anode and cathode; electrode potentials can be measured against a reference electrode that are requirements to be incorporated in the electrode compartment (Bard and Faulkner 2001). Electricity is calculated using Ohm's law ($I = E_{cell}/R$) given the measured voltage.

Further comprehensive understanding of the (bio-) electrochemical system can be attained using a potentiostat (e.g., Eco Chemie, the Netherlands; Princeton Applied Research, United States; Gamry Scientific, United States). The electrochemical response of the electrode at that specific condition can be studied with a potentiostat. It can be also controlled either by the potential or by the current of an electrode. The potentiostat is normally operated in a three-electrode setup, which consists of a working electrode (anode or cathode), a reference electrode, and a counter electrode (Bard and Faulkner 2001). The potentiostatic mode of this instrument is often utilized for voltammetry tests in which the potential of the working electrode (anode or cathode) is varied at a certain scan rate (expressed in $V\ s^{-1}$). If the scan goes only in one way, it is referred as linear sweep voltammetry; if the scan continues on the reverse track and comes back to the starting potential, it is referred as cyclic voltammetry. Voltammetry could be employed to evaluate the electrochemical activity of microbial strains or consortia (Aelterman et al. 2008; Fricke et al. 2008; Richter et al. 2009), determining the standard redox potentials of redox-active components (Rabaey et al. 2005a) and testing the performance of novel cathode materials (Zhao et al. 2005). A potentiostat can also be operated in a two-electrode setup to obtain polarization curves or to determine the ohmic resistance using the current interrupt technique as described in the following. In the two-electrode setup, the working electrode connector is connected to the cathode (positive terminal), and both the counter electrode and reference electrode connectors are connected to the anode.

More advanced measurements can be done when the potentiostat is equipped with a frequency response analyzer, allowing electrochemical impedance spectroscopy (EIS) measurements (Bard and Faulkner 2001). In EIS, a sinusoidal signal with small amplitude is superimposed on the applied potential of the working electrode. By varying the frequency of the sinusoidal signal over a wide range (typically 104–106 Hz) and plotting the measured electrode impedance, detailed information can be obtained from the electrochemical system. EIS can be used to measure the ohmic and internal resistances of an MFC (He et al. 2006; Min et al. 2005), as well as to provide additional insight into the operation of an MFC. The interpretation of EIS data can be rather complex, and therefore EIS techniques are not further discussed here.

25.2.4.1 Electrode Potential

The electrode (anode or cathode) potential can be calculated by determining the voltage against an electrode with a known potential (reference electrode). The most popular reference electrode is the silver–silver chloride (Ag/AgCl), which is simple, stable, and nontoxic. In a saturated potassium chloride solution at 25°C, the Ag/AgCl reference electrode develops a potential of +0.197 V against the normal hydrogen electrode (NHE).

25.2.4.2 Power

The overall presentation of the MFC is assessed by several ways, but principally, it is assessed through power output and coulombic efficiency. The power is determined by assessing the voltage and current. The voltage is measured across a fixed external resistor (R_{ext}), while the current is calculated from Ohm's law as $P = E_{cell} \times I$. This is the direct measure of the electric power. The maximum power is calculated from the polarization curve.

$$P = \frac{E_{cell}^2}{R_{ext}} \tag{25.4}$$

25.2.4.3 Power Density

The power density (P_{An}, W/m^2) is therefore calculated on the basis of the area of the anode (A_{An}).

25.2.4.4 Polarization Curves

Polarization curves represent a powerful tool for the analysis and characterization of MFCs. A polarization curve represents the voltage as a function of the current (density). Polarization curves can be measured for the anode, the cathode, or the whole MFC using a potentiostat. The polarization curve should be measured for both up and down (i.e., from high to low external resistance) and vice versa. When a variable external resistance is used to obtain a polarization curve, the current and potential values need to be taken only when pseudo-steady-state conditions have been established. The power density in MFC depends on the factors such as the nature of the substrate, mediator type, type of exoelectrogens, reactor configuration, nature of anode material and cathode material, and physical condition like temperature and pH value.

25.3 Marine MFC

Various forms of energy-producing technology are presently being explored, but almost all power supplies for marine sensor networks come in the form of one or many batteries. Although the upfront cost of batteries is relatively minimal, their necessary pressure housings and lifetime maintenance make it costly. MFCs have shown great promise for novel energy-harvesting technology, and they can provide consistent, maintenance-free power for long periods of time, well beyond the lifetimes of sensor and communication hardware. MFCs were first installed in marine sediments about 12 years ago (Reimers et al. 2001). In 2007, these MFCs were established to be viable power sources for undersea sensor and communications systems (Logan et al. 2006).

25.3.1 Benthic MFC

The majority of the MFC investigations are paying attention to the performance and characterization of laboratory reactors, benthic MFCs (BMFCs), or sediment fuel cells. MFCs rely on a potential gradient at a sediment–water interface and thus have played a significant role in the evolution of MFC technology. Laboratory MFCs allow study of the basic attributes of MFC performance and microbial physiology, for example, ohmic resistance, current density, and mechanisms of electron transfer by laboratory-cultivated microbes. But BMFCs offer a unique opportunity to study (1) how energy transfers through microbial communities, (2) the effectiveness of harvesting electricity from natural systems, (3) how BMFCs generate power and are also involved in the bioremediation of the natural world.

BMFCs produce electricity from the electropotential difference between oxic seawater and anoxic sediments (Reimers et al. 2001, 2006; Tender et al. 2002; Whitfield 1972). Electrons are transferred to the anode by microbes either directly from organic material or indirectly from inorganic products degraded from organic matter (Reimers et al. 2007). At the cathode, these electrons reduce dissolved oxygen to form water. Since microorganisms cause organic matter oxidation in natural anoxic environments to proceed a variety of interactive electron acceptors (Burdige 2006; Froelich et al. 1979), the processes at the anode can be especially complex and critical to BMFC performance.

The BMFC also uses the same basic MFC principles described earlier, whereby ocean sediment acts like nutrient-rich anodic media, the inoculum, and the PEM. Sediments naturally contain the complex community of microbes, including the electrogenic microbes required for MFCs and complex sugars and other nutrients accumulated by the degradation of animal material. Additionally, the aerobic microbes present in the top few centimeters of the sediment act as an oxygen filter, and they cause the redox potential of the sediment to decrease with greater depth, typically leveling at approximately −0.4 V vs. standard hydrogen electrode (SHE). While the level of dissolved oxygen within a body of water does generally decrease with greater depth, continuous water circulation ensures that water above the seafloor retains a high enough level of dissolved oxygen to maintain proper cathode function even at extreme depths. It has been shown that cathodes may still exhibit potentials of 0.4 V vs. SHE at depths in excess of 950 m (Nielsen et al. 2008).

Early BMFC models occupied the simple architecture of having a graphite-based anode plate hidden within the ocean sediment, while a graphite-based cathode lay fixed in the overlying water. In this design, new nutrients arrive at the anode via simple diffusion through the sediment pore water (Figure 25.4). This diffusion is sustained by natural seafloor seepage. BMFCs normally display power densities of 30 mW m^{-2} of footprint continuously, depending on sediment conditions (Nielsen et al. 2007).

In general, BMFCs generate relatively low levels of electricity due to low rates of diffusion in sediments, low concentrations of labile organic matter, and passivation of anode surfaces (Finkelstein et al. 2006; He et al. 2007; Holmes et al. 2004; Lowy et al. 2006; Reimers et al. 2001, 2006; Ryckelynck et al. 2005). Among the different strategies attempted to improve electricity generation from BMFCs, the most successful have employed carbon cloth anodes enriched with organic substrates (Farzaneh et al. 2007), using a rotating cathode (He et al. 2007). There are several developments in wastewater-fueled MFC technology, which involves the pumping of waste fluids through porous anode materials contained in chambers (Cheng et al. 2006b; He et al. 2005; Jang et al. 2004; Moon et al. 2006). Based on these developments, electricity generation from BMFCs has been improved further by removing the anode from the sediments and enclosing it in a benthic chamber.

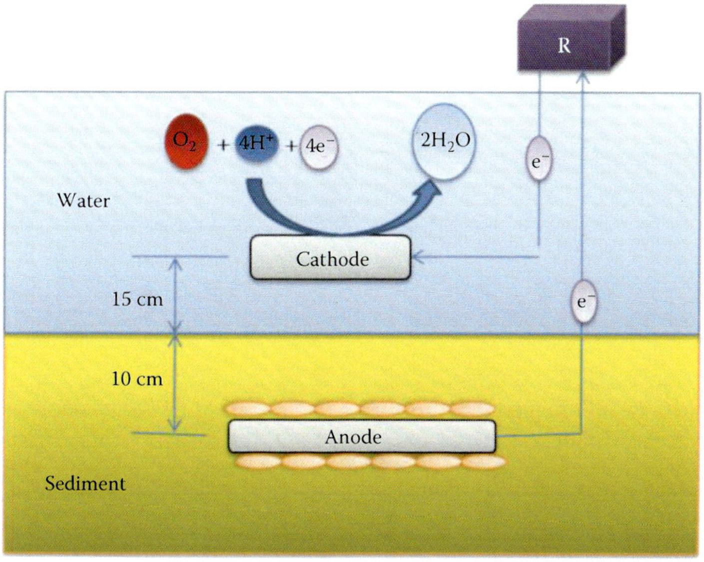

FIGURE 25.4
BMFC.

Integration of the benthic chamber would facilitate the use of high-surface-area anodes and the advection of reductant-rich pore water to anode materials.

A novel benthic chamber MFC (BC-MFC) is designed to tackle the restrictions of nutrient diffusion. The BC-MFC utilizes a big semienclosed chamber that stays strong on the seafloor (Figure 25.5). Through the use of a one-way check valve, the chamber uses the tidal fluctuations in water pressure to passively pump pore water from the underlying sediment into the chamber. This water is not only rich in nutrients but also uses oxygen due to the prosperous group of aerobic microbes present within the sediment. These microbes filter oxygen from the water before it reaches the chamber. The carbon fiber brushes that are set inside the chamber serve as anodes. The nutrient-rich water flows through the brushes and comes into contact with the thick films of electrogenic microbes that develop on their surface, before exiting the chamber. The one-way check valve situated at the top of the chamber permits only outflow from the chamber, preventing the overlying, oxygen-laden water from reaching the anode. The cathode also contains a carbon fiber brush at the top of the chamber, and it floats freely in the water with the aid of syntactic foam floats. With this new design, the negative effect of diffusion limitation on power generation is nullified and increases the power densities to 380 mW m^{-2} of footprint (Nielsen et al. 2008).

The most relevant evaluation procedure is to conduct an extended discharge experiment under field conditions with some program of controlled cell potential (E_{cell}), current (I), or external load resistance(s) (R_{ext}). The system can be characterized by using Ohm's law ($E_{cell} = IR_{ext}$), and internal resistance terms may be assigned to each of the processes that cause a drop in potential from the open circuit voltage ($E_{cell} = OCV - IR_{int}$) (Logan et al. 2006). As described by Logan et al. (2006), the three main loss categories that contribute to the steady-state internal resistance (R_{int}) are ohmic losses (R_o), activation losses (R_a), and mass-transfer losses (R_{mt}); by giving the relationship

$$R_{int} = R_o + R_a + R_{mt} \tag{25.5}$$

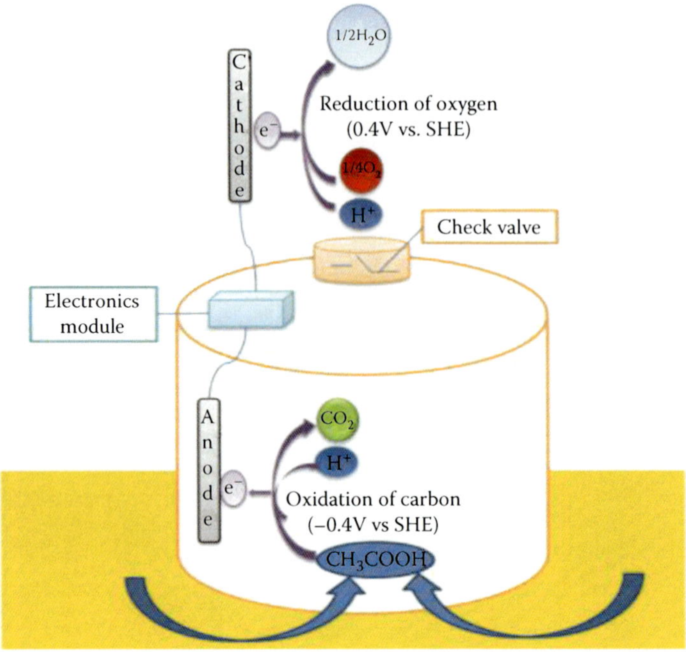

FIGURE 25.5
BC-MFC.

R_o includes resistances to the flow of current through connections in the device and the resistance to counterion flow through the electrolyte, which is seawater in ocean BMFCs (there are no PEMs). R_a refers to losses due to the activation energy required by oxidation/reduction reactions. R_{mt} are losses that arise from limitations in the rate of delivery of electroactive species to the electrodes. The rate of delivery is controlled by the concentration of electroactive species in the vicinity of the electrodes and by factors that determine mass transport. The terms R_a and R_{mt} may also be separated into anode and cathode components and will vary depending on the load placed on the system (Logan et al. 2006).

BMFCs have been mainly examined with the aim of using low-power-consuming marine instrumentation, such as oceanographic sensors, monitoring devices, and telemetry systems (Shantaram et al. 2005). Tender et al. (2008) explained the first demonstration of a marine MFC as a substitute to batteries for a low-power-consuming application. To produce adequate power for the telemetry system, energy generated by the MFC was stored in a capacitor and used in small intervals when required. The specific application reported was a meteorological buoy (about 18 mW average consumption) measuring air temperature, pressure, relative humidity, and water temperature, which was configured for real-time telemetry of data. The prototype sustained 36 mW power equivalent to 26 alkaline D-cells per year at 25°C.

25.3.2 Marine Floating MFC

To reduce the ohmic loss due to the anode–cathode spacing, a new type of marine MFC has been designed based on the floating MFC concept (An et al. 2009; Huang et al. 2012).

The marine floating MFC is a device better suited to field application, especially in contaminated aqueous environments. The cathode of the floating-type MFC is placed on top of the device and opens to the air. The anode is located anaerobically underwater, the position of which is adjustable according to the specific operating condition. Marine floating MFC is effective with a stable interelectrode spacing of 15 cm, and also it varies according to the water column (tidal phenomena). In addition, a cell working with a floating system can benefit from the *natural* agitation created by the movement of the waves, which can promote mass transfer within the anodic compartment. A prototype of floating MFC was tested in Atlantic coastal waters during a 6-month campaign. The microbial anodes were prepared in the laboratory from wild marine biofilm, while the aerobic cathodes were built directly on site in open seawater. A test size was used in order to (1) assess the relevance of the design and (2) evaluate the robustness of this first-generation of floating MFCs in real conditions. Owing to the materials used and the gas accumulation under PEM generated by microorganism, the floating-type MFC can operate on the surface of aqueous environments singly or in series. The maximum power density of this floating-type MFC was 8 mW m^{-2} (An et al. 2009; Chang and An 2009).

25.3.3 Phototrophic MFC

A sediment-type self-sustained phototrophic MFC has been designed (Figure 25.6) to produce electricity by the synergistic interaction between photosynthetic microorganisms and heterotrophic bacteria. Under lighting, the MFC continuously generates electricity

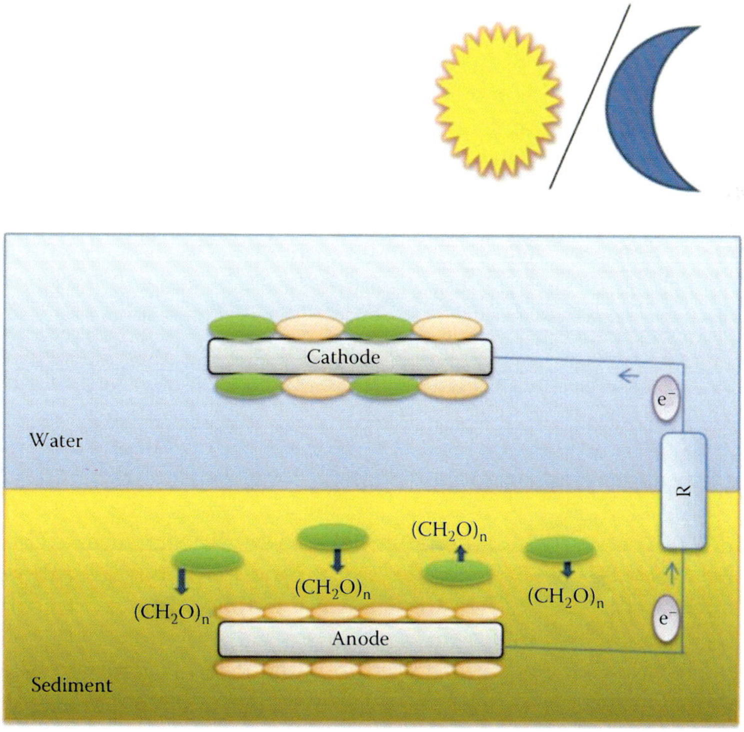

FIGURE 25.6
Sediment-type phototrophic MFC.

without the external input of exogenous organics or nutrients. The generation of the current increased in the dark and decreased with the light, possibly due to the negative effect of the oxygen produced via photosynthesis. Continuous lighting inhibited current production, whereas the continuous dark period inspired electricity generation. Complete darkness resulted in a decrease of current, probably due to the consumption of the organics accumulated during the light phase. Usage of color filters or increasing the thickness of the sediment resulted in a decline of the oxygen-induced inhibition (He et al. 2007; Zou et al. 2009). Molecular taxonomic analysis showed that photosynthetic microorganisms such as cyanobacteria and microalgae predominated in the water phase, adjacent to the cathode and on the surface of the sediment. In contrast, the sediments were enriched by heterotrophic bacteria, becoming low with increasing depth. Algae and some bacterial groups present in marine sediment can supply organic matter (e.g., excreted polysaccharides) to heterotrophic bacteria via photosynthesis and thus maintain synergistic communities. Marine microbiota was also tested in a sediment-type photo MFC with a microbial anode and cathode (Malik et al. 2009). Here, photosynthetic microorganisms in the overlaying water layers produced oxygen for the cathodic oxygen reduction and organic matter, which could be utilized as carbon source at the anode in the anaerobic sediment, resulting in a self-maintained synergistic bioelectrochemical system with light input and electricity output.

25.4 Microbes Involved in Marine MFCs

Microbial community analysis of marine MFC biofilms reveals that there is a group of bacterial communities that develop on the anode. Different bacteria are capable of producing electricity due to the different range of operating conditions, system architectures, electron donors, and electron acceptors. Additionally, a part of the bacterial community can be sustained by alternative metabolisms such as fermentation, methanogenesis, and use of terminal electron acceptors that do not result in electricity generation, as evidenced by a low coulombic efficiency. The electrochemically active bacteria in MFCs are thought to be iron-reducing bacteria such as *Shewanella* and *Geobacter* species (Bond et al. 2002; Kim et al. 1999).

Shewanella putrefaciens is a Gram-negative, facultative anaerobe that has the ability to reduce iron and manganese metabolically; that is, it can use iron and manganese as the terminal electron acceptor in the electron transport chain (in contrast to obligate aerobes that must use oxygen for this purpose). It is also one of the organisms associated with the odor of rotting fish, as it is a marine organism that produces trimethylamines (Sharma and Kundu 2010).

Holmes et al. (2004) investigated the potential role of Geobacteraceae, *Geopsychrobacter electrodiphilus* gen. nov., sp. nov., in an electricity producing marine sediment fuel cell. Microorganisms from the anode surface of a marine sediment fuel cell were enriched and isolated with Fe(III) oxide. Two unique marine isolates were recovered, strains A1T and A2. They are Gram-negative, nonmotile rods, with abundant c-type cytochromes. Phylogenetic analysis of the 16S rRNA, *recA*, *gyrB*, *fusA*, *rpoB*, and *nifD* genes indicated that strains A1T and A2 represented a unique phylogenetic cluster within the Geobacteraceae. Both strains were able to grow with an electrode serving as the sole electron acceptor and transferred ca. 90% of the electrons available in their organic electron donors to the electrodes.

These organisms were the first psychrotolerant members of the Geobacteraceae and grown at temperatures between 4°C and 30°C, with an optimum temperature of 22°C. They utilized a wide range of traditional electron acceptors, including all forms of soluble and insoluble Fe(III) and anthraquinone 2,6-disulfonate. In addition to acetate, both strains utilize a number of other organic acids, amino acids, long-chain fatty acids, and aromatic compounds to support growth with Fe(III) nitrilotriacetic acid as an electron acceptor. The metabolism of these organisms differed in that only strain A1T used acetoin, ethanol, and hydrogen as electron donors, whereas only strain A2 used lactate, propionate, and butyrate. The name *G. electrodiphilus* gen. nov., sp. nov., was proposed for strains A1T. Strains A1T and A2 (ATCC BAA-770; JCM 12470) represented the first organisms recovered from anodes that effectively coupled the oxidation of organic compounds to an electrode.

Marine sediment microorganisms were also investigated by Bond et al. (2002) to generate electricity. Electricity was generated from marine sediments by inserting a graphite electrode (the anode) in the anoxic zone and linking it to a graphite cathode in the overlying aerobic water. A specific enrichment of microorganisms of the family Geobacteraceae on the energy-harvesting anode showed that these microorganisms conserved energy to support their growth by oxidizing organic compounds with an electrode helping as the sole electron acceptor. This was helpful in promoting the bioremediation of organic contaminants in subsurface environments. The highly structured, multilayered biofilms on the anode surface of an MFC developed by *Geobacter sulfurreducens* was studied by Reguera et al. (2006). Cells at a distance from the anode remained viable, and there was no decrease in the efficiency of current production as the thickness of the biofilm increased. Genetic studies showed that efficient electron transfer through the biofilm required the presence of electrically conductive pili. These pili represented an electronic network permeating the biofilm that promoted long-range electrical transfer in an energy-efficient manner.

Molecular analyses of microbial biofilms formed on the anode surface exposed changes in microbial community composition along the anode as a function of sediment depth. Within the 20–29 cm sediment horizon, the anodic biofilm was dominated by microorganisms phylogenetically associated with *Desulfuromonas acetoxidans* (Reimers et al. 2006). It suggests that the availability of Fe(III) in the surrounding sediments or other geochemical factors restricted their distribution and abundance so as to enrich only in the top section of the anode. *D. acetoxidans* growth was also supported by the oxidation of organic compounds such as acetate and transferring these electrons to electrodes (Holmes et al. 2004). Phylotypes within the Geobacteraceae, like *D. acetoxidans,* have also been shown to transfer electrons to Fe(III) oxides via a membrane-bound Fe(III) reductase (Childers et al. 2002; Magnuson et al. 2001; Nevin and Lovley 2000). The middle (46–55 cm depth) and bottom (70–76 cm depth) sediment horizons have enriched for other more diverse communities on the anodic surface. The middle sediment horizon contains δ- and ε-Proteobacteria, whereas the bottom horizon was dominated by the phylotypes, *Syntrophus aciditrophicus* and ε-Proteobacteria (Reimers et al. 2006).

25.5 Conclusion and Future Perspectives

Fossil fuels should not supply the growing energy requirement in the future. The development of sustainable and renewable energy resources and technologies is necessary for sustainable development. Wireless sensor networks are a mounting requirement in today's information-driven world. The collection and propagation of pervasive data in difficult operating environments such as on the ocean floor is becoming increasingly critical for a range of services, from monitoring ship traffic and environmental conditions to enhancing communications. The devices used for these tasks have a critical limitation due to the lack of a robust, long-term power source. Wired connections, batteries, solar panels, and ROV services are all impractical for long-term deployments of field instrumentation due to their operational costs and poor reliability. The capital and maintenance savings, long lifetime, redundancy features, and covert installment make marine MFCs as a promising technology for a wide variety of remote marine applications.

Organic matter stored in the marine sediment environment and waste represents an immense energy resource. The capacity of the MFC to exploit energy from waste and the natural environment holds the promise of MFCs applied for environmental management or as sustainable energy technology. Marine MFC research has provided critical insights into the geochemical and microbiological factors that contribute and may ultimately limit power production in complex systems. To date, marine MFCs have been shown to be reliable, long-term power sources suitable for many marine applications. As further improvements are still required in this nascent technology and power generation capability continues to climb, further market opportunities will undoubtedly present themselves.

Successful commercial development of marine MFCs requires both an engineer and a scientist. The growing pressure on our environment and the call for renewable energy sources will further stimulate the development of this technology, leading soon we hope to its successful implementation. Necessary elements in the marine MFC include temperature variation, pressure, fluid flow, mass transport, and reactant conversion. Once these elements are considered, it should become possible to design a biofuel cell as a unit operation that can be employed as a part of a larger process. In the near future, the development of MFCs to generate useful power for these applications will be limited by the efficiency and cost of materials, physical architecture, and chemical limitations, such as solution conductivity and pH. However, to render MFCs as an economical and reliable technology, more systematic and multidisciplinary research must be undertaken, for example, by setting standards and protocols, by conducting statistically designed experiments, and (most of all) by increasing the degree of transversal interaction among diverse disciplines, such as microbiology, electrochemistry, chemical engineering, electrical engineering, materials science, and nanotechnology so that higher power densities can be achieved one day by overcoming the limitations in MFCs.

References

Aelterman, P., Freguia, S., Keller, J., Verstraete, W., and Rabaey, K. 2008. The anode potential regulates bacterial activity in microbial fuel cells. *Appl Microbiol Biotechnol* 78:409–418.

Aelterman, P., Rabaey, K., Pham, T. H., Boon, N., and Verstraete, W. 2006. Continuous electricity generation at high voltages and currents using stacked microbial fuel cells. *Environ Sci Technol* 40:3388–3394.

An, J., Kim, D., Chun, Y., Lee, S., Ng, H. Y., and Chang, I. S. 2009. Floating type microbial fuel cell for treating organic contaminated water. *Environ Sci Technol* 43:1642–1647.

Atsumi, S., Hanai, T., and Liao, J. C. 2008. Non-fermentative pathways for synthesis of branched-chain higher alcohols as biofuels. *Nature* 451:86–89.

Bard, A. J. and Faulkner, L. R. 2001. *Electrochemical Methods: Fundamentals and Applications*. New York: John Wiley & Sons.

Bentley, R. W. 2002. Global oil and gas depletion: An overview. *Energy Policy* 30:189–205.

Bergel, A., Feron, D., and Mollica, A. 2005. Catalysis of oxygen reduction in PEM fuel cell by seawater biofilm. *Electrochem Commun* 7:900–904.

Bond, D. R., Holmes, D. E., Tender, L. M., and Lovley, D. R. 2002. Electrode-reducing microorganisms that harvest energy from marine sediments. *Science* 295:483–485.

Bond, D. R. and Lovley, D. R. 2003. Electricity production by *Geobacter sulfurreducens* attached to electrodes. *Appl Environ Microbiol* 69:1548–1555.

Burdige, D. J. 2006. *Geochemistry of Marine Sediments*. Princeton, NJ: Princeton University Press.

Chang, I. S. and An, J. 2009. Floating-type microbial fuel cell. WO patent 2,009,134,035.

Chaudhuri, S. K. and Lovley, D. R. 2003. Electricity generation by direct oxidation of glucose in mediatorless microbial fuel cells. *Nat Biotechnol* 21:1229–1232.

Cheng, S., Liu, H., and Logan, B. E. 2006a. Power densities using different cathode catalysts (Pt and CoTMPP) and polymer binders (Nafion and PTFE) in single chamber microbial fuel cells. *Environ Sci Technol* 40:364–369.

Cheng, S., Liu, H., and Logan, B. E. 2006b. Increased power generation in a continuous flow MFC with advective flow through the porous anode and reduced electrode spacing. *Environ Sci Technol* 40:2426–2432.

Childers, S. E., Ciufo, S., Lovley, D. R. 2002. *Geobacter metallireducens* accesses insoluble Fe(III) oxide by chemotaxis. *Nature* 416:767–769.

Dylan, S., Parot, S., Delia, M. L., and Bergel, A. 2007. Electroactive biofilms: New means for electrochemistry. *J Appl Electrochem* 37:173–179.

Faaij, A. P. C. 2006. Bio-energy in Europe: Changing technology choices. *Energy Policy* 34:322–342.

Farzaneh, R., Richard, T. L., Brennan, R. A., and Logan, B. E. 2007. Substrate enhanced microbial fuel cells for improved power generation from sediment based systems. *Environ Sci Technol* 41:4053–4058.

Finkelstein, D. A., Tender, L. M., and Zeikus, J. G. 2006. Effect of electrode potential on electrode-reducing microbiota. *Environ Sci Technol* 40:6990–6995.

Fortman, J. L., Chhabra, S., Mukhopadhyay, A., Chou, H., Lee, T. S., Steen, E., and Keasling, J. D. 2008. Biofuels alternatives to ethanol: Pumping the microbial well. *Trends Biotechnol* 26:375–381.

Fricke, K., Harnisch, F., and Schroder, U. 2008. On the use of cyclic voltammetry for the study of anodic electron transfer in microbial fuel cells. *Energy Environ Sci* 1:144–147.

Froelich, P. N., Klinkhammer, G. P., Bender, M. L., Luedtke, N. A., Heath, G. R., Cullen, D., Dauphin, P., Hammond, D. E., Hartman, B., and Maynard, V. 1979. Early oxidation of organic matter in pelagic sediments of the eastern equatorial Atlantic: Suboxic diagenesis. *Geochim Cosmochim Acta* 43:1075–1090.

Gil, G. C., Chang, I. S., Kim, B. H., Kim, M., Jang, J. K., Park, H. S., and Kim, H. J. 2003. Operational parameters affecting the performance of a mediator-less microbial fuel cell. *Biosens Bioelectron* 18:327–334.

Gorby, Y. A. and Beveridge, T. J. 2005. Composition, reactivity, and regulation of extracellular metal-reducing structures (nanowires) produced by dissimilatory metal reducing bacteria. Presented at *DOE/NABIR Meeting*, April 20, 2005, Warrenton, VA.

Gorby, Y. A., Yanina, S., McLean, J. S., Rosso, K. M., Moyles, D., Dohnalkova, A. et al. 2006. Electrically conductive bacterial nanowires produced by *Shewanella oneidensis* strain MR-1 and other microorganisms. *PNAS* 103:11358–11363.

Habermann, W. and Pommer, E. H. 1991. Biological fuel cells with sulphide storage capacity. *Appl Microbiol Biotechnol* 35:128–133.

He, Z., Haibo, S., and Angenent, L. T. 2007. Increased power production from a sediment microbial fuel cell with a rotating cathode. *Biosens Bioelectron* 22:3252–3255.

He, Z., Minteer, S. D., and Angenent, L. T. 2005. Electricity generation from artificial wastewater using an upflow microbial fuel cell. *Environ Sci Technol* 39:5262–5267.

He, Z., Wagner, N., Minteer, S. D., and Angenent, L. T. 2006. The upflow microbial fuel cell with an interior cathode: Assessment of the internal resistance by impedance spectroscopy. *Environ Sci Technol* 40:5212–5217.
Holmes, D. E., Bond, D. R., O'Neill, R. A., Reimers, C. E., Tender, L. R., and Lovley, D. R. 2004. Microbial communities associated with electrodes harvesting electricity from a variety of aquatic sediments. *Microb Ecol* 48:178–190.
Huang, Y., He, Z., Kan, J., Manohar, A. K., Nealson, K. H., and Mansfeld, F. 2012. Electricity generation from a floating microbial fuel cell. *Bioresour Technol* 114:308–313.
Jang, J. K., Pham, T. H., Chang, I. S., Kang, K. H., Moon, H., Cho, K. S., and Kim, B. H. 2004. Construction and operation of a novel mediator and membrane less microbial fuel cell. *Process Biochem* 39:1007–1012.
Kerr, R. A. 2007. Global warming is changing the world. *Science* 316:188–190.
Kim, B. H., Park, D. H., Shin, P. K., Chang, I. S., and Kim, H. J. 1999. Mediator-less biofuel cell. U.S. patent 5,976,719.
Kim, H. J., Park, H. S., Hyun, M. S., Chang, I. S., Kim, M., and Kim, B. H. 2002. A mediator-less microbial fuel cell using a metal reducing bacterium, *Shewanella putrefaciens*. *Enzyme Microb Technol* 30:145–152.
Kim, N., Choi, Y., Jung, S., and Kim, S. 2000. Development of microbial fuel cell using *Proteus vulgaris*. *Bull Korean Chem Soc* 21:44–49.
Liu, H., Ramnarayanan, R., and Logan, B. E. 2004. Production of electricity during wastewater treatment using a single chamber microbial fuel cell. *Environ Sci Technol* 38:2281–2285.
Liu, J. L., Lowy, D. A., Baumann, R. G., and Tender, L. M. 2007. Influence of anode pretreatment on its microbial colonization. *J Appl Microbiol* 102:177–183.
Logan, B. E. 2004. Extracting hydrogen and electricity from renewable resources. *Environ Sci Technol* 38:160a–167a.
Logan, B. E., Murano, C., Scott, K., Gray, N. D., and Head, I. M. 2005. Electricity generation from cysteine in a microbial fuel cell. *Water Res* 39:942–952.
Logan, B., Aelterman, P., Hamelers, B., Rozendal, R., Schroder, U., Keller, J., Freguia, S., Verstraete, W., and Rabaey, K. 2006. Microbial fuel cells: Methodology and technology. *Environ Sci Technol* 40:5181–5192.
Lovley, D. R. 2006. Bug juice: Harvesting electricity with microorganisms. *Nat Rev Microbiol* 4:497–508.
Lovley, D. R. 2008. The microbe electric: conversion of organic matter to electricity. *Curr Opin Biotechnol* 19:564–571.
Lowy, D. A., Tender, L. M., Zeikus, J. G., Park, D. H., and Lovley, D. R. 2006. Harvesting energy from the marine sediment-water interface II—Kinetic activity of anode materials. *Biosens Bioelectron* 21:2058–2063.
Magnuson, T. S., Isoyama, N., Hodges M. A. L., Davidson, G., Maroney, M. J., Geesey, G. G., Lovley, D. R. 2001. Isolation, characterization and gene sequence analysis of a membrane- associated 89 kDa Fe(III) reducing cytochrome c from *Geobacter sulfurreducens*. *Biochem J* 359:147–152.
Malik, S., Drott, E., Grisdela, P., Lee, J., Lee, C., Lowy, D. A. et al. 2009. A self-assembling self-repairing microbial photoelectrochemical solar cell. *Energy Environ Sci* 2:292–298.
Min, B., Cheng, S., and Logan, B. E. 2005. Electricity generation using membrane and salt bridge microbial fuel cells. *Water Res* 39:1675–1686.
Min, B. and Logan, B. E. 2004. Continuous electricity generation from domestic wastewater and organic substrates in a flat plate microbial fuel cell. *Environ Sci Technol* 38:5809–5814.
Mo, Y. H., Liang, P., Huang, X., Wang, H., and Cao, X. 2009. Enhancing the stability of power generation of single-chamber microbial fuel cells using an anion exchange membrane. *J Chem Technol Biotechnol* 84:1767–1772.
Mohan, S. V., Raghavulu, S. V., and Sarma, P. N. 2008. Biochemical evaluation of bioelectricity production process from anaerobic wastewater treatment in a single chambered microbial fuel cell (MFC) employing glass wool membrane. *Biosens Bioelectron* 23:1326–1332.

Moon, H., Chang, I. S., and Kim, B. H. 2006. Continuous electricity production from artificial wastewater using a mediator-less microbial fuel cell. *Bioresour Technol* 97:621–627.

Nevin, K. P. and Lovley, D. R. 2000. Lack of production of electron shuttling compounds or solubilization of Fe(III) during reduction of insoluble Fe(III) oxide by *Geobacter metallireducens*. *Appl Environ Microbiol* 66:2248–2251.

Nielsen, M. E., Reimers, C. E., and Stecher, H. A. 2007. Enhanced power from chambered benthic microbial fuel cells. *Environ Sci Technol* 41:7895–7900.

Nielsen, M. E., Reimers, C. E., White, H. K., Sharma, S., and Girguis, P. R. 2008. Sustainable energy from ocean cold seeps. *Energy Environ Sci* 1:584–593.

Niessen, J., Schroder, U., Rosenbaum, M., and Scholz, F. 2004. Fluorinated polyanilines as superior materials for electro catalytic anodes in bacterial fuel cells. *Electrochem Commun* 6:571–575.

Oh, S. and Logan, B. E. 2006. Proton exchange membrane and electrode surface areas as factors that affect power generation in microbial fuel cells. *Appl Microbiol Biotechnol* 70:162–169.

Oh, S., Min, B., and Logan, B. E. 2004. Cathode performance as a factor in electricity generation in microbial fuel cells. *Environ Sci Technol* 38:4900–4904.

Park, D. H. and Zeikus, J. G. 1999. Utilization of electrically reduced neutral red by *Actinobacillus succinogenes*: Physiological function of neutral red in membrane-driven fumarate reduction and energy conservation. *J Bacteriol* 181:2403–2410.

Park, D. H. and Zeikus, J. G. 2003. Improved fuel cell and electrode designs for producing electricity from microbial degradation. *Biotechnol Bioeng* 81:348–355.

Phung, N. T., Lee, J., Kang, K. H., Chang, I. S., Gadd, G. M., and Kim, B. H. 2004. Analysis of microbial diversity in oligotrophic microbial fuel cells using 16S rDNA sequences. *FEMS Microbiol Lett* 233:77–82.

Rabaey, K., Boon, N., Hofte, M., and Verstraete, W. 2005a. Microbial phenazine production enhances electron transfer in biofuel cells. *Environ Sci Technol* 39:3401–3408.

Rabaey, K., Boon, N., Siciliano, S. D., Verhaege, M., and Verstraete, W. 2004. Biofuel cells select for microbial consortia that self-mediate electron transfer. *Appl Environ Microbiol* 70:5373–5382.

Rabaey, K., Clauwaert, P., Aelterman, P., and Verstraete, W. 2005b. Tubular microbial fuel cells for efficient electricity generation. *Environ Sci Technol* 39:8077–8082.

Rabaey, K., Lissens, G., Siciliano, S. D., and Verstraete, W. 2003. A microbial fuel cell capable of converting glucose to electricity at high rate and efficiency. *Biotechnol Lett* 25:1531–1535.

Reguera, G., McCarthy, K. D., Mehta, T., Nicoll, J. S., Tuominen, M. T., and Lovley, D. R. 2005. Extracellular electron transfer via microbial nanowires. *Nature* 435:1098–1101.

Reguera, G., Nevin, K. P., Nicoll, J. S., Covalla, S. F., Woodard, T. L., and Lovley, D. R. 2006. Biofilm and nanowire production leads to increased current in *Geobacter sulfurreducens* fuel cells. *Appl Environ Microbiol* 72:7345–7348.

Reimers, C. E., Girguis, P., Stecher, H. A., Tender, L. M., and Ryckelynck, N. 2006. Microbial fuel cell energy from an ocean cold seep. *Geobiology* 4:123–136.

Reimers, C. E., Stecher, H.A., Westall, J. C., Alleau, Y., Howell, K. A., Soule, L., White, H. K., and Girguis, P. R. 2007. Substrate degradation kinetics, microbial diversity, and current efficiency of microbial fuel cells supplied with marine plankton. *Appl Environ Microbiol* 73:7029–7040.

Reimers, C. E., Tender, L. M., Fertig, S., and Wang, W. 2001. Harvesting energy from the marine sediment-water interface. *Environ Sci Technol* 35:192–195.

Ren, H., Lee, H., and Chae, J. 2012. Miniaturizing microbial fuel cells for potential portable power sources: Promises and challenges. *Microfluid Nanofluid* 13:353–381.

Rhoads, A., Beyenal, H., and Lewandowski, Z. 2005. Microbial fuel cell using anaerobic respiration as an anodic reaction and bio-mineralized manganese as a cathodic reactant. *Environ Sci Technol* 39:4666–4671.

Richter, H., Nevin, K. P., Jia, H., Lowy, D. A., Lovley, D. R., and Tender, L. M. 2009. Cyclic voltammetry of biofilms of wild type and mutant *Geobacter sulfurreducens* on fuel cell anodes indicates possible roles of OmcB, OmcZ, type IV pili, and protons in extracellular electron transfer. *Energy Environ Sci* 2:506–516.

Rodrigo, M. A., Canizares, P., Lobato, J., Paz, R., Saez, C., and Linares, J. J. 2007. Production of electricity from the treatment of urban waste water using a microbial fuel cell. *J Power Sources* 169:198–204.
Rozendal, R. A., Hamelers, H. V. M., and Buisman, C. J. N. 2006. Effects of membrane cation transport on pH and microbial fuel cell performance. *Environ Sci Technol* 40:5206–5211.
Ryckelynck, N., Stecher, H. A., and Reimers, C. E. 2005. Understanding the anodic mechanism of a seafloor fuel cell: Interactions between geochemistry and microbial activity. *Biogeochemistry* 76:113–139.
Schroder, U., Niessen, J., and Scholz, F. 2003. A generation of microbial fuel cells with current outputs boosted by more than one order of magnitude. *Angew Chem Int Ed* 42:2880–2883.
Sell, D., Kramer, P., and Kreysa, G. 1989. Use of an oxygen gas diffusion Cathode and a three-dimensional packed bed anode in a bioelectrochemical fuel cell. *Appl Microbiol Biotechnol* 31:211–213.
Shantaram, A., Beyenal, H., Raajan, R., Veluchamy, A., and Lewan-dowski, Z. 2005. Wireless sensors powered by microbial fuel cells. *Environ Sci Technol* 39:5037–5042.
Sharma, V. and Kundu, P. P. 2010. Biocatalysts in microbial fuel cells. *Enzyme and Microb Tech* 47:179–188.
Sorrel, S., Speirs, J., Bentley, R., Brandt, A., and Miller, R. 2010. Global oil depletion: A review of the evidence. *Energy Policy* 38:5290–5295.
Stephanopoulos, G. 2007. Challenges in engineering microbes for biofuels production. *Science* 315:801–804.
Tender, L. M., Gray, S. M., Groveman, E., Lowy, D. A., Kauffman, P., Melhado, J. et al. 2008. The first demonstration of a microbial fuel cell as a viable power supply: Powering a meteorological buoy. *J Power Sources* 179:571–575.
Tender, L. M., Reimers, C. E., Stecher, H. A., Holmes, D. E., Bond, D. R., Lowy, D. A. et al. 2002. Harnessing microbially generated power on the seafloor. *Nat Biotechnol* 20:821–825.
Tsuneo, H. 2001. Research and development of international clean energy network using hydrogen energy. *Int J Hydrogen Energy* 27:115–129.
Voiland, A. 2009. 2009: Second warmest year on record; end of warmest decade. NASA Goddard Institute for Space Studies. http://www.giss.nasa.gov/research/news/20100121 (accessed November 14, 2013).
Whitfield, M. 1972. The electrochemical characteristics of natural redox cells. *Limnol Oceanogr* 17:383–393.
Zhao, F., Harnisch, F., Schroder, U., Scholz, F., Bogdanoff, P., and Herrmann, I. 2005. Application of pyrolysed iron (II) phthalocyanine and CoTMPP based oxygen reduction catalysts as cathode materials in microbial fuel cells. *Electrochem Commun* 7:1405–1410.
Zou, Y., Pisciotta, J., Billmyre, R. B., and Baskakov, I. V. 2009. Photosynthetic microbial fuel cells with positive light response. *Biotechnol Bioeng* 104:939–946.
Zuo, Y., Cheng, S., Call, D., and Logan, B. E. 2008. Ion exchange membrane cathodes for scalable microbial fuel cells. *Environ Sci Technol* 42:6967–6972.

Section VI

Marine Waste for Bioenergy

26

Waste-Derived Bioenergy Production from Marine Environments

Kyoung C. Park and Patrick J. McGinn

CONTENTS

26.1 Introduction

The question of land availability and costs for bioenergy production brings us to the concept of *bioenergy production using waste sources*. When we produce bioenergy using selected feedstocks, a more targeted biofuel type can be produced, but the cost will be high. However, when we use waste streams originating from sewage treatment and fishery production, it may be less productive for the targeted biofuel type but may well be more cost-effective overall.

Marine environments were once considered a major disposal site for various types of wastes derived from both land and ocean. Such waste disposal practices caused many environmental issues, especially in estuaries and coastal waters, such as degraded or contaminated water quality, red tides, reduced tourism, and other problems. In recent years, the regulations on waste disposal to marine environments have been strengthened, especially in economically advanced countries. This caused the question: *How can we properly manage wastes for marine environments?* The main sources of disposal into marine environments are sewage sludge, industrial wastes, and dredged material. Additionally, industrial and municipal effluents are also eventually introduced into marine environments.

In the past two decades, world aquaculture production has dramatically increased and accounted for more than 40% of total (capture and aquaculture) fishery production with 148 million metric tons (MT) in 2010 (FAO 2013). Global aquaculture and fishery activities produce tremendous wastes such as fecal matter, uneaten feed pellets, and solid and liquid wastes from fish-processing operations. These wastes also cause a lot of environmental problems in estuaries and coastal settings. To minimize aquaculture and fishery wastes, many waste management programs were developed and implemented.

In recent years, we have seen a huge interest and support for producing biofuels from many types of biomass including waste material and effluents. The main bioenergy products converted from waste sources are in the form of liquid biofuels (e.g., biodiesel) and biogases (e.g., methane), although bioethanol and biocrude oil also have been studied.

In practical terms, the most intensively studied area is methane production using municipal wastewater sludge, while biodiesel production using microalgae grown in municipal wastewater has more often been studied in the laboratory or at the pilot scale. Unfortunately, bioenergy production from marine waste sources has not been much studied, although it has huge potential. Thus, in this review, we discuss the potential of marine waste for producing bioenergy as well as bioenergy production using wastes commonly released to marine environments.

26.2 Bioenergy Production Using Wastewater

Most highly populated cities are located near coastlines, meaning that treated and untreated municipal and industrial wastes are eventually discharged into sensitive coastal areas. Although waste discharge into marine environments is becomingly reduced compared to common practice 10–20 years ago, large amounts of different types of wastes continue to be released into marine environments. These wastes include municipal wastewater effluents and sludge, industrial effluents, and dredged material. Municipal wastes are generated in two main forms: solids and treated wastewaters. Solid wastes are produced when highly liquefied sludge is dewatered, which produces a liquid with a high concentration of nutrients (called *centrates*). Municipal solid wastes (MSWs) are produced in tremendous quantities globally. In the United States alone, an amount between 200 and 400 million MT per year is produced (Tonjes and Greene 2012). This has been traditionally sent to landfills or used for methane gas and electricity production or for fertilizer. It is estimated that the volume of liquid wastewater effluents generated in North America, Europe, and Latin America is approximately 70, 63, and 47 km^3/year, respectively (UNO 2009). Conventional treatment can remove only a fraction of the total nitrogen (40% removal) and phosphorous (12% removal) present in these effluents, and wastewater treatment plants (WWTPs) commonly show N and P values around 20–70 and 4–12 mg/L, respectively (Carey and Migliaccio 2011). These levels are still high for discharge at the legislated nutrient levels, even though the recommended levels vary from country to country. To further reduce nutrient levels, biological systems are often considered to be the ideal means (Rawat et al. 2011). Algae-based domestic wastewater treatment has been practiced, as a tertiary treatment, for a long time for cost-effective treatment and nutrient recovery (Oswald et al. 1957; Pittman et al. 2011). Microalgae-based systems have shown a high potential to assist with nutrient removal (Oswald et al. 1957; Beneman et al. 1980; Liang 2013). Many species of microalgae are able to grow well in municipal (Li et al. 2011; Zhou et al. 2011; McGinn et al. 2012; Park et al. 2012b), agricultural (Gonzalez et al. 1997; Wilkie and Mulbry 2002; An et al. 2003), and industrial wastewaters (Chinnasamy et al. 2010) through their ability to utilize abundant organic carbon and inorganic N and P residuals. The commonly isolated and used genera in microalgae-based wastewater treatment systems were species from the genera *Actinastrum, Coelastrum, Dictyosphaerium, Micractinium, Pediastrum,* and *Scenedesmus* (Oswald 1988; Banat et al. 1990; Beneman 2003; Wells 2005; Heubeck et al. 2007; Park and Craggs 2010). Representative species from these genera are known to be able to grow also mixotrophically (Park et al. 2011, 2012b, 2014; Zhou et al. 2011). Microalgae have been shown to be very efficient at removing N and P from wastewater. For example, various species of *Chlorella* and *Scenedesmus* spp. completely removed ammonia, nitrate, and phosphorous from wastewater (Martinez et al. 2000; Zhang et al. 2008; Ruiz-Marin et al. 2010; McGinn et al. 2012; Park et al. 2012b).

Microalgae as a source of fuel and/or energy have been intensively studied over the past two decades. There are four types of biofuels that can potentially be produced from algal biomass: biodiesel, biomethane, bioalcohols, and hydrogen. Many promising results suggesting the potential for biodiesel production using wastewater-grown microalgae have been obtained (Sheehan et al. 1998; Oswald 2003; Kong et al. 2010; Park et al. 2011). Algal lipids extracted from biomass grown on waste CO_2 from the combustion of fossil fuels, and wastewater have been suggested as a strategy to lower the costs of microalgae biofuels. Although microalgae systems show high potential, the main bottleneck toward an effective application in the energy sector is the costs associated with the primary production of massive quantities of microalgal biomass and the associated costs of processing the biomass to fuel products (Wijffels and Barbosa 2010; Singh and Olsen 2011).

26.2.1 Alternative Approaches for Mass Microalgal Cultivation: Mixotrophy

One approach to reduce biomass production cost is to maximize biomass or lipid productivity in the growth system of choice. To date, previous research has shown that autotrophic cultivation using secondary wastewater commonly employed is limited by light and nutrient availability. Compared to primary wastewater that generally contains higher levels of organic substrates and nutrients, secondary wastewater has very low levels of organic carbon and has characteristics more akin to autotrophic growth media (Park et al. 2012b). Much previous research has been done using different characteristics of primary and secondary wastewater under different culturing conditions. A variety of approaches that have been taken have made it difficult to determine which culturing condition is better suited for higher productivity. However, it seems that mixotrophic-based systems, in which photosynthetic cultures are supplemented with organic C substrates, have higher productivity than autotrophic cultivation (Ogawa and Aiba 1981; Ogbonna and Tanaka 1998; Cerón et al. 2000; Perez-Garcia et al. 2001; Liang et al. 2009; Kong et al. 2010; Sydney et al. 2011; Tanoi et al. 2011; Wan et al. 2011; Park et al. 2012b), especially in raceway pond systems where sunlight is the common limiting factor for maximum production.

Many researchers have tried to increase lipid productivity (biomass × lipid content/day) by mixotrophic cultivation using primary wastewater or secondary wastewater with the addition of external carbon sources. Other carbon sources for mixotrophic cultivation with secondary wastewater include acetate, glycerol, anaerobic digestates, and centrate from sludge dewatering (Li et al. 2011; Zhou et al. 2011; Park et al. 2012a,b, 2014; Bjornsson et al. 2013). Finding cheap supplemental carbon sources for algal cultivation is an important factor for reducing biomass production costs. Thus, the use of centrates or anaerobic digestates could be a cost-effective choice for external carbon sources as well as a feasible strategy for inorganic nutrient supplementation.

Primary wastewater contains high levels of organic carbon as well as other dissolved inorganic nutrients such as N and P. Some researchers have studied the potential for biofuel production using a combination of primary wastewater and mixotrophic microalgae (Table 26.1). Based on previous research conducted at laboratory scales, maximum biomass (2000 mg/L/day) and lipid productivity (505 mg/L/day) reported in the literature were obtained from *Chlamydomonas reinhardtii* cultivated in a biocoil photobioreactor using municipal wastewater amended with centrate (Table 26.1). Even though growth conditions were not optimized, *Dictyosphaerium* sp. cultured in 250 mL Erlenmeyer flasks in the laboratory displayed rates of biomass and lipid productivity of 456 and 162 mg/L/day, respectively, autotrophically using secondary wastewater. According to Cabanelas et al. (2013), a microalgae production system treating wastewater at a scale of 200 m^3/day for 240 days/year could potentially produce 2–6 MT of biomass/year, depending on the strain used.

Even though mixotrophic cultivation has many potential advantages such as higher biomass and/or lipid productivity and the tendency under certain conditions to induce the formation of specific fatty acids more favorable for biodiesel production, such as oleic acids (Knothe 2006, Park et al. 2012b), there are problems with this approach. One such problem is related to the management of contaminating microorganisms such as bacteria, fungi, and zooplankton predators. Under laboratory conditions, these contaminants can be controlled to a certain extent. At pilot or production scales in the field, however, this is much more difficult, and these contaminants can seriously compromise system productivity. This can be partially mitigated through careful screening and selection of microalgae strains more resistant to invasion by foreign organisms.

Other factors affecting the efficiency and stability of mixotrophic growth are growth conditions such as the external carbon source used, the CO_2 concentrations, and light intensity. In our laboratory, we have observed that different strains exert different preferences for external carbon sources. Interestingly, the addition of external carbon sources such as glycerol and centrates did not enhance lipid productivity in the presence of high CO_2 and high light conditions (Park et al. 2012b, 2014).

The efficient growth of microalgae in wastewater depends on a number of variables including wastewater characteristics such as N and P concentrations and ratios, the presence of toxic chemicals, and the presence of other microorganisms such as pathogenic bacteria or predators. These effects will vary depending on the local wastewater treatment technology used and therefore serve to reemphasize the importance of using local microalgae strains isolated in situ from the wastewater that one wishes to remediate.

The high biomass productivity of wastewater-grown microalgae suggests that this cultivation method offers real potential as a viable approach to use for the production of sustainable and renewable energy.

The combination of algal biomass production with wastewater treatment and flue gas remediation can have a significant impact in the wastewater treatment and carbon mitigation sectors as well as in the biofuel industry.

Land-based production of algal biofuels has focused more intensively on microalgae, whereas macroalgae have dominated research in natural marine settings. In the early 1970s, marine microalgae were first tested for their capacity for growth and remediation in seawater–wastewater mixtures and found to be effective for nutrient removal and biomass production (Ryther et al. 1972; Goldman and Stanley 1974; Goldman 1975). These approaches were focused on aquaculture aspects. In the 1990s, two studies in particular showed that marine microalgae were excellent for wastewater treatment (Craggs et al. 1995, 1997). In those papers, 102 marine microalgae were tested, and it was found that *Phaeodactylum tricornutum* was an excellent species for mass culture using wastewater–seawater mixtures. In more recent years, a variety of novel systems combining

TABLE 26.1

Comparison of Biomass and Lipid Productivity between Autotrophic and Mixotrophic Cultivation of Microalgae Using Municipal Wastewater

Species	NH_4 (mg/L)	PO_4 (mg/L)	Auto-/Mixotrophic Cultivation	Biomass Productivity (mg/L/Day)	Lipid Productivity (mg/L/Day)	Cultivation System	References
Chlorella vulgaris	39.4–43.9	6.7–7.1	MW-1st[a] and glycerol	118	18	1 L Erlenmeyer flask	Cabanelas et al. (2013)
Botryococcus terribilis	46.1–47.4	11.4–11.5	MW-1st and glycerol	282	35	1 L Erlenmeyer flask	Cabanelas et al. (2013)
Chlamydomonas reinhardtii	67.0/128.6 (TKN)	120.6 (TP)	MW-2nd[b] and centrate	2000	505	Biocoil–bioreactor	Kong et al. (2010)
Scenedesmus sp.-AMDD	79.3	38.29	MW-2nd and centrate	140	N.D.	250 mL Erlenmeyer flask	Park et al. (2012a)
Scenedesmus obliquus	27.4	11.8	MW-2nd	26	8 (FAMEs)	1 L jacketed cylindrical bioreactors	Martínez et al. (2000)
Botryococcus braunii	15.0	11.5	MW-2nd	346 ($mg/dm^3/h$)	62 ($mg/dm^3/h$)	Bioreactor	Órpez et al. (2009)
Dictyosphaerium sp.	20.8	2.15	MW-2nd	456	162 (FAMEs)	250 mL Erlenmeyer flask	Park et al. (2014)
Scenedesmus sp.-AMDD	20–30	2.2–3.5	MW-2nd	267	32 (FAMEs)	Continuous chemostat	McGinn et al. (2012)

TKN, total nitrogen; TP, total phosphate; FAMEs, fatty acid methyl esters.
[a] MW-1st: primary treated municipal wastewater.
[b] MW-2nd: secondary treated municipal wastewater.

microalgae–wastewater remediation with seawater have been discussed and proposed for biofuel production (Li et al. 2008; Jiang et al. 2011). More recently, applied research at NASA has supported pilot-scale projects like the offshore membrane enclosures for growing algae (OMEGA) system with US$10 million, launched in 2012. In this project, production of microalgae takes place just offshore in floating membrane enclosures using wastewater as a nutrient source. As the cultures grow and remediate the wastewater, the low-salinity wastewater tends to cross the diffusible membrane enclosure as the system reaches osmotic equilibrium with the surrounding seawater. This has the effect of passively concentrating the microalgae biomass in situ and avoids the energetically costly step of postcultivation algae dewatering. Such concepts as the OMEGA system are obviously innovative but still have many technical hurdles to surmount such as biofouling, sunlight attenuation, wave action, and other physical forces in the ocean that could damage the rather delicate membranes.

26.3 Bioenergy Production Using Fisheries and Aquaculture Wastes

26.3.1 Fishery Wastes: Biodiesel Production

Biodiesel production, manure composting, biogas production, and burning of fish/shellfish wastes to produce energy are different ways to utilize these materials (Yahyaee et al. 2013). According to the Food and Agriculture Organization (FAO), in 2010, global fish production was around 148 million MT (FAO 2013). Approximately 75% of this production was used for direct consumption, and the remainder processed to nonfood products (FAO 2006). It has been suggested that approximately 50% (by weight) of total processed fish ends up as a waste product, and of this amount, the oil content varies from 40% to 65% (Arruda et al. 2007). About 250,000 MT of fish-processing waste such as viscera, fins, eyes, and tails is produced worldwide and could be converted into useful products such as fish meal, oil, and fertilizer. When fish oils are used directly or blended with other commercial oils, problems could emerge, such as changes in viscosity, lubricity, and acidity. Thus, converting fish oils into a fuel such as biodiesel could be a better approach. This has been more intensively studied using salmon and anchovy oils. Biodiesel is commonly produced from vegetable oils and animal fats. Haas et al. (2006) compared the biodiesel production costs of salmon and soybean oils, using a computer model, and found that the cost of biodiesel production from salmon oil was almost twice that produced from soybean due to the price of feedstock. However, the relatively low cost of waste fish oil of US$0.25 per gallon compared to the current (but erratic) price US$1.19 per gallon of diesel fuel provides an economic incentive to regard these waste oils as practical feedstocks to produce biodiesel fuels competitively with fossil sources (Yahyaee et al. 2013).

Oils extracted from marine fish have some advantages compared to vegetable oils. They contain a higher percentage of long-chain, omega-3 polyunsaturated fatty acids (PUFAs) and require minimal processing to be made usable as a fuel. Marine fishes such as cod, tuna, and squid generally contain high levels of PUFAs. These may be beneficial for diesel engine performance and may represent a reduction in pollutant emissions (Lin and Li 2009). Fish oil is produced in large quantities by the fish-processing industry as a byproduct in many countries. Several studies have been carried out for using waste fish oil to produce biofuel. Eleven percent of total fish wastes were extracted fish oil, and 0.9 L of biodiesel was produced from each liter of extracted fish oil. The highest fatty acid ratio (40.3%) was oleic acid (C18:1) followed by palmitic acid (C16:0) with 20.9% and linoleic acid (C18:2) with 13.9% (Yahyaee et al. 2013).

Biodiesels produced from fish oils have higher viscosity, lower lubricity, more acidity, and higher flash point compared to petroleum diesel. This was one of the concerns of early research (Steigers 2002). Recent studies have shown that, despite the differences in the composition of salmon oil from that of corn oil, salmon oil methyl esters had comparable viscosity, volatility, low-temperature properties, heating value, acid value, and specific gravity and better oxidative stability than that of corn oil methyl esters (Chiou et al. 2006). Engine tests have revealed that biodiesel obtained from fish oils such as anchovy can be used in place of petroleum diesel in diesel engines with little or no modification (Alberta et al. 2009; Öner and Altun 2009; Anderson and Weinbach 2010). Even though there is a great potential of waste fish oil as a sustainable source for production of biodiesel, there is a substantial need for more research work to study the other economic issues related to biodiesel production from these oils (Yahyaee et al. 2013). As other biofuels, fish oil also requires catalysts to efficiently produce biodiesel. Recent studies showed fishery wastes from shrimp and mollusk shell contain a variety of efficient catalysts for effective biodiesel production (Yang et al. 2009; Viriya-empikul et al. 2010).

26.3.2 Fishery Wastes: Biogas Production

Literature on methane production from waste fish biomass is scarce (Callaghan et al. 1999; Mshandete et al. 2004; Bouallagui et al. 2009; Eiroa et al. 2012; Kafle et al. 2013).

It has been shown that anaerobic codigestion of multiple waste streams improves their digestibility and biogas yields. In an early study, Callaghan et al. (1999) tested codigestion of fish waste and cattle manure slurry. The ratio of codigestion was 70% (w/w) cattle slurry, 20% fish offal, and 10% digester inoculum. The production was significantly increased from 280 mL CH_4/g $VS_{removed}$ with cattle manure alone to 380 mL CH_4/g $VS_{removed}$. Bouallagui et al. (2009) also found that fish waste codigests well with fruit and vegetable waste. The addition of fish waste increased biomethane production by 8.1%, compared to fruit/vegetable waste alone. Mshandete et al. (2004) also showed positive results from codigestion trials. When sisal pulp (from *Agave sisalana*) and fish waste alone (consisting of viscera, scales, gills, and washing water) were digested, methane production was 0.32 and 0.39 m^3 CH_4/kg VS, respectively, at a total solid (TS) level of 5%; the production was increased to 0.62 m^3 CH_4/kg VS_{added} when it was codigested as 33% fish waste and 67% sisal pulp at 16.6% TS level.

Eiroa et al. (2012) and Kafle et al. (2013) tested fish waste alone as a feedstock for the production of biomethane and biogas. Eiroa et al. (2012) evaluated biomethane potential using solid fish wastes from tuna, sardine, needle fish, and mackerel. Biomethane production was highest from mackerel waste at 0.59 g COD–CH_4/g COD_{added} compared to other fish wastes with around 0.47 g COD–CH_4/g COD_{added}. Kafle et al. (2013) used Pacific saury fish waste to make fish silage and produced biogas and methane at 671–763 and 441–482 mL/g VS, respectively. These biomethane production rates were slightly higher than methane production from digestion of microalgae (300–400 mL/g TVS) (Frigon et al. 2013). In some countries, jellyfish are a major problem to swimmers as well as fishermen. Kim et al. (2012) tested the potential of biogas production using moon jellyfish (*Aurelia aurita*) waste. The highest hydrogen and methane gas production were 121.35 and 870.12 mL/g, respectively, under optimized conditions.

26.3.3 Aquaculture Wastes

The decline of marine fisheries worldwide and the global increase in the demand for seafood have stimulated rapid growth in aquaculture production. Based on total world fishery production in 2010, aquaculture production contributed 60 million MT (40.3%) to the total production of 148 million MT (FAO 2013). This highly intensified production of seafood by aquaculture has impacted significantly on marine environments, as well as producing tremendous amounts of fish. In particular, the environmental impacts of finfish and shrimp aquaculture operated mainly along coastal zones are becoming a matter of greater concern in many parts of the world. The release of mainly organic solid wastes and dissolved wastes (e.g., nitrogen and phosphorous) by aquaculture operations has resulted in highly eutrophic marine coastal zones (Bureau and Hua 2010). Even though, in recent years, improvements to feed formulations and the development of efficient feeding strategies have reduced waste outputs by fish culture operations, the uncertainty of the fate of these wastes is still one of the most significant factors that limit the implementation of sustainable aquaculture practices.

A number of strategies have been developed to reduce negative impacts of fish and shrimp aquaculture on local marine environments. These include integrated multitrophic aquaculture (IMTA), land-based aquaculture, and dredging systems to collect waste sediments. IMTA technology is gaining interest as a strategy to reduce solid wastes and as a means of remediating dissolved nitrogenous and phosphorous-enriched effluents (Chopin 2011). IMTA is an integrated aquaculture strategy wherein organisms that complement each other are cocultivated in order to optimize nutrient utilization and reduce solid waste that goes to sediments. In most IMTA farms, seaweed has been grown near fish aquaculture sites, and this approach has shown highly positive results for seaweed production as well as a means of bioremediating excess nutrients (Neori et al. 2004; Schneider et al. 2005; Chávez-Crooker and Obreque-Contreras 2010; Skriptsova and Miroshnikova 2011; Ren et al. 2012).

However, data for macroalgae biomass productivity linked to potential bioenergy production are limited, even though there are some species that integrate well with fish aquaculture installations as the waste nutrient remediation agent (Shpigel et al. 1993b; Krom et al. 1995; Buschmann et al. 1996; Neori 1996, Neori et al. 2000; Schneider et al. 2005; Skriptsova and Miroshnikova 2011). These studies have shown that it is possible to produce *Ulva lactuca* (sea lettuce) biomass at between 3.3 and 7 times (wet weight) the quantity of fish biomass in IMTA systems. For the IMTA systems under study in those reports, *Ulva* spp. and *Gracilaria* spp. were the most intensively studied macrophytes. The relative quantities of fish or seaweed produced in IMTA systems in marine environments depend on many factors including the type of fish and seaweed grown, fish feed composition, location, and climate conditions, among others. Therefore, it is not easy to calculate possible production for a given pairing of fish and seaweed. However, in land-based systems, it is easier to calculate possible production. Shpigel et al. (1993b) suggested, based on certain assumptions, that a 1 ha land-based integrated seabream–shellfish–seaweed farm could produce 25 MT of fish, 50 MT of bivalves, and 30 MT of seaweeds (all fresh weight) annually. Another 1 ha farm model could produce 55 MT of seabream or 92 MT of salmon, with 385 or 500 MT of seaweed, respectively, without pollution (Shpigel et al. 1993a).

In the Bay of Fundy in eastern Canada, IMTA systems have been more intensively studied in North America during the last 20 years (Barrington et al. 2009). In this area, annual production of Atlantic salmon in sea-cage systems was estimated to be 25,000 MT in 2012.

Based on the findings of previous studies that have shown that roughly 3 MT of fresh seaweed biomass could be produced through IMTA for every MT of fish biomass, this corresponds to 75,000 MT of seaweed (fresh weight) (Shpigel et al. 1993b; Krom et al. 1995; Buschmann et al. 1996; Neori 1996, Neori et al. 2000; Schneider et al. 2005; Skriptsova and Miroshnikova 2011). Using *Laminaria* as the seaweed model, it is possible to calculate the biomethane production potential if 75,000 MT of biomass were anaerobically digested to biogas. At 10% dry weight (relative to fresh weight) and a specific methane yield of 260–280 m^3 CH_4/MT VS (Chynoweth 2005), this would correspond to 2.1×10^7 m^3 of CH_4, equivalent to 80×10^5 GJ of bioenergy.

Based on salmonid aquaculture operations, releases of solid wastes equivalent to 25% of the weight of fish produced are common (Bureau and Hua 2010). In sea-cage systems, mollusks have been tested for efficient use of solid waste in IMTA systems, but the recycling of these wastes using other organisms highly depends on other factors such as hydro physics, including current speed and direction. In recent years, land-based IMTA systems are becoming more favored in terms of sustainability. In this case, environmental impacts are minimized by taking aquaculture operations onto land where waste products can be accessed more readily. For flow-through systems, the gain is marginal as large volumes of wastewater are still discharged into a receiving water body. However, for recirculating systems, the volume of the waste streams becomes more manageable, and various treatment options can be considered (Chávez-Crooker and Obreque-Contreras 2010). Solid and dissolved wastes can be efficiently collected and more efficiently used for bioenergy production. About 3–4 million MT of farmed shrimp is produced globally each year, which is increasing at a rate of 10% per year (FAO 2008) with most of this production taking place in China and Thailand. High-intensity shrimp ponds operated near coastal water columns are facing major environmental challenges, requiring development of best management practices, especially in the United States, Australia, and the Mediterranean (Castine et al. 2013). This requires more effective management of solid and dissolved nutrient waste streams, essentially the same objectives as municipal WWTPs, yet in saline water. The characteristics of land-based aquaculture wastewater such as in shrimp farms are different from municipal wastewater and intensive recirculating aquaculture wastewater (Castine et al. 2013) in terms of nutrient compositions and solid constituents, which require both modified wastewater treatment systems and selective strains for those wastewater characters. In pond systems such as shrimp farms, feeding is not efficient, and 50%–80% of protein can be released to water columns as wastes (Briggs and Funge-Smith 1994; Karakassis et al. 2005). Even though there is no report on biofuel production using shrimp farm wastes so far, it has high potentials because tremendous amount of free nutrients and solid wastes are produced from shrimp farms where less expansive lands are available for microalgae and biofuel production.

26.3.4 Marine Natural Wastes

Proliferation of macroalgae blooms is a worldwide environmental problem in coastal ecosystems (Morand et al. 1990; Fletcher 1996; Blomster et al. 2002; Liu et al. 2009; Yabe et al. 2009). Based on methane production potential, *Laminaria* and *Ulva* have been considered as possible candidate feedstocks for bioenergy applications (Morand et al. 1990). *Ulva* sp. as a drifting and therefore easily harvestable seaweed has attracted more attention as a bioenergy feedstock than any other macroalgae. On the coast of Brittany, France, 50,000 m^3 of drift seaweeds was harvested per year, (Briand 1989) and in Italy, 1–1.2 million MT was collected in a Venice lagoon (Orlandini 1988). Migliore et al. (2012) collected mixtures of *Gracilariopsis longissima* and *Chaetomorpha linum* and sediment from the Orbetello lagoon (Tuscany, Italy). The aim of their study was to determine biogas production through anaerobic digestion in the area where opportunistic seaweeds bloom as an environmental problem. Without pretreatment of sample, they could produce methane with 380 dm^3 kg/VS_{added}. In recent years, there were reports of the world's largest macroalgal bloom in China. The bloom was caused by one species, *Ulva prolifera*, successively from 2007 to 2011 (Liu et al. 2009; Zhao et al. 2013). In late June 2008, a massive green tide by *Ulva prolifera* covered about 400 km^2 of the Qingdao, China coastline and over 1×10^6 MT of wet macroalgae was removed for the cleanup. This corresponded to about 1.4×10^5 MT dry weight based on the calculation of Kamer and Fong (2000). It has been reported that *Ulva* biomass can be converted into methane through anaerobic digestion within the range of 180–330 L CH_4/kg (Habig et al. 1984; Briand and Morand 1997; Bruhn et al. 2011). Based on previous research, thus, one time of cleanup on the Qingdao coast could produce methane with the volume of 25.1–47.1 billion liters. These results may indicate that anaerobic digestion of waste macroalgae can be a useful approach in local applications for eutrophic coast lagoons where seasonal or more frequent algal blooms typically occur. The use of locally produced bioenergy from marine waste would be beneficial for maritime countries. The potential for biofuel from fish waste is a function of the location and abundance of the waste material. Based on the world production in 2010 (FAO 2013), China accounted for 35% of total fishery production and 61% of aquaculture production. In these aspects, China has huge potential for bioenergy production using captured fisheries and aquaculture wastes.

Based on current technologies, algal cultivation for biofuel production alone is unlikely to be economically viable or provide a positive energy return. Dual-use microalgal cultivation for wastewater treatment coupled with biofuel generation is therefore a more attractive option in terms of reducing the energy cost, GHG emissions, and the nutrient (fertilizer) and freshwater resource costs of biofuel generation from microalgae.

26.4 Conclusions

Marine environments have been continuously challenged by human activities such as waste disposal and fishery and aquaculture activities. In recent years, we have seen a huge interest in biofuel production due to increased concern over the economic and environmental problems associated with the combustion of fossil fuels. To lower production costs and to minimize environmental impacts, biofuel production using waste streams has been attracting more supporters. Traditionally, bioenergy production using common wastewater was seen as a more practical strategy, which served the dual function of fuel production and environmental remediation. However, bioenergy production using marine waste streams has not attracted much serious study, even though this sector has tremendous potential. Algae have attracted a great deal of interest and investment in recent years as a potential feedstock for biofuel and bioenergy production, due to their high productivity and ease of cultivation on marginal lands using waste inputs. Marine wastes obtained from the global fishery and aquaculture industries present a tremendous opportunity to produce bioenergy as well as huge environmental challenges. Solid wastes from fishery industries have been tested for biogas and biodiesel production. To reduce marine environmental effects caused by aquaculture, IMTA has been developed at a pilot scale and is operating on a commercial scale in some sectors. In this system, seaweed coproduction occurs through uptake and assimilation of excess nutrients that drift with pervading currents from finfish production pens. The cultivated seaweed can then potentially be used as a feedstock for the production of biomethane or bioethanol. Marine natural wastes such as macroalgal blooms may also be a useful source of biomass for bioenergy production, even though this is a highly seasonal phenomenon. Based on current reviews, bioenergy production using waste sources could be of great benefit to the overall health of marine environments, as well as to the fisheries and aquaculture sectors looking for additional revenue streams.

Acknowledgment

The authors wish to acknowledge the support from the National Research Council of Canada through the Algal Carbon Conversion Flagship Program.

References

Alberta NA, Aryee R, van de Voort F, Simson BK (2009) FTIR determination of free fatty acids in fish oils intended for biodiesel production. *Process Biochem* 44:401–405.

An JY, Sim SJ, Lee JS, Kim BW (2003) Hydrocarbon production from secondary treated piggery wastewater by the green alga *Botryococcus braunii*. *J Appl Phycol* 15:185–191.

Anderson O, Weinbach JE (2010) Residual animal fat and fish for biodiesel production. Potential in Norway. *Biomass Biotechnol* 34:1183–1188.

Arruda LF, Borghesi R, Oetterer M (2007) Use fish waste as silage—A review. *Braz Arch Biol Technol* 50:879–886.

Banat I, Puskas K, Esen I, Daher RA (1990) Wastewater treatment and algal productivity in an integrated ponding system. *Biol Waste* 32:265–275.

Barrington K, Chopin T, Robinson S (2009) Integrated multi-trophic aquaculture (IMTA) in marine temperate waters. In D Soto (ed.) *Integrated Mariculture: A Global Review*. FAO Fisheries and Aquaculture Technical Paper No. 529. FAO, Rome, Italy, pp. 7–46.

Beneman JR (2003) Biofixation of CO_2 and greenhouse gas abatement with microalgae—Technology roadmap. Report No. 7010000926 prepared for the U.S. Department of Energy National Energy Technology Laboratory, Morgantown, WV.

Beneman JR, Miyamoto K, Hallenbeck PC (1980) Bioengineering aspects of biophotolysis. *Enzyme Microbial Technol* 2:103–111.

Bjornsson WJ, Nicol RW, Dickinson KE, McGinn PJ (2013) Anaerobic digestates are useful nutrient sources for microalgae cultivation: Functional coupling of energy and biomass production. *J Appl Phycol* 25:1523–1528.

Blomster J, Back S, Fewer DP et al. (2002) Novel morphology in *Enteromorpha* (Ulvophyceae) forming green tides. *Am J Bot* 89:1756–1763.

Bouallagui H, Lahdheb H, Ben Romdan E, Rachdi B, Hamdi M (2009) Improvement of fruit and vegetable waste anaerobic digestion performance and stability with co-substrates addition. *J Environ Manage* 90:1844–1849.

Briand X (1989) Prolifération de l'algue verte Ulva sp. en Baie de Lannion par fermentation anaérobie. *Th. 3e cycle: Biologie et Physiologie végétale*. Lille, France, 210pp.

Briand X, Morand P (1997) Anaerobic digestion of *Ulva* sp. 1. Relationship between *Ulva* composition and methanisation. *J Appl Phycol* 9:511–524.
Briggs MRP, Funge-Smith SJ (1994) A nutrient budget of some intensive marine shrimp ponds in Thailand. *Aquacult Fish Manag* 25:789–811.
Bruhn A, Dahl J, Nielsen HB, Nikolaisen L, Rasmussen MB, Markager S, Olsen B, Arias C, Jensen PD (2011) Bioenergy potential of *Ulva lactuca*: Biomass yield, methane production and combustion. *Bioresour Technol* 102:2595–2604.
Bureau DP, Hua K (2010) Towards effective nutritional management of waste outputs in aquaculture, with particular reference to salmonid aquaculture operations. *Aquacult Res* 41:777–792.
Buschmann AH, Troell M, Kautsky N, Kautsky L (1996) Integrated tank cultivation of salmonids and *Gracilaria chilensis* (Gracilariales Rhodophyta). *Hydrobiologia* 326/327:75–82.
Cabanelas ITD, Arbib Z, Chinalia FA, Souza CO, Perales JA, Almeida PF, Druzian JI, Nascimento IA (2013) From waste to energy: Microalgae production in wastewater and glycerol. *Appl Energy* 109:283–290.
Callaghan FJ, Wase DAJ, Thayanithy K, Forster CF (1999) Co-digestion of waste organic solids: Batch studies. *Bioresour Technol* 67:117–122.
Carey RO, Migliaccio KW (2011) Contribution of wastewater treatment plant effluents to nutrient dynamics in aquatic systems: A review. *Environ Manag* 44:205–217.
Castine SA, McKinnon AD, Paul NA, Trott LA, Nys R (2013) Wastewater treatment for land-based aquaculture: Improvements and value-adding alternatives in model systems from Australia. *Aquacult Environ Interact* 4:285–300.
Cerón García MC, Fernández Sevilla JM, Acién Fernández AG, Molina Grima E, García Camacho F (2000) Mixotrophic growth of *Phaeodactylum tricornutum* on glycerol: Growth rate and fatty acid profile. *J Appl Phycol* 12:239–248.
Chavez-Crooker P, Obreque-Contreras J (2010) Bioremediation of aquaculture wastes. *Curr Opin Biotechnol* 21:313–317.
Chinnasamy S, Bhatnagar A, Hunt RW, Das KC (2010) Microalgae cultivation in a wastewater dominated by carpet mill effluents for biofuel applications. *Bioresour Technol* 101:3097–3105.
Chiou B, El-Mashad HM, Avena-Bustilos RJ et al. (2006) Rheological and thermal properties of salmon processing by-products. ASABE Paper No. 066157.
Chopin T (2011) Progression of the integrated multi-trophic aquaculture (IMTA) concept and upscaling of IMTA systems towards commercialization. *Aquacult Eur* 36(4):5–12.
Chynoweth DP (2005) Renewable biomethane from land and ocean energy crops and organic wastes. *Hortscience* 40:283–286.
Craggs RJ, McAuley PJ, Smith VJ (1997) Wastewater nutrient removal by marine microalgae grown on a corrugated raceway. *Water Res* 31:1701–1707.
Craggs RJ, Smith VJ, McAuley PJ (1995) Wastewater nutrient removal by marine microalgae cultured under ambient conditions in mini-ponds. *Water Sci Technol* 31:151–160.
Eiroa M, Costa JC, Alves MM, Kennes C, Veiga MC (2012) Evaluation of the biomethane potential of solid fish waste. *Waste Manage* 32:1347–1352.
FAO—Food and Agriculture Organization (2006) The state of world fisheries and aquaculture. Food and Agriculture Organization, Rome, Italy.
FAO—Food and Agriculture Organization (2008) Food outlook, global market analysis. Food and Agriculture Organization, Rome, Italy.
FAO—Food and Agriculture Organization (2013) Statistics-fisheries and aquaculture. Food and Agriculture Organization, Rome, Italy.
Fletcher RL (1996) The occurrence of "green tides": A review. In W Schramm, PH Nienhuis (eds.) *Marine Benthic Vegetation: Recent Changes and the Effects of Eutrophication*. Springer, Berlin, Germany, pp. 7–43.
Frigon JC, Matteau-Lebrun F, Abdou RH, McGinn PJ, O'Leary SJB, Guiot SR (2013) Screening microalgae strains for their productivity in methane following anaerobic digestion. *Appl Energy* 108:100–107.

Goldman JC (1975) Nutrient transformations in mass cultures of marine algae. *Journal of the Environmental Engineering Division* 101:351–364.
Goldman JC, Stanley HI (1974) Relative growth of different species of marine algae in wastewater-seawater mixtures. *Mar Biol* 28:17–25.
Gonzalez LE, Canizares RO, Baena S (1997) Efficiency of ammonia and phosphorous removal from a Colombian agroindustrial wastewater by the microalgae *Chlorella vulgaris* and *Scenedesmus dimorphus*. *Bioresour Technol* 60:259–261.
Haas MJ, McAloon AJ, Yee WC, Fioglia TA (2006) A process model to estimate biodiesel production costs. *Bioresourc Technol* 97:671–678.
Habig C, DeBusk TA, Ryther JH (1984) The effect of nitrogen content on methane production by the marine algae *Gracilaria tikvahiae* and *Ulva* sp. *Biomass* 4:239–251.
Heubeck S, Craggs RJ, Shilton A (2007) Influence of CO_2 scrubbing from biogas on the treatment performance of a high rate algal pond. *Water Sci Technol* 55:193–200.
Jiang L, Luo S, Fan X, Yang Z, Guo R (2011) Biomass and lipid production of marine microalgae using municipal wastewater and high concentration of CO_2. *Appl Energy* 88:3336–3341.
Kafle GK, Kim SH, Sung KI (2013) Ensiling of fish industry waste for biogas production: A lab scale evaluation of biochemical methane potential (BMP) and kinetics. *Bioresour Technol* 127:326–336.
Kamer K, Fong P (2000) A fluctuating salinity regime mitigates the negative effects of reduced salinity on the estuarine macroalga, *Enteromorpha intestinalis* (L) link. *J Exp Mar Biol Ecol* 254:53–69.
Karakassis I, Pitta P, Krom MD (2005) Contribution of fish farming to the nutrient loading of the Mediterranean. *Sci Mar* 69:313–321.
Kim JY, Lee SM, Lee JW (2012) Biogas production from moon jellyfish (*Aurelia aurita*) using of the anaerobic digestion. *J Ind Eng Chem* 18:2147–2150.
Knothe G (2006) Analyzing biodiesel: Standards and other methods. *J Am Oil Chem Soc* 83:823–833.
Kong QX, Li L, Martinez B, Chen P, Ruan R (2010) Culture of microalgae *Chlamydomonas reinhardtii* in wastewater for biomass feedstock production. *Appl Biochem Biotechnol* 160:9–18.
Krom MD, Ellner S, van Rijin J, Neori A (1995) Nitrogen and phosphorous cycling and transformations in a prototype "non-polluting" integrated mariculture system, Eilat, Israel. *Mar Ecol Prog Ser* 118:25–36.
Li Y, Horsman M, Wu N, Lan CQ, Dubois-Calero N (2008) Biofuels from microalgae. *Biotechnol Prog* 24:815–820.
Li Y, Zhou W, Hu B, Min M, Chen P, Ruan RR (2011) Integration of algae cultivation as biodiesel production feedstock with municipal wastewater treatment: Strains screening and significance evaluation of environmental factors. *Bioresour Technol* 102:10861–10867.
Liang Y (2013) Producing liquid transportation fuels from heterotrophic microalgae. *Appl Energy* 104:860–868.
Liang Y, Sarkany N, Cui Y (2009) Biomass and lipid productivities of *Chlorella vulgaris* under autotrophic, heterotrophic and mixotrophic growth conditions. *Biotechnol Lett* 31:1043–1049.
Lin CY, Li RJ (2009) Fuel properties of biodiesel produced from the crude fish oil from the soapstock of marine fish. *Fuel Process Technol* 90:130–136.
Liu D, Keesing JK, Xing Q, Shi P (2009) World's largest macroalgal bloom caused by expansion of seaweed aquaculture in China. *Mar Pollut Bull* 58:888–895.
Martinez ME, Sanchez S, Jimenez JM, El Yousfi F, Munoz L (2000) Nitrogen and phosphorus removal from urban wastewater by the microalga *Scenedesmus obliquus*. *Bioresour Technol* 73:263–272.
McGinn PJ, Dickinson KE, Park KC, Whitney CG, MacQuarrie SP, Black FJ, Frigon JC, Guiot SR, O'Leary SJ (2012) Assessment of the bioenergy and bioremediation potentials of the microalga *Scenedesmus* sp. AMDD cultivated in municipal wastewater effluent in batch and continuous mode. *Algal Res* 1:155–165.
Migliore G, Alisi C, Sprocati AR, Massi E, Ciccoli R, Lenzi M, Wang A (2012) Anaerobic digestion of macroalgal biomass and sediments sourced from the Orbetello Lagoon, Italy. *Biomass Bioenergy* 42:69–77.

Morand P, Charlier RH, Mazé J (1990) European bioconversion projects and realizations for macroalgal biomass: Saint-Cast-Le-Guildo (France) experiment. *Hydrobiologia* 204/205:301–308.
Mshandete A, Kivais A, Rubindamayugi M, Mattiasson B (2004) Anaerobic batch co-digestion of sisal pulp and fish wastes. *Bioresour Technol* 95:19–24.
Neori A (1996) The form of N-supply (ammonia or nitrate) determines the performance of seaweed biofilters integrated with intensive fish culture. *Isr J Aquac Bamidgeh* 48:19–27.
Neori A, Chopin T, Troell M, Buschmann AH, Kraemer GP, Halling C, Shpigel M, Yarish C (2004) Integrated aquaculture: Rationale, evolution and state of the art emphasizing seawater biofiltration in modern mariculture. *Aquaculture* 231:361–391.
Neori A, Shpigel M, Ben-Ezra D (2000) A sustainable integrated system for culture of fish, seaweed and abalone. *Aquaculture* 186:279–291.
Ogawa T, Aiba S (1981) Bioenergetic analysis of mixotrophic growth in *Chlorella vulgaris* and *Scenedesmus acutus*. *Biotechnol Bioeng* 23:1121–1132.
Ogbonna JC, Tanaka H (1998) Cyclic autotrophic/heterotrophic cultivation of photosynthetic cells: A method of achieving continuous cell growth under light/dark cycles. *Bioresour Technol* 65:65–72.
Öner C, Altun S (2009) Biodiesel production from inedible animal tallow and an experimental investigation of its use as alternative fuel in a direct injection diesel engine. *Appl Energy* 27:173–181.
Orlandini M (1988) Harvesting of algae in polluted lagoons of Venice and Orbetello and their effective and potential utilization. In J de Waart, PH Nienhuis (eds.) *Aquatic Primary Biomass (Marine macroalgae): Biomass Conversion, Removal and Use Nutrients. II. Proceedings of the Second Workshop of the COST 48 Sub-Group 3*. Zeist and Yerseke, the Netherlands. October 25–27, 1988. E.E.C., Brussels, Belgium, pp. 20–23.
Órpez R, Martínez E, Hodaifa G, Yousfi FE, Jbari N, Sánchez S (2009) Growth of the microalga *Botryococcus braunii* in secondarily treated sewage. *Desalination* 246:625–630.
Oswald WJ (1988) Micro-algae and waste-water treatment. In MA Borowitzka, LJ Borowitzka (eds.) *Micro-Algal Biotechnology*. Cambridge University Press, Cambridge, U.K., pp. 305–328.
Oswald WJ (2003) My sixty years in applied algology. *J Appl Phycol* 15:99–106.
Oswald WJ, Gotaas HB, Golueke CG, Kellen WR (1957) Algae in waste treatment. *Sewage Ind Wastes* 29:437–455.
Park JBK, Craggs RJ (2010) Wastewater treatment and algal production in high rate algal ponds with carbon dioxide addition. *Water Sci Technol* 61:633–639.
Park JBK, Craggs RJ, Shilton AN (2011) Wastewater treatment high rate algal ponds for biofuel production. *Bioresour Technol* 102:35–42.
Park KC, McGinn P, Kadota P (2012a) The potential for integrating algal carbon capture, wastewater nutrient removal and the production of bioenergy and bioproducts by testing municipal wastewater from the Annacis Island Wastewater Treatment Plant. *62th Canadian Chemical Engineering Conference*, Vancouver, British Columbia, Canada, October 14–17, 2012.
Park KC, Whitney C, Kozera C, O'Leary S, McGinn P (2014) Seasonal isolation of microalgae from municipal wastewater for remediation and biofuel applications. *Appl Microbiol Biotechnol* (submitted).
Park KC, Whitney C, McNichol J, Dickinson K, MacQuarrie S, Skrupski BP, Zhou J, Wilson K, O'Leary SJB, McGinn PJ (2012b) Mixotrophic and photoautotrophic cultivation of fourteen microalgae isolates from Saskatchewan, Canada: Potential applications for wastewater remediation for biofuel production. *J Appl Phycol* 24:339–348.
Perez-Garcia O, Bashan Y, Puente ME (2001) Organic carbon supplementation of sterilized municipal wastewater is essential for heterotrophic growth and removing ammonium by the microalga *Chlorella vulgaris*. *J Phyol* 47:190–199.
Pittman JK, Dean AP, Osundeku O (2011) The potential of sustainable algal biofuel production using wastewater resources. *Bioresour Technol* 102:17–25.
Rawat I, Kumar RR, Mutanda T, Bux F (2011) Dual role of microalgae: Phycoremediation of domestic wastewater and biomass production for sustainable biofuels production. *Appl Energy* 88:3411–3424.

Ren J, Stenton-Dozey J, Plew DR, Fang J, Gall M (2012) An ecosystem model for optimizing production in integrated multitrophic aquaculture systems. *Ecol Model* 246:34–46.
Ruiz-Marin A, Mendoza-Espinosa LG, Stephenson T (2010) Growth and nutrient removal in free and immobilized green algae in batch and semi-continuous cultures treating real wastewater. *Bioresour Technol* 101:58–64.
Ryther JH, Debusk TA, Blakeslee M (1984) Cultivation and conversion of marine macroalgae (*Gracilaria* and *Ulva*). In SERI/STR-231-2360, Solar Energy Research Institute. Golden, CO, pp. 1–88.
Ryther JH, Dunstan WM, Tenore KR, Huguenin JE (1972) Controlled eutrophication-increasing food production from the sea by recycling human wastes. *Biosciences* 22:144–152.
Schneider O, Sereti V, Eding EH, Verreth JAJ (2005) Analysis of nutrient flows in integrated intensive aquaculture systems. *Aquacult Eng* 32:379–401.
Sheehan J, Dunahay T, Benemann J, Roessler P (1998) A look back at the U.S. Department of Energy's Aquatic Species Program—Biodiesel from algae. National Renewable Energy Program. NREL TP-580-24190, National Renewable Energy Laboratory, Golden, CO.
Shpigel M, Lee J, Soohoo B, Fridman R, Gordin H (1993a) The use of effluent water from fish ponds as a food source for the Pacific oyster *Crassostrea gigas* Tunberg. *Aquac Fish Manage* 24:529–543.
Shpigel M, Neori A, Popper DM, Gordin H (1993b) A proposed model for "environmentally clean" land-based culture of fish, bivalves and seaweeds. *Aquaculture* 117:115–128.
Singh A, Olsen SI (2011) A critical review of biochemical conversion, sustainability and life cycle assessment of algal biofuels. *Appl Energy* 88:3548–3555.
Skriptsova AV, Miroshnikova N (2011) Laboratory experiment to determine the potential of two macroalgae from the Russian Far-East as biofilters for integrated multi-trophic aquaculture (IMTA). *Bioresour Technol* 102:3149–3154.
Steigers JA (2002) Demonstrating the use of fish oil as fuel in a large stationary diesel engine. In PJ Bechtel (ed.) *Advances in Sea Food Byproducts: 2002 Conference Proceedings*. Alaska Sea Grant, Fairbank, AK, pp. 187–200.
Sydney EB, da Silva TE, Tokarski A et al. (2011) Screening of microalgae with potential for biodiesel production and nutrient removal from treated domestic sewage. *Appl Energy* 88:3291–3294.
Tanoi T, Kawachi M, Watanabe MM (2011) Effects of carbon source on growth and morphology of *Botryococcus braunii*. *J Appl Phycol* 23:25–33.
Tonjes DJ, Greene KL (2012) A review of national municipal solid waste generation assessments in the USA. *Waste Manag Res* 30:758–771.
UNO (2009) *The United Nations World Water Development Report 3. Water in a Changing World*, Istanbul, Turkey.
Viriya-empikul N, Krasae P, Puttasawat B, Yoosuk B, Chollacoop N, Faungnawakij K (2010) Waste shells of mollusk and egg as biodiesel production catalysts. *Bioresour Technol* 101: 3765–3767.
Wan M, Liu P, Xia J, Rosenberg JN, Oyler GA, Betenbaugh MJ, Nie Z, Qiu G (2011) The effect of mixotrophy on microalgal growth, lipid content, and expression levels of three pathway genes in *Chlorella sorokiniana*. *Appl Microbiol Biotechnol* 91:835–844.
Wells CD (2005) Tertiary treatment in integrated algal ponding systems. Master of Science thesis, Department of Biotechnology, Rhodes University, Grahamstown, South Africa.
Wijffels R, Barbosa M (2010) An outlook on microalgal biofuels. *Science* 329:796–799.
Wilkie AC, Mulbry WW (2002) Recovery of dairy manure nutrients by benthic freshwater algae. *Bioresour Technol* 84:81–91.
Yabe T, Sshii Y, Amano Y et al. (2009) Green tide formed by free-floating *Ulva* spp. at Yatsu tidal flat. *Jpn Limnol* 10:239–245.
Yahyaee R, Ghobadian B, Najafi G (2013) Waste fish oil biodiesel as a source of renewable fuel in Iran. *Renew Sust Energ Rev* 17:312–319.
Yang L, Zhang A, Zheng X (2009) Shrimp shell catalyst for biodiesel production. *Energy Fuels* 23:3859–3865.

Zhang ED, Wang B, Wang QH, Zhang SB, Zhao BD (2008) Ammonia-nitrogen and orthophosphate removal by immobilized *Scenedesmus* sp isolated from municipal wastewater for potential use in tertiary treatment. *Bioresour Technol* 99:3787–3793.

Zhao J, Jiang P, Liu ZY, Wei W, Lin HZ, Li FC, Wang JF, Qin S (2013) The Yellow Sea green tides were dominated by one species, *Ulva* (*Enteromorpha*) *prolifera*, from 2007 to 2011. *Chin Sci Bull* 58:2298–2302.

Zhou W, Li Y, Min M, Hu B, Chen P, Ruan R (2011) Local bioprospecting for high-lipid producing microalgal strains to be grown on concentrated municipal wastewater for biofuel production. *Bioresour Technol* 102:6909–6919.

27

Algal Biosorption of Heavy Metals from Wastes

Ozcan Konur

CONTENTS

27.1 Introduction

27.1.1 Issues

Global warming, air pollution, and energy security have been some of the most important public policy issues in recent years (Jacobson 2009; Wang et al. 2008; Yergin 2000). With increasing global population, food security has also become a major public policy issue (Lal 2004). The development of biofuels generated from biomass has been a long awaited solution to these global problems (Demirbas 2007; Goldemberg 2007; Lynd et al. 1991). However, the development of early biofuels produced from agricultural plants such as sugar cane (Goldemberg 2007) and agricultural wastes such as corn stovers (Bothast and Schlicher 2005) has resulted in a series of substantial concerns about food security (Godfray et al. 2010). Therefore, the development of algal biofuels as a third-generation biofuel has been considered as a major solution for the problems of global warming, air pollution, energy security, and food security (Chisti 2007; Demirbas 2007; Kapdan and Kargi 2006; Spolaore et al. 2006; Volesky and Holan 1995).

On the other hand, besides the development of algal biofuels, the sorption of heavy metals from wastes has also been a major research area in recent years (Bailey et al. 1999; Cervantes et al. 2001; Davis et al. 2003; Fein et al. 1997; Guibal 2004; Kapoor et al. 1999; Kratochvil and Volesky 1998; Mohan and Singh 2002; Reddad et al. 2002; Volesky 2001; Volesky and Holan 1995). Bailey et al. (1999) report some of the highest adsorption capacities for cadmium, chromium, lead, and mercury as 1587 mg Pb/g lignin, 796 mg Pb/g chitosan, 1123 mg Hg/g chitosan, 1000 mg Hg/g CPEI cotton, 92 mg Cr(III)/g chitosan, 76 mg Cr(III)/g pear, 558 mg Cd/g chitosan, and 215 mg Cd/g seaweed.

As Volesky and Holan (1995) point out, the potential of metal biosorption by biomass materials has been well established. For economic reasons, of particular interest are abundant biomass types generated as standard waste by-product of large-scale industrial fermentations or certain metal-binding algae found in large quantities in the sea. These biomass types serve as a basis for newly developed metal biosorption processes foreseen particularly as a very competitive means for the detoxification of metal-bearing industrial effluents. The algae have also been important in the absorption of the heavy metals and other materials from the wastes and dyes (e.g., Davis et al. 2003; Juhasz and Naidu 2000; Volesky and Holan 1995).

Although there have been many reviews on the use of marine algae as biosorbents (e.g., Davis et al. 2003; Juhasz and Naidu 2000; Volesky and Holan 1995) and a number of scientometric studies on algal biofuels (Konur 2011), there has not been any study on the citation classics in the algal biosorption of heavy metals from wastes as in the other research fields (Baltussen and Kindler 2004a,b; Dubin et al. 1993; Gehanno et al. 2007; Konur 2012a,b, 2013; Paladugu et al. 2002; Wrigley and Matthews 1986).

As North's new institutional theory suggests, it is important to have up-to-date information about current public policy issues to develop a set of viable solutions to satisfy the needs of all the key stakeholders (Konur 2000, 2002a,c, 2006a,b, 2007a,b, 2012c; North 1994).

Therefore, brief information on a selected set of citation classics on the algal biosorption of heavy metals from wastes is presented in this chapter to inform key stakeholders relating to the global problems of warming, air pollution, food security, and energy security about the use of the algae as biosorbents for the solution of these problems in the long run, thus complementing a number of recent scientometric studies on the biofuels and global energy research (Konur 2011, 2012d–p).

27.1.2 Methodology

A search was carried out in the Science Citation Index Expanded (SCIE) and Social Sciences Citation Index (SSCI) databases (version 5.11) in September 2013 to locate the papers relating to the algal biosorption of heavy metals from wastes using the keyword set of Topic = (biosorpt* or sorpt* or "bio* sorpt*" or waste* or bioremoval or "bio* removal" or bioremediation or "bio* remediation" or Sorbent* or biosorbent* or "bio* sorbent*") AND Topic = (alga* or "macro* alga*" or "micro* alga*" or macroalga* or microalga* or cyanobacter* or seaweed* or diatoms or sea* weed* or reinhardtii or braunii or chlorella or sargassum or gracilaria or spirulina) in the abstract pages of the papers. For this chapter, it was necessary to embrace the broad algal search terms to include diatoms, seaweeds, and cyanobacteria as well as to include many spellings of biosorption rather than the simple keywords of biosorption and algae.

It was found that there were 5700 papers indexed between 1980 and 2013. The key subject categories for the algal research were environmental sciences (2251 references, 38.9%), biotechnology applied microbiology (1248 references, 21.6%), environmental engineering (1187 references, 20.5%), and water resources (862 references, 14.9%), altogether comprising 66% of all the references on the algal biosorption of heavy metals from wastes. It was also necessary to focus on the key references by selecting articles and reviews.

The located highly cited 50 papers were arranged by decreasing number of citations for two groups of papers: 14 and 36 papers for nonexperimental research and experimental research, respectively. In order to check whether these collected abstracts correspond to the larger sample on these topical areas, new searches were carried out for each topical area.

The summary information about the located citation classics is presented under two major headings of nonexperimental and experimental researches in the order of the decreasing number of citations for each topical area.

The information relating to the document type, the affiliation and location of the authors, journal, indexes, subject area of the journal, concise topic, total number of citations received, and total average number of citations received per year are given in the tables for each paper.

27.2 Nonexperimental Studies on Algal Biosorption of Heavy Metals from Wastes

27.2.1 Introduction

The nonexperimental research on algal biosorption of heavy metals from wastes has been one of the most dynamic research areas in recent years. Fifteen citation classics in the field with more than 153 citations were located, and the key emerging issues from these papers were presented by decreasing order of the number of citations (Table 27.1). These papers give strong hints about the determinants of efficient algal biosorption of heavy metals from wastes and emphasize that marine algae are efficient sorbents in comparison with the traditional sorbents.

The papers were dominated by the researchers from 10 countries, usually through the intracountry institutional collaboration, and they were multiauthored. Canada (six papers) and Australia and India (two papers each) were the most prolific countries. Similarly, McGill University and Volesky were the most prolific institution and author, respectively.

Similarly, all these papers were published in the journals indexed by the Science Citation Index (SCI) and/or SCIE. There were papers indexed by the SSCI or Arts & Humanities Citation Index (A&HCI). The number of citations ranged from 153 to 924, and the average number of citations per annum ranged from 8.5 to 57.9. Five of the papers were articles, while nine of them were reviews.

27.2.2 Nonexperimental Research

Volesky and Holan (1995) discuss the biosorption of heavy metals in an early review paper with 924 citations. They note that algal and other biomass types serve as a basis for newly developed metal biosorption processes foreseen particularly as a very competitive means for the detoxification of metal-bearing industrial effluents. In this context, lead and cadmium have been effectively removed from very dilute solutions by the dried biomass of some ubiquitous species of brown marine algae such as *Ascophyllum* and *Sargassum*, which accumulate more than 30% of biomass dry weight in the metal. Biosorption isotherm curves, derived from equilibrium batch sorption experiments, are used in the evaluation of metal uptake by different biosorbents. Further studies are focusing on the assessment of biosorbent performance in dynamic continuous-flow sorption systems, while new methodologies are being developed that are aimed at mathematical modeling of biosorption

TABLE 27.1

Nonexperimental Research on Algal Biosorption

No.	Paper References	Year	Document	Affiliation	Country	No. of Authors	Journal	Index	Subjects	Topic	Total No. of Citations	Total Average Citations per Annum	Rank
1	Volesky and Holan	1995	R	McGill Univ.	Canada	2	*Biotechnol. Prog.*	SCI, SCIE	Biot. Appl. Microb., Food Sci. Tech.	Algal biosorption	924	48.6	1
2	Davis et al.	2003	R	McGill Univ.	Canada	3	*Water Res.*	SCI, SCIE	Eng. Env., Env. Sci., Water. Res.	Algal biosorption	637	57.9	2
3	Juhasz and Naidu	2000	R	CSIRO Land and Water	Australia	2	*Int. Biodeterior. Biodegrad.*	SCIE	Biot. Appl. Microb., Env. Sci.	Algal biosorption	319	22.8	4
4	Wang and Chen	2009	R	Tsinghua Univ.	China	2	*Biotechnol. Adv.*	SCI, SCIE	Biot. Appl. Microb.	Algal biosorption	263	52.6	8
5	Ahluwalia and Goyal	2007	R	Thapar Inst. Engn. Technol.	India	2	*Bioresour. Technol.*	SCI, SCIE	Biot. Appl. Microb., Ener. Fuels	Algal biosorption	263	37.6	9
6	Malik	2004	R	Utsunomiya Univ.	Japan	1	*Environ. Int.*	SCI, SCIE	Env. Sci.	Algal biosorption	238	23.8	11
7	Volesky	1994	A	BV Sorbex, Inc.	Canada	1	*FEMS Microbiol. Rev.*	SCI, SCIE	Microb.	Algal biosorption	230	11.5	13
8	Wilde and Benemann	1993	R	Westinghouse Savannah River Co.	United States	2	*Biotechnol. Adv.*	SCI, SCIE	Biot. Appl. Microb.	Algal biosorption	220	10.5	14

(*Continued*)

TABLE 27.1 (*Continued*)
Nonexperimental Research on Algal Biosorption

No.	Paper References	Year	Document	Affiliation	Country	No. of Authors	Journal	Index	Subjects	Topic	Total No. of Citations	Total Average Citations per Annum	Rank
9	Schiewer and Volesky	1995	A	McGill Univ.	Canada	2	*Environ. Sci. Technol.*	SCI, SCIE	Eng. Env., Env. Sci.	Algal biosorption	186	9.8	19
10	Figueira et al.	2000	A	McGill Univ., Univ. Fed. Minas Gerais, RMIT	Canada, Brazil, Australia	4	*Water Res.*	SCI, SCIE	Eng. Env., Env. Sci., Water. Res.	Algal biosorption	172	12.3	22
11	Chong and Volesky	1995	A	McGill Univ.	Canada	2	*Biotechnol. Bioeng.*	SCI, SCIE	Biot. Appl. Microb.	Algal biosorption	161	8.5	26
12	Chojnacka et al.	2005	A	Wroclaw Univ. Technol.	Poland	3	*Chemosphere*	SCI, SCIE	Env. Sci.	Algal biosorption	157	17.4	27
13	Mehta and Gaur	2005	R	Banaras Hindu Univ.	India	2	*Crit. Rev. Biotechnol.*	SCI, SCIE	Biot. Appl. Microb.	Algal biosorption	155	17.2	29
14	Munoz and Guieysse	2006	R	Lund Univ., Univ. Valladolid	Sweden, Spain	2	*Water Res.*	SCI, SCIE	Eng. Env., Env. Sci., Water. Res.	Algal biosorption	153	19.1	32

SCI, Science Citation Index; SCIE, Science Citation Index Expanded; SSCI, Social Sciences Citation Index; A, article; R, review.

systems and their effective optimization. They argue that "elucidation of mechanisms active in metal biosorption is essential for successful exploitation of the phenomenon and for regeneration of biosorbent materials in multiple reuse cycles." They focus on the composition of marine algae polysaccharide structures, which are instrumental in metal uptake and binding.

Davis et al. (2003) discuss the biochemistry of heavy metal biosorption by brown algae in a review paper with 637 citations. They note that brown algae have been the most effective and promising substrates for metal biosorption. They argue that it is the properties of cell wall constituents, such as alginate and fucoidan, which are chiefly responsible for heavy metal chelation. They outline the biochemical properties of brown algae. They describe the macromolecular conformation of the alginate biopolymer in order to explain the heavy metal selectivity displayed by brown algae. They evaluate the role of cellular structure, storage polysaccharides, cell wall, and extracellular polysaccharides in terms of their potential for metal sequestration. They then discuss the binding mechanisms including the key functional groups involved and the ion-exchange process. They argue that "quantification of metal-biomass interactions is fundamental to the evaluation of potential implementation strategies and they review sorption isotherms, ion-exchange constants, as well as models used to characterize algal biosorption." They finally evaluate the sorption behavior (i.e., capacity, affinity) of brown algae with various heavy metals and their relative performance.

Juhasz and Naidu (2000) discuss the bioremediation of high-molecular-weight polycyclic aromatic hydrocarbons (PAHs) with a focus on the microbial degradation of benzo[a] pyrene (BaP) in a review paper with 319 citations. They focus on the high-molecular-weight PAH BaP. There is concern about the presence of BaP in the environment because of its carcinogenicity, teratogenicity, and toxicity. BaP has been observed to accumulate in marine organisms and plants that could indirectly cause human exposure through food consumption. They "provide an outline of the occurrence of BaP in the environment and the ability of bacteria, fungi and algae to degrade the compound, including pathways for BaP degradation by these organisms."

Wang and Chen (2009) discuss biosorbents for heavy metal removal and their future in a review paper with 263 citations. They review biosorbents widely used for heavy metal removal mainly focusing on their cellular structure, biosorption performance, pretreatment, modification, regeneration/reuse, modeling of biosorption (isotherm and kinetic models), evaluation, potential application, and future and the development of novel biosorbents. They evaluate the pretreatment and modification of biosorbents aiming to improve their sorption capacity. They argue that molecular biotechnology is a potent tool to elucidate mechanisms at the molecular level and to construct engineered organisms with higher biosorption capacity and selectivity for the objective metal ions. They then argue that "although the biosorption application is facing a great challenge, there are two trends for the development of the biosorption process for metal removal. One trend is to use hybrid technology for pollutants removal, especially using living cells. Another trend is to develop the commercial biosorbents using immobilization technology, and to improve the biosorption process including regeneration/reuse, making the biosorbents just like a kind of ion exchange resin, as well as to exploit the market with great endeavor."

Ahluwalia and Goyal (2007) discuss microbial and plant-derived biomass for removal of heavy metals from wastewater in a review paper with 263 citations. They note that due to unique chemical composition, biomass sequesters metal ions by forming metal complexes from solution and obviates the necessity to maintain special growth-supporting conditions.

They argue that "biomass of *Aspergillus niger, Penicillium chrysogenum, Rhizopus nigricans, Ascophyllum nodosum, Sargassum natans, Chlorella fusca, Oscillatoria angustissima, Bacillus firmus* and *Streptomyces* sp. have highest metal adsorption capacities ranging from 5 to 641 mg g^{-1} mainly for Pb, Zn, Cd, Cr, Cu and Ni." Biomass generated as a by-product of fermentative processes offers great potential for adopting an economical metal-recovery system. They review the available information on various attributes of utilization of microbial and plant-derived biomass, and they explore the possibility of exploiting them for heavy metal remediation.

Malik (2004) discusses metal bioremediation through growing cells in a review paper with 238 citations. She notes that the strains (bacteria, yeast, and fungi) isolated from contaminated sites possess excellent capability of metal scavenging. She argues that some bacterial strains possess high tolerance to various metals and may be potential candidates for their simultaneous removal from wastes. Evidently, the stage has already been set for the application of metal-resistant growing microbial cells for metal harvesting. She focuses on the applicability of growing bacterial/fungal/algal cells for metal removal and the efforts directed toward cell/process development to make this option technically/economically viable for the comprehensive treatment of metal-rich effluents.

Volesky (1994) discusses advances in biosorption of metals with a focus on the selection of biomass types in an early paper with 230 citations. He notes that ions of lead and cadmium have been found to be bound very efficiently from very dilute solutions by the dried biomass of some ubiquitous brown marine algae such as *Ascophyllum* and *Sargassum* that accumulate more than 30% of biomass dry weight in the metal. The common yeast *Saccharomyces cerevisiae* is a *mediocre* metal biosorbent. He argues that the "construction of biosorption isotherm curves serves as a basic technique assisting in evaluation of the metal uptake by different biosorbents. The methodology is based on batch equilibrium sorption experiments extensively used for screening and quantitative comparison of new biosorbent materials." Experimental methodologies used in the study of biosorption and selected recent research results demonstrate the route to novel biosorbent materials, some of which can even be repeatedly regenerated for reuse.

Wilde and Benemann (1993) discuss the bioremoval of heavy metals by the use of microalgae in a review paper with 220 citations. They note that although microalgae are not unique in their bioremoval capabilities, they offer advantages over other biological materials in some conceptual bioremoval process schemes. They argue that "selected microalgae strains, purposefully cultivated and processed for specific bioremoval applications, have the potential to provide significant improvements in dealing with the worldwide problems of metal pollution." In addition to strain selection, significant advances in the technology appear possible by improving biomass containment or immobilization techniques and by developing bioremoval process steps utilizing metabolically active microalgae cultures. They argue that the latter approach is especially attractive in applications where extremely low levels of residual metal ions are desired. They summarize the current literature, highlighting the potential benefits and problems associated with the development of novel algal-based bioremoval processes for the abatement of heavy metal pollution.

Schiewer and Volesky (1995) model proton–metal ion exchange in biosorption in an early paper with 186 citations. They perform metal ion–binding experiments with continuously controlled pH. They model the metal ion and proton binding at equilibrium as a function of pH and metal ion concentration using a modified multicomponent Langmuir sorption model. They consider both the exchange of metal ions for protons from functional groups in their acidic form and the sorption of metal ions on ionized groups.

They argue that the "model is applicable to adsorption by biomass with free or protonated metal binding sites as well as to metal ion desorption with acids since the direction of the reaction depends simply on the given initial conditions." They incorporate the model parameters into the MINEQL+ equilibrium program, leading to a prediction of the equilibrium, for example, of metal ion–laden biosorbent desorption performance for given initial conditions.

Figueira et al. (2000) discuss the biosorption of metals in brown seaweed biomass in a paper with 172 citations. Biosorption of Cd by biomass of the brown seaweeds *Durvillaea, Laminaria, Ecklonia,* and *Hormosira* presaturated with Ca, Mg, or K was coupled with the release of these light ions. They evaluate the feasibility of biomass pretreatment to develop a better biosorbent by its biosorption performance, by the degree of its component leaching (measured by the weight loss and TOC), as well as by the number of ion-exchange sites remaining in the biomass after the pretreatment. Multicomponent Langmuir and ion-exchange models applied to the equilibrium sorption data for pH 4.5 confirmed the ion-exchange mechanism involved in the biosorption of metals. They find that "both models fitted well the experimental data and their parameters can be used in the derivation of dimensionless ion-exchange isotherms which are instrumental in predicting the behavior of the biosorbents in dynamic flow-through biosorption systems." They assert that the sequence of biomass affinities established for the selected heavy metals can be correlated with the chemical pretreatment of the biomass.

Chong and Volesky (1995) describe the 2-metal biosorption equilibria by Langmuir-type models in an early paper with 165 citations. They examine a biosorbent prepared from *A. nodosum* seaweed biomass, FCAN2, for its sorption capacity. They discussed equilibrium batch sorption studies that were performed using two-metal systems containing either (Cu + Zn), (Cu + Cd), or (Zn + Cd). In the evaluation of the two-metal sorption system performance, simple isotherm curves had to be replaced by 3D sorption isotherm surfaces. In order to describe the isotherm surfaces mathematically, they evaluate three Langmuir-type models. They use the apparent one-parameter Langmuir constant (b) to quantify FCAN2 *affinity* for one metal in the presence of another one. They find that the uptake of Zn decreased drastically when Cu or Cd was present, while the uptake of Cd was much more sensitive to the presence of Cu than to that of Zn. They argue that the "presence of Cd and Zn alter the 'affinity' of FCAN2 for Cu the least at high Cu equilibrium concentrations" and assert that the "mathematical model of the two-metal sorption system enabled quantitative estimation of one-metal (bio)sorption inhibition due to the influence of a second metal."

Chojnacka et al. (2005) discuss the biosorption of Cr^{3+}, Cd^{2+}, and Cu^{2+} ions by blue-green algae *Spirulina* sp. with a focus on the kinetics, equilibrium, and the mechanism of the process in a paper with 157 citations. They investigate quantitatively the potential binding sites present at the surface of *Spirulina* sp., using both potentiometric titrations and adsorption isotherms. The kinetic experiments showed that the process equilibrium was reached quickly, in less than 5–10 min. They find that the equilibrium dependence between biosorption capacity and bulk metal ion concentration could be described with the Langmuir equation. They argue that the mechanism of biosorption is rather chemisorption than physical adsorption and it was further confirmed by the low surface area associated with physical adsorption and by the presence of cations that appeared in the solution after biosorption. The maximum contribution of physical adsorption in the overall biosorption process was evaluated as 3.7%. They propose that "functional groups on the cell surface contributed to the binding of metal ions by a biosorbent via equilibrium reaction" as three functional groups capable of cation exchange were identified on the cell surface. The biomass was described as weak acidic ion exchanger. Since deprotonation of each functional group depends on pH, the process of biosorption is strongly pH dependent. This was confirmed in the biosorption experiments carried out at different pH. The contribution of functional groups in the biosorption process was confirmed by chemical modification of the groups. They find that chemically blocked groups did not show neither biosorption nor ion-exchange capabilities. They assert that "growth conditions can affect the metal adsorption properties of microalgae."

Mehta and Gaur (2005) discuss the use of algae for removing heavy metal ions from wastewater with a focus on the progress and prospects in a review paper with 155 citations. They note that algae can effectively remove metals from multimetal solutions. Dead cells sorb more metal than live cells. Various pretreatments enhance metal sorption capacity of algae. They argue that "$CaCl_2$ pretreatment is the most suitable and economic method for activation of algal biomass." Algal periphyton has great potential for removing metals from wastewaters. An immobilized or granulated biomass-filled column can be used for several sorption/desorption cycles with unaltered or slightly decreased metal removal. They argue that Langmuir and Freundlich models cannot precisely describe metal sorption since they ignore the effect of pH, biomass concentration, etc. For commercial application of algal technology for metal removal from wastewaters, they recommend that emphasis should be given to "(i) selection of strains with high metal sorption capacity, (ii) adequate understanding of sorption mechanisms, (iii) development of low-cost methods for cell immobilization, (iv) development of better models for predicting metal sorption, (v) genetic manipulation of algae for increased number of surface groups or over expression of metal binding proteins, and (vi) economic feasibility."

Munoz and Guieysse (2006) discuss algal–bacterial processes for the treatment of hazardous contaminants in a review paper with 153 citations. They note that when proper methods for algal selection and cultivation are used, it is possible to use microalgae to produce the O_2 required by acclimatized bacteria to biodegrade hazardous pollutants such as PAHs, phenolics, and organic solvents. Well-mixed photobioreactors with algal biomass recirculation are recommended to protect the microalgae from effluent toxicity and optimize light utilization efficiency. The optimal biomass concentration depends mainly on the light intensity and the reactor configuration: At low light intensity, they argue that "the biomass concentration should be optimized to avoid mutual shading and dark respiration whereas at high light intensity, a high biomass concentration can be useful to protect microalgae from light inhibition and optimize the light/dark cycle frequency." Photobioreactors can be designed as open (stabilization ponds or high rate algal ponds) or enclosed (tubular, flat plate) systems. The latter are generally costly to construct and operate but more efficient than open systems. The best configuration to select will depend on factors such as process safety, land cost, and biomass use.

27.2.3 Conclusion

The nonexperimental research on algal biosorption of heavy metals from wastes has been one of the most dynamic research areas in recent years. Fourteen citation classics in the field of algal biosorption of heavy metals from wastes with more than 153 citations were located, and the key emerging issues from these papers were presented by the decreasing order of the number of citations. These papers give strong hints about the determinants of the efficient algal biosorption of heavy metals and other materials from wastes and dyes and emphasize that marine algae are efficient biosorbents in comparison to the traditional sorbents.

27.3 Experimental Studies on Algal Biosorption of Heavy Metals from Wastes

27.3.1 Introduction

The experimental research on algal biosorption of heavy metals and other materials from wastes and dyes has been one of the most dynamic research areas in recent years. Thirty-six experimental citation classics in the field with more than 105 citations were located, and the key emerging issues from these papers were presented by decreasing order of the number of citations (Table 27.2). These papers give strong hints about the determinants of efficient biosorption of heavy metals and other materials from wastes and dyes and emphasize that marine algae are efficient biosorbents in comparison with the traditional sorbents.

The papers were dominated by the researchers from the 15 countries, usually through the intracountry institutional collaboration, and they were multiauthored. Canada (eight papers); Turkey and India (six papers each); Australia, South Korea, and the United States (three papers each); and Brazil and China (two papers each) were the most prolific countries. Similarly, McGill University (seven papers), Hacettepe University and Indian Institute of Technology (five papers each), and Griffith University and Pohang University of Science and Technology (three papers each) were the most prolific institutions.

Similarly, all these papers were published in the journals indexed by the SCI and/or SCIE. There were no papers indexed by the A&HCI or SSCI. The number of citations ranged from 105 to 340, and the average number of citations per annum ranged from 6.4 to 34.6. All of these papers were articles.

27.3.2 Experimental Research

Sheng et al. (2004) study the sorption of lead, copper, cadmium, zinc, and nickel by marine algal biomass (*Sargassum* sp., *Padina* sp., *Ulva* sp., and *Gracilaria* sp.) with a focus on the characterization of biosorptive capacity and investigation of mechanisms in a paper with 340 citations. They find that the biosorption capacities were significantly affected by solution pH, with higher pH favoring higher metal-ion removal. They carry out kinetic and isotherm experiments at the optimal pH: at pH 5.0 for lead and copper and at pH 5.5 for cadmium, zinc, and nickel. They find that the "metal removal rates were rapid, with 90% of the total adsorption taking place within 60 min." Furthermore, "*Sargassum* sp. and *Padina* sp. showed the highest potential for the sorption of the metal ions, with the maximum uptake capacities ranging from 0.61 to 1.16 mmol/g for *Sargassum* sp. and 0.63 to 1.25 mmol/g for *Padina* sp." The general affinity sequence for *Padina* sp. was Pb > Cu > Cd > Zn > Ni, while that for *Sargassum* sp. was Pb > Zn > Cd > Cu > Ni. XPS and Fourier transform infrared (FTIR) analysis of *Sargassum* sp. and *Padina* sp. revealed the chelating character of the ion coordination to carboxyl groups. They assert that "carboxyl, ether, alcoholic, and amino groups are responsible for the binding of the metal ions."

Holan et al. (1993) study the biosorption of cadmium by biomass of marine algae (dried brown marine algae *S. natans*, *Fucus vesiculosus*, and *A. nodosum*) in an early paper with 290 citations. They evaluate the metal uptake by these materials using sorption isotherms. They find that "biomass of *A. nodosum* accumulated the highest amount of cadmium exceeding 100 mg Cd^{2+}/g (at the residual concentration of 100 mg Cd/L and pH 3.5), outperforming a commercial ion exchange resin." They obtain a new biosorbent material based on

TABLE 27.2

Experimental Research on Algal Biosorption

No.	Paper References	Year	Document	Affiliation	Country	No. of Authors	Journal	Index	Subjects	Topic	Total No. of Citations	Total Average Citations per Annum	Rank
1	Sheng et al.	2004	A	Natl. Univ. Singapore	Singapore	4	*J. Colloid Interface Sci.*	SCI, SCIE	Chem. Phys.	Algal biosorption	340	34.0	3
2	Holan et al.	1993	A	McGill Univ.	Canada	3	*Biotechnol. Bioeng.*	SCI, SCIE	Biot. Appl. Microb.	Algal biosorption	290	13.8	5
3	Fourest and Volesky	1996	A	McGill Univ.	Canada	2	*Environ. Sci. Technol.*	SCI, SCIE	Eng. Env., Env. Sci.	Algal biosorption	285	15.8	6
4	Aksu	2001	A	Hacettepe Univ.	Turkey	1	*Sep. Purif. Technol.*	SCIE	Eng. Chem.	Algal biosorption	265	20.4	7
5	Gupta et al.	2001	A	Univ. Roorkee	India	3	*Water Res.*	SCI, SCIE	Eng. Env., Env. Sci., Water. Res.	Algal biosorption	244	18.8	10
6	Holan and Volesky	1994	A	McGill Univ.	Canada	2	*Biotechnol. Bioeng.*	SCI, SCIE	Biot. Appl. Microb.	Algal biosorption	238	11.9	12
7	Aksu	2002	A	Hacettepe Univ.	Turkey	"	*Process Biochem.*		Bioch. Mol. Biol., Biot. Appl. Microb., Eng. Chem.	Algal biosorption	219	18.3	15
8	Kratochvil et al.	1998	A	McGill Univ.	Canada	3	*Environ. Sci. Technol.*	SCI, SCIE	Eng. Env., Env. Sci.	Algal biosorption	216	13.5	16
9	Donmez et al.	1999	A	Hacettepe Univ., Ankara Univ., Nigde Univ.	Turkey	4	*Process Biochem.*	SCI, SCIE	Bioch. Mol. Biol., Biot. Appl. Microb., Eng. Chem.	Algal biosorption	201	13.4	17
10	Davis et al.	2000	A	McGill Univ., Fed. Univ. Ceara	Canada, Brazil	3	*Water Res.*	SCI, SCIE	Eng. Env., Env. Sci., Water. Res.	Algal biosorption	188	13.4	18

(Continued)

TABLE 27.2 (*Continued*)

Experimental Research on Algal Biosorption

No.	Paper References	Year	Document	Affiliation	Country	No. of Authors	Journal	Index	Subjects	Topic	Total No. of Citations	Total Average Citations per Annum	Rank
11	Yun et al.	2001	A	Pohang Univ. Sci. Technol., McGill Univ.	South Korea, Canada	4	*Environ. Sci. Technol.*	SCI, SCIE	Eng. Env., Env. Sci.	Algal biosorption	184	14.2	20
12	Gupta and Rastogi	2009	A	Indian Inst. Technol. Roorkee	India	2	*J. Hazard. Mater.*	SCI, SCIE	Eng. Env., Eng. Civil, Env. Sci.	Algal biosorption	173	34.6	21
13	Leusch et al.	1995	A	McGill Univ.	Canada	3	*J. Chem. Technol. Biotechnol.*	SCI, SCIE	Biot. Appl. Microb., Chem. Mult., Eng. Chem.	Algal biosorption	171	9.0	23
14	Gupta and Rastogi	2008a	A	Indian Inst. Technol.	India	2	*J. Hazard. Mater.*	SCI, SCIE	Eng. Env., Eng. Civil, Env. Sci.	Algal biosorption	169	28.2	24
15	Matheickal et al.	1999	A	Griffith Univ.	Australia	3	*Water Res.*	SCI, SCIE	Eng. Env., Env. Sci., Water. Res.	Algal biosorption	161	10.7	25
16	Matheickal and Yu	1999	A	Griffith Univ.	Australia	2	*Bioresour. Technol.*	SCI, SCIE	Biot. Appl. Microb., Ener. Fuels	Algal biosorption	156	10.4	28
17	Gupta and Rastogi	2008b	A	Indian Inst. Technol. Roorkee	India	2	*J. Hazard. Mater.*	SCI, SCIE	Eng. Env., Eng. Civil, Env. Sci.	Algal biosorption	154	25.7	30
18	Donmez and Aksu	2002	A	Ankara Univ., Hacettepe Univ.	Turkey	2	*Process Biochem.*	SCI, SCIE	Bioch. Mol. Biol., Biot. Appl. Microb., Eng. Chem.	Algal biosorption	154	12.8	31
19	Gupta and Rastogi	2008c	A	Indian Inst. Technol.	India	2	*J. Hazard. Mater.*	SCI, SCIE	Eng. Env., Eng. Civil, Env. Sci.	Algal biosorption	151	25.2	33

(*Continued*)

TABLE 27.2 (*Continued*)

Experimental Research on Algal Biosorption

No.	Paper References	Year	Document	Affiliation	Country	No. of Authors	Journal	Index	Subjects	Topic	Total No. of Citations	Total Average Citations per Annum	Rank
20	Aksu and Tezer	2005	A	Hacettepe Univ.	Turkey	2	*Process Biochem.*	SCI, SCIE	Bioch. Mol. Biol., Biot. Appl. Microb., Eng. Chem.	Algal biosorption	150	16.7	34
21	Jalali et al.	2002	A	Atom. Energy Org. Iran, Univ. Alzahra	Iran	5	*J. Hazard. Mater.*	SCI, SCIE	Eng. Env., Eng. Civil, Env. Sci.	Algal biosorption	148	12.3	35
22	Hashim and Zhu	2004	A	Univ. Canterbury, Univ. Malaya	New Zealand, Malaysia	2	*Chem. Eng. J.*	SCI, SCIE	Eng. Env., Eng. Chem.	Algal biosorption	143	14.3	36
23	Park et al.	2005	A	Pohang Univ. Sci. Technol., Chonbuk Natl. Univ.	South Korea	3	*Chemosphere*	SCI, SCIE	Env. Sci.	Algal biosorption	141	15.7	37
24	Crist et al.	1994	A	Messiah Coll., Georgetown Univ.	United States	6	*Environ. Sci. Technol.*	SCI, SCIE	Eng Env., Env. Sci.	Algal biosorption	141	7.1	38
25	Gupta and Rastogi	2008d	A	Indian Inst. Technol.	India	2	*Colloid Surf. B Biointerfaces*	SCI, SCIE	Biophys., Chem. Phys., Mats. Sci. Biomats.	Algal biosorption	138	23.0	39
26	Axtell et al.	2003	A	Univ. Minnesota	United States	3	*Bioresour. Technol.*	SCI, SCIE	Biot. Appl. Microb., Ener. Fuels	Algal biosorption	133	12.1	40
27	Yee et al.	2004	A	Univ. Leeds, Univ. Toronto	England, Canada	4	*Environ. Sci. Technol.*	SCI, SCIE	Eng. Env., Env. Sci.	Algal biosorption	131	13.1	41
28	Klimmek et al.	2001	A	Tech. Univ. Berlin	Turkey	5	*Environ. Sci. Technol.*	SCI, SCIE	Eng. Env., Env. Sci.	Algal biosorption	127	9.8	42
29	Nourbakhsh et al.	1994	A	Hacettepe Univ., Firat Univ.	Turkey	6	*Process Biochem.*	SCI, SCIE	Bioch. Mol. Biol., Biot. Appl. Microb., Eng. Chem.	Algal biosorption	127	6.4	43

(Continued)

TABLE 27.2 (*Continued*)

Experimental Research on Algal Biosorption

No.	Paper References	Year	Document	Affiliation	Country	No. of Authors	Journal	Index	Subjects	Topic	Total No. of Citations	Total Average Citations per Annum	Rank
30	Cruz et al.	2004	A	Univ. Estado Rio De Janeiro	Brazil	4	*Bioresour. Technol.*	SCI, SCIE	Biot. Appl. Microb., Ener. Fuels	Algal biosorption	126	12.6	44
31	Gong et al.	2005	A	Nanjing Univ., Anhui Normal Univ.	China	5	*Chemosphere*	SCI, SCIE	Env. Sci.	Algal biosorption	125	13.9	45
32	Park et al.	2004	A	Pohang Univ. Sci. Technol., Chonbuk Natl. Univ.	South Korea	3	*Environ. Sci. Technol.*	SCI, SCIE	Eng. Env., Env. Sci.	Algal biosorption	115	11.5	46
33	Lodeiro et al.	2006	A	Univ. Coruna	Spain	4	*Environ. Pollut.*	SCI, SCIE	Env. Sci.	Algal biosorption	111	13.9	47
34	Yu et al.	1999	A	Griffith Univ.	Australia	4	*Water Res.*	SCI, SCIE	Eng. Env., Env. Sci., Water. Res.	Algal biosorption	109	7.3	48
35	Roden and Urrutia	1999	A	Univ. Alabama	United States	2	*Environ. Sci. Technol.*	SCI, SCIE	Eng. Env., Env. Sci.	Algal biosorption	108	7.2	49
36	Schiewer and Wong	2000	A	Hong Kong Baptist Univ.	China	2	*Chemosphere*	SCI, SCIE	Env. Sci.	Algal biosorption	105	7.5	50

SCI, Science Citation Index; SCIE, Science Citation Index Expanded; SSCI, Social Sciences Citation Index; A, article; R, review.

A. nodosum biomass by reinforcing the algal biomass by formaldehyde (FA) cross-linking. They find that the "prepared sorbent possessed good mechanical properties, chemical stability of the cell wall polysaccharides and low swelling volume." Furthermore, "desorption of deposited cadmium with 0.1–0.5 M HCl resulted in no changes of the biosorbent metal uptake capacity through five subsequent adsorption/desorption cycles." There was no damage to the biosorbent that retained its macroscopic appearance and performance in repeated metal uptake/elution cycles.

Fourest and Volesky (1996) study the contribution of sulfonate groups and alginate to heavy metal biosorption by the dry biomass of *Sargassum fluitans* in an early paper with 285 citations. Simultaneous potentiometric and conductometric titrations gave some information concerning the amount of strong and weak acidic functional groups in the biomass (0.25 ± 0.05 mequiv/g and 2.00 ± 0.05 mequiv/g, respectively). Those results were confirmed by the chemical identification of sulfonate groups (0.27 mequiv/g ± 0.03) and alginate (45% of the dry weight) corresponding to 2.25 mmol of carboxyl groups/g of biomass. Modification of these functional groups by methanolic hydrochloride or propylene oxide demonstrated the predominant role of alginate in the uptake of cadmium and lead. However, they argue that "sulfonate groups can also contribute, to a lower extent, to heavy metal binding, particularly at low pH." Eventually, FTIR spectrophotometry on protonated or cadmium-loaded alginate and *S. fluitans* biomass physically demonstrated that "cadmium binding arises by bridging or bidentate complex formation with the carboxyl groups of the alginate."

Aksu (2001) studies the equilibrium and kinetic modeling of cadmium(II) biosorption by *Chlorella vulgaris* in a batch system with a focus on the effect of temperature in a paper with 265 citations. She finds that the "algal biomass exhibited the highest cadmium(II) uptake capacity at 20°C, at the initial pH value of 4.0 and at the initial cadmium(II) ion concentration of 200 mg L^{-1}." Furthermore, "biosorption capacity decreased from 85.3 to 51.2 mg g^{-1} with an increase in temperature from 20 to 50°C at this initial cadmium(II) concentration." She finds that the "equilibrium data fitted very well to both Freundlich and Langmuir isotherm models in the studied concentration range of cadmium(II) ions at all the temperatures studied." She also applies "the pseudo first- and pseudo second-order kinetic models to experimental data assuming that the external mass transfer limitations in the system can be neglected and biosorption is sorption controlled." She asserts that the "cadmium(II) uptake process followed the second-order rate expression and adsorption rate constants decreased with increasing temperature."

Gupta et al. (2001) study the biosorption of chromium(VI) from aqueous solutions by biomass of filamentous algae *Spirogyra* species in a paper with 244 citations. Biosorption of heavy metals is an effective technology for the treatment of industrial wastewaters. They conduct batch experiments to determine adsorption properties of biomass, and they observe that the "adsorption capacity of the biomass strongly depends on equilibrium pH." They also obtain equilibrium isotherms and find that "maximum removal of Cr(VI) was around 14.7×10^3 mg metal/kg of dry weight biomass at a pH of 2.0 in 120 min with 5 mg/L of initial concentration." They assert that the "biomass of *Spirogyra* species is suitable for the development of efficient biosorbent for the removal and recovery of Cr(VI) from wastewater."

Holan and Volesky (1994) study the biosorption of lead and nickel by biomass of marine algae in an early paper with 238 citations. Screening tests of different marine algae biomass types revealed "a high passive biosorptive uptake of lead up to 270 mg Pb/g of biomass in some brown marine algae. Members of the order Fucales performed particularly well in this descending sequence: Fucus > Ascophyllum > Sargassum." They argue that "although decreasing the swelling of wetted biomass particles, their reinforcement by crosslinking may significantly affect the biosorption performance." They observe "lead uptakes up to 370 mg Pb/g in crosslinked *Fucus vesiculosus* and *Ascophyllum nodosum*," while they find that "at low equilibrium residual concentrations of lead in solution, however, ion exchange resin Amberlite IR-120 had a higher lead uptake than the biosorbent materials." Then they observe that an order-of-magnitude lower uptake of nickel in all of the sorbent materials was examined.

Aksu (2002) determines the equilibrium, kinetic, and thermodynamic parameters of the batch biosorption of nickel(II) ions onto *C. vulgaris* with respect to temperature and initial metal ion concentration in a paper with 219 citations. She finds that "algal biomass exhibited the highest nickel(II) uptake capacity at 45°C at an initial nickel(II) ion concentration of 250 mg L^{-1} and an initial pH of 4.5 whilst biosorption capacity increased from 48.1 to 60.2 mg g^{-1} with an increase in temperature from 15 to 45°C at this initial nickel(II) concentration." She then finds that "equilibrium data fitted very well to all the Freundlich, Langmuir and Redlich-Peterson isotherm models in the studied concentration range of nickel(II) ions at all the temperatures studied." She then applies the saturation-type kinetic model to experimental data at different temperatures changing from 15°C to 45°C to describe the batch biosorption kinetics assuming that the external mass transfer limitations in the system can be neglected and biosorption is chemical sorption controlled. She determines the "activation energy of biosorption (EA) as 25.92 kJ $mole^{-1}$ using the Arrhenius equation."

Kratochvil et al. (1998) study the removal of trivalent and hexavalent chromium by seaweed biosorbent in an early study with 216 citations. They find that "protonated or Ca-form *Sargassum* seaweed biomass bound up to 40 mg/g of Cr(Ill) by ion exchange at pH 4." An ion-exchange model assuming that the only species taken up by the biomass was $Cr(OH)^{2+}$ successfully fitted the experimental biosorption data for Cr(III). They explain the "maximum uptake of Cr(VI) by protonated *Sargassum* biomass at pH 2 by simultaneous anion exchange and Cr(VI) to Cr(III) reduction." They then find that "at pH <2.0, the reduction of Cr(VI) to Cr(III) dominated the equilibrium behavior of the batch systems, which was explained by the dependence of the reduction potential of $HCrO_4^-$ ions on the pH." They argue that "at pH >2.0, the removal of Cr(VI) was linked to the depletion of protons in equilibrium batch systems via an anion-exchange reaction." They argue that "the optimum pH for Cr(VI) removal by sorption lies in the region where the two mechanisms overlap, which for *Sargassum* biomass is in the vicinity of pH 2." They then explain the "existence of the optimum pH for the removal of Cr(VI) by taking into account (a) the desorption of Cr(III)from biomass at low pH and (b) the effect of pH on the reduction potential of Cr(VI) in aqueous solutions. They assert that 70% of Cr(VI) bound to the seaweed at pH 2 can be desorbed with 0.2 M H_2SO_4 via reduction to Cr(III)."

Donmez et al. (1999) study the heavy metal biosorption characteristics of some algae in a paper with 201 citations. They test the biosorption of copper(II), nickel(II), and chromium(VI) from aqueous solutions on dried (*C. vulgaris*, *Scenedesmus obliquus*, and *Synechocystis* sp.) algae under laboratory conditions as a function of pH, initial metal ion, and biomass concentrations. They find that "optimum adsorption pH values of copper(II), nickel(II) and chromium(VI) as 5.0, 4.5 and 2.0. respectively, for all three algae whilst at the optimal conditions, metal ion uptake increased with initial metal ion concentration up to 250 mg L^{-1}." They also find the influence of the alga concentration on the metal uptake for all the species. They argue that "both the Freundlich and Langmuir adsorption models were suitable for describing the short-term biosorption of copper(IT), nickel(II) and chromium(VI) by all the algal species."

Davis et al. (2000) study *Sargassum* seaweed as biosorbent for heavy metals in a paper with 188 citations. They compare six different species of nonliving *Sargassum* biomass on the basis of their equilibrium Cd and Cu uptake in order to evaluate potential variability in the sorption performance of different *Sargassum* species. They find that "biosorption uptakes for Cd at the optimal pH of 4.5 ranged from $q(max) = 0.90$ mmol/g for *Sargassum* sp. 1 to 0.66 mmol/g for *S. filipendula* I representing a 36% difference." They evaluate "three species for their Cu uptake where $q(max) = 0.93$ mmol/g for *S. vulgare*; 0.89 mmol/g for *S. filipendula* I, and 0.80 for *S. fluitans*, representing a 16% difference between the lowest and highest values." They carry out potentiometric titrations on *S. vulgare*, *S. fluitans*, and *S. filipendula* I and find the "similar results of 1.5 mmol/g weakly acidic sites for *S. vulgare* and *S. fluitans*, and 1.6 mmol/g for *S. filipendula* I." They then obtain estimates of 0.3 mmol/g of strongly acidic sites for *S. fluitans* and *S. filipendula* I and 0.5 mmol/g for *S. vulgare*. The total number of active sites averaged 1.9 ± 0.1 mmol/g. They determine the "elution efficiency for Cu-desorption from *S. filipendula* for $CaCl_2$, $Ca(NO_3)_2$, and HCl at various concentrations and solid:liquid ratios (S/L)." They find that the "highest elution efficiency was >95% for Cu for all elutants at $S/L = 1$ g/L and decreased for both calcium salts with increasing S/L to less than 50% at $S/L = 10$ g/L as a new batch sorption equilibrium was reached quickly." They argue that $CaCl_2$ was the most suitable metal-cation desorbing agent.

Yun et al. (2001) study the biosorption of trivalent chromium on protonated brown alga *Ecklonia* biomass in a paper with 184 citations. Titration of the biomass revealed that it contains at least three types of functional groups. The FTIR spectrometry showed that the "carboxyl group was the chromium-binding site within the pH range (pH 1–5), where chromium does not precipitate." They estimate the "pK value and the number of carboxyl groups to be 4.6 +/− 0.1 and 2.2 +/− 0.1 mmol/g, respectively." The equilibrium sorption isotherms determined at different solution pH indicated that the "uptake of chromium increased significantly with increasing pH." The model was able to predict the equilibrium sorption experimental data at different pH values and chromium concentrations. In addition, they predict the "speciation of the binding site as a function of the solution pH using the model in order to visualize the distribution of chromium ionic species on the binding site."

Gupta and Rastogi (2009) study the biosorption of hexavalent chromium by raw and acid-treated green alga *Oedogonium hatei* from aqueous solutions in a paper with 173 citations. They conduct batch experiments to determine the biosorption properties of the biomass. They find the "optimum conditions of biosorption as a biomass dose of 0.8 g/L, contact time of 110 min, pH and temperature 2.0 and 318 K respectively." They then find that both Langmuir and Freundlich isotherm equations could fit the equilibrium data. Under the optimal conditions, they find that the "biosorption capacities of the raw and acid-treated algae were 31 and 35.2 mg Cr(VI) per g of dry adsorbent, respectively." Then "thermodynamic parameters showed that the adsorption of Cr(VI) onto algal biomass was feasible, spontaneous and endothermic under studied conditions." The pseudo first-order kinetic model adequately describes the kinetic data in comparison to second-order model, and the process involving rate-controlling step is much complex involving both boundary layer and intraparticle diffusion processes. They determine the physical and chemical properties of the biosorbent and evaluate the nature of biomass–metal ion interactions by FTIR analysis, which showed "the participation of –CCOH, –OH, and $-NH_2$ groups in the biosorption process." Biosorbents could be regenerated using 0.1 M NaOH solution, with up to 75% recovery. Thus, they assert that the algal biomass was effective materials for the treatment of chromium-bearing aqueous solutions.

Leusch et al. (1995) study the biosorption of heavy metals (Cd, Cu, Ni, Pb, Zn) by chemically reinforced biomass of marine algae in an early paper with 171 citations. Particles of two different sizes (0.105–0.295 and 0.84–1.00 mm diameter) of two marine algae, *S. fluitans* and *A. nodosum*, were cross-linked with FA and glutaraldehyde (GA) or embedded in polyethylene imine (PEI), followed by GA cross-linking. They were used for equilibrium sorption uptake studies with cadmium, copper, nickel, lead, and zinc. They find that the "metal uptake by larger particles (0.84–1.00 mm) was higher than that by smaller particles (0.105–0.295 mm) whilst the order of adsorption for *S. fluitans* biomass particles was Pb > Cd > Cu > Ni > Zn, for A. nodosum copper and cadmium change places." Further, they find that "uptakes of metals range from $q_{max}=378$ mg Pb/g for *S. fluitans* (FA, big particles) to $q_{max}=89$ mg Zn g^{-1} for *S. fluitans* (FA, small particles) as the best sorption performance for each metal." Generally, they argue that "*S. fluitans* is a better sorbent material for a given metal, size and modification, although there were several exceptions in which metal sorption by A. nodosum was higher whilst the metal uptake for different chemical modifications showed the order GA > FA > PEI."

Gupta and Rastogi (2008a) study the biosorption of lead from aqueous solutions by green algae *Spirogyra* species with a focus on the kinetics and equilibrium studies in a paper with 169 citations. Biosorption is the effective method for the removal of heavy metal ions from wastewaters. They find that "the maximum adsorption capacity of Pb(II) ion was around 140 mg metal/g of biomass at pH 5.0 in 100 min with 200 mg/L of initial concentration. Temperature change in the range 20–40°C affected the adsorption capacity and the nature of the reaction was endothermic in nature." Uptake kinetics follows the pseudo second-order model, and equilibrium is well described by Langmuir isotherm. FTIR analysis of algal biomass revealed the "presence of amino, carboxyl, hydroxyl and carbonyl groups, which are responsible for biosorption of metal ions." They assert that "the biomass of *Spirogyra* sp. is an efficient biosorbent for the removal of Pb(II) from aqueous solutions."

Matheickal et al. (1999) study the biosorption of cadmium(II) from aqueous solutions by pretreated biomass of marine alga *Durvillaea potatorum* in a paper with 169 citations. They find that "pretreatment of the native biomass with calcium chloride and subsequent thermal treatment considerably improved the swelling properties and physical stability of the biomass granules." They then find that "the adsorption capacity of the biomass strongly depends on equilibrium solution pH where at solution pH of 5, the maximum adsorption capacity of the pretreated biomass is 1.1 mmol/g." The kinetics of cadmium adsorption was fast with 90% of adsorption taking place within 30 min. They assert that the "pretreated biomass of *D. potatorum* can be used as an efficient biosorbent for the treatment of cadmium bearing waste streams."

Matheickal and Yu (1999) study the biosorption of lead(II) and copper(II) from aqueous solutions by pretreated biomass of Australian marine algae *D. potatorum*–and *Eucalyptus radiata*–based biosorbents (DP95Ca and ER95Ca) in a paper with 156 citations. They find that "two stage modification processes substantially improved the leaching characteristics of the biomass whilst batch equilibrium experiments showed that the maximum adsorption capacities of DP95Ca for lead and copper were 1.6 and 1.3 mmol/g, respectively." The corresponding values for ER95Ca were 1.3 and 1.1 mmol/g. They argue that these capacities are comparable with those of commercial ion-exchange resins and are much higher than those of natural zeolites and powdered activated carbon. They then find that the "heavy metal uptake process was rapid with 90% of the adsorption completed in about 10 min in batch conditions. Heavy metal adsorption was observed at pH values as low as 2.0 and maximum adsorption was obtained approximately at a pH of 4.5."

They assert that "both biosorbents were effective in removing lead and copper in the presence of chelating agents and other light metal ions in waste water."

Gupta and Rastogi (2008b) study the equilibrium and kinetic modeling of cadmium(II) biosorption by nonliving algal biomass *Oedogonium* sp. from aqueous phase with respect to initial pH, algal dose, contact time, and the temperature in a paper with 154 citations. They find that the "algal biomass exhibited the highest cadmium(II) uptake capacity at 25°C, at the initial pH value of 5.0 in 55 min and at the initial cadmium(II) ion concentration of 200 mg L^{-1}." They then find that "biosorption capacity decreased from 88.9 to 80.4 mg g^{-1} with an increase in temperature from 25 to 45°C at this initial cadmium(II) concentration." Uptake kinetics follows the pseudo second-order model, and equilibrium is well described by Langmuir isotherm. FTIR analysis of algal biomass showed the "presence of amino, carboxyl, hydroxyl and carbonyl groups, which are responsible for biosorption of metal ions." Next, they find that "acid pretreatments did not substantially increase metal sorption capacity but alkali like NaOH pretreatment slightly enhanced the metal removal ability of the biomass whilst during repeated sorption/desorption cycles at the end of fifth cycle, Cd(II) sorption decreased by 18%, with 15–20% loss of biomass." They assert that "*Oedogonium* sp. is a good sorbent for removing metal Cd(II) from aqueous phase."

Donmez and Aksu (2002) study the removal of chromium(VI) from saline wastewaters by *Dunaliella* species as a function of pH, initial metal ion, and salt (NaCl) concentrations in a batch system in a paper with 154 citations. They find that the "biosorption capacity of both *Dunaliella* strains strongly depends on solution pH and maximum as Chromium(VI) sorption capacities of both sorbents were obtained at pH 2.0 in the absence and in the presence of increasing concentrations of salt." They then find that "equilibrium uptakes of chromium(VI) increased with increasing chromium(VI) concentration up to 250–300 mg L^{-1} and decreased sharply by the presence of increasing concentrations of salt for both the sorbents." They then find that "*Dunaliella* 1 and *Dunaliella* 2 biosorbed 58.3 and 45.5 mg g^{-1} of chromium(VI), respectively, in 72 It at 100 mg L^{-1} initial chromium(VI) concentration without salt medium. When salt concentration increased to 20% (w/v), these values dropped to 20.7 and 12.2 mg g^{-1} of chromium(VI) at the same conditions." Both the Freundlich and Langmuir adsorption models were suitable for describing the biosorption of chromium(VI) individually and in salt-containing medium by both algal species. The pseudo second-order kinetic model was successfully applied to single chromium(VI) and chromium(VI)–salt mixture biosorption data.

Gupta and Rastogi (2008c) study the sorption and desorption studies of chromium(VI) from nonviable cyanobacterium (*Nostoc muscorum*) biomass in a paper with 151 citations. They find that "sorption interaction of chromium on to cyanobacterial species obeyed both the first and the second-order rate equation and the experimental data showed good fit with both the Langmuir and Freundlich adsorption isotherm models." The maximum adsorption capacity was 22.92 mg/g at 25°C and pH 3.0. The adsorption process was endothermic, and the values of thermodynamic parameters of the process were calculated. They observe sorption–desorption of chromium into inorganic solutions and distilled water, and this indicated the "biosorbent could be regenerated using 0.1 M HNO_3 and EDTA with up to 80% recovery." The biosorbents were reused in five biosorption–desorption cycles without a significant loss in biosorption capacity. Thus, they assert that the cyanobacterial biomass *N. muscorum* could be used as an efficient biosorbent for the treatment of chromium(VI) bearing wastewater.

Aksu and Tezer (2005) study the biosorption of reactive dyes (three vinyl sulfone–type reactive dyes [Remazol Black B (RB), Remazol Red RR (RR), and Remazol Golden Yellow

RNL (RGY)]) on the green alga *C. vulgaris* in a paper with 150 citations. They find that the "algal biomass exhibited the highest dye uptake capacity at the initial pH value of 2.0 for all dyes. The effect of temperature on equilibrium sorption capacity indicated that maximum capacity was obtained at 35°C for RB biosorption and at 25°C for RR and RGY biosorptions. Biosorption capacity of alga increased with increasing initial dye concentration up to 800 mg/L for RB and RR dyes and up to 200 mg L^{-1} for RGY dye. Among the three dyes, RB was adsorbed most effectively by the biosorbent to a maximum of approximately 419.5 mg g^{-1}." They then find that "equilibrium data of RB biosorption fitted very well to Freundlich, Langmuir, Redlich-Peterson and Koble-Corrigan adsorption models except that the Langmuir model, while this model was most suitable for describing the biosorptions of RR and RGY dyes in the studied concentration and temperature ranges." They assert that the "dye uptake process followed the pseudo second-order and saturation type rate expressions for each dye–*C. vulgaris* system."

Jalali et al. (2002) study the removal and recovery of lead using nonliving biomass of marine algae in a paper with 148 citations. They find that "biosorption of lead was rapidly occurred onto algal biosorbents and most of the sorbed metal was bound in <30 min of contact. Three species of brown algae, namely *Sargassum hystrix, S. natans* and *Padina pavonia,* removed lead most efficiently from aqueous solution, respectively." They show an increasing uptake of the metal by biosorbents with increasing pH. Desorption of the adsorbed lead on biosorbent was conducted by decreasing the pH values to lower than 1.0. "Removal of lead from *Sargassum* biomass was successfully achieved by eluting with 0.1 M HNO_3 for 15 min and a high degree of metal recovery was observed (95%)." For optimum operation in the subsequent metal uptake cycle, "regeneration of the *Sargassum* biomass was efficiently performed by 0.1 M $CaCl_2$ for 15 min that was total and reversible." In repeated use of biomass experiment, the lead uptake capacity of *Sargassum* biomass was constantly retained (98%), and no significant biomass damage took place after 10 sorption–desorption cycles.

Hashim and Chu (2004) study the biosorption of cadmium by brown, green, and red seaweeds in a paper with 143 citations. Although all the seaweed types investigated were capable of binding appreciable amounts of cadmium, considerable variability in their biosorption performance was observed. They find that "maximum cadmium uptake capacities at pH 5 ranged from the highest value of 0.74 mmol/g for the brown seaweed *Sargassum baccularia* to the lowest value of 0.16 mmol/g for the red seaweed *Gracilaria salicornia,* representing a 363% difference." In general, "brown seaweeds exhibited the best overall cadmium ion removal." The equilibrium uptakes of cadmium were similar within the pH 3–5 range but decreased significantly when the solution pH was reduced to pH 2. They then find that "the presence of background cations such as sodium, potassium, and magnesium and anions such as chloride, nitrate, sulphate, and acetate up to a concentration of 3.24 mmol/L had no significant effect on the equilibrium uptake of cadmium. However, the biosorbent uptake of cadmium was markedly inhibited in the presence of calcium ions at 3.24 mmol/L." Kinetic studies revealed that cadmium uptake was fast with 90% or more of the uptake occurring within 30–40 min of contact time.

Park et al. (2005) describe the studies on hexavalent chromium (Cr(VI)) biosorption by chemically treated biomass of *Ecklonia* sp. to enhance the Cr(VI)-reducing capacity of the biomass using various chemical treatments and to elucidate the mechanisms governing Cr(VI) reduction in a paper with 141 citations. They find that "acid-treatment showed the best performance with regards the improvement of Cr(VI) removal from the aqueous phase, while organic solvent-treatment significantly improved the removal efficiency of total Cr in the equilibrium state." They then find that "methylation of the amino group significantly decreased the Cr(VI) removal rate, but amination of the carboxyl group significantly increased the Cr(VI) removal rate meanwhile, esterification of the carboxyl group and carboxylation of the amino group decreased the Cr(VI) removal rate, but the former showed a more negative effect than the latter." They assert that the "amino and carboxyl groups take part in the Cr(VI) removal from the aqueous phase."

Crist et al. (1994) study the interaction of metals and protons with algae with a focus on the ion exchange versus adsorption models and a reassessment of Scatchard plots with a focus on the ion-exchange rates and equilibria compared with calcium alginate in an early paper with 141 citations. They use an experimental ion-exchange constant for Zn displacing Ca from Rhizoclonium to calculate concentrations over a wide enough range to assess interpretations given to Langmuir and Scatchard plots. They argue that while such plots may be convenient to describe maximum sorption and systematize data, misinterpretations can occur at low metal concentrations in an ion-exchange system. They find that "values of K_{ex} for seven metals displacing Ca from Vaucheria correlated with formation constants of the metal acetates and with K_{ex} of the metals on calcium alginate, a model of cell wall components, indicating the bonding of metals to carboxylate groups of algal cell walls." They then find that "rates of metal desorption from Vaucheria by EDTA-Li are inversely related to binding strengths, and rates of Cd desorption from calcium alginate justify assumptions in the K-ex expression."

Gupta and Rastogi (2008d) study the biosorption of lead(II) from aqueous solutions by nonliving algal biomass (*Oedogonium* sp. and *Nostoc* sp.) under varying range of pH (2.99–7.04), contact time (5–300 min), biosorbent dose (0.1–0.8 g/L), and initial metal ion concentrations (100 and 200 mg/L) in a paper with 138 citations. They find that the "optimum conditions for lead biosorption are almost same for the two algal biomass *Oedogonium* sp. and *Nostoc* sp. (pH 5.0, contact time 90 and 70 min, biosorbent dose 0.5 g/L and initial Pb(II) concentration 200 mg/L)." However, they argue that the "biomass of *Oedogonium* sp. was more suitable than *Nostoc* sp. for the development of an efficient biosorbent for the removal of lead(II) from aqueous solutions, as it showed higher values of q_e adsorption capacity (145.0 mg/g for *Oedogonium* sp. and 93.5 mg/g for *Nostoc* sp.)." Furthermore, the equilibrium data fitted well in the Langmuir isotherms than the Freundlich isotherm, thus proving monolayer adsorption of lead on both the algal biomass. They argue that the process involves second-order kinetics and thermodynamic treatment of equilibrium data shows endothermic nature of the adsorption process. The spectrum of FTIR confirms that the amino and carboxyl groups on the surface of algal biomass were the main adsorption sites for lead removal. Both the biosorbents could be regenerated using 0.1 mol/L HCl solution, with up to 90% recovery. The biosorbents were reused in five biosorption–desorption cycles without a significant loss in biosorption capacity. They assert that both the algal biomass could be used as an efficient biosorbents for the treatment of lead(II)-bearing wastewater streams.

Axtell et al. (2003) study the lead and nickel removal using *Microspora* and *Lemna minor* in a paper with 133 citations. *Microspora* was tested in a batch and semibatch process for lead removal. *L. minor* was tested in a batch process with lead and nickel to examine the potential competition between metals for adsorption. The *Microspora* was exposed to 39.4 mg/L of lead over 10 days. They find that "up to 97% of the lead was removed in the batch process and 95% in the semi-batch process. Initial concentrations below 50 mg/l (a dose that kills the algae) had no effect on the final concentration." The *L. minor* was exposed to lead and nickel using a full 3(2) factorial experimental design (nine experiments, plus replications). Initial lead concentrations were 0.0, 5.0, and 10.0 mg/L, and nickel concentrations were 0.0, 2.5, and 5.0 mg/L in the experiment. "Overall, *L. minor* removed 76% of the lead, and 82% of the nickel."

Yee et al. (2004) study the characterization of metal–cyanobacteria sorption reactions with a focus on a combined macroscopic and infrared spectroscopic investigation in a paper with 131 citations. Infrared spectra were collected with samples in solution for intact cyanobacterial filaments and separated exopolymeric sheath material to examine the deprotonation reactions of cell surface functional groups. The infrared spectra of intact cells sequentially titrated from pH 3.2 to 6.5 display an "increase in peak intensity and area at 1400 cm^{-1} corresponding to vibrational COO^- frequencies from the formation of deprotonated carboxyl surface sites." Similarly, bulk acid–base titration of cyanobacterial filaments and sheath material indicates that the "concentration of proton-active surface sites is higher on the cell wall compared to the overlying sheath." A three-site model provides an excellent fit to the titration curves of both intact cells and sheath material with corresponding pK_a values of 4.7 ± 0.4, 6.6 ± 0.2, and 9.2 ± 0.3 and 4.8 ± 0.3, 6.5 ± 0.1, and 8.7 ± 0.2, respectively. The modeling indicates that "metal ions (Cu^{2+}, Cd^{2+}, and Pb^{2+}) are partitioned between the exopolymer sheath and cell wall and that the carboxyl groups on the cyanobacterial cell wall are the dominant sink for metals at near neutral pH." They assert that the cyanobacterial surfaces are complex structures that contain distinct surface layers, each with unique molecular functional groups and metal binding properties.

Klimmek et al. (2001) carry out a comparative analysis of the biosorption of cadmium, lead, nickel, and zinc by 30 strains of algae in a paper with 127 citations. They find that the "cyanophyceae *Lyngbya taylorii* exhibited high uptake capacities for the four metals. The algae showed maximum capacities according to the Langmuir Adsorption Model of 1.47 mmol lead, 0.37 mmol cadmium, 0.65 mmol nickel, and 0.49 mmol zinc per gram of dry biomass. The optimum pH for *L. taylorii* was between pH 3 and 7 for lead, cadmium, and zinc and between pH 4 and 1 for nickel." Studies with the algae indicated a preference for the uptake of lead over cadmium, nickel, and zinc in a four-metal solution. They argue that the "metal binding abilities of *L. taylorii* could be improved by phosphorylation of the biomass as the modified biosorbent demonstrated maximum capacities of 2.52 mmol cadmium, 3.08 mmol lead, 2.79 mmol nickel, and 2.60 mmol zinc per gram of dry biomass." Investigations with phosphated *L. taylorii* indicated high capacities for the four metals also at low pH.

Nourbakhsh et al. (1994) carry out a comparative study of various biosorbents (nonliving biomass of *C. vulgaris, Cladophora crispata, Zoogloea ramigera, Rhizopus arrhizus*, and *S. cerevisiae*) for the removal of chromium(VI) ions from industrial wastewaters in an early paper with 127 citations. They find that the "initial pH of the metal ion solution effected metal uptake capacity of the biomass and the optimum initial pH was found as 1.0–2.0 for all microorganisms. Maximum adsorption rates of metal ions to microbial biomass were obtained at temperatures in the range 25–35°C." They then find that the "adsorption rate increased with increasing metal ion concentration for *C. vulgaris, C. crispata, R. arrhizus, S. cerevisiae* and *Z. ramigera* up to 200, 200, 125, 100 and 75 mg/litre, respectively."

Cruz et al. (2004) study the kinetic modeling and equilibrium studies during cadmium(II) biosorption by dead *Sargassum* sp. biomass with a focus on the influence of different experimental parameters: initial pH, shaking rate, sorption time, temperature, and initial concentrations of cadmium ions on cadmium uptake in a paper with 126 citations. They argue that cadmium uptake could be described by the Langmuir adsorption model, being the monolayer capacity negatively affected with an increase in temperature. Analogously, they find that the "adsorption equilibrium constant decreased with increasing temperature. The kinetics of the adsorption process followed a second-order adsorption, with characteristic constants increasing with increasing temperature.

Activation energy of biosorption could be calculated as equal to 10 kcal/mol." They assert that the biomass used was suitable for the removal of cadmium from dilute solutions as its maximum uptake capacity was 120 mg/g. Thus, they argue that *Sargassum* sp. has great potential for removing cadmium ions especially when concentration of this metal is low in samples such as wastewater streams.

Gong et al. (2005) study the lead biosorption and desorption by intact and pretreated *Spirulina maxima* biomass with a focus on the effects of operational conditions (e.g., pH, contact time, biomass concentration) on lead biosorption in a paper with 125 citations. They find that the "biosorption was solution pH dependent and the maximum adsorption was obtained at a solution pH of about 5.5 whilst the adsorption equilibrium was reached in 60 min." The biosorption followed the Freundlich isotherm model. They then find that "maximum removal ratios of lead were about 84% in intact biomass and 92% in pretreated biomass whilst the lead adsorbed could be desorbed effectively by 0.1 M nitric acid, EDTA and hydrochloric acid." They assert that pretreated biomass of *S. maxima* was a promising candidate for removing lead from wastewater.

Park et al. (2004) study the reduction of hexavalent chromium (Cr(VI)) with the brown seaweed *Ecklonia* biomass in a paper with 115 citations. They find that "when synthetic wastewater containing Cr(VI) was placed in contact with the biomass, the Cr(VI) was completely reduced to Cr(III). The converted Cr(III) appeared in the solution phase or was partly bound to the biomass. The Cr(VI) removal efficiency was always 100% in the pH range of this study (pH 1 similar to 5)." Furthermore, the Cr(VI) reduction was independent of the Cr(III) concentration, the reaction product, suggesting that the reaction was an irreversible process under test conditions. They then find that "proton ions were consumed in the ratio of 1.15 +/− 0.02 mol of protons/mol of Cr(VI), and the rate of Cr(VI) reduction increased with decreasing the pH. An optimum pH existed for the removal efficiency of total chromium (Cr(VI) plus Cr(III)), but this increased with contact time, eventually reaching approximately pH 4 when the reaction was complete. The electrons required for the Cr(VI) reduction also caused the oxidation of the organic compounds in the biomass. One gram of the biomass could reduce 4.49 +/− 0.12 mmol of Cr(VI)." They assert that the abundant and inexpensive *Ecklonia* biomass could be used for the conversion of toxic Cr(VI) into less toxic or nontoxic Cr(III).

Lodeiro et al. (2006) study the brown seaweed *Cystoseira baccata* as biosorbent for cadmium(II) and lead(II) removal with a focus on the kinetic and equilibrium studies in a paper with 111 citations. Kinetic experiments demonstrated rapid metal uptake, and kinetic data were satisfactorily described by a pseudo second-order chemical sorption process. Temperature change from 15°C to 45°C showed small variation on kinetic parameters. Langmuir–Freundlich equation was selected to describe the metal isotherms and the proton binding in acid–base titrations. They then find that the "maximum metal uptake values were around 0.9 mmol g^{-1} (101 and 186 mg g^{-1} for cadmium(II) and lead(II), respectively) at pH 4.5 (raw biomass), while the number of weak acid groups were 2.2 mmol g^{-1} and their proton binding constant, K-H, 10(3.67) (protonated biomass)." FTIR analysis confirmed the participation of carboxyl groups in metal uptake. The metal sorption increased with the solution pH reaching a plateau above pH 4, while calcium and sodium nitrate salts in solution affected considerably the metal biosorption.

Yu et al. (1999) study the heavy metal (cadmium, copper, and lead) uptake capacities of a group of nine marine macroalgal biomass in a paper with 109 citations. Equilibrium isotherms for each biomass–heavy metal system were obtained from batch adsorption experiments as they find that the "maximum uptake capacities of the biomass ranged from around 0.8 to 1.6 mmol/g (dry), which were much higher than those of other types of biomass."

They assert that the biomass of marine algae is suitable for the development of efficient biosorbents for the removal and recovery of heavy metals from wastewater.

Roden and Urrutia (1999) study the effects of the ferrous iron (Fe(II)) removal on the microbial reduction of crystalline iron(III) oxides by *Shewanella* alga strain BrY in a paper with 108 citations. They find that "aqueous phase replacement in semicontinuous cultures (average residence time of 9 or 18 days) resulted in a 2–3-fold increase in the cumulative amount of Fe(II) produced from synthetic goethite reduction over a 2-month incubation period, compared to parallel batch cultures." A more modest (maximum 30%) but significant stimulation of natural subsoil Fe(III) oxide reduction was observed. The extended Fe(III) reduction resulted from enhanced generation of aqueous Fe(II), which was periodically removed from the cultures. A concomitant stimulation of bacterial protein production was detected, which suggested that "Fe(II) removal also promoted bacterial growth." A simulation model in which Fe(II) sorption to the solid phase resulted in blockage of surface reduction sites captured the contrasting behavior of the batch versus semicontinuous Gt reduction systems. They assert that elimination of Fe(II) via advective transport could play a significant role in governing the rate and extent of microbial Fe(III) oxide reduction in sedimentary environments.

Schiewer and Wong (2000) study the ionic strength effects in biosorption of metals by marine algae in a paper with 105 citations. The green alga *Ulva fascia* and the brown seaweeds *S. hemiphyllum*, *Petalonia fascia*, and *Colpomenia sinuosa* were characterized in terms of their number of binding sites, their charge density, and intrinsic proton binding constant (pK_a) using pH titrations at different ionic strengths. They find that the "determined number of binding sites decreased in the order Petalonia greater than or equal to *Sargassum* > *Colpomenia* > *Ulva*." They argue that due to their high number of binding sites, *Sargassum* and *Petalonia* are most promising for biosorption applications. They argue that the decrease of proton binding with increasing ionic strength and pH as well as the increase of Cu and Ni binding with increasing pH and decreasing ionic strength could be described by the "Donnan model in conjunction with an ion exchange biosorption isotherm."

27.3.3 Conclusion

The experimental research on the algal biosorption of heavy metals from wastes has been one of the most dynamic research areas in recent years. Thirty-six experimental citation classics in the field of algal biodiesel with more than 105 citations were located, and the key emerging issues from these papers were presented by decreasing order of the number of citations. These papers give strong hints about the determinants of the efficient algal biosorption of heavy metals and other materials from wastes, dyes, and other materials and emphasize that marine algae are efficient sorbents in comparison with the traditional sorbents.

27.4 Conclusion

The citation classics presented under the two main headings in this chapter confirm the predictions that the marine algae have a significant potential to serve as a major solution for the global problems of warming, air pollution, energy security, and food security in the form of algal biosorption of heavy metals from wastes.

Further research is recommended for the detailed studies in each topical area presented in this chapter including scientometric studies and citation classic studies to inform the key stakeholders about the potential of marine algae for the solution of the global problems of warming, air pollution, energy security, and food security in the form of algal biosorption of heavy metals from wastes.

References

Ahluwalia, S.S. and D. Goyal. 2007. Microbial and plant derived biomass for removal of heavy metals from wastewater. *Bioresource Technology* 98:2243–2257.

Aksu, Z. 2001. Equilibrium and kinetic modelling of cadmium(II) biosorption by *C. vulgaris* in a batch system: Effect of temperature. *Separation and Purification Technology* 21:285–294.

Aksu, Z. 2002. Determination of the equilibrium, kinetic and thermodynamic parameters of the batch biosorption of nickel(II) ions onto Chlorella vulgaris. *Process Biochemistry* 38:89–99.

Aksu, Z. and S. Tezer. 2005. Biosorption of reactive dyes on the green alga *Chlorella vulgaris*. *Process Biochemistry* 40:1347–1361.

Axtell, N.R., S.P.K. Sternberg, and K. Claussen. 2003. Lead and nickel removal using *Microspora* and *Lemna minor*. *Bioresource Technology* 89:41–48.

Bailey, S.E., T.J. Olin, R.M. Bricka, and D.D. Adrian. 1999. A review of potentially low-cost sorbents for heavy metals. *Water Research* 33:2469–2479.

Baltussen, A. and C.H. Kindler. 2004a. Citation classics in anesthetic journals. *Anesthesia & Analgesia* 98:443–451.

Baltussen, A. and C.H. Kindler. 2004b. Citation classics in critical care medicine. *Intensive Care Medicine* 30:902–910.

Bothast, R.J. and M.A. Schlicher. 2005. Biotechnological processes for conversion of corn into ethanol. *Applied Microbiology and Biotechnology* 67:19–25.

Cervantes, C., J. Campos-Garcia, S. Devars et al. 2001. Interactions of chromium with microorganisms and plants. *FEMS Microbiology Reviews* 25:335–347.

Chisti, Y. 2007. Biodiesel from microalgae. *Biotechnology Advances* 25:294–306.

Chojnacka, K., A. Chojnacki, and H. Gorecka. 2005. Biosorption of Cr^{3+}, Cd^{2+} and Cu^{2+} ions by blue-green algae *Spirulina* sp.: Kinetics, equilibrium and the mechanism of the process. *Chemosphere* 59:75–84.

Chong, K.H. and B. Volesky. 1995. Description of 2-metal biosorption equilibria by Langmuir-type models. *Biotechnology and Bioengineering* 47:451–460.

Crist, R.H., J.R. Martin, D. Carr, J.R. Watson, H.J. Clarke, and D.L.R. Crist. 1994. Interaction of metals and protons with algae.4. Ion-exchange vs. adsorption models and a reassessment of Scatchard plots—Ion-exchange rates and equilibria compared with calcium alginate. *Environmental Science and Technology* 28:1859–1866.

Cruz, C.C.V., A.C.A. da Costa, C.A. Henriques, and A.S. Luna. 2004. Kinetic modeling and equilibrium studies during cadmium biosorption by dead *Sargassum* sp biomass. *Bioresource Technology* 91:249–257.

Davis, T.A., B. Volesky, and A. Mucci. 2003. A review of the biochemistry of heavy metal biosorption by brown algae. *Water Research* 37:4311–4330.

Davis, T.A., B. Volesky, and R.H.S.F. Vieira. 2000. Sargassum seaweed as biosorbent for heavy metals. *Water Research* 34:4270–4278.

Demirbas, A. 2007. Progress and recent trends in biofuels. *Progress in Energy and Combustion Science* 33:1–18.

Donmez, G. and Z. Aksu. 2002. Removal of chromium(VI) from saline wastewaters by *Dunaliella* species. *Process Biochemistry* 38:751–762.

Donmez, G.C., Z. Aksu, A. Ozturk, and T. Kutsal. 1999. A comparative study on heavy metal biosorption characteristics of some algae. *Process Biochemistry* 34:885–892.

Dubin, D., A.W. Hafner, and K.A. Arndt. 1993. Citation-classics in clinical dermatological journals—Citation analysis, biomedical journals, and landmark articles, 1945–1990. *Archives of Dermatology* 129:1121–1129.

Fein, J.B., C.J. Daughney, N. Yee, and T.A. Davis. 1997. A chemical equilibrium model for metal adsorption onto bacterial surfaces. *Geochimica et Cosmochimica Acta* 61:3319–3328.

Figueira, M.M., B. Volesky, V.S.T. Ciminelli, and F.A. Roddick. 2000. Biosorption of metals in brown seaweed biomass. *Water Research* 34:196–204.

Fourest, E. and B. Volesky. 1996. Contribution of sulfonate groups and alginate to heavy metal biosorption by the dry biomass of *Sargassum fluitans*. *Environmental Science and Technology* 30:277–282.

Gehanno, J.F., K. Takahashi, S. Darmoni, and J. Weber. 2007. Citation classics in occupational medicine journals. *Scandinavian Journal of Work, Environment and Health* 33:245–251.

Godfray, H.C.J., J.R. Beddington, I.R. Crute et al. 2010. Food security: The challenge of feeding 9 billion people. *Science* 327:812–818.

Goldemberg, J. 2007. Ethanol for a sustainable energy future. *Science* 315:808–810.

Gong, R.M., Y. Ding, H.J. Liu, Q.Y. Chen, and Z.L. Liu. 2005. Lead biosorption and desorption by intact and pretreated *Spirulina maxima* biomass. *Chemosphere* 58:125–130.

Guibal, E. 2004. Interactions of metal ions with chitosan-based sorbents: A review. *Separation and Purification Technology* 38:43–74.

Gupta, V.K. and A. Rastogi. 2008a. Biosorption of lead from aqueous solutions by green algae *Spirogyra* species: Kinetics and equilibrium studies. *Journal of Hazardous Materials* 152:407–414.

Gupta, V.K. and A. Rastogi. 2008b. Equilibrium and kinetic modelling of cadmium(II) biosorption by nonliving algal biomass *Oedogonium* sp from aqueous phase. *Journal of Hazardous Materials* 153:759–766.

Gupta, V.K. and A. Rastogi. 2008c. Sorption and desorption studies of chromium(VI) from nonviable cyanobacterium *Nostoc muscorum* biomass. *Journal of Hazardous Materials* 154:347–354.

Gupta, V.K. and A. Rastogi. 2008d. Biosorption of lead(II) from aqueous solutions by non-living algal biomass *Oedogonium* sp and *Nostoc* sp—A comparative study. *Colloids and Surfaces B—Biointerfaces* 164:170–178.

Gupta, V.K. and A. Rastogi. 2009. Biosorption of hexavalent chromium by raw and acid-treated green alga *Oedogonium hatei* from aqueous solutions. *Journal of Hazardous Materials* 163:396–402.

Gupta, V.K., A.K. Shrivastava, and N. Jain. 2001. Biosorption of chromium(VI) from aqueous solutions by green algae *Spirogyra* species. *Water Research* 35:4079–4085.

Hashim, M.A. and K.H. Chu. 2004. Biosorption of cadmium by brown, green, and red seaweeds. *Chemical Engineering Journal* 97:249–255.

Holan, Z.R. and B. Volesky. 1994. Biosorption of lead and nickel by biomass of marine-algae. *Biotechnology and Bioengineering* 43:1001–1009.

Holan, Z.R., B. Volesky, and I. Prasetyo. 1993. Biosorption of cadmium by biomass of marine-algae. *Biotechnology and Bioengineering* 41:819–825.

Jacobson, M.Z. 2009. Review of solutions to global warming, air pollution, and energy security. *Energy and Environmental Science* 2:148–173.

Jalali, R., H. Ghafourian, Y. Asef, S.J. Davarpanah, and S. Sepehr. 2002. Removal and recovery of lead using nonliving biomass of marine algae. *Journal of Hazardous Materials* 92:253–262.

Juhasz, A.L. and R. Naidu. 2000. Bioremediation of high molecular weight polycyclic aromatic hydrocarbons: A review of the microbial degradation of benzo[a]pyrene. *International Biodeterioration and Biodegradation* 45:57–88.

Kapdan, I.K. and F. Kargi. 2006. Bio-hydrogen production from waste materials. *Enzyme and Microbial Technology* 38:569–582.

Kapoor, A., T. Viraraghavan, and D.R. Cullimore. 1999. Removal of heavy metals using the fungus *Aspergillus niger*. *Bioresource Technology* 70:95–104.

Klimmek, S., H.J. Stan, A. Wilke, G. Bunke, and R. Buchholz. 2001. Comparative analysis of the biosorption of cadmium, lead, nickel, and zinc by algae. *Environmental Science and Technology* 35:4283–4288.

Konur, O. 2000. Creating enforceable civil rights for disabled students in higher education: An institutional theory perspective. *Disability and Society* 15:1041–1063.

Konur, O. 2002a. Assessment of disabled students in higher education: Current public policy issues. *Assessment and Evaluation in Higher Education* 27:131–152.

Konur, O. 2002b. Access to employment by disabled people in the UK: Is the Disability Discrimination Act working? *International Journal of Discrimination and the Law* 5:247–279.

Konur, O. 2002c. Access to nursing education by disabled students: Rights and duties of nursing programs. *Nursing Education Today* 22:364–374.

Konur, O. 2006a. Participation of children with dyslexia in compulsory education: Current public policy issues. *Dyslexia* 12:51–67.

Konur, O. 2006b. Teaching disabled students in higher education. *Teaching in Higher Education* 11:351–363.

Konur, O. 2007a. A judicial outcome analysis of the Disability Discrimination Act: A windfall for the employers? *Disability and Society* 22:187–204.

Konur, O. 2007b. Computer-assisted teaching and assessment of disabled students in higher education: The interface between academic standards and disability rights. *Journal of Computer Assisted Learning* 23:207–219.

Konur, O. 2011. The scientometric evaluation of the research on the algae and bio-energy. *Applied Energy* 88:3532–3540.

Konur, O. 2012a. 100 citation classics in energy and fuels. *Energy Education Science and Technology Part A—Energy Science and Research* 2012(si):319–332.

Konur, O. 2012b. What have we learned from the citation classics in energy and fuels: A mixed study. *Energy Education Science and Technology Part A* 2012(si):255–268.

Konur, O. 2012c. The gradual improvement of disability rights for the disabled tenants in the UK: The promising road is still ahead. *Social Political Economic and Cultural Research* 4:71–112.

Konur, O. 2012d. Prof. Dr. Ayhan Demirbas' scientometric biography. *Energy Education Science and Technology Part A—Energy Science and Research* 28:727–738.

Konur, O. 2012e. The evaluation of the research on the biofuels: A scientometric approach. *Energy Education Science and Technology Part A—Energy Science and Research* 28:903–916.

Konur, O. 2012f. The evaluation of the research on the biodiesel: A scientometric approach. *Energy Education Science and Technology Part A—Energy Science and Research* 28:1003–1014.

Konur, O. 2012g. The evaluation of the research on the bioethanol: A scientometric approach. *Energy Education Science and Technology Part A—Energy Science and Research* 28:1051–1064.

Konur, O. 2012h. The evaluation of the research on the microbial fuel cells: A scientometric approach. *Energy Education Science and Technology Part A—Energy Science and Research* 29:309–322.

Konur, O. 2012i. The evaluation of the research on the biohydrogen: A scientometric approach. *Energy Education Science and Technology Part A—Energy Science and Research* 29:323–338.

Konur, O. 2012j. The evaluation of the biogas research: A scientometric approach. *Energy Education Science and Technology Part A—Energy Science and Research* 29:1277–1292.

Konur, O. 2012k. The scientometric evaluation of the research on the production of bio-energy from biomass. *Biomass and Bioenergy* 47:504–515.

Konur, O. 2012l. The evaluation of the global energy and fuels research: A scientometric approach. *Energy Education Science and Technology Part A—Energy Science and Research* 30:613–628.

Konur, O. 2012m. The evaluation of the biorefinery research: A scientometric approach, *Energy Education Science and Technology Part A—Energy Science and Research* 2012(si):347–358.

Konur, O. 2012n. The evaluation of the bio-oil research: A scientometric approach, *Energy Education Science and Technology Part A—Energy Science and Research* 2012(si):379–392.

Konur, O. 2012o. What have we learned from the citation classics in energy and fuels: A mixed study. *Energy Education Science and Technology Part A—Energy Science and Research* 2012(si):255–268.

Konur, O. 2012p. The evaluation of the research on the biofuels: A scientometric approach. *Energy Education Science and Technology Part A—Energy Science and Research* 28:903–916.

Konur, O. 2013. What have we learned from the research on the International Financial Reporting Standards (IFRS)? A mixed study. *Energy Education Science and Technology Part D: Social Political Economic and Cultural Research* 5:29–40.

Kratochvil, D., P. Pimentel, and B. Volesky. 1998. Removal of trivalent and hexavalent chromium by seaweed biosorbent. *Environmental Science and Technology* 32:2693–2698.

Kratochvil, S. and B. Volesky. 1998. Advances in the biosorption of heavy metals. *Trends in Biotechnology* 16:291–300.

Lal, R. 2004. Soil carbon sequestration impacts on global climate change and food security. *Science* 304:1623–1627.

Leusch, A., Z.R. Holan, and B. Volesky. 1995. Biosorption of heavy-metals (Cd, Cu, Ni, Pb, Zn) by chemically-reinforced biomass of marine-algae. *Journal of Chemical Technology and Biotechnology* 62:279–288.

Lodeiro, P., J.L. Barriada, R. Herrero, and M.E.S. De Vicente. 2006. The marine macroalga *Cystoseira baccata* as biosorbent for cadmium(II) and lead(II) removal: Kinetic and equilibrium studies. *Environmental Pollution* 142:264–273.

Lynd, L.R., J.H. Cushman, R.J. Nichols, and C.E. Wyman. 1991. Fuel ethanol from cellulosic biomass. *Science* 251:1318–1323.

Malik, A. 2004. Metal bioremediation through growing cells. *Environment International* 30:261–278.

Matheickal, J.T. and Q.M. Yu. 1999. Biosorption of lead(II) and copper(II) from aqueous solutions by pre-treated biomass of Australian marine algae. *Bioresource Technology* 69:223–229.

Matheickal, J.T., Q.M. Yu, and G.M. Woodburn. 1999. Biosorption of cadmium(II) from aqueous solutions by pre-treated biomass of marine alga *Durvillaea potatorum*. *Water Research* 33:335–342.

Mehta, S.K. and J.P. Gaur. 2005. Use of algae for removing heavy metal ions from wastewater: Progress and prospects. *Critical Reviews in Biotechnology* 25:113–152.

Mohan, D. and K.P. Singh. 2002. Single- and multi-component adsorption of cadmium and zinc using activated carbon derived from bagasse—An agricultural waste. *Water Research* 36:2304–2318.

Munoz, R. and B. Guieysse. 2006. Algal-bacterial processes for the treatment of hazardous contaminants: A review. *Water Research* 40:2799–2815.

North, D. 1994. Economic-performance through time. *American Economic Review* 84:359–368.

Nourbakhsh, M., Y. Sag, D. Ozer, Z. Aksu, T. Kutsal, and A. Caglar. 1994. A comparative-study of various biosorbents for removal of chromium(vi) ions from industrial-waste waters. *Process Biochemistry* 29:1–5.

Paladugu, R., M.S. Chein, S. Gardezi, and L. Wise. 2002. One hundred citation classics in general surgical journals. *World Journal of Surgery* 26:1099–1105.

Park, D., Y.S. Yun, and J.M. Park. 2004. Reduction of hexavalent chromium with the brown seaweed *Ecklonia* biomass. *Environmental Science and Technology* 38:4860–4864.

Park, D., Y.S. Yun, and J.M. Park. 2005. Studies on hexavalent chromium biosorption by chemically-treated biomass of *Ecklonia* sp. *Chemosphere* 60:1356–1364.

Reddad, Z., C. Gerente, Y. Andres, and P. Le Cloirec. 2002. Adsorption of several metal ions onto a low-cost biosorbent: Kinetic and equilibrium studies. *Environmental Science and Technology* 36:2067–2073.

Roden, E.E. and M.M. Urrutia. 1999. Ferrous iron removal promotes microbial reduction of crystalline iron(III) oxides. *Environmental Science and Technology* 33:1847–1853.

Schiewer, S. and B. Volesky. 1995. Modeling of the proton-metal ion-exchange in biosorption. *Environmental Science and Technology* 29:3049–3058.

Schiewer, S. and M.H. Wong. 2000. Ionic strength effects in biosorption of metals by marine algae. *Chemosphere* 41:271–282.

Sheng, P.X., Y.P. Ting, J.P. Chen, and L. Hong. 2004. Sorption of lead, copper, cadmium, zinc, and nickel by marine algal biomass: Characterization of biosorptive capacity and investigation of mechanisms. *Journal of Colloid and Interface Science* 275:131–141.

Spolaore, P., C. Joannis-Cassan, E. Duran, and A. Isambert. 2006. Commercial applications of microalgae. *Journal of Bioscience and Bioengineering* 101:87–96.

Volesky, B. 1994. Advances in biosorption of metals—Selection of biomass types. FEMS *Microbiology Reviews* 14:291–302.

Volesky, B. 2001. Detoxification of metal-bearing effluents: Biosorption for the next century. *Hydrometallurgy* 29:203–216.

Volesky, B. and Z.R. Holan. 1995. Biosorption of heavy-metals. *Biotechnology Progress* 11:235–250.

Wang, B., Y.Q. Li, N. Wu, and C.Q. Lan. 2008. CO(2) bio-mitigation using microalgae. *Applied Microbiology and Biotechnology* 79:707–718.

Wang, J.L. and C. Chen. 2009. Biosorbents for heavy metals removal and their future. *Biotechnology Advances* 27:195–226.

Wilde, E.W. and J.R. Benemann. 1993. Bioremoval of heavy-metals by the use of microalgae. *Biotechnology Advances* 11:781–812.

Wrigley, N. and S. Matthews. 1986. Citation-classics and citation levels in geography. *Area* 18:185–194.

Yee, N., L.G. Benning, V.R. Phoenix, and F.G. Ferris. 2004. Characterization of metal-cyanobacteria sorption reactions: A combined macroscopic and infrared spectroscopic investigation. *Environmental Science and Technology* 38:775–782.

Yergin, D. 2006. Ensuring energy security. *Foreign Affairs* 85:69–82.

Yu, Q.M., J.T. Matheickal, P.H. Yin, and P. Kaewsarn. 1999. Heavy metal uptake capacities of common marine macro algal biomass. *Water Research* 33:1534–1537.

Yun, Y.S., D. Park, J.M. Park, and B. Volesky. 2001. Biosorption of trivalent chromium on the brown seaweed biomass. *Environmental Science and Technology* 35:4353–4358.

28

CDM Potential through Phycoremediation of Industrial Effluents

K. Sankaran, C. Naveen, K.K. Vasumathi, M. Premalatha, M. Vijayasekaran, and V.T. Somasundaram

CONTENTS

28.1 Introduction

The Clean Development Mechanism (CDM) is a market-based trading mechanism originated by the Kyoto Protocol. It functions to provide subsidy to the developing world for greenhouse gas (GHG) emission reduction, thereby encouraging developing countries to emit less than otherwise they would. It is designed with insight in such a way that the marginal cost of emission reduction especially in rapidly developing countries like India would be less than the developed ones. The reason for the lower marginal cost in developing countries is due to the high potential of CDM availability. The higher emissions in developing countries are due to scale of operation, efficiency of the technology adopted, fluctuation in plant load factor, and high air and water pollution loads. CDM can also be viewed as market and political mechanisms. It is a market mechanism because its subsidy is provided through the creation of certified emission reduction (CER), and carbon credit values will be calculated accordingly. It is political because it induces developing countries to come forward and participate in the Kyoto Protocol and benefit from the GHG emission reduction. The global warming potential of six major GHGs is given in Table 28.1.

TABLE 28.1

Global Warming Potential of Six Major GHGs

Gas	Global Warming Potential	Atmospheric Life (Years)
CO_2	1	5–200
CH_4	21	12
N_2O	310	114
HFC	140–11,700	1.4–260
PFC	6,500–9,200	10,000–50,000+
SF_6	23,900	3,200

Sources: IPCC, *Climate Change 2001—Mitigation. The Third Assessment Report of the Intergovernmental Panel on Climate Change*, B. Metz, O. Davidson, R. Swart, and J. Pan (eds.), Cambridge University Press, Cambridge, U.K., 2001a; IPCC, *Climate Change 2001. The Third Assessment Report of the Intergovernmental Panel on Climate Change*, Cambridge University Press, Cambridge, U.K., 2001b.

Industry contributes maximally to the development of any nation. It serves societies through its products/services. The size and number of industries are creeping up to meet worldwide demand. However, on the other hand, the wastewater being generated from industries possesses various threats to the environment and human health. The extent to which the industries concentrate well on the production process is not that much focused on wastewater treatment. The scenario could be changed if effective treatment techniques are employed, which can generate additional revenue to industry. The treatment should be chosen in such a way that it is not only minimizing/eradicating the waste pollution load but also producing some product demand out of it. Hence, biological treatment is one that can be adopted. However, it requires a lot of insight to be observed and checked for suitability for a longer run. Among various biological treatments, microalgae treatment is quite attractive as it possesses huge advantages over the other processes, one of which is its suitability for large scale. Employing microalgae for treatment studies results in multifold benefits such as reduced pollution load and carbon dioxide mitigation besides its regular benefit in terms of product and by-product or the whole cell.

Phycoremediation qualifies for CDM in many ways, which includes (1) microalgae sequestration of CO_2 from flue gas emitted due to combustion of fuel for thermal/electrical power generation, (2) nutrient removal from wastewater, (3) biogas production through anaerobic digestion of algae biomass, and (4) industry repowering by methane-rich biogas so obtained. Hence, phycoremediation has a very large scope and is truly a green technology, which generates revenues through all stages. This chapter explores the CDM potential of industrial wastewater treatment with microalgae.

28.2 Wastewater Characteristics

Understanding wastewater characteristics plays a vital role in phycoremediation. In general, industrial wastewater can be categorized as shown in Figure 28.1. It has physical, chemical, and biological characteristics (Figure 28.2). Microalgae grow well in wastewater in which nitrogen, potassium, and phosphorus are readily available. Characteristic highlights of model industrial wastewaters are shown in Table 28.2.

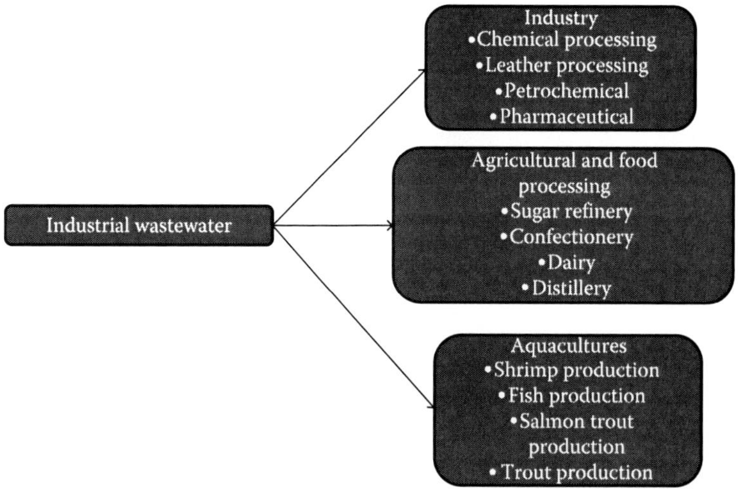

FIGURE 28.1
Industrial wastewater categories. (Adapted from Rawat, I. et al., Phycoremediation by high-rate algal ponds (HRAPs), in: *Biotechnological Applications of Microalgae: Biodiesel and Value Added Products*, ed. Bux, F., CRC Press, London, U.K., 2013, pp. 179–200.)

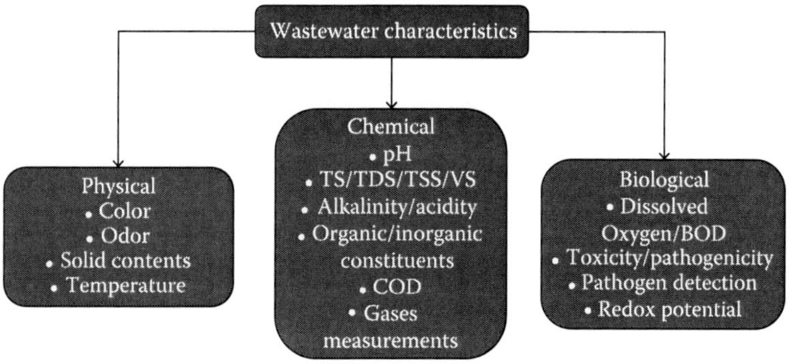

FIGURE 28.2
Wastewater characteristics. (Adapted from Rawat, I. et al., Phycoremediation by high-rate algal ponds (HRAPs), in: *Biotechnological Applications of Microalgae: Biodiesel and Value Added Products*, ed. Bux, F., CRC Press, London, U.K., 2013, pp. 179–200.)

28.3 Microalgae Species and Their Cultivation for Phycoremediation

Microalgae have been extensively used for various wastewater treatments since the 1950s (Oswald et al., 1953; Oswald and Gotaas, 1957; Fallowfield and Garrett, 1985; Lincoln and Earle, 1990; Ghosh, 1991; Oswald, 1991, 2003; Borowitzka, 1999; Hanumantha Rao et al., 2011). Wastewater treatment using microalgae (phycoremediation) would be the major long-term sustainable method among various biological methods employed since microalgae possess the strongest ability to fix CO_2 and take up nutrients from

TABLE 28.2

Characteristics of Model Industrial Wastewaters

Parameters	Values (mg/L)[a]	Values (mg/L)[b]	Values (mg/L)[c]
TDS	1,200	35,000	16,000
Total hardness	300	3,000	400
Sodium	—	250	600
Potassium	—	12,000	100
Chloride	—	8,000	400
Sulfate	—	4,000	250
Phosphate	12	360	10
BOD	250	10,000	5,000
COD	700	30,000	11,000

[a] Characteristics of dairy effluent (with dilution).
[b] Characteristics of biomethanated distillery effluent.
[c] Characteristics of oil drilling industrial effluent.

pollutant sites with less energy consumption and capital investment (Hirata et al., 1996; Murakami and Ikenouchi, 1997). Microalgae genera such as *Chlamydomonas, Chlorella,* and *Scenedesmus* are known for CO_2 biofixation and *Chlorella, Dunaliella, Botryococcus, Scenedesmus, Oscillatoria, Nostoc,* and *Phormidium* are known for wastewater treatment studies. Microalgae can be cultivated in extensive or open ponds, intensive or raceway ponds, closed photobioreactors in many designs, and closed fermentor systems. However, a closed fermentor system is not applicable for industrial wastewater treatment since microalgae do not degrade compounds in dark conditions. HRAPs exhibit better performance when compared to others in terms of economics. Photobioreactor cultivation can be encouraged for higher productivity of algae biomass since it provides a controlled environment.

28.4 CDM Potential through CO_2 Biofixation by Microalgae

One of the key challenges for environmental sustainability is global warming. It gets attention due to increased level of CO_2 in the atmosphere. Furthermore, the carbon credits trading scheme was created by the United Nations in conjunction with the Kyoto Protocol. Hence, the development of CO_2 capture and sequestration technology is vital today. Emission regulations are already framed for industries and vehicles for transportation in all countries. However, with present understanding and maturity, no single technology is available for CO_2 capture in a sustainable way. Microalgae CO_2 fixation is largely discussed and practiced by researchers all over the world since it has a tremendous ability to fix CO_2 in a sustainable route. Microalgae biomass has a rich fuel content, which can further be used to obtain bioethanol and biodiesel as alternatives to existing fossil fuels. Its growth is 100 times faster than terrestrial plants and doubling time ranges in hours (Tredici, 2010). It was reported that 1.8 kg of CO_2 is fixed by producing 1 kg of microalgae biomass and that microalgae cells contain approximately 50% carbon (Yusuf, 2007). Hence, this method would be more technically feasible

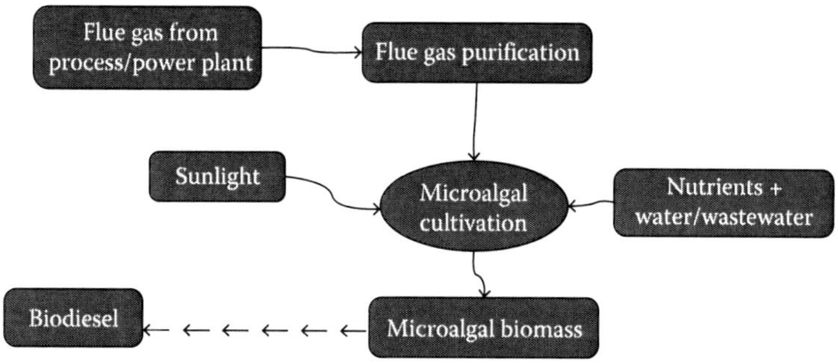

FIGURE 28.3
Basic diagram for CO_2 capture and biodiesel production route from microalgae.

TABLE 28.3
Growth Parameters for Different Algae Strains

Species	Temperature	pH	CO_2 (%)	Doubling Time (h)
Chlorococcum sp.	15–27	4–9	Up to 70	8
Chlorella sp.	15–45	3–7	Up to 60	2.5–8
Euglena gracilis	23–27	3.5	Up to 100	24
Goldieria sp.	Up to 50	1–4	Up to 100	13
Viridiella sp.	15–42	2–6	Up to 50	2.9
Synechococcus lividus	40–55	Up to 8.2	Up to 70	8

Source: Li, Y. et al., *Biotechnol. Progr.*, 24(4), 815, 2008.

because microalgae can biofix CO_2 from the atmosphere while producing renewable fuels like bioethanol and biodiesel. CO_2 balance is positive before biomass generation and becomes negative after biodiesel production. Hence, research works are highly in need to find various alternative routes of obtaining biodiesel/products from algae and to maintain the CO_2 balance as overall positive. The basic block diagram of CO_2 capture and route for CDM analysis is given in Figure 28.3. The important microalgae used in CO_2 fixation are listed in Table 28.3.

28.4.1 Calculation for CDM Potential through Microalgae CO_2 Mitigation and Biodiesel Production

Assume that 10 million liters of algal oil is the target for a 1-year scale. HRAP cultivation is considered. The calculation is done with the assumed conditions given in Table 28.4 with the close reference work of Davis et al. (2011) and Soares et al. (2013) for *Scenedesmus* species. The cost such as fixed cost, overheads, maintenance, insurance, tax, and lifetime of plant and other indirect costs are adopted from the National Renewable Energy Laboratory (NREL (2002)) and Soares et al. (2013) (Table 28.5). With reference to Xu et al. (2011), Soares et al. (2013) mentioned the biomass conversion yield as

$$1 \text{ ton of dry biomass} \rightarrow 370 \text{ kg of biodiesel.}$$

TABLE 28.4

Conditions Assumed for HRAP Cultivation

Assumed Conditions	Value
Scale of production	10 million L/year of algal oil
Algal productivity	30 g/m^2/day
Water evaporation	550 L/L algal oil
Volume of water for culturing	Approx. 166 × 10^6 L/year
Cell density	0.5 g/L
Operation	300 days

TABLE 28.5

Scale and Cost of Biodiesel Production through HRAP Cultivation

Scale and Cost	Value
Production	
Algal oil	10 million L/year
Biodiesel with 93% efficient extraction	9.3 million L/year
Land requirement	
Pond area	608 acre
Water demand	10,000 million L/year
CO_2 demand	1,45,000 tons/year
Urea	Approx. 5,000 tons/year
Phosphate (DAP)	Approx. 4,000 tons/year
Energy	80 (M kWh/year)
System cost	
Capital cost	Approx. 400 million US dollar

TABLE 28.6

Carbon Credit Calculation for CO_2 Capture by Microalgae

Particulars	Value	Unit
Density of biodiesel	0.88	kg/L
Calorific value	9,500	kcal/kg
Biodiesel production	9.3	Millions liters/year
Biodiesel production	8,184,000	kg/year
	8,184	Tons/year
Biodiesel yield from 1 ton dried biomass	0.37	Ton
Biomass requirement to produce 8184 tons of biodiesel	22,118.92	Tons/year
Productivity of biomass	30	g/m^2/day
	0.164	g/L/day
Cell density	0.5	g/L
CO_2 capture	0.2952	g/L/day
CO_2 capture daily	132.7	Tons/day
CO_2 capture for a year	39,810	Tons/year
Market value of carbon credit	12	US dollar/ton
Carbon credit value for annual biomass production	4,77,720	US dollar/year

Li et al. (2008) stated that 1 kg biomass of microalgae will capture 1.9 kg of CO_2. With these assumptions and data, the CDM potential can be calculated for CO_2 mitigation by *Scenedesmus* sp. with the route to produce biodiesel. The market value for carbon credit per ton of CO_2 would be US dollar 12/ton. Table 28.6 gives the carbon credit values for CO_2 capture.

28.5 CDM Potential by Methane Recovery through Anaerobic Digestion of Algal Biomass Produced from Wastewater Treatment

Effluent treatment with algae species is not a new technique and it has been practiced for the past five decades. Productivity and carbohydrate, lipid, and protein contents of microalgae depend on the type of algae used and type and load of effluent discharged on the microalgae cultivation site (Table 28.7). Studies have revealed that the use of wastewater as a substrate for biofuel production may make the process economically viable (Brennan and Owende, 2010; Boelee et al., 2011; Cho et al., 2011).

Focusing the growth of microalgae on biomass productivity rather than lipid productivity may be beneficial as higher amounts of biomass improve the viability of conversion to alternate fuels (Pittman et al., 2011). The yields of biomass from HRAPs depend on the type of effluent being treated with a specific regard to nutrient content. Figure 28.4a and b shows the treatment efficiency of microalgae used in dairy and oil drilling industrial effluent treatment. Preliminary studies were conducted for the phycoremediation of the recalcitrant compound melanoidin present in biomethanated distillery wastewater and it resulted in 20% COD reduction.

28.5.1 Calculation for CDM Potential through Power Generation from Biogas Produced in Treating Industrial Wastewater

Assume that microalgae have the productivity of 0.089 g/L/day (dry basis) and dry biomass of 20 g/m²/day. Operation days are taken as 300 for a year. One kilogram of biomass producing 0.5 m³ biogas has been taken as basis (Craggs et al., 2013) and calculation was worked out accordingly. Table 28.8 shows a detailed calculation for carbon

TABLE 28.7

Biomass, Lipid Content, and Productivity of Various Microalgae on Carpet Mill Industrial Effluent

Wastewater Type	Microalgae Species	Biomass Productivity (Dry Weight) (mg/L/Day)	Lipid Content (Dry Weight)%	Lipid Productivity (mg/L/Day)
Carpet mill	*Botryococcus braunii*	34	13.2	4.5
	Chlorella saccharophila	23	18.1	4.2
	Dunaliella tertiolecta	28	15.2	4.3
	Pleurochrysis carterae	33	12.0	4.0

Source: Adapted from Rawat, I. et al., *Appl. Energ.*, 88, 3411, 2011.

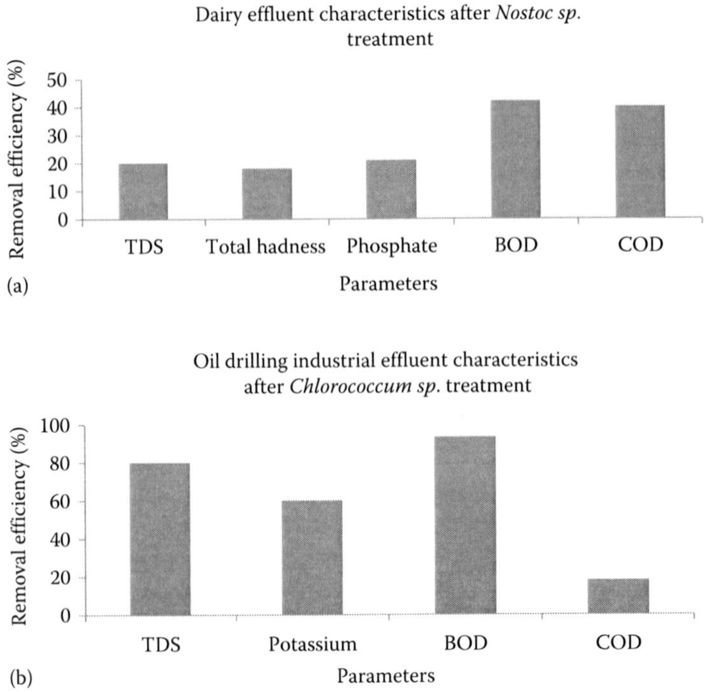

FIGURE 28.4
(a) Characteristics of dairy effluent after microalgae *Nostoc sp.* treatment. (b) Characteristics of oil drilling industrial effluent after *Chlorococcum sp.* treatment. (Adapted and calculated from Kotteswari, M. et al., *Int. J. Environ. Res. Dev.*, 2(1), 35, 2012; Sivasubramanian, V. and Muthukumaran, M., *J. Algal Biomass Utiliz.*, 3(4), 5, 2012.)

credit due to power generation through phycoremediation of industrial effluent. Carbon credit calculation is based on replacing the equivalent power from fossil- or coal-based power plants.

28.6 Concluding Remarks

Among various biological treatment techniques, microalgae are feasible for scale-up and profit maximization. The number of applications using microalgae is steadily increasing. Though phycoremediation has been practiced for several decades, research is highly needed to identify the best microalgae for treatment studies and mechanism through which these can nullify toxicity. The calculation worked out in this chapter was based on arriving at carbon credit values for CO_2 sequestration and power generation from algal biomass. However, detailed economics have to be done for the entire growth and product cycle of microalgae. If alternative upcoming energy-efficient technologies could be followed for downstream processing, algal technology will become a highly viable one.

TABLE 28.8

Carbon Accounting for Phycoremediation of Industrial Wastewater

Land for Cultivation (Acre)	Biomass (Tons/Day)	Biogas Generation (m^3/Day)	Energy (kcal/Day)	Power Generation (Megawatt)	Carbon Credit due to Power Generation/ Day (Minimum)[a]	Annual Carbon Credit (Minimum)	Revenue Generated by Carbon Credit (Minimum) US Dollar
200	16.19	8,095	43,227,300	2.09	32	9,600	1,15,200
400	32.37	16,185	86,427,900	4.19	64	19,200	2,30,400
600	48.56	24,280	129,655,200	6.28	96	28,800	3,45,600
800	64.75	32,375	172,882,500	8.38	128	38,400	4,60,800
1,000	80.94	40,470	216,109,800	10.47	160	48,000	5,76,000

[a] 1 kWh power generation is equivalent to 0.667 kg of CO_2 (minimum).

Acknowledgments

The author thanks CEESAT, NIT Tiruchirappalli for providing the necessary facilities to carry out the research. The author also thanks the industrial partner TDCL, Tiruchirappalli for the research work on microalgae treatment of distillery wastewater, Science and Engineering Research Board (SERB), and Confederation of Indian Industry (CII) for the financial support under Prime Minister's Fellowship Scheme for Doctoral Research.

References

Boelee, N.C., Temmink, H., Janssen, M., Buisman, C.J.N., and Wijffels, R.H. 2011. Nitrogen and phosphorus removal from municipal wastewater effluent using microalgal biofilms. *Water Research* 45:5925–5933.

Borowitzka, M.A. 1999. Commercial production of microalgae: Ponds, tanks, tubes and fermenters. *Journal of Biotechnology* 70:313–321.

Brennan, L. and Owende, P. 2010. Biofuels from microalgae—A review of technologies for production, processing, and extractions of biofuels and co-products. *Renewable and Sustainable Energy Reviews* 14:557–577.

Cho, S., Luong, T.T., Lee, D., Oh, Y.K., and Lee, T. 2011. Reuse of effluent water from a municipal wastewater treatment plant in microalgae cultivation for biofuel production. *Bioresource Technology* 102:8639–8645.

Craggs, R.J., Lundquist, T.J., and Benemann, J.R. 2013. Wastewater treatment and algal biofuel production. In *Development in Applied Phycology 5: Algae for Biofuels and Energy*, eds. M.A. Borowitzka and N.R. MoheiMani, pp. 153–164. Springer, Dordrecht, the Netherlands.

Davis, R., Aden, A., and Pienko, P.T. 2011. Techno-economic analysis of autotrophic microalgae for fuel production. *Applied Energy* 88:3524–3531.

Fallowfield, H.J. and Garrett, M.K. 1985. The treatment of wastes by algal culture. *Journal of Applied Bacteriology—Symposium Supplement* 59:187S–205S.

Ghosh, D. 1991. Ecosystem approach to low-cost sanitation in India. In *Ecological Engineering for Wastewater Treatment*, eds. C. Etnier and B. Guterstam, pp. 63–79. Bokeskogen, Gothenburg, Sweden.

Gutierrez, R., Gutierrez-Sanchez, R., and Nafidi, A. 2008. Trend analysis using non-homogeneous stochastic diffusion processes. Emission of CO_2; Kyoto protocol in Spain. *Stochastic Environmental Research and Risk Assessment* 22:57–66.

Hanumantha Rao, P., Ranjith Kumar, R., Raghavan, B.G., Subramanian, V.V., and Sivasubramanian, V. 2011. Application of phycoremediation technology in the treatment of wastewater from a leather-processing chemical manufacturing facility. *Water South Africa* 37:7–14.

Hirata, S., Hayashitani, M., Taya, M., and Tone, S. 1996. Carbon dioxide fixation in batch culture of *Chlorella* sp. using a photobioreactor with a sunlight-collection device. *Journal of Fermentation Bioengineering* 81:470–472.

IPCC, 2001a, *Climate Change 2001—Mitigation. The Third Assessment Report of the Intergovernmental Panel on Climate Change*. B. Metz, O. Davidson, R. Swart, and J. Pan (eds.). Cambridge University Press, Cambridge, U.K.

IPCC, 2001b, *Climate Change 2001. The Third Assessment Report of the Intergovernmental Panel on Climate Change*. Cambridge University Press, Cambridge, U.K.

Kondili, E.M. and Kaldellis, J.K. 2007. Biofuel implementation in East Europe: Current status and future prospects. *Renewable and Sustainable Energy Reviews* 11:2137–2151.

Kotteswari, M., Murugesan, S., and Ranjith Kumar, R. 2012. Phycoremediation of dairy effluent by using the microalgae *Nostoc* sp. *International Journal of Environmental Research and Development* 2(1):35–43.
Li, Y., Horsman, M., Wu, N., Lan, C.Q., and Dubois-Calero, N. 2008. Biofuels from microalgae. *Biotechnology Progress* 24(4):815–820.
Lincoln, E.P. and Earle, J.F.K. 1990. Wastewater treatment with microalgae. In *Introduction to Applied Phycology*, ed. I. Akatsuka, pp. 429–446, SPB Academic Publications, The Hague, the Netherlands.
Murakami, M. and Ikenouchi, M. 1997. The biological CO_2 fixation and utilization project by RITE (2). *Energy Conversion and Management* 38:S493–S497.
NREL (National Renewable Energy Laboratory). Energy analysis report, Golden, CO, 2002.
Oswald, W.J. 1991. Introduction to advanced integrated wastewater ponding systems. *Water Science Technology* 24:1–7.
Oswald, W.J. 2003. My sixty years in applied algology. *Journal of Applied Phycology* 15:99–106.
Oswald, W.J. and Gotaas, H. 1957. Photosynthesis in sewage treatment. *Transactions of the American Society for Civil Engineering* 122:73–105.
Oswald, W.J., Gotaas, H.B., Ludwig, H.F., and Lynch, V. 1953. Algal symbiosis in oxidation ponds. *Sewage and Industrial Waste* 25:692–704.
Pittman, J.K., Dean, A.P., and Osundeko, O. 2011. The potential of sustainable algal biofuel production using wastewater resources. *Bioresource Technology*, 102:17–25.
Rawat, I., Ranjith Kumar, R., and Bux, F. 2013. Phycoremediation by high-rate algal ponds (HRAPs). In *Biotechnological Applications of Microalgae: Biodiesel and Value Added Products*, ed. F. Bux, pp. 179–200. CRC Press, London, U.K.
Rawat, I., Ranjith Kumar, R., Mutanda, T., and Bux, F. 2011. Dual role of microalgae: Phycoremediation of domestic wastewater and biomass production for sustainable biofuels production. *Applied Energy* 88:3411–3424.
Sivasubramanian, V. and Muthukumaran, M. 2012. Large scale phycoremediation of oil drilling effluent. *Journal of Algal Biomass Utilization* 3(4):5–17.
Soares, F.R., Martins, G., and Seo, E.S.M. 2013. An assessment of the economic aspects of CO_2 sequestration in a route for biodiesel production from microalgae. *Environmental Technology* 34:13–14, 1777–1781.
Tredici, M.R. 2010. Photobiology of microalgae mass cultures: Understanding the tools for the next green revolution. *Biofuels* 1:143–162.
Usui, N. and Ikenouchi, M. 1997. The biological CO_2 fixation and utilization project by RITE(1): Highly-effective photobioreactor system. *Energy Conversion and Management* 38(Supplement 1): S487–S492.
Xu, L., Brilman, D.W.F., Withag, J.A.M., Brem, G., and Kersten, S. 2011. Assessment of a dry and wet route for the production of biofuels from microalgae: Energy balance analysis. *Bioresource Technology* 102:5113–5122.
Yusuf, C. 2007. Biodiesel from microalgae. *Biotechnology Advances* 25:294–206.

Section VII

Commercialization and Global Market

29

Commercialization of Marine Algae-Derived Biofuels

Anoop Singh, Poonam Singh Nigam, and Dheeraj Rathore

CONTENTS

29.1 Introduction

Microalgae feedstock is gaining attention in the present energy scenario due to its fast growth potential with very little land use coupled with high lipid, carbohydrate, protein, nucleic acid, and bioactive metabolite contents and capacity to reduce pollution levels by utilizing nutrients from wastewater and carbon dioxide from flue gases (Amaro et al. 2013; Nigam and Singh 2011; Singh and Olsen 2011; Singh et al. 2011a,b,c). All of these properties render microalgae an excellent source for biofuels such as biodiesel, bioethanol, biohydrogen, and biomethane, as well as a number of other valuable pharmaceutical and nutraceutical products (Rathore and Singh 2013; Singh and Gu 2010; Singh et al. 2012).

The commercial viability of algae-based biofuel production depends on economics of the technology and sources of inputs required during algal cultivation. The principal investment for an algae biomass project may be split into the costs associated with algal cultivation (algal biomass growth, harvesting, dewatering) and algal oil extraction systems. Singh and Gu inferred that regardless of whatsoever advances might come in terms of technological and biological innovations, the hard fact remains that the commercial marketplace shall have to have an enthusiasm for funding capital intensive energy projects by ensuring the risk–return ratio to the acceptable stratum for debt and equity financiers (Singh and Gu 2010).

Algal biofuel plant cost can be fractioned into plant establishment cost (e.g., engineering, permitting, infrastructure preparation, balance of plant, installation and integration, land or leasing, and contractor fees) and operation and management (O&M) costs (e.g., nutrients [generally N–P–K], CO_2 distribution, and water replenishment due to evaporative losses, utilities, components replacement, and labor costs) (Table 29.1).

TABLE 29.1

Costs Involved in an Algal Biofuel Plant

Plant Establishment Cost	Operation and Management Cost
Engineering	Nutrients
Permitting	CO_2 distribution
Infrastructure preparation	Water replenishment
Plant, installation, and integration	Utilities
Contractor fees	Components replacement
Land or leasing	Labor

Source: Adapted from Singh, J. and Gu, S., *Renew. Sustain. Energy Rev.*, 14, 2596, 2010.

Singh and Gu (2010) pointed out that significant speculation in research would be required to assure high levels of productivity that can match commercial-scale requirements. Worldwide, a number of companies and government organizations (Table 29.2) have previously assessed different methodologies as well as designs and prepared cost estimates for the production of algal biomass at commercial scale. Many of these investigations recommend that an algal biofuel plant may be effectively developed on land adjacent to power stations (Singh and Gu 2010), so that CO_2 from exhausts can be utilized for cultivation of algal biomass that could effectively reduce the input cost involved in the cultivation.

TABLE 29.2

Algae Production Companies around the World

Company	Region	Cultivation Method
Live Fuels, Menlo Park, California	United States	Open pond
OriginOil Inc., Los Angeles, California	United States	Open pond
PetroSun, Scottsdale, Arizona	United States	Open pond
Kelco, San Diego, California	United States	Natural settings
Neptune Industries, Boca Raton, Florida	United States	Natural settings
Blue Marble Energy, Seattle, Washington	United States	Natural settings
A2BE Carbon Capture, Boulder, Colorado	United States	Closed systems
GreenFuel Technologies, Cambridge, Massachusetts	United States	Closed systems
Solazyme, Inc., San Francisco, California	United States	Closed systems
Algenol Biofuels, Fort Myers, Florida	United States	Closed systems
Sapphire Energy, San Diego, California	United States	Closed systems
Inventure Chemical Technology, Seattle, Washington	United States	Closed systems
Solena, Washington State, DC	United States	Closed systems
Solix Biofuels, Fort Collins, Colorado	United States	Closed systems
XL Renewables, Phoenix, Arizona	United States	Closed systems
Aurora Biofuels, Alameda, California	United States	Closed systems
Bionavitas, Snoqualmie, Washington	United States	Closed systems
Cellana, Kailua-Kona, Hawaii	United States	Closed systems
Neste Oil, Helsinki, Finland	Europe	Open pond
Ingrepo, Borculo, the Netherlands	Europe	Open pond
Biofuel Systems, Alicante, Spain	Europe	Natural settings
Seambiotic, Ashkelon, Israel	Mediterranean	Open pond
Aquaflow Bionomics, Richmond, New Zealand	New Zealand	Natural settings

Source: Singh, J. and Gu, S., *Renew. Sustain. Energy Rev.*, 14, 2596, 2010.

The algal products market is estimated to be of the order of billions of dollars per year with more than 20 different commercial products, especially food products, including nutraceuticals and functional foods (Hudek et al. 2014). This chapter aims to provide a holistic view of commercialization of algal biofuels, limitations, and key strategy for commercialization of algal biofuels.

29.2 Sustainability and Life Cycle Assessment

The World Commission on Environment and Development defined the term *sustainability* as "the development that meets the needs of the present without compromising the ability of future generations to meet their own needs" (UNCED 1992). Energy conversion, utilization, and access underlie many of the great challenges associated with sustainability, environmental quality, security, and poverty (Korres et al. 2010, 2011; Singh and Olsen 2012). Sustainability assessment of products or technologies is normally seen as encompassing impacts in three dimensions, that is, social, environmental, and economic (Elkington 1998). These three dimensions form the backbone of sustainability standards. To replace the fossil fuels with biofuels, there is a need to maximize the environmental and social value of biofuels that is also important for the future of biofuels industry and market potential depends on being cost competitive with fossil fuels. These three interrelated goals must stay in balance for biofuels to remain sustainable (Singh and Olsen 2012).

The sustainability of biofuels production depends on the net energy gain fixed in the biofuels, which depends on the production process parameters, such as land type where the biomass is produced; the amount of energy-intensive inputs; and the energy input for harvest, transport, and running the processing facilities (Haye and Hardtke 2009), emissions, and their production cost. The most used indicators to measure the biofuels sustainability includes life cycle energy balance, quantity of fossil energy substituted per hectare, coproduct energy allocation, life cycle carbon balance, and changes in soil utilization (Silva Lora et al. 2011). Gnansounou suggested that due to the multidimensional impact of biofuels, the sustainability impact assessment of policies is as relevant as the sustainability assessment of production pathways and regulatory impact assessment (Gnansounou 2011). The interactions among various sustainability issues make the assessment of biofuel development difficult and complicated. The complexity during the whole biofuel production chain generates significantly different results due to the differences in input data, methodologies applied, and local geographical conditions (Singh and Olsen 2011; Yan and Lin 2009). The life cycle assessment (LCA) can be used as an effective tool for addressing environmental sustainability issues.

LCA is a tool to define the environmental burdens from a product, process, or activity by identifying and quantifying energy and materials usage, as well as waste discharges, assessing the impacts of these wastes on the environment, and it also evaluates the opportunities for environmental improvements over the whole life cycle (Singh and Olsen 2012). LCA considers all attributes or aspects of natural environment, human health, and resources (Korres et al. 2010) and can be defined as a method for analyzing and assessing environmental impacts of a material, product, or service along its entire life cycle (ISO 2005). The life cycle of biofuels includes each and every step involved in production, for example, raw material production and extraction, processing, transportation, manufacturing, storage, distribution, and utilization of the biofuel. In addition, the life cycle stages can have harmful effects or benefits of different environmental, economical, and social dimensions; due to these facts, the complete fuel chains from different perspectives is of crucial importance in order to achieve sustainable biofuels (Markevičius et al. 2010).

LCA has been the method of choice in recent years for various kinds of new technologies for bioenergy and carbon sequestration. LCA is a universally accepted approach of determining the environmental consequences of a particular product over its entire production cycle (Pant et al. 2011). Various scientists have employed LCA for biofuel production systems (Gnansounou et al. 2009; Reinhard and Zah 2011; Singh and Olsen 2013), and some useful results considering the factors (e.g., biomass, technologies, use, system boundary, allocation, reference system) affecting the outcome of the analysis have been obtained (Singh et al. 2010). Gnansounou et al. also stated that monitoring reduction of greenhouse gas (GHG) emissions and estimations of substitutional efficiency with respect to fossil fuels is subject to significant uncertainty and inaccuracy associated with the LCA approach (Gnansounou et al. 2009).

In spite of a lot of thrust on algal biofuel development, an adequate LCA algal biofuel production study is still not available. LCA studies reported by researchers lack in one aspect or other. Some of these studies present a comparative evaluation of biofuel production LCA from various biomass feedstocks and some have carried out LCA on the basis of lab scale data after extrapolation. Some studies have not included all the factors required for an exact LCA (Singh and Gu 2010). Yang et al. examined the life cycle of water and nutrients usage of microalgae-based biodiesel production. The results indicated that using seawater or wastewater can reduce the freshwater usage life cycle by as much as 90%. However, a significant amount of freshwater (about 400 kg/kg biodiesel) must be used for culture whether or not sea/wastewater serves as the culture medium or how much harvested water is recycled. They also analyzed the life cycle usages of nitrogen, phosphorous, potassium, magnesium, and sulfur and reported that when the harvest water is 100% recycled, the usage of these nutrients decreases by approximately 55%. Using sea/wastewater for algal culture can reduce nitrogen usage by 94% and eliminate the need of potassium, magnesium, and sulfur. Overall, the water footprint of microalgae-based biodiesel production gradually decreases from north to south as solar radiation and temperature increase (Yang et al. 2011). Clarens et al. have compared the life cycle of algae feedstocks with three traditional terrestrial biocrops, namely, corn, canola, and switchgrass. The authors present a *cradle to gate* analysis, which included all the products and process upstream of delivered dry biomass, rather than *cradle to grave* approach that would include all processes up to oil utilization (Clarens et al. 2010). Lardon et al. reported in an LCA study of algal biodiesel production that drying and hexane extraction accounted for up to 90% of the total process energy (Lardon et al. 2009). Levine et al. concluded that drying algal biomass and treating it as a substitute for terrestrial oilseeds in traditional solvent extraction and subsequent transesterification processes is not likely to be a net energy positive route (Levine et al. 2010). A comparative LCA study of an algal biodiesel facility has been undertaken by Lardon et al. to assess the energetic balance and the potential environmental impacts of the whole process chain, from the biomass production to biodiesel combustion. The outcome confirms the potential of microalgae as an energy source but highlights the imperative necessity of decreasing energy and fertilizer consumption (Lardon et al. 2009). Therefore, control of nitrogen stress during the culture and optimization of wet extraction seem to be valuable options.

Kim and Dale have modeled the algae production process using an open-pond raceway configuration and reported that energy production from the four crops is net positive, while terrestrial crops were found to have significantly lower energy use as well as gas emission and water use than algae. These results concluded that algae require more fossil-based carbon to produce the same amount of energy; however, the land use offers a clear-cut and appreciable improvement over corn, canola, and switchgrass (Kim and Dale 2004). An analysis of the energy life cycle for production of biomass using the oil-rich microalgae *Nannochloropsis* sp. was performed by Jorquera et al., which included raceway ponds and tubular and flat-plate photobioreactors (PBRs) for algal cultivation. The net energy ratio (NER) for each process was calculated. The NER of a system has been defined as the ratio of the total energy produced (energy content of the oil and residual biomass) over the energy content of PBR construction and material plus the energy required for all plant operations. The results showed that the use of horizontal tubular PBRs is not economically feasible (NER < 1) and that the estimated NERs for flat-plate PBRs and raceway ponds is >1. The NER for ponds and flat-plate PBRs could be raised to significantly higher values if the lipid content of the biomass were increased to 60% dw/cwd (Jorquera et al. 2010). Campbell et al. conducted a comparative LCA study of a national production system designed for Australian conditions to compare biodiesel production from algae (with three different scenarios for carbon dioxide supplementation and two different production rates) with canola and ultralow sulfur (ULS) diesel. Comparisons of GHG emissions (g CO_2-eq per t km) and costs (¢ per t km) are given. Algae GHG emissions (27.6–18.2) compare very favorably with canola (35.9) and ULS diesel (81.2). Costs are not so favorable, with algae ranging from 2.2 to 4.8, compared with canola (4.2) and ULS diesel (3.8). This highlights the need for a high production rate to make algal biodiesel economically attractive (Campbell et al. 2011). In another study, Collet et al. conducted an LCA of biogas production from the microalgae *Calluna vulgaris* and the results are compared to algal biodiesel and to first-generation biodiesels. These results suggest that the impacts generated by the production of methane from microalgae are strongly correlated with electricity consumption. Progress can be achieved by decreasing the mixing costs and circulation between different production steps or by improving the efficiency of the anaerobic process under controlled conditions (Collet et al. 2011). Singh and Olsen in an LCA study compare six different biodiesel production pathways (three different harvesting techniques, i.e., aluminum as flocculent, lime flocculent, and centrifugation, and two different oil extraction methods, i.e., supercritical CO_2 [sCO_2] and press and cosolvent extraction) with conventional diesel in a EURO 5 passenger car used for transport purpose. The cultivation of *Nannochloropsis* sp. is considered in a flat-panel PBR (FPPBR). Results of the study depicted that impacts of algal biodiesel on climate change were far better than conventional diesel, but impacts on human health, ecosystem quality, and resources were higher than conventional diesel. Singh and Olsen recommended more practical data at a pilot-scale production plant with maximum utilization of by-products generated in production to produce a sustainable algal biodiesel (Singh and Olsen 2013).

The majority of the reported LCAs of algal biofuel does not sound very positive about the advantage of algal biofuel over the biofuels from terrestrial crops. However, none of these studies has been carried out using any commercial plant data due to nonavailability of data from a commercial plant. The land use impact of microalgae offers a significant advantage over other oil crops. Thus, it may be possible to utilize microalgae-based fuels cost-effectively, with a judicious processing. However, a detailed LCA for a larger algal biofuel production system could further clarify.

29.3 Commercial Products

Algal biomass production, from past practical experience, is for production of other useful products, not for biofuel production. The largest algal biomass–based market is food products, including nutraceuticals and functional foods. There are multiple product streams associated with the separation process. Production occurs in freshwater, marine water, and harvests from natural areas of seas and oceans. Because of the diversity of growth environments and species, many more potential products are possible, and additional research is to be encouraged (Hudek et al. 2014). This is depicted from studies reported that algal biofuel production alone is not very economically attractive; therefore, the concept of multiproduct delivery (biorefinery) system could add value to make the system sustainable and increase the economic viability of algal biofuel production system. The by-products generated during the production of biofuels from the algal biomass can be utilized for production of other useful products, enhance the income, and reduce the cost of algal biofuel.

Algae have great diversity and production of algal products have great promise that algal features may serve to enhance a large variety of current products and produce new ones. The current expansion of market for algae-based goods indicates a bright future for the research and development of algal products (Hudek et al. 2014). There are many potential useful products that can be discovered and developed through future research (De Luca et al. 2012).

TABLE 29.3

Refined Commercial Products from Algae

Carotenoids	**Proteins and Amino Acids**	**Polysaccharides**
Astaxanthin	Allophycocyanin	Agar
Beta-carotene	Kainic acid	Alginates
Bixin	Mycosporine-like amino acids	Ascophyllan
Fucoxanthin		Carrageenan
Lutein	Phycocyanins	Fucoidans
Lycopene	Phycoerythrins	Furcellaran
		Polyuronides
Polyunsaturated Fatty Acids	**Sterols**	**Isotopically Labeled Compounds**
Arachidonic acid (AA)	Desmosterol	Carbon 13 isotopes
Docosahexaenoic acid (DHA)	Fucosterol	Nitrogen 15 isotopes
Eicosapentaenoic acid (EPA)		Hydrogen 2 isotopes
Gamma-linolenic acid (GLA)		

Sources: Bhatia, S.K. et al., *Pharmacogn. Res.*, 2, 45, 2010; Bixler, H.J. and Porse, H., *J. Appl. Phycol.* 23, 321, 2011; Burja, A.M. and Radianingtyas, H., Nutraceuticals and functional foods from marine microbes: An introduction to a diverse group of natural products isolated from marine macroalgae, microalgae, bacteria, fungi, and cyanobacteria, in Barrow, C.J. and Shahidi, F. (eds.), *Marine Nutraceuticals and Functional Foods*, CRC Press, Boca Raton, FL, 2008, pp. 367–403; Fitton, J.H. et al., Marine algae and polysaccharides with therapeutic applications, in Barrow, C.J. and Shahidi, F. (eds.), *Marine Nutraceuticals and Functional Foods*, CRC Press, Boca Raton, FL, 2008, pp. 345–365; Hudek, K., et al., Commercial products from algae, in Bajpai, R., Prokop, A., and Zappi, M. (eds.), *Algal Biorefineries*, Vol. 1, Cultivation of Cells and Products, Springer Science+Business Media, Dordrecht, the Netherlands, 2014, pp. 275–295; Miyashita, K. and Hosokawa, M., Beneficial health effects of seaweed carotenoid, fucoxanthin, in Barrett, C.J. and Shahidi, F. (eds.), *Marine Nutraceuticals and Functional Foods*, CRC Press, Boca Raton, FL, 2008, pp. 297–319; Spolaore, P. et al., *J. Biosci. Bioeng.*, 101, 87, 2006; Venugopal, V., *Marine Products for Healthcare*, CRC Press, Boca Raton, FL, 2009.

Hudek et al. summarized refined products from algal biomass having commercial use in their study (Table 29.3). The largest market for algae is the food industry, and commercial products include algal powders, polyunsaturated fatty acids, polysaccharides, carotenoids, and food colorants (Capelli et al. 2010; Delgado-Vargas and Paredes-Lopez 2003; Fitton et al. 2008; Henrikson 2009; Miyashita and Hosokawa 2008; Olaizola 2007; Shahidi 2008; Venugopal 2009; Yuan 2008). Algae are also used for animal feed, especially in aquaculture (Spolaore et al. 2006; Venugopal 2009). Hydrocolloids (e.g., agars, alginates, and carrageenans) are marketed for their functional use in foods as well as their use in many industrial products (Bixler and Porse 2011; Venugopal 2009). Products from algae are also used in sunscreen, cosmetics, and pharmaceutical products. Substances from algae also have many applications in scientific research, for example, many cyanobacteria produce phycobiliproteins, which are used in analytical chemistry (Hudek et al. 2014).

29.4 Limitations in Commercialization

Commercial-scale cultivation of algal biomass is mainly for nutritional products by small- and medium-scale production systems and producing huge biomass annually. The main algal genera currently cultivated at commercial scale for the production of nutritional product include *Spirulina*, *Chlorella*, *Haematococcus*, and *Dunaliella*. The algal biomass production mainly takes place in China, Japan, United States, Australia, India, and some other countries. The commercial production of *Spirulina* has been continuing since 1982 at Earthrise Farms in California (Henrikson 2009). Ponds with liners are used to grow *Spirulina* at a high pH, with CO_2 bubbled into the water to supply carbon and nitrogen; potassium and other needed nutrients are supplied from sterilized sources to avoid contamination. *Spirulina* grow optimally at 35°C–38°C; however, it is not possible to control temperature in the open-pond system. Commercial operations have been set up in the sunny California desert, in Hawaii, and in other locations where sun and temperature are appropriate for good production. The Colorado River is used as the source of freshwater at Earthrise Farms (Hudek et al. 2014). Filtration is used to separate the *Spirulina* from the water and drying is done by spray drying technique, as it is a quick process and preserves heat-sensitive nutrients (Henrikson 2009; Hu 2004).

The most commonly used technique for commercial-scale production of algal biomass is raceway ponds (Chisti 2012; Terry and Raymond 1985) with an exceedingly low productivity. Whether raceways can be used to produce sufficient algal biomass inexpensively for making biofuels is debatable (Chisti 2012). Closed culture systems such as PBRs are more productive but are expensive and require a lot of energy to operate. A high energy expenditure for producing algal biomass resulted in low NER, which makes the algal biofuel unsustainable and unacceptable (Singh et al. 2012; Wongluang et al. 2013). Novel biomass production methods that rely on sunlight, industrial flue gases, marine water, and wastewater could be inexpensive and energy efficient. Zijffers et al. reported that a high-density algae culture is attainable only at high light intensities with sufficient nutrient supply and only in a growth system with a shallow depth (Zijffers et al. 2010). Algal biomass productivity can be substantially enhanced not only through engineering but also by addressing issues of algal biology. In terms of the biomass and oil productivity, the existing culture systems do not come close to the biological limits of productivity. This is mainly because an algal culture is inevitably light limited due to mutual shading by the cells and is susceptible to photoinhibition at a light level that is only about 10% of the peak midday sunlight level in a tropical region (Chisti 2013). Turbulence-induced light–dark cycling of the algal cells is claimed to increase the photosynthetic efficiency (Camacho Rubio et al. 2003; Grobbelaar 2009; Mussgnug et al. 2007), but attaining the suitable intensity of turbulence is impractical in a commercial culture device in view of the energy expenditure (Chisti 2013).

Chisti pointed out that the main constraint in commercial production of algal biomass is that a low-cost point supply of concentrated carbon dioxide collocated with the other essential resources is necessary for sustainable production of algal fuels. Sustainability of algal biofuel production requires the ability to almost fully recycle the nutrients that are necessary for algal culture. The requirement for nitrogen biofixation in algal strains ought to be an important long-term objective to support production of algal fuels. The limited supply of freshwater could pose a significant limitation for commercial-scale production even if marine algae are used. Processes for recovering energy or useful products from algal biomass left after the extraction of oil are required for achieving a net positive energy balance. Genetic and metabolic engineering of microalgae to boost production of fuel oil and ease its recovery are essential for commercialization of algal fuels (Chisti 2013). Hudek et al. reported that in order to produce a commercial product, the algae must be able to be grown at reasonable cost and with appropriate quality control to have a product with sufficient purity to serve customer needs. The processing required to produce a consistent product with desired functional properties must be at an economical cost to compete with other existing products. The value of products from algae must exceed the cost of production to have good sales and a growing market (Hudek et al. 2014).

29.5 Key Strategies for Commercialization

Algae can be grown to produce useful products with less land use. However, commercial-scale cultivation of algal biomass for the production of commercial goods must be carried out with appropriate environmental management. In a study, Thurmond has identified five key strategies: fatter, faster, cheaper, easier, and fractionation marketing approaches to help producers to reduce costs and accelerate the commercialization of algae biodiesel, biocrude, and drop-in fuels. A primary strategy for most algae biofuels producers is to identify algae species having high oil content and growth rates to produce biodiesel, biocrude, and drop-in fuels. Algae producers are interested in selecting algal species with a high triglyceride (TAG) oil content for biodiesel and biocrude production. The utilization of fatter algae with 60% oil content can significantly reduce the size and footprint of algae biofuels production systems, resulting in significant reduction in capital and operating costs and savings for systems. The significant economic challenge for commercial-scale production of algal biomass is identifying low-cost oil extraction and harvesting methods. Cost reductions in algae production systems are essential for algae producers to establish economically sustainable and profitable enterprises. Algae production systems are a composite of several subsets of systems (e.g., production, harvesting, extraction, drying systems); reducing the steps in algae biofuel production is essential to make the production system easier, better, and low cost. The free fatty acids can produce DHA (fish oil equivalents), omega 3 and omega 6 (heart healthy oils), as well as valuable products, for example, beta-carotene and other nutraceutical and pharmaceutical supplements from carotenoids. The other fractions of the algal biomass contain valuable chemicals or compounds that can be used to produce green plastics, green detergents, cleaners, and polymers that are biodegradable, nontoxic, and can be sold at a premium price. These biomass coproduct marketing strategies will be critical to the success of aspiring algae biodiesel, biocrude, renewable diesel, aviation fuel, and drop-in fuel producers (Thurmond 2009). Singh and Gu concluded that a hybrid biofuel refinery concept can be implemented profitably for microalgae-based biofuels. CO_2 and nutrients may be recycled for microalgae culture and thus help in carbon sequestration. Apart from fuels, other valuable products can be cogenerated to make the commercialization process a profitable venture (Singh and Gu 2010).

29.6 Conclusion

Microalgae have been explored as sources of fuels and chemicals from the past few decades. Different algal production systems are being considered for a sustainable cultivation of algal biomass at a commercial scale. The barriers in the commercialization of algal fuels include impossibly high demands of resources, higher production cost, and the need to achieve a higher NER that will be well above unity. The commercial production of algal biofuel can be sustainable by combining the production of other useful products from the by-products generated during the production of biofuel and utilization of all waste in the production cycle. A successful biorefinery requires not only a set of efficient and productive processing technologies but also an effective system of collection and transportation of raw materials, economic analysis of integrated systems, public education, and favorable policy environment.

References

Amaro HM, Barros R, Guedes AC, Sousa-Pinto I, and Malcata FX. 2013. Microalgal compounds modulate carcinogenesis in the gastrointestinal tract. *Trends Biotechnol* 31:92–98.

Bhatia SK, Sharma AG, Namdeo BB, Chaugule M, and Kavales N. 2010. Broad-spectrum sun-protective action of porphyra-334 derived from *Porphyra vietnamensis*. *Pharmacogn Res* 2:45–49.

Bixler HJ and Porse H. 2011. A decade of change in the seaweed hydrocolloids industry. *J Appl Phycol* 23:321–335.

Burja AM and Radianingtyas H. 2008. Nutraceuticals and functional foods from marine microbes: An introduction to a diverse group of natural products isolated from marine macroalgae, microalgae, bacteria, fungi, and cyanobacteria. In: Barrow CJ and Shahidi F (eds.), *Marine Nutraceuticals and Functional Foods*, CRC Press, Boca Raton, FL. pp. 367–403.

Camacho Rubio F, García Camacho F, Fernández Sevilla JM, Chisti Y, and Molina Grima E. 2003. A mechanistic model of photosynthesis in microalgae. *Biotechnol Bioeng* 81:459–473.

Campbell PK, Beer T, and Batten D. 2011. Life cycle assessment of biodiesel production from microalgae in ponds. *Bioresour Technol* 102:50–56.

Capelli B, Keily S, and Cysewski GR. 2010. *The Medical Research of Astaxanthin*, Cyanotech Corporation, Kailua-Kona, HI.

Chisti Y. 2012. Raceways-based production of algal crude oil. In: Posten C and Walter C (eds.), *Microalgal Biotechnology: Potential and Production*, de Gruyter, Berlin, German. pp. 113–146.

Chisti Y. 2013. Constraints to commercialization of algal fuels. *J Biotechnol* 167:201–214.

Clarens AF, Resurreccion EP, White MA, and Colosi LM. 2010. Environmental life cycle comparison of algae to other bioenergy feedstocks. *Environ Sci Technol* 44(5):1813–1819.

Collet P, Hélias A, Lardon L, Ras M, Goy R-A, and Steyer J-P. 2011. Life-cycle assessment of microalgae culture coupled to biogas production. *Bioresour Technol* 102:207–214.

De Luca Y, Salim V, Atsumi SM, and Yu F. 2012. Mining the biodiversity of plants: A revolution in the making. *Science* 236:1658–1661.

Delgado-Vargas F and Paredes-Lopez O. 2003. *Natural Colorants for Food and Nutraceutical Uses*, CRC Press, Boca Raton, FL.

Elkington J. 1998. *Cannibals with Forks—The Triple Bottom Line of 21st Century Business*, New Society Publishers, Gabriola, British Columbia, Canada.

Fitton JH, Irhimeh MR, and Teas J. 2008. Marine algae and polysaccharides with therapeutic applications. In: Barrow CJ and Shahidi F (eds.), *Marine Nutraceuticals and Functional Foods*, CRC Press, Boca Raton, FL. pp. 345–365.

Gnansounou E. 2011. Assessing the sustainability of biofuels: A logic-based model. *Energy* 36:2089–2096.

Gnansounou E, Dauriat A, Villegas J, and Panichelli L. 2009. Life cycle assessment of biofuels: Energy and greenhouse gas balances. *Bioresour Technol* 100:4919–4930.

Grobbelaar JU. 2009. Upper limits of photosynthetic productivity and problems of scaling. *J Appl Phycol* 21:519–522.

Haye S and Hardtke CS. 2009. The roundtable on sustainable biofuels: Plant scientist input needed. *Trends Plant Sci* 14(8):409–412.

Henrikson R. 2009. Earth Food Spirulina, Ronore Enterprises, Inc., Hana, Maui, HI, http://www.spirulinasource.com/PDF.cfm/EarthFoodSpirulina.pdf (accessed on December 14, 2012).

Hu Q. 2004. Industrial production of microalgal cell-mass and secondary products-major industrial species: *Arthrospira (Spirulina) platensis*. In: *Handbook of Microalgal Culture: Biotechnology and Applied Phycology*, Blackwell, Oxford, U.K. pp. 264–272.

Hudek K, Davis LC, Ibbini J, and Erickson L. 2014. Commercial products from algae. In: Bajpai R, Prokop A, and Zappi M (eds.), *Algal Biorefineries*, Vol. 1, Cultivation of Cells and Products, Springer Science+Business Media, Dordrecht, the Netherlands. pp. 275–295.

ISO. 2005. 14040-Environmental management—Life cycle assessment—Requirements and guidelines. International Standard Organisation, Geneva, Switzerland, p. 54.

Jorquera O, Kiperstok A, Sales EA, Embirucu M, and Ghirardi M. 2010. Comparative energy life-cycle analyses of microalgal biomass production in open ponds and photobioreactors. *Bioresour Technol* 101:1406–1413.

Kim S and Dale BE. 2004. Cumulative energy and global warming impact from the production of biomass for biobased products. *J Ind Ecol* 7:147–162.

Korres NE, Singh A, Nizami AS, and Murphy JD. 2010. Is grass biomethane a sustainable transport biofuel? *Biofuels Bioprod Biorefining* 4:310–325.

Korres NE, Thamsiriroj T, Smyth BM, Nizami AS, Singh A, and Murphy JD. 2011. Grass bio-methane for agriculture and energy. In: Lichtfouse E (ed.), *Sustainable Agriculture Reviews*, Vol. 7, Genetics, Biofuels and Local Farming Systems, Springer Science+Business Media B.V. *Amsterdam*, the Netherlands, pp. 5–49.

Lardon L, Helias A, Sialve B, Steyer J-P, and Bernard O. 2009. Life-cycle assessment of biodiesel production from microalgae. *Environ Sci Technol* 43:6475–6481.

Levine RB, Pinnarat T, and Savage PE. 2010. Biodiesel production from wet algal biomass through in situ lipid hydrolysis and supercritical transesterification. *Energy Fuels* 24:5235–5243.

Markevičius A, Katinas V, Perednis E, and Tamasauskiene M. 2010. Trends and sustainability criteria of the production and use of liquid biofuels. *Renew Sustain Energy Rev* 14:3226–3231.

Miyashita K and Hosokawa M. 2008. Beneficial health effects of seaweed carotenoid, fucoxanthin. In: Barrett CJ and Shahidi F (eds.), *Marine Nutraceuticals and Functional Foods*, CRC Press, Boca Raton, FL. pp. 297–319.

Mussgnug JH, Thomas-Hall S, Rupprecht J, Foo A, Klassen V, McDowall A, Schenk PM, Kruse O, and Hankamer B. 2007. Engineering photosynthetic light capture: Impacts on improved solar energy to biomass conversion. *Plant Biotechnol J* 5(6):802–814.

Nigam PS and Singh A. 2011. Production of liquid biofuels from renewable resources. *Prog Energ Combust Sci* 37:52–68.

Olaizola M. 2008. The production and health benefits of astaxanthin. In: *Marine Nutraceuticals and Functional Foods*, CRC Press, Boca Raton, FL pp. 321–343.

Pant D, Singh A, Bogaert GV, Gallego YA, Diels L, and Vanbroekhoven K. 2011. An introduction to the life cycle assessment (LCA) of bioelectrochemical systems (BES) for sustainable energy and product generation: Relevance and key aspects. *Renew Sustain Energy Rev* 15:1305–1313.

Rathore D and Singh A. 2013. Biohydrogen production from microalgae. In: Gupta VK and Tuohy MG (eds.), *Biofuel Technologies*, Springer-Verlag, Berlin, Germany. pp. 317–333.

Reinhard J and Zah R. 2011. Consequential life cycle assessment of the environmental impacts of an increased rapemethylester (RME) production in Switzerland. *Biomass Bioenergy* 35(6):2361–2373.

Shahidi F. 2008. Omega-3 oils: Sources, applications, and health effects. In: Barrow CJ and Shahidi F (eds.), *Marine Nutraceuticals and Functional Foods*, CRC Press, Boca Raton, FL. pp. 23–61.

Silva Lora EE, Escobar Palacio JC, Rocha MH, Grillo Renó ML, Venturini OJ, and del Olmo OA. 2011. Issues to consider, existing tools and constraints in biofuels sustainability assessments. *Energy* 36:2097–2110.

Singh A, Nigam PS, and Murphy JD. 2011a. Renewable fuels from algae: An answer to debatable land based fuels. *Bioresour Technol* 102:10–16.

Singh A, Nigam PS, and Murphy JD. 2011b. Mechanism and challenges in commercialisation of algal biofuels. *Bioresour Technol* 102:26–34.

Singh A and Olsen SI. 2011. Critical analysis of biochemical conversion, sustainability and life cycle assessment of algal biofuels. *Appl Energy* 88:3548–3555.

Singh A and Olsen SI. 2012. Key issues in life cycle assessment of biofuels. In: Gopalakrishnan K, van Leeuwen JH, and Brown RC (eds.), *Sustainable Bioenergy and Bioproducts Value Added Engineering Applications*, Springer-Verlag, London, U.K. pp. 213–228.

Singh A and Olsen SI. 2013. Comparison of algal biodiesel production pathways using life cycle assessment tool. In: Singh A, Pant D, and Olsen SI (eds.), *Life Cycle Assessment of Renewable Energy Sources*, Springer-Verlag, London, U.K. pp. 145–168.

Singh A, Olsen SI, and Nigam PS. 2011c. A viable technology to generate third generation biofuel. *J Chem Technol Biotechnol* 86:1349–1353.

Singh A, Pant D, Korres NE, Nizami AS, Prasad S, and Murphy JD. 2010. Key issues in life cycle assessment of ethanol production from lignocellulosic biomass: Challenges and perspectives. *Bioresour Technol* 101(13):5003–5012.

Singh A, Pant D, Olsen SI, and Nigam PS. 2012. Key issues to consider in microalgae based biodiesel. *Energy Educ Sci Technol A: Energy Sci Res* 29:687–700.

Singh J and Gu S. 2010. Commercialization potential of microalgae for biofuels production. *Renew Sustain Energy Rev* 14:2596–2610.

Spolaore P, Joannis-Cassan C, Duran E, and Isambert A. 2006. Commercial applications of microalgae. *J Biosci Bioeng* 101:87–96.

Terry KL and Raymond LP. 1985. System design for the autotrophic production of microalgae. *Enzyme Microb Technol* 7:474–487.

Thurmond W. 2009. Five key strategies for algae biofuels commercialisation. In: *Algae 2020, Emerging Market Online* (EMO), Houston, TX, http://www.emerging-markets.com/algae/Algae%2020205%20Strategies%20For%20Commercialization%20Emerging%20Markets%20Online.pdf (accessed on 27th December 2014).

UNCED. 1992. Promoting sustainable human settlement development. The United Nation Conference on Environment and Development Agenda 21, United Nations Division for Sustainable Development, Rio de Janeiro, Brazil.

Venugopal V. 2009. *Marine Products for Healthcare*, CRC Press, Boca Raton, FL.
Wongluang P, Chisti Y, and Srinophakun T. 2013. Optimal hydrodynamic design of tubular photobioreactors. *J Chem Technol Biotechnol* 88:55–61.
Yan J and Lin T. 2009. Bio-fuels in Asia. *Appl Energy* 86:S1–S10.
Yang J, Xu M, Zhang X, Hu Q, Sommerfeld M, and Chen Y. 2011. Life-cycle analysis on biodiesel production from microalgae: Water footprint and nutrients balance. *Bioresour Technol* 102:159–165.
Yuan YV. 2008. Marine algal constituents. In: Barrow CJ and Shahidi F (eds.), *Marine Nutraceuticals and Functional Foods*, CRC Press, Boca Raton, FL. pp. 259–296.
Zijffers J-WF, Schippers KJ, Zheng K, Janssen M, Tramper J, and Wijffels RH. 2010. Maximum photosynthetic yield of green microalgae in photobioreactors. *Mar Biotechnol* 12:708–718.

30

Algal High-Value Consumer Products

Ozcan Konur

CONTENTS

30.1 Introduction

30.1.1 Issues

Global warming, air pollution, and energy security have been some of the most important public policy issues in recent years (Jacobson 2009, Wang et al. 2008, Yergin 2006). With increasing global population, food security has also become a major public policy issue (Lal 2004). The development of biofuels generated from biomass has been a long-awaited solution to these global problems (Demirbas 2007, Goldember 2007, Lynd et al. 1991). However, the development of early biofuels produced from agricultural plants such as sugarcane (Goldemberg 2007) and agricultural wastes such as corn stovers (Bothast and Schlicher 2005) has resulted in a series of substantial concerns about food security (Godfray et al. 2010). Therefore, the development of algal biofuels as a third-generation biofuel has been considered as a major solution for the problems of global warming, air pollution, energy security, and food security (Chisti 2007, Demirbas 2007, Kapdan and Kargi 2006, Spolaore et al. 2006, Volesky and Holan 1995).

Besides the great success of algal biomass for the production of biofuels, the success of the production of algal medicines for a number of life-limiting diseases such as cancer (An and Carmicael 1994, Ohta et al. 1994) and HIV (Boyd et al. 1997, Witvrouw and DeClercq 1997) has relatively been unknown in the public. Furthermore, algal biomass has also been used in the production of pigments (Johnson and An 1991, Lorenz and Cysewski 2000) and biomaterials (Gombotz and Wee 1998) as well as in other commercial areas (Grima et al. 2003, Herrero et al. 2006, Pulz and Gross 2004, Spolaore et al. 2006).

Although there have been many reviews on the use of marine algae for the production of high-value consumer products (e.g., Burja et al. 2001, Gombotz and Wee 1998, Spolaore et al. 2006) and there have been a number of scientometric studies on the algal biofuels (Konur 2011), there has not been any study on the citation classics on algal high-value consumer products as in the other research fields (Baltussen and Kindler 2004a,b, Dubin et al. 1993, Gehanno et al. 2007, Konur 2012a,b, 2013, Paladugu et al. 2002, Wrigley and Matthews 1986).

The development of efficient drugs for cancer and HIV patients has been one of the major public policy initiatives in recent years aiming to reduce the public burden from these life-limiting diseases to the society at large (Calman 1984, Chesney et al. 2000, Dunkelschetter 1984, Gray et al. 2001, Hall et al. 2008, Hammer et al. 2008, Kamb et al. 1998, Mathers and Loncar 2006, Portenoy et al. 1994, Potosky et al. 1993, Shepard et al. 2005, Wlodawer and Erickson 1993).

As North's new institutional theory suggests, it is important to have up-to-date information about current public policy issues to develop a set of viable solutions to satisfy the needs of key stakeholders (Konur 2000, 2002a–c, 2006a,b, 2007a,b, 2012c, North 1994).

Therefore, brief information on a selected set of citation classics on algal high-value consumer products is presented in this chapter to inform key stakeholders about the production of algal medicines, pigments, biomaterials, and other commercial products, thus complementing a number of recent scientometric studies on the biofuels and global energy research (Konur 2011, 2012d–p).

30.1.2 Methodology

A search was carried out in the Science Citation Index Expanded (SCIE) and Social Sciences Citation Index (SSCI) databases (version 5.11) in September 2013 to locate the papers relating to the algal high-value consumer products using the keyword set of (biotech* or "bio* tech*" or commercial or pharmaceutical* or "high* value* product*" or astaxanthin or cosmetic* or antibiotic* or antiviral or "anti* viral*" or antineoplastic or drug* or "aids virus" or "Human* Immunodeficiency* Virus" or cancer) AND (alga* or "macro* alga*" or "micro* alga*" or macroalga* or microalga* or cyanobacter* or seaweed* or diatoms or "sea* weed*" or reinhardtii or braunii or chlorella or sargassum or gracilaria or spirulina) in the abstract pages of the papers. For this chapter, it was necessary to embrace the broad algal search terms to include diatoms, seaweeds, and cyanobacteria as well as to include many spellings of high-value consumer products rather than the simple keywords of algae and biotechnology.

It was found that there were 6800 papers indexed between 1980 and 2013. The key subject categories for research were marine and freshwater biology (1253 references, 18.4%), biotechnology and applied microbiology (1202 references, 17.6%), biochemistry and molecular biology (721 references, 10.6%), and plant sciences (632 references, 9.3%), altogether comprising

50% of all the references on the algal high-value consumer products. It was also necessary to focus on the key references by selecting articles and reviews.

The located highly cited 50 papers were arranged by decreasing number of citations for two groups of papers: 27 and 23 papers for nonexperimental research and experimental research, respectively. In order to check whether these collected abstracts correspond to the larger sample on these topical areas, new searches were carried out for each topical area.

The summary information about the located citation classics is presented under two major headings of nonexperimental and experimental research in the order of the decreasing number of citations for each topical area.

The information relating to the document type, the affiliation and location of the authors, the journal, the indexes, the subject area of the journal, the concise topic, the total number of citations received, and the total average number of citations received per year is given in the tables for each paper.

30.2 Nonexperimental Studies on Algal High-Value Consumer Products

30.2.1 Introduction

The nonexperimental research on algal high-value consumer products has been one of the most dynamic research areas in recent years. Twenty-seven citation classics in the field of algal high-value consumer products with more than 123 citations were located, and the key emerging issues from these papers are presented in the decreasing order of the number of citations in Table 30.1. These papers give strong hints about the determinants of efficient algal high-value consumer product design and emphasize that it is possible in the light of the biotechnological advances in recent years to meet the demands of the consumers in several fields such as medicine, medical aids, pigments, and health foods.

The papers were dominated by researchers from 16 countries, usually through the intra-country institutional collaboration, and they were multiauthored. The United States (six papers), Germany (four papers), France (three papers), and Belgium, Brazil, Japan, and Spain (two papers each) were the most prolific countries.

Similarly, all these papers were published in journals indexed by the SCI and/or SCIE. There were no papers indexed by the SSCI or Arts & Humanities Citation Index (A&HCI). The number of citations ranged from 123 to 471 and the average number of citations per annum ranged from 5.9 to 58.1. Only 3 of the papers were articles, while 24 of them were reviews.

The development of algal drugs has been one major research area among the nonexperimental studies with 14 papers. Two and three of them focus on cancer and HIV, respectively. There were seven papers on general studies, while there were three papers each for the areas of algal biomaterials and algal pigments.

30.2.2 Nonexperimental Research

30.2.2.1 Algal Drugs

30.2.2.1.1 General Algal Drugs

Burja et al. (2001) discuss pharmacological chemistry of natural products from marine cyanobacteria with a focus on the metabolic pathways in a review paper with 277 citations.

TABLE 30.1

Nonexperimental Research on Algal High-Value Consumer Products

No.	Paper References	Year	Document	Affiliation	Country	No. of Authors	Journal	Index	Subjects	Topic	Total No. of Citations	Total Average Citations per Annum	Rank
1	Gombotz and Wee	1998	R	Immunex Res. Dev. Corp.	United States	2	*Adv. Drug Deliv. Rev.*	SCI, SCIE	Phar. Phar.	Biomats—alginate matrices	471	29.4	1
2	Spolaore et al.	2006	R	Ecole Cent. Paris, Evaflor	France	4	*J. Biosci. Bioeng.*	SCI, SCIE	Biotech. Appl. Microb., Food Sci. Tech.	Commercial applications	465	58.1	2
3	Burja et al.	2001	R	Heriot Watt Univ., Univ. Perpignan	Scotland, France	5	*Tetrahedron*	SCI, SCIE	Chem. Org.	Algal drugs—pharmacological chemistry	277	21.3	6
4	Johnson and An	1991	R	Hai Tai Confectionery Co.	S. Korea	2	*Crit. Rev. Biotechnol.*	SCI, SCIE	Biotech. Appl. Microb.	Pigments	266	11.6	8
5	Pulz and Gross	2004	R	IGV Inst Getreideverarbeitung	Germany	2	*Appl. Microbiol. Biotechnol.*	SCI, SCIE	Biotech. Appl. Microb.	Commercial applications	264	26.4	9
6	Lorenz and Cysewski	2000	A	Cyanotech Corp.	United States		*Trends Biotechnol.*	SCI, SCIE	Biotech. Appl. Microb.	Pigments	253	18.1	10
7	Grima et al.	2003	R	Massey Univ., Univ. Almeria	New Zealand, Spain	5	*Biotechnol. Adv.*	SCI, SCIE	Biotech. Appl. Microb.	Metabolites	251	22.8	11
8	Witvrouw and DeClercq	1997	R	Univ Catholique L.	Belgium	2	*Gen. Pharmacol.*[a]	SCI, SCIE	Phar. Phar.	Drugs—HIV	225	13.2	12
9	Herrero et al.	2006	R	CSIC	Spain	3	*Food Chem.*	SCI, SCIE	Chem. Appl., Food Sci. Tech.	Food	211	26.4	13

(*Continued*)

TABLE 30.1 (*Continued*)

Nonexperimental Research on Algal High-Value Consumer Products

No.	Paper References	Year	Document	Affiliation	Country	No. of Authors	Journal	Index	Subjects	Topic	Total No. of Citations	Total Average Citations per Annum	Rank
10	Proksch et al.	2002	R	Univ. Dusseldorf	Germany	3	*Appl. Microbiol. Biotechnol.*	SCI, SCIE	Biotech. Appl. Microb.	Algal drugs—microbial diversity	194	16.2	14
11	Simmons et al.	2005	R	Oregon St. Univ.	United States	5	*Mol. Cancer Ther.*	SCIE	Oncol.	Drugs—cancer	186	20.7	16
12	Huneck	1999	R	na	Germany	1	*Naturwis-senschaften*	SCI, SCIE	Mult. Sci.	Algal drugs—lichens	170	11.3	20
13	De Philippis and Vincenzini	1998	R	Univ Florence, CNR	Italy	2	*Fems Microbiol. Rev.*	SCI, SCIE	Microbial.	Drugs—HIV	169	10.6	21
14	De Clercq	2000	A	Univ Catholique L.	Belgium	1	*Medicinal. Res. Rev.*	SCI, SCIE	Chem. Med., Phar. Phar.	Drugs—HIV	166	11.9	22
15	Clardy et al.	2006	R	Harvard Univ.	United States	3	*Nat. Biotechnol.*	SCI, SCIE	Biotech. Appl. Microb.	Algal drugs—antibiotics	161	20.1	23
16	Smit	2004	R	Univ Cape Town	S. Africa	1	*J. Appl. Phycol.*	SCI, SCIE	Biotech. Appl. Microb., Mar. Fresh. Biol.	Algal drugs	159	15.9	27
17	Tan	2007	R	Nanyang Technol. Univ.	Singapore	1	*Phyto-chemistry*	SCI, SCIE	Bioch. Mol. Biol., Plant Sci.	Drugs—cancer	159	22.7	28
18	Boussiba	2000	R	Ben Gurion Univ.	Israel		*Physiol. Plant.*	SCI, SCIE	Plant Sci.	Pigments	156	11.1	29
19	Borowitzka	1995	R	Murdoch Univ.	Australia	1	*J. Appl. Phycol.*	SCI, SCIE	Biotech. Appl. Microb., Mar. Fresh. Biol.	Algal drugs—prospect	154	8.1	32

(*Continued*)

TABLE 30.1 (*Continued*)

Nonexperimental Research on Algal High-Value Consumer Products

No.	Paper References	Year	Document	Affiliation	Country	No. of Authors	Journal	Index	Subjects	Topic	Total No. of Citations	Total Average Citations per Annum	Rank
20	Jonas and Farah	1998	R	Desen-volvimento Biotechnol, GBF	Brazil, Germany	2	*Polym. Degrad. Stabil.*	SCI, SCIE	Poly. Sci.	Biomats—microbial cellulose	151	9.4	35
21	Rinaudo	2008	R	Univ. Grenoble 1	France	1	*Polym. Int.*	SCI, SCIE	Poly. Sci.	Biomats—polysaccharides	151	25.2	36
22	Cardozo et al.	2007	R	Univ. Sao Paulo, Univ. Cruzeiro Sul	Brazil	11	*Comp. Biochem. Physiol. C Toxicol. Pharmacol.*	SCI, SCIE	Bioch. Mol. Biol., Endocr. Metabol., Toxic., Zool.	Metabolites	140	20.0	39
23	Apt and Behrens	1999	R	Martek Biosci. Corp.	United States	2	*J. Phycol.*	SCI, SCIE	Plant Sci., Mar. Fresh. Biol.	Commercial applications	137	9.1	41
24	Moore	1996	A	Univ. Hawaii	United States	1	*J. Indust. Microbiol.*	SCI, SCIE	Bioch. Mol. Biol.	Cyclic peptides	137	7.6	42
25	Hussein et al.	2006	R	Int. Res. Ctr. Tradit. Med., Toyama Univ.	Japan	5	*J. Nat. Prod.*	SCI, SCIE	Plant. Sci., Chem. Med., Phar. Phar.	Algal drugs—astaxanthin	125	15.6	46
26	Belay et al.	1993	R	Dainippon Ink Chem. Inc.	Japan	3	*J. Appl. Phycol.*	SCI, SCIE	Biotech. Appl. Microb., Mar. Fresh. Biol.	Algal drugs—spirulina	124	5.9	47
27	Higuera-Ciapara et al.	2006	R	Ctr. Invest. Alimentac. Desarrollo AC	Mexico	3	*Cr. Rev. Food Sci. Nutr.*	SCI, SCIE	Food Sci. Tech., Nutr. Diet.	Algal drugs—astaxanthin	123	15.4	50

SCI, Science Citation Index; SCIE, Science Citation Index Expanded; SSCI, Social Sciences Citation Index; A, article; R, review.

[a] General pharmacology continues as vascular pharmacology.

They argue that natural products will become even more important in the future. Marine cyanobacteria have been proven to biosynthesize secondary metabolites of unlimited structural diversity that can further be enlarged by structure modification by applying strategies of combinatorial chemistry. In the future, they argue that progress would depend substantially on methods and techniques that accelerate validation of new target genes potentially related to human disorders. Validation can be provided either by functional gene expression in animal models or by immediate transfer to high-throughput screening in order to get rapid access to a low-molecular mass effector that gives rise to both a drug candidate and target validation.

Proksch et al. (2002) discuss marine drugs with a focus on the current status and microbiological implications in a review paper with 194 citations. They note that numerous natural products from marine invertebrates show striking structural similarities to known metabolites of microbial origin, suggesting that microorganisms (bacteria, microalgae) are at least involved in their biosynthesis or are in fact the true sources of these respective metabolites. They argue that this assumption is corroborated by several studies on natural products from sponges that proved these compounds are localized in symbiotic bacteria or cyanobacteria. Recently, molecular methods have successfully been applied to study microbial diversity in marine sponges and to gain evidence for an involvement of bacteria in the biosynthesis of the bryostatins in the bryozoan *Bugula neritina*.

Huneck (1999) discusses the significance of lichens and their metabolites in a review paper with 170 citations. He notes that lichens, symbiotic organisms of fungi and algae, synthesize numerous metabolites, the "lichen substances," which comprise aliphatic, cycloaliphatic, aromatic, and terpenic compounds. He argues that lichens and their metabolites have a manifold biological activity such as antiviral, antibiotic, antitumor, allergenic, plant growth inhibitory, antiherbivore, and enzyme inhibitory. Usnic acid, a very active lichen substance, is used in pharmaceutical preparations. Similarly, large amounts of *Pseudevernia furfuracea* and *Evernia prunastri* are processed in the perfume industry, and some lichens are sensitive reagents for the evaluation of air pollution.

Clardy et al. (2006) discuss new antibiotics from bacterial natural products in a review paper with 161 citations. They note that advances in technology have sparked resurgence in the discovery of natural product antibiotics from bacterial sources. In particular, efforts have refocused on finding new antibiotics from old sources (e.g., streptomycetes) and new sources (e.g., other actinomycetes, cyanobacteria, and uncultured bacteria). This has resulted in several newly discovered antibiotics with unique scaffolds and/or novel mechanisms of action, with the potential to form a basis for new antibiotic classes addressing bacterial targets that are currently underexploited.

Smit (2004) discusses the medicinal and pharmaceutical uses of seaweed natural products in a review paper with 159 citations. They search the literature for natural products from marine macroalgae in the Rhodophyta, Phaeophyta, and Chlorophyta with biological and pharmacological activity. Substances that receive the most attention from pharmaceutical companies for use in drug development or from researchers in the field of medicine-related research included sulfated polysaccharides as antiviral substances, halogenated furanones from *Delisea pulchra* as antifouling compounds, and kahalalide F from a species of *Bryopsis* as a possible treatment for lung cancer, tumors, and AIDS. Other substances such as macroalgal lectins, fucoidans, kainoids, and aplysiatoxins are routinely used in biomedical research, and a multitude of other substances have known biological activities.

Borowitzka (1995) discusses microalgae as sources of pharmaceuticals and other biologically active compounds in an early review paper with 154 citations. He notes that many cyanobacteria produce antiviral and antineoplastic compounds. A range of pharmacological activities have also been observed with extracts of microalgae; however, the active principles are as yet unknown in most cases. He argues that several of the bioactive compounds may find application in human or veterinary medicine or in agriculture, while others should find application as research tools or as structural models for the development of new drugs. He asserts that microalgae are particularly attractive as natural sources of bioactive molecules since they have the potential to produce these compounds in culture that enable the production of structurally complex molecules that are difficult or impossible to produce by chemical synthesis.

Hussein et al. (2006) discuss astaxanthin, a carotenoid with potential in human health and nutrition, in a review paper with 125 citations. They note that Compound 1 has considerable potential and promising applications in human health and nutrition. The recent scientific literature is covered on the most significant activities of Compound 1, including its antioxidative and anti-inflammatory properties; its effects on cancer, diabetes, the immune system, and ocular health; and other related aspects. They also discuss the green microalga *Haematococcus pluvialis*, the richest source of natural Compound 1, and its utilization in the promotion of human health, including the antihypertensive and neuroprotective potentials of Compound 1, emphasizing the experimental data on the effects of dietary astaxanthin on blood pressure, stroke, and vascular dementia in animal models.

Belay et al. (1993) discuss the potential health benefits of *Spirulina* in an early paper with 124 citations. They note that *Spirulina* is a microscopic filamentous alga that is rich in proteins, vitamins, essential amino acids, minerals, and essential fatty acids like gamma-linolenic acid. It is produced commercially and sold as a food supplement in health food stores around the world. However, numerous people are looking into the possible therapeutic effects of *Spirulina*. Many preclinical studies and a few clinical studies suggest several therapeutic effects ranging from reduction of cholesterol and cancer to enhancing the immune system, increasing intestinal lactobacilli, and reducing nephrotoxicity by heavy metals and drugs and radiation protection. They present a critical review of some data on therapeutic effects of *Spirulina*.

Higuera-Ciapara et al. (2006) discuss the chemistry and applications of astaxanthin in a review paper with 123 citations. They note that astaxanthin plays a key role as an intermediary in reproductive processes and that synthetic astaxanthin dominates the world market, but recent interest in natural sources of the pigment has increased substantially. Common sources of natural astaxanthin are the green algae *H. pluvialis*, the red yeast (*Phaffia rhodozyma*), as well as crustacean by-products. Astaxanthin possesses an unusual antioxidant activity that has caused a surge in the nutraceutical market for the encapsulated product. Also, health benefits such as cardiovascular disease prevention, immune system boosting, bioactivity against *Helicobacter pylori*, and cataract prevention have been associated with astaxanthin consumption. They argue that research on the health benefits of astaxanthin is very recent and has mostly been performed in vitro or at the preclinical level with humans.

30.2.2.1.2 Algal Cancer Drugs

Simmons et al. (2005) discuss marine natural products as anticancer drugs in a review paper with 186 citations. These compounds range in structural class from simple linear peptides, such as dolastatin 10, to complex macrocyclic polyethers, such as halichondrin B; equally as diverse are the molecular modes of action by which these molecules impart their biological activity. They highlight several marine natural products and their synthetic derivatives that are currently undergoing clinical evaluation as anticancer drugs.

Tan (2007) discusses bioactive natural products from marine cyanobacteria for drug discovery in a review paper with 159 citations. They note that the prokaryotic marine cyanobacteria continue to be an important source of structurally bioactive secondary metabolites. A majority of these molecules are nitrogen-containing compounds biosynthesized by large multimodular nonribosomal polypeptide (NRP) or mixed polyketide-NRP enzymatic systems. He presents a total of 128 marine cyanobacterial alkaloids, published in the literature between 2001 and 2006 with emphasis on their biosynthesis and biological activities. In addition, he identifies a number of highly cytotoxic compounds such as hectochlorin, lyngbyabellins, apratoxins, and aurilides as potential lead compounds for the development of anticancer agents. They discuss the distribution of natural product biosynthetic genes as well as the mechanisms of tailoring enzymes involved in the biosynthesis of cyanobacterial compounds.

30.2.2.1.3 Algal HIV/AIDS Drugs

Witvrouw and DeClercq (1997) discuss sulfated polysaccharides extracted from marine algae as potential antiviral drugs in an early review paper with 225 citations. They note that heparin and other sulfated polysaccharides are potent and selective inhibitors of HIV-1 replication in cell culture. Since 1988, the activity spectrum of sulfated polysaccharides has been shown to extend to various enveloped viruses, including viruses that emerge as opportunistic pathogens (e.g., herpes simplex virus and cytomegalovirus) in immunosuppressed patients. As potential anti-HIV drug candidates, sulfated polysaccharides offer a number of promising features. They are able to block HIV replication in cell culture at concentrations as low as 0.1–0.01 μg/mL without toxicity to the host cells at concentrations up to 2.5 mg/mL. Some polysulfates show a differential inhibitory activity against different HIV strains, suggesting that marked differences exist in the target molecules with which polysulfates interact. They argue that these polysulfates not only inhibit the cytopathic effect of HIV but also prevent HIV-induced syncytium (giant cell) formation. Furthermore, experiments carried out with dextran sulfate samples of increasing molecular weight and with sulfated cyclodextrins of different degrees of sulfation have shown that antiviral activity increases with increasing molecular weight and degree of sulfation. These polysulfates lead very slowly to virus drug resistance development, and they show activity against HIV mutants that have become resistant to reverse transcriptase inhibitors, such as AZT and tetrahydro-imidazo [4,5,1-jk] [1,4] benzodiazepin 2 (1H) thione (TIBO). They conclude that polysulfates exert their anti-HIV activity by shielding off the positively charged sites in the V3 loop of the viral envelope glycoprotein (gp120). The V3 loop is necessary for virus attachment to cell surface heparan sulfate, a primary binding site, before more specific binding occurs to the CD4 receptor of $CD4^+$ cells. This general mechanism also explains the broad antiviral activity of polysulfates against enveloped viruses. Variations in the viral envelope glycoprotein region may result in differences in the susceptibility of different enveloped viruses to compounds that interact with their envelope glycoproteins. They assert that polysulfates may be considered as potentially effective in a vaginal formulation to protect against HIV infection.

De Philippis and Vincenzini (1998) discuss exocellular polysaccharides from cyanobacteria and their possible applications and provide an overview of the current knowledge on both released exocellular polysaccharides (RPS)-producing cyanobacterial strains (including the possible roles of the exopolysaccharides) and chemical characteristics of the cyanobacterial RPSs given, with particular emphasis on RPS properties and possible industrial applications in a review paper with 169 citations. On the whole, cyanobacterial RPSs are characterized by a great variety in both number (from 2 to 10) and type of constitutive monosaccharides (various arrangements of acidic and neutral sugars). All cyanobacterial RPSs so far tested showed a pseudoplastic behavior, but with marked differences in both viscosity values and shear thinning. In terms of RPS production, the responses of cyanobacteria to changes of culture conditions are strain dependent. RPS productivities shown by some cyanobacteria are well comparable with those reported for other photosynthetic microorganisms proposed for polysaccharide production, but very low in comparison with those of heterotrophic microorganisms. Nevertheless, they argue that cyanobacteria may be regarded as a very abundant source of structurally diverse polysaccharides, some of which may possess unique properties for special applications, not fulfilled by the polymers currently available.

De Clercq (2000) discusses leading natural products for the chemotherapy of HIV infection in a paper with 166 citations. Cyanovirin-N (CV-N), a 11 kDa protein from cyanobacterium (blue-green alga), irreversibly inactivates HIV and also aborts cell-to-cell fusion and transmission of HIV, due to its high-affinity interaction with gp120. Various sulfated polysaccharides extracted from seaweeds (i.e., *Nothogenia fastigiata, Agardhiella tenera*) inhibit the virus adsorption process, while ingenol derivatives may inhibit virus adsorption at least in part through downregulation of CD4 molecules on the host cells. Inhibition of virus adsorption by flavanoids such as (−)epicatechin and its 3-*O*-gallate has been attributed to an irreversible interaction with gp120. For the triterpene glycyrrhizin, the mode of anti-HIV action may at least in part be attributed to interference with virus–cell binding. The mannose-specific plant lectins from *Galanthus, Hippeastrum, Narcissus, Epipactis helleborine*, and *Listera ovata* and the *N*-acetylglucosamine-specific lectin from *Urtica dioica* would primarily be targeted at the virus–cell fusion process. Various other natural products qualify as HIV-cell fusion inhibitors: the siamycins (siamycin I [BMY-29304], siamycin II [RP 71955, BMY 29303], and NP-06 [FR901724]), which are tricyclic 21-amino-acid peptides isolated from *Streptomyces* spp. that differ from one another only at position 4 or 17 (valine or isoleucine in each case), the betulinic acid derivative RPR 103611, and the peptides tachyplesin and polyphemusin that are highly abundant in hemocyte debris of horseshoe crabs. Both T22 and T134 block T-tropic X4 HIV-1 strains through a specific antagonism with the HIV coreceptor CXCR4. A number of natural products interact with the reverse transcriptase, that is, baicalin, avarol, avarone, psychotrine, phloroglucinol derivatives, and, in particular, calanolides and inophyllums. The natural marine substance ilimaquinone would be targeted at the RNase H function of the reverse transcriptase. Curcumin, dicaffeoylquinic and dicaffeoyltartaric acids, L-chicoric acid, and a number of fungal metabolites have all been proposed as HIV-1 integrase inhibitors. Yet, he argues that L-chicoric acid owes its anti-HIV activity to a specific interaction with the viral envelope gp120 rather than integrase. A number of compounds would be able to inhibit HIV-1 gene expression at the transcription level: the flavonoid chrysin (through inhibition of casein kinase II), the antibacterial peptides melittin and cecropin, and EM2487, a novel substance produced by *Streptomyces*.

30.2.2.2 Algal Biomaterials

Gombotz and Wee (1998) discuss protein release from alginate matrices in a review paper with 471 citations. They note that alginate has several unique properties that have enabled it to be used as a matrix for the entrapment and/or delivery of a variety of biological agents. Alginate polymers are a family of linear unbranched polysaccharides that contain varying amounts of 1,4′-linked beta-D-mannuronic acid and alpha-L-guluronic acid residues. The relatively mild gelation process has enabled not only proteins but cells and DNA to be incorporated into alginate matrices with retention of full biological activity. Furthermore, by selection of the type of alginate and coating agent, the pore size, degradation rate, and ultimately release kinetics can be controlled. Gels of different morphologies can be prepared including large block matrices, large beads (>1 mm in diameter), and microbeads (<0.2 mm in diameter). In situ gelling systems have also been made by the application of alginate to the cornea or on the surfaces of wounds. Alginate is a bioadhesive polymer that can be advantageous for site-specific delivery to mucosal tissues. All of these properties, in addition to the nonimmunogenicity of alginate, have led to an increased use of this polymer as a protein delivery system. They discuss the chemistry of alginate, its gelation mechanisms, and the physical properties of alginate gels with an emphasis on applications in which biomolecules have been incorporated into and released from alginate systems.

Rinaudo (2008) discusses applications of some polysaccharides in the domain of biomaterials and bioactive polymers in a review paper with 151 citations. The main polysaccharides currently used in the biomedical and pharmaceutical domains are chitin and its derivative chitosan, hyaluronan, and alginates. Alginates are well known for their property of forming a physical gel in the presence of divalent counterions (Ca, Ba, Sr), whereas carrageenans form a thermoreversible gel. These seaweed polysaccharides are mainly used to encapsulate different materials (cells, bacteria, fungi). Other promising systems are the electrostatic complexes formed when an anionic polysaccharide is mixed with a cationic polysaccharide (e.g., alginate/chitosan or hyaluronan/chitosan). She argues that an important development of the applications of polysaccharides would be in relation to their intrinsic properties such as biocompatibility and biodegradability in the human body for some of them as they are also renewable and have interesting physical properties (film-forming, gelling, and thickening properties). In addition, they are easily processed in different forms such as beads, films, capsules, and fibers.

Jonas and Farah (1998) discuss the production and application of microbial cellulose from algae in a paper with 151 citations. Cellulose is also produced by green algae (*Valonia*) and some bacteria, principally of the genera *Acetobacter*, *Sarcina*, and *Agrobacterium*. Acetobacter strains oxidize alcohols to acids and ketones, especially for the production of vinegars using ethanol, wine, or cider as carbon sources. The formation of the cellulose pellicle occurs on the upper surface of the supernatant film. Cellulose production was stimulated by addition of lactic acid, methionine, tea infusion, and corn steep liquor. They argue that although for the production process new nonconventional bioreactors have been developed, static cultures are still preferred. A large surface area is important for a good productivity. Relatively low glucose concentrations also gave better productivity and yields than higher ones. New applications were described as thickener to maintain viscosity in food, cosmetics, etc., as nonwoven fabric or paper for old document repair and as food additives and others. They argue that we could make use of cellulose films as a temporary substitute for human skin in the case of burns, ulcers, decubitus, and others. Biofill®, Bioprocess®, and Gengiflex® are products of microbial cellulose that now have wide applications in surgery and dental implants.

30.2.2.3 General Studies on Algal Biotechnology

Spolaore et al. (2006) discuss the commercial applications of microalgae in a review paper with 465 citations. They note that there are numerous commercial applications of microalgae where they (1) can be used to enhance the nutritional value of food and animal feed owing to their chemical composition, (2) play a crucial role in aquaculture, and (3) can be incorporated into cosmetics. Moreover, they are cultivated as a source of highly valuable compounds. For example, polyunsaturated fatty acid oils are added to infant formulas, and nutritional supplements and pigments are important as natural dyes. Stable-isotope biochemicals help in structural determination and metabolic studies. They assert that future research should focus on the improvement of production systems and the genetic modification of strains to ensure that microalgal products would become even more diversified and economically competitive.

Pulz and Gross (2004) discuss the production of high-value products from biotechnology of microalgae in a review paper with 264 citations. They note that these products range from simple biomass production for food and feed to valuable products for ecological applications. They argue that for most of these applications, the market is still developing and the biotechnological use of microalgae is expanding into new areas. They represent one of the most promising sources for new products and applications. They assert that with the development of sophisticated culture and screening techniques, microalgal biotechnology can already meet the high demands of both the food and pharmaceutical industries.

Grima et al. (2003) discuss the recovery of microalgal biomass and metabolites with a focus on process options and economics in a review paper with 251 citations. They argue that commercial production of intracellular microalgal metabolites requires (1) large-scale monoseptic production of appropriate microalgal biomass, (2) recovery of biomass from a relatively dilute broth, (3) extraction of metabolite from the biomass, and (4) purification of crude extract. They examine the options available for recovery of biomass and the intracellular metabolites from the biomass. They finally discuss the economics of monoseptic production of microalgae in photobioreactors and the downstream recovery of metabolites using eicosapentaenoic acid recovery as a representative case study.

Herrero et al. (2006) discuss sub- and supercritical fluid extraction of functional ingredients from algae and microalgae in a review paper with 211 citations. Although they discuss the extraction of some compounds with antibacterial, antiviral, or antifungical activity, they focus on the extraction of antioxidant compounds, due to their important role in food preservation and health promotion.

Cardozo et al. (2007) discuss metabolites from algae with economical impact in a review paper with 140 citations. They describe the main substances biosynthesized by algae with potential economical impact in food science, pharmaceutical industry, and public health. They focus on fatty acids, steroids, carotenoids, polysaccharides, lectins, mycosporine-like amino acids, halogenated compounds, polyketides, and toxins.

Apt and Behrens (1999) discuss the commercial developments in microalgal biotechnology in a review paper with 137 citations. They note that new products are being developed for use in mass commercial markets as opposed to the "health food" markets. These include algal-derived long-chained polyunsaturated fatty acids, mainly docosahexaenoic acid, for use as supplements in human nutrition and animals. They argue that large-scale production of algal fatty acids is possible through the use of heterotrophic algae and the adaptation of classical fermentation systems providing consistent biomass under highly controlled conditions that result in a very high-quality product. New products have also been developed for use in the development of pharmaceutical and research products. These include stable-isotope biochemicals produced by algae in closed-system photobioreactors and extremely bright fluorescent pigments. They argue that cryopreservation has also had a tremendous impact on the ability of strains to be maintained for long periods of time at low cost and maintenance while preserving genetic stability.

Moore (1996) discusses cyclic peptides and depsipeptides from cyanobacteria in a paper with 137 citations. He notes that an elaborate array of structurally novel and biologically active cyclic peptides and depsipeptides are found in blue-green algae (cyanobacteria). Several of these compounds possess structures that are similar to those of natural products from marine invertebrates. Most of these cyclic peptides and depsipeptides, such as the microcystins and the lyngbyatoxins, will probably only be useful as biochemical research tools. A few, however, he argues, have the potential for development into useful commercial products. For example, cryptophycin-1, a novel inhibitor of microtubule assembly from *Nostoc* sp. GSV 224, shows impressive activity against a broad spectrum of solid tumors implanted in mice, including multidrug-resistant ones, and majusculamide C, a microfilament-depolymerizing agent from *Lyngbya majuscula*, shows potent fungicidal activity and may have been used in the treatment of resistant fungal-induced diseases of domestic plants and agricultural crops.

30.2.2.4 Algal Pigments

Johnson and An (1991) discuss astaxanthin from microbial sources including algae. Astaxanthin (3,3′-dihydroxy-beta, beta-carotene-4,4′-dione) is the principal carotenoid pigment of salmonids and gives attractive pigmentation in eggs, flesh, and skin in an early review paper with 266 citations. The alga *Haematococcus* has a high concentration of astaxanthin (0.2%–2%), but its industrial application is limited by the lengthy autotrophic cultivation in open freshwater ponds and the requirement for disruption of the cell wall to liberate the carotenoid. They evaluate the biological sources of astaxanthin and compare them with the synthetic compound presently being used in animal feeds.

Lorenz and Cysewski (2000) discuss the commercial potential for *Haematococcus* microalgae as a natural source of astaxanthin in a paper with 253 citations as they note that astaxanthin produced by *Haematococcus* is a product that has become a commercial reality through novel and advanced technology. They argue that cultivation methods have been developed to produce *Haematococcus* containing 1.5%–3.0% astaxanthin by dry weight, with potential applications as a pigment source in aquaculture, in poultry feeds, and in the worldwide nutraceutical market.

Boussiba (2000) discusses carotenogenesis in the green alga *H. pluvialis* with a focus on cellular physiology and stress response in a review paper with 156 citations. Astaxanthin accumulation in *Haematococcus* is induced by a variety of environmental stresses that limit cell growth in the presence of light. This is accompanied by a remarkable morphological and biochemical *transformation* from green motile cells into inert red cysts. In recent years, they have studied this transformation process from several aspects: defining conditions governing pigment accumulation, working out the biosynthetic pathway of astaxanthin accumulation, and questioning the possible function of this secondary ketocarotenoid in protecting *Haematococcus* cells against oxidative damage. He finds that astaxanthin synthesis proceeds via canthaxanthin and that this exceptional stress response is mediated by reactive oxygen species through a mechanism that is not yet understood. He asserts that astaxanthin is the by-product of a defense mechanism rather than the defending substance itself, although at this stage one cannot rule out other protective mechanisms, and argues that *Haematococcus* may serve as a simple model system to study response to oxidative stress and mechanisms involved to cope with this harmful situation.

30.2.3 Conclusion

The nonexperimental research on algal high-value consumer products has been one of the most dynamic research areas in recent years. Twenty-seven citation classics in the field with more than 123 citations were located, and the key emerging issues were presented in the decreasing order of the number of citations. These papers give strong hints about the determinants of efficient algal high-value consumer product design and emphasize that such design is possible in the light of the biotechnological advances in recent years to meet the demands of the consumers in several fields such as medicines, pigments, and health foods.

The development of algal drugs has been one major research area among the nonexperimental studies with 14 papers. Two and three of them focus on cancer and HIV, respectively. There were also seven papers on general studies, while there were three papers each for the areas of algal biomaterials and algal pigments.

30.3 Experimental Studies on Algal High-Value Consumer Products

30.3.1 Introduction

The experimental research on algal high-value consumer products has been one of the most dynamic research areas in recent years. Twenty-three experimental citation classics in the field with more than 124 citations were located, and the key emerging issues were presented in the decreasing order of the number of citations in Table 30.2. These papers give strong hints about the determinants of efficient algal high-value consumer product design and emphasize that the efficient design is possible in light of the recent biotechnological advances in recent years to meet the demands of the consumers in several fields such as medicines, pigments, and health foods.

The papers were dominated by researchers from 12 countries, usually through the intracountry institutional collaboration, and they were multiauthored. The United States (12 papers), Japan (7 papers), and Australia (2 papers) were the most prolific countries.

Similarly, all these papers were published in journals indexed by the SCI and/or SCIE. There were no papers indexed by the A&HCI or SSCI. The number of citations ranged from 124 to 437 and the average number of citations per annum ranged from 4.4 to 22.9. All of these papers were articles.

The development of algal drugs has been one major research area among nonexperimental studies with 20 papers. Eleven and three of them focus on cancer and HIV, respectively. There were also three papers on algal pigments.

30.3.2 Experimental Research

30.3.2.1 Algal Drugs

30.3.2.1.1 General Algal Drugs

Dittmann et al. (1997) study the insertional mutagenesis of a peptide synthetase gene that is responsible for hepatotoxin production in the cyanobacterium *Microcystis aeruginosa* PCC 7806 in a paper with 193 citations. They identify putative peptide synthetase genes in the microcystin-producing strain *M. aeruginosa* PCC 7806. Nonhepatotoxic strains of *M. aeruginosa* lack these genes as strain PCC 7806 was transformed to chloramphenicol resistance. They find that the "antibiotic resistance cassette insertionally inactivated a peptide synthetase gene of strain PCC 7806 as revealed by southern hybridization and DNA amplification as a first report of genetic transformation and mutation, by homologous recombination, of a bloom-forming cyanobacterium." They further find the "inability of derived mutant cells to produce any variant of microcystin while maintaining their ability to synthesize other small peptides." The disrupted gene therefore encodes a peptide

TABLE 30.2

Experimental Research on Algal High-Value Consumer Products

No.	Paper References	Year	Document	Affiliation	Country	No. of Authors	Journal	Index	Subjects	Topic	Total No. of Citations	Total Average Citations per Annum	Rank
1	Nishiwaki-Matsushima et al.	1992	A	Natl. Canc Ctr., Jap. Fdn. Canc. Res., Univ. Tokyo, Wright St. Univ.	Japan, United States	8	*J. Cancer Res. Clin. Oncol.*	SCI, SCIE	Oncol.	Algal drugs—cancer promotion	437	19.9	3
2	Ueno et al.	1999	A	Tokyo Metr. Res. Lab. P.H., Shinshu Univ., Shanghai Med. Univ., Haimen City. Hlth. Anti Epidem. Stn.	Japan, China	9	*Carcinogenesis*	SCI, SCIE	Oncol.	Algal drugs—cancer promotion	323	21.5	4
3	Boyd et al.	1997	A	NCI, Frederick Canc. Res. Dev. Ctr., NIDDKD	United States	17	*Antimicrob. Agents Chemother.*	SCI, SCIE	Microbiol., Phar. Phar.	Algal drugs—HIV, CV-N	305	17.9	5
4	An and Carmichael	1994	A	Wright St. Univ.	United States	2	*Toxicon*	SCI, SCIE	Phar. Phar., Toxic.	Algal drugs—cancer promotion	269	13.5	7
5	Dittmann et al.	1997	A	Univ. New S Wales	Australia	5	*Mol. Microbiol.*	SCI, SCIE	Bioch. Mol. Biol., Microbiol.	Algal drugs—peptide synthetase gene	193	11.4	15
6	Schwartz et al.	1990	A	Merck Sharp Dohme Ltd	United States	10	*J. Indust. Microbiol.*	SCI, SCIE	Biotech. Appl. Microb.	Algal drugs—development	183	8.3	17
7	Boussiba and Vonshak	1991	A	Ben Gurion Univ.	Israel	2	*Plant Cell Physiol.*	SCI, SCIE	Plant Sci., Cell Biol.	Pigments—astaxanthin	176	7.7	18
8	Ohta et al.	1994	A	Saitama Canc. Ctr., Aichi Canc. Ctr., NCI, Wright St. Univ.	Japan, United States	10	*Cancer Res.*	SCI, SCIE	Oncol.	Algal drugs—cancer promotion	176	8.8	19

(*Continued*)

TABLE 30.2 (*Continued*)

Experimental Research on Algal High-Value Consumer Products

No.	Paper References	Year	Document	Affiliation	Country	No. of Authors	Journal	Index	Subjects	Topic	Total No. of Citations	Total Average Citations per Annum	Rank
9	Cumashi et al.	2007	A	Russ. Acad. Sci., Scottish Assoc. Marine Sci., Univ. Buenos Aires, Univ G. DAnnunzio, Ist. Ric. Farmacol M.N.	Russia, Scotland, Argentina, Italy	18	*Glycobiology*	SCI, SCIE	Bioch. Mol. Biol.	Algal drugs—fucoidans	160	22.9	24
10	Namikoshi and Rinehart	1996	A	Univ. Ill.	United States	2	*J. Ind. Microbiol. Biotechnol.*	SCI, SCIE	Biotech. Appl. Microb.	Algal drugs—bioactive compounds	160	8.9	25
11	Kobayashi et al.	1993	A	Hiroshima Univ.	Japan	3	*Appl. Environ. Microbiol.*	SCI, SCIE	Biotech. Appl. Microb., Microb.	Pigments—astaxanthin	159	7.6	26
12	Trimurtulu et al.	1994	A	Univ. Hawaii, Wayne State Univ.	United States	7	*J. Am. Chem. Soc*	SCI, SCIE	Chem. Mult.	Algal drugs—cryptophycins	156	7.8	30
13	Romay et al.	1998	A	CNIC, Univ. Havana	Cuba	5	*Inflamm. Res.*	SCI, SCIE	Cell Biol., Immun.	Algal drugs—C-phycocyanin	155	9.7	31
14	Gustafson et al.	1989	A	NCI, Univ. Hawaii	United States	8	*J. Natl. Cancer Inst.*	SCI, SCIE	Oncol.	Algal drugs—AIDS antivirals	153	6.1	33
15	Kotake-Nara et al.	2001	A	Hokkaido Univ., Natl. Food Res. Inst.	Japan	6	*J. Nutr.*	SCI, SCIE	Nutr. Ditet.	Algal drugs—carotenoids	152	11.7	34
16	Smith et al.	1994	A	Univ. Hawaii M.	United States	4	*Cancer Res.*	SCI, SCIE	Oncol.	Algal drugs—cryptophycin	151	7.6	37
17	Walsh et al.	1997	A	Univ. Alabama	United States	3	*FEBS Lett.*	SCI, SCIE	Bioch. Mol. Biol., Biophys., Cell Biol.	Algal drugs—fostriecin	143	8.4	38

(Continued)

TABLE 30.2 (*Continued*)

Experimental Research on Algal High-Value Consumer Products

No.	Paper References	Year	Document	Affiliation	Country	No. of Authors	Journal	Index	Subjects	Topic	Total No. of Citations	Total Average Citations per Annum	Rank
18	Neilan et al.	1999	A	Univ. New S Wales, Humboldt Univ., Univ. Helsinki	Australia, Germany, Finland	7	*J. Bacteriol.*	SCI, SCIE	Microbiol.	Algal drugs—peptide synthesis	138	9.2	40
19	Koyanagi et al.	2003	A	Fukuoka Univ.	Japan	5	*Biochem. Pharmacol.*	SCI, SCIE	Phar. Phar.	Algal drugs—fucoidans	136	12.4	43
20	Verdier-Pinard et al.	1998	A	NCI, Univ. Texas, Oregon St. Univ., Univ. Pittsburgh	United States	12	*Mol. Pharmacol.*	SCI, SCIE	Phar. Phar.	Algal drugs—curacin A	132	8.3	44
21	Chang et al.	2004	A	Univ. Michigan, Oregon St. Univ.	United States	8	*J. Nat. Prod.*	SCI, SCIE	Plant Sci., Chem. Med., Phar. Phar.	Algal drugs—curacin A	125	12.5	45
22	Kobayashi et al.	1991	A	Osaka Univ.	Japan	3	*J. Ferment. Bioeng*	SCI, SCIE	Biotech. Appl. Microb., Food Sci. Tech.	Pigments—astaxanthin	124	5.4	48
23	Tsai et al.	2004		Univ. Wash., St George Hosp, NCI, Univ. S Alabama	United States	9	*Aids Res. Hum. Retrovir.*	SCI, SCIE	Immun., Infect. Dis., Virol.	Algal drugs—AIDS, CV-N	124	12.4	49

SCI, Science Citation Index; SCIE, Science Citation Index Expanded; SSCI, Social Sciences Citation Index; A, article; R, review.

synthetase (microcystin synthetase) that is specifically involved in the biosynthesis of microcystins. They assert that "microcystins are synthesized non-ribosomally and that a basic difference between toxic and nontoxic strains of *M. aeruginosa* is the presence of one or more genes coding for microcystin synthetases."

Schwartz et al. (1990) discuss the development of pharmaceuticals from cultured algae (cultured cyanobacteria and cultured eukaryotic microalgae) in an early paper with 183 citations. They develop the methods for the isolation, purification, preservation, and cultivation of axenic cyanobacteria and eukaryotic cultures. They find that screening of these groups for biologically active components has led to the "isolation of pachydictyol and caulerpenyne from cultured macroalgae, while a series of hapalindoles and an antifungal depsipeptide have been isolated from cyanobacteria."

Namikoshi and Rinehart (1996) study bioactive compounds produced by cyanobacteria in an early paper with 160 citations. Cyanobacteria produce a large number of compounds with varying bioactivities including toxins such as hepatotoxins (such as microcystins and nodularins) and neurotoxins (such as anatoxins and saxitoxins). Cytotoxicity to tumor cells has been demonstrated for other cyanobacterial products, including 9-deazaadenosine, dolastatin 13, and analogs. A number of compounds in cyanobacteria are inhibitors of proteases (micropeptins, cyanopeptolins, oscillapeptin, microviridin, aeruginosins) and other enzymes, while still other compounds have no recognized biological activities. They argue that in general cyclic peptides and depsipeptides are the most common structural types, but "a wide variety of other types are also found such as linear peptides, guanidines, phosphonates, purines and macrolides." They assert that the "close similarity or identity in structures between cyanobacterial products and compounds isolated from sponges, tunicates and other marine invertebrates suggests the latter compounds may be derived from dietary or symbiotic blue-green algae."

Trimurtulu et al. (1994) study the total structures of cryptophycins, potent antitumor depsipeptides, from the blue-green alga *Nostoc* sp. strain GSV 224 in an early paper with 156 citations. They note that cryptophycin (A, 1), the major cytotoxin in the blue-green alga (cyanobacterium) *Nostoc* sp. GSV 224, shows excellent activity against solid tumors implanted in mice. This cyclic depsipeptide had previously been isolated from *Nostoc* sp. ATCC 53789 as an antifungal agent and its gross structure determined by researchers. Six minor cryptophycins (B-G, 2–7) have also been isolated from GSV 224 and their total structures and cytotoxicities determined. They find that "two types of cryptophycins are present in this alga, the major series possessing a monochlorinated L-O-methyltyrosine unit and the minor series possessing a nonchlorinated D-O-methyltyrosine unit." They describe structure–activity relationship studies of the cryptophycins and several derivatives and degradation products (8–14) and present preliminary in vivo results for cryptophycin A against six solid tumors.

Romay et al. (1998) study the antioxidant and anti-inflammatory properties of C-phycocyanin from blue-green algae in a paper with 155 citations. They evaluate phycocyanin as a putative antioxidant in vitro by using (1) luminol-enhanced chemiluminescence generated by three different radical species (O_2^-, OH·, RO·) and by zymosan-activated human polymorphonuclear leukocytes (PMNLs), (2) deoxyribose assay, and (3) inhibition of liver microsomal lipid peroxidation induced by Fe^{2+}–ascorbic acid. They also assay the antioxidant activity in vivo in glucose oxidase (GO)–induced inflammation in mouse paw. They find that "phycocyanin is able to scavenge OH· (IC50 = 0.91 mg/mL) and RO· (IC50 = 76 μg/mL) radicals, with activity equivalent to 0.125 mg/mL of dimethyl sulphoxide (DMSO) and 0.038 μg/mL of trolox, specific scavengers of those radicals respectively." They then find that, in the deoxyribose assay, the "second-order rate constant was 3.56×10^{11} M^{-1} S^{-1}, similar to that obtained for some non-steroidal anti-inflammatory drugs." They then argue that "phycocyanin also inhibits liver microsomal lipid peroxidation (IC50 = 12 mg/mL), the CL response of PMNLs ($p < 0.05$) as well as the edema index in GO-induced inflammation in mouse paw ($p < 0.05$)."

Neilan et al. (1999) study nonribosomal peptide synthesis and toxigenicity of cyanobacteria in a paper with 138 citations. They note that cyanobacteria produce a myriad array of secondary metabolites, including alkaloids, polyketides, and nonribosomal peptides, some of which are potent toxins. They address the molecular genetic basis of nonribosomal peptide synthesis in diverse species of cyanobacteria. They achieve amplification of peptide synthetase genes by use of degenerate primers directed to conserved functional motifs of these modular enzyme complexes. They show specific detection of the gene cluster encoding the biosynthetic pathway of the cyanobacterial toxin microcystin for both cultured and uncultured samples. They find "a broad distribution of peptide synthetase gene orthologues in cyanobacteria." The results demonstrate a molecular approach to assessing preexpression microbial functional diversity in uncultured cyanobacteria. They assert that the "nonribosomal peptide biosynthetic pathways detected may lead to the discovery and engineering of novel antibiotics, immunosuppressants, or antiviral agents."

30.3.2.1.2 Algal Cancer Drugs

Nishiwaki-Matsushima et al. (1992) study tumor promotion by the cyanobacterial cyclic peptide toxin microcystin-LR in an early paper with 437 citations. They note that the inhibition of protein phosphatase type 1 (PP1) and type 2A (PP2A) activities by microcystin-LR is similar to that of the known protein phosphatase inhibitor and tumor promoter okadaic acid. Microcystin-LR, applied below the acute toxicity level, dose-dependently increases the number and percentage area of positive foci for the placental form of glutathione S-transferase in rat liver, which was initiated with diethylnitrosamine. The result was obtained independently through two animal experiments. They argue that "microcystin-LR is a new liver tumor promoter mediated through inhibition of protein phosphatase type 1 and type 2A activities." They assert that the "okadaic acid pathway is a general mechanism of tumor promotion in various organs, such as mouse skin, rat glandular stomach and rat liver."

Ueno et al. (1996) study the detection of microcystins, a blue-green algal hepatotoxin, in drinking water sampled in Haimen and Fusui, endemic areas of primary liver cancer in China, by a highly sensitive immunoassay in a paper with 323 citations. The first survey in September 1993 found that three out of 14 ditch water specimens were positive for microcystine (MC), with a range of 90–460 pg/mL. Several toxic algae such as *Oscillatoria agardhii* were present in some of the ditches. In the second trial, samples were collected from five ponds/ditches, two rivers, two shallow wells, and two deep wells monthly for the whole year of 1994. These data showed that MC was highest in June to September, with a range of 62–296 pg/mL. A third trial on the 989 different water samples collected from the different types of water sources in July 1994 revealed that 17% of the pond/ditch water, 32% of the river water, and 4% of the shallow-well water were positive for MC, with averages of 101, 160, and 68 pg/mL, respectively. No MC was detected in deep well water. A similar survey on 26 drinking water samples in Fusui, Guangxi province, demonstrated a high contamination frequency of MC in the water of ponds/ditches and rivers but no MC in shallow and deep wells. They argue that the "blue-green algal toxin MC in the drinking water of ponds/ditches and rivers, or both, is one of the risk factors for the high incidence of PLC in China" and they propose "an advisory level of MC in drinking water to below 0.01 μg/L."

An and Carmichael (1994) study the use of a colorimetric protein phosphatase inhibition assay and enzyme-linked-immunosorbent assay for the study of microcystins and nodularins in an early paper with 269 citations. Using a rabbit anti-microcystin-LR polyclonal antibody preparation, they test the cross-reactivity with 18 microcystin and nodularin variants. They find essential a "hydrophobic amino acid, 3-amino-9-methoxy-10-phenyl-2,6,8-trimethyl-deca-4(E),6(E)-dienoic acid (Adda), which has the (E) form at the C-6 double bond in both microcystin and nodularin," for these toxins to express antibody specificity. Modification of –COOH in glutamic acid of microcystin and nodularin did not alter their antigenicity. They compare antibody cross-reactivity of these toxins with their ability to inhibit PP1. They find that "the concentration of microcystin-LR causing 50% inhibition of recombinant PP1 activity (IC50) was about 0.3 nM, while that of two modified microcystins had a significantly higher IC50. Microcystin-LR and nodularin with the (z) form of Adda at the C-6 double bond or having the monoester of glutamic acid did not inhibit PP1." These three toxins were also nontoxic in the mouse bioassay. They stress the importance of Adda and glutamic acid in toxicity of these cyclic peptides and that PPI inhibition is related to the toxins' mechanism of action.

Ohta et al. (1994) study nodularin, a potent inhibitor of protein phosphatase-1 and phosphatase-2a, which is a new environmental carcinogen in male f344 rat liver in a paper with 176 citations. They find that nodularin is also a new tumor promoter in rat liver. A two-stage carcinogenesis experiment in rat liver initiated with diethylnitrosamine and without partial hepatectomy revealed that nodularin stimulated glutathione S-transferase placental form–positive foci in rat liver more effectively than did microcystin-1R and that nodularin alone induced glutathione S-transferase placental form–positive foci as well as did diethylnitrosamine alone. Thus, "nodularin itself is a new liver carcinogen, and microcystin-LR is a tumor promoter rather than a carcinogen." They find that "nodularin induced hyperphosphorylation of cytokeratin peptides 8 and 18 in primary cultured rat hepatocytes 20% more effectively than did microcystin-LR, suggesting that nodularin penetrates more easily into the hepatocytes than does microcystin-LR." Nodularin upregulated induction of c-jun, jun-B, jun-D, c-fos, fos-B, and fra-1 mRNA transcripts in a rat liver after i.p. administration, and the accumulation of the mRNA transcripts was sustained for over 9 h after treatment.

Cumashi et al. (2007) carry out a comparative study of the anti-inflammatory, anticoagulant, antiangiogenic, and antiadhesive activities of nine different fucoidans from brown seaweeds to examine the influence of fucoidan origin and composition on their biological activities in a paper with 160 citations. They find that all fucoidans inhibited leucocyte recruitment in an inflammation model in rats, and neither the content of fucose and sulfate nor other structural features of their polysaccharide backbones significantly affected the efficacy of fucoidans in this model. In vitro evaluation of P-selectin-mediated neutrophil adhesion to platelets under flow conditions revealed that "only polysaccharides from *Laminaria saccharina, L. digitata, Fucus evanescens, F. serratus, F. distichus, F. spiralis,* and *Ascophyllum nodosum* could serve as P-selectin inhibitors." Furthermore, "all fucoidans, except that from *Cladosiphon okamuranus* carrying substantial levels of 2-O-alpha-D-glucuronopyranosyl branches in the linear (1 → 3)-linked poly-alpha-fucopyranoside chain, exhibited anticoagulant activity as measured by activated partial thromboplastin time whereas only fucoidans from *L. saccharina, L. digitata, F. serratus, F. distichus,* and *F. evanescens* displayed strong antithrombin activity in a platelet aggregation test." The last fucoidans potently inhibited human umbilical vein endothelial cell (HUVEC) tubulogenesis in vitro, and this property correlated with decreased levels of plasminogen-activator inhibitor-1 in HUVEC supernatants, suggesting a possible mechanism of fucoidan-induced inhibition of tubulogenesis. Finally, they find that "fucoidans from *L. saccharina*, *L. digitata*, *F. serratus*, *F. distichus*, and *F. vesiculosus* strongly blocked MDA-MB-231 breast carcinoma cell adhesion to platelets, an effect which might have critical implications in tumor metastasis." They provide a new rationale for the development of potential drugs for thrombosis, inflammation, and tumor progression.

Kotake-Nara et al. (2001) study the effect of 15 kinds of carotenoids on the proliferation of three lines of human prostate cancer cells (PC-3, DU 145, and LNCaP) in a paper with 152 citations. They find that "when the prostate cancer cells were cultured in a carotenoid-supplemented medium for 72 h at 20 μmol/L, 5,6-monoepoxy carotenoids, namely, neoxanthin from spinach and fucoxanthin from brown algae, significantly reduced cell viability to 10.9 and 14.9% for PC-3, 15.0 and 5.0% for DU 145, and nearly zero and 9.8% for LNCaP, respectively." Acyclic carotenoids such as phytofluene, zeta-carotene, and lycopene, all of which are present in tomato, also significantly reduced cell viability. On the other hand, they find that "phytoene, canthaxanthin, beta-cryptoxanthin and zeaxanthin did not affect the growth of the prostate cancer cells." They detect DNA fragmentation of nuclei in neoxanthin- and fucoxanthin-treated cells by in situ TdT-mediated dUTP nick end labeling assay. They argue that "neoxanthin and fucoxanthin reduce cell viability through apoptosis induction in the human prostate cancer cells." They assert that "ingestion of leafy green vegetables and edible brown algae rich in neoxanthin and fucoxanthin might have the potential to reduce the risk of prostate cancer."

Smith et al. (1994) study cryptophycin as a new antimicrotubule agent active against drug-resistant cells in an early paper with 151 citations. They find that incubation of L1210 leukemia cells with cryptophycin resulted in dose-dependent inhibition of cell proliferation in parallel with increases in the percentage of cells in mitosis (half-maximal effects at <10 pM). They next find that the "treatment of A-10 vascular smooth muscle cells with cryptophycin results in marked depletion of cellular microtubules and reorganization of vimentin intermediate filaments, similar to the effects of vinblastine. Cytochalasin B caused the depolymerization of microfilaments in these cells, while neither vinblastine nor cryptophycin affected this cytoskeletal component." Pretreatment of cells with taxol prevented microtubule depolymerization in response to either vinblastine or cryptophycin. While microtubule depolymerization in response to vinblastine was rapidly reversed by removal of the drug, they find that "cells treated with cryptophycin remained microtubule depleted for at least 24 h after removal of the compound." Combinational treatments with vinblastine and cryptophycin resulted in additive cytotoxicity. They argue that "ovarian carcinoma and breast carcinoma cells which are multiply drug resistant due to overexpression of P-glycoprotein are markedly less resistant to cryptophycin than they are to vinblastine, colchicine, and taxol." Therefore, cryptophycin is a new antimicrotubule compound that is a poorer substrate for P-glycoprotein than are the vinca alkaloids, and they assert that "this property may confer an advantage to cryptophycin in the chemotherapy of drug-resistant tumors."

Walsh et al. (1997) study fostriecin, an antitumor antibiotic with inhibitory activity against serine/threonine PP1 and PP2A in a paper with 143 citations. They note that fostriecin, an antitumor antibiotic produced by *Streptomyces pulveraceus*, is a strong inhibitor of type 2A (PP2A; IC50 3.2 nM) and a weak inhibitor of type 1 (PP1; IC50 131 μM) serine/threonine protein phosphatases. They find that "fostriecin has no apparent effect on the activity of PP2B, and dose-inhibition studies conducted with whole cell homogenates indicate that fostriecin also inhibits the native forms of PP1 and PP2A whilst the studies with recombinant PP1/PP2A chimeras indicate that okadaic acid and fostriecin have different binding sites."

Koyanagi et al. (2003) study the effect of the oversulfation of fucoidan on its antiangiogenic and antitumor activities in a paper with 136 citations. They demonstrate earlier that fucoidan can inhibit tube formation following migration of HUVEC and that its chemical oversulfation enhances the inhibitory potency. They test the hypothesis that fucoidan may suppress tumor growth by inhibiting tumor-induced angiogenesis. They find that "both natural and oversulfated fucoidans (NF and OSF) significantly suppressed the mitogenic and chemotactic actions of vascular endothelial growth factor 165 ($VEGF_{165}$) on HUVEC by preventing the binding of $VEGF_{165}$ to its cell surface receptor." The suppressive effect of OSF was more potent than that of NF, suggesting an important role for the numbers of sulfate groups in the fucoidan molecule. They argue that consistent with its inhibitory actions on $VEGF_{165}$, "OSF clearly suppressed the neovascularization induced by $Sarcoma_{180}$ cells that had been implanted in mice." The inhibitory action of fucoidan was also observed in the growth of Lewis lung carcinoma and B16 melanoma in mice. They assert that the "antitumor action of fucoidan is due, at least in part, to its anti-angiogenic potency and that increasing the number of sulfate groups in the fucoidan molecule contributes to the effectiveness of its anti-angiogenic and antitumor activities."

Verdier-Pinard et al. (1998) study the structure–activity analysis of the interaction of curacin A, the potent colchicine site antimitotic agent, with tubulin and effects of analogs on the growth of MCF_7 breast cancer cells in a paper with 132 citations. They prepare a series of curacin A analogs to determine the important structural features of the molecule. These modifications include reduction and E-to-Z transitions of the olefinic bonds in the 14-carbon side chain of the molecule; disruption of and configurational changes in the cyclopropyl moiety; disruption, oxidation, and configurational reversal in the thiazoline moiety; configurational reversal and substituent modifications at C_{13}; and demethylation at C_{10}. They find that the "most important portions of curacin A required for its interaction with tubulin are the thiazoline ring and the side chain at least through C_4, the portion of the side chain including the C_{9-10} olefinic bond, and the C_{10} methyl group." Only two modifications totally eliminated the tubulin–drug interaction. The inactive compounds were a segment containing most of the side chain, including its two substituents, and analogs in which the methyl group at the C_{13} oxygen atom was replaced by a benzoate residue. They find that "antiproliferative activity comparable with that observed with curacin A was only reproduced in compounds that were potent inhibitors of the binding of colchicine to tubulin." The most active analogs overlapped extensively with curacin A but failed to provide an explanation for the apparent structural analogy between curacin A and colchicine.

Chang et al. (2004) study the biosynthetic pathway and gene cluster analysis of curacin A, an antitubulin natural product from the tropical marine cyanobacterium *L. majuscule* in a paper with 125 citations. They feed a series of stable-isotope-labeled precursors to cultures of curacin A–producing strains and allowed determination of the metabolic origin of all atoms in the natural product (1 cysteine, 10 acetate units, 2 S-adenosyl methionine-derived methyl groups) as well as several unique mechanistic insights. Moreover, these incorporation experiments facilitated an effective gene cloning strategy that allowed identification and sequencing of the approximately 64 kb putative curacin A gene cluster. They find that the "metabolic system is comprised of a nonribosomal peptide synthetase (NRPS) and multiple polyketide synthases (PKSs) and shows a very high level of colinearity between genes in the cluster and the predicted biochemical steps required for curacin biosynthesis." Unique features of the cluster include "(1) all but one of the PKSs are monomodular multifunctional proteins, (2) a unique gene cassette that contains an

HMG-CoA synthase likely responsible for formation of the cyclopropyl ring, and (3) a terminating motif that is predicted to function in both product release and terminal dehydrative decarboxylation."

30.3.2.1.3 Algal HIV/AIDS Drugs

Boyd et al. (1997) discovered CV-N, a novel HIV-inactivating protein that binds viral surface envelope glycoprotein gp120 with a focus on the potential applications to microbicide development in a paper with 305 citations. They isolate and sequence a novel 11 kDa virucidal protein, named CV-N, from cultures of the cyanobacterium (blue-green alga) *Nostoc ellipsosporum*. They also produce CV-N recombinantly by expression of a corresponding DNA sequence in *Escherichia coli*. They find that "low nanomolar concentrations of either natural or recombinant CV-N irreversibly inactivate diverse laboratory strains and primary isolates of human immunodeficiency virus (HIV) type 1 as well as strains of HIV type 2 and simian immunodeficiency virus." In addition, CV-N aborts cell-to-cell fusion and transmission of HIV-1 infection. They then find that "continuous, 2-day exposures of uninfected CEM-SS cells or peripheral blood lymphocytes to high concentrations (e.g., 9,000 nM) of CV-N were not lethal to these representative host cell types." They assert that the "antiviral activity of CV-N is due, at least in part, to unique, high-affinity interactions of CV-N with the viral surface envelope glycoprotein gp120." They further assert that the "biological activity of CV-N is highly resistant to physicochemical denaturation, further enhancing its potential as an anti-HIV microbicide."

Gustafson et al. (1989) study AIDS-antiviral sulfolipids from cyanobacteria in an early paper with 153 citations. A recently developed tetrazolium-based microculture assay was used to screen extracts of cultured cyanobacteria (blue-green algae) for inhibition of the cytopathic effects of the HIV (HIV-1), which is implicated as a causative agent of AIDS. They find a "number of extracts to be remarkably active against the AIDS virus and discover a new class of HIV-1-inhibitory compounds, the sulfonic acid-containing glycolipids, through the use of the microculture assay to guide the fractionation and purification process." They next find that the "pure compounds were active against HIV-1 in cultured human lymphoblastoid CEM, MT-2, LDV-7, and C3-44 cell lines in the tetrazolium assay as well as in p24 viral protein and syncytium formation assays."

Tsai et al. (2004) study the inhibition of AIDS virus infections in vaginal transmission models by CV-N in a paper with 124 citations. The cyanobacterial protein CV-N potently inactivates diverse strains of HIV-1 and other lentiviruses due to irreversible binding of CV-N to the viral envelope glycoprotein gp120. They show that "recombinant CV-N effectively blocks HIV-1(Ba-L) infection of human ectocervical explants." Furthermore, they demonstrate the "in vivo efficacy of CV-N gel in a vaginal challenge model by exposing CV-N-treated female macaques (*Macaca fascicularis*) to a pathogenic chimeric SIV/HIV-1 virus, SHIV89.6P. All of the placebo-treated and untreated control macaques (8 of 8) became infected. In contrast, 15 of 18 CV-N-treated macaques showed no evidence of SHIV infection." Further, CV-N produced no cytotoxic or clinical adverse effects in either the in vitro or in vivo model systems. They assert that "CV-N is a good candidate for testing in humans as an anti-HIV topical microbicide."

30.3.2.2 Algal Pigments

Boussiba and Vonshak (1991) study astaxanthin accumulation in the green alga *H. pluvialis* in an early paper with 176 citations. Cells of the green microalga *H. pluvialis* were induced to accumulate the ketocarotenoid pigment astaxanthin. This induction was achieved by the application of light intensity (170 μmol/[m^2/s]), phosphate starvation, and salt stress (NaCl 0.8%). They find that "these conditions retarded cell growth as reflected by a decrease in cell division rate, but led to an increase in astaxanthin content per cell." Accumulation of astaxanthin required nitrogen and was associated with a change in the cell stage from biflagellate vegetative green cells to nonmotile and large resting cells. They assert that "environmental or nutritional stresses, which interfere with cell division, trigger the accumulation of astaxanthin. Indeed, when a specific inhibitor of cell division was applied, a massive accumulation of astaxanthin occurred."

Kobayashi et al. (1993) study enhanced carotenoid biosynthesis by oxidative stress in acetate-induced cyst cells of a green unicellular alga *H. pluvialis* in an early paper with 159 citations. In a green alga, *H. pluvialis*, a morphological change of vegetative cells into cyst cells was rapidly induced by the addition of acetate or acetate plus Fe^{2+} to the vegetative growth phase. They find that accompanied by cyst formation, "algal astaxanthin formation was more enhanced by the addition of acetate plus Fe^{2+} than by the addition of acetate alone. Encystment and enhanced carotenoid biosynthesis were inhibited by either actinomycin D or cycloheximide. However, after cyst formation was induced by the addition of acetate alone, carotenoid formation could be enhanced with the subsequent addition of Fe^{2+} even in the presence of the inhibitors." The Fe^{2+}-enhanced carotenogenesis was inhibited by potassium iodide, a scavenger for hydroxyl radical, suggesting that hydroxyl radical formed by an iron-catalyzed Fenton reaction may be required for enhanced carotenoid biosynthesis. Moreover, they demonstrate that "four active oxygen species, singlet oxygen, superoxide anion radical, hydrogen peroxide, and peroxy radical, were capable of replacing Fe^{2+} in its role in the enhanced carotenoid formation in the acetate-induced cyst." They assert that "oxidative stress is involved in the posttranslational activation of carotenoid biosynthesis in acetate-induced cyst cells."

Kobayashi et al. (1991) study astaxanthin production by a green alga *H. pluvialis* accompanied with morphological changes in acetate media in an early paper with 124 citations. They develop two acetate-containing media for astaxanthin production by a green unicellular alga *H. pluvialis*. They find that the "basal medium, a vegetative growth medium facilitated algal cell growth, whereas the modified medium was likely to induce morphological changes with the formation of large cysts and bleached cells which seemed to consequently enhance the carotenoid biosynthesis." They argue that in the "two-stage culture, the injection of ferrous ion with acetate into the basal medium on the fourth day, was greatly stimulative for both the algal cell growth and the astaxanthin formation at a high light intensity." In addition, carotenoid precursors, mevalonate, and pyruvate were effective on the carotenoid formation in the modified medium. They assert that "pyruvate was an especially good carbon source both for the algal cell growth and the carotenoid synthesis."

30.3.3 Conclusion

The experimental research on algal high-value consumer products has been one of the most dynamic research areas in recent years. Twenty-three experimental citation classics in the field with more than 124 citations were located, and the key emerging issues were presented in decreasing order of the number of citations. These papers give strong hints about the determinants of efficient algal high-value consumer product design and emphasize that such design is possible in the light of the recent biotechnological advances in recent years to meet the demands of the consumers in several fields such as medicines, pigments, and health foods.

The development of algal drugs has been one major research area among the nonexperimental studies with 20 papers. Eleven and three of them focus on cancer and HIV, respectively. There were also three papers on algal pigments.

30.4 Conclusion

The citation classics presented under the two main headings in this chapter confirm the predictions that marine algae have a significant potential to serve as a major solution for the global problems of health in life-limiting diseases like cancer and HIV as well as other commercial areas like algal pigments in addition to the well-known solution of problems relating to the warming, air pollution, energy security, and food security through the design of efficient algal high-value consumer products.

Further research is recommended for the detailed studies in each topical area presented in this chapter including scientometric studies and citation classic studies to inform key stakeholders about the potential of marine algae for the solution of the global problems of health in addition to the well-known solution of problems relating to warming, air pollution, energy security, and food security in the form of algal high-value consumer products.

References

An, J.S. and W.W. Carmichael. 1994. Use of a colorimetric protein phosphatase inhibition assay and enzyme-linked-immunosorbent-assay for the study of microcystins and nodularins. *Toxicon* 32:1495–1507.

Apt, K.E. and P.W. Behrens. 1999. Commercial developments in microalgal biotechnology. *Journal of Phycology* 35:215–226.

Baltussen, A. and C.H. Kindler. 2004a. Citation classics in anesthetic journals. *Anesthesia & Analgesia* 98:443–451.

Baltussen, A. and C.H. Kindler. 2004b. Citation classics in critical care medicine. *Intensive Care Medicine* 30:902–910.

Belay, A., Y. Ota, K. Miyakawa, and H. Shimamatsu. 1993. Current knowledge on potential health benefits of spirulina. *Journal of Applied Phycology* 5:235–241.

Borowitzka, M.A. 1995. Microalgae as sources of pharmaceuticals and other biologically-active compounds. *Journal of Applied Phycology* 7:3–15.

Bothast, R.J. and M.A. Schlicher. 2005. Biotechnological processes for conversion of corn into ethanol. *Applied Microbiology and Biotechnology* 67:19–25.

Boussiba, S. 2000. Carotenogenesis in the green alga *Haematococcus pluvialis*: Cellular physiology and stress response. *Physiologia Plantarum* 108:111–117.

Boussiba, S. and A. Vonshak. 1991. Astaxanthin accumulation in the green-alga haematococcus-pluvialis. *Plant and Cell Physiology* 32:1077–1082.

Boyd, M.R., K.R. Gustafson, J.B. McMahon et al. 1997. Discovery of cyanovirin-N, a novel human immunodeficiency virus-inactivating protein that binds viral surface envelope glycoprotein gp120: Potential applications to microbicide development. *Antimicrobial Agents and Chemotherapy* 41:1521–1530.

Burja, A.M., B. Banaigs, E. Abou-Mansour, J.G. Burgess, and P.C. Wright. 2001. Marine cyanobacteria—A prolific source of natural products. *Tetrahedron* 57:9347–9377.

Calman, K.C. 1984. Quality of life in cancer-patients—An hypothesis. *Journal of Medical Ethics* 10:124–127.

Cardozo, K.H.M., T. Guaratini, M.P. Barros et al. 2007. Metabolites from algae with economical impact. *Comparative Biochemistry and Physiology C—Toxicology & Pharmacology* 146:60–78.

Chang, Z.X., N. Sitachitta, J.V. Rossi et al. 2004. Biosynthetic pathway and gene cluster analysis of curacin A, an antitubulin natural product from the tropical marine cyanobacterium *Lyngbya majuscule*. *Journal of Natural Products* 67:1356–1367.

Chesney, M.A., J.R. Ickovics, D.B. Chambers et al. 2000. Self-reported adherence to antiretroviral medications among participants in HIV clinical trials: The AACTG Adherence Instruments. *Aids Care—Psychological and Socio-Medical Aspects of AIDS/HIV* 12:255–266.

Chisti, Y. 2007. Biodiesel from microalgae. *Biotechnology Advances* 25:294–306.

Clardy, J., M.A. Fischbach, and C.T. Walsh. 2006. New antibiotics from bacterial natural products. *Nature Biotechnology* 24:1541–1550.

Cumashi, A., N.A. Ushakova, M.E. Preobrazhenskaya et al. 2007. A comparative study of the anti-inflammatory, anticoagulant, antiangiogenic, and antiadhesive activities of nine different fucoidans from brown seaweeds. *Glycobiology* 17:541–552.

De Clercq, E. 2000. Current lead natural products for the chemotherapy of human immunodeficiency virus (HIV) infection. *Medicinal Research Reviews* 20:323–349.

De Philippis, R. and M. Vincenzini. 1998. Exocellular polysaccharides from cyanobacteria and their possible applications. *FEMS Microbiology Reviews* 22:151–175.

Demirbas, A. 2007. Progress and recent trends in biofuels. *Progress in Energy and Combustion Science* 33:1–18.

Dittmann, E., B.A. Neilan, M. Erhard, H. vonDohren, and T. Borner. 1997. Insertional mutagenesis of a peptide synthetase gene that is responsible for hepatotoxin production in the cyanobacterium Microcystis aeruginosa PCC 7806. *Molecular Microbiology* 26:779–787.

Dubin, D., A.W. Hafner, and K.A. Arndt. 1993. Citation-classics in clinical dermatological journals—Citation analysis, biomedical journals, and landmark articles, 1945–1990. *Archives of Dermatology* 129:1121–1129.

Dunkelschetter, C. 1984. Social support and cancer—Findings based on patient interviews and their implications. *Journal of Social Issues* 40:77–98.

Gehanno, J.F., K. Takahashi, S. Darmoni, and J. Weber. 2007. Citation classics in occupational medicine journals. *Scandinavian Journal of Work, Environment & Health* 33:245–251.

Godfray, H.C.J., J.R. Beddington, I.R. Crute et al. 2010. Food security: The challenge of feeding 9 billion people. *Science* 327:812–818.

Goldemberg, J. 2007. Ethanol for a sustainable energy future. *Science* 315:808–810.

Gombotz, W.R. and S.F. Wee. 1998. Protein release from alginate matrices. *Advanced Drug Delivery Reviews* 31:267–285.

Gray, R.H.., W.J. Wawer, R. Brookmeyer et al. 2001. Probability of HIV-1 transmission per coital act in monogamous, heterosexual, HIV-1-discordant couples in Rakai, Uganda. *Lancet* 357:1149–1153.

Grima, E.M., E.H. Belarbi, F.G.A. Fernandez, A.R. Medina, and Y. Chisti. 2003. Recovery of microalgal biomass and metabolites: Process options and economics. *Biotechnology Advances* 20:491–515.

Gustafson, K.R., J.H. Cardellina, R.W. Fuller et al. 1989. Aids-antiviral sulfolipids from cyanobacteria (blue-green-algae). *Journal of the National Cancer Institute* 81:1254–1258.

Hall, H.I., R.G. Song, P. Rhodes et al. 2008. Estimation of HIV incidence in the United States. *Journal of the American Medical Association* 300:520–529.

Hammer, S.M., J.J. Eron, P. Reiss et al. 2008. Antiretroviral treatment of adult HIV infection—2008 recommendations of the International AIDS Society USA panel. *Journal of the American Medical Association* 300:555–570.

Herrero, M., A. Cifuentes, and E. Ibanez. 2006. Sub- and supercritical fluid extraction of functional ingredients from different natural sources: Plants, food-by-products, algae and microalgae—A review. *Food Chemistry* 98:136–148.

Higuera-Ciapara, I., L. Felix-Valenzuela, and F.M. Goycoolea. 2006. Astaxanthin: A review of its chemistry and applications. *Critical Reviews in Food Science and Nutrition* 46:185–96.

Huneck, S. 1999. The significance of lichens and their metabolites. *Naturwissenschaften* 86:559–570.

Hussein, G., U. Sankawa, H. Goto, K. Matsumoto, and H. Watanabe. 2006. Astaxanthin, a carotenoid with potential in human health and nutrition. *Journal of Natural Products* 69:443–449.

Jacobson, M.Z. 2009. Review of solutions to global warming, air pollution, and energy security. *Energy & Environmental Science* 2:148–173.

Johnson, E.A. and G.H. An. 1991. Astaxanthin from microbial sources. *Critical Reviews in Biotechnology* 11:297–326.

Jonas, R. and L.F. Farah. 1998. Production and application of microbial cellulose. *Polymer Degradation and Stability* 59:101–106.

Kamb, M.L., M. Fishbein, J.M. Douglas et al. 1998. Efficacy of risk-reduction counseling to prevent human immunodeficiency virus and sexually transmitted diseases—A randomized controlled trial. *Journal of the American Medical Association* 280:1161–1167.

Kapdan, I.K. and F. Kargi. 2006. Bio-hydrogen production from waste materials. *Enzyme and Microbial Technology* 38:569–582.

Kobayashi, M., T. Kakizono, and S. Nagai. 1991. Astaxanthin production by a green-alga, *Haematococcus pluvialis* accompanied with morphological-changes in acetate media. *Journal of Fermentation and Bioengineering* 71:335–339.

Kobayashi, M., T. Kakizono, and S. Nagai. 1993. Enhanced carotenoid biosynthesis by oxidative stress in acetate-induced cyst cells of a green unicellular alga, *Haematococcus pluvialis*. *Applied and Environmental Microbiology* 59:867–873.

Konur, O. 2000. Creating enforceable civil rights for disabled students in higher education: An institutional theory perspective. *Disability & Society* 15:1041–1063.

Konur, O. 2002a. Assessment of disabled students in higher education: Current public policy issues. *Assessment and Evaluation in Higher Education* 27:131–152.

Konur, O. 2002b. Access to employment by disabled people in the UK: Is the Disability Discrimination Act working? *International Journal of Discrimination and the Law* 5:247–279.

Konur, O. 2002c. Access to nursing education by disabled students: Rights and duties of nursing programs. *Nursing Education Today* 22:364–374.

Konur, O. 2006a. Participation of children with dyslexia in compulsory education: Current public policy issues. *Dyslexia* 12:51–67.

Konur, O. 2006b. Teaching disabled students in higher education. *Teaching in Higher Education* 11:351–363.

Konur, O. 2007a. A judicial outcome analysis of the Disability Discrimination Act: A windfall for the employers? *Disability & Society* 22:187–204.

Konur, O. 2007b. Computer-assisted teaching and assessment of disabled students in higher education: The interface between academic standards and disability rights. *Journal of Computer Assisted Learning* 23:207–219.

Konur, O. 2011. The scientometric evaluation of the research on the algae and bio-energy. *Applied Energy* 88:3532–3540.

Konur, O. 2012a. 100 citation classics in energy and fuels. *Energy Education Science and Technology Part A—Energy Science and Research* 2012(si):319–332.

Konur, O. 2012b. What have we learned from the citation classics in energy and fuels: A mixed study. *Energy Education Science and Technology Part A* 2012(si):255–268.

Konur, O. 2012c. The gradual improvement of disability rights for the disabled tenants in the UK: The promising road is still ahead. *Social Political Economic and Cultural Research* 4:71–112.

Konur, O. 2012d. Prof. Dr. Ayhan Demirbas' scientometric biography. *Energy Education Science and Technology Part A—Energy Science and Research* 28:727–738.

Konur, O. 2012e. The evaluation of the research on the biofuels: A scientometric approach. *Energy Education Science and Technology Part A—Energy Science and Research* 28:903–916.

Konur, O. 2012f. The evaluation of the research on the biodiesel: A scientometric approach. *Energy Education Science and Technology Part A—Energy Science and Research* 28:1003–1014.

Konur, O. 2012g. The evaluation of the research on the bioethanol: A scientometric approach. *Energy Education Science and Technology Part A—Energy Science and Research* 28:1051–1064.

Konur, O. 2012h. The evaluation of the research on the microbial fuel cells: A scientometric approach. *Energy Education Science and Technology Part A—Energy Science and Research* 29:309–322.

Konur, O. 2012i. The evaluation of the research on the biohydrogen: A scientometric approach. *Energy Education Science and Technology Part A—Energy Science and Research* 29:323–338.

Konur, O. 2012j. The evaluation of the biogas research: A scientometric approach. *Energy Education Science and Technology Part A—Energy Science and Research* 29:1277–1292.

Konur, O. 2012k. The scientometric evaluation of the research on the production of bio-energy from biomass. *Biomass and Bioenergy* 47:504–515.

Konur, O. 2012l. The evaluation of the global energy and fuels research: A scientometric approach. *Energy Education Science and Technology Part A—Energy Science and Research* 30:613–628.

Konur, O. 2012m. The evaluation of the biorefinery research: A scientometric approach. *Energy Education Science and Technology Part A—Energy Science and Research* 2012(si):347–358.

Konur, O. 2012n. The evaluation of the bio-oil research: A scientometric approach. *Energy Education Science and Technology Part A—Energy Science and Research* 2012(si):379–392.

Konur, O. 2012o. What have we learned from the citation classics in energy and fuels: A mixed study. *Energy Education Science and Technology Part A—Energy Science and Research* 2012(si):255–268.

Konur, O. 2012p. The evaluation of the research on the biofuels: A scientometric approach. *Energy Education Science and Technology Part A—Energy Science and Research* 28:903–916.

Konur, O. 2013. What have we learned from the research on the International Financial Reporting Standards (IFRS)? A mixed study. *Energy Education Science and Technology Part D: Social Political Economic and Cultural Research* 5:29–40.

Kotake-Nara, E., M. Kushiro, H. Zhang, T. Sugawara, K. Miyashita, and A. Nagao. 2001. Carotenoids affect proliferation of human prostate cancer cells. *Journal of Nutrition* 131:3303–3306.

Koyanagi, S., N. Tanigawa, H. Nakagawa, S. Soeda, and H. Shimeno. 2003. Oversulfation of fucoidan enhances its anti-angiogenic and antitumor activities. *Biochemical Pharmacology* 65:173–179.

Lal, R. 2004. Soil carbon sequestration impacts on global climate change and food security. *Science* 304:1623–1627.

Lorenz, R.T. and G.R. Cysewski. 2000. Commercial potential for *Haematococcus* microalgae as a natural source of astaxanthin. *Trends in Biotechnology* 18:160–167.

Lynd, L.R., J.H. Cushman, R.J. Nichols, and C.E. Wyman. 1991. Fuel ethanol from cellulosic biomass. *Science* 251:1318–1323.

Mathers, C.D. and D. Loncar. 2006. Projections of global mortality and burden of disease from 2002 to 2030. *PLoS Medicine* 3:e442.

Moore, R.E. 1996. Cyclic peptides and depsipeptides from cyanobacteria: A review. *Journal of Industrial Microbiology* 16:134–143.

Namikoshi, M. and K.L. Rinehart. 1996. Bioactive compounds produced by cyanobacteria. *Journal of Industrial Microbiology & Biotechnology* 17:373–384.

Neilan, B.A., E. Dittmann, L. Rouhiainen et al. 1999. Nonribosomal peptide synthesis and toxigenicity of cyanobacteria. *Journal of Bacteriology* 181:4089–4097.

Nishiwaki-Matsushima, R., T. Ohta, S. Nishiwaki et al. 1992. Liver-tumor promotion by the cyanobacterial cyclic peptide toxin microcystin-lr. *Journal of Cancer Research and Clinical Oncology* 118:420–424.

North, D. 1994. Economic-performance through time. *American Economic Review* 84:359–368.

Ohta, T., E. Sueoka, N. Iida et al. 1994. Nodularin, a potent inhibitor of protein phosphatase-1 and phosphatase-2a, is a new environmental carcinogen in male f344 rat-liver. *Cancer Research* 54:6402–6406.

Paladugu, R., M.S. Chein, S. Gardezi, and L. Wise. 2002. One hundred citation classics in general surgical journals. *World Journal of Surgery* 26:1099–1105.
Portenoy, R.K., H.T. Thaler, A.B. Kornblith et al. 1994. Symptom prevalence, characteristics and distress in a cancer population. *Quality of Life Research* 3:183–189.
Potosky, A.L., G.F. Riley, J.D. Lubitz et al. 1993. Potential for cancer-related health-services research using a linked medicare-tumor registry database. *Medical Care* 31:732–748.
Proksch, P., R.A. Edrada, and R. Ebel. 2002. Drugs from the seas—Current status and microbiological implications. *Applied Microbiology and Biotechnology* 59:125–134.
Pulz, O. and W. Gross. 2004. Valuable products from biotechnology of microalgae. *Applied Microbiology and Biotechnology* 65:635–648.
Rinaudo, M. 2008. Main properties and current applications of some polysaccharides as biomaterials. *Polymer International* 57:397–430.
Romay, C., J. Armesto, D. Remirez, R. Gonzalez, N. Ledon, and I. Garcia. 1998. Antioxidant and anti-inflammatory properties of C-phycocyanin from blue-green algae. *Inflammation Research* 47:36–41.
Schwartz, R.E., C.F. Hirsch, D.F. Sesin et al. 1990. Pharmaceuticals from cultured algae. *Journal of Industrial Microbiology* 5:113–123.
Shepard, C.W., L. Finelli, and M. Alter. 2005. Global epidemiology of hepatitis C virus infection. *Lancet Infectious Diseases* 5:558–567.
Simmons, T.L., E. Andrianasolo, K. McPhail, P. Flatt, and W.H. Gerwick. 2005. Marine natural products as anticancer drugs. *Molecular Cancer Therapeutics* 4:333–342.
Smit, A.J. 2004. Medicinal and pharmaceutical uses of seaweed natural products: A review. *Journal of Applied Phycology* 16:245–262.
Smith, C.D., X.Q. Zhang, S.L. Mooberry, G.M.L. Patterson, and R.E. Moore. 1994. Cryptophycin—A new antimicrotubule agent active against drug-resistant cells. *Cancer Research* 54:3779–3784.
Spolaore, P., C. Joannis-Cassan, E. Duran, and A. Isambert. 2006. Commercial applications of microalgae. *Journal of Bioscience and Bioengineering* 101:87–96.
Tan, L.T. 2007. Bioactive natural products from marine cyanobacteria for drug discovery. *Phytochemistry* 68:954–979.
Trimurtulu, G., I. Ohtani, G.M.L. Patterson et al. 1994. Total structures of cryptophycins, potent antitumor depsipeptides from the blue-green-alga *Nostoc* sp. strain Gsv-224. *Journal of the American Chemical Society* 116:4729–4737.
Tsai, C.C., P. Emau, Y.H. Jiang et al. 2004. Cyanovirin-N inhibits AIDS virus infections in vaginal transmission models. *AIDS Research and Human Retroviruses* 20:11–18.
Ueno, Y., S. Nagata, T. Tsutsumi et al. 1996. Detection of microcystins, a blue-green algal hepatotoxin, in drinking water sampled in Haimen and Fusui, endemic areas of primary liver cancer in China, by highly sensitive immunoassay. *Carcinogenesis* 17:1317–1321.
Verdier-Pinard, P., J.Y. Lai, H.D. Yoo et al. 1998. Structure–activity analysis of the interaction of curacin A, the potent colchicine site antimitotic agent, with tubulin and effects of analogs on the growth of MCF-7 breast cancer cells. *Molecular Pharmacology* 53:62–76.
Volesky, B. and Z.R. Holan. 1995. Biosorption of heavy-metals. *Biotechnology Progress* 11:235–250.
Walsh, A.H., A.Y. Cheng, and R.E. Honkanen. 1997. Fostriecin, an antitumor antibiotic with inhibitory activity against serine/threonine protein phosphatases types 1 (PP1) and 2A (PP2A), is highly selective for PP2A. *FEBS Letters* 416:230–234.
Wang, B., Y.Q. Li, N. Wu, and C.Q. Lan. 2008. CO(2) bio-mitigation using microalgae. *Applied Microbiology and Biotechnology* 79:707–718.
Witvrouw, M. and E. DeClercq. 1997. Sulfated polysaccharides extracted from sea algae as potential antiviral drugs. *General Pharmacology* 29:497–511.
Wlodawer, A. and J.W. Erickson. 1993. Structure-based inhibitors of HIV-1 protease. *Annual Review of Biochemistry* 62:543–585.
Wrigley, N. and S. Matthews. 1986. Citation-classics and citation levels in geography. *Area* 18:185–194.
Yergin, D. 2006. Ensuring energy security. *Foreign Affairs* 85:69–82.

31

Enhancing Economics of Spirulina platensis *on a Large Scale*

E.M. Nithiya, K.K. Vasumathi, and M. Premalatha

CONTENTS

31.1 Introduction

Spirulina platensis is a mesophilic planktonic filamentous cyanobacterium forming massive populations in freshwater and brackish lakes and some marine environment (Vonshak, 1997). Spirulina is extraordinarily richer in nutrition than in any food item. It contains high content of protein (70%), extremely high concentration of beta-carotene, vitamin B12, essential fatty acids, essential amino acids, antioxidants, and minerals (Belay et al., 1993; Vonshak, 1997). Spirulina is highly potent in providing nourishment and mental clarity, assisting in cancer recovery. It is rich in beta-carotene, a source of antioxidant that helps in effectively fighting against diseases. Spirulina is rich in vitamin B that prevents cardiovascular diseases. It is a good anti-inflammatory agent, which enhances the immune system and contains antiaging properties. Due to the potential use of spirulina as a nutritional supplement, commercial production has profound international demand. Large-scale production of biomass is an intricate process that involves optimization of numerous parameters for profitable yield. Among the various constraints in a large-scale production, physical, physiological, and economical limitations play an inevitable role (Mostert and Grobbelaar, 1987; Raoof, 2002). In developing countries such as India, focus should be given to minimize the production cost (Gupta and Changwal, 1988) to overcome the economical limitation.

31.2 Media for Cultivation of Spirulina

The Zarrouk medium is the only conventional medium used for spirulina production on a large scale. A number of conventional media are experimented and biomass productivities corresponding to different media are reported (Celeki and Yavuzatmaca, 2009; Pandey et al., 2010; Zeng et al., 2012). The cost for commercial media reported in literature is compared to produce 1 g of dry biomass and shown in Table 31.1. Both the Zarrouk and modified Schlosser medium have good productivity, and the cost of the medium to produce 1 g of dry biomass is comparatively much lesser than other media, although for large-scale cultivation, only addition of media is not economical. CO_2 emitted from thermal power plants or from point sources added to media can make large scale cultivation economical.

31.2.1 Supply of CO_2 in Spirulina Cultivation Other Than Nutrients

Global energy consumption releases abundant CO_2 emissions, which is a major cause of greenhouse gas (Chen et al., 2011; Matondo et al., 2004). Among the CO_2 reduction techniques, microalgal sequestration is the most effective (Morais and Costa, 2007; Zeng et al., 2012), which reduces CO_2 by converting it into organic compounds through photosynthesis (Ho et al., 2012; Jacob Lopes et al., 2009). Microalgae convert solar energy into chemical energy. Apart from CO_2 mitigation, algal biomass generated from this source leads to various high-value products, human food, animal feed, cosmetics, and medicine (Chen et al., 2010; Ho et al., 2010, 2011; Yeh et al., 2012). Zeng et al. (2012) reported that the spirulina grown in 14 h of light period with illumination intensity of 200 μmol/(m^2 s) in the presence of CO_2 exhibited significant change in the yield.

Maximum algal productivity on a large scale ranges from 10 to 30 g/m^2/day (Jimenez et al., 2003a,b; Putt, 2007). Assuming an average microalgal productivity as 10 g/m^2/day and the functional days in a year as 300, the annual yield of spirulina for 2.5 acres is 30.35 tons. One gram of dry algal biomass consumes 1.83 g of CO_2. The cost of 1 kg of spirulina dry powder is Rs. 1000, and the cost of 1 ton of carbon credit is Rs. 735 ($12). The investment cost for 2.5 acres is Rs. 400 lakhs (inclusive of land, equipment, and utility cost) (KSIDC, 2012). The addition of CO_2 apart from nutrients increases the yield up to

TABLE 31.1

Price Comparison of Different Reported Media and Their Biomass Yield

Media	Light Condition	Dry Weight (g/L)	Approximate Cost of Media for 1 g Dry Weight (Rs.)	Approximate Cost of Media for 1000 L (Merck)[a] (Rs.)
Zarrouk medium	200 μmol/(m^2 s)	3.2	3.0025	9,608
Modified Schlosser medium	2 Klux	3.497	2.92	10,245.2
Revised Medium 6	5 Klux	1.3	6.174	8,027.4
Oferr medium	5 Klux	1.2	4.5	5,393.7
Rao medium	5 Klux	0.84	9.847	8,271.87
CFTRI medium	5 Klux	0.72	6.852	4,933.7
Bangladesh medium	5 Klux	0.112	10.33	7,794.2

[a] The cost is calculated based on the Merck catalogue.

TABLE 31.2

Anticipated Revenue Generation for Spirulina Production in Various Conditions

Parameters	Spirulina Production without CO_2	Spirulina Production with CO_2
Annual biomass productivity (tons)	30.35	56.14
Revenue generated from algal biomass yield (Rs.)	303.52 Lakhs	561.48 Lakhs
Revenue generated from sequestered CO_2 (Rs.)	22,307	41,262.9
Total revenue generation (Rs.)	303.22	561.89
Investment cost for 2.5 acres (land, equipment, and utilities) (Rs.)	400 Lakhs	400 Lakhs
Payback period (exclusive of operating cost) (years)	1.32	0.712

1.85 times (Zeng et al., 2012). The revenue generated from algal biomass yield, subsequent CO_2 sequestered, and payback period is calculated and compared with the corresponding conditions in the presence and absence of CO_2 as given in Table 31.2.

31.3 Influence of Optical Fibers on Biomass Yield

Open pond cultivation requires less capital investment than photobioreactors; still contamination is the potential risk associated with nutraceutical compounds extracted from algae. Other impediments in open pond reactors are the requirement of a huge land area for large-scale cultivation. One gram of algal dry biomass fixes 1.83 g of CO_2 (Weissman et al., 1988). The maximum biomass yield is 30 g/day/m^2. Hence, to produce 1 ton of biomass, it requires 0.033 km^2 land area. Therefore, the land requirement is very high and also the maintenance and operation is highly difficult. Apart from these constraints, culture cells present near the surface receive light and the remaining microalgal cells are under shading effect. Thus, designing an efficient photobioreactor that reduces the land area, as well as providing maximum biomass productivity, is necessary.

Use of optical fibers in photobioreactors maximizes the depth and minimizes land area. Optical fibers also provide uniform light distribution throughout the system during the daytime. They are robust and flexible, which are made up of plastic or glass fibers used in transmitting the light to the photobioreactor from the light collector. Due to enormous light accumulation in one point of the optical fiber, the selection of optical fibers should be based on the ability to withstand very high temperature. Optical fibers should be designed to transmit only light but not heat. Optical fiber provided as a source of illumination to microalgal cells also restricts UV and IR radiations enhancing photoautotrophic cultivation (An and Kim, 2000). Optical fibers were first used as an axially fixed light source in bubble column photobioreactor (Mori, 1986). Optical fibers with various diameters and lengths are available, which can be chosen according to the dimension of photobioreactor and the light required by the microalgal species to produce maximum biomass yield. Xue et al. (2013) reported 38% increase in yield using optical fibers in 130 L airlift photobioreactors.

31.4 Influence of Intermittent Illumination (Light/Dark Cycle)

Light requirement is one of the essential parameters required for microalgal cultivation. Providing light at optimum intensity, wavelength, and duration gives outstanding photosynthetic efficiency enabling higher biomass yields (Carvalho et al., 2011). Flashing light effect increases the efficiency of photosynthesis. To enhance optimized light to dark cycle ratio, various techniques are practiced to modify the configuration of photobioreactor as well as to impose the suitable turbulent flow (Carvalho et al., 2011).

In stirred photobioreactors during sunlight conditions, photosynthetic ability of microalgae keeps fluctuating dominantly between compensation light intensity (region limited by light intensity where the microalgal cells grow) and light saturation region (no further growth of microalgae in spite of increasing light intensity, where the excess light is released as heat) and rarely in photoinhibition region (where the excessive light becomes injurious). An effective alternative to the continuous supply of light for extended periods can be replaced by the use of intermittent light; in fact photosynthetic efficiency is increased for short flashes of light separated by an extended dark period, which still has not been explored in practice for large-scale cultivation (Carvalho et al., 2011). Xue et al. (2011) reported under intermittent illumination, when L/D frequency increased from 0.01 to 20 Hz, specific growth rate and light efficiency were enhanced. Higher light efficiency is obtained under light-limited region, whereas efficiency decreases with increasing light intensity.

31.5 Influence of Agitation

Agitation in a microalgal culture improves mass transfer and uniform distribution of light to the microalgal cells. Among the mixing techniques involved, bubbling air in the column offered maximum biomass productivity (Ravelonandro et al., 2011). A bubble column induces optimum shear stress in the reactor, also highly minimizing the shading condition in algal culture. Hence, agitation through use of a bubble column seemed to be effective in large-scale spirulina cultivation. A high L/D frequency in the microalgal system needs a very high extent of agitation, thus leading to abundant mixing or pumping input. Solar light intensity is low in the morning or evening and is high in the afternoon. Therefore, especially during noon at high light intensity, the algal culture needs to be agitated under high turbulence favoring high L/D frequency than morning or evening (Xue et al., 2011).

31.6 Influence of Photobioreactors in Large-Scale Cultivation

Increasing population accompanied with exorbitant cost of land and seasonal temperature variation, solar light fluctuation throughout the year has created high demand for the development of large-scale photobioreactors (Ugwv et al., 2008). Photobioreactors have received more attention in providing a highly controlled environment than open systems that are easily prone to contamination. With closed photobioreactors, optimum parameters are easily maintained, facilitating higher biomass productivity and value-added products (Najafour, 2007). Although many photobioreactors are operated on a laboratory scale, very few are scaled up due to constraints in maintaining optimum process parameters (Ugwv et al., 2008). Carlozzi (2003) reported 1.90 g/L/day biomass productivity in *Arthrospira platensis* using a undular–tubular photobioreactor. Zeng et al. (2012) reported a 1.27-fold increase in biomass yield in a custom designed photobioreactor under optimized bioprocess conditions over an open system.

31.7 Semicontinuous Mode of Operation

Microalgal growth exhibits an autocatalytic process. In a batch experiment of microalgae, an optimum biomass concentration is reached in a certain period of time. If this optimum biomass concentration is maintained, it leads to rapid increase in biomass. Hence, harvesting microalgae systematically sustains the maximum growth rate and provides profitable yield (Vasumathi et al., 2013). Ravelonandro et al. (2011) reported the biomass productivity of 0.21–0.23 g/L/day in the semicontinuous mode conducted with light 1200 lux, green light, and 1% CO_2 in *S. platensis*, where harvesting is done four times throughout the experiment. In the semicontinuous mode of experiment, the productivity is high compared with open pond cultivation (Radmann et al., 2007; Ravelo, 2001) and in lab-scale aerated photobioreactors (Colla et al., 2007; Wang et al., 2007), whose productivity ranges from 0.02 to 0.05 g/L/day.

31.8 Downstream Processing of Microalgae

Filtration, centrifugation, and flotation are conventionally used downstream techniques for harvesting microalgal cells. Microalgae such as spirulina, anabaena, and microcysts move in a vertical position through gas vesicles (Oliver and Ganf, 2000), facilitating harvesting by filtration. Flotation is a cheap and cost-effective method performed without any physical damage of harvested algal biomass. The floating cells of microalgae are easily harvested from the surface by skimming. The quality of biomass in a stationary phase is low and thus monitoring optimum harvesting time is highly essential. The optimum harvesting time of *S. platensis* can be predicted by the cellular protein to carbon ratio, and the biomass productivity was found to be maximum at day 6 (Kim et al., 2005). On a commercial scale, spirulina is harvested by filtration in three stages. The first stage of filtration is the elimination of pond debris (Belay, 1997). The second stage is the harvesting of microalgae and the final stage is to further concentrate the spirulina cells. These cells are subjected to sun drying and further various potential compounds are extracted. Products generated from spirulina are given in Table 31.3.

31.9 Concluding Remarks

Among the three commercial algal genera *Spirulina, Dunaliella,* and *Chlorella, Spirulina* has the maximum possibility of establishing efficient large-scale production due to high profit generated from value-added products and also achieving carbon credits for

TABLE 31.3

Products Generated from Spirulina Cultivation

Commercial Products of Spirulina	List of Websites
Capsules, nutrition drinks and bars, beauty soap, fairness cream, powder, pasta, skin care products such as toning lotion, active moisturising gel, beauty, cream mask and Eye zone repair gel, noodles, shampoo, and other therapeutical products.	http://www.biopathica.co.uk/page4.html http://www.jaffexim.com/products.html http://www.algaeindustrymagazine.com/special-report-spirulina-part-6/. http://www.tradeindia.com/suppliers/spirulina-capsule.html

mitigating CO_2. The common problems encountered in large-scale production are low productivity per unit area and inconsistency of product quality throughout the production period. Various strategies for achieving the quality and economics of cultivation on a large scale are discussed, which can effectively overcome the current problems to considerable extent, developing spirulina industry in a sustained and efficient way. More pilot-scale studies need to be carried out in research organizations before commercializing the product in the industry. Further, the interaction between research institutes and industry can improve the methods employed for cultivation, production technology, quality control, and screening of value-added products.

Acknowledgment

We express our sincere gratitude to the Department of Science and Technology for providing an opportunity to research and explore CO_2 sequestration using microalgae (DST/IS-STAC/CO2-SR-12/07).

References

An, J.Y. and Kim, B.W. 2000. Biological desulfurization in an optical-fiber photobioreactor using an automatic sunlight collection system. *Journal of Biotechnology* 80:35–44.

Belay, A. 1997. Mass culture of *Spirulina* outdoors: The earthrise farms experience. In *Spirulina Platensis (Arthrospira): Physiology, Cell-Biology and Biotechnology*. ed. A. Vonshak, pp. 131–158. Taylor & Francis Ltd., London, U.K.

Belay, A., Ota, Y., Miyakawa, K., and Shimamatsu, H. 1993. Current knowledge on potential health benefits of *Spirulina*. *Journal of Applied Phycology* 5:235–241.

Carlozzi, P. 2003. Dilution of solar radiation through culture lamination in photobioreactor rows facing south–north: A way to improve the efficiency of light utilization by cyanobacteria (*Arthrospira platensis*). *Biotechnology and Bioengineering* 81:305–315.

Carvalho, A.P., Silva, S.O., Baptista, J.M., and Malcata, F.X. 2011. Light requirements in microalgal photobioreactors: An overview of biophotonic aspects. *Applied Microbiology and Biotechnology* 89:1275–1288.

Celekli, A. and Yavuzatmaca, M. 2009. Predictive modeling of biomass production by *Spirulina platensis* as function of nitrate and NaCl concentrations. *Bioresource Technology* 100:1847–1851.

Chen, C.Y., Yeh, K.L., Aisyah, R., Lee, D.J., and Chang, J.S. 2011. Cultivation, photobioreactor design and harvesting of microalgae for biodiesel production: A critical review. *Bioresource Technology* 102:71–81.

Chen, C.Y., Yeh, K.L., Su, H.M., Lo, Y.C., Chen, W.M., and Chang, J.S. 2010. Strategies to enhance cell growth and achieve high-level oil production of a *Chlorella vulgaris* isolate. *Biotechnology Progress* 26:679–686.

Colla, L.C., Reinehr, C.O., Reichert, C., and Viera Costa, J.A. 2007. Production of biomass and nutraceutical compounds by *Spirulina platensis* under different temperature and nitrogen regimes. *Bioresource Technology* 98:1489–1493.

Gupta, R.S. and Changwal, M.L. 1988. Biotechnology of mass production of *Spirulina* and *Arthrospira* in fresh water. In *Spirulina: Proceedings of the ETTA National Symposium*, ed. Seshadri, C.V. and Jeeji Bai, N, pp. 125–128. MCRC: Madras.

Ho, S.H., Chen, C.Y., and Chang, J.S. 2012. Effect of light intensity and nitrogen starvation on CO_2 fixation and lipid/carbohydrate production of an indigenous microalga *Scenedesmus obliquus* CNW-N. *Bioresource Technology* 113:244–252.

Ho, S.H., Chen, C.Y., Lee, D.J., and Chang, J.S. 2011. Perspectives on microalgal CO_2-emission mitigation systems—A review. *Biotechnology Advances* 29:189–198.

Ho, S.H., Chen, W.M., and Chang, J.S. 2010. *Scenedesmus obliquus* CNW-N as a potential candidate for CO_2 mitigation and biodiesel production. *Bioresource Technology* 101(22):8725–8730.

KSIDC (Kerala State Industrial Development Corporation). 2012. Project profile on commercial production of Spirulina algae. http://www.emergingkerala2012.org/.

Jacob Lopes, E., Revah, S., Hernandez, S., Shirai, K., and Franco, T.T. 2009. Development of operational strategies to remove carbon dioxide in photobioreactors. *Chemical Engineering Journal* 153:120–126.

Jimenez, C., Cossıo, B.R., Labella, D., and Niell, F.X. 2003a. The feasibility of industrial production of *Spirulina* (*Arthrospira*) in southern Spain. *Aquaculture* 217:179–190.

Jimenez, C., Cossıo, B.R., and Niell, X. 2003b. Relationship between physicochemical variables and productivity in open ponds for the production of *Spirulina*: A predictive model of algal yield. *Aquaculture* 221:331–345.

Kim, S.G., Choi, A., Ahn, C.Y., Park, C.S., Park, Y.H, and Oh, H.M. 2005. Harvesting of *Spirulina platensis* by cellular flotation and growth stage determination. *Letters in Applied Microbiology* 40:190–194.

Matondo, J.I., Peter, G., and Msibi, K.M. 2004. Evaluation of the impact of climate change on hydrology and water resources in Swaziland. *Physics and Chemistry of the Earth, Part B: Hydrology, Oceans and Atmosphere* 29:1193–2002.

Morais, M.G. and Costa, J.A.V. 2007. Biofixation of carbon dioxide by *Spirulina* sp. and *Scenedesmus obliquus* cultivated in a three-stage serial tubular photobioreactor. *Journal of Biotechnology* 129:439–445.

Mori, K. 1986. Photoautotrophic bioreactor using visible solar rays condensed by Fresnel lenses and transmitted through optical fibers. *Biotechnology and Bioengineering Symposium* 15:331–345.

Mostert, E.S. and Grobbelaar, J.U. 1987. The influence of nitrogen and phosphorus on algal growth and quality in outdoor mass algal cultures. *Biomass* 13:219–233.

Najafour, D.G. 2007. *Biochemical Engineering and Biotechnology*. Oxford, U.K.: Elsevier.

Oliver, R. and Ganf, G. 2000. Freshwater blooms. In *The Ecology of Cyanobacteria: Their Diversity in Time and Space*, eds. B. Whitton and M. Potts, pp. 149–194. Dordrecht, the Netherlands: Kluwer Academic Publishers.

Pandey, J.P., Tiwari, A., and Mishra, R.M. 2010. Evaluation of biomass production of *Spirulina maxima* on different reported media. *Journal of Algal Biomass Utilization* 1(3):70–81.

Putt, R. 2007. *Algae as a Biodiesel Feedstock: A Feasibility Assessment*. Department of Chemical Engineering, Auburn University, Auburn, AL.

Radmann, E.M., Reinehr, C.O., and Costa, J.A.V. 2007. Optimization of the repeated batch cultivation of microalga *Spirulina platensis* in open raceway ponds. *Aquaculture* 265:118–126.

Raoof, B. 2002. Standardization of growth parameters for outdoor biomass production of *Spirulina* sp. PhD thesis, Division of Microbiology, Indian Agricultural Research Institute, New Delhi, India.

Ravelo, V. 2001. Bio-écologie, valorisation du gisement naturel de *Spiruline* de Belalanda (Toliara, sud-ouest de Madagascar) et technologie de la culture. PhD thesis, University of Toliara, Tuléar, Madagascar.

Ravelonandro, P.H., Ratianarivo, D.H., Joannis-Cassan, C., Isambert, A., and Raherimandimbya, M. 2011. Improvement of the growth of *Arthrospira* (*Spirulina*) *platensis* from Toliara (Madagascar): Effect of agitation, salinity and CO_2 addition. *Food and Bioproducts Processing* 89:209–216.

Ugwv, C.U., Aoyagi, H., and Uchiyama, H. 2008. Photobioreactors for mass cultivation of algae. *Bioresource Technology* 99:4021–4028.

Vasumathi, K.K., Premalatha, M., and Subramanian, P. 2013. Experimental studies on the effect of harvesting interval on yield of *Scenedesmus arcuatus* var. capitatus. *Ecological Engineering* 58:13–16.

Vonshak, A. 1997. *Spirulina platensis* (*Arthrospira*): Physiology, cell biology and biotechnology. *Journal of Applied Phycology* 9(3):295–296.

Wang, C.Y., Fu, C.C., and Liu, Y.C. 2007. Effects of using light-emitting diodes on the cultivation of *Spirulina platensis*. *Journal of Biochemical Engineering* 37:21–25.

Weissman, J.D., Goebel, R.P., and Benemann, J.R. 1988. Photobioreactor design: Mixing, carbon utilization and oxygen accumulation. *Biotechnology and Bioengineering* 3:336–344.

Xue, S., Su, Z., and Cong, W. 2011. Growth of *Spirulina platensis* enhanced under intermittent illumination. *Journal of Biotechnology* 151:271–277.

Xue, S., Zhang, Q., Wu, X., Yan, C., and Cong, W. 2013. A novel photobioreactor structure using optical fibers as inner light source to fulfill flashing light effects of microalgae. *Bioresource Technology* 138:141–147.

Yeh, K.L., Chen, C.Y., and Chang, J.S. 2012. pH-stat photoheterotrophic cultivation of indigenous *Chlorella vulgaris* ESP-31 for biomass and lipid production using acetic acid as the carbon source. *Biochemical Engineering Journal* 64:1–7.

Zeng, X., Danquah, M.K., Zhanga, S., Zhang, X., Wu, M., Chen, X.D., Ng, I.-S., Jing, K., and Lu, Y. 2012. Autotrophic cultivation of *Spirulina platensis* for CO_2 fixation and phycocyanin production. *Chemical Engineering Journal* 183:192–197.

32

Algal Economics and Optimization

Ozcan Konur

CONTENTS

32.1 Introduction

32.1.1 Issues

Global warming, air pollution, and energy security have been some of the most important public policy issues in recent years (Jacobson 2009; Wang et al. 2008; Yergin 2006). With the increasing global population, food security has also become a major public policy issue (Lal 2004). The development of biofuels generated from biomass has been a long-awaited solution to these global problems (Demirbas 2007; Goldember 2007; Lynd et al. 1991). However, the development of early biofuels produced from agricultural plants such as sugarcane (Goldemberg 2007) and agricultural wastes such as corn stovers (Bothast and Schlicher 2005) has resulted in a series of substantial concerns about food security (Godfray et al. 2010). Therefore, the development of algal biofuels as a third-generation biofuel has been considered as a major solution for global warming, air pollution, energy security, and food security (Chisti 2007; Demirbas 2007; Kapdan and Kargi 2006; Spolaore et al. 2006; Volesky and Holan 1995).

The economics of production of algal biofuels and algal biocompounds has been an important research area in recent years, complementing technological aspects of the production of algal biofuels and biocompounds (e.g., Cardozo et al. 2007; Grima et al. 2003; Pulz 2001). Based on the results from studies on algal economics, the issues relating to the optimization of the production of the algal biofuels and compounds through LCA have also emerged as an important research area (e.g., Clarens et al. 2010; Jorquera et al. 2010; Lardon et al. 2009).

For example, Borowitzka (1999) argues that the "main problem facing the commercialization of new microalgae and microalgal products is the need for closed culture systems and the fact that these are very capital intensive. The high cost of microalgal culture systems relates to the need for light and the relatively slow growth rate of the algae." Pulz (2001) argues that the technical design of photobioreactors (PBRs) is the "most important issue for economic success in the field of phototrophic biotechnology."

Li et al. (2008) argues that the production of algal biofuels can be coupled with flue gas CO_2 mitigation, wastewater treatment, and the production of high-value chemicals. They assert that the developments in microalgal cultivation and downstream processing (e.g., harvesting, drying, and thermochemical processing) would further "enhance the cost effectiveness of the biofuel from microalgae strategy." Greenwell et al. (2010) argue that "simulations that incorporate financial elements, along with fluid dynamics and algae growth models, are likely to be increasingly useful for predicting reactor design efficiency and life cycle analysis to determine the viability of the various options for large-scale culture." They argue that the greatest potential for cost reduction and increased yields most probably lies within closed or hybrid closed–open production systems.

Similarly, relating to the optimization of the processes for the production of algal biofuels and algal biocompounds, Lardon et al. (2009) confirm the "potential of microalgae as an energy source but highlight the imperative necessity of decreasing the energy and fertilizer consumption" and assert that the "control of nitrogen stress during the culture and optimization of wet extraction are valuable options." They also stress the potential of anaerobic digestion of oil cakes as a way to reduce external energy demand and to recycle some of the mineral fertilizers. Clarens et al. (2010) find that "conventional crops have lower environmental impacts than algae in energy use, greenhouse gas emissions, and water regardless of cultivation location. Only in total land use and eutrophication potential do algae perform favorably." They argue that the large environmental footprint of algal cultivation is driven predominantly by upstream impacts, such as the demand for CO_2 and fertilizer. To reduce these impacts, they argue that "flue gas and, to a greater extent, wastewater could be used to offset most of the environmental burdens associated with algae."

Jorquera et al. (2010) find that the "use of horizontal tubular PBRs is not economically feasible (net energy ratio, [NER] <1) and that the estimated NERs for flat-plate PBRs and raceway ponds is >1." They argue that the net energy ratio (NER) for ponds and flat-plate PBRs could be raised to significantly higher values "if the lipid content of the biomass were increased to 60% dw/cwd." Stephenson et al. (2010) find that if the future target for the productivity of lipids from microalgae, such as *Chlorella vulgaris*, of around 40 tons ha^{-1} year^{-1} could be achieved, "cultivation in typical raceways would be significantly more environmentally sustainable than in closed air-lift tubular bioreactors." They next find that "if air-lift tubular bioreactors were used, the GWP of the biodiesel would be significantly greater than the energetically equivalent amount of fossil-derived diesel." They finally find that the global warming potential (GWP) and fossil-energy requirement in this operation are particularly sensitive to "(i) the yield of oil achieved during cultivation, (ii) the velocity of circulation of the algae in the cultivation facility, (iii) whether the culture media could be recycled or not, and (iv) the concentration of carbon dioxide in the flue gas."

Although there have been many studies on the economics and optimization of algal biofuels and algal biocompounds (e.g., Cardozo et al. 2007; Clarens et al. 2010; Grima et al. 2003; Jorquera et al. 2010; Lardon et al. 2009; Pulz 2001) and there have been a number of scientometric studies on the algal biofuels (Konur 2011), there has not been any study on the citation classics on algal economics and optimization of algal biofuels and algal biocompounds as in other research fields (Baltussen and Kindler 2004a,b; Dubin et al. 1993; Gehanno et al. 2007; Konur 2012a,b, 2013; Paladugu et al. 2002; Wrigley and Matthews 1986).

As North's new institutional theory suggests, it is important to have up-to-date information about current public policy issues to develop a set of viable solutions to satisfy the needs of key stakeholders (Konur 2000, 2002a,b,c, 2006a,b, 2007a,b, 2012c; North 1994).

Therefore, brief information on a selected set of 50 citation classics in the field is presented in this chapter to inform the key stakeholders about the economic and optimal production of algal biofuels and algal biocompounds for the solution of these problems in the long run, thus complementing a number of recent scientometric and review studies on biofuels and global energy research (Konur 2011, 2012d,e,f,g,h,i,j,k,l,m,n,o,p).

32.1.2 Methodology

A search was carried out in the Science Citation Index Expanded (SCIE) and Social Sciences Citation Index (SSCI) databases (version 5.11) in September 2013 to locate papers relating to the algal economics and algal optimization using the keyword set of Topic = (econ* or *life* cycle** or commercialization or commercialization) AND Topic = (alga* or *macro* alga** or *micro* alga** or macroalga* or microalga* or cyanobacter* or seaweed* or diatoms or *sea* weed** or reinhardtii or braunii or chlorella or sargassum or gracilaria) in the abstract pages of the papers. For this chapter, it was necessary to embrace the broad algal search terms to include diatoms, seaweeds, and cyanobacteria as well as to include many forms of economics rather than the simple keywords of economics and algae.

It was found that there were 3400 papers indexed between 1980 and 2013. The key subject categories for the algal research were marine and freshwater biology (1072 references, 30.1%), biotechnology and applied microbiology (524 references, 15.1%), plant sciences (478 references, 13.8%), and environmental sciences (455 references, 13.1%), altogether comprising 61.2% of all the references on algal economics and algal optimization. It was also necessary to focus on the key references by selecting articles and reviews.

The located highly cited 50 papers were arranged by decreasing number of citations for 2 groups of papers: 31 and 19 papers for the research on algal economics and algal optimization, respectively. In order to check whether these collected abstracts correspond to the larger sample on these topical areas, new searches were carried out for each topical area.

The summary information about the located citation classics is presented under two major headings of algal economic and algal optimization research in the order of the decreasing number of citations for each topical area.

The information relating to the document type, affiliation and location of the authors, journal, indexes, subject area of the journal, concise topic, total number of citations received, and total average number of citations received per year were given in the tables for each paper.

32.2 Studies on Algal Economics

32.2.1 Introduction

The research on algal economics has been one of the most dynamic research areas in recent years. Thirty-one citation classics in the field with more than twenty-one citations were located, and the key emerging issues from these papers are presented in the decreasing order of the number of citations in Table 32.1. These papers give strong hints about the determinants of the economic production of algal biofuels and biocompounds and emphasize that marine algae are efficient feedstocks in comparison with first- and second-generation biofuel feedstocks.

The papers were dominated by researchers from 20 countries, usually through the intra-country institutional collaboration, and they were multiauthored. The United States (10 papers), England and the Netherlands (3 papers each), and Wales, Malaysia, India, and New Zealand (2 papers each) were the most prolific countries.

Similarly, all these papers were published in journals indexed by the Science Citation Index (SCI) and/or SCIE. There was also one paper indexed by the SSCI. The number of citations ranged from 21 to 272 and the average number of citations per annum ranged from 1.0 to 29.7. Twenty of the papers were articles, while eleven of them were reviews.

32.2.2 Research on Algal Economics

Borowitzka (1999) discusses commercial production of microalgae with a focus on ponds, tanks, tubes, and fermenters in a paper with 272 citations. He notes that production of microalgae for aquaculture is generally on a much smaller scale and in many cases is carried out indoors in 20–40 L carboys or in large plastic bags of up to approximately 1000 L in volume. More recently, a helical tubular PBR system, the BIPCOIL™, has been developed that allows algae to be grown reliably outdoors at high cell densities in a semicontinuous culture. Other closed PBRs such as flat panels are also being developed. He argues that the "main problem facing the commercialization of new microalgae and microalgal products is the need for closed culture systems and the fact that these are very capital intensive. The high cost of microalgal culture systems relates to the need for light and the relatively slow growth rate of the algae." He asserts that "although this problem has been avoided in some instances by growing the algae heterotrophically, not all algae or algal products can be produced this way."

Grima et al. (2003) discuss the recovery of microalgal biomass and metabolites with a focus on the process options and economics in a paper with 251 citations. They argue that commercial production of intracellular microalgal metabolites requires: "(1) large-scale monoseptic production of the appropriate microalgal biomass; (2) recovery of the biomass from a relatively dilute broth; (3) extraction of the metabolite from the biomass; and (4) purification of the crude extract." They examine the options available for recovery of the biomass and the intracellular metabolites from the biomass. They finally discuss the economics of monoseptic production of microalgae in PBRs and the downstream recovery of metabolites using eicosapentaenoic acid recovery as a representative case study.

Pulz (2001) discusses PBRs with a focus on the production systems for phototrophic microorganisms in a paper with 205 citations. He argues that the design of the technical and technological basis for PBRs for the production of algal biofuels and biocompounds is the "most important issue for economic success in the field of phototrophic biotechnology." For future applications, open pond systems for large-scale production have a lower innovative potential than closed systems. He asserts that for high-value products in particular, "closed systems of photobioreactors are the more promising fields for technical developments despite very different approaches in design."

Li et al. (2008) discuss biofuels from microalgae in a paper with 178 citations. They note that microalgae can potentially be employed for the production of biofuels in an economically effective and environmentally sustainable manner. Microalgae have been investigated

TABLE 32.1
Research on Algal Economics

No.	Paper References	Year	Document	Affiliation	Country	No. of Authors	Journal	Index	Subjects	Topic	Total No. of Citations	Total Average Citations per Annum	Rank
1	Borowitzka	1999	A	Murdoch Univ.	Australia	1	*J. Biotechnol.*	SCI, SCIE	Biot. Appl. Microb.	Algal economics	272	18.1	1
2	Grima et al.	2003	R	Massey Univ., Univ. Almeria	New Zealand, Spain	5	*Biotechnol. Adv.*	SCI, SCIE	Biot. Appl. Microb.	Algal economics	251	22.8	2
3	Pulz	2001	R	IGV Inst. Cereal Proc.	Germany	1	*Appl. Microbiol. Biotechnol.*	SCI, SCIE	Biot. Appl. Microb.	Algal economics	205	15.8	3
4	Li et al.	2008	A	Univ. Ottawa, S China Normal Univ., AirScience Biofuels	Canada, China	5	*Biotechnol. Prog.*	SCI, SCIE	Biot. Appl. Microb., Food Sci. Tech.	Algal economics	178	29.7	5
5	Cardozo et al.	2007	R	Univ. Sao Paulo, Univ. Cruzeiro Sul	Brazil	11	*Comp. Biochem. Physiol. C-Toxicol. Pharmacol.*	SCI, SCIE	Bioch. Mol. Biol., Endoc. Met., Toxic., Zool.	Algal economics	140	20.0	7
6	Greenwell et al.	2010	R	Univ. Durham, NREL, Univ. Swansea	Wales England United States	5	*J. R. Soc. Interface*	SCI, SCIE	Mult. Sci.	Algal economics	112	28.0	8
7	Williams and Laurens	2010	R	Univ. Bangor	Wales	2	*Energy Environ. Sci.*	SCI, SCIE	Chem. Mult., Ener. Fuels, Eng. Chem., Env. Sci.	Algal economics	106	26.5	9
8	Pienkos and Darzins	2009	A	NREL	United States	2	*Biofuels Bioprod. Biorefining*	SCIE	Biot. Appl. Microb., Ener. Fuels	Algal economics	94	18.8	10
9	Borowitzka	1992	A	Murdoch Univ.	Australia	1	*J. Appl. Phycol.*	SCI, SCIE	Biot. Appl. Microb., Mar. Fresh. Biol.	Algal economics	85	3.9	11
10	Pittman et al.	2011	A	Univ. Manchester	England	3	*Bioresour. Technol.*	SCI, SCIE	Biot. Appl. Microb., Agr. Eng., Ener. Fuels	Algal economics	83	27.7	12
11	Benemann	2000	A	Univ. Calif. B.	United States	1	*J. Appl. Phycol.*	SCI, SCIE	Biot. Appl. Microb., Mar. Fresh. Biol.	Algal economics	72	5.1	14

(*Continued*)

TABLE 32.1 (*Continued*)

Research on Algal Economics

No.	Paper References	Year	Document	Affiliation	Country	No. of Authors	Journal	Index	Subjects	Topic	Total No. of Citations	Total Average Citations per Annum	Rank
12	Singh and Cu	2010	A	Univ. Southampton, Indian Inst. Petr.	England, India	2	*Renew. Sust. Energ. Rev.*	SCI, SCIE	Ener. Fuels	Algal economics	70	17.5	15
13	Harun et al.	2010	R	Monash Univ., Univ. Pertanian Malaysia	Australia, Malaysia	4	*Renew. Sust. Energ. Rev.*	SCI, SCIE	Ener. Fuels	Algal economics	69	17.3	17
14	Matthijs et al.	1996	A	Univ. Amsterdam	Netherlands	6	*Biotechnol. Bioeng.*	SCI, SCIE	Biot. Appl. Microb.	Algal economics	69	3.8	18
15	Gao and McKinley	1994	A	Univ. Hawaii Manoa	United States	2	*J. Appl. Phycol.*	SCI, SCIE	Biot. Appl. Microb., Mar. Fresh. Biol.	Algal economics	67	3.4	19
16	Park et al.	2011	A	NIWA, Massey Univ.	New Zealand	3	*Bioresour. Technol.*	SCI, SCIE	Biot. Appl. Microb., Agr. Eng., Ener. Fuels	Algal economics	66	22.0	20
17	Khan et al.	2009	R	Indian Agr Res Inst, NIPER	India	4	*Renew. Sust. Energ. Rev.*	SCI, SCIE	Ener. Fuels	Algal economics	64	12.8	21
18	Norsker et al.	2011	R	Wageningen Univ.	Netherlands	4	*Biotechnol. Adv.*	SCI, SCIE	Biot. Appl. Microb.	Algal economics	54	18.0	23
19	Davis et al.	2011	A	NREL	United States	3	*Appl. Energy*	SCI, SCIE	Eng. Chem., Ener. Fuels	Algal economics	52	17.3	25
20	Uduman et al.	2010	R	Monash Univ.	Australia	5	*J. Renew. Sustain. Energy*	SCIE	Ener. Fuels	Algal economics	52	13.0	27
21	Singh et al.	2011	A	Univ. Ulster, Univ. Coll. Cork	Ireland, N. Ireland	3	*Bioresour. Technol.*	SCI, SCIE	Biot. Appl. Microb., Agr. Eng., Ener. Fuels	Algal economics	51	17.0	28
22	Um and Kim	2009	R	Univ. Maine, Dankook Univ.	United States, S. Korea	2	*J. Ind. Eng. Chem.*	SCIE	Chem. Mult., Eng. Chem.	Algal economics	37	7.4	34

(*Continued*)

TABLE 32.1 (*Continued*)

Research on Algal Economics

No.	Paper References	Year	Document	Affiliation	Country	No. of Authors	Journal	Index	Subjects	Topic	Total No. of Citations	Total Average Citations per Annum	Rank
23	Brune et al.	2009	A	Clemson Univ., Calif. Polytech. St. Univ. S.L.O., Inst. Environm. Management Inc.	United States	33	*J. Environ. Eng.-ASCE*	SCI, SCIE	Eng. Env., Env. Sci., Eng. Civil	Algal economics	33	6.6	36
24	Zamalloa et al.	2011	A	Univ. Ghent	Belgium	4	*Bioresour. Technol.*	SCI, SCIE	Biot. Appl. Microb., Agr. Eng., Ener. Fuels	Algal economics	31	10.3	37
25	Pokoo-Aikins et al.	2010	A	Texas A&M Univ., McMaster Univ.	United States, Canada	4	*Clean Technol. Environ. Policy*	SCIE	Eng. Env., Env. Sci.	Algal economics	29	7.3	38
26	Danquah et al.	2009	A	Monash Univ.	Australia	5	*J. Chem. Technol. Biotechnol.*	SCI, SCIE	Chem. Mult., Eng. Chem.	Algal economics	29	5.8	39
27	Kadam	1997	A	NREL	United States	1	*Energy Conv. Manag.*	SCIE	Therm., Ener. Fuels, Mechs., Phys, Nucl. Ener.	Algal economics	29	1.7	40
28	Pizarro et al.	2006	A	USDA, Univ. Maryland	United States	4	*Ecol. Eng.*	SCIE	En. Env., Env. Sci.	Algal economics	25	3.1	41
29	Tapie and Bernard	1988	A	Cen. Cadarache	France	2	*Biotechnol. Bioeng.*	SCI, SCIE	Biot. Appl. Microb.	Algal economics	25	1.0	42
30	Lam and Lee	2012	R	Univ. Sains Malaysia	Malaysia	2	*Biotechnol. Adv.*	SCI, SCIE	Biot. Appl. Microb.	Algal economics	23	11.5	44
31	Lee	2011	A	Natl. Taiwan Ocean Univ.	Taiwan	1	*Bioresour. Technol.*	SCI, SCIE	Biot. Appl. Microb., Agr. Eng., Ener. Fuels	Algal economics	21	7.0	46

SCI, Science Citation Index; SCIE, Science Citation Index Expanded; SSCI, Social Sciences Citation Index; A, article; R, review.

for the production of a number of different biofuels including biodiesel, bio-oil, biosyngas, and biohydrogen, while he notes that the production of these biofuels can be coupled with flue gas CO_2 mitigation, wastewater treatment, and the production of high-value chemicals. Microalgal farming can also be carried out with seawater using marine microalgal species as the producers. They assert that the developments in microalgal cultivation and downstream processing (e.g., harvesting, drying, and thermochemical processing) would further "enhance the cost effectiveness of the biofuel from microalgae strategy."

Cardozo et al. (2007) discuss metabolites from algae with an economical impact in a paper with 140 citations. They note that recent trends in drug research from natural sources have shown that algae are promising organisms to furnish novel biochemically active compounds. They describe the main substances biosynthesized by algae with potential economical impact in food science, pharmaceutical industry, and public health. They focus on "fatty acids, steroids, carotenoids, polysaccharides, lectins, mycosporine-like amino acids, halogenated compounds, polyketides, and toxins."

Greenwell et al. (2010) discuss the placement of microalgae on the biofuel priority list with a focus on technological challenges in a recent paper with 112 citations. They critically review current designs of algal culture facilities, including PBRs and open ponds, with regard to photosynthetic productivity and associated biomass and oil production, and include an analysis of alternative approaches using models, balancing space needs, productivity, and biomass concentrations, together with nutrient requirements. They also evaluate the options for potential metabolic engineering of the lipid biosynthesis pathways of microalgae. They conclude that although significant literature exists on microalgal growth and biochemistry, "significantly more work needs to be undertaken to understand and potentially manipulate algal lipid metabolism." Furthermore, with regard to chemical upgrading of algal lipids and biomass, they describe alternative fuel synthesis routes and discuss and evaluate the application of catalysts traditionally used for plant oils. They assert that "simulations that incorporate financial elements, along with fluid dynamics and algal growth models, are likely to be increasingly useful for predicting reactor design efficiency and life cycle analysis to determine the viability of the various options for large-scale culture." They argue that the greatest potential for cost reduction and increased yields most probably "lies within closed or hybrid closed-open production systems."

Williams and Laurens (2010) discuss microalgae biodiesel and biomass feedstocks with a focus on biochemistry, energetics, and economics in a paper with 106 citations. They analyze a number of aspects of large-scale lipid and overall algal biomass production from a biochemical and energetic standpoint. They show that the maximum conversion efficiency of total solar energy into primary photosynthetic organic products falls in the region of 10%. Biomass biochemical composition further conditions this yield as 30% and 50% of the primary product mass is lost on producing cell protein and lipid, while obtained yields are 1/3 to 1/10 of the theoretical ones. They argue that "wasted energy from captured photons is a major loss term and a major challenge in maximizing mass algal production." They produce a simple model of algal biomass production and its variation with latitude and lipid content. An economic analysis of algal biomass production considers a number of scenarios and the effect of changing individual parameters. They conclude that "(i) the biochemical composition of the biomass influences the economics, in particular, increased lipid content reduces other valuable compounds in the biomass; (ii) the 'biofuel only' option is unlikely to be economically viable; and (iii) among the hardest problems in assessing the economics are the cost of the CO_2 supply and uncertain nature of downstream processing."

Pienkos and Darzins (2009) discuss the promise and challenges of microalgal biofuels in a paper with 94 citations. They note that the interest in algal biofuels has been growing recently due to increased concern over peak oil, energy security, greenhouse gas (GHG) emissions, and the potential for other biofuel feedstocks to compete for limited agricultural resources. The high productivity of algae suggests that much of the U.S. transportation fuel needs can be met by algal biofuels at a production cost competitive with the cost of petroleum seen during the early part of 2008. The development of algal biomass production technology, however, they argue remains in its infancy. They discuss past algal research sponsored by the DOE, the potential of microalgal biofuels, and the "technical and economic barriers that need to be overcome before production of microalgal-derived diesel-fuel substitutes can become a large-scale commercial reality."

Borowitzka (1992) discusses algal biotechnology products and processes with a focus on matching science and economics in an early paper with 85 citations. He provides the outline for a rational approach in evaluating which algae and which algal products are the most likely to be commercially viable. This approach involves "some simple market analysis followed by economic modeling of the whole production process." It also permits an evaluation of "which steps in the production process would have the greatest effect on the final production cost of the alga or algal product," thus providing a guide as to what area the research and development effort should be directed to. The base model gives a "production cost of microalgal biomass at about A$ 14–15 kg^{-1}, excluding the costs of further processing, packaging and marketing." The model also shows that "some of the key factors in microalgal production are productivity, labor costs and harvesting costs." He asserts that given the existing technology, "high value products such as carotenoids and algal biomass for aquaculture feeds have the greatest commercial potential in the short term."

Pittman et al. (2011) discuss the potential of sustainable algal biofuel production using wastewater resources in a recent paper with 83 citations. They note that if microalgal biofuel production is to be economically viable and sustainable, further optimization of mass culture conditions is needed. They argue that "wastewaters derived from municipal, agricultural and industrial activities potentially provide cost-effective and sustainable means of algal growth for biofuels." In addition, "there is also potential for combining wastewater treatment by algae, such as nutrient removal, with biofuel production." They discuss the potential benefits and limitations of using wastewaters as resources for cost-effective microalgal biofuel production.

Benemann (2000) discusses hydrogen production by microalgae in a paper with 72 citations. Based on a preliminary engineering and economic analysis, he argues that "biophotolysis processes must achieve close to an overall 10% solar energy conversion efficiency to be competitive with alternatives sources of renewable hydrogen, such as photovoltaic-electrolysis processes as such high solar conversion efficiencies in photosynthetic CO_2 fixation could be reached by genetically reducing the number of light harvesting (antenna) chlorophylls and other pigments in microalgae." Another challenge is "to scale-up biohydrogen processes with economically viable bioreactors." He asserts that "solar energy driven microalgae processes for biohydrogen production are potentially large-scale, but also involve long-term and economically high-risk R&D whilst in the nearer-term, it may be possible to combine microalgal H_2 production with wastewater treatment."

Singh and Cu (2010) discuss the commercialization potential of microalgae for biofuel production in a recent paper with 70 citations. They provide a critical appraisal of the commercialization potential of microalgal biofuels. They discuss the available literature on various aspects of microalgae, for example, its cultivation, LCA, and conceptualization of an algal biorefinery. They assert that "the economic viability of the process in terms of minimizing the operational and maintenance cost along with maximization of oil-rich microalgae production is the key factor, for successful commercialization of microalgal biofuels."

Harun et al. (2010) discuss the bioprocess engineering of microalgae to produce a variety of consumer products in a paper with 69 citations. They argue that for most algal products, the "production process is moderately economically viable and the market is developing." Considering the enormous biodiversity of microalgae and recent developments in genetic and metabolic engineering, they argue that "this group of organisms represents one of the most promising sources for new products and applications." With the development of detailed culture and screening techniques, they assert that "microalgal biotechnology can meet the high demands of food, energy and pharmaceutical industries."

Matthijs et al. (1996) discuss the application of light-emitting diodes (LEDs) in bioreactors with a focus on the flashing light effects and energy economy in algal culture (*C. pyrenoidosa*) in an early paper with 69 citations. They use LEDs as the sole light source in continuous culture of the green alga *C. pyrenoidosa*. They find that at standard voltage in continuous operation, the "light output of the diode panel was more than sufficient to reach maximal growth. Lower frequencies fell short of sustaining the maximal growth rate. However, the light flux decrease resulting from lowering of the flash frequency reduced the observed growth rates less than in the case of a similar flux decrease with light originating from LEDs in continuous operation. Flash application also showed reduction of the quantum requirement for oxygen evolution at defined frequencies." They argue that "LEDs may open interesting new perspectives for studies on optimization of mixing in mass algal culture via the possibility of separation of interests in the role of modulation on light energy conversion and saturation of nutrient supply." They assert that the "use of flashing LEDs in indoor algal culture yielded a major gain in energy economy in comparison to luminescent light sources."

Gao and McKinley (1994) discuss the use of macroalgae for marine biomass production and CO_2 remediation in an early paper with 67 citations. They note that environmental considerations will increasingly determine product and process acceptability and drive the next generation of economic opportunity as some countries, including Japan, are actively promoting green technologies that will be in demand worldwide in the coming decades. They show that macroalgae have great potential for biomass production and CO_2 bioremediation. They focus on recent data on productivity, photosynthesis, nutrient dynamics, optimization, and economics. They argue that microalgal biomass "promises to provide environmentally and economically feasible alternatives to fossil fuels" provided "the techniques and technologies for growing macroalgae on a large-scale and for converting feedstocks to energy carriers fully developed."

Park et al. (2011) discuss the wastewater treatment of high-rate algal ponds (HRAPs) for biofuel production in a recent paper with 66 citations. They argue that a niche opportunity may exist where "algae are grown as a by-product of high rate algal ponds (HRAPs) operated for wastewater treatment." In addition to significantly better economics, they argue that "algal biofuel production from wastewater treatment HRAPs has a much smaller environmental footprint compared to commercial algal production HRAPs which consume freshwater and fertilizers." They discuss critical parameters that limit algal cultivation, production, and harvest as well as the practical options that may enhance the net harvestable algal production from wastewater treatment HRAPs including CO_2 addition, species control, control of grazers and parasites, and bioflocculation.

Khan et al. (2009) discuss the prospects of biodiesel production from microalgae in India in a paper with 64 citations. They argue that renewable and carbon neutral biodiesels are necessary for environmental and economic sustainability. They note that microalgae are photosynthetic microorganisms that convert sunlight, water, and CO_2 to sugars, from which macromolecules, such as lipids and triacylglycerols (TAGs), can be obtained. These TAGs are the promising and sustainable feedstocks for biodiesel production. They then argue that "microalgal biorefinery approach can be used to reduce the cost of making microalgal biodiesel whilst microalgal-based carbon sequestration technologies cover the cost of carbon capture and sequestration."

Norsker et al. (2011) discuss microalgal production with a focus on the economics in a recent paper with 54 citations. They calculate microalgal biomass production costs for three different microalgal production systems operating at commercial scale today, open ponds, horizontal tubular PBRs, and flat panel PBRs. For these three systems, they find that "resulting biomass production costs including dewatering, were 4.95, 4.15 and 5.96 Euros per kg, respectively. The important cost factors are irradiation conditions, mixing, photosynthetic efficiency of systems, medium- and carbon dioxide costs." Optimizing production with respect to these factors, they find a price of €0.68 kg^{-1}. They assert that "at this cost level microalgae become a promising feedstock for biodiesel and bulk chemicals whilst the photobioreactors may become attractive for microalgal biofuel production."

Davis et al. (2011) discuss the technoeconomic analysis of autotrophic microalgae for biofuel production in a recent paper with 52 citations. They note that there is a wide lack of public agreement on the near-term economic viability of algal biofuels, due to uncertainties and speculation on process scale-up associated with the nascent stage of the algal biofuel industry. They establish baseline economics for two microalgal pathways, by performing a comprehensive analysis using a set of assumptions for what can plausibly be achieved within a 5-year time frame. Specific pathways include autotrophic production via both open pond and closed tubular PBR systems. The production scales were set at 10 million gallons per year of raw algal oil, subsequently upgraded to a green diesel blend stock via hydrotreating. They find that the "cost of lipid production to achieve a 10% return was $8.52/gal for open ponds and $18.10/gal for PBRs. Hydrotreating to produce a diesel blend stock added onto this marginally, bringing the totals to $9.84/gal and $20.53/gal of diesel, for the respective cases." They recommend that the "near-term research should focus on maximizing lipid content as it offers more substantial cost reduction potential relative to an improved algae growth rate."

Uduman et al. (2010) discuss the dewatering of microalgal cultures as a major bottleneck to algal biofuels in a recent paper with 52 citations. They note that the dilute nature of harvested microalgal cultures creates a huge operational cost during dewatering, thereby rendering algal biofuels less economically attractive. They argue that "technique that may result in a greater algal biomass may have drawbacks such as a high capital cost or high energy consumption as the choice of which harvesting technique to apply will depend on the species of microalgae and the final product desired." They assert that "algal properties such as a large cell size and the capability of the microalgae to autoflocculate can simplify the dewatering process."

Singh et al. (2011) discuss the mechanism and challenges in the commercialization of algal biofuels in a recent paper with 51 citations. They review up-to-date literatures on the composition of algae, mechanism of oil droplets, TAG production in algal biomass, research and development made in the cultivation of algal biomass, harvesting strategies, and recovery of lipids from algal mass. They then discuss the "economical challenges in the production of biofuels from algal biomass in view of the future prospects in the commercialization of algal biofuels."

Um and Kim (2009) discuss a chance for South Korea to advance algal biodiesel technology in a paper with 37 citations. They note that renewable and recycled energy, which constituted 2.3% of S. Korea's total energy resources in 2006, will be required to reach 5% in 2011 and 9% in 2030, while biodiesel, which is currently only 1% of diesel oil consumed, will be required to be 3% in 2012. S. Korea emitted 591 million tons of carbon dioxide in 2005, which is a 98.7% increase from 1990. They note that the "pilot scale studies evolve practices to produce algal biodiesel and obtain optimal harvest of such aquatic algae with anthropogenic CO_2." They recommend that work should be initiated to "establish a multilateral network, taking into consideration institutional infrastructure, scientific capabilities, and cost effectiveness."

Brune et al. (2009) discuss microalgal biomass for GHG reductions with a focus on the potential for replacement of fossil fuels and animal feeds in a paper with 33 citations. They present an initial analysis of the potential for GHG avoidance using a proposed algal biomass production system coupled to recovery of flue gas CO_2 combined with waste sludge and/or animal manure utilization. They construct a model around a 50 MW natural-gas-fired electrical generation plant operating with 50% capacity as a semibase-load facility. This facility is projected to produce 216 million kW h/240-day season while releasing 30.3 million kg-C/season of GHG–CO_2. They find that an "algal system designed to capture 70% of flue-gas CO_2 would produce 42,400 t (dry wt) of algal biomass/season and requires 880 ha of high-rate algal ponds operating at a productivity of 20 g-dry-wt/m^2-day." This algal biomass is assumed to be fractionated into 20% extractable algal oil, useful for biodiesel, with the 50% protein content providing animal feed replacement and 30% residual algal biomass digested to produce biomethane gas, providing gross GHG avoidances of 20%, 8.5%, and 7.8%, respectively. They then find that the "total gross GHG avoidance potential of 36.3% results in a net GHG avoidance of 26.3% after accounting for 10% parasitic energy costs." At CO_2 utilization efficiencies predicted to range from 60% to 80%, net GHG avoidances are estimated to range from 22% to 30%. To provide nutrients for algal growth and to ensure optimal algal digestion, importation of 53 t day^{-1} of wastepaper, municipal sludge, or animal manure would be required. They caution that "although theoretically promising, successful integration of waste treatment processes with algal recovery of flue-gas CO_2 will require pilot-scale trials and field demonstrations to more precisely define the many detailed design requirements."

Zamalloa et al. (2011) study the technoeconomic potential of renewable energy (biomethane) through the anaerobic digestion of microalgae in a recent paper with 31 citations. They examine the potential of straightforward biomethanation, which includes "preconcentration of microalgae and utilization of a high rate anaerobic reactor based on the premises of achievable upconcentration from 0.2–0.6 kg m^{-3} to 20–60 kg dry matter (DM) m^{-3} and an effective biomethanation of the concentrate at a loading rate of 20 kg DM m^{-3} day^{-1}." They calculate the "costs of biomass available for biomethanation under such conditions in the range of 86–124 Euros ton^{-1} DM whilst the levelized cost of energy by means of the process line 'algae biomass - biogas - total energy module' would be in the order of 0.170–0.087 Euro kW h^{-1}, taking into account a carbon credit of about Euros 30 ton^{-1} $CO_{2(eq)}$."

Pokoo-Aikins et al. (2010) discuss the design and technoeconomic analysis of an integrated system for the production of biodiesel from algal oil produced via the sequestration of carbon dioxide from the flue gas of a power plant in a recent paper with 29 citations. They carry out a process simulation using Aspen Plus to model a two-stage alkali-catalyzed transesterification reaction for converting microalgal oil of *Chlorella* species to biodiesel. They find that for the algal oil to biodiesel process, "factors such as choosing the right algal species, using the appropriate pathway for converting algae to algal oil, selling the resulting biodiesel and glycerol at a favorable market selling prices, and attaining high levels of process integration can collectively render algal oil to be a competitive alternative to food-based plant oils."

Danquah et al. (2009) explore the performance and economic viability of chemical flocculation and tangential flow filtration for the dewatering of *Tetraselmis suecica* microalgal culture in a paper with 29 citations. They find that "TFF concentrates the microalgae feedstock up to 148 times by consuming 2.06 kWh m^{-3} of energy while flocculation consumes 14.81 kWh m^{-3} to concentrate the microalgae up to 357 times." They then find that "even though TFF has higher initial capital investment than polymer flocculation, the payback period for TFF at the upper extreme of microalgae revenue is similar to 1.5 years while that of flocculation is similar to 3 years." They assert that "improved dewatering levels can be achieved more economically by employing TFF."

Kadam (1997) discusses plant flue gas as a source of CO_2 for microalgal cultivation with a focus on the economical impact of different process options in an early paper with 29 citations. He develops an economic model for CO_2 recovery from power-plant flue gas and its delivery to microalgal ponds, while he devises a design basis for recovering CO_2 from flue gas emitted by a typical 500 MW power plant located in the southwestern United States. For the standard process, which included monoethanolamine extraction, compression, dehydration, and transportation to the ponds, he estimates a "delivered CO_2 cost of \$40/t." He also uses the model to evaluate the efficacy of directly using the flue gas, and he finds this option more expensive. He argues that the economics of microalgal cultivation using power-plant flue gas can be evaluated by integrating this model for CO_2 recovery with a previously developed model for microalgal cultivation. He finds that the model predictions for a long-term process are "a lipid cost of \$1.4/gal (unextracted) and a mitigation cost of \$30/t CO_2 (CO_2 avoided basis)." He asserts that these costs are economically attractive and demonstrate the promise of microalgal technology.

Pizarro et al. (2006) discuss the economic assessment of algal turf scrubber technology at the farm scale for the treatment of dairy manure effluent for a hypothetical 1000-cow dairy in a paper with 25 citations. The costs were developed for farms with and without anaerobic pretreatment. They find that the "majority of capital costs were due to land preparation, installation of liner material, and engineering fees whilst the majority of operational costs were due to energy requirements for biomass drying, pumping water, and repayment of capital investment." On farms using anaerobic pretreatment, waste heat from the burning of biogas could be used to offset the energy requirements of biomass drying. In addition, they argue that biogas combustion exhaust gas could then be recycled back to the algal system to supply dissolved inorganic carbon for optimal algal production and pH control. Under the best case (algal system coupled with anaerobic digestion pretreatment), they find that the "yearly operational costs per cow, per kg N, per kg P, and per kg of dried biomass were \$454, \$6.20, \$31.10, and \$0.70, respectively. Without anaerobic digestion pretreatment, the yearly operational costs were 36% higher, amounting to \$631 per cow, \$8.70 per kg N, \$43.20 per kg P, and \$0.97 per kg of dried biomass."

Tapie and Bernard (1988) discuss microalgal production with a focus on technical and economic evaluations in an early paper with 25 citations. They find that the total production costs of a nonprocessed biomass range from US\$0.15 to US\$4.0 kg^{-1}, according to various authors. They present a cost analysis for a tubular bioreactor system that shows that, assuming a productivity of 60 tons ha^{-1} year, "production costs would range from FF24 to FF29 kg^{-1} for such a system." They note that operating costs as well as fixed charges account for approximately 50% of the cost. They finally analyze the parametric sensitivity of these costs and find that "if productivity would be 30, 45, or 90 tons ha^{-1} yr, total cost would be around FF48, FF33, and FF19 kg^{-1}."

Lam and Lee (2012) discuss the issues, problems, and the way forward relating to algal biofuels in a recent paper with 23 citations. Based on the maturity of current technology, they argue that the "true potential of microalgal biofuel towards energy security and its feasibility for commercialization are still questionable." They depict the practical problems that are facing the microalgal biofuel industry, covering upstream to downstream activities by accessing the latest research reports and critical data analysis. Additionally, they suggest "several interlink solutions to the problems with the purpose to bring current microalgae biofuel research into a new dimension and consequently, to revolutionize the entire microalgae biofuel industry towards long-term sustainability."

Lee (2011) discusses the algal biodiesel economy and competition among biofuels in a recent paper with 21 citations as he focuses on the possible results of policy support in developed and developing economies for developing algal biodiesel through to 2040. He adopts the Taiwan General Equilibrium Model-Energy for Bio-fuels (TAIGEM-EB) to predict competition among the developments of algal biodiesel, bioethanol, and conventional crop-based biodiesel. He finds that "algal biodiesel will not be the major energy source in 2040 without strong support in developed economies. In contrast, algal bioethanol enjoys a development advantage relative to both forms of biodiesel. Finally, algal biodiesel will almost completely replace conventional biodiesel."

32.2.3 Conclusion

The research on algal economics has been one of the most dynamic research areas in recent years. Thirty-one citation classics in the field with more than twenty-one citations were located, and the key emerging issues were presented in the decreasing order of the number of citations. These papers give strong hints about the determinants of the economic production of algal biofuels and biocompounds and emphasize that marine algae are efficient feedstocks in comparison with first- and second-generation biofuel feedstocks.

32.3 Studies on Algal Optimization

32.3.1 Introduction

The research on algal LCA has been one of the most dynamic research areas in recent years. Nineteen citation classics in the field of algal optimization with more than nineteen citations were located, and the key emerging issues from these papers are presented in the decreasing order of the number of citations in Table 32.2. These papers give strong hints about the determinants of efficient algal LCA processes and emphasize that marine algae are efficient feedstocks in comparison with first- and second-generation biofuel feedstocks.

The papers were dominated by researchers from the 10 countries, usually through the intercountry institutional collaboration, and they were multiauthored. The United States (nine papers) and England and France (two papers each) were the most prolific countries.

Similarly, all these papers were published in journals indexed by the SCI and/or SCIE. There were no papers indexed by the Arts & Humanities Citation Index (A&HCI) or SSCI.

TABLE 32.2

Research on Algal Optimization

No.	Paper References	Year	Document	Affiliation	Country	No. of Authors	Journal	Index	Subjects	Topic	Total No. of Citations	Total Average Citations per Annum	Rank
1	Lardon et al.	2009	A	INRA, Montpellier SupAgro, INRIA	France	5	*Environ. Sci. Technol.*	SCI, SCIE	Eng. Env., Env. Sci.	Algal optimization	193	38.6	4
2	Clarens et al.	2010	A	Univ. Virginia	United States	4	*Environ. Sci. Technol.*	SCI, SCIE	Eng. Env., Env. Sci.	Algal optimization	163	40.8	6
3	Jorquera et al.	2010	A	NREL, Univ. Fed. Bahia	United States, Brazil	5	*Bioresour. Technol.*	SCI, SCIE	Biot. Appl. Microb., Agr. Eng., Ener. Fuels	Algal optimization	76	19.0	13
4	Stephenson et al.	2010	A	Univ. Cambridge	England	6	*Energy Fuels*	SCI, SCIE	Ener. Fuels, Eng. Chem.	Algal optimization	70	17.5	16
5	Yang et al.	2011	A	Georgia Inst. Technol., Arizona State Univ.	United States	6	*Bioresour. Technol.*	SCI, SCIE	Biot. Appl. Microb., Agr. Eng., Ener. Fuels	Algal optimization	56	18.7	22
6	Kadam	2002	A	NREL	United States	1	*Energy*	SCI, SCIE	Therm., Ener. Fuels	Algal optimization	54	4.5	24
7	Sander and Murthy	2010	A	Oregon State Univ.	United States	2	*Int. J. Life Cycle Assess.*	SCIE	Eng. Env., Env. Sci.	Algal optimization	52	13.0	26
8	Campbell et al.	2011	A	CSIRO	Australia	3	*Bioresour. Technol.*	SCI, SCIE	Biot. Appl. Microb., Agr. Eng., Ener. Fuels	Algal optimization	50	16.7	29
9	Aresta et al.	2005	A	Univ. Bari	Italy	3	*Fuel Process. Technol.*	SCI, SCIE	Chem. Appl., Ener. Fuels, Eng. Chem.	Algal optimization	47	5.2	30
10	Collet et al.	2011	A	INRA, Montpellier SupAgro, Naskeo	France	6	*Bioresour. Technol.*	SCI, SCIE	Biot. Appl. Microb., Agr. Eng., Ener. Fuels	Algal optimization	46	15.3	31

(Continued)

TABLE 32.2 (*Continued*)
Research on Algal Optimization

No.	Paper References	Year	Document	Affiliation	Country	No. of Authors	Journal	Index	Subjects	Topic	Total No. of Citations	Total Average Citations per Annum	Rank
11	Batan et al.	2010	A	Colorado State Univ.	United States	4	*Environ. Sci. Technol.*	SCI, SCIE	Eng. Env., Env. Sci.	Algal optimization	41	10.3	32
12	Singh and Olsen	2011	A	Tech. Univ. Denmark	Denmark	2	*Appl. Energy*	SCI, SCIE	Ener. Fuels, Eng. Chem.	Algal optimization	40	13.3	33
13	Mulder	2003	A	Amecon Environmental Consultancy	Netherlands	1	*Water Sci. Technol.*	SCI, SCIE	Eng. Env., Env. Sci., Water Res.	Algal optimization	37	3.4	35
14	Khoo et al.	2011	A	Inst. Chem. & Engn. Sci.	Singapore	6	*Bioresour. Technol.*	SCI, SCIE	Biot. Appl. Microb., Agr. Eng., Ener. Fuels	Algal optimization	24	8.0	43
15	Soratana and Landis	2011	A	Univ. Pittsburgh	United States	2	*Bioresour. Technol.*	SCI, SCIE	Biot. Appl. Microb., Agr. Eng., Ener. Fuels	Algal optimization	23	7.7	45
16	Luo et al.	2010	A	Georgia Inst. Technol., Algenol Biofuels	United States	6	*Environ. Sci. Technol.*	SCI, SCIE	Eng. Env., Env. Sci.	Algal optimization	21	5.3	47
17	Clarens et al.	2011	A	Univ. Virginia	United States	5	*Environ. Sci. Technol.*	SCI, SCIE	Eng. Env., Env. Sci.	Algal optimization	20	6.7	48
18	Razon and Tan	2011	A	De La Salle Univ.	Philippines	2	*Appl. Energy*	SCI, SCIE	Ener. Fuels, Eng. Chem.	Algal optimization	20	6.7	49
19	Shirvani et al.	2011	A	Univ. Oxford	England	5	*Energy Environ. Sci.*	SCI, SCIE	Chem. Mult., Ener. Fuels, Eng. Chem., Env. Sci.	Algal optimization	19	6.3	50

SCI, Science Citation Index; SCIE, Science Citation Index Expanded; SSCI, Social Sciences Citation Index; A, article; R, review.

The number of citations ranged from 19 to 193, and the average number of citations per annum ranged from 3.4 to 40.8. All of these papers were articles.

32.3.2 Research on Algal Optimization

Lardon et al. (2009) discuss the LCA of potential environmental impacts of biodiesel production from microalgae in a paper with 193 citations. They undertake a comparative LCA study of a virtual facility to assess the energetic balance and the potential environmental impacts of the whole process chain, from biomass production to biodiesel combustion. They test two different culture conditions, nominal fertilizing or nitrogen starvation, as well as two different extraction options, dry or wet extraction in comparison to first-generation biodiesel and oil diesel. They confirm the "potential of microalgae as an energy source but highlight the imperative necessity of decreasing the energy and fertilizer consumption." Therefore, they argue that "control of nitrogen stress during the culture and optimization of wet extraction are valuable options." They also stress the potential of anaerobic digestion of oil cakes as a way to reduce external energy demand and to recycle some mineral fertilizers.

Clarens et al. (2010) discuss the environmental life-cycle comparison of algae to other bioenergy feedstocks in a recent paper with 163 citations. They determine the impacts associated with algal production using a stochastic life-cycle model and compared with switchgrass, canola, and corn farming. They find that "these conventional crops have lower environmental impacts than algae in energy use, greenhouse gas emissions, and water regardless of cultivation location. Only in total land use and eutrophication potential do algae perform favorably." They argue that the large environmental footprint of algal cultivation is driven predominantly by upstream impacts, such as the demand for CO_2 and fertilizer. To reduce these impacts, they argue that "flue gas and, to a greater extent, wastewater could be used to offset most of the environmental burdens associated with algae." They expand the model to demonstrate the benefits of algal production coupled with wastewater treatment, the model to include three different municipal wastewater effluents as sources of nitrogen and phosphorus. They find that "each effluent provided a significant reduction in the burdens of algae cultivation."

Jorquera et al. (2010) carry out comparative energy LCAs of microalgal biomass production in open ponds and PBRs in a recent paper with 76 citations. They perform an analysis of the energy life cycle for production of biomass using the oil-rich microalgae *Nannochloropsis* sp., which included both raceway ponds and tubular and flat-plate PBRs for algal cultivation as they calculate the NER for each process. They find that the "use of horizontal tubular photobioreactors (PBRs) is not economically feasible ([NER] < 1) and that the estimated NERs for flat-plate PBRs and raceway ponds is >1." They argue that the NER for ponds and flat-plate PBRs could be raised to significantly higher values "if the lipid content of the biomass were increased to 60% dw/cwd."

Stephenson et al. (2010) discuss the LCA of potential algal biodiesel production in the United Kingdom to investigate the GWP and fossil-energy requirement of a hypothetical operation in which biodiesel is produced from the freshwater alga *C. vulgaris*, grown using flue gas from a gas-fired power station as the carbon source with a focus on a comparison of raceways and airlift tubular bioreactors in a recent paper with 70 citations. They consider cultivation using a two-stage method, whereby the cells were initially grown to a high concentration of biomass under nitrogen-sufficient conditions, before the supply of nitrogen was discontinued, whereupon the cells accumulated triacylglycerides. They find that if the future target for the productivity of lipids from microalgae, such as *C. vulgaris*, of approximately 40 tons ha^{-1} $year^{-1}$ could be achieved, "cultivation in typical raceways would be significantly more environmentally sustainable than in closed air-lift tubular bioreactors." While biodiesel produced from microalgae cultivated in raceway ponds would have a GWP similar to 80% lower than fossil-derived diesel (on the basis of the net energy content), they next find that "if air-lift tubular bioreactors were used, the GWP of the biodiesel would be significantly greater than the energetically equivalent amount of fossil-derived diesel." They finally find that the GWP and fossil-energy requirement in this operation are particularly sensitive to "(i) the yield of oil achieved during cultivation, (ii) the velocity of circulation of the algae in the cultivation facility, (iii) whether the culture media could be recycled or not, and (iv) the concentration of carbon dioxide in the flue gas."

Yang et al. (2011) discuss the LCA of biodiesel production from microalgae with a focus on water footprint and nutrient balance in a recent paper with 56 citations. They confirm the competitiveness of microalgae-based biofuels and highlight the necessity of recycling harvested water and using sea/wastewater as the water source. They find that "to generate 1 kg biodiesel, 3726 kg water, 0.33 kg nitrogen, and 0.71 kg phosphate are required if freshwater used without recycling whilst the recycling harvest water reduces the water and nutrients usage by 84% and 55%." Furthermore, "using sea/wastewater decreases 90% water requirement and eliminates the need of all the nutrients except phosphate." They recommend suitable microalgal biofuel implementation pathways and identify potential bottlenecks.

Kadam (2002) discusses the environmental implications of power generation via coal–microalgal cofiring in a paper with 54 citations. An LCA was conducted to compare the environmental impacts of electricity production via coal firing versus coal/algal cofiring. It was found that "there are potentially significant benefits to recycling CO_2 toward microalgae production as it reduces CO_2 emissions by recycling it and uses less coal, there are concomitant benefits of reduced greenhouse gas emissions." However, there are also other energy and fertilizer inputs needed for algal production, which contribute to key environmental flows. This researcher observed "lower net values for the algae cofiring scenario for SO_x, NO_x, particulates, carbon dioxide, methane, and fossil energy consumption using the direct injection process (in which the flue gas is directly transported to the algae ponds)." Also observed were lower values for the algal cofiring scenario for greenhouse potential and air acidification potential. However, "impact assessment for depletion of natural resources and eutrophication potential showed much higher values."

Sander and Murthy (2010) discuss the well-to-pump LCA of algal biodiesel to investigate the overall sustainability and net energy balance of an algal biodiesel process in a paper with 52 citations. The functional unit was 1000 MJ of energy from algal biodiesel using existing technology. Primary data for this study were obtained from the U.S. LCI Database and the Greenhouse Gases, Regulated Emissions, and Energy Use in Transportation (GREET) model. Carbohydrates in coproducts from algal biodiesel production were assumed to displace corn as a feedstock for ethanol production. For every 24 kg of algal biodiesel produced (1000 MJ algal biodiesel), 34 kg of coproducts is also produced. The total energy input without solar drying is 3292 and 6194 MJ for the process with filter press and centrifuge as the initial filtering step, respectively. Net CO_2 emissions are −20.9 and 135.7 kg/functional unit for a process utilizing a filter press and centrifuge, respectively. In addition to the −13.96 kg of total air emissions per functional unit, 18.6 kg of waterborne wastes, 0.28 kg of solid waste, and 5.54 Bq are emitted. They find that the "largest energy input (89%) is in the natural gas drying of the algal cake." Although the net energies for both filter press and centrifuge processes are −6670 and −3778 MJ/functional unit, respectively, CO_2 emissions are positive for the centrifuge process, while they are negative for the filter press process. Additionally, 20.4 m^3 of wastewater per functional unit is lost from the growth ponds during the 4-day growth cycle due to evaporation. They argue that there is one major obstacle in algal technology: the need to efficiently process the algae into its usable components. Thermal dewatering of algae requires high amounts of fossil fuel–derived energy (3556 kJ kg^{-1} of water removed) and consequently presents an opportunity for significant reduction in energy use. They assert that there is a need for new technologies to make algal biofuels a sustainable, commercial reality.

Campbell et al. (2011) discuss the LCA of biodiesel production from microalgae in ponds with a focus on the potential environmental impacts and economic viability of producing biodiesel in a recent paper with 50 citations. They conduct a comparative LCA study of a national production system designed for Australian conditions to compare biodiesel production from algae (with three different scenarios for carbon dioxide supplementation and two different production rates) with canola and ultralow sulfur diesel. They give comparisons of GHG emissions (g $CO_2^{-e/t}$ km) and costs (not subset of/t km). They find that "algal GHG emissions (−27.6 to 18.2) compare very favorably with canola (35.9) and ULS diesel (81.2)." However, they find that "costs are not so favorable, with algae ranging from 2.2 to 4.8, compared with canola (4.2) and ULS diesel (3.8)." They recommend for a high production rate to make algal biodiesel economically attractive.

Aresta et al. (2005) discuss the utilization of macroalgae for enhanced CO_2 fixation and biofuel production with a focus on the development of computing software for an LCA study in a paper with 47 citations. They test macroalgae obtained from the Adriatic and Ionian seas. They utilize different techniques (supercritical CO_2, organic solvents, and pyrolysis) for the extraction of biofuel. They find that "supercritical CO_2 is the most effective."

Collet et al. (2011) discuss the LCA of microalgal culture (of *C. vulgaris*) coupled to biogas production (biomethane) in comparison to algal biodiesel and to first-generation biodiesels in a recent paper with 46 citations. They find that the "impacts generated by the production of methane from microalgae are strongly correlated with the electric consumption." They argue that improvements can be "achieved by decreasing the mixing costs and circulation between different production steps, or by improving the efficiency of the anaerobic process under controlled conditions." They assert that this new bioenergy generating process strongly competes with other biofuel productions.

Batan et al. (2010) discuss the net energy and GHG emission evaluation of biodiesel derived from microalgae in a recent paper with 41 citations. They propose a detailed, industrial-scale engineering model for the species *Nannochloropsis* using a PBR architecture. They integrate this process-level model with a life-cycle energy and GHG emission analysis compatible with the methods and boundaries of the Argonne National Laboratory GREET model, thereby ensuring comparability to preexisting fuel-cycle assessments. They evaluate the NER and net GHG emissions of microalgal biodiesel in comparison to petroleum diesel and soybean-based biodiesel with a boundary equivalent to *well to pump*. They find that the "resulting NER of the microalgae biodiesel process is 0.93 MJ of energy consumed per MJ of energy produced and in terms of net GHGs, microalgae-based biofuels avoids 75 g of CO_2-equivalent emissions per MJ of energy produced." They assess the scalability of the consumables and products of the proposed microalgae-to-biofuel processes in the context of 150 billion liters (40 billion gallons) of annual production.

Singh and Olsen (2011) provide a critical review of biochemical conversion, sustainability, and LCA of algal biofuels in a recent paper with 50 citations. They note that the cultivation of algal biomass provides dual benefit as it provides biomass for the production of biofuels and it also saves our environment from air and water pollution. They find that "algal biofuels are environmentally better than the fossil fuels but economically it is not yet so attractive."

Mulder (2003) discusses the quest for sustainable nitrogen removal technologies in a paper with 37 citations. For the assessment of the sustainability, he uses five indicators: sludge production, energy consumption, resource recovery, area requirement, and N_2O emission. For the evaluation of the position of the individual nitrogen removal systems in the anthropogenic nitrogen cycle, he presents a broad outline for an LCA.

Khoo et al. (2011) discuss the life-cycle energy and CO_2 analysis of microalgae to biodiesel to analyze various biofuel production technologies from *cradle to gate* in a paper with 24 citations. They carry out energy and CO_2 balances for a hypothetical integrated PBR-raceway microalgae-to-biodiesel production in Singapore. Based on a functional unit of 1 MJ biofuel, they find that the "total energy demands are 4.44 MJ with 13% from biomass production, 85% from lipid extraction, and 2% from biodiesel production." They carry out the sensitivity analysis for adjustments in energy requirements, percentage lipid contents, and lower/higher heating product value. They project an *optimistic case* "with estimates of: 45% lipid content, reduced energy needs for lipid extraction (1.3 MJ per MJ biodiesel), and heating value of biodiesel (42 MJ kg^{-1})" where the "life cycle energy requirements dropped significantly by about 60%."

Soratana and Landis (2011) evaluate industrial symbiosis and algal cultivation from a life-cycle perspective in a paper with 23 citations as they conduct a comparative LCA on 20 scenarios of microalgal cultivation. These scenarios examined the utilization of nutrients and CO_2 from synthetic sources and waste streams as well as the materials used to construct a PBR. They use a "0.2-m^3 closed PBR of *Chlorella vulgaris* at 30%-oil content by weight with the productivity of 25 g m^{-2} × day was used as a case study." They find that the "utilization of resources from waste streams mainly avoided global warming potential (GWP) and eutrophication impacts." They find that the "impacts from the production of material used to construct the PBR dominate total impacts in acidification and ozone depletion categories, even over longer PBR lifetimes," as the choice of PBR construction materials is important.

Luo et al. (2010) discuss the life-cycle energy and GHG emissions for a bioethanol production process based on blue-green algae in a recent paper with 21 citations. They calculate the life-cycle energy and GHG emissions for the three different system scenarios for the proposed bioethanol production process, using process simulations and thermodynamic calculations. They find that the "energy required for bioethanol separation increases rapidly for low initial concentrations of ethanol, and, unlike other biofuel systems, there is little waste biomass available to provide process heat and electricity to offset those energy requirements." They argue that the "ethanol purification process is a major consumer of energy and a significant contributor to the carbon footprint." With a lead scenario based on a natural-gas-fueled combined heat and power system to provide process electricity and extra heat and conservative assumptions around the ethanol separation process, they find that the "net life cycle energy consumption, excluding photosynthesis, ranges from 0.55 MJ/MJ_{EtOH} down to 0.20 MJ/MJ_{EtOH}, and the net life cycle greenhouse gas emissions range from 29.8 g CO_2e/MJ_{EtOH} down to 12.3 g CO_2e/MJ_{EtOH} for initial ethanol concentrations from 0.5 wt% to 5 wt%." They argue that in comparison to gasoline, these predicted values represent 67% and 87% reductions in the carbon footprint for this bioethanol on an energy equivalent basis. They assert that energy consumption and GHG emissions can be further reduced via employment of higher-efficiency heat exchangers in ethanol purification and/or with use of solar thermal for some of the process heat.

Razon and Tan (2011) discuss the net energy analysis of the production of biodiesel and biogas from the microalgae (*Haematococcus pluvialis* and *Nannochloropsis*) in a paper with 20 citations. Even with very optimistic assumptions regarding the performance of processing units, they find a "large energy deficit for both systems, due mainly to the energy required to culture and dry the microalgae or to disrupt the cell." They argue that "some energy savings may be realized from eliminating the fertilizer by the use of wastewater or in the case of *H. pluvialis*, recycling some of the algal biomass to eliminate the need for a photobioreactor, but these are insufficient to completely eliminate the deficit." They make recommendations to develop wet extraction and transesterification technology to make microalgal biodiesel systems viable from an energy standpoint.

Clarens et al. (2011) discuss the environmental impacts of algae-derived biodiesel and bioelectricity for transportation in a paper with 20 citations. They undertake a *well-to-wheel* LCA to evaluate algae's potential use as a transportation energy source for passenger vehicles. They assess four algal conversion pathways resulting in combinations of bioelectricity and biodiesel for several relevant nutrient procurement scenarios. They find that "algae-to-energy systems can be either net energy positive or negative depending on the specific combination of cultivation and conversion processes used." They then argue that "conversion pathways involving direct combustion for bioelectricity production generally outperformed systems involving anaerobic digestion and biodiesel production, and they generate four and fifteen times as many vehicle kilometers traveled (VKT) per hectare as switchgrass or canola, respectively." Despite this, they assert that algal "systems exhibited mixed performance for environmental impacts (energy use, water use, and greenhouse gas emissions) on a 'per km' basis relative to the benchmark crops." They recommend that "both cultivation and conversion processes must be carefully considered to ensure the environmental viability of algae-to-energy processes."

Shirvani et al. (2011) discuss the life-cycle energy and GHG analysis for algae-derived biodiesel in a recent paper with 19 citations. Through a life-cycle approach, they evaluate whether algal biodiesel production can be a viable fuel source once the energy and carbon intensity of the process is managed accordingly. Currently, they find that "algal biodiesel production is 2.5 times as energy intensive as conventional diesel and nearly equivalent to the high fuel-cycle energy use of oil shale diesel." They argue that "biodiesel from advanced biomass can realize its inherent environmental advantages of GHG emissions reduction once every step of the production chain is fully optimized and decarbonized. This includes smart co-product utilization, decarbonization of the electricity and heat grids as well as indirect energy requirements for fertilizer, transport and building material." Only if all these factors are taken into account, they assert that the cost of heat and electricity is reduced and GHG emissions are fully mitigated.

32.3.3 Conclusion

The research on algal LCA has been one of the most dynamic research areas in recent years. Nineteen citation classics in the field of algal optimization with more than nineteen citations were located, and the key emerging issues from these papers were presented in the decreasing order of the number of citations. These papers give strong hints about the determinants of efficient algal LCA processes and emphasize that marine algae are efficient feedstocks in comparison with first- and second-generation biofuel feedstocks.

32.4 Conclusion

The citation classics presented under the two main headings in this chapter confirm the predictions that the marine algae have a significant potential to serve as a major solution for the global problems of warming, air pollution, energy security, and food security in the form of algal biofuels and algal biocompounds. However, the issues relating to the optimal and economic production of algal biofuels and biocompounds have to be considered to realize these production targets.

Further research is recommended for the detailed studies in each topical area presented in this chapter including scientometric studies and citation classic studies to inform key stakeholders about the potential of marine algae for the solution of the global problems of warming, air pollution, energy security, and food security in the form of algal biofuels and algal biocompounds through interdisciplinary studies such as economic and optimization studies.

References

Aresta, M., A. Dibenedetto, and G. Barberio. 2005. Utilization of macro-algae for enhanced CO_2 fixation and biofuels production: Development of a computing software for an LCA study. *Fuel Processing Technology* 86:1679–1693.

Baltussen, A. and C.H. Kindler. 2004a. Citation classics in anesthetic journals. *Anesthesia and Analgesia* 98:443–451.

Baltussen, A. and C.H. Kindler. 2004b. Citation classics in critical care medicine. *Intensive Care Medicine* 30:902–910.

Batan, L., J. Quinn, B. Willson, and T. Bradley. 2010. Net energy and greenhouse gas emission evaluation of biodiesel derived from microalgae. *Environmental Science and Technology* 44:7975–7980.

Benemann, J.R. 2000. Hydrogen production by microalgae. *Journal of Applied Phycology* 12:291–300.

Borowitzka, M.A. 1992. Algal biotechnology products and processes—Matching science and economics. *Journal of Applied Phycology* 4:267–279.

Borowitzka, M.A. 1999. Commercial production of microalgae: Ponds, tanks, tubes and fermenters. *Journal of Biotechnology* 70:313–321.

Bothast, R.J. and M.A. Schlicher. 2005. Biotechnological processes for conversion of corn into ethanol. *Applied Microbiology and Biotechnology* 67:19–25.

Brune, D.E., T.J. Lundquist, and J.R. Benemann. 2009. Microalgal biomass for greenhouse gas reductions: Potential for replacement of fossil fuels and animal feeds. *Journal of Environmental Engineering—ASCE* 135:1136–1144.

Campbell, P.K., T. Beer, and D. Batten. 2011. Life cycle assessment of biodiesel production from microalgae in ponds. *Bioresource Technology* 102:50–56.

Cardozo, K.H.M., T. Guaratini, M.P. Barros et al. 2007. Metabolites from algae with economical impact. *Comparative Biochemistry and Physiology C—Toxicology and Pharmacology* 146:60–78.

Chisti, Y. 2007. Biodiesel from microalgae. *Biotechnology Advances* 25:294–306.

Clarens, A.F., H. Nassau, E.P. Resurreccion, M.A. White, and L.M. Colosi. 2011. Environmental impacts of algae-derived biodiesel and bioelectricity for transportation. *Environmental Science and Technology* 45:7554–7560.

Clarens, A.F., E.P. Resurreccion, M.A. White, and L.M. Colosi. 2010. Environmental life cycle comparison of algae to other bioenergy feedstocks. *Environmental Science and Technology* 44:1813–1819.

Collet, P., A. Helias, L. Lardon, M. Ras, R.A. Goy, and J.P. Steyer. 2011. Life-cycle assessment of microalgae culture coupled to biogas production. *Bioresource Technology* 102:207–214.
Danquah, M.K., L. Ang, N. Uduman, N. Moheimani, and G.M. Fordea. 2009. Dewatering of microalgal culture for biodiesel production: Exploring polymer flocculation and tangential flow filtration. *Journal of Chemical Technology and Biotechnology* 84:1078–1083.
Davis, R., A. Aden, and P.T. Pienkos. 2011. Techno-economic analysis of autotrophic microalgae for fuel production. *Applied Energy* 88:3524–3531.
Demirbas, A. 2007. Progress and recent trends in biofuels. *Progress in Energy and Combustion Science* 33:1–18.
Dubin, D., A.W. Hafner, and K.A. Arndt. 1993. Citation-classics in clinical dermatological journals—Citation analysis, biomedical journals, and landmark articles, 1945–1990. *Archives of Dermatology* 129:1121–1129.
Gao, K. and K.R. Mckinley. 1994. Use of macroalgae for marine biomass production and CO_2 remediation—A review. *Journal of Applied Phycology* 6:45–60.
Gehanno, J.F., K. Takahashi, S. Darmoni, and J. Weber. 2007. Citation classics in occupational medicine journals. *Scandinavian Journal of Work, Environment and Health* 33:245–251.
Godfray, H.C.J., J.R. Beddington, I.R. Crute et al. 2010. Food security: The challenge of feeding 9 billion people. *Science* 327:812–818.
Goldemberg, J. 2007. Ethanol for a sustainable energy future. *Science* 315:808–810.
Greenwell, H.C., L.M.L. Laurens, R.J. Shields, R.W. Lovitt, and K.J. Flynn. 2010. Placing microalgae on the biofuels priority list: A review of the technological challenges. *Journal of the Royal Society Interface* 7:703–726.
Grima, E.M., E.H. Belarbi, F.G.A. Fernandez, A.R. Medina, and Y. Chisti. 2003. Recovery of microalgal biomass and metabolites: Process options and economics. *Biotechnology Advances* 20:491–515.
Harun, R., M. Singh, G.M. Forde, and M.K. Danquah. 2010. Bioprocess engineering of microalgae to produce a variety of consumer products. *Renewable and Sustainable Energy Reviews* 14:1037–1047.
Jacobson, M.Z. 2009. Review of solutions to global warming, air pollution, and energy security. *Energy and Environmental Science* 2:148–173.
Jorquera, O., A. Kiperstok, E.A. Sales, M. Embirucu, and M.L. Ghirardi. 2010. Comparative energy life-cycle analyses of microalgal biomass production in open ponds and photobioreactors. *Bioresource Technology* 101:1406–1413.
Kadam, K.L. 1997. Plant flue gas as a source of CO_2 for microalgae cultivation. Economic impact of different process options. *Energy Conversion and Management* 38:S505–S510.
Kadam, K.L. 2002. Environmental implications of power generation via coal-microalgae cofiring. *Energy* 27:905–922.
Kapdan, I.K. and F. Kargi. 2006. Bio-hydrogen production from waste materials. *Enzyme and Microbial Technology* 38:569–582.
Khan, S.A., M.Z. Hussain, S. Prasad, and U.C. Banerjee. 2009. Prospects of biodiesel production from microalgae in India. *Renewable and Sustainable Energy Reviews* 13:2361–2372.
Khoo, H.H., P.N. Sharratt, P. Das, R.K. Balasubramanian, P.K. Naraharisetti, and S. Shaik. 2011. Life cycle energy and CO_2 analysis of microalgae-to-biodiesel: Preliminary results and comparisons. *Bioresource Technology* 102:5800–5807.
Konur, O. 2000. Creating enforceable civil rights for disabled students in higher education: An institutional theory perspective. *Disability and Society* 15:1041–1063.
Konur, O. 2002a. Assessment of disabled students in higher education: Current public policy issues. *Assessment and Evaluation in Higher Education* 27:131–152.
Konur, O. 2002b. Access to employment by disabled people in the UK: Is the Disability Discrimination Act working? *International Journal of Discrimination and the Law* 5:247–279.
Konur, O. 2002c. Access to nursing education by disabled students: Rights and duties of nursing programs. *Nursing Education Today* 22:364–374.
Konur, O. 2006a. Participation of children with dyslexia in compulsory education: Current public policy issues. *Dyslexia* 12:51–67.

Konur, O. 2006b. Teaching disabled students in higher education. *Teaching in Higher Education* 11:351–363.

Konur, O. 2007a. A judicial outcome analysis of the Disability Discrimination Act: A windfall for the employers? *Disability and Society* 22:187–204.

Konur, O. 2007b. Computer-assisted teaching and assessment of disabled students in higher education: The interface between academic standards and disability rights. *Journal of Computer Assisted Learning* 23:207–219.

Konur, O. 2011. The scientometric evaluation of the research on the algae and bio-energy. *Applied Energy* 88:3532–3540.

Konur, O. 2012a. 100 Citation classics in energy and fuels. *Energy Education Science and Technology Part A: Energy Science and Research* 2012(si):319–332.

Konur, O. 2012b. What have we learned from the citation classics in energy and fuels: A mixed study. *Energy Education Science and Technology Part A: Energy Science and Research* 2012(si):255–268.

Konur, O. 2012c. The gradual improvement of disability rights for the disabled tenants in the UK: The promising road is still ahead. *Social Political Economic and Cultural Research* 4:71–112.

Konur, O. 2012d. Prof. Dr. Ayhan Demirbas' scientometric biography. *Energy Education Science and Technology Part A: Energy Science and Research* 28:727–738.

Konur, O. 2012e. The evaluation of the research on the biofuels: A scientometric approach. *Energy Education Science and Technology Part A: Energy Science and Research* 28:903–916.

Konur, O. 2012f. The evaluation of the research on the biodiesel: A scientometric approach. *Energy Education Science and Technology Part A: Energy Science and Research* 28:1003–1014.

Konur, O. 2012g. The evaluation of the research on the bioethanol: A scientometric approach. *Energy Education Science and Technology Part A: Energy Science and Research* 28:1051–1064.

Konur, O. 2012h. The evaluation of the research on the microbial fuel cells: A scientometric approach. *Energy Education Science and Technology Part A: Energy Science and Research* 29:309–322.

Konur, O. 2012i. The evaluation of the research on the biohydrogen: A scientometric approach. *Energy Education Science and Technology Part A: Energy Science and Research* 29:323–338.

Konur, O. 2012j. The evaluation of the biogas research: A scientometric approach. *Energy Education Science and Technology Part A: Energy Science and Research* 29:1277–1292.

Konur, O. 2012k. The scientometric evaluation of the research on the production of bio-energy from biomass. *Biomass and Bioenergy* 47:504–515.

Konur, O. 2012l. The evaluation of the global energy and fuels research: A scientometric approach. *Energy Education Science and Technology Part A: Energy Science and Research* 30:613–628.

Konur, O. 2012m. The evaluation of the biorefinery research: A scientometric approach, *Energy Education Science and Technology Part A: Energy Science and Research* 2012(si):347–358.

Konur, O. 2012n. The evaluation of the bio-oil research: A scientometric approach, *Energy Education Science and Technology Part A: Energy Science and Research* 2012(si):379–392.

Konur, O. 2012o. What have we learned from the citation classics in energy and fuels: A mixed study. *Energy Education Science and Technology Part A: Energy Science and Research* 2012(si):255–268.

Konur, O. 2012p. The evaluation of the research on the biofuels: A scientometric approach. *Energy Education Science and Technology Part A: Energy Science and Research* 28:903–916.

Konur, O. 2013. What have we learned from the research on the International Financial Reporting Standards (IFRS)? A mixed study. *Energy Education Science and Technology Part D: Social Political Economic and Cultural Research* 5:29–40.

Lal, R. 2004. Soil carbon sequestration impacts on global climate change and food security. *Science* 304:1623–1627.

Lam, M.K. and K.T. Lee. 2012. Microalgae biofuels: A critical review of issues, problems and the way forward. *Biotechnology Advances* 30:673–690.

Lardon, L., A. Helias, B. Sialve, J.P. Steyer, and O. Bernard. 2009. Life-cycle assessment of biodiesel production from microalgae. *Environmental Science and Technology* 43:6475–6481.

Lee, D.H. 2011. Algal biodiesel economy and competition among bio-fuels. *Bioresource Technology* 102:43–49.

Li, Y., M. Horsman, N. Wu, C.Q. Lan, and N. Dubois-Calero. 2008. Biofuels from microalgae. *Biotechnology Progress* 24:815–820.

Luo, D.X., Z.S. Hu, D.G. Choi, V.M. Thomas, M.J. Realff, and R.R. Chance. 2010. Life cycle energy and greenhouse gas emissions for an ethanol production process based on blue-green algae. *Environmental Science and Technology* 44:8670–8677.

Lynd, L.R., J.H. Cushman, R.J. Nichols, and C.E. Wyman. 1991. Fuel ethanol from cellulosic biomass. *Science* 251:1318–1323.

Matthijs, H.C.P., H. Balke, U.M. VanHes, B.M.A. Kroon, L.R. Mur, and R.A. Binot. 1996. Application of light-emitting diodes in bioreactors: Flashing light effects and energy economy in algal culture (*Chlorella pyrenoidosa*). *Biotechnology and Bioengineering* 50:98–107.

Mulder, A. 2003. The quest for sustainable nitrogen removal technologies. *Water Science and Technology* 48:67–75.

Norsker, N.H., M.J. Barbosa, M.H. Vermue, and R.H. Wijffels. 2011. Microalgal production—A close look at the economics. *Biotechnology Advances* 29:24–27.

North, D. 1994. Economic-performance through time. *American Economic Review* 84:359–368.

Paladugu, R., M.S. Chein, S. Gardezi, and L. Wise. 2002. One hundred citation classics in general surgical journals. *World Journal of Surgery* 26:1099–1105.

Park, J.B.K., R.J. Craggs, and A.N. Shilton. 2011. Wastewater treatment high rate algal ponds for biofuel production. *Bioresource Technology* 102:35–42.

Pienkos, P.T. and A.L. Darzins. 2009. The promise and challenges of microalgal-derived biofuels. *Biofuels Bioproducts and Biorefining—BIOFPR* 3:431–440.

Pittman, J.K., A.P. Dean, and O. Osundeko. 2011. The potential of sustainable algal biofuel production using wastewater resources. *Bioresource Technology* 102:17–25.

Pizarro, C., W. Mulbry, D. Blersch, and P. Kangas. 2006. An economic assessment of algal turf scrubber technology for treatment of dairy manure effluent. *Ecological Engineering* 26:321–327.

Pokoo-Aikins, G., A. Nadim, M.M. El-Halwagi, and V. Mahalec. 2010. Design and analysis of biodiesel production from algae grown through carbon sequestration. *Clean Technologies and Environmental Policy* 12:239–254.

Pulz, O. 2001. Photobioreactors: Production systems for phototrophic microorganisms. *Applied Microbiology and Biotechnology* 57:287–293.

Razon, L.F. and R.R. Tan. 2011. Net energy analysis of the production of biodiesel and biogas from the microalgae: *Haematococcus pluvialis* and *Nannochloropsis*. *Applied Energy* 88:3507–3514.

Sander, K. and G.S. Murthy. 2010. Life cycle analysis of algae biodiesel. *International Journal of Life Cycle Assessment* 15:704–714.

Shirvani, T., X.Y. Yan, O.R. Inderwildi, P.P. Edwards, and D.A. King. 2011. Life cycle energy and greenhouse gas analysis for algae-derived biodiesel. *Energy and Environmental Science* 4:3773–3778.

Singh, A., P.S. Nigam, and J.D. Murphy. 2011. Mechanism and challenges in commercialisation of algal biofuels. *Bioresource Technology* 102:26–34.

Singh, A. and S.I. Olsen. 2011. A critical review of biochemical conversion, sustainability and life cycle assessment of algal biofuels. *Applied Energy* 88:3548–3555.

Singh, J. and S. Cu. 2010. Commercialization potential of microalgae for biofuels production. *Renewable and Sustainable Energy Reviews* 14:2596–2610.

Soratana, K. and A.E. Landis. 2011. Evaluating industrial symbiosis and algae cultivation from a life cycle perspective. *Bioresource Technology* 102:6892–6901.

Spolaore, P., C. Joannis-Cassan, E. Duran, and A. Isambert. 2006. Commercial applications of microalgae. *Journal of Bioscience and Bioengineering* 101:87–96.

Stephenson, A.L., E. Kazamia, J.S. Dennis, C.J. Howe, S.A. Scott, and A.G. Smith. 2010. Life-cycle assessment of potential algal biodiesel production in the United Kingdom: A comparison of raceways and air-lift tubular bioreactors. *Energy and Fuels* 24:4062–4077.

Tapie, P. and A. Bernard. 1988. Microalgae production—Technical and economic evaluations. *Biotechnology and Bioengineering* 32:873–885.

Uduman, N., Y. Qi, M.K. Danquah, G.M. Forde, and A. Hoadley. 2010. Dewatering of microalgal cultures: A major bottleneck to algae-based fuels. *Journal of Renewable and Sustainable Energy* 2:012701.

Um, B.H. and Y.S. Kim. 2009. Review: A chance for Korea to advance algal-biodiesel technology. *Journal of Industrial and Engineering Chemistry* 15:1–7.

Volesky, B. and Z.R. Holan. 1995. Biosorption of heavy-metals. *Biotechnology Progress* 11:235–250.

Wang, B., Y.Q. Li, N. Wu, and C.Q. Lan. 2008. CO(2) bio-mitigation using microalgae. *Applied Microbiology and Biotechnology* 79:707–718.

Williams, P.J.L. and L.M.L. Laurens. 2010. Microalgae as biodiesel & biomass feedstocks: Review & analysis of the biochemistry, energetics & economics. *Energy and Environmental Science* 3:554–590.

Wrigley, N. and S. Matthews. 1986. Citation-classics and citation levels in geography. *Area* 18:185–194.

Yang, J., M. Xu, X.Z. Zhang, Q.A. Hu, M. Sommerfeld, and Y.S. Chen. 2011. Life-cycle analysis on biodiesel production from microalgae: Water footprint and nutrients balance. *Bioresource Technology* 102:159–165.

Yergin, D. 2006. Ensuring energy security. *Foreign Affairs* 85:69–82.

Zamalloa, C., E. Vulsteke, J. Albrecht, and W. Verstraete. 2011. The techno-economic potential of renewable energy through the anaerobic digestion of microalgae. *Bioresource Technology* 102:1149–1158.

33

Microalgal Hydrothermal Liquefaction: A Promising Way to Sustainable Bioenergy Production

Dong Ho Seong, Choul-Gyun Lee, and Se-Kwon Kim

CONTENTS

33.1 Introduction

In the past, humans derived their daily energy needs mainly from bioenergy, such as firewood, through simple biomass burning. Thus far, a large part of energy consumption for heating and cooking in undeveloped regions is still similar to the ways of the past. The biomass productivity per land area is low, so the forests in those regions have become bare and barren. On the other hand, in developed countries, most of the energy for transportation, plant operation, lighting, home heating and cooking, etc., originates from nonrenewable sources, such as fossil oil, coal, and natural gas. Therefore, a more effective use of renewable energy is needed for sustainability and environment issues. "Sustainable energy is the sustainable provision of energy that meets the needs of the present without compromising the ability of future generations to meet their needs" (Freya and Linke 2002). Sustainable energy should satisfy the following: First, all the raw materials needed to produce the energy will not be depleted because they are infinite or involve recycling, regardless of long-term use. Second, an energy production process should have high energy production efficiency, that is, it should have a high energy returned on energy invested (EROEI). Third, the manufacturing process and results of energy use should not damage the global environment, and it must be environmentally friendly. In addition, it is a superior sustainable energy if the energy is a fuel source and can provide chemical substances for human life.

Microalgae are photosynthetic microorganisms that demand only light, carbon dioxide, water, and a small amount of nutrients for cell growth and biosynthesis of organic materials, and they have an excellent growth rate and high lipid productivity. Microalgae are promising feedstock for the production of sustainable energy.

Microalgal fuel has become an important subject of research into the production of future biofuel. Thus far, most research of microalgal bioenergy has concentrated on the culturing of microalgae, but the processing technology of biomass is also important for the practical realization of microalgal energy (Kim et al. 2013). Technologies that convert microalgal biomass to a useful form of fuel are also important for the success of the future practical application of the energy from microalgae.

The high lipid content microalgal cells are very beneficial for making biofuels because of the convenient lipid extraction and simple purification. Generally, high lipid content microalgae grow slowly, and fast-growing cells contain small amounts of lipids. Biomass productivity and the lipid content of cells during cultivation are normally different. To achieve more effective biofuel production with microalgae, the microalgae need to accumulate lipid at the same time of cell growth, but there are time differences between cell growth and lipid accumulation. Therefore, a biofuel manufacturing method with a low lipid content biomass is needed for the economical production of microalgal energy.

Among the harvesting steps of biomass from a microalgal culture broth, the drying process of wet microalgae is the most energy-consuming step (Xu et al. 2011). Nevertheless, dry biomass is essential for converting it to useful fuel forms in many processes. A bioenergy production method using the wet cells themselves is more effective in terms of the EROEI.

Hydrothermal liquefaction (HTL) is a bioenergy production process, where the biomass is processed in water at high temperatures and high pressures, and its main product is biocrude, a type of crude liquid oil. The energy production efficiency of HTL is good because it uses wet biomass as a feedstock. During the reaction, water is used as both a solvent and a reagent for reacting with biomass. The HTL process is performed at high temperatures and pressures in water, that is, subcritical water condition under the critical point temperature. Biocrude, the main product, has high energy density. The aqueous phase is easily separated, and it contains many dissolved nutrients, such as phosphate and nitrogen compounds, for microalgal growth (Biller and Ross 2012). As a feedstock of HTL, the microalgal whole cell and lipid defatted algae (LEA) can be a good material (Yu et al. 2011a,b). The HTL process using microalgae is an efficient biofuel manufacturing method and an economical way of producing bioenergy (Delrue et al. 2013; Zhu et al. 2013).

Microalgal HTL is an HTL process using microalgal biomass as a feedstock, it produces bioenergy efficiently, and all raw materials used for the process can be recycled. In addition, it can supply its product for chemical substance production. Microalgae, which are the feedstock of HTL, have a high cell growth rate and high lipid content, and microalgal cultivation does not compete with land use for food production. Therefore, they have attracted considerable interest for future energy production (Patil et al. 2008). Microalgal biomass harvested as wet cells can be a feedstock in the HTL process without a drying step. Microalgal HTL produces good-quality fuel, namely, liquid oil with an excellent higher heating value (HHV) in high yield. The HTL process is a desirable processing method for producing biofuel from microalgal biomass (Shuping et al. 2010; Valdez et al. 2012; Xu et al. 2012). As the microalgal cell structure and biochemical composition differ according to the strains and culturing methods, the production yield and biocrude properties of microalgal HTL can vary.

Up to now, microalgal HTL biocrude has not been commercialized, but it has many advantages. If critical technologies are developed, microalgal HTL will be one of the future sustainable fuel production systems.

33.2 Biofuel Production Processes Using Microalgae

Many processing methods have been developed for the practical use of microalgal biofuel. A representative biofuel of microalgae is biodiesel, which is a liquid fuel with short-chain alkyl ester of fatty acids that is produced through the transesterification or esterification reaction of lipids with alcohol. Although the chemical structure of biodiesel is basically different from fossil diesel fuel, the fuel properties of the two are similar. Biodiesel is being used as a mixed-type fuel of petroleum diesel for diesel engines, which is a representative internal combustion engine, without special structural changes to it.

Among the many components in microalgae, only lipids are the raw material for biodiesel production, but the residue of microalgae after the extraction of lipids is a useful protein source for animal and fish feed as well as a raw material producing sugars to ferment for bioethanol production. Hence, all parts of the microalgal cell can be used to produce useful products.

Microalgae consume carbon dioxide, a representative greenhouse gas, during photosynthesis for the biosynthesis of organic matter, which offsets the carbon dioxide released during the combustion of biodiesel. Microalgal energy can be real *carbon neutral* if the energy consumed for processing and transportation is excluded.

Pyrolysis and gasification are also microalgal biofuel production methods. Fast pyrolysis is a thermal cracking method using dry biomass as a feedstock to produce pyrolysis oil at high temperatures around 500°C, under a relatively low pressure and anaerobic condition for seconds. Gasification is conducted at temperatures higher than 700°C without combustion by the controlled supply of oxygen or steam for the production of hydrogen and carbon monoxide gases. The gases produced, hydrogen and carbon monoxide, can be purified and used as the main raw materials for the Fischer–Tropsch process to prepare synthetic hydrocarbon fuels, such as jet fuel, diesel, and gasoline. All these processes require dry feedstock, but the drying step in the microalgal harvesting process is the most energy-consuming step. Therefore, to make an economical microalgal biofuel, it is helpful to convert wet microalgae directly to fuel without a cell drying step, namely, using wet biomass.

Hydrothermal treatment using biomass under aqueous conditions under high temperatures and high pressures is the representative process of biofuel production using wet biomass and can be classified according to its main product type. *Hydrothermal carbonization* (HTC) produces mainly solid fuel, HTL makes liquid fuel, and *hydrothermal gasification* (HTG) manufactures gas phases. HTC is conducted at approximately 200°C under pressures less than 2 MPa. HTL reacts under subcritical water conditions at 280°C–374°C at 10–25 MPa. HTG converts biomass under supercritical water conditions at 400°C–700°C and 25–30 MPa (López Barreiro et al. 2013a).

Microalgal HTL uses wet microalgal biomass as a feedstock in water under extremely high temperatures and pressures; water acts as both the solvent and reagent. The major reactions are hydrolysis and various types of decompositions and condensations. The main products are liquid-phase oil called *biocrude* or *bio-oil*.

As desirable downstream methods of microalgal fuel, the following are expected: (1) Microalgal biomass regardless of the lipid content should be available for feedstock. (2) The wet biomass itself can be used directly as a feedstock. (3) The process should have high EROEI. (4) The product has high HHV. (5) The product is a liquid fuel compatible with petroleum oil. (6) Every raw material used in the production process must be recyclable. (7) The process and product are environmentally friendly.

33.3 Basic Knowledge of Hydrothermal Liquefaction

33.3.1 Characteristics of Subcritical and Supercritical Water

A substance changes its phase (solid, liquid, and gas) depending on temperature and pressure. Unique phases that exist under high temperatures and pressures are different in character from normal liquid or gas and have intermediate properties of liquid and gas. This state of a substance is called a *supercritical fluid*, and the supercritical fluid starting point of temperature and pressure is called the *critical point*. The supercritical point of water is 374°C and 22.064 MPa. Water becomes a supercritical fluid under harsher conditions than this point. Under these conditions, supercritical water exists as a compressed steam state, but it is not steam; it has intermediate properties of water and steam. Supercritical water mixes well with nonpolar substances, such as lipids, and dissolves large amounts of gases, such as hydrogen, oxygen, and nitrogen. On the other hand, inorganic salts, such as sodium chloride, are insoluble in supercritical water (Franck 1968).

Subcritical water, which is water under the critical point, has different characteristics not only with those of supercritical water but also normal water. Subcritical water has a lower viscosity, density, and permittivity; higher solubility of hydrophobic compounds; and a higher concentration of ionic products than normal water. As an example, at 350°C and 25 MPa subcritical water, the ionic product of water pKw 10^{-12} is two orders higher than the 10^{-14} of normal water at 25°C, 0.1 MPa. Namely, subcritical water has high concentration of H^+ and OH^- ions, so hydrolysis reactions are easily induced in subcritical water (López Barreiro et al. 2013a). This means that the concentration of protons and hydroxide ions in subcritical water is approximately 100 times higher than that of normal water, so acid and alkali reactions can be conducted without catalysts to hydrolyze biomass in subcritical water (Toor et al. 2011). Subcritical water at 200°C–374°C is a unique substance for organic chemical reactions; water serves as both a solvent and a reagent (Akiya and Savage 2002; Hunter and Savage 2004). This means that the biomass can be converted easily to new substances through acid and alkali hydrolysis reactions under subcritical water conditions without the addition of acid and alkali reagents. Microalgal constituents are hydrolyzed and reconstructed to produce new compounds easily. For example, solid microalgal cells are converted easily to liquid biocrude by an HTL reaction.

On the other hand, the elevation of water reactivity at subcritical water conditions means an increase in the corrosion activity of water. Normal stainless steel reactors are corroded during HTL, even though no acid catalysts are added in the reaction.

33.3.2 HTL Equipment

Until now, the equipment used for HTL studies, batch-type autoclaves, is the major type of research reactor, but there is a gap time to reach the planned reaction temperature, which is a problem that cannot be controlled precisely in research factors. The reaction conditions in semicontinuous (stop flow type) and continuous-type reactors, such as temperature and reaction time, are easier to control (Marcilla et al. 2013; Mørup et al. 2012). As an example, continuous HTL equipment was established with a 2 L reaction volume of reactor at the University of Sydney. They tested a pilot-scale microalgal HTL process with *Chlorella* and *Spirulina* microalgal biomass (Jazrawi et al. 2013).

The downstream process of microalgal fuel needs to have good equipment productivity to make an economical biocrude. Because of the low price of biofuel, the equipment productivity is more important. Batch-type reactors need to be stopped intermittently, which impedes the productivity making them inadequate for the industrial-scale production of microalgal biofuels. Continuous-type reactors work without stopping, so they have high equipment productivity. On the other hand, many developments related to the device should be achieved before continuous HTL systems can be used for industrial-scale microalgal HTL, such as a material resistant to the corrosion of subcritical water, device parts, equipment scale-up, and operation. In particular, the devices need to operate correctly during the long-term operation under high temperatures and pressures.

Glass reactors are durable to chemical corrosion, but they are weak to physical pressure. Therefore, they cannot be used because of the high-pressure operation of HTL. Metal materials are strong but can decay due to a chemical reaction, and the generally used industrial equipment alloy, stainless steel, is corroded easily by subcritical water (Watanabe et al. 2004). Nickel alloys, such as Inconel 625 or Hastelloy C-276, are more durable to the corrosion of subcritical water (Sridharan et al. 2009; Zhang et al. 2009). On the other hand, a material more durable to corrosion, stronger to physical stress, and device parts, such as a valve, should be developed. A pump system that can supply a microalgal feedstock paste to a reactor is needed for industrial-scale continuous HTL because microalgal feedstock mixed with water is too viscous to pump at a constant flow rate.

33.4 HTL Reaction Characteristics of the Microalgal Component

33.4.1 HTL of Lipids

Lipid is one of the major components of microalgae. Triacylglycerol (TAG), which consists of a single molecule of glycerol bonded to three molecules of fatty acids by an ester bond, is the major kind of lipid. Fat is solid and oil is liquid at room temperature. The characteristics of TAG are determined by the composition of fatty acid. When TAG has a high saturated fatty acid content, it becomes fat, whereas it will become oil if it has a high unsaturated fatty acid content. Fat and oil are immiscible in water at room temperature, but they mix easily in subcritical water because of its low dielectric constant and high nonpolar affinity (Carr et al. 2011). Subcritical water hydrolyzes a vegetable oil rapidly. As an example, more than 97% of oil is hydrolyzed to free fatty acids at 260°C–280°C and 330°C–340°C over a 15–20 min period (Holliday et al. 1997) and over a 10–15 min period, respectively, without a catalyst, but small amount of *trans*-fatty acids can also appear (King et al. 1999).

TAG is hydrolyzed more rapidly than other major biological components, protein and a carbohydrate, under the subcritical water conditions and can be separated into fatty acids and glycerol. Fatty acids enhance the hydrolysis activity of subcritical water, and the fatty acids produced from TAG in HTL reaction exhibit the autocatalytic activity. Therefore, it accelerates the generation of free fatty acids (Milliren et al. 2013).

Fatty acids are relatively stable to heat. Therefore, most fatty acids produced by HTL exist without change. A small amount of *trans*-fatty acids can form during the HTL process of TAG, whereas only the *cis*-form fatty acids exist in raw TAG (King et al. 1999; Kocsisová et al. 2006). Stearic acid ($C_{17}H_{35}COOH$), a saturated fatty acid, is reasonably stable, even in supercritical water, which exists at a higher temperature than subcritical water. When alkali hydroxides, such as sodium hydroxide or potassium hydroxide, are used as a catalyst, the major products of stearic acid in supercritical water reaction are C_{17} alkane and carbon dioxide because the decarboxylation reaction of fatty acids is enhanced. Metal oxides, particularly zirconium hydroxide ($ZrO_2 \cdot xH_2O$), also enhance the decarboxylation reaction of fatty acids under supercritical water conditions, but the major products of the decarboxylation of stearic acid are different when using alkali hydroxide as a catalyst, that is, C_{16} alkene and carbon dioxide. Alkene can be converted to alkane via a hydrogen addition reaction (Watanabe et al. 2006). The hydrocarbons, *n*-alkanes, produced from fatty acids by supercritical water with a catalyst are the same as the major components of petro diesel and jet fuel. In addition, *n*-alkanes can be produced in high yield from vegetable oil, through oil hydrolysis to fatty acids in subcritical water followed by hydrogenation with a Pd/C catalyst at 300°C (Kubičková et al. 2005; Wang et al. 2012).

The biocrude production yield of HTL using high TAG content microalgal feedstock is high, because fatty acids are reasonably stable under subcritical water conditions and TAG is decomposed to fatty acids and glycerol under HTL conditions. Therefore, many fatty acids accumulate in biocrude. Glycerol is decomposed to acetaldehyde, formaldehyde, acrolein, methanol, propionaldehyde, and allyl alcohol under HTL conditions (Bühler et al. 2002). Glycerol is the primary hydrolysis product of TAG, but the glycerol produced through HTL converts to its derivative during the HTL process. Glycerol dissolves in the water phase rather than in biocrude, but many derivatives of glycerol exist in the biocrude part. Glycerol is transformed to acrolein by a dehydration reaction and then to acrylic acid through oxidation. Acrylic acid is a raw material for the production of poly(acrylic acid) that is used to produce superabsorbent resin, cement admixtures, water treatment agent, adhesives, etc. Acrolein is produced when glycerol is reacted with zinc sulfate as a catalyst in subcritical water (Ott et al. 2006).

33.4.2 HTL of Proteins

Protein is the main component of microalgae and is comprised of amino acids connected continuously by a peptide bond. The carboxylic acid and amino group in the amino acids bond by a peptide bond successively to form a protein. The amino acids have different side chains according to the amino acid type; some have a ring structure, and the others have aliphatic side functional group. Therefore, there is a minimum of one nitrogen atom in a peptide bond per amino acid unit in the protein. Asparagine, glutamine, lysine, and tryptophan have another nitrogen atom; histidine has two; and arginine has three nitrogen atoms in the side chain. Therefore, proteins have a nitrogen content of 15%–19% that differs according to the composition of amino acids. Another set of biomolecules containing nitrogen are chlorophylls, which are involved in photosynthesis, and pyrimidine and purine bases in nucleotides. Biocrude, the HTL main product, contains nitrogen mainly from protein. The nitrogen components in microalgal feedstock are distributed after HTL in biocrude, aqueous phase, gas, and solid. The nitrogen in biocrude can form NOx, an serious cause of air pollution, and the strange smell of biocrude. Therefore, a lower nitrogen content in biocrude is generally desirable. The nitrogen dissolved in the water phase can be recycled using as a nutrient source for cultivation medium of microalgae or microorganisms.

The peptide bond in proteins is more stable than the glycoside bond in starch or cellulose and ester bond between fatty acid and glycerol in TAG. Therefore, it is hydrolyzed slowly to amino acids during the HTL process. Carbon dioxide in subcritical water acts as a catalyst and elevates the reaction rate of peptide bond hydrolysis. Because the amino acids undergo a decomposition process, the amino acid concentration in HTL reactant is not high, even though amino acids and peptides are produced during HTL. Among the amino acids, glycine and alanine are relatively stable, so relatively large amounts of them with very small amounts of other amino acids are detected in the HTL reactant (Brunner 2009). The major decomposition products of a protein are acetaldehyde, acetaldehyde hydrate, diketopiperazine, ethylamine, methylamine, formaldehyde, lactic acid, propionic acid, etc. (Klingler et al. 2007).

Amino acids can be decomposed via two main routes in HTL. One is the *deamination* of amino acid to ammonia and organic acid, and the other is *decarboxylation* producing carbonic acid and amines. By deamination, alanine is converted to lactic acid by deamination with one molecule of H_2O addition, or to pyruvic acid with a 1/2 molecule O_2 addition, finally decomposing to carbon dioxide. When alanine, leucine, phenylalanine, serine, and aspartic acid are degraded by subcritical water at 200°C–340°C and 20 MPa, the amino acid decomposition rate was a first-order function of temperature within the experimented conditions. The decomposition rate of amino acids is different and rapid in the order of aspartic acid, serine, phenylalanine, leucine, and alanine. Alanine decomposes mainly to ammonia, carbonic acid, lactic acid, pyruvic acid, and acetic acid by hydrothermal degradation at 300°C and 20 MPa (Sato et al. 2004). In the hydrothermal degradation of proteins, the temperature and reaction time are the most influential factors for amino acid generation. The amino acid production yield is 3% and 4% based on the starting protein bovine serum albumin (BSA) at 310°C for 30 s and 270°C for 90 s, respectively, so the amino acid production yield is very low. The amino acid production yield decreases sharply when the reaction temperature is elevated. This means that the amino acids produced decompose rapidly under hydrothermal treatment condition.

Carbon dioxide addition to the HTL hydrolysis of proteins increases the amino acid production yield by enhancing peptide bond degradation via an acid hydrolysis reaction (Rogalinski et al. 2005). During the deamination reaction of amino acids, the nitrogen in amino acids is degraded to ammonia gas, which results in a decrease in the nitrogen content in biocrude. Because biomass is a complex material, the HTL reaction of biomass produces not only amino acid but also monosaccharides. The Maillard reaction, a representative browning reaction between amino acid and sugar, occurs. As a result, N-containing heterocyclic compounds, such as pyrimidines and pyrroles, are produced and exist in the biocrude of HTL (Kruse et al. 2005, 2007).

33.4.3 HTL of Carbohydrates

Starch is a representative organic compound produced by a photosynthesis reaction with carbon dioxide and is accumulated in microalgal cells. Its constitution unit is glucose, which produces a polymer with a straight chain via an α-(1 → 4) linkage and branched chain via an α-(1 → 6) linkage. Hydrolysis and degradation reactions take place when starch is hydrolyzed under subcritical water conditions without a catalyst. In the experiment using potato starch as a raw material, the maximum glucose production yield was 63% at 200°C for a 30 min reaction. The glucose production yield was decreased in the hydrothermal reaction at higher temperatures due to the decomposition of the glucose produced. The starch in biomass converts to glucose in a subcritical water reaction; the glucose epimerizes to fructose; and the fructose decomposes faster than glucose. The decomposition products of glucose reacted in subcritical water are fructose, saccharinic acids, erythrose, glyceraldehyde, 1,6-anhydroglucose, dihydroxyacetone, pyruvaldehyde, hydroxymethylfurfural (HMF), 1,6-anhydroglucose, and furfural, etc. (Kabyemela et al. 1997; Miyazawa et al. 2006; Nagamori and Funazukuri 2004).

33.5 Effects of the HTL Reaction Conditions

The representative reaction temperature for biomass is a subcritical temperature, 300°C–374°C, and the biocrude production yield is elevated as the reaction temperature is increased under subcritical water conditions (Garcia Alba et al. 2012; Ross et al. 2010). As the reaction temperature is increased, the oxygen content in biocrude is decreased, resulting in a high HHV. On the other hand, biocrude has a higher nitrogen content from protein degradation, and solid has a lower production yield. When the reaction temperature is elevated over the critical point to the supercritical region, the biocrude yield is decreased and gas production is increased.

The results of the microalgal HTL process using a continuous reactor with *Chlorella* and *Spirulina* microalgal biomass were as follows (Jazrawi et al. 2013). The microalgal concentration reaction temperature and residence times were 1%–10% (w/w), 250°C–350°C, and 3–4 min, respectively. The biocrude yield was improved at a higher biomass loading and higher temperature. Harsher reaction conditions produced biocrude with a lower oxygen content, and a higher reaction temperature induced a higher nitrogen content in biocrude, because more bio-oil originated from proteins. The maximum biocrude production yield was 41.7% (w/w) when the loading of *Chlorella* cells was 10% (w/w) at 350°C for 3 min. The production condition of high-quality biocrude did not correspond to that of the maximum yield. The gas composition was more than 90 mol% carbon dioxide, and the amount of gas increased with increasing strength of the reaction condition. The solid yield decreased with increasing reaction temperature and residence time (Jazrawi et al. 2013).

The biocrude production yield using microalgal biomass was highest near the critical point. The biocrude yield according to the microalgal strains is affected by the reaction temperature. When the reactions were conducted at a relatively low temperature of 250°C for 5 min using an autoclave, the yield was 17.7%–44.8%, but the biocrude production yield near the critical point 375°C was high, and the yield difference depending on the strains was small, 45.6%–58.1% (López Barreiro et al. 2013b,c).

Catalysts improve the HTL process, such as increasing the biocrude production yield. The catalysts used in the HTL process can be divided into homogeneous catalysts and heterogeneous catalysts; homogeneous alkali catalysts are often used in the microalgal HTL process. Homogeneous alkali catalysts have a tendency to increase the biocrude yield during the HTL process and inhibit the production of char. When an empty palm fruit bunch was treated in subcritical water without a catalyst at 270°C for 20 min, the liquid oil yield was 30.4% (w/w), whereas under the same conditions except using 1.0 M K_2CO_3, the yield was 68% (w/w) (Akhtar et al. 2010). In the other case, corn stalk was treated by HTL with 1.0% (w/w) Na_2CO_3, and the biocrude yield was higher than that without a catalyst from 33.4% to 47.2%, and the gas production was decreased (Song et al. 2004). Potassium salts are a more effective catalyst than sodium salts.

Potassium carbonate catalyzes the *water–gas shift reaction* in the HTL process; it consumes water and CO and produces H_2 and CO_2. The reaction equations are as follows (Elliott et al. 1983; Sinag et al. 2003):

$$K_2CO_3 + H_2O \rightarrow KHCO_3 + KOH$$

$$KOH + CO \rightarrow HCOOK$$

Formic acid salt reacts with water to produce hydrogen and potassium bicarbonate:

$$HCOOK + H_2O \rightarrow KHCO_3 + H_2$$

Potassium bicarbonate is converted to CO_2 and K_2CO_3:

$$2KHCO_3 \rightarrow H_2O + K_2CO_3 + CO_2$$

Therefore, the net reaction is

$$H_2O + CO \leftrightarrow HCOOH \leftrightarrow H_2 + CO_2$$

CO gas produced during HTL is consumed, so the CO content in HTL gas is low and H_2 and CO_2 are produced. H_2 gas participates in the reductive reaction of biocrude, so it improves the quality and HHV of biocrude. The major gas product of HTL is CO_2, and it can be recycled in the photosynthesis reaction of microalgae.

Heterogeneous catalysts can overcome the difficulty of recovery and environmental pollution of dissolved catalysts, which are the disadvantages of homogeneous catalysts. They can be recovered easily, so they do not behave as a pollutant but can undergo fouling by the reactant that would deteriorate its activity. Many kinds of heterogeneous catalysts, such as nickel, ruthenium, zirconia, and zeolite, are being used for research (Azadi and Farnood 2011; Azadi et al. 2013; Duan and Savage 2011a; Perego and Bosetti 2011; Shi et al. 2012; Yoosuk et al. 2012; Zhang et al. 2013a). In some studies using heterogeneous catalyst, the biocrude yield was elevated, and the biocrude quality, HHV, was improved (Biller et al. 2011; Duan and Savage 2011a). Microalgal biocrude characteristics can be upgraded by a heterogeneous catalyst. When the HTL biocrude was reacted with a Pt/C catalyst under high-pressure hydrogen under supercritical water conditions, the HHV of oil was elevated, the acidity was decreased, and the HHV became similar to that of petro oil (Duan and Savage 2011b).

Methanol and ethanol can be used as an HTL solvent. In the experiment, ethanol was used for HTL of low lipid content *Chlorella pyrenoidosa*. The oil yield was 51.1% compared to 40.5% in the case of using water for HTL, and the thermal stability of oil was also improved (Zhang et al. 2013b; Zhou et al. 2012).

33.6 HTL Products

33.6.1 Biocrude

The major phases of HTL products are oil, water, gas, and a solid phase. The oil called *biocrude* is the main product produced in large quantities in the microalgal HTL process. Biocrude is a black liquid substance with a high heating value, organic solvent extractable, and high viscosity. A higher production yield of microalgal biocrude that is 50%–60% was reported (Biller and Ross 2011). The typical biocrude composition is 70%–75% carbon, 10%–16% oxygen, 4%–6% nitrogen, and 0.5%–1% sulfur, and it has a high energy content, that is, higher HHV of 33.4–39.9 MJ kg^{-1}. The main components of biocrude are aromatic hydrocarbon, nitrogen heterocycles, long-chain fatty acids, alcohols, etc. Biocrude is a complex of compounds constituting with various molecules with a range of molecular weights (Brunner 2009).

The oxygen content of biocrude is lower than that of its raw material, microalgal biomass. Therefore, its HHV is high and it has high value as a fuel. The chemical composition and physical characteristics of biocrude are affected significantly by the composition of feed microalgae, HTL temperature, reaction time, catalyst, and separation method of biocrude. When the same feedstock is used, the reaction temperature is the most influencing factor. The biocrude produced by HTL is more suitable as a fuel; it has a lower oxygen, water content, and higher storage stability than those of pyrolysis oil, which is the oil produced from biomass by fast pyrolysis under high temperatures and low pressures for a short time without water (Jena and Das 2011).

Although biocrude contains a lower nitrogen content than feed substance microalgae, it contains considerable amounts of nitrogen and sulfur. If it is used without upgrading, large quantities of NOx and SOx, which are representative air pollutants, will be released. Therefore, the production method for biocrude with low nitrogen and sulfur contents is necessary. The main components of microalgal cells are protein, carbohydrate, lipid, nucleic acid, etc. In particular, the lipid content has a significant effect on biocrude production; a higher lipid content results in a higher biocrude yield. Biocrude from TAG does not contain nitrogen and sulfur, and its major product, fatty acids, can be converted to more qualified fuel components, that is, long-chained aliphatic hydrocarbons (Kubičková et al. 2005; Wang et al. 2012).

33.6.2 Aqueous Phase

The aqueous phase in microalgal HTL is important for recycling resources, and most inorganic materials exist as an aqueous phase. Among them, phosphate is one of the most essential nutrients for microalgal cultivation. If phosphate is only used for the mass culture of microalgae without recycling, the phosphate resources that come from the limited phosphate rocks will be depleted. For sustainable bioenergy production, phosphate recycling in feedstock biomass is essential. The phosphate in microalgal cell accumulates in the aqueous phase during the HTL process. Many inorganic elements and organic substances, such as NH_4^+, PO_4^-, acetate, NO_3^-, phenol, K, Na, and Ni, are dissolved in aqueous phases. These can be recycled by reusing in the cultivation of microalgae (Biller et al. 2012; Garcia Alba et al. 2013a,b).

When the aqueous phase is used as a nutrient source for a microalgal culture, some substances existing in the water phase, such as phenol, inhibit microalgal growth.

The aqueous phase can be diluted to avoid inhibiting cell growth as a culture medium. In other aspects, the acetate existing large quantities in the water phase can be a good nutrient for the *heterotrophic growth* of microalgae. Heterotrophic nutrition is one of the microalgal organic substance obtaining methods that use an organic carbon source. This can be compared with *autotrophic nutrition* that uses carbon dioxide through photosynthesis. Acetate is a good carbon source for microalgal heterotrophic growth.

In addition, the aqueous phase of microalgal HTL can be a growth media for normal microorganisms, such as *Escherichia coli* and *Pseudomonas putida*, which grow well using the HTL aqueous phase of *Nannochloropsis oculata* as the sole carbon, nitrogen, and phosphate sources. Moreover, *Saccharomyces cerevisiae* requires only a glucose supplement in the aqueous phase of microalgal HTL as the growth media (Nelson et al. 2013).

33.6.3 Gas Phase

In general, the most abundant gas in the HTL gas phase is carbon dioxide, which is more than 90 mol%. Hydrogen, carbon monoxide, methane, nitrogen gas, etc., are also present in the gas phase. The type of gas and contents are changed according to the feedstock components, reaction temperature, reaction time, type of catalyst, and other reaction conditions. Generally, the amount of gas produced from microalgal HTL is approximately 20% based on the organic substance in feed microalgae (Brown et al. 2010; Garcia Alba et al. 2012).

Organic nitrogen in biomass is degraded mainly through a deamination reaction. Generally, microalgal biomass has a nitrogen content of approximately 8%, and biocrude contains 4%–6% nitrogen. This is transformed to a gas through HTL and converted to N_2, HCN, NH_3, and N_2O type gas. The gas production differs according to the catalyst and reaction conditions even when the same biomass is used. Normally, in the case of HTL using an organic acid, the nitrogen content is decreased in the aqueous phase. As a result, NH_3 and HCN increase in the gas phase. The reason for the small CO content in the gas phase is the conversion of CO produced during the reaction to CO_2 and H_2 through a water–gas shift reaction (Elliott and Sealock 1983).

33.6.4 Solid Phase

The solid residue in microalgal HTL has a low residual yield; less than 10% of organic substance is used as the feedstock. The harsher reaction conditions result in less solid in microalgal HTL. The main components of the solid are ash, which might be useful for soil modification.

33.7 Summary

Microalgal HTL is a biofuel production method that uses microalgae as a feedstock, reacts under subcritical water conditions, and produces biocrude as the main product. Biocrude along with an aqueous phase, gas, and solid is produced by microalgal HTL. Biocrude is a black liquid with a high energy value that is a potential alternative to petro oil as a raw fuel and chemical substance. The aqueous phase is dissolved with phosphates, nitrogen, various inorganic salts, and organic substance and can be recycled as a nutrient source for microalgal cultivation.

Microalgae grow and produce organic materials using only sunlight, carbon dioxide, water, and a few nutrients. The microalgal wet biomass is processed through HTL to produce a high energy value fuel, biocrude. Phosphate and trace mineral substances used to culture microalgae can be recycled easily using the aqueous phase of HTL as a nutrient source for microalgal cultivation.

The HTL process uses wet feedstock, microalgal biomass, as a wet cell, so it does not require a drying step, which is the most energy-consuming process in microalgal cell production. HTL is high EROEI process for bioenergy production.

The main product of microalgal HTL, biocrude, has high HHV, and its properties can be upgraded, so it is a favorable alternative to crude petroleum oil. For the practical production of microalgal HTL bioenergy, many technologies must be developed, such as microalgal feedstock production, HTL processing device, HTL operation condition, product upgrading, and substance recycling. Although there are no practical uses so far, microalgal HTL is a promising way of producing sustainable bioenergy.

References

Akhtar, J., S. K. Kuang, and N. S. Amin. 2010. Liquefaction of empty palm fruit bunch (EPFB) in alkaline hot compressed water. *Renewable Energy* 35(6):1220–1227.

Akiya, N. and P. E. Savage. 2002. Roles of water for chemical reactions in high-temperature water. *Chemical Reviews* 102(8):2725–2750.

Azadi, P., E. Afif, H. Foroughi, T. Dai, F. Azadi, and R. Farnood. 2013. Catalytic reforming of activated sludge model compounds in supercritical water using nickel and ruthenium catalysts. *Applied Catalysis B: Environmental* 134–135:265–273.

Azadi, P. and R. Farnood. 2011. Review of heterogeneous catalysts for sub- and supercritical water gasification of biomass and wastes. *International Journal of Hydrogen Energy* 36(16):9529–9541.

Biller, P., R. Riley, and A. B. Ross. 2011. Catalytic hydrothermal processing of microalgae: Decomposition and upgrading of lipids. *Bioresource Technology* 102(7):4841–4848.

Biller, P. and A. B. Ross. 2011. Potential yields and properties of oil from the hydrothermal liquefaction of microalgae with different biochemical content. *Bioresource Technology* 102(1):215–225.

Biller, P. and A. B. Ross. 2012. Hydrothermal processing of algal biomass for the production of biofuels and chemicals. *Biofuels* 3(5):603–623.

Biller, P., A. B. Ross, S. C. Skill, A. Lea-Langton, B. Balasundaram, C. Hall, R. Riley, and C. A. Llewellyn. 2012. Nutrient recycling of aqueous phase for microalgae cultivation from the hydrothermal liquefaction process. *Algal Research* 1(1):70–76.

Brown, T. M., P. Duan, and P. E. Savage. 2010. Hydrothermal liquefaction and gasification of *Nannochloropsis* sp. *Energy and Fuels* 24(6):3639–3646.

Brunner, G. 2009. Near critical and supercritical water. Part I. Hydrolytic and hydrothermal processes. *Journal of Supercritical Fluids* 47(3):373–381.

Bühler, W., E. Dinjus, H. J. Ederer, A. Kruse, and C. Mas. 2002. Ionic reactions and pyrolysis of glycerol as competing reaction pathways in near- and supercritical water. *Journal of Supercritical Fluids* 22(1):37–53.

Carr, A. G., R. Mammucari, and N. R. Foster. 2011. A review of subcritical water as a solvent and its utilisation for the processing of hydrophobic organic compounds. *Chemical Engineering Journal* 172(1):1–17.

Delrue, F., Y. Li-Beisson, P. A. Setier, C. Sahut, A. Roubaud, A. K. Froment, and G. Peltier. 2013. Comparison of various microalgae liquid biofuel production pathways based on energetic, economic and environmental criteria. *Bioresource Technology* 136:205–212.

Duan, P. and P. E. Savage. 2011a. Hydrothermal liquefaction of a microalga with heterogeneous catalysts. *Industrial and Engineering Chemistry Research* 50(1):52–61.
Duan, P. and P. E. Savage. 2011b. Upgrading of crude algal bio-oil in supercritical water. *Bioresource Technology* 102(2):1899–1906.
Elliott, D. C., R. T. Hallen, and L. J. Sealock Jr. 1983. Aqueous catalyst systems for the water-gas shift reaction. 2. Mechanism of basic catalysis. *Industrial and Engineering Chemistry Product Research and Development* 22(3):431–435.
Elliott, D. C. and L. J. Sealock Jr. 1983. Aqueous catalyst systems for the water-gas shift reaction. 1. Comparative catalyst studies. *Industrial and Engineering Chemistry Product Research and Development* 22(3):426–431.
Franck, E. U. 1968. Supercritical water. *Endeavour* 27(101):55–59.
Freya, G. W. and D. M. Linke 2002. Hydropower as a renewable and sustainable energy resource meeting global energy challenges in a reasonable way. *Energy Policy* 30(14):1261–1265.
Garcia Alba, L., C. Torri, D. Fabbri, S. R. A. Kersten, and D. W. F. Wim Brilman. 2013a. Microalgae growth on the aqueous phase from Hydrothermal Liquefaction of the same microalgae. *Chemical Engineering Journal* 228:214–223.
Garcia Alba, L., C. Torri, C. Samorì, J. Van Der Spek, D. Fabbri, S. R. A. Kersten, and D. W. F. Brilman. 2012. Hydrothermal treatment (HTT) of microalgae: Evaluation of the process as conversion method in an algae biorefinery concept. *Energy and Fuels* 26(1):642–657.
Garcia Alba, L., M. P. Vos, C. Torri, D. Fabbri, S. R. A. Kersten, and D. W. F. Brilman. 2013b. Recycling nutrients in algae biorefinery. *ChemSusChem* 6(8):1330–1333.
Holliday, R. L., J. W. King, and G. R. List. 1997. Hydrolysis of vegetable oils in sub- and supercritical water. *Industrial and Engineering Chemistry Research* 36(3):932–935.
Hunter, S. E. and P. E. Savage. 2004. Recent advances in acid- and base-catalyzed organic synthesis in high-temperature liquid water. *Chemical Engineering Science* 59(22–23):4903–4909.
Jazrawi, C., P. Biller, A. B. Ross, A. Montoya, T. Maschmeyer, and B. S. Haynes. 2013. Pilot plant testing of continuous hydrothermal liquefaction of microalgae. *Algal Research* 2(3):268–277.
Jena, U. and K. C. Das. 2011. Comparative evaluation of thermochemical liquefaction and pyrolysis for bio-oil production from microalgae. *Energy and Fuels* 25(11):5472–5482.
Kabyemela, B. M., T. Adschiri, R. M. Malaluan, and K. Arai. 1997. Kinetics of glucose epimerization and decomposition in subcritical and supercritical water. *Industrial and Engineering Chemistry Research* 36(5):1552–1558.
Kim, J., G. Yoo, H. Lee, J. Lim, K. Kim, C. W. Kim, M. S. Park, and J. W. Yang. 2013. Methods of downstream processing for the production of biodiesel from microalgae. *Biotechnology Advances* 31(6):862–876.
King, J. W., R. L. Holliday, and G. R. List. 1999. Hydrolysis of soybean oil: In a subcritical water flow reactor. *Green Chemistry* 1(6):261–264.
Klingler, D., J. Berg, and H. Vogel. 2007. Hydrothermal reactions of alanine and glycine in sub- and supercritical water. *Journal of Supercritical Fluids* 43(1):112–119.
Kocsisová, T., J. Juhasz, and J. Cvengroš. 2006. Hydrolysis of fatty acid esters in subcritical water. *European Journal of Lipid Science and Technology* 108(8):652–658.
Kruse, A., A. Krupka, V. Schwarzkopf, C. Gamard, and T. Henningsen. 2005. Influence of proteins on the hydrothermal gasification and liquefaction of biomass. 1. Comparison of different feedstocks. *Industrial and Engineering Chemistry Research* 44(9):3013–3020.
Kruse, A., P. Maniam, and F. Spieler. 2007. Influence of proteins on the hydrothermal gasification and liquefaction of biomass. 2. Model compounds. *Industrial and Engineering Chemistry Research* 46(1):87–96.
Kubičková, I., M. Snåre, K. Eränen, P. Mäki-Arvela, and D. Yu Murzin. 2005. Hydrocarbons for diesel fuel via decarboxylation of vegetable oils. *Catalysis Today* 106(1–4):197–200.
López Barreiro, D., W. Prins, F. Ronsse, and W. Brilman. 2013a. Hydrothermal liquefaction (HTL) of microalgae for biofuel production: State of the art review and future prospects. *Biomass and Bioenergy* 53:113–127.

López Barreiro, D., F. Ronsse, and W. Prins. 2013b. Influence of microalgae strain-specific parameters in hydrothermal liquefaction of microalgae for biofuels production. *Communications in Agricultural and Applied Biological Sciences* 78(1):167–172.

López Barreiro, D., C. Zamalloa, N. Boon, W. Vyverman, F. Ronsse, W. Brilman, and W. Prins. 2013c. Influence of strain-specific parameters on hydrothermal liquefaction of microalgae. *Bioresource Technology* 146:463–471.

Marcilla, A., L. Catalá, J. C. García-Quesada, F. J. Valdés, and M. R. Hernández. 2013. A review of thermochemical conversion of microalgae. *Renewable and Sustainable Energy Reviews* 27:11–19.

Milliren, A. L., J. C. Wissinger, V. Gottumukala, and C. A. Schall. 2013. Kinetics of soybean oil hydrolysis in subcritical water. *Fuel* 108:277–281.

Miyazawa, T., S. Ohtsu, Y. Nakagawa, and T. Funazukuri. 2006. Solvothermal treatment of starch for the production of glucose and maltooligosaccharides. *Journal of Materials Science* 41(5):1489–1494.

Mørup, A. J., P. R. Christensen, D. F. Aarup, L. Dithmer, A. Mamakhel, M. Glasius, and B. B. Iversen. 2012. Hydrothermal liquefaction of dried distillers grains with solubles: A reaction temperature study. *Energy and Fuels* 26(9):5944–5953.

Nagamori, M. and T. Funazukuri. 2004. Glucose production by hydrolysis of starch under hydrothermal conditions. *Journal of Chemical Technology and Biotechnology* 79(3):229–233.

Nelson, M., L. Zhu, A. Thiel, Y. Wu, M. Guan, J. Minty, H. Y. Wang, and X. N. Lin. 2013. Microbial utilization of aqueous co-products from hydrothermal liquefaction of microalgae *Nannochloropsis oculata*. *Bioresource Technology* 136:522–528.

Ott, L., M. Bicker, and H. Vogel. 2006. Catalytic dehydration of glycerol in sub- and supercritical water: A new chemical process for acrolein production. *Green Chemistry* 8(2):214–220.

Patil, V., K. Q. Tran, and H. R. Giselrød. 2008. Towards sustainable production of biofuels from microalgae. *International Journal of Molecular Sciences* 9(7):1188–1195.

Perego, C. and A. Bosetti. 2011. Biomass to fuels: The role of zeolite and mesoporous materials. *Microporous and Mesoporous Materials* 144(1–3):28–39.

Rogalinski, T., S. Herrmann, and G. Brunner. 2005. Production of amino acids from bovine serum albumin by continuous sub-critical water hydrolysis. *Journal of Supercritical Fluids* 36(1):49–58.

Ross, A. B., P. Biller, M. L. Kubacki, H. Li, A. Lea-Langton, and J. M. Jones. 2010. Hydrothermal processing of microalgae using alkali and organic acids. *Fuel* 89(9):2234–2243.

Sato, N., A. T. Quitain, K. Kang, H. Daimon, and K. Fujie. 2004. Reaction kinetics of amino acid decomposition in high-temperature and high-pressure water. *Industrial and Engineering Chemistry Research* 43(13):3217–3222.

Shi, F., P. Wang, Y. Duan, D. Link, and B. Morreale. 2012. Recent developments in the production of liquid fuels via catalytic conversion of microalgae: Experiments and simulations. *RSC Advances* 2(26):9727–9747.

Shuping, Z., W. Yulong, Y. Mingde, I. Kaleem, L. Chun, and J. Tong. 2010. Production and characterization of bio-oil from hydrothermal liquefaction of microalgae *Dunaliella tertiolecta* cake. *Energy* 35(12):5406–5411.

Sinag, A., A. Kruse, and V. Schwarzkopf. 2003. Key compounds of the hydropyrolysis of glucose in supercritical water in the presence of K_2CO_3. *Industrial and Engineering Chemistry Research* 42(15):3516–3521.

Song, C., H. Hu, S. Zhu, G. Wang, and G. Chen. 2004. Nonisothermal catalytic liquefaction of corn stalk in subcritical and supercritical water. *Energy and Fuels* 18(1):90–96.

Sridharan, K., Y. Chen, L. Tan, X. Ren, A. Kruizenga, M. Anderson, and T. Allen. 2009. Perspectives on corrosion in supercritical water environment: Materials and novel treatments. In NACE—International Corrosion Conference Series. NACE International, Houston, TX.

Toor, S. S., L. Rosendahl, and A. Rudolf. 2011. Hydrothermal liquefaction of biomass: A review of subcritical water technologies. *Energy* 36(5):2328–2342.

Valdez, P. J., M. C. Nelson, H. Y. Wang, X. N. Lin, and P. E. Savage. 2012. Hydrothermal liquefaction of *Nannochloropsis* sp.: Systematic study of process variables and analysis of the product fractions. *Biomass and Bioenergy* 46:317–331.

Wang, W. C., N. Thapaliya, A. Campos, L. F. Stikeleather, and W. L. Roberts. 2012. Hydrocarbon fuels from vegetable oils via hydrolysis and thermo-catalytic decarboxylation. *Fuel* 95:622–629.

Watanabe, M., T. Iida, and H. Inomata. 2006. Decomposition of a long chain saturated fatty acid with some additives in hot compressed water. *Energy Conversion and Management* 47(18–19):3344–3350.

Watanabe, Y., H. Abe, Y. Daigo, R. Fujisawa, and M. Sakaihara. 2004. Effect of physical property and chemistry of water on cracking of stainless steels in sub-critical and supercritical water. *Key Engineering Materials* 261–263:1031–1036.

Xu, L., D. W. F. Wim Brilman, J. A. M. Withag, G. Brem, and S. Kersten. 2011. Assessment of a dry and a wet route for the production of biofuels from microalgae: Energy balance analysis. *Bioresource Technology* 102(8):5113–5122.

Xu, Y., H. Yu, L. Zhu, K. Wang, Z. Cui, and X. Hu. 2012. Preparation of bio-fuel from *Chlorella pyrenoidosa* by hydrothermal catalytic liquefaction. *Nongye Gongcheng Xuebao/Transactions of the Chinese Society of Agricultural Engineering* 28(19):194–199.

Yoosuk, B., D. Tumnantong, and P. Prasassarakich. 2012. Amorphous unsupported Ni-Mo sulfide prepared by one step hydrothermal method for phenol hydrodeoxygenation. *Fuel* 91(1):246–252.

Yu, G., Y. Zhang, L. Schideman, T. L. Funk, and Z. Wang. 2011a. Hydrothermal liquefaction of low lipid content microalgae into bio-crude oil. *Transactions of the ASABE* 54(1):239–246.

Yu, G., Y. Zhang, L. Schideman, T. Funk, and Z. Wang. 2011b. Distributions of carbon and nitrogen in the products from hydrothermal liquefaction of low-lipid microalgae. *Energy and Environmental Science* 4(11):4587–4595.

Zhang, J., W. T. Chen, P. Zhang, Z. Luo, and Y. Zhang. 2013a. Hydrothermal liquefaction of *Chlorella pyrenoidosa* in sub- and supercritical ethanol with heterogeneous catalysts. *Bioresource Technology* 133:389–397.

Zhang, J., Z. Luo, and Y. Zhang. 2013b. Hydrothermal liquefaction of *Chlorella pyrenoidosa* in water and ethanol. *Transactions of the ASABE* 56(1):253–259.

Zhang, Q., R. Tang, C. Li, X. Luo, C. Long, and K. Yin. 2009. Corrosion behavior of Ni-base alloys in supercritical water. *Nuclear Engineering and Technology* 41(1):107–112.

Zhou, D., S. Zhang, H. Fu, and J. Chen. 2012. Liquefaction of macroalgae *Enteromorpha prolifera* in sub-/supercritical alcohols: Direct production of ester compounds. *Energy and Fuels* 26(4):2342–2351.

Zhu, Y., K. O. Albrecht, D. C. Elliott, R. T. Hallen, and S. B. Jones. 2013. Development of hydrothermal liquefaction and upgrading technologies for lipid-extracted algae conversion to liquid fuels. *Algal Research* 2(4):455–464.

Index

B

D

M

R

S

Z